# The Cambridge Encyclopedia of **Human Growth and Development**

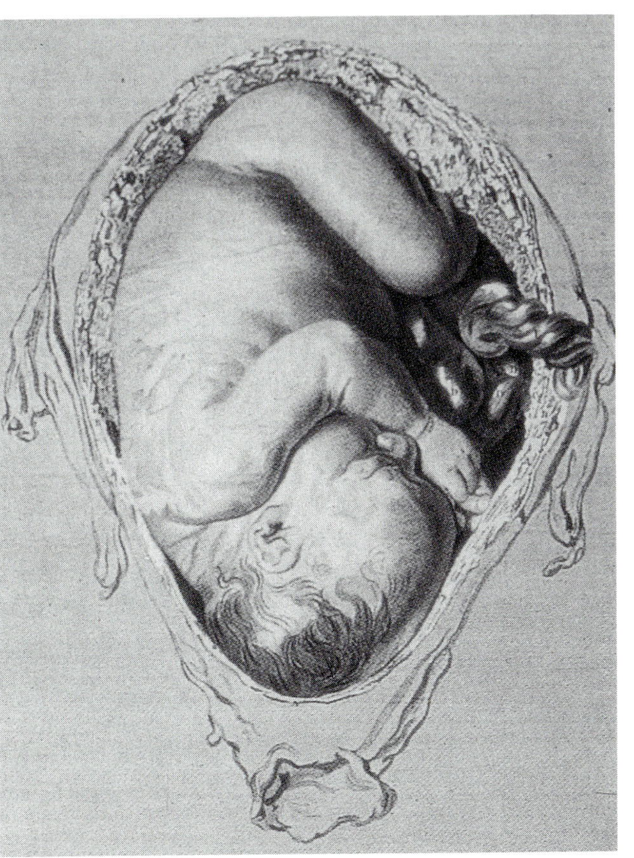

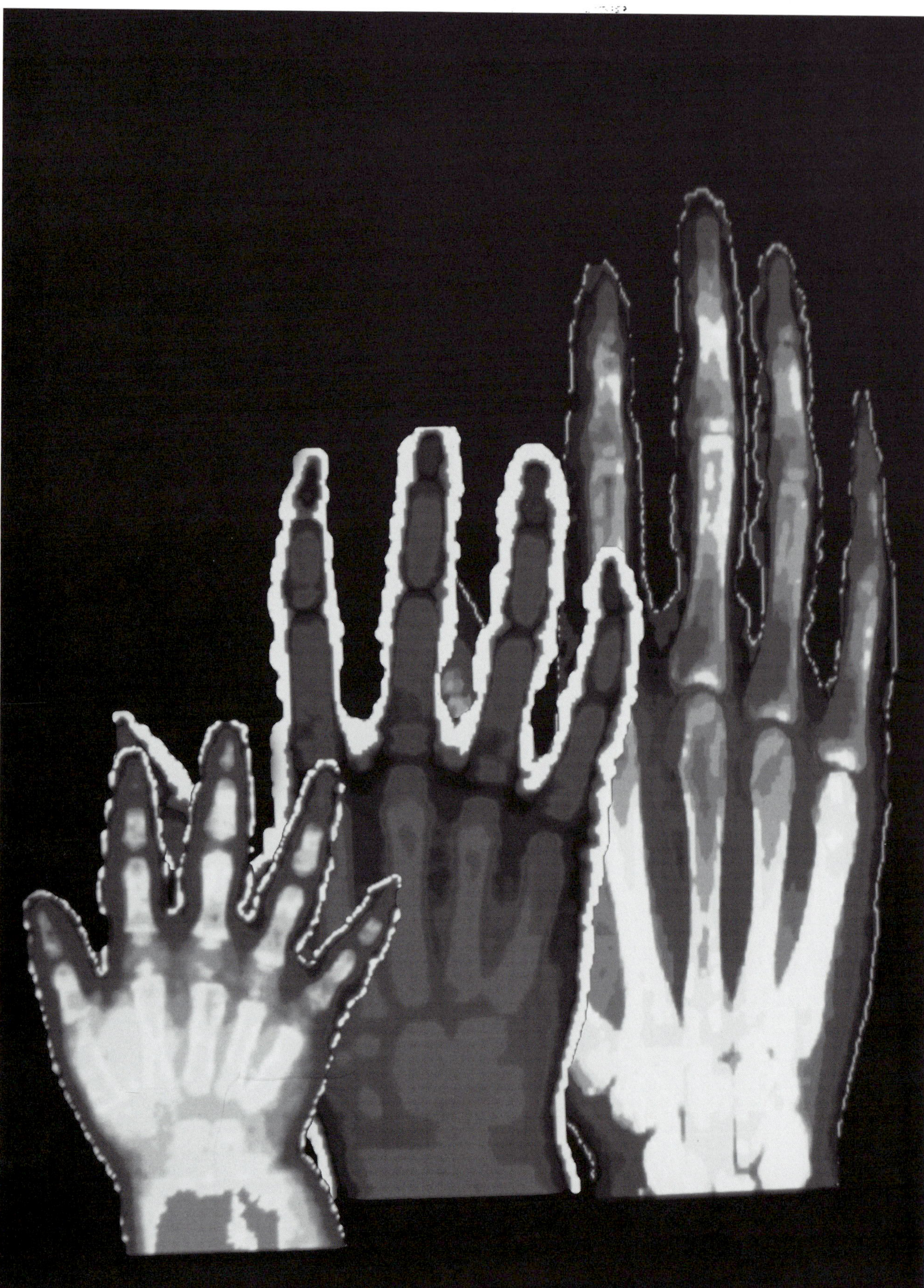

The Cambridge
Encyclopedia of

# Human Growth and Development

Edited by **Stanley J. Ulijaszek, Francis E. Johnston and Michael A. Preece**

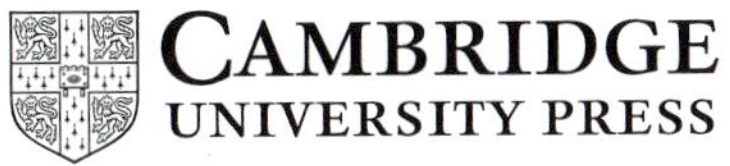

PUBLISHED BY THE PRESS SYNDICATE OF THE UNIVERSITY OF CAMBRIDGE
The Pitt Building, Trumpington Street, Cambridge CB2 1RP, United Kingdom

CAMBRIDGE UNIVERSITY PRESS
The Edinburgh Building, Cambridge CB2 2RU, United Kingdom
40 West 20th Street, New York, NY 10011–4211, U.S.A.
10 Stamford Road, Oakleigh, Melbourne 3166, Australia



First published 1998

Printed in the United Kingdom at the University Press, Cambridge

Typeset in Adobe Joanna 10.25/12.5pt, in QuarkXpress™ [DS]

Layout and design by David Seabourne

Artwork: David Gregson Associates

*A catalogue record for this book is available from the British Library*

*Library of Congress Cataloguing in Publication data*
The Cambridge encyclopedia of human growth and development /
edited by Stanley Ulijaszek, Francis E. Johnston, Michael Preece.
p. cm.
Includes bibliographical references and index.
ISBN 0-521-56046-2 (hardbound)
1. Human growth – encyclopedias. 2. Developmental psychology – encyclopedias.
I. Ulijaszek, Stanley J.
II. Johnston, Francis E., 1931– .
III. Preece, M. A.
[DNLM: 1. Human Development – encyclopedias. 2. Growth – encyclopedias. 3. Fetal Development – encyclopedias.
4. Anthropometry – encyclopedias.
WS 13 C177 1997]
QP84.C26 1997
612.6'03 – dc21
DNLM/DC 97–14707
for Library of Congress CIP

ISBN 0 521 56046 2 hardback

*Half-title*: From William Hunter's *Anatomy of the Human Gravid Uterus* (1774), by Jan van Rymsdyk. Courtesy of the Wellcome Institute, London.
*Frontispiece*: Bone growth at 2, 6 and 19 years. Photograph: Scott Camazine, courtesy of Science Photo Library.

# Contents

Part Five

## Post-natal growth and maturation Robert M. Malina

176



Part Six

## Behavioural and cognitive development Paolo Parisi

236

*For Jeff, Susan, Todd, Jan, Michael, Alexandra and Peter.*

**Stanley J. Ulijaszek** is a nutritional anthropologist who trained at the University of London, has conducted fieldwork in Papua New Guinea, South and South-East Asia, and the Pacific Islands, and has taught in Australia, the United Kingdom and the United States. He is currently Head of the Department of Nutrition, Curtin University School of Public Health, Australia, and Research Associate in Biological Anthropology at the University of Cambridge. He has been prominent in extending the use of physiological methods to the study of human growth and variation.

**Francis E. Johnston** is a biological anthropologist and Professor of Anthropology at the University of Pennsylvania. Johnston's research specialization is in the growth and development of children, especially in relation to nutritional status and health. He has worked extensively throughout Latin America, especially Guatemala (since 1966) but also in Mexico, Peru, Ecuador, and Cuba. In addition he is involved in projects in Papua New Guinea on the biosocial effects of modernization, and in West Philadelphia as part of the University of Pennsylvania Center for Community Partnerships.

**Michael A. Preece** is a growth physiologist, who trained in Medicine at Guy's Hospital Medical School and is currently the Professor of Child Health and Growth at the Institute of Child Health of University College London Medical School. His interests are in human growth regulation with particular reference to the genetic control of fetal growth and development.

# Contributors

**Professor R. McNeill Alexander**
University of Leeds

**Dr Imelda T. Angeles**
University of Indonesia

**Professor Jesús Argente**
Hospital Nino Jesús, Madrid

**Professor D. J. P Barker**
University of Southampton

**Dr Pawel Bergman**
Polish Academy of Sciences, Warsaw

**Professor Tadeusz Bielicki**
Polish Academy of Sciences, Warsaw

**Professor C. W. Binns**
Curtin University, Perth

**Professor Alan H. Bittles**
Edith Cowan University, Western Australia

**Professor Barry Bogin**
University of Michigan-Dearborn

**Professor Dr Ingeborg Brandt**
Zentrum für Kinderheilkunde Universitatskinderklinik und Poliklinik, Bonn

**Dr André Briend**
Hôpital St Lazare, Paris

**Professor C. G. D. Brook**
University College Medical School and Royal Free Hospital of Medicine, London

**Dr G. Brush**
University of Oxford

**Professor R. G. Burwell**
University of Nottingham

**Professor Noel Cameron**
Loughborough University of Technology and University of the Witwatersrand

**Leslie Carlin**
University of Durham

**Dr Julie A. Chowen**
Hospital Nino Jesús, Madrid

**Dr T. J. Cole**
Medical Research Council, Cambridge

**Dr Douglas E. Crews**
Ohio State University

**Dr Peter H. Dangerfield**
University of Liverpool

**Dr M. J. Dauncey**
The Babraham Institute, Cambridge

**Professor Peter S. W. Davies**
Queensland University of Technology

**Professor Arto Demirjian**
Université de Montreal

**Dr Marinos Elia**
Medical Research Council, Cambridge

**Professor P. T. Ellison**
Harvard University

**Dr Phyllis B. Eveleth**

**Professor Roderick Floud**
London Guildhall University

**Kristine Formica**
University of London

**Dr Diana M. Gibb**
University of London

**Professor Elena Godina**
Moscow State University

**Dr Gail Goldberg**
Medical Research Council, Cambridge

**Professor Michael H. N. Golden**
University of Aberdeen

**Professor Alan H. Goodman**
Hampshire College, Amherst

**Dr Rainer Gross**
University of Indonesia

**Professor A. Guarino**
Università degli Studi di Napoli

**Professor Jere D. Haas**
Cornell University

**Gillian J. Harper**
Ohio State University

**Professor Edward F. Harris**
University of Tennessee

**Professor G. A Harrison**
University of Oxford

**Professor R. C. Hauspie**
Vrije Universiteit Brussel

**Professor David A. Hay**
Curtin University, Perth

**Professor Anthony R. Hayward**
University of Colorado School of Medicine

**Professor C. J. K. Henry**
Oxford Brookes University

**Dr Michael Hermanussen**

**Professor Manuel Hernández**
Hospital Nino Jesús, Madrid

**Dr C. M. Hill**
University of Durham

**Professor John H. Himes**
University of Minnesota

**Dr Peter Hindmarsh**
University College London Medical School and Royal Free Hospital School of Medicine

**Professor Ieuan A. Hughes**
University of Cambridge

**Dr Carol Jenkins**
Institute of Medical Research, Papua New Guinea

**Professor Francis E. Johnston**
University of Pennsylvania

**Professor D. A. Jones**
University of Birmingham

**Professor Peter R. M. Jones**
Loughborough University of Technology

**Dr Enamul Karim**
Institute of Epidemiology, Dhaka

**Professor Johan Karlberg**
University of Hong Kong

**Dr N. F. Kember**

**Dr Kyra Marie Landzelius**
Center College, Kentucky

**Dr Michelle Lampl**
Emory University, Atlanta

**Dr P. C. Lee**
University of Cambridge

**Dr Lynette Leidy**
University of Massachusetts

**Dr Florence Levy**
Royal South Sydney Hospital

**Professor Gunilla Lindgren**

**Professor Michael A. Little**
State University of New York

**Dr Ann L. Magennis**
Colorado State University

**Professor Robert M. Malina**
Michigan State University

**Dr Diane Markowitz**
Rowan University, New Jersey

**Dr C. G. N. Mascie-Taylor**
University of Cambridge

**Dr Ishiro Miyata**
Vanderbilt University, Nashville

**Anna Molesworth**
Public Health Laboratory Service, London

**Professor Stephen T. McGarvey**
Brown University School of Medicine, Providence

**Professor Hyton B. Meire**
King's College Hospital, London

**Marilee Monnot**
University of Cambridge

**Professor William H. Mueller**
University of Texas

**Dr F. Müller**

**Lola Nathanail**
Save the Children Fund, London

**Dr Angus Nicoll**
Public Health Laboratory Service, London

**Dr N. G. Norgan**
Loughborough University of Technology

**Dr R. O'Rahilly**

**Dr C. Panter-Brick**
University of Durham

**Professor Paolo Parisi**
Instituto della Enciclopedia Italiana, Rome

**Dr Jay D. Pearson**
National Institute on Aging, Baltimore

**Dr J. M. Pell**
The Babraham Institute, Cambridge

**Dr Jean Peters**
University of Sheffield

**Dr Stewart Petersen**
University of Leicester

**Professor John A. Phillips**
Vanderbilt University, Nashville

**Professor E. M. E. Poskitt**
Medical Research Council Laboratories, The Gambia

**Professor Michael A. Preece**
University of London

**Professor Dr Michael B. Ranke**
Eberhard-Karls University, Tübingen

**Dr Martin Richards**
University of Cambridge

**Dr L. Rees**
Royal Free Hospital, London

**Dr Simon P. Robins**
Rowett Research Institute, Aberdeen

**Professor Derek Roberts**

**Aileen Robertson**
World Health Organization, Copenhagen

**Dr Marie-Françoise Rolland-Cachera**
Institut National de la Santé et de la Recherche Médicale, Paris

**Dr Roberto Rona**
St Thomas' Hospital, London

**Dr Rosanna Rooney**
Curtin University, Australia

**Dr Joan M. Round**
University of Birmingham

**Dr Soemilah Sastroamidjojo**
University of Indonesia

**Dr Lawrence M. Schell**
State University of New York at Albany

**Dr Theresa O. Scholl**
UMDNJ School of Medicine, New Jersey

**Dr Werner Schultinck**
University of Indonesia

**Dr John Seaman**
Save the Children Fund, London

**Professor Brian T. Shea**
Northwestern University Medical School, Chicago

**Professor P. S. Shetty**
University of London

**Professor Roger M. Siervogel**
Wright State University School of Medicine, Ohio

**Professor David H. Skuse**
University of London

**Professor James L. Smart**
St Mary's Hospital, Manchester

**Professor G. B. Spurr**
Medical College of Wisconsin

**Dr Virginia A. Stallings**
Children's Hospital of Philadelphia

**Professor A. T. Steegman Jr**
State University of New York at Buffalo

**Dr Robin Stevens**
University of Nottingham

**Professor William A. Stini**
University of Arizona

**Dr David W. Stock**
Pennsylvania State University

**Dr Simon S. Strickland**
University of London

**Professor C. Susanne**
Vrije Universiteit Brussel

**Professor James M. Tanner**

**Dr Jennifer L. Thompson**
University of Toronto

**Dr Bradford Towne**
Department of Community Health, Yellow Springs, Ohio

**Professor Stanley J. Ulijaszek**
Curtin University, Perth
University of Cambridge

**Dr Johannes D. Veldhuis**
University of Virginia

**Dr M. Vercauteren**
Vrije Universiteit Brussel

**Dr Mike Wailoo**
University of Leicester

**Professor J. A. Walker-Smith**
Royal Free Hospital, London

**Professor J. O. Warner**
University of Southampton School of Medicine

**Dr Tessa Webb**
University of Birmingham

**Professor Kenneth M. Weiss**
Pennsylvania State University

**Professor Robin Winter**
University of London

**Dr Carol M. Worthman**
Emory University, Atlanta

**Dr M. Zavattaro**
Vrije Universiteit Brussel

**Dr Babette S. Zemel**
Children's Hospital of Philadelphia

**Dr Zhiyong Zhao**
Pennsylvania State University

# General introduction

The formal study of human growth and development has over 300 years of history. As Pilbeam has said, 'humans are fascinated by humanity', and one of the most palpable aspects of humanity is the change in shape, form and function from helpless baby, to fully formed adult. From the point of view of population size, humans are a very successful species. Much of this success stems from the human design, which involves having a large body-size, a brain that is disproportionately large relative to that body-size, and an extended period of childhood. This gives us the metabolic and ecological advantages of greater size relative to other species, the ability to think our way through problems, and the time to develop behaviours through play and learning that will make us successful, social, problem-solving animals.

Another feature of our success as a species is the flexibility (plasticity) of growth in response to environmental or ecological stress. Thus, we are able to reduce or stop growth altogether in the face of famine, and to varying degrees catch up during times of plenty. The characterization of the patterns of human growth, the ways in which they differ from other species, vary within- and between-populations, and how they are regulated genetically and endocrinologically forms one theme of this encyclopedia.

Another theme is concerned with the cultural, behavioural, economic and environmental factors that are related to growth and their relationships with health and well-being. The context is global: humans as populations and individuals, past and present. Of the future? In geological time, perhaps extinction. In the sort of time that is measured in hours, days, weeks, years, decades and perhaps centuries, there is still much to learn and much to do.

The study and understanding of human growth in its many facets is important intellectually as well as practically, and while, for example, studies of the genetic mechanisms of growth are in their infancy, enough is perhaps known about patterns of growth to be able to use this as a diagnostic tool and in the monitoring of health of populations. However, relating one to the other is one of the continuing challenges in improving monitoring, screening and treatment procedures; furthermore, the understanding of these relationships and how they affect the well-being of populations is also a fundamental aim of human biologists. This encyclopedia is for the health professional, the biologist, anthropologist and educationist; indeed it is for everyone interested in growth and development.

*Stanley J. Ulijaszek, Francis E. Johnston, Michael A. Preece*

INTRODUCTION

# A brief history of the study of human growth

Looking back through the 300-year history of auxology, we can make out three distinct strands of interest, running sometimes separately, sometimes intertwined. We may call them the social, the medical/educational and the intellectual/scientific, after the impulses that gave rise to them: growth study in the service of social reform or social-economic history; growth study in the service of individual children, both in nurturing good development and in treating disorder; and growth study in the service of 'truth', to understand the form, the mechanisms and the evolution of the human growth curve.

The first time we hear the voice of genuine observation of children's growth – and a stentorian voice it is indeed – is in Guarinoni's immense scrap-book (see Appendix 1, 1), inveighing against the harm done to a boy's growth by anxiety at school, and commending the peasant girls of his beloved countryside for their late development (menarche at 17, 18 or even 20 years of age), in contrast to that of the too juicily nourished city girls. A little later we see the word *Anthropometria* for the first time, in a Paduan thesis describing the first known instrument for measuring humans.

Then, in 1729, the story really begins with the first textbook on human growth (see Appendix 1, 3). Comprehensive and detailed, it was dedicated to the irascible and giant-guardsman-collecting Frederick William I of Prussia, with the claim that the usefulness and efficacy of the subject was proved by the fact that the Crown Prince (later Frederick the Great), formerly abhorrently puny and delicate had been cured 'by nothing else than a speedy growth in length'. Stöller confused post-illness catch-up with the normal pubertal growth spurt, a confusion that persisted right through the time of Quetelet (see Appendix 1, 7).

Stöller's work was theoretical-clinical, after the manner of the time: he measured no children. But 25 years later, in 1754, the first thesis derived from a true growth study was presented in Halle (see Appendix 1, 3). Only in hindsight can we fully appreciate Jampert's long-lost work, with its growth tables, its clear comprehension of the difference between longitudinal and cross-sectional methods of study, and its understanding of the problems of variation and sampling. So here we see the first children of whom we have measurements: the first and the smallest.

Jampert does not say how he measured height – surprisingly, for he gives details about all the other measurements he took. Probably already the system for measuring army recruits (see p. 2) was in use. The sketch of the measuring was made in 1779 by no less a person than Johann Wolfgang von Goethe who, to his intense disgust, had to include recruiting as part of his duties for the Duke of Sachsen-Weimar. The method of measurement is absolutely modern, and much better than school measurements 200 years later: head correctly aligned with upward pressure upon it, block sliding down the back-board, shoes removed and a

**Opposite: Goethe's drawing of the measurement of recruits for the Duke of Sachsen-Weimar's army in 1779. Reproduced by courtesy of the Nationale Forschungs- und Gedenkstatten der Klassischen Deutschen Literatur in Weimar, Goethe Nationale-Museum.**

recorder recording. The earliest military data retrieved and analysed go back to 1741, in Norway, but evidently measurements were made, if not recorded, somewhat earlier than this. Already, in 1724, the Reverend Joseph Wasse had demonstrated that stature decreased during the day, and he had used that information to help would-be soldiers refused 'for being a little under the standard' persuade recruiting officers they would be acceptable in the morning.

## The artistic tradition

When it came to representing the growth of children, the artistic community was well ahead of the medical one. In ancient times, and again at the Renaissance, canons or rules of proportion were developed for the use of painters and sculptors. The Vitruvius-Leonardo da Vinci canon of man inscribed in a square and circle is universally remembered; later Albrecht Dürer (1528) gave rules for drawing peasants and burghers, and illustrated geometrical methods for transforming faces that became in D'Arcy Thompson's hands (see Appendix 1, 14) the method of transformed co-ordinates.

The first book dealing specifically with change of proportions during childhood, published in 1723, was by Bergmüller, Professor of Painting in Augsberg: it was also called *Anthropometria*,

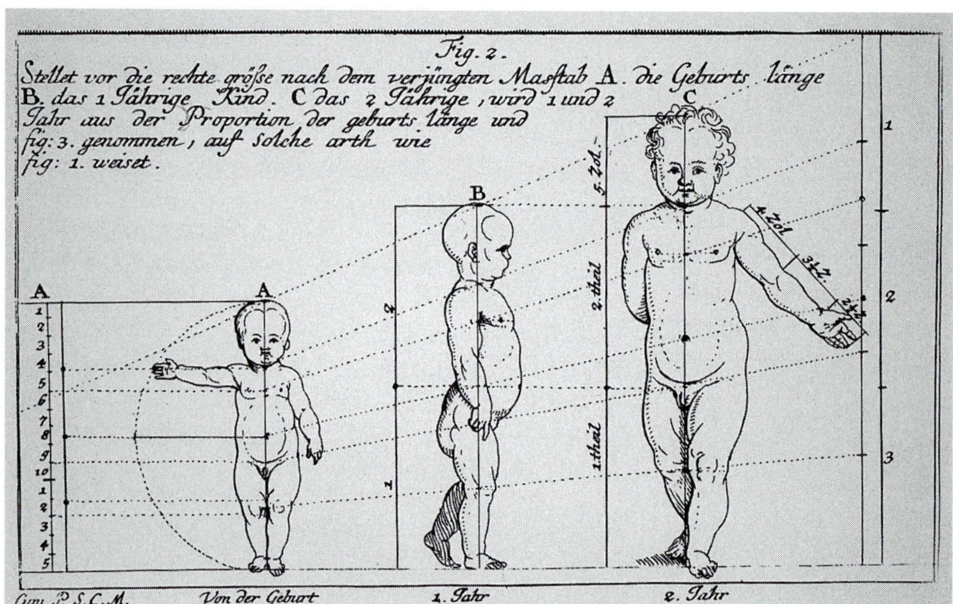


**Right: Bergmüller's (1723) drawings of a child at birth, age 1 and age 2. Courtesy of the Countway Library, Boston.**

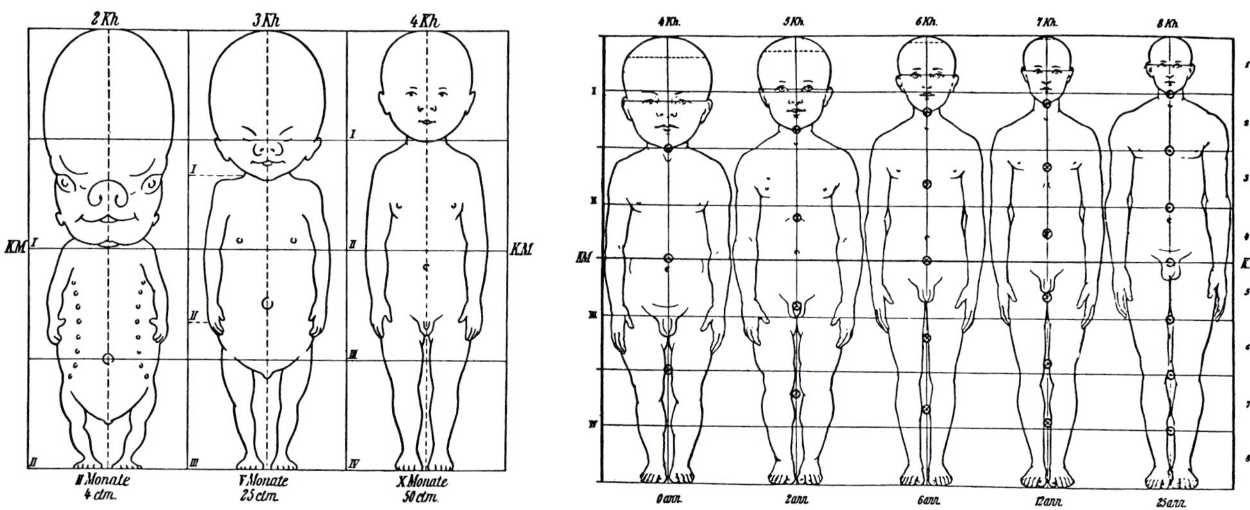


**Below: Stratz's (1909) diagram of change in body proportions during (a) fetal (left) and (b) post-natal (right) growth.**

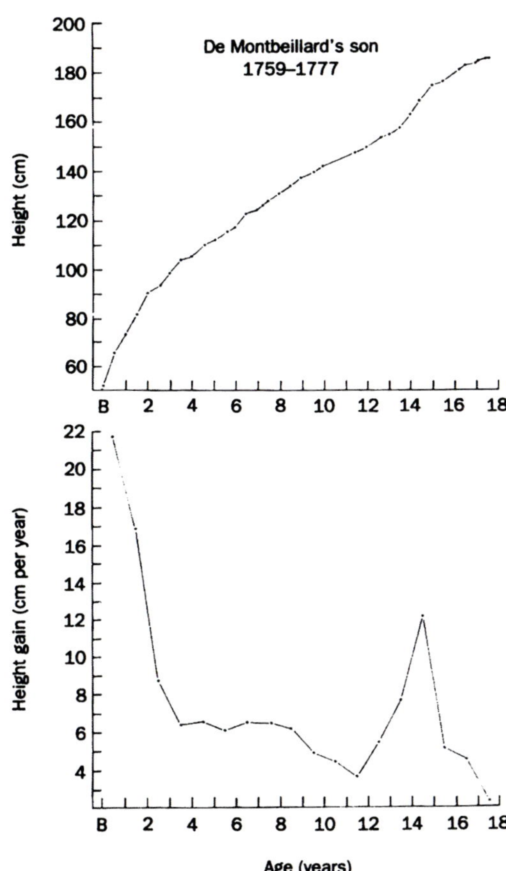


Height (above) and height velocity (below) curves of de Montbeillard's son.

like Elsholtz' thesis. It gave a strict geometrical rule which generated a height curve, though not a very naturalistic one; it omitted any pubertal spurt. Not till 100 years later did more accurate rules for children appear, enabling later artists to escape the criticism of Carl Heinrich Stratz in his Edwardian era textbook on child care that 'the cupids of Boucher, Fragonard and even Rubens and Titian are lusty little men-about-town and certainly no infants' (1903).

## The first longitudinal growth study

Very few growth studies have been made in the spirit of pure inquiry to establish the 'truth' of how things are. But the first ever longitudinal study, instigated by Buffon (see Appendix 1, 5) and made by de Montbeillard on his son in the years 1759–77 was just such a one. It was made in the spirit of the times; the years of the European Enlightenment and the French Encyclopédists. It established the existence of the pubertal growth spurt, and seasonal changes in growth-rate, and confirmed the occurrence of shrinkage during the day. It is a record that has never been surpassed, and seldom equalled, in elegance and presumed accuracy.

## Quetelet's fitted growth curve

Sixty years and two major Revolutions later science was firmly established and astronomers and mathematicians ruled the roost. Adolphe Quetelet (see Appendix 1, 7), a star of the first magnitude, made two cross-sectional studies of Belgian children, in 1831 and 1832. His interest was in the form of the human growth curve – but as an example of Natural Beauty rather than Natural History. He fitted, for the first time, a mathematical curve to the empirical values, but the elegance of its shape disguised from him its falsity. It showed no pubertal spurt, and such was his authority that it took nearly 50 years for the existence of the spurt to be re-established, by the longitudinal studies of Bowditch (see Appendix 1, 10) in Boston and Pagliani (see Appendix 1, 11) in Turin.

## Growth as a mirror of the conditions of society: the beginning of auxological epidemiology

In the early part of the 19th century a new tradition of growth studies appeared, born of the reaction of humanitarians to the appalling conditions of the poor and their children. This activity has been called *auxological epidemiology*, 'the use of growth data to search out, and later to define suboptimal conditions of health' (Tanner, 1981). It monitors *groups* of children rather than individuals, which distinguishes it from the medical/educational tradition. The upshot of the monitoring is not individual treatment but public health measures or even political reforms, targeted at those groups in a population whose diminished growth betrays the signs of neglect: neglect, that is, of education, amenity, housing, work, and money.

Villermé (see Appendix 1, 6) and Chadwick (see Appendix 1, 8) were the pioneers in this field. The illustration of mean height of boys working in factories in the 1830s is that of Chadwick plotted on modern British standards; they were smaller than practically any third-world population nowadays.

In the 1870s Roberts and Galton (see Appendix 1, 9) were able to contrast the growth of working children with those being educated in private schools. Heights of sons of non-manual workers were greater than those of manual workers. A little later Key (1885) in Sweden showed a similar difference between children in elementary schools (mostly working-class) and those in schools providing a longer period of education (mostly upper middle-class).

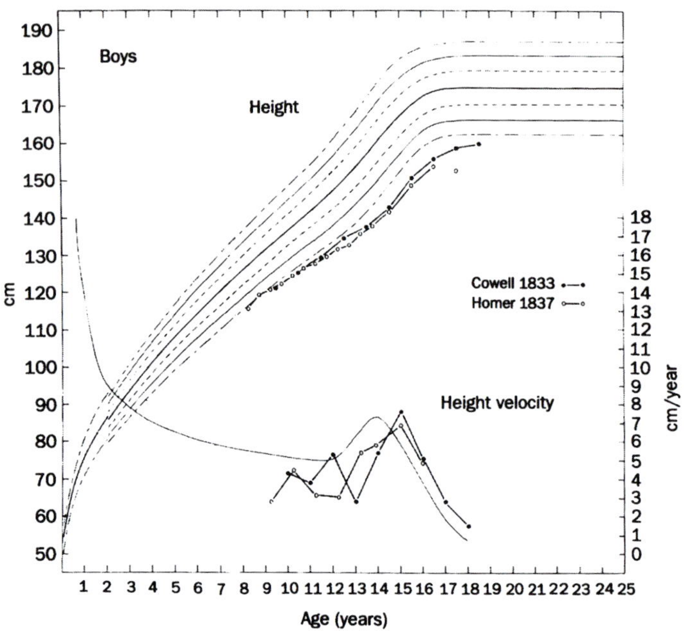


**Mean height of boys working in factories in the Manchester/Leeds area in 1833 and 1837.**

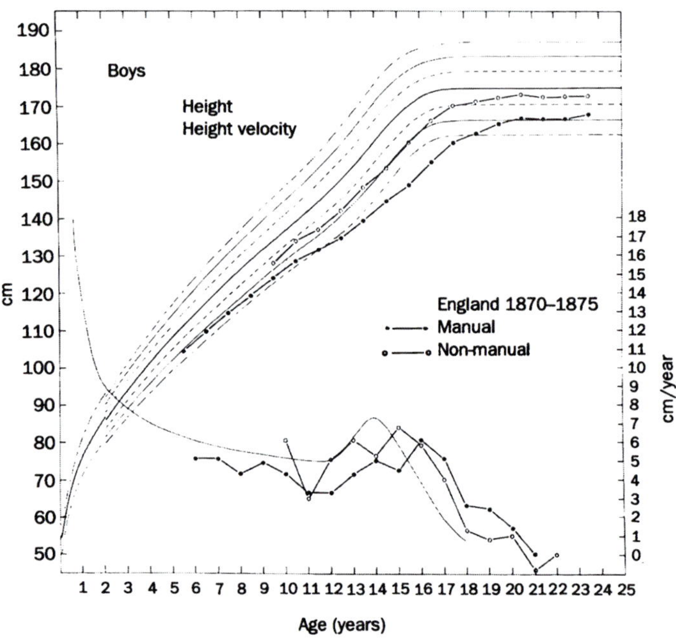


**Heights of sons of manual workers and non-manual workers, mostly professional class, in about 1870 in England.**

Social class differences have been much investigated since. They are not inevitable, as Lindgren showed in 1976: in Sweden no differences were present in children born in 1955. But worsening conditions later caused the differences to return (see Part 11).

From the 1870s to the present there has been an uninterrupted history of child growth surveys. Bowditch and Pagliani both made large-scale surveys of schoolchildren and soon school committees everywhere demanded surveys to monitor school conditions and especially to investigate the supposed effects of 'over-pressure' of work in the factory-like secondary schools of the day. In 1893 William Porter, later a colleague of Bowditch, showed in St Louis that pupils who achieved above average grades were taller than pupils of identical age who received below average grades. He thus started a series of investigations on social mobility, social class, height and achievement, which still continue.

Comprehensive surveys of a country's whole population of children, however, or even of clearly defined parts of it, are of rather recent origin. Periodical cross-sectional surveys have been made in, for example Holland in 1955, 1965 and 1980; Czechoslovakia in 1951, 1961, 1971 and 1981; Cuba in 1972–4; Hungary in 1981–5, and by the United States National Center for Health Statistics intermittently from the 1960s to the present (HES II and III and NHANES I, II and III). In the United Kingdom, the alternative system of continuous *surveillance* was set up in 1972. Such a permanently operating system has the advantage that it can respond rapidly to social changes by altering its target population: currently schoolchildren aged 5 to 11 are monitored, and changes in sampling were introduced in the 1980s so that the representation of groups believed to be most disadvantaged and at risk of growth failure (inner city and ethnic minorities) was up-weighted. In a few countries, chiefly Scandinavian, the routine measurements taken by school authorities have been used for surveillance purposes; but in most either such measurements were too seldom and/or too badly done to be useful.

On an international scale, agencies such as the World Bank began in the 1970s to use height of children as a proxy for national well-being, and change of height over time as a criterion of the success of economic aid. As part of the International Biological Programme of 1967–72, statistics on growth and maturation, many unpublished, were collected by Eveleth and Tanner and published as *Worldwide Variation in Human Growth* (1976, second edition 1990). From these surveys, and related data on age at menarche, knowledge of the so-called secular trend (**11.3**) (or shift) towards larger size and faster maturation accumulated and began to receive increasing attention from about the time of the Second World War. War-time measurements, in the First World War in Germany and in the Second World War in Norway under Nazi occupation, showed the trend could be negative under bad conditions and dispelled the idea, once held, that the trend was somehow connected with global climate changes or long-term shifts in world ecology. The largest positive secular shift was seen in the period 1950–80 in Japan, associated with swift and intensive economic development. Here, as also in China and perhaps generally, the trend was associated with increased leg growth rather than growth of trunk (Tanner, Hayashi et al., 1982), reminding one of Dürer's difference between peasants and burghers.

## The human growth curve

To define the human growth curve, to study *growth* itself, longitudinal studies are a *sine qua non*; the curves obtained by fitting cross-sectional means are different in shape from individuals' curves, something not universally realized till the time of Boas (see Appendix 1, 13), Baldwin (see Appendix 1, 15) and Shuttleworth (see Appendix 1, 16) in the 1930s. The figure below is from Tanner (1962) but drawn exactly after the manner of Shuttleworth (1937) and Boas (1930). The difference between longitudinal and cross-sectional curves has implications for growth standards, but here we are concerned with the growth curve as a sign of underlying processes and as a characteristic subject to evolution.

As soon as numbers of individual growth curves began to be examined, and particularly *velocity* curves, as advocated by D'Arcy Thompson (see Appendix 1, 14), it was evident some children reached their 'age of maximal growth' (Boas) or 'peak height velocity' (Tanner) earlier than others. Boas coined the expression *tempo of growth*, drawing the analogy with classical music: some children are marked *allegro* (fast), others *lento* (slow), and Boas' empirical finding was that their final adult height did not, on average, differ. Thus there were two independent parameters of growth – size and maturation rate.

There were two great periods of longitudinal studies, the first in America *c.*1930–70, and the second in Europe *c.*1950–80. Before these, however, there were a number of individual enthusiasts for using serially taken measurements to monitor individuals' health and school progress. Chief amongst them was Paul Godin (see Appendix 1, 12) in France, who in 1919 introduced the word *auxology*, which soon came into common usage in the Latin languages and was introduced to the Anglo-Saxon literature by Tanner (see Appendix 1, 22) and others in the 1970s. The International Association of Human Auxologists was founded in 1977 and since then International Congresses of Auxology have been held every 3 years. The proceedings of the Fifth Congress, held in 1988 in Exeter, England, entitled *Auxology 88: perspectives in the*

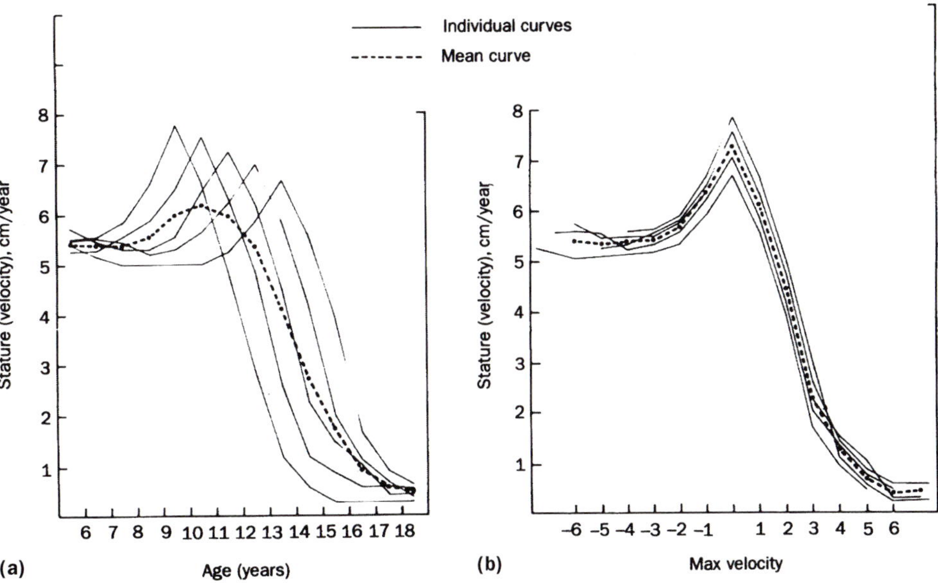


(a) Relation between individual velocity curves and mean increment curve (heavy broken line) during the pubertal growth spurt. (b) The same individual curves plotted against years before and after age of maximum growth velocity. Mean again shown by the heavy broken line.

*science of growth and development* (Tanner, 1989) were deliberately designed, through invited and contributed papers, to give a complete conspectus of the scope and content of auxology at that time. It is thus a valuable historical source.

## The North American longitudinal studies

In the 1920s a powerful child welfare movement in North America spawned a series of studies which shaped the whole pattern of human auxology in the period 1935–55. All were devoted to the 'whole child', though some had educationists as directors (Baldwin, Dearborn), some psychologists (Bayley, McFarlane), some medical men (Todd, Washburn, Stuart, Sontag) and one, in Philadelphia, a physical anthropologist (Krogman, see Appendix 1, 18). Each study had its special characteristics and its own particular triumphs. At Iowa, the first to be established, there was Baldwin and later Meredith (see Appendix 1, 15) concentrating on physical measurements and their analysis. At the Brush Foundation Todd developed the first *Atlas of Skeletal Maturation*, later the *Greulich–Pyle Atlas*. At Boston efforts, still continuing, were made to examine their devoted cohort of subjects into middle- and old age. At Denver two generations of pediatricians were taught growth and development in a hands-on mode, and the mathematical fitting of growth curves was initiated. Of the three California studies, one was oriented towards pubertal physiological change (something almost unique at the time), one to behavioural development, and the third, the Berkeley Growth Study, was a birth-to-maturity cohort study recruited in 1928 by Nancy Bayley (see Appendix 1, 17), which produced landmark analyses both of physical growth and intellectual development.

The largest and longest-lived of the American studies, however, was the Fels Research Institute. The history and results of this study have been admirably described by Roche (1992). Begun in 1929, there were about 600 participants in the first generation, then some 350 who were offspring of a Fels parent, and lastly 90 enrolled 'grandchildren' (in 1991). Privately funded and located on the campus of a Midwestern progressive liberal arts college it boasted in its heyday separate departments of physical growth, biochemistry, social psychology and psychophysiology, all housed in a large 80-room building, an Institute in itself. It was directed first (1929–70) by one of its creators, Lester Sontag, part pediatrician, part psychiatrist, and second (1970–8) by Frank Falkner (see Appendix 1, 21), who before emigrating to the States had set up both the London and Paris teams of the International Children's Centre studies (see below). Falkner brought his experience as a pediatrician and an auxologist, as well as administrator, to the Fels. He was editor, with Tanner, of the massive three-volume *Human Growth: a Comprehensive Treatise* (two editions 1978 and 1986). In 1977 the funding and location changed and the Fels Longitudinal Study became part of Wright State University Department of Community Health. Beside its directors, the Fels was blessed with three successive, outstanding physical anthropologists: Earle Reynolds, Stanley Garn (see Appendix 1, 24) and Alex Roche (see Appendix 1, 23), around whose numerous papers much of American auxology coalesced.

### NORTH AMERICAN LONGITUDINAL GROWTH STUDIES

**University of Iowa Child Welfare Research Station**
1917–70 Baldwin, Meredith

**Harvard Growth Study (School of Education)**
1922–34 Dearborn, Shuttleworth

**University of California Institute of Child Welfare**
- Berkley Growth Study 1928–54 Bayley
- Child Guidance Study 1930–50 MacFarlane
- Adolescent Growth Study 1932–39 Stolz, Jones, Shock

**Harvard School of Public Health Study**
1929–54 Stuart, Valadian

**Fels Research Institute. Yellow Springs**
1929– Sontag, Reynolds, Garn, Roche, Falkner

**Child Research Council, University of Colorado**
1930–71 Washburn, Maresh, Deming

**Brush Foundation. Western Reserve University**
1930–71 Todd, Greulich, Broadbent

**Philadelphia Center for Research in Growth**
1948– Krogman, Johnston

*James M. Tanner*

## The European longitudinal studies

Though most of the European longitudinal studies had an element of psychometry and some study of behaviour, their primary orientation was towards medicine. Effectively the first was the *Harpenden Growth Study*, located in a children's home, directed by Tanner (see Appendix 1, 22), with Whitehouse as anthropometrist. Before setting it up Tanner had visited all the major American studies, and learned the measuring techniques from Meredith. When the International Children's Centre (ICC) Co-ordinated Studies began, first in London under Frank Falkner, initiated by Alan Moncrieff, Professor of Child Health, they all adopted the Harpenden techniques. Thus there was a close historical link between the European studies and the American ones.

Whitehouse designed a new range of anthropometric instruments for Harpenden, and the Harpenden rack-and-pinion wall-mounted stadiometer became the world standard. A new system for assessing skeletal maturity, the so-called Tanner–Whitehouse method (1975, 1983), was introduced. Also at Harpenden the Tanner pubertal stages (**1.10**), now used worldwide, were developed (1955), drawing on the previous work of Earle Reynolds at Fels. During puberty the subjects were seen every 3 months, and this permitted analyses of the pubertal staging data and also rather accurate curve-fitting to anthropometric measurements. This made possible the construction of the first practical longitudinal, or course-of-growth, reference curves and charts. Charts of growth velocity as well as 'distance' were produced and, following Nancy Bayley's (see Appendix 1, 17) (1956) original suggestion, specific curves for early, mid- and late maturers.

The difference between cross-sectional population standards suitable for screening and for surveys, and longitudinal course-of-growth standards suitable for following individuals in school or clinic was a constant theme of Tanner and his colleagues. From 1956 the Harpenden Study and the ICC London Study were part of the Department of Growth and Development of the Institute of Child Health. Set up by the Nuffield Foundation, and self-consciously a mini-Fels, it also had sections of infant ethology, chemical endocrinology, experimental animal auxology, and a Growth Disorder Clinic located at The Hospital for Sick Children. It attracted a steady stream of post-doctoral students from overseas (Hauspie, Lindgren and Falkner, 1995) but, less lucky than Fels, it was dissolved entirely in the early 1990s.

The biometrical expertise necessary for this work was supplied from 1950 onwards by Michael Healy (see Appendix 1, 20) then working under Frank Yates in the famous statistical department of Rothamsted Experimental Station, itself located in Harpenden. Healy's abiding interest in the manipulation of growth data brought about a considerable transformation of human auxology. One of his students, Harvey Goldstein, working in Tanner's department, became the statistical adviser and teacher to the whole group of ICC studies.

**Structural average curves of height velocity of the 25% ultimately tallest and 25% ultimately shortest (dashed) subjects of Zürich First Longitudinal Growth Study. Boys above, girls below. From Gasser et al. (1990). The height difference between these two groups arises in the period age 2 years to puberty. This is the generalized version of de Montbeillard's son.**

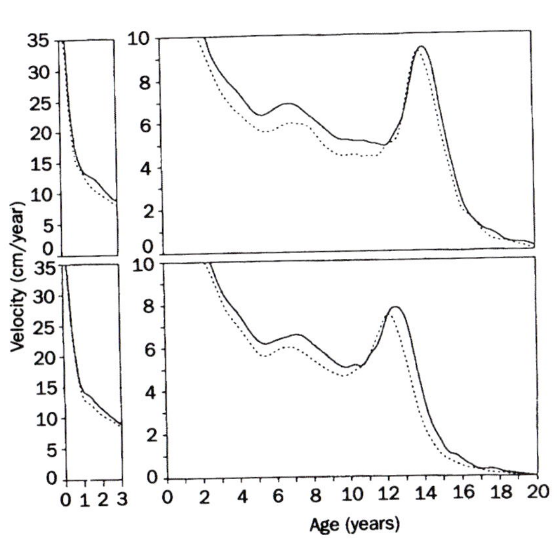


The ICC studies were all of birth-to-maturity cohorts. All were co-ordinated by Natalie Masse, director of the ICC, so that exactly the same techniques were used in every place. They were to serve as replicates, since no single longitudinal study boasted more than 400 children, reducing to about half that number followed to maturity. Every 18 months or so the workers of all studies met to compare techniques and results (Falkner, 1960).

Like the American studies, each had its particular interests and triumphs, and in 1977 an invaluable bibliography of all the teams' publications was published (ICC, 1977). The triumph of the ICC itself was that the meetings increasingly drew researchers from other European countries and stimulated them to embark on similar studies.

The medical orientation of these studies meant that there was much greater feed-back into pediatric departments than in most of the American ones. The prime figure in this was Andrea Prader (see

**EUROPEAN LONGITUDINAL GROWTH STUDIES**

**Oxford Child Health Survey**
1944–64 Ryle, Stewart, Acheson

**Harpenden Growth Study**
1948–71 Tanner, Whitehouse, Marshall, Hughes, Cameron

**International Children's Centre Co-ordinated Studies**
- **London**
  1949–69 Falkner, Tanner
- **Paris**
  1953–75 Masse, Sempé
- **Zürich**
  1954–80 Prader, Gasser
- **Stockholm**
  1955–75 Karlberg, Taranger
- **Brussels**
  1955–75 Graffar
- **Louisville**
  1962– Falkner, Wilson
- **Dakar**
  1954–75 Sénécal, Massé

**Stockholm School of Education Study**
1954–66 Ljung, Lindgren

**Helsinki Growth Study**
1955–75 Backström-Järvinen

**Prague Growth Study**
1956–81 Prokopec

**Brno Growth Study**
1961–81 Bouchalova

**Lublin Growth Study**
1964–80 Chrzastek-Spruch

**Wroclaw Growth Study**
1961–71 Bielicki

**Budapest Growth Study**
1970–88 Eiben

**Edinburgh Growth Study**
1972–95 Ratcliffe

*James M. Tanner*

Appendix 1, 19), a greatly respected pediatric endocrinologist, Professor of Pediatrics in Zürich and director of the Zürich longitudinal study, itself perhaps the most successful of the ICC group.

## Economic history

One of the most striking recent developments in auxology has been its impact on economic history and even economic theory. The chief impacting instrument, the Chicago economist Robert Fogel, has recounted the history of the collision. He and his colleagues found thousands of records of heights of children of Black slaves in American archives, and in 1976, in collaboration with Tanner, used the height curve to estimate average age of menarche, a statistic they needed for economic and demographic reasons. Once started, these diggers in government archives unearthed more data on growth in the 19th and even 18th centuries than auxologists dreamed existed. Here was hard data throwing light on the old politico-historical debate as to whether the Industrial Revolution first lowered the conditions of life of the working class, or whether, to the contrary, it provided a rescue by urban immigration from the Malthusian imperative in the countryside, over-population producing starvation.

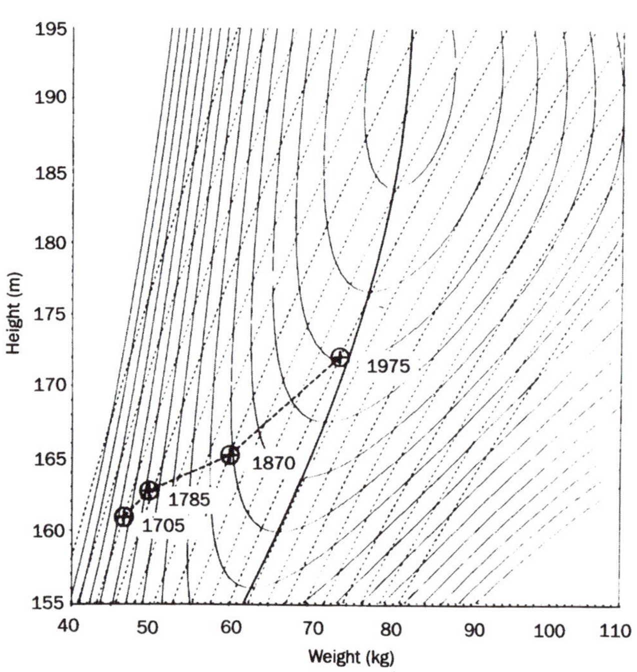


**Iso-mortality curves of relative risks for height and weight among Norwegian men aged 50 to 64 years. Average values of height and weight for the French population at four dates plotted, giving at each date the expected mortality rate of the population. From Fogel (1994).**

This is one of the fastest growing areas linked to auxology at present. In 1984 Waaler published data on Norwegians showing that at all ages up to 80 years short men and women had a higher overall mortality rate than tall men and women. Height was a proxy for living conditions, especially in the infant years, and evidently poor conditions had long-lasting effects. Those with high, or very low weight-for-height also had a higher rate of mortality. Fogel (1994) developed these data into so-called Waaler surfaces, in which contours representing the mortality rate for each combination of height and weight were plotted. The combination of height and weight giving the lowest rate at ages 50 to 64 is shown by the continuous solid line. Fogel has used such surfaces in historical stud-

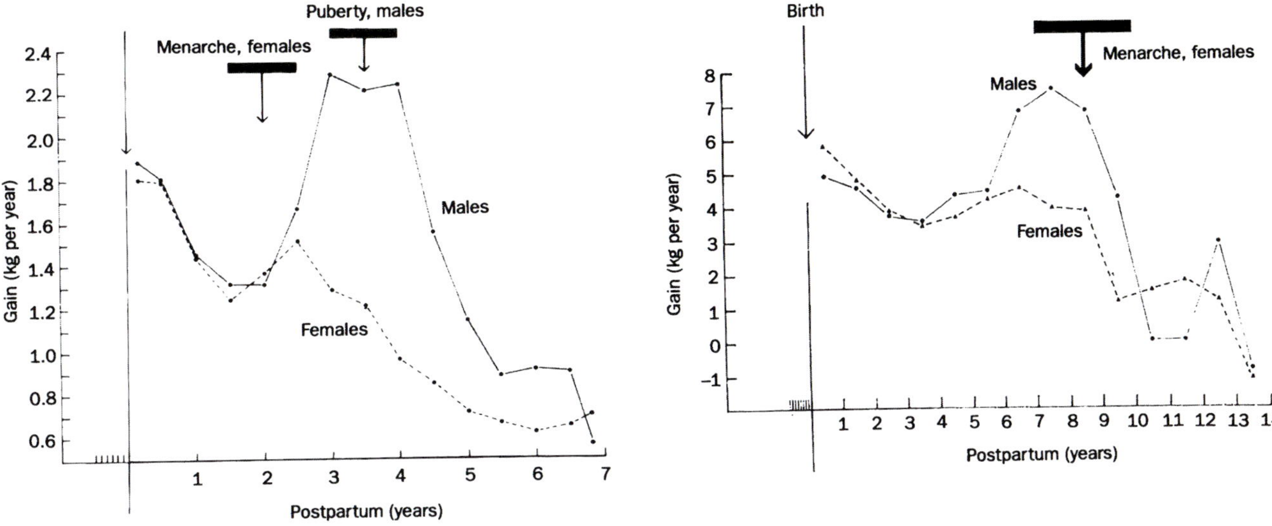


Weight-velocity curves for rhesus (above left), chimpanzee (above right) and humans (right). From Tanner (1962).

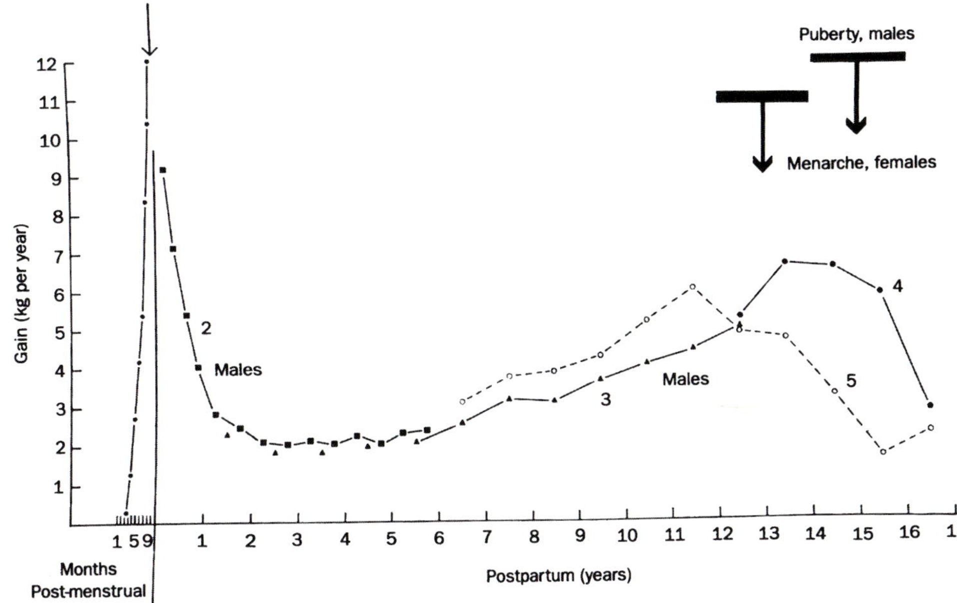


ies: the change in mortality rate in France, for example, until the very recent era, can be closely associated with the increase in height and weight. Further work in this field can be accessed through the collection of papers edited by Komlos (1994, 1995).

## The evolution of the human growth curve

Lastly, we return to the intellectual/scientific strand of growth study. On the mechanistic side, this means the physiology of growth, but this would take us beyond the space allocated to this brief history. Suffice to say that the discovery that growth hormone was species-specific and that growth hormone deficiency could be successfully treated (in 1958) gave an enormous, if temporary, boost to the study of auxology, as clinicians vied with each other to demonstrate successful diagnoses and outcomes. Methods for height prediction burgeoned, the assessment of skeletal maturity was briefly automated, course-of-growth standards became *de rigeur*. By 1990 the effect had largely worn off however, and most clinicians were back to assessing individuals' velocities by looking at cross-sectional, height-attained charts.

But since Darwin, the scientific impulse has centred on the light that growth studies of humankind can shed on the mechanism and history of evolution (**2.3**). The figures on the previous page show the growth curves of weight of Rhesus, Chimpanzee and Humans. They are all similar, but the interval between infancy and puberty has progressively lengthened. The physiology was determined in the 1980s: neurones of the gonadotrophin-releasing hormone (GnRH) nucleus, secreting at birth, are turned off soon after, and stay turned off for a short time in Rhesus, longer in Chimpanzee, longest in Man. Evolution proceeds by alterations in timing and intensity of growth and development. Changes in adult form are produced by increased growth at particular sites (the Gibbon's limbs come to mind) and the cellular physiology of this is in process of becoming clear. Change in timing, or heterochronic development, is crucial: one part, such as the GnRH nucleus, slowing up so that another, the cognitive part of the brain, has the possibility of developing more than before. These are the ways of evolution, and it is along this path that the history of auxology will perhaps proceed furthest.

*James M. Tanner*

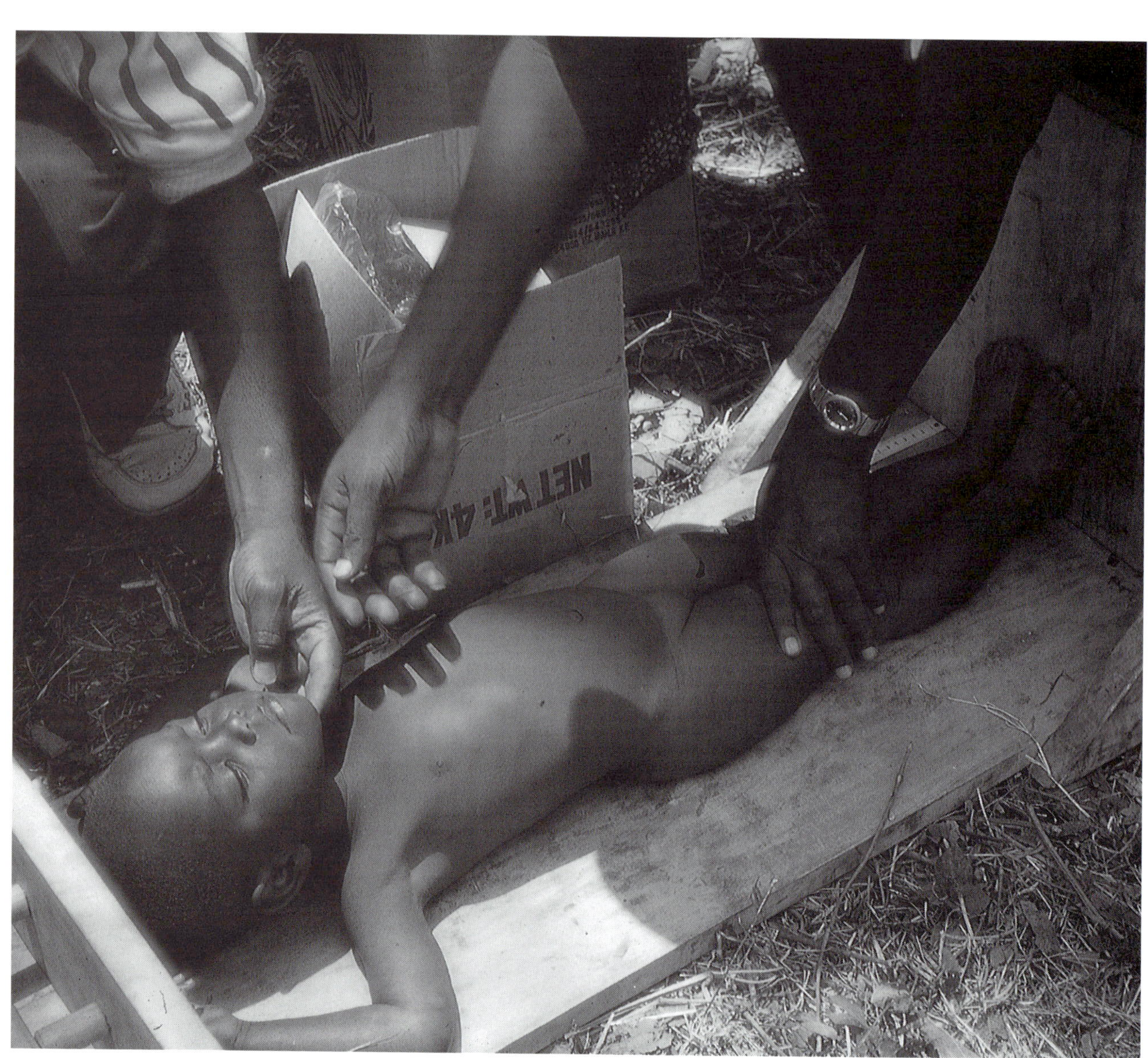
NET WT: 4K

## PART ONE

# Measurement and assessment

The measurement and assessment of human growth and development involves monitoring structure, and structure as a reflection of function, from conception to maturity. In other words, the interpretation of the measurements taken to describe human growth can be simply descriptive, such as height or weight at particular ages, or functional, such as implying that the structure denotes a functional ability. Secondary sexual development, for instance, is assessed through the appearance of specific characteristics, such as breasts, genitalia and pubic hair, but the interpretation of completed sexual maturation is that the individual has the functional ability to procreate. The main problem facing those wishing to assess human growth and development is that their subject, the child, is in a constant state of change. Thus it is impossible to repeat a measurement or assessment with any major time-lag and expect to obtain the same result. The vast majority of measurements are therefore cross-sectional in nature, i.e., they describe the child at one particular moment in time. Repeated measurements describe the changes that occur during growth and thus provide a longitudinal appraisal of change over time. Both approaches are essential to properly understand the process of human growth and development, and both approaches involve an understanding of the nature of variation both within and between subjects. In addition, human growth and development is assessed in a variety of contexts. The individual may be assessed because of concern over abnormal growth, while the sample or population may be measured because a comparison is required with other samples or populations to examine the differences in morphology between groups. This section details not only the techniques, such as anthropometry and radiography, used to assess morphology or structure, but also the research designs and statistical methods needed to appropriately collect and analyse such data, such as cross-sectional and longitudinal designs and the rationale that forms the basis for growth references.

Measurement and assessment naturally begin from conception with the staging of embryonic development (**1.1**). At about 8 to 10 post-fertilization weeks, when the embryo is 'recognizably human', it is described as a fetus and ultrasonic evaluation is carried out while it is within the protective environment of the uterus. Prior to the availability of ultrasonic evaluation, data on fetal growth was based on examination of the results of spontaneous, 'social' and clinical abortions or still-births. As a result of damage to the fetus during such procedures, data on fetal growth was relatively inaccurate. However, ultrasonic measurement is a relatively

**Opposite: Measuring a child with a portable device. Photograph by Lola Nathanail.**

non-invasive process applied to the living fetus (**1.2**) and thus has been able to provide accurate information not only on total body growth but also on the growth of different structures, such as the kidney, spleen, stomach and bowel. Ultrasound methods are mainly two-dimensional and are thus hampered by the technological problems involved in describing a three-dimensional structure in two dimensions. However, three-dimensional analysis is becoming a technical possibility that should expand our knowledge of fetal growth in the near future. The embryo can be identified by ultrasound at 4 post-conception weeks, and such methods can be used to assess prenatal age (**1.3**). While crown–rump length is the initial measurement, by 10 post-conception weeks it is possible to measure both head-size and the length of limb bones. The head, abdomen and femur form the focus for clinically monitoring fetal growth during mid- and late pregnancy. These measurements are applied to growth charts which allow the normality of the pattern of human growth to be established prior to birth, and allow intervention methods to be applied if that pattern demonstrates abnormalities.

Body measurement is the basis for the vast majority of available information on human growth (**1.4**). The techniques are relatively simple and have changed little in the last century, but the availability of more sophisticated electronic equipment that incorporates high degrees of accuracy with ease of use is allowing auxologists to better understand the growth process. Fundamental to that understanding is the realization that measurement error impedes the accurate description of that process and thus the statistical evaluation of measurement error is an important consideration. While the concept of 'reliability' has been accepted by auxologists, its mathematical or statistical representation has been the source of considerable debate. This section provides equations to calculate the most commonly used statistics of the Technical Error of Measurement (TEM) and the Reliability Coefficient (R). It must be appreciated that neither of these statistics incorporates a correction for bias – the systematic over- or under-estimation of a dimension. Bias can be calculated form the average deviation between sets of repeated measurements, i.e., the sum of deviations divided by the number of subjects. If this is not zero, but is consistently positive or negative, then bias exists and must be considered during subsequent data analysis.

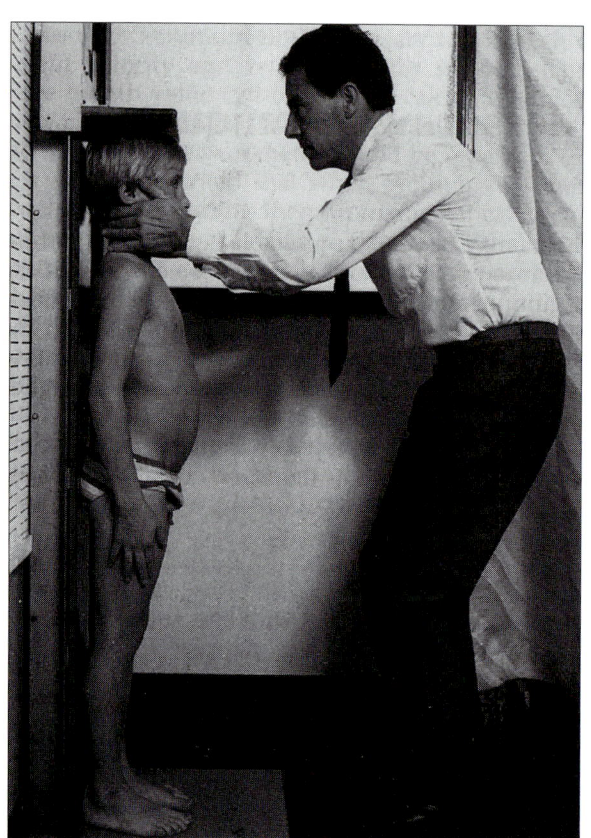

**Measurement of stature with a stadiometer. Photograph by permission of Castlemead.**

The majority of anthropometric measurements are undertaken on subjects who can assume a standing posture. However, the very young and very old are often unable to maintain the positions required for normal anthropometric measurement and thus techniques for recumbent anthropometry have been developed and are discussed in this section. At other times, and particularly when dealing with subjects suffering from skeletal disorders, the normally fundamental measurement of stature is not possible. At these times dimensions of limb segments assume greater importance, and growth charts to assess normality have been developed. Imaging techniques (**1.5**) allow surface morphology to be examined either descriptively, or to be used in clinical diagnosis.

Beneath the surface of the skin the composition of the body, in terms of the relative amounts of fat and lean tissue, changes during growth (**1.6**). The identification of these changes has important implications for physiology and homeostasis. Direct analysis of body composition has seldom been undertaken because of the problems involved in analysing post-mortem or preserved material and thus indirect methods of analysing body composition in the living have been developed. These methods analyse the body at different levels of complexity; from the level of tissues (fat, muscle, and bone) down to the level of basic elements (carbon, oxygen, and hydrogen) and provide important information on changes during growth and maturation, and the relationship of those changes to maturation, diet, heredity, environment, and disease. Biochemical markers of collagen metabolism (**1.7**) are recently developed research tools which allow short-term bone growth and bone turnover to be studied.

Body composition, in terms of the relative amounts of fat, muscle, and bone apparent in anterio-posterior radiographs of the limbs, has also been assessed, but the invasive nature of this approach has resulted in radiographs being used only for the assessment of skeletal maturity in children suspected of having growth disorders. Most countries have made radiographic assessment for research purposes on normal children illegal, although data collected prior to such legislation was fundamental to the understanding of skeletal maturation, and to the consequent development of techniques to assess skeletal or bone 'age'. These 'atlas' and 'bone-specific scoring' techniques (**1.8**) are an essential part of the clinician's armamentarium to assess the relationship between growth and maturation in a child presenting with a suspected growth disorder. Radiographic techniques can also be used to assess dental development (**1.9**) but, once again, the invasive nature of this approach has meant that for research purposes ages of tooth eruption are the most common assessment method.

Pubertal, or secondary sexual development, is most easily assessed using the staging techniques developed in America and England in the 1960s. In a clinical setting such techniques form the basis of physical examination (**1.10**), but in a research setting it is becoming more common to use a 'self-assessment' technique in which adolescents, aided by suitable illustrations, appraise their own development. Such self-assessment techniques are becoming used more commonly because the results so obtained have been shown to be not significantly different from objective observer assessments and because they avoid the problem of subject compliance. Obesity is a health problem in the industrialized nations and is becoming increasingly prevalent in modernizing populations. From a clinical and public health perspective, the early detection of obesity is of considerable importance, and the use of decrease in fatness from the age of 1 year across childhood is useful in the prediction of fatness-onset in older children (**1.11**).

At the population or national level, it has been suggested that growth differentials can be used to determine the extent of social and economic stratification (**1.12**). This may be of public health significance if it is found that such differences also associate with differentials in health parameters.

Growth studies require appropriate sampling procedures, designs and analytical methods (**1.13** and **1.14**) according to the research questions in hand. Furthermore, specific markers of the development process, such as the onset of puberty by estimation of age at menarche (**1.15**), can give population-level indicators, which require different types of analysis.

Many countries have undertaken large-scale cross-sectional and longitudinal growth surveys to assess the health and well-being of their children. The resulting data have often been used to make charts of height and weight for age and sex (**1.16**) that are thought to be a more sensitive indicator of the growth status of local children than international reference charts. The creation of such reference charts is often difficult, and each dimension being investigated

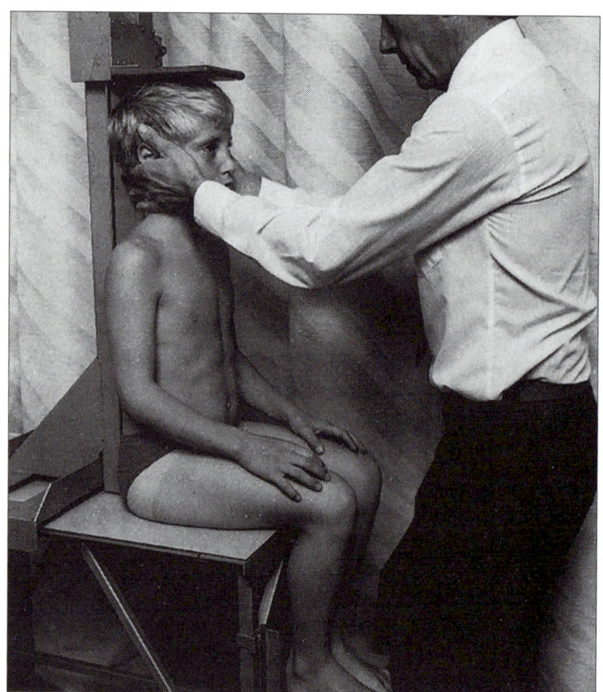

**Measurement of sitting-height. Photograph by permission of Castlemead.**

presents, to some extent, a different growth curve and thus interpretation. The use of such charts is widespread in both the developed and developing world (**1.17**). An important consideration is the difference between a growth *reference* and a growth *standard*. References reflect the growth patterns of a particular sample of children that have been selected as a reference group because they have particular characteristics.

Ideally, and to reflect 'normal' growth, one would require a source sample of healthy, well-nourished children. The American National Center for Health Statistics references, for example, have a source sample of children from the Fels Longitudinal Growth Study to represent growth during infancy, and a nationally representative sample of children from throughout America to represent growth from childhood to adulthood. These children are nationally representative in that they do not reflect the growth of any particular group. The use of these charts in a country other than America would be as references for the comparison of groups of children. The British Clinical Longitudinal Standards, on the other hand, include source samples of British children followed longitudinally, and within the United Kingdom are viewed as standards that accurately reflect the growth to be expected from British children. Because of their longitudinal source sample they are used as standards to assess the growth of individuals rather than groups. Fundamental to such charts is the appreciation that they are sample-dependent, i.e., that their accuracy in reflecting the growth of any child or group of children depends upon the proximity of that individual or group to the source sample on which the charts were based. Thus the identification of an appropriate reference sample is vitally important. Once such a sample has been identified and the dimensions of interest appropriately measured, the statistical processes required to display the data in a format that can be used with equal accuracy and confidence by both the expert research auxologist and the less expert field-worker must be applied to the data. A variety of statistical approaches are possible, from the less sophisticated averaging of data at each age to more complex curve-fitting procedures.

An additional consideration is the variables that should be used in different contexts to assess growth status and growth-rate. Traditionally height- and weight-for-age have been the dimensions of major interest. But in the context of assessing nutritional status, weight-for-height, as a reflection of wasting, assumes importance. The growth of a child from time $t$ to time $t + 1$ is to some extent dependent on the growth status of the child at time t. In addition, the growth status of the child is dependent on the genetic contribution of his/her parents. Charts which incorporate these concepts are known as *conditional* references and, while few have been developed to date, their usefulness in certain clinical contexts is important. The use of growth references and standards takes place in a variety of contexts such as developed and developing nations during times of major social change and environmental stress; the results of such screening and surveillance procedures are vital in assessing the impact of social change on the health and well-being of children.

*Noel Cameron*

# Embryonic staging

The need for standardized stages in the embryonic development of various organisms for the purpose of accurate description of normal development and also for use in experimental work has long been recognized, particularly for amphibian and chick embryos. Embryos can be arranged as a series of selected individual examples, numbered in the presumed order of their development. In such a series, however, any given embryo may be advanced in one respect while being retarded in another, and hence it may prove impossible to match a new embryo exactly with any one of the illustrated norms. This disadvantage is eliminated by using staging rather than seriation.

## Embryonic stages

An embryonic stage is an arbitrarily cut section through the developmental time axis, equivalent to a frame taken from a ciné film. Despite this, staging is necessary in assessing the sequence and timing of events. The idea of using a standard system of enumeration of stages throughout the vertebrate sub-phylum was proposed by Witschi in 1956. The rate of development of individual organs, however, varies from one species to another, so that comparative staging is best confined to an individual organ, such as the brain. Nevertheless, an interesting atlas for staging mammalian and chick embryos has been produced by Butler and Juurlink, and is based on an extension of the Carnegie system to other species.

## Somitic count

The somitic count is an important feature used in all systems of embryonic staging. Traditionally the somitic period has been thought to extend from 3 to 4 post-fertilizational weeks. On the basis of dating from ultrasonography *in vivo*, however, it seems more likely that the somitic period begins and ends several days later. Counting more than 30 pairs of somites in

*Developmental stages in human embryos*

| Pairs of somites | Stage | Size (mm) | Approximate age (days) | Features |
|---|---|---|---|---|
| | 1 | 0.1–0.15 | 1 | Fertilization |
| | 2 | 0.1–0.2 | 2–3 | From 2 to about 16 cells. 'Morula.' Compaction |
| | 3 | 0.1–0.2 | 4–5 | Free blastocyst } 'Hatching' |
| | 4 | 0.1–0.2 | 6 | Attaching blastocyst } 'Hatching' |
| | 5 | 0.1–0.2 | | Implanted although previllous |
| | 5a | 0.1 | 7–8 | Solid trophoblast |
| | 5b | 0.1 | 9 | Trophoblastic lacunae |
| | 5c | 0.15–0.2 | 11–12 | Lacunar vascular circle |
| | 6 | 0.2 | 17 | Chorionic villi: primitive streak may appear |
| | 6a | | | Chorionic villi |
| | 6b | | | Primitive streak |
| | 7 | 0.4 | 19 | Notochordal process |
| | 8 | 1.0–1.5 | 23 | Primitive pit; notochordal and neurenteric canals; neural folds may appear |
| 1–3 | 9 | 1.5–2.5 | 25 | Somites first appear |
| 4–12 | 10 | 2–3.5 | 28 | Neural folds begin to fuse; 2 pharyngal bars; optic sulcus |
| 13–20 | 11 | 2.5–4.5 | 29–30 | Rostral neuropore closes; optic vesticle |
| 21–29 | 12 | 3–5 | 30 | Caudal neuropore closes; 3–4 pharyngal bars; upper limb buds appearing |
| 30–? | 13 | 4–6 | 32 | Four limb buds; lens disc; otic vesticle |
| | 14 | 5–7 | 33 | Lens pit and optic cup; endolymphatic appendage distinct |
| | 15 | 7–9 | 36 | Lens vesticle; nasal pit; antitragus beginning; hand plate; trunk relatively wider; future cerebral hemispheres distinct |
| | 16 | 8–11 | 38 | Nasal pit faces ventrally; retinal pigment visible in intact embryo; auricular hillocks beginning; foot plate |
| | 17 | 11–14 | 41 | Head relatively larger; trunk straighter; nasofrontal groove distinct; auricular hillocks distinct; finger rays |
| | 18 | 13–17 | 44 | Body more cuboidal; elbow region and toe rays appearing; eyelid folds may begin; tip of nose distinct; nipples appear; ossification may begin |
| | 19 | 16–18 | 46 | Trunk elongating and straightening |
| | 20 | 18–22 | 49 | Upper limbs longer and bent at elbows |
| | 21 | 22–24 | 51 | Fingers longer; hands approach each other, feet likewise |
| | 22 | 23–28 | 53 | Eyelids and external ear more developed |
| | 23 | 27–31 | 56 | Head more rounded; limbs longer and more developed |

From R. O'Rahilly and F. Müller, *Human Embryology and Teratology*, Wiley-Liss- New York, 2nd edition, 1996.

**The approximate ages (in post-fertilizational days) are those currently assigned to each stage, taking ultrasonic findings into account. Post-menstrual days would be greater by about 14.**

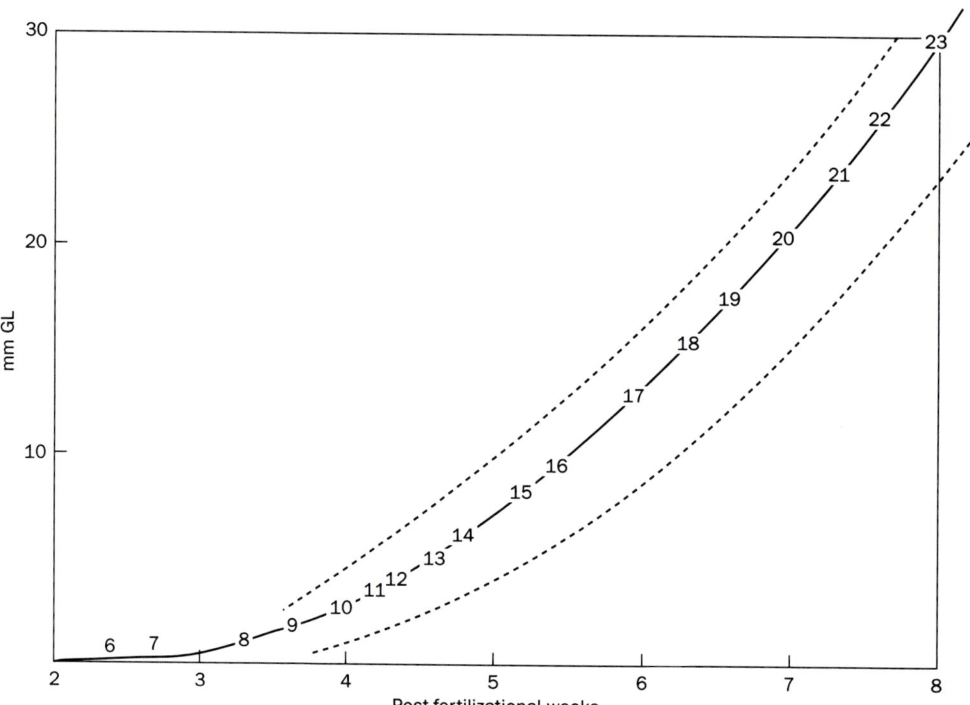


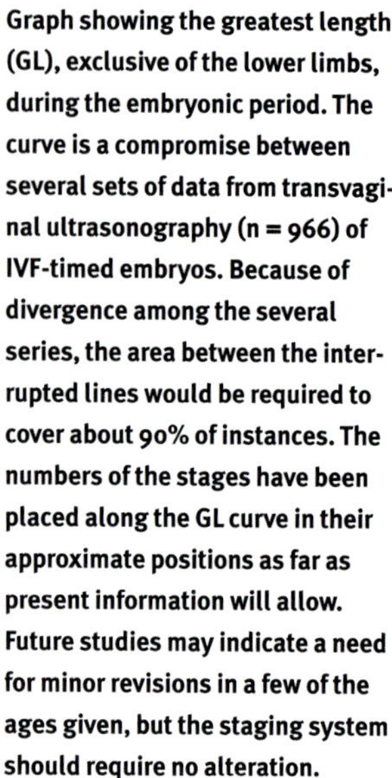
**Graph showing the greatest length (GL), exclusive of the lower limbs, during the embryonic period. The curve is a compromise between several sets of data from transvaginal ultrasonography (n = 966) of IVF-timed embryos. Because of divergence among the several series, the area between the interrupted lines would be required to cover about 90% of instances. The numbers of the stages have been placed along the GL curve in their approximate positions as far as present information will allow. Future studies may indicate a need for minor revisions in a few of the ages given, but the staging system should require no alteration.**

the human embryo becomes difficult and is seldom undertaken. The full complement is about 38 or 39 (and not 42–44 as formerly thought), although they are never all visible at one time. The somitic count is generally a good indication of the level of development, but embryos having the same number of somites may show some slight variations in differentiation, for example, of the nervous and vascular systems.

## Carnegie stages

Embryonic staging was first applied to the human by Mall in 1914. The Carnegie system, now accepted internationally, was established by Streeter for the later stages and by O'Rahilly for the earlier stages. The combined Streeter–O'Rahilly scheme was given the name 'Carnegie Stages' by O'Rahilly. These concern the embryonic period (the first 8 post-fertilizational weeks) only and the scheme consists of 23 developmental stages based on external and internal morphological appearances instead of directly on either size or age. The stages depend mainly on features that change rapidly, such as the number of paired somites, the early appearance of the eye, or the form of the emerging limbs.

Because the Carnegie staging system is not based on age, the morphological criteria characteristic of a given stage remain intact even when newer information concerning embryonic age becomes available. Thus, as a result of recent ultrasonic studies, slightly older ages are now assigned to certain stages. Nevertheless, it needs to be emphasized that the stages depend on both external and internal morphology, some features of which are not accessible in ultrasonographic images.

In terms of embryonic staging, the most detailed information available concerns the brain, an atlas of which has been published. In contrast, no similarly comprehensive work on the heart is to be found, and staged data are unfortunately scattered throughout the literature.

## Incorrect usage

The term *stage* is used in embryology for a specific level of external and internal morphological development, and should be restricted to one of the recognized staging systems, which vary from one species to another. Designations such as 'at the 18-millimetre stage' should read 'at a length of 18 millimetres', because 18 millimetres is a measurement and not a morphological stage. The statement 'at the 150-millimetre stage' contains an additional error, because no morphological staging system has (as yet) been devised for the fetal period. This lack occurs because changes are then neither sufficiently rapid nor adequately spectacular. Furthermore, a bad habit has arisen in the literature on human embryology, that of assigning a Carnegie stage merely on the basis of a measurement or a supposed age. In point of fact, however, an embryo of 20 millimetres, for example, may, depending on the degree of structural development, belong to any one of three stages. Similarly an embryo of 40 post-fertilizational days may belong to any one of several stages.

In summary, although morphological staging is necessary in embryology for correct assessment of the sequence and timing of developmental events, care must be taken in both determination and reporting.

*R. O'Rahilly and F. Müller*

See also 'Growth of the human embryo' (4.3) and 'Developmental morphology of the embryo and fetus' (4.5)

# Ultrasound measurement of prenatal growth

Ultrasound has been used for the assessment of fetal growth for over 30 years. Throughout this time there has been enormous technological development which shows no sign of abating. The ease with which increasingly accurate measurements can be obtained is progressively improving and the number of structures for which normal growth curves have been identified continues to expand. However, there is still a great deal to be learned about patterns of intra-uterine growth and the mechanisms which control and influence it. In addition, the ability to measure fetal volume or weight accurately remains rather poor as all measurements currently made are only linear ones. Three dimensional data acquisition and analysis is just becoming a technical possibility and may permit the development of more sophisticated measurement algorithms in the future.

## Early pregnancy

Using modern transvaginal ultrasound scanners, the embryo can be identified reliably less than 4 weeks after conception and its length measured. The subsequent rapid and accelerating growth of the fetal crown–rump length can be measured by either transvaginal or transabdominal ultrasound scans up to about 10 weeks post-conception, 12 weeks after the last menstrual period. It is now also possible to measure the dimensions of individual embryonic structures such as head size and limb bone lengths.

## Mid and late pregnancy

After 12 weeks, a rapidly increasing number of fetal structures can be seen with clarity and can be measured. Normal growth charts exist for many of them. The measurements which have proved of greatest clinical value are those of the head, abdomen and femur.

In addition to measuring these individual structures, the ratios between some, such as the head and abdomen circumference, or head circumference and femur length, can be computed and are of particular value in identifying growth-retarded fetuses. In growth retardation, soft tissue growth, such as that of the abdomen, tends to be affected early, while limb growth, and especially head growth, are preserved until late in the disorder.

*Fetal structures which can be measured after 12 weeks post-conception, and for which normal growth charts are available.*

| | |
|---|---|
| Biparietal diameter | Orbital diameter |
| Head circumference | Interocular distance |
| Occipito-frontal diameter | Cerebellum |
| Cephalic index | Kidney |
| Abdominal circumference | Spleen |
| Femur length | Stomach |
| Other limb bone lengths | Bowel |
| Scapula | Ear length |
| Foot length | |

## Growth patterns

It is seldom appreciated by the users of ultrasound measurements of the fetus that these are merely linear measurements of a complex three-dimensional structure. All obstetricians are familiar with the graphs showing the growth rate of structures such as the fetal head, abdomen and femur, and much has been written about the significance of the shapes of these curves. What many authors have overlooked is the mathematical relationship between these linear dimensions and the weight or volume of the structures from which the measurements are taken. For example, while there is an apparent slow-down in the rate of growth of both head and abdominal circumferences towards the end of pregnancy, the volumes represented by these linear measurements show a linear rate of increase in the

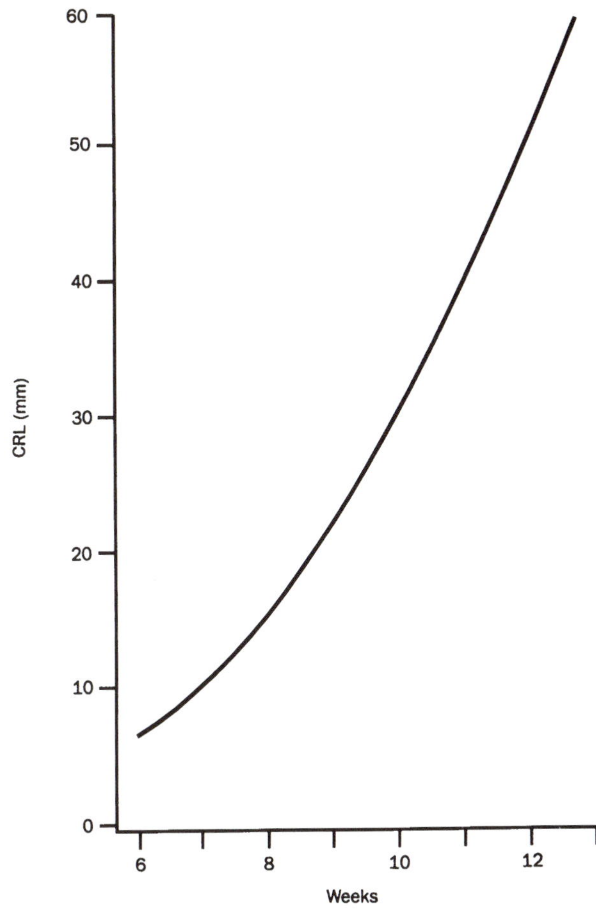


**Growth in crown–rump (CRL) length between 6 and 12 weeks post-conception**

case of the abdomen, and only a gentle decrease in head growth late in pregnancy. The apparent fall-off in growth seen in the standard charts is thus a mathematical artifact.

In recent years it has been shown that fetal growth patterns can be influenced by socio-economic factors and that the sizes and proportions of fetuses from different ethnic groups may show small but significant differences. For example, the limb bones of Japanese fetuses are shorter that Western fetuses and certain ethnic groups from the Indian subcontinent have longer legs and smaller abdomens than European fetuses. It has been interesting to study these ethnic groups outside their original geographic locations, and it is now clear that while some South Asian babies are genetically small, others achieve normal Western birth-weights if the mothers are part of a modernized population.

## Abnormal growth patterns

The main impetus for the development of ultrasound fetal measurements has been the study of fetal growth. These studies have shown that when fetal nutrition is compromised liver growth is initially affected, with other organs following in sequence. Interestingly, brain growth appears to be affected last. This 'brain-sparing effect' appears to be peculiar to the human species. Perhaps there is a genetically determined mechanism to preserve the intellect at the expense of other organs.

There are currently few treatments available for the disorders of fetal growth. However, if maternal overwork is the cause, bed-rest and mild sedation may result in *catch-up* growth, a period of unusually rapid growth until the fetus is

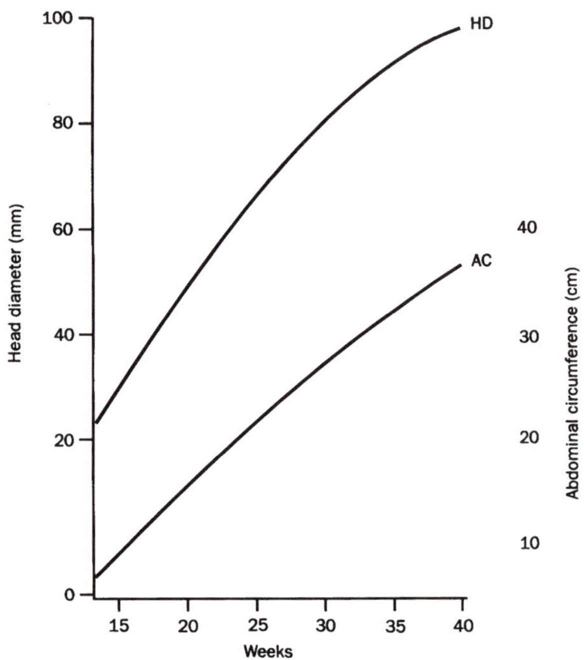


**Growth of head diameter and abdominal circumference between 15 and 40 weeks post-conception. Both appear to slow down in rate towards the end of the pregnancy.**

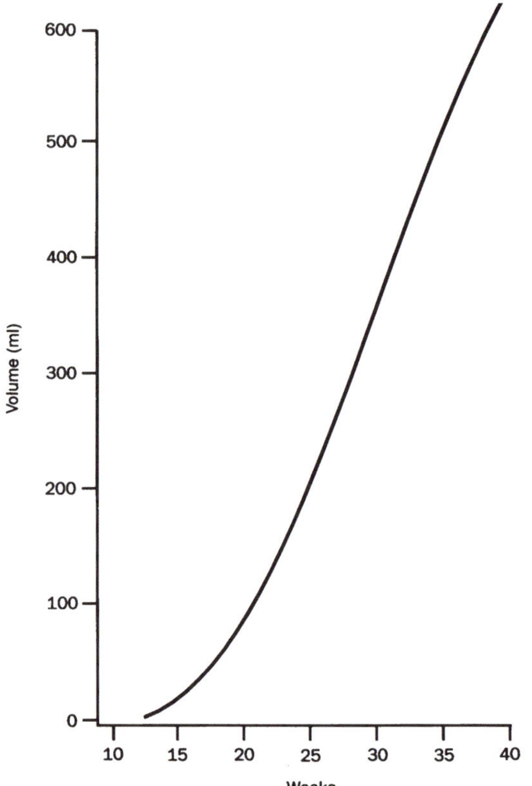


**Volumes represented by linear measurement of fetal head and abdomen, respectively.**

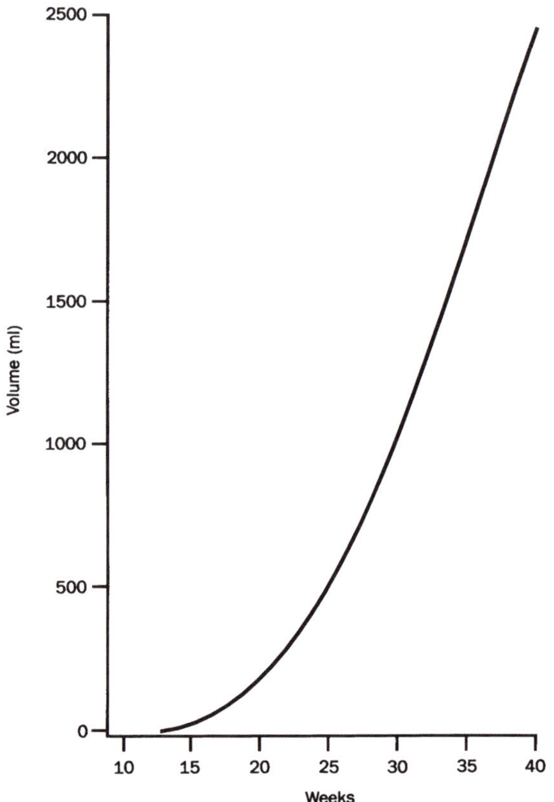


**There is a linear rate of increase in abdominal volume during the last half of pregnancy and only a gentle decrease in head growth. The apparent fall-off in growth represented by linear measurement is thus an artifact caused by not taking into consideration volumetric relationships.**

once again appropriate in size for the gestational age. The same phenomenon has been seen after birth with post-natal catch-up growth (**9.6**, **9.7**) for a minority of fetuses who have been growth-retarded *in utero*.

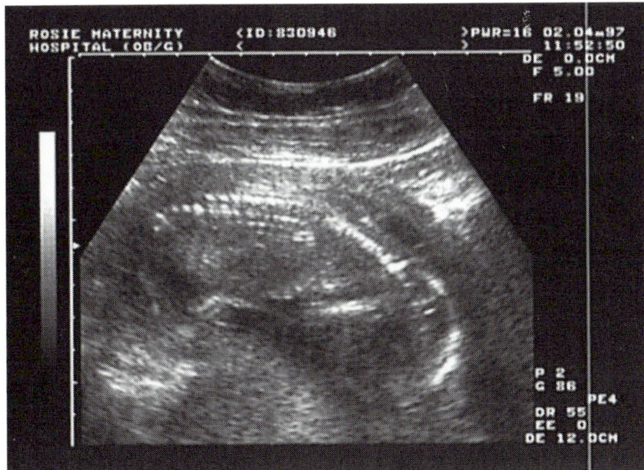


**Ultrasound of fetus at 18 weeks with spine and skull prominent.**

## Factors controlling growth

The ability to make these limited observations of fetal growth is enabling attempts to understand some of the controlling mechanisms for intra-uterine growth. For example, the adverse effects on fetal growth of maternal smoking (**8.3**), over-work and stress, hypertension and diabetes can now be observed. The beneficial effects of correcting those causes which can be removed or reversed can also be seen and are beginning to be of value in the management of a limited number of cases of intra-uterine growth retardation.

Ultrasound is almost the only tool in existence for the assessment of intra-uterine growth of the fetus and has permitted enormous advances in the understanding of fetal growth in recent years. Despite these advances there is still considerable ignorance about the mechanisms which control normal fetal growth.

*Hylton B. Meire*

See also 'Standards and references for the assessment of fetal growth and development' (**4.9**) and 'Body-size at birth' (**10.3**)

# Prenatal age

Prenatal age is, by definition, the time elapsed since fertilization. It is best expressed as post-fertilizational weeks and/or days because a month is a variable interval. Ovulation, as elicited by ultrasonography or as estimated from the basal body temperature, is sufficiently close to fertilization as to be serviceable. Hence, for all practical purposes, post-ovulatory age and post-fertilizational age can be regarded as contemporaneous.

The word 'conception' is best avoided in scientific writing as a synonym for fertilization because any one of two or more starting points may have been used: 1) the last menstrual period; 2) ovulation/fertilization; or 3) implantation. Similarly, the starting point of 'pregnancy' is not free of ambiguity.

## Post-menstrual interval

In clinical practice the duration from the first day of the last menstrual period (LMP) is commonly and conveniently used. This interval is expressed as post-menstrual (or simply menstrual) weeks and/or days. During the first 2 weeks of this interval no embryo exists, so that the incorrect term menstrual 'age' is best avoided.

'Gestational age' is frequently considered to be a synonym for the post-menstrual interval but it also is unsatisfactory because that interval is not age and because the starting point of 'gestation' or pregnancy has not received agreement.

## Assessment of prenatal age

Knowledge of prenatal age is very important in obstetrics. Menstrual histories, however, are frequently unreliable, so that ultrasonic examination (**1.2**) *in vivo* is undertaken. The morphology is studied and various measurements are taken. The greatest length (GL) exclusive of the lower limbs is the most valuable single measurement (and is generally what is measured ultrasonically under the rubric 'crown–rump length'). Other measurements are also used, and then compared with those in standard age tables.

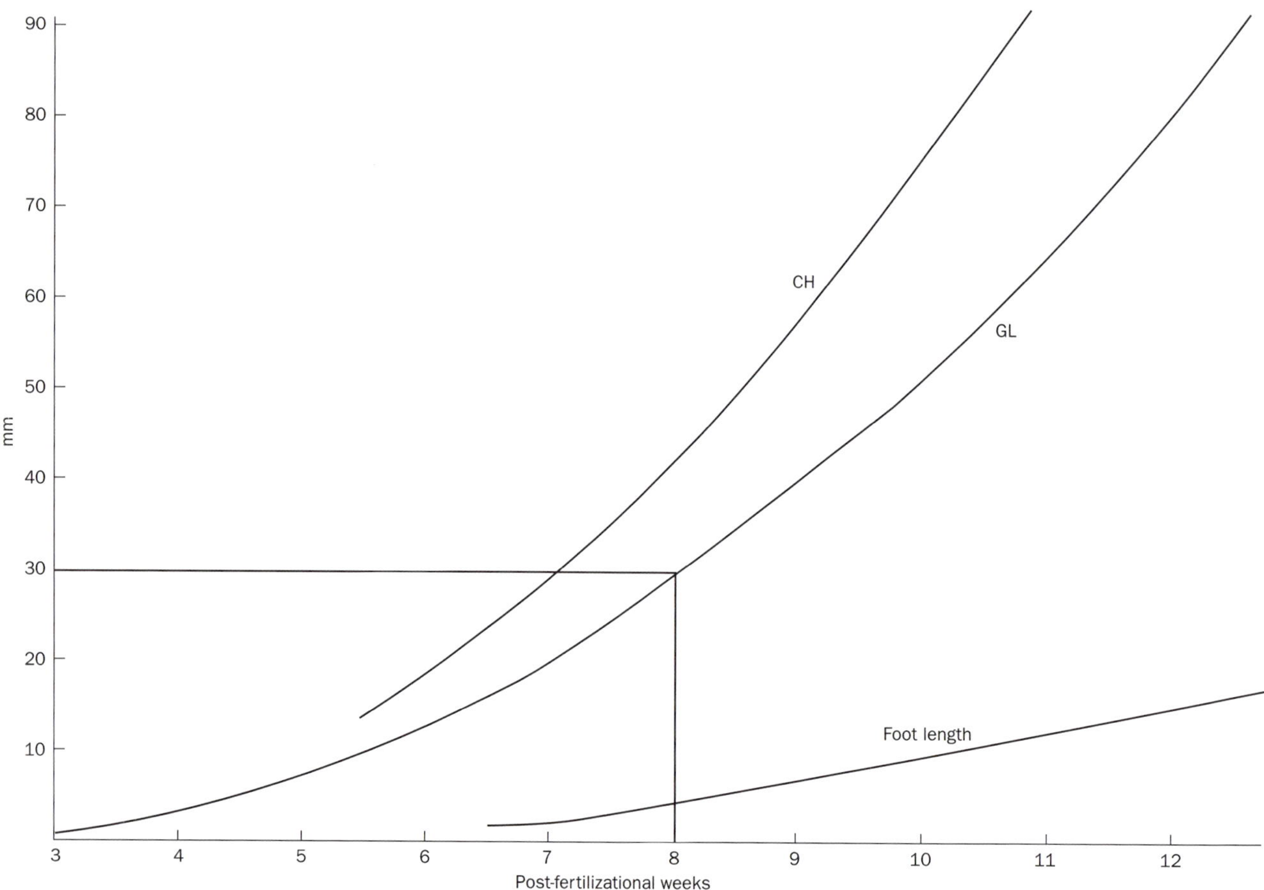


**Prenatal age during the first trimester as related to crown–heel (CH), greatest length (GL), exclusive of the lower limbs, and foot length. These are representative figures, but the range of variation is not included. The rectangle indicates the embryonic period.**

## The embryonic period

In the embryonic period proper (the first 8 weeks), an unjustified comparison with the ages of macaque embryos was soon abandoned as it became clear that a 30-millimetre embryo would be 8 post-fertilizational weeks of age. This was later confirmed by ultrasonographic studies *in vivo* and has long since reached general acceptance both in embryology and in obstetrics. The introduction of transvaginal ultrasonography has resulted in greater precision and the possibility of finding the greatest length of very young embryos. More reliable data for establishing a standard have been obtained by the use of that technique for embryos of known age, namely in instances of *in vitro* fertilization or gamete intra-tubal transfer. In embryos between about 0.2 and 10 millimetres in greatest length, it is now believed that the ages previously used in embryology were too low and should be increased in some instances by about 4 days according to one study, about 6 days according to another.

In embryos with a GL of 3 to 30 millimetres an excellent approximation of age in days can readily be obtained by adding the figure 27 (26 was suggested originally) to the embryonic length. Thus a 10-millimetre embryo may be expected to have an age of about 10 + 27 = 37 days. In terms of post-menstrual days, the figure 41 should be added to the length instead of 27. Variation is found, as would be expected, but during the embryonic period the time interval between the 10th and the 90th percentile for a given embryonic length is believed not to exceed one week, and to be less in the earlier stages. At birth, however, the variation in age for a given fetal length may be of the order of several weeks.

During the embryonic period, particular attention is paid to the morphology (for example, the chorionic sac, and the limb buds) and several measurements can be taken in addition to the GL: the maximum diameter of the chorionic cavity, the amniotic cavity, and the umbilical vesicle (the so-called yolk sac).

*Embryonic age related to greatest length, according to current opinion*

| Greatest length (mm) | Embryonic age (post-fertilizational weeks) |
|---|---|
| 0.15 | 1 |
| 0.2 | 2 |
| 1 | 3 |
| 3 | 4 |
| 8 | 5 |
| 15 | 6 |
| 20 | 7 |
| 30 | 8 |

## The fetal period

The fetal period is frequently and conveniently considered in terms of trimesters, as exemplified in reference to the GL. The transition from the embryonic to the fetal period occurs at 8 post-fertilizational weeks, when the GL is approximately 30 millimetres. At the end of the first trimester, some 90 days, the GL is about 90 millimetres. At the middle of prenatal life it has increased to approximately 190 millimetres, and to about 250 millimetres at the junction between the second and third trimesters. At birth it is most frequently of the order of 335 millimetres.

A much more precise assessment of prenatal age is needed in obstetrics, where ultrasonography *in vivo* is used. Although morphological features are taken into consideration, the main emphasis is on various linear and circumferential measurements, chiefly the head circumference, GL, abdominal circumference, biparietal diameter, foot length, and (ossified) femoral length. In addition, fetal weight can be calculated from the biparietal diameter and the abdominal circumference by a formula. The post-menstrual interval is then read from elaborate tables of several key measurements and their range, which are listed for each week.

The duration of prenatal life is the time period from fertilization to birth. The mean duration is 38 weeks or, more precisely, 264 days, with a range of 254–274 days.

In summary, estimates of prenatal age are important in both embryology and obstetrics, and can be assessed from standardized tables arranged week by week according to post-menstrual intervals. *R. O'Rahilly and F. Müller*

See also 'Growth of the human embryo' (**4.3**), 'Embryonic development of teeth' (**4.4**) and 'Developmental morphology of the embryo and fetus' (**4.5**)

## 1.4

# Anthropometry

The basic technique for describing the growth at the level of the individual is anthropometry, the measurement of the body. Its roots may be traced to the Greeks, who first considered the mathematical relationship of body segments and regions to each other (i.e. proportionality), and to artists of the Renaissance and subsequent periods whose concern with representing the body led them to even more systematic investigations. However, as a science, anthropometry blossomed in the 19th century when it was brought, largely by German investigators, into anthropology as a formal method of conducting a reliable, quantitative description of the human body. It has since been further refined by the development of a battery of precise and specialized instruments, the formulation of standard protocols for collecting data, and the construction of specific measurements appropriate to the purpose of the study.

Anthropometric data are basic to the design of workspaces and their components; to the analyses of biological variability within and between populations (**2.5**, **10.5**, **10.6**); to the assessment of the health status of individuals and groups, and to the manufacture of clothing that both fits and is appropriate to the task at hand. But nowhere is it more fundamental than it is to the study of physical growth, for nothing captures growth as a dynamic process of change better than does the measurement of an individual child through time, or the aggregation of measurements of some defined sample of children and youth.

As does any scientific technique, anthropometry requires well-defined procedures that produce data which are both precise and accurate, which are valid estimates of the biological structure of concern, and which may be replicated as part of the process of generating scientific knowledge. Adherence to these principles necessitates sensitivity to the dynamic interplay between the need to stay with established protocols of measurement – to ensure maximum comparability – and the need to develop new protocols which reflect not only new knowledge of growth, but also the hypotheses forming the basis of a particular study. An investigator should be neither a slave to past traditions nor someone whose data cannot be generalized beyond the immediate investigation.

With respect to the analysis of human growth, anthropometric data should reflect underlying developmental processes, i.e. they should make biological sense. Thus weight, the most widely taken measure of growth, is an estimate of the total mass of the body, its cells, its fluids, and its stores and, while not a particularly accurate indicator of any single tissue or segment, is none the less a useful measurement for a range of purposes. Stature, or height, is another global measure which, while lumping together a range of individual growth processes, still provides valuable data on the whole child.

Central to anthropometry is the landmark, a defined anatomical point from which a particular dimension is mea-

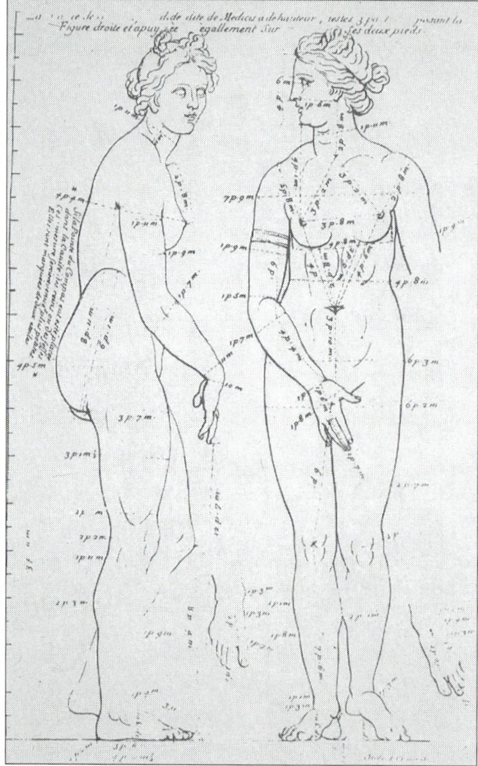

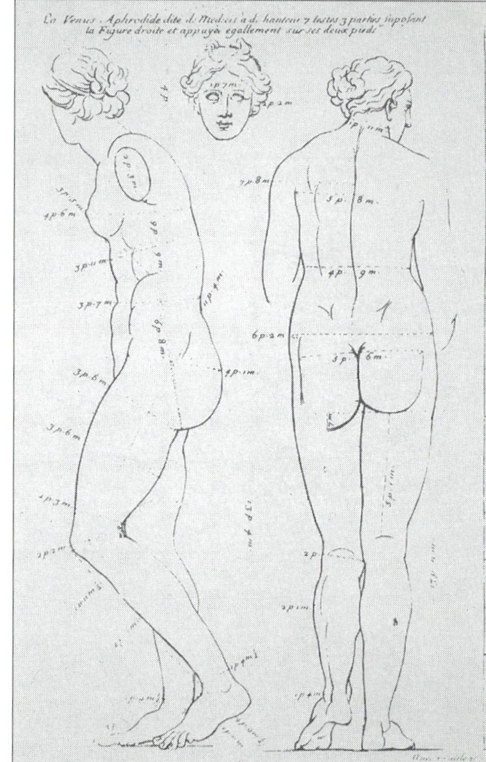

**Audran's measurements of the Medici statue of Venus. From Audran (1683).**

sured, which identifies the measurement and allows it to be replicated by others. Thus the vertex is the most superior point in the mid-line of the head (when the head is oriented in a precisely defined plane called the Frankfurt Horizontal). Stature (or supine length) is defined as the distance between the vertex and the surface on which the child is standing (in the standard erect position).

Anthropometry should be carried out as carefully as any laboratory method, with due attention to all aspects of the procedure. When this is done, measurement error will be held to the absolute minimum, making possible the calculation of growth increments in individuals, reliable descriptive statistics of samples, and analysis of the components of variance in these growth indicators.

## Advantages and disadvantages of anthropometry

Anthropometry is a reliable method for analysing growth. It can produce valid, objective information when applied rationally to biologically meaningful constructs. It is non-invasive and the instruments used are relatively inexpensive. It can be used with large, national samples or it can be applied to the measurement of growth over the interval of a few days or even less. Anthropometric data may be used to study the growth of different regions or body segments, or in describing body proportions and their changes during growth (**10.6**). It can be applied to children and youth in a variety of circumstances: the non-ambulatory; those who have various disorders; infants of very low birth-weight, and adolescents who may be morbidly obese. And even though indirect, anthropometric data may provide reliable estimates, at the population level, of the growth of different body tissues and compartments, for example, lean body mass.

At the same time, anthropometry can only take measurements from the surface of the body, and while these can indicate the state of underlying growth in cells, tissues, and organs, they can only provide estimates of amount. It is impossible to measure growth at a single locus using this technique. The understanding of growth and its quantification at individual and population level requires the integration of data from a range of sources and the employment of a broad set of methods and techniques. Anthropometry is one essential component of the study of growth useful at many levels and in many settings.

*Francis E. Johnston*

See also 'The use of growth references' (**1.17**), 'The human growth curve' (**2.4**), 'Within-population variation in growth patterns (**2.5**), 'Morphology' (**5.5**), 'Growth cyclicities and pulsatilities' (**5.13**), 'Fat and fat-patterning (**5.16**), 'Identification of abnormal growth' (**7.1**), 'Growth in chronic diseases' (**7.8**), 'Maternal anthropometry and birth outcome' (**8.1**), and 'Migration and changing population characteristics' (**11.6**)

**Anthropometry training session, Dhaka, Bangladesh. Photograph by Stanley Ulijaszek.**

## MEASUREMENT ERROR

Anthropometric measurement errors are of three sorts: imprecision, undependability and inaccuracy. Imprecision is due to intra- and inter-observer measurement differences. Undependability is due to non-nutritional factors that influence the reproducibility of the measurement – such as differences in height of an individual across the day as a consequence of compression of the spinal column due to standing against gravity. Inaccuracy is a function of instrument error. All potential anthropometrists should receive adequate training from an expert or supervisor to reach a measurable level of expertise prior to survey, and maintain this level of expertise throughout. Targets for anthropometric assessment have been put forward by Zerfas (1985), using a repeat-measures protocol. The trainee and trainer measure the same subjects until the difference between the two of them is good or, at the very least, fair. However, these values should not be used uncritically, since differences between trainer and trainee at the upper level of 'goodness' for height, weight, arm circumference and skinfolds represent different proportions of the absolute measure according to the size of the measurement. Thus, although Zerfas (1985) gives values for differences that are possible given the techniques available, a 5-millimetre difference in height measurement is more accurate than the same difference in arm circumference.

The most commonly used measures of this are the technical error of measurement (TEM) and reliability (R). The TEM is obtained by carrying out a number of repeat measurements on the same subject, either by the same observer, or by two or more observers, taking the differences and entering them into an appropriate equation. For intra-observer TEM, and inter-observer TEM involving two measurers only, the equation is:

$$\text{TEM} = \sqrt{\frac{\Sigma D^2}{2N}}$$

where D is the difference between measurements, and N is the number of individuals measured. When more than two observers are involved, the equation is more complex:

$$\text{TEM} = \sqrt{\left(\Sigma_1^N\left[\Sigma_1^K M(n)^2 - \frac{\Sigma_1^K M(n)^2}{K}\right]N(K-1)\right)}$$

where N is the number of subjects, K is the number of determinations of the variable taken on each subject, and M(n) is the difference between replicates of the measurement, where n varies from 1 to K. The units of TEM are the same as the units of the anthropometric measurement in question. Although acceptable maximum TEMs have been recommended as reference values for a variety of measures by Frisancho (1990), these ignore the age dependence of TEM, and fail to give values for height. The coefficient of reliability, R, ranges from 0 to 1, and can be calculated using the following equation:

$$R = 1-[(\text{TEM})^2/(\text{SD})^2]$$

where SD is the total inter-subject variance for the study, including measurement error. This coefficient reveals what proportion of the between subject variance in a measured population is free from measurement error. In the case of a measurement with an R of 0.9, 90 per cent of the variance is due to factors other than measurement error. Measures of R can be used to compare the relative reliability of different anthropometric measurements, and to estimate sample size requirements in anthropometric surveys.

*Stanley Ulijaszek*

*Evaluation of measurement error among trainees (after Zerfas, 1985)*

| | Difference between trainee and trainer | | | |
|---|---|---|---|---|
| Measurement | Good | Fair | Poor | Gross error |
| Height/length (cm) | 0–0.5 | 0.6–0.9 | 1.0–1.9 | 2.0 or > |
| Weight (kg) | 0–0.1 | 0.2 | 0.3–0.4 | 0.5 or > |
| Arm circumference (cm) | 0–0.5 | 0.6–0.9 | 1.0–1.9 | 2.0 or > |
| Skinfolds (any) (mm) | 0–0.9 | 1.0–1.9 | 2.0–4.9 | 5.0 or > |

*Upper limits for total technical error of measurement at two levels of reliability (males)*

| Age group (years) | Height (cm) | Sitting-height (cm) | arm circumference (cm) | Triceps skinfold (mm) | Subscapular skinfold (mm) |
|---|---|---|---|---|---|
| *Reliability = 0.95* | | | | | |
| 1–4.9 | 1.03 | 0.40[a] | 0.31 | 0.61 | 0.43 |
| 5–10.9 | 1.30 | 0.35 | 0.52 | 0.97 | 0.87 |
| 11–17.9 | 1.69 | 0.30 | 0.75 | 1.45 | 1.55 |
| 18–64.9 | 1.52 | 0.30 | 0.73 | 1.38 | 1.79 |
| 65+ | 1.52 | 0.30 | 0.74 | 1.29 | 1.74 |
| *Reliability = 0.99* | | | | | |
| 1–4.9 | 0.46 | 0.18[a] | 0.14 | 0.28 | 0.19 |
| 5–10.9 | 0.58 | 0.16 | 0.23 | 0.43 | 0.39 |
| 11–17.9 | 0.76 | 0.13 | 0.33 | 0.65 | 0.69 |
| 18–64.9 | 0.68 | 0.13 | 0.33 | 0.62 | 0.80 |
| 65+ | 0.68 | 0.13 | 0.33 | 0.58 | 0.78 |

[a] 2–4.9 years.

*Upper limits for total technical error of measurement at two levels of reliability (females)*

| Age group (years) | Height (cm) | Sitting-height (cm) | arm circumference (cm) | Triceps skinfold (mm) | Subscapular skinfold (mm) |
|---|---|---|---|---|---|
| *Reliability = 0.95* | | | | | |
| 1–4.9 | 1.04 | 0.34[a] | 0.30 | 0.65 | 0.47 |
| 5–10.9 | 1.38 | 0.36 | 0.54 | 1.05 | 1.08 |
| 11–17.9 | 1.50 | 0.29 | 0.78 | 1.55 | 1.74 |
| 18–64.9 | 1.39 | 0.31 | 0.98 | 1.94 | 2.39 |
| 65+ | 1.35 | 0.32 | 0.98 | 1.86 | 2.27 |
| *Reliability = 0.99* | | | | | |
| 1–4.9 | 0.47 | 0.15[a] | 0.13 | 0.29 | 0.21 |
| 5–10.9 | 0.62 | 0.16 | 0.24 | 0.47 | 0.48 |
| 11–17.9 | 0.67 | 0.13 | 0.35 | 0.69 | 0.78 |
| 18–64.9 | 0.62 | 0.14 | 0.44 | 0.87 | 1.07 |
| 65+ | 0.60 | 0.14 | 0.44 | 0.83 | 1.02 |

[a] 2–4.9 years.

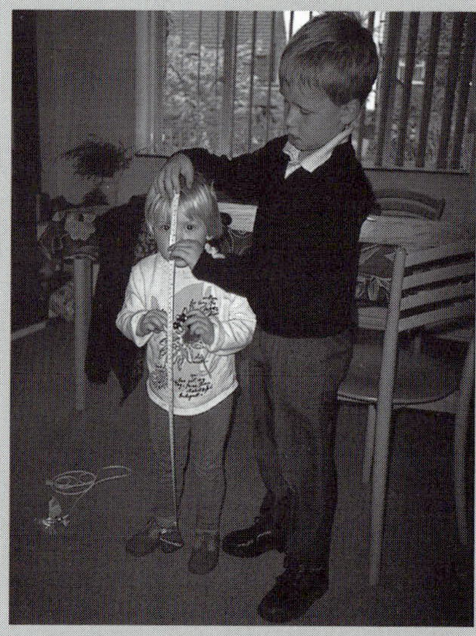

**How not to measure stature. Photograph by Stanley Ulijaszek.**

## RECUMBENT ANTHROPOMETRY

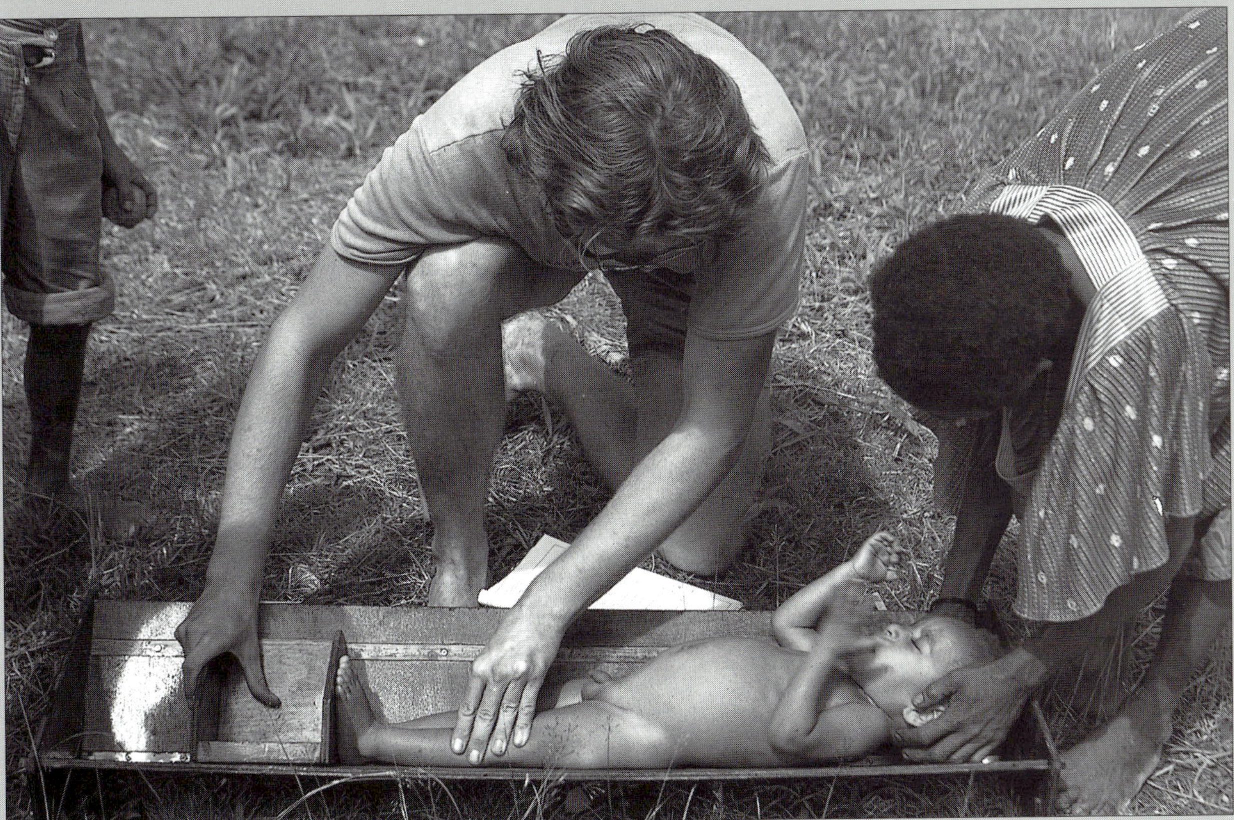

**Measuring recumbent length of a young child with the modified AHRTAG length measurer. Photograph by John Lourie.**

Recumbent anthropometry is used for measurement of individuals unable to stand. Accordingly, in the past it has been of greatest use for the assessment of infants and small children, but is increasingly applied to adults or the elderly who are unable to stand or remain standing. Recumbent measurement of the young principally aims to monitor growth. Recent findings which suggest that greatest permanent growth deficits occur under the age of 2 years will probably spur even greater use of early growth monitoring for population-based, international public-health efforts. Relevant measurements include total length and crown–rump length, which correspond to standing- and sitting-heights respectively. Recumbent lengths have been shown to be somewhat more reliable than the standing equivalents. That children under the age of 3 years are measured supine accounts for the curious disjunction in reference growth curves at that age, in which children appear to shrink suddenly. This measurement artifact occurs because a prone position does not subject the spinal column to the vertical compression exerted under erect posture; compression has also been invoked as the cause of diurnal variation in individual stature, amounting to as much as 1 centimetre. Thus, recumbent length is systematically greater than the corresponding standing-height. Accurate measurement of recumbent total and crown–rump lengths requires two persons, one to position and hold the child's head and shoulders, and the other to manipulate the instrument and take the reading. Each measurement simply requires a calibrated board with a fixed head-board and a sliding-board, which is brought to the feet or the rump for taking the reading.

Measurements of incapacitated adults are used to assess nutritional status, body-size and fat distribution: recumbent measurements of girths (waist, hip, mid-thigh, lower thigh, calf); diameter (abdominal saggital); length (knee height), and skinfold (triceps, forearm, subscapular, abdominal, suprailiac, anterior thigh, medial calf) have been described. Instrumentation for these measurements is narrow steel tape-measure, Kahn abdometer, anthropometer or Mediform calliper, and skinfold calliper, respectively. The girths, skinfolds and abdominal diameter allow assessment of nutritional status and body-fat patterning, while knee height has been shown to yield a fairly accurate estimate of overall stature when a formula appropriate for body proportions of the target population is used. Precision of recumbent measurements and their correlation with standing equivalents is generally good. Notable exceptions include the problems of hip circumference yielding lower values, and suprailiac skinfold being less precise in the recumbent than standing position. Precision of abdominal diameter is much higher than related girths and skinfolds, and thus may provide a better index of abdominal fat. Indeed, increased abdominal diameter is associated with elevated risk for cardiovascular events. Adult recumbent measurements assist in studies of normal ageing processes, risk factors for chronic disease and nutritional risk among the elderly or ill. *Carol M. Worthman*

**Portable infant measuring device used in Bangladesh. Photograph by Stanley Ulijaszek.**

# Body imaging by three-dimensional surface anthropometry

Three-dimensional (3-D) surface anthropometry describes the size and shape of the human form from any point on the body surface using measurements in 3-D and includes measures of length, breadth and depth.

The study of human size and shape is not new. Hippocrates, in 400 BC, classified people as one of two physical types: *phthisic habitus* or *apoplectic habitus* – while Sheldon more recently defined individual shape, or somatotype, as intermediate between three extremes: endomorph, mesomorph and ectomorph. However, such studies relied on traditional methods of recording lengths and circumferences using instruments such as tape-measures and callipers (**1.4**). The resultant information described size rather than shape.

## Development of 3-D measurement

### *Early systems*

The first study, in 1973, of the human body as a 3-D object used light-sectioning. Many of the early systems, including optical and opto-electronic techniques, back-lighting or structured light, suffered from limitations, including being labour-intensive, time-consuming, or unable to deal with re-entrant shapes. The moiré fringe method was successful in measuring a human torso. The technique involves illumination and observation of the object through a 25 lines-per-inch ruled grating. Interference fringes produced by the grating are photographed using a conventional camera and the pictures analysed.

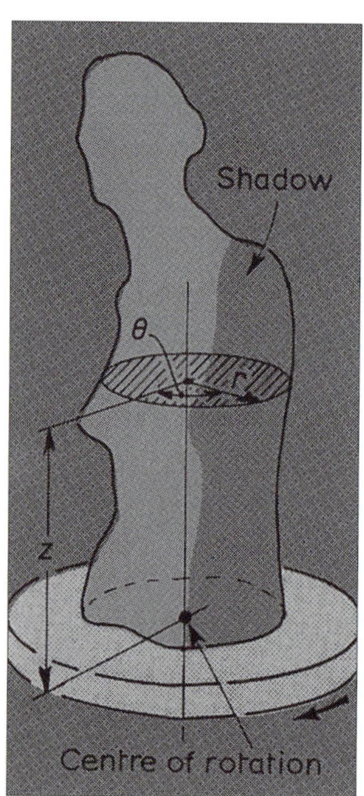


**The LASS principle involves the rotation of the subject being measured so that the radii and angles of the body can be measured in conjunction with height. The subject being measured is placed on a turntable which is rotated through 360° in measured angular increments (θ). Four narrow strips of light are projected onto the body in vertical planes which pass through the centre of rotation of the turntable, from four banks of four special light sources. All points where the edge of the light falls on the body define the horizontal radii (r) at points in the vertical plane (z).**

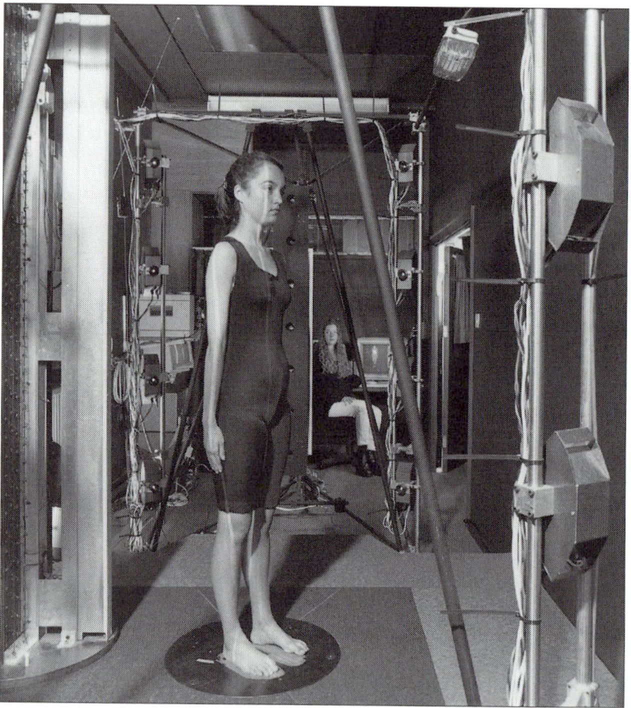

**The Loughborough Anthropometric Shadow Scanner in operation.**

### *Computer aided systems*

The first system to use computer rather than human interpretation of the measurements appeared in 1986. A pattern of dots was projected to define particular points on a body surface and viewed by two cameras simultaneously. Each spot on the image was computer-matched with a particular spot from the projector so that, by triangulation, a position in space could be calculated. Limitations with this system, and a Body Information Modelling System, were that neither could produce a complete 360-degree view, although the latter was successfully used to measure the shape and contours of the human back.

### *Body imaging systems in use*

Four fully automated whole-body systems are currently in use:

- A British development, the Loughborough Anthropometric Shadow Scanner (LASS).
- An American system, the Cyberware scanner.
- A Japanese scanner.
- The Scoliosis Image Processing System (SIPS), Liverpool, England.

### *LASS*

The general principle employed in LASS is that a whole body can be measured in terms of radii and angles in conjunction with height. The radii are measured using two banks of seven

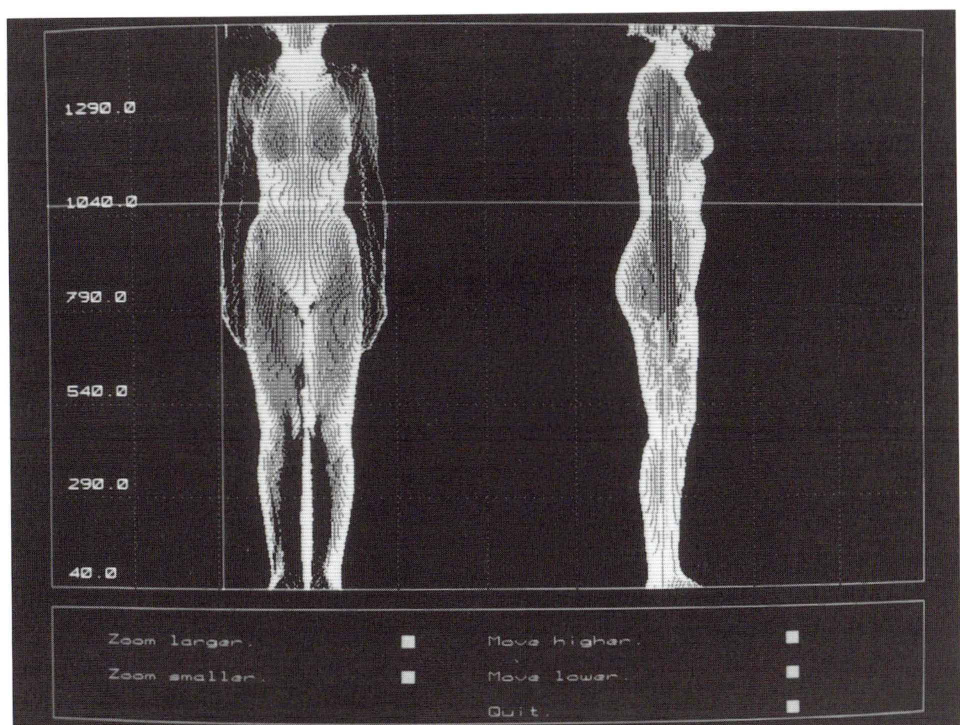


**The computerized whole-body image after scanning.**

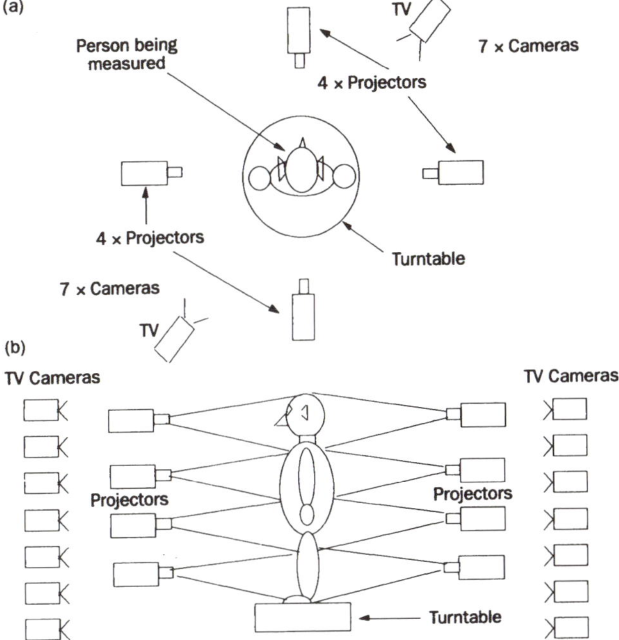


**Plan (a) and elevation (b) of the configuration of the Loughborough Anthropometric Shadow Scanner.**

television cameras which are aligned so that some known point at the light-slit edge of the field of view also coincides with the centre of rotation of the turntable. Because the plane of the slit of light passes through the centre of the turntable, all points where the edge of light falls on the body define the horizontal radii at points in the vertical plane. If one horizontal line of the scene viewed by the television camera is considered, the distance from the reference point, which coincides with the centre of the turntable, to the shadow is accurately related to the radius of the body at that point. Further television lines measure radii further down the body and information from the transducer mounted on the turntable gives the angle at which the radii are measured. The full shape of the body is, therefore, defined in cylindrical co-ordinates. The television cameras are mounted at 60 degrees to the plane of the shadow as a compromise between a good viewing angle to allow the radii to be measured with sufficient resolution and reduction in obscured re-entrant areas by other parts of the body.

Video signals from the camera are processed and converted by a computer programme into a graphical display, either in the form of vertical planes at any given angle, or in horizontal planes viewed from above, at any level of the body. A print-out of the displayed information provides a permanent record.

## Applications of 3-D anthropometry

### *Medical*

Distortion of body-shape due to natural or pathological processes has been studied, in for example, respiratory mechanics, scoliosis before and after corrective surgery, craniofacial abnormalities related to fetal alcohol syndrome, and the identification of malignant breast lesions from asymmetry. Three-dimensional scanning at specific sites also has medical applications. For example, optic disc measurements can be used for early diagnosis of glaucoma, wax dental imprints are useful for orthodontic preparations, and body-shape pre- and post-surgery is useful to examine changes made, especially in the maxillofacial region, and for cranioplasty and ear prosthetics. Other applications include for the manufacture and fitting of customized prosthetics in dentistry, orthopedics, and mastectomy. Accurate information on body volumes and/or surface areas can provide guidance for maintenance of homeostatic control, drug administration and medical management in, for example, children with liver disease. Accurate measurement of burn surface areas, arthritic swellings, and expansible tumours could aid treatment decisions and planning of radiation therapy.

*Human Systems Engineering*

This relates to the environment in which people live and work, including everything used and worn. Body-size and shape in relation to unique environments, such as space and mines, have attracted special attention, but on a more mundane level, an operator's geometric and physical properties need to be incorporated into car-seat or aircraft control design. Clothing and manufacturing industries require body dimension information for incorporation into product design and manufacture. This includes not only fashion wear but anti-gravity suits and personal protective clothing including helmets, face-masks, gloves and shoes, all of which need to be shaped for optimal fit and therefore maximum protection in, for example, chemical warfare or hazardous environments.

Both the Japanese scanner and LASS are being used for population screening. Such information will contribute to the assessment of long-term changes in body-shape and relative rates of change in dimensions such as soft and skeletal tissue.

*Comparative morphology*

Studies of human morphology have the potential for improving early diagnosis in a large number of diseases. In addition, with special focus on growth, speciation and sexual dimorphism, they are central to evolutionary research. Three-dimensional imaging can be used in reconstruction of fossil specimens, facial features and skulls. Dynamic studies of human motion are using 3-D measurements at individual joint level and in planning intra- and extra-vehicular space activities.

## GRATING PROJECTION METHODS: THE LIVERPOOL EXPERIENCE

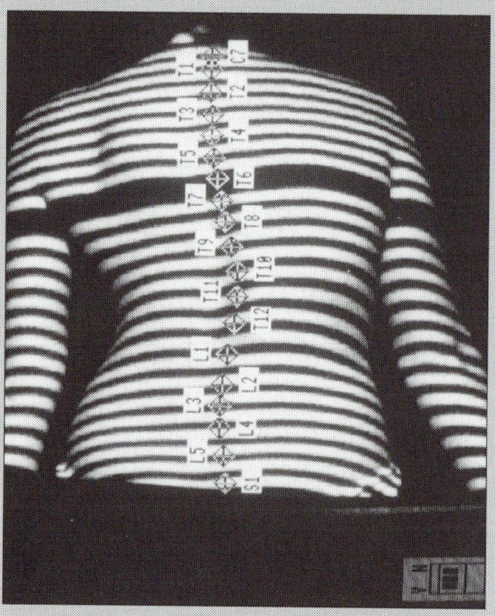


**Automatic location of the features of the vertebral column on a scoliosis patient's back.**

The technique of grating projection (sometimes called raster stereography) uses structured light to generate fringe patterns on the surface of the object under view. Simple in concept, the technique employs a 35-millimetre slide of a number of fine horizontal lines which form a special grating. Using a suitable light source, the grating pattern is projected onto the surface of an object to create a series of fringes represented by parallel shadows or lines. The surface is then viewed by a video camera mounted horizontally at a known distance from the surface and at a fixed distance below the projector. The camera acquires an image of the back surface and a frame grab/store device digitizes the image, enabling a computer to process the signal. Image-processing algorithms are then employed by the computer which extract three-dimensional information allowing reconstructions of the shape of the object. This method of image analysis has been employed by researchers at the University of Liverpool and the Liverpool John Moores University to develop a method of recording and evaluating the deformity of spinal curvatures called scoliosis (7.3). Scoliosis is a deformity of the vertebral column of the body in which the normal straight spine develops a right- or left-sided curvature accompanied by the rotation of the individual vertebral bones. The result is a visible deformity of the back and trunk called a rib-hump.

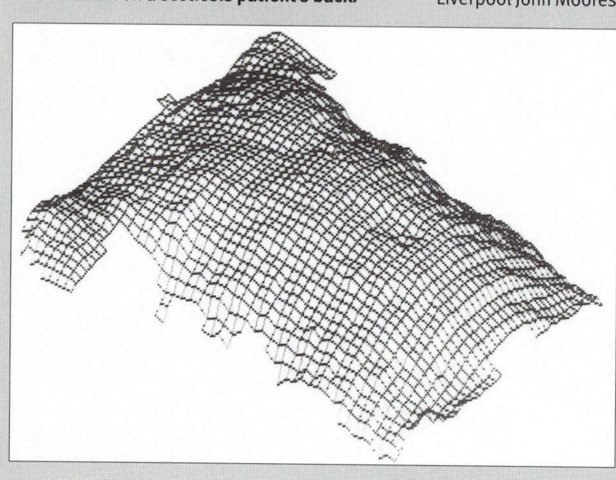

**Three dimensional reconstruction s of the shape of the patient. This is the chest and abdomen of the subject.**

The development of the imaging system (Scoliosis Image Processing System: SIPS) has allowed the research team to measure the deformity as it first is seen by orthopedic doctors and allow a critical evaluation of the condition as it is treated. The technique is entirely non-invasive, unlike X-rays, and thus avoids all the risks of X-ray imaging. Many patients have been assessed and the results have contributed to the knowledge and understanding of the condition of scoliosis.

Future development of the SIPS system will inevitably include a continuing programme of upgrading the computer system, permitting faster data collection and more detailed images to be analysed. Phase-measuring profilometry is currently under investigation whereby the object's shape is converted into a phase distribution, as in interferometry, and is analysed by digital-phase measuring techniques: accuracy to less than 1 millimetre is possible without the need for reference plane images. This technique has been used to measure the three-dimensional shape of the trunk, face and breasts, and has obvious applications in plastic surgery.

Further developments under investigation include three-dimensional real-time reconstructions of body-shape, including dynamic movement recording. These techniques require very fast computing and high-speed image frame grabbers. Combination with magnetic resonance imaging (MRI) is also being studied. The goal is to be able to view the body in movement, such as walking, in conjunction with gait- and movement-analysis methodologies, leading to application in

### *Virtual reality and communications*

Special effects in science fiction movies have been achieved with Cyberware's system and computers are being used to synthesize 3-D digital animated images of human models. Three-dimensional facial-image processing techniques are being developed in Japan for visual telecommunications.

## Limitations of current scanning systems

### *Human*

The human body, being a living organism, is in constant motion and subject to variation in shape from external (gravity) and internal factors. These small but ever-present movements and shape changes contribute an error factor in the accuracy of the scanned measurement. Most optical sensing techniques underestimate skin elevation because human skin is quite transparent, although the degree of underestimation varies with skin pigmentation. Additional problems may arise from variation in distribution and density of body hair and its associated pigmentation.

### *Equipment*

Speed of scanning is one of the most important limiting factors for body imaging. Access to equipment may be difficult, as portable devices have yet to be developed. Finally, no current surface imaging system is able to digitize 100 per cent of the body's surface, especially if several postures need to be recorded, as the human shape is too complex and possesses too many degrees of freedom to devise a practical 100 per cent surface collection digitizing system. Irrespective of such limitations, 3-D imaging is capable of producing useful information for a multitude of differing applications and extending way beyond that derived from simple anthropometric techniques.

Peter R. M. Jones and Jean Peters

See also 'Regional growth disorders' (7.3)

the field of movement science and sports analysis. Growth of children could be accurately recorded by such techniques without any problems of keeping a young child still. Dynamic movement is the normal state of the human body at any age and the challenge of being able to record these changing shapes accurately at low cost is the long-term aim of this research programme.

*Peter H. Dangerfield*

Grating projection onto a subject's back. Note the scoliosis deformity produces asymmetry of the shoulders and the chest.

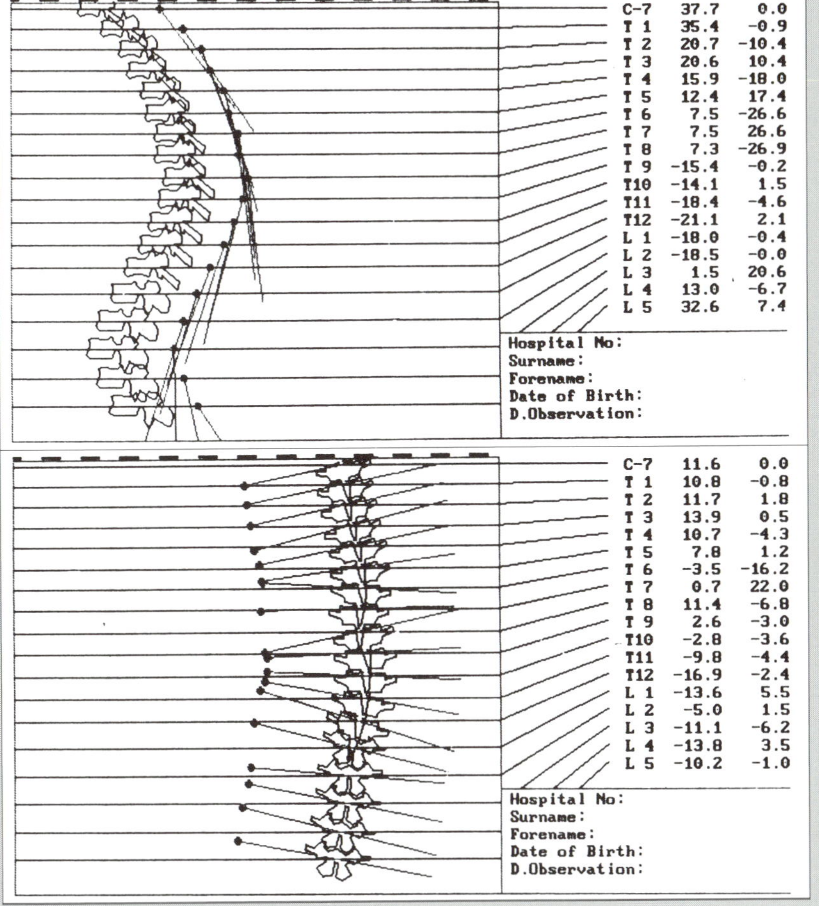


Image analysis permits the presentation of measurements of the body in any chosen profile. The examples shown here indicate shape changes running down the length of the vertebral column, seen in a side view (sagittal plane) or back (posterior) view.

# The assessment of body composition

Through the life-cycle, the human body changes in its chemical composition. As growth, maturation and ageing proceed, the fat, muscle and bone compartments of the body undergo alterations in the absolute amounts and relative proportions of lipid, protein, water and minerals. Understanding these changes is important because they are an integral part of the biological changes of the life-cycle (**5.11**) . For example, there are considerable changes in fatness across the course of growth and development. Variability within and between populations in body composition during growth and development is due to differences in biological maturation, diet, heredity, environment, and disease-related factors (**10.6**). In addition, body composition is a major determinant of the nutritional needs associated with the support of normal growth, maturation and health.

Body composition methods evaluate the body's components at numerous levels of biological complexity, from basic elements (for example, carbon, oxygen and hydrogen), molecules (water, lipids, protein), to whole tissue compartments (fat, muscle and bone) of the human body. Each level of complexity provides different kinds of information about how the body changes and matures during growth and development. Research is limited by the ability to measure various compartments safely, accurately and reliably *in vivo*. This is an important consideration in studies of infants and young children for whom certain methods widely used in adults, such as hydrodensitometry (underwater weighing), are not feasible. In addition, the cost, availability and risks associated with various methods impose further limitations.

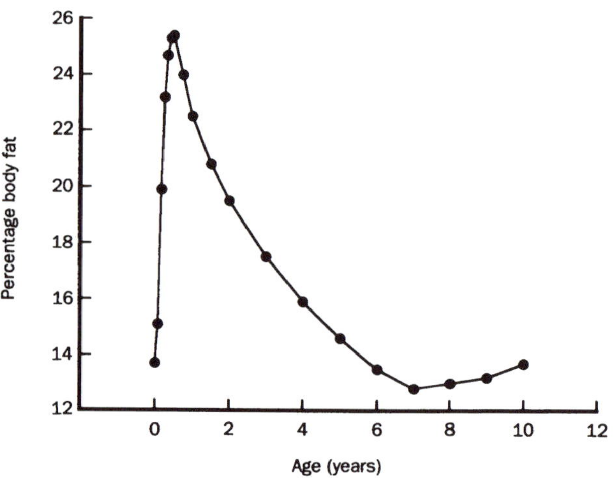


**Changes in percentage body fat from birth to 10 years of age in a male child. Adapted from: Fomon S.J., Haschke F., Ziegler E.E. and Nelson SE. Body composition of reference children from birth to age 10 years (1982).**

Most methods available for use during growth and development measure whole tissue compartments using either a two compartment (fat and fat-free mass) or three compartment (fat, lean body mass and bone) model. The most commonly used methods are described below.

### Isotope dilution methods

Stable isotopes are used to estimate the size of various compartments of the body using the classic dilution principle. The stable isotopes, deuterium oxide ($^2H_20$) or oxygen-18 ($0^{18}$), are naturally occurring isotopes and are used safely and effectively to measure the size of the total body water pool (TBW) in infants and children. A concentrated dose of isotope is administered and, after an equilibration period, its concentration is determined in a body fluid such as urine. It is estimated that $^2H_20$ overestimates TBW by 4 per cent, whereas $0^{18}$ overestimates TBW by 1 per cent due to the mixing of these isotopes with non-aqueous fractions of the body.

Hydration factors which estimate the fraction of the total body water in fat-free mass are used to calculate fat-free mass, fat mass and per-cent body fat. Sodium bromide can be used in a similar manner to estimate the extracellular water compartment so that the distribution of TBW in the intra- and extra-cellular water compartments can be determined.

### Bioelectrical methods

The water and electrolytes in the body have electrical properties which can be measured for estimation of TBW or FFM

*Comparison of the time, expense practicality and accuracy of body composition methods in children*

| Method | Time | Expense | Practicality in young children | Practicality in older children | Accuracy |
|---|---|---|---|---|---|
| Anthropometry | + | ++ | +/– | + | +/– |
| TOBEC | ++ | – – | +/– | ++ | + |
| BIA | ++ | + | – | + | – |
| Hydrodensitometry | – – | – | – – – | + | + |
| Isotope dilution | + | +++ | +++ | +++ | +++ |
| DXA | + | +++ | + | ++ | +++ |
| K-40 | – – – | – – – | – – – | – – – | ? |

+ represents relative advantage – represents relative disadvantage

Estimated hydration of fat-free mass during growth and development for males and females

| Males | | | | Females | | | |
|---|---|---|---|---|---|---|---|
| Fomon[1] | | Lohman[2] | | Fomon[1] | | Lohman[2] | |
| Age group | Hydration factor | Age group | Hydration factor | Age group | Hydration factor | Age group | Hydration factor |
| Birth | 80.6 | | | Birth | 80.6 | | |
| 1 month | 80.5 | | | 1 month | 80.5 | | |
| 2 months | 80.3 | | | 2 months | 80.2 | | |
| 3 months | 80.0 | | | 3 months | 79.9 | | |
| 4 months | 79.9 | | | 4 months | 79.7 | | |
| 5 months | 79.7 | | | 5 months | 79.5 | | |
| 6 months | 79.6 | | | 6 months | 79.4 | | |
| 9 months | 79.3 | | | 9 months | 79.0 | | |
| 12 months | 79.0 | 1 year | 79.0 | 12 months | 78.8 | 1 year | 78.8 |
| 18 months | 78.5 | 1–2 years | 78.6 | 18 months | 78.4 | 1–2 years | 78.5 |
| 24 months | 78.1 | | | 24 months | 78.2 | | |
| 3 years | 77.5 | | | 3 years | 77.9 | | |
| 4 years | 77.0 | 3–5 years | 77.8 | 4 years | 77.7 | 3–5 years | 78.3 |
| 5 years | 76.6 | | | 5 years | 77.6 | | |
| 6 years | 76.3 | 5–6 years | 77.0 | 6 years | 77.5 | 5–6 years | 78.0 |

Anthropometric prediction equations for estimating fat-free mass, fat mass and percentage body fat

**1. For skinfold prediction equations estimating body density (D):**

Ages 1 to 11 years[1]:
Males: $D = 1.1690 - 0.0788 \cdot \log_{10}\Sigma$ (biceps triceps subscapular suprailiac
Females: $D = 1.2063 - 0.0999 \cdot \log_{10}\Sigma$ (biceps triceps subscapular suprailiac

Ages 12 to 16 years[2]:
Males: $D = 1.1533 - 0.0643 \cdot \log_{10}\Sigma$ (biceps triceps subscapular suprailiac
Females: $D = 1.1369 - 0.0598 \cdot \log_{10}\Sigma$ (biceps triceps subscapular suprailiac

Ages 17 to 19 years[3]:
Males: $D = 1.1620 - 0.0630 \cdot \log_{10}\Sigma$ (biceps triceps subscapular suprailiac
Females: $D = 1.1549 - 0.0678 \cdot \log_{10}\Sigma$ (biceps triceps subscapular suprailiac

Percentage fat = $\{4.95/D - 4.5\} \cdot 100$
Fat mass (kg) = Percentage fat · body-weight (kg)
Fat-free mass (kg) = body weight (kg) − fat mass (kg)

**2. Two skinfold prediction equations estimating percentage fat (% fat)[4]:**

Prepubescent Males:
White: % fat = 1.21 (triceps + subscapular) − 0.008 (triceps + subscapular)$^2$ − 1.7
Black: % fat = 1.21 (triceps + subscapular) − 0.008 (triceps + subscapular)$^2$ − 3.2

Pubescent Males:
White: % fat = 1.21 (triceps + subscapular) − 0.008 (triceps + subscapular)$^2$ − 3.4
Black: % fat = 1.21 (triceps + subscapular) − 0.008 (triceps + subscapular)$^2$ − 5.2

Post-pubescent Males:
White: % fat = 1.21 (triceps + subscapular) − 0.008 (triceps + subscapular)$^2$ − 5.5
Black: % fat = 1.21 (triceps + subscapular) − 0.008 (triceps + subscapular)$^2$ − 5.2

All Females:
White: % fat = 1.33 (triceps + subscapular) − 0.013 (triceps + subscapular)$^2$ − 2.5

For sum of triceps and subscapular > 35 mm:
Males: % fat = 0.783 (triceps + subscapular) + 1.6
Females: % fat = 0.546 (triceps + subscapular) + 9.7
Fat mass (kg) = Percentage fat · body weight (kg)
Fat-free mass (kg) = body weight (kg) − fat mass (kg)

(fat-free mass). Total body electrical conductivity (TOBEC) devices for infants and children provide accurate, rapid, non-invasive estimates of fat-free mass, fat mass and per-cent body fat. As the body passes through a low-energy electromagnetic coil, alterations in the conductance caused by the water and electrolytes are measured and the signal is converted to body composition estimates by computerized prediction equations developed for this method. Bioelectrical impedance analysers measure the impedance of a low-energy electrical signal as it passes through the body. As with the TOBEC device, this method uses a calibration equation to convert the resistance signal to estimates of body composition. Care should be taken to use the prediction equations devised for children.

## Anthropometric methods

Fat-free mass, fat mass and per-cent body fat can be estimated from prediction equations that use skinfold thickness measurements and other body dimensions. (**1.4**) It is important that these prediction equations be used on samples of the appropriate age. Ethnicity, obesity and puberty status are other factors that have been included in some prediction equations.

## Hydrodensitometry

Underwater weighing is a commonly used method for determining adult body composition and is used in children (aged 6 years or more) and adolescents. The density of the body is determined from measurement of body-weight in air and while immersed in water. Corrections are needed for the volume of air in the lungs and intestines, and for the density of air and water. Given the known densities of fat (0.900 g/cc) and fat-free mass (1.100 g/cc) and the measured density of the body, the fat mass, fat-free mass and per-cent fat can be calculated. In children and adolescents, the chemical composition of the body is changing, in terms of decreasing hydration and increasing mineralization of the fat-free mass. Therefore, multicompartment approaches that combine hydrodensitometry with other body composition

[1] From: Brook CGD (1971): Determination of body composition of children from skinfold measurements. *Archives of Diseases in Childhood* 46:182–184.
[2] From: Durnin JVGA and Rahaman MM (1967): The assessment of the amount of fat in the human body from measurement of skinfold thickness. *British Journal of Nutrition* 21:681–689.
[3] From: Durnin JVGA and Womersley J (1974): Body fat assessed from total body density and its estimation from skinfold thickness: measurements on 481 men and women aged from 16 to 72 years. *British Journal of Nutrition* 32:77–97.
[4] From: Slaughter MH, Lohman TG, Boileau RA, Horswill CA, Stillman RJ, Van Loan MD, Bembem DA (1988): Skinfold prediction equations for estimation of body fatness in children and youth. *Human Biology* 60(5):709–723.

methods that target these components of the body, such as TBW or dual energy absorptiometry, greatly improve the accuracy of body composition estimates.

### Potassium-40

A naturally occurring stable isotope of potassium ($^{40}K$) is present in the body and emits a strong gamma ray. The $^{40}K$ content of the body represents 0.0118 per cent of total body potassium. Total body potassium is representative of the body cell mass and is highly correlated with fat-free mass. Potassium-40 counters consist of a lead-shielded room with a gamma ray detector for determination of the whole body content of $^{40}K$.

### Absorptiometry methods

Dual photon absorptiometry (DPA) and dual energy X-ray absorptiometry (DXA) are techniques that measure three compartments of the body: the bone mass, lean body mass and fat mass. Because of their varying densities, bone, lean tissue and fat attenuate the energy beams differentially. By using dual energy beams, it is possible to solve for three tissue compartments. DXA is becoming increasingly available and provides more accurate measurements than DPA. The radiation exposure of DXA is extremely low (3.5 mrad) and whole-body estimates of body composition for infants, children and adolescents can be obtained in less than five minutes. In addition, since bone mineral density is one of the sources of variability contributing to errors in estimating the density of fat-free mass, measurement of bone mineral improves the accuracy in estimating fat-free mass in a two-compartment model. Compared to bioelectrical and anthropometric methods of body composition assessment described above, it has the added advantage of being independent of sample-based prediction equations.

Numerous body composition techniques are available for use in infants children and adolescents. However, there is no single method that is ideal. Methods vary in their accuracy, availability, and appropriateness depending on the age of the subject and the setting (field, clinic or laboratory). The goals for which body composition is being assessed should be carefully considered in the choice of a technique to maximize the information obtained and minimize costs, inconvenience, and discomfort for the child and parents.

*Babette S. Zemel*

See also 'Body composition' (**5.11**)

### BODY SCANNING, IMAGING AND MEASUREMENT

Before the discovery of X-rays, plaster casts, mechanical devices and photographs were used in a range of crude body-scanning and shape-recording techniques. Imaging and measurement techniques in use today represent a development of traditional anthropometric techniques. They employ a range of investigative scanning and imaging techniques based around photography, X-rays, microwaves, ultrasound and nuclear magnetic resonance technology (MRI).

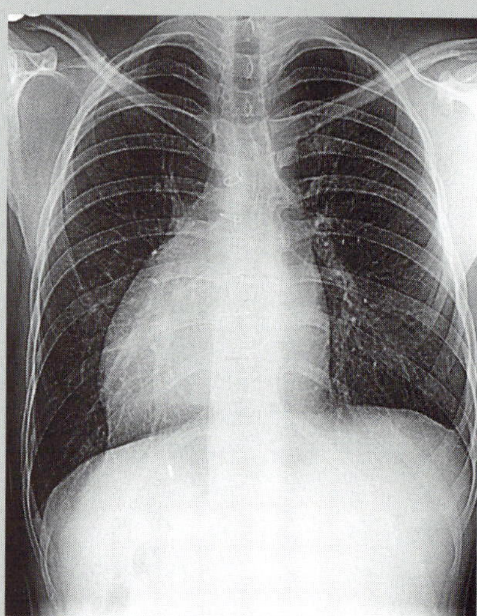

**A digital radiograph of the chest. This is a posterior–anterior view of a patient who has undergone heart surgery. Note the details of the bones and the lung markings that are revealed by the technique.**

Imaging and scanning range from basic techniques with easy-to-interpret images to more complex, and highly expensive, methods. The interaction of the components of skeletal anatomy is best understood by employment of imaging methods based around X-rays. Technical developments strive to reduce dosage, so that, with the newest digital systems, the dosage of radiation is reduced by a factor of at least 10 times. Computer analysis, on-screen real-time imaging and other advances improve the images and view of structures examined. However, the risks from ionizing radiation remain, which raises ethical issues for anyone employing such methods in a pure research capacity. The combining of radiography and computer technology has resulted in the development of computerized tomography scanning (CT-scanning). This technique offers the opportunity to create cross-sectional and three-dimensional reconstructions of skeletal and soft-tissue structures of the body and limbs which now have important medical diagnostic applications.

Magnetic resonance imaging (MRI) produces highly detailed images of soft tissue and hard tissue. It exploits the property of application of a powerful magnetic field to the body, measuring the processing movement of hydrogen and other atoms in the magnetic field when they are disturbed from a stationary state. Advanced image-processing methods already allow three-dimensional reconstruction of images of tissues to be achieved. Furthermore, the ionic contents of tissues can be quantified using the decay times of molecular movement activated by the magnetic field (T1 and T2 images). Acquiring geometric joint data in functional motion studies and procuring static and dynamic images will become feasible as the design of equipment advances and costs fall.

Very recently, a completely new magnetic resonance imaging system has been developed which employs polarized helium gas. The helium is polarized using an initiator of rubidium vapour, locally magnetized by polarized light. When inhaled into the lungs, the helium's magnetic field can be detected and used to create images of the air-filled lung anatomy, something conventional MRI cannot do. Further work suggests that xenon may also act in a similar fashion but, in this instance, be absorbed into fatty body tissue, opening up the opportunity to image the circulatory system or ner-

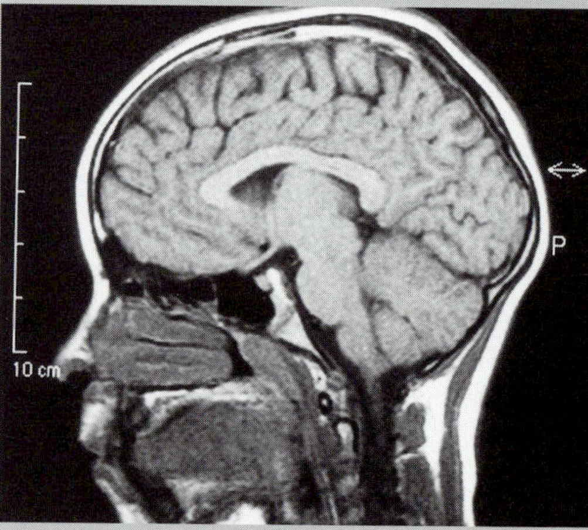


**MRI image of a sagittal section through the head. The details of the brain, nose and mouth can easily be seen.**

## CHEMICAL IMMATURITY

Body-composition assessment of children depends heavily on the measurement and understanding of body density relationships with age. Body density is the usual criterion of assessing fatness, but such measurements give overestimations in children. This is because the density of fat-free mass is lower than the value of 1.1 kilogram per litre used in equations developed for the assessment of body fatness in adults, using, for example, the Siri formula (1956). In addition, the use of fat-free mass assessments using potassium-40 dilution will also give overestimations of body fatness if the constant adult value of 2.66 grams of potassium per kilogram of fat-free mass is used. The low density of fat-free mass in childhood is due to: 1) changing contributions of the individual constituents of fat-free mass; 2) reduced hydration; and 3) increased bone mineral content, in the course growth and development. The low levels of bone mineralization and body potassium relative to adult values is known as chemical immaturity. Estimates of body fatness from body density assessments should therefore use age-specific values for fat-free mass density. *Stanley Ulijaszek*

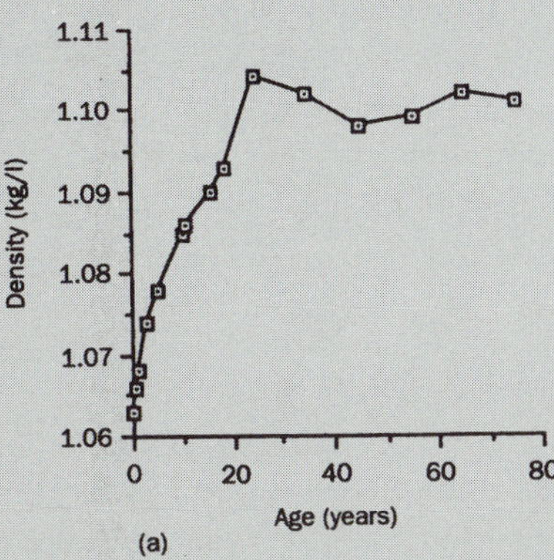


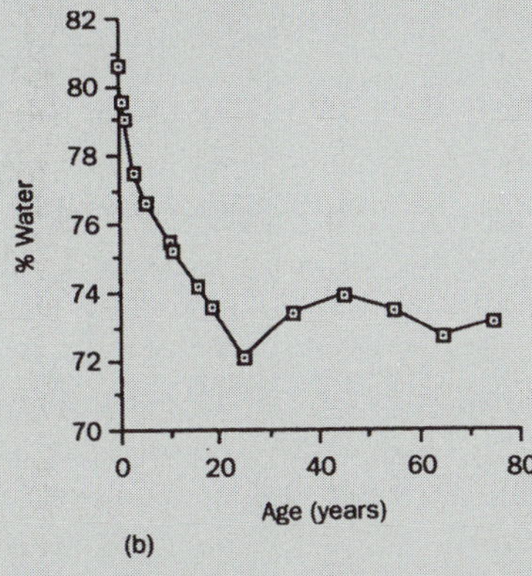


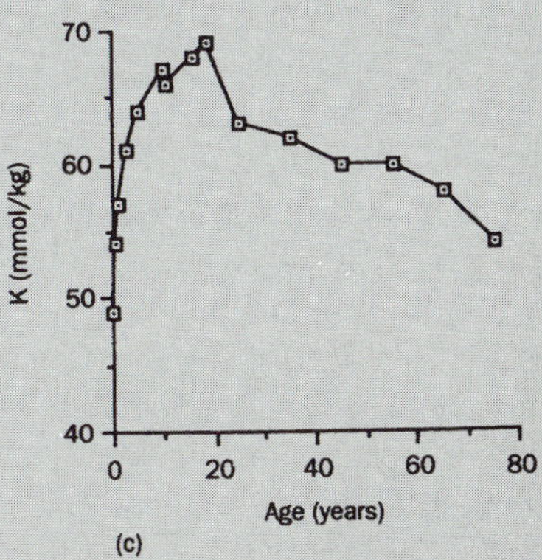


**Changes in the composition of fat-free mass with age, in European males (a) body density, (b) body water, (c) total body potassium. From Jebb and Elia, 1995.**

vous system. These new developments offer the opportunity for low-cost imaging systems without the need for expensive, powerful magnets.

In the early 1970s, von Ramm in America developed the conventional two-dimensional ultrasound technologies which are now routinely used in hospitals and clinics around the world for medical imaging during pregnancy for antenatal screening of the fetus, to ensure it is growing correctly within the uterus of the mother. It is also employed to assess the structure of internal organs and structures. Real-time images offer a high degree of detail to skilled observers but have a big limitation as they can only scan a small thickness of the body at a time in a slice just 2 millimetres wide. Recent ultrasound developments are likely to change this difficulty with the advent of three-dimensional ultrasound images.

This new 3-D ultrasound process uses parallel computing to analyse instantly a myriad of reflected sound waves produced from a wand in a similar way to military phased-array radar. Many hundreds of miniature ceramic crystals are used to emit high-frequency pulses of sound so that an entire volume of the body is swept simultaneously. Other crystals receive the returning echoes, which are converted and processed into digital images by the computer. The resulting images are presented as slices which can be at different angles and can be made thicker or thinner as required. Since images are captured at high speed, the technique allows accurate 3-D viewing of moving objects, such as the heart or moving fetus, in 'real time' for the first time.

Light-based scanning has also been applied to body imaging (**1.5**). Still-film cameras, electronic cameras from Kodak and other manufacturers, computer frame-grabbers and video-based techniques all offer research concepts to be developed towards goals of 3-D image reconstruction and movement analysis in medical, sporting and biological fields. Progress is very rapid due to the ever-increasing speed of computer processors. Lasers, ultrasound and microwaves are all finding application in specialist biological and medical research with the goal of replacing invasive touch techniques which are labour-intensive with fast, accurate and safe methods. However, it must be constantly remembered that biological and medical fieldwork often takes place in places where simple equipment is required and it may well be a long time before traditional anthropometric equipment is superseded by high-tech electronics. *Peter H. Dangerfield*

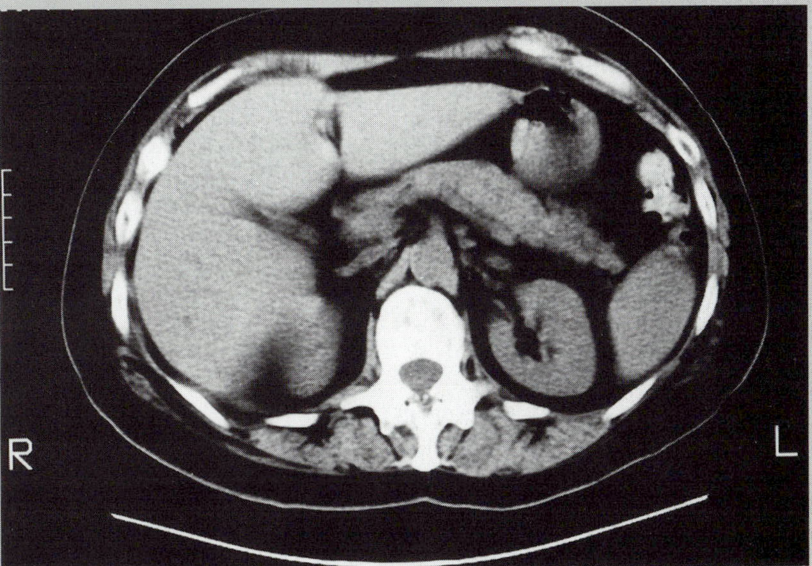


**CT image of the upper abdomen. This is a cross-sectional view. Note the large grey shadow representing the liver on the right (labelled R) and the white outline of the bones of the vertebral column and ribs. The kidney, spleen and pancreas may also be identified.**

## MODELS OF BODY COMPOSITION

The assessment of human body composition has improved significantly since Hippocrates postulated that we were made up of four constituents: blood, phlegm, choler, and melancholy. Nevertheless, the majority of work in this area still divides the body into two components: fat mass and fat-free mass. In this definition fat consists of all fat in the body, i.e. that deposited in subcutaneous stores, deep stores (e.g. surrounding organs such as the kidneys, or in the greater omentum of the abdomen) fat in the nervous system and brain, and in cell walls. The term *lean body mass* is often used interchangeably with fat-free mass, but lean-body mass represents the body-weight minus storage fat, and thus fat in cell walls and in the brain would be included in lean body mass.

As with all biological variables there are differences between individuals in the changes in fat and fat-free masses during growth but the general trend can be described. At birth an average infant has about 14 per cent body fat – girls having a percentage more or so than boys, a difference which will increase as growth proceeds. The amount of fat in the body increases fourfold in the first 6 months of life, leading to the average infant being about 25–20 per cent body fat at the end of this time. From then on levels begin to decline in percentage terms, although in absolute terms fat continues to be laid down, but at a slower rate. By the age of 7–8 years the average child is carrying about 3 kilograms of fat – again slightly more in girls than in boys – representing percentages of about 13 per cent in boys and 17 per cent in girls. From this point on there is a gradual increase in percentage body fat throughout adolescence, leading to a young adult woman being, again, on average 25 per cent body fat, and the young male being about 15 per cent body fat. Levels then fluctuate throughout adulthood, dictated by fashion and leisure activity as much as biology.

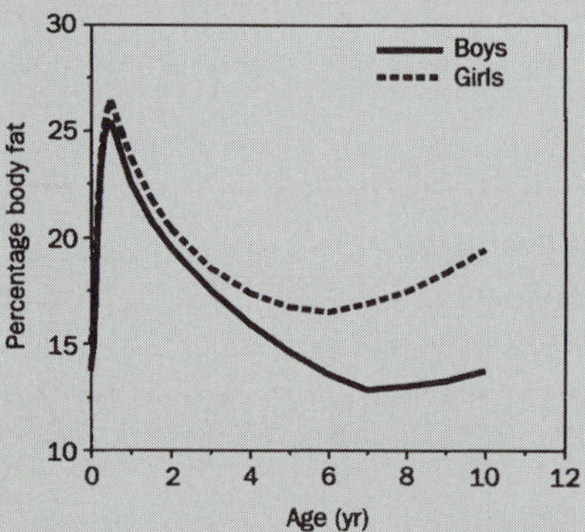


**The average percentage body fat in boys and girls over the first 10 years life. Note the rapid increase in both sexes following birth. By the age of about 1 year, girls have a slightly higher percentage fat than boys, and this difference becomes more apparent throughout childhood. There is some evidence emerging from both the United States and the United Kingdom that in recent years the decline in percentage body fat after about 1 year of age is less dramatic, leading to concerns relating to the potential for a major upturn in childhood obesity in future years. Also the point at which the percentage body fat begins to increase again (around 7 years in this example) has been associated with obesity. It has been suggested that if the upturn is early in childhood there is a greater probability of later obesity.**

### *DIVIDING THE BODY INTO MULTIPLE COMPONENTS*

The division of the body into two components has produced useful data, but while this approach is appealing because of its relative simplicity it fails to address the important fact that the composition of the fat-free mass itself changes throughout growth. As with all mammals, the fat-free mass becomes 'drier' with age, with the percentage of fat-free mass that is water falling from about 81 per cent at birth to about 73 per cent at maturity. Equally, protein levels rise consistently from birth, as do mineral levels – notably bone mineral – but to a lesser extent. Thus, in recent years it has been the aim of researches in the field of body composition to assess body composition using three-, four- or five-compartment models in order to obtain more accurate and meaningful information.

A three-compartment model of body composition divides the human into fat, water and protein plus mineral. This model, therefore, requires that total body water be measured. This is relatively easily achieved using a dilution principle, usually employing isotopes of water such as deuterium or tritium. The latter isotope cannot, however, be used in children or pregnant woman due to its radioactive nature.

Additional compartmental models require that body protein be measured, usually via the assessment of body nitrogen stores, or total body calcium be measured to obtain an estimate of bone mineral content. Dual X-ray absorptiometry has been a recent advance which can produce detailed estimates of mineral content, both in the whole body and in differing regions.

### *CHANGES IN BODY COMPOSITION IN DISEASE STATES*

There can be marked deviation form 'normal' body composition in a number of diseases, including obesity, renal disease, liver disease and many hormonal imbalances. One of the challenges that faces researchers in body composition is to be able to adapt models to accommodate the physiological changes that accompany many diseases in order to be able to assess body composition accurately. For example, in renal disease there are often major disturbances in the amount of total body water. Thus, it would be inappropriate to assume that an individual with renal disease has the same fat-free mass composition as a healthy individual.

*Peter S.W. Davies*



**The principle of measuring total body water, a key body composition parameter, revolves around the ability to measure precisely a known quantity and concentration of a suitable 'tracer' before oral administration and then obtaining a small sample of physiological fluid in which the concentration of the tracer is again measured. Mass spectrometry is often used for these analyses, limiting its use to specialized research centres.**

# Biochemical Markers

Growth and development is accompanied by a great deal of tissue remodelling, involving degradation of existing structures and replacement with new conformations. Skeletal growth (**5.6**) may act as a primary drive for growth of soft tissue and many of the biochemical markers of growth have been developed specifically to reflect bone growth and metabolism. Bone is, however, a unique tissue in that maintenance of its integrity requires continual remodelling to remove microfractures resulting from mechanical stress. This remodelling process occurs throughout life but its effects in terms of biochemical markers are superimposed on the much larger effects of modelling during growth. The latter process reflects changes in the architecture of the skeleton with changes in size.

Over 90 per cent of the matrix of bone is collagen, a protein that constitutes about one third of the total body protein. Not surprisingly, therefore, many of the biochemical markers of growth are based on collagenous constituents. Collagens comprise a family of about 17 genetically distinct types, consistent with the burgeoning knowledge about the diverse functions of this protein. By far the most common form, collagen type I, exists as a fibrillar component providing the main structural framework for bone, skin, tendons and most other tissues. The other primary fibrillar collagens are type II, which is mainly present in cartilage, and collagen type III, which is abundant in soft tissues, particularly skin and blood vessels, in close association with collagen type I. Collagen type III is, however, virtually absent from bone and metabolites of this component therefore provide useful markers for soft tissue growth and metabolism.

## Synthesis and degradation

Many of the markers have not been developed specifically for monitoring growth but to assess changes in the balance between synthesis and degradation in adults. This is particularly true for bone markers where the formative and resorptive processes are so responsive to hormonal influences. Markers for measuring synthesis are usually directed at particular cellular activities or the concentrations of specific secreted proteins, whereas those for degradation usually detect protein fragments released from tissue, but may also

*A summary of currently available biochemical markers of bone growth and metabolism*

| Marker | Fluid | Assay type | Specificity | Main advantages | Main drawbacks |
|---|---|---|---|---|---|
| Hydroxyproline | Urine | Colorimetric, HPLC | Collagens from all tissues and other components such as C1q | Gives overall estimate of the growth and turnover of all tissues | Requires gelatin-free diet |
| Galactosyl-hyroxylysine | Urine | HPLC | Mainly derived from bone collagen | Good marker of bone modelling and remodelling | Technically demanding assay; may reflect degradation of newly synthesized collagen |
| Deoxypyridinoline | Urine | ELISA, HPLC | Almost exclusively derived from bone | Direct immunoassay available; good marker of bone modelling and remodelling; correlates with growth | |
| Pyridinium crosslinks | Urine | ELISA, HPLC | Bone and skeletal tissue | Direct immunoassay available; good marker of bone modelling and remodelling; correlates with growth | |
| N-telopeptide (NTx) | Urine | ELISA | Degradation of collagen type I | Direct immunoassay available; correlates well with growth-rate | |
| C-telopeptide (CTx) | Urine | LISA | Degradation of collagen type I | Direct immunoassay available | |
| Osteocalcin | Serum | ELISA, RIA | Synthesized by osteoblasts | Gives estimate of bone formation | Immunochemical heterogeneity; dependent on serum collection and storage conditions |
| Procollagen type I C-propeptide (PICP) | Serum | ELISA, RIA | Synthesis of all collagen type I but mainly from bone | Direct immunoassay; correlates well with growth-rate | Variable clearance rates may affect serum concentrations |
| Procollagen type III N-propeptide (PIIINP) | Serum | ELISA, RIA | Synthesis of all collagen type III in soft tissues | Direct immunoassay; good marker for soft tissue growth | Variable clearance rates may affect serum concentrations |
| Bone-specific alkaline phosphatase | Serum | RIA, microtitre plate immuno-activity assay | Synthesized by osteoblasts | Good marker for bone formation rate | |

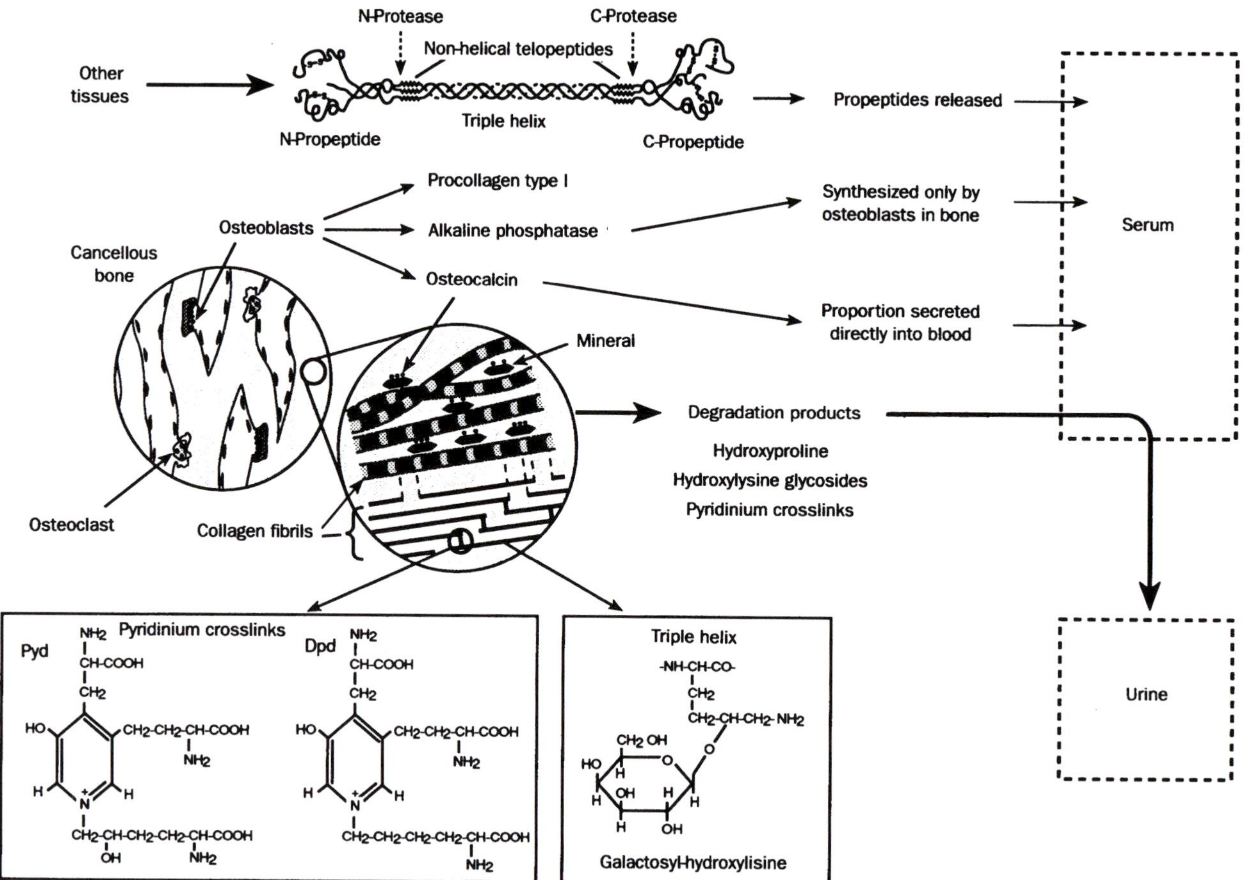


**Serum and urinary markers for bone formation and resorption.**

include activity measurements of degradative enzymes; both types of marker are useful for monitoring growth. Collagen markers fall into both categories because of the extensive biochemical processing of this protein during synthesis and maturation of the fibrils. Thus, the fibrillar collagens are synthesized as precursors that are 50 per cent larger than the molecules finally incorporated into fibrils; globular regions at both the N- and C-terminal ends of the molecule are removed during biosynthetic processing and these procollagen fragments can be measured in serum as indicators of the rate of collagen synthesis.

For measuring tissue degradation rates a number of different fragments derived from collagen have been used. Urinary hydroxyproline is the archetypal assay for growth and collagen degradation, but this procedure is known to have many drawbacks. Most important is the fact that hydroxyproline is present in several proteins other than collagens, including serum components with rapid turnover. Also, not all collagen precursors synthesized are processed and incorporated into fibrils; a proportion (15–40%) are degraded intra- and extracellularly, contributing to the hydroxyproline pool. Another major disadvantage of urinary hydroxyproline as a marker is the large contribution from dietary sources if a gelatin-free diet is not observed.

Some of the problems with hydroxyproline as a marker are overcome by using galactosyl-hydroxylysine, a metabolite of collagen largely present in bone. More recently, however, measurements of collagen cross-links has become the method of choice for measuring collagen degradation, with significant advantages over other collagen degradation markers. Pyridinoline (Pyd) and deoxypyridinoline (Dpd) are products of the lysyl oxidase-mediated cross-linking system of collagen. As these cross-links are formed extracellularly during the final stages of maturation of collagen fibrils, the compounds act as markers only of insoluble collagen, avoiding any contribution from degradation of newly synthesized material. During tissue degradation the cross-links are released and are not metabolized further so that their concentration in urine or serum indicates directly the amount of bone resorbed. Because of its restricted location in mineralized tissue, Dpd is essentially a specific marker for bone; both of the cross-links are absent from skin, thus providing more tissue specificity than for hydroxyproline measurements. More recently, another type of immunoassay has been developed based on measurements of peptide sequences of collagen type I around the sites of the cross-link formation, either at the N- or C-terminal end: these are referred to as NTx and CTx assays respectively. Because the presence of the pyridinium cross-link is not necessary, these assays are not specific for bone but for collagen type I from any tissue: this does not, however, affect their utility as growth markers.

## Non-collagen markers

There is a large number of protein components that have been suggested as markers, but relatively few that have been used widely. Again, most of the well-utilized markers relate to the skeleton. Osteocalcin (or Bone Gla Protein) has been studied extensively as a bone formation marker. This 49-amino acid residue protein is characterized by its content of 3 γ-carboxyglutamic acid residues added post-ribosomally through a vitamin K-dependent process. Synthesized almost exclusively by osteoblasts, the protein is largely taken up by the mineralizing bone matrix but a proportion, assumed to be constant, is secreted directly into blood. To act as a true bone formation marker, therefore, assays should recognize only intact osteocalcin and not fragments released into the blood from resorbing bone. Osteocalcin expression is dependent on 1,25-vitamin $D_3$. In addition to the dependence on vitamins D and K status, the assay is susceptible to any hemolysis during blood collection as this leads to degradation of the protein. Nevertheless, serum osteocalcin values are about fourfold higher in the first decade of life than in adults and the pubertal growth spurt is detectable.

Bone-specific alkaline phosphatase provides an additional bone formation marker as the concentration of this enzyme in blood reflects the activity of osteoblasts. The main difficulty in measurement is distinguishing between the liver and bone forms of the enzyme in blood. These isotypes differ in their carbohydrate composition and can be separated using lectins, but more recently direct immunoassays for the bone form have become available.

Most of the biochemical markers developed to monitor bone metabolism are also suitable markers to monitor growth, with three- to twentyfold higher values during growth compared with adult values. Most of the markers also correlate with linear growth velocities and are useful, therefore, in monitoring conditions of impaired growth, such as growth-hormone deficiency and chronic steroid treatment for asthma.

Simon P. Robins

See also 'Radiographic assessment' (**1.8**), 'Skeletal development' (**5.6**), 'The human growth plate' (**5.7**), 'Adulthood and developmental maturity' (**5.17**), 'Identification of abnormal growth' (**7.1**), 'Catch-up growth in height' (**9.6**) and 'Ageing as part of the development process' (**12.3**)

# Radiographic assessment

The ability of radiographs to distinguish between hard and soft tissue allows such procedures to be used to monitor the growth and development of a child *beneath* the superficial aspect provided by anthropometry. Radiographic assessment procedures were originally, and are still most commonly used, to assess skeletal maturation (**5.17**), but have also been used to assess dental maturity, and somatic growth through changes in bone, muscle and fat, and growth in a number of organs, such as the heart and lungs.

## Somatic growth

A number of major growth studies in America (Harvard, FELS, Philadelphia, and Denver) and England (Harpenden, Oxford and Forsythe) in the middle of this century subjected their participants to radiographic procedures. As a result it was possible to analyse the changes occurring in soft and hard tissue, notably of the limbs, during growth. The procedure in the Harpenden Longitudinal Growth Study involved radiographing the limb in either a lateral (arm), or anterio-posterior (calf) position using a constant anode-film distance of 2.5 metres to minimize magnification error. At all ages the central vertical plane of the limb was positioned a set distance from the film (5 cm for the arm and 10 cm for the calf) so that the percentage magnification remained constant from one age to the next. Safety measures included the coning of the radiographic field and the requirement for participants to wear protective lead shields. Measurements on the radiographs were made with callipers, measuring to 0.1 millimetres, at a horizontal plane identified by a radio-opaque marker that had been put on the limb prior to radiation. While in some studies the placement of the radio-opaque marker was arbitrary and fixed for convenience it usually corresponded with the middle of the 'upper arm', i.e., the midpoint between the lateral end of the acromion and the olecranon, and the maximum muscle diameter of the calf. Changes in the proportions of fat, muscle, and bone were similar for all studies.

## Skeletal maturity

The major use of radiography in measurement and assessment has been to estimate skeletal maturity. Two techniques to estimate skeletal maturity using the hand and wrist have emerged since 1937 to become the most widely used and frequently researched techniques available. These techniques are known as (*Atlas* and *bone-specific scoring* techniques. The former has its origins in the work of the American T. Wingate Todd who published an *Atlas of Skeletal Maturity* in 1937. Todd's *Atlas* was updated and modified by W.M. Greulich and I. Pyle in 1950. The latter technique was due to the work in Britain of J.M. Tanner who, with his colleagues R.H. Whitehouse and M.J.R. Healy, published their first system, known as TW1 in 1959, which was later superseded in 1983 by an improved version called TW2. Other attempts to quantify skeletal maturity have been published using the knee and foot and ankle, but only the *Atlas* and bone-specific techniques for the hand and wrist have survived in general usage.

### *Atlas of Skeletal Maturity*

Prior to the *Atlas* technique, attempts to formulate practical methods, using planimetry, to assess skeletal maturity had used the times of appearance of ossification centres, the time of appearance of the 'newest' ossification centre, or the area occupied by bones. Milo Hellman's serialization in 1928 of changes in the growth cartilage plate during fusion prompted Wingate Todd to realize that observing sequential changes that had been identified through serial studies of maturing bones could lead to a practical system for assessing maturity. These changes were called *maturity indicators* and were defined as features of individual bones that can be seen in a roentgenogram (radiograph) and which, because they

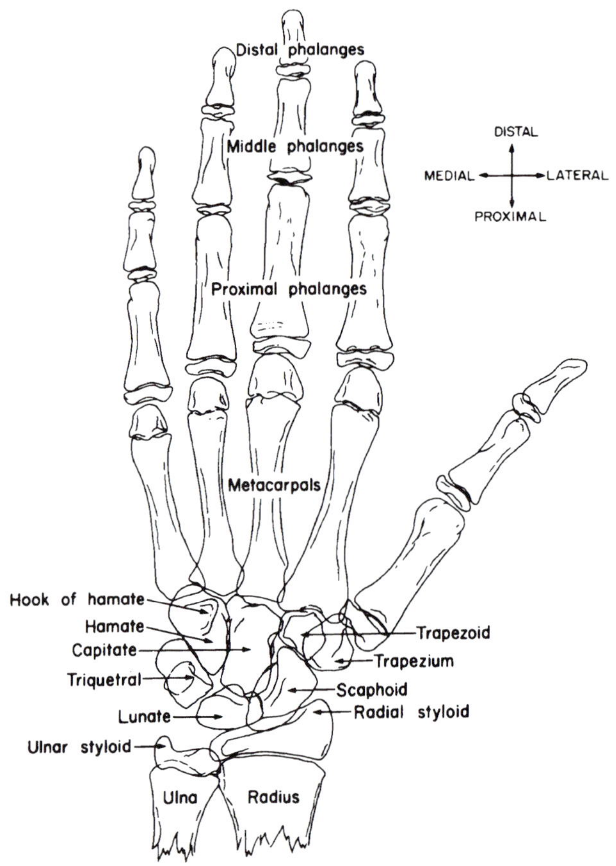


**Bones of the hand and wrist. From Tanner, Whitehouse, Cameron et al., 1983.**

tend to occur regularly and in a definite and irreversible order, mark their progress towards maturity.

The isolation of a particular stage (maturity indicator) in the continuum of maturity from birth to adulthood made it theoretically possible to compare the temporal relationships between children if standards reflecting the sequential changes occurring in normal children could be constructed. Thus Todd required 1) the pattern of change in each particular bone, and 2) an integrated picture of the changes that would be found occurring in a child of known chronological age and normal rate of maturation. These data were obtained from 1000 participants in the Brush Foundation Study of Human Growth and Development at the Western Reserve University School of Medicine in Cleveland, Ohio. While the majority of participants had been studied longitudinally, the data set was expanded at adolescent ages with the addition of cross-sectional data from various 'public schools' and 'social agencies' in the Cleveland area. The 'Todd Atlas' that resulted from this work in 1937 was later modified by William Walter Greulich and Idell Pyle and resulted in the *Greulich–Pyle Atlas* in 1950. Greulich and Pyle's modifications included increasing the number of Brush Foundation subjects at adolescence, reducing the number of photographic standards that constituted the *Atlas*, and adding a graphical method to describe skeletal status (known as Pyle's Red Graph) that disclosed any increase or decrease in the maturity differences between the least mature and most mature bones in response, so it was thought, to insults to an individual's health. The second edition of the *Greulich–Pyle Atlas* in 1959 improved the quality of the original standard photographic plates and the drawings of the maturity indicators, and added a method of adult height prediction developed by Nancy Bayley in 1952.

The *Atlas* contains a series of photographic plates of radiographs representing skeletal 'age' from birth to 18 years for males and females. Each facing page contains the skeletal ages for each bone. Following the male and female standards are detailed descriptions of the maturity indicators in each bone in addition to the number of the standard(s) in which they appear, for example, Ulna indicator II: 'The epiphyseal bone nodule is rounded and its margins are smooth. Male standards 14 and 15, Female standards 14 and 15'.

The recommended method for using the *Atlas* is that the radiograph is compared to the standards to orientate the assessor as to the likely development of the child. Each bone is then compared to the maturity indicators and a standard number assigned to each of the 30 bones. The median of these 30 numbers is used to determine the skeletal age. In practice, however, most assessors simply compare the radiograph to the standards and assign the skeletal age or interpolated skeletal age if the radiograph is intermediate between standards. The use of this sort-cut method is obviously open to erroneous interpretation and ignores the maturity indicators on which the *Atlas* system is based.

### *Bone-specific scoring techniques*

Bone-specific scoring techniques were developed to overcome the two main disadvantages of the *Atlas* techniques: the concept of the 'evenly maturing skeleton' and the difficulty of using 'age' in a system measuring maturity. The first attempt was the *Oxford Technique* of R.M. Acheson. Using radiographs on 500 preschool children from Oxford, UK, Acheson used the simple expedient of applying the scores 1, 2, 3, et cetera. to the first, second, third, et cetera, maturity indicators. This strategy, however, did not account for the fact that the passage from one stage to the next may involve very different changes in maturity according to which bone and which stages were involved. The rationale employed by J.M. Tanner (see Appendix 1, 22)and his colleagues in 1962 was that the development of each single bone reflected a single process which they defined as maturation. Therefore each of the scores from each of the bones in a particular individual should be the same. This common score, with suitable standardization,

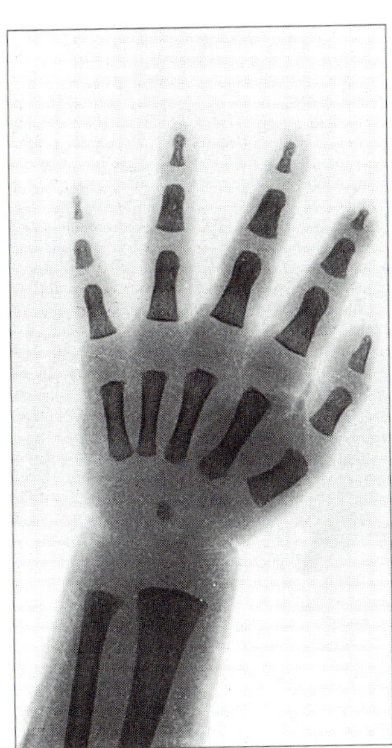
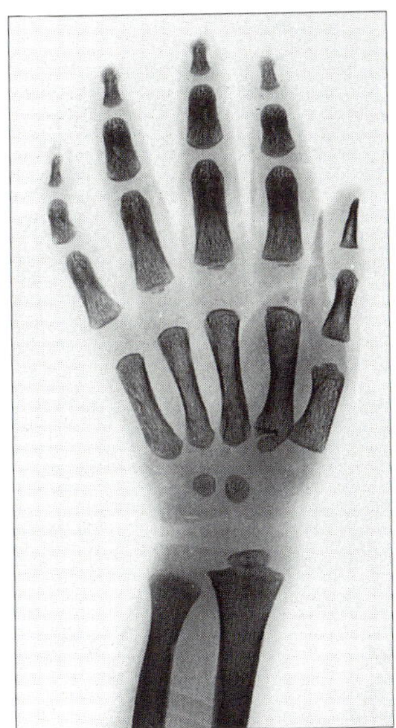
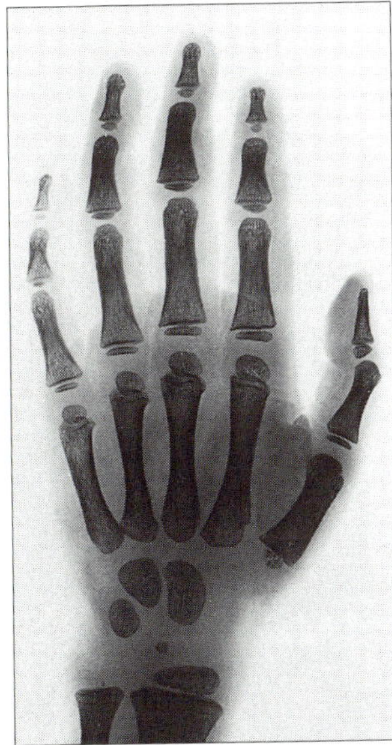
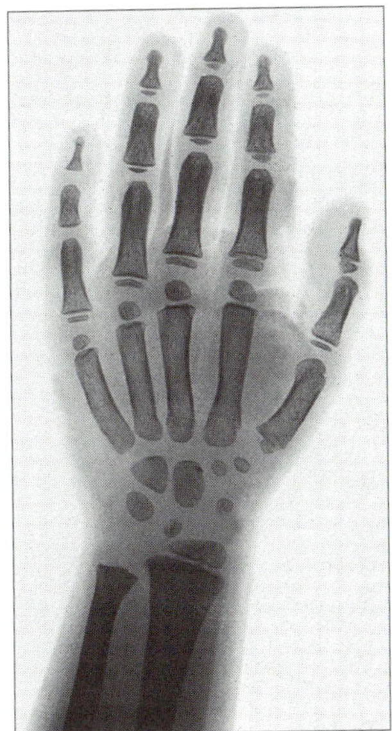

**Radiographs of the hand and wrist of males aged (from left to right): 3 months, 2 years, 4 years, 6 years.**

would be the individual's maturity. Practically, the scores would not be the same because, for example, there were large gaps between successive events in a single bone. The scores were therefore defined in a such a way as to minimize the overall disagreement between the different bones of an individual. The disagreement was measured by the sum of the squares of the deviations of an individual's bone score about their mean value. In addition, the constraints that each bone score should extend from 0 to 100 and that each bone should start from 0 were applied. In order that the final estimate of maturity should not be biased in favour of the more numerous metacarpal and phalangeal bones, the second and fourth rays were omitted from the calculations. Thus, in effect, the total maturity was a combination of 10 per cent for the radius, ulna and the three metacarpal-phalangeal rays and 7 per cent for each of the seven carpal bones. This TW1 system was updated in 1975 to create TW2. TW2 featured:

- A combination of certain end stages, for example, Stage 1 in the capitate, because they were difficult to rate.
- The removal of the constraint of a 0 to 100 score for each bone to be replaced by a constraint that the whole scale, rather than each bone, should be 'anchored' to pass from 0 to 1000.
- The introduction of sex-specific scores.
- An increase in the source sample.
- A change in the technique of plotting Bone Maturity Score against age for greater convenience.
- A choice of systems to allow skeletal maturity to be based on either the Carpus (CARPAL) or the Radius, Ulna, and Short bones (RUS) or a combination of the 13 RUS bones and the 7 CARPAL bones called 'TW2 (20)'.

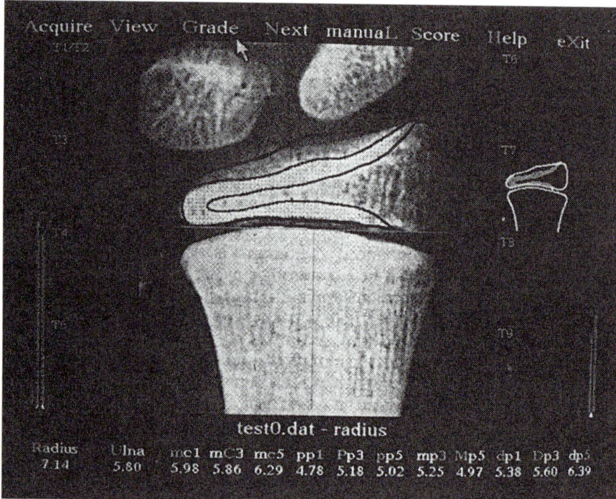


**The developing radius (wrist) at stage G of the TW2 system of skeletal maturity assessment**

The bone-specific scoring technique requires the grading of maturity indicators in 20 bones of the hand and wrist. The manual contains written criteria, drawings and photographs of the maturity indicators. Each bone has a series of eight or nine stages identified by the letters A to H or I. Having decided on the maturity indicator reflected in the radiograph a score is assigned to the bone depending on the sex and system being used, i.e., TW2 (20), RUS or CARPAL. These scores are summed to provide a Bone Maturity Score (BMS) which may then be used to assign a 'bone age' or plotted on a standard graph to compare to other children of the same sex.

*Noel Cameron*

See also 'Biochemical markers' (**1.7**), 'Skeletal development' (**5.6**), 'The human growth plate' (**5.7**), 'Adulthood and developmental maturity' (**5.17**), 'Identification of abnormal growth' (**7.1**) and 'Bone dysplasias' (**7.5**)

# Dental maturation

The timing of tooth formation and tooth eruption is strongly regulated by the individual's genotype. The environment can accelerate or delay the rate of development to a measurable extent, but most of the control seems to be inherent in the person's genetic make-up. Just as other genetic characteristics vary among individuals (like eye colour and stature) so the *tempo* of growth differs among people. The public relies on chronological age (CA) for social and legal definitions of how old a person is. An individual needs to be so many years old to begin school, drive a car, or vote. But, it is obvious that all children at a particular CA are not equally mature (**5.17**), either emotionally or developmentally, so it can be more useful to use some measure of a person's degree of biological maturity. Such measures yield a *physiological age* in contrast to CA. By definition, most children in a population are 'average' growers because their CA and biological ages are similar to one another. At one extreme, though, are the early maturers whose biological age appreciably exceeds their CA. Conversely, slow-growing late-maturers are those in whom biological age lags substantially behind CA.

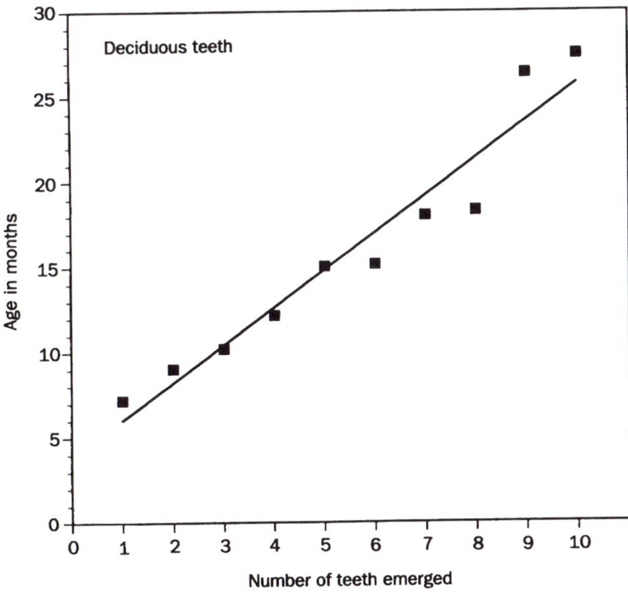


**The 20 deciduous teeth erupt into occlusion between about 8 months and 2.5 years. Left and right teeth are combined in this plot. Emergence typically begins with the mandibular central incisors, and the second molars are the last to emerge.**

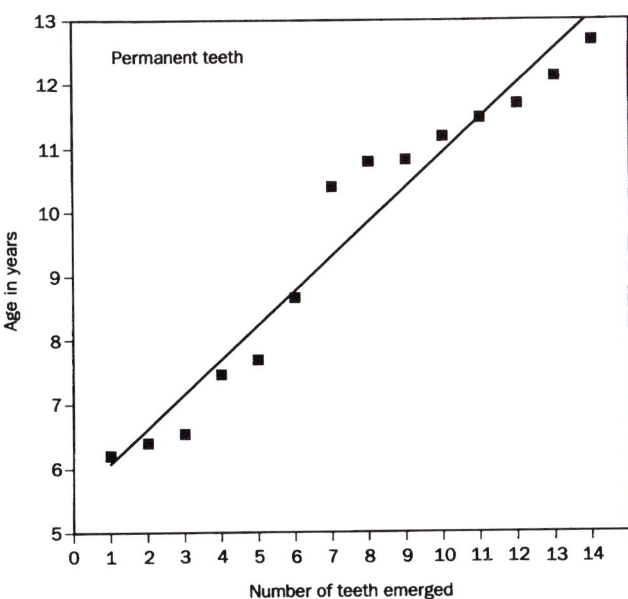


**Emergence of the permanent teeth is only roughly linear with time. Left and right sides are pooled here, and the highly variable third molars are excuded. The first grouping of teeth, from about 6 to 9 years of age, consists initially of the first molars, followed by the four permanent incisor tooth types. The second grouping (ages 10–12 years) begins with emergence of the canines and premolars and ends with emergence of the second molars.**

Several key events of maturity occur around puberty (**5.15**), such as deepening of the voice and growth of facial hair in boys, and the onset of menstruation and breast development in girls, and these are useful in defining one particular stage of development. What is needed, however, is a way of measuring biological age that is uniformly applicable across most or all of the growth span between birth and adulthood. There are just two such tissue systems that change progressively throughout this long interval of about 20 years: one is bone formation and the other is tooth formation. The intent in assessing tooth formation is to quantify the degree of dental development in an individual, their *dental age*, and use that directly as a measure of biological age or compare it to CA.

## Tooth eruption

Various kinds of information can be collected to arrive at dental age. The simplest approach is to count the number of teeth erupted into the mouth. This can be done using either the primary (deciduous) teeth, which begin emerging through the gingiva about 8 months of age and are all in occlusion by about 30 months, or the permanent teeth, which begin about 6 years of age and conclude with emergence of the wisdom teeth around the age of 20 years. 'Emergence' has to be interpreted with caution because there are at least three operational definitions. One is the appearance of a tooth as it breaks through the gingiva and is visible on oral examination. Another is to count those teeth that have completed eruption and are in functional occlusion. When dealing with radiographs of teeth and jaws and with skeletal material, there is no gingival boundary, so researchers often use *alveolar* emergence, denoting that at least part of a tooth's crown has erupted beyond the bony margin of the jaw. These differences in definition can lead to conflicts in comparing data based on different criteria.

The number of teeth can be used in assessment, but this is not a linear function of time; instead, teeth erupt in clusters. Left and right antimeres erupt at nearly identical ages, with no side preference. There is a small but systematic arcade difference, with mandibular teeth erupting a few months ahead of their maxillary analogues. A more pronounced difference is between the sexes; as a statistical average, the teeth of females emerge at earlier chronological ages than those of males. Sex differences can be up to about 1 year, and the emergence of the canine tooth typically shows the greatest difference.

The 8 incisors and the first molars in the deciduous dentition emerge within a few months of one another. The canines erupt at about a year and a half, followed by a pause until the second molars emerge at about 28 months of age. There are two clusters or transitions of emergence in the permanent dentition, one from about 6 to 8 years of age, when the incisors and the first molars emerge, and the second from about 10 to 12 years, when the canines, premolars, and second molars emerge. This leaves the third molar (M3), which in many regards is the most variable tooth. The median age for alveolar emergence of M3 in American whites is about 18 years. A unique feature of the third molar is that it typically erupts at an earlier age in boys than girls. All other permanent teeth typically emerge at earlier chronological ages in girls. Precocious emergence of M3 in males may relate to the fact that this is the only tooth to erupt post-pubertally; its tempo may be influenced by androgenic hormones, but this is speculative.

## Tooth mineralization

Most events of tooth formation occur prior to eruption (which begins roughly when the root is half-formed), and much more data are available if the stages of crown-root formation can be imaged. Such stages are obvious on skeletal material where the degree of formation can be viewed directly. Most interest focuses on the living, however, and standards have been developed for use with radiographs of the teeth and jaws.

Tooth formation proceeds in an invariant sequence. The first radiographic evidence is formation of a bony crypt, followed by mineralization of the crown tips. Some researchers discuss tooth *calcification*, but calcium is not even a major constituent of most dental tissues; *mineralization* is a more appropriate term. Mineralization proceeds from the crown tips down the sides of the presumptive tooth. Root mineralization does not begin until crown formation is complete, and root formation ceases with the reduction of the apical foramen. This reduction is called apexification. Tooth mineralization is a continuous process, but it can be partitioned into a set of ordinal-scale grades against which the extent of formation can be scored.

Since the mineralization of the deciduous dentition begins *in utero*, early in the second trimester, and the root formation of M3 may not be complete until after 20 years of age, tooth mineralization as an indicator of biological age is applicable over a broader span of life than any other measure. Dental age (DA) is assessed by comparing a child's tooth-formation status with normative data. The degree of completeness of each forming tooth is compared against an ordinal scheme, and the most advanced stage that the tooth has achieved is scored. It is not desirable to interpolate to the closest stage since that can overestimate the child's degree of biological maturity. Reference tables are then consulted to find the average chronological age at which the normative sample attained that stage. This is done for each scorable tooth, and the average is the person's dental age. In practice, it can be useful to calculate the DA–CA discrepancy for each tooth as well as the composite for the whole dentition to test whether some tooth types are affected disproportionately.

## Modulation of dental maturation

Most adverse conditions can slow down the formation and emergence of the teeth, and it seems that factors that modulate tooth formation affect emergence in the same manner. Well-documented examples of such modulation include various genetic diseases, including Down syndrome, cleidocranial dysostosis and ectodermal dysplasia, as well as growth-hormone deficiency, chronic undernutrition, recurrent diseases, and low socio-economic status. Conditions that cause somatic growth to slow down, including slow mitotic activity and protein synthesis also slow the rate of tooth formation and eruption.

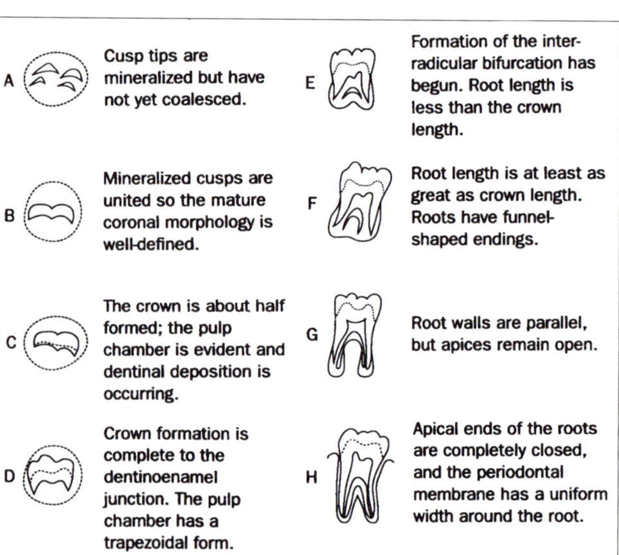


**An eight-grade scheme showing how the continuum of crown–tooth formation can be partitioned into visually assessable morphological grades. A similar system can be used for the anterior, single-rooted teeth. Tables of normative values are available in the literature that provide the race- and sex-specific average chronological ages at which each stage occurs in normal children.**

Factors enhancing growth rates are the opposite of those causing growth to slow down, and include genotypic normalcy, adequate diet, freedom from disease, and high SES status (which probably enhances the likelihood of these other factors occurring). Local increases in dental maturation that affect just some teeth can be brought on by conditions such as tumours and hemifacial hypertrophy that proliferate the vascular supply to some teeth, but such conditions are rare.

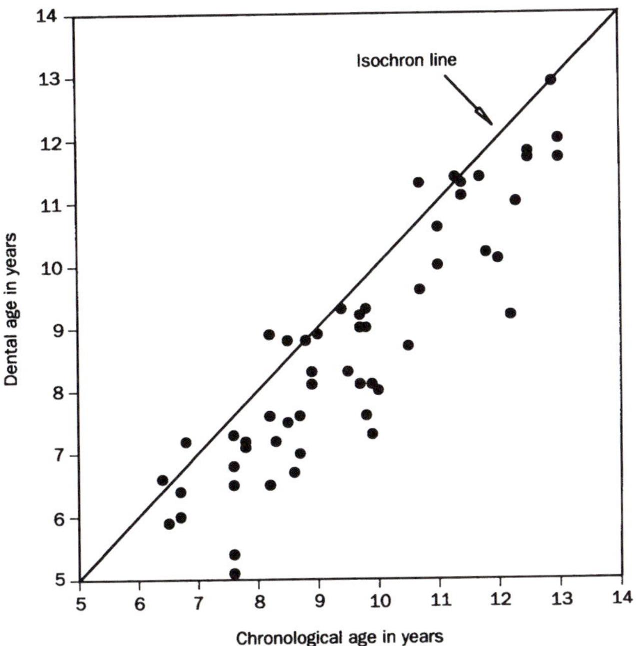


Chronological and dental ages are approximately the same in normal-growing children, so the disribution of cases should be centred on the isochron line. In this sample of children with left lip and palate, 90% (48/54) of the children have dental ages that lag behind their chronological ages and are clustered below the isochron line.

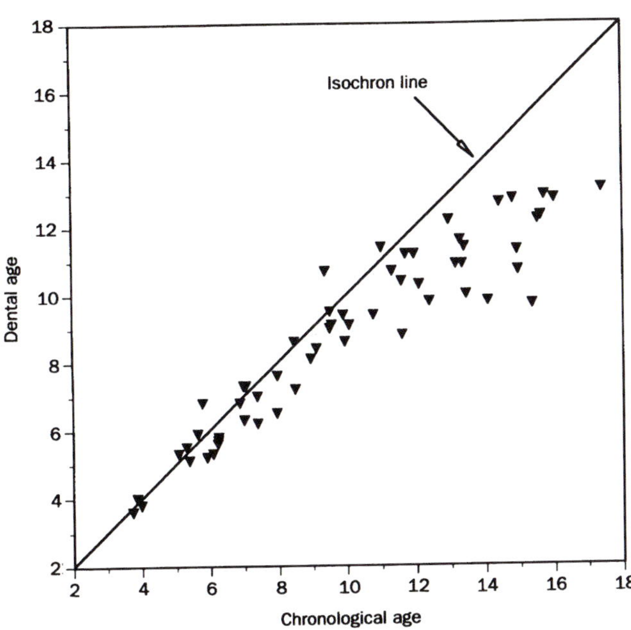


Averaging across the scorable (i.e., incompletely formed) permanent teeth is a sample of American children with sickle cell disease shows that the majority have delayed dental ages, and the delay is disproportionately large in the older, teenage cases.

There is evidence of a secular trend in dental maturation. In Westernized countries (where there are long-term records), teeth now form and erupt at earlier chronological ages than in the past. This trend probably is tied to improved nutrition and lower morbidity. Perhaps the greatest influence on the tempo of dental maturation is ethnicity. Ethnic differences, which include differences in genes regulating the tempo of growth, generally exceed sex differences in magnitude. Europeans are rather slow maturers. It has long been documented that sub-Saharan Africans (including African-Americans) develop their dentitions sooner than whites. Although African-Americans are about 12 per cent of the US population, there are few dentition standards for them, and almost nothing is known about other segments of the population, such as Mexican-Americans or groups from Southeast Asia. What data exist for Asian populations suggests that their teeth mature at rates intermediate between those of Europeans and those of Africans.

## Applications

Dental ageing has a number of applications, two of which are described here. The first is from the study of children with isolated cleft lip and palate. The majority of these have delayed

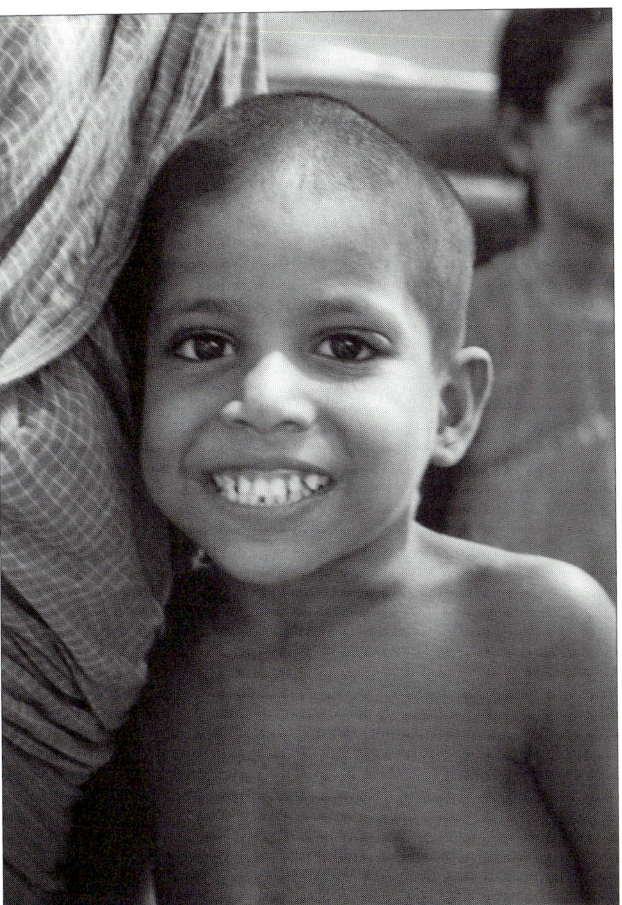

Indian child. Photograph by P. van Someren.

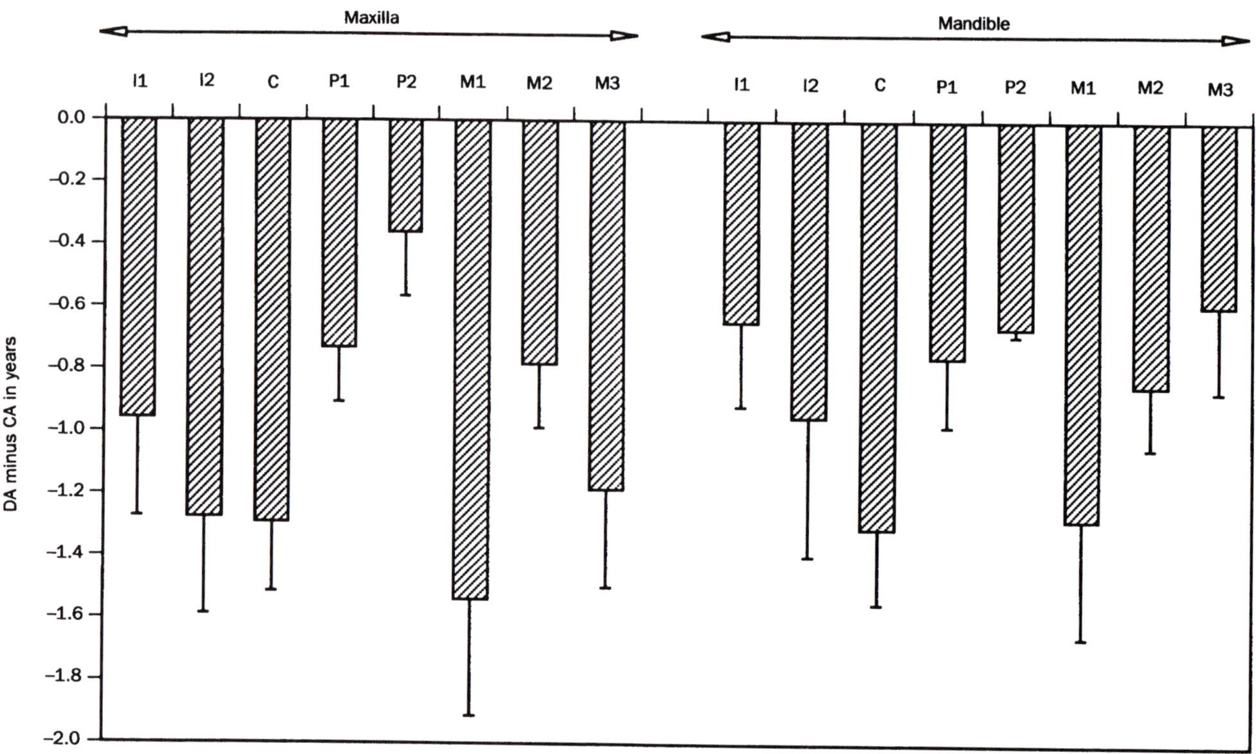


**This plot of the average DA–CA difference for each permanent tooth type shows that children with sickle cell disease are appreciably delayed in their tempo of growth. In each instance, the mean difference is negative and significantly below zero (where DA = CA).**

dental development, so that DA–CA is always negative. The presumed causes of such delay involve the debilitating post-natal effects of feeding problems, recurrent middle ear and upper respiratory infections, repeated surgeries, and, perhaps, psycho-social stress in children with cleft lip and palate.

The second example is from a study of American black children with sickle cell disease (SCD), a genetic abnormality of the beta-hemoglobin chain that reduces its capacity to transport oxygen and causes it to polymerize, leaving the child with chronic hemolytic anemia. Children with SCD have dental ages significantly delayed behind their chronological ages compared to American black children without SCD. Averaging over all scorable teeth in each individual shows that the majority of children have delayed DA, with older individuals being affected disproportionately more.

*Edward Harris*

See also 'Embryonic development of teeth' (**4.4**), 'Post-natal craniofacial growth' (**5.9**) and 'Variation in time of tooth formation and eruption' (**5.10**)

# Physical examination

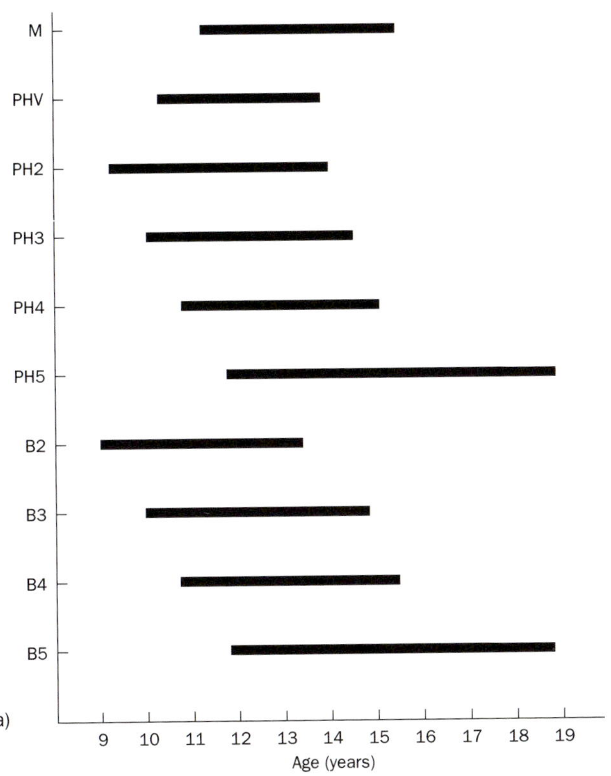


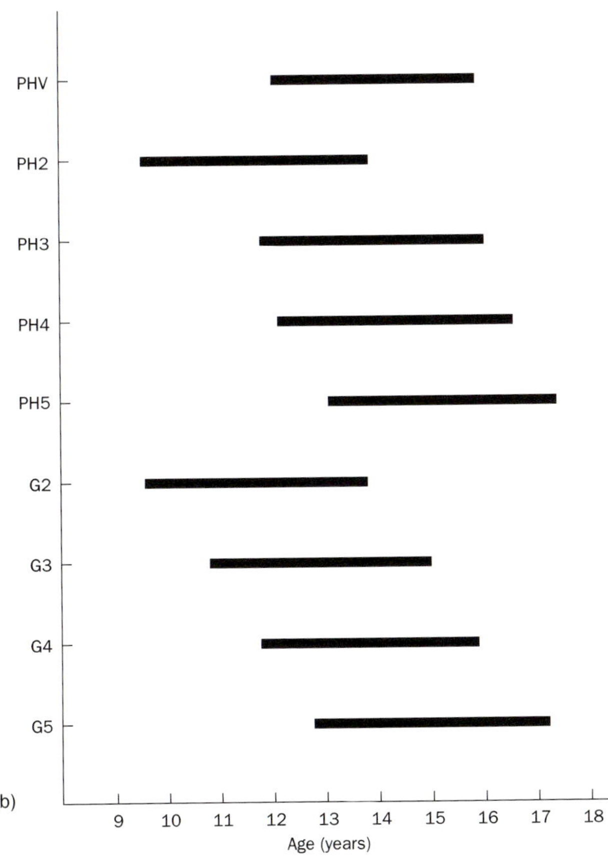


Physical examination associated with growth and development during infancy and childhood is dominated by measurements of weight, length and head circumferences as indices of normal development. Body length is measured in the supine position until about 2 years of age when the child can thereafter be measured standing. The relationship of weight to height is important to record as a sign of a possible disease process. For example, a child underweight for length may have a malabsorption problem, whereas the converse may suggest an endocrinopathy.

Across childhood, physical examination is less important than body measures (**1.4**) relative to previous measures and/or reference values. However, it becomes more relevant again at the time of puberty. Distinct morphological changes become apparent at puberty (**5.5**, **5.15**) in response to the reactivation of the hypothalamo-pituitary-gonadal network already imprinted during fetal and early post-natal life. Puberty may be defined as the development of secondary sexual characteristics arising from increased gonadal steroid secretion which ultimately contribute to the acquisition of reproductive capacity. It is a continuation of physical changes which takes place over a number of years. The starting age is highly variable, but by convention, onset of signs can vary from 8–14 and 9–15 years in normal girls and boys, respectively.

## Puberty in the female

The first sign of puberty in the female is breast-budding due to an elevation of the breast and papilla and an increase in diameter of the areola. This occurs on average at 11–13 years of age, and is termed stage B2 by convention. Full development takes about 4 years, through to maturity at stage B5. Almost concomitant with breast-budding is the onset of pubic hair growth (stage PH2) which usually starts first on the labia. Again there are 5 stages until the inverse triangular distribution of hair over the mons pubis is reached. In some women, there is further hair growth on the medial aspect of the thighs and along the linea alba. Axillary hair growth starts slightly later, at about 12.5 years, and is complete within 18 months. Physical changes occur to the vagina, vulva, uterus and cervix in response to increased estrogen. There is also increased fat deposition especially around the hips and mons pubis.

**Range of age of onset of different signs of puberty: (a) females, (b) males. Onset of menarche (M), peak height velocity (PHV), pubic hair stages 2, 3, 4 and 5 (PH2, PH3, PH4, PH5), breast stages 2, 3, 4 and 5 (B2, B3, B4, B5) and genital stages 2, 3, 4 and 5 (G2, G3, G4, G5). Reproduced with permission from *Clinical Paediatric Endocrinology*, 3rd edition. Ed C.G.D. Brook, Blackwell Scientific Publications.**

The growth spurt starts early in puberty and peak height velocity occurs around 12 years of age. By this stage, about 90 per cent of adult height has been achieved. Menarche occurs quite late in puberty and when the growth-rate is declining. The mean age of menarche in Britain is currently 12.9 years. Normal puberty is characterized by the harmonious interplay of the various stages.

## Male puberty

A similar 5-stage process is assigned to quantitating physical signs in the male. The first sign of puberty in boys is an increase in testicular size from a prepubertal volume of 2–3 millilitres to a size of 4 millilitres. This can be quantitated by comparing the ovoid shape of the testis with a set of volumetric standards ranging from 1–25 millilitres, the Prader orchidometer. With the onset of this sign there is also a change in the texture and pigmentation of the scrotal skin and then a gradual increase in the size of the penis, firstly in length and later accompanied by an increase in breadth as the glans penis develops. Pubic hair growth starts slightly later than genital development, with the first hair appearing either at the base of the penis or in the scrotal skin. The 5 stages are similar to those described for the female, but many adult males have hair spreading up the abdominal wall. Axillar hair development is quite late, starting around 14 years. Facial hair growth starts initially at the lateral ends of the upper lip area, then in the side-burns and much later over the chin when genital development is fully complete. Voice-breaking is often referred to by lay persons as the sign of puberty starting in boys. In fact it occurs quite late in the pubertal process but fairly abruptly between stages G3 and G4. This coincides with a decrease in the fundamental voice frequency and a lengthening of the vocal cords, as shown by ultrasound. There is also some breast development in the male at puberty which is quite easily palpable in more than 70 per cent of boys. It then usually regresses spontaneously. The growth spurt starts late in puberty with the peak height velocity occurring around 14 years of age and between stages G4 and G5. Boys are taller than girls at the respective onset of the growth spurt because of their extra 2 years of prepubertal growth achieved. This, together with the slightly greater magnitude of the growth increment at puberty, explains why males are generally taller than females in adulthood.

*Comparative physical features of female and male puberty*

| Physical features | Female | | Male | | |
|---|---|---|---|---|---|
| | Age (years) | Stage | Age | Stage (years) | Testis volume (ml) |
| Onset of puberty | 11.3 | B2 | 11.8 | G2 | 4 |
| Peak height velocity (cm/yr) | 12 | B3 | 14 | G4–5 | 12 |
| Axillary hair | 12.5 | AH2 | 14.3 | AH2 | 12 |
| Menarche | 12.9 | B4 | – | – | – |
| Spermarche | – | | 13 | G3 | 8–10 |
| Voice 'breaking' | – | – | 14.5 | G3–4 | 10–12 |
| Total height increment (cm) | | 22 | | 28 | |

B, breast; G, gental; AH, axillary hair: stages according to Tanner

## A comparison of male and female pubertal development

It is a misconception that girls mature much earlier than boys. The onset of genital development in boys occurs only about 6 months later than the age of onset of breast-budding in girls; clearly, such a sign is more overt than testicular enlargement. The main difference in pubertal development between males and females is the much earlier growth spurt in girls. Females also experience menarche (**1.15**). There is no such equivalent in the male, although the term 'spermarche' has been coined to define the onset of spermatogenesis which can be documented by the appearance of sperm in an early morning urine specimen. Interestingly, this is a relatively early event which occurs before the growth spurt when the testicular volume is 8–10 millilitres. Musculoskeletal differences are quite marked between the sexes at puberty. Changes are more profound in the male with respect to trunk and limb length, breadth of shoulders and hips, and muscle mass. The total growth achieved during puberty in the male exceeds that in the female by about 6 centimetres. The spurt in sitting length, as an indicator of trunk length, continues for about 6 months longer in the male. The accentuated muscle strength is more profound in the upper body muscle mass.

*Ieuan A. Hughes*

See also 'Assessment of age at menarche' (**1.15**), 'Hormonal regulation of growth in childhood and puberty' (**5.14**), 'Sexual maturation' (**5.15**), 'Chromosome aberrations and growth' (**7.4**), 'Teenage pregnancy' (**8.6**) and 'Between-population differences in adolescent growth' (**10.5**)

# Adiposity rebound and prediction of adult fatness

During growth, weight increases regularly both by age and by height. This mainly reflects an increase in body-size. Conversely, fatness development, assessed by either skinfolds or by the weight/height[2] body mass index (BMI) (**1.16**) shows ascending and descending phases. For the prediction of adult status from childhood measurements, absolute changes are of less concern than changes across percentiles of the reference distribution. Tracking in a measurement implies maintenance of the same rank order among a group of individuals across age. Tracking of fatness is well documented, most studies showing that the degree of continuity is greater between childhood and adolescence than between infancy and childhood. These differences between successive periods of growth can be clearly understood by following the BMI curves. Like skinfold thickness, the BMI curve rises during the first year of life, so that by the age of 1 year children look chubby. They then slim down, so that by the age of 6 years, a child on the 50th centile looks thin. If, at this age, a child is on the 10th centile, he or she looks severely underweight, but on the 90th centile they look just normal or slightly overweight. Subsequently, fatness increases, this increase being termed the *adiposity rebound*. Since the duration of fatness decrease after age 1 year varies from child to child, the adiposity rebound can occur at between 4 and 8 years of age. In general, the earlier the rebound, the higher is the adiposity at the end of growth. However, in most cases, an early or late rebound regulates transient under- or overweight of the first years of life. This explains why before the age of 8 years most individual curves cross percentiles: the majority of fat children do not stay fat as they join the average after a late rebound, and thin children can join the average or occasionally become fat after an early rebound. The adiposity rebound usually occurs before the age of 8 years, and thereafter most children follow the same percentile until the end of growth. The child–adult correlation in BMI values reaches its highest level at this time, after a steep increase from age 1 year.

It is often reported that adolescence is a critical period for the development of obesity. However, changes observed at

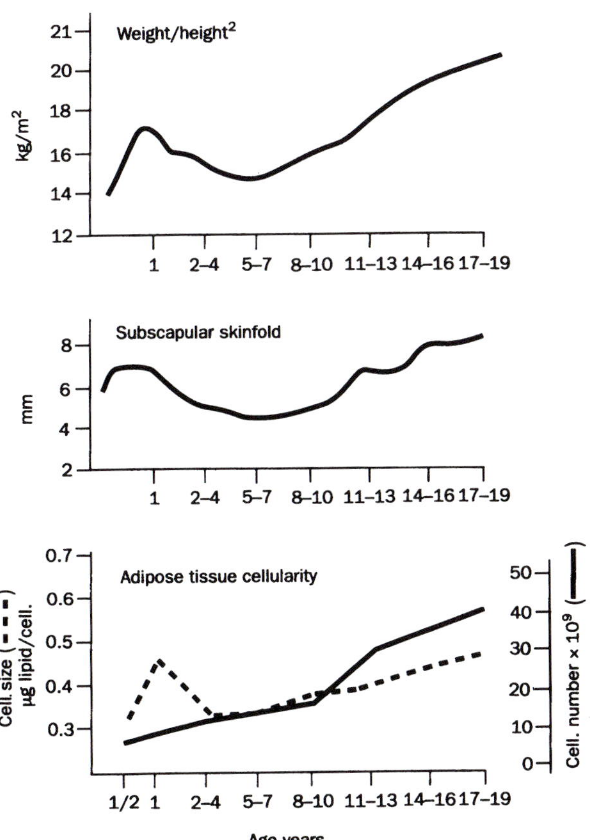


**Development of fatness as assessed by BMI and skinfolds and corresponding changes of adipose tissue cellularity.**

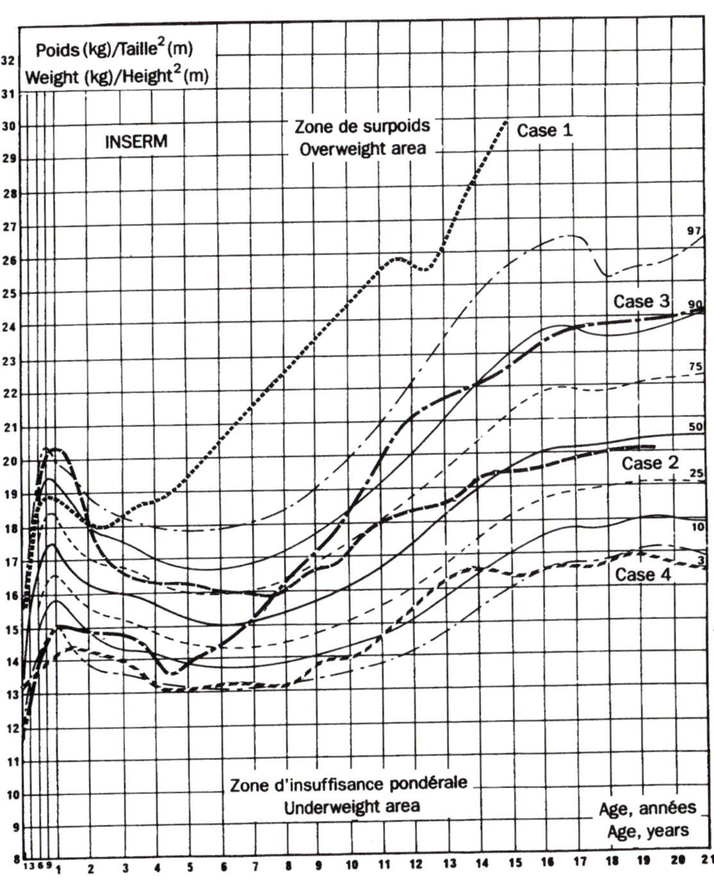


**Four examples of BMI development: Case 1, fat child at 1 year, who remained fat after an early adiposity rebound (2 years); case 2, fat child at 1 year, who did not stay fat after a late adiposity rebound (8 years); case 3, lean child at 1 year, who did not stay lean after an early adiposity rebound (4.5 years); case 4, lean child at 1 year, who remained lean after a late adiposity rebound (8 years).**

adolescence usually have their origins many years before. For example, many individuals have an early adiposity rebound at a time when they are lean, but their BMI curve rises relative to reference values from this time on, obesity revealing itself formally only many years later. Such unfavourable development cannot be detected using traditional weight- and height-for-age curves.

### Rate of maturation

An early adiposity rebound is a strong characteristic of BMI changes relative to reference values in almost all obese children. Among children who become obese, the adiposity rebound occurs at about 3 years of age, as compared to about 6 years for children who do not develop obesity. Age at adiposity rebound is associated with bone age: the earlier the rebound, the older the bone age. This is consistent with the well-known accelerated growth recorded in the obese.

### Adipose tissue cellularity

The pattern of BMI by age reflects the development of adipocyte cellularity: cell size increases during the first year of life and then decreases. Subsequently, cell number starts to increase from the age of 8 years. On the basis of individual BMI patterns, it has been suggested that transient obesity at the beginning of life could involve the increase of cell size, while persistent obesity starting with an early adiposity rebound would be associated with early (at 3 instead of 8 years of age) cell multiplication.

### Hormonal status

Up to the age of 6–8 years, levels of insulin-like growth factor-1 (IGF-I) are low, subsequently reaching adult values. The period of low IGF-I corresponds to the period of low adipocyte number, while the period of high IGF-I corresponds to the period of high multiplication of adipocytes. This observation is consistent with observations that IGF-I is involved in adipocyte multiplication. Adiposity rebound might reflect adipocyte multiplication and the early adiposity rebound recorded in the obese might correspond to precocious IGF-I production. The changes in hormonal status point out that growth takes place in clearly different periods: a period of maturation of the fat-free mass, with low adipocyte multiplication and steep decrease of height velocity which take place during the first 6–8 years of life (the *maturation phase*) is followed by a period where the chemical maturation of the fat-free mass is achieved, and adipocyte number and height velocity increase (the *multiplication phase*). An early adiposity rebound could reflect an accelerated growth but a shortened maturation phase. This mechanism could explain some aspects of immaturity in the body composition of the obese, such as low density of fat-free mass.

### Comparisons between countries

Differences in age at adiposity rebound between countries are consistent with differences in the prevalence of obesity. A late rebound is recorded in Senegal (8 years), while it occurs at age 6.2 years in France and 5.2 years in the United States. By the age of 2 years, the lowest BMI value is observed in the United States and the highest in Senegal, while by the age of 10 years, the highest BMI value is observed in the United States and the lowest in Senegal. These differences can be explained by a shift of the BMI curves towards younger ages in industrialized countries, reflecting accelerated growth.

Early-life measurements of fatness are poor predictors of adult fatness. The duration of fatness decrease after the age of 1 year seems to be a better predictor than fatness in early child-

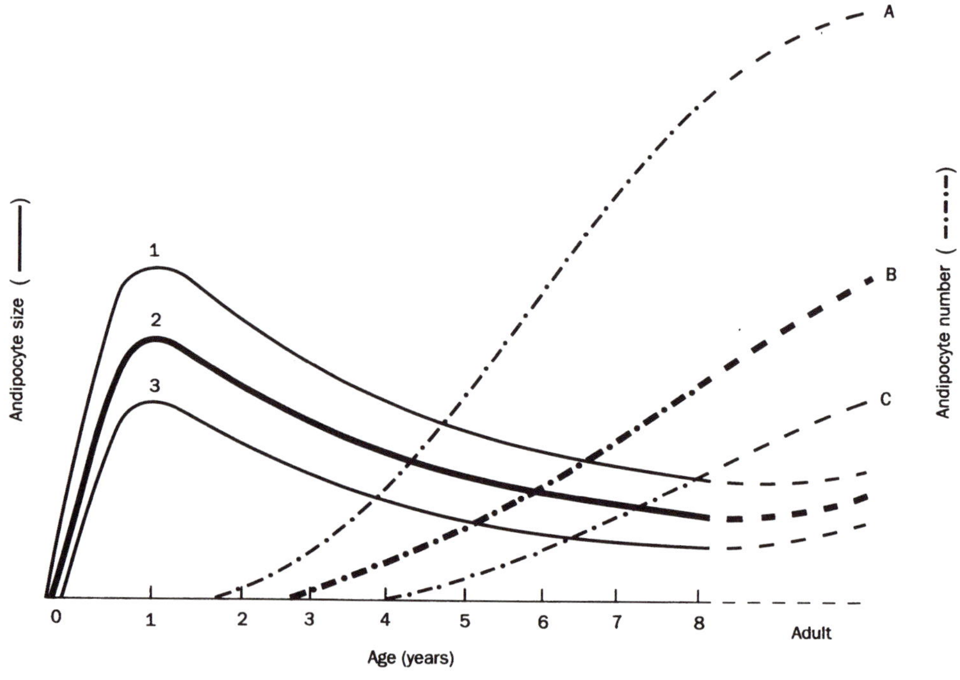


**An hypothesis for explaining different body mass index (BMI) patterns during growth, according to adipocyte size and number. For example, size (2) letter (B) correspond to a median BMI pattern, with a mean age at adiposity rebound (6 years). Size (1), letter (A) correspond to a precocious multiplication of large adipocytes after an early adiposity rebound (4.5 years).**

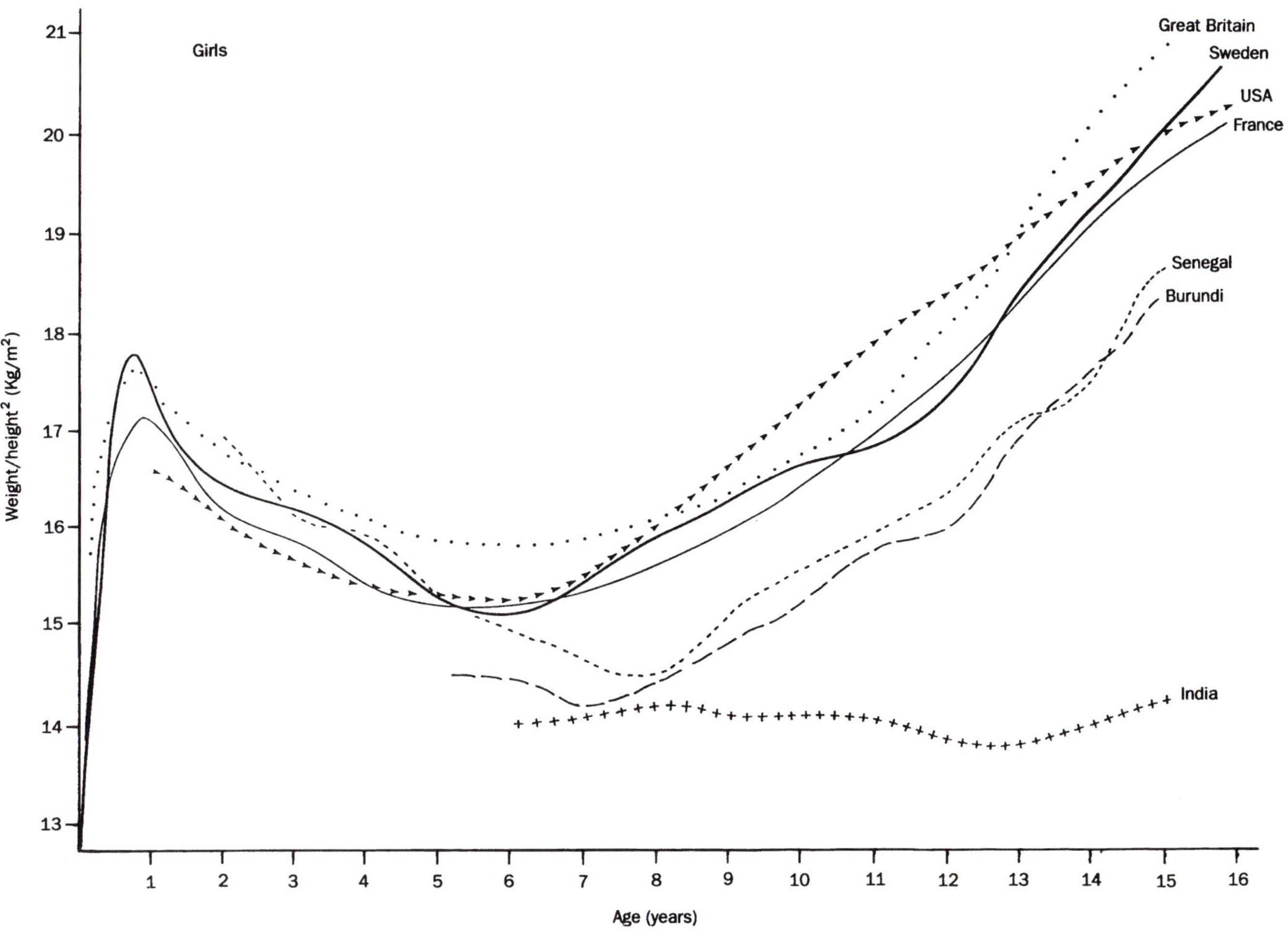


**BMI change across age in several countries.**

hood. This knowledge might help to improve growth surveillance by avoiding pointless treatment of obesity in early childhood, and in identifying early abnormal development.

*Marie-Françoise Rolland-Cachera*

See also 'Creation of growth references' (**1.16**), 'Body composition' (**5.11**), 'Fat and fat-patterning' (**5.16**), 'Endocrine growth disorders' (**7.7**) 'Nutrition' (**9.2**), 'The secular trend' (**11.3**) and 'Obesity, fatness and modernization' (**11.8**)

# Growth as an indicator of social inequalities

It has long been recognized that data on physical growth and maturation of children can provide important insights into the extent of social inequalities within a society, as well as the temporal changes in the economic condition of that society as a whole or of its various sub-groups. These social uses of growth data are made possible by two circumstances. First, many parameters of growth – including the two most widely used for such purposes: height-for-age (in both sexes) and age at menarche (in girls) (**1.16**) – are traits for which variation is both genetic and environmental in origin. Second, among the environmental factors contributing to this variation there are two whose contribution is particularly significant: the adequacy of nutrition and the rate of morbidity. Both constitute important elements of what is termed 'the standard of living' or 'the economic well-being' of individuals, families, and societies.

As a consequence, growth and maturation show two conspicuous patterns of socially induced variation:

1. *Social gradients.* In most societies children from upper strata, whether these are defined in terms of parental education or occupational rank or income, tend to be taller than their peers from lower strata. Also, they tend to mature physiologically, and to experience the pubertal growth spurt, earlier. As a result, social-class differences in height (**11.4**) are most accentuated at adolescence; they later decline, but they do not disappear, they persist into adulthood. It is noteworthy that when the social criterion used to distinguish between the 'upper' or 'better-off' and 'lower' or 'less well-off' strata is expressed not as a simple dichotomy but as a series of grades on an ordering scale, the relationship between that scale and indices of growth often proves monotonic; for example, menarcheal age increases gradually with decreasing parental education or increasing number of children in the family.
2. *Intergenerational changes.* These are usually referred to as secular trends (**11.3**) towards taller stature and accelerated maturation. During the 20th century in most industrial societies the mean statures of young adults have increased, while mean menarcheal age has decreased, at the rate of about 1–2 centimetres and 0.2–0.6 years per decade, respectively. Despite much discussion as to the causes of these trends, there is strong evidence that they constitute a phenotypic response to improvements in living conditions, rather than a reflection of any presumed shifts in the genetic make-up of populations.

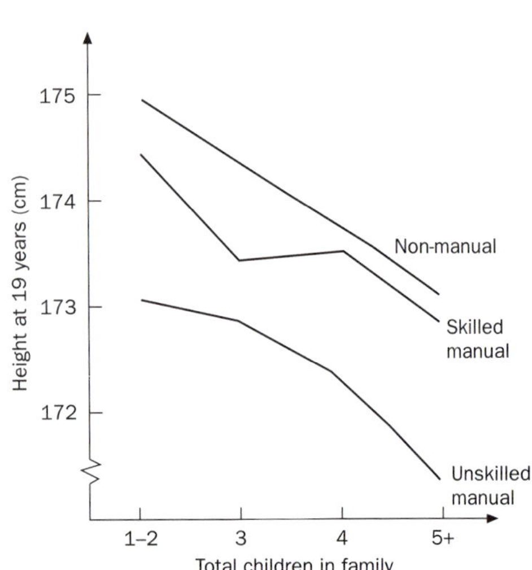


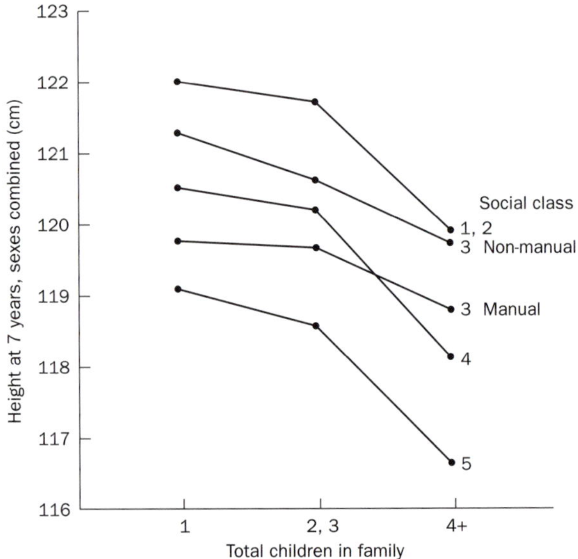


**The independent influence of two different factors of the social environment, parental social class and number of children in the family, on mean heights among 19-year-old Polish conscripts measured in 1976 (above) and among 7-year-old British children measured in 1965 (above right). In each of the two populations heights decrease with increasing size of sibship, and this gradient appears within each of the several social groups considered separately. Conversely, heights decrease with decreasing position on the social-class scale, even when sibship size is held constant. The striking similarity in the pattern of variation between these British and Polish data is all the more significant in that the comparison involves two populations 1) ethnically different, 2) living at the time under radically different political and economic systems (a free-market versus a centrally planned economy), and 3) represented here by subjects measured at two very different stages of development: the beginning of school age versus young adulthood (from Tanner 1981, and Bielicki 1986).**

**Right; Dramatic secular increases in height, but no tendency for the social contrasts to decline, can be seen from a comparison of two national surveys of Polish 19-year-old conscripts, carried out in 1965 and 1986. Each of the two cohorts is represented here by seven selected groups defined by a combination of three variables: urbanization, father's education and occupational status, and number of children in family. For example, group 1– large cities, father college education, 1–2 children; group 4 – medium cities, father skilled worker, three children; group 7 – rural communities, father peasant, five children. Note that gains in height, although very intense, were roughly parallel in all groups; hence, not only the order of the groups on the statural scale but also the gaps between them remained unchanged. The only exception is group 1, the large-city intelligentsia, in which the gain was smaller, perhaps because in that social stratum, compared to other strata, many more young adults have, by the age of 19 years, already approached the upper limit of their genetically determined potential for growth (from Bielicki and Szklarska, unpublished).**

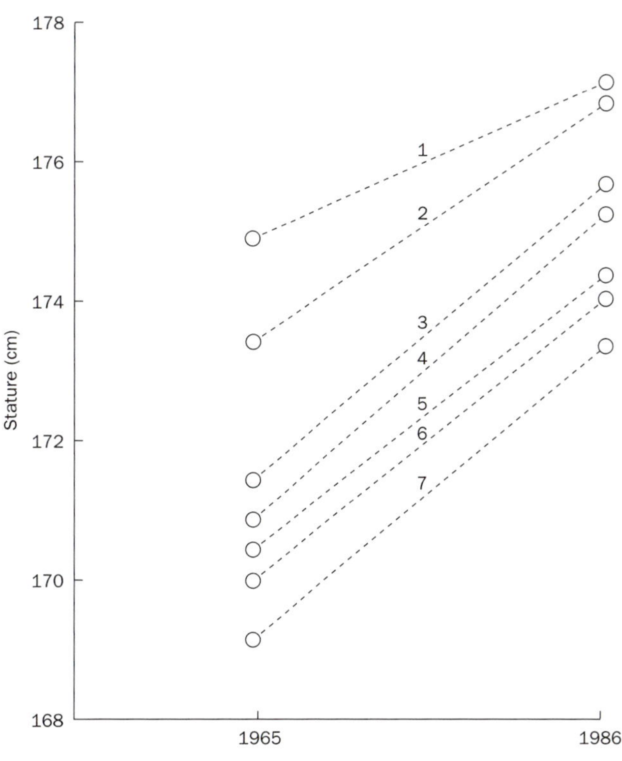


**The great sensitivity of children's growth to changes in a society's living conditions is spectacularly illustrated below by mean heights of Norwegian (Oslo) school-children measured regularly since 1920. The long-term trend towards greater tallness in the successive birth cohorts, clearly seen in these data, was briefly reversed in the early 1940s, the period of Nazi occupation of Norway during World War II. The effect is unmistakable: dips in the curves of height-for-age appear at that time in both sexes and at each age level (from Brundtland et al., 1980). Temporary reversals of the trend, closely coinciding with periods of economic hardship, have also been recorded in a number of other populations, notably Germany, France, Russia and Japan.**

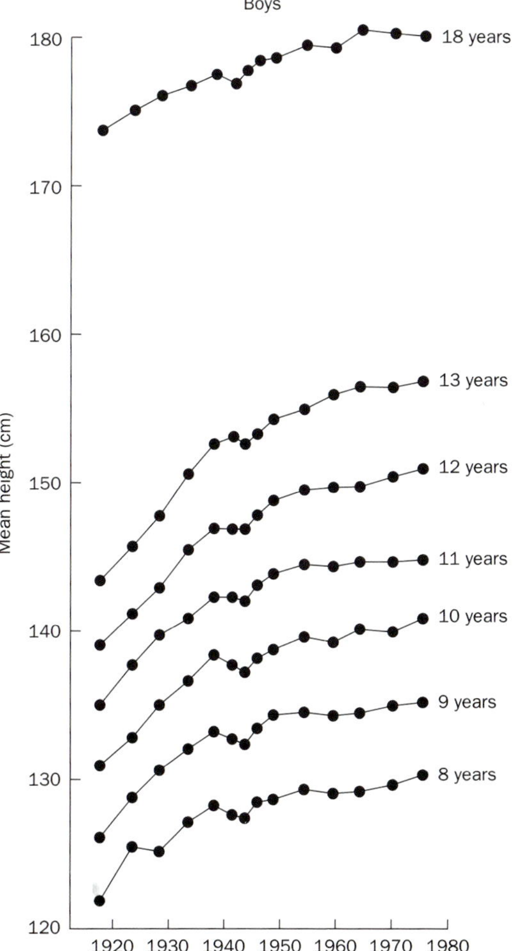


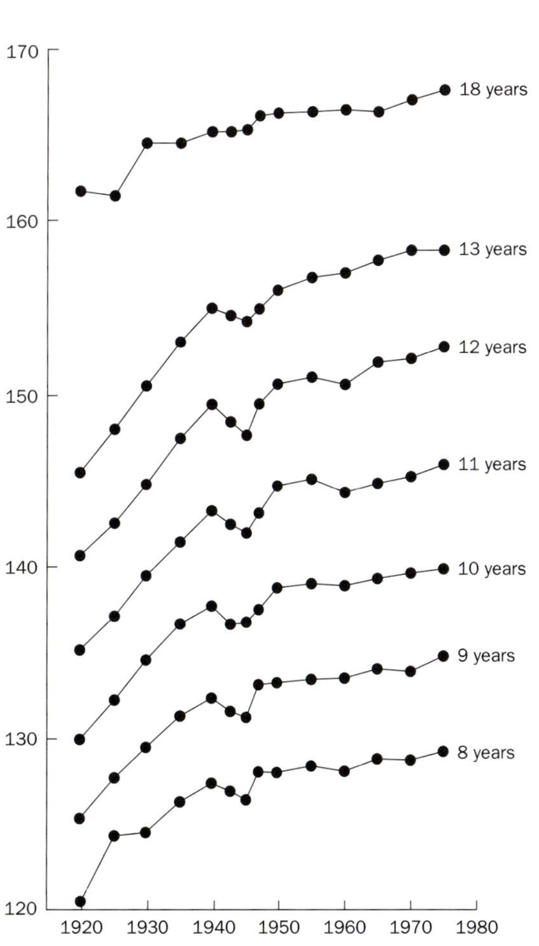

Growth surveys which combine anthropometric with sociological data and are carried out in a given population periodically, for example at 10-year intervals, are particularly useful in identifying social inequality and/or social change, in that they enable secular trends to be traced separately for different social groups. Data of this sort permit one to see whether the gaps between such groups in children's growth (and therefore, by inference, in the groups' living standards) have narrowed, widened or remained unchanged. An even more powerful method of extracting socially important information from growth surveys consists in sorting out statistically the 'net impact' on growth of each of several specific factors of the family's social situation, such as parental education, family size and urbanization. It is important to note that in many societies such factors are intercorrelated. For example, family size tends to decrease with greater duration of urban life and the type of urban centre lived in. A question arises: is the relative shortness and delayed maturation of rural and small-town children due solely to the fact that they come, on average, from larger (and, also, less well educated) families than do their large-city peers, or are some growth-inhibiting pressures associated with the rural, or small-town environment as such? Is there an effect on growth of the factor 'degree of urbanization' *per se*? In some populations this is indeed the case. Statistical analyses have shown that a) urbanization, b) family size, c) parental status and d) income are factors – each of which may retain a significant, residual effect on children's growth even after the effects of the other factors have been partialled out. Furthermore, such analytical techniques, when applied to data from periodically conducted growth surveys, can give answers to important questions such as: how does the impact on growth of a particular social factor change in time; and, for example, how is the impact on children's growth of parental education changing as compared with that of family size, or of urbanization.

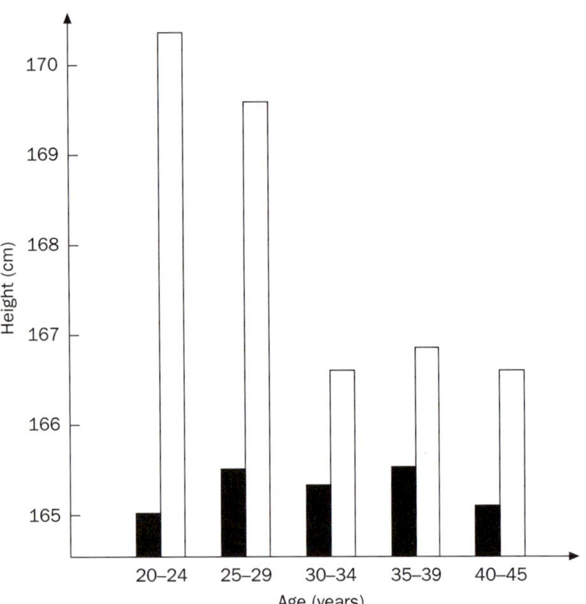


**Part of the 20th-century social history of Russia as reflected in anthropometric data. The vertical bars depict mean heights of factory workers recorded in two surveys carried out at a half-a-century interval in the same factories in ethnic central Russia: first in 1925–28 (filled bars), and again in 1972–75 (open bars). Note the conspicuous stability of heights across all age categories in the earlier survey, an indication that there was little or no secular trend towards increased height among the successive birth cohorts maturing during the last decades of the Tsarist era and World War I. Note also the contrastingly different pattern of height-for-age variation in the later survey: dramatically increased heights of the 20–29 year olds compared to the 30–45 year olds, suggesting an abrupt take-off of the secular trend among the birth cohorts which were entering puberty during the late 1950s and early 1960s, i.e., at the beginning of the period of political and social liberalization, the post-Stalinist 'thaw' (drawn on the basis of data in Table 1 of Volkova, 1979).**

One of the most thought-provoking contributions of authors of growth studies to the understanding of social change was the finding that in a few economically prosperous modern societies lower social strata have, in children's growth and maturation-rate terms, recently caught up with the uppermost ones. Cases of such obliteration of social differences among urban children were first reported in Sweden and Norway in the 1970s. These findings suggest the attainment, probably for the first time in history, of living standards of the society as a whole being such as to make any environmental deficits inhibiting children's physical development equally rare among lower-class families as among upper-class ones.

*Tadeusz Bielicki*

See also 'Within-population variation in growth patterns' (**2.5**), 'Skeletal growth and time of agricultural intensification' (**11.1**), 'The secular trend' (**11.3**), 'Social and economic class' (**11.4**), 'Modernization and growth' (**11.5**) and 'Urbanism and growth' (**11.7**)

# Cross-sectional studies

Cross-sectional studies are used in analyses of growth and development to provide information at one point in time. They differ from longitudinal studies (**1.14**) where a person is re-measured on a number of occasions. In observational studies one of the main problems is selection of the subjects for study. Since it is usually impossible and impractical to study everyone (the population) the survey is limited to a subset (sample) of the population (**1.16**). In random sampling each subject in the population has an equal chance of being chosen for the study and this approach maximizes the likelihood that the results can be generalized to the entire population. However, circumstances arise when it may be necessary to use a different sampling strategy, for instance a stratified random sample in which the sample is divided into strata or levels, such as educational levels or occupational groups (**11.4**). Subjects are then sampled randomly from the stratum into which they fall. It is common to make use of multi-stage cluster sampling where the cluster might be the household or family. However, the people in the household or family will share the same diet and environment, and so outcomes are likely to be correlated. Studies that use cluster sampling usually need larger sample sizes than investigations which make use of strictly random sampling.

Measurements obtained as part of a research design involve knowledge of the required sample size and of the way in which the sample is selected. The calculation of the sample size required depends on the complexity of the study design but in simple terms it incorporates an hypothesis (null or alternate), and the levels of Type I ($\alpha$) and Type II ($\beta$) errors. The power of the test is equal to ($1-\beta$). In addition an unknown population parameter $\sigma^2$ has to be estimated from a pilot sample or other source.

*Types of cross-sectional statistical analysis*

| DEPENDENT VARIABLE | | | |
|---|---|---|---|
| **CONTINUOUS** e.g. blood pressure, skinfold thickness, egg count etc. | | **CATEGORICAL** e.g. infection (yes, no), egg count classes, mother's education etc. | |
| INDEPENDENT | TEST PROCEDURE | TEST PROCEDURE | INDEPENDENT |
| A. Dependent variable follows normal distribution | | CHI-SQUARE | One or more categorical variables |
| Categorical variable with 2 levels | T-TEST | LOGISTIC REGRESSION | One or more categorical variables |
| Categorical variable with 3 or more levels | ONEWAY | | |
| One or more categorical variables with or without continuous independent variable | ANOVA | | |
| One or more categorical or continuous variable | REGRESSION | | |
| One continuous variable | CORRELATION | | |
| B. Dependent variable does not follow normal distribution | | | |
| Categorical variable with 2 levels | MANN-WHITNEY | | |
| Categorical variable with 3 or more levels | KRUSKAL-WALLIS | | |
| One continuous variable | SPEARMAN RANK CORRELATION | | |

| | Categorical variable | Categorical variable following normal distribution | Categorical variable not following normal distribution |
|---|---|---|---|
| Paired observation | MCNEMAR (2 x 2 table) | PAIRED T-TEST | WILCOXON MATCHED PAIRS SIGNED RANKS TEST |
| Three or more related observations | | MANOVA | FRIEDMAN TWO-WAY ANALYSIS OF VARIANCE |

If anthropometry is used, cross-sectional analysis is usually straightforward, since body measurement variables are usually continuous, quantitative characters. However, 'parametric' statistical tests assume that the data approximate to a normal distribution with no marked skewness or kurtosis. There are a number of statistical tests available for testing for normality. Some tests, such as the Kolmogorov–Smirnoff test, examine the cumulative distribution, while others, such as the Cox test, examine the extent of skewness and kurtosis separately. In general, skewness is more constraining than kurtosis. If an anthropometric variable does show significant skewness then it is quite likely that a simple logarithmic transformation ($\log_{10}$ or $\log_e$) will normalize the distribution, in which case the geometric mean, rather than arithmetic mean, should be used. Caution should be exercized with large or very large sample sizes when there is always a danger of significant departures from normality even though the magnitude of the departures may be trivial.

Cross-sectional surveys are usually easier to analyse than are data from longitudinal studies because they do not involve repeated measures. If the dependent variable is continuous, then analyses can include the testing of the difference between two means (using a t-test with equal or unequal sample variances), or analysis of variance with one categorical variable with three or more levels (such as might be carried out in a comparison of mean heights in relation to four parental educational levels). Analysis of variance involves partitioning sums of squares between and within groups. If the F-test is significant, an *a posteriori* test, such as the Scheffé test, can be applied to see which means are different. More complex analyses of variance (ANOVAs) which examine the relationship between a dependent variable and a number of categorical variables (for example, sex of child, parental educational level, morbidity status) with different numbers of levels can be constructed. These analyses also test for non-additivity of categorical variables through two-way and higher-order interactions. Most computer packages also provide a multiple classification analysis that gives unadjusted and adjusted (taking into account the other categorical variables) differences from the grand mean. The coefficient of determination ($R^2$) provides a measure of how much of the variance has been explained by the categorical variables. ANOVAs can be extended to include covariates (for example, weight). Various methods exist for decomposing sums of squares and the order of entry of covariates and main effects (categorical variables) can be changed depending on the objectives of the study.

Correlation and regression analysis examine the bivariate relationship between two continuous variables. Regression is used when there is either a known dependent (Y axis) – independent (X axis) relationship or when the best fitting linear relationship needs to be plotted; correlation analysis provides a measure of the association. The best fitting line is described by the equation $Y = \pm a \pm b X$, where a is the intercept on the Y axis and b is the regression coefficient or slope. More complex curvilinear relationships can be examined (quadratic, cubic, quartic and so on); the equation including a quadratic function is $Y = \pm a \pm b X \pm cX^2$.

## Prediction

One aim of statistical analysis, when anthropometric data is used in auxological epidemiology (see Introduction), is for prediction. An example of this is how well one can predict a persons height, given various background information, such as age, sex and educational level. For continuous characters the technique of multiple regression analysis is used to see how much of the variation can be explained by a large number of independent continuous and categorical variables. Categorical variables are coded as 0 and 1, or if it has more than two levels, dummy variables are used. The model can be expressed as $Y_i = \beta_0 + \beta_2 X_{1i} + \beta_1 X_{2i} + \ldots \beta_p X_{pi} + e_i$. The notation

*Calculation of Relative Risk and Odds Ratio. An example from Bangladesh. Mothers with a weight of less than 40 kg are 2.27 times more likely to have a child death than mothers with body-weight greater than, or equal to, 40 kg.*

| Maternal weight | Died | Survived | Total |
|---|---|---|---|
| <40 kg | 60 (a) | 292 (b) | 352 (a + b) |
| ≤40 kg | 285 (c) | 2513 (d) | 2798 (c + d) |
| Total | 345 (a + c) | 3805 (b + d) | 4150 (t) |

$$\text{Relative Risk} = \frac{\text{Incidence rate of death in exposed group}}{\text{Incidence rate of death in unexposed group}}$$

$$= \frac{a/(a+b)}{c/(c+d)} = \frac{60/352}{285/3798} = \frac{0.170}{0.075} = 2.27$$

$$\text{Odds Ratio} = \frac{a/b}{c/d} = \frac{60/292}{295/3513} = \frac{0.205}{0.081} = 2.53$$

*Confidence limits for the Odds Ratio*

1. Calculate $\log_e OR = \text{Log}_e 2.53 = 0.928$
2. Calculate the estimated standard error

$$SE(\log_e OR) = \left\{\frac{1}{a} + \frac{1}{b} + \frac{1}{c} + \frac{1}{d}\right\}^{1/2}$$

$$= \left\{\frac{1}{60} + \frac{1}{292} + \frac{1}{285} + \frac{1}{3513}\right\}^{1/2} = 0.154$$

3. The 95% confidence limits of OR

$$\log_e OR \pm 1.96 = \left\{\frac{1}{a} + \frac{1}{b} + \frac{1}{c} + \frac{1}{d}\right\}^{1/2}$$

$\log_e OR \pm 1.96 = 0.928 \pm 0.154$

95% confidence limits on $\log_e OR = 0.774$ and $1.082$

95% confidence limits for $OR = e^{0.774}$ and $e^{1.082} = 2.17$ and $2.95$

*Calculation of Sensitivity and Specificity, using maternal weight as a screening device for infant mortality*

| Maternal weight | Died | Survived | Total |
|---|---|---|---|
| <40kg | 60 (a) | 292 (b) | 352 (a + b) |
| ≤40 kg | 285 (c) | 3513 (d) | 3798 (c + d) |
| Total | 345 (a + c) | 3805 (b + d) | 4150 (t) |

| | True Condition | | |
|---|---|---|---|
| **Maternal weight** | **Died** | **Survived** | **Total** |
| <40 kg | True positive | False positive | Indicator Positive |
| ≤40 kg | False negative | True negative | Indicator Negative |
| Total | Total positive | Total negative | Total |

The true positive rate = a/a + c = 60/345 = 17.4. This rate is also called the sensitivity of the test.

The true negative rate = d/b + d = 3513/3805 = 92.3. This rate is also called the specificity of the test.

Positve predictive value = the percentage of the indicator positives who are true positives = a/a + b = 60/352 = 17.0

$X_{pi}$ indicates the value of the pth independent variable for case I. The predicted value can be obtained from using the partial regression coefficients. In generating a model, independent variables can be entered by forward selection, backward elimination and step-wise regression although there is no simple 'best' way.

Frequently researchers decide to split a continuous distribution at a particular cut-off. For example, in Bangladesh a maternal weight of less than 40 kilograms is treated as low and attempts have been made to relate low maternal weight to increased risk of child deaths. A simple way of testing whether there is any relationship between maternal weight and child death is to draw up a 2 × 2 matrix of maternal weight (less than 40 kg and greater than or equal to 40 kg) by child status (survived and died) and to calculate the chi-square statistic. To measure the strength of the association between the risk factor (less than 40 kg) and the condition (child death) the relative risk and/or the odds ratio can be calculated. A value of 0 indicates no association; a value greater than 1 (positive association) signifies that the risk or odds of death are greater when exposed to the risk factor while a value less than 1 (negative association) indicates reduced risk or odds of death with exposure to the risk factor. Numerical differences in the relative risk or odds ratio may occur but they always point in the same direction.

Another way of examining predictability is in terms of sensitivity and specificity. The true positive rate is also called the sensitivity of the test while the true negative rate refers to the specificity of the test. A good predictor is one which has a high sensitivity and high specificity; in addition a low false positive rate is important. Sensitivity and specificity are dependent on one another. High sensitivity is required for the identification of all the child deaths. Lowered specificity and a high false positive rate leads to incorrectly identifying women as high-risk. A high false positive rate is not as serious as a high false negative rate (failure to identify women as high-risk) but it will burden any screening programme.

An alternative which has a number of advantages to the sensitivity/specificity approach is to use the multivariate technique of logistic regression which is designed to test for the relationship between a binary dependent variable (for example child died/survived) and a number of independent variables which can be categorical and/or continuous. Thus the logistic regression analysis directly estimates the probability of an event occurring.

Finally, discriminant analysis can be used to classify cases into two or more groups. The concept underlying discriminant analysis is that linear combinations of the independent (predictor) variables are formed. Assumptions are that each group must be a sample of a multivariate normal population and the population covariance matrices must all be equal.

*C.G.N. Mascie-Taylor*

See also 'The use of growth references' (**1.17**) and 'Maternal anthropometry and birth outcome' (**8.1**)

# Longitudinal analysis

The nature of longitudinal data presents a number of analytical challenges. They are a time-series data set which, due to the realities of most study exigencies, consist of data collected at unequal time points. Longitudinal studies are a repeated-measures design and provide serial data that are dependent and correlated. Finally, each data point consists of potential systematic and stochastic error components and these may or may not be correlated through time.

There are two general classes of methods for analysing longitudinal data directly: serial and incremental analysis. *Serial analysis* can be carried out in a number of ways. One is to consider the feature assessed on the child (such as body-height) as a function of time, employing the raw empirical data in analyses. This approach is statistically the least complex because there is no *a priori* reason to assume that the uncertainty of measurement of a parameter on one day will be significantly correlated with the uncertainty of measurement on a subsequent day. Traditional approaches to the identification of growth and developmental change through time have involved the curvilinear smoothing of serial data points by fitting equations based on the assumption that the best fit of a continuously increasing mathematical function is a good descriptor (**1.16**). Both linear and non-linear regression models have been employed. A number of such equations have been fit to different data series, and it has been found that different functions best fit serial data encompassing different developmental ages.

A primary problem in mathematical modelling has been to find a curve that can describe the entire growth period from birth to adulthood. The characteristic growth-rate changes that typify development, with high growth velocities during infancy and adolescence by comparison with the intervening childhood years, demand a flexible mathematical function to capture these trends. In general, work has been oriented towards identifying functions that permit a shift from a general exponential function (a good descriptor for growth during infancy) to a polynomial function (for childhood growth-rates) and finally a sigmoidal curve, approximated by logistic functions, to describe the acceleration and final deceleration characteristic of the adolescent growth spurt. This problem has been most successfully approached by the Preece–Baines and Karlberg models, both aiming to accommodate these developmental shifts in their equations.

A critique of these model-fitting techniques is that there is no clear theory of growth process directing choice of model assumptions, models being based on assumptions of relatively constant growth-rates that do not permit the identification of shorter-term velocity shifts, and the results of these analyses produce mathematical constants of unclear biological meaning. Thus, other approaches have included incorporating aspects of smoothing methods into the analytical procedure, aiming to identify more specific characteristics of acceleration and deceleration changes in individual growth processes within the general trends. Techniques aiming to provide this opportunity include spline functions, kernel estimation and moving average techniques. In general, these methods average data over short sequences and derive a growth curve based on these averaged values. This is an attempt to capture shorter-term velocity variability that is missed in the large-scale developmental equations previously discussed, where short-term variability is a residual rather than contributing to the dynamics of the curve itself. This type of technique was used on biannual data to identify the presence of a 'mid-childhood' growth spurt and was applied to weekly and monthly lower-leg growth data to identify the presence of *mini growth spurts* in the longitudinal data of some children. All of these approaches are useful for individual-based analyses. Analysis of individual longitudinal data has identified the importance of growth-rate differences between children and recent reference standards incorporate these observations.

Additional approaches to the analysis of longitudinal data have included the examination of correlations and non-parametric relationships between specific characteristics of acceleration and deceleration changes in individual growth curves as derived from the curve-fitting methods and principal components analyses. Less often employed due to insufficiently frequent measurement protocols is *time-series analysis*. Traditional time-series approaches required equally spaced data points, but newer methods can accommodate unequally spaced data as well. Traditional Fourier methods pose some problems in that they are designed to identify sine waveforms in data, and will do so whether or not the original data are sinusoidal. This can introduce spurious wave characteristics to non-wave-form data and requires caution in the use of Fourier analytical methods. By contrast, Baysean methods do not share these assumptions and are useful approaches to delineate the temporal characteristics of longitudinal data, focusing on local changes in the time-series process.

The second general approach to longitudinal data is to analyse growth increments. *Incremental analyses* present unique problems. When first differences of sequential data are taken to create growth increments or velocities, a significant autocorrelation is introduced into the resulting incremental data series that requires statistical consideration and treatment. These features are often ignored in velocity presentations, but are critical aspects of longitudinal analysis. Serial velocity or incremental differences are negatively dependent and a significance level must be set that includes consideration of measurement error at three points. A statistically rigorous approach is a must for sequential differences that sets a confidence interval at a selected probability level, considers the size of the data set and the order effects, such that a significant difference between sequential increments is defined as that greater than [$t_{(\alpha/2,\ df)} \times \sqrt{6} \times$ the technical error of measurement]. The square root of 6 reflects the fact that serial increments result from a minimum of three measurements, a factorial experimental design.

Other recent approaches to longitudinal data analysis have been developed for the specific requirements of short-term data analysis. These methods incorporate the identification of significant incremental changes using raw serial data, and are modifications of pulse detection techniques developed for similar data in other biological sciences. All of the methods that analyse original data series can be compared between one another for goodness-of-fit by F-ratios of weighted residuals and autocorrelation to identify which model assumptions more closely approximate the process underlying the longitudinal data.

*Michelle Lampl*

See also 'Creation of growth references' (**1.16**), 'The use of growth references' (**1.17**), 'The human growth curve' (**2.4**), 'Within-population variation in growth patterns' (**2.5**), 'Growth cyclicities and pulsatilities' (**5.13**), 'Identification of abnormal growth' (**7.1**), 'Growth in infancy and pre-adolescence' (**10.4**), 'Growth modelling and growth references: the future' (**13.1**) and 'Growth patterns associated with new problem complexes' (**13.3**)

# Assessment of age at menarche

There are three methods for collecting menarche data and it is important to recognize the advantages and disadvantages of each. The three methods are: 1) *status quo*, 2) prospective, 3) retrospective. In the *status quo* method each subject is asked two questions: her date of birth and whether she has begun to menstruate. The sample must be sufficiently large and representative of the population and the age range broad enough to include young girls who have not yet menstruated. An age range of 8 to 16 years is suitable for a study of American girls. However, since mean age of menarche varies among different populations, the age range sampled needs to be adjusted accordingly when studying other populations.

From these data a table is constructed giving the percentage of 'yes' and 'no' answers at each age. Using either probit or logit transformation, the median age and standard error and standard deviation of the age of menarche in the population is then estimated. This method does not give the age of menarche of any one individual. The percentage of post-menarcheal girls at each age is shown, and the probits corresponding to those percentages are plotted. A line is fitted to these probits, and the age corresponding to 50 per cent is the median.

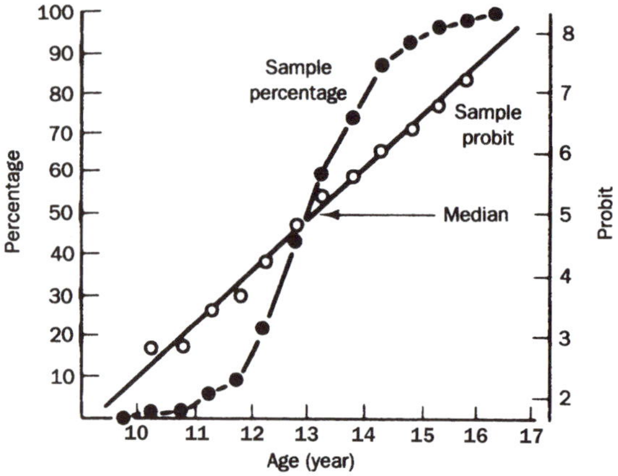


**Determination of age at menarche by the status quo method. The sigmoid curve indicates the percentage of girls in the sample (made in London, 1966) who had experienced menarche by each age. From this, the probits are determined and a straight line fitted through them (from Marshall and Tanner, 1986).**

The prospective method can be employed in longitudinal growth studies (**1.14**, **1.16**). Girls are asked at each visit whether they have begun to menstruate. This method gives reliable results if the interval between visits is short (not over 1 year) or if a diary is kept. This is the most accurate method for determining age at menarche in individuals. It requires a large and representative sample to give a reliable estimate of the population mean. In addition to the mean it provides a frequency distribution curve. From these studies we know that age at menarche follows a normal distribution curve (**1.13**).

Retrospective or recall studies involve asking each subject the age when she began to menstruate. This method is fraught with problems of poor memory and even deliberate falsification. It may be more reliable closer to the time of the event than many years after. A second problem concerns the sample taken. If the sample includes many girls who have not yet begun to menstruate, and these subjects are simply excluded, the results will be biased downwards, i.e. toward earlier mean menarcheal age. If the sample is one of older women, it frequently is carried out in a prenatal clinic. Since infertile women would be excluded, this also probably biases the mean downwards. Another bias comes from the practice in the Western world of giving one's age as that of the preceding birthday. For example, a girl who reached menarche at 12.75 years would report that she was 12. In a large sample this kind of error would lead to the mean being underestimated by 0.5 years. In a smaller sample, the bias is less consistent. In sum, the recall method is clearly unsatisfactory, although it was used in nearly all the older surveys and still has some proponents.

*Phyllis B. Eveleth*

See also 'Physical examination' (**1.10**), 'Hormonal regulation of growth in childhood and puberty' (**5.14**), 'Sexual maturation' (**5.15**), 'The development of sexuality' (**6.6**), 'Chromosome aberrations and growth' (**7.4**), 'Teenage pregnancy' (**8.6**), 'Between-population differences in adolescent growth' (**10.5**), 'The secular trend' (**11.3**) and 'Menopause' (**12.2**)

# Creation of growth references

A growth reference is a graphical summary of the relationship between two variables, for example, weight and age, as represented by a series of centile curves. The data are obtained from a suitable reference sample. The first stage in creating a growth reference is to identify the reference sample (**1.13**), ensuring that it is representative of the parent population. Secondly the required measurements need to be collected from the sample subjects, using trained observers to ensure accurate measurements.

The third stage is the statistical process of fitting smooth centile curves to the data. The growth reference has two important requirements: it should faithfully represent the reference data, and it should at the same time be visually convincing. This implies that each of the reference centiles should be smooth, and ideally should also be suitably spaced relative to its neighbours. This is partly a matter of visual simplicity, but it is also justified statistically in terms of reduced bias. Unfortunately, data fidelity and curve smoothness are directly contradictory. If the centiles are to follow the data closely then they cannot also be smooth. What is required is a compromise between the two, which in the past has been achieved by drawing each centile curve by eye. There is much to commend this method, as the necessary trade-off between goodness of fit and smoothness is provided by the artist.

The alternative is to fit the centiles objectively, using some form of statistical smoothing. If the age trend is not too complicated in shape, simple parametric curves like a polynomial in age can be used, but they are relatively inflexible. Recently, computer-intensive statistical methods have been developed to provide a realistic alternative to the artist and drawing board, in that they generate smooth curves which convincingly follow the age trends in the data. Cubic spline smoothers and kernel smoothers are two such techniques. There are also some relatively complex parametric functions available to model height during childhood, for example the eight-parameter Jolicoeur–Pontier–Abidi model 2:

$$\text{Height} = A\left\{1 - \frac{1}{1 + \left(\frac{t+E}{D_1}\right)^{C_1} + \left(\frac{t+E}{D_2}\right)^{C_2} + \left(\frac{t+E}{D_3}\right)^{C_3}}\right\}$$

where t is age, A is adult height, and the other seven parameters define three phases of growth. However this type of function is normally used to model individual growth curves rather than to summarize a whole sample.

Whichever method of smoothing is used, it can be applied either to the mean of the reference data or to selected centiles like the median or the 90th centile, and in this way a set of smooth centile curves can be generated. However, this is an inefficient process, as the fitting of each centile ignores information from neighbouring centiles. A better solution is to fit all the centiles at the same time. This involves specifying how far apart neighbouring centiles ought to be, i.e. what the underlying distribution of the measurement is. This distribution then needs to change smoothly with age, to ensure that the centiles remain smooth. For measurements like height or

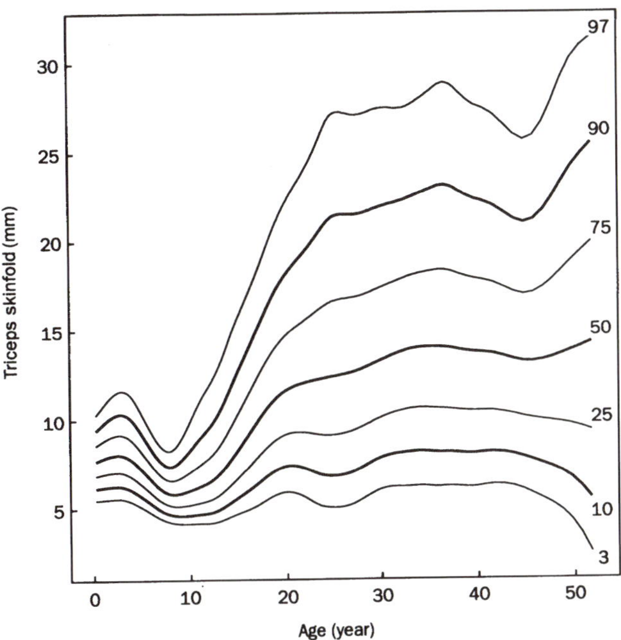


**Over- (left) and under-smoothed (above) centiles. From Cole TJ, Green PJ. Smoothing reference centile curves: the LMS method and penalized likelihood. *Statistics in Medicine* 1992; 11: 1305–1319.**

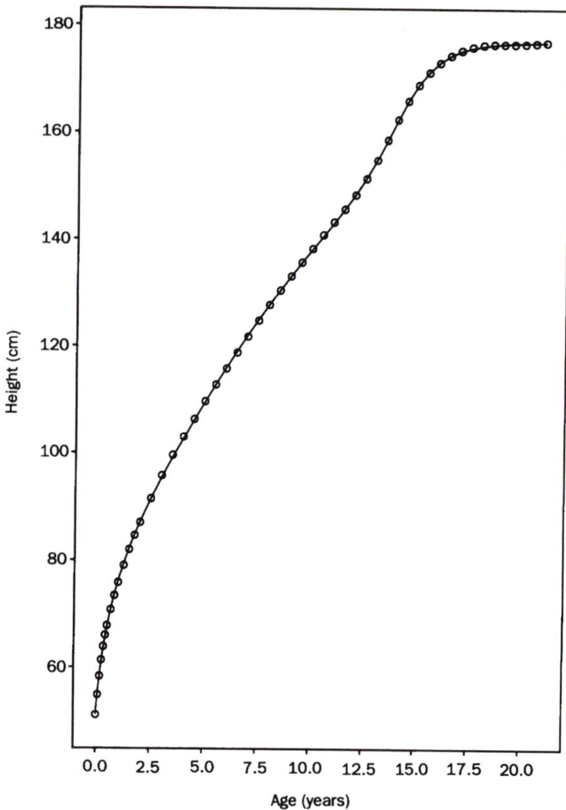


Graph of British boys' (1990) median height fitted by the Jolicoeur-Pontier-Abidi 2 model.

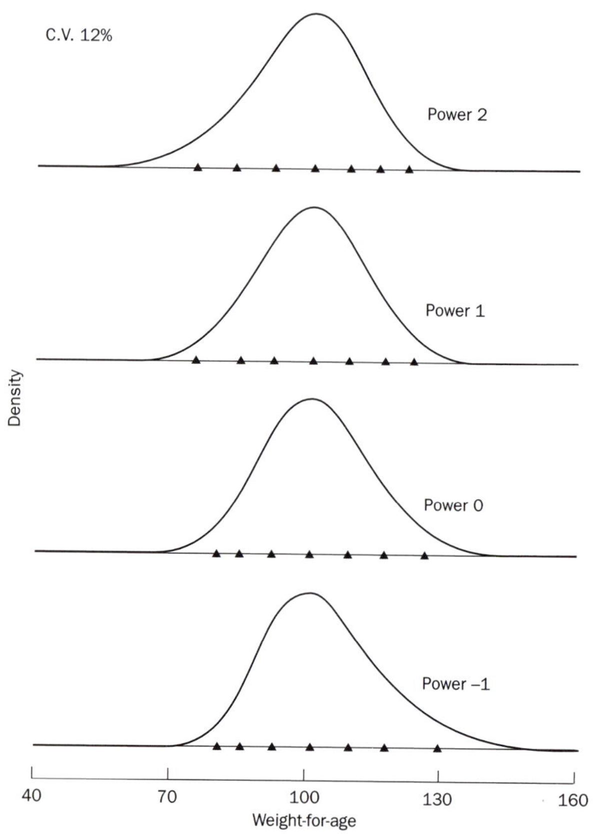


Using a power transformation to adjust for differing degrees of skewness in weight. Fig.1. From Cole TJ. The LMS method for constructing normalized growth standards. *European Journal of Clinical Nutrition* 1990 44: 45–60.

head circumference, which are normally distributed, smooth centiles can be obtained simply by smoothing the mean and standard deviation of the data.

However, skew measurements (**1.13**) such as weight or skinfold thickness cannot be handled this way. Two methods that do handle skewness have been described in detail, one using a power transformation of the normal distribution to summarize the data at each age (the LMS method), and the other fitting parametric functions (such as polynomials in age) to each of the selected centiles, and then relating the coefficients of the functions to those of neighbouring centiles (the HRY method). Both approaches specify the form of the distribution in a way that allows for skewness.

The LMS method converts measurements for a child of known age and sex (Y) to an SDS using the formula

$$SDS = \frac{(Y/M)^{L} - 1}{LS}$$

where the values of L, M and S are read off smooth sex-specific curves plotted against age. The M curve defines the median of the measurement, the S curve is the measurement's coefficient of variation, and the L curve is the power

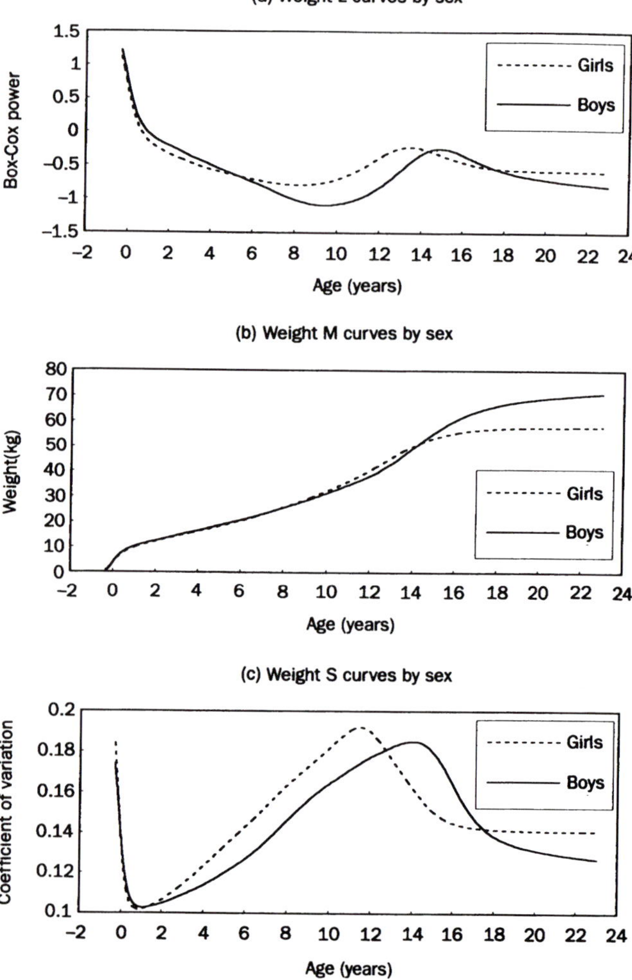


LMS curves for weight of British children (1990) by sex.

of the Box–Cox transformation needed to make the measurement normally distributed. Each curve is estimated as a cubic spline using maximum penalized likelihood.

The LMS method has been used to fit British, French, Swedish and Hong Kong growth reference curves. Once the LMS curves are known, then any required centile curve can be constructed based on them, using the formula:

$$C_{100\alpha} = M\,(1 + L\,S\,Z_{\alpha})^{1/L}$$

where $Z_{\alpha}$ is the SDS of the required centile. For example for the third centile, $100\alpha = 3$, $\alpha = 0.03$, $Z_{\alpha} = -1.88$ and $C_3 = M\,(1 - 1.88\,L\,S)^{1/L}$.

The HRY method does not provide such a simple conversion between measurement centile and SDS and back again, but unlike the LMS method it summarizes the growth reference very compactly, in terms of a few coefficients defining the general shape of the centile curves, and a few coefficients that define how this shape changes as the centile changes. The LMS method needs to tabulate the values of L, M and S by age. Other methods have been described for smooth centile fitting, but they all rely on the same general principle of allowing the distribution of the measurement to change smoothly with age.

An advantage of the LMS method is that it can be used to combine national growth references to make an international reference which is both valid and politically acceptable. One way to combine growth references such as the British, American and Dutch national weight references which have clearly similar shapes for the LMS functions would be to average the curves. This offers advantages over the more obvious process of fitting centiles to the combined reference samples, which leads to centiles that are too far apart – the national differences in the mean increase the variability as well. Using the averaged LMS curves to define an international reference means that every nation is equally represented, yet the reference centiles are no wider apart than for the individual national references.

T.J. Cole

See also 'Adiposity rebound and prediction of adult fatness' (**1.11**), 'Longitudinal analysis' (**1.14**), 'The use of growth references' (**1.17**), 'The human growth curve' (**2.4**), 'Within-population variation in growth patterns' (**2.5**), 'Standards and references for the assessment of fetal growth and development' (**4.9**), 'Identification of abnormal growth' (**7.1**), 'Catch-up growth in height' (**9.6**), 'Catch-up weight-gain' (**9.7**) and 'Growth modelling and growth references: the future' (**13.1**)

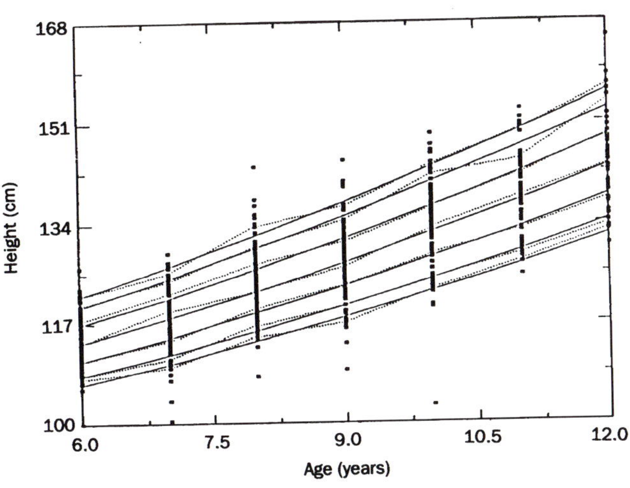


**Centiles of height by age constructed by the method of Healy et al. (1988). From Ayatollahi SMT. Growth of school children of southern Iran in relation to the NCHS standard.** ***Annals of Human Biology 1991*** **18: 515–522.**

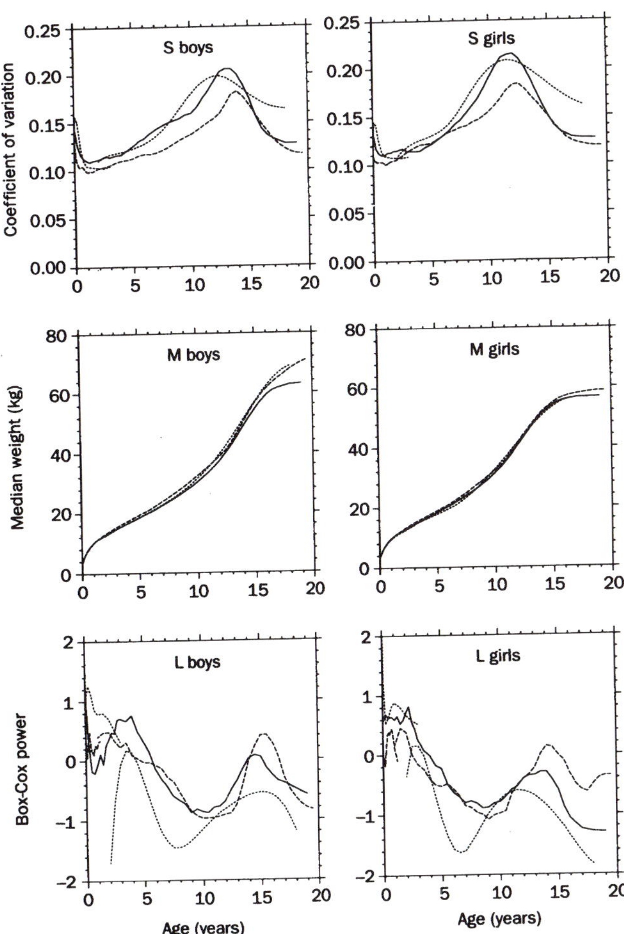


**LMS curves for three national weight references. From Cole T.J. The British, American NCHS and Dutch weight standards compared using the LMS method.** ***American Journal of Human Biology 1989*** **1: 387–408.**

## THE THREE-IN-ONE INFANT WEIGHT-MONITORING CHART

The well-known weakness of the conventional weight-growth chart is its failure to quantify poor growth velocity. An extended infant weight chart is described which, in addition to the usual weight centiles, identifies children who are growing slowly. The centiles are augmented with a series of thrive lines, whose slopes define a rate of marginal weight-gain appropriate for the child's age and weight centile. If the child gains weight at a slower rate than the slope of the nearest thrive line, then their weight-gain is below the fifth centile. The chart is thus a combination of distance, velocity and conditional charts, and it makes true weight monitoring possible for the first time.

Growth charts have been used for over 30 years to monitor nutritional status. Yet the term 'growth chart' is actually a misnomer, because the chart is constructed from cross-sectional data and contains no direct information about growth. Despite this, charts are used all over the world as a tool to monitor weight-gain in infancy. The assumption is that normally-growing infants track along their chosen weight centile, while growth failure manifests itself as centile crossing downwards. However this raises the question: how much crossing is needed to constitute growth failure?

The weight-distance chart (to use Tanner's terminology) is not calibrated for centile crossing, and so cannot monitor weight-gain. The weight velocity chart was developed for this purpose, but it too has major disadvantages. Firstly it involves plotting on two charts rather than one; secondly, weight velocity has to be calculated for each pair of measurements and the time interval between them; and thirdly, the chart is designed for measurements a fixed time interval apart (e.g. 1 month or 3 months). For these reasons velocity charts are not used widely during infancy.

The weight-velocity chart suffers from a further problem, that of regression to the mean. Light infants, with weights below the median, tend over time to catch up (or regress) towards median weight, while the opposite occurs for relatively heavy infants. So, the smaller the initial weight centile is at a particular age, the greater the mean weight velocity. This is a statistical phenomenon, and a conditional reference of weight adjusted for previous weight is the way to adjust for it. However, conditional references also require extra charts and a fixed time interval between measurements.

A new form of growth chart has been devised which combines distance, velocity and conditional references in a single format. It solves most of the problems, and for the first time provides a way of monitoring weight-gain in infancy.

#### *METHODS*

The chart consists of nine weight centiles from the British 1990 growth reference, augmented with extra lines whose slopes define marginal growth. They are called *thrive lines*, as they define the cut-off for failure to thrive. The amount of centile crossing that occurs over time depends on the time interval between measurements, which has to be specified at the outset: an interval of 4 weeks is chosen here.

If weight is expressed as a standard deviation score (SDS), the mean and standard deviation (SD) of the change in weight SDS over 4 weeks depends only on the correlation between the two measurements, which in turn depends on the child's age. Cole has developed a formula for this correlation using data from the Cambridge Infant Growth Study: 223 infants measured at birth and then 4-weekly during the first year. In addition, the correlation between birth and 4 weeks is calculated as 0.8. The formula allows the fifth centile of weight-gain to be calculated for a series of ages and weight centiles. If the gain is calculated from two weight SDSs 4 weeks apart, called, say, $SDS_1$ and $SDS_2$, and the correlation between them is $r$, then they are related as follows:

$$SDS_2 = SDS_1 + \text{Mean gain} + Z_{0.05} \times \text{SD of gain}$$

$$= r.SDS_1 - 1.645\sqrt{1-r^2}$$

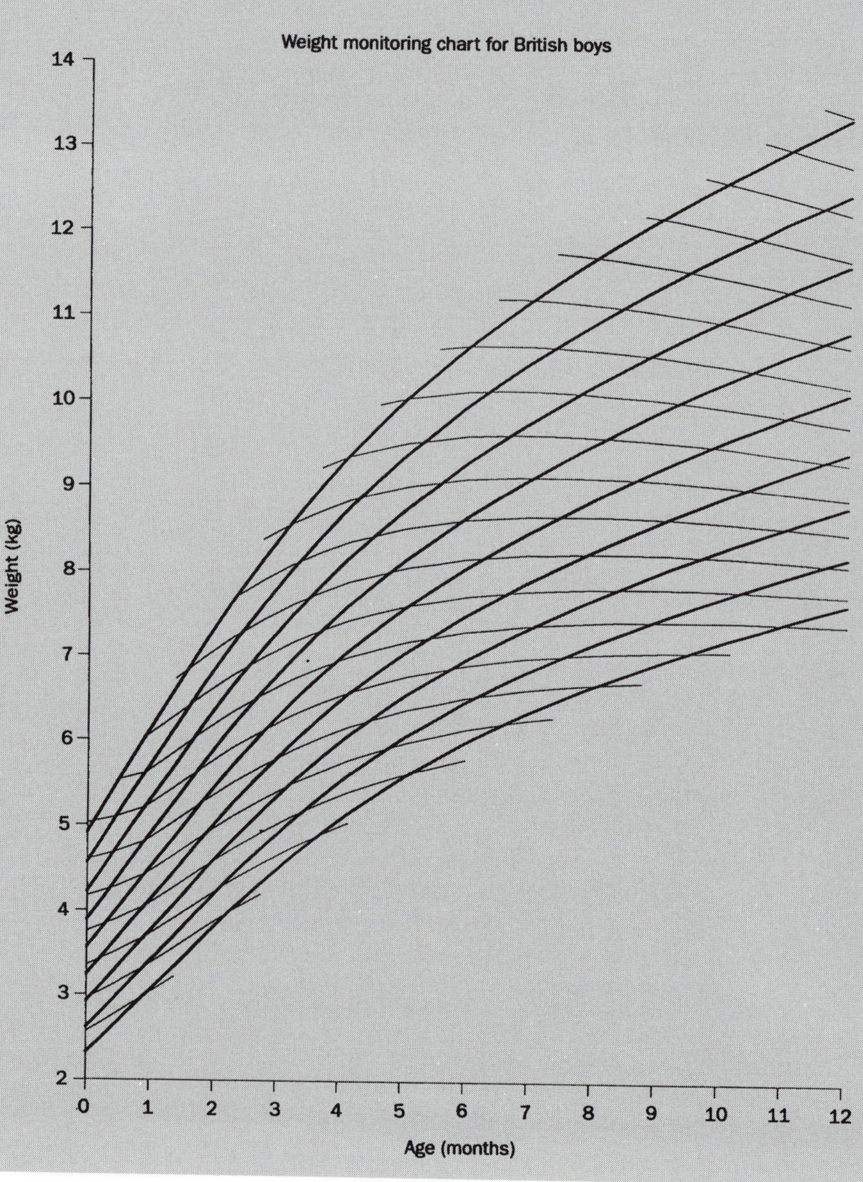

where $Z_{0.05}$ is the Normal equivalent deviate for the fifth centile. The weight centiles $SDS_1$ and $SDS_2$ are plotted on the chart at the appropriate ages 4 weeks apart, and the slope of the line joining them defines the fifth weight gain centile. The lines are spaced to cover the range of ages and weight centiles for the first year.

The way the chart is used is as follows: the child's most recent weight is plotted on the chart and joined up to the previous weight, ideally about 4 weeks earlier. The slope of the resulting line is then compared with the slope of the nearest thrive line. If the child's slope is the smaller, i.e. the rate of weight-gain is less than that shown by the thrive line, then the child's weight-gain is below the fifth centile.

### *RESULTS*

The graphs show the weight-monitoring charts for British boys and girls respectively during the first year of life. The nine weight centiles are in bold, while the thrive lines are lighter and appear as traces of downward centile crossing. The rate of weight-gain they represent is greatest at 4 weeks, after birth-weight has been regained, then they flatten off around 6 months and go negative. In addition the gain is greater (or the loss less) on lower weight centiles, indicating the effect of regression to the mean. This is particularly obvious at birth, where infants near the second centile need to put on about 0.4 kilograms in the first 4 weeks, while on the 99.6th centile they need gain only 0.1 kilogram.

The thrive lines near the 99.6th centile level off at about 6 months, but are 3–4 months later near the 0.4th centile. Past these ages normal growth includes no growth, so that many normally growing infants fail to put on weight.

The chart combines weight distance (nine centiles), weight velocity (the thrive line slopes depend on age) and conditional weight (the slopes vary with weight centile at each age). So it is a three-in-one chart, and its properties make it potentially valuable for weight-monitoring. The conditional adjustment for regression to the mean recognizes the fact that light infants cannot afford to lose weight, while heavy infants often can.

The new chart addresses all the problems mentioned earlier except for one. For it to work properly the measurements need to be 4 weeks apart, and this is unrealistic as infants are often weighed irregularly. However, in practice the chart is robust to a range of measurement intervals, so long as the interpretation is modified accordingly. For intervals of 2–3 weeks, weight-gain is more variable and a larger proportion than 5 per cent of infants will appear to be growing slowly. For intervals longer than 4 weeks the opposite holds, and the proportion will be less than 5 per cent. In the extreme case of an 8-week interval, i.e. two 4-week intervals side-by-side, the chance of failing to thrive is 5 per cent of 5 per cent, or 0.25 per cent. Thus a child who grows parallel to or slower than the thrive line for 2 months or more is clearly failing to thrive.

The 5 per cent cut-off chosen for the chart is conservative, and should not be viewed as a firm definition of growth failure. It corresponds to a 1 in 20 chance, and many infants will dip below it occasionally without necessarily being ill. Poor quality measurements will also make this more likely, and they should obviously be guarded against. As the data come from a research study with a relatively small measurement error, the cut-off will in practice pick up slightly more than the nominal 5 per cent of infants.

It would be straightforward to construct a chart based on the 95th rather than the fifth weight-gain centile, to monitor for excessive weight-gain. However, including both centiles in the same chart could lead to information overload. It would also be simple to combine height distance and annual height velocity charts in the same way, so that the three-in-one chart should improve the quality of infant weight-monitoring worldwide.

*T. J. Cole*

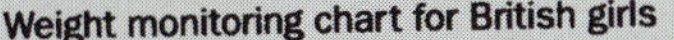
Weight monitoring chart for British girls

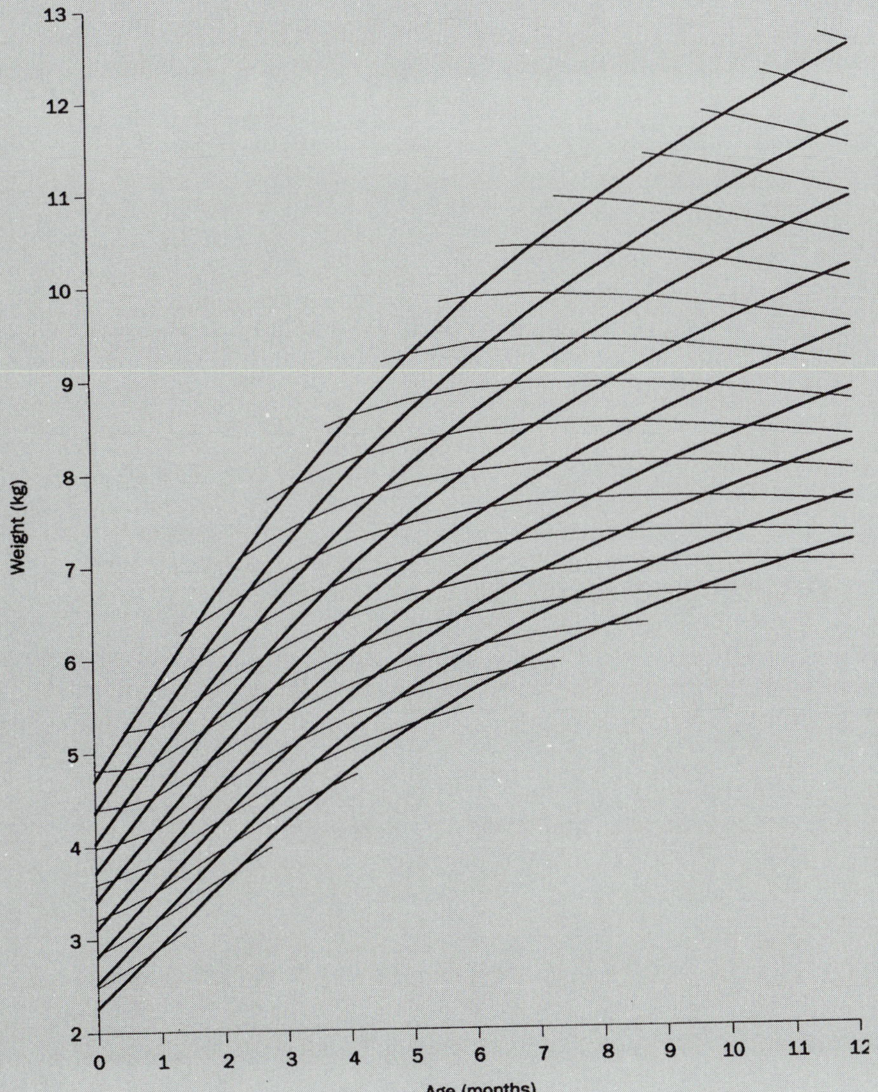

## BODY MASS INDEX REFERENCES

### CHOICE OF METHODS

During growth, the choice of method for assessing nutritional status depends on the reference charts available. Weight-for-age and height-for-age charts are the most widely used, and it is current practice to assess nutritional status on this basis. A child who has the same standard deviation (SD) value or who is on the same percentile range for height and weight is often considered normal. This is equivalent to using the formula: weight $_{SD}$ – height $_{SD}$ (a). However, this calculation assumes that the correlation between weight and height is perfect (i.e. = 1), when it is about 0.7 until puberty (**1.10, 5.15**) and declines thereafter. Consequently, the degree of under- and overweight in light or heavy children, are considerably underestimated using this method.

Weight-for-height charts allow a more precise assessment of the degree of under- and overweight, but these charts are less available, particularly after the age of 10 years, because this method is questionable during puberty. Like in the weight-for-age chart, the rising curve mainly reflects changes in height rather than in fatness. Finally, age is not taken into account, giving no information on the time interval between two measurements.

### INDICES COMBINING WEIGHT AND HEIGHT

Weight and height are each measures of the size of an individual. Conversely, the weight/height indices are measures of shape, closely associated with body composition and nutritional status. The validity of such indices is based on the following criteria: a high correlation with body-fatness and low correlation with height. The weight/height$^2$ or body mass index (BMI) has a high correlation with weight and fatness and a low correlation with height except during the first year of life and at adolescence, when weight/height$^3$ has a lower correlation with height. For this reason some authors propose that the value of $n$, the power of height, should change with age. However although weight/height$^3$ is better at adolescence with its low height correlation, it is also poorly correlated with weight. The weight/height$^2$ index is better in this respect. In spite of some limitations, it is now admitted that the advantage of using a single index throughout life is sufficient to justify weight/height$^2$ as an indicator of fatness at all ages.

### COMPARISON OF THE DIFFERENT METHODS

Comparing different methods, weight according to height and weight/height$^2$ according to age, give similar results, while the use of height-for-age and weight-for-age separately gives an underestimation of overweight. In addition, weight changes appear more clearly and earlier using BMI charts than other traditional charts. Since it takes age into account, the BMI chart improves the estimation of weight-for-height at adolescence.

As the BMI is largely independent of height, the BMI chart is the only method based on weight and height that shows ascending and descending phases reflecting the development of body-shape and fatness. This curve gives more information on the child's fatness changes during growth than the previous weight–age and weight–height charts, and the surveillance of individual BMI pattern permits early prediction of subsequent development.

*Marie Françoise Rolland-Cachera*

**Right: Comparison of methods for assessing weight development: (1) weight- and height-for-age, (2) weight-for-height, (3) weight/height$^2$-for-age. For example, in a girl aged 6.5 years whose weight = 29 kg (+ 3.2 SD) and height = 1.2 m (+ 1.3 SD), her overweight as assessed by method (1) = +1.9 SD, by method (2) = +3.2 SD, and by method (3) = +3,3 SD. The last two methods (Wt according to Ht and Wt/Ht$^2$ according to age) give similar results (approximately +3.2 SD) and show that this child is obese, while using method (1) she is 1.9 SD overweight only (after Rolland-Cachera, 1995).**

## SKINFOLDS

Reference data for skinfolds are needed in the assessment of fatness (**5.16**), most often in cases of overnutrition and obesity(**9.2**), but also where there is undernutrition and wasting. The procedures followed for constructing the reference values are generally the same as those for height and weight, except for the fact that the distributions of skinfold thicknesses tend to be markedly skewed to the left. In addition there is much greater fluctuation of the values, both across percentiles at the same age, and along a particular percentile at different ages. Smoothed curves

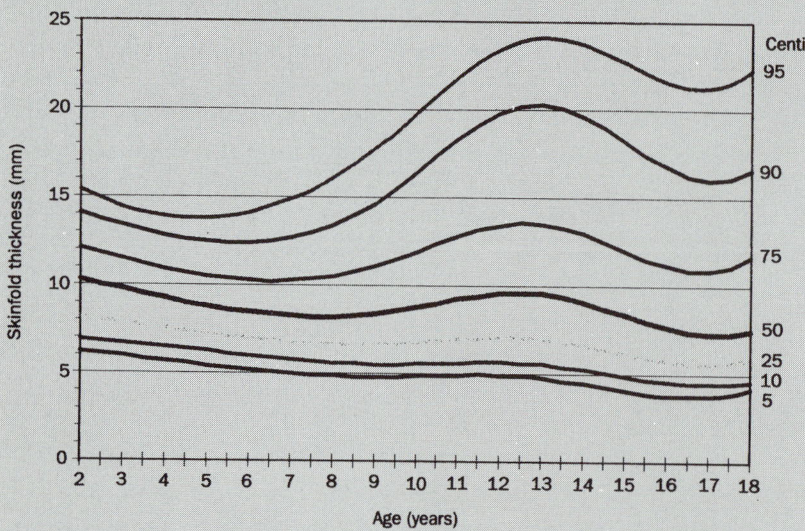


**Distribution of triceps skinfold thickness in American White males, NHANES-1 (Johnston et al., 1981).**

of percentiles 5, 50, and 95 for the triceps skinfold of American males from the first National Health and Nutrition Examination Survey (NHANES-1) can be examined for skewness by comparing the distances between them. While the distribution is generally symmetrical around the median in the earlier years, the deviation in the upper half of the distribution is readily apparent by age 6 and becomes increasingly marked thereafter.

Other skinfolds show even greater skewness. In general, those sites where there is greater accumulation of fat display greater degrees of skewness. The skewness may be so great that it cannot be corrected by a simple mathematical transformation. For example, various transformations of the abdominal skinfold in 485 youths of 12–17 years of age, from Chandigarh in northern India, show a variety of distributions. All nine transformations are skewed. These transformations are: 1) cube of the value; 2) square of the value; 3) the raw value; 4) square root of the value; 5) log e of the value; 6) reciprocal of the square root; 7) reciprocal of the value; 8) reciprocal of the square; and 9) reciprocal of the cube. The natural logarithm comes closest to a normal distribution.

Recognizing that in most cases a logarithmic transformation gives the closest to a normal

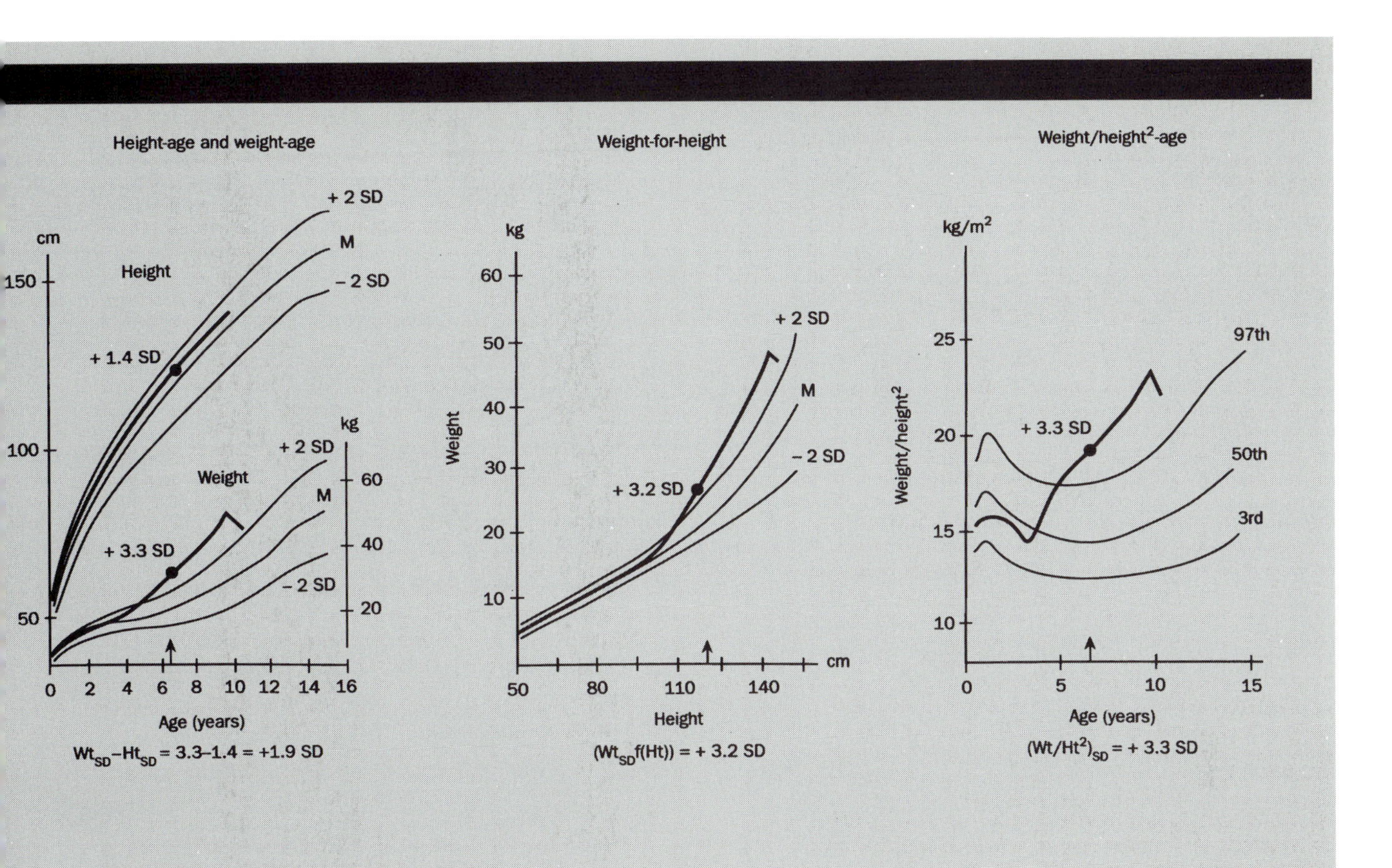

distribution(**1.13**), Hammond et al. (1955) published a formula that includes such a transform and that produces a normal distribution. This formula has been widely used in the past but much less so in more recent years. More commonly, investigators and clinicians have turned to reference standards for skinfolds which allow individual values to be expressed as centiles of the reference of the appropriate age and sex.

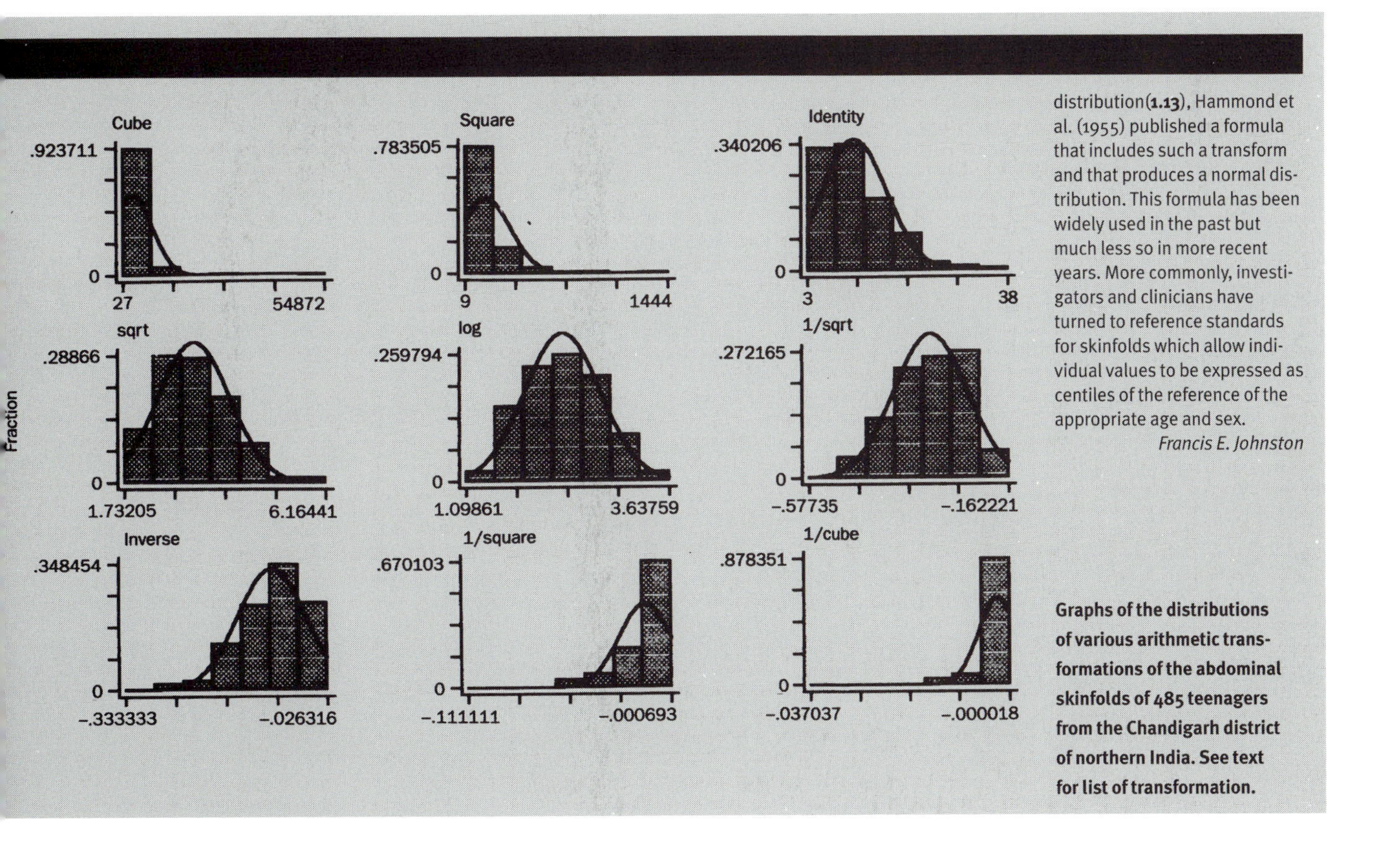

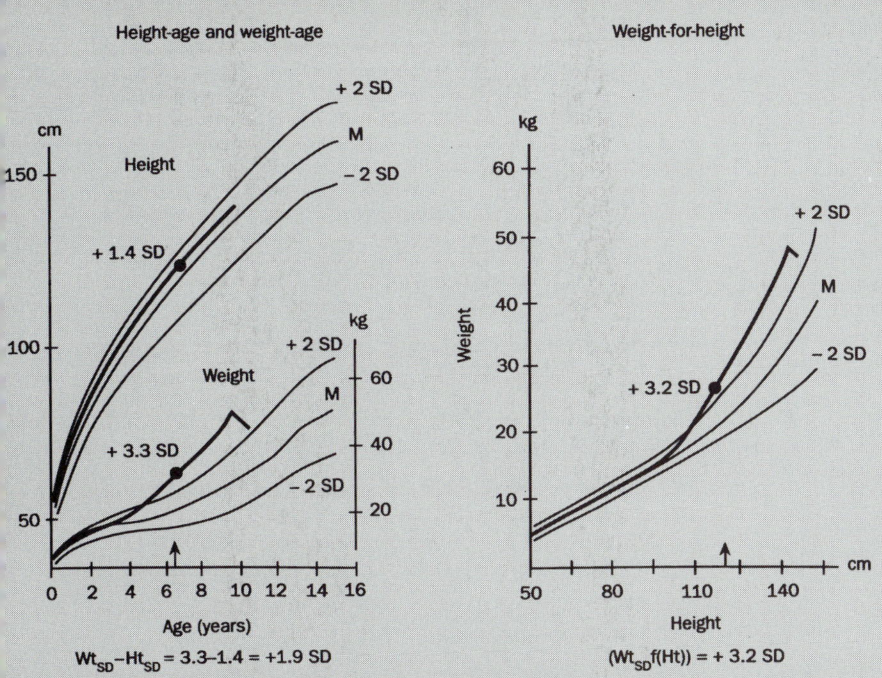

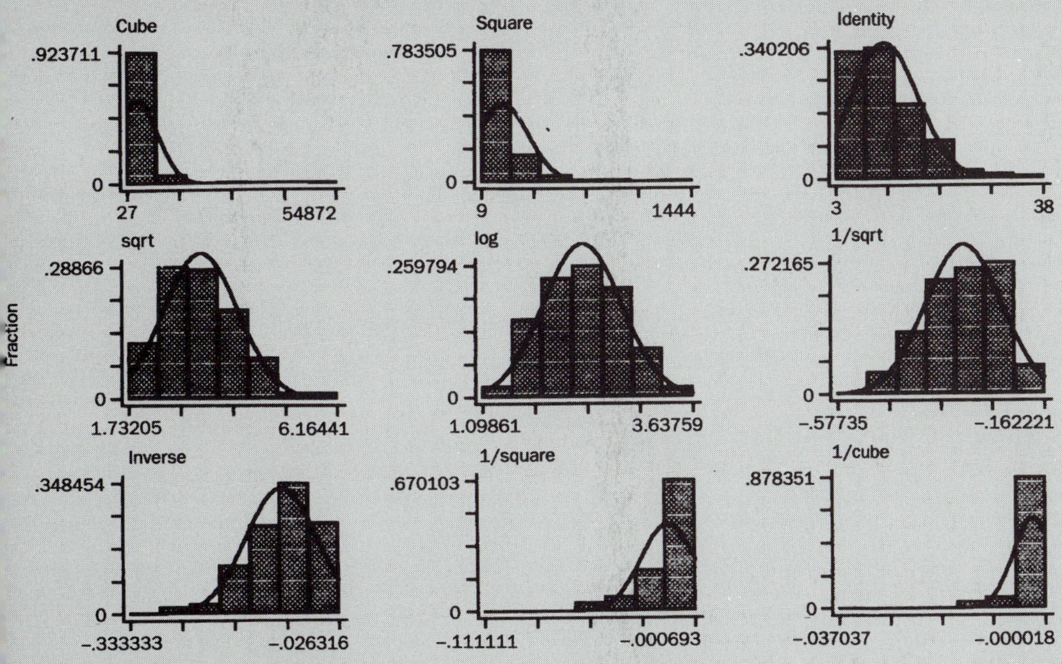

*Francis E. Johnston*

**Graphs of the distributions of various arithmetic transformations of the abdominal skinfolds of 485 teenagers from the Chandigarh district of northern India. See text for list of transformation.**

## DENTAL DEVELOPMENT

Dental development is accepted as one of the maturity indicators of the growing child (**1.9**); height, skeletal development and the secondary sexual characteristics being the others (**1.10**). Dental age (DA) is used and taught now frequently along with the skeletal age, in different fields of medical and paramedical sciences such as human biology, physical anthropology, endocrinology, nutrition, physical education and forensic medicine. Its most frequent use, however, is naturally in general and pediatric dentistry and orthodontics.

For the evaluation of dental age, preference should be given to the use of dental development assessed from radiographs. This is a better measure of physiological maturity and ultimately of dental-age determination, rather than dental emergence (wrongly defined as eruption). One should be aware that formation (mineralization) and emergence involve two essentially separate processes that may be influenced differently by genetic, environmental and hormonal factors. Furthermore, the clinical appearance of a tooth defined as emergence is a fleeting event and its precise time is very difficult to determine. However, its mineralization is a continuous process that can be assessed by permanent records, such as X-ray films.

The evaluation of a dental maturity system should have two general properties: 1) maturity scores must reflect the continuous nature of biological development and succeed each other in a smooth, logical progression; and 2) the variability in the maturity score at each age should be sufficient to reflect the natural variability of the general population (**5.10**). The assessment of dental maturity, like skeletal maturity, should be based on biological criteria rather than on length or width measurements alone. The latter are really only estimates of growth, as distinct from maturity. Like the bones of the hand–wrist area, teeth undergo different sequences of maturational stages. The first stage is the actual formation of the crypt, and the final stage is the fully mature tooth, defined as the closure of the apex. During this maturational process, one sees continuous changes in the size and shape of the teeth. Each tooth follows the same sequence; in order to study the entire process, arbitrary stages must be selected and fully described that trace the entire developmental process from beginning to end. These stages must 1) describe the major developmental stages of the tooth; 2) be clearly defined (not merely on the basis of length increases); and 3) be objective enough to be reproducible. Unlike the skeletal system, the dental system has two overlapping developmental periods for two sets of teeth. The developmental period for the deciduous teeth extends from the third month of intra-uterine life to the third year of post-natal life. For the permanent teeth, it is from the post-natal age of 6 months to 14–15 years, excluding the third molars.

Because the radiographic view of the mandibular teeth is so clear, it has been proposed that, for purposes of standardization, all mandibular teeth (excluding the third molars) on the left side, be selected for study. Once a group of teeth has been agreed on for assessment a maturity scale can be constructed. The concept of a maturity scale is different from that of a scale of height, for example, in that all subjects pass through the same series of points on the scale, starting with a stage designated as 0 for complete immaturity and concluding with 100, corresponding to complete maturity. Scores between these two extremes are assigned to the degree of development of the specific teeth under consideration. Then the total score is transformed to dental age.

The dental literature is rich with articles on different factors affecting dental development, such as, sex, race, ethnic origin, and socio-economic conditions. However, the study of secular trend in dental maturity should be studied further in order to elucidate the early formation and emergence pattern of the dentition in the today's populations.

With the advance of information technologies in the last decade, the evaluation of dental age may also be automated. The clinician can assign the developmental stage for each tooth, and the total of scores as well as the dental age are calculated instantly by the computer. This is part of a recent computer program, where the tutorial and the standardization of dental age evaluation, as well examples of longitudinal dental X-rays are on the same CD-ROM. *Arto Demirjian*

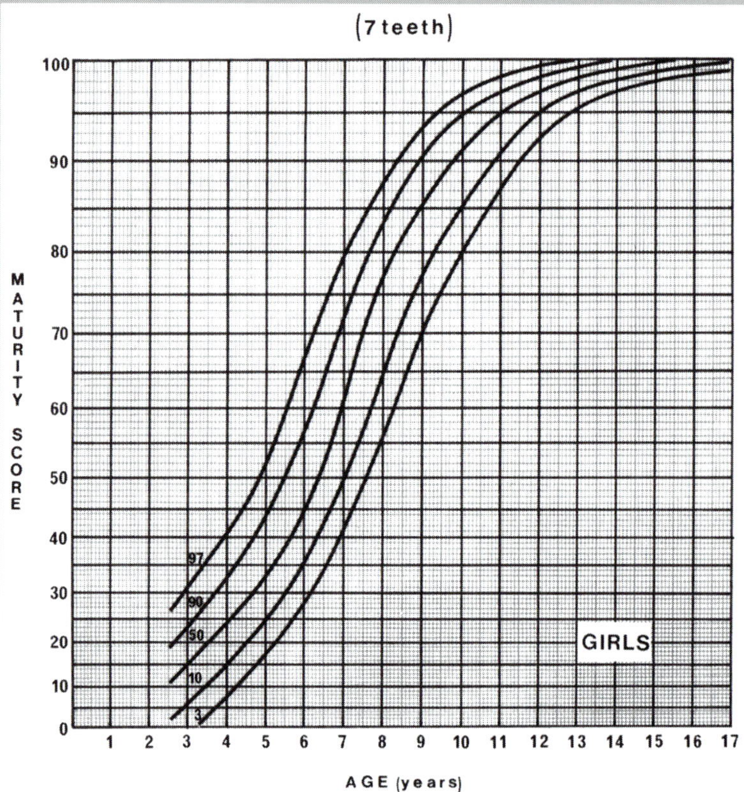


**Dental maturity percentiles.**

*Median ages of attainment of each developmental stage and emergence for girls and boys*

| | | A (1) | B (2) | C (3) | D (4) | E (5) | F (6) | G (7) | H (8) | EMERG. |
|---|---|---|---|---|---|---|---|---|---|---|
| $M_2$ | GIRLS | 3.5 | 4.0 | 4.6 | 5.9 | 7.9 | 9.9 | 11.5 | 14.9 | 11.3 |
| | *diff. | 0.0 | 0.0 | 0.3 | 0.4 | 0.6 | 0.5 | 0.5 | 0.4 | 0.3 |
| | BOYS | 3.5 | 4.0 | 4.9 | 6.3 | 8.5 | 10.4 | 12.0 | 15.3 | 11.6 |
| $M_1$ | GIRLS | - | - | - | - | 3.7 | 5.2 | 6.3 | 9.5 | 6.1 |
| | *diff. | - | - | - | - | 0.4 | 0.2 | 0.4 | 0.7 | 0.2 |
| | BOYS | - | - | - | - | 4.1 | 5.4 | 6.7 | 10.2 | 6.3 |
| $PM_2$ | GIRLS | 3.8 | 4.1 | 4.7 | 5.6 | 7.1 | 9.3 | 11.1 | 13.6 | 11.2 |
| | *diff. | 0.0 | 0.0 | 0.0 | 0.0 | 0.5 | 0.3 | 0.5 | 0.6 | 0.4 |
| | BOYS | 3.8 | 4.1 | 4.7 | 5.9 | 7.6 | 9.6 | 11.6 | 14.2 | 11.6 |
| $PM_1$ | GIRLS | - | - | 3.5 | 4.2 | 6.0 | 8.6 | 10.1 | 12.7 | 10.3 |
| | *diff. | - | - | 0.1 | 0.3 | 0.5 | 0.5 | 0.7 | 0.7 | 0.4 |
| | BOYS | - | - | 3.6 | 4.5 | 6.5 | 9.1 | 10.8 | 13.4 | 11.6 |
| C | GIRLS | - | - | - | 2.9 | 4.9 | 7.6 | 9.6 | 12.2 | 9.6 |
| | *diff. | - | - | - | 0.4 | 0.5 | 0.9 | 1.0 | 1.2 | 0.9 |
| | BOYS | - | - | - | 3.3 | 5.4 | 8.5 | 10.6 | 13.4 | 10.5 |
| $I_2$ | GIRLS | - | - | - | - | 3.7 | 6.1 | 7.3 | 9.2 | 7.1 |
| | *diff. | - | - | - | - | 0.7 | 0.4 | 0.4 | 0.4 | 0.3 |
| | BOYS | - | - | - | - | 4.4 | 6.5 | 7.7 | 9.6 | 7.4 |
| $I_1$ | GIRLS | - | - | - | - | 3.5 | 5.3 | 6.5 | 8.1 | 6.0 |
| | *diff. | - | - | - | - | 0.4 | 0.4 | 0.3 | 0.4 | 0.4 |
| | BOYS | - | - | - | - | 3.9 | 5.7 | 6.8 | 8.5 | 6.4 |

*The figures in italic represent the differences between the sexes (boys minus girls).

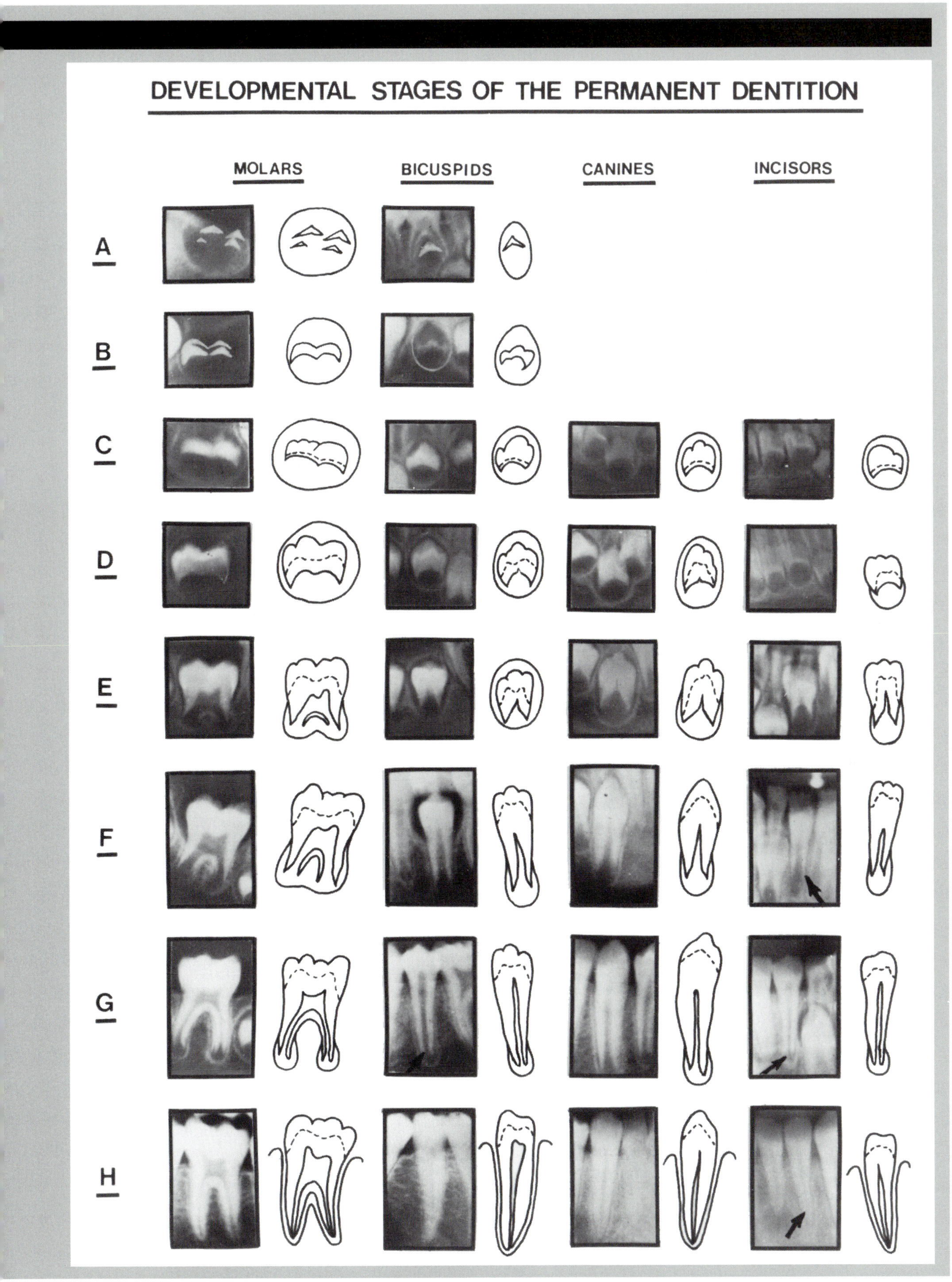

DEVELOPMENTAL STAGES OF THE PERMANENT DENTITION

MOLARS BICUSPIDS CANINES INCISORS

A

B

C

D

E

F

G

H

## SKELETAL DEVELOPMENT

While human somatic growth can be assessed and monitored with reference to changes in height, the ongoing process of maturation requires the use of specific indicators that reflect the complex array of physical changes that constitute maturity (**1.10, 5.17**). While secondary sexual development is an appropriate maturity indicator during the adolescent years it has little use prior to this age range. Only changes in dentition (**1.9**) and the skeleton (**1.8**) accurately record the maturation of an individual from birth to adulthood. The normality of a child's growth status is assessed using growth references or standards. Similarly, the normality of a child's maturation may be assessed using references or standards that relate specifically to the process of dental or skeletal maturation. Concern over the exposure of children to radiation has resulted in skeletal maturity being the most commonly used method to assess the maturational status and rate of maturity of children.

Skeletal maturation poses considerable problems in the development of an appropriate reference chart or standard. Maturity is linked to a genetic rather than a chronological clock; one year of chronological time may not represent one year of skeletal maturity. This fact was little understood in the early *Atlas* techniques used to assess skeletal maturity. Thus a growth 'reference' or 'standard' for skeletal maturity was depicted as a diagonal line passing through 0 years, in which each year of chronological age corresponded to a single year of skeletal 'age'. Because the rationale behind the Atlas technique was that a perfectly healthy child would present with all bones at the same level of maturity, variability in the skeletal ages of different bones *within* a child's hand–wrist would reflect 'general bodily health'. According to Greulich and Pyle, 1950, as this variability increased or decreased in subsequent radiographs the physician was theoretically able to, '...determine to what extent, if any, the child has corrected the imbalance in his skeletal development.'

It was clearly understood that the rate of skeletal development of the source sample may not be equivalent to that of another group of children. If the Atlas standards were found to be too advanced or retarded to adequately fit another group of children, Greulich and Pyle (1950) recommended that, '...the age equivalents of the standards can be scaled upward or downward in order to better conform to the developmental rate...' of the other group.

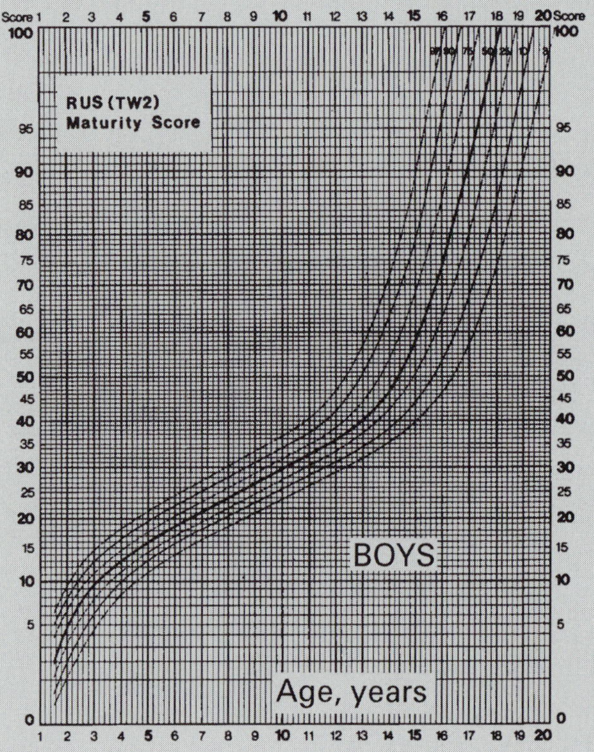


**Standards for RUS skeletal maturity score: boys.**

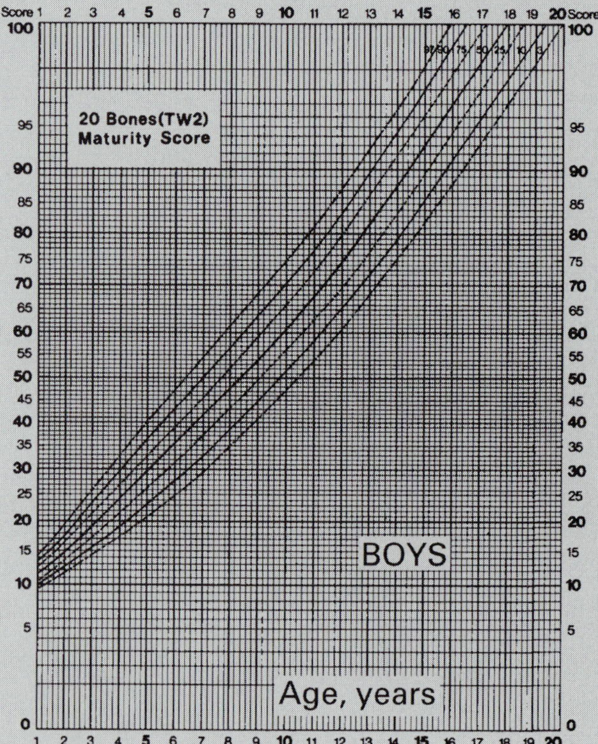


**Standards for 20-bone TW2 skeletal maturity score: boys.**

The later bone-specific scoring techniques did not use *skeletal age* on the Y-axis of skeletal maturity standards but instead used the *bone maturity score* resulting from the technique of assessment. These 'standards' were calculated in the same way as standards for height and weight. The vertical, or Y-axis scale, however was not a linear one but was transformed according to $y=sin^{-1}\sqrt{p}$) where $p$ is the score and $y$ is the transform. Centiles for the TW2 (20) bone system were almost straight, those for CARPAL maturity were straight from 5 years onwards, but those for the Radius, Ulna and Short bone (RUS) score demonstrated a marked curvature.

These centile 'standards' were cross-sectional and thus did not accurately reflect the pattern of change in skeletal maturity of an individual child who was more likely to demonstrate a steeper increase in TW2 (20) and RUS scores at adolescence. Because bone 'ages' are widely used to express maturity the specific system included tables of bone ages corresponding to given maturity scores. *Noel Cameron*

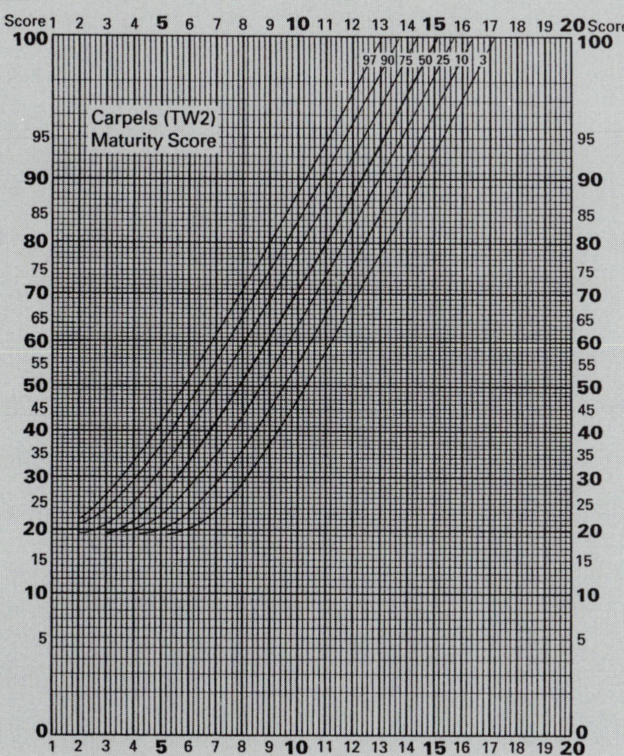


**Standards for carpal skeletal maturity score: boys.**

## GROWTH SOFTWARE

Computers play an important role in the collection, use and interpretation of growth information, and this is likely to increase as they become more widespread and software more sophisticated. The ability to store large quantities of data, and then to analyse and/or display them on demand makes computers essential for many different growth-related tasks.

The variety of these tasks is such that no single piece of software can be expected to cover all requirements. Clinicians in their growth clinics (see Part seven) have different needs from the statisticians charged with developing new growth charts. The situation is also fluid, with new software being developed all the time. This summary of available software can only hope to be a snapshot at one moment in time, and incomplete at that.

A basic requirement of growth software is to express anthropometric data as centiles using a reference chart. Two examples are the epidemiological package Epi-Info, which calculates weight-for-age, height-for-age and weight-for-height based on the international growth reference of the World Health Organization, and the British 1990 growth reference, which is available as a worksheet and a set of built-in functions for the spreadsheet package Microsoft Excel.

Spreadsheets like Excel are powerful for calculating and displaying data, but on their own they lack a database facility. The Castlemead Growth Program provides these features in a single package, which makes it useful for longitudinal growth studies. Data can be stored and then displayed in a variety of formats, using one of the growth standards provided, and others can be added.

If the aim is to construct new growth references, as opposed to using existing ones, then more specialized software is needed. Again, general-purpose packages can do the job in a limited way, but for the latest methodology tailor-made programs work best. The program Grostat II fits growth references by the HRY method, and there is a public domain program to fit the LMS method. Both programs require specialist knowledge to get the best out of them.

Auxal is another program for fitting growth references, but designed more for the individual than the population. Unlike the HRY and LMS methods it is parametric in form, fitting the JPA2 model to individual height growth curves, and it uses Bayesian methods to compensate for missing data. Individual curves can then be amalgamated to give population centiles as required.

### SOFTWARE SOURCES

1. Epi-Info 6. From: ftp.cdc.gov, directory /pub/epi/epiinfo, by anonymous ftp (PC, public domain).
2. British 1990 growth reference (MS Excel add-ins). From: Child Growth Foundation, 2 Mayfield Avenue, London W4 lPW, UK (PC, Mac, handling charge).
3. Castlemead Growth Program. From: Castlemead Publications, 12 Little Mundells, Welwyn Garden City, Herts AL7 lEW, U.K. (PC, commercial).
4. Grostat II. From: Maternal and Child Health Division, World Health Organisation, Geneva, Switzerland (PC, commercial).
5. LMS program. From: lib.stat.cmu.edu, file /general/lms, by anonymous ftp (Unix, public domain).
6. Auxal 3. From: SSI, 1525 East 53rd Street, Suite 530, Chicago, IL 60615-4530, U.S.A. (PC, commercial).

*T.J.Cole*

## ALTERNATIVE MEASURES FOR ASSESSING LINEAR GROWTH

The measurement of body segments (**1.4, 10.6**) offers the opportunity to evaluate their growth and to assess linear growth when length or stature measurements are not feasible. Upper arm length and lower leg length are two measures which serve both these purposes and for which good reference data are available.

### *MEASUREMENT TECHNIQUE*

*Infants (0 to 24 months)*: Sliding callipers (0–200mm) are used for young infants. For older infants, a standard anthropometer is used. From birth to 24 months, the lower leg length measure consists of a heel to knee measurement. With the infant supine, the right leg should be elevated and the knee bent at a 90-degree angle. The callipers are placed at the centre of the heel pad and the superior surface of the knee. The upper arm length is measured from the elbow to the shoulder. The right arm is flexed at the elbow at a 90-degree angle. The measurement is taken from the superior lateral surface of the acromion to the inferior surface of the elbow (olecranon). If deformities or asymmetries exist, the least affected side should be measured and the side noted.

*Children (3 to 18 years)*: Upper arm and lower leg length in older children are taken using an anthropometer. For the upper arm length measurement the subject should be positioned with the arm extended and relaxed at the side of the body. The measurement is taken from the superior lateral surface of the acromion to the head of the radius (radiale). The lower leg-length measurement is taken while sitting with the leg relaxed and the medial surface exposed (the right leg crossed horizontally over the left knee). The measurement is taken from the lower border of the medial maleolus (sphyrion) to the medial tip of the tibia (tibiale). The right side, or least affected side when asymmetry exists, should be measured and the side noted.

### *USE AND INTERPRETATION OF REFERENCE DATA FOR UPPER ARM LENGTH AND LOWER LEG LENGTH*

Reference charts for growth in length of the upper arm and lower leg are adapted from measurements of 230 infants (0–24 months) and 1239 children (2–19 years) measured in the survey of The Highway Safety Research Institute. For girls, the reference data did not indicate detectable growth after the age of 16 years, whereas for boys, growth continues until age 18. Children whose growth in upper arm or lower leg length fails outside the range given for their age and sex can be considered to have an unusual growth pattern. If growth is below the fifth percentile, this may be indicative of a nutritional deficiency, marked growth or developmental delay or some other pathology. In non-ambulatory children, it is not uncommon for the growth of the lower leg to be more retarded than the upper arm, relative to the reference standards.

*Babette S. Zemel and Virginia A. Stallings*

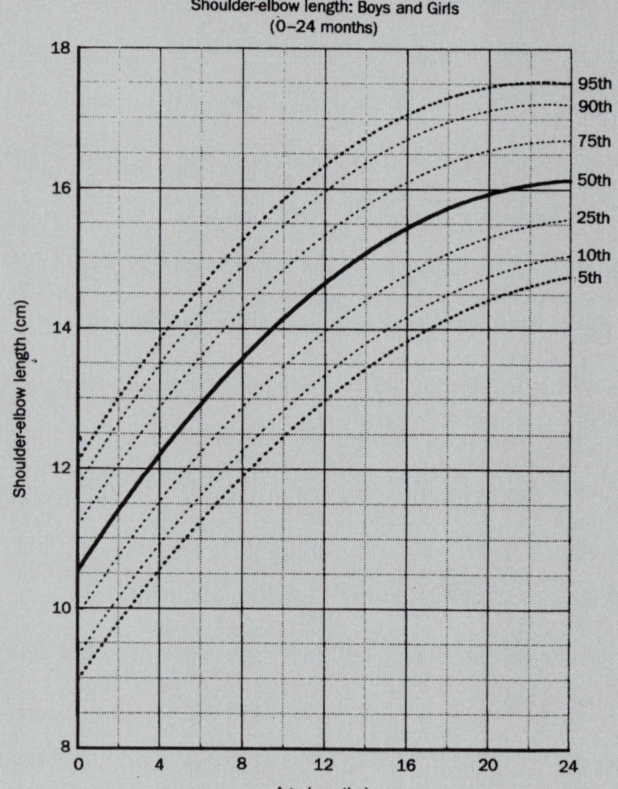


**Reference data for body segments.**

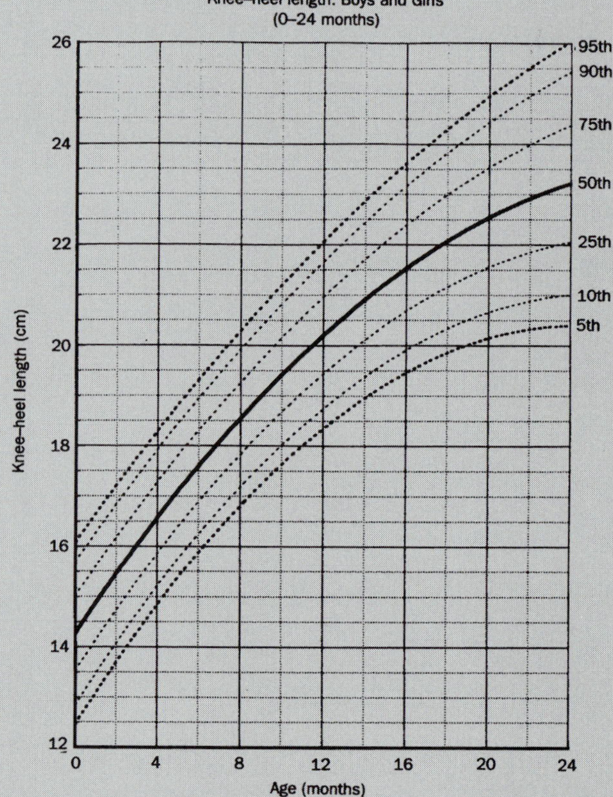


**Reference data for body segments.**

[1] **Reprinted with permission from Stallings, V.A. and Zemel, B.S. Nutritional assessment of the disabled child. In: *Clinics in Developmental Medicine: Feeding the disabled Child*. Edited by Peter Sullivan and Lewis Rosenbloom. Mac Keith Press, pp. 62–76, 1996.**

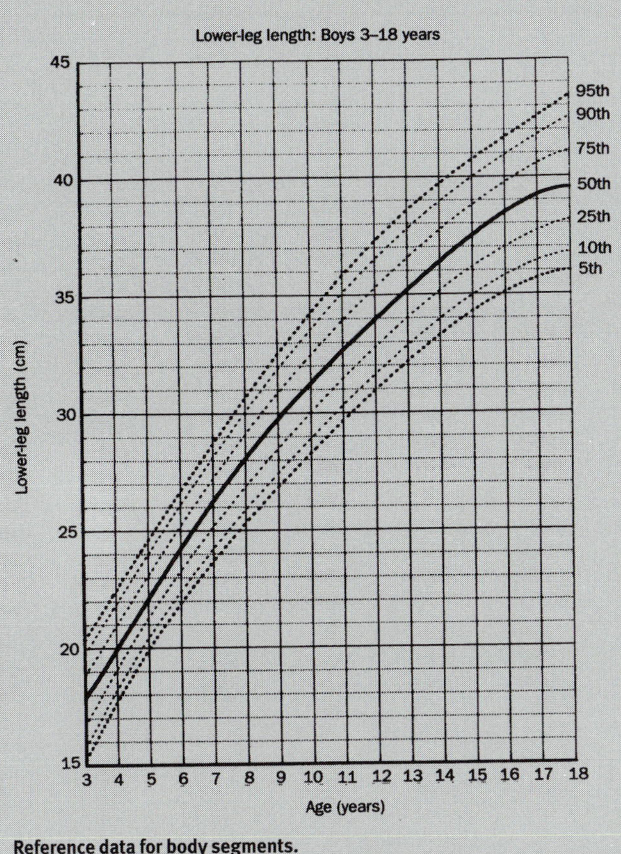


Reference data for body segments.

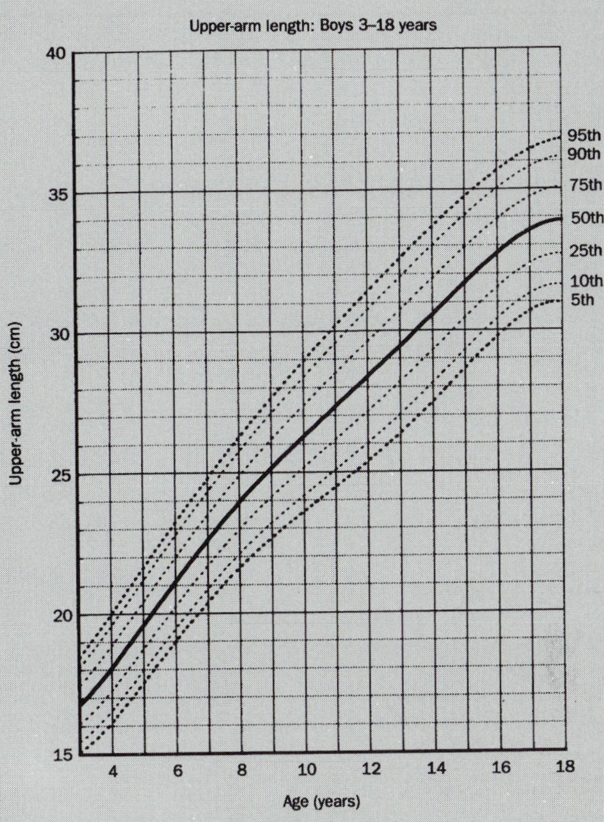


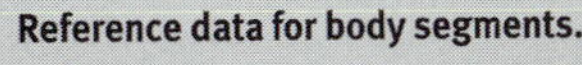
Reference data for body segments.

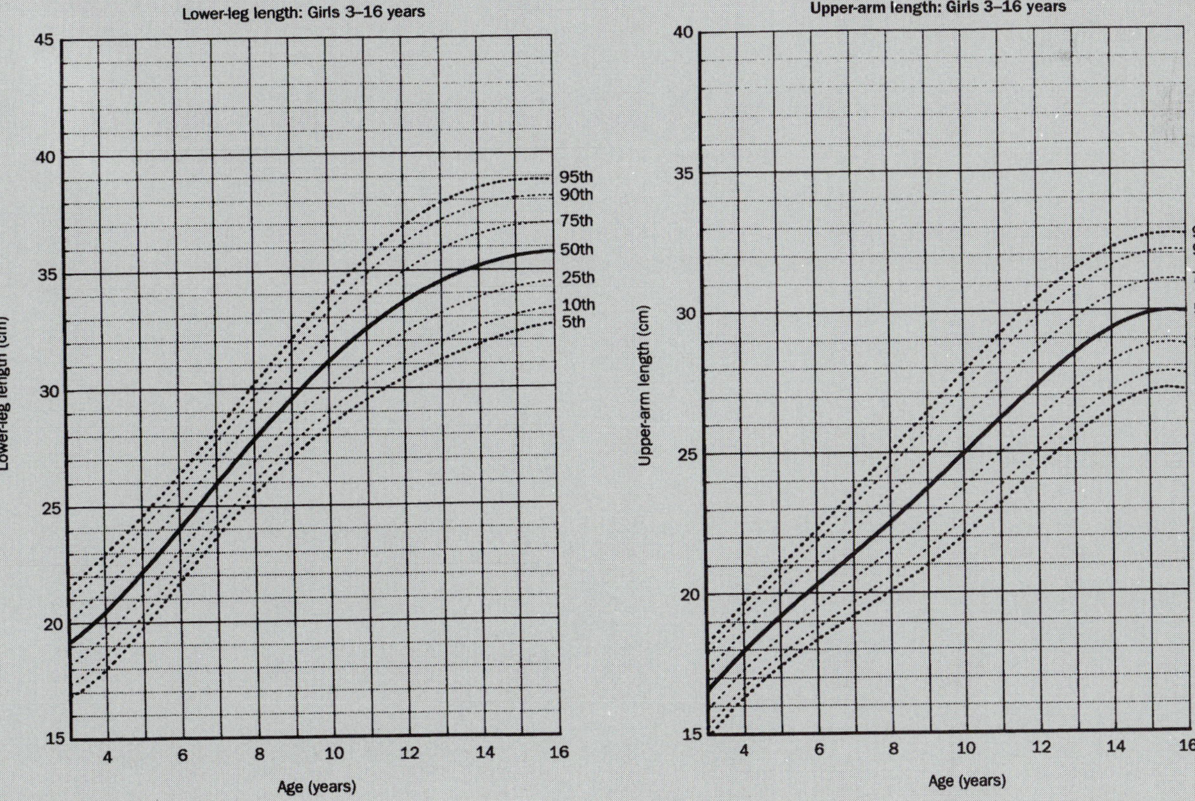


Reference data for body segments.

Reference data for body segments.

## CONDITIONAL GROWTH REFERENCES

The term *conditional reference* refers to a growth reference or growth chart where the measurement's distribution is conditional on some other variable. It is to some extent a misnomer as all growth references are by their nature conditional – even the simplest reference of weight plotted against age is a chart of weight conditional on, i.e. adjusted for, age. The more complete definition is that the reference is conditional on an extra variable over and above the age–sex adjustment. For example the Tanner–Whitehouse conditional reference of height on parental height adjusts the child's height for age and in addition for the height of his or her parents. Inevitably, introducing a second conditioning variable to the reference makes it more complex to use than the simpler reference conditioned just on age.

The purpose of a conditional reference is to exploit the fact that the measurement is highly correlated with the extra variable. If the variable is easy to measure, this provides a more precise assessment of the child's growth status. However, to be useful its advantage has to be balanced against the greater complexity of the chart.

Another example is the Tanner–Whitehouse clinical longitudinal or tempo-conditional height standard, which is conditional on the timing of puberty. The process of constructing a cross-sectional height reference blunts the intensity of the pubertal spurt, and as a result the centiles rise less steeply at this time than the heights of individual subjects plotted on the chart. In an attempt to get round this, Tanner and Whitehouse used as their median curve the height trace of a child of median height, who in addition had his/ her growth spurt at the mean age for all boys/ girls. Thus the chart is conditional on age at peak height velocity. In practice this form of chart is difficult to use, as the timing of the growth spurt is not known accurately except in retrospect, which reduces its value for predictive purposes.

Many conditional references are *regression* references, where the relationship with the conditioning variable is assumed to be linear and is estimated by linear regression. The predictive equation is of the form:

$$\text{Measurement} = \text{constant} + \beta.\ \text{Conditioning variable}$$

where the regression coefficient $\beta$ indicates by how much the measurement changes for a unit change in the conditioning variable. When height is the measurement, and height at an earlier age is the conditioning variable, this is a conditional reference to assess height velocity. If for any particular age group the slope $\beta$ is less than 1, this indicates that short boys grow faster at this age than tall boys. Hence the conditional reference is preferable to a velocity reference, which assumes that height-gain is unrelated to starting height.

The Tanner–Whitehouse parental height reference is a regression reference, and similar references have been described involving weight conditional on height, height conditional on height one year earlier, and birthweight conditional on sibling birth-weight.

Regression references are greatly simplified if they are expressed in terms of the SDS of the measurement and the SDS of the conditioning variable. The SDS of the measurement is by definition adjusted for age and sex, and the predictive equation simplifies to:

$$\text{Measurement SDS} = r.\ \text{Conditioning variable SDS}$$

where the constant in the previous equation is now zero and the regression coefficient $\beta$ is equal to r, the correlation between the measurement and conditioning variable SDSs. The residual standard deviation is known to be $\sqrt{(1-r^2)}$, and this allows the conditional SDS for residual standard deviation is known to be the measurement to be calculated explicitly. As an example, take weight SDS conditional on height SDS, expressed as an SDS of weight-for-height. The conditional SDS is given by:

$$\frac{\text{Weight SDS} - r.\text{Height SDS}}{\sqrt{(1-r^2)}}$$

This is a novel index of weight-for-height which depends only on the correlation between weight and height at different ages. Before puberty the correlation is fairly constant at about 0.7, and this leads to the simple weight-for-height index:

$$1.4 \times \text{Weight SDS} - \text{Height SDS}$$

A similar regression reference can be used to adjust height for mid-parent height, where mid-parent height is defined as

$$\text{Mid-parent SDS} = \frac{\text{Mother SDS} + \text{Father SDS}}{\sqrt{(2(1+r_{mf})}}$$

where $r_{mf}$ is the correlation between parents' heights. In practice this is about 0.3, due to assortative mating.

*T.J. Cole*

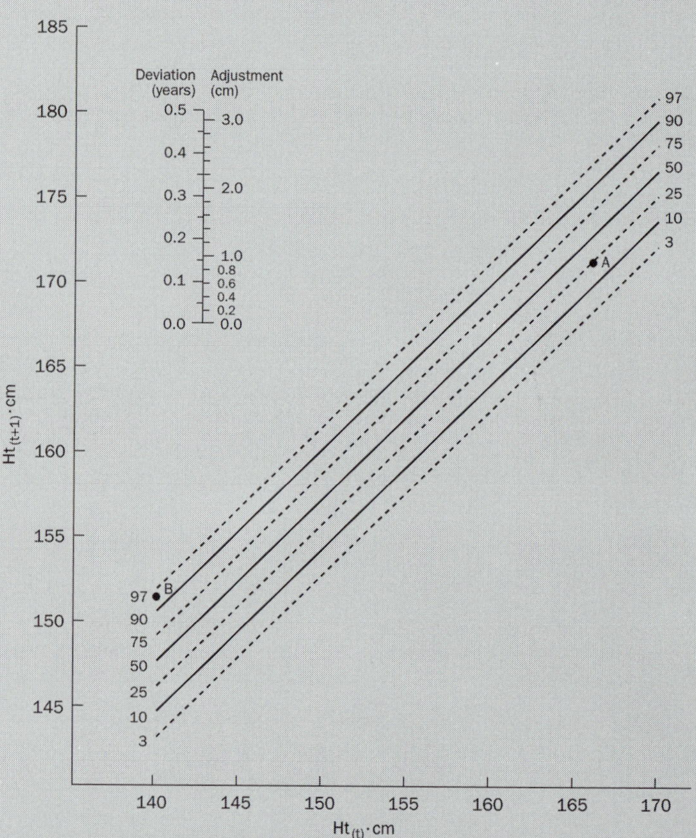


**Reference for boys' height conditional on height 1 year earlier. From Cameron N. Conditional standards for growth in height of British children from 5.0 to 15.99 years of age. *Annals of Human Biology* 1980 7: 331–337.**

## HEIGHT AND WEIGHT REFERENCES

A reference standard for growth is not just a set of means and standard deviations, no matter how large the sample might be. Much of the available data may be useful as describing the means (**1.13**), perhaps the standard deviations, of various samples from around the world and of allowing investigators to compare different groups. However, these descriptive samples cannot serve as true reference standards.

An acceptable reference standard should conform to a set of criteria which specify the characteristics of the population which has been sampled and the methods by which the data have been aggregated and presented. The following criteria were developed by an expert committee of the World Health Organization:

- the population should be well-nourished
- each age/sex group should contain at least 200 individuals
- the sample should be cross-sectional
- sampling procedures should be defined and reproducible
- measuring procedures should be optimal
- measurements should include all variables to be used in nutritional evaluation
- raw data and smoothing procedures should be available

The most common variables for evaluating growth status are height and weight. They are global measurements which are universally understood by professionals and non-professionals alike. Height and weight are useful indicators of the overall adequacy of growth and are sensitive to a range of environmental factors, even if they are not particularly specific as to the cause of any deviation. Finally, height and weight are relatively easily to measure under a variety of conditions, and their technical errors of measurement are acceptably low.

Except for the circumpubertal years, height is distributed normally so that conventional statistics may be used to describe the distribution. Weight on the other hand is usually positively skewed and the standard deviation may not be accurate as an indicator of dispersion about the mean, especially at the more distant centiles.

In constructing the shape of growth curves for height and weight at varying distances from the mean, a variety of smoothing techniques have been used. These techniques involve different mathematical functions, parametric as well as non-parametric, for smoothing along a particular centile, and other techniques for smoothing across the centiles at a particular age. The growth curves for American children as constructed by the United States National Center for Health Statistics (NCHS) involved the fitting of third order polynomials to relatively short age ranges along specific centiles, and the curves generated by the polynomials joined sequentially by a splining technique. These data have been adopted widely by a range of public and private organizations and have become accepted as international reference standards. However, the sample was measured some 20 years ago and the cubic spline technique has been superseded by newer and better methods of smoothing. While the NCHS standards have been valuable for a variety of purposes, they are being updated with newer data and smoothing techniques.

Why be concerned about sampling and smoothing? Very simply, if reference data are to be valid indicators of the population from which they were derived, and useful for clinical (see Part Seven) and epidemiological purposes, then they must portray accurately the distributions of height and weight through the growing years. Given that height and weight are distributed normally (or near normally), it is a relatively simple matter to estimate their means and nearby percentiles. However, considerable effort is required to estimate more distant centiles, for example below the 10th and above the 90th. This is crucial since it is these extreme centiles which are used in screening for growth failure or for obesity. Consequently, the construction of reference standards for height and weight is not a simple task and requires considerable planning and execution, from the selection of the sample to the collection of the data to the development and presentation of the reference data.

*Francis E. Johnston*

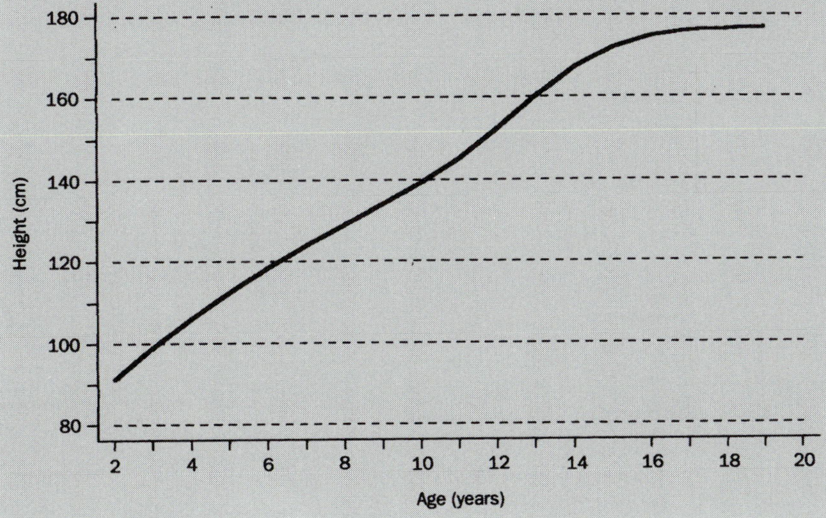


**Smoothed growth curve of height of American males (from Hamill et al., 1977).**

## BIBLIOGRAPHY OF SOME CURRENT HEIGHT- AND WEIGHT-BY-AGE REFERENCES

Listed here are references to selected samples which provide reference data on height- and weight-by-age for local regions of the world. In choosing from the large number of data reported in the literature several criteria were employed. First, references for growth velocity, body mass index, skinfolds, arm circumference and other measures were excluded, for the sake of simplicity. Second, studies required adequate sample size. The third criterion was geographical area, with an attempt made to cover the major population groups and continents, and to avoid an overemphasis on any one area, such as Europe. The fourth was the age of the data, with every effort made to include the most recently published studies.

Ideally, reference data covers a broad cross-section of a country or region, so as to avoid presenting statistics on children who are disproportionately malnourished, or who are drawn only from the élite strata of society. Also, the data should be truly cross-sectional, and smoothed. These criteria are not always satisfied fully, leading to noticable gaps in the data, especially with respect to some geographical areas, such as the African continent.


- Algeria
  Chamla M-C, Demoulin F. (1976). *Croissance des Algèriens de l'Enfance à l'Age Adulte*. Paris: Centre National de la Recherche Scientifique.
- Belgium
  Wachholder A, Hauspie RC. (1986). 'Clinical standards for growth in height of Belgian boys and girls, aged 2 to 18 years'. *International Journal of Anthropology* 1: 327–38.
- China
  Zhang X and Qin ZT. 'Physical growth and development'. In *Practical Pediatrics* ed. FT Zhu. Beijing: People's Health Publishing House.
- Cuba
  Jordan JR. (1979). *Desarrollo Humano en Cuba*. Cientifico-Tecnica, La Habana.
- Germany
  Danker-Hopfe H, Thiel P, Tsiakalos G. (1981). 'Untersuchungen zur Bevolkerungsbiologie Norddeutschlands'. *Anthropologischer Anzeiger* 39: 82–88.
- Hong Kong
  Leung SF. (1995). *A Simple Guide to Childhood Nutritional Assessment*. Hong Kong: Maeil.
- Hungary
  Eiben OG, Barabás A, Pantó E. (1991). *The Hungarian National Growth Study*. Budapest: Humanbiologia Budapestinensis.
- India
  Indian Council of Medical Research. (1972). *Growth and Development of Indian Infants and Children*. New Delhi: Medical Enclave.
- Ireland
  Hoey HMCV, Tanner JM and Cox LA. (1987). 'Clinical growth standards for Irish children'. *Acta Paediatrica Scandinavica*, suppl. 338: 1–3.
- Japan
  Kikuta F, Takaishi M. (1987). 'Studies on physical growth standards for school-children in Japan. Part I. Centile curves for height and weight based on cross-sectional data and a consideration of secular trend of the centile curves'. *Journal of Child Health* (Japan) 46: 27–33.
- Mexico
  Faulhaber JA and Sáenz F. (1994). *Terminando de Crecer en México. Antropometría de Subadultos*. Mexico: Universidad Nacional Autónoma de México.
- The Netherlands
  Roede MJ and van Wieringen JC. (1985). 'Growth diagrams, 1980'. *Tijdshrift voor Sociale Gezondheidszorg* 63 (suppl. 1985): 1–34.
- Poland
  Hulanicka B, Brajczewski C, Jedlinska W, Slawinska T, Waliszko A. (1990). *City-Town-Village, Growth of Children in Poland in 1988*. Monograph 7, Institute of Anthropology. Wroclaw: Polish Academy of Sciences.
- Spain
  Hernández M, Castellett J, Narvaíza, Rincón JM, Ruiz I, Sanchez E, Sobradillo B and Zurimendi A. (1988). *Curvas y Tablas de Crecimiento*. Bilbao: Fundación F. Orbegozo.
- Sweden
  Lindgren GW and Strandell A. (1986). *Fysisk Utveckling och Halsa. Rapport 4. Institutionen för Pedogogik*. Högskolan för Lärarutbildning, Stockholm.
- United Kingdom
  Cole TJ, Freeman JV and Preece MA. 'British 1990 reference centiles for height, weight, body mass index and head cricumference fitted by maximum penalized likelihood'. *Statistics in Medicine* 1996.
- United States of America
  Hamill PVV, Johnson CL, Reed RB, Roche AF. (1977). *NCHS Growth Curve for Children, Birth–18 Years, United States*. DHEW Publication (PHS) 78–1650. Washington DC: U.S. Government Printing Office.
- Venezuela
  Project Venezuela (1994). *Proyecto Venezuela 1981–1987. Resultados Nacionales*. Caracas: Fundacredesa.


*Francis E. Johnston*

## WEIGHTS DEFINING SMALL-FOR-GESTATIONAL-AGE IN TWINS

Although intra-uterine growth retardation has long been recognized as an issue, no set of norms has been published outlining what constitutes small-for-gestational-age (SGA) in twins in any other than American data. The data here come from a group of 1972 twin pairs aged between 4 and 12 who were participants in the Australian Twin Attention Deficit Hyperactivity Disorder project in 1991. Full details of the sample are provided in Levy et al. (1996). Children with significant congenital abnormalities have been excluded. In the absence of major effects of zygosity, data from monozygotic (MZ) and dizygotic (DZ) twins were pooled. Information assessing birth-weight and gestation length were collected by mailed questionnaire from parents, along with wider behavioural data. The SGA norms were calculated from twins who were born from 34 to 40 weeks gestation, as prior to 34 weeks and after 40 weeks there were insufficient numbers in the sample to calculate accurate SGA percentile ranks at the extreme centiles. While being extremely growth-restricted has often been described as being below the third percentile, to maximize subject numbers the extremely growth-restricted category included those babies at, as well as below, the third percentile in weight for their gestational age and gender. Similarly, small for gestational age was taken as being at or below the 10th percentile, rather than simply below the 10th percentile. The birth-weights in grams at the third (WGA3) and the tenth (WGA10) percentile for male and female twins born from 34 to 40 weeks gestation inclusive, are given in the table. The table also shows the percentage of males and females under singleton WGA3 and WGA10 percentiles. The singleton means with which the twin means are compared come from singleton norms in Western Australia published by Blair and Stanley (1985).

*Rosanna Rooney, David A. Hay and Florence Levy*

*Weight at the third (WGA3) and the tenth (WGA10) percentile, and percentage under singleton WGA3 and WGA10 percentiles for male and female twins born from 34 to 40 weeks gestation*

| gestation age (weeks) | g | MALE TWINS % below singleton male mean | g | FEMALE TWINS % below singleton female mean |
|---|---|---|---|---|
| **WGA3** | | **WGA3** | | **WGA3** |
| 34 | 1443 | 17 | 1490 | 24 |
| 35 | 1580 | 28 | 1588 | 33 |
| 36 | 1765 | 26 | 1800 | 30 |
| 37 | 2030 | 25 | 1726 | 21 |
| 38 | 2155 | 45 | 2000 | 21 |
| 39 | 2300 | 18 | 2075 | 20 |
| 40 | 2292 | 30 | 2245 | 18 |
| **WGA10** | | **WGA10** | | **WGA10** |
| 34 | 1789 | 33 | 1760 | 52 |
| 35 | 1810 | 50 | 1856 | 55 |
| 36 | 2042 | 54 | 1970 | 50 |
| 37 | 2240 | 51 | 2160 | 50 |
| 38 | 2377 | 45 | 2235 | 45 |
| 39 | 2547 | 46 | 2296 | 41 |
| 40 | 2496 | 51 | 2468 | 49 |

# The use of growth references

A growth reference is a table or chart summarizing how an anthropometric measurement such as height, weight or head circumference changes during childhood. The growth reference is based on a defined reference sample, and the chart consists of selected centiles (the median, 10th, 98th centile etc.) of the measurement at different ages during childhood. Broadly speaking, the purpose of a growth reference is to account for age and sex differences in anthropometry. In practice it is used in two distinct contexts: as a clinical sign, to monitor the growth status of individual children, and as a public health tool, to summarize and compare the anthropometry of groups of children.

There is an important distinction between a growth *reference* and a growth *standard*. A standard represents 'healthy' growth, whereas a reference makes no claims about the health or otherwise of its reference sample. Thus in principle a standard is better than a reference for diagnosing growth disorders, always assuming that the standard is appropriate for the child being assessed. In practice a reference is easier to construct than a standard, as there is no need to define 'good' growth or to exclude individuals from the reference sample who fail to achieve it.

Growth charts are used not only in growth clinics, but also by obstetricians to monitor fetal size, in well-baby clinics, by school nurses for height screening, and in general practice. The child's measurement is plotted on the chart, and the centile read off. If the centile is sufficiently extreme, or if it has changed materially since the previous measurement, then the child may need referral for more detailed investigations.

In the developing world, growth charts are also used for education purposes. The 'Road to Health' chart is based on a developed-world reference, using just two centiles to provide

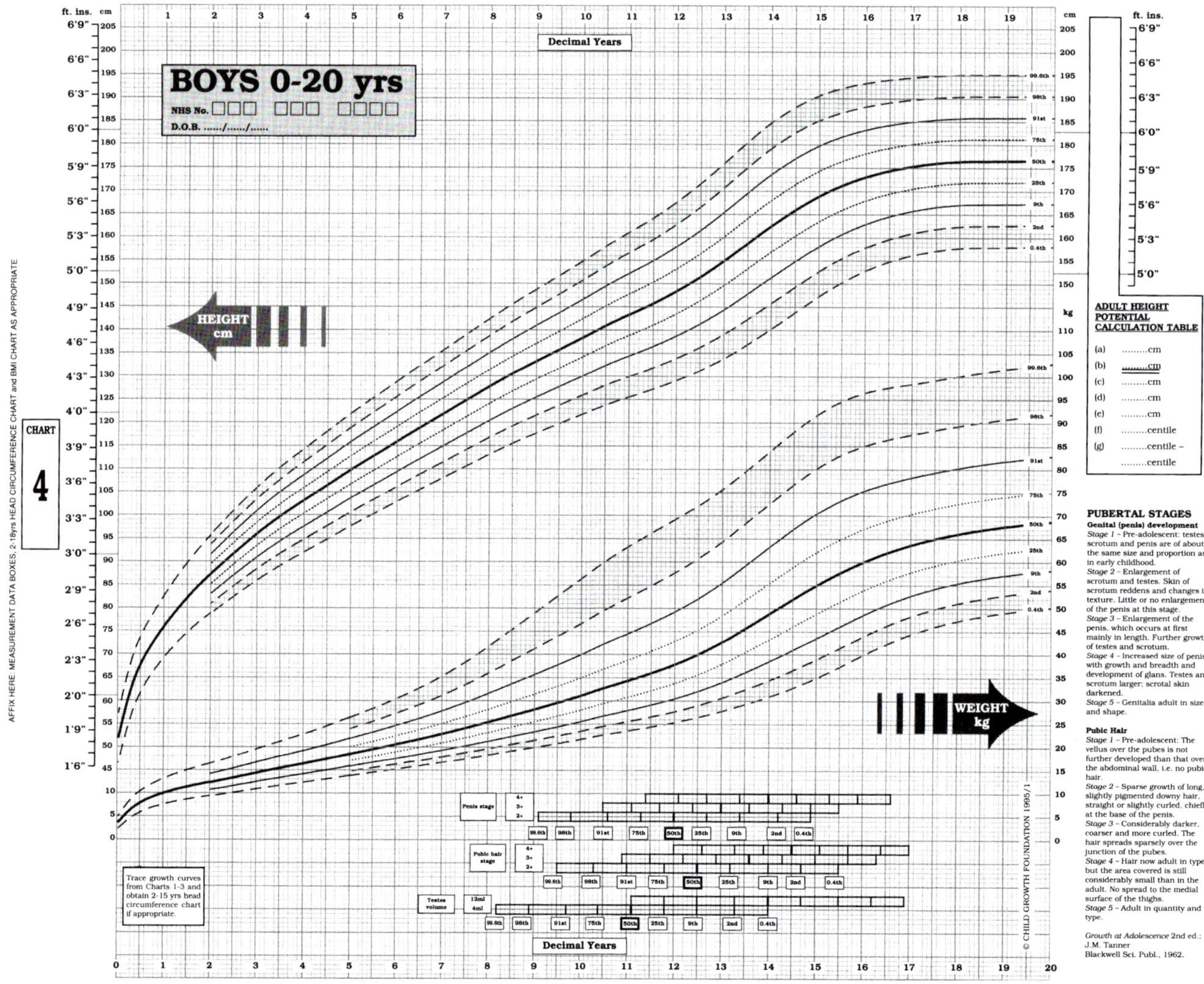


The 1990 British growth reference for height and weight, showing nine centiles from birth to 20 years of age.

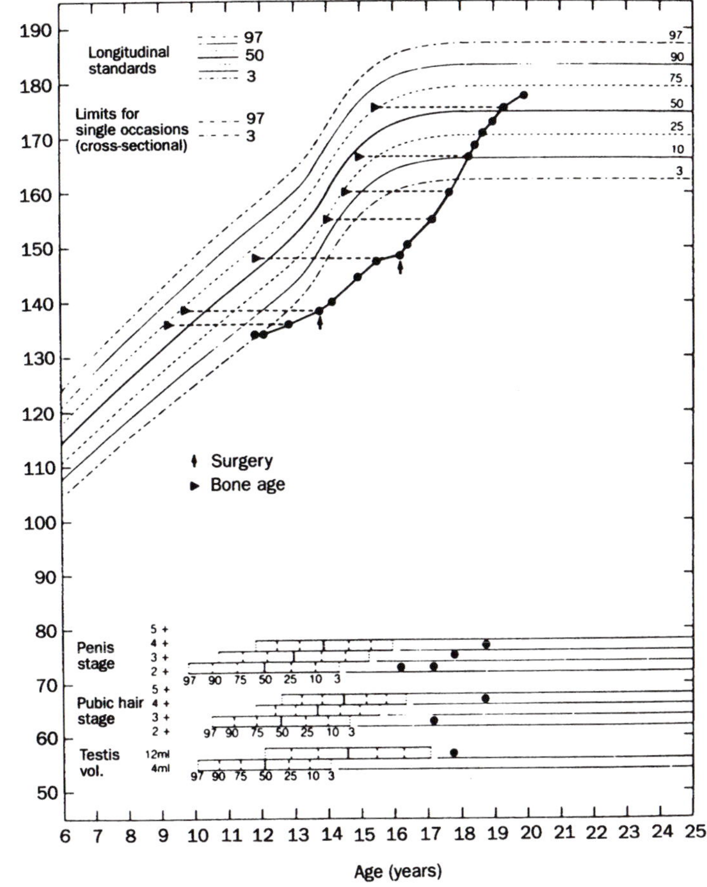


**Height chart of a child with Crohn's disease: the effect of the two bowel resections on his growth pattern. After the second resection, height velocity increases dramatically.**

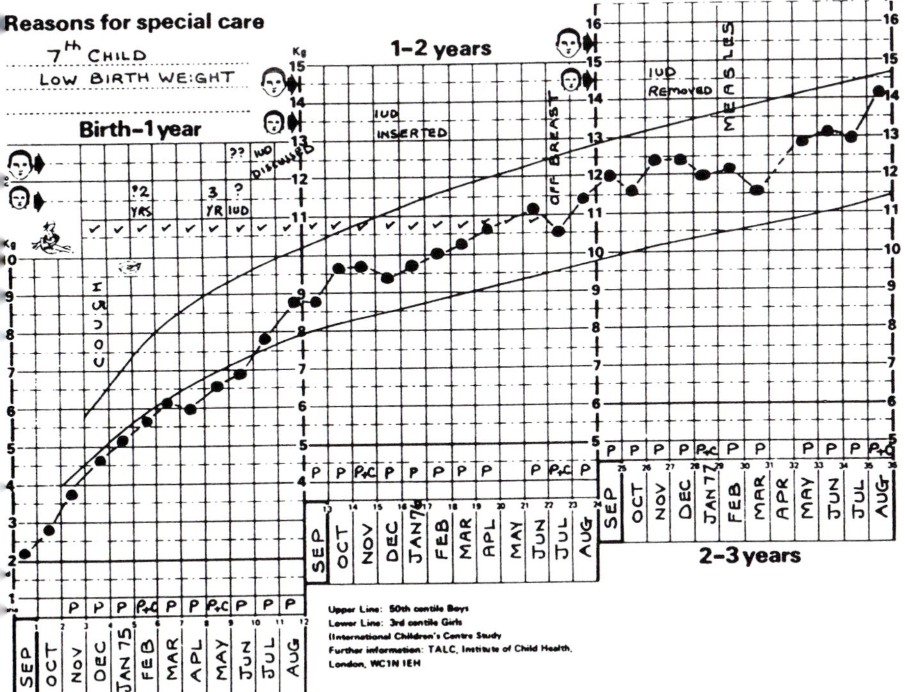


**A 'Road to Health' chart typical of those used in maternal and child health clinics across the developing world. Weights are represented as dots.**

a channel or 'road' along which children should grow. The chart may also provide other relevant information for primary care, such as the timing of weaning or immunizations.

Screening for growth uses an extreme centile on the chart to identify children who are unusually small or large. The fifth or third centile is commonly used for this purpose, but it screens in a large proportion of children, i.e. about 5 per cent and 3 per cent respectively of the normal population. Only a small minority of such children are likely to have a growth disorder, and to avoid such a high false positive rate a more extreme centile can be used. The British growth reference provides a four per 1000 centile, which substantially increases the proportion of screened-in children with a growth disorder. At the same time it reduces the number of short normal children who are investigated unnecessarily.

Most growth references are based on cross-sectional data, i.e. each child in the reference sample contributes a single measurement at one particular age. Such references are often used for growth monitoring, i.e. following a child's centile on the chart over a period of time, on the basis that normally growing children should remain close to their chosen centile. However, the chart tells only part of the story, because the reference contains no information about growth as such, so the range of variability about the child's chosen centile cannot be determined. Velocity references or conditional references are more appropriate for monitoring growth over time.

Another difficulty with cross-sectional references arises during puberty. The pubertal peak in height and weight velocity occurs over a wide range of ages in the population, and the reference smooths out and hence flattens the slope of the median curve. As a result no child stays on the same centile throughout puberty – they all grow faster around the time of their growth spurt, and slower afterwards. The graph shows the difference between cross-sectional and longitudinal median height and height velocity during puberty for boys on the Tanner–Whitehouse chart.

A growth reference chart is a powerful tool in clinical practice – it is visually effective, simple to use and easy to understand. For public health purposes though, the visual element is less important than the ability of the reference to adjust measurements mathematically for age and sex. On the chart the measurement appears as a centile, but for statistical purposes this is better expressed as an SD score, Z-score or SDS. Some growth references give the distribution of the measurement at each age in terms of the mean and standard deviation as well as selected centiles, to simplify the adjustment. Thus the heights and weights of groups of children can be sum-

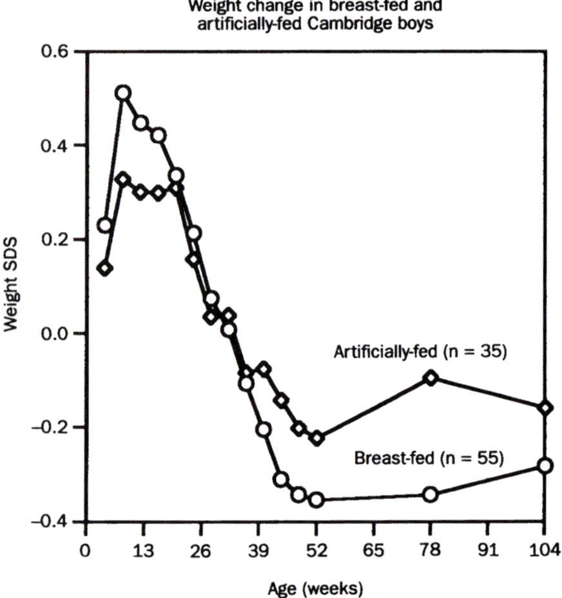


**Differences in weight SDS (Z-score) between groups of breast-fed and formula-fed infants during the first year of life, where the formula-fed group grows faster. The adjustment for age removes the age trend of increasing weight, so that normally growing children appear to grow along a horizontal line. The patterns of weight-gain are very different from the straight-line growth predicted by the reference. The mismatch arises from the reference being out of date, and not taking into account recent changes in infant feeding practice.**

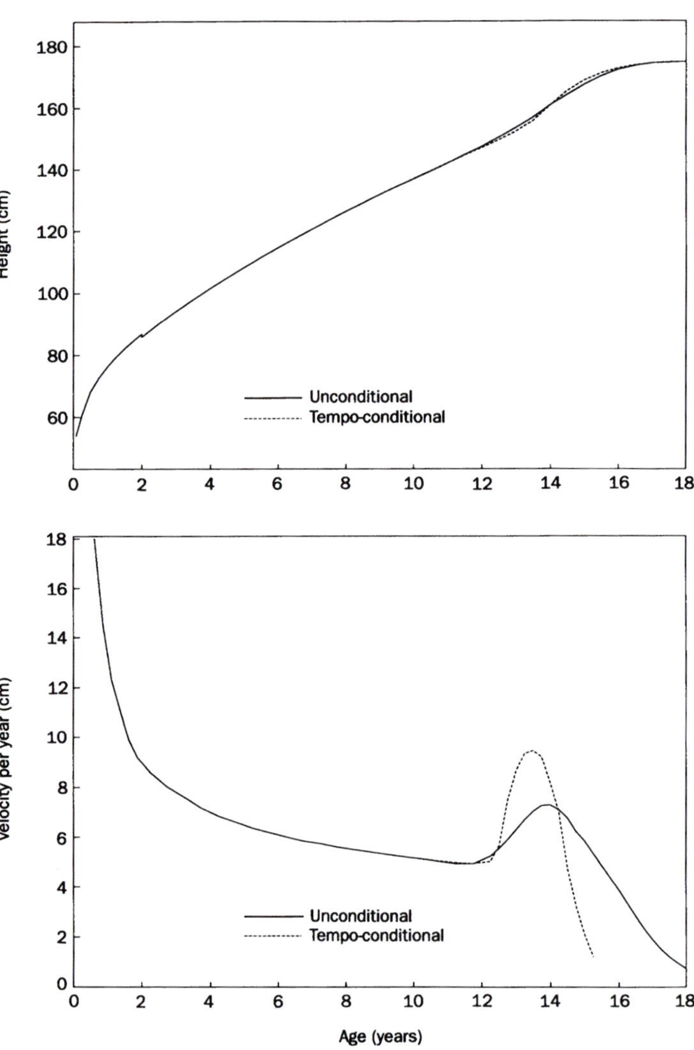


**The difference between cross-sectional and longitudinal median height, and height velocity during puberty, on the Tanner–Whitehouse 1966 chart. Comparison of tempo-conditional and unconditional height distance and height velocity curves (from Cole, 1993).**

marized in terms of their means and standard deviations of height SDS and weight SDS.

The validity of a growth reference depends critically on its underlying reference sample. National references in the Western world, such as Sweden, The Netherlands, Britain or America, are based on samples representative of the national population. In the developing world there is a need, both health-wise and politically, to distinguish between the whole population, where many children grow relatively poorly, and selected élite groups, where the growth pattern matches that of developed-world children. This highlights an important dilemma, whether the sample should be representative of all children or just the élite.

For group comparisons, the choice of sample does not matter too much – the growth reference acts purely as a neutral baseline. However for clinical use the reference needs to represent the pattern of growth that monitored children can reasonably expect to achieve, and the reference sample should be representative of such 'local' children. Secular trends in height and weight make it advisable to revise national references every 10 to 15 years. Children with genetic defects known to affect growth, for example, cystic fibrosis, or Down's, Turner's or Prader–Willi Syndrome, are special cases of children requiring their own growth references. Taken to extremes, the need for a local reference suggests that there should be separate growth references for every regional group within a country. This is clearly impractical. A simpler alternative is to have a single national reference and to express distinct geographical or ethnic groups relative to it in terms of their mean offset in SDS (Z-score) units. For example, regional differences in height and weight exist for white children in Britain; those from the north and west tend to be slightly shorter and lighter than those from the south-east, although the differences amount to no more than 5 centile points.

For international comparisons the concept of a local reference is irrelevant, and a different strategy is needed. The current World Health Organization international reference is based on the American National Center for Health Statistics reference, but this is to be replaced by a reference that is representative of several countries rather than just one. *T.J. Cole*

See also 'Longitudinal analysis' (**1.14**), 'Creation of growth references' (**1.16**), 'Standards and references for the assessment of fetal growth and development' (**4.9**), 'Extra-uterine growth after premature birth' (**5.2**), 'Identification of abnormal growth' (**7.1**), 'Endocrine growth disorders' (**7.7**), 'Growth in chronic diseases' (**7.8**) and 'Growth modelling and growth references: the future' (**13.1**)

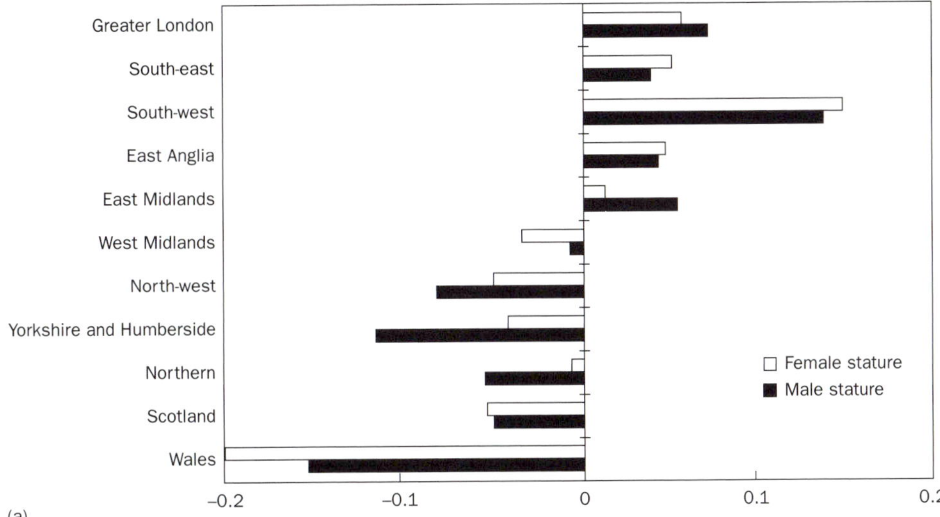


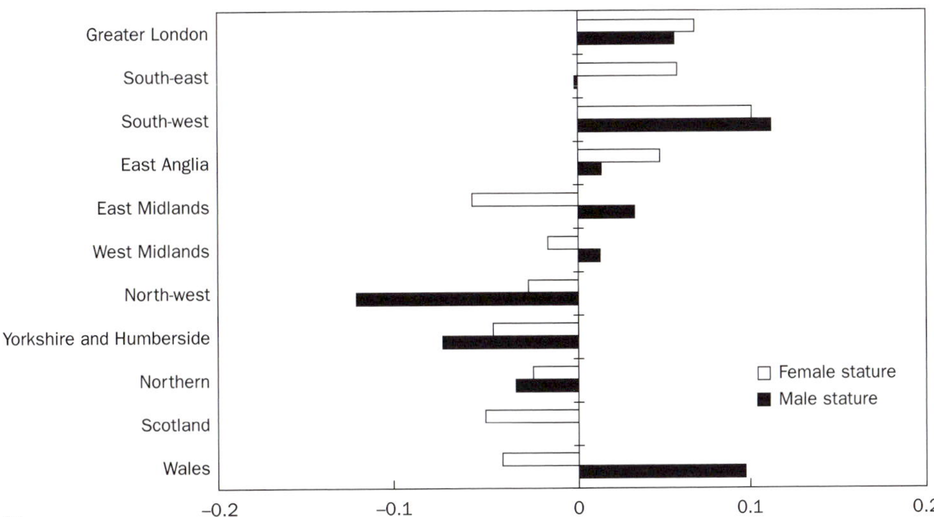


**Regional differences in height and weight for White children in Britain, expressed as SDS units (Z-scores).**

## MONITORING OF TREATMENT FOR GROWTH FAILURE

The role of monitoring during treatment of a growth disorder is to ensure that treatment is optimal, achieving maximal sustained catch-up growth. To ensure that this is achieved it is necessary that the child is observed at appropriate intervals, measured accurately and the measurements compared to growth reference data that is appropriate.

During catch-up growth (**9.6**) the child is, by definition, growing at a velocity that is greater than normal for that age. Equally, the error of measurement of growth velocity is inversely proportional to the amount of growth that there has been between adjacent measurements. During the bulk of childhood, between the ages of 2 and 10 years, normal growth velocity is such that measurements more frequent than 3–4 monthly are seldom worthwhile. That is assuming that the measurements are made with the highest practical level of precision with a Standard Error of Measurement ($S_{meas}$) of 0.2–0.3 centimetre. In the presence of catch-up growth with higher velocities this time interval may be reduced, but as treatment for growth failure is always long-term, shorter measurement intervals seldom add significant clinical information.

Accuracy of measurement is discussed elsewhere, but it is particularly important in this clinical monitoring role. It clearly ties in with the issues of measurement frequency. Values of $S_{meas}$ significantly greater than 0.3 centimetre rapidly degrade the quality of the data so that clinically useful monitoring becomes impossible.

The issue of appropriate standards is more one of interpretation than the nature of the reference data, assuming that they are properly constructed in the first place. The point is that the patient's progress is judged partly by comparison to the standards, but partly by comparison with the child's previous growth pattern. In mid-childhood there is little difficulty, but at puberty things can get very difficult, particularly if the patient has significant growth delay or advance. The former is the most common occurrence. A problem arises because the patient is being compared to age-group peers, but they are not maturity peers and they are showing a spontaneous acceleration in growth due to puberty. This can give the impression that the short patient (who is also delayed) is not showing catch-up growth despite treatment. If the growth curve was compared to prepubertal individuals of the same age it would be clear that catch-up growth was taking place.

Growth monitoring may also play a role in the treatment of chronic childhood illnesses where the growth failure is secondary to the main disease (**7.8**). The use of careful measurements of height and weight can add greatly to the clinical management of a child in this situation and form a powerful adjunct to the care of the underlying disease.

The final issue that requires supervision, rather than formal monitoring, is the decision of when to stop treatment. In the case of treatments, like androgens, for simple delayed adolescence then treatment is stopped as soon as it is clear that the individual is in well-established puberty. On the other hand, treatments for disorders such as juvenile hypothyroidism are lifelong and will be continued long after growth has ceased, as thyroid hormone is required for functions other than growth. *Michael Preece*

## Z-SCORES

*Approximate percentiles and proportion of a normal distribution falling above or below selected values of Z*

| Z | Percentile | Approximate proportion below | Z | Percentile | Approximate proportion above |
|---|---|---|---|---|---|
| 0.0 | 50 | 0.500 | 0.0 | 50 | 0.500 |
| −1.0 | 16 | 0.160 | 1.0 | 84 | 0.160 |
| −2.0 | 2 | 0.020 | 2.0 | 98 | 0.020 |
| −3.0 | 0.1 | 0.001 | 3.0 | 99.9 | 0.001 |

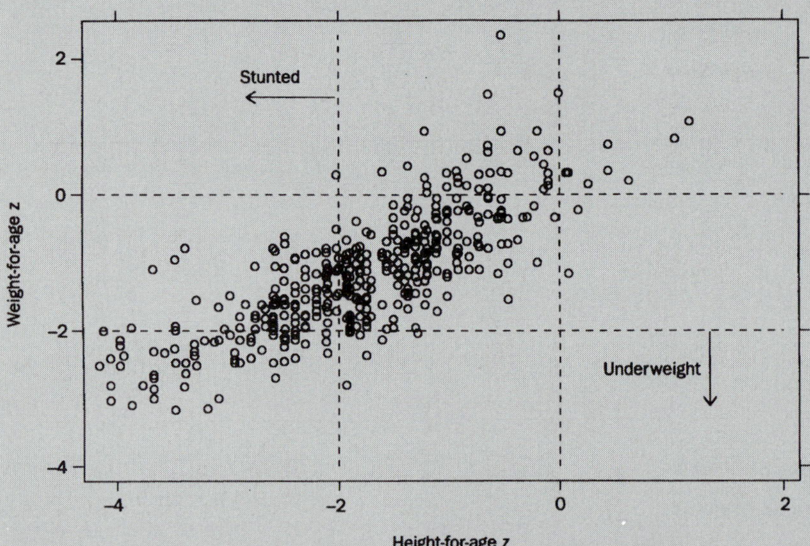


**Distributions of Z-scores of weight and height of 421 7-year-old boys and girls from a poor community on the periphery of Guatemala. The raw measurements were converted to Z-scores using the means and standard deviations of the reference data from the U.S. National Center for Health Statistics (NCHS). The individual points represent the height and weight Z-scores of individual children. The arrow in the upper left part of the graph indicates those whose height-for-age Z-scores fall below −2 (less than −2 SD), a common cut-off value for stunting. The arrow in the lower right indicates those with weight-for-age Z-scores below −2, a measure of underweight.**

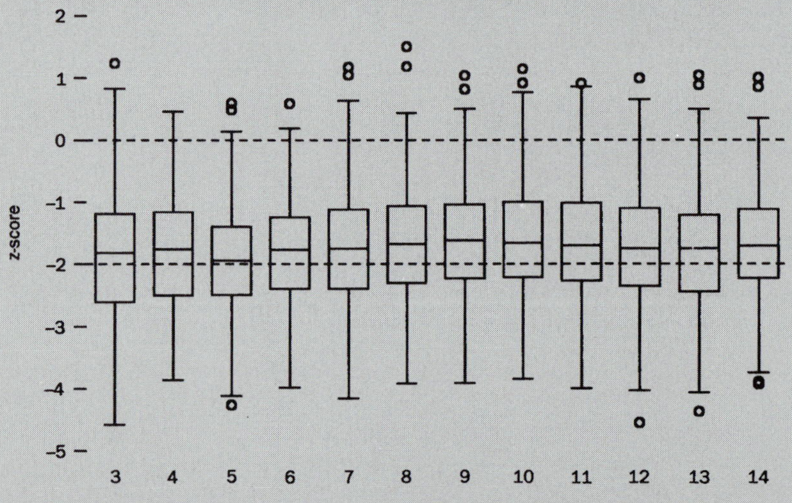


**Another use of Z-scores using data from the same study in Guatemala. Box plots of height Z-scores (from the U.S. National Center for Health Statistics) by age show how change in relative height status can be visualized across a 12-year age span. The box indicates the interquartile range (between percentiles 25 and 75), and the horizontal line in each box, the median.**

Alternative names for Z-scores are standard deviation units, and standard deviation scores (SDS). The formula for a Z-score is:

$$Z = \frac{\text{Measurement} - \text{Mean}}{\text{Standard Deviation}}$$

where Measurement equals the value of a particular measurement, Mean equals the mean of the distribution, and Standard Deviation equals the standard deviation of the distribution.

Because of Central Limit Theorem, the Z-score is the basis of inferential statistics. This theorem states that the distribution of sample means of a specified size will approach the normal as the number of samples increase, regardless of the form of the original distribution of the variable. If, in the formula, Measurement refers to a sample mean, and Mean and Standard Deviation refer to a sample drawn for analysis, the resulting Z defines the area of a normal curve (or the proportion of samples) with scores either below or above the value for the sample.

In studies of growth, a Z-score provides a standard indicator of either a sample from some reference value, or an individual within a sample from an appropriate reference. And, with measurements of different units, or scales of the same unit, Z-scores provide a way of comparing deviations from the reference. When the distribution of the raw data is normal, then, by using a cumulative normal frequency distribution, a particular Z-score indicates the proportion of the distribution above and below that value. For example, the proportion of a population below −2 Z-scores is 2 per cent, while the proportion below −1 Z-score is 16 per cent.

When the distribution is not normal, Z-scores are not equivalent either to percentiles or to proportions under the curve, though they may be useful as standardized measures of distance from a mean.

Z-scores permit comparison of different measurements of growth. For example, in a Guatemalan sample, height Z-scores were found to be lower than those for weight, a finding borne out by the mean Z-score, −1.75 for height and −1.13 for weight, indicating a greater retardation of growth in height, relative to the reference, than in weight. The percentile equivalent for the mean height Z-score is 4. Since the distribution of weight was skewed, there is not an exact equivalent of percentile for Z-score. However, the mean percentile rank for the sample is 7.

Z-scores transform a measurement into a standardized, relative measure. They are widely used in studies of growth, especially when the purpose is to evaluate its adequacy.

*Francis E. Johnston*

## SCREENING FOR GROWTH FAILURE IN INDUSTRIALIZED COUNTRIES

Screening is the identification of preclinical disease by a relatively simple test. In the case of growth failure the aim of screening is to uncover disease at an early stage so that appropriate management can be implemented. The prerequisites of screening were listed by Wilson and Jungner in 1968. Screening techniques must be able to identify the severity of a growth problem, if it exists. Primarily, they must be able to identify children with growth failure. However, their utility must be seen in the context of organizational and sociological issues in which growth may, or may not, be seen as an important public health issue, and where treatment of growth failure may be controversial.

*List of prerequisites for an effective screening programme, according to Wilson and Jungner (1968)*

1. The condition of interest should be an important health problem.
2. There should be an accepted treatment for the condition.
3. Facilities for treatment and diagnosis should be available.
4. There should be a recognizable latent or early symptomatic stage.
5. There should be a suitable test.
6. The test should be acceptable to the population
7. The natural history of the condition should be adequately understood.
8. There should be an agreed policy on whom to treat.
9. The cost should be balanced in relation to medical expenditure as a whole.
10. The screening process should be a continuing activity.

### *Severity of the problem*

The initial problem is that growth failure by itself is not a condition *per se*, but it is the result of a large number of conditions. These will include several conditions that are not treatable, and others that will be detected before the child participates in the screening programme, for example, congenital malformations of the heart. Among the conditions amenable to treatment, or to which palliative treatment can be offered in a developed country, are growth-hormone deficiency, Turner's syndrome, thyroid deficiency, coeliac disease, chronic renal failure (see Part seven) and emotional deprivation (**9.4**). Most of these conditions are of low frequency. Congenital hypothyroidism can be detected more appropriately using a specific test soon after birth, and many developed countries have already implemented a screening programme. Approximately 50 per cent of children with Turner's syndrome are above the cut-off point of height, defined as the third percentile, when screening takes place below age 5. The frequency of growth-hormone deficiency has been estimated between 1 in 4000 to 1 in 10,000, and the frequency of severe emotional deprivation is unknown, but in studies which have evaluated growth failure in populations, the number of cases with this condition is very low or non-existent. Renal failure is usually detected by other means.

### *Identifying growth failure*

The use of attained height as the screening test has been shown to be sufficiently accurate in community studies. The sensitivity and specificity of the test, in this case height, are usually not given in the evaluation of screening programmes of growth failure. The positive predictive value is more frequently given. This can be construed as assessing the probability that a child with growth failure has a condition amenable to treatment as defined above, but usually is defined as the percentage of children with growth failure that have an organic condition. In the assessment of growth failure the cut-off point is an important consideration. Studies have focused on the assessment of positive predictive value using cut-offs of less than −2SD or below the third percentile. The positive predictive value reported with these cut-off values have been approximately 20 per cent increasing to 40 per cent if the cut-off point is changed to less than −2.5SD. In the studies carried out a large proportion of the children identified by screening were already known to health services. Of the newly identified children few were amenable to treatment. A further problem in implementing a screening programme in preschool children is that only two thirds of the target group will participate in the programme. On the other hand, a large percentage of parents with a short normal child or with familial short child can be reassured. In regard to this, it is worthwhile to keep in mind a surprising recent finding which shows that only a third of the short children without an organic disorder identified at age 7 will become a short adult. In rare cases, an added bonus for parents whose short child has a genetic disorder is that they can be given genetic counselling, especially if planning another child.

Some have advocated the use of height velocity, instead of attained height, as the screening measure to use. This is totally inappropriate in a community programme as the child has to be measured accurately twice in a year, and the power to discriminate between normal growth and short children based on height velocity is low, while the resource implications are high.

### *Organizational issues*

There is little scientific support for implementing a screening programme of growth failure using traditional criteria for screening. The available economic appraisals of the value of a population screening programme for growth failure are technically unsound. In some developed countries the measurement of height has been put at the bottom of the list of priorities. However, the systematic measurement of height has some benefits that should not be forgotten, such as the opportunity to meet every child in the population in an non-controversial environment. Growth information also provides the opportunity for parents to discuss the health of their child. It also puts the community services in a position to provide basic core data about the child to health services. In Britain the Report of the Joint Working Party on Child Health Surveillance recommended that all children should be measured at age 3, and between 4 and 5 years.

### *Sociological issues*

The potential for increasing height of short normal children with biosynthetic growth hormone can provide a different perspective to growth-failure screening as the bulk of children below −2SD would be then amenable to treatment. The unknown long-term effectiveness, the side-effects of treatment, the commercial pressures from the drug manufacturers, and the sheer high cost of treatment makes this intervention highly contentious. In addition, the treatment of these children could be considered merely cosmetic and difficult to include as a health priority in times of increasing demand and mounting costs in the health services.

*Roberto Rona*

*The assessment of validity and predictive values of the screening test*

| | | Disease | |
|---|---|---|---|
| | | Present | Absent |
| The test (For example height) | Positive | a | b |
| | Negative | c | d |

Sensitivity = a/(a + c)
Specificity = d/(b + d)
Positive predictive value = a/(a + b)
Negative predictive value = d/(c + d)

## SCREENING FOR GROWTH FAILURE IN DEVELOPING COUNTRIES

Growth failure is most often defined as either a low height-for-age or a low weight-for-height. In the first case this is referred to as stunting and assessed as a value for height that is greater than 2 standard deviations below the reference mean for age (i.e., a Z-score that is <−2). This corresponds to a height-for-age that is approximately 92 per cent of the reference median and approximately the second centile. A low weight-for-height (for age) is called wasting and is generally defined as a value that is less than −2SD of the reference value. Stunting may or may not be accompanied by wasting. In some instances, weight-for-age may be used as the criterion of failure, in which cases a Z-score of <−2 is termed underweight.

In terms of etiology, stunting is usually interpreted as a chronic condition caused by the cumulative effects of mild-to-moderate nutritional deficiency and/or diarrhoeal disease. In the absence of wasting, stunting indicates an overall, (more or less) proportional faltering of growth as the effects of environmental factors accumulate through time. Stunted children may appear morphologically normal (**5.5**) until one learns their age, discovering that they are several years older than might otherwise be thought.

Wasting, on the other hand, is more likely to be an acute condition, where the body is depleted of its nutrient reserves. Weight-loss occurs and height may or may not be affected, depending on the specifics of the condition, including its duration and severity. There are two different reasons for screening for growth failure. The first is clinical and refers to the identification of individuals who are targeted for follow-up and possible treatment. Such was the case during the civil war in Nigeria, when a massive screening campaign was carried out in Biafra to identify children who had become clinically malnourished. What is required is an acceptable reference standard and appropriate cut-off points, for example, the fifth percentile. In Biafra investigators used the ratio of arm circumference to height. Since individual ages weren't readily available, arm circumference provided an estimate of the body's fat (**5.16**) and muscle stores, and height became a proxy for general growth status.

The second reason for screening is to evaluate the extent and type of growth failure. This is usually part of a general scheme of nutritional monitoring. In contrast to the identification of individuals for follow-up, nutritional monitoring is aimed at quantifying the prevalence of growth failure. Viewed in this sense, the prevalence of growth failure is a valuable indicator of underlying public health problems and of use in the following settings:

- It provides a practical way of describing the problem.
- It provides one of the best general proxies for constraints to human welfare of the poorest.
- It provides feasible predictors, at individual and population levels, of subsequent ill health, functional impairment, and/or mortality.
- It can provide an appropriate indicator of the success or failure of interventions aimed at the underlying causal agents.

Stunting is widespread among the world's developing countries, wasting less so.

For example, low body-weight is much less common among these same children. Of the 158 represented, only 96 (60.8%) fall into this category.

Victoria (1992) has reviewed stunting and wasting in 175 samples of 1–2-year-old children. He reports that while the prevalence of stunting shows little variation between Asian, African, Latin American, and Eastern Mediterranean samples, the variation in wasting was sevenfold:

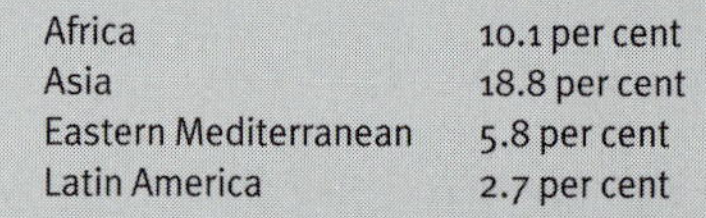

| | |
|---|---|
| Africa | 10.1 per cent |
| Asia | 18.8 per cent |
| Eastern Mediterranean | 5.8 per cent |
| Latin America | 2.7 per cent |

It is clear that the type of growth failure, and by extension the ecological conditions responsible, vary among the developing nations of the world.

*Francis E. Johnston*

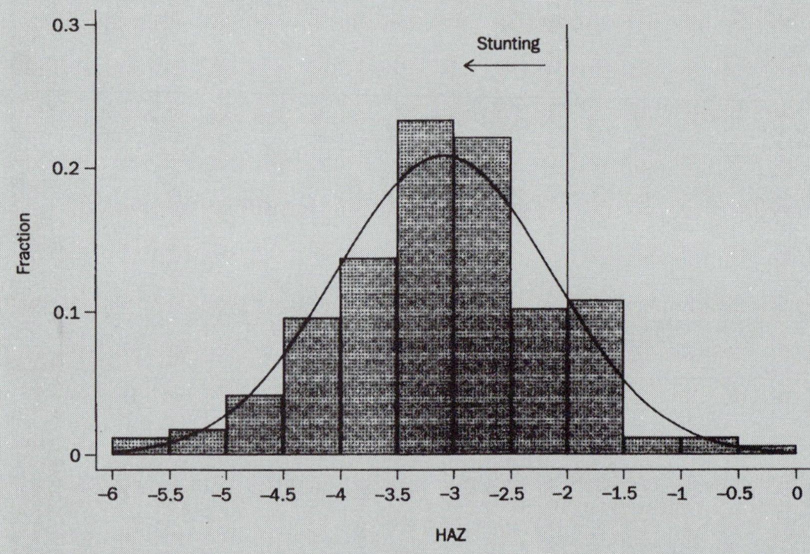


**The extent of stunting in the Guatemalan population. Distribution of height-for-age Z-scores in 158 rural Guatemalan preschool children 87% of them have Z-scores below −2, indicating a high degree of stunting.**

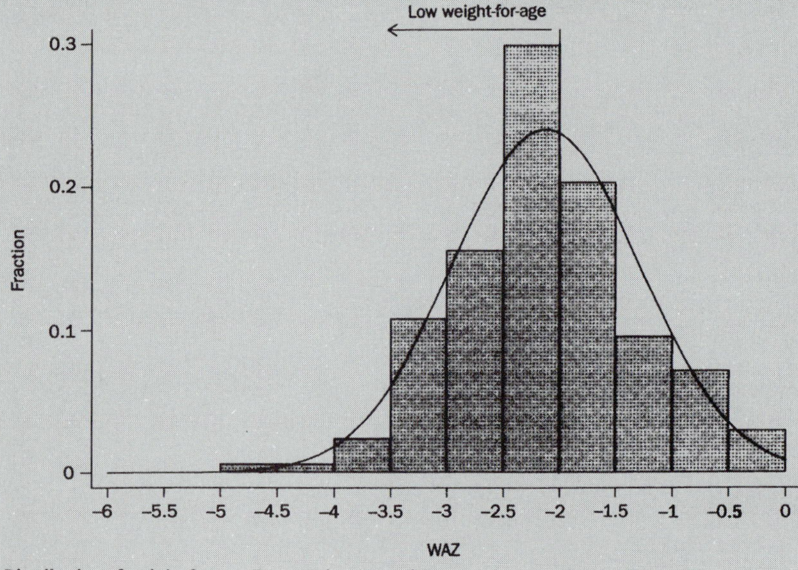


**Distribution of weight-for-age Z-scores in 158 rural Guatemalan preschool children. Fewer of them are wasted than are stunted, 61% have weight-for-age below −2 Z-scores.**

## SCREENING AND SURVEILLANCE DURING FAMINE

The word famine has no generally accepted definition. The word is used loosely, commonly to describe any situation of possible or actual starvation. In practice, it encompasses a wide range of events, including military siege, mass rural starvation, drought and, most typically, the migration of populations in search of food. In the case of starvation – or more typically, a high proportion of severe malnutrition (**9.2**) in children – death may result as much from disease (**9.3**) and poor food quality as from an absolute shortage of food.

As the physical nature of famine varies, so does its pattern of onset. In a world where large rural populations often cannot now reliably meet their food needs from their own production and increasingly rely on obtaining cash (from paid employment, petty trade, the sale of livestock or cash crops) to buy food, famine may be comparatively slow in onset. For example, famine may develop from a succession of years of bad production, culminating in a collapse of the relative prices of assets and grain. Rarely, famine may be provoked by a sudden rise in the price of grain beyond the ability of the poor to buy – unrelated to local events or through the obstruction of access to markets by war.

### *ASSESSING NEED*

In providing relief, a difficulty common to all famine crises is that not all of those affected will be in equal need of food. In each situation some people, often a large majority, will retain some capacity to find food for themselves. The central difficulty of relief is to accurately and quickly allocate relief resources within large and disrupted populations where administrative control is slight. In this context, where the circumstances of each individual cannot be known, the anthropometric assessment (**1.4**) of nutritional status is one of the most important indicators of the needs of individuals and of public health in general.

Anthropometry provides a practical, reasonably cheap and, with good technique, reproducible means of discriminating between those in more and less need. It is of most value where malnutrition mostly affects children, as it is in this group where the relationship between undernutrition and health is best understood. It is of less value for the assessment of adults where the importance of different levels of thinness is less clear and is of no practical use for children at the age of puberty. In famine relief, anthropometry is used at both the individual and population level.

### *ANTHROPOMETRIC ASSESSMENT*

There is no standard methodology for screening or surveillance. Approaches vary, depending on location, resources, the expertise available, the specific objectives, time and the nature of the emergency. The most widely used nutritional indicator is weight-for-height (or weight-for-length in younger children), as this provides an estimate of a child's thinness and does not require knowledge of age, which can be difficult to establish reliably. Other anthropometric indices used include weight-for-age, height-for-age, mid-upper-arm-circumference (MUAC) or MUAC-for-height. For adults, body mass index (weight [kg]/ height-squared [m$^2$]) or MUAC-for-height are sometimes used.

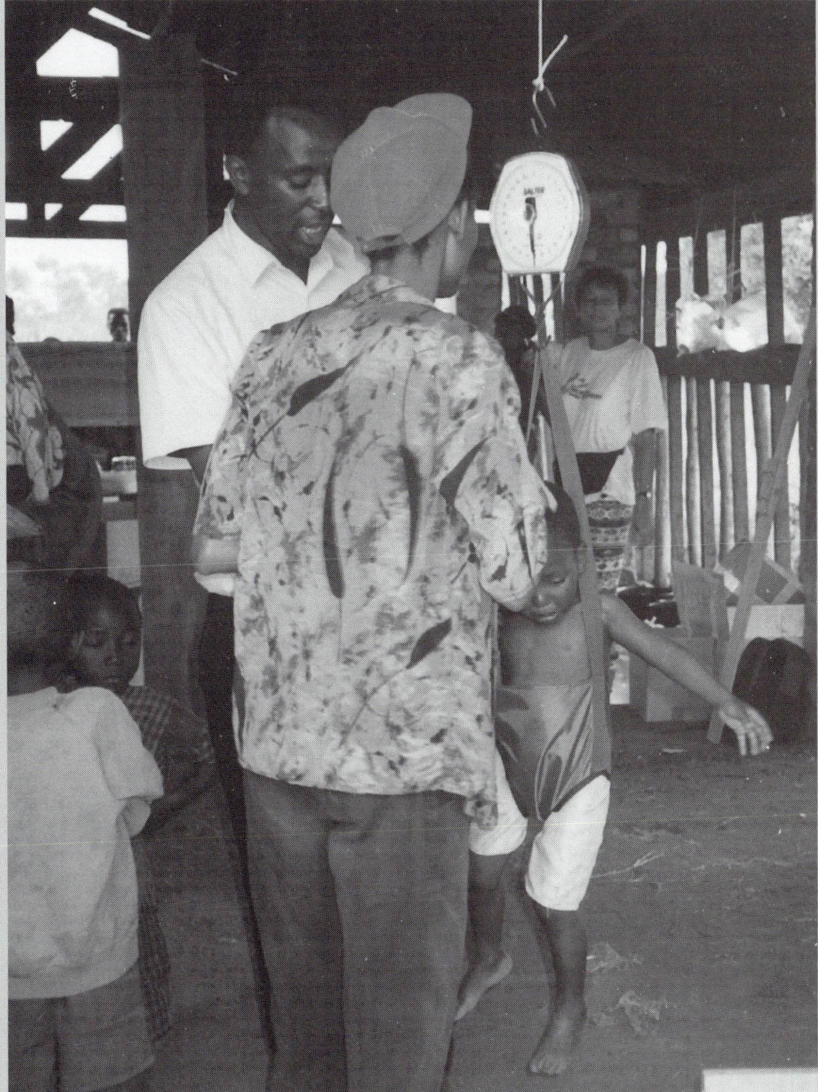

**Anthropometric assessment during famine: weight measurement by hanging scale. Photograph by Lola Nathanail.**

### *SCREENING*

Anthropometry can be used to screen populations to select those individuals requiring additional food or health care. This usually occurs as refugees enter a camp, or as children attend a health clinic. In feeding centres, anthropometry is used to monitor, through regular measurements, the recovery of individuals from malnutrition.

### *SURVEILLANCE*

Anthropometric population surveys, usually using standardized random cluster samples (or some other method of randomly selecting individuals from a population), allow reasonably accurate estimates to be made of the proportion of malnutrition in different population groups – whether in sections of a camp or in a rural population. This provides a basis for the planning of relief services. Repeated surveys (nutritional surveillance) allow changes in the nutritional status of a population to be monitored over time, which provides an objective estimate of change in the general circumstances of the population.

### *FAMINE EARLY WARNING*

Repeated anthropometric surveys, when used in conjunction with other indicators, such as crop production and food prices, have been proposed, and to a limited extent used, as a means of famine early warning. The theory is that famine could be predicted by the extrapolation of an observed decline in nutritional status. In practice, it has been found that declines in nutritional status are often abrupt and that anthropometry has limited predictive value. Nevertheless, such systems have been useful in identifying current crises and allowing relief to be provided more promptly.

*Lola Nathanail and John Seaman*

## SCREENING AND SURVEILLANCE DURING WARFARE

Very often countries at war have no nutrition information system and any pre-war systems have collapsed because of active warfare. Warfare or not, if famine is suspected the purpose of nutrition screening and surveillance is to:

- Confirm or refute that a nutritional emergency exists.
- If a nutritional emergency exists estimate its magnitude.
- Assess the impact on mortality, morbidity and nutritional status.
- Identify the most effective measures to minimize the impact of a nutritional emergency.
- Monitor the effectiveness of nutrition response over time.

Civilians who are displaced because of war are particularly vulnerable. Those groups of the population who are the most nutritionally vulnerable are often used as a proxy for the overall nutritional situation. For example, during wars in developing countries, traditionally only the children below the age of 5 years were measured (**9.1**). In the countries of Central and Eastern Europe (CCEE) and former Yugoslavia, the children do not always provide a good nutritional proxy for the population. The elderly (those over 60 years) living without any adult support and/or adult women may provide a more objective overview of the situation in the CCEE.

*Trends in nutritional status of the elderly: December 1993–May 1994, Bosnia*

| | Elderly (60+ years) | | | | | |
|---|---|---|---|---|---|---|
| | Dec | Jan | Feb | Mar | Apr | May |
| Sample size | 550 | 567 | 553 | 510 | 504 | 486 |
| Mean BMI* | 23.0 | 22.8 | 22.8 | 23.0 | 23.1 | 23.6 |
| SD | 4.14 | 4.20 | 4.14 | 4.08 | 4.11 | 4.13 |
| %BMI<18.5 | 14.2 | 15.0 | 14.6 | 11.8 | 11.5 | 9.5 |

Among the elderly, demispan was used as a proxy for height; height = demispan × 2 × 0.73 + 0.43 (Golden, M. Personal communication).

Data from Bosnia collected during the war in 1993 help to illustrate the naiveté of earlier thinking in surveillance during warfare: the body mass index (BMI, weight [kg]/ height [m²]) of the mothers, the distribution and the trends, gave a far better perspective of energy intake in Bosnia compared with monitoring only the children. When compared with pre-war data, a substantial weight-loss was indicated and a large number of adult females had BMIs just above the critical cut-off point of 18.5 for undernutrition. Moreover, a much higher percentage (12%) of residents living in cities (Sarajevo and Zenica), compared with rural mothers, were classified as energy deficient.

Having established that the rural community was relatively self-sufficient in food, we recommended that the urban population receive priority for food aid distribution. The percentage of the elderly with BMIs below 18.5 was much higher than expected and peaked at 15 per cent in January 1994. Elderly people living in the Old People's Homes had higher levels of undernutrition when compared with the elderly living with families or alone.

Monitoring trends can help to evaluate the effectiveness of nutrition interventions. Prior to the cease-fire of February 1994, the downward trend in body-weight among adults and the elderly suggested that the nutritional situation was deteriorating. After the cease-fire a steady weight-gain was noted in all groups.

Surveillance during warfare requires the development of rapid, objective screening methods. These methods should include the nutritional monitoring of adults and the elderly as well as children. In order to develop this novel approach the benefits of linking fieldworkers with major academic centres should be pursued. The impact of ruthlessly applied scientific thinking on nutritional surveillance and screening should continue to influence politicians and change policy in the interest of the innocent victims of war.

*Aileen Robertson*

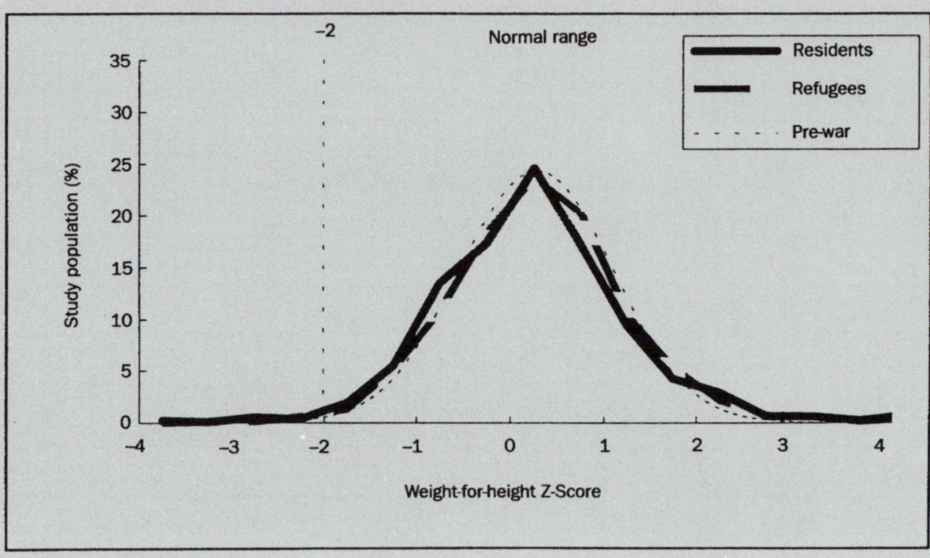


**Distribution of weight-for-height in Bosnian children, 1993.**

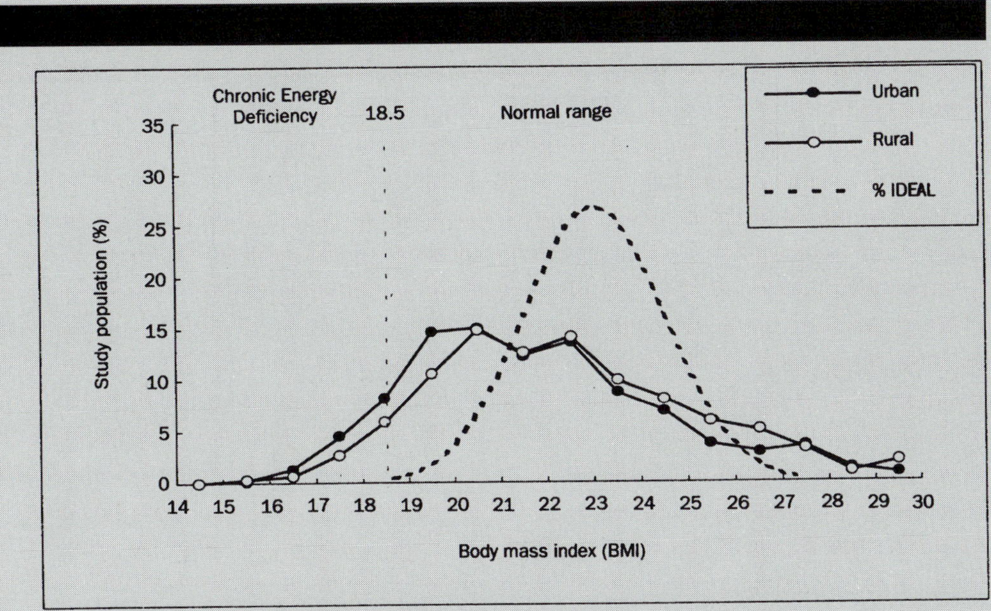


**Distribution of BMI of Bosnian mothers, 1993.**

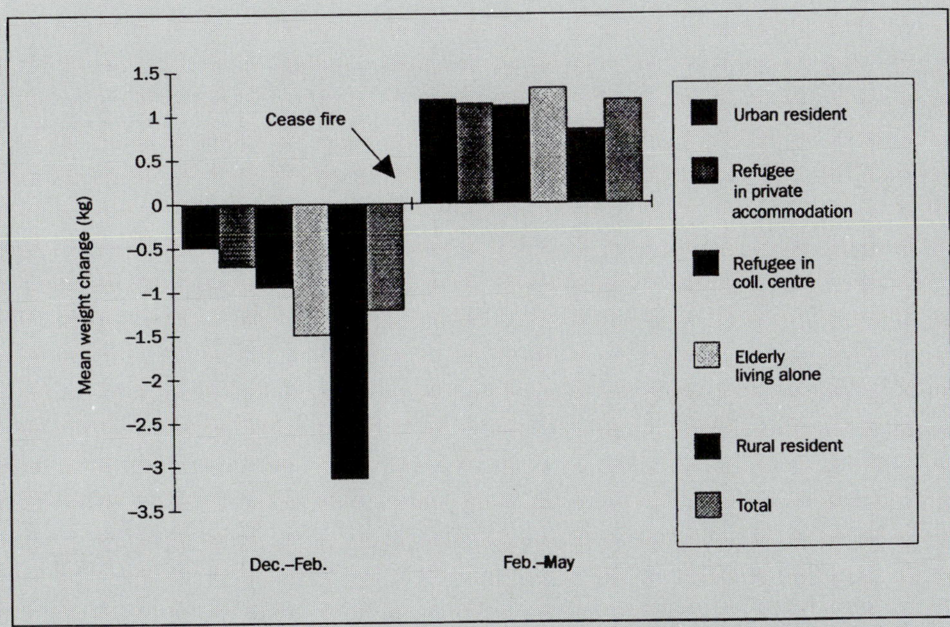


**Weight change in adults and the elderly by household group in Bosnia during December 1993–May 1994.**

# Part Two

# Patterns of human growth

Opposite: A family, from pregnancy to adulthood. Photograph by Stanley Ulijaszek.

Anthropologists have become increasingly interested in explaining the significance of the human life-cycle (also known as life-history). This is because it stands in sharp contrast to other species of social mammals, even other primates. Theory needs to explain how humans successfully combined a vastly extended period of offspring dependency and delayed reproduction with helpless newborns, a short duration of breast-feeding, an adolescent growth spurt, and menopause. Are these characteristics a package or a mosaic?

## Life-history and stages of the life-cycle

Brian Shea has broadly defined life-history as 'including not only the traditional foci such as age-related fecundity and mortality rates, but also the entire sequence of behavioural, physiological, and morphological changes that an organism passes through during its development from conception to death'. Recent work in mammalian life-history (**2.1** and **2.2**), and its evolution (**2.3**), focuses on the post-natal to adulthood period of the life-cycle. One way to define the stages of the life-cycle is by biological characteristics. Changes in the rate of growth and the onset of sexual maturation (puberty) are two such characteristics. The majority of mammals progress from infancy to adulthood seamlessly, without any intervening stages, and puberty occurs while their growth-rates are in decline. However, highly social mammals, such as wolves, wild dogs, lions, elephants, and the primates, postpone puberty by inserting a period of juvenile growth and behaviour between infancy and adulthood. In these animals puberty occurs while the rate of growth is still decelerating.

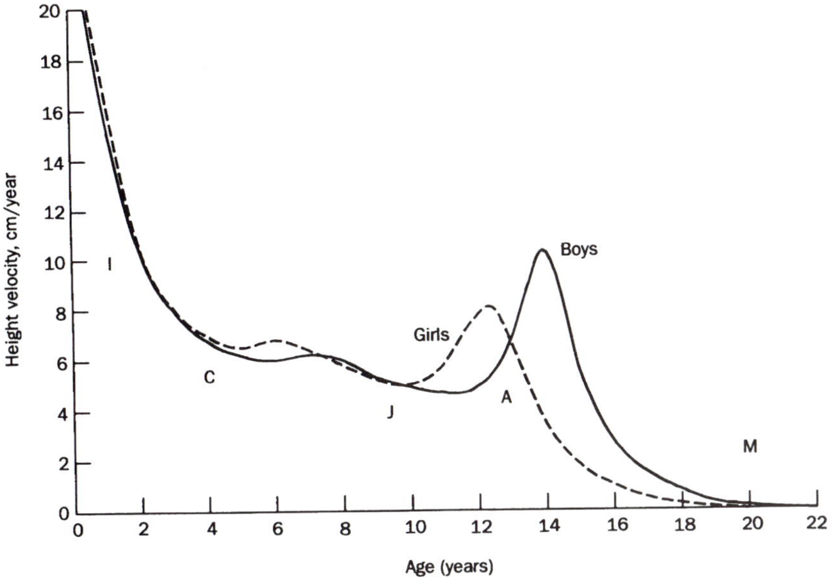

The pattern of human growth after birth may be characterized by five stages: 1) infancy, 2) childhood, 3) juvenile, 4) adolescence, and 5) adulthood. Changes in growth-rate and behaviour are associated with each

Idealized mean velocity curves of growth in height for healthy girls and boys show in the post-natal stages of the pattern of human growth. Note the spurts in growth-rate at mid-childhood and adolescence. Both spurts occur earlier, on average, for girls than for boys. The childhood and adolescence stages of growth, and their growth spurts, are unique to the human species. I-infancy, C-childhood, J-juvenile, A-adolescence, M-mature adult.

stage, and an important area of study with implications for anthropologists, comparative zoologists as well as clinical practitioners and public health workers is the characterization of the human growth curve (**2.4**). As for all mammals, human infancy is the period when the mother provides all or some nourishment to her offspring via lactation. Human infancy ends when the child is weaned from the breast, which in pre-industrialized societies occurs at a median age of 36 months.

Childhood is defined as the period following weaning, when the youngster still depends on older people for feeding and protection. Children require specially prepared foods due to the immaturity of their dentition and digestive tracts, and the rapid growth of their brain. These constraints necessitate a diet low in total volume but dense in energy, fat, and protein. Children are also especially vulnerable to predation and disease and thus require protection. There is no society in which children survive if deprived of this care provided by older individuals.

Important developments that allow children to progress to the next stage are the eruption of the first permanent molars and completion of growth of the brain (in weight). First molar eruption takes place, on average, between the ages of 5.5 and 6.5 years in most human populations. Recent morphological and mathematical investigation shows that brain growth in weight is complete at a mean age of 7 years. At this stage of development the child becomes much more capable dentally of processing an adult type diet. Furthermore, nutrient requirements for brain and body growth diminish, and cognitive capacities mature to new levels of self-sufficiency, for example, shifting from the preoperational to concrete operational stage, using the terminology of Piaget.

The child then progresses to the juvenile stage. Juveniles have been defined by Pereira and Altmann (1985) as, '...prepubertal individuals that are no longer dependent on their mothers (parents) for survival'. This definition is derived from ethological research with social mammals, especially non-human primates, but applies to the human species as well. Ethnographic research shows that juvenile humans have the physical and cognitive abilities to provide much of their own food and to protect themselves from predation and disease. In girls, the juvenile period ends, on average, at about the age of 10, 2 years before it usually ends in boys, the difference reflecting the earlier onset of puberty in girls.

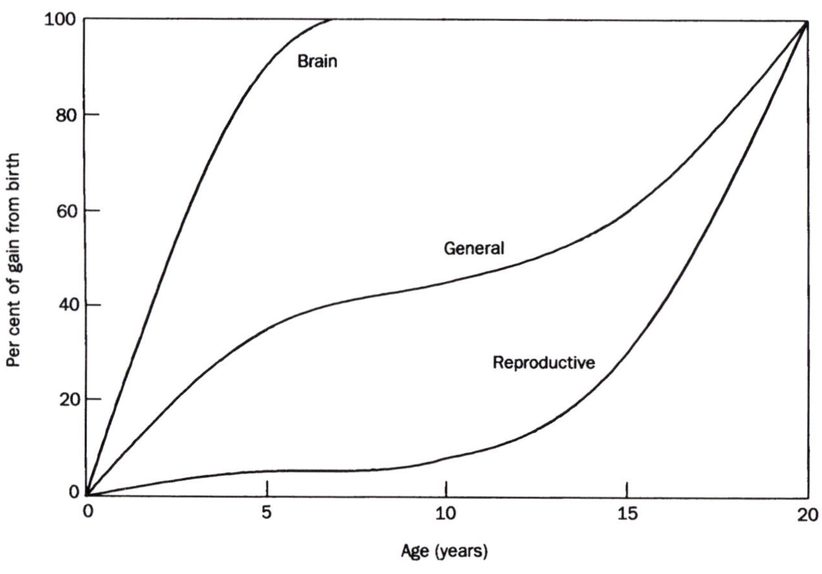


**Growth curves for different body tissues. The 'general' curve represents growth in stature or total bodyweight. The 'brain' curve is for total weight and the 'reproductive' curve represents the weight of the gonads and primary reproductive organs. Note that brain growth ends by about age 7 years, allowing a diversion of nutrients to supply growth of the body and reproductive system (from Scammon, 1930).**

Human adolescence begins with puberty, marked by some visible sign of sexual maturation. The adolescent stage also includes development of the secondary sexual characteristics and the onset of adult patterns of sociosexual and economic behaviour.

These physical and behavioural changes occur in many species of social mammals. What makes human adolescence different is that during this life-stage both boys and girls experience a rapid acceleration in the growth of virtually all skeletal tissue – the adolescent growth spurt. Other primate species may show a rapid acceleration in soft tissue growth, such as muscle mass in many male monkeys and apes. However, in contrast to humans other primate species either have no acceleration in skeletal growth, or an increase in growth-rate that is very small. The human skeletal growth spurt is unequalled by other species and, when viewed graphically, the growth spurt defines human adolescence.

Adolescence ends and early adulthood begins with the completion of the growth spurt, the attainment of adult stature, and the achievement of full reproductive maturity. The latter includes both physiological, socio-economic, and psychobehavioural attributes which all coincide, on average, by about age 19 in women and 21 to 25 years of age in men.

## Why do new life-stages evolve?

In the book *Size and Cycle*, J.T. Bonner develops the idea that the stages of the life-cycle of an individual organism, a colony, or a society are '...the basic unit[s] of natural selection'. Bonner's focus on life-cycle stages follows in the tradition of many of the 19th-century embryologists who proposed that speciation is often achieved by altering rates of growth of existing life-stages and by adding or deleting stages. Bonner shows that the presence of a stage, and its duration, in the life-cycle relate to such basic adaptations as locomotion, reproductive rates, and food acquisition. From this theoretical perspective it is profitable to view the evolution of human childhood and adolescence as adaptations for both feeding and reproduction.

The evolution of human childhood is a feeding and reproductive adaptation because childhood frees the mother from the demands of nursing an infant, and the inhibition of ovulation related to continuous nursing. This, in turn, decreases the interbirth interval and increases reproductive fitness. In comparison with living apes (**2.2**), people have a shorter infancy and shorter birth interval, which in apes and traditional human societies are virtually coincident. The net result is that humans have the potential for greater lifetime fertility than any ape, but also the problem of how to take care of the children, who are still dependent on older individuals for feeding and protection.

The peoples of traditional societies solved the problem of childcare by spreading the responsibility among many individuals. The child must be given foods that are specially chosen and prepared and these may be provided by older juveniles, adolescents, or adults. In Hadza society (African hunters and gatherers), for example, grandmothers and great-aunts are observed to supply a significant amount of food and care to children. In Agta society (Philippine hunter-gathers) women with children hunt large-game animals. They accomplish this by living in extended family groups – two or three brothers and sisters, their spouses, children and parents – and sharing the childcare.

Among the Maya of Guatemala (horticulturalists and agriculturalists), many people live together in extended family compounds. Women of all ages work together in food preparation, manufacture of clothing, and childcare. Juvenile girls associate with these working groups and the girls provide much of the direct care and feeding of children, but always under the guidance of adolescents and adults. In some societies fathers provide significant childcare, including the Agta, who take their children on hunting trips, and the Aka Pygmies, a hunting-gathering people of central Africa. No other primate or social mammal does all of this.

An adolescent stage of human growth may have evolved to provide the time needed to practice complex social skills needed to be an effective parent. The evolution of childhood afforded hominid females the opportunity to give birth at shorter intervals, but producing offspring is only a small part of reproductive fitness. Rearing the young to their own reproductive maturity is a more sure indicator of success. Studies of yellow baboons, toque macaques, and chimpanzees show that between 50 and 60 per cent of first-born offspring die in infancy. By contrast, in hunter-gather human societies between 44 per cent (!Kung) and 39 per cent (Hadza) of offspring die in infancy. Studies of wild baboons show that while the infant mortality rate for the first-born is 50 per cent, mortality for second-born drops to 38 per cent, and for third- and fourth-born reaches only 25 per cent. The difference in infant survival is, in part, due to experience and knowledge gained by the mother with each subsequent birth. Such maternal information is mastered by human women during adolescence which gives the women a reproductive edge. The initial human advantage may seem small, but it means that up to 21 more people than baboons or chimpanzees survive out of every 100 first-born – more than enough over the vast course of evolutionary time to make the evolution of human adolescence an overwhelmingly beneficial adaptation.

Adolescent girls gain knowledge of sexuality, reproduction, and infant care because they look mature sexually, and are treated as such, several years before they actually become fertile. The adolescent growth spurt serves as a signal of maturation. Early in the spurt, before peak height velocity is reached, girls develop pubic hair and fat deposits on breasts, buttocks and thighs. In essence they appear to be maturing sexually. About a year after peak height velocity girls experience menarche, an unambiguous external signal of internal reproductive system development. However, most girls experience one to three years of anovulatory menstrual cycles following menarche. Fertility is not achieved until near the end of the adolescent growth stage. Nevertheless, the dramatic changes of adolescence stimulate both the girls and adults around them to participate in adult social, sexual, and economic behaviour. For the adolescent girls this participation is 'risk-free' in terms of pregnancy, but does allow them to learn and practice behaviours that lead to increased reproductive fitness in later life.

Even though healthy girls may become fertile at the age of 15 years, the cross-cultural evidence shows that age at first marriage, and childbirth, clusters around 19 years for women from such diverse cultures as the Kikuyu of Kenya, Mayans of Guatemala, Copper Eskimo of Canada, and both the Colonial-period and contemporary United States. A major anatomical reason for this is that the female pelvis has its own unique slow and continuous pattern of growth, which continues for several years even after adult stature is achieved. For this very fundamental biological reason girls must wait up to a decade from the time of menarche to reach full reproductive maturity. Thus, both biology and culture work together to help ensure the reproductive success of the first-time human mother.

The adolescent development of boys is quite different from girls. Boys become fertile well before they assume adult size and the physical characteristics of men. Analysis of urine samples from boys age 11 to 16 years old show that they begin producing sperm at a median age of 13.4 years. Yet the cross-cultural evidence is that few boys successfully father children until they are into their third decade of life. In the United States, for example, only 3.09 per cent of infants live-born in 1990 were fathered by men under 20 years of age. Traditional Kikuyu men (East African pastoralists) do not marry and become fathers until about age 25 years, although they become sexual active following their own circumcision rite at around the age of 18 years.

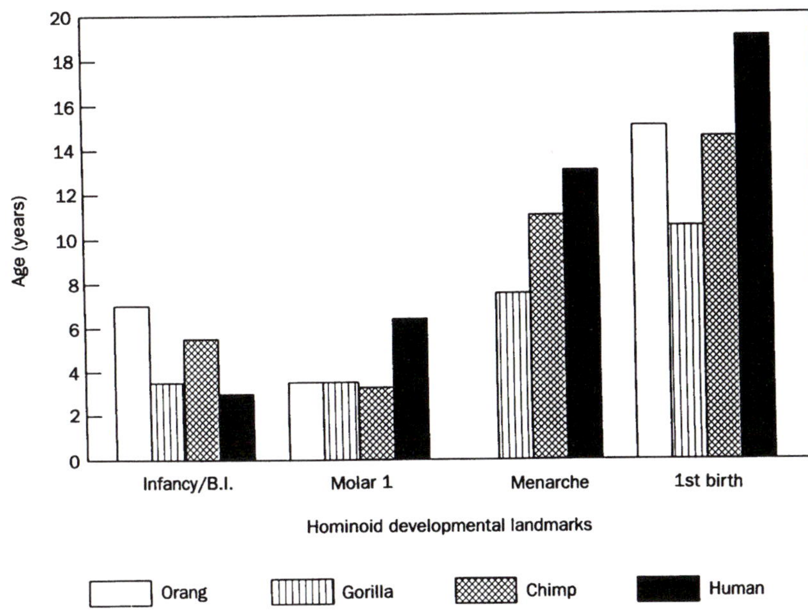


**Hominoid developmental landmarks. Data based on observations of wild-living individuals or, for humans, healthy individuals from various cultures. Note that, compared with apes, people experience developmental delays in eruption of the first permanent molar, age at menarche, and age at first birth. However, people have a shorter infancy and shorter birth interval, which in apes and traditional human societies are virtually coincident. The net result is that humans have the potential for greater lifetime fertility than any ape. Species abbreviations are: Orang = *Pongo pygmaeus*, Gorilla = *Gorilla gorilla*, Chimp = *Pan troglodytes*, Human = *Homo sapiens*. Developmental landmarks are: Infancy/B.I. = period of dependency on mother for survival, usually coincident with mean age at weaning and/or a new birth (B.I. = birth interval); Molar 1 = mean age at eruption of first permanent molar; Menarche = mean age at first estrus/menstrual bleeding, 1st birth = mean age of females at first offspring delivery. From Bogin and Smith, 1996.**

A probable reason for the lag between sperm production and fatherhood is that the average boy of 13.4 years is only beginning his adolescent growth spurt. In terms of physical appearance, physiological status, psychosocial development and economic productivity he is still more of a child than an adult. Few women, and more importantly from a cross-cultural perspective few prospective in-laws, view the teenage boy as a biologically, economically, and socially viable husband and father.

The delay between sperm production and reproductive maturity is not wasted time in either a biological or social sense. The obvious and the subtle psychophysiological effects of testosterone, and other androgen hormones, that are released following gonadal maturation and during early adolescence may 'prime' boys to be receptive to their future roles as men. Alternatively, it is possible that physical changes provoked by the endocrines provide a social stimulus toward adult behaviours. Whatever the case, early in adolescence sociosexual feelings intensify, including guilt, anxiety, pleasure, and pride. At the same time boys become more interested in adult activities, adjust their attitude to parental figures, and think and act more independently. In short, they begin to behave like men.

However – and this is where the survival advantage may lie – they still look like boys. Because their adolescent growth spurt occurs late in sexual development, young males can practice behaving like adults before they are actually perceived as adults. The sociosexual antics of young adolescent boys are often considered to be more humorous than serious. Yet, they provide the experience to fine-tune their sexual and social roles before either their lives, or those of their offspring, depend on them. For example, competition between men for women favours the older, more experienced man. Since such competition may be fatal, the child-like appearance of the immature, but hormonally and socially primed, adolescent male may be life-saving, as well as educational.

Biologists usually measure evolutionary success in reproductive terms and by most criteria the human species is highly successful. The evolution of the childhood and adolescent growth stages, as well as the human brain, language and culture, facilitate unique styles of human social organization, cultural learning, and reproductive behaviour that are part of the reason for the evolutionary success of our species. Furthermore, the variation observed within groups or populations (**2.5**) may reflect differentials in survivorship and reproductive success both of which are important in the understanding of human evolution, and in public health screening of populations at risk.

*Barry Bogin*

# Comparative growth and development of mammals

Mammals vary enormously in life-history pattern and adult body-size. The interspecific range of mature body mass in mammals varies in the order of 100 million. If primate species, including humans, are excluded, then the length of different life-history periods is greater with larger body-size. The relationship between adult weight and gestation length is strong, while those between adult body weight and the age at which 50 per cent and 98 per cent of adult weight is attained are less so. When the comparison includes primate species, all three relationships are greatly weakened. Between species, the duration of growth and development is related to the state of development of the young at birth, with slower-developing mammals producing precocial young who are reasonably independent from birth. The skeletal muscle of altricial mammals is close to the end of the primary stages of differentiation. However, in precocial mammals, development beyond this stage has taken place *in utero*, allowing mobility soon after birth. Humans are in an ambiguous position, with long gestation period but dependent young.

*Proportion of lifespan spent in growth*

| | % | | % |
|---|---|---|---|
| Elephant | 15 | Human | 25 |
| Pig | 15 | Chimpanzee | 29 |
| Rat | 6 | Gorilla | 24 |
| Mouse | 6 | Orangutan | 26 |
| | | Gibbon | 20 |

*Growth characteristics of different mammalian species (females)*

| | Adult weight (kg) | Length of gestation (months) | Post-natal age (months) at which different proportions of adult weight are attained | |
|---|---|---|---|---|
| | | | 50% | 98% |
| Horse | 678 | 11.0 | 8.5 | 47.7 |
| Cow | 420 | 9.4 | 12.6 | 71.6 |
| Pig | 200 | 4.0 | 11.0 | 63.0 |
| Sheep | 50 | 5.0 | 3.3 | 20.3 |
| Rabbit | 3.9 | 1.0 | 4.6 | 25.6 |
| Guinea pig | 0.3 | 2.2 | 4.1 | 29.0 |
| Mouse | 0.02 | 0.7 | 1.0 | 5.0 |
| Human | 56 | 9.5 | 106 | 198 |
| Chimpanzee | 43 | 8.4 | 67 | 180 |

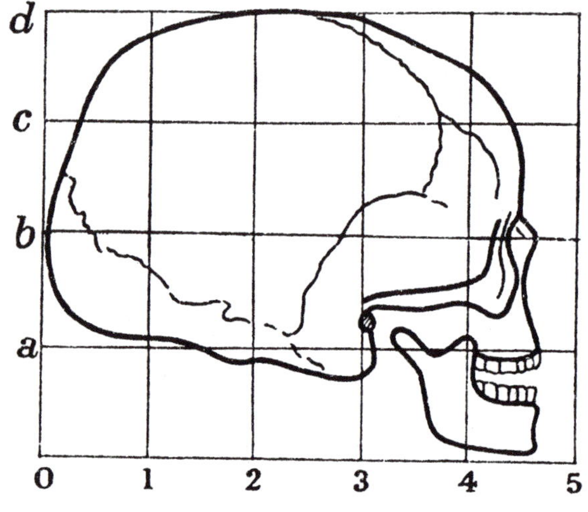


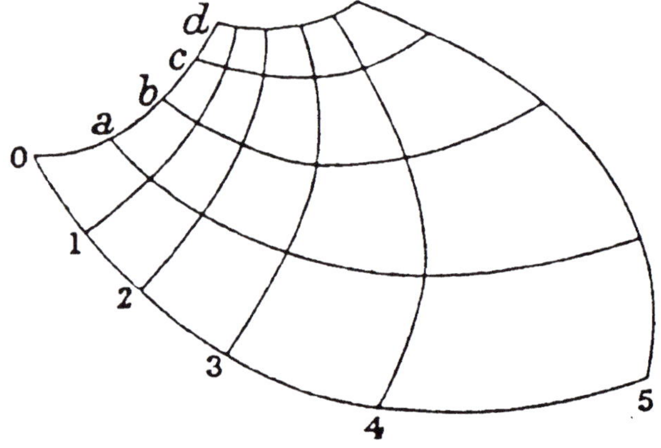


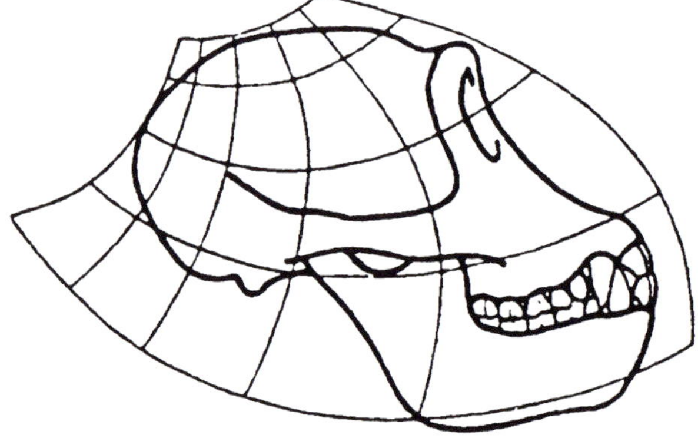

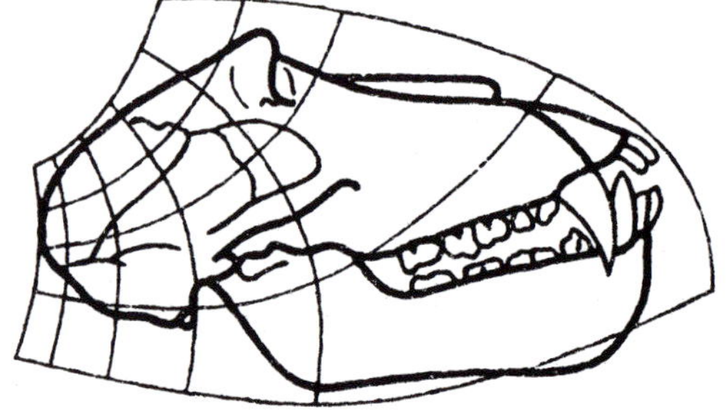

**The method of transformed co-ordinates depicting, according to D'Arcy Thompson (see Appendix 1, 14), the relationships between human, chimpanzee and baboon skulls**

An important feature of human evolution (**2.3**) and that of other primates has been the increase in the time between birth and sexual maturity, or of extended childhood relative to other species, although it is clearly greatest in human beings. This is true also of the rates of growth, which in most mammalian species is fast, with 50 per cent of adult body mass being attained at an early age in less than 20 per cent of their total period of post-natal growth and development. In contrast, humans attain 50 per cent of their adult mass at about half way through the course of post-natal growth, while chimpanzees attain the same proportion at about one third of their total growing period. Humans and some other primates make up for this slowness of growth with a period of accelerated growth in puberty (**5.15**).

A major milestone in growth is the attainment of reproductive performance. Rats and elephants reach sexual matu-

## THE ALLOMETRY OF BASAL METABOLIC RATE

Allometric relationships proposed for basal metabolic rate (BMR) (i.e. weight$^{0.75}$) have been widely accepted, and the term 'metabolic body size' (MBS) is commonly used to describe it. It is widely held that among adult mammals Kleiber's generalization that BMR is related to weight to the power of 0.75 is the best method of scaling BMR for size. The relationship holds for most species and provides values that permit inter-specific comparison. Furthermore, it is widely agreed that maintenance energy intake also varies at the $W^{0.75}$ level and is approximately 1.5 BMR, i.e. $105W^{0.75}$. In addition, voluntary food intake in adults of different species also varies in the same way, establishing that maximal energy metabolism in both poikilotherms and homeotherms is proportional to the power function $W^{0.75}$. Thus, BMR (kcal/day) = $70W^{0.75}$ (where W = kg).

Despite this, the precise allometry of BMR is still debatable (**9.2**). However, there are certain points of agreement. Firstly, it is broadly agreed that allometric generalizations about BMR are least confusing when they are restricted to adult animals in good nutritional state, and that size changes associated with growth or with malnutrition do not fit to any simple adult relationship. Secondly, it is generally accepted that among eutherian mammals of different species, BMR rises as weight rises and BMR per unit weight falls. It is also agreed that among birds, fish and reptiles, the same is true. However, where weights overlap, passerine birds have a higher BMR than mammals, and placental mammals have a higher BMR than marsupials. In addition, mammals have a higher BMR than fish and reptiles. *C.J.K. Henry*

*Allometric equations of basal metabolism in different species*

| Species | Unit of measurement | Body-weight | Coefficient b | Exponent a |
|---|---|---|---|---|
| Amphibians and reptiles | kcal/day | kg | 1.79 | 7.28 |
| Birds | | | | |
| Nonpasserine | $ml0_2$/min | g | 0.078 | 0.723 |
| Passerine | $ml0_2$/min | g | 0.126 | 0.724 |
| Dassyurid marsupials | $cm^3 0_2$/(g.h) | g | 2.45 | −0.261 |
| Hibernating animals | | | | |
| Summer | kcal/day | g | 63.6 | 0.62 |
| Winter | kcal/day | g | 2.09 | 0.69 |
| Homeotherms | kcal $h^{-1}$ | g | $2 \times 10^{-2}$ | 0.751 |
| Mammals | ml $0_2$/min | g | 0.057 | 0.75 |
| | | | 0.061 | 0.76 |
| | | | 0.064 | 0.734 |
| Placental mammals | $cm^3 0_2$/(g.h) | g | 3.8 | −0.265 |
| Poikilotherms | kcal.$h^{-1}$ | g | $6.9 \times 10^{-4}$ | 0.751 |
| Unicellular organisms | kcal.$h^{-1}$ | g | $8.43 \times 10^{-4}$ | 0.751 |

Based on $y = bx^a$

## BIRTH-WEIGHT, GESTATION LENGTH AND RELATIVE MATURITY

The gestation length and birth-weight (**10.3**) of mammals varies enormously, but cross-species comparison shows them to be strongly related to adult weight. However, the latter relationship is less good for primates than other mammals. The anabolic drive *in utero* appears to enable a rate of protein deposition in the immature mammal which is as close as possible to the genetic maximum. The body protein to fat-free mass ratio is an index of chemical maturity; human birth-weight lies midway in the range of both birth-weight–mature-weight ratio and chemical maturity index, despite such infants being relatively altricial, and dependent at birth.

Although there is no clear relationship between the magnitude of fat stores in the newborn and its size relative to adult size, this is not the case for the fetal load imposed on the gestating mother. Across species, the nutritional and energetic cost of fetal growth and development is inversely related to neonatal fatness. Fat deposition therefore reflects the extent to which energy intake exceeds the energy expenditure associated with fetal maintenance and protein deposition. Since humans have a low fetal load, they are born as fat mammals. *Stanley Ulijaszek*

*Neonatal body composition, chemical maturity and fetal load in different mammalian species*

| | Birth-weight/ mature weight (%) | Protein/ fat-free weight (%) | Fetal load[(a)] | Fat/ body-weight (%) |
|---|---|---|---|---|
| Human | 5.0 | 14.5 | 0.06 | 16.0 |
| Guinea pig | 15.8 | 18.7 | 0.2 | 10.0 |
| Rabbit | 1.9 | 11.2 | 0.7 | 4.0 |
| Pig | <1.0 | 11.7 | 0.4 | 1.1 |
| Cat | 3.4 | 13.1 | 1.0 | 1.8 |
| Rat | 1.6 | 10.0 | 2.8 | 2.0 |
| Mouse | 6.6 | 14.4 | 13.4 | 3.2 |

(a) Fetal load calculated as (birth-weight × litter size) divided by ([0.33 × gestation length] × maternal weight)

## LOCOMOTION

Humans are the only mammals to use bipedal walking and running as their usual gait. However, our close relatives the apes occasionally stand and walk bipedally. Chimpanzees and gorillas usually walk on all fours, but rise on their hind legs in situations in which they appear to be expressing dominance. Gibbons spend their time in trees, swinging by their arms from one small branch to another, but sometimes walk bipedally on larger branches.

The quadrupedal gaits of apes are unusual in several respects. They do not place the palms of their hands on the ground but walk on the knuckles of their flexed fingers. Their hind feet are plantigrade; that is to say, the whole sole of the foot from heel to toe is placed on the ground. Many small mammals are plantigrade, but among larger ones only humans, apes, monkeys and bears are plantigrade. Other mammals are digitigrade, like dogs and cats, standing on the toes of their fore and hind feet while keeping the rest of the foot off the ground; or unguligrade, like horses and cattle, standing on hooves at the tips of their toes. Another peculiarity of apes and monkeys is that when standing and walking quadrupedally, they support about 60 per cent of their weight on their hind legs and only 40 per cent on the fore legs. In other mammals, such as dogs and sheep, these proportions are reversed.

Humans have much shorter toes than apes, whose feet are capable of functioning like hands, grasping branches as they climb. The big toe of apes resembles the human thumb.

Adult humans normally walk at speeds below 2 metres per second (4.5 mph) and run at higher speeds. Measurements of oxygen consumption show that below 2 metres per second walking requires less energy than running whereas, above that speed, the reverse is true. In running, each foot is on the ground for less than half the time and we travel in a series of leaps. The Achilles tendon and the ligaments of the arch of the foot serve as springs, saving energy by the principle of the bouncing ball.

In walking, each foot is on the ground for more than half the time (so there are stages when both are on the ground). While a foot is on the ground the knee is kept almost straight – much straighter than in other mammals. This makes human beings bob up and down when walking, raising the body highest when the leg is vertical. Energy is saved by the principle of the pendulum; we slow down as we rise and speed up as we fall. Even when walking bipedally, apes move very differently from humans. They keep their legs much less straight. They place the whole length of the sole of the foot on the ground more or less simultaneously, whereas humans set down the heel well before the fore-foot. Also, apes walk with the trunk sloping forward. The combination of bent knees and sloping trunk makes the angles between trunk and thigh quite similar to those of quadrupeds. In contrast, the thighs of a standing human are parallel to the trunk. This seems to explain why the pelvis of humans is shaped very differently from those of other mammals, including apes.

Human babies start crawling at about 10 months, moving their arms and legs in the same sequence as walking quadrupeds. They do not generally learn to walk until they are about 15 months old. This is exceptionally late, in comparison with other mammals. Dogs walk at 3 weeks of age, and sheep and deer within a few hours of birth.

*R. McNeill Alexander*

## MAMMALIAN LIFE-HISTORY

There are three major groups of mammals; the monotremes, the marsupials and the eutherians. The first two of these are primitive groups with distinctive life-histories. The monotremes (the duck-billed platypus and the spiny ant-eaters) lay eggs instead of giving birth. The marsupials (kangaroos, opossums etc.) give birth to tiny young at a very early stage of development, which spend their early life firmly attached to one of the mother's nipples, usually inside a pouch. For example, 30-kilogram red kangaroos give birth, about one month after copulation, to young weighing only about 0.7 grams.

The vast majority of mammals including humans belong to the third group, the eutherians. All of these have qualitatively similar life-histories, but there are marked differences in rates of development and other quantitative characteristics. Among eutherians, there are striking differences between altricial species, which are born at a relatively early stage of development, at which they are helpless, and precocial species which are born better developed and more competent. Carnivores and most rodents are altricial, born with their eyes closed and with little hair. They remain incapable of supporting themselves on their legs until some time after birth. In contrast, precocial mammals, such as deer, antelopes and horses, are born with their eyes open and with a good coat of hair, and start walking within a few hours of birth.

*Some life-history characteristics of eutherian mammals with adult weight between 40 and 60 kg*

| | Adult mass (kg) | Newborn mass (kg) | Litter size | Gestation period (months) | Age at weaning (months) | Age of female sexual maturity (years) |
|---|---|---|---|---|---|---|
| **Carnivores** | | | | | | |
| Wolf | 50 | 0.4 | 4 | 2 | 1 | 2 |
| Puma | 48 | 0.4 | 3 | 3 | 3 | 2.5 |
| Sun bear | 45 | 0.3 | 1 | 3 | ? | ? |
| **Ruminants** | | | | | | |
| Mule deer | 57 | 3 | 2 | 7 | 4 | 1.5 |
| Impala | 44 | 4.5 | 1 | 7 | 5–7 | 1 |
| Sheep | 50 | 4.9 | 1* | 5 | 3–6 | 1 |
| **Apes** | | | | | | |
| Chimpanzee | 56 | 1.8 | 1 | 8 | 49 | 10 |
| Orangutan | 40 | 1.5 | 1 | 9 | 37 | 7 |
| Gorilla | 60 | 1.8 | 1 | 9 | 52 | 7 |
| **Human** | 56 | 3.2 | 1 | 9 | 24 | 13 |

* Modern breeds have been selected artifically to produce twins.

There are differences in the pace of life, between similar animals of different sizes. Lions are about 100 times as heavy as their close relative the domestic cat, and both species have altricial young. The gestation period is about 110 days for lions and only 63 days for cats. This is an example of the general tendency for everything to happen faster for smaller animals. Not only do small animals develop faster than large ones and live less long, but their heart-beats and their breathing are faster.

A comparison of human life-history with the life-histories of other eutherian mammals of similar body mass, between 40 and 60 kilograms, highlights differences between humans, carnivores, ruminants and apes, while minimizing any effects of size differences on life-history. In general, carnivores give birth to tiny, altricial young after a short gestation period. After birth they develop rapidly and are weaned while still relatively young. Most carnivores produce litters of several pups. In this respect the sun bear, which commonly gives birth to a single pup, is unusual. In contrast, ruminant mammals give birth to large, precocial calves after a long gestation period. They produce single calves or at most twins; more than two large calves would be too big a burden for the mother. The young are weaned after a few months and mature at an early age. The apes, the closest relative of human beings, produce single babies after long gestation periods. Though large, these babies are helpless for their first few months. They are cared for by their mothers for much longer than the other mammals in the table and are not weaned until they are several years old. Human life-history is broadly similar.

*R. McNeill Alexander*

rity when their weight is about 40 times greater than at birth, while mice and pigs reach the same stage after about 30- and 100-fold increases, respectively. Humans and chimpanzees are exceptions, reaching sexual maturity at 15 and 21 times birth-weight, respectively. Another feature which distinguishes primates from other mammals is the proportion of their total lifespan spent in growth (see Introduction). Again, this is longest for the primates, including human beings. Thus the growth of humans and other moderately sized primates can be largely distinguished from those of other mammalian species on the basis of a long period of growth and development in absolute terms and as a proportion of total lifespan, slower growth in the earliest stages of post-natal growth, with low adult body mass and weight at sexual maturity relative to birth-weight.

*Stanley Ulijaszek*

See also 'Growth in non-human primates' (**2.2**), 'Adulthood and developmental maturity' (**5.17**) and 'Ageing as part of the development process' (**12.3**)

## ALLOMETRY: A BRIEF HISTORY

The study of allometry has its origins in attempts to quantitatively compare the anatomy and physiology of animals of different sizes. Size differences in mammals are enormous, ranging from 2 grams body-weight for an adult shrew to 150 tonnes ($10^8$g) for the blue whale. Despite these enormous weight ranges, it is evident that the physiology and function of these mammals show great qualitative similarity. In order to describe these similarities in quantitative terms it is necessary to derive a scaling factor that will negate the effects of body-size on metabolic activity. This is allometric analysis, a technique which essentially investigates Brody's (1945) proposition that 'organisms change geometrically to remain the same physiologically'.

The fundamental principles of allometry can be found in the treatise of Huxley (1932) *Problems of Relative Growth*. Huxley showed that plotting gross organ measurement against body-weight yields essentially two types of relationships. The first is a simple linear proportionality, evident when the organ- and body-weight measurements are plotted on an arithmetic scale. Huxley called these relationships *isometric*. Furthermore, he pointed out that the second class of relationships, where organ-size varies curvilinearly with body-size, generally yields a straight-line relationship in a log : log-plot graph. Huxley coined the term *allometry* (from Greek *allometron*, meaning that by which anything is measured) to describe such relationships, and his generalized allometric equation was of the form:

$$y = b(x)a$$

where $y$ represented the size of an organ, $x$ the size of the total body, and $a$ and $b$ are numerical constants for a given data series. This equation may be rewritten as:

$$\log y = \log b + a \log x$$

which is a linear equation of the form

$$y = ax + b$$

In the first equation, $a$ is the slope of the log-transformed plot and $b$ is the value of $y$ when $x = 1$. The coefficient $a$ is the ratio of specific rate of increase of $y$ with $x$. If $a$ is greater than 1 (what Huxley called positive allometry) there is a differential increase of $y$ relative to $x$, as $x$ increases. If $a$ is less than 1 (negative allometry), this indicates that $y/x$ ratio decreases with increasing value of $x$. If $a = 1$, geometrical similarity is maintained with increasing size, i.e., isometry, a special presentation of an allometric relationship. In this equation, $a$ is sometimes called the mass exponent, and $b$ the allometric coefficient.

After Huxley, the allometric relationship between body-weight and metabolic rate was investigated by both Brody (1932) and Kleiber (1932), independently. Both authors sought to define the exact mass exponent relating body-weight to Basal Metabolic Rate (BMR). Brody (1932) and Kleiber (1932) did not arrive at the same conclusion as to what this relationship was, or should be. Both investigators, however, agreed that it was *not* $W^{0.67}$ but an exponent between 0.73 and 0.76. There is no clear anatomical meaning to be attached to this finding. Kleiber appears to have been the first to go beyond the graphical (illustrative) approach of Huxley and attempt to analyse the allometric relationship by a more rigorous mathematical technique of linear regression analysis. Subsequent investigations of allometry employing this approach include relationships between body-weight and urine output, heart-beat, tidal volume, water-intake, as well as the weight of the kidneys, heart, lungs, and liver.

*C.J.K. Henry*

*Examples of some allometric analyses relating quantitative properties with body-weight*

| Investigator | Measurement | Year |
|---|---|---|
| Adolph | Water uptake | 1949 |
| Anderson, Rahn, Prange | Supportive tissue | 1979 |
| Bartels | Oxygen transport | 1964 |
| Brody | Organ weights, metabolic rate | 1945 |
| Bertalanffy | Oxygen consumption | 1957 |
| Edwards | Renal function | 1975 |
| Hemmingsen | Energy metabolism | 1960 |
| Hunt and Giles | Body composition | 1956 |
| Kleiber | Metabolic rate | 1932 |
| Martin | Brain-size | 1981 |
| Milton and May | Home range area | 1976 |
| Prothero | Heart, blood and liver weight | 1979, 1980, 1982 |
| Prange, Anderson, Rahn | Skeletal muscle | 1979 |
| Stahl | Respiratory variables organ weight | 1965, 1967 |
| Zeuthen | Oxygen uptake | 1953 |

*Allometric equations relating body-weight to some quantitative properties*

| | | |
|---|---|---|
| Urine output ml/hr | = U = | $0.0064B^{0.82}$ |
| Heart-beat (hr) | = H = | $0.0000119B^{0.27}$ |
| Breath duration (hr) | = Q = | $0.000047B^{0.28}$ |
| Ventilation-rate (ml/hr) | = V = | $120B^{0.74}$ |
| Tidal volume (ml) | = T = | $0.0062B^{1.01}$ |
| Water-intake (ml/hr) | = I = | $0.010B^{0.88}$ |
| Kidney-weight (g) | = K = | $0.0212B^{0.85}$ |
| Heart-weight (mg) | = J = | $0.0066B^{0.98}$ |
| Lung-weight (g) | = F = | $0.0124B^{0.99}$ |
| Liver-weight (g) | = L = | $0.082B^{0.87}$ |
| Gut-weight (stomach and intestine, g) | = G = | $0.112B^{0.74}$ |
| Blood-weight (g) | = $B_L$ = | $20.55B^{0.99}$ |

B = body-weight

# Growth in non-human primates

Biologists have come to appreciate that it is the entire developmental life-cycle, and not merely the terminal adult stage, which undergoes evolutionary transformation over time. Therefore primate growth patterns should be viewed as adaptations within the broad context of the evolution of mammalian life-history strategies. Relevant components of a species' life-history profile include not only growth-timing, -rates and -duration, but such features as basic reproductive strategy, age at weaning, and age at sexual maturity and/or first reproduction. Broad comparative analysis of life-history and growth in mammals reveals a dual pattern – first a basic one of consistency and constraint linked to phylogeny, body-size and other general factors, and then an overlying one of variability and specialization determined by extrinsic ecological influences.

The eutherian mammals range along a continuum from highly *altricial* species, which are generally short-lived, small in body and relative brain-size, with large litters, rapid growth, and early sexual maturity; to very *precocial* species, which are typically long-lived, larger in body and relative brain-size, with small litters, slower growth, and later reproductive maturity. Primates are perhaps the archetypal precocial mammalian order, generally producing very few well-developed offspring, which grow slowly, reach sexual maturity only after a prolonged period of socialization and learning, and are highly encephalized. A number of these features are undoubtedly linked, and in fact the biologist Eric Charnov has recently argued that the long average adult lifespans and small number of offspring per year that characterize primates as a group (when corrected for body-size and compared to other mammals) may in fact be related to their very low growth-production rates, characteristic of juveniles in particular. Why primates have such slow individual growth-rates compared to other mammals of comparable size is a key unanswered question, but it may be linked to the energetic demands of having a relatively large brain. Slower growth-rates may also minimize chances of juvenile mortality, a viable strategy especially in social primates subject to lower predation pressures. In any case, the slow growth-rates of primates are part of a matrix of retardation – birth-rates, death-rates and individual growth-rates are all only a fraction of the values found in other comparably sized mammals.

This pattern was quite possibly found in even the earliest primates. In fact, the unusual combination of developing a highly precocial reproductive strategy at the small body-size characteristic of early primates may very well have been an important life-history complex shared by the primates, colugos (flying lemurs), tree shrews and bats, four mammalian orders often phylogenetically united in the superorder *Archonta*. These groups may share aspects of an ancient basal adaptation produced by selection for a relatively precocial life-history pattern in an arboreal setting within fairly stable tropical-forest environments.

While the primates share aspects of a basic pattern of growth- and life-history, they also exhibit considerable variation in specific components of that pattern. For example, there is considerable diversity found in simple plots of growth-in-time among different groups of primates, ranging from linear relationships with age for *Macaca* species, to simple decelerating exponential curves in *Callitrichidae*, and sigmoid and complex curves for lemurs, and chimpanzees, respectively. There is also some evidence that overall growth-rate decreases across the major primate taxonomic groups, i.e., from prosimians to New World monkeys to Old World monkeys and apes to humans. Additional variance in growth

**Primate species show striking differences in patterns of somatic growth. In the examples above, the marmosets show no growth spurt, the baboons show considerable sexual dimorphism in growth spurts, and colobus monkeys show growth spurts in both sexes.**

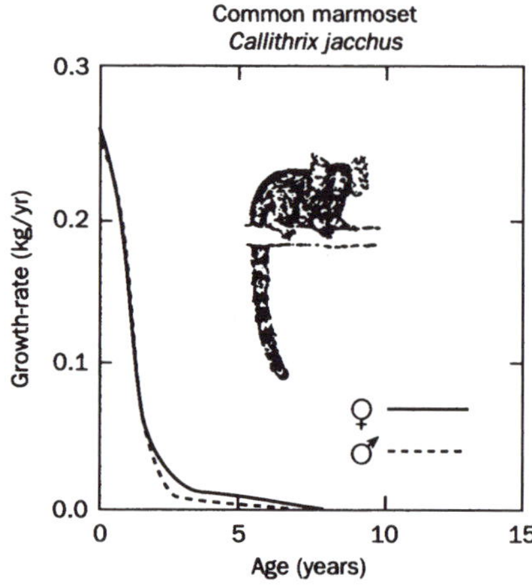


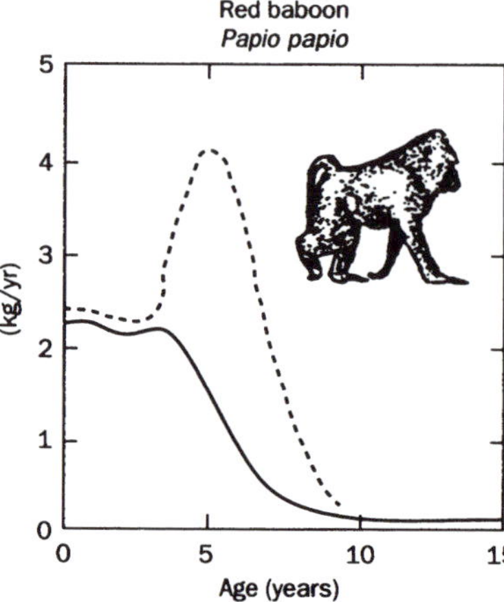


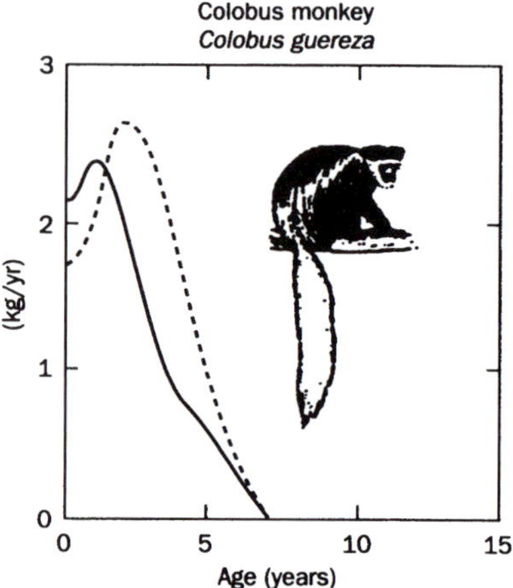

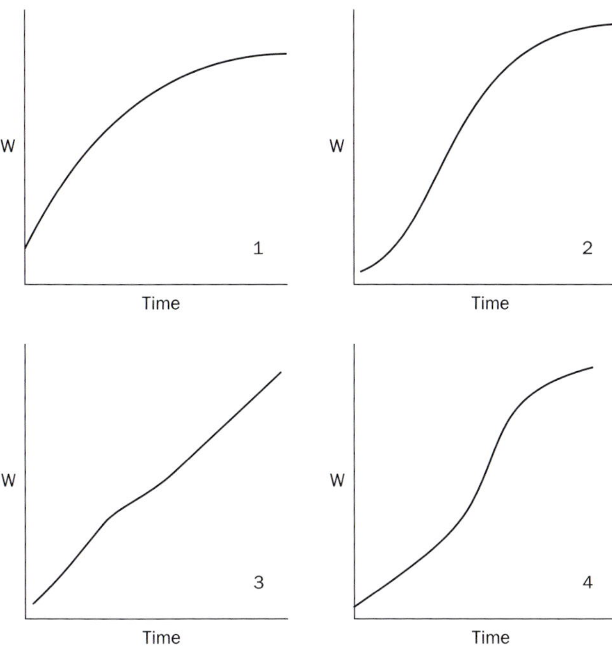


**Patterns of growth of body-weight (W) with age (time) in primates. 1. Simple decelerating exponential curve; this pattern is seen in Callitrichidae. 2. Sigmoid curve; seen in *Lemur* and *Gorilla*. 3. Linear; *Macaca* species. 4. Complex curves; curves of this shape are seen in *Pan* and *Homo*. It is not intended that growth patterns can be meaningfully classified into such groups. These curves are shown merely to demonstrate that there are considerable differences in growth patterns between species of primates.**

**Variation in body-size of very young rhesus monkeys (macaca mulatta) of the same age. Photograph by Stanley Ulijaszek.**

patterns appears linked to particular ecological influences. Such growth variations are best considered as novel patterns which have evolved within phylogenetically nested groups of species.

Biological anthropologist Steven Leigh has recently shown that folivorous primates generally differ from comparably sized frugivorous or omnivorous relatives in exhibiting more rapid rates of growth that are established early in ontogeny and tend to cease earlier. This is true of the folivorous indriids relative to other lemurs, the folivorous colobine monkeys relative to other Old World monkeys, the folivorous howler monkey (*Alouatta caraya*) relative to other New World monkeys, the folivorous siamang (*Hylobates syndactylus*) relative to other lesser apes, and the folivorous gorilla (*Gorilla gorilla*) relative to chimpanzees. Leigh explains this pattern in terms of a growth and life-history model proposed by the ecologists Charles Janson and Carel van Schaik, which predicts that the reduced levels of (especially juvenile) intraspecific feeding competition characteristic of folivorous species allow for more rapid growth and truncated growth periods. Leaves often have moderately high protein content, which may also account for the relatively accelerated and truncated growth of other primate species, such as the patas monkey (*Cercopithecus patas*), which eats large amounts of animal protein derived from insects and invertebrates.

Growth-rates appear to be associated with ecological contexts in other primate species as well, for reasons which require additional clarification. The marmosets and tamarins have relatively accelerated growth-rates among the New World monkeys, associated with a more rapid reproductive turnover, and perhaps linked to their inhabitation of forest fringes and secondary forest environments. A similar contrast could be made between the rapid-developing lesser bush-baby (*Galago moholi*) in wooded savanna regions, and the tropical rain forest dwelling Allen's bush-baby (*Galago alleni*), which has slower rates of growth, development and reproductive turnover.

The evolution of primate sexual dimorphism, or marked differences in phenotypic characters between males and females, is also elucidated through a focus on growth patterns and processes. This approach leads away from a focus on static adult end-points, and towards questions such as 'how and why do males and females grow and mature differently?' An observed pattern of adult dimorphism may in fact be produced in multiple ways in terms of underlying growth trajectories. For example, a derived increase in the degree of adult size (and allometrically correlated shape) dimorphism could be produced by increasing growth-rate and terminal size in males, decreasing growth-rate and terminal size in females, by combining both these changes, by having both sexes increase ancestral rates of growth and terminal size but more strongly so in males, and so forth. Furthermore, since a larger size may be generated by extending the duration of the ancestral growth period in time, as well as by altering growth-rates, the possibilities underlying a particular adult pattern of dimorphism are very complex. Quite distinct ecological contexts and selective scenarios may be linked to these various ontogenetic pathways, as is indicated by variations in sexually dimorphic growth patterns in the large-bodied hominoids. Here the reduced degree of adult dimorphism seen in the common chimpanzee (*Pan troglodytes*) as compared to the gorilla (*Gorilla gorilla*) and perhaps even the pygmy chimpanzee (*Pan paniscus*) is produced by slowed growth, delayed maturation and increased body-size in females. This may relate to increased levels of intrasexual female competition in frugivorous animals specializing in small, unevenly distributed patches of food resources. By comparison, the marked body-size and shape dimorphism seen in adult orangutans (*Pongo pygmaeus*) appears related to strong sexual selection for elevated and prolonged rates of growth in males, who compete vigorously for females and territories.

## PRIMATE LIFE-HISTORY

The life-history of a species is a mathematical description of the frequency and timing of events from conception to death (**2.1, 10.1**). The terms were originally derived from models of population growth, and classically differentiated into two basic types; species where the timing of events and frequency maximize the population's net rate of growth (r-strategists) and those species living close to the population's mathematically defined carrying capacity (K-strategists), where a premium is placed on quality of individuals and competitive ability in an environment containing scarce resources.

Primates are considered to be extreme K-strategists, in that they reproduce at a late age, produce a single offspring at each reproductive event, and allocate substantial time to the production and care of each infant, thus reproducing only infrequently over a lifespan. Infant quality, its survival and competitive ability, has been enhanced at the expense of infant numbers. Associated with these traits are long gestations, extended periods of infancy and adolescence, relatively slow growth to puberty, long lifespan, a relatively large body-size, and an expanded brain-size for body-weight. In a sense, energy used for growth and reproduction has been 'traded-off' against time, enhancing infant survival as well as reproductive chances.

Traits can be quantitatively compared, and used to describe species and populations. Recent theoretical work has extended the concept of life-history variance into the level of the individual, to facilitate comparisons between individuals of different social status or living under different ecological conditions. Life-history theory is thus a powerful tool for describing adaptation at the level of the species and also for understanding alternative behavioural strategies.

Are all primates the same? The strepsirhine (or prosimian) primates do not show the extensive degree of K-adaptations found among the higher primates. Strepsirhines, typified by the galagos, lorises, and the lemurs of Madagascar tend to be small in size and live life at a faster rate, reproducing earlier, more frequently and often producing twins or even triplets. This 'grade' difference underscores the importance of controlling for evolutionary heritage in interspecific comparisons. Among the haplorhine (higher) primates,

**Japanese macaques (*Macaca fuscata*). Courtesy of Orion Press/Natural History Photographic Agency.**

**A male silverback gorilla beats his chest during a threat display. Courtesy of A.H. Harcourt/Anthro-Photo File.**

Thus, an approach which focuses on growth and life-history features as adaptations in diverse socio-ecological settings is likely to yield the greatest insights into an understanding of the evolution of developmental patterns in our primate relatives.

*Brian T. Shea*

See also 'Comparative growth and development of mammals' (**2.1**), 'Human growth from an evolutionary perspective' (**2.3**), 'Morphology' (**5.5**), 'Sexual maturation' (**5.15**) and 'Growth and natural selection' (**10.1**)

**Chimpanzee (*Pan troglodytes*). Photograph by Nick Newton-Fisher**

the apes and humans lie at a further extreme, with extraordinarily slow reproduction and long periods of infant dependence. In the largest apes, the gorillas, chimpanzees and orang-utans, successive births occur between 3–6 years apart and the onset of reproduction is delayed until 10–12 years of age for females, and considerably later for males. Even taking their size into account, apes live life in the slow-lane. The monkeys tend to have a medium body-size, produce infants at intervals of 1–2 years, and reproduce at 3–5 years of age. The richness and stability of the habitat account for some differences between monkey species, with those in more variable habitats following an r-strategy, while those in stable habitats are somewhat more K-strategists. Even within this markedly K-selected group of species, a trend to specialize towards the extremes can be found, following the early ecological predictions.

There are some interesting social traits which also influence primate life histories. Among the monkeys and apes, those with access to help in rearing infants tend to have higher reproductive rates than do those lacking parenting assistance. Underlying much of these differences are infant growth-rates: with help, infants can grow faster and become independent at an earlier age, allowing the mother to reproduce again more quickly. Mothers lacking help must expend more energy in providing milk or transport, and thus are less able to reproduce again themselves.

Life-history traits are essentially descriptive; can they be used to make statements about cause and effect in adaptation? The separation of cause from consequence in life-history theory is an area of intense research. The primates, with their well-studied lives and extremes of size and sociality, are important in such comparative work. Of critical interest are the energetics of growth (**9.2**), sources of infant mortality, and individual variation in social behaviour. Relating these factors to body-size, food availability and reproductive rates (**10.1**) should provide insights into the evolution of life-history traits.

*P. C. Lee*

# Human growth from an evolutionary perspective

The stages of the life-cycle may be studied directly only for living species. However, there is evidence of the nature of life-cycle variables of extinct species. Such inferences for the hominids come from comparative anatomy, comparative physiology, comparative ethology, and archaeology. These methods may be used to study patterns of brain and body growth in apes, humans, and their ancestors.

Apes have a pattern of brain growth that is rapid before birth and relatively slower after birth. In contrast, humans have rapid brain growth both before and after birth. Relative to body-size, human adult brain-size is 3.5 times larger than the chimpanzee. The rate of human brain growth exceeds that of most other tissues of the body during the first few years after birth. Human neonates also have remarkably large brains (corrected for body-size) compared with other primate species. Together, relatively large neonatal brain-size and the high post-natal growth-rate give adult humans the largest degree of encephalization (a scaling of brain to body-size) of all higher primates.

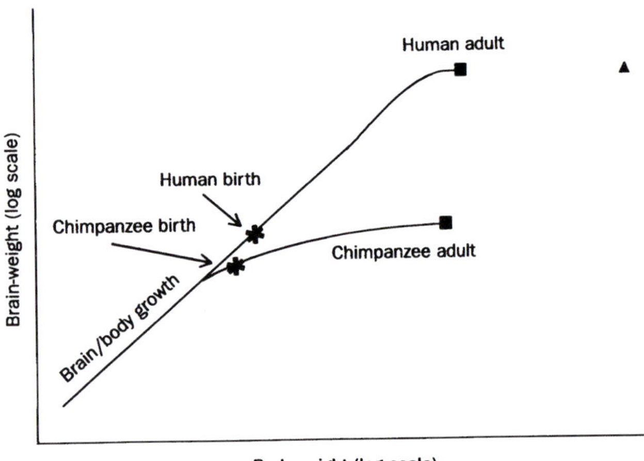


**Growth curve for human brain and body compared with the chimpanzee. The length of the human fetal phase, in which brain and body grow at the same rate for both species, is extended for humans. Chimpanzee brain growth slows after birth, but humans maintain the high rate of brain growth during the post-natal phase. In contrast, the rate of human body growth slows after birth. If human brain/body growth-rate were equal to the chimpanzee rate, then adult humans would weight 454 kg and stand nearly 3.1 metres tall (indicated by the ▲ symbol).**

Robert Martin argues that a 'human-like' pattern of brain and body growth becomes necessary once adult hominid brain-size reaches about 850 cubic-centimetres. This biological marker is based on an analysis of cephalo-pelvic dimensions of fetuses and their mothers across a wide range of social mammals, including cetaceans, extant primates and fossil hominids. Given the mean rate of post-natal brain growth for living apes, an 850 cubic-centimetre adult brain-size may be achieved by all hominoids, including extinct hominids, by lengthening the fetal stage of growth. At brain-sizes above 850 cubic-centimetres the size of the pelvic inlet of the fossil hominids, and living people, does not allow for sufficient fetal growth. Thus, a period of rapid post-natal brain growth and slow body growth – the human pattern – is needed to reach adult brain-size.

Against this background, it is possible to attempt a summary of the evolution of the human pattern of growth and development. This must be considered tentative, as only data for extant species are known with some certainty. Although *Australopithecus afarensis* is a hominid, it shares many anatomical features with non-hominid species including an adult brain-size of about 400 cubic-centimetres. Analysis of its dentition

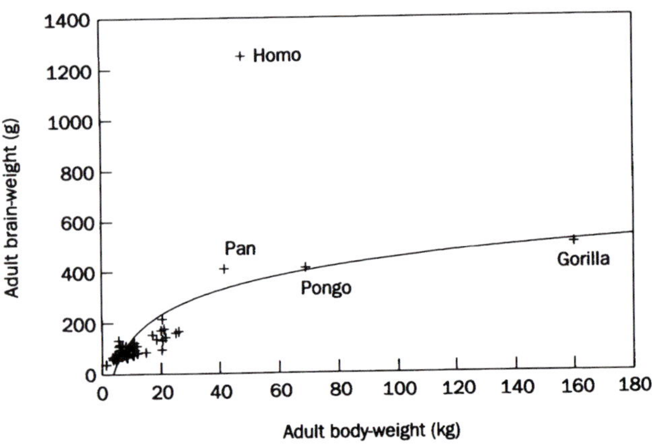


**Adult body-weight and brain-weight plotted for 61 species of Cercopithcidae (Old World monkeys, apes, and people). The curve is a logarithmic regression function, fit to the data for all species.**

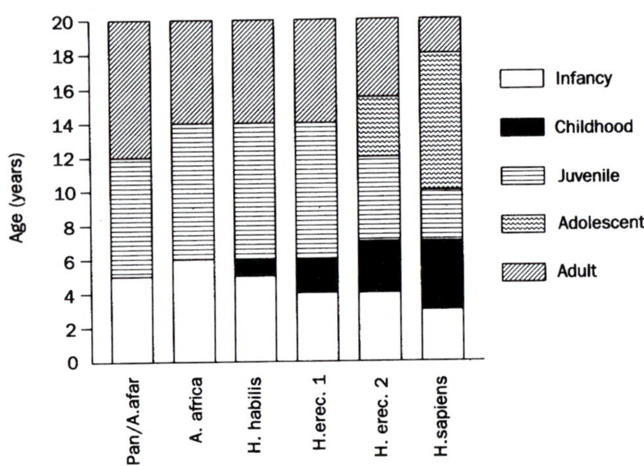


**The evolution of hominid life-history during the first 20 years of life. Abbreviated nomenclature as follows: A. afar. – *Australopithecus afarensis*, A. africa – *Australopithecus africanus*, H. habilis – *Homo habilis*, H. erec. 1 – early *Homo erectus*, H. erec. 2 – late *Homo erectus*, H. sapiens – *Homo sapiens*.**

indicates a rate of dental development indistinguishable from extant apes. Therefore, the chimpanzee and *A. afarensis* are depicted as sharing the typical tripartite stages of post-natal growth of social mammals – infant, juvenile, and adult. To achieve the larger adult brain-size of *A. africanus* (442 cc) may have only required an addition to the length of the fetal or, possibly, the infant stage.

The rapid expansion of adult brain-size during the time of *Homo habilis* (650–800 cc) might have been achieved with expansion of the fetal, the infant and, even the juvenile periods. However, further extension of infancy may have placed a severe demographic constraints on *Homo habilis* populations. Female primates, including humans, cannot reproduce a new infant successfully if they are still nursing their current infant. Chimpanzees, for example, average 5.5 years between successful births in the wild and young chimpanzees are infants, dependent on their mothers, for about 5 years. Actuarial data collected on wild-living animals indicate that between 35 per cent and 38 per cent of all live-born chimpanzees survive to their mid-20s. Although this is a significantly greater percentage of survival than for most other species of animals, the chimpanzee is at a reproductive threshold. Jane Goodall reports that for the period 1965 to 1980 there were 51 births and 49 deaths in one community of wild chimpanzees at the Gombe Stream National Park, Tanzania. During a 10-year period at the Mahale Mountains National Park, Tanzania, Toshisada Nishida observed 74 births and deaths as well as 14 immigrations and 13 emigrations in one community. Chimpanzee population growth is, by these data, effectively equal to zero. Extending infancy and birth intervals beyond the chimpanzee range may not have been possible for early hominids such as *Homo habilis*.

The insertion of a brief childhood stage into life history could have reduced the reproductive strain. Children no longer nurse, so they may be fed and cared for by individuals other than the mother. This frees the mother for further reproduction. The archaeological evidence suggests that *Homo habilis* intensified stone tool manufacture and use to scavenge animal carcasses, especially bone marrow. This behaviour may be interpreted as a strategy to feed children. Such scavenging may have been needed to provide the essential amino acids, some of the minerals and, especially, the fat (dense source of energy) that children require for rapid brain growth.

Brain-size increased further during the time of *H. erectus*. The earliest adult specimens have brain-sizes of 850 to 900 cubic-centimetres. This places *H. erectus* at Martin's adult brain-size marker, and may justify an expansion of the childhood period to provide the high quality foods needed for the rapid, human-like, pattern of brain growth. Note also that infancy is depicted as decreasing in duration as childhood expands. Hypothetically, this gives *H. erectus* a reproductive advantage over all other hominoids. With this advantage it is easier to understand why population size and the geographic range of *H. erectus* expands beyond that of all prior hominids.

There is evidence that early *H. erectus* did not have an adolescent growth spurt. Holly Smith analysed the skeleton and dentition of the fossil specimen KMN-WT 15000, or the 'Turkana boy' a 1.6-million-year-old (early) *H. erectus* skeleton. Using the data for skeletal growth and maturation of living apes and humans Smith developed a model of *H. erectus* skeletal growth and development. The 'Turkana boy,' who was probably less than 11 years old at death, was 'too advanced' in skeletal development to have experienced an adolescent growth spurt.

Late *H. erectus* (c.400,000 BP), with adult brain-sizes up to 1100 cubic-centimetres, is depicted here with further expansion of childhood and the insertion of the adolescent stage. Along with bigger brains, late *H. erectus* shows increased complexity of technology (tools, fire, and shelter) and social organization that were likely to require an adolescent stage of development in order to become a successful adult member of society. The transition to archaic and finally modern *H. sapiens* expands the childhood and adolescent stages to their current dimensions.

*Barry Bogin*

See also 'Comparative growth and development of mammals' (**2.1**), 'Growth in non-human primates' (**2.2**), 'The human growth curve' (**2.4**), 'Homeobox genes' (**3.3**), 'Post-natal craniofacial growth' (**5.9**) and 'Growth and natural selection' (**10.1**)

## NEANDERTHAL GROWTH AND DEVELOPMENT

When the modern human rate and pattern of growth and development first appeared is a key question in paleobiological studies. Studies of growth and development in Plio-Pleistocene hominids have demonstrated that early hominids followed an ape-like rate and pattern of growth and development, but that later hominids became more similar to modern humans over the course of evolution. Of particular importance are hominids of the Mid-Upper Pleistocene as this is a time when modern humans first evolved and, by examining fossils from this time-period, we may be able to establish the first appearance of the modern rate and pattern of growth and development.

The Neanderthals lived in Europe and the Middle East between 180,000–30,000 years ago and overlapped in time with modern humans. Comparing the rate and pattern of growth and development of this population to modern humans may reveal: 1) the ancestry of the modern condition; and 2) the relationship between ourselves and Neanderthals. Given that studies of Neanderthal ontogeny have the potential to answer such important questions in paleoanthropology, what do we know about their rate and pattern of growth and development?

In studying Neanderthal ontogeny, researchers have focused on the rate of growth, with a number of authors arguing that Neanderthals had a greatly accelerated dental growth-rate (**5.10**) relative to modern humans.

### RATE OF GROWTH

Recent work by Tompkins (1996) has revealed that the duration of Neanderthal dental growth may be somewhat shorter than that of populations of European descent.The duration of dental growth was probably more similar to (though possibly of slightly shorter duration) than that seen in recent human populations from southern Africa, since the third molar formed more rapidly in Neanderthals. Tompkins' work also demonstrated contrasts between Neanderthals and modern humans in the rate of speed of growth of certain teeth. For instance, compared with recent humans, Neanderthals (and other Mid-Upper Pleistocene hominid fossils) show: 1) a relative delay of the first incisor and third premolar formation; and 2) a relative advance in their second molar formation compared with a modern sample of European descent; and 3) relative advance in their third molar formation compared with modern southern Africans. These contrasts indicate a difference in the growth-rate between Neanderthals and populations of modern humans. The fact that samples of early modern human fossils also demonstrate contrasts with recent modern humans means that the rate of dental growth probably changed from the Upper Pleistocene onwards.

### PATTERN OF GROWTH

The pattern of growth relates to the relative sequence of certain events over the course of life. Neanderthal teeth erupted in the same sequence as ours do, so in this respect they were like modern humans. The fact that growth events occur in a certain order allows researchers to assess how mature an individual was at the time of death. Dentally, they examine the number of teeth formed and erupted in the jaws. Skeletally, the degree of maturation is assessed by examining how ossified a bone is, or if the epiphyses have or have not yet fused to the shaft. This information tells us how close an individual was to the adult form, allows an age at death to be assigned, and sometimes can tell us something about the nature of growth and development of a particular species. For example, most of the teeth of the Le Moustier 1 Neanderthal were fully

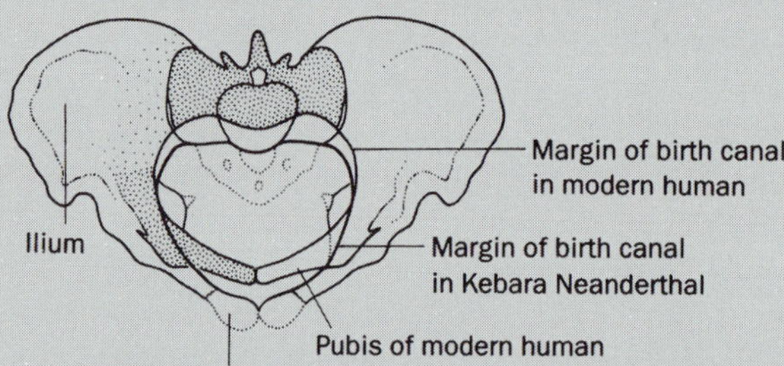


**The Kebara Neanderthal pelvis compared to a modern human pelvis. The superior pubic ramus of the Kebara Neanderthal is longer and thinner than that of the modern human.**

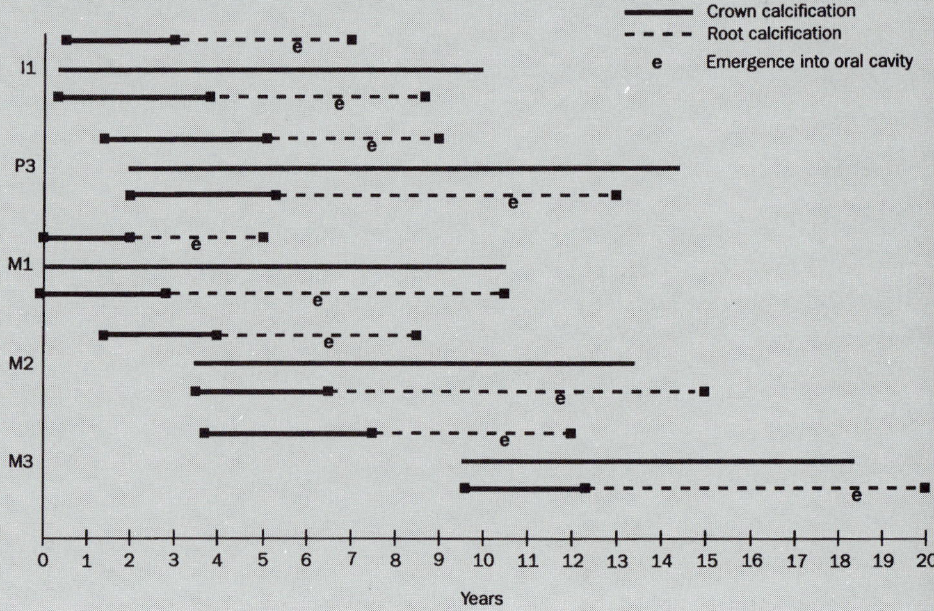


**Provisional rate of Neanderthal dental development compared with the developing dentition in chimpanzees and modern humans, showing differences in rate and pattern of growth (information on humans and chimpanzees redrawn from chart in Anemone et al., 1996). In Neanderthals, the teeth developed at a different rate than in humans. Neanderthal I1s and P3s developed somewhat slower, while M2s and M3s developed somewhat faster, than some human populations (see text). Since the exact calcification and emergence times of Neanderthal teeth are not yet known, only the total duration of tooth development can be estimated as indicated in this tentative figure.**

**Order of samples is the same within each tooth category: chimpanzees at top, Neanderthals in the middle, humans at the bottom.**

formed and erupted, except for the third molars, indicating a dental age of approximately 15.5 years (based on modern human standards). Skeletally, none of the long bones seems to have fused, which indicates that the specimen was younger than 14 years. Further analysis by Thompson and Nelson (1997), based on the length of the long bones, indicates an age of 11 for this individual. This lack of agreement between the dental and skeletal ages indicates that growth and development in Neanderthals differed somewhat from our own.

Another way to assess growth pattern is to examine the points at which typical adult Neanderthal features appeared over the course of their growth and development. Adult Neanderthals are characterized by a number of distinctive features, including a gap between the third permanent molar and the ascending ramus of the mandible, known as the retromolar space; a projection of the mid-facial region; no chin; a handle-bar shaped eyebrow region, accompanied by a rather flat forehead; and a projection of the rear of the skull, known as an occipital bun. In the postcranial skeleton, there are several other ways in which Neanderthals contrast with modern humans, including having had shorter forearms and lower legs, and a pelvis with a longer superior pubic ramus. These features did not appear fully-fledged in infants; instead they appeared at different times during growth. For example, at 10 months of age, a Neanderthal infant lacked a chin and had several other features typical of adults. By 1.7 years of age, a Neanderthal child showed an elongation of the superior pubic ramus and, from the age of 3 or more, achieved roughly 90 per cent of adult brain-size. Later still, after the eruption of the second molars, the features of the retromolar space, mid-facial prognathism, and other craniofacial features complete their growth.

### *DEVELOPMENT*

Modern human development is characterized by: 1) secondary altriciality; an extreme delay in motor development during infancy; 2) the childhood stage of development; and, 3) the adolescent growth spurt (**2.1, 2.4**). Based on estimates of the ratio of adult/newborn brain-weight, the first feature was seemingly already established in Neanderthals. The stage of infancy ended with weaning (**9.1**). It is said that while there is an isometric relationship between birth-weight and weaning weight, the timing of this threshold event varies according to a number of factors, including food availability and quality. This relationship between birth-weight and weaning was confirmed by Dettwyler who found that in humans, weaning (complete cessation of breastfeeding) takes place at around 3 years of age in well-nourished populations, but later (between 3 and 4 years of age) in less well-nourished ones. While the birth-weight of Neanderthal infants is yet unknown (although they were likely to have been similar in size to modern human infants (**10.3**) due to the constraints of the mother's pelvic dimensions), the weight of the mother can be used to estimate weaning age. Given a generous range of estimates (50–65 kg) of Neanderthal female body-weight, weaning ages for Neanderthals fall between 3 and 4 years of age. Even if Neanderthals experienced a somewhat accelerated rate of growth relative to modern human infants, this might be offset by the relative scarcity of resources in the Ice Age environment of the later Upper Pleistocene of Europe and the Middle East (compared to well-nourished modern human populations of today). Thus the length of infancy in Neanderthals was likely to have fallen within the modern human range of variability.

It is not known if the childhood stage of development was present in Neanderthals. None the less, since Neanderthal dental specimens show a period of stasis between the eruption of the lateral incisors and the second molars, a delay seen in humans but not apes, this is seemingly evidence that a childhood stage of development was present in Neanderthals. However, the duration of this stage, relative to that seen in modern humans, and relative to the Neanderthal stages of infancy and adolescence, is unknown.

During adolescence, modern humans experience a characteristic growth spurt (**5.17**). Whether or not this stage of development was present in Neanderthals is yet unknown. Smith and Tompkins predicted that if the growth spurt occurred in Neanderthals, then 'juveniles of any age will show concordance or random discordance between dental development and stature'. While work on Le Moustier 1 reveals a juvenile with dental (15.5 years) and skeletal ages (11 years) which disagree, work in progress will test the hypothesis that Neanderthals experienced a growth spurt like that seen in modern human populations.

*Jennifer L. Thompson*

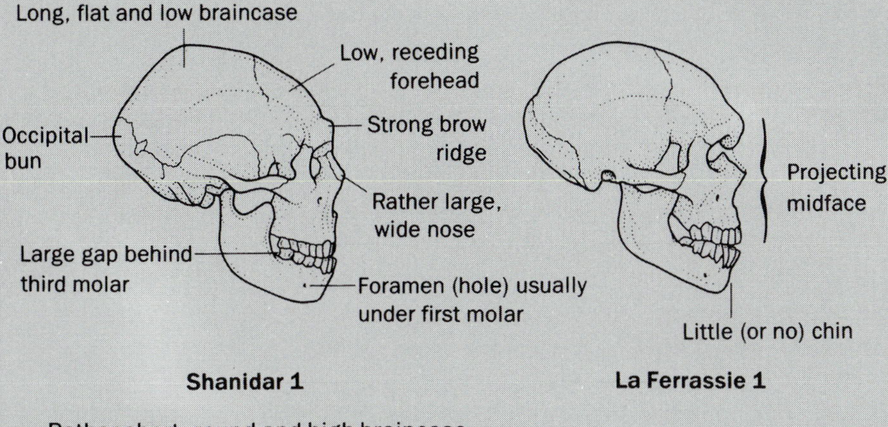


Rather short, round and high braincase

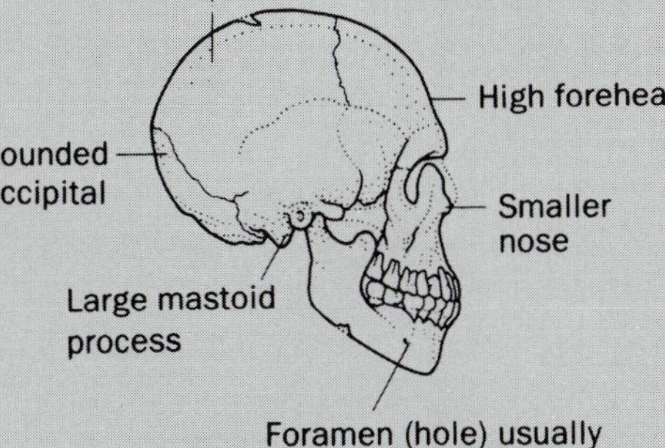


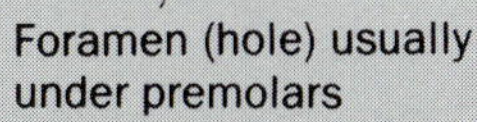


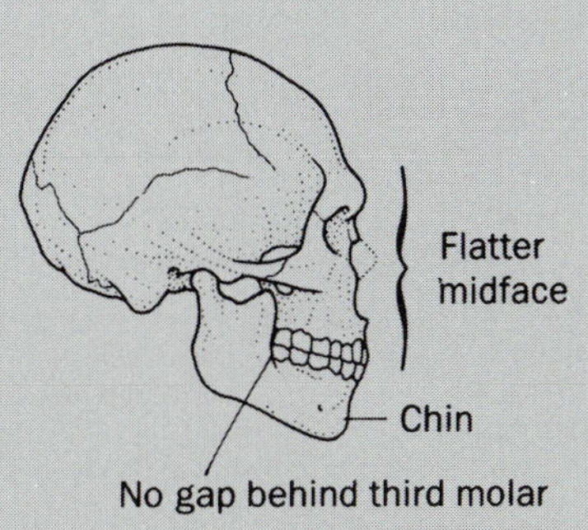


**Reconstructed skulls of two Neanderthals from Shanidar in Iraq and La Ferrassie in France, showing some of the features that distinguish Neanderthals from early modern *Homo sapiens*, represented here by skulls from Qafzeh in Israel and Předmosti in the Czech Republic.**

# The human growth curve

Growth of the physical dimensions of the body is a highly regulated process. The pattern of post-natal growth is well documented, with a high growth-rate immediately after birth, and rapid deceleration until 3 years of age. This is followed by a period with lower, slowly decelerating velocity until the onset of puberty. Puberty starts with an increased growth-rate and after the period of peak velocity, there is deceleration until growth ceases.

## Hormonal regulation of linear growth

The growth process is under the control of the endocrine system. Not only hormones are involved, but also hormone-binding proteins, growth factors and their binding proteins, and the hormone- and growth-factor receptors on the target cells. The secretion of hormones, such as growth hormone (GH), is pulsatile with higher peaks during the night as well as during puberty (**5.13**). A further complicating factor is that some growth can take place without involvement of central control, thus catch-up growth can be regulated locally at the tissue level and not under the influence of serum growth factors. For these reasons, the measurement of hormone or growth factor in a single serum, urine or tissue sample will demonstrate only a small part of the growth-regulatory puzzle.

However, it is generally agreed that there are at least three distinct endocrine phases of linear growth. How fetal linear growth is regulated is not precisely defined and no key circulating hormone has been identified. Uterus size, nutritional support and oxygen level in conjunction with insulin-like growth factors and insulin are believed to be involved in regulating fetal growth (**4.1**). During fetal life, the serum GH level is high and GH receptors have also been detected. However, fetal linear growth is known to behave almost independently of GH. A lack of growth response to GH during fetal life may be due to immature GH-specific receptors in the growth plate, as noted in the rabbit. For instance, GH-deficient children are on average 1 to 2 centimetres, or 2 to 4 per cent shorter at birth than normal infants. Whether this minor deviation is a secondary effect due to insufficiency of the influential metabolic action of GH, or to the lack of a direct effect of GH on the cartilage is currently not known.

It is also generally accepted that GH is responsible for growth during childhood given a normal thyroid hormone secretion. The exact age when it begins to control linear growth significantly in humans is still uncertain. However, the majority of isolated GH-deficient children grow normally or close to normal during the initial six months of life, but not thereafter. Growth during adolescence is related both to GH and sex steroids; testosterone in males and estrogens in females. Both GH and sex hormones are needed for normal pubertal growth although the presence of either one alone is associated with some growth during this period. It is not clear whether GH and sex steroids interact or act independently of each other.

It is thus reasonable to conclude that linear growth, from birth to maturity, is regulated by at least three different growth-promoting systems. Two simultaneously active systems are known to be involved in the adolescent growth spurt (**1.10**). Similarly, a post-natal continuation of the nutritionally driven fetal growth in conjunction with the GH-dependent phase of childhood growth characterizes the growth in the first year of life.

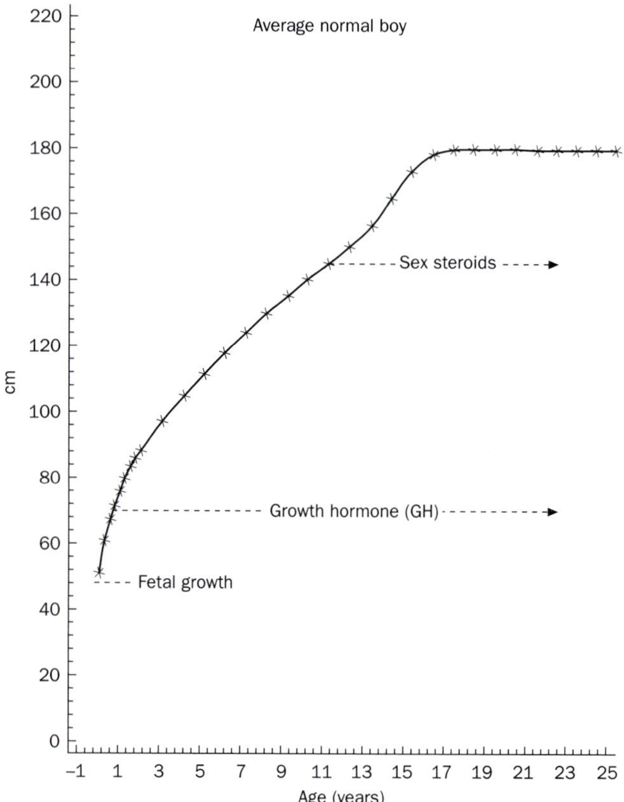


**Stature for a typical male from birth to maturity. The essential hormones involved in the regulation of linear growth are also included.**

## The infancy–childhood–puberty (ICP) growth model

The ICP growth model breaks down linear growth mathematically into three additive and partly superimposed components – infancy, childhood and puberty. This model represents linear growth during the first 3 years of life by a combination of a sharply decelerating infancy component and a slowly decelerating childhood component, the latter starting from the second half of the first post-natal year. From about 3 years of age to maturity, linear growth is represented by the sum of the infancy and childhood components and a sigmoid-shaped puberty component operating throughout adolescence.

The mid-growth spurt, a short-lived acceleration that has been observed in about two-thirds of healthy children, occurs, when detectable, at about 7 to 8 years of age and may be emphasized by a dip in velocity before puberty. For the ICP model it was deliberately smoothed and absorbed into the individual childhood function. This was done because of its very small influence on the mean curve, especially for females.

## Biological interpretation of the modelled components

Empirical observations have shown that the three components of the ICP model can be observed in isolation and that they are additive. Each component of this model is therefore assumed to represent a separate biological phase of the growth process. The *infancy component*, tentatively starting in mid-gestation and continuing up to 3 to 4 years of age, is believed to represent the post-natal continuation of fetal growth. It is assumed that the *childhood component*, slowly decelerating during childhood and adolescence, corresponds basically to the effect of GH. The sigmoid-shaped *puberty component*, the size of which is found to be independent of its

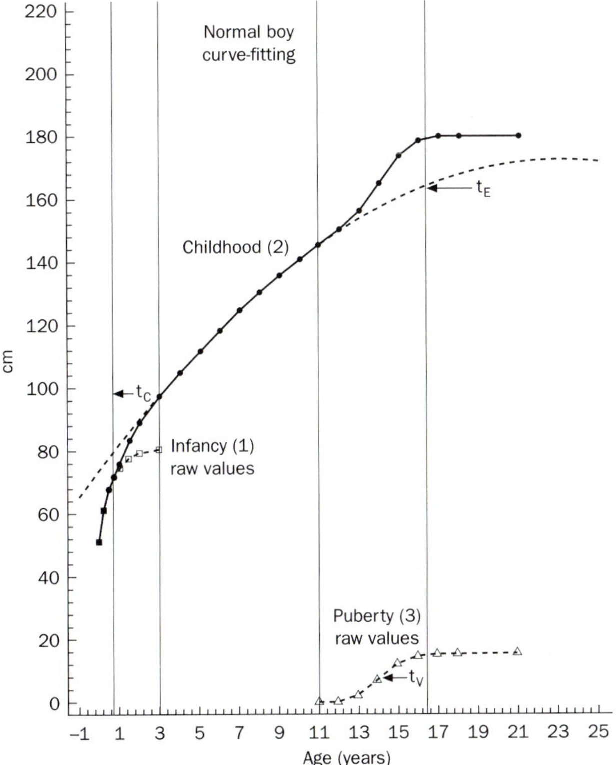


**The ICP model fitted to the growth curve for height for a typical male. In view of the stability of the growth process it seems reasonable to represent the period of slowly decelerating growth from 3 years of age up to the onset of puberty by a single component, which we have termed 'childhood'. A simple quadratic function was found to fit growth perfectly during this period. After subtracting the extrapolated values of this childhood component from the observed values during the periods before and after this age, two further components were extracted indirectly and modelled. As a result of the modelling process, three distinct components have been isolated: an *infancy component*, assumed to start during fetal life with a rapidly decelerating course ceasing at 3–4 years of age, which was explained by an *exponential function*; a *childhood component*, starting during the first year of life with a slowly decelerating course and continuing until end of growth, which was explained by a *second degree polynomial function*; and a *puberty component*, representing the additional growth induced by puberty and accelerating up to age at peak velocity (PV), then decelerating until the end of growth, which was represented by a *logistic function*.**

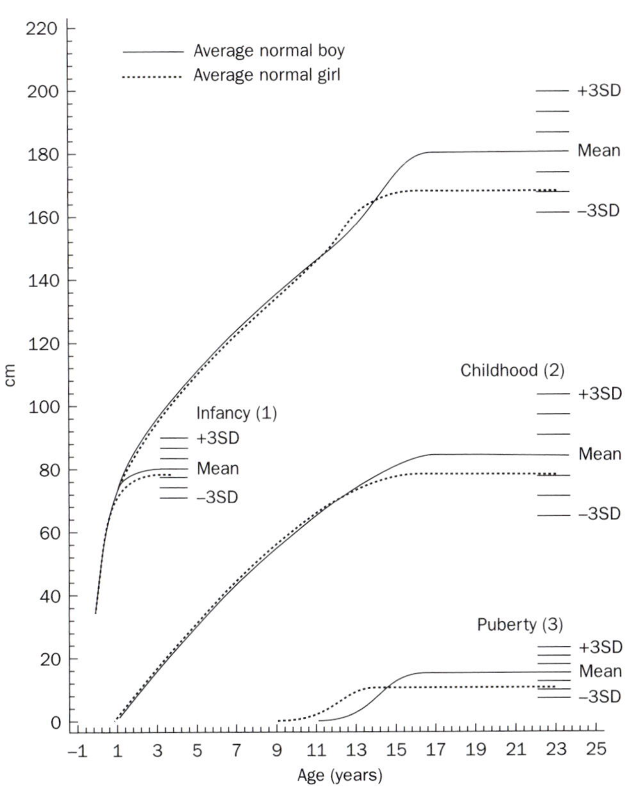


| | | Normal boy | Normal girl | Difference to normal boy |
|---|---|---|---|---|
| Infancy | (cm) | 80.4 | 78.3 | –2.1 |
| Childhood | (cm) | 84.3 | 78.6 | –5.7 |
| Puberty | (cm) | 15.5 | 11.0 | –4.5 |
| Total | (cm) | 180.2 | 167.9 | –12.3 |
| Childhood | | | | |
| onset | (yrs) | 0.8 | 0.7 | –0.1 |
| ceases | (yrs) | 17.6 | 15.8 | –1.8 |
| duration | (yrs) | 16.8 | 15.1 | –1.7 |

**The ICP growth model for normal boys and girls including mean values for each component and the combined growth. Reference values of 1, 2, 3 SD about the mean are also given for boys for each individual component and for the combined growth.**

timing, most likely describes the part of adolescent linear growth stimulated by sex steroids.

What mediates the onset of the childhood phase of growth has not been elucidated because information about the serum level of growth factors and hormones, as well as their interaction with their respective receptors during this critical period, is still limited. However, the most plausible explanation is that it represents the age at which GH begins to influence normal human linear growth significantly.

The final adult height of an individual is an additive result of the three underlying components. There is individual variation around the mean values of each component of the ICP model and the standard deviation (SD) values for each are 2 to 4 centimetres. It is not only the magnitude of each component that is important for the final height, but also the duration of the childhood phase; a late onset reduces height and a late end point increases height.

## The infancy–childhood–puberty model versus the triple logistic model

Other researchers have proposed models, either on theoretical or empirical grounds, which comprise three distinct functions. The ICP model shares some features with one of them, the triple logistic model. In contrast to the ICP model, Bock and Thissen preferred to think of the first two components of their model as an amalgam representing prepubertal activity, rather than as two components each with its own separate identity. Furthermore, the triple logistic model was applied to individual growth data for height from 1 year of age to maturity, whereas the ICP model covers the entire post-natal period. By using a logistic function instead of a second degree polynomial (like ICP) to describe the second component the two remaining components are affected in size and shape. A different rationale is also used by the two models concerning onset of the second component. A further distinction is that the ICP model views length as the sum of the separately modelled segments, leg length and sitting-height, and so is able to deal with growth of height more realistically than the triple logistic model, especially with the behaviour of growth towards completion.

## Some biological conclusions from the infancy–childhood–puberty model

The three separate and additive components of the ICP model are in agreement with known endocrinological regulation of growth and seem to be individually biologically applicable. This indicates the existence of different growth-promoting systems for each component. However, the onset of a component might be dependent on the state of previous components. The onset of the childhood component is positively related to the magnitude of the infancy component and the onset of the puberty component is negatively correlated with attained size prior to puberty. Low height-gain during the infancy component seems thus to be compensated for by an early start of the childhood component, and low attained size before puberty seems to be compensated for by late onset of the puberty component, which allows for a longer duration of the childhood component before adult size is reached. By adapting the onset of one component to the gain of the previous one the variation in final size in a healthy population seems to be reduced.

Although the shape of each of the three measurement components is similar, their gains, rates and timing are quite different for the two segments of the body. The two segments of the body are differently affected by each of the three components. The gain in sitting-height is associated more with the infancy and puberty components, while the gain in leg length is associated more with the childhood component. Different tissue, represented here by short bones in the vertebra and long bones in the legs, seems to be specifically sensitive to substances governing each component. The well-known dramatic change in body proportions during the first years of life is certainly due to a greater influence of the childhood component on leg length than on sitting-height, while after puberty the legs stop growing sooner than the upper segment of the body. The childhood component thus has little effect on leg length after cessation of the puberty component, but it

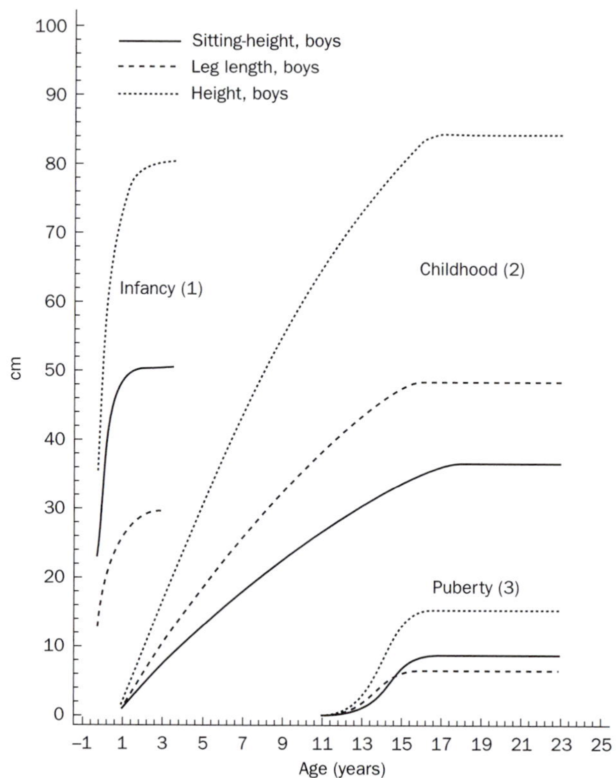


**The ICP model for the two segments of the body – mean values of sitting-height and leg-length for each of three components. The sum of the two segments gives the mean height ICP values.**

might continue to affect sitting-height. Early pubertal maturation, as seen for females in comparison with males will, for this reason, result in shorter legs in relation to the upper part of the body.

Males are on average about 12 centimetres taller than females at maturity. Only a small part (2 cm) of this difference develops during fetal life or is due to post-natal gain during the infancy component. Persistent differences are due to males having a longer duration of the childhood component (6 cm) and a larger puberty component (4 cm). The childhood component exerts a greater influence on females than males, in the sense that, for females, acceleration and deceleration of this component are both more rapid. At about 12 years of age the gains in height for the two sexes due to this component are about equal, but because of the longer duration for males they show a greater final gain.

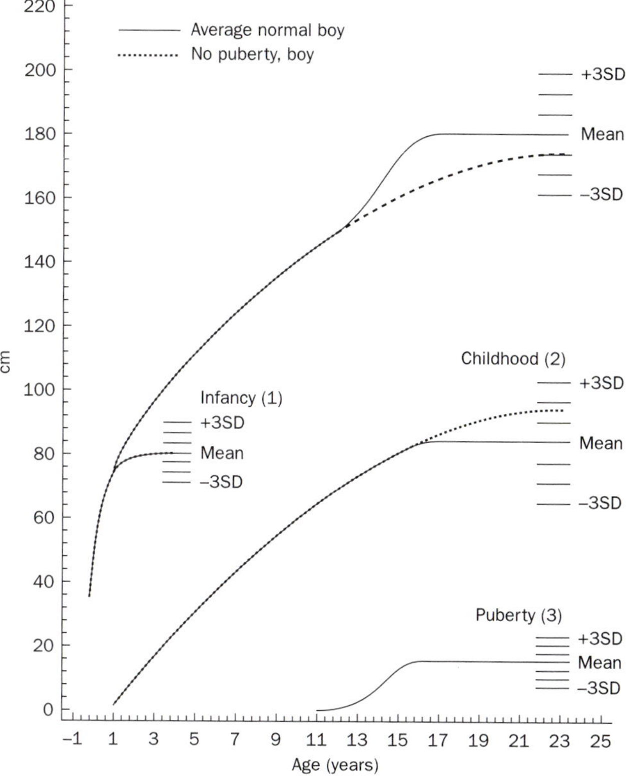


| | | Normal boy | No puberty boy | Difference to normal boy |
|---|---|---|---|---|
| Infancy | (cm) | 80.4 | 80.4 | 0.0 |
| Childhood | (cm) | 84.3 | 94.1 | 9.8 |
| Puberty | (cm) | 15.5 | 0.0 | −15.5 |
| Total | (cm) | 180.2 | 174.5 | −5.7 |
| Childhood | | | | |
| onset | (yrs) | 0.8 | 0.8 | 0.0 |
| ceases | (yrs) | 17.6 | 24.0 | 6.4 |
| duration | (yrs) | 16.8 | 23.2 | 6.4 |

**The ICP growth model for normal boys and girls with no puberty component, including mean values for each component and the combined growth. Reference values of 1, 2, 3 SD about the mean are also given for boys for each individual component and for the combined growth.**

Adolescent growth due to the puberty component alone explains only a small part of the final height – 7 per cent for females and 9 per cent for males. Children lacking pubertal maturation could theoretically almost reach normal values for final height (4 cm below the mean for females and 6 cm below the mean for males), in spite of the absence of a pubertal growth spurt, if they continued to grow up to about 21 years of age for females and to about 24 years for males. These ages represent the point when the maxima of the parabolas representing the childhood values for height are reached. The puberty component may thus be regarded more as a regulator of the cessation of growth than a factor influencing the size of the gain. This is also indicated by the fact that children with early onset of puberty are taller in childhood than children with late onset, but are equal in final size. A further point indicated by the model is that early pubertal maturation is associated with a higher peak velocity than late pubertal maturation. Such variation in the pubertal gain is related to the decelerating shape of the childhood component alone, because the size of the puberty component was found to be independent of the time of its occurrence.

## Growth patterns constructed from the infancy–childhood–puberty growth model

### *Precocious puberty*

Early pubertal maturation is one reason for adult shortness. For example, the growth pattern of a boy with precocious puberty may have a normal magnitude for all three components, but with an early pubertal onset and consequently a short duration of the childhood component, 10 years shorter than for normal boys. This boy would end up with a final height of 138.0 centimetres, or 42 centimetres below the mean.

### *Small-for-gestational-age without catch-up growth*

The reason that some babies are small at birth can be genetic or disease-related, caused by chronic malnutrition and/or oxygen deprivation, or simply being at the lower extreme of normal. More than 80 per cent of all small-for-gestational-age (SGA) infants catch-up during the first years of life. One reason that the remaining infants lack this catch-up growth may be that they have a magnitude of the infancy component in the lower extreme of the normal. For example, an infant may have an infancy magnitude of −3 SDS, but otherwise with a normal pattern of growth. The result is a final height of 170, or 10 centimetres below the mean. The timing of puberty is usually normal or even in some cases early for such children.

*Stunting in early life*

About 25 per cent of all children ending up short in final height are born small and have no or little catch-up growth post-natally (**9.6**). Another 25 per cent of the short adult population is born with normal body-size, but have growth impairment, or growth faltering during the first months of life. This growth pattern is also the major reason behind shortness in developing countries. In some populations more than 80 per cent of the children end up short in final height, mostly because of growth stunting in early life. An extreme example of this pattern might be where final height is 17 centimetres below the normal, this being solely explained by a delayed onset of the childhood phase (at three years of age). The cause of the childhood-onset delay in developing countries is, however, still unknown, although it is clear that the mechanism is related to this delay.

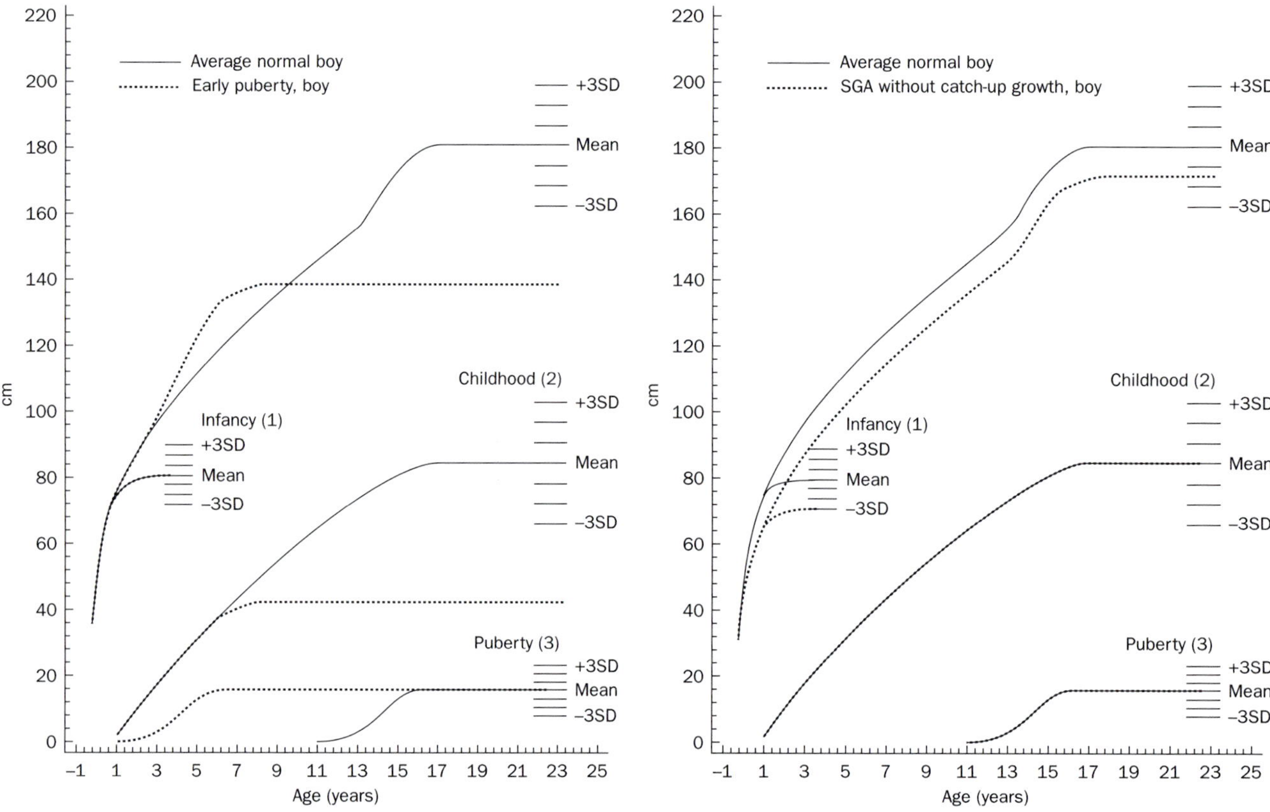


| | | Normal boy | Early puberty boy | Difference to normal boy |
|---|---|---|---|---|
| Infancy | (cm) | 80.4 | 80.4 | 0.0 |
| Childhood | (cm) | 84.3 | 42.1 | −42.2 |
| Puberty | (cm) | 15.5 | 15.5 | 0.0 |
| Total | (cm) | 180.2 | 138.0 | −42.2 |
| Childhood | | | | |
| onset | (yrs) | 0.8 | 0.8 | 0.0 |
| ceases | (yrs) | 17.6 | 7.6 | −10.0 |
| duration | (yrs) | 16.8 | 6.8 | −10.0 |

**The ICP growth model for normal boys and boys with early puberty; mean values for each component and the combined growth. Reference values of 1, 2, 3 SD about the mean are also given for boys for each individual component and for the combined growth.**

| | | Normal boy | SGA boy | Difference to normal boy |
|---|---|---|---|---|
| Infancy | (cm) | 80.4 | 70.6 | −9.8 |
| Childhood | (cm) | 84.3 | 84.3 | 0.0 |
| Puberty | (cm) | 15.5 | 15.5 | 0.0 |
| Total | (cm) | 180.2 | 171.6 | −9.8 |
| Childhood | | | | |
| onset | (yrs) | 0.8 | 0.8 | 0.0 |
| ceases | (yrs) | 17.6 | 17.6 | 0.0 |
| duration | (yrs) | 16.8 | 16.8 | 0.0 |

**The ICP growth model for normal boys and small-for-gestational-age boys with no post-natal catch-up growth, including mean values for each component and the combined growth. Reference values of 1, 2, 3 SD about the mean are also given for boys for each individual component and for the combined growth.**

*Tallness and shortness*

At two extremes, a comparison of a very short boy (149 cm) and a very tall one (215 cm) in final height, is instructive. The short boy has low magnitude of all three components (−2 SDS), late onset of the childhood component and early pubertal onset (both ages equal to 2 SD from the mean). The tall boy has high magnitude of all three components (+2 SDS), early onset of the childhood component and late pubertal onset (both ages equal to 2 SD from the mean). Normally, there is no correlation between the magnitude of the three components, but in very extreme cases such combinations can lead to tallness or shortness despite close to normal behaviour of the three individual components. *Johan Karlberg*

See also 'Anthropometry' (**1.4**), 'Longitudinal analysis' (**1.14**), 'Creation of growth references' (**1.16**), Use of growth references' (**1.17**), 'Human growth from an evolutionary perspective' (**2.3**), 'Within-population variation in growth patterns' (**2.5**), 'Endocrinological regulation of post-natal growth' (**5.1**), 'Growth cyclicities and pulsatilities' (**5.13**), 'Hormonal regulation of growth in childhood and puberty' (**5.14**) and 'Growth modelling and growth references: the future' (**13.1**)

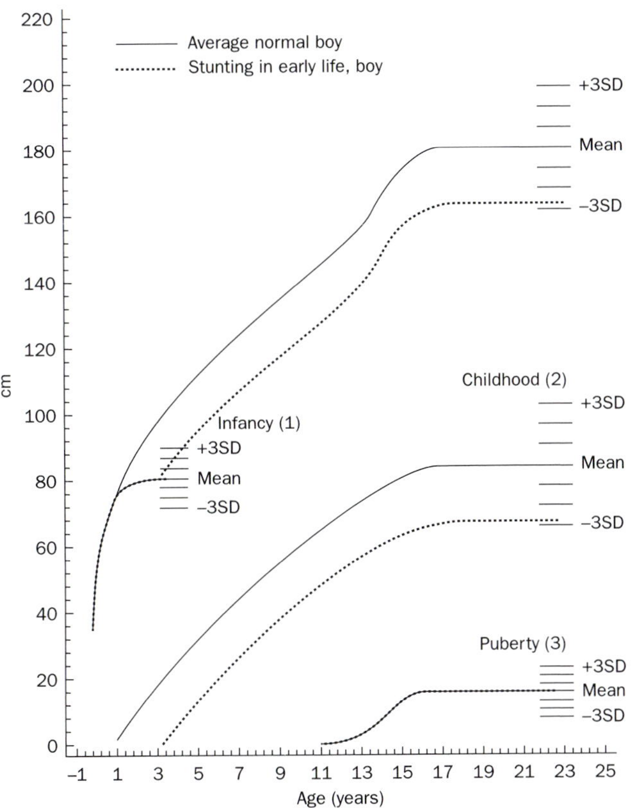


| | | Normal boy | Stunting boy | Difference to normal boy |
|---|---|---|---|---|
| Infancy | (cm) | 80.4 | 80.4 | 0.0 |
| Childhood | (cm) | 84.3 | 67.6 | −16.7 |
| Puberty | (cm) | 15.5 | 15.5 | 0.0 |
| Total | (cm) | 180.2 | 163.5 | −16.7 |
| Childhood | | | | |
| onset | (yrs) | 0.8 | 3.0 | 2.2 |
| ceases | (yrs) | 17.6 | 17.6 | 0.0 |
| duration | (yrs) | 16.8 | 14.6 | −2.2 |

**The ICP growth model for normal boys and and boys experiencing stunting in early life, including mean values for each component and the combined growth. Reference values of 1, 2, 3 SD about the mean are also given for boys for each individual component and for the combined growth.**

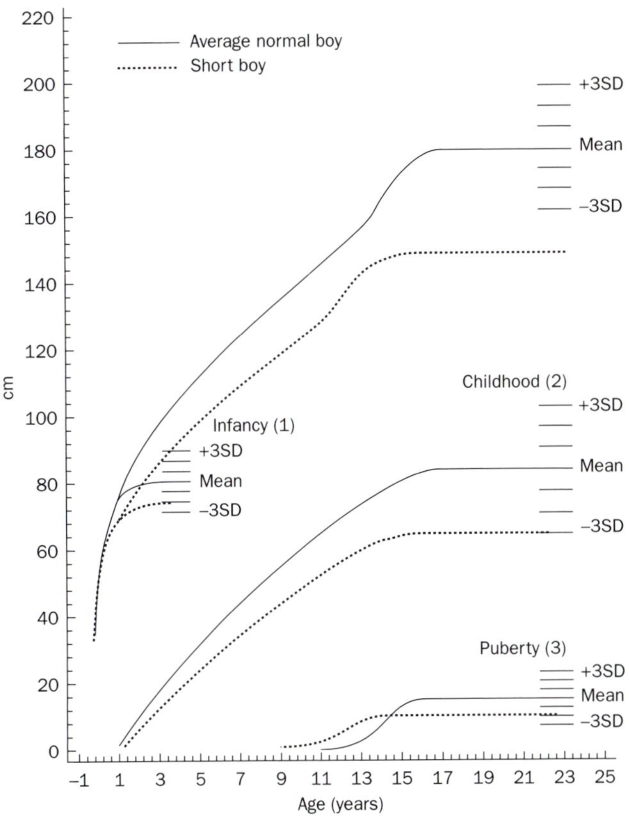


| | | Normal boy | Short boy | Difference to normal boy |
|---|---|---|---|---|
| Infancy | (cm) | 80.4 | 73.9 | −06.5 |
| Childhood | (cm) | 84.3 | 65.2 | −19.1 |
| Puberty | (cm) | 15.5 | 10.2 | −5.3 |
| Total | (cm) | 180.2 | 149.3 | −30.9 |
| Childhood | | | | |
| onset | (yrs) | 0.8 | 1.1 | 0.3 |
| ceases | (yrs) | 17.6 | 15.6 | −2.0 |
| duration | (yrs) | 16.8 | 14.5 | −2.3 |

**The ICP growth model for normal boys and boys with low magnitude of all three components (−2 SDS), late onset of the childhood component and early pubertal onset (both ages equal to 2 SD from the mean), including mean values for each component and the combined growth. Reference values of 1, 2, 3 SD about the mean are also given for boys for each individual component and for the combined growth.**

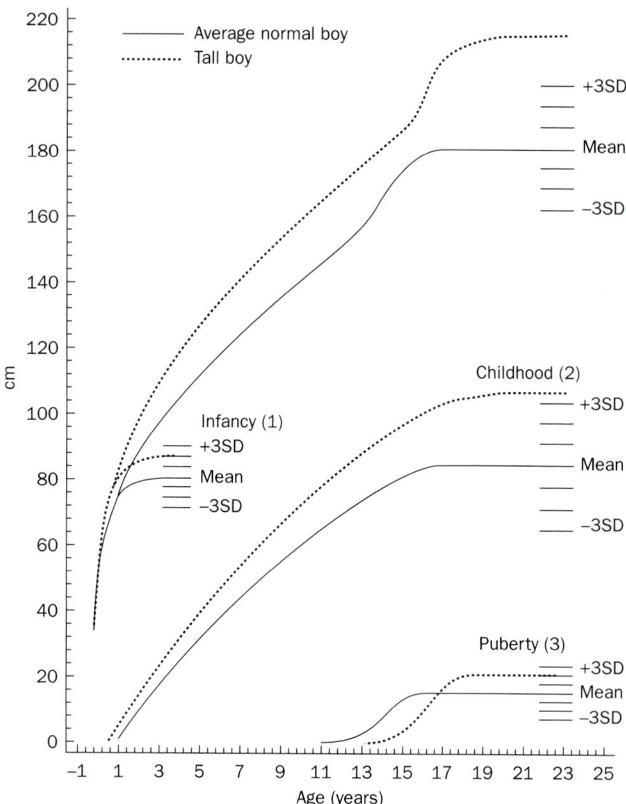


| | | Normal boy | Tall boy | Difference to normal boy |
|---|---|---|---|---|
| Infancy | (cm) | 80.4 | 87.1 | 6.5 |
| Childhood | (cm) | 84.3 | 106.7 | 22.4 |
| Puberty | (cm) | 15.5 | 20.8 | 5.3 |
| Total | (cm) | 180.2 | 214.5 | 34.2 |
| Childhood | | | | |
| onset | (yrs) | 0.8 | 0.5 | −0.3 |
| ceases | (yrs) | 17.6 | 19.6 | 2.0 |
| duration | (yrs) | 16.8 | 19.1 | 2.3 |

**The ICP growth model for normal boys and boys with high magnitude of all three components (+2 SDS), early onset of the childhood component and late pubertal onset (both ages equal to 2 SD from the mean), including mean values for each component and the combined growth. Reference values of 1, 2, 3 SD about the mean are also given for boys for each individual component and for the combined growth.**

## CURVE-FITTING

Human growth is the process of change in size and shape which occurs between conception and full maturity. Most of our knowledge of children's growth relates to the post-natal period and comes from sequential measurements of size (stature, for instance), taken at regular intervals on the same subject (longitudinal data) (**1.14**). These serial data allow the determination of the underlying continuous process, i.e. to produce a smooth growth curve which fits our observations, and which can be used to estimate growth between measurement occasions. Various mathematical models have been proposed to estimate this continuous growth process, starting from a set of discontinuous growth measurements. These growth models have variable success in describing the pattern of human growth depending on the type of growth variable (height, leg length, head circumference, or other) used, the precision of the measurements, the frequency and age range of the observations, the ability of the model to describe either a part (childhood or adolescence, for instance) or the whole of the human growth curve, the flexibility of the model to cope with the wide variations in the overall shape of normal human growth patterns and with particular features such as a prepubertal growth spurt, for instance. Adjusting a mathematical model to a set of growth data is called growth curve-fitting or growth modelling.

The fit of a model to serial growth data is a regression technique which consists of finding the set of values for the function parameters that satisfies some 'best-fitting' criterion. The oldest and most widely used approach in curve-fitting is the 'least-squares' method, which provides the values of the function parameters that minimize the sum of squared deviations of the observed values from those predicted by the equation. In some instances, other parameter estimation techniques such as 'maximum likelihood' may be a more appropriate approach.

There are basically two families of growth models, 'non-structural' and 'structural'. Non-structural models merely provide a good description of the growth process as shown by the empirical data. They mainly consist of polynomials of varying degrees which provide a smoothing of the individual serial growth data in order to suppress measurement error. The general form is:

$$y = b_0 + b_1 t + b_2 t^2 + \ldots + b_n t^n$$

where $y$ is size, $t$ is age, and $b_0, \ldots, b_n$, are the polynomial coefficients. Polynomials are easy to fit and are essentially useful in describing growth over short time intervals, i.e. over periods of a few years only. High-order polynomials would be required to describe long-term growth, but this approach has several limitations including: 1) the large number of parameters to be estimated; 2) a lack of functional interpretation of the parameters; 3) lack of flexibility of polynomials to describe sudden changes in growth rate; 4) instability of high-order polynomials in the extremities of the data, and 5) no asymptote to a final value, rendering polynomials entirely unsuitable for extrapolation beyond the observation range or for prediction of final size. Some of the disadvantages of high-order polynomials have been overcome by using smoothing spline functions. These models consist of series of polynomials (cubic, for example), which are fitted over only a small age range and which are connected by constraints of continuity, i.e. equality of the first and second derivative at the points of transition or 'knots'. Further sophistication of non-structural curve-fitting is the *kernel method*, which is similar to smoothing splines, but uses weighted averages of the measurements. Although the latter approaches yield much better fits than simple polynomials they still require the estimation of a large number of parameters and do not tend asymptotically to a final value.

Structural models are based on the idea that the growth pattern has a basic functional form to which a direct 'biological' interpretation can be attributed. Structural models are essentially non-linear functions and provide a good description of the growth pattern in a relatively small number of parameters. Structural models sometimes have the drawback of not being able to represent certain features of growth (such as a prepubertal spurt) or they may impose a too rigid shape upon the growth pattern. An example of a structural model is the JPA2 model of Jolicoeur and colleagues. The model has the following mathematical expression:

$$y = A\left\{1 - \frac{1}{1 + \left[\frac{t+E}{D_1}\right]^{C_1} + \left[\frac{t+E}{D_2}\right]^{C_2} + \left[\frac{t+E}{D_3}\right]^{C_3}}\right\}$$

with $y$ = size (stature), $t$ = age, and $A, D_1, D_2, D_3, C_1, C_2, C_3, E$, the eight function parameters.

Growth modelling is often used to estimate biological parameters of the growth curve such as age, size and velocity at the start of the adolescent growth spurt (age at 'take-off') and at the age of maximal velocity during the adolescent growth spurt. Such biological parameters include age at take-off, velocity at take-off, age at peak height velocity, and peak height velocity. Several other quantities, characterizing some aspect of the shape of the growth pattern, can be derived from the fitted curve. These biological parameters form the basis for studies comparing growth patterns between individuals or between groups of individuals.

Curve-fitting is a technique which allows the estimation of smooth growth curves based on a set of discrete growth data. In addition it can be used to summarize growth data with a limited number of constants (function parameters or biological parameters) which have the same meaning for all subjects and which can easily be used for further analysis of the shape of the growth pattern. *R.C. Hauspie*

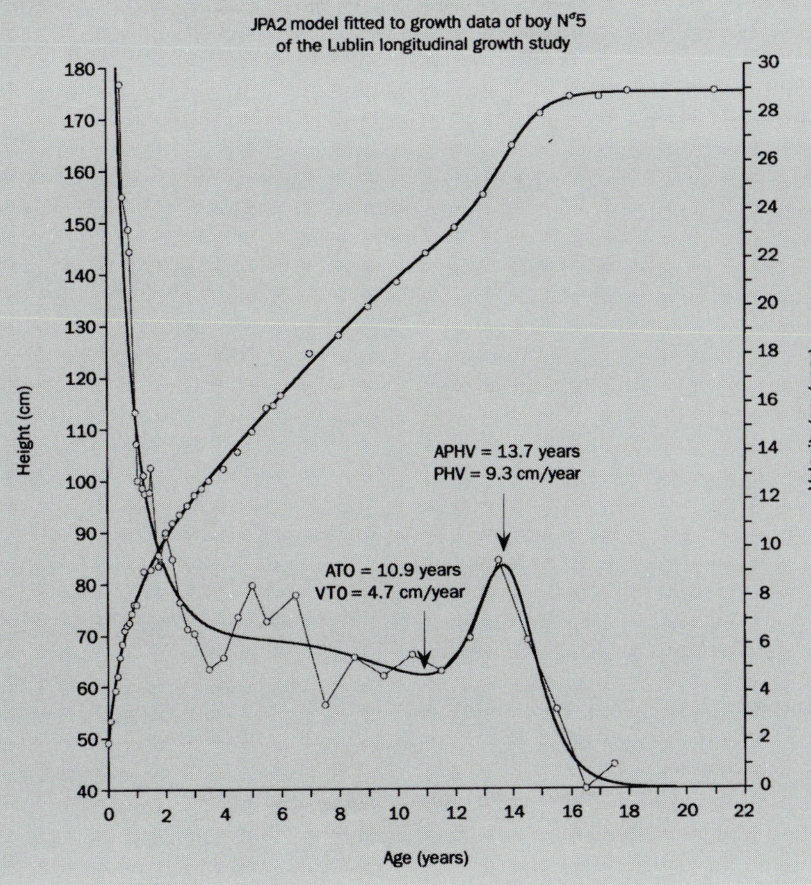


**An example of the JPA2 model of Jolicoeur and colleagues, fitted to the growth data of an individual child measured from birth to full maturity. The upper part of the figure depicts attained height together with the fitted curve; the lower part shows the whole-year increments in height together with the instantaneous growth velocity curve obtained as the mathematical first derivative of the curve fitted to the distance data. Although the model fits the data fairly well, it cannot describe the prepubertal growth spurt seen in the plot of the yearly increments in height around ages 5 to 7 years. On the other hand, it fits the adolescent cycle very well. ATO = age at take-off; VTO = velocity at take-off; APHV = age at peak height; PHV = peak height velocity.**

# Within-population variation in growth patterns

It is the nature of science to aggregate data, presenting them as mean, median, variance, and so on. While such an approach is necessary as a descriptive device, all too often one can lose sight of the wide range of individual variability that exists within any population group, for any variable. For example, the size of a dimension usually shows differences in variation across and within age groups. In addition, the growth increment shows great variability within age groups. There is also great variation in adolescent parameters. For example, in the third Harvard Growth Study the age of attainment of peak velocity and the peak height velocity itself show great variation.

## Sources of variation

Individual variability in body-size and growth increment reflects two different features of the human growth curve. First is that feature related to adult size. Other things being equal, larger adults will tend to have been larger during their growing years, just as smaller adults will have been smaller children. This 'canalization' of the growth process is well known and is at the base of techniques used to predict the adult size of children. The second is related to the tempo of development, the rate at which individual children move through the various stages of maturation. Early maturers, other things being equal, will tend to be larger at a particular chronological age than late maturers. Variability is present for height as well as for other measures of growth, body composition, and maturation.

*Francis E. Johnston*

See also 'Growth as an indicator of social inequalities' (**1.12**), 'Longitudinal analysis' (**1.14**), 'Creation of growth references' (**1.16**), Growth in infancy and pre-adolescence' (**10.4**), 'Between population differences in adolescent growth' (**10.5**) and 'Migration and changing population characteristics' (**11.6**)

*Distribution of height increments from 7.0 to 8.0 years in 405 Guatemalan children, showing clear variation*

| Increment (cm) | Number of children |
|---|---|
| 2 | 4 |
| 3 | 9 |
| 4 | 41 |
| 5 | 134 |
| 6 | 153 |
| 7 | 54 |
| 8 | 9 |
| 9 | 1 |

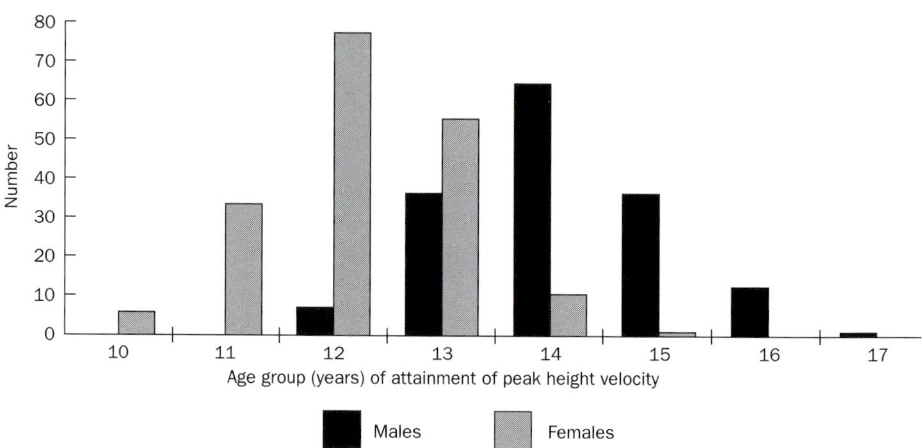


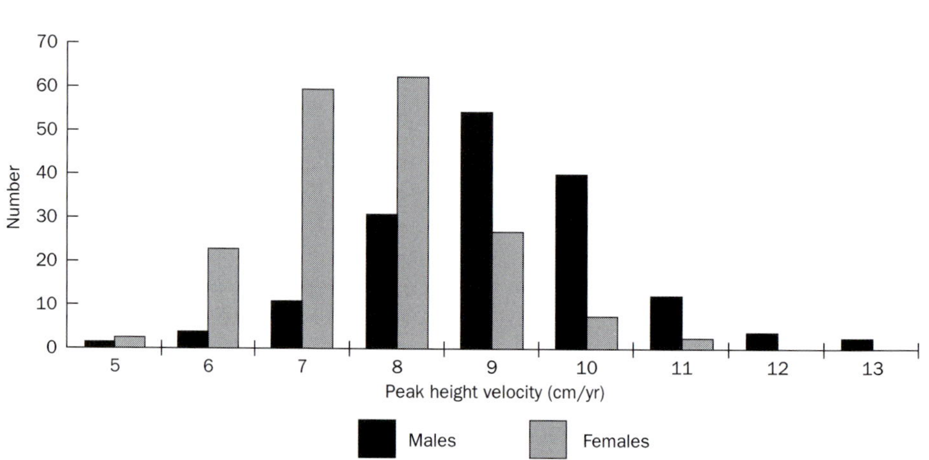


**Variability in the patterns of growth over time of individual children from the Third Harvard Growth Study. The children of this sample were from schools in the Boston area and were measured annually between 1923 and 1934. Using a non-linear model (Preece–Baines fit) the heights of the 338 individuals were regressed on age at examination. A separate fit was carried out for each child and, again for each, biological parameters (i.e., developmental landmarks) were estimated from the equations. Both have been rounded off to the nearest whole year or centimetre/year. Age of attainment of peak velocity and peak height velocity (PHV) show great variation. The estimated values are of course dependent upon the success of the non-linear model at achieving convergence. And the expected sex differences, especially in age at PHV, are obvious.**

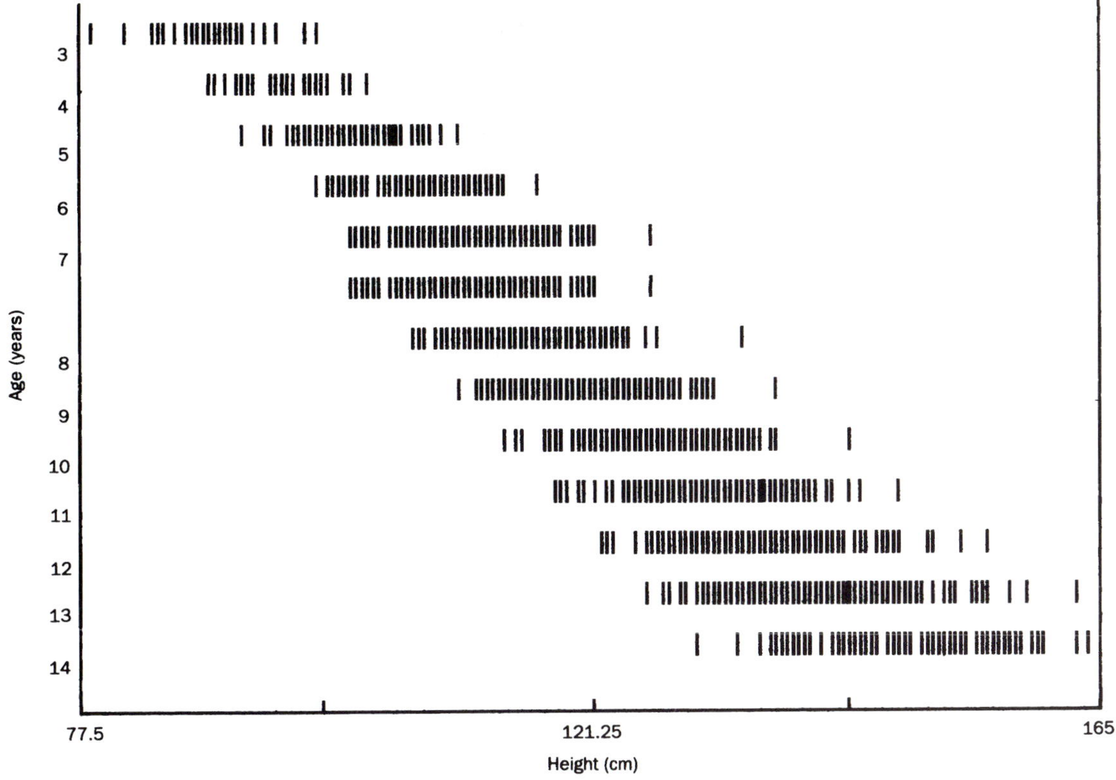


**The range of variation in height of a mixed-longitudinal sample of Guatemalan children from 3 to 14 years of age. The numbers in each age group range from 272 to 421. Since measurements were made within a week of each child's birthday, there was no increased variability due to age grouping. While the means increase steadily, from 88cm at 3 years of age to 149cm at 14, the amount of variability among individuals within each age group is substantial and the distributions of adjacent age groups show considerable overlap.**

## A TECHNIQUE FOR ANALYSING THE GROWTH OF POPULATION SUB-GROUPS

When it is of interest to examine the growth of children from different populations (**10.4**, **10.5**), or even sub-populations (**1.12**, **11.2**, **11.4**), it is convenient to compare average growth at corresponding ages. The use of the average curve is, however, potentially misleading. Even when two curves are very close, they represent only the way the population averages change with time and not necessarily the specific shapes of individual growth curves.

Two methods have been explored which investigate the relationship between individual and average growth patterns. The first of these, called *multilevel modelling*, proceeds from fitting individual growth curves. The second makes use of variance and correlation to examine the effect that growth has on the distribution of a trait in a population. It takes part of its inspiration from classical quantitative genetics and the partitioning of the phenotypic variance into genetic and environmental components.

Both methods depend on longitudinal data. In multilevel modelling, the growth curves are fitted to individual data, measured at intervals. For the *distributional method*, the mean and variance of the trait are calculated at each interval and the relation between the two examined. Different measures of distance and increment are analysed to provide complementary information on patterns of variance. Correlation is used to examine movements within the distribution of the trait over time. In all cases both the magnitude and pattern of change is of interest.

There are clear differences in growth pattern which emerge when variances in growth are considered compared to the use of means. In comparing Hong Kong and Khartoum infants, there is a difference in body-length variance between the two groups which was already present at birth. To consider whether differences can also arise after birth, another measure of growth is used. This measure, called the *accumulated increment*, is computed simply as the length gained from birth onwards (e.g. $L_1-L_0, L_2-L_0, \dots, L_n-L_0$, where $L_0$ is birth-length, and $L_1, \dots, L_n$ are lengths at 1 month, 2 months, etc.). Although birth is used in the present context, accumulated increments can be calculated from any age of interest.

A property of accumulated increments is that their variance is initially zero (birth in the present context) and non-zero thereafter. It is therefore possible to observe how variation in a trait is generated by growth. Again with the Khartoum –Hong Kong infants' comparison, the variance of length increases steadily with accumulated growth after birth in both populations of infants. It is striking that the Khartoum infants are more variable than the Hong Kong infants in any length-gain from birth. Not only are the Khartoum infants more variable at birth, their length growth is also more variable thereafter.

Because accumulated increments display the variance of a trait generated through growth, attempts have been made to determine how much of the variance is due to environmental, genetic and other causes of variation. This is possible because some sources of variation have different theoretical relationships with average growth. Consider a purely genetic component of variance, and a theoretical population of growth curves. The only variation present in growth is that determined by variation in genes. The variance of the accumulated increments increases at a non-linear rate relative to mean growth. This in fact is a quadratic function of the mean, which can be used to estimate the genetic component of growth variance in the population, and is called the quadratic component of the variance.

Environmental sources of variance can be conveniently if somewhat arbitrarily divided into those that persist and those that are more transient in nature. Persistent environmental influences are most important to growth variance if they are distributed differently among subgroups of a population. An example is poor and well-off groups experiencing contrasting levels of nutritional intake and exposure to pathogens. Because the prolonged effect of such environmental stratification on growth is similar to the effect of genes, the two variances may be indistinguishable and thus commingled in the quadratic component. This will lead to overestimates

*(continued over page)*

## A TECHNIQUE FOR ANALYSING THE GROWTH OF POPULATION SUB-GROUPS (CONTINUED)

of the genetic variance unless the strata can be identified and analysed separately.

Transient environmental effects are those which effect growth variation only for a short period. Discrete episodes of disease (**9.3**), for example, generate variation in a population by perturbing the growth of some children at different times. When any one child recovers, the effect of the disease on length is commonly removed by the well-known spurt of catch-up growth (**9.6**). This process, which has been termed the homeostatic cycle by Harrison and Brush, does not allow variation arising from acute challenges to accumulate indefinitely. With children in various stages of the cycle a dynamic constant of variance is established, termed the *homeostatic constant*.

The homeostatic constant is related to the number of children in the homeostatic cycle, the extent to which individuals are deflected from their growth targets, and the efficiency of the homeostatic response. The constant is of potential use for measuring the quality of a population's environment, with larger constants representing poorer conditions. Its theoretical independence from the rate of growth means that assessments of environmental quality will not be confounded

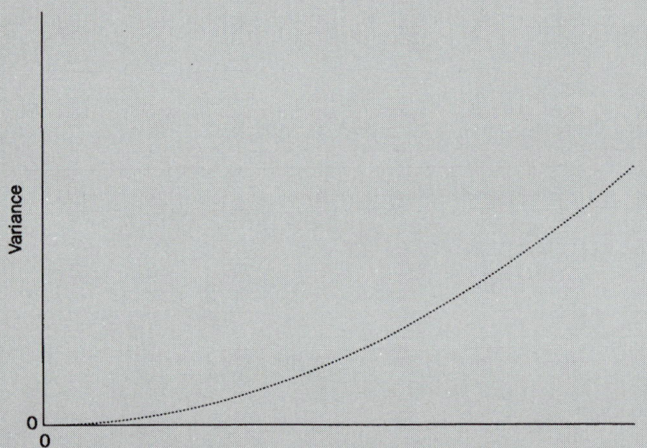


**The pattern of variation generated in relation to mean growth when growth-rate is determined solely by genetic variation. The variation is expected to increase as a quadratic function of mean growth.**

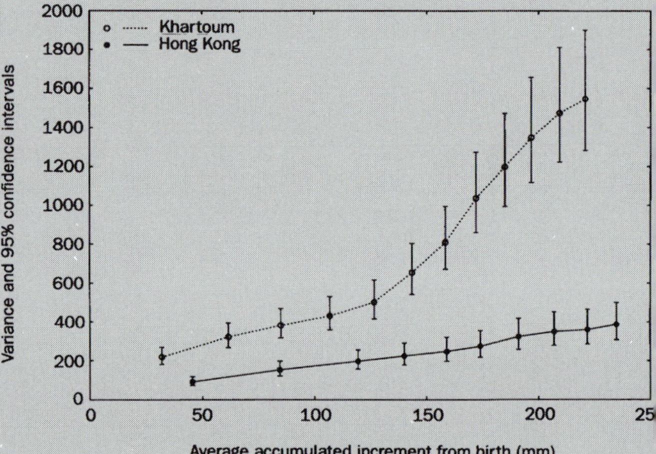


**Variance in relation to mean of length; accumulated increments, taken monthly from birth, in Khartoum and Hong Kong infants. The error bars represent 95% confidence intervals.**

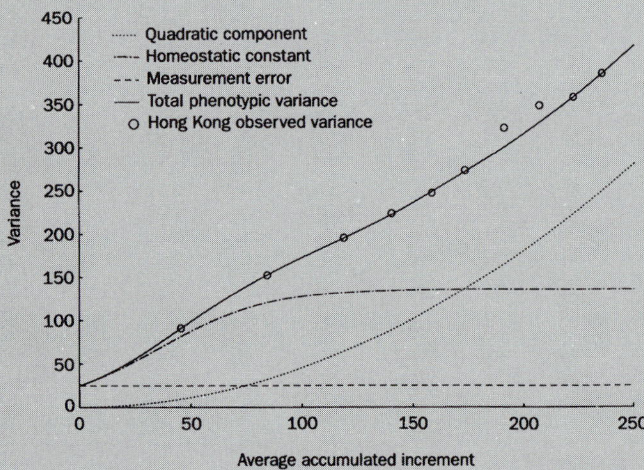


**Fitted model of the quadratic component of variance and the homeostatic constant to the phenotypic variances of the Hong Kong infants. The x axis shows the average amount of length gained after birth. When length-gain is small, the phenotypic variances are dominated by measurement error and the homeostatic constant, as these are large relative to the quadratic component. However, as growth continues, the quadratic component of variance increases rapidly, first overtaking measurement error and then, eventually, the homeostatic constant.**

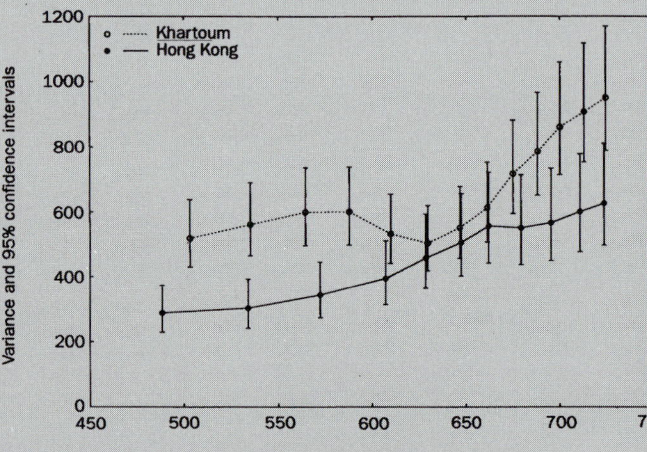


**Changes in variance of body-length with average growth over the first 10 months of life in Khartoum and Hong Kong infants. The Khartoum variances are significantly greater than the Hong Kong variances except for the dip midway through the period. A different picture of growth emerges from that given by comparing only the means; as a group, the Khartoum infants are more dissimilar in length than are the infants from Hong Kong.**

by genetic differences in growth-rate among populations. The estimation and practical usability of the constant remain to be more fully explored.

Another source of variation is measurement error. The simplest assumption about its variance is it is constant and unrelated to growth. It is therefore commingled with the homeostatic constant, leading to a potential overestimation of the latter. The logistic curve seems to model this build up to the constant adequately, although other curves may also be appropriate. The measurement error variance, being constant from the outset, is assumed to be at the point where the logistic curve intercepts the *y* axis. The phenotypic variance is the sum of these components, which fits the observed variation in the Hong Kong infants very well. For this group, when length-gain is small, the phenotypic variance is dominated by measurement error and the homeostatic constant, as these are large relative to the quadratic component. However, as growth continues, the quadratic component of the variance increases rapidly, first overtaking measurement error and then eventually the homeostatic constant. With increasing growth, the phenotypic variance is expected to increasingly resemble the quadratic component in shape. In magnitude, the phenotypic variance will be greater than the quadratic component as a result of it containing the additional constant components of variation.

This model of growth variance represents the simplest pattern of variation arising from a set of unchanging conditions. When complicated patterns occur these can be interpreted as deviations from the basic pattern. An example comes from the Khartoum data. At about 150 days after birth there is a rapid increase in variance that is not predicted by the model, and in fact the model cannot be fitted to these data.

Complexities such as this can be investigated through correlation. Consider a set of correlations between birth-length and lengths at subsequent measurement intervals. In a stable situation in which infants retain their sizes relative to one another over time, correlations are expected to be positive and constant. In the Khartoum infants, however, these correlations are not constant, but decrease smoothly from a correlation of 0.8 between birth-length and length at 1 month, to 0 between birth-length and length at 10 months. This suggests that the Khartoum growth curves do not produce a simple pattern but are in some way more complicated.

To complement the analysis, correlations between birth-length and subsequent monthly increments can also be examined. These indicate how short periods of growth relate to an initial size, birth-length in this case. Rather than the correlations being positive and constant as predicted, in the Khartoum data they are all negative. The negative correlations indicate that some infants large at birth are growing slowly in comparison to their birth-size, and likewise some infants small at birth are growing rapidly for their initial size. The implication is that over these 10 months of post-natal growth, some infants are moving up and others moving down within the length distribution. The re-arrangement of the infants is so complete as to make length at birth unrelated to length 10 months later.

Some movement of this kind is to be expected in post-natal growth (see Part five). It is well recognized that a period of adjustment follows birth, and that these adjustments can result in negative correlations between birth-length and subsequent length increments. However, the Khartoum sample is complicated in at least one other way: in a stratification based on the occupational status of the mother (whether she holds a job or is a housewife). Infants of working mothers in this sample are larger at birth, but from 2 months onwards they begin to grow more slowly than do the infants of the non-housewives. As a result, after the first 6 months of growth the infants of the working mothers have become shorter than the infants of the housewives.

This shift in growth-rate at 2 months and a crossover in length at 6 months complicates the pattern of length variance generated after birth. The Khartoum variance increases at a slow rate when the two groups are converging in average length, and then suddenly increases after they cross each other and diverge. This crossing over in length is also visible in the variance of length distance. The dip in variation corresponds to the time of the crossing over, when the infants of the Khartoum housewives and working mothers are most similar in length. It is thus not surprising that the variance model does not fit the Khartoum data, as it could not anticipate such complexities. However, the unusual pattern of variation in Khartoum highlights a more complicated pattern of growth which leads eventually to identifying the relationship between growth and maternal occupational status. The relationship is very important in the Khartoum situation, and seems to reflect conflicting demands between household income and infant care.

When evidence of stratification arises it is useful to analyse the strata separately, subject to constraints on sample size. For the Khartoum data, the variance model can be fitted separately to the infants of the housewives and working mothers, suggesting simpler patterns of growth within each of the two groups.

*G. Brush and G.A. Harrison*

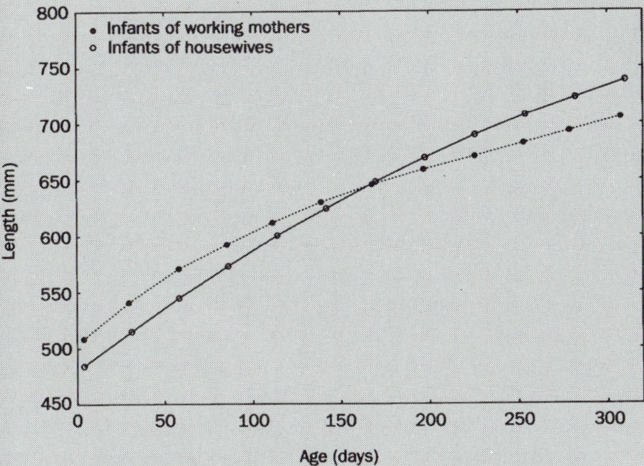


**Average length of infants of Khartoum working mothers and housewives, from birth to 10 months of age.**

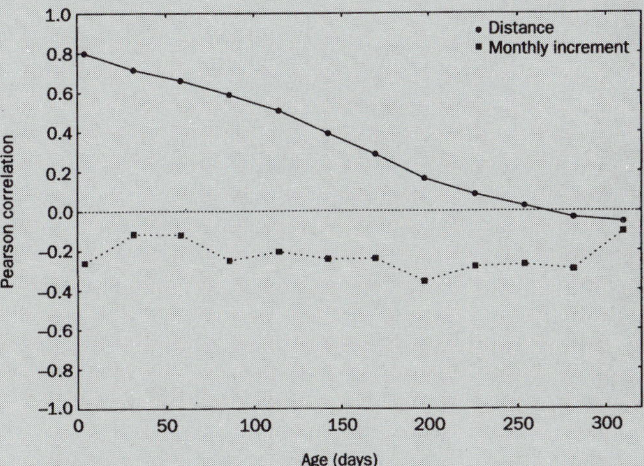


**Correlations between birth-length on the one hand and subsequent length distances, and one-month length increments on the other, in Khartoum infants.**

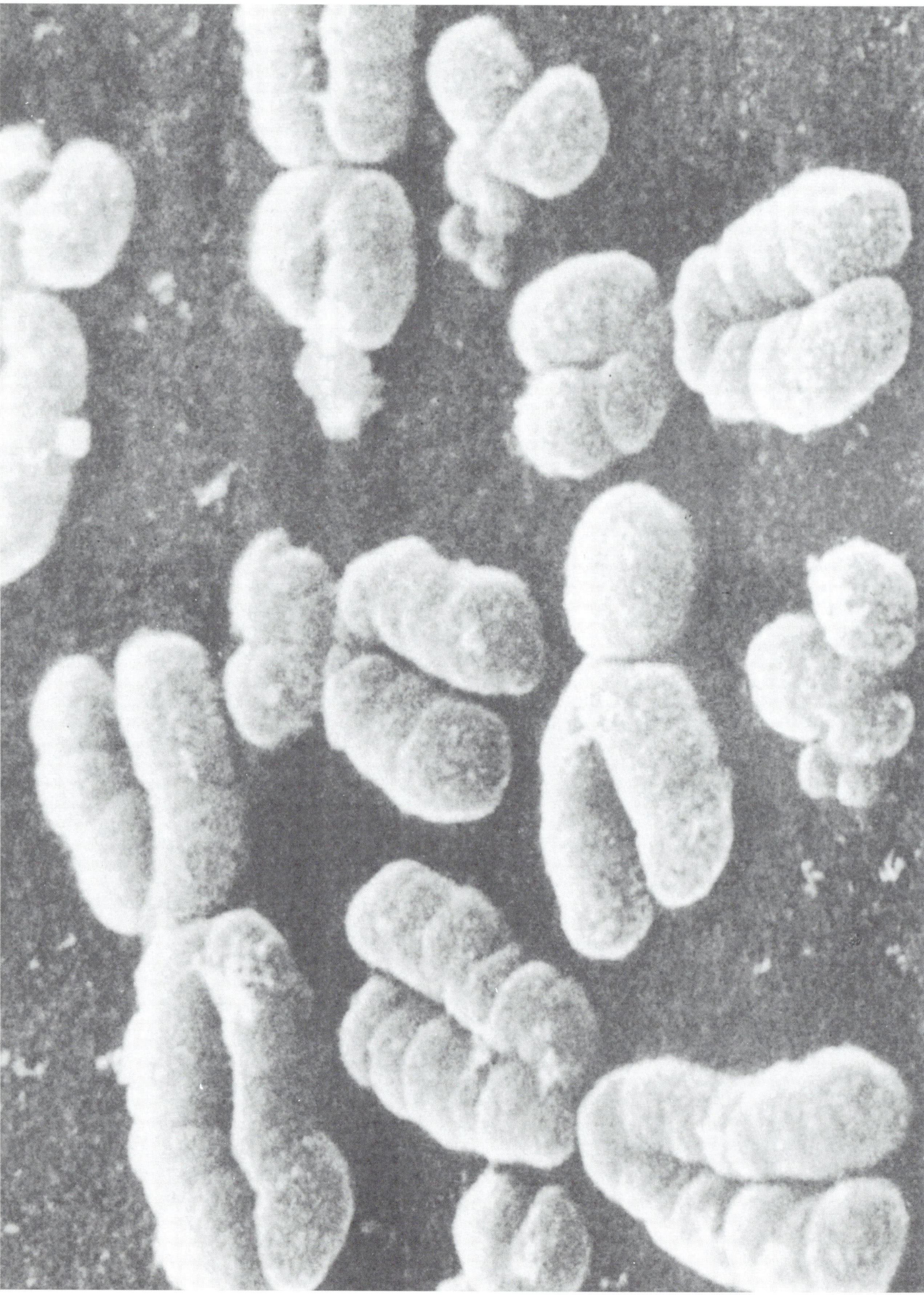

## PART THREE

# The genetics of growth

Although the familiality of growth and the development of physical characteristics is commonly acknowledged, the study of the genetics of growth has a fairly recent history. In 1918, Fisher proposed a model of multifactorial inheritance in which he postulated that quantitative traits, such as height, were determined by a large number of segregating genes, and by environmental factors. This model has remained central to the study of genetic and environmental factors influencing growth to the present day. Quantitative genetic analysis has been important in attempting to determine the extent to which genetic factors influence the observed variation in phenotypic traits, such as height and other skeletal measurements, as well as soft tissue measurements such as skinfold thicknesses. In particular, familial studies, including studies of the growth of twins, have been useful in identifying the extent of the genetic contribution to growth under different conditions. However, such studies can at best only give approximations of the genetic component to growth, since there are many reasons why aggregation of phenotypic traits happens in families. These include behavioural, cultural and economic factors which to some extent create a common environment. Thus, the similar growth and body-size of two brothers within a family might be as much a function of a shared nutritional environment as of genetics. However, a variety of study designs, including those examining relationships in body-size characteristics between: 1) sibs; 2) parents and offspring; and 3) twins, both monozygotic and dizygotic, suggests a strong genetic component to growth and development.

A number of these designs have been used to study the genetics of prenatal and post-natal growth, maturation, energetics, and body fatness (**3.1**). With respect to fetal growth (**3.2**), the study of the genetic contribution to growth is limited by the methods available to measure growth phenotypes. Despite this, considerable ingenuity has been employed in developing methods based on studies of fetal loss, gestational duration, familial aggregation of birth measurements, and the use of biochemical markers of fetal development.

More recently, the genetics of growth has been studied using molecular and biochemical approaches in an attempt to understand developmental mechanisms. These studies are in their early stages, but have generated considerable excitement because they hold the key to understanding fundamental biological processes. Common structures and mechanisms of development are found among different species of animals, and even between vertebrates and non-vertebrates. These include gene structures, types of gene regulation and interactions between cells. The identification of growth factors and their form and function has been of prime importance in the understanding of the genetic and environmental interactions which influence growth. This is because of their ubiquity, the peptide nature of most of them

**Opposite: Pairs of human chromosomes. Courtesy of Biophoto Associates/Science Photo Library.**

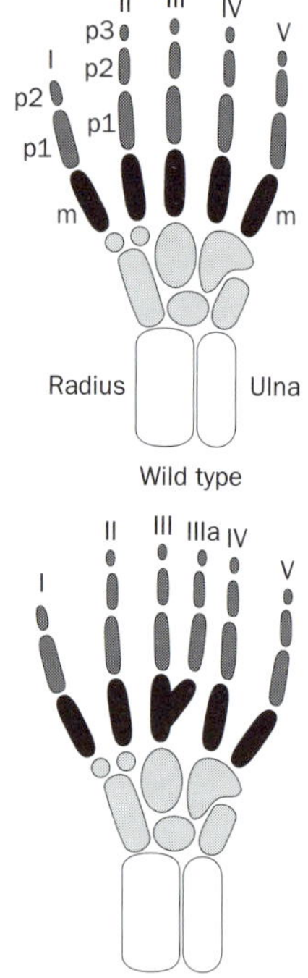


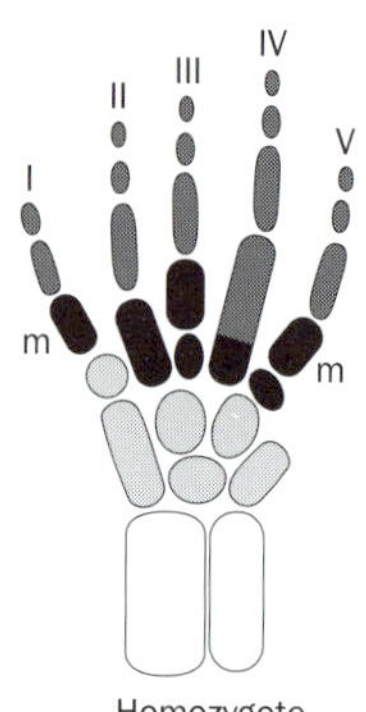


**Examples of hand skeletons from individuals heterozygous or homozygous for the *Hox D13* mutation responsible for synpolydactyly compared with a normal hand. Carpal bones are lightly shaded, metacarpals are black, phalanges (denoted by p1, p2 and p3) are heavily shaded. In the heterozgote, metacarpal III (black) is branched and gives rise to an extra digit IIIa. In the more severe phenotype of the homozygote, the metacarpals are fully or partially replaced by carpal-like bones. (Adapted from Muragaki et al. 1996)**

(which means they are directly produced by the genome), and because of the environmental sensitivity of the genome and of some of the hormones which regulate transcription of parts of the genome-coding for growth factors.

At another level, the study of individual genes with powerful and universally acting products like growth hormone is important for the understanding of gene structure and regulation, as is the role of hormones themselves on gene expression. Recently, there has been considerable progress in the identification of transcription factors and regulatory genes associated with growth (**3.3**). The genetic regulation of growth and development involves the coordinated activation of a variety of genes, but in order for a gene to be identified as being involved in growth regulation, it must be shown to lead to measurable phenotypic traits. Transcription factors are proteins coded by genes involved in the regulation of other genes, including those that produce the structural and processual components of the fully differentiated adult. Studies of the relationships between regulatory genes and transcription factors are improving the understanding of how, for example, the developmental architecture in the embryo is achieved and regulated. Furthermore, the study of the phylogeny of transcription factors has allowed them to be seen in an evolutionary context; of the many transcription factors in the human genome, several have arisen by gene-duplication events from a common ancestral gene, and have related functions.

Of particular importance is the homeobox gene family of transcription factors, which are key players in the development of most organ systems. Pioneering work on the homeotic genes of fruit flies by 1995 Nobel Prize winner, Edward Lewis, lead him to conclude that *Hox* genes, which are types of homeobox genes, have important roles to play in the control of morphogenesis during embryonic development. Mammals have 38 *Hox* genes, organized as 4 small clusters in on different chromosomes, and these share many features of the same type of gene in the fruit fly. Hox genes have been identified as being involved in specifying the regional identity of important skeletal and neurological structures in the developing organism. In vertebrate species, these include the vertebral column, hind-brain segments, gut, limbs, and genital structures. The *Hox* and other homeobox genes operate within a regulatory hierarchy, their expression being induced by other genes, including homeobox genes and genes responding to growth or transcription factors. Complex interactions among these various factors induce homeobox genes, and allow them to induce tissue-specific functions. These functions include cell division, cell adhesion and migration to create structures such as the neural crest and bronchial arches, and to stimulate morphological differentiation in the vertebrae, digits, limbs, and brain. While mutations in *Hox* genes have been considered strong candidates for causing a number of human developmental disorders, attempts to identify associations between *Hox* gene mutations and disease have met with little success. One exception, however, is the finding by Muragaki and colleagues in 1996 that mutations in the *Hox D13* gene of humans are associated with synpolydactyly, a rare, inherited abnormality of limb development.

The genetic control of development has two basic components: the regulation of genes by way of their control elements, and the action of those genes by way of protein products which recognize the control elements of other genes and either activate or repress their expression according to circumstance. The discovery that genes with growth-promoting protein products may be regulated by nutritional factors is an important one which provides a mechanism (perhaps the first of many) whereby genetic and environment factors interact in growth and development (**3.4**). The key to this process is thought to be by way of mediation of growth hormone action by the growth hormone receptor (GHR) and the gene coding for it.

Differential regulation of GHR gene expression can occur in a tissue-specific fashion, possibly mediated in part by the changes in thyroid hormone status induced by nutrition. Nutrition affects GHR gene transcription, and tissue-specific responses of GHR to nutrition suggest that a low food intake would result in reduced insulin-like growth factor-I (IFG-I) production in the liver, and hence reduced growth. Regulation of IGF-I gene expression may also be regulated by nutrition, as may other growth-promoting factors. In addition, recent studies have identified leptin, the gene product of the obese gene, as potentially playing an important role in the regulation of body-weight by signalling the size of the adipose tissue mass. Although this mechanism is being questioned collectively, such work is of great interest because it points to mechanisms which may control adaptation and responsiveness to the external environment (developmental plasticity), and provide additional fine-tuning of genetically programmed development.

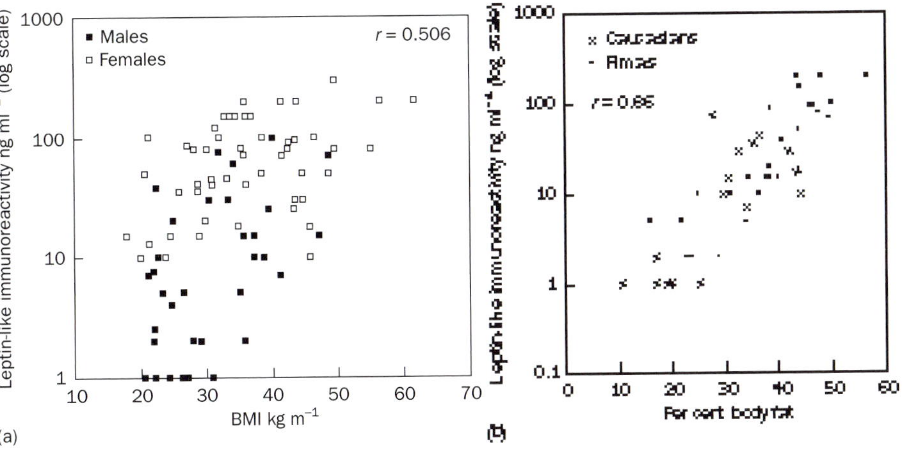


Leptin levels in humans. Above: Leptin levels in lean and obese humans. Leptin levels were measured in 87 lean and obese men and women and plotted relative to BMI (kg m$^{-2}$). There was a linear correlation between leptin concentration and BMI ($r$ = 0.506, $P$ < 0.001). The subjects included a sample of mixed ethnic origin (n = 59) and Pima Indians (n = 28). The sex of the subjects is indicated. At each BMI the levels were higher on average in women than men. This sexual dimorphism was not observed when the level was correlated with per cent body fat. Above right: Leptin levels vs. per cent body fat. The leptin levels were compared in 46 individuals with per cent body fat as measured by underwater weighing and there was a linear correlation of 0.86 ($P$ < 0.0001). (×) refers to Europeans and (●) refers to Pima Indians. Data from Maffei et al. (1995).

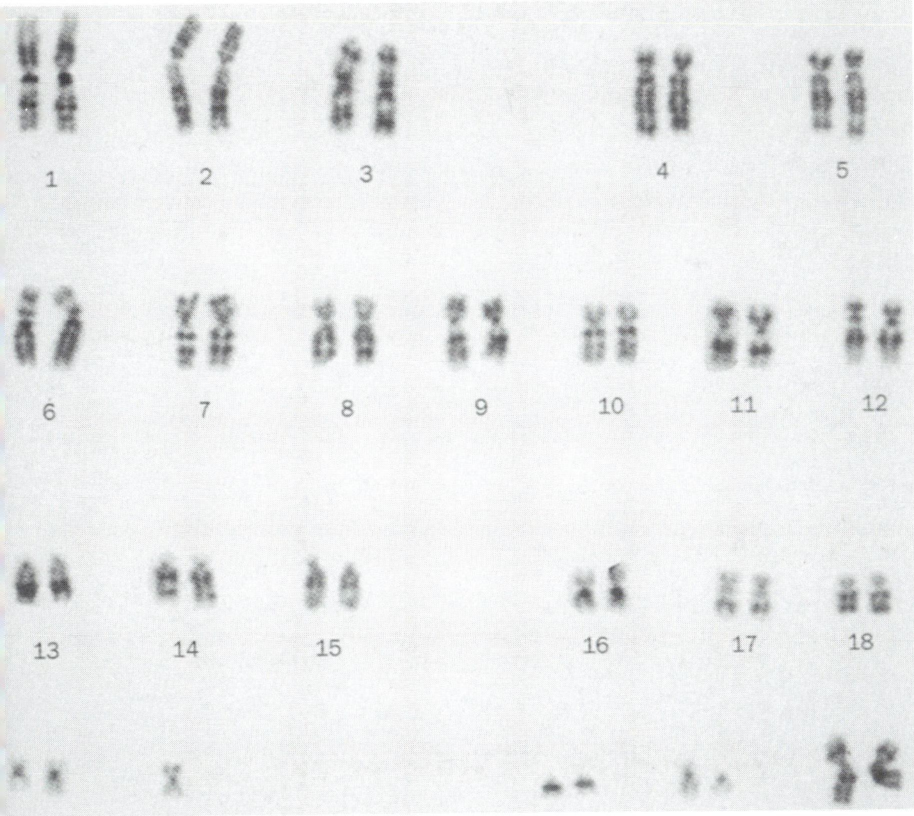


Above: A normal human female karyotype of 46 chromosomes prepared by the G-banding method and arranged according to size and position of the centromeres.

The study of genetics and growth, with its two approaches, has identified morphological and physiological traits which aggregate along familial lines, and has identified some of the mechanisms that underlie the growth process and its regulation. Furthermore, intriguing possibilities exist for the study of human growth plasticity at the genetic level. The gap between studies of relatedness in growth traits and those of genetic mechanisms of growth regulation is at first sight enormous. However, it is no wider than that which exists between cancer epidemiology and the genetics of cancer. In this case, the relatively new discipline of genetic epidemiology has proved promising in relating population-level characteristics with mechanisms of disease. One can hope that a new field, genetic auxology, might arise, and do the same for the study of human growth and development.

*Stanley Ulijaszek*

# Genetics of child growth

Human growth generally refers to changes in body-size and form, which take place from conception until attainment of final size, i.e. by approximately the end of the second decade of life. However, this is more a practical than a realistic definition of the human growth-cycle since several characteristics of the body may continue increasing slightly in size beyond the second decade of life, or decrease at a later age, as a consequence of senescence (**12.3**). Measures of growth, such as stature and weight, show continuous variation with every grade of intermediate. These are so-called quantitative traits for which the frequency distribution in the population often conforms more or less to the 'normal' or Gaussian distribution (**1.13**), as demonstrated by Quetelet (see Appendix 1, 7)in 1870.

When studying familial correlations of quantitative traits, Fisher, in 1918, proposed a model of multifactorial inheritance in which he postulated that quantitative traits were determined by a large number of segregating genes (with small, additive and equally important effects on the phenotype), and by environmental factors, also assumed to be numerous and of small effect. Mendelian laws for single-gene inheritance are valid for each of the contributing genes of a polygenic trait, but variation in allelic combinations (i.e. variation in genotypes), together with variation in environmental factors, result in practically indistinguishable phenotypic differences between genotypic classes. While it can be expected that a character, influenced by many segregating loci, will show a continuous distribution, the opposite is not necessarily true. Any feature that is controlled by a small number of genes with small differences between the genotypic means, but with relatively large environmental variation, will also show continuous variation of the phenotypes. As few as five or six genes suffice to produce such a distribution.

**Three generations of women in a Guatemalan family. Photograph by Barry Bogin.**

Although nowadays Fisher's work still lies at the basis of genetic theory of quantitative traits, it has sometimes proved to be of limited practical use because of various assumptions which are often not met in human familial data or which cannot be satisfactorily verified. However, recent knowledge of the genetics of growth and development is not solely based on the findings about familial aggregation of growth variables, but also comes from progress made in molecular genetics, physiology, biochemistry, analysis of single-gene polymorphisms, and the study of pathological growth in genetic disorders (chromosomal aberrations, single-gene mutations), as well as population comparisons of growth characteristics.

## The polygenic model and heritability

According to Fisher's model for inheritance of quantitative traits, the phenotype (P) of an individual can be considered as the sum of all that person's genetic factors (G), and the influence of a number of environmental factors (E), plus any interaction (GE) between G and E:

$$P = G + E + GE$$

The genetic component (G) reflects the combination of genes that a child gets from the parents, while the environmental component (E) relates to factors such as nutrition, hygienic situation, general psychosocial environment, climatic conditions, and cultural influences. Interaction between G and E occur if particular genotypes are favoured in specific environments. For instance, better nutrition may produce a greater increase in stature in genetically tall than in genetically small children. However, for many traits, the interaction GE is often considered as negligible and ignored in the model. Hence, the phenotypic variance of a trait in the population can be written as:

$$V_P = V_G + V_E$$

The genetic variance ($V_G$) can further be decomposed into the additive genetic variance ($V_A$), the dominance genetic variance ($V_D$), and the epistatic genetic variance ($V_I$), while the environmental variance ($V_E$) can be decomposed into the variance between means of families ($V_{EC}$), and the within-family variance ($V_{EW}$). The phenotypic variance thus becomes:

$$V_P = V_A + V_D + V_I + V_{EC} + V_{EW}$$

Dominance genetic variance ($V_D$) contributes to $V_P$ if heterozygotes are not exactly intermediate between homozygotes, while epistasis is the non-additive interaction among alleles at different loci also resulting in additional so-called *epistatic* genetic variance. Epistasis occurs if the effect of a gene at a particular locus depends on the alleles present at other loci. Although there is evidence that some traits may have other than purely additive genetic variance, it is often assumed that only additive genetic variance is in play. It is also known that many genes have an effect on more than one organizational level (pleiotropism).

In quantitative genetic analysis, the question is not whether genes actually play a role in the development of a trait, but how much of the observed variation of a trait is influenced by genes. The latter is quantified by the coefficient of heritability, i.e. the ratio of genetic variance over the total phenotypic variance:

$$H^2 = \frac{V_G}{V_P}$$

$H^2$ is also denoted *heritability in the broad sense* which is of interest in human population analysis, and should not be confused with $h^2 = V_A/V_P$, *heritability in the narrow sense*, which is useful in plant- and animal-breeding experiments. Heritability is not a property of a given character, but a quantity that depends on the genetic and environmental variation of the population in which it is estimated. Therefore, estimates of heritability are typical of a specific population, examined at a specific time under the specific environmental conditions of the examination period. The usefulness of heritability estimates lies rather in setting up rank orders of heritability of various traits within a population; it cannot be extrapolated from one population and set of environments to another.

Estimates of heritability coefficients can be obtained from familial data, in particular from simple correlations calculated between pairs of mono- and dizygotic twins, between parents and their offspring, between sibs, and between nephews and nieces. However, in humans, these estimates tend to be more or less inflated by the effect of varying degrees of similarities in the environment of these relatives (cultural inheritance), and by the effect of assortative mating. Parent-offspring correlations may also be diminished as a result of genotype–environment interactions, particularly when environments differ between generations. It is therefore very difficult to determine to what extent a particular trait is heritable or familial. Statistical techniques, such as path analysis and structural equation modelling, for instance, give opportunities to better separate the effects of genetic and environmental factors than classical familial correlations.

*Estimations of heritability coefficients from correlations* (**r**) *between relatives*

| | Formulae | Comments |
|---|---|---|
| Twins | | |
| | $H^2 = r_{MZ}$ | Less inflated by familial environment affected by assortative mating |
| | $H^2 = \frac{\sigma^2_{DZ} - \sigma^2_{MZ}}{\sigma^2_{DZ}}$ | |
| | $H^2 = \frac{r_{MZ} - r_{DZ}}{1 - r_{DZ}}$ | |
| Parent–child (p.c.) | | |
| | $H^2 = b_{mp.c}$ | Regression midparent–child (mp.c.) inflated by familial environment |
| | $H^2 = 2b_{p.c}$ | twice regression parent–child, inflated by familial environment and assortative mating |
| | $H^2 = \frac{2r_{p.c}}{1 + m_p}$ | m = correlation between spouses |
| Siblings | | |
| | $H^2 = 2r_{sibs}$ | Inflated by familial environment and by assortative mating |
| Nephews and nieces | | |
| | $H^2 = 4r_{nephews/nieces}$ | |

## Genetic control and single-gene action

The exact path from DNA replication to the development of a polygenic character is not known because the nature and the number of the genes, interacting with each other to produce an integrated whole, are unknown. The expression of an inherited trait also depends on an appropriate internal (physiological, metabolic, immunological, hormonal) and external (nutritional, physical) environment. Moreover, not all genes are active during the whole growth process. The sequence of different hemoglobins in humans, adapted to early and later fetal life, is just one example of the switch mechanism that turns on and off the synthesis of particular gene products at specific periods of development. Another example is the activation of genes responsible for the secretion of androgenic hormones at puberty.

The role of specific genes in human growth has partly become understood from the study of polymorphisms, and from single-gene diseases (achondroplasia, for instance). Most polymorphic genetic markers are in the blood, and few of these seem directly relevant to the growth process. However, some variants of glucose 6-phosphate dehydrogenase, an enzyme of the carbohydrate metabolism, have a slightly reduced physiological efficiency and may therefore affect growth-rate and contribute to normal variation in growth.

## Genetics of prenatal growth

Most knowledge about the genetics of prenatal growth does not come from serial ultrasonic measurements taken during pregnancy, but is obtained indirectly from studying the variation in birth-weight, the culmination of prenatal development. Important evidence for genetic control of prenatal growth has been provided by the observed effects of genetic defects (single-gene anomalies and chromosomal abnormalities) on the development of prematurely aborted embryos and fetuses, as well as on birth-weight.

Variations in the number of chromosomes (aneuploidy) and partial translocations of chromosomal material usually results in specific malformations and severely retarded overall growth of the fetus (trisomy 21 = Down syndrome; trisomy 18 = Edward's syndrome; trisomy 13 = Patau syndrome). Exceptions to that are the mosaic trisomy 20. Additional Y chromosomes (47, XYY males) have no noticeable effect on the size of newborns, but supernumerary X chromosomes do affect prenatal growth in both males and females (XXY = Klinefelter syndrome; 47, XXX = triple X females). It was found that the rate of cell replication in cultures of human fibroblasts was higher in cells with a 45, X genotype, than in cells with a normal 46, XX genotype, while the rate of cell division in 47, XXX cells was slower than in 46, XX. Although the second and subsequent X chromosome(s) are inactivated in early embryonic development, it is suggested that, at least in humans, control of the rate of cell division could be mediated by heterochromatin.

Although prenatal growth is clearly influenced by a child's genes, it is also, to a considerable extent, controlled by maternal factors such as maternal weight (**8.1**), age, health, and blood pressure, intra-uterine constraints, pregnancy spacing, parity, smoking habits (**8.3**) and alcohol intake (**8.5**) of the mother, and twinning. Only 20–40 per cent of the variability in birth-weight can be explained by fetal genes. The often greater discrepancy in birth-weight within pairs of monozygotic twins than within pairs of dizygotic twins typically reflects the strong effect of a variable intra-uterine environment on prenatal development. Indeed, the generally higher degree of vascular anastomoses between the placentas of monozygotic twins than of dizygotic twins often results in unequal sharing of placental blood and unbalanced distribution of oxygen and nutrients in the former. This explains the lower within-pair correlations at birth in monozygotic twins than in dizygotic twins. The correlation between parent's stature and the child's length at birth is

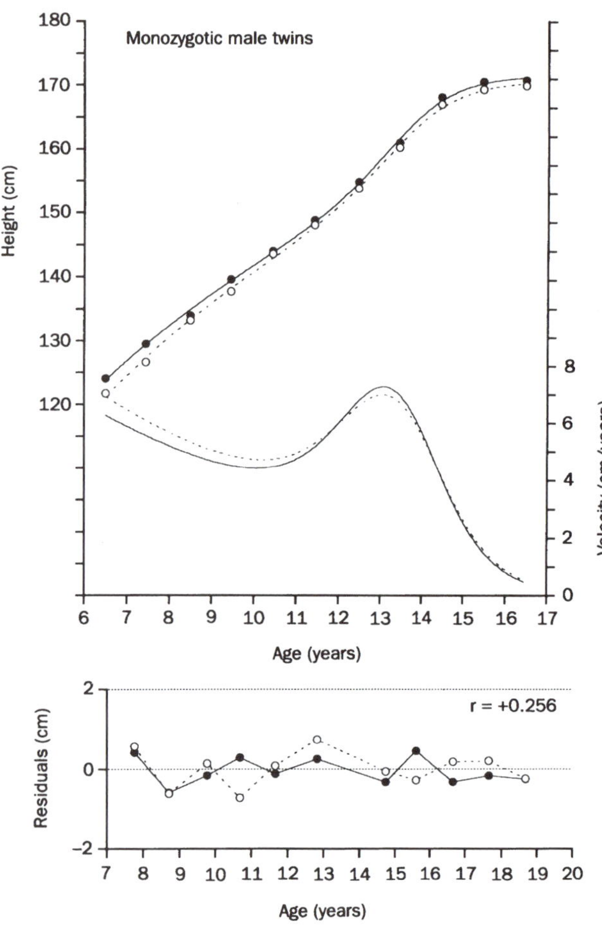


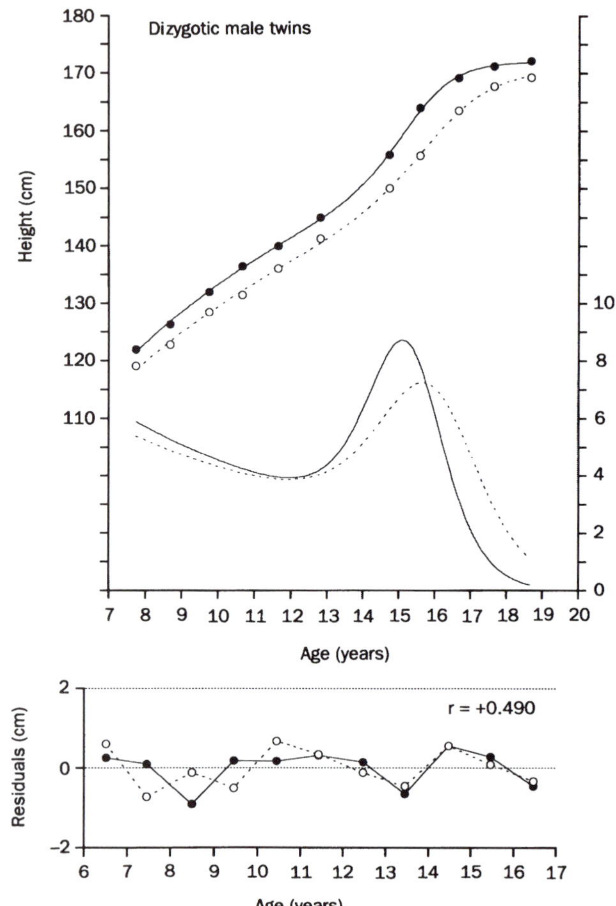


**Examples of growth patterns of monozygotic and dizygotic twins.**

**The monozygotic twins typically show a great similarity in the overall shape of their growth curve, attained height at each age as well as the timing of the adolescent growth spurt. Dizygotic twins typically show greater intra-pair differences in attained height and in the timing as well as the intensity of the adolescent growth spurt. Results obtained on samples of twins indicate a that there is a genetic component in the variability of the shape of the human growth curve (upper-part).**

**Short-term variations in growth are reflected in the residuals obtained after fitting a mathematical model to the observed serial growth data (lower part). The examples shown illustrate the greater similarity in the pattern of residuals among the monozygotic than among the dizygotic twin pairs. Based on larger samples, there is strong evidence for genetic control of short-term variations in growth velocity as well.**

also very low as a result of the variable intra-uterine constraints, but increases rapidly during the first 2–3 years of post-natal life, i.e. when the child grows to a channel determined by his or her genotype.

## Genetics of post-natal growth

Genetics of post-natal growth is mainly documented through studies of twins, familial data, and population comparisons. Intra-pair differences in height indicate that monozygotic twins are more concordant than dizygotic twins throughout the whole post-natal period. The convergence of the growth curves of monozygotic twins during the first few years after birth illustrates the strong genetic control on post-natal growth. The pattern of parent–offspring correlations during post-natal growth is also a clear demonstration of the gradually increasing impact of the child's own genotype on his or her growth. Shortly after birth these correlations are low, due to the relatively strong effect of the intra-uterine environment on growth, but then rise sharply during the first 2 years of post-natal life. Thereafter, these correlations increase further more slowly and even slightly decrease during adolescence, due to the variation in timing of the adolescent growth spurt.

The effect of genetic control on adult stature is considerable: the 95 per cent limits of normal variation around the mean for adult stature in males is 25 centimetres for unrelated individuals, 16 centimetres for brothers, and only 1.6 centimetres for monozygotic twins. Longitudinal growth data on twins also indicates that the overall shape of the post-natal growth curve as well as short-term variations in growth velocity are under genetic control. There is substantial evidence that children tend to grow along a genetically predetermined growth channel. Adverse environmental conditions (intra-uterine constraints (**4.8**), disease (**9.3**), malnutrition (**9.2**), psychosocial deprivation (**9.4**)) may temporarily push a child from its growth trajectory. However, abolition of the growth-disturbing conditions (release from the restrictions of the womb, appropriate medical treatment, better nutrition, improved psychosocial climate) usually result in a period of accelerated growth, or catch-up growth (**9.6**, **9.7**), until the original growth channel is reached again. Similarly, children suffering from overgrowth syndromes may show a catch-down after appropriate treatment. Growth is a target-seeking or self-stabilizing process.

The growth of a child seems to be very sensitive to quantitative changes of genetic material. Additional autosomal chromosomes not only result in reduced prenatal development, but also in reduced post-natal growth. It is believed that the short stature and the increased variation in developmental channels, seen in Down syndrome (trisomy 21), is the result of an abnormality of developmental homeostasis or canalization caused by the extra chromosome 21. Additional Y chromosomes, and to a much lesser extent extra X chromosomes, result in taller individuals, while the lack of sex chromosomes stunts growth. An adult male with a supernumerary Y chromosome (47, XYY) may be as tall as 200 centimetres, but adult Turner syndrome patients (45, X) are on average only 140 centimetres tall. Adult stature of males with two Y chromosomes is on average 13 centimetres above that of their fathers. This difference equals about the normal sex difference in adult stature noticed in most populations, which suggests that genes for this characteristic are on the Y chromosome. Tallness of individuals with supernumerary sex chromosomes is associated with long-leggedness because a significant proportion of the extra growth is acquired during childhood

**Brothers, sisters and cousins from one family. Photograph by I. Prince.**

**Siblings in one Australian family. Photograph by S. Scott.**

when the growth in legs is more important. Size and shape seem to be controlled separately, both by genetic and environmental factors. The genetic control of shape is much more rigorous than that of size, presumably because shape represents chiefly the number of cells in particular spaces, while size represents more the size of the various cells.

## Genetics of maturation

Children not only vary in size attained at all ages during the growth-cycle, but also in the time needed to complete their growth process, i.e. in tempo of growth. Children with a high tempo of growth, associated with an early onset of the adolescent growth spurt and a relatively shorter growth period, are said to have an early somatic maturation, and vice versa. Somatic maturation (**1.10**, **5.17**), as well as other maturational events (pubertal development, ossification of the bones, tooth formation and eruption, for instance) are all under varying degrees of genetic control, which seems, furthermore, to be independent of the genetic control of adult size, and, to a large extent, of final shape.

Familial studies and twin data clearly show that there is a strong genetic component in the timing and the shape of the adolescent growth spurt, one of the best studied features of growth tempo. The best-documented evidence for genetic involvement in sexual maturation comes from sister–sister and mother–daughter correlations for age at menarche (first menses), which is estimated between 0.25 and 0.5. Heritability of age at menarche is of the order of 0.80. In monozygotic twin sisters there is, on average, only 2 months difference in the age at menarche, while in dizygotic sisters and ordinary sisters this difference on average is 12 months. Twin data also show a greater similarity among monozygotics than among dizygotics for breast and testicular development, and to a lesser extent in the timing of hair appearance and voice change.

Skeletal maturation is also partly under genetic control. Variation in ossification sequence (**1.8**, **5.7**) occurs between individuals and between populations, possibly reflecting genetic polymorphism. Also tooth formation, eruption and eruption sequence show inter-individual and inter-population variations which are largely genetic. While it seems that sexual and skeletal maturation are to some extent correlated, and thus probably depend on common genetic factors, it appears that dental maturation, though under strong genetic influence itself, is fairly independent from skeletal maturation.

*R.C. Hauspie and C. Susanne*

See also 'Genetic and environmental influences on fetal growth' (**3.2**), 'Variation in time of tooth formation and eruption' (**5.10**), 'Hormonal regulation of growth in childhood and puberty' (**5.14**), 'Sexual maturation' (**5.15**), 'Adulthood and developmental maturity' (**5.17**) and 'Chromosome aberrations and growth' (**7.4**)

### GROWTH OF TWINS

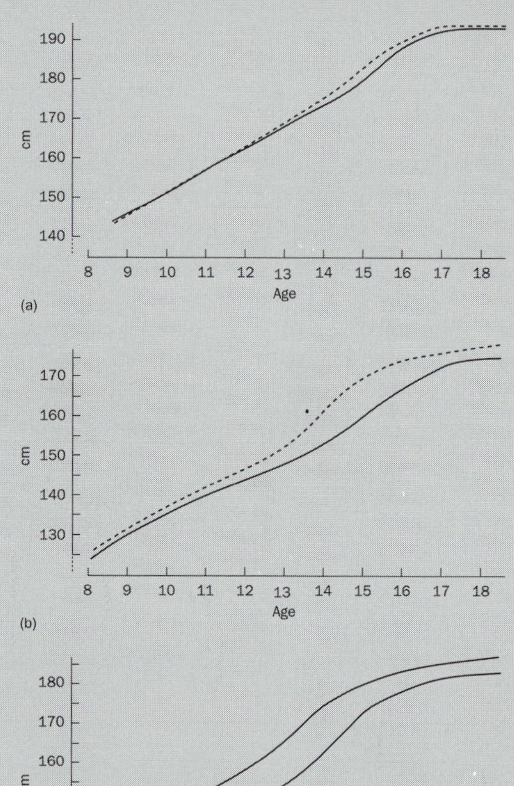

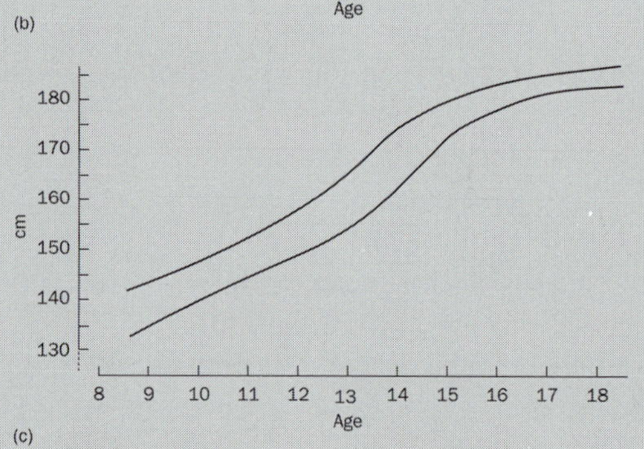


**Various patterns of similarity and dissimilarity growth trajectories of co-twins, as illustrated by individual curves of growth in height in three pairs of male twins.**

**Pair (a) Dizygotic twins differing both in stature (elevation of the curve) and in tempo of maturation (shape of the curve). The dotted partner remains taller at all ages but in addition he happens to be an earlier maturer: he experienced the pubertal spurt earlier and his spurt was more intense. The two effects are, in this case, superimposed, so that the statural differences between the twins become, temporarily, quite dramatic; they decline as the two partners approach adulthood. Evidence from a variety of growth studies indicates that height on the one hand and the timing and intensity of the spurt on the other are two uncorrelated features, probably controlled by two independent sets of genes.**

**Pair (b) Monozygotic twins whose growth curves closely coincide, although a very small transient difference appears during adolescence. This effect appears often in MZ pairs. It suggests a slight, temporary increase of the ecosensitivity of growth at the time when 'the second set of genes', that initiating the spurt, becomes active.**

**Pair (c) A very rare example of monozygotic twins whose growth curves differ markedly. The shorter twin had a much smaller birth-weight, during his childhood suffered a head injury, and went through several prolonged infectious diseases. Note, however, that these environmental onslaughts, although they depressed his stature, did not affect significantly the shape of the growth curve.**
**(Data from the Polish Longitudinal Twin Study).**

**The strong genetic determination of the development of various features of facial morphology is illustrated by serial photographs of two pairs of monozygotic twins. Each pair photographed at ages 9, 13 and 16 years. Photographs from the Polish Longitudinal Twin Study.**

Studies of twins shed much light on the strength of genetic versus environmental influences on growth processes. From the fact that monozygotic (MZ) twins are genetically identical, whereas dizygotic (DZ) twins have, on average, only one half of their genes in common it follows that growth trajectories, to the extent that such trajectories are genetically conditioned, should be more similar within MZ than within DZ pairs. This indeed proves to be the rule. For example, the mean intra-pair difference in menarcheal age (**1.15**), as recorded in a number of longitudinal studies of twins, is 8–10 months in DZ pairs but only 2–3 months in MZ pairs. The latter were found to be markedly more similar in many other aspects of growth as well, notably in skeletal maturity at any age, in age of attaining the successive developmental stages of secondary sex characters, in height for age, in the timing of the adolescent growth spurt (age at take-off and at peak velocity of growth) and in the intensity of the spurt (the magnitude of the statural gain during the spurt). These findings demonstrate that each of these features is under strong genetic control. On the other hand, the fact that differences between MZ partners in all these respects, although small, are in most cases non-zero is proof that developmental trajectories are determined not solely by genes but are also susceptible to modifications by certain environmental 'stimulants' or 'inhibitors'.

A number of techniques have been developed to utilize data on twins to numerically assess the relative importance of genetic versus environmental influences on various aspects of growth. Generally speaking, such techniques consist in comparing the magnitude of intra-pair differences, or intra-pair correlations, among MZ twins with those among like-sexed DZ twins drawn from the same population. The smaller such differences, or the stronger the correlations, for MZ pairs relative to those for DZ pairs, the higher the 'heritability' of the trait, i.e. the higher the contribution of genetic diversity (compared to that of environmental diversity) to the variation of the trait observed in the population. This method has certain limitations, two of which are mentioned here. First, heritability, as defined above, cannot be thought of as an absolute *property* of a trait: other things being equal, heritability will be *directly* related to the amount of genetic diversity (in the genes affecting a given aspect of growth) present in the population from which the twins were drawn; and it will be *inversely* related to whatever sociocultural factors are at work in the population which facilitate divergencies in some aspects of living conditions to develop between sibs in the family. Therefore, heritability estimates may vary between populations. Second, heritability may be affected by genotype–environment correlations, i.e., by the tendency for genetically identical siblings to adopt similar life styles either spontaneously, because of similar inborn predispositions, or as a result of social expectations. Correlations of this sort will tend to further augment the similarity in physical growth between MZ partners, though not between DZ partners, and thus to inflate heritability estimates.

*Pawel Bergman and Tadeusz Bielicki*

## SEX STEROIDS AND GROWTH HORMONE-RELEASING HORMONE (GHRH) GENE EXPRESSION

It has been over a decade since the discovery of GHRH, a hypothalamic peptide known to play an important role in controlling the synthesis and secretion of growth hormone (GH) (**5.1, 7.6**). Released into the hypophyseal-portal system, GHRH stimulates GH release from the anterior pituitary. The human GHRH gene has been mapped to chromosome 20q11.2. The human GHRH receptor has been cloned and characterized recently and mapped to chromosome 7p14. The secretion of GH is regulated by the interplay between the stimulatory effects of GHRH and the inhibitory effects of somatostatin (SS). The neurons producing GHRH are located mainly in the arcuate (ARCN) and ventromedial (VMH) nuclei of the hypothalamus, while the somatostatin neurons involved in GH secretion are located in the hypothalamic periventricular nucleus. Three different molecular forms of GHRH have been isolated from human hypothalami (11–44, 1–40 and 1–37) and all three possess biological activity.

### *GHRH AND SEX STEROIDS IN THE LABORATORY RAT*

Systemic growth and the GH secretory pattern are sexually dimorphic in the post-pubertal laboratory rat. This is partially due to the sexual dimorphism of the GHRH system. Adult male rats have significantly more hypothalamic GHRH peptide and GHRH mRNA than females, indicating a greater capacity for the production and secretion of this peptide. The differences in GHRH synthetic capacity become significant after the onset of puberty, suggesting the involvement of gonadal steroids.

Both the neonatal and adult steroid environments modulate the GH secretory pattern and systemic growth of pubertal and adult mammals. These effects are manifest at specific times during development and are mediated, at least in part, through modulation of the number, synthetic capacity and sensitivity of the hypothalamic neurons controlling GH secretion, including the GHRH neurons. Castration of adult male rats results in a significant decline in GHRH mRNA levels in the ARCN and VMH. These levels are restored to normal with testosterone. This could be attributed to an effect of estrogen produced by local aromatization of the testosterone. However, an androgen effect is more likely, at least in the ARCN, since the non-aromatizable androgen, dihydrotestosterone, restores these levels to normal. Estrogen did not

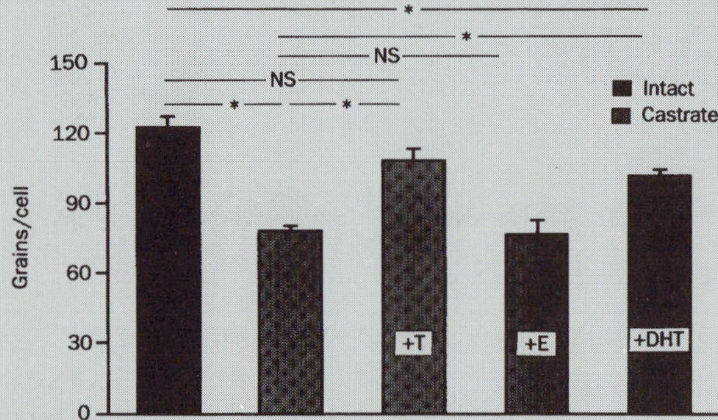


**Castration of adult male rats results in a significant decline in GHRH mRNA levels in the hypothalamic arcuate nucleus. Replacement with physiological levels of testosterone (T) prevents this decline. The non-aromatizable androgen, dihydrotestosterone (DHT), is capable of mimicking the effects of testosterone, but estradiol (E) has no significant effect on GHRH mRNA levels.**

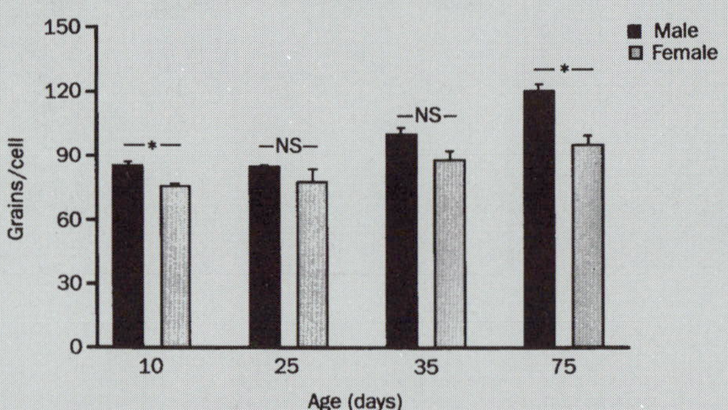


**Hypothalamic growth-hormone-releasing-hormone (GHRH) mRNA levels, as reflected in autoradiographic silver grains (per cell) throughout development of male and female rats. GHRH mRNA levels increase throughout maturation in both sexes. Male rats have significantly higher levels than females at 10 and 75 days of age.**

## GENETICS AND BODY FATNESS

Being overweight and having a centralized or truncal fat pattern are cardiovascular disease (CVD) risk factors which are associated to varying degrees with other CVD risks including dislipidemia, hypertension, insulin resistance and glucose intolerance. Furthermore, it is now well established that there are significant genetic influences on the amount and distribution of body fat. Although the genetic and physiologic links between different CVD risk factors have yet to be fully examined, it is reasonable to hypothesize that genetic predispositions for a number of adult onset CVD risks that have excess adiposity as part of their etiology likely begin to emerge during childhood (**12.1**). Increasing rates of adolescent obesity, especially in industrialized nations (**1.11**), provide impetus for studies elucidating the genetic and environmental sources of variation in the amount and distribution of body fat during growth and development. The primary practical goals of such studies are to identify as early as possible children at high risk for developing obesity and to devise appropriate intervention strategies.

Evidence suggesting that genes influence the amount and distribution of body fat (**5.16**) during childhood is provided by studies comparing members of different racial or ethnic groups living in the same locale. Such studies have generally found that significant racial or ethnic differences in measures of adiposity persist in spite of individuals being exposed to many of the same key environmental features, such as diet and level of physical activity. Numerous studies of patterns of familial correlations also suggest a genetic component to childhood adiposity, but such studies are not able to definitively separate genetic from shared familial environmental influences. More direct evidence for genetic involvement in the deposition of body fat during childhood comes from adoption studies, twin studies, or a combination of the two. The adoption and/or twin studies almost all concur that the variation in body mass attributable to shared familial environment is small compared to the effects attributable to the unique genetic constitution of individuals.

There are at present only a few studies, however, that use body-fat data collected during childhood to explicitly quantify genetic influences on childhood adiposity, and most of these are twin studies from which such quantitative genetic parameter estimates need to be viewed in light of potential limitations of the method used. None

affect GHRH mFINA levels. However, estrogen receptors have been reported in a subset of GHRH neurons, suggesting that this sex steroid does have a direct action on GHRH neurons.

Adult male rats have more GHRH neurons than adult females, neonatal castration of male rats results in a reduction in the number of GHRH neurons in the adult animal. Likewise, injection of a female with testosterone during the neonatal period increases the number of detectable GHRH neurons in the hypothalamus of the adult animal. Although adult sex steroids do not affect the number of GHRH neurons, testosterone stimulates GHRH mRNA levels in both adult females and males. Furthermore, the response of individual GHRH neurons to the adult steroid is greater if they had been exposed to sex steroids during the neonatal period.

#### *GHRH AND SEX STEROIDS IN THE HUMAN*

Circulating GHRH peptide levels increase throughout puberty in humans. Furthermore, both estradiol and testosterone are capable of increasing serum GHRH levels, suggesting that sex steroids are also capable of stimulating GHRH gene expression in humans.

*Jesús Argente and Julie A. Chowen*

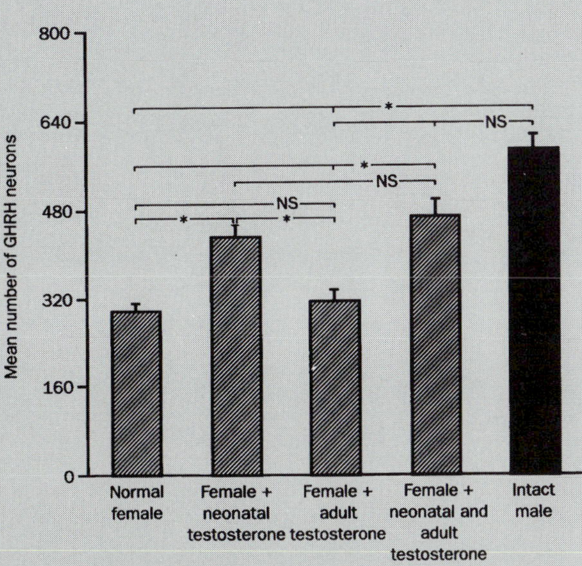


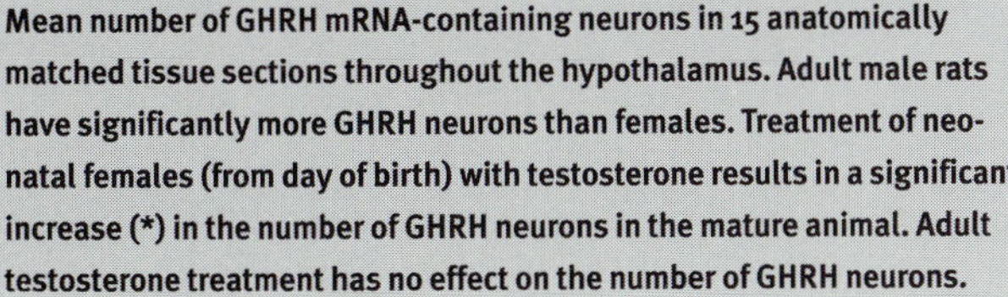

Mean number of GHRH mRNA-containing neurons in 15 anatomically matched tissue sections throughout the hypothalamus. Adult male rats have significantly more GHRH neurons than females. Treatment of neonatal females (from day of birth) with testosterone results in a significant increase (*) in the number of GHRH neurons in the mature animal. Adult testosterone treatment has no effect on the number of GHRH neurons.

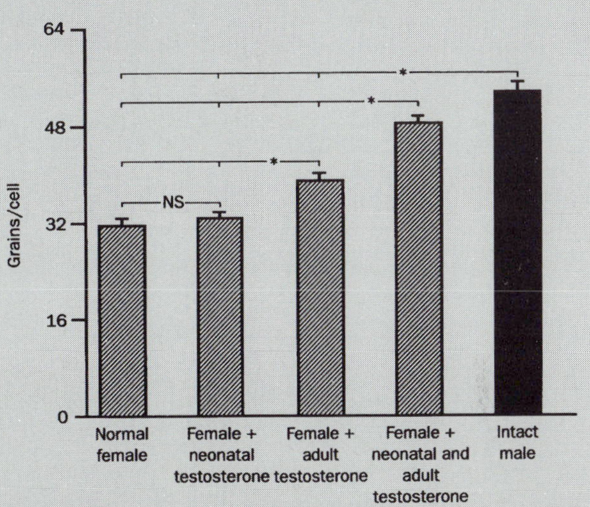


**GHRH mRNA levels in the hypothalamus. Growth-hormone-releasing hormone (GHRH) mRNA levels (as reflected in grains/cell) in the hypothalamus. Normal adult females have significantly lower levels of GHRH mRNA per cell than normal males (*). Neonatal treatment of females (day of birth) with testosterone has no effect on GHRH mRNA levels in the adult animal. Adult testosterone treatment of normal females stimulates GHRH mRNA levels. However, this effect is significantly greater if the animal had previously received neonatal testosterone.**

the less, the findings of these studies are in general accord with those from other types of studies and suggest that there are significant genetic effects on different measures of adiposity that are manifested to one degree or another throughout childhood. Brook and colleagues (1975) examined a large sample of twins between the ages of 3 and 15 years and derived heritability estimates (the proportion of the observed variance in a trait accounted for by the genetic relationships among family members) of 0.98 and 0.46, respectively, for trunk and limb subcutaneous fat measured by skinfolds. Cardon (1994) examined early- and middle-childhood data from twins and found that genetic influences on the body mass index (BMI) (**1.11**) are expressed during these stages of development. Bodurtha et al. (1990) examined a large sample of 11-year-old twins and obtained heritability estimates of 0.75 for triceps skinfold, 0.77 for subscapular skinfold, and 0.87 for the BMI. Using longitudinal data from the Medical College of Virginia Twin Study sample of Bodurtha and colleagues (1990), Meyer and colleagues (1996) found that the high heritability of the BMI seen in the 11-year-olds remains high throughout their adolescence up to the age of 17 years. Similarly, in a family study of 325 children from over 100 kindreds in the Fels Longitudinal Study, Towne and colleagues (1996) found significant heritabilities of 0.64 for both the pre- and post-pubertal growth spurt BMI.

Specific challenges for future genetic epidemiological studies of the accumulation of body fat during childhood will involve using new and more precise measures of adipose tissue deposition in more sophisticated statistical genetic analyses. These analyses will incorporate sex differences and age changes in body-fat accumulation during growth and development, as well as quantifications of pertinent environmental factors, such as nutrient intake and energy expenditure. Combined with increasingly refined molecular marker maps of the human genome, the potential exists to identify and locate specific major genes affecting body-fat accumulation during childhood and to determine the magnitude and timing of their effects.

*Bradford Towne*

## GENETICS AND ENERGY METABOLISM

Energy expenditure reflects aspects of human physiological function, many of which are related to physical growth. Human energy expenditure can be broken down into a number of components, including the energy cost of maintenance, thermic effect of food (TEF), physical activity, thermoregulation, growth, and reproduction. These components vary between individuals and populations, and differ according to the extent to which they can be modified by human behaviour.

Intraclass correlation coefficients for BMR (kJ/kg FFM) in MZ and DZ twins

| | MZ | DZ |
|---|---|---|
| Henry et al. 1990 | 0.85 | −0.04 |
| Bouchard et al. 1989 | 0.77 | 0.30 |
| Fontaine et al. 1985 | 0.45 | 0.21 |

Evidence for genetic adaptation in energy expenditure comes from studies of parent–offspring relationships and from twin study designs. A familial dependence of resting metabolic rate has been shown in Pima Indians in Arizona, while various studies have shown significant heritability of basal metabolic rate (BMR). Intraclass correlation coefficients for BMR expressed per kilogram of fat-free mass largely remove associations due to twin similarities in body size (**5.5**) and composition (**5.11**), and show substantially lower values in dizygotic than monozygotic twins, suggesting an important genetic component to maintenance metabolism.

Genetic characteristics influencing various components of energy expenditure have been inferred from the study of Canadian twins. For physical activity, a heritability coefficient of up to 25–30 per cent of total variance has been found, while the mechanical efficiency of physical work has a significant genetic component at levels of exercise below six times resting metabolic rate. Furthermore, there is a significant genetic contribution to the TEF with intraclass correlations coefficients of 0.35 and 0.52 for dizygotic and monozygotic twin pairs respectively. Bouchard and colleagues (1991) have concluded that the genotype accounts for a significant fraction of the individual differences in resting metabolic rate, TEF and energy cost of exercise, even after age, sex, body mass and composition are taken into account. They also conclude that body energy-gains during overfeeding vary among individuals, and are probably genotype dependent. If this is the case, then it is not unreasonable to suggest that variation in response to underfeeding may also have a genetic component. It is also possible that variation in growth-rates of children according to nutritional circumstance may also have a genetic component, although this remains untested.

*Stanley Ulijaszek*

# Genetic and environmental influences on fetal growth

Understanding genetic and environmental influences on the fetus is limited by the methods used to study human growth and development. These methods include: 1) fetal loss and birth defects, particularly the involvement of aneuploidy (unusual numbers of chromosomes) and inborn errors of metabolism; 2) epidemiologic studies of factors affecting intra-uterine growth and gestational duration; 3) biometric studies of familial aggregation of birth measurements, and 4) biochemical development of the fetus with particular regard to genetic polymorphisms.

## Fetal loss and birth defects

The prevalence of aneuploidy is as high as 50 per cent in spontaneous abortions in the first 3 months of pregnancy. This reduces to 8 per cent by six months and is about 2 per cent in newborns. Spontaneous abortions are quite frequent in humans occurring in about 30–50 per cent of all conceptions, many of these going undetected as very early abortions or as delayed menstrual periods. This natural biological mechanism for selection for normal development works less well in women of very young childbearing age and also in those of 35 years or older. After the age of 35, the risk for a live-born child with trisomy (an extra chromosome – usually #21 or others of very small size) increases significantly.

Monosomy of the X-chromosome is also relatively common. Children with trisomy #21 have lower weights-for-gestational-age so there is a general effect on growth. Infant death and birth defects are more prevalent in children of consanguineous marriages, suggesting an effect of recessive genes. However, most of these deaths occur after birth when the newborn can no longer depend upon maternal metabolism.

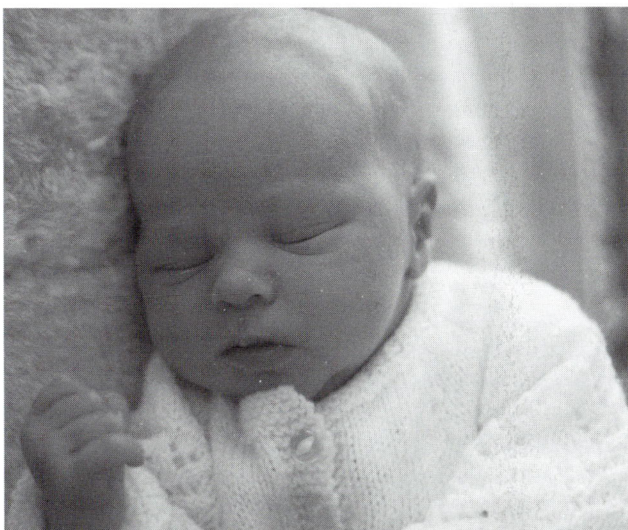

**A recently born baby. Photograph by Stanley Ulijaszek.**

## Epidemiologic studies of intra-uterine growth

The importance of factors affecting intra-uterine growth may vary according to the general level of economic development of the society: low birth-weight in industrial countries is largely due to short gestation (**4.9**); while in agricultural and developing societies most of it is related to intra-uterine growth retardation (**4.8**). There are 21 separate factors shown to be consistently important in influencing intra-uterine growth and gestational duration, out of a total of 43 factors studied in nearly 900 publications. In general, effects of each factor on birth-weight are in the range of 100 to 300 grams reduction, somewhat less than 10 per cent of the mean. Causal, important and modifiable factors include prepregnancy weight (**8.1**), gestational weight-gain, caloric intake, cigarette smoking and tobacco chewing (**8.3**), as well as young maternal age (**8.6**). Factors of potential importance which have not been thoroughly studied are occupational fatigue, psychosocial stress and the availability and use of medical care. There are fewer factors associated with gestational duration than birth-weight. A limitation of epidemiologic methods is the large error in gestational age evaluation.

*Established Determinants of Intra-uterine Growth and Gestational Duration*

| Intra-uterine growth | Gestational duration |
|---|---|
| **Direct** | **Direct** |
| Infant sex | Prepregnancy weight |
| Racial/ethnic origin | Prior prematurity |
| Maternal height | Prior spontaneous abortion |
| Prepregnancy weight | *In utero* diethylpstilbestrol exposure |
| Paternal height and weight | Cigarette-smoking |
| Maternal birth-weight | |
| Parity | |
| Prior low-birth-weight infant | |
| Gestational weight-gain | |
| Caloric intake | |
| General morbidity | |
| Malaria | |
| Cigarette-smoking | |
| Alcohol consumption | |
| Tobacco-chewing | |
| Altitude | |
| **Indirect** | **Indirect** |
| Young maternal age | Young maternal age |
| Socio-economic status | Socio-economic status |

From Kramer, M.S. (1987). *Pediatrics* 80:502–511.

## Biometric studies of familial aggregation of birth-size and gestational age

Such studies are almost always concerned with birth-weight because of the availability of data. Gestational age has been studied as well, though not nearly as much as size. Genetic influences on gestational age are very low or have not been evaluated adequately. Restrictions in most countries on the routine use of ultrasound have precluded this as a research tool to get at other aspects of fetal body composition (brain, skeletal, muscle, and fat tissue). Fetal weight is increasing in the last trimester of pregnancy, so that it is most likely to be affected by factors present at this time. Fetal length (skeletal mass) grows fastest in mid-pregnancy, and the fetal brain has two peaks of rapid growth-rate: one early in pregnancy, which reflects the growth of neurons; and another near birth, which reflects the growth of the supporting structures of the brain. These rapid periods of growth are, in a quantitative sense, potentially sensitive periods when interference from the environment could be important. Thus, severe mental retardation resulted from *in utero* exposure early in pregnancy to atomic bomb radiation. On the other hand, the effect of smoking on birth-weight is more apparent in mothers who smoked during the last 6 months of pregnancy. Women exposed to the famine of 1944–45 in war-torn Holland had babies with reduced birth-weight when exposure was in the last 3 months of pregnancy. These babies grew up to have no increased risk of having low birth-weight infants. When exposure was in early pregnancy, birth-weights of children were normal, yet these same children grew up to have a greater risk for lower birth-weight babies. This suggests that windows of vulnerability, organ and tissue specific, are in effect early in pregnancy and may be significant predictors of later health.

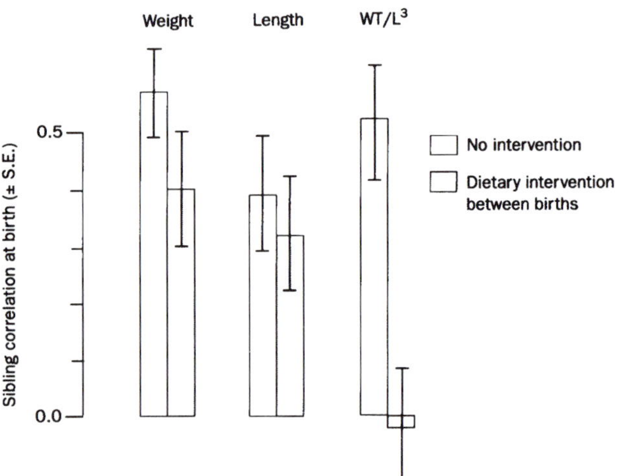


**Sibling correlation in birth measurements in two groups: one in which there has been dietary intervention during pregnancy of second child, the other no such intervention. Sibling correlations are depressed in the experimental group relative to controls, suggesting that the height-to-length relationship at birth is more reflective of environmental factors than either weight or length alone. Data from W.H. Mueller and E. Pollitt. (1982). *Human Biology* 54:455–468**

Sibling correlations of birth-weight are commonly reported. Correlations for weight, length and weight/length[3] (ponderal index) at birth for two groups of siblings (one exposed to nutritional stress, the other in which mothers were given an energy supplement during pregnancy) from a

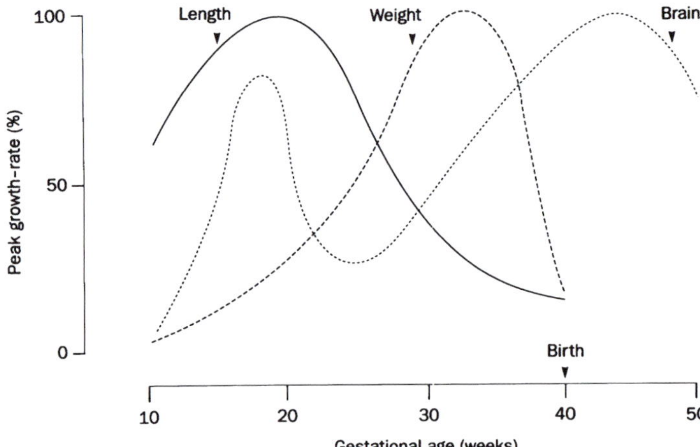


**Variation in growth-rates of three components of fetal development. Adapted from J.M. Tanner, *Foetus into Man*, and J. Dobbing and J. Sands (1970). *Nature* 226:639.**

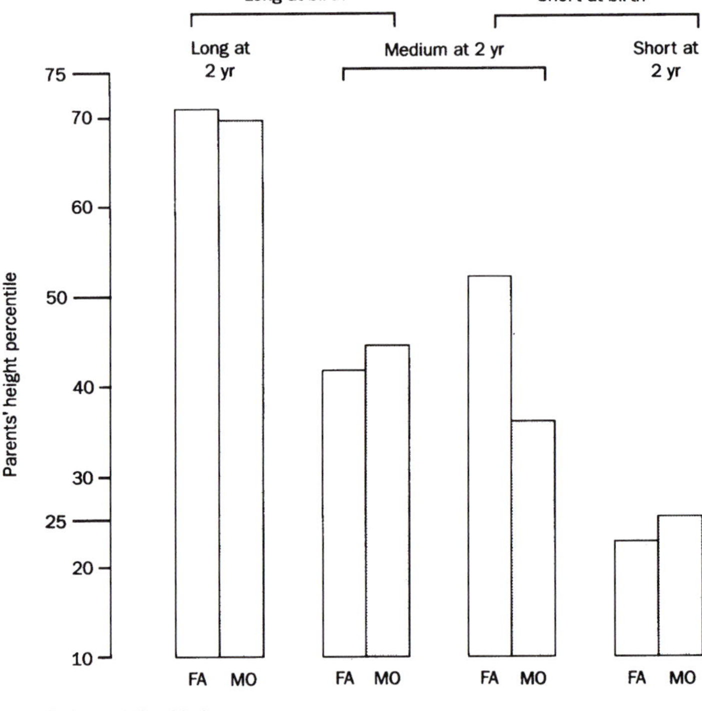


**Child's length status at birth and at 2 years, relative to parents' height percentile. The child's birth-length linearly follows the mother's height percentile, but not the father's. However, both the mother's and the father's heights are important in a child's height at 2 years, especially in children who started off small or large, but end up of average size at age 2 years (middle 2 bars). These data suggest that maternal body-size to some extent regulates the growth of the fetus and, after birth, genetic factors of the child play an increasingly important role in length variation. Data from D.W. Smith et al. (1976). *Pediatrics* 89:225–230.**

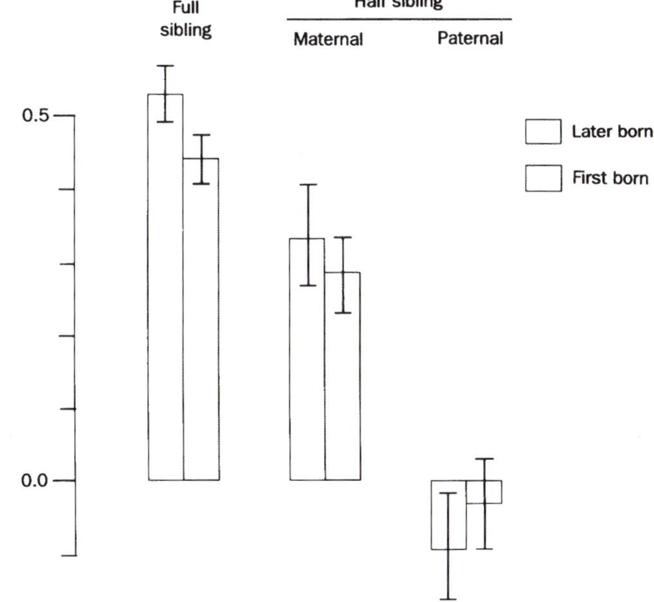


**Birth-weight correlations for sibling pairs, involving a first-born and pairs involving only later born sibs. Both fetal genetic and maternal factors are evident in the data; genetic because half-sibling correlations are lower than full-sibling correlations. Full siblings have the same mother, but some effects of parity on the correlations are evident. Correlations between pairs of siblings involving a first-born are less than those involving only pairs of later-borns. There seems to be something physiologically unusual about a first birth. This fits with the observation that birth-weight tends to increase with birth order up until about five children. Half-siblings may either have the same mother or the same father. Correlations are evident for maternal half-siblings but not paternal, suggesting the inheritance of intra-uterine factors. What these are specifically it is impossible to say, but they must have to do with maternal size and health. Data from Nance et al. (1983). *American Journal of Human Genetics* 35:1211–1223.**

rural population at risk of mild seasonal malnutrition have identified interesting relationships between genetic and environmental factors. Sibling correlations were depressed in the supplemented group compared to controls, as expected on the supposition of a genotype and environment interaction. This was particularly notable for the weight/length$^3$ index, much less so for either weight or length. This suggests that the relation of weight to length at birth is more reflective of environmental factors than weight or length alone. The reduced sibling correlation in weight/length$^3$ has been linked to birth season, suggesting seasonal nutritional factors in the pattern.

In most mammals, including humans, the intra-uterine environment and maternal body-size are important regulators of fetal growth, but after birth, particularly in the first 2 years, the genetic factors of the child play an increasingly important role in variation of body-size. Of several studies dating from the 1950s, there is considerable variability in the estimates of the influences of fetal genes, maternal factors and random environmental effects on fetal growth. The role of fetal genes varies from practically nil to 51 per cent, maternal factors from 27–50 per cent, and random environment from 8–43 per cent in the variation in birth-weight. Most of the studies indicate a maternal effect which may reflect maternal genetic as well as intra-uterine environmental factors.

## Genetic polymorphisms

Genetic polymorphisms may manifest in antigenic substances (e.g. blood groups), proteins of various functions and DNA polymorphisms. The rhesus (Rh) blood group polymorphism was the second one discovered because of its role in hemolytic disease of the newborn which results in severe anemia and usually still-birth. Potentially, this

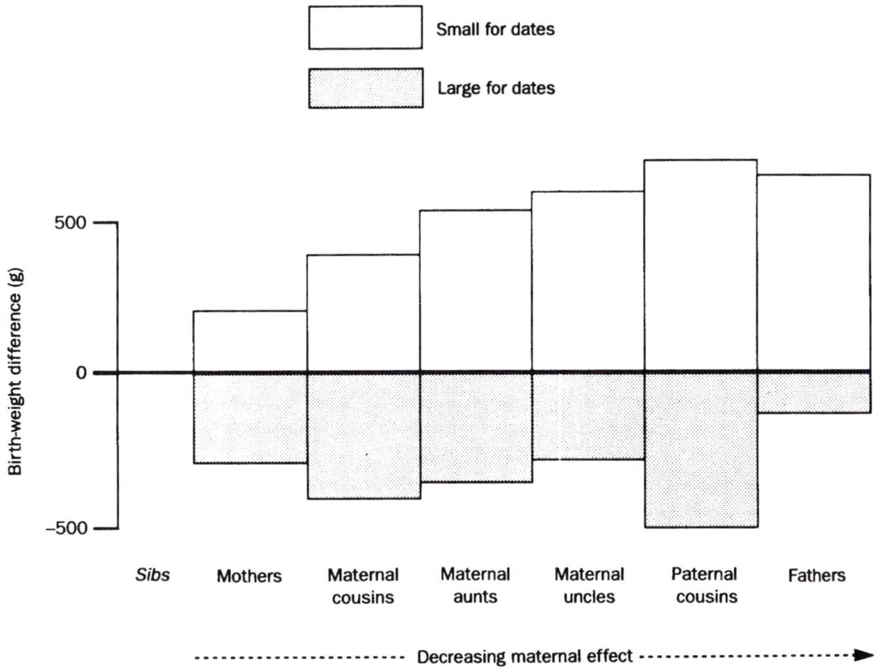


**Average birth-weight difference between various relationships and siblings of small- or large-for-gestational-age infants. As one moves from left to right, the involvement of maternal genes decreases with type of relationship. If maternally transmitted factors are important, one should see differences increase from left to right. In fact this happens only where growth was restricted *in utero* (small), but not in large-for-dates infants. Thus both maternal genetic and fetal influences are seen in this figure at different levels of birth-size. Fetal genetic factors are evident in the small difference between father's birth-weight and birth-weights of siblings of large-for-dates infants. Finally, environmental factors, such as smoking, may alter the familial correlations in birth-weight, but this is rarely taken into account. Data from M. Ounsted (1974). *British Medical Journal* 4:163.**

can occur when an Rh negative mother carries an Rh positive fetus and when the mother's immune system becomes sensitized to the presence of her baby's positive antigens, usually from previous births. The Rh blood group system is a known genetic polymorphism with affects on fetal health. Beyond this, few of the polymorphisms have demonstrable direct affects on fetal growth. Some protein polymorphisms come in different forms that vary through fetal life. The hemoglobin molecule is an example. Hemoglobin carries oxygen in the red blood cells, from which it is distributed to all cells. Fetal forms of this hemoglobin exist which favour the attraction of oxygen from the maternal circulation; the principal fetal hemoglobin is Hemoglobin F(HbF). HbF rises in fetal life and gradually falls off near and after birth to be replaced gradually by adult hemoglobin (HbA). This provides a model for molecular changes that occur in physiologically important substances in fetal life which we know are polymorphic in some adult populations. Promising variants are the family of esterase enzymes, of which the qualitative changes during fetal life seem to correlate with central nervous system development, particularly the development of myelin.

*William H. Mueller*

See also 'Growth of the human embryo' (**4.3**), 'Developmental morphology of the embryo and fetus' (**4.5**), 'Fetal growth retardation' (**4.8**), 'Standards and references for the assessment of fetal growth and development' (**4.9**), 'Chromosome aberrations and growth' (**7.4**), 'Maternal anthropometry and birth outcome' (**8.1**), 'Nutrition' (**8.2**) and 'Body-size at birth' (**10.3**)

# Homeobox genes

Complex physiological, histological, and morphological traits arise during development by the co-ordinated activation of context-specific sets of genes, in a process called *gene regulation*. Among many factors that are involved in gene regulation are genes called transcription factors. The proteins coded by *transcription factor* genes activate or repress the expression of other genes, including those that produce the physical and chemical components of the differentiated adult structure or system.

There are many transcription factors in the human genome. Many are members of gene families, having arisen by gene-duplication events from a common ancestral gene and that have related functions. The transcription factors work by recognizing various types of short DNA sequence or *control elements*, usually located near the coding region of the gene whose expression is being regulated. The regulatory proteins physically bind to the control element DNA sequence and, in combination with regulatory proteins bound to other elements, cause the nearby regulated gene to be expressed (transcribed into messenger RNA), or repressed, in that cell.

One important family of transcription factors is the *homeobox* group of genes. These genes code for proteins that contain a 60 amino-acid DNA-binding region, known as the *homeodomain*. Homeobox genes play various roles in development, but the family derives its name from a subset involved in specifying the morphological identity of segmental morphological systems. Mutations in these genes cause *homeotic* changes of segment identity in the embryo. The first known were flies with legs in the place of antennae or wings on thoracic segments that do not normally have them; similar natural mutations had long attracted the attention of evolutionary biologists. Similar effects have been observed in mammals, such as morphological shifts of elements of the vertebral column or limb, although the mutations involved are generally not known.

*Selected multimember homeobox classes and some of their known roles in invertebrate development*

| Gene class | Aspects of expression pattern or function |
|---|---|
| Hox | Hind-brain, limb, vertebral column, blood, gut |
| Dlx | Limb, mid- and forebrain, teeth |
| En | Hind-brain, craniofacial region, pharynx |
| Msx | Limb, eye, brain, teeth |
| POU | Pituitary, CNS, general cellular response growth hormones |
| Pax | Brain, eye, kidney, teeth |
| Otx | Brain |
| Emx | Brain |
| Evx | Central axis, caudal region |
| Nkx | Brain, teeth |
| Barx | Gut, teeth |
| Csx | Heart muscle |

There are tens of classes of homeobox genes, descendants of an ancestral gene that first contained a homeobox region. These are found in all animals tested to date and in many plants. Vertebrate homeobox gene classes consist of varying numbers of known members. Each class has characteristic conserved amino acid variants in its homeodomain as well as other parts of the protein, and the members often have related function.

## *Hox* genes in axial specification

The role of homeobox genes can be illustrated by the subclass of about 40 related genes known as *Hox genes* in vertebrates. The *Hox* genes are involved in specifying regional identity in several important skeletal (**5.6**) and neurological structures, and are arranged in a contiguous linear array on a given chromosome. This arrangement is conserved in animals as diverse as chordates, arthropods, nematodes, and molluscs.

The pattern of expression of *Hox* genes is similarly conserved across taxa. The expression domains of *Hox* genes along the main body axis of a developing embryo are co-linear with their position along their respective chromosome. The genes on a given chromosome form nested domains of expression such that each has a unique anterior boundary of expression but overlaps in its posterior expression with other genes. Moving along the chromosome from one end, genes have progressively more posteriorly restricted domains of expression. Thus, anterior parts of the body express a few *Hox* genes, while more posterior regions express more. In addition to this spatial co-linearity, there is also a temporal co-linearity. More posteriorly restricted genes are expressed later in development than their more anteriorly expressed chromosomal neighbours. In general, the genes control the morphological identity of regions near their anterior expression boundary. This phenomenon is known as *posterior prevalence*; in more posterior regions of the body, several *Hox* genes from a given chromosome may be expressed, but the identity of a structure is dependent on the most posteriorly-restricted gene expressed there. Although posterior prevalence appears to explain much experimental data, in some cases the results are more consistent with a combinatorial model in which the combination of genes expressed at a given axial level determine its morphology (**5.5**).

Existing knowledge of the function of *Hox* genes in vertebrate development is derived from 1) detailed experimental manipulations of their homologues in the fruit fly *Drosophila*; 2) experiments to reveal the spatial and temporal patterns of *Hox* expression in a particular vertebrate species; and 3) transgenic animals in which altered *Hox* genes have been introduced to (a) 'knockout' the function of a gene, (b) expand its expression to ectopic regions, or (c) study the expression patterns of 'reporter' genes fused to the control elements of a *Hox* gene (such a reporter produces a compound that can be visually stained in all cells in which the gene is normally expressed).

In vertebrates, *Hox* genes appear to be responsible for determining regional identities along a number of body axes. These include the vertebral column, hindbrain segments, branchial (gill) arches, gut, limbs, and genital structures. For example, knockout or overexpression experiments have changed the identity of, or entirely eliminated, vertebrae, limb bones (like the radius and ulna), or digits. In most cases, patterning is along an anterior–posterior (A–P) axis, but in the limbs, *Hox* genes appear to specify pattern along both the A–P and proximo-distal (P–D) axes. They are also involved in the differentiation of hematopoietic (blood-forming) cell lineages.

Vertebrates differ from most other animal taxa in possessing four separate *Hox* clusters, each on a different chromosome. In the hindbrain, genes occupying similar positions on different chromosomes (members of paralogue groups) exhibit similar restrictions of expression along the A–P axis. Expression of paralogue group members along the A–P axis of the vertebral column is also similar, but may be somewhat more divergent than in the hindbrain. In the limbs, nested expression domains are characteristic of the members of the *Hox A, C,* and *D* clusters. The expression domains of *Hox D* cluster members tend to be oriented along the A–P axis and those of the *Hox A* complex more along the P–D axis, although both change over time during development. The expression of *Hox C* genes is somewhat more complex and has not been as well studied.

Even in cases where the A–P expression limits are similar among paralogues, there may be differences in expression within tissue layers or along the dorso-ventral (D–V) axis. It is likely that paralogues exhibit partial functional redundancy, because of their evolutionarily related sequence similarities, but are also responsible for distinct functions. Knockout experiments of paralogues tend to result in overlapping but distinct phenotypes and the phenotypes of double (two-gene) knockouts are more severe than either mutation alone.

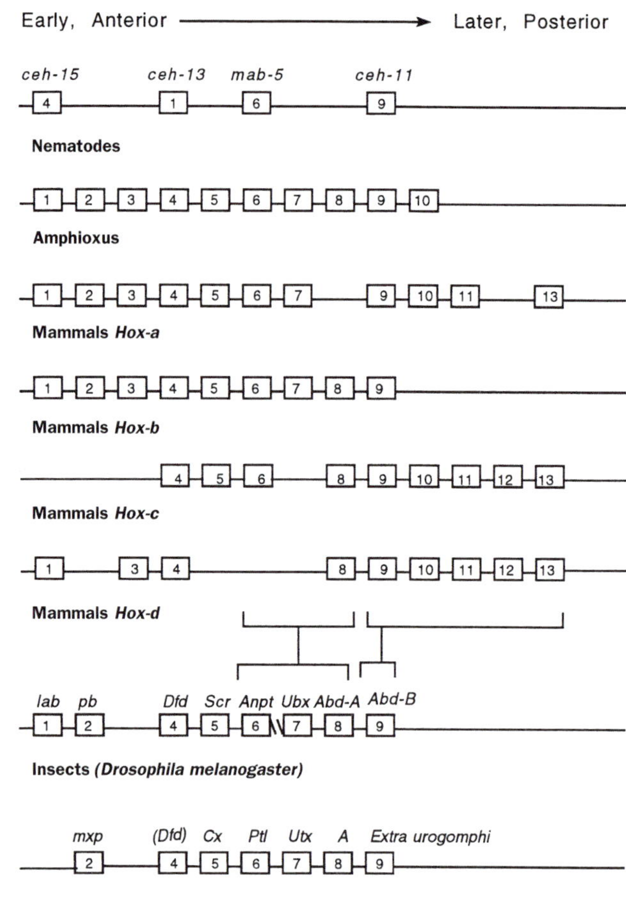


**Phylogeny of the *Hox* gene cluster as found in selected species. Genes (boxes) are arrayed in their known chromosome order. Numbers in boxes refer to paralogue groups; that is, genes in the clusters that descend from the same gene in a common ancestral cluster, based on the amino acid sequence of their homeodomain regions. Conventional gene names are given for the corresponding invertebrate genes. Multiple descendants of single genes are indicated by the brackets relating to insects or to vertebrates. The top line shows genes expressed more anteriorly, or earlier in embryogenesis.**

## Mechanisms for complex regulation

The *Hox* and other homeobox genes operate within a regulatory hierarchy. Their expression is induced by other genes, including homeobox genes and genes responding to growth or transcription factors like the zinc-finger gene Krox-20, the secreted signalling proteins and growth factors like *EGF, FGF, BMP, SHH* and *WNT*; and genes responding to small molecules like retinoic acid. In turn, homeobox genes activate other target genes like cell adhesion (*N-CAM*) factors, and may even activate themselves.

Complex interactions among these various factors induce homeobox genes, and allow them to induce tissue-specific functions. These functions include cell division (blood cells), cell adhesion and migration (neural crest, branchial arches), programmed cell death (interdigital area), and morphological differentiation (vertebrae, digits, limbs, brain). However, to date few downstream target genes have been identified in mammals.

At least some of the regulation of homeobox genes occurs through the recognition of short DNA control elements within 1000 or so base pairs of the genes themselves. Experimental manipulation of these regions in *Msx, Dlx, Hox* and other homeobox genes shows that different elements are responsible for different aspects of a gene's expression; experimental deletion of one element removes only some aspects of the gene's expression. These regions can be highly conserved among taxa, so that upstream regulatory regions from an amphibian homeobox gene can properly regulate the corresponding gene in mammals, or mammal regulatory regions those in flies.

There is overlap in function among paralogue groups of homeobox genes, which may protect the animal by functional redundancy. However, the expression of paralogous genes does not overlap completely, so that each gene has its own distinct patterns. It appears that it may be the expression as much as the protein function *per se* of paralogous genes that is responsible for their different functions. As an example, mice have two En genes, both of which are expressed in the embryonic brain. However, En-1 is turned on earlier than En-2, and mice with an En-1 knockout mutation have several defects of the brain that are lethal. Experimentally replacing the En-1 protein-coding region with that of En-2 so that En-2 is expressed in a manner similar to the natural pattern of En-1 resulted in normal mice.

Thus, it is proper to think of development has having two basic components: the regulation of genes in terms of their control elements, and the action of those genes' coded protein products to recognize other genes' control elements to activate or repress their expression context-specifically. However, our understanding of the genetic mechanisms behind the development – and hence, the evolution – of complex phenotypes is still young and incomplete. One major question is the origin of organ-system differentiation, including specification of serially homologous segmental systems. Homeobox genes are involved in both systems.

*Kenneth M. Weiss, David W. Stock and Zhiyong Zhao*

See also 'Human growth from an evolutionary perspective' (**2.3**), 'Growth of the human embryo' (**4.3**) and 'Developmental morphology of the embryo and fetus' (**4.5**)

## HOMEOBOX GENES: EVOLUTIONARY ISSUES

The Hox cluster arrangement evolved early in animal evolution. This arrangement and its consequent temporal and spatial co-linearity of expression with structural differentiation along anatomic axes has been conserved for hundreds of millions of years as a tool for specification of position and/or time during development. However, the details of what develops in a given position vary tremendously among species.

Evolution has used a highly conserved tool to regulate very malleable downstream function. These points can be illustrated with some examples from experimental studies. *Hox* genes from paralogue group 5 demarcate the position of the forelimbs or pectoral fins, while genes from group 9 demarcate the thoracic-trunk boundary, but the number and morphology of vertebrae, and the appendicular structures within each region, vary greatly. Aspects of pentadactyly in tetrapods appear to relate to the regional expression patterns of five members of the *Hox D* cluster in the distal part of developing limbs, although the situation is complex and not yet clearly understood.

A pair of *Distal-less* (*Dlx*) genes is chromosomally linked to each of three mammal *Hox* clusters. We do not know the reason for this conservation. However, the *Dlx* genes are all expressed in the developing branchial arches, and/or teeth, in somewhat differing temporal and spatial patterns, and *Dlx* genes along with members of other classes seem to be segmentally expressed in the developing forebrain. Thus, these classes of homeobox genes are thought to have been important in the evolution of the vertebrate head, where Hox genes are not expressed.

Almost incredibly, many homologous homeobox genes are expressed in physiologically similar structures in invertebrates and vertebrates. *Dlx*, *Otx*, and *Emx* genes are expressed in head, mouth, and limb sites in flies and vertebrates. The similar homeobox gene expression, and regulatory interactions, found in mammal nervous system, heart, eye, limb, gut, kidney, and blood-forming systems also have somewhat histologically related function in invertebrates. *Pax-6* gene has similar roles in the development of eyes in both classes of animal. These patterns reveal a deep level of 'homology' in animal evolution in the toolkit for organ-system development.

Studies of the homeobox gene expression patterns is still rather recent, but experiments have already confirmed ideas derived from paleontology and comparative embryology about the evolution of the head skeleton, segmentation of skeletal and dental system and brain (**10.1**), and the evolution of the tetrapod limb. For example, experimental mutations have generated 'atavistic' structures, that resemble evolutionarily earlier states of these traits. These genes clearly have been of major importance in mammalian evolution.

A number of human diseases have already been identified that seem to be caused at least in part by mutations in homeobox genes. Mutations in *Msx* genes can affect cranial suture closure and palatofacial development. Anomalies in Pax genes produce developmental disorders of the kidney and (in both vertebrates and invertebrates) the eye, are involved in some neural and muscle cancers, and cause craniofacial and pigmentation anomalies and deafness in Waardenburg syndrome. Experimental *Hox* mutations in mice are associated with developmental anomalies in the bony structures of the ear. In humans, benign conditions such as extra cervical or lumbar ribs mimic *Hox* experimental mutations, and traits of digit number or formation, such as polydactyly or brachydactyly exist and mimic changes generated experimentally by *Hox*-gene mutations in transgenic mice.

*Kenneth M. Weiss, David Stock and Zhiyong Zhao*

## EVOLUTION OF A GENE SYSTEM

Different animal phyla have related sets of Hox genes. These genes are thought to have evolved by cluster duplication, such that the *paralogue groups* among taxa or clusters within taxa, are descendants from the corresponding gene in an ancient Hox cluster. Within arthropods there are varying numbers of these genes, in one or two different clusters of adjacent genes. Likely modern-day representatives of basal chordates, like *Amphioxus* and tunicates, have a single *Hox* cluster. Lampreys appear to have two or more clusters, while bony fish and land vertebrates have four, with each cluster on a different chromosome.

Because of their conserved chromosomal arrangement and expression pattern along the major anterior–posterior morphological axis, the *Hox* gene expression system has been viewed as a fundamental characteristics of all animals, known as the *zootype*. There is a time in the embryological development of members of each animal phylum, termed the *phylotypic stage*, when all of the major body parts are represented as undifferentiated tissue condensations. As a general rule, the stage of development in which the zootype in an animal is manifest – all *Hox* genes expressed in their proper relative positions in the embryo (**4.5**) – is also the phylotypic stage.

The other homeobox gene classes have fewer known members and less conserved chromosomal arrangements. However, two members of the *Dlx* ('distal-less') class, and one *Evx* ('even-skipped') gene are chromosomally linked to several Hox clusters. It is thought that an ancestral cluster had four Hox genes, and perhaps three other homeobox genes (including *Dlx* and *Evx*) were also at one time part of such a cluster. A number of conserved arrangements that include other developmental or growth-related genes are linked to at least some of the *Hox* clusters.

*Kenneth M. Weiss, David W. Stock and Zhiyong Zhao*

## GROWTH HORMONE GENE DELETIONS

Other categories of genes are also important in growth and development and considerable progress has been made in characterizing genes coding for growth hormone (**7.6**). The growth hormone (GH) and chorionic somatomammotropin (CSH) genes reside in a 50 kilobase gene cluster located on human chromosome 17q22–q24. The cluster includes (5' to 3') the *GH1* gene, chorionic somatomammotropin pseudogene-1 (*CSHP1*); *CSH1*, *GH2* and *CSH2* genes, as well as 48 *Alu* repeats. These genes all consist of five *exons* and four *introns*. The *GH1* gene encodes normal pituitary GH, while *GH2* encodes a protein that is expressed in the placenta and differs from the primary sequence of GH by 13 amino acids. The *CSH1* and *CSH2* genes encode proteins of identical sequences, whereas *CSHP1* encodes a protein differing by 13 amino acids. A high degree of sequence homology (92~98%) exists within the immediate flanking, intervening, and coding sequences of these five genes, suggesting that this multigene family arose through a series of gene duplication events. These duplications are thought to have arisen from successive unequal recombinations between *Alu* family repeats.

One form of familial isolated GH deficiency (IGHD) called IGHD IA, has an autosomal recessive mode of inheritance and is caused by deletion of the *GH1* genes. Affected individuals may have subtle intra-uterine growth retardation manifest by reduced height compared to weight. Hypoglycemia occasionally occurs in infancy and severe dwarfism is apparent by six months of age. Typical facial features include a bulging forehead and a small nose with retracted nasal bridge. Many IGHD IA subjects initially respond well but within a few months poorly to exogenous GH replacement therapy due to their formation of anti-GH antibodies because of their complete lack of endogenous GH.

Initially, genomic DNA deletions of ~ 6.7, 7.0, or 7.6 kilobases including the *GH1* gene have been detected by Southern blot analysis in individuals with IGHD IA. Most subsequent subjects (~ 75%) have deletions of ~ 6.7 kilobases (1). Homologous recombination between sequences flanking *GH1* has been proposed as the common underlying mechanism for these various sizes of GH1 deletions. Interestingly, IGHD IA associated with a double deletion of the *GH* and *CSH* gene cluster has also been reported, but the mechanism of its origin is unknown.

Currently, *GH1* deletions are detected by polymerase chain reaction (PCR) amplification of the homologous regions flanking *GH1* and detection of the fusion fragments associated with *GH1* deletions by restriction enzyme digestion. Published studies suggest that ~ 15 per cent of subjects with severe IGHD (less than −4.5 SD in height) have *GH1* deletions. Recently, frameshift and nonsense mutations of the *GH1* gene have been found in some subjects with the IGHD IA phenotype. These mutations also cause complete GH deficiency and, in some cases, affected subjects have developed anti-GH antibodies. Finally, a compound heterozygote with a 6.7-kilobase deletion including one *GH1* gene and a two-base deletion in exon III of the other *GH1* gene has been reported.

Heterogeneous *GH1* deletions can be detected by Southern blot or PCR analysis. Homozygosity for these or heterozygosity in combination with other *GH1* gene mutations that prevent any *GH1* expression cause a severe form of IGHD called IGHD IA. Clinically, such affected individuals present with 1) severe growth retardation, 2) absence of GH secretory response to provocative stimuli and 3) good initial growth response to exogenous GH often followed by poor growth due to anti-GH antibodies. In the cases which become refractory to exogenous GH, treatment with insulin-like growth factor I (IGF-I) (**4.2**) has proven to be one effective alternative therapy.

*Ichiro Miyata and John A. Phillips III*

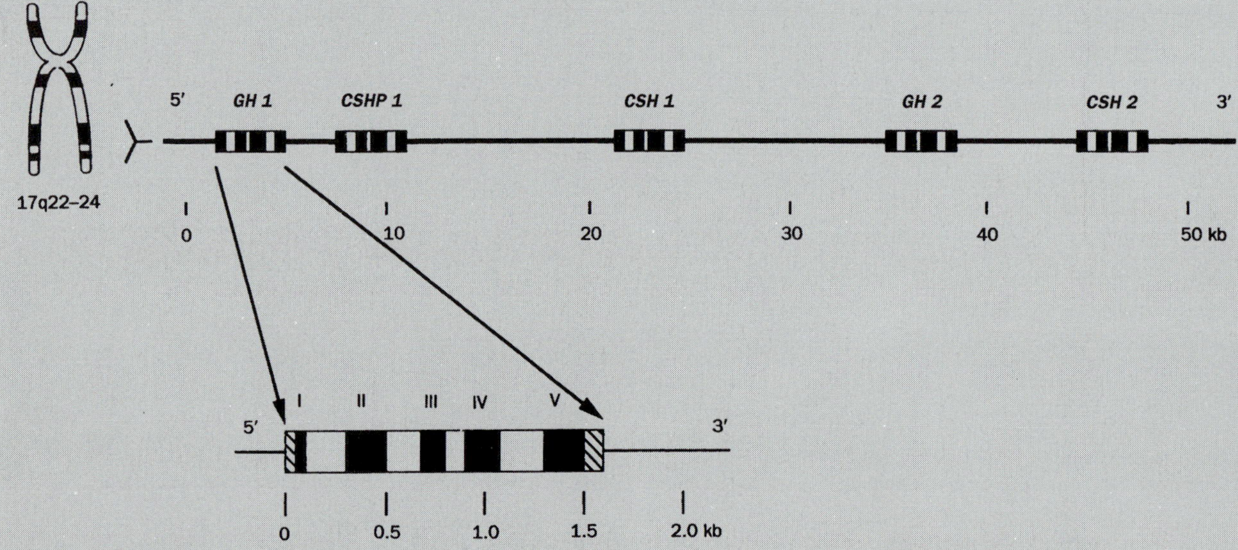


**Schematic representation of the *GH* gene cluster and its localization on human chromosome 17. Exons, introns, and non-translated sequences are depicted by solid, open, and shaded rectangles.**

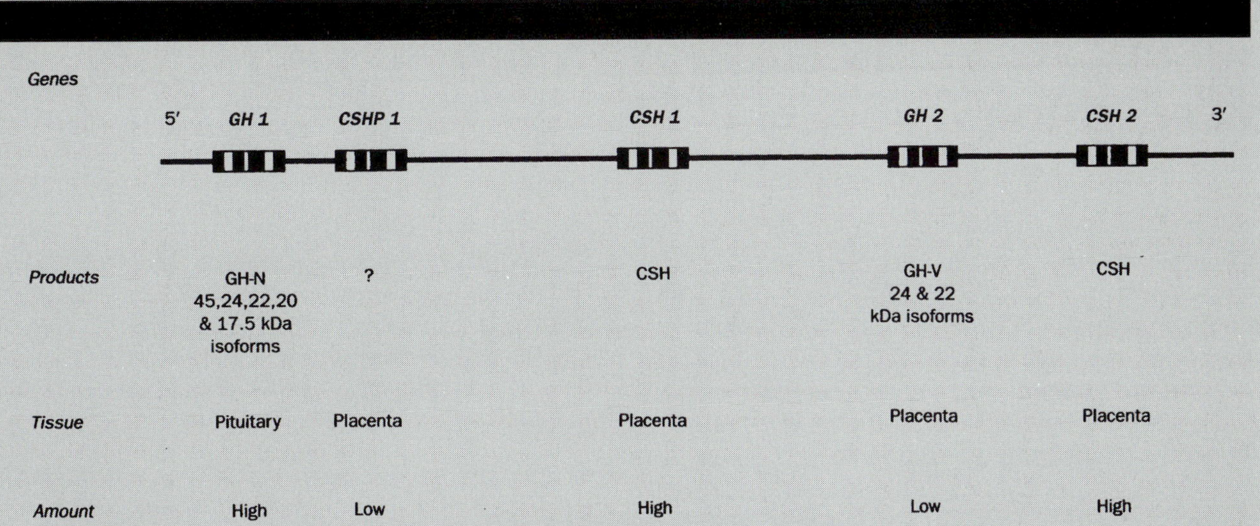


The hGH gene cluster is shown at the top and the corresponding protein products and the sites and relative amounts of their expression are shown below it.

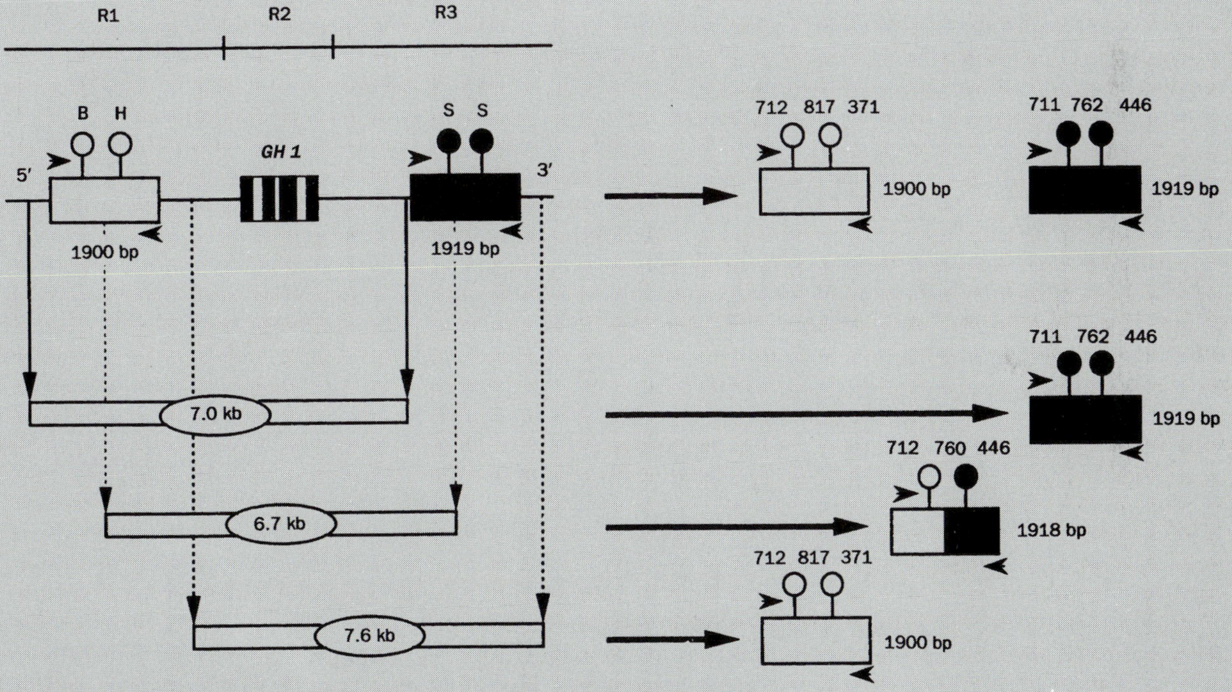


Schematic representation of the three *EcoR* I-derived fragments that flank (R1 and R3) or contain (R2) the *GH-1* gene. The sizes and relative locations of 6.7, 7.0, and 7.6 kb deletions are shown. The locations of sequences homologous to oligonucleotide primers used in Polmerase Chain Reaction (PCR) detection of the various deletions are indicated by small rightward and leftward arrows. The PCR products and fragment sizes obtained following restriction endonuclease digestion (B: *Bgl* I, H: *Hae* II and S: *Sma* I) are shown on the right from a normal control (top) and 7.0, 6.7 and 7.6 kb deletion homozygotes (bottom).

# Genetic regulation of growth-promoting factors by nutrition

Numerous factors are required for promoting growth and development during fetal and post-natal life, and particularly important is the inherited genotype and its interactions with nutritional status (**9.1**, **9.2**) and the external environment. Growth promoting factors (GPFs) act not only by classical endocrine mechanisms but also by paracrine, juxtacrine and autocrine mechanisms. Major actions of GPFs are to increase cell division (hyperplasia), and hence the total number of cells, and to increase cell size (hypertrophy) and modify cell composition by altering nutrient uptake and utilization. In addition, GPFs regulate cell lineage and commitment, helping to programme exit from the proliferative cycle to differentiated function. Binding proteins and receptors both play key roles in determining the actions of GPFs by altering their tissue availability and ability to act on target tissues. Regulation of the action of GPFs can therefore be exerted on the factors themselves and also on their binding proteins and receptors. Moreover, regulation of gene expression can occur at the levels of gene transcription, mRNA processing, mRNA stability and mRNA translation.

Knowledge of nutrient–gene interactions during critical stages of development is central to an understanding of the mechanisms which regulate growth and the precise role of nutrition in health and disease. The following sections highlight the mechanisms by which nutrition (**9.2**), and especially the energy and protein content of the diet, regulate gene expression during growth. A brief outline of the many GPFs so far identified will be followed by specific examples of the role of diet in regulating gene expression of these factors and their receptors in a tissue-specific manner.

## Growth-promoting factors (GPFs)

GPFs include both the classical endocrine hormones and the more recently discovered growth factors. Endocrine hormones which play a central role in growth regulation include growth hormone (GH) (**5.1**), thyroid hormones (TH) and insulin. Not only do each of these hormones have specific growth-promoting actions but interactions occur between them and other hormones such as glucocorticoids at the level of genetic regulation. These hormones act in concert to signal nutrient availability and determine the tissue abundance and activity of nutrient transporters and rate-limiting enzymes. In addition to the classical hormones, a large number of growth factor protein families have been and continue to be identified. Unlike the more classical hormones, these growth factors are often synthesized ubiquitously by many tissues of the body, but in a highly regulated manner. They have highly specific actions in potentiating or inhibiting cell proliferation and differentiation in a wide variety of tissues. Of the growth factors introduced here, the best characterized in terms of the nutritional and endocrine regulation of growth, are the insulin-like growth factors (IGFs), which have both endocrine and local actions.

The potency of many growth factors is such that they are not found in the free form; instead, binding proteins in

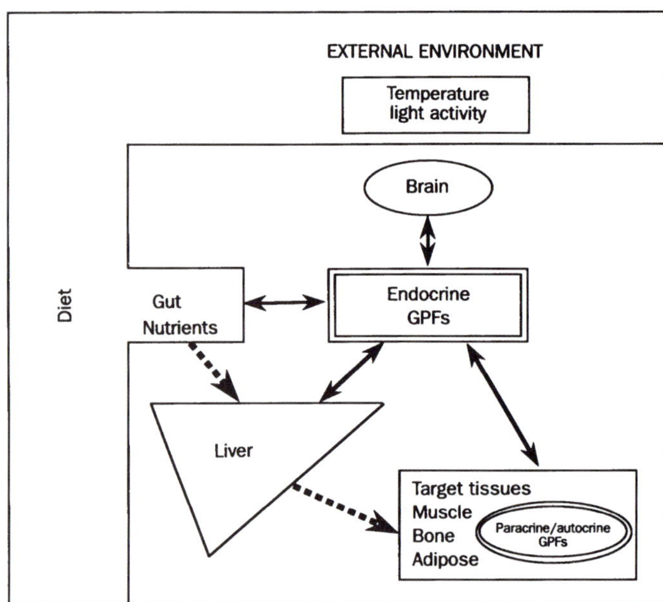


**Schematic representation of the interactions between endogenous factors, including nutrition and thermal environment, and the regulation of growth by growth-promoting factors (GPFs).**

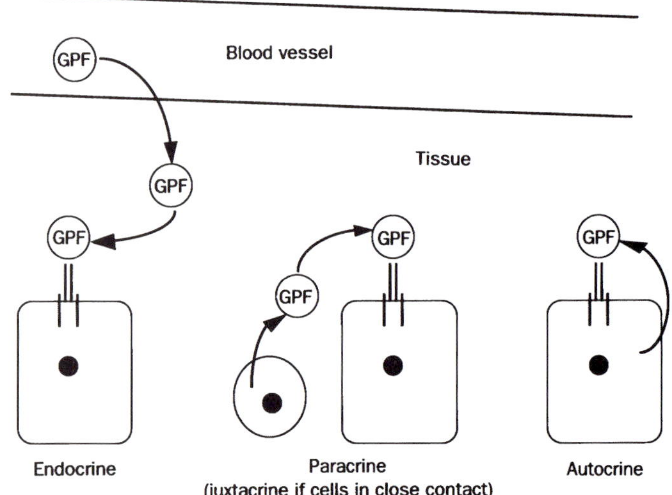


**Comparison between endocrine, paracrine, juxtacrine and autocrine mechanisms for the action of growth-promoting factors (GPFs).**

plasma and tissues serve to transport them and modulate their function, thus providing an additional tier of regulation. For example, IGFs can be complexed with one of six high-affinity binding proteins (IGFBPs) each of which is regulated differently and exerts different functions. Although most studies on the function of IGFBPs have been carried out *in vitro*, evidence suggests that they can probably also influence IGF-I bioactivity *in vivo*. The tissue receptors for GPFs have been identified in numerous sites including the cell membrane, cytoplasm and nucleus, and their functions include transport into and within the cell, and ligand-binding and associated transcriptional regulation of specific genes.

## Nutritional regulation

Changes in nutrition can alter the cell surface availability and activity of pre-existing proteins or influence synthesis of new proteins by influencing gene transcription, promoter usage or post-translational processing of the gene products. In most adult tissues, when an individual is in energy balance, there will be an equilibrium between gene products with anabolic actions and those with catabolic actions. During development, by contrast, regulatory mechanisms will differ because net growth is essential. Growth is in turn dependent on the amount of energy available for growth after the demands of thermoregulation and activity have been met, and therefore energy supply in relation to requirement must be considered, rather than simply the energy intake. For example, plasma levels of IGF-I are affected not just by energy intake but by net energy availability and there is a strong positive correlation between plasma IGF-I concentration and growth-rate.

*Major growth-promoting factors*

| **A. Classical endocrine hormones** |
|---|
| Androgens<br>Growth hormone (GH)<br>Insulin<br>Estrogens<br>Thyroid hormones (TH) |
| **B. Growth factor families** |
| Epidermal growth factor (EGF)/Transforming growth factor α (TGFα)<br>Fibroblast growth factors (FGFs)<br>Hepatocyte growth factor/scatter factor (HGF/SF)<br>Insulin-like growth factors (IGFs)<br>Nerve growth factors (NGFs)<br>Transforming growth factor β family (TGFβ)<br>Vascular endothelial growth factors (VEGFs) |

GH is essential for normal growth and development and is an obligatory stimulus for IGF-I synthesis; indeed, IGF-I mediates many of the anabolic actions of GH in addition to the independent and directly stimulated effects of GH. Studies in a wide variety of species have demonstrated that dietary energy and protein intakes regulate GH secretion from the pituitary gland. However, circulating IGF-I concentrations are not correlated simply with blood GH levels because the GH receptor (GHR) is a further important step in mediating GH action. Therefore, activity through the GH/IGF axis is subject to several key stages of regulation. One way to assess each of these stages individually is to examine the tissue-specific expression of the genes responsible for encoding the GHR and IGF-I.

## Regulation of growth hormone receptor (GHR) gene expression

The GHR is a member of the superfamily of GH/prolactin/cytokine receptor transmembrane proteins. The mature pro-

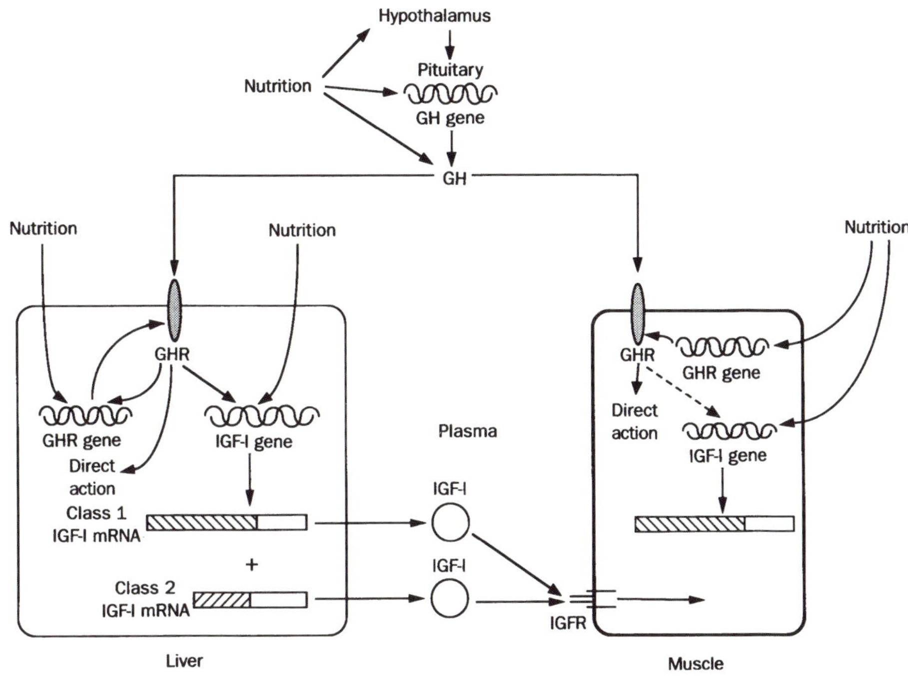


**Mechanisms in the genetic regulation of growth hormone (GH), insulin-like growth factor-I (IGF-I) and their receptors (GHR and IGFR) by nutrition.**

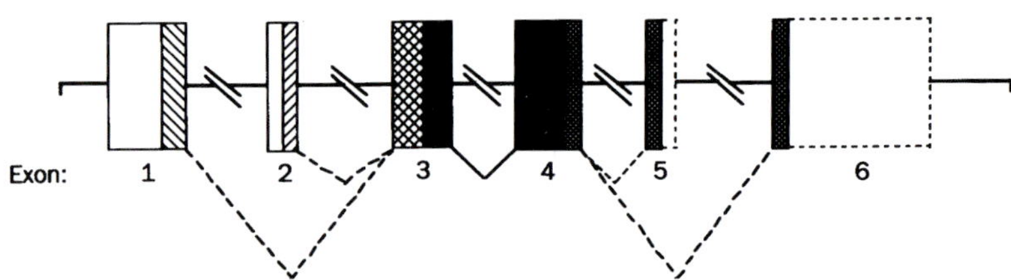


**Structure of mammalian IGF-I gene. Regions of exons are shaded to depict the following:**

**= mature IGF-I protein**

**= common leader sequence**

**and = specific leader sequence, for the class 1 or class 2 IGF-I mRNA transcripts, respectively**

**= terminal peptide sequence.**

**Alternative splicing between leader and terminal exons is demonstrated by dashed lines.**

tein of 620 amino acids comprises a long intracellular domain, a 24 amino acid transmembrane region, and a 246 amino acid extracellular GH-binding domain. The latter is virtually identical to the soluble short-form GHR present in blood which acts as a specific GH binding protein (GHBP). In humans and most other non-rodent mammalian species a single 4.5 kilobases mRNA encodes the full-length GHR; the GHBP is produced by proteolysis of the mature GHR. By contrast, in rats and mice, alternative splicing of a single primary transcript results in two distinct mRNAs; these encode the full-length GHR and the short-form GHBP.

Nutritional deprivation is accompanied by resistance of the liver to GH, and protein-energy malnutrition results in high plasma levels of GH but a reduction in plasma IGF-I concentration. Differing responses of the hepatic GHR in fasting and protein restriction have been suggested from studies at both the tissue and molecular levels: a low energy intake results in defective liver GHR whereas a low protein intake results in defects at the post-receptor level. Thus, not only is there a reduction in GH binding to liver membranes and hepatocytes after fasting, but this decline concurs with reductions in hepatic GHR gene expression and IGF-I mRNA. By contrast, liver GH binding is normal after protein restriction and the reduction in plasma IGF-I levels is probably due to an intracellular defect in GH action.

Differential regulation of GHR gene expression can occur in a tissue-specific fashion. By contrast with the down-regulation of hepatic GHR in undernutrition, a reduction in energy status due to either a low intake or a high metabolic demand results in an up-regulation of GHR mRNA in muscle. It has been hypothesized that these changes are mediated in part by the changes in thyroid hormone status induced by nutrition. Whether nutrition affects not only gene transcription but also the stability and translation rate of GHR mRNA is unknown. Nevertheless, these tissue-specific responses of the GHR to nutrition suggest that a low food intake would result in reduced hepatic IGF-I production and hence reduced growth, but increased muscle GHR and hence an increase in the direct metabolic actions of GH which would conserve glucose and increase fat utilization.

## Regulation of insulin-like growth factor-I (IGF-I) gene expression

Given the apparent simplicity of mature IGF-I protein, with only 70 amino acids and encoded only by parts of exons 3 and 4, it has a surprisingly complex gene structure. However, its potential importance is indicated by the conservation of this arrangement across many species. Exons 1 and 2 encode leader sequence and are differentially spliced to exon 3 to produce class 1 and class 2 transcripts respectively.

Studies in normal animals have demonstrated that hepatic levels of class 2 transcripts are more sensitive to dietary protein and energy status than are class 1 transcripts, and that the ratio of class 1 : class 2 IGF-I mRNA expression in liver is related to growth potential. Moreover, the IGF-I mRNA derived from exon 2 is poorly expressed in many extra-hepatic tissues, such as muscle, and the liver has very low levels of IGF receptors. Therefore, it has been hypothesized that exon 2-derived hepatic IGF-I may have special significance, either in its sensitivity to certain transcription factors, such as those stimulated via the GHR, or that the precursor IGF-I encoded by exon 2 is conferred with specific properties, for example efficient export into blood for action at peripheral tissues. It should be emphasized, however, that these types of studies are in their infancy and that much remains to be confirmed and discovered.

In addition to the few specific examples given here, many other GPFs are likely to be affected by nutrition and the level of genetic regulation is currently being investigated in detail. It is certain that GPFs act in concert to co-ordinate and orchestrate the underlying mechanisms of mammalian growth, control adaptation and responsiveness to the external environment, and provide additional fine-tuning of genetically programmed development.

*M.J. Dauncey and J.M. Pell*

See also 'Hormonal regulation of fetal growth' (**4.1**), 'Growth factors and development' (**4.2**) 'Endocrinological regulation of post-natal growth' (**5.1**) 'Gene–environment interactions' (**10.2**) and 'Long-term consequences of early enviromental influences' (**12.1**)

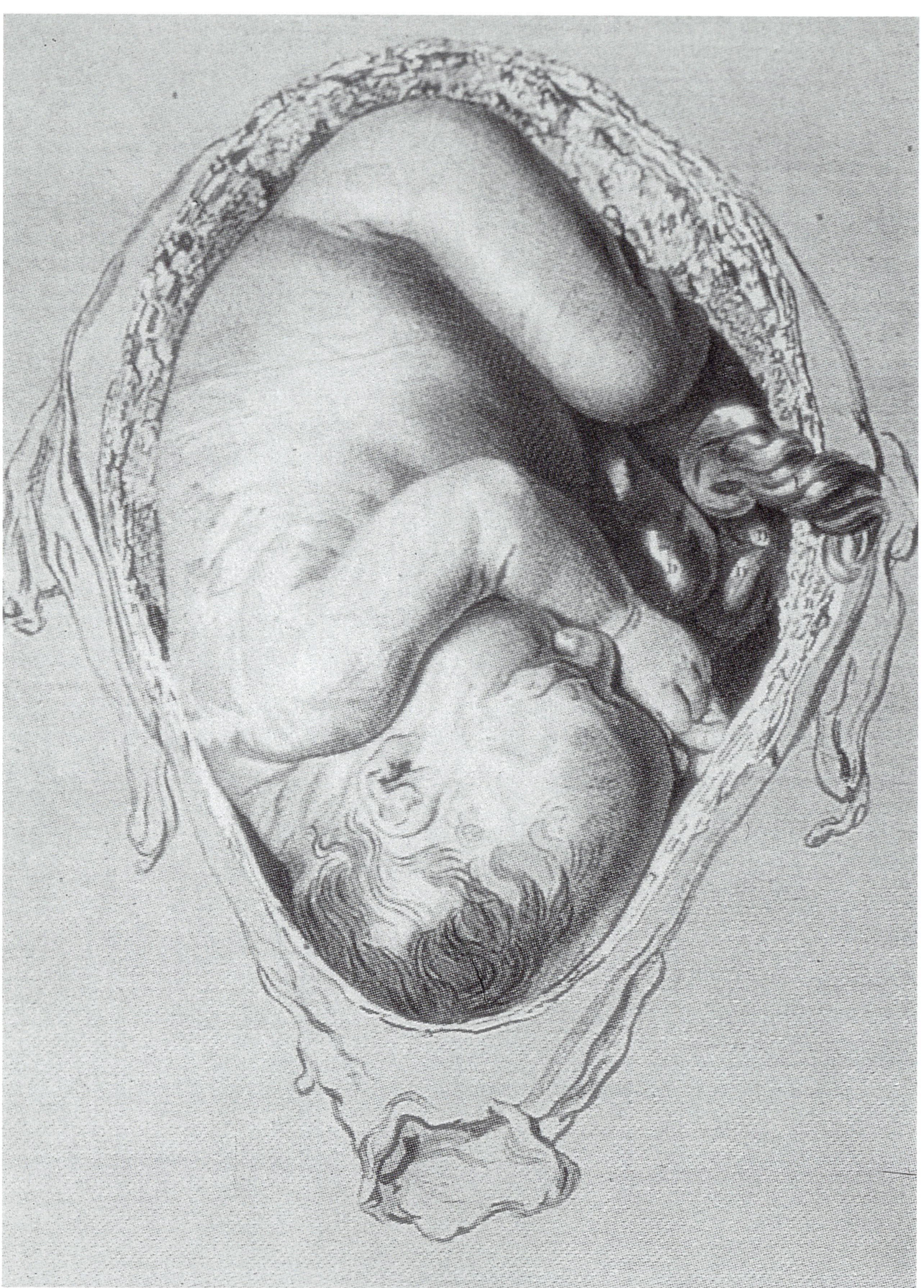

## Part Four

# Fetal growth

Growth of the embryo (**4.3, 4.5**) and fetus is the result of a complex array of genetic and environmental factors which interplay in a co-ordinated and longitudinal manner. Knowledge about some of the physiological processes involved is inferred from examination of the consequences of pathogenetic mechanisms. For example, abnormal placental structure and function can vividly demonstrate the importance of an adequate vascular and nutrient supply to promote growth of the fetus, especially after early pregnancy (**4.5**). Placentation requires the establishment of an efficient blood supply to link maternal and fetal circulations. Trophoblast invasion of the spiral arterioles in the placental bed initially blocks these blood vessels, thereby protecting the embryo from the effects of high ambient oxygen concentrations. Later, the plugged vessels are re-canalized but without muscular walls so that a low-resistance

**Opposite: William Hunter's *Anatomy of the Human Gravid Uterus* (1774) contained this exceptionally life-like illustration (opposite) by Jan van Rymsdyk. Courtesy of the Wellcome Institute, London.**

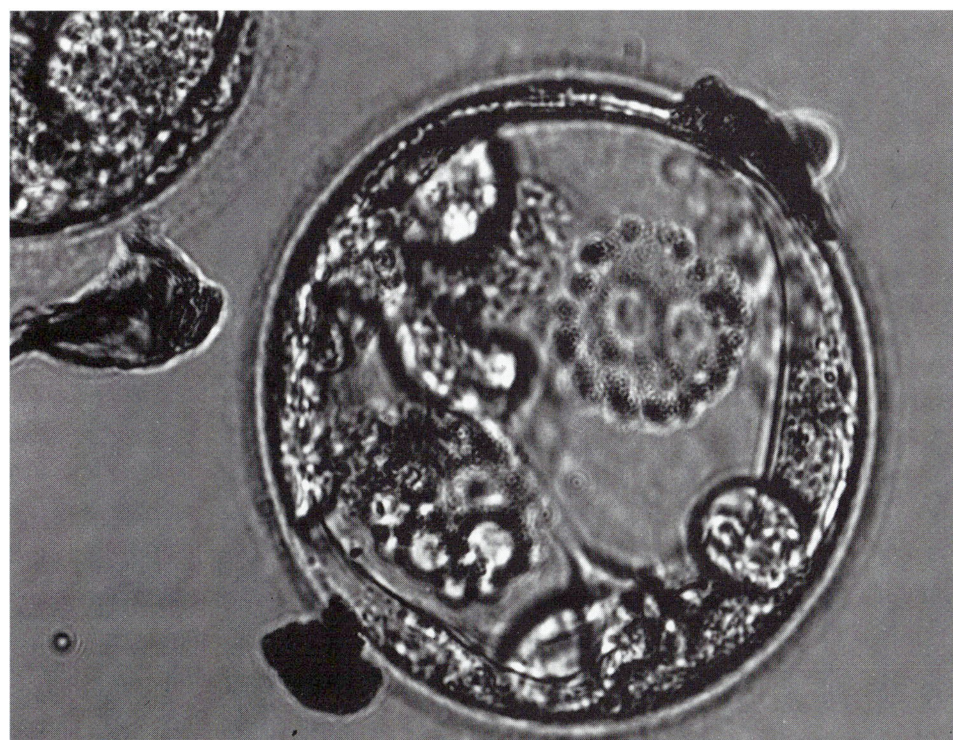

**A human blastocyst at 5.5 days post-insemination. Courtesy of Dr Simon Fishel.**

blood-flow system operates to supply the embryo. Growth of the placenta is clearly dependent on an adequate blood supply but also the state of maternal nutrition during early placentation. This in turn affects growth of the embryo so that, in general, fetal weight in late gestation is directly related to placental weight. Early maternal nutritional status may have longer lasting effects (**4.8**), as evidenced by maternal undernutrition during the Dutch famine in World War II adversely affecting fetal growth, not so much in the first but in the second generation thereafter.

The utero-placental unit provides the fetus with oxygen in a rate-limiting manner. This is dependent on the difference in partial pressures between maternal and fetal blood as well as the differences in oxygen affinity between maternal and fetal hemoglobin. The latter has a higher affinity for oxygen in order to accommodate the metabolic activity generated by a rapidly growing fetus. The placenta has a high rate of oxygen consumption, equalling, at least, that of the brain and liver. This requirement is needed to meet the metabolic demands of adequate uptake and transfer of glucose, amino acids, and lipids across the placenta. It appears that glucose availability is an important determinant of fetal growth and, together with the high oxygen demand, the placenta is a dominant consumer of maternal fuel supply.

These metabolic processes do not occur independent of a range of hormones acting in both an endocrine and paracrine manner (**4.1, 4.2**). There is a close interdependence between nutritional and hormonal factors in the regulation of fetal growth, which is in turn influenced greatly by the tricompartmental maternal-placental-fetal link. A dominant influence is exercised by the insulin-like growth factors (IGFs). IGF-I, IGF-II and the numerous IGF binding proteins are ubiquitously expressed in fetal tissues. There is preferential expression of IGF-II in early gestation which is probably constitutive in nature. IGF-I becomes more dominant in later gestation, and this developmental switch in IGFs appears to be nutrition-dependent as the maternal-placental-fetal unit becomes anatomically and functionally firmly established.

There is a positive correlation between birth-weight and umbilical cord blood concentrations of IGF-I. Direct evidence for the role of IGFs in fetal growth is based on gene disruption studies in experimental animals. Mice null mutants for IGF-I and IGF-II are both growth-

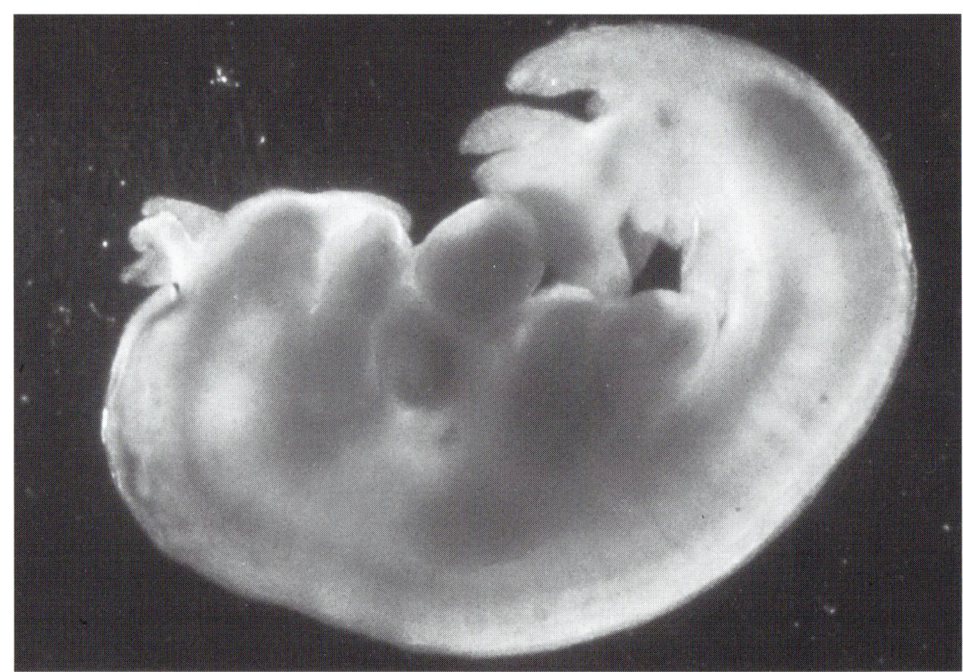

**A human embryo after 28 days development, seen in profile with the head on the left. Most of the primitive organ systems are formed, and the embryonic heart is already beating. The embryo's curved appearance is due to the symmetrical growth of the brain, which now comprises five divisions. The spinal cord becomes thickened on the underside, forming clusters of nerve cells. These unite with the upper and lower sides of the neural canal to form spinal nerves which connect with other organs as they are formed. Lungs bud in a branching manner, forming bronchi, bronchioles and finally air sacs. Glands, such as the thyroid, pancreas, thymus and parathyroid, arise from the primitive gut. Courtesy of Science Photo Library, London.**

retarded, but the onset of fetal growth retardation occurs earlier and is less persistent in the IGF-II deficient mice. Disruption of the gene for the Type I IGF receptor (IGF-I receptor) as well as double null mutants for IGF-I and IGF-II result in severe growth retardation, thereby demonstrating the roles of both IGFs acting via the IGF-I receptor in controlling fetal growth. A rare example of human congenital IGF-I deficiency due to an IGF-I gene deletion has been reported. The infant was severely growth-retarded at birth and was extremely short by the time of investigation. There was a positive growth response to treatment with recombinant IGF-I. Developmental delay was also present in keeping with the function of IGFs along with other growth factors in the control of neuroglial development.

Insulin is also a growth-promoting hormone, as illustrated by macrosomia in the presence of insulin excess. However, under physiological conditions insulin acts in a more permissive role to regulate IGF's dependence on nutrient supply, especially for glucose and amino acids. Growth hormone is not a major determinant of fetal growth even though circulating concentrations are high in mid-gestation. This apparent unresponsiveness is partly explained by delayed fetal expression of growth-hormone receptors, particularly in the liver. Fetal growth-hormone deficiency is associated with a smaller size at birth but not as pronounced as in the case of IGF-I deficiency. The placenta is also a source of growth hormone, producing a peptide which differs by 13 amino acids from pituitary growth hormone. This molecule has an effect on the maternal side of the circulation whereby pulsatile maternal growth hormone secretion is suppressed during pregnancy. Furthermore, growth-hormone-dependent IGF-I and IGF binding protein levels are normalized during pregnancy in the case of maternal pituitary deficiency. Placental lactogen has similar actions to growth hormone and together with the placental growth hormone variant, raises levels of glucose and free fatty acids in the maternal circulation. These increased substrates are important for nutrient delivery to the fetus. There is no evidence for a direct biological role for placental growth hormone and lactogen on the fetus, although placental lactogen may influence fetal anabolism.

Other hormones traditionally associated with growth effects post-natally, such as thyroid, glucocorticoid, vitamin D and sex hormones, are all produced in significant quantities by the

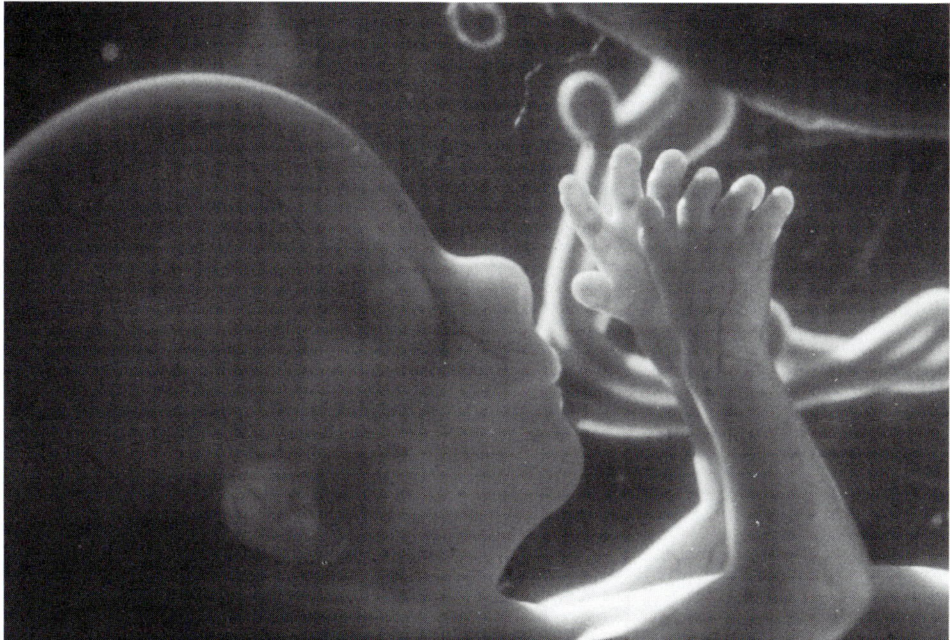

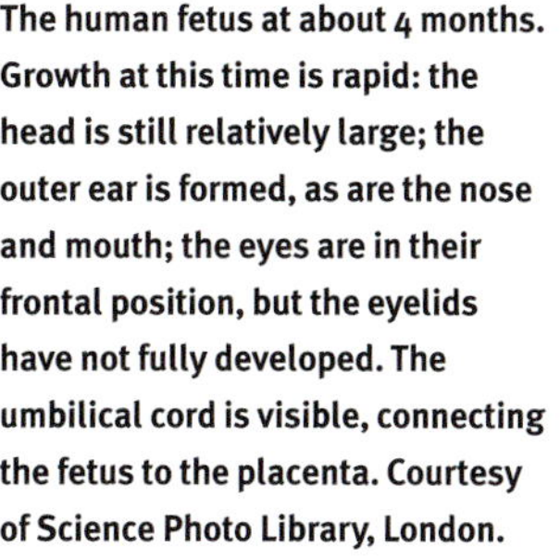

**The human fetus at about 4 months. Growth at this time is rapid: the head is still relatively large; the outer ear is formed, as are the nose and mouth; the eyes are in their frontal position, but the eyelids have not fully developed. The umbilical cord is visible, connecting the fetus to the placenta. Courtesy of Science Photo Library, London.**

fetus but are not directly targeted towards fetal growth. Neural development (**4.6**) is critically dependent on thyroid hormones which probably mediate their effects via stimulation of neurotrophic growth factors. A fetus with thyroid agenesis is spared significant neurological deficit at birth because of some transplacental passage of maternal thyroid hormones and an active deiodinase system in the brain to ensure sufficient concentrations of the more biologically active thyroid hormone, tri-iodothyronine. This dependence is illustrated by the poor outcome for an infant of a mother with pituitary deficiency of growth hormone, prolactin and thyroid-stimulating hormone resulting from mutation in pit-l, a transcription factor active in early embryogenesis (**4.3**). Equally complex mechanisms operate in the embryonic development of teeth (**4.4**) and the ontogeny of the immune system (**4.7**).

The maternal-placental-fetal unit is involved in the control of glucocorticoid exposure to the fetus. Glucocorticords are essential for the maturation of several fetal organs such as the liver, pancreas, gastrointestinal tract and especially, the lungs. However, maternal glucocorticoids readily cross the placenta and in high concentrations would be growth-inhibiting to the fetus. During most of gestation, the fetus is protected from excess cortisol by the action of placental 11 β-hydroxysteroid dehydrogenase, which inactivates cortisol to cortisone. The reaction is reversible as befits an enzyme shuttle system and the 11-oxidoreductase component becomes more active towards the end of gestation to ensure sufficient cortisol to induce critical maturational events before delivery.

An over-riding influence on fetal size at birth is maternal constraint which has its most dominant effect in late gestation (**4.8**). The capacity of the uterus for growth must obviously influence fetal size, particularly in the case of multiple fetuses. The role of uterine environment and thus maternal phenotype was demonstrated by the use of embryo transfer experiments and cross-breeding experiments using different sized breeds. As an example, an embryo from a small breed transplanted to a uterus from a large breed grows larger than its non-transplanted control. The effect of maternal constraint is also illustrated by the results of the converse experiment. Thus fetal growth is controlled by a complex interplay of factors whose effects are not consistently influential throughout gestation. The major influence on fetal growth overall appears to be the capacity for the placenta to deliver adequate nutrients (including oxygen) to enable appropriate IGFs to effect growth of the fetus optimally within the physical constraints of the uterus. Adequate fetal development is critical to post-natal health, and standards and references for its assessment (**4.9**) are vitally important.

*Ieuan A. Hughes*

# Hormonal regulation of fetal growth

The intra-uterine period of human growth is a complex process during which a single undifferentiated cell, the zygote, is transformed into the most complex organism of the animal kingdom: the newborn human. The first weeks of intra-uterine growth are dominated by the processes of hyperplasia, differentiation and morphogenesis which results in the formation of the endodermal, mesodermal and ectodermal layers, and subsequently the precursors of the principal organs are formed (**1.1**). During the second trimester, one of the most dramatic processes is that of longitudinal growth, which approaches its maximal velocity during the 20th week of gestation. Finally, during the last 3 months, longitudinal growth slows while body-weight increases mainly due to the hypertrophy or increase in cell size, the accumulation of fat and the incorporation of new molecules into the extracellular space.

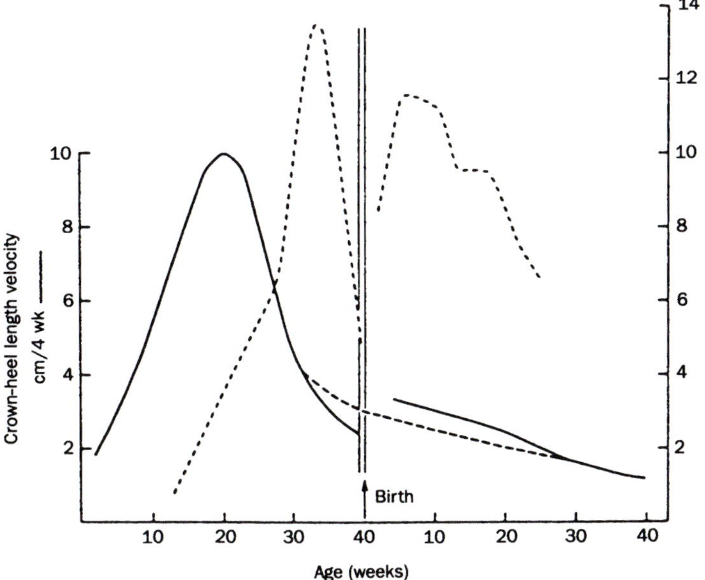


**Rate of linear growth- (——) and weight-gain (---) *in utero* and during first 40 weeks after birth. Data from Tanner J.M., *Foetus into Man*. Ware: Castlemead Publications, 1989.**

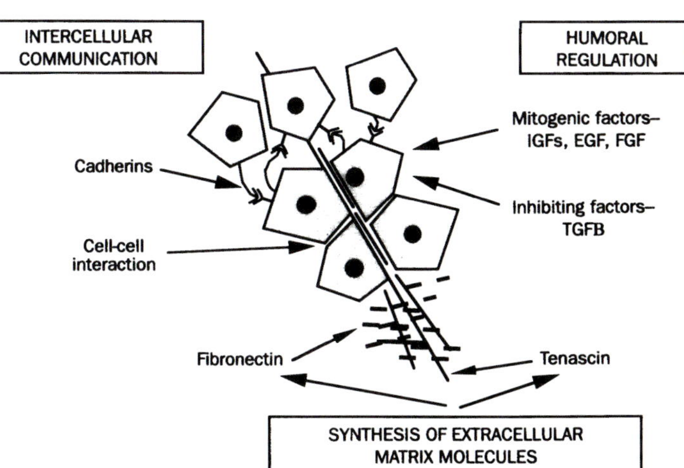


**Basic mechanisms of differentiation, morphogenesis and fetal growth. Cell–cell interaction, synthesis, and secretion of molecules to the extracellular space and regulation of the cellular multiplication through stimulating factors (IGF-I, IGF-II, EGF, PDGF, FGF) and inhibitors (TGF β, TNF).**

## Basic mechanisms of the differentiation, morphogenesis and growth of the fetus

The basic mechanisms which result in the differentiation, morphogenesis and growth of the fetus consist, in essence, of the synthesis and secretion of molecules into the extracellular space, remodelling of cellular morphology as a result of the direct contact between them or due to cell recognition molecules, and the expression of hormones or intracellular messengers that accelerate or retard the rhythm of cell division, as well as cellular metabolism.

## Hormonal regulation of fetal growth

In contrast to what occurs during extra-uterine life, the majority of the hormones that play a role in the regulation of fetal growth do not act via systemic or endocrine systems, but are synthesized locally. These hormones act directly on the cells that produce them, or on those cells in close proximity, hence, in an autocrine or paracrine fashion (**3.4**). The most important of these hormones are the peptide growth factors.

## Peptide growth factors

This group is comprised of peptides that, similar to traditional hormones, have a common mechanism of action through the use of receptors in the cell membrane. Their union with the receptor provokes physical modifications of the membrane which result in changes in the velocity of transport of certain ions ($K^+$) and metabolic precursors (glucose, amino acids, nucleotides). The resulting biochemical changes in the cell interior culminate in an increase in cell size, the beginning of cell division or the synthesis and secretion of certain molecules.

Among these factors, those which play one of the most dramatic roles are the somatomedins or insulin-like growth factors (IGF-I and IGF-II). Both of these factors are present during the first phases of development, and in addition to their mitogenic activities, they are important for the differentiation of some cell types (e.g., myoblasts and adipocytes) and act upon cells that are already differentiated to stimulate

*Peptide growth factors*

**Somatomedins**
IGF I
IGF II

**Epidermal growth factor (EGF) family**
EGF
TGF α
Amphiregulin

**Fibroblast growth factors (heparin-binding growth factors)**
Acidic
Basic
Oncogenes (e.g., v-int-2, v-hst)

**Platelet-derived growth factors**
Beta, beta (v-sis oncogene)
Alpha, beta
Alpha, alpha

**Transforming growth factor β family***
TGF β types 1, 2 and 3
Inhibins
Activins
Antimüllerian hormone
Oncogenes (vg 1)
Bone morphogenic proteins
DPP-C (decapentaplegic transcript) (pattern formation in Drosophila)

* Growth inhibiting factors

the production of various substances, such as type-I collagen in chondrocytes and glycogen in hepatocytes.

The relative importance of IGF-I and IGF-II is still undetermined. However, the available data suggest that IGF-II plays a predominant role during the early stages of embryonic development, while once organogenesis has finished, IGF-I becomes more important and its plasma concentrations correlate positively with the weight of the newborn. The activity of IGF-I is modulated by the transfer of nutrients, which are the principle regulators of its expression, and the equilibrium between its binding proteins (IGFBPs), of which six have been identified. Although the mechanisms by which these binding proteins modulate the activity of IGF-I are not well understood, it is generally accepted that they act to increase or decrease IGF-I's effects on the target cell. Indeed, there is a positive correlation between IGFBP-3 and fetal weight, while IGFBP-1 and IGFBP-2 are increased during intra-uterine growth retardation. Other growth factors of lesser importance include epidermal, fibroblast, platelet-derived and transforming growth factors.

## The role of other hormones and growth factors

Of the traditional hormones, those which are synthesized in specialized cells and act at a distance from their place of origin, the only one which plays an important role in the regulation of fetal growth is the insulin produced by the fetal endocrine pancreas, since insulin of maternal origin cannot pass the placental barrier. Insulin's mechanism of action is two-fold: on the one hand it plays a permissive function regulating the transport of nutrients across the cell membrane; and on the other, it regulates the synthesis of the peptide growth factors and the binding proteins.

Other hormones have little influence over intra-uterine growth. Thyroid hormones are important for the development of the nervous system, while androgens and estrogens play a role in skeletal maturation and sexual differentiation, but have little influence on longitudinal growth. Pituitary growth hormone, which is the most important hormone in the control of post-natal growth, plays only a secondary role during the last weeks of gestation. This is most likely

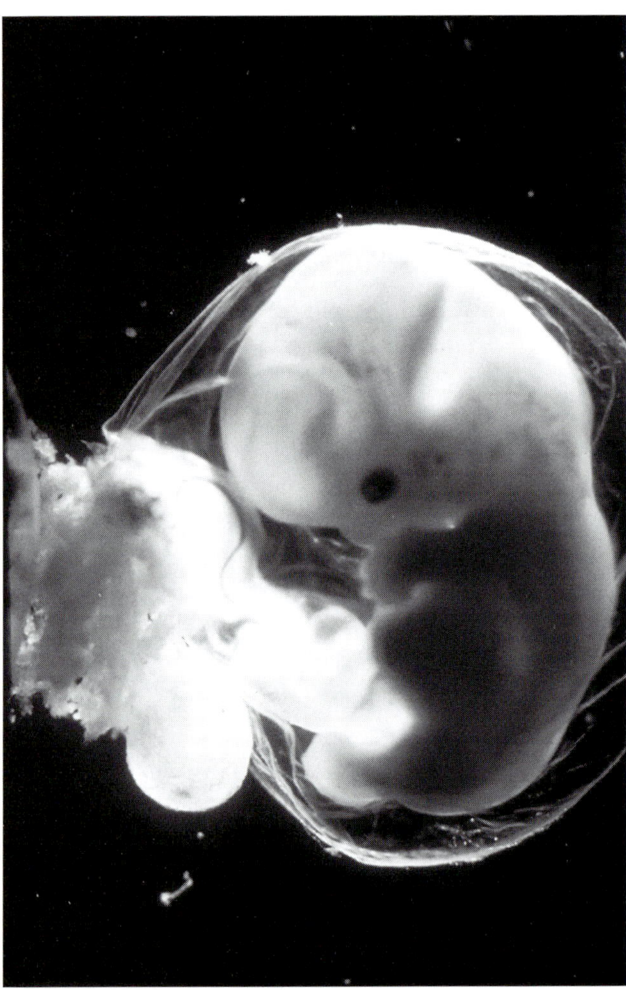

**The human embryo at 5–6 weeks. The umbilical cord (carrying two arteries and one vein to the placenta) connects the embryo with the maternal circulation, supplying oxygen and nourishment and removing waste materials. Primitive organ systems have formed and are developing rapidly. The void at the back of the head awaits partial filling by a further division of the brain (at present in three parts, the fore-, mid- and hind-brain), leaving the fourth ventricle under the cerebellum. The retina of the eye is visible (dark spot), the arms and legs have budded. The heart has formed and has been beating since the third week. At this stage the embryo is about 15 mm long and the umbilical cord is formed. All the internal organs, such as the liver, heart, stomach and sex organs, have begun to form. It is in this period that the embryo is most vulnerable to substances (alcohol, medication etc.) consumed by the mother which may cause defects. The smaller sac at the bottom left is the yolk sac which provides nourishment for the early embryo. Courtesy of Science Photo Library, London.**

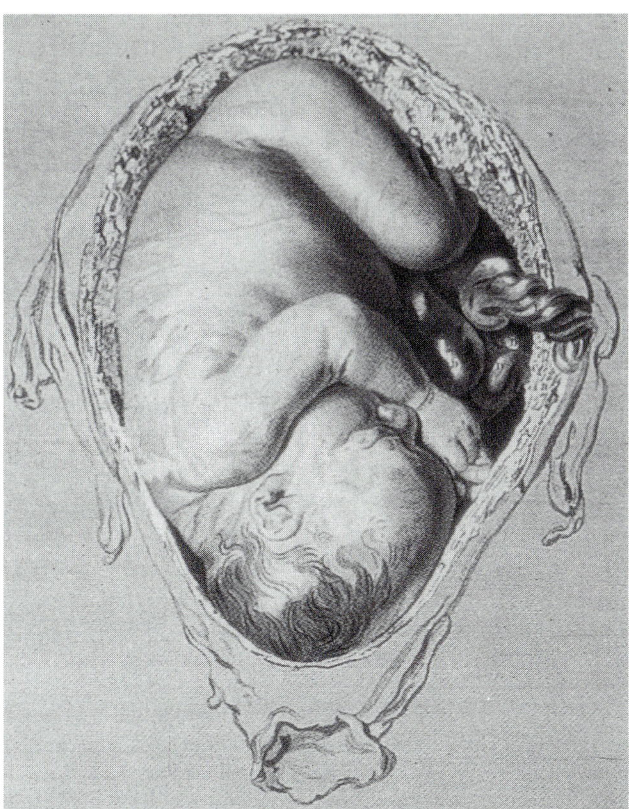

**A baby at term, by Jan van Rymsdyk. It was published in William Hunter's *Anatomy of the Human Gravid Uterus* (1774). Courtesy of the Wellcome Institute, London.**

due to the low level of receptors for this hormone in the fetus. The importance of other hormones such as placental growth hormone or placental lactogen is currently a topic of discussion.

## The relationship between hormones, peptide growth factors and nutrition

The activity of IGF-I, as well as that of its binding proteins, is modulated by the nutrient environment, especially the availability of glucose in the fetal circulation and the subsequent secretion of insulin by the fetal pancreas. At the same time, the circulating levels of IGF-I in the fetus is a determining factor for the distribution of nutrients between the fetus and placenta, aiding the transference of nutrients to the fetal circulation. In addition, the maternal IGF-I favours the drainage of nutrients from the mother to the placenta. In addition to being the main factor regulating fetal growth, IGF-I plays a fundamental role in the optimal usage of nutrients, resulting in the equilibrated growth of the fetus and placenta.

*Manuel Hernández*

See also 'The human growth curve' (**2.4**), 'Genetic regulation of growth-controlling factors by nutrition' (**3.4**), 'Growth of the human embryo' (**4.3**), 'Developmental morphology of the embryo and fetus' (**4.5**), 'Endocrinological regulation of post-natal growth' (**5.1**) and 'Skeletal development' (**5.6**)

# Growth factors and development

Communication among cells plays an essential role in embryonic development. Intercellular communication is achieved by an elaborate signalling system, the extracellular constituents of which include relatively small organic molecules (e.g. steroids and vitamin A and D) and polypeptides. Many of the polypeptide-signalling molecules have been grouped under the imprecise term 'growth factor' because of their initial isolation based on an ability to promote cell multiplication in culture. Rita Levi-Montalcini and Stanley Cohen won the Nobel Prize in 1986 for their discovery of growth factors in the 1960s. It is now known that growth factors have diverse effects on cells including promotion of differentiation, cell death, and locomotion. Some growth factors have been referred to by a variety of other terms depending on the

## PEPTIDE-SIGNALLING FACTORS IN DEVELOPMENT

There have been hundreds of peptide-signalling factors identified to date (**4.1**). Listed here are ones that have been extensively studied in development. Each factor has various functions in different types of cells and tissues, at different developmental stages, under different conditions, and when interacting with different co-factors. Therefore, only the major functions for each family are mentioned below.

### GROWTH FACTORS

- *Epidermal growth factor (EGF) and transforming growth factor α (TGF-α)*: these two factors are members of the EGF family. They are similar in both structure and function and bind to the same (EGF) receptor. Both factors are involved in skin and hair formation. EGF also plays a role in early embryonic development, for example, egg cleavage.
- *TGF-β superfamily*: this is the largest growth-factor family, containing at least 24 members to date. Members of this family function as dimers. Receptors for this superfamily are serine/threonine kinases. Most of them are associated with bone formation and wound repair. Some function in early embryonic development (e.g. BMPs, Vg-1, and nodal), others in later cell differentiation (TGF-βs).
- *TGF-β family*: this family includes TGF-βl, TGF-β2, and TGF-β3. Growth inhibition may be the main activity of the family, although these factors can also stimulate some types of cells to divide. Correlation with cell death has been reported for TGF-β1. TGF-β3 is an important factor in secondary palate fusion.
- *Bone morphogenetic proteins (BMP)*: eight members have been discovered.

| Protein | Other names |
|---|---|
| BMP-1 | |
| BMP-2 | BMP-2A |
| BMP-3 | osteogenin |
| BMP-4 | BMP-2B |
| BMP-5 | |
| BMP-6 | Vgr-1 |
| BMP-7 | OP-1 |
| BMP-8 | |

- *Activin and inhibin*: activin is composed of two types of β chains, which combine to form activin A ($\beta_A\beta_A$), AB ($\beta_A\beta_B$), and B ($\beta_B\beta_B$), while inhibin is made up of α and β chains (inhibin A ($\alpha\beta_A$) and B ($\alpha\beta_B$). An activin- and inhibin-binding protein, follistatin, regulates their activities and plays an important role in embryonic development. In addition to functioning in craniofacial development during embryogenesis, activins and inhibins modulate hormone production in the pituitary and gonads.
- *Vg-1, nodal and dorsalin*: these factors are important in embryonic development.
- *Fibroblast growth factor (FGF) family*: this has nine members to date. Four types of receptor (FGFR) have been identified.

| Protein | Gene or other names |
|---|---|
| FGF-1 | acidic FGF |
| FGF-2 | basic FGF |
| FGF-3 | int-2 |
| FGF-4 | Oncogene hst; hstf-l; K-FGF |
| FGF-5 | |
| FGF-6 | hst-2 |
| FGF-7 | Keratinocyte growth factor (KGF) |
| FGF-8 | Androgen-induced growth factor (AIGF) |
| FGF-9 | Glia-activating factor (GAF) |

FGFs are potent mitogens. They participate in angiogenesis, wound-healing, and tissue regeneration. In embryonic development they are involved in various processes including mesoderm formation, gastrulation, primordial outgrowth and organ formation. Mutations in FGFs and FGFRs are associated with human skeletal dysplasias and craniosynostotic syndromes, for example, Crouzon and Apert syndromes.

- *Platelet-derived growth factors (PDGF)*: PDGF consists of A and B chains (AA, AB, and BB). Their receptors are also dimers (αα, αβ, and ββ). PDGFs are potent mitogens associated with various cancers. They also participate in wound-healing, various inflammatory joint diseases, and fibrosis.
- *Nerve growth factor (NGF)*: receptors are TrkA, TrkB and TrkC, and p75. NGF acts to control neuronal survival and neurite outgrowth.
- *Insulin-like growth factors (IGF)*: IGF-I and IGF-II. There are two type of receptors, one of which has a tyrosine kinase domain and the other one of which does not. Besides their mitogenetic activity, IGFs also promote cell differentiation in chondroblasts, myoblasts, osteoblasts, neuroblasts, the lens, and hematopoietic cells. In embryogenesis, IGF expression has been found in early development and limb development.

### OTHER FACTORS

Some recently identified peptide-signalling factors function very similarly to growth factors and play important roles in embryonic development.

- *The hedgehog family*: the mammalian hedgehog (HH) family consists of three members, Sonic hedgehog (SHH), Desert hedgehog (DHH), and Indian hedgehog (IHH). Receptors for the HHs have not yet found. A transmembrane protein, known as Patched, is involved in the HH signalling system in both insects and vertebrates. SHH is expressed in various embryonic structures that play inductive or organizing roles in patterning and growth, such as the notochord and floor plate in the spinal cord, the ZPA in the limb bud, and the enamel knot in the tooth germ. Experiments have shown that SHH is necessary and sufficient to mimic the functions of the embryonic organizers in the control of pattern formation.
- *The Wnt family*: Wnt molecules are secreted glycoproteins. Their receptors have not been identified. Developmental genetic studies showed that they are required for various developmental processes including central nervous system and limb formation.
- *Noggin and chordin*: these two unrelated factors are secreted proteins expressed in the Spemann's organizer that are believed to serve as dorsalizing signals.
- *Steel factor*: its receptor is a tyrosine kinase encoded by the *c-kit* gene. This signalling system is required for the normal development of germ cells, pigment cells, and hematopoietic cells.
- *Delta-Serrate-Lag-2 family*: this family is the ligands for the Notch receptor family. These signalling systems may participate in cell-fate determination in the nervous system and other tissues such as hair and teeth.

*Zhiyong Zhao, Kenneth M. Weiss and David W. Stock*

## MODES OF ACTION OF PEPTIDE-SIGNALLING FACTORS

There are four main ways in which peptide-signalling factors act, endocrine, autocrine, paracrine and juxtacrine.

- *Endocrine*. Factors (solid triangles in the figure) are carried by the blood stream to remote target cells (cell B in the figure, which has receptors). This is the pathway for hormones, but also for some growth factors.
- *Autocrine*. Factors act on the cells that produce them (cell A in the figure).
- *Paracrine*. Factors act on the cells adjacent to the producing cells.
- *Juxtacrine*. Factors are presented on the cell surface but not secreted. They act on adjacent cells that possess the receptors (e.g. EGF and TGF-α precursors).

In addition, some maternally produced growth factors can penetrate through the placenta to the fetus (e.g. TGF-β1).

*Zhiyong Zhao, Kenneth M. Weiss and David W. Stock*

Endocrine Autocrine Paracrine Juxtacrine

**Modes of action of peptide-signalling factors.**

type of study involved, for example, interleukins, cytokines, lymphokines, and monokines by immunologists and colony-stimulating factors by hematologists.

Growth factors are not a single class of related molecules, but rather comprise several distinct families. They are generally differentiated from classical endocrine hormones such as insulin by their site of synthesis (multiple types of cells and tissues in the former case versus specialized glands in the latter) and mode of action. More recently, additional polypeptide-signalling molecules resembling growth factors have been identified by developmental genetic methods in both vertebrates and the fruit fly. The common feature of the polypeptide signalling factors is that they function through binding to specific receptors on the cell surface.

A growth factor is usually synthesized in a precursor form, as a long polypeptide. The precursor is processed to become a short mature molecule by proteolytic cleavage. For example, EGF precursor has 1200 amino acid (aa) residues, while the mature molecule is only 53 amino acids long. The mature, or activated growth factor binds to its receptor on the surface of either the cell that produces it or other cells to trigger a series of biochemical reactions in the process of signal transduction.

Receptors for growth factors are transmembrane proteins having extracellular, transmembrane, and intracellular (cytoplasmic) domains. The extracellular domain is used for recognizing and binding to ligands. The intracellular domain is often a protein kinase, either a tyrosine or serine/threonine kinase. The binding between a growth factor and its receptor activates the kinase domain to phosphorylate its substrates in the cytoplasm. The substrates are often other kinases, and when activated by phosphorylation, phosphorylate their substrates in turn. The series of phosphorylation reactions transfer signals to the nucleus and eventually regulates transcription of specific genes, by either activation or suppression. Activities of growth factors are also regulated by non-signalling binding proteins, in addition to the transmembrane receptors.

The multitude of cellular activities occurring in embryonic development and adult physiological functioning requires a diversity of signals. The number of possible signalling activities generated by the limited number of growth factors may be increased in a number of ways:

- *Dimerization*. A growth factor may function as either a single peptide (monomer) or as a dimer of identical (homodimer) or different (heterodimer) subunits. Receptors may also function as dimers.
- *Interactions of multiple ligands and multiple receptors*. For each growth-factor family there is usually more than one receptor. Each ligand can bind to different receptors and in turn each receptor can bind to different ligands with different affinities.
- *Concentration*. Many growth factors diffuse through tissues and may form a concentration gradient from the producing cell or cells to a certain distance. Different concentrations in the gradient may have different biological effects on target cells.

The development of molecular techniques in the 1980s and 1990s provided powerful tools for investigating developmental

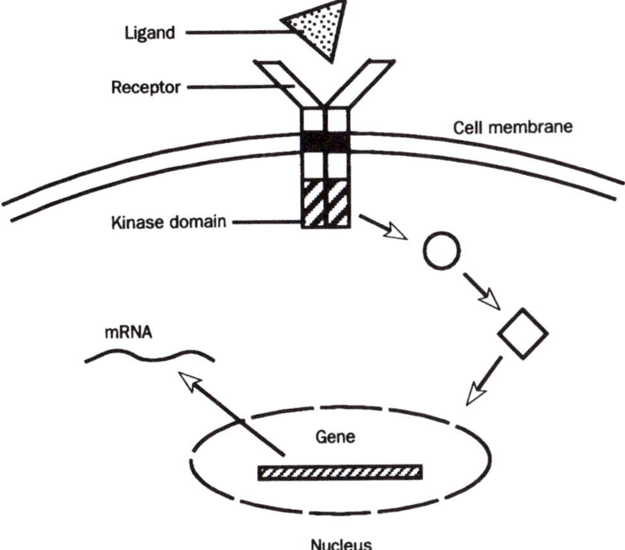


**Schematic representation of growth factor signal transduction pathway (see the text for details). The small circle and square represent the components in the signal transduction cascade.**

mechanisms. Accumulating data on molecular developmental biology indicate that fundamental mechanisms in development are conserved among species of animals and even between vertebrates and invertebrates, for example, gene structures, gene regulation, cell–cell interactions and the factors involved in the processes. In developmental biology, model systems are used to understand fundamental mechanisms. Growth factors once considered as mitogenetic agents in tumorigenesis have been found to be essential factors in normal embryonic development. They control pattern formation of the body and organs, regulate cell proliferation and tissue growth, induce cell differentiation and death, and influence cell migration.

*Models for the study of skeletal development*

*Embryo:*
*Xenopus*, chick, rodent, zebrafish

*Post-natal:*
Rodent, chick, bovine, rabbit, deer antler

*Models of induced bone formation in vivo:*
Subcutaneous implantation of demineralized bone matrix.
Bone and cartilage formation in intraperitoneal diffusion chambers.

*In vitro systems:*
Cell culture: stromal cells, osteoblasts, oesteoclasts, growth plate chondrocytes, periosteal fibroblasts etc.
Organ culture: developing mandible and limb bud, rodent long bone, fetal or post-natal rodent calvariae etc.

*Genetic manipulations:*
Gene disruption: *in vivo* and *in vitro* models.

From Price et al. (1994). The cell biology of bone growth. *European Journal of Clinical Nutrition*, Supplement 1, S131–S149.

## The Xenopus model

The egg of the frog provides a good model for early embryonic pattern formation. At the gastrulation stage, the fundamental body plan is formed with anterior–posterior (A–P) and dorsal–ventral (D–V) axes and three germ layers that later will differentiate to form various types of tissues and cells. Many aspects of this process are controlled by cell–cell interactions. Signals, such as Vg-1 and Wnt-11, from the vegetal hemisphere induce mesoderm formation, and the signals are relayed or amplified by FGFs and TGF-βs (activins, BMPs). The dorsal blastopore, known as Spemann's organizer, produces noggin, chordin, and follistatin to control A–P patterning and differentiation along the axis. The activities of the organizer factors are antagonized by the ventralizing factors, such as BMP-4, to make ventral tissues.

## The vertebrate limb model

Initiation of the outgrowth of limb primordia appears to be controlled by regional expression of growth factors, such as FGF-8. The proximal–distal growth of the limb is regulated by interaction between the apical ectodermal ridge (AER) and the underling mesenchyme. The AER expresses a number of growth factors, such as FGFs and BMPs. Experiments have demonstrated that some of the growth factors can mimic the AER in limb development. The A–P pattern (number and order of the digits) is governed by signals from a posterior mesenchymal region known as the zone of polarizing activity (ZPA). One such signalling molecule is SHH. The AER and the ZPA signals also interact with each other either directly or indirectly to control the limb patterns. D–V pattern formation involves Wnt-7a (dorsal) and BMP-4. *Zhiyong Zhao, Kenneth Weiss and David W. Stock*

See also 'Genetic regulation of growth-promoting factors by nutrition' (3.4), 'Hormonal regulation of fetal growth' (4.1), 'Growth of the human embryo' (4.3) and 'Developmental morphology of the embryo and fetus' (4.5)

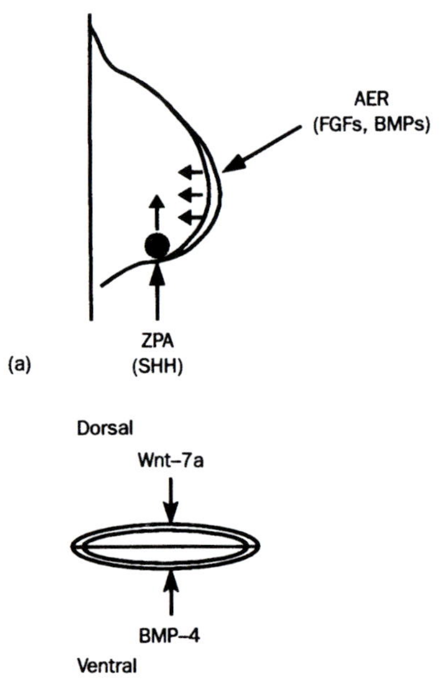


**Diagrammatic representation of signalling systems in vertebrate limb patterning. (a) Side view of a limb bud. (b) Distal view of a limb bud.**

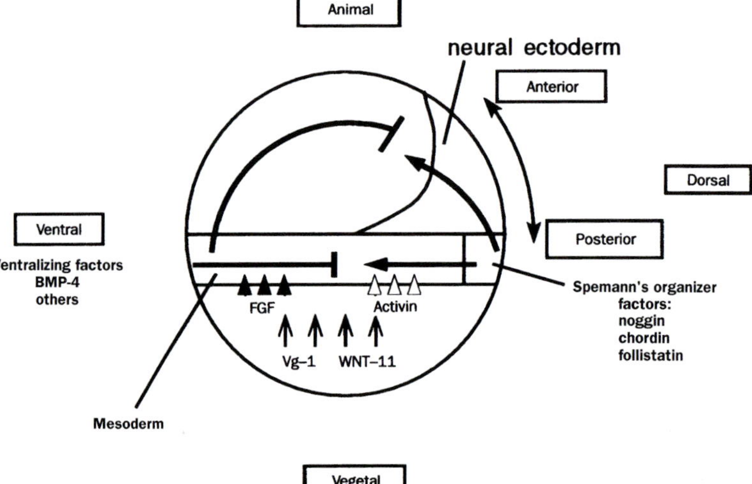


**Diagrammatic representation of signalling systems in the Xenopus embryo. Polarities of embryo are indicated by the boxed words.**

# Growth of the human embryo

The embryonic period, the first 8 post-fertilizational weeks, is the time during which most of the thousands of named structures of the body first become visible. It is also the time during which most congenital anomalies first become apparent. The embryonic period begins with fertilization of an oocyte. The technical criterion for the end of the embryonic period is the initial formation of marrow in the humerus.

## Overall growth

The most useful single measurement throughout prenatal life is the greatest length (GL) of the body, exclusive of the lower limbs, which are greatly flexed. It corresponds to the sitting-height post-natally. The GL is close to the commonly mentioned crown–rump (CR) length, which is not as satisfactory because both points C and R are difficult to determine consistently and with precision. The GL is about 3 millimetres at 4 weeks, 8 millimetres at 5 weeks, 15 millimetres at 6 weeks, 20 millimetres at 7 weeks, and 30 millimetres at 8 post-fertilizational weeks. That a 30-millimetres embryo may be expected to be 8 post-fertilizational weeks of age has been confirmed repeatedly by ultrasonography *in vivo*.

The crown–heel (CH) length involves the addition of three separate measurements prenatally. It corresponds to the standing-height post-natally. The CH length can also be calculated from the GL according to Noback's formula

$$\frac{CH = (3 \times GL) - 3}{2}$$

The CH length, which is about 44 millimetres at 8 weeks, is mostly of interest during the fetal period.

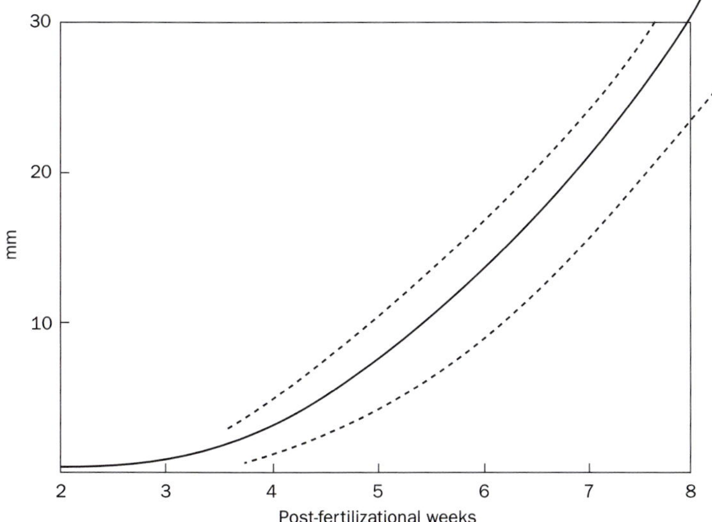

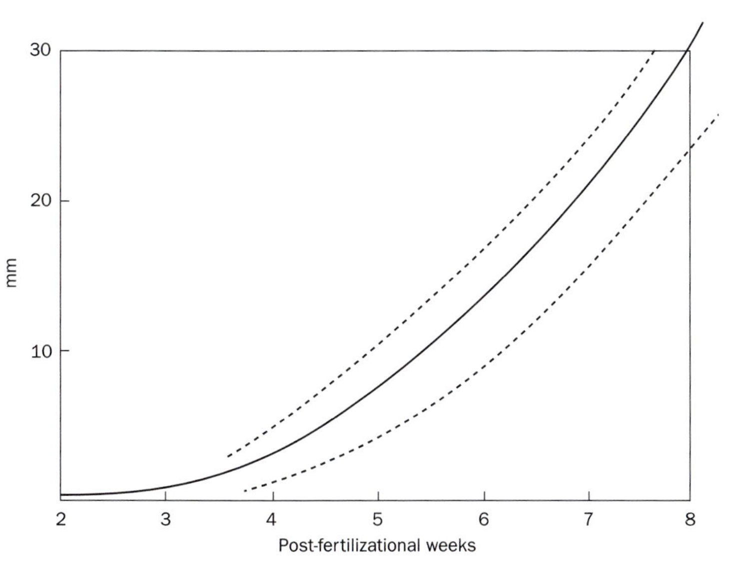

**Graph showing the greatest length (GL), exclusive of the lower limbs, during the embryonic period. The curve is a compromise between several sets of data from transvaginal ultrasonography (n=966) of IVF-timed embryos. Because of divergence among the several series, the area between the interrupted lines would be required to cover about 90% of instances.**

A few other measurements are relevant to the embryonic period. The allantoic diverticulum increases in length from about 0.1 millimetre at 2 weeks to 0.5–1 millimetre at 3 weeks. The umbilical vesicle (or so-called secondary yolk sac) is about 2 millimetres in diameter at 3 weeks and 5 millimetres at 7 weeks. The maximum chorionic diameter increases from about 25 millimetres at 4 weeks to approximately 65 millimetres at 8 weeks, when the chorionic sac is 40–50 millimetres. The placental diameter is then about 50 millimetres. The foot length (FL) can be measured near the end of the embryonic period. It reaches about 4.5 millimetres at 8 weeks. The body-weight at this time is 2–3 grams. Reliable information on the weights of individual organs, such as has been recorded for the fetal period, is not available for the embryonic period.

## Regional growth

### *Head and vertebral column*

The maximum diameter of the head increases from about 2 millimetres at 4 weeks to approximately 13 millimetres at 8 weeks, and the biparietal diameter reaches 30 millimetres at 12 weeks. The maximum circumference of the head is about 40 millimetres at 8 weeks and reaches 110 millimetres at 12 weeks.

The vertebral centra and the basioccipital begin to chondrify early (*stage* 17) and are followed rapidly by the laminae and ribs (*stage* 18) and by the pedicles (*stage* 19). The presellar part of the chondrocranium then develops (*stage* 20). Ossification begins in the mandible (*stages* 18–20) and in the maxilla (*stages* 19 *and* 20). The precursors of most 'membrane' bones of the skull are appearing at 8 weeks (*stage* 23), when the tectum posterius begins to ossify. The vertebral column is about 23 millimetres in length at that time and consists of 33–34 cartilaginous vertebrae. A normal spina bifida occulta totalis is characteristic.

### *Brain and eyes*

It should be stressed that the three major divisions of the brain (prosencephalon, mesencephalon, rhombencephalon) become distinguishable in the completely open neural tube at 3–4 weeks (*stage* 9), before any so-called cerebral vesicles could have appeared. At this time the brain occupies four fifths of the length of the neural tube, and the rhombencephalon remains the longest portion for the next fortnight. At 4 weeks (*stage* 10) telencephalon medium can already be distinguished from diencephalon. At this time the prosencephalon begins its rapid growth, and it occupies one third of the length of the brain. The rostral and caudal neuropores should have closed by 4–5 weeks, resulting in a completely closed neural tube. At this time, when metencephalon and myelencephalon can be discerned in the rhombencephalon, the five major subdivisions of the brain become distinguishable (*stage* 13). The cerebellum also makes its appearance and has its own rate of growth. The growth of the forebrain is again noticeable when the future cerebral hemispheres become distinct at 5 weeks (*stages* 14 *and* 15). The diameter of the eye increases from about 0.2 millimetres at 4 weeks to approximately 1.6 millimetres at 8 weeks.

### *Limbs*

Limb length increases from about 2 millimetres at 5 weeks to approximately 13 millimetres at 8 weeks. Differences between the lengths of the upper and lower limbs are negligible during the first trimester. The skeleton of the limbs, which appears first in mesenchyme (*stages* 15 *and* 16), soon begins to chondrify (*stages* 16–18). Ossification begins in the clavicle (*stages* 18–20) and is found later in the major long bones (*stages* 21–23). Chondrific centres in the hand and foot appear in a definite sequence (*beginning at stages* 17 *and* 18) and their number increases until the total reaches 26 in the foot and 28 in the hand at 8 weeks (*stage* 23).

In summary, at the end of the embryonic period (*stage* 23), 8 weeks after fertilization, the greatest length of the embryo is about 30 millimetres, the crown–heel length 44 millimetres, the foot length 4.5 millimetres, and the body-weight is 2–3 grams.

*R. O'Rahilly and F. Müller*

See also 'Embryonic staging' (**1.1**), 'Prenatal age' (**1.3**), 'Genetic and environmental influences on fetal growth' (**3.2**), 'Neurological development' (**4.6**) and 'Standards and references for the assessment of fetal growth and development' (**4.9**)

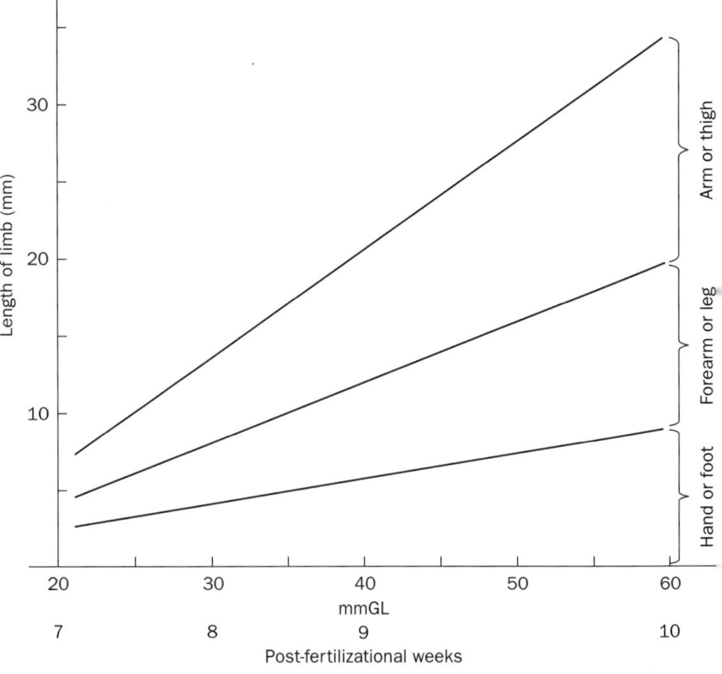


**The increase in length of the upper and lower limbs during the first 10 weeks. Based on data provided by Bossy and Katz.**

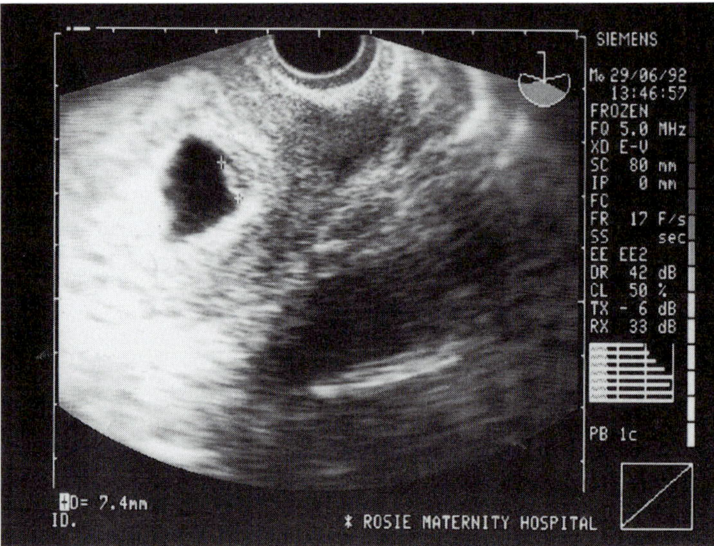


**Embryo at 5 post-fertilizational weeks, with greatest length of 7.4mm.**

# Embryonic development of teeth

Mammalian teeth, regardless of shape, consist of enamel, dentine, pulp, root, the periodontal ligament, and the cementum. Enamel is produced by ameloblasts, which develop from oral epithelial cells. Dentine is produced by odontoblasts, which develop from mesenchymal cells derived from the cranial neural crest, from which cells migrate into the first branchial arch.

The stages in the development of a mammalian tooth are denoted as *initiation*, *morphogenesis*, *differentiation*, and *eruption*. Tooth initiation is characterized by thickening of the oral epithelium at the sites of future teeth. This, using the first lower molar as an example, takes place at about 11 days of gestation (embryonic day 11 or E11) in the mouse, and at about 28–30 days of pregnancy in humans. The thickened epithelial band soon invaginates into the mesenchyme forming a dental lamina at E12, and a tooth bud at E13, and this period is therefore named the *bud stage*. At the same time, the mesenchymal cells condense around the epithelial invagination forming the dental mesenchyme. Interactions between the epithelium and the dental mesenchyme further control tooth morphogenesis and differentiation.

Differential proliferation of the cells in the tooth bud causes morphological changes leading to a cap-shaped structure at E14–E15. The epithelial structure, now called the enamel organ, differentiates into the inner enamel epithelium (IEE), which will become ameloblasts, the outer enamel epithelium (OEE), which will form root structures, the stellate reticulum, the enamel knot, the enamel stratum, and the enamel navel, which are believed to play a role in the regulation of tooth shape. The dental mesenchyme differentiates into the dental papilla and the dental follicle, the latter surrounds the enamel organ and the dental papilla. The dental follicle cells will give rise to the periodontal ligament and the cementum.

At the bell stage (E16–E17) the molar morphology has been established, and cells begin differentiation. Components of predentine and enamel proteins are deposited into the gap between ameloblast and odontoblast layers at about E19–20 and 1–2 days of post-natal development respectively. The enamel proteins and the predentine will later be mineralized to become mature enamel and dentine, as the teeth undergo crown pattern formation and root formation.

By 18 post-natal days in the mouse, the first molar is ready to erupt. Tooth eruption involves tooth movement and resorption of the surrounding alveolar bone. A number of growth factors play a role in this process. Dramatic cytological changes occur during dental cell differentiation. The IEE and the dental papilla cells that directly contact the basement membrane elongate and polarize perpendicularly to the basement membrane, becoming preameloblasts and preodontoblasts respectively. The nuclei move toward the basal part of the cells, and the protein synthesizing and processing machinery takes up the major space of the cell. Within the enamel organ, a layer of cells, the stratum intermedium, develops between the stellate reticulum and the IEE, and produces paracrine stimulating factors (e.g. growth factors) to regulate ameloblast and odontoblast differentiation.

Dentine consists of collagen types I, V, and VI, which form a matrix for the deposition of crystals of carbonate apatite. Some non-collagenous proteins are also found in dentine. Among them, dentine sialoprotein and phosphophoryn are restricted to teeth.

Secretion of enamel proteins occurs about 24–36 hours later than that of predentine, and correlates with the onset of mineralization of the predentine. Amelogenins constitute about 90 per cent of proteins secreted by ameloblasts, the remaining proteins include enamelin, tuftelin, and tuf proteins. Amelogenins serve as a nucleus around which very dense, highly organized hydroxyapatite crystals form. Amelogenins also regulate crystal growth, size, and orientation.

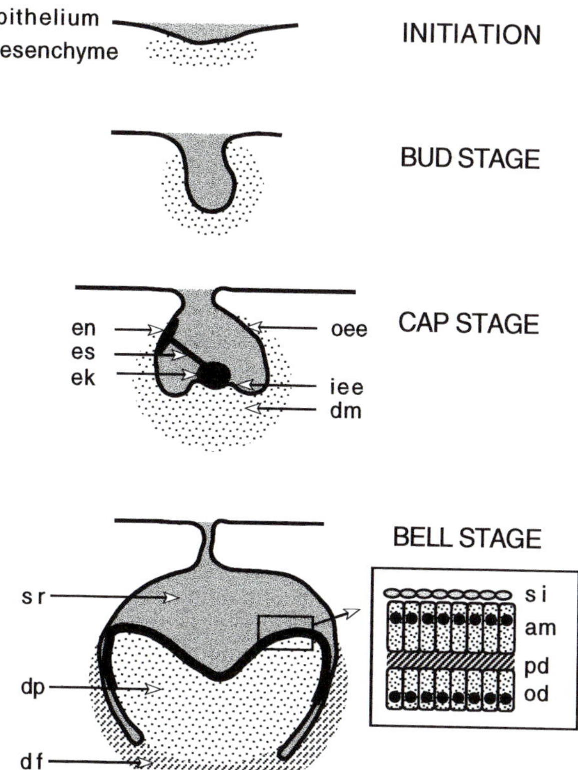


**Diagrammatic representation of molar development: am = ameloblast; df = dental follicle; dm = dental mesenchyme; dp = dental papilla; ek = enamel knot; en = enamel navel; es = enamel stratum; iee = inner enamel epithelium; od = odontoblast; oee = outer enamel epithelium; pd = predentine; si = stratum intermedium; sr = stellate reticulum.**

During enamel maturation they will be removed by proteinases. Dental mineralization is a result of an accumulation of calcium and phosphate ions in predentine and enamel.

A number of regulatory factors, including paracrine/autocrine signalling factors (e.g. Shh, growth factors), receptor kinases, transcription factors (e.g. homeodomain proteins, HMG proteins, nuclear receptors) (**3.3**), and cell adhesion molecules are expressed in the first branchial arches and/or tooth germs. Experimental inactivation of a number of these genes either by *in vitro* antisense, neutralizing antibody inhibitions, or homologous recombination alters tooth initiation and morphogenesis.

## Control of tooth shape and position

The mammalian dentition, unlike that of most fish, amphibians, and reptiles, is heterodont, namely having different-shaped teeth along the jaws. From mesial to lateral (i.e. front to back) of each side of the jaw, specific tooth types (incisor, canine, premolar, molar) are located at specific positions. During embryonic development, the enamel knot may play a role in folding the IEE that creates the enamel gloves, which will lead to different cusps within a tooth. In molars, secondary enamel knots appear on the tip of each cusp. The enamel knot expresses FGF-4, SHH, BMP-3,BMP-4, BMP-7 and MSX-2. FGF-4 a mitogenic molecule, has been demonstrated to stimulate enamel epithelial cell proliferation *in vitro*.

The mechanism responsible for differentiating tooth type along the jaw is not known, although two major hypotheses have been advanced. A field model invokes a chemical gradient of a morphogen along the jaw. Cells at different positions along the jaw react to different concentrations of the morphogen, and give rise to different types of teeth. In contrast, a clone or cell-lineage model proposes that teeth are formed from individual cell clones that are genetically programmed to produce different tooth types before they divide, separate, and form the clones of cells that are the progenitors of each tooth. A so-called progress zone model, based on the observations of alligator dentition, suggests that the number and position of teeth are dependent on the differential growth of the embryonic jaw. Each tooth has a zone of inhibition surrounding it that defines its own position and that of others beside it. However, none of these models has been convincingly established.

## Evolution of dentition

Among the major characteristics of vertebrate evolution was the development of the neural crest. Cells from the neural crest migrate away from the mid-line of the early embryo to interact with ectodermal cells to form various structures in the body. It is thought that teeth originated from exoskeletal sensory pores that evolved protective covering in the form of scales. With the evolution of the jaws, scales in the head and pharyngeal region were recruited for feeding purposes. The structure of teeth (tubular dentine surrounding a pulp cavity and covered with enamel) is found in the scales of some primitive fish, but true teeth exist today only in jawed vertebrates. Other chordates have structures that may be homologous. In conodonts, the earliest chordates in the fossil record, the feeding apparatus consists of an enamel-like covering of lamellar apatite. Teeth in cyclostomes (hagfishes and lampreys) are composed of horny material instead of dentine.

The dentition is a segmental system composed of individual units descended from scales, but in most earlier jawed vertebrates the dentition is homodont, that is, each tooth has a similar, single-cusped morphology. Mammals and some other fossil taxa, have a heterodont dentition consisting of morphologically different types of teeth. An additional distinctive feature of the mammalian dentition is that the teeth are basically fixed in number and only generated one or two times during life, whereas in other vertebrates the number of teeth is larger, much more variable, and teeth are typically regenerated continuously. Tooth attachment has also become more firm and flexible (the cementum is unique to mammals). How heterodonty evolved from homodonty, and how multi-cusped teeth evolved from single-cusped ancestral teeth is not known. However, an early characteristic of fossil mammals was a simple three-cusped molar, which led to the tri-tubercular theory that this structure consisted of three homologues of the earlier single-cusped primitive tooth. It seems likely that answers to how this may have occurred at the gene level will be found with modern developmental genetic methods. *Zhiyong Zhao, David W. Stock and Kenneth M. Weiss*

See also 'Dental maturation' (**1.9**), 'Developmental morphology of the embryo and fetus' (**4.5**), 'Post-natal craniofacial growth' (**5.9**) and 'Variation in time of tooth formation and eruption' (**5.10**)

# Developmental morphology of the embryo and fetus

Changes are by far more rapid and more marked during the embryonic than during the fetal period.

*The human as a chordate is characterized by four features in the embryo.*

| Structure | Morphological appearance (Carnegie stages) |
|---|---|
| Notochord | 10–11 |
| Dorsally situated neural tube | 10–12 |
| Pharyngeal arches | 10–13 |
| Subpharyngeal (thyroid) gland | 10–13 |

## The embryonic period

The first 8 weeks after fertilization are considered in 23 morphological stages. Some of the changes observed are summarized very briefly below.

### *First half of the embryonic period*

The unicellular embryo (*stage* 1) formed at fertilization is a genetically distinct human organism, although the embryonic genome does not become activated until the next stage. The cleaving embryo (*stage* 2) proceeds along the uterine tube. It acquires a cavity within the cellular mass, is then termed a blastocyst (*stage* 3), lies in the uterine cavity, and possesses an embryonic disc that shows dorsal and ventral surfaces. The blastocyst becomes attached to the uterine lining (*stage* 4) and implantation within the endometrium begins. The solid trophoblast (*stage 5a*) acquires lacunae (*stage 5b*) and a vascular circle (*stage 5c*) is formed. The amniotic cavity and the umbilical vesicle (so-called yolk sac) appear (*stage* 5). Chorionic villi develop (*stage* 6) and axial features (such as the primitive streak) become recognizable. The embryo now has right and left sides, as well as rostral and caudal ends. Twinning (other than certain conjoined types) arises before the appearance of axial features. The embryo has now attained determinate individuality, a stable ontological human identity.

A further axial structure, the notochordal process, develops (*stage* 7) and is followed by the appearance of the neural groove (*stage* 8), the first morphological indication of the nervous system. As the embryo lengthens (*stage* 9) the three major divisions of the brain (forebrain, mid-brain, hind-brain) can be detected in the open neural folds. The optic primordium is visible, the heart develops, and somites commence their appearance in pairs. The neural folds then begin their fusion to form the neural tube (*stage* 10) and the optic primordium can be recognized. Cardiac activity can be shown by ultrasound.

### *Second half of the embryonic period*

The ends of the neural tube become closed, first at the rostral neuropore (*stage* 11) and then at the caudal neuropore (*stage* 12). A complicated process termed secondary neurulation begins. The notochord has developed by 4 weeks. Three pairs of pharyngeal arches are now present. The pharyngeal grooves, arches, and pouches are not branchial, i.e., this region does not form gills in mammals. All four limb buds have appeared (*stage* 13). The heart changes from 'in series' to 'in parallel' and atrial septation begins. The future cerebral hemispheres develop (*stage* 14) and the optic cups have formed. The metanephros becomes recognizable. Lobar buds can be distinguished in the bronchial tree (*stage* 15). Transverse subdivisions of the embryonic brain known as neuromeres (called rhombomeres in hind-brain) reach 16 in number at 5 weeks. Pigment is visible in the retina (*stage* 16). Primordial germ cells have migrated to the gonad and the normal umbilical hernia appears. The caudal tip of the body is tapered but is non-vertebrated, i.e., it is in no sense a tail. Embryonic movements can be detected by ultrasound.

Cartilaginous (later replaced by bony) vertebrae arise from somites and develop around the notochord; to follow these complicated changes it is necessary to distinguish the central material (perichordal sheath) from the lateral component (sclerotomes). The four occipital somitic pairs give rise to the basioccipital part of the skull. Chondrification begins in some vertebrae and in the upper limbs (*stage* 17) and the mesonephros begins to secrete urine. The semicircular ducts begin to develop (*stage* 18). Ossification begins in the clavicle and mandible (*stages* 18–20). The upper and lower limb buds are practically parallel, the region of the thumb and big toe being situated on the pre-axial borders (*stage* 19). The upper limbs become longer and bent at the elbows (*stage* 20). Testes and ovaries can be distinguished, and the second interventricular foramen of the heart becomes obliterated. The cortical plate can be seen in the cerebral hemispheres (*stage* 21), and the eyelids and external ears become more developed (*stage* 22). At the end of the embryonic period proper the embryo is about 30 millimetres in length and is 8 post-fertilizational weeks in age. The soles more or less face each other ('praying feet').

## The fetal period

The embryonic period has been studied in much greater detail than the fetal one, and the morphology of the fetus needs much further investigation. The fetal period is

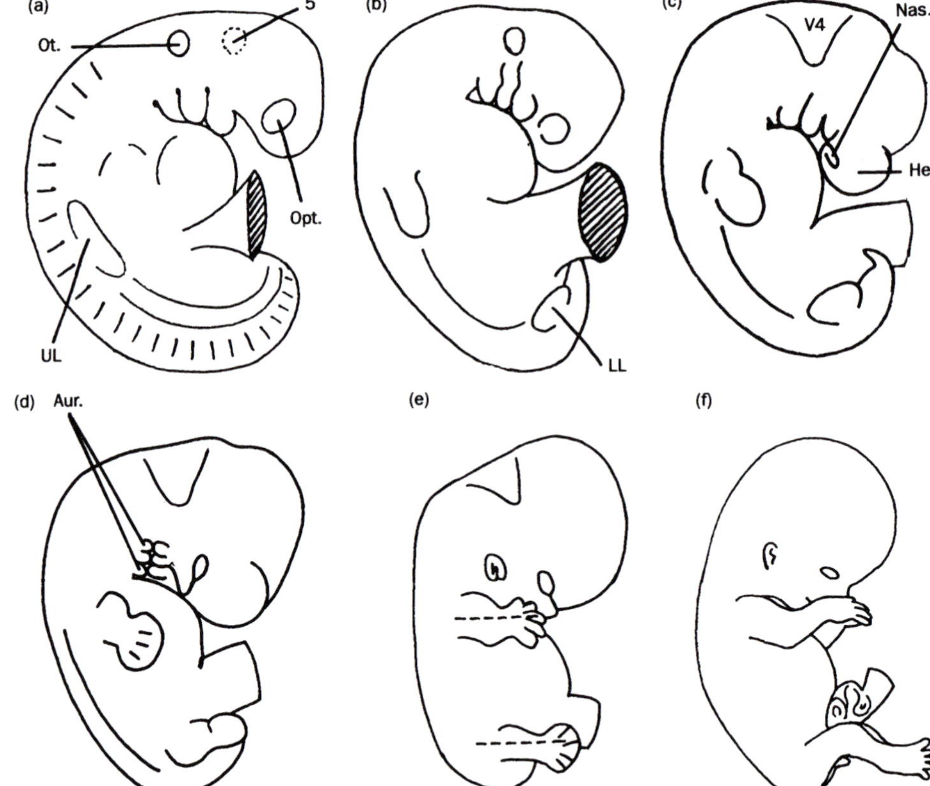


**Right lateral views of six human embryos showing the external morphology at Carnegie stages 12, 13, 15, and 17, 19, 23. They range in greatest length from 4 to 31 mm, but here the outlines have been drawn to the same total height.**

**(a) This embryo has 28 pairs of somites, most of which can be seen from the surface, as can the otic vesicle also, trigeminal ganglion, optic vesicle, and three pharyngeal arches (which are not branchial). The upper limb buds have appeared.**

**(b) Four pharyngeal arches are visible and all four limb buds have formed.**

**(c) The future cerebral hemisphere, the nasal pit, and the hand plate are evident. The tapered caudal end of the body is non-vertebrated and in no sense a tail.**

**(d) Six auricular hillocks, finger rays, and the foot plate are discernible.**

**(e) Toe rays have appeared, and the upper and lower limbs are more or less parallel. Their longitudinal axes are indicated by interrupted lines. The thumb and the big toe are pre-axial in position.**

**(f) This view shows the appearance at the end of the embryonic period. The physiological umbilical hernia is evident.**

***Abbreviations*: Aur. = auricular hillocks; Hem. = future cerebral hemisphere; LL = lower limb bud; Nas. = nasal pit; Opt. = optic vesicle; Ot. = otic vesicle; UL = upper limb bud; V4 = fourth ventricle; 5 = trigeminal ganglion.**

frequently and conveniently considered in trimesters. Many ossific centres appear during all three trimesters.

*Trimester 1*

During the 4–5 weeks following the embryonic period, a number of important features become noticeable. The palatal shelves become fused together either at the end of the embryonic or at the beginning of the fetal period. The eyelids also become fused together, although only temporarily. The fetal uterus can be defined. The normal umbilical hernia becomes reduced. The external genitalia become distinguished as to sex. The fusion of the paramesonephric ducts is completed and the metanephros begins to function. The ciliary body and the iris gradually become distinguishable in the retinal coat. Hair follicles make their appearance. The cerebral hemispheres gradually cover the diencephalon and the mesencephalon.

*Trimester 2*

The following are representative of the changes that occur. The respiratory tree develops acini, and myelination begins in the spinal cord. All 10 layers of the retina can be detected and myelination begins in the brain. Sulci appear on the surface of the cerebral hemispheres at about the middle of prenatal life. The eyelids become gradually separated and the testis enters the inguinal canal. Alveoli develop and the fetus becomes viable during trimester 2.

*Trimester 3*

Dramatic morphological changes are not observed during trimester 3. However, respiratory alveoli continue to develop, the testis reaches the scrotum, although full descent may be delayed until during the first post-natal year. The pulmonary blood flow increases during trimester 3, in readiness for the essential circulatory changes at birth.

In summary, morphological changes are very numerous and very rapid during the embryonic period (the first 8 post-fertilizational weeks), during which most of the thousands of named structures of the body first become visible. It is also the time during which most congenital defects appear. Subsequently, changes continue and some new features arise, but the main characteristic is rather an elaboration of existing structures in preparation for birth.

*R. O'Rahilly and F. Müller*

See also 'Embryonic staging' (**1.1**), 'Ultrasound measurement of prenatal growth' (**1.2**), 'Prenatal age' (**1.3**), 'Genetic and environmental influences on fetal growth' (**3.2**), 'Growth of the human embryo' (**4.3**) and 'Standards and references for the assessment of fetal growth and development' (**4.9**)

## FETAL BODY COMPOSITION

The composition of the fetal body varies enormously across gestation. In healthy individuals, about 180 grams of fat accumulates between 36 and 38 weeks of gestation, about half the total amount between 33 and 40 weeks of gestation. At birth, fat constitutes about 14 per cent of the neonate's body mass, with protein comprising 12 per cent. The vast majority of protein accretion occurs in the first two trimesters of pregnancy, with over 50 per cent of term proportion of body-weight as protein being acquired in the first 16 weeks of fetal life.

With respect to fat-free mass, the proportion of water declines, and the proportion of protein increases steadily across the course of gestation. The fall in the total water proportion is due largely to reductions in intracellular fluid. In parallel with this change is an increase in the predominantly intracellular mineral constituent potassium and a decrease in extracellular sodium and chloride. The rates of calcium accretion in the fetus are greatest in the early stages of development, but continue steadily across pregnancy. This reflects the process of bone mineralization, as does the steady accretion of phosphorus, magnesium and zinc. Zinc is associated with alkaline phosphatase at calcification sites and is deposited within the inorganic matrix, while magnesium may play a role in the control of crystal formation and stability.

Measurements of fetal organ size suggest that the growth of liver, heart and kidneys is similar in form to the exponential increase of fetal body-weight across pregnancy. Although of exponential character, the growth curve of the fetal brain is more complex than that of other tissues.

*Stanley Ulijaszek*

*Body composition of the human fetus (calculated from Metcoff, 1986)*

| Fetal age (weeks) | 12 | 16 | 20 | 24 | 28 | 32 | 36 | 40 |
|---|---|---|---|---|---|---|---|---|
| Weight | 0.02 | 0.1 | 0.3 | 0.75 | 1.35 | 2.0 | 2.7 | 3.4 |
| Percentage weight due to: | | | | | | | | |
| Fat | 0.5 | 0.7 | 0.9 | 1.7 | 3.2 | 5.1 | 8.5 | 14.0 |
| Protein | 5.5 | 6.3 | 7.5 | 8.7 | 9.1 | 9.4 | 10.3 | 11.8 |
| Total minerals | 1.1 | 1.3 | 1.5 | 1.6 | 1.7 | 1.7 | 1.8 | 1.8 |
| Water | 92.9 | 91.7 | 91.1 | 88.0 | 86.0 | 83.8 | 79.4 | 72.4 |

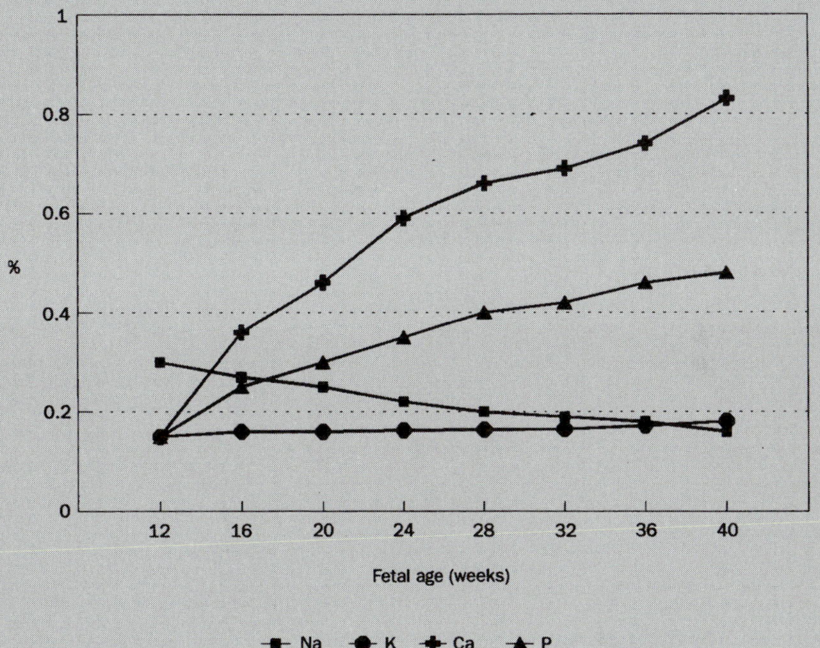


**Proportion of total fetal weight as minerals, across pregnancy. Data from Metcoff, 1986.**

4.6

# Neurological development

There is now greater understanding of the hitherto secret intra-uterine life of the fetus. Today the fetus can be watched from all sides with continuous real-time ultrasound observations.

## Fetal movements

Up to 20 weeks, ultrasound examination is facilitated by the fact that the entire fetus can be visualized within the fields of one ultrasound transducer (**1.2**). The intra-uterine movements of the fetus are similar to movements after birth when the advantageous effects of the amniotic fluid are taken into account. Thus the fetus moves like an astronaut.

Active fetal movements and responses develop earlier and resemble post-natal behaviour much more than previously realized. At an early age the fetus shows a series of movement patterns which do not have a specific function before post-natal life, for example, breathing movements, eye movements, yawns, hiccups, and hand–face contact.

## When does embryo-fetal motor activity start?

*At 7–8.5 post-menstrual weeks* (post-menstrual age is defined as the time from the first day of the mother's last menstrual period) the first and just discernible movements can be seen. The crown–rump length (CRL) of the 8 week-old fetus is only 14.7 millimetres. It is growing at a very high velocity; from 8 to 12 post-menstrual weeks it nearly quadruples its crown–rump length from 14.7 to 51.7 millimetres.

Since the fetal body floats in the amniotic fluid, its specific weight being only slightly above that of the fluid, very little effort is needed for the fetus to move. *At 10 to 12 weeks*, general movements become forceful. They are of large amplitude, frequently lead to changes in the fetal position, and are on average, 10 per hour.

*After 12 weeks* general fetal movements are more variable in speed and amplitude, and may last from about 1 to 4 minutes but wax and wane during this period. The gracefulness of the early fetal movements resembles that of a young female Balinese temple dancer. *By 14 post-menstrual weeks* 16 different movement patterns can be seen. Among these are general movements, isolated movements of the extremities, breathing movements, sucking and swallowing, head retroflexion, rotation and anteflexion, hand–face contacts.

Since fetal motor behaviour represents the output of the central nervous system, a limited neurological examination of the young fetus is possible by this time. *At 13 to 15 post-menstrual weeks*, the number of positional changes of the fetus shows a peak: 10 to 20 changes per hour, including rotations around the sagittal or transverse axis and somersaults. However, the supine position exceeds all other positions during this time. De Vries (1982) has described complex movements of the fetus at this point. These include a complete change in position around the transverse axis with a backwards somersault, achieved by a general movement with alternating leg movements resembling neonatal stepping, and rotations around the longitudinal axis occurring by leg

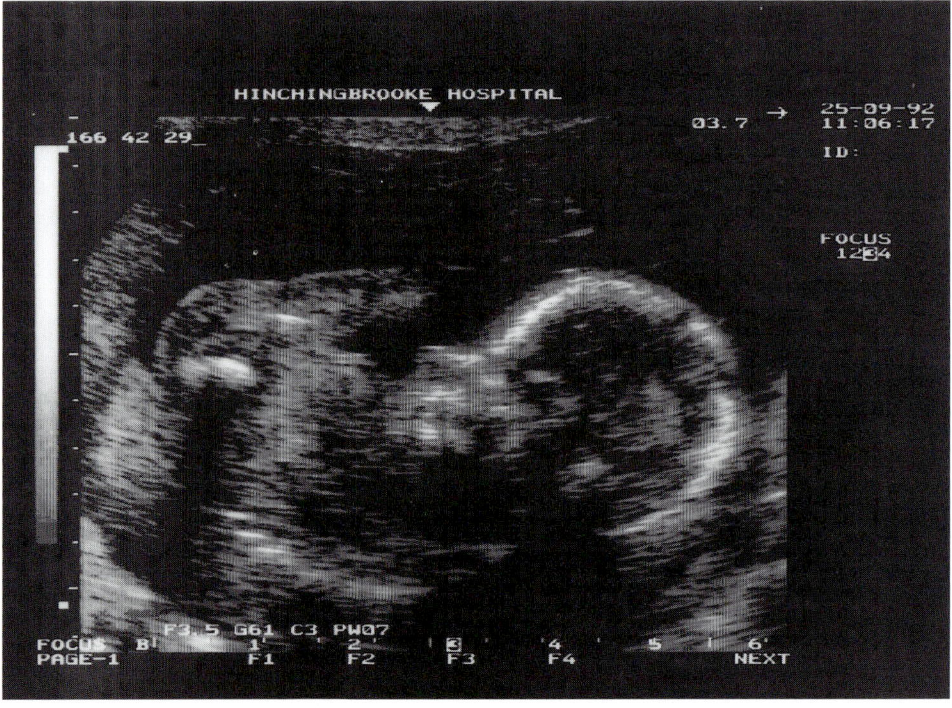


**Fetus at 15 post-menstrual weeks.**

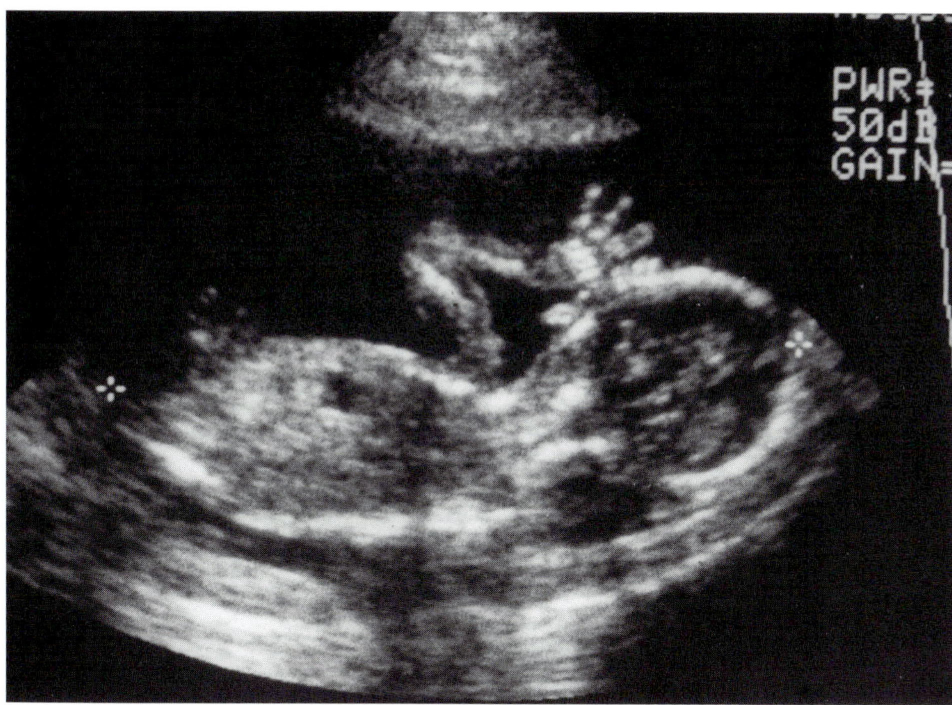


**Hand–face contact in a 17 post-menstrual weeks old fetus.**

movements with hip rotation or by rotation of the head, followed by trunk rotation.

*At 15 weeks* the crown–rump length of the fetus has increased to 101.1 millimetres. The duration of movements increases; they last about 5 minutes with pauses up to 14 minutes. Isolated movements of the arms are more frequent than those of the legs. Rhythmic flexion and extension movements of the legs are seen, unequivocally the stepping movements of the newborn; they are considered a specific intra-uterine adaptation, and continue after birth for a few weeks. *After 15 weeks*, the rate of changes in the fetal position decreases to about 10 per hour on average. This is not the result of a decrease of motor activity but might be attributable to the spatial conditions in the uterus for the rapidly growing fetus. The extent is also dependent upon the amount of amniotic fluid; the findings are variable and are not appropriate diagnostic criteria. The figure shows a 17 post-menstrual-weeks-old fetus with hand–face contact.

*At about 20 post-menstrual weeks*, fetal movements become more vigorous and are felt by the mother as quickening because the wall of the uterus is deformed. *From 24 weeks*, the number and rate of somersaults and loopings decreases, and the fetus, when turning, exhibits marked motility of the extremities. Hand opposition to mouth and repetitive mandible movements are seen, probably thumb-sucking. *From 26 post-menstrual weeks to term* fetal movements have been studied by continuous recording with a tocodynamometer. Four types of movement have been distinguished:

- rolling
- simple movements, i.e., short and easily palpable, possibly originating from an extremity
- high-frequency movement
- respiratory movement.

Many primary reflexes and reactions have intra-uterine functions and remain for some time after birth. Of all five senses the sense of touch seems to be of greatest importance for the fetus. By this it can avoid obstacles and therefore is prevented from entangling itself with the umbilical cord. After intra-uterine deaths, secondary umbilical cord entanglements are frequently observed. Active fetal movements and responses develop earlier and resemble post-natal behaviour much more than previously assumed. The central nervous system is not just switched on at birth – the neonate is well prepared for extra-uterine life, with a continuity of neural functions from prenatal to post-natal life. Thus, according to J.W. Ballantyne (1902): 'Truly, birth marks not a beginning but a stage in life's journey'. *Ingeborg Brandt*

See also 'Embryonic staging' (**1.1**) 'Prenatal age' (**1.3**), 'Growth of the human embryo' (**4.3**), 'Standards and references for the assessment of fetal growth and development' (**4.9**), 'Extra-uterine development after premature birth' (**5.3**), 'Motor development and performance' (**6.2**) and 'Language development' (**6.3**)

# Ontogeny of the immune system

Before birth, growth and development in the specific immune system comprises the production of lymphocytes and the cells responsible for presenting antigen to them. This phase, which continues throughout life, is characterized by the elimination of lymphocytes which could cause autoimmune disease (by virtue of self-reactivity). The two populations of antigen-specific lymphocytes, T cells and B cells, undergo these generation and selection procedures in the thymus and bone marrow respectively. T lymphocytes go through an additional selection step in the thymus by which cells which could not bind antigen are eliminated.

Antigens encountered after birth drive the specific immune system through additional adaptive phases in which cells providing protection are stimulated to multiply, while cells which start to respond to self-antigens have to become tolerant to them. The mechanisms which regulate these two phases of development are incompletely understood.

Most experimental data on development of lymphocytes before birth derive from animals with a much shorter incubation period than humans (mice) or with a differently constructed placenta (sheep). Nevertheless, most of the critical immunity-development steps of mice have their counterparts in humankind.

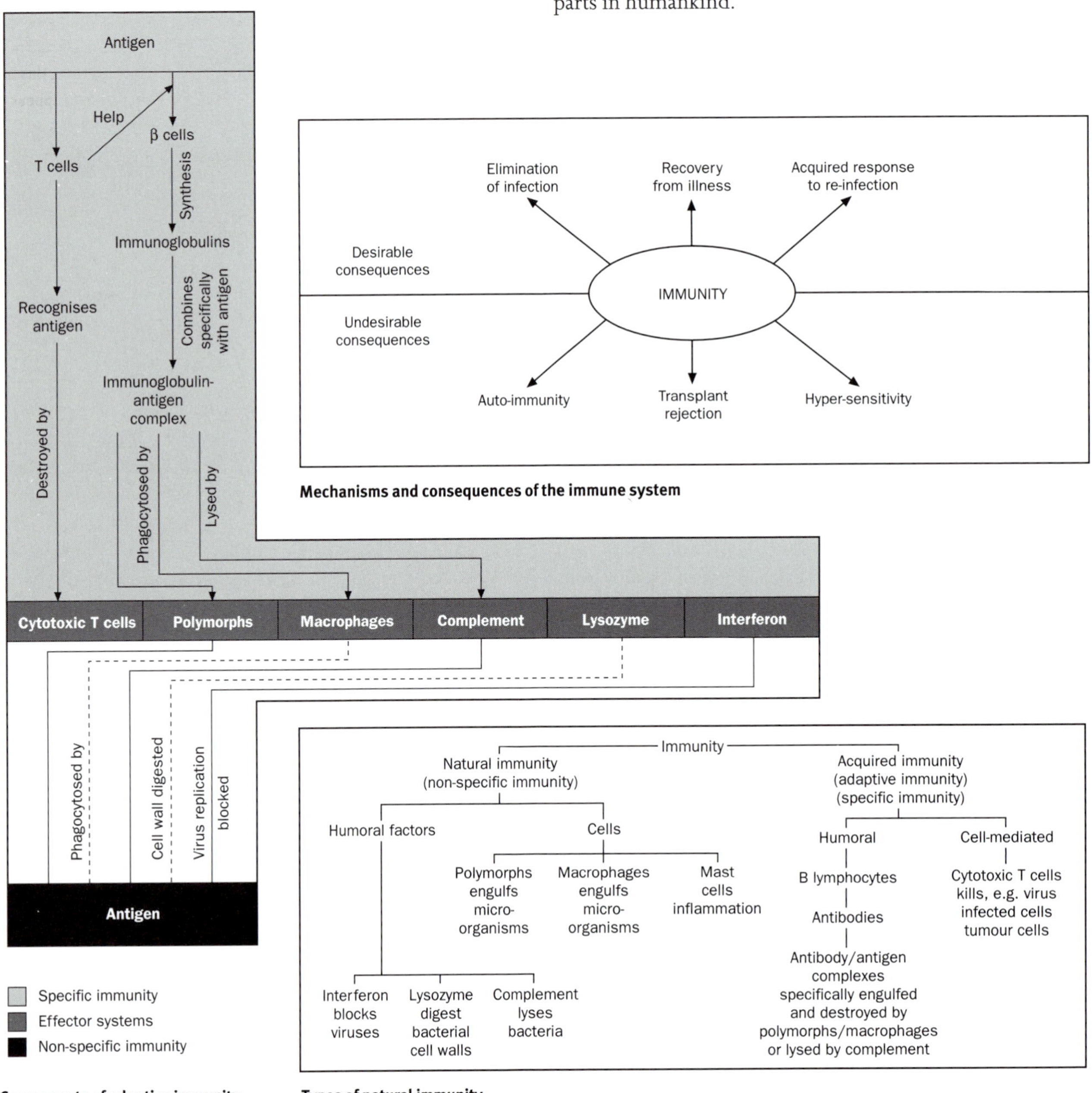


**Mechanisms and consequences of the immune system**

**Components of adaptive immunity**

**Types of natural immunity**

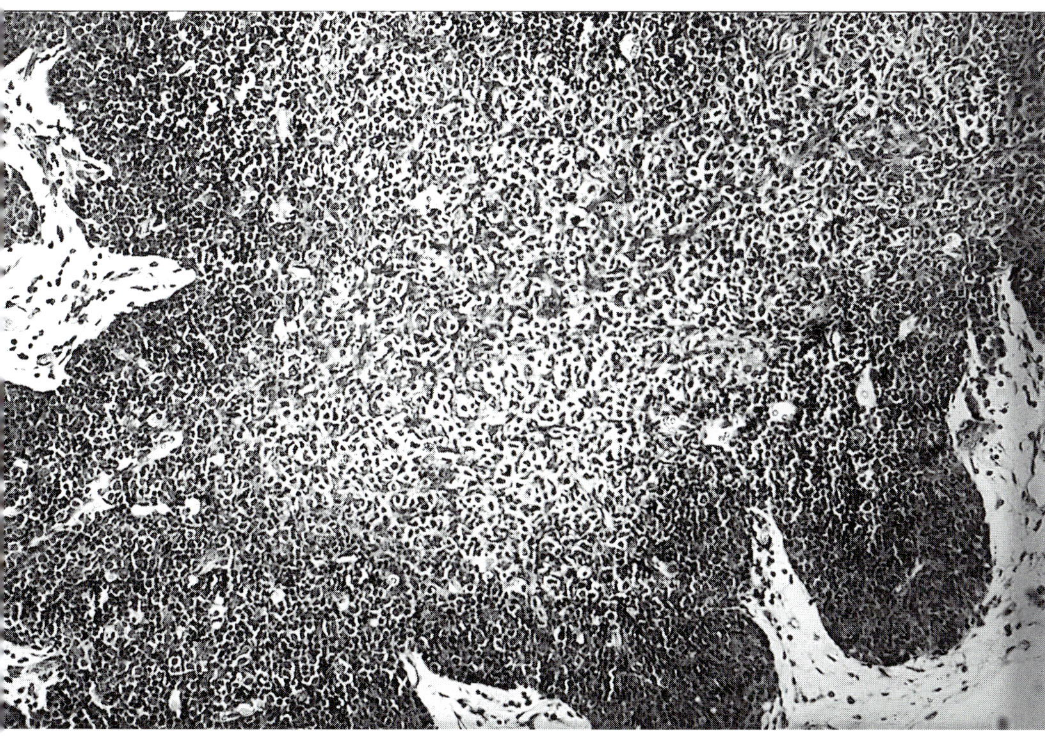

**Human fetal thymus at 10 weeks gestation. Cortical lymphopoiesis has started, but the number of lymphocytes in the medulla is still low. This is an age at which T cell receptor rearrangements are underway, but the thymus has not started to export many selected cells. Hassals corpuscles do not usually appear until 13–14 weeks. Photograph courtesy of Anthony R. Haywood.**

## Stem cells

The lymphocytes, and the specialized antigen-presenting and phagocytic cells which together mediate specific immunity, derive from pluripotential stem cells. Functionally, these cells are defined by unequal divisions, which yield a stem cell and a daughter cell which will differentiate down a myeloid, lymphoid or erythroid pathway. Phenotypically, stem cells are characterized by the presence of CD34 on the cell surface, without the lineage-specific antigens (CD2, CD3, CD4, CD8 or CDw52). Stem cells are present in the human yolk sac at 21 days of gestation and produce erythroid and myeloid lineages from the fifth week. As the fetus develops, the primary location of these cells changes, to the liver around 60 days and to the marrow after 90 days. The differentiation of stem cells to the erythroid or lymphoid series is regulated by a variety of DNA binding, zinc finger proteins (**3.3**) such as the GATA family and, for lymphoid development, the 'Ikaros' family of proteins.

Cytokines such as G-CSF, GM-CSF and stem cell factor are likely to affect the induction and/or suppression of individual regulatory proteins (**4.2**) and hence the production of blood cells. This is currently an area of intense investigation because the stress of birth (or cytokine treatment in adults) is accompanied by the release of marrow stem cells into the blood, from which they can be recovered and used for transplantation.

## B lymphocyte development

CD34+ stem cells from the bone marrow will grow into lymphoid colonies in the presence of bone marrow stromal cells and the cytokine, interleukin (IL) 7. The lymphoid colonies which enter the B cell pathway are defined as pro-B cells when they start to express CD19 on the cell surface. They start to make an enzyme, terminal deoxytransferase (TdT), which contributes to receptor diversification but have not started to make immunoglobulin heavy chains at this point. Further differentiation along the B pathway to a pre-B cell follows the activation of RAG genes and the rearrangement of the V and D gene segments of an immunoglobulin heavy chain to a J segment. If the first rearrangement fails to yield a usable transcript then the VDJ regions on the second chromosome are rearranged. Cells which transcribe and translate heavy chain genes are called pre-B cells before they make light chains. Pre-B cells are present in the fetal liver of mice at 16 days' gestation.

Human fetuses have pre-B cells in the liver, lungs and kidneys from 14 weeks gestation, and the bone marrow takes over as the principal site for B cell production from about 19 weeks. The κ and κ light chains are rearranged 48 or more hours after the heavy chains and, as they become available, a complete immunoglobulin molecule can be expressed on the cell surface. The differentiation of stem cells through the pro- and pre-B cell stages is affected, at least in tissue culture, by the cytokines IL1α, IL4 and TNFα (**5.6**). At this point the developing B cell acquires the ability to bind antigen: either an autoantigen, if the random rearrangement process has yielded a self-reactive cell, or a non-self-antigen. Cells reactive to self-antigens which might be available in the marrow (such as serum proteins) probably activate endogenous endonucleases which result in their own destruction by programmed cell death (or apoptosis **12.3**). Those which do not undergo apoptosis enter the circulation, where they form the pool of naive B cells with the potential for proliferating and differentiating further in the event that their cell surface immunoglobulin binds to antigen.

There are probably at least two pathways of B cell differentiation. In adults the major pathway is in the bone marrow and yields a B2 cell population which can subsequently differentiate to make IgG, IgA or IgE. B1 cells of mice have CD5 on their cell surface and they are thought to differentiate primarily in the peritoneal cavity. B1 cells make IgM following stimulation, without differentiating to other isotypes. The highest concentration of CD5+ B cells in the fetus is in the spleen. Most of the B cells in cord blood are B1 cells and B2 cells do not predominate in blood for several months of life.

## T lymphocytes

The thymus, where T cells will differentiate, is initially formed by the migration of epithelial cells from the third and fourth pharyngeal pouches into the superior mediastinum. In man these anlage fuse in the mid-line around 8 weeks of gestation and the first lymphocytes are seen in the thymus in the end of the ninth week of gestation. Almost all understanding of T cell development comes from mice, in which epithelial migration is complete around day 10 of the 21 day gestation period and lymphocytes are seen in the thymus around day 11.

The critical step which characterizes a cell as developing along a T cell pathway is the rearrangement of the T cell receptor genes. This process starts with the expression of specific recombinase enzymes and a looping of the DNA so that a selected D (diversity) region gene is approximated to J (joining) and C (constant) region genes. Terminal deoxytransferase TdT) can act on the nucleotides as the intervening DNA is excised to introduce additional variability into the receptor diversity. Rearrangement starts with the genes for β, γ and δ chains of T cell receptors on day 13–14. Successful β rearrangements are transcribed and translated to yield a polypeptide which binds to an invariant a chain called pre-T-α. This resulting heterodimer is transported to the cell surface and binds to CD3, where it could presumably be subject to a selection process. The variable α gene DNA, which will ultimately provide the other variable component of the T cell receptor (TcR) is rearranged about 2 days after the β chain. Cells with γδ polypeptides on their cell surface appear on days 15–16 and αβ-expressing cells are found from day 17. Not all rearrangements are successful and a cell may rearrange genes on first one and then the other chromosome, and may follow a γ rearrangement with a β rearrangement until a successful combination can be transcribed and translated.

Although TdT-mediated diversification is important in adult animals, two studies of T cell receptor rearrangements in human fetal T cells at 8 and 12 weeks gestation suggested that diversity was achieved more by germ line recombination than by nucleotidase action. It is possible that TdT-mediated diversity only develops after 16 weeks gestation, when the enzyme becomes detectable in the fetal thymus by immunofluorescent staining.

The T cell receptor for antigen binds to foreign peptides which are held in the grove of major histocompatibility complex (MHC) molecules. TcR rearrangement and the diversity provided by nucleotidase modification at the joins generates binding specificities at random. A selection process (positive selection) is therefore required to prevent the accumulation in the body of T cells lacking binding affinity for the individual's own MHC. Selection is also required to prevent or minimize the production of lymphocytes capable of initiating auto-immune disease. The details of the two selection processes are still incompletely understood. Low affinity binding by developing T cells to any cell expressing MHC molecules in the thymus seems sufficient for positive selection, and cells which either fail to bind, or bind with an affinity high enough to give them auto-immune potential, enter a programmed cell-death pathway in which the nucleus undergoes apoptosis.

The thymus generates at least two major populations of T cells differing in the HLA molecules they bind to. CD4 cells bind to antigenic peptides associated with HLA DR or DQ molecules and CD8 cells bind to antigen peptides on HLA A and B molecules. CD4 expression may be the default pathway, with CD8 expression occurring only if there is positive affinity and triggering by the HLA A or B antigens. Most developing T cells in the thymus express CD4 and CD8 proteins, have low levels of CD3 and express the R0 isoform of CD45, along with CD2 and other integrins. These cells cannot reconstitute irradiated mice and it seems likely that they comprise the population which is destined to die in the thymus, presumably because they had no potential for function. A smaller subset of thymocytes has either CD4 or CD8, high levels of CD3 and the RA or RB isoform of CD45. These cells can function in irradiated hosts and are presumably the cells which survived the selection process.

## NK cells

Cells with CD56+, CD3 surface phenotype of NK cells appear in the fetal liver between 6 and 8 weeks of gestation. Subsequently they are present in the spleen and blood, but not the thymus. Fetal NK cells have the δ and ε chains of CD3 in their cytoplasm but they do not rearrange or express a TcR. Several studies suggest that the NK cells of fetuses and newborns are not as cytolytically active as those of adults. The difference may reflect the cells' exposure to stimulatory cytokines such as IL2 and α-IFN.

## Accessory cells

Although the TcR suffices for antigen binding by a T cell it is not sufficient to trigger T cells to make cytokines, to divide and express effector function. T cell activation requires that the cell makes IL2, and this is contingent on the cross-linking of additional T cell surface molecules. Examples of triggering molecules include CD28 (binding to BB7), CD27 and CD2. CD28 is present on the surface of specialized antigen-presenting cells such as dendritic cells, B lymphocytes and macrophages. Much of the post-natal development of memory T cells must depend on the development and migration of specialized antigen-presenting cells but there is little information on the process in human fetuses. Fetal mononuclear cells can stimulate unrelated T cells in mixed lymphocyte cultures, arguing for at least some maturation of accessory cell function before birth. Dendritic cells purified from cord blood at birth have lower levels of accessory and MHC molecules, and stimulated adult T cells less efficiently than the corresponding adult dendritic cells.

## Immune cells at birth

T and B cell numbers in blood at birth approximate to those found in infants and adults, though the vast majority are 'naive' in that they have not yet been stimulated by antigen. Healthy newborns do not have palpable lymph nodes and their tissues have few or no plasma cells, presumably because maternal antigens are not presented to the fetus in a stimulatory context. The potential for a response is nevertheless present before birth. Perhaps the strongest evidence for this view comes from the presence of plasma cells (implying that B cells have been driven to effector function) in fetuses aborting around 20 weeks of gestation because of congenital syphilis. By the time that infants with other congenital infections (rubella or toxoplasmosis) are born, they usually have IgM antibody in their serum. Since IgM does not cross the placenta it seems likely that the IgM antibody originates from the fetus. It is more difficult to assess IgG production by fetuses because of the maternal IgG antibody which actively transported across the placenta.

## Antigen-driven immune responses after birth

The 'naive' T cells of experimental animals are clearly slower to respond to an antigen stimulus, and make a smaller range of cytokines, than do memory T cells. When human naive T lymphocytes from cord blood are non-specifically stimulated with a mitogen they make IL2 in amounts similar to the memory T cells of adults, but they make much less γ-interferon. A second cycle of stimulation however results in γ-IFN production, but still relatively little IL4 production. The differences in cytokine production appear to be determined by upstream regulators of the γ-IFN gene in naive T cells. Whether a T cell differentiates towards the production of type 1 cytokines (IL2 and γ-IFN) or type 2 cytokines (IL4 and IL10) appears to be determined by the cytokine environment in which a cell is stimulated. Thus, IL4 favours differentiation towards the Th2 phenotype while interferon and IL12 favour the production of the Th1 phenotype. Both pathways of differentiation have been reproduced in tissue cultures of purified newborn CD4 and CD8 cells stimulated through their TcR by anti-CD3.

The delay (compared with adults) in cytokine production by naive T cells in newborn and immature animals is likely to affect responses to pathogens. For example, the differences in the cytokines made by naive and memory T cells in humans may account for the weaker inflammatory responses to viruses made by human newborns compared with infants aged 3 months or more. Sluggish responses to activation by newborn's T cells have been suggested to contribute to the predominance of IgM and slowness to switch to IgG of newborns' B cells. Immunoglobulin class switching depends on the expression of the 39 kDa ligand for CD40 (CD40L) on the surface of activated T cells. Adult T cells express CD40L efficiently after activation by agents which bypass normal activation pathways, such as phorbol and ionomycin. These agents are poor stimulators of CD40L expression by newborns' T cells. Nevertheless, both adult and newborns' T cells have recently been shown to have equivalent CD40L expression when stimulated by antibody to CD3.

*Anthony R. Hayward*

See also 'Maternal HIV infection' (**8.4**) and 'Infection' (**9.3**)

4.8

# Fetal growth retardation

Growth retardation in the fetus has been shown to be related to numerous factors associated with maternal constitution and health and the quality of the maternal internal and external environments. Fetuses which suffer growth retardation have been shown to have greater risks than normally grown fetuses of morbidity and mortality during the perinatal period. They also have been shown to suffer more from physical and mental disabilities and certain chronic diseases later in life. These adverse consequences of fetal growth retardation may be due to the physiologic and metabolic disturbances resulting from the same intra-uterine factors that cause the growth retardation, or they may result from post-natal exposure to the same harmful environment that stressed the mother during pregnancy. A challenge of research on the consequences of fetal growth retardation is to distinguish these various prenatal and post-natal causes of disability, illness and premature death later in life. The assessment of fetal growth retardation is a useful clinical tool in

## INTRA-UTERINE GROWTH RETARDATION: PERSPECTIVES FROM THE BONN STUDY

There is increasing interest in how very low birth-weight (VLBW) infants (infants with birth-weight less than 1500 g) with severe intra-uterine growth retardation develop in subsequent life. In the Bonn Longitudinal Study (BLS), growth patterns of VLBW preterm infants born between 1970 and 1975 were analysed, with special attention paid to those who were born small for their gestational age (SGA), which means with a birth-weight below the 10th percentile of intra-uterine growth curves, but excluding infants with malformations, chromosomal aberrations and prenatal infections. These children represent an interesting and heterogeneous group.

The main cause for intra-uterine growth retardation (IUGR) was maternal pre-eclampsia of the mother. The 51 VLBW SGA preterm infants of the BLS had a mean birth-weight of 1200 grams, and a mean post-menstrual age of 33 weeks. In these infants, nutrition and the amount of energy given immediately after birth are important in allowing them to catch up on the prenatal retardation. Infants were classified according to whether or not catch-up in head growth occurred, and if the neuromotor and psychological outcome was favourable – i.e., whether their developmental quotient (DQ) and intelligence quotient (IQ) was similar to those of the appropriate for gestational age (AGA) preterm infants and the full-term controls of the BLS. This gave rise to two groups: group A (n = 28, mean post-menstrual age [PMA] 32.5 weeks, mean birth-weight 1130 g), with complete catch-up of head circumference and with a favourable development, and group B (n=23, mean PMA 33.2 weeks, mean birth-weight 1230 g), with incomplete or no catch-up growth (**9.6**) of head circumference and mostly with a less favourable development. Not only weight but also growth of head circumference and supine length were retarded although to a lesser degree. Growth retardation started in the third trimester of pregnancy. In the heterogeneous group of SGA preterm infants, the post-natal growth patterns of weight, length and head circumference differed much more than in AGA preterm infants.

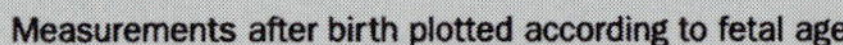


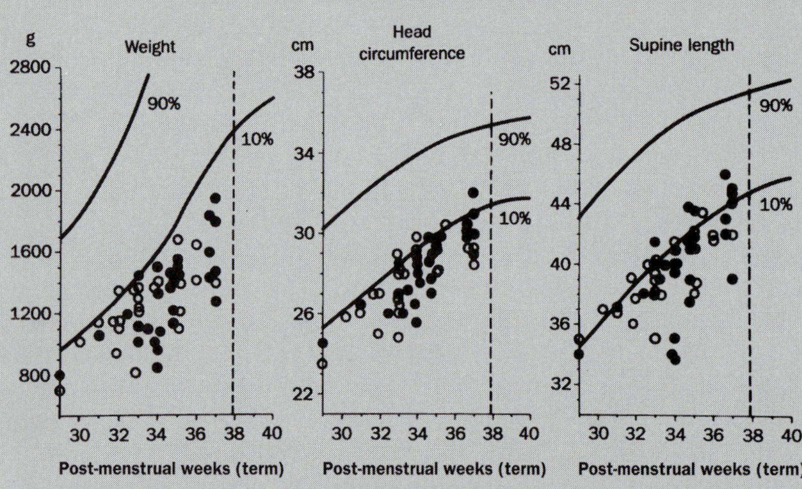


**Birth measurements of the 51 SGA preterm infants, plotted against the Lubchenco curves. Circles with an asterisk: SGA infants with catch-up growth of head circumference, Group A, n=28; open circles: SGA infants without catch-up growth of head circumference, Group B, n=23.**

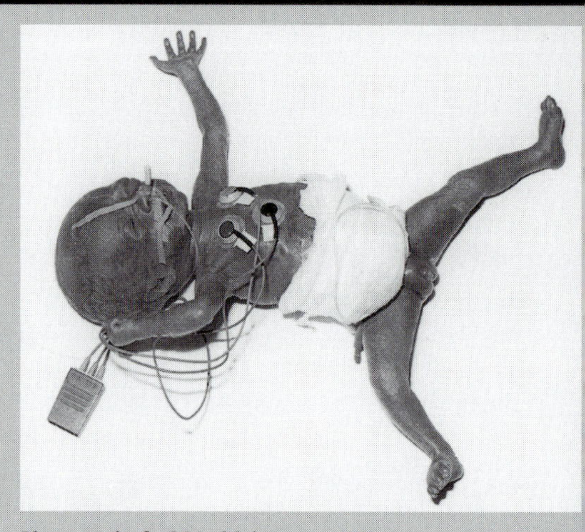

**Photograph of a SGA girl, born at 29 PMW, birth-weight 800g, at the age of 5 days.**

Weight development is most unpredictable. Most of the SGA preterm infants remain lighter than controls during the first years of life. However, at later ages individual children catch up in weight and some of them become even heavier than the controls.

Growth in supine length/height on average remained retarded during the first 2 years, catch-up growth of supine length/height in individual children taking place at later ages than that of head circumference. Furthermore it may occur independently, i.e., whether or not catch-up growth in head circumference was seen in the first months of life.

### *HEAD CIRCUMFERENCE*

Because of the rapidly decreasing growth velocity of head circumference after 6 months of corrected age, the chances for catch-up growth after intra-uterine or early post-natal growth retardation diminish with increasing age. The growth spurt of head circumference extends from the 31st post-menstrual week (PMW) until the sixth month after term with a peak velocity of 4.3 centimetres per month in the 34th PMW. At 6 months of corrected age, head circumference velocity has decreased to one-fifth of its peak, i.e. 0.8 centimetres per evaluating the health and developmental status of the fetus and newborn infant so that appropriate medical, nutritional and social interventions can be prescribed.

Growth retardation implies a process of inadequate growth in size between two measurement periods. Reasonable assessment of this process requires that multiple measurements be taken over time, that gestational age be known, and that there is a standard of growth to which the changes in size can be compared. When assessing growth retardation in the human fetus all of these conditions are problematic. Since it is not possible to remove the growing fetus from the uterus in order to perform the relevant measurements, much of what is known about fetal growth is based on indirect assessment of the process.

In order to describe the normative pattern of achieved fetal growth at various stages of gestation, measurements have been obtained from either aborted or still-born fetuses or from live-born infants delivered preterm. The use of any of these to infer normal growth assumes that the growth trajectories and achieved growth of the dead or prematurely delivered fetus followed the same course of growth as a fetus who survived to the same stage of gestation. It is reasonable to ask how accurately a fetal growth curve based on these cross-sectional measures actually reflects normal fetal growth. There is evidence that live-born preterm infants are actually smaller than fetuses of the same gestational age who remain in the uterus.

The only direct safe method of measuring the growing fetus *in utero* is through the use of ultrasound. However, this

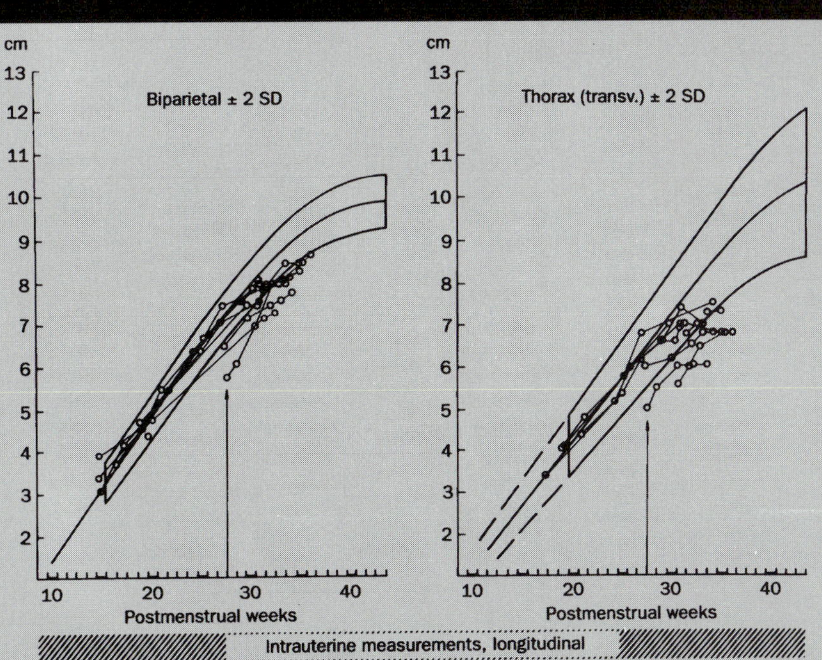


**Intra-uterine ultrasound (US) measurements of the biparietal and thorax diameter of the SGA infants in Group A, who had repeated US-examinations in the second and third trimester, plotted against the standards of Hansmann (1976). The increasing retardation from 30 PWM ('late flattening') is clearly to be seen.**

month, and at 9 months to one-tenth, i.e. only 0.4 centimetres per month. The brain, however, continues to grow at a higher rate, and a velocity decrease to one-tenth of the peak value is observed at 19 months only. In most of the SGA infants of group A, the catch-up in head circumference was accomplished before the corrected age of 9 months.

Since head circumference catch-up growth is closely related to brain development (**2.3, 5.9**), it is strongly dependent on energy supply immediately after birth. After the introduction of early high-energy feeding (from 2 to 3 hours after birth) in 1972 a catch-up of head circumference could be observed provided that there was no other disadvantageous condition which might cause microcephaly. Catch-up growth of head circumference, and thus presumably of the brain, following intra-uterine growth retardation had not been reported before the 1970s. Also, the SGA infants of group B do not exhibit catch-up. At 35 post-menstrual weeks the mean head circumference difference to the AGA preterm infants amounted to 2.3 centimetres, and remained at this level in the following months, being 2.0 centimetres at 6 years. Their mental development, on average, was less favourable than that of the full-term controls. There are still few reports on catch-up growth of head circumference after intra-uterine growth retardation.

In all of the SGA preterm infants of Group A, the catch-up growth of head circumference was completed for the greatest part at the age of 6 months, after a period of very rapid growth velocity. The mean head circumference difference relative to the AGA preterms was 2.1 centimetres at birth, diminishing to 0.5 centimetres at the age of 6 months and to only 0.2 centimetres at 5 years of age.

In the first months of life, and during a period of very high growth velocity, there seems to be a sensitive phase when catch-up growth can take place if environmental and nutritional conditions are favourable. The children of the BLS have now been examined as adults. The SGA infants of group A who experienced catch-up growth of head circumference have realized their genetic potential. Their head circumferences (HC) resemble those of their parents in the same way as do the measurements of the AGA preterms and the full-term controls.

This has been confirmed by a comparison with the target head circumference (THC). THC is derived from father's head circumference plus mother's head circumference divided by 2, adding 0.7 centimetre for boys and subtracting 0.7 centimetre for girls (the basis for this addition or subtraction being to allow for the mean sex difference of head circumference of 1.4 cm). In addition, the mental development of these SGA preterm infants does not differ significantly from that of the full-term controls.

### *A CASE REPORT*

D.G., was a girl, born at 29 PMW with a birth-weight of 800 grams, below the 10th percentile, due to gestosis of the mother. Because of post-natal asphyxia she was on the respirator for about 6 weeks. Oral-feeding started at the third day and there was no initial weight loss. Up to the age of 18 months she exhibited complete catch-up growth of head circumference. At 17 years, with a head circumference of 55.4 centimetres, she nearly reached target head circumference of 55.8 centimetres. Furthermore, there was complete catch-up growth of height. At 17 years (164.2 cm) she is taller than her mother (160.4 cm) and taller than target height of 163.2 centimetres. Her mental development is above average with an IQ of 140. Thus the results of the BLS are encouraging. This case example suggests that intra-uterine growth retardation need not be a disadvantage, if post-natal care and nutrition are good.

*Ingeborg Brandt*

method is restricted to measures of skeletal dimension, such as femur length and biparietal diameter. Estimates of fetal weight from these skeletal measurements are not sufficiently accurate or precise to serve as the basis for fetal weight standards. Estimation of fetal size from measurements of the uterine (symphysis-fundus) height is less accurate than ultrasound but has been used to assess fetal growth retardation with serial measurements in the latter stages of gestation.

*Intra-uterine growth retardation* (IUGR) is the most common term used to refer to the results of inadequate growth in the fetus. Occasionally, *fetal growth retardation* (FGR) is also used. Retardation implies delayed growth which should result in a smaller fetus at the end of the period of observation. which is usually at delivery. Therefore, one often refers to the IUGR infant as being *small-for-gestational age* (SGA), or *small-for-dates*. Not all SGA infants, however, are growth-retarded. Since a small percentage of the population of normal fetuses used to create fetal-growth reference standards are naturally small, and the criteria for diagnosing SGA is a weight or linear measurement below a statistical cut-off value for that population sample (usually the 10th percentile or minus two standard deviations, which is expressed as a −2.0 Z-score), there will be some non-growth-retarded infants falling below that cut-off value. Likewise, fetuses who are larger than the cut-off value may be growth-retarded, but not identified. These would be fetuses whose intra-uterine growth was genetically programmed to result in a larger than average size, but was constrained by intra-uterine factors that slowed its growth-rate and inhibited its size, although not excessively.

Cases of fetal growth retardation can also be missed by this procedure when the growth retardation was experienced early in gestation and compensatory growth later in gestation allowed catch-up growth to take place. However, the probability that a fetus is growth-retarded increases as the measurement of interest deviates more and more below the cut-off value.

Retrospective inference to the timing of fetal growth retardation has been made by examining several different measures in the newborn infant. It is assumed that growth in recumbent length has its maximum growth-rate in mid-gestation while weight-gain is maximum during the last third of the gestation period. By 26 to 28 weeks of gestation, the normal fetus has attained about 70 per cent of its total fetal growth in length but only about 32 per cent of its full-term birth-weight. Therefore, full-term newborn infants with low weights and short recumbent lengths are described

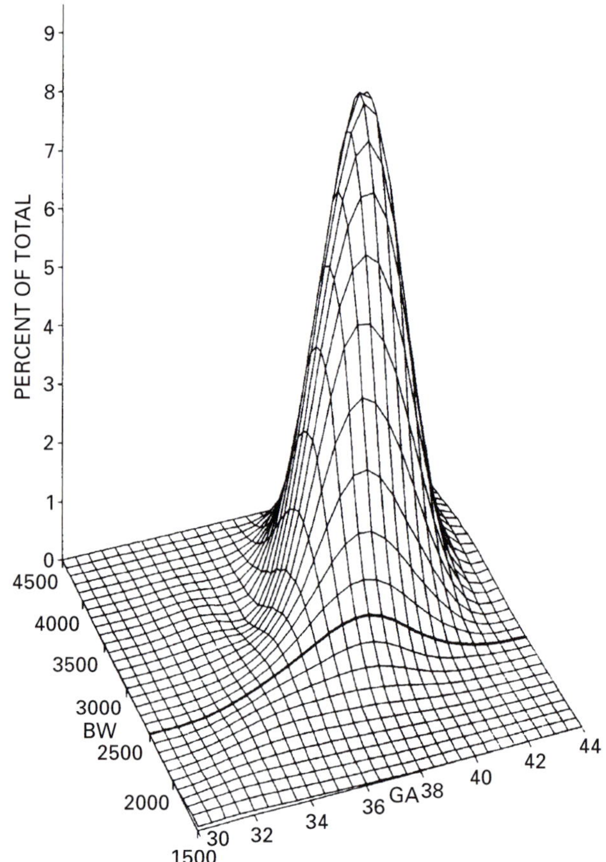


**Three-dimensional relationship between birth-weight (g) gestational age (week) and frequency (%). A cut-off level of 2500 g is indicated. Note the mixture with respect to gestational age in the group < 2500 g.**

as *proportionately* or *symmetrically* growth-retarded and are believed to have suffered growth retardation for much of the gestation period. In contrast, infants with low weights and near normal lengths are called *disproportionately* or *asymmetrically* growth-retarded and believed to have suffered a more recent or acute type of growth retardation. This designation is often made based on the interpretation of *Rohrer's ponderal index* (weight/length$^3$), where a low value represents a disproportionately grown infant.

*Jere D. Haas*

See also 'Genetic and environmental influences on fetal growth' (**3.2**), 'Standards and references for the assessment of fetal growth and development' (**4.9**), 'Maternal anthropometry and birth outcome' (**8.1**), 'Growth in high-altitude populations' (**9.9**) and 'Body-size at birth' (**10.3**)

# Standards and references for the assessment of fetal growth and development

Standards and references are a prerequisite for understanding the patterns of intra-uterine growth. They are indispensable for judging the timing and extent of fetal growth retardation and of post-natal catch-up growth. In addition, they are necessary to allow comparison of intra-uterine growth with the post-natal extra-uterine growth of preterm infants until their expected date of delivery.

Normal growth is one measure of fetal well-being, and intra-uterine growth curves are necessary to quantify normal fetal growth. Numerous attempts have been made to create standards. There are two different methods of assessing intra-uterine growth: one based on extra-uterine measurements of fetuses of different post-menstrual ages; the other based on intra-uterine ultrasound measurement of fetuses at different post-menstrual ages.

## Assessment based on extra-uterine measurements of fetuses of different post-menstrual ages (PMA).

The PMA is calculated from the first day of the last menstrual period, and fetal weight curves have been created by Brenner and colleagues (1976): (a) from 12 to 20 post-menstrual weeks (PMW) based on fetuses aborted with prostaglandins; and (b) from 24 PMW, based on post-natal cross-sectional measurements of preterm and full-term infants born at different PMA.

The maximum percentage increase in weight is in early pregnancy, when weight-gain (growth-rate) is expressed as the percentage increase in weight over the previous week. It progressively decreases throughout pregnancy, from 78 per cent at 12 PMW, to 40 per cent at 16 PMW, and 20 per cent at 22 PMW. Such 'intra-uterine curves' can only be considered as estimates of fetal growth because preterm birth may be related to unphysiological states of variable duration in either mother or fetus, and a comparison with the intra-uterine fetus is not feasible.

## Assessment based on intra-uterine ultrasound measurements of fetuses at different post-menstrual ages.

In the earliest stages of development up to 20 PMW the crown–rump length (CRL) is taken as a growth parameter. With increasing refinement of ultrasound biometry techniques it has become possible to assess the growth pattern of the fetal crown–rump length during the first half of pregnancy. Pioneers of this method were Robinson and Fleming (1975) from Great Britain. Sonar growth data of the developing fetus differ from the anatomical values published by

*Growth of crown–rump length (CRL) in the first weeks of preganacy*

| | Crown–rump length in mm | |
|---|---|---|
| **Post-menstrual age in weeks** | **Streeter (1920)** | **Robinson and Fleming (1975)** |
| 8 | 23 | 14.7 |
| 9 | 31 | 21.9 |
| 10 | 40 | 30.5 |
| 11 | 50 | 40.4 |
| 12 | 61 | 51.7 |
| 13 | 74 | 64.3 |
| 14 | 87 | 78.3 |

Streeter (1920) in that they are smaller. That means that length growth in early pregnancy – like weight – is less rapid as has been assumed formerly.

Measurements of fetal CRL are the most reliable method for gestational age assessment. The reason for the high accuracy of the CRL is the excellent correlation between length and age in early pregnancy, when growth is minimally affected by pathological disorders. Between 11 and 14 PMW, velocity of CRL amounts to 10–12 millimetres per week. This rapid growth velocity enables an accurate estimation of gestational age; one single measurement yields an accuracy of ± 4.7 days, and three independent measurements give an accuracy of + 2.7 days, with a reliability of 95 per cent.

Ultrasound measurements of the biparietal diameter (BPD) are possible from 10 PMW. This was the first parameter used to assess the PMA. Its accuracy is maximal between 12 and 20 PMW. Ultrasound measurements of the BPD, before it exceeds 78 millimetres, corresponding to 29 PMW, enable the obstetrician to estimate fetal age with confidence

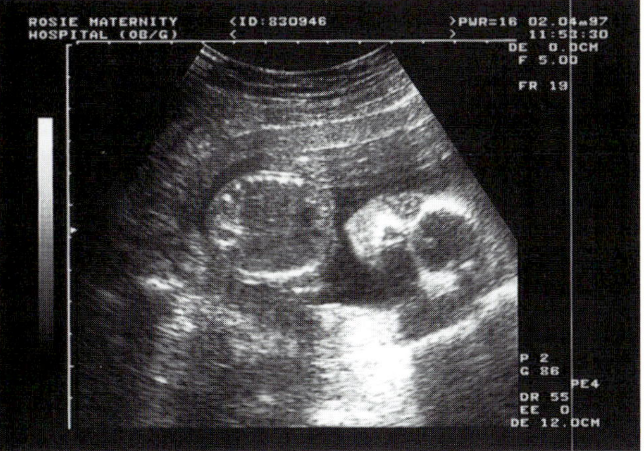


**Fetus at 18 post-menstrual weeks, clearly showing the biparietal diameter and fronto-occipital diameter of the skull.**

limits ± 10 days in 96 per cent of cases. Determination of fetal age in the third trimester by a biparietal diameter greater than 90 millimetres, corresponding to 35 PMW, is clinically no longer reliable.

Standards of the BPD are necessary to judge fetal growth. For the diagnosis of intra-uterine growth retardation, however, knowledge of the exact PMA is indispensable. From measurements of the biparietal and fronto-occipital diameter (FOD) calculation of fetal head circumference (HC) is possible by using the formula of Jeanty and Romero (1986):

$$\text{Head perimeter} = 1.62\ (\text{BPD} + \text{FOD}).$$

Another possibility is to use a modified elliptic formula:

$$\text{HC} = 2.325 \times \sqrt{(\text{BPD}^2 + \text{FOD}^2)}.$$

The 50th percentile of this curve is similar to the 50th percentile of 'intra-uterine curves' constructed from post-natal measurements.

Lubchenco and colleagues (1963) were one of the first groups to publish 'intra-uterine' weight charts in percentiles. In 1966 this was followed by a publication of charts for length and head circumference for practical use. Her standards are still widely used. The 10th percentile for weight serves for separation of appropriate-for-gestational-age (AGA) and small-for-gestational-age (SGA) infants. This differentiation is important for immediate post-natal care and nutrition, and because the prognosis of SGA preterm infants may be less favourable.

Between 33 and 37 post-menstrual weeks the 10th percentiles of Lubchenco and colleagues (1963), Hosemann (1949), Gruenwald (1966), and Hohenauer (1973) agree quite well, whereas the curves of Thomson and colleagues (1968) and Babson (1970) run at a higher level. From 28 to 32 post-menstrual weeks the 10th percentile of the American standards of Brenner and colleagues (1976) agrees with the corresponding Lubchenco curve until 40 weeks; subsequently it runs at a slightly higher level. Until the age of 37 post-menstrual weeks, the 10th percentile of Lubchenco also shows good agreement with the standards of Hohenauer (1980) for the Deutschen Sprachraum. Beyond 38 post-menstrual weeks the Lubchenco curves flatten; this may be due to an intra-uterine growth retardation caused by the high altitude of Denver (1600 m), where the infants were measured.

The mean weight curve of Keen and Pearce (1988) agrees up to 34 PMW with the 50th percentile of Lubchenco and thereafter runs at an increasingly higher level. At 28, 34 and 39 post-menstrual weeks the 50th percentile of Lubchenco agrees with the mean of the Bristol Perinatal Growth Chart of Dunn (1979). With respect to other 'intra-uterine growth curves' which are used as dividing lines between SGA and AGA infants, the 10th percentile of Lubchenco agrees well with the curves of minus two standard devia-

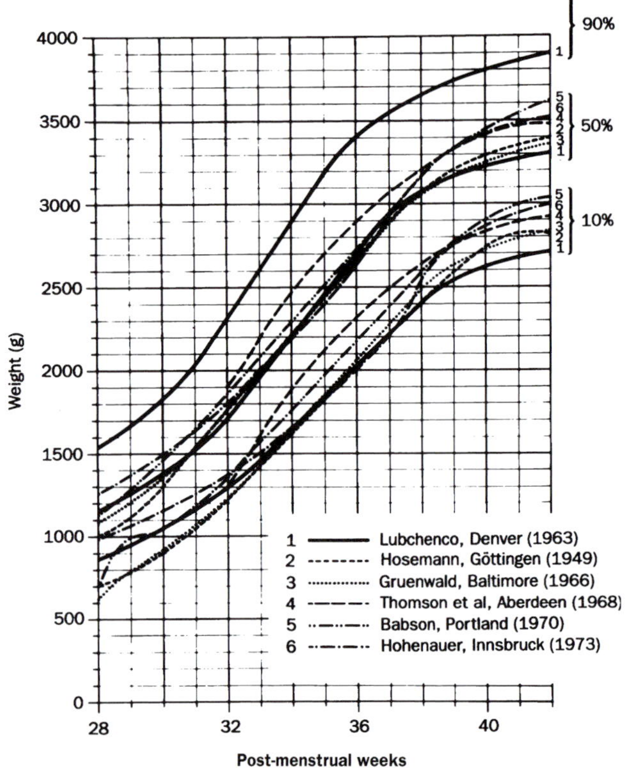


**'Intra-uterine' weight curves of different authors from 28 to 42 PMW, based on birth-weight measurements of infants born at different PMA (cross-sectional).**

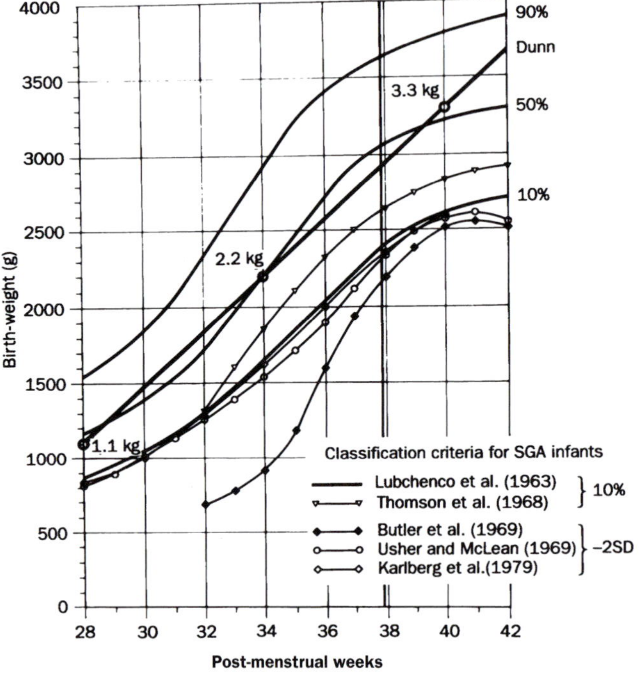


**Classification criteria for SGA infants by 'intra-uterine' growth curves. The 10th percentile of Thomson et al. (1968), as well as minus two standard deviations of Usher and McLean (1969) and Karlberg et al. (1979), are plotted against the percentiles of Lubchenco et al. (1966) and the mean of the Bristol Perinatal Growth Chart of Dunn (1979).**

tions of Usher and McLean (1969) and of Karlberg and colleagues (1979). The 10th percentile of Thomson and colleagues (1968) runs at a slightly higher level from 32 weeks. From that it follows that the widely used 10th percentile of Lubchenco (1963) may well be used for separation of AGA and SGA infants with known PMA, and international comparisons are then possible.

The criterion of 'below the 10th percentile' as a classification for SGA corresponds to international practice over the last 25 years, and the 10th percentile of Lubchenco is used by many authors. The *ponderal index* (PI), or weight–supine-length ratio, was introduced by Rohrer in 1921 for assessment of nutritional status and for comparisons among groups of school children:

$$PI = \frac{100 \times \text{weight in g}}{(\text{supine length in cm})^3}$$

This index can also be used to evaluate the nutritional status of newborn infants. Standards have been set up by Lubchenco and colleagues (1966), among others. Between 30 and 40 PMW the PI of Rohrer increases from 2.33 to 2.62, meaning that the fetus becomes heavier for length during the last trimester of pregnancy. This is in accordance with the growth spurt of weight from 32.5 to 36.5 PMW, when the fat stores are laid down. There is a maximum weight velocity of 1.015 grams per month at 34.5 weeks.

The PI is recommended for classification between different types of intra-uterine growth retardation. In the literature, SGA infants are classified according to the value of Rohrer's ponderal index (1921) into those who are *wasted* and those who are *non-wasted*, or malnourished and well-nourished, using the 10th percentile of the PI as a dividing line.

The use of the term 'small for gestational age' only has its limitations in differentiating between types of intra-uterine growth retardation. In SGA infants, a low PI demonstrates a retardation mainly in weight, and thus a lack of energy reserves; this enhances the risk of hypoglycemia. Therefore the additional use of the ponderal index for classification is recommended.

*Ingeborg Brandt*

See also 'Ultrasound measurement of prenatal growth' (**1.2**), 'Prenatal age' (**1.3**), 'Creation of growth references' (**1.16**), 'Genetic and environmental influences on fetal growth' (**3.2**), 'Growth of the human embryo' (**4.3**), 'Developmental morphology of the embryo and fetus' (**4.5**), 'Fetal growth retardation' (**4.8**), 'Extra-uterine growth after premature birth' (**5.2**), 'Extra-uterine development after premature birth' (**5.3**), 'The development of patterns of deep body temperature in infancy' (**5.4**), 'Maternal anthropometry and birth outcome' (**8.1**) and 'Body-size at birth' (**10.3**)

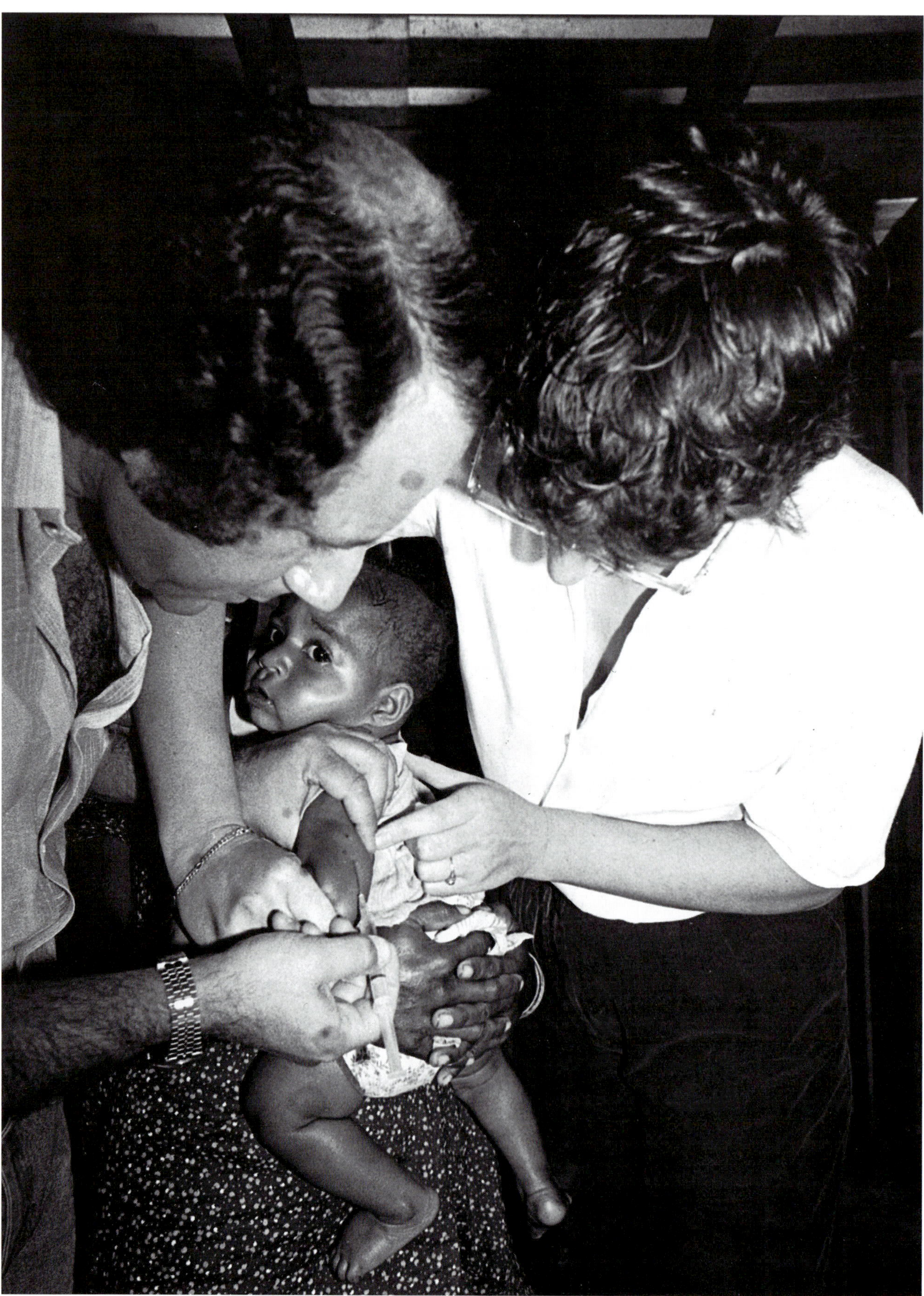

## Part Five

# Post-natal growth and maturation

Growth is the increase in the size of the body or its parts (size attained), while maturation refers to the timing and tempo of progress towards the mature biological state, which varies with the specific system considered, whether endocrine, reproductive, skeletal, digestive or immunological. The processes underlying growth and maturation are cellular: increase in cell number or cell division (hyperplasia), increase in cell size (hypertrophy) and accretion of intercellular substances. The study of growth and maturation is based, on the other hand, upon measurement (e.g., stature, mass, body composition) and observation (skeletal maturation, secondary sex characteristic development) of the outcomes of the cellular processes. The outcomes are the result of complex interactions among genes, endocrine secretions, energy and nutrients, and environmental stresses. The nature of the interactions and resulting outcomes vary with age (timing), beginning prenatally and continuing post-natally.

In Part 5, various aspects of post-natal growth and maturation are considered. Many insights and generalizations about post-natal growth are limited to, or derived from, height (stature) and weight (body mass), and to a lesser extent other body dimensions, tissues and systems, and body composition. However, there is considerable information on changes in other anthropometric dimensions during growth (**5.5**), and progress in other measurable aspects of growth and maturation is enormous, as evidenced in the discussions of endocrine regulation (**5.1**), extra-uterine growth after premature birth (**5.2**), skeletal development (**5.6**), the development of skeletal muscle (**5.8**) the growth plate and cell kinetics (**5.7**), and the development of normal temperature rhythms in infants (**5.4**). Furthermore, the study of growth and development of early-born infants (**5.2**, **5.3**) helps to shed light on growth across the prenatal-postnatal continuum.

**Opposite: Growth and maturation outcomes are the result of complex interactions among genes, endocrine output, nutrition, and environmental stresses; among them, infection. Immunization of infants enhances their resistance to infections and may promote growth directly and indirectly. Photograph courtesy of OK Tedi Mining Ltd.**

Height and weight are the two most widely used, and perhaps mis-used, dimensions in growth studies. Both are complex dimensions, height being a composite of linear dimensions contributed by the lower extremities, trunk, neck and head, and weight being a composite of independently varying tissues. The classic observations of Richard Scammon (1930), including his curves of systemic growth – neural, genital, lymphoid and general – illustrate the differential nature of post-natal growth; i.e., between birth and adulthood, different dimensions, tissues, organs and systems grow at different rates and at different times. This is especially apparent in the growth of the craniofacial complex (**5.9**) and in dental differentiation and maturation (**5.10**). The upper part of the face, for example, follows an overall pattern of growth similar to Scammon's neural curve (see Introduction), while the lower face and mandible follow a pattern similar to that for the body as a whole. Similarly, dental mineralization and eruption tend to proceed independently of the commonly used indicators of biological maturation, i.e., skeletal and sexual maturation, and age at peak height velocity.

An aspect of the study of growth and maturation that has experienced significant progress is the area of body composition (**5.11**). The study of body composition has historically been driven by technology, i.e., what can be measured in contrast to what is desired, and by models developed on adults and animals. Care must be used in applying the latter to infants, children and youth, so that a good deal of the presently available data provide only estimates of body composition. Of particular importance is the chemical composition of the fat-free mass (FFM). It is generally assumed that the FFM is constant, which is not the case. The chemical composition of the FFM changes during prenatal and post-natal growth, and it is not until late adolescence that the FFM reaches 'chemical maturity'. There is a need for more specific data on the chemical composition of the FFM and of specific tissues during infancy, childhood and adolescence.

**Changing physique across childhood and adolescence. Boy of European origin.**

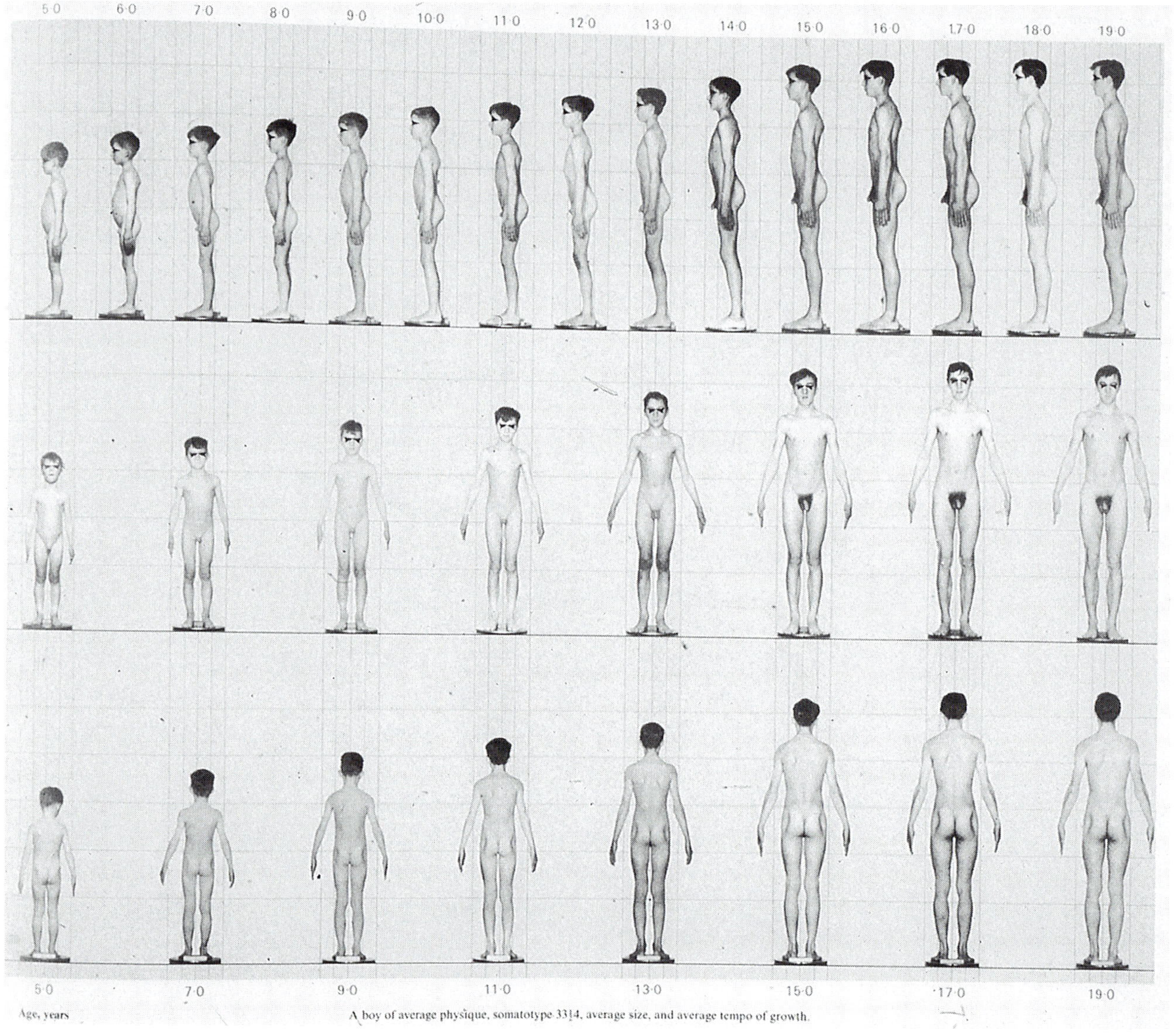

Advances in technology have provided several non-invasive methods to assess skeletal mineral status regionally or in the total body (e.g., single and dual photon absorptiometry and dual energy X-ray absorptiometry), which have contributed to a better understanding of the FFM. Visceral and subcutaneous abdominal adipose tissue are currently of major public health significance, specifically in the context of several metabolic complications in adulthood. Computed tomography is the primary tool in the study of visceral and subcutaneous adipose tissue in the abdominal region. Because computed tomography requires a radiation dose, its application to clinically normal children is unlikely. The development of magnetic resonance imaging has provided some information on the development of abdominal visceral and subcutaneous adipose tissue in children and youth, and in turn to further understanding of relative adipose tissue distribution (**5.16**). Relative fat distribution in growth studies is commonly assessed with ratios of skinfolds and principal-components analysis in cross-sectional data. The pattern of change in individual skinfolds on the trunk and extremities needs further longitudinal study, especially during the adolescent growth spurt and sexual maturation.

**Girl of African parentage. Changing physique across childhood and adolescence.**

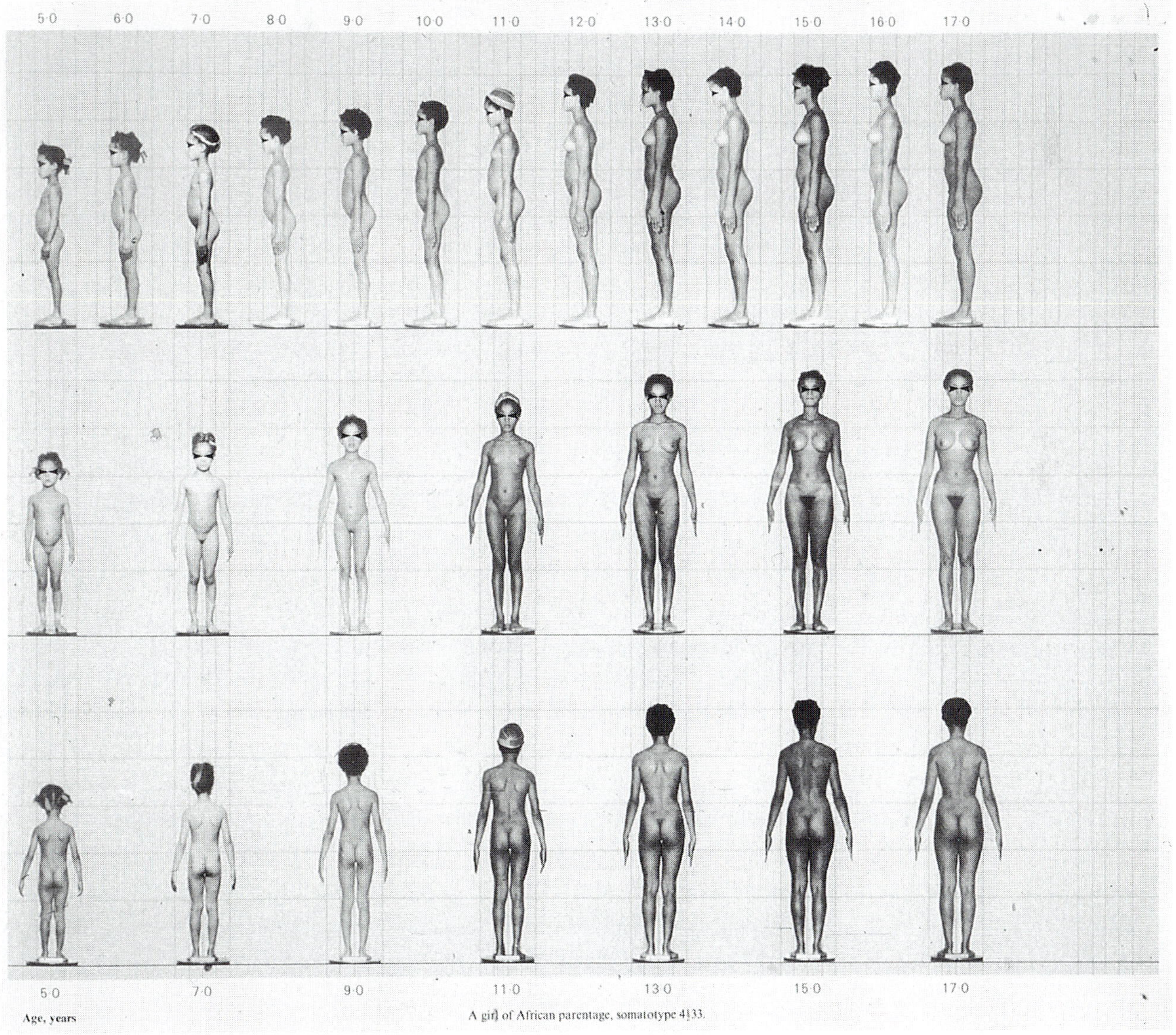

In the assessment of sexual maturation (**5.15**), the literature is dominated by the terms 'Tanner Scale' or 'Tanner Stages'. The criteria described by Tanner in 1962 are used for breast, genital and pubic hair maturation, and it is more appropriate to describe the stages specific to each characteristic, i.e., breast stage 3 or genital 3, rather than 'pubertal stage 3' or 'Tanner stage 3'.

Epiphyseal union represents the attainment of skeletal maturity. Epiphyseal union also marks the cessation of physiological growth of a long bone in length. The process of union as observed on radiographs of the hand–wrist is well described, due in part to the widespread use of hand–wrist radiographs in growth studies. Note, however, that skeletal maturation begins prenatally and continues into the third decade of life, and variation in ossification and epiphyseal union in other regions of the body is considerable. It is important to extend observations beyond the hand and wrist to the entire skeleton, and to study aspects of developmental maturity beyond the age of epiphyseal fusion (**5.17**).

Growth of the individual is not a linear or continuous process. There is cyclicity or seasonality of post-natal growth and maturation, as well as pulsatility of linear growth (**5.13**). The adolescent growth spurts in height and weight are well studied, as is the endocrinological regulation of that growth (**5.14**), but information on adolescent spurts in specific dimensions, tissues and functions is less extensive. The mid-growth spurt in height is less well studied. Apparently, not all children show a mid-growth spurt, which probably reflects real biological variation among individuals and the frequency with which height is measured. The debate about the existence of mini growth spurts and saltations (**5.13**) highlight the importance of issues related to the identification and interpretation of short-term growth, for example, total body-length or stature, or lengths of individual bones or segments; measurement variability; diurnal and seasonal variation; and methods of mathematically fitting the data. Nevertheless, the evidence suggests that growth in length or height, or of individual bones is a non-linear process, which highlights the limitation of attempting to predict long-term growth from short-term observations.

There is currently some confusion about the role of physical activity as a factor which may influence growth and especially sexual maturation (**5.12**). This confusion arises in part from a loose or imprecise use of the terms 'physical activity', 'exercise' and/or 'training' (physical activity or exercise are not equivalent to training); the lack of adequate longitudinal observations in which physical activity/training are controlled; a tendency to make inferences from athletes in several sports with highly select criteria to the general population; erroneous statistical inferences (correlation does not imply a cause–effect relationship); and a failure to consider other factors that are known to influence indicators of growth and maturation.

Regular physical activity and training for sport are, however, associated with presumably favourable changes in body composition, i.e., decrease in fatness, increase bone mineralization, and possibly an increase in fat-free and muscle mass. Changes in body composition tend to be specific to the type of training program, i.e., strength or resistance versus endurance, weight-bearing versus non-weight-bearing, while changes in fat and fat-free masses depend upon continued training for their maintenance. Both skeletal and muscular tissues normally increase with growth and maturation in both sexes so that it is difficult to partition effects of training upon these tissues from expected changes associated with normal growth and sexual maturation during puberty. Furthermore, there is a need for documentation of just how much activity is needed to maintain changes in body composition associated with training.

*Domains of growth, maturation, and development*

| Growth | Maturation | Development |
|---|---|---|
| Size | Skeletal | Cognitive |
| Physique | Sexual | Emotional |
| Composition | Somatic | Social |
| Systemic | Neuromuscular | Motor |
| | Self-concept | |
| | Perceived competence | |

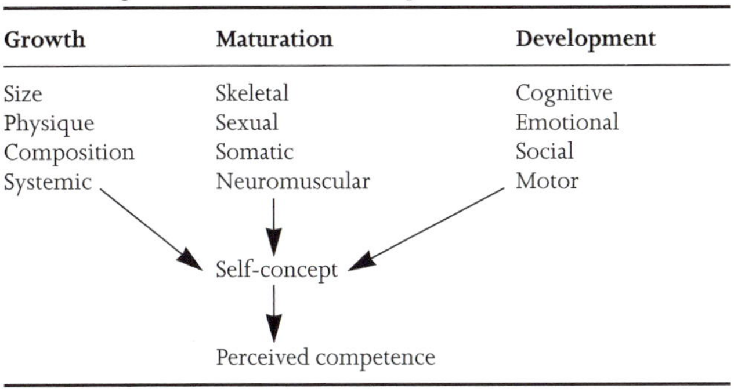

Although different aspects of post-natal growth and maturation have been considered by various researchers, one danger of reductive agendas in studies of growth and development is that interactions among growth, maturation and development are often forgotten. Development refers to behavioural competence in a variety of interrelated domains: social, emotional, cognitive and moral competence, as the child adapts to the behavioural demands of his/her culture. Motor competence may be included (**6.2**). In discussions of sexual maturation (**5.15**) an 'increased competitive ability' associated with size and maturity in males has been identified. Males show adolescent spurts in strength, power, speed and maximal aerobic power, but it is difficult to relate these spurts to 'competitive ability', which is influenced by the cultural context. There is thus a need to consider biological and behavioural interactions that occur during childhood and adolescence, from a biocultural perspective.

The term 'development' is also used in the context of functional competence, including sexual (**5.17**), and muscular, including aerobic power (**5.7**). The context, however, is usually biological and often does not include the behavioural correlates of biological maturation. Discussions of aerobic power highlight the problem of normalizing the expression of maximal aerobic capacity per unit body mass or fat-free mass. The limitation of this type of analysis has been recognized in auxology for some time, and a variety of approaches have been used with data for maximal oxygen uptake, for example, allometric regression, multilevel analysis and covariance analysis. This emphasizes the need for further study of allometry (**2.1**), i.e., differences in proportions correlated with changes in absolute magnitude of the total organism or part of it, during post-natal growth and maturation.

*Robert M. Malina*

5.1

# Endocrinological regulation of post-natal growth

Post-natal growth arises predominantly as a result of growth hormone secretion and the final component of post-natal growth puberty, from an interaction between growth hormone and the gonadal steroids. Growth hormone (GH) is a 22 kilodalton (KD) protein comprising of 191 amino acids which is secreted in a pulsatile fashion from the somatotrophs in the anterior pituitary gland. Growth hormone in turn is regulated by two hypothalamic peptides, growth-hormone-releasing hormone and somatostatin. Growth-hormone-releasing hormone, a 40–44-amino-acid protein, causes release and synthesis of growth hormone whereas somatostatin, which inhibits growth hormone release is largely involved in the modulation of the frequency of GH pulses. The hypothalamic peptides cycle at 180 degrees out of phase with each other, so that a GH pulse is generated by a reduction in somatostatin levels within the hypophyseal portal system, coupled with a rise in growth-hormone-releasing hormone concentrations.

Growth-hormone-releasing hormone promotes GH synthesis and secretion by binding to a specific receptor. The growth-hormone-releasing hormone receptor is a member of a family of G protein receptors, the characteristic feature of which is the presence of seven hydrophobic membrane-spanning helices. Ligand-receptor binding leads to activation of a G protein secondary message system and cyclic AMP generation. A complex series of interactions then take place involving protein kinase A and phosphorylation of cyclic AMP response-element-binding protein leading to increased transcription of the growth hormone gene. Secretion arises as a result of interaction with calcium channels.

In humans the pulsatile pattern of GH secretion is associated with the growth-rate of the individual. There is a dose-dependent relationship between the amount of growth hormone secreted over a 24 hour period and the growth-rate of an individual. The system appears to be pulse-amplitude rather than -frequency modulated.

There are receptors for GH on just about every single cell in the mammalian body. The receptor induces a wide range of actions, including long-bone and soft-tissue growth, fat metabolism and insulin action. Signalling to the target organ takes place through a specific growth hormone receptor and arises as a result of dimerization of two growth hormone receptors with one growth hormone molecule. The receptor has a single transmembrane domain, an extracellular hormone-binding domain of 246 amino acids and a cytoplasmic domain of 350 amino acids. The dimerization process leads to tyrosine phosphorylation of the Janus kinase family and of proteins of the signal transducers and activators of transcription family. These transcription factors then migrate to the nucleus and bind to DNA elements responsible for the modulation of gene transcription (**4.2**). In many cells the end result is the local generation of insulin-like growth factor I (IGF-I).

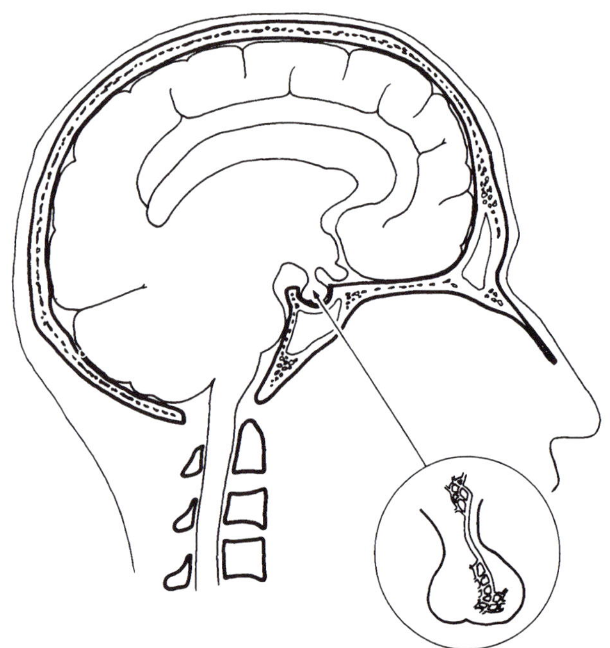

Diagram showing position of pituitary gland immediately below, and joined to the brain. The hypothalamus is the area of the brain just above the pituitary. The circled inset shows the system of blood vessels going from the hypothalmus to the anterior part of the pituitary.

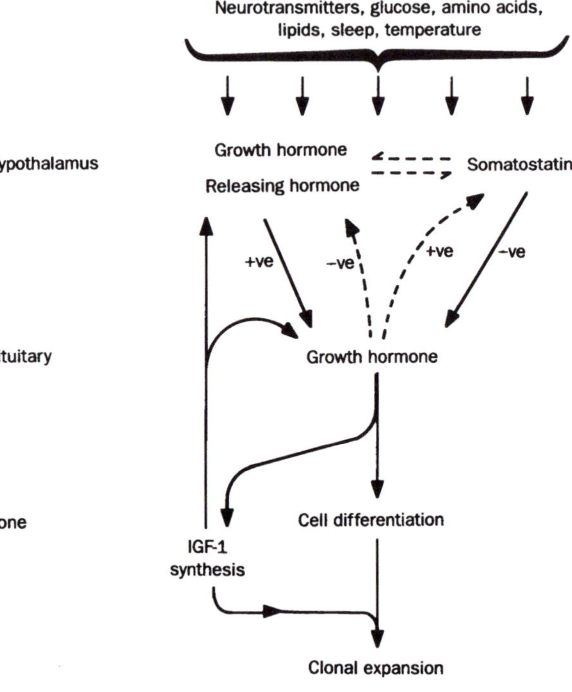


Regulation of growth hormone secretion, and growth hormone action.

This peptide acts predominantly in an a paracrine fashion although it does have some endocrine function.

In the circulation, IGF-I is complexed to a binding protein (IGFBP-3) and an acid labile subunit. This prolongs the half-life of the peptide. Other binding proteins, particularly those in the extracellular space, alter the delivery of IGF-I to the tissue. IGFBP-1, for example, is also negatively regulated by insulin, leading to the reduced availability of IGF-I during starvation; when insulin levels are low, IGFBP-1 levels are high. Teleologically this is useful as this slows growth during a period of substrate unavailability.

At the growth plate, GH appears to be important in allowing differentiation of the chondrocyte precursors to take place. Local generation of IGF-I follows and it is the effect of local production of IGF-I which leads to clonal expansion of the chondrocyte cell line and ultimately growth.

Sex steroids modulate the final component of post-natal growth, namely the pubertal growth spurt. These agents influence growth hormone secretion at hypothalamic and pituitary levels. Both testosterone and estradiol increase growth hormone pulse amplitude without altering pulse frequency. Aromatization of testosterone to estradiol (**5.15**) appears to be the important step, rather than the conversion of testosterone to its more active metabolite dihydrotestosterone. Estrogens have been shown to be important modulators of growth hormone gene transcription and are more efficient at this process than testosterone.

*Peter Hindmarsh*

See also 'Genetic regulation of growth-promoting factors by nutrition' (**3.4**), 'Hormonal regulation of fetal growth' (**4.1**), 'Genetic disorders of the growth hormone axis' (**7.6**), 'Endocrine growth disorders' (**7.7**), 'Treatment of growth disorders' (**7.9**) and 'Gene–environment interactions' (**10.2**)

# Extra-uterine growth after premature birth

There are numerous studies on post-natal growth of very low birth-weight (VLBW) (< 1500 g) preterm infants. However, there are only a few among them where the infants were measured longitudinally at short intervals from birth. Short intervals between measurements are important if velocity curves are to be calculated. Growth velocity is very high and decelerates quickly during the first months of post-natal life, and frequent measurements are essential for evaluation of growth at a time when it is so rapid can easily be negatively affected by disadvantageous environmental factors and disease. On the other hand, growth may be promoted by favourable conditions, such as optimal nutrition and care, when even catch-up growth may be possible.

In the past, there have been contradictory findings with respect to the post-natal growth of preterm infants. In many studies, it has been reported that preterm infants remain below the growth standards of full-term infants. However, in some other studies authors have reported no growth differences between preterm and full-term infants. Some of the more recent publications on VLBW infants consider birth-weight only, disregarding the post-menstrual age at birth and making no differentiation between appropriate for gestational age (AGA) and small for gestational age (SGA) infants. In some studies, the growth of preterm infants is reported to be less favourable than that of full-term controls.

Normal intra-uterine growth is considered a good marker of fetal well-being; the same is valid for post-natal growth of preterm infants. In VLBW preterm infants, especially SGA, nutrition and the amount of energy given immediately after birth play an important role for the prevention of growth failure.

## The Bonn study

The growth patterns presented here are based on a longitudinal study of growth and development from birth of VLBW preterm infants. Between 1967 and 1975, all preterm infants with a birth-weight of 1500 grams and below who were admitted to the Bonn Universitats-Kinderklinik were included in the study. By the criteria of post-menstrual age and birth-weight, the 116 preterm infants of the Bonn study are classified into 65 AGA infants, and 51 SGA infants. This report is based on the group of the 65 AGA infants, mean birth-weight 1350 grams and mean post-menstrual age 30.5 weeks. The study also includes 85 full-term infants as a control group.

All the body measurements were made by the author, at weekly to fortnightly intervals before term, at monthly intervals during the first year, at 3-month intervals in the second year, and 6-month intervals thereafter. Since the growth velocity in the last 10 post-menstrual weeks and in the first months after term is very high and changes quickly, frequent examinations at short intervals are essential for setting up velocity curves (**1.14**).

*Head circumference velocity (in cm/month), for boys and girls: AGA preterm infants.*

| | | | Percentiles (%) | | |
|---|---|---|---|---|---|
| Post-menstrual age (weeks) | Mean | SD | 10 | 50 | 90 |
| 30.5 | 3.32 | 1.04 | 1.96 | 3.30 | 4.80 |
| 31.5 | 3.58 | 1.01 | 2.35 | 3.60 | 5.05 |
| 32.5 | 3.89 | 0.89 | 2.86 | 3.90 | 5.30 |
| 33.5 | 4.33 | 0.80 | 3.20 | 4.30 | 5.49 |
| 34.5 | 4.15 | 0.79 | 3.00 | 4.20 | 5.09 |
| 35.5 | 3.87 | 0.85 | 2.86 | 3.95 | 4.74 |
| 36.5 | 3.71 | 0.70 | 2.70 | 3.70 | 4.50 |
| 37.5 | 3.46 | 0.63 | 2.58 | 3.49 | 4.22 |
| 38.5 | 3.25 | 0.60 | 2.39 | 3.32 | 3.98 |
| 39.5 | 3.00 | 0.60 | 2.25 | 3.00 | 3.80 |

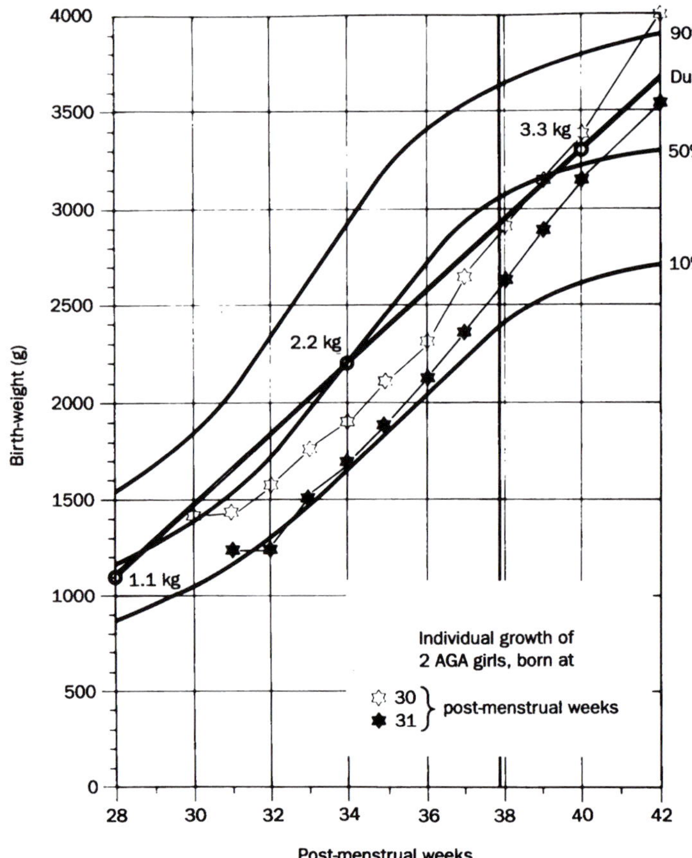


**Growth curves of weight (line of stars) of two AGA girls, born at 30 and 31 post-menstrual weeks with a birth-weight of 1420 g and 1250 g respectively. They are plotted against the percentiles of Lubchenco (1966) and the Bristol Perinatal Growth Chart of Dunn (1979). The 50th percentile of Lubchenco is surpassed around term.**

From these data, percentiles, means, and standard deviations of growth in weight, length, and head-circumference were calculated. In the preterm infants, a correction for post-menstrual age has always been made, i.e., discounting the time of prematurity from chronological age.

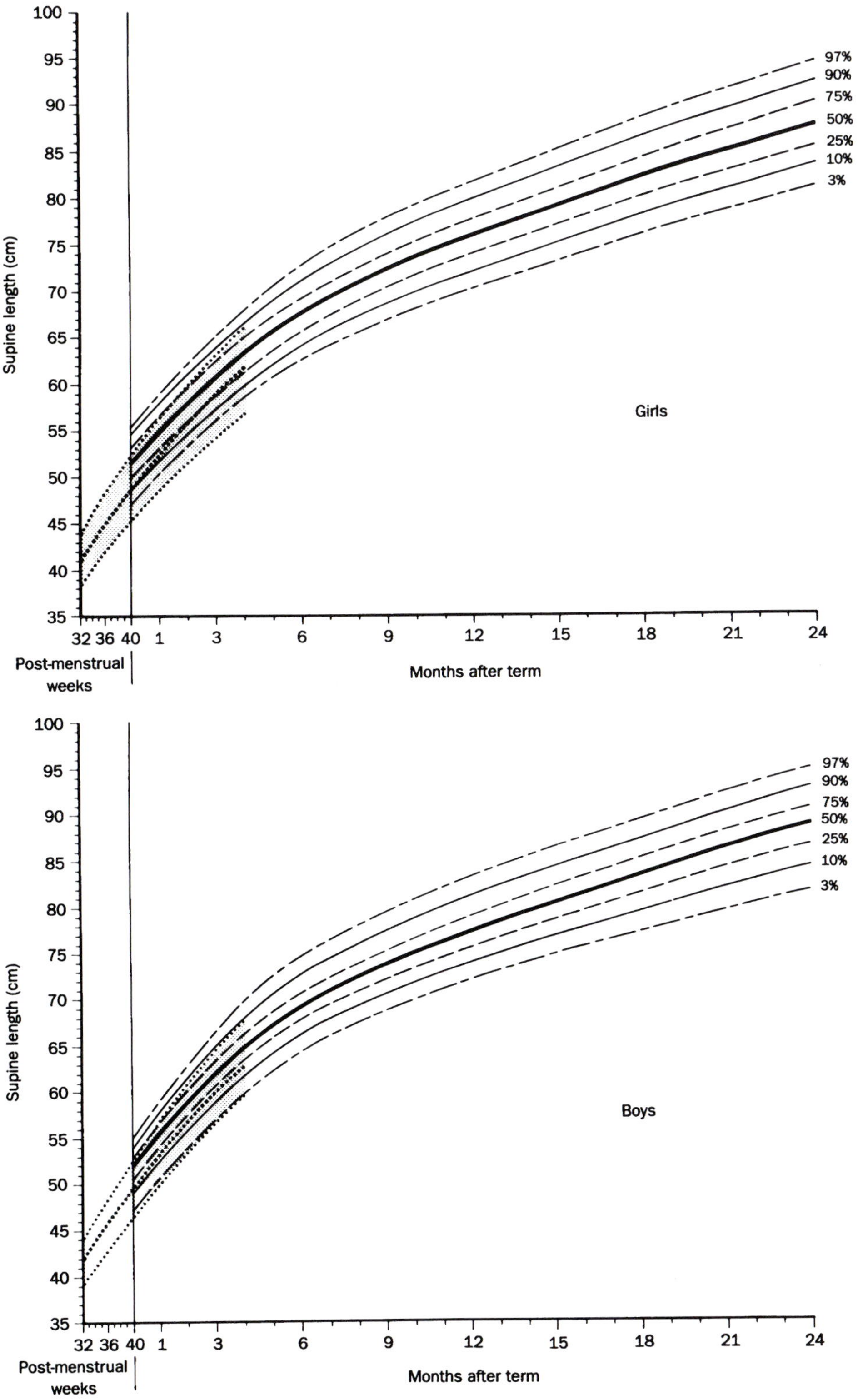

The post-natal growth of weight, length and head circumference differ in their patterns. Post-natal weight-gain is only somewhat smaller than that suggested for the intra-uterine fetus. This is demonstrated in the top figure by the growth curves (line of stars) of two AGA girls, born at 30 and 31 post-menstrual weeks with a birth-weight of 1420 grams and 1250 grams, plotted against the percentiles of Lubchenco and colleagues (1963) and the Bristol Perinatal Growth Chart of Dunn (1979). The 50th percentile of Lubchenco is crossed around term. Early and high-energy feeding allows the preterm infants to grow similarly to a fetus. Already from the age of 2 months their weight curve agrees with that of the full-term controls.

With respect to length and height, post-natal growth is retarded up to the age of about 21 months. From two years of age, the growth curves of preterm infants show no statistically significant difference to the full-term controls. Preterm infants grow faster during the first 2 years, and level up their initial post-natal retardation. In the bottom figure, the third, 50th, and 97th percentiles of the preterm infants (line of dots, shaded area) from 32 post-menstrual weeks to 4 months after term are compared with the percentiles of the full-term control infants. At term, the 50th percentile of the preterm infants corresponds to the 10th percentile of the full-term control infants, and at 4 months it corresponds to the 25th.

Interestingly, the initial lag of supine length of the preterms is mainly due to a significantly smaller crown–rump length. At 40 post-menstrual weeks (term) the difference of supine length is 2.8 centimetres and that of crown–rump length 2.4 centimetres, and at 2 years the difference decreases to 0.7 centimetres. There are only small differences (0.1–0.5 cm) in leg length between the AGA preterm and full-term infants from term to 18 months.

In addition, the later development of the VLBW AGA preterm infants is as favourable as that of the full-term controls. For example, in the figure overleaf, the growth curve of an AGA preterm girl from term to 5 years is

**Percentile curves (3rd, 50th, and 97th) for supine length of AGA preterm infants from 32 post-menstrual weeks to 4 months after term (line of dots, shaded area) compared with percentiles (3rd to 97th) of the full-term controls from 0 to 18 months (continuous and broken lines).**

shown, plotted against the percentiles ofTanner and colleagues (1966). She was born at 31 post-menstrual weeks with a birth-weight of 1340 grams. Her growth curve, starting at the 10th percentile, reaches the 90th percentile by 12 months of corrected age, where she stays and continues to grow.

Head-circumference is an important growth parameter, its close relationship to brain-weight in the first 2 years being shown in many studies. At term, 63 per cent of the adult head-circumference value is achieved, while at 6 years, 93 per cent of the adult value is attained. The post-natal growth of the AGA preterm infants until their expected date of delivery is similar to that of the intra-uterine fetus. From term, head-circumference growth of the preterms is identical with that of the full-term control infants.

The difference between the age-corrected and age-uncorrected head-circumference growth curve diminishes from 6.3 centimetres at term to 0.4 centimetres at 18 months, and is no longer relevant (as can be seen in the fourth figure). Therefore, the answer to the question 'how long should age correction be carried out for preterm infants?', is: up to the age of 18 months.

The approximation of the age-uncorrected curve to the age-corrected curve which is only due to the rapidly decreasing growth velocity, has led, in some earlier publications, to the erroneous assumption that catch-up growth takes place in normal AGA preterm infants.

From 30.5 to 39.5 post-menstrual weeks (PMW) the extra-uterine growth velocity of head circumference is significantly higher than after term. It amounts to 3.3 centimetres per month in the 31st PMW, reaches a peak of 4.3 centimetres per month in the 34th PMW, and decreases to 3.2 centimetres per month in the 39th PMW. After term, velocity decreases radiply, and it amounts to 2.5 centimetres per month in the first month and is similar to that of the full-term control infants, at 2.4 centimetres per month. A period of rapid head-circumference growth, the spurt, extends from the 31st PMW until the sixth month after term. The curve in the bottom figure clearly shows the considerable decrease of velocity from 4.3 centimetres per month at its peak in the 34th PMW to 0.8 centimetres per month in the sixth month after

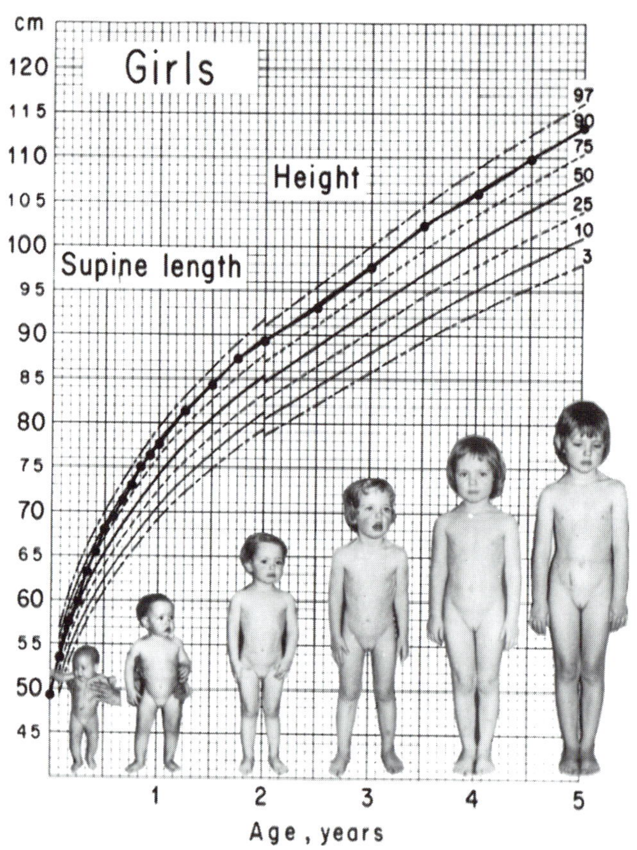


**Growth of supine length/height of an AGA preterm girl from the expected date of delivery (term) to 5 years. The curve starts near the 10th percentile of Tanner et al. (10), and reaches the 90th percentile already at 1 year of corrected age, where the girl continues to grow.**

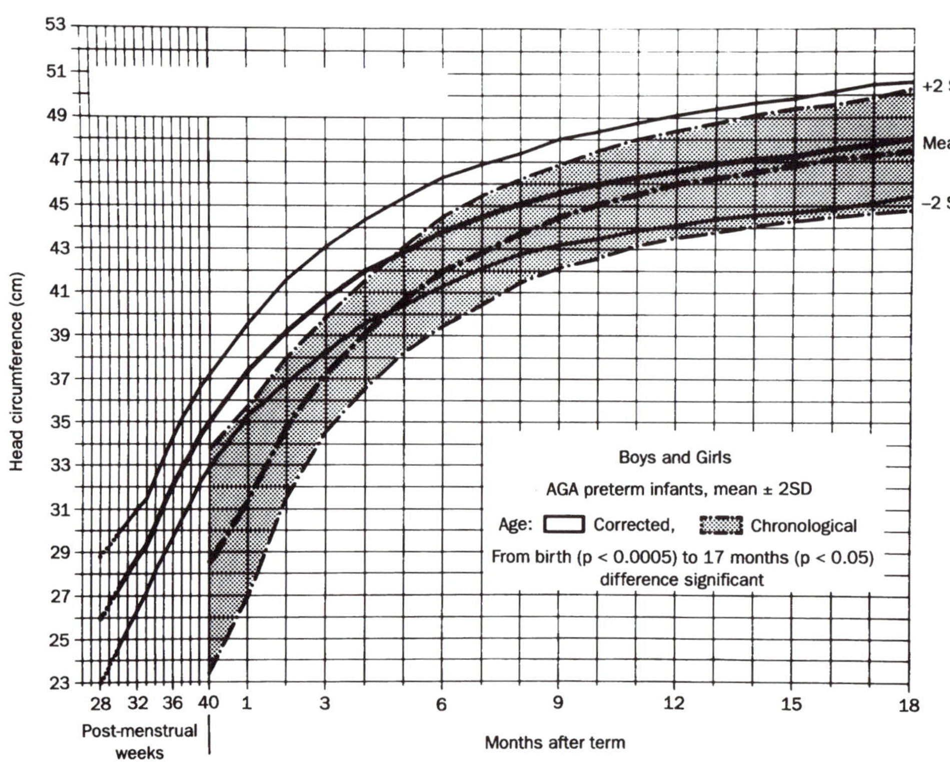


**Growth curve of head circumference of AGA preterm infants from birth to 18 months, mean ± 2 standard deviations. Comparison of uncorrected (chronological) age (dots and broken lines, shaded area) and corrected age (continuous line).**

term, to 0.4 centimetres per month in the 10th month and to 0.04 centimetres per month in the fifth year. This high velocity of head circumference in the perinatal period corresponds to a phase of rapid brain development, and suggests a high energy requirement for preterm infants.

The results of the Bonn Study demonstrate that optimal nutrition and favourable environmental conditions enable even the VLBW AGA preterm infant to develop in a way similar to that of the intra-uterine fetus and the full-term infant after birth.

*Ingeborg Brandt*

See also 'Neurological development' (**4.6**), 'Fetal growth retardation' (**4.8**), 'Standards and references for the assessment of fetal growth and development' (**4.9**) and 'Body-size at birth' (**10.3**)

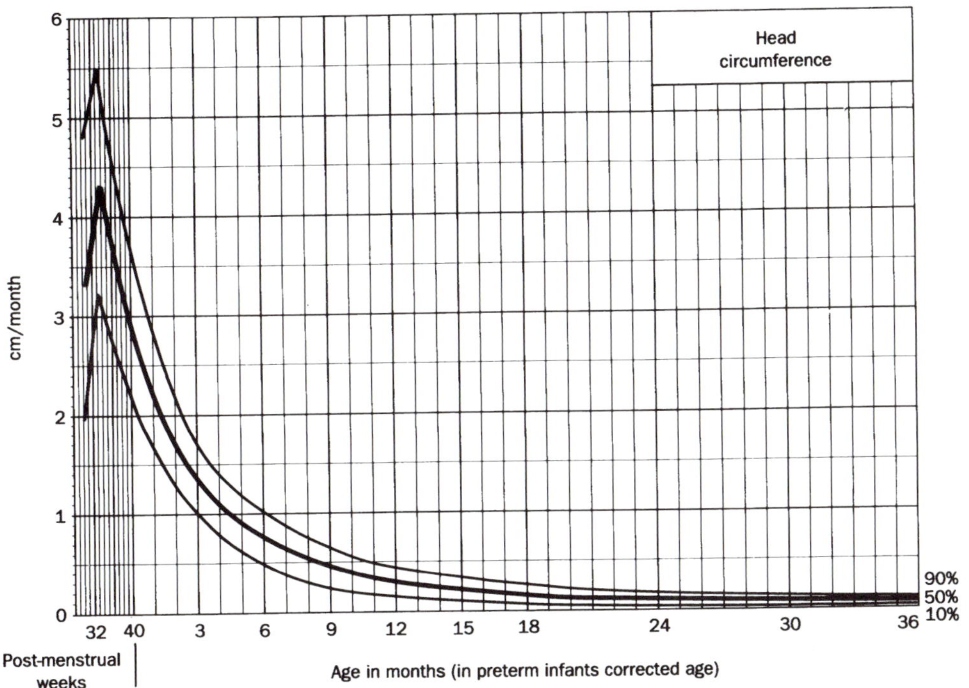


**Post-natal head circumference velocity, calculated in cm/month, the 10th, 50th, and 90th percentile, based on longitudinal measurements of AGA preterm infants (see table on p. 152).**

# Extra-uterine development after premature birth

The movements of the preterm infant resemble those of the intra-uterine fetus because brain development (**2.3**) is independent whether the individual grows inside or outside the uterus. However, the preterm infant's motor activity is influenced by gravity, and the dietary energy requirements are higher as a consequence. Prematurity displaces the time of birth, but it does not thereby dislocate the normal sequences of development when the sequences are estimated from the fundamental baseline of conception. There is no indication that preterm infants benefit from the prolonged exposure to extra-uterine life which is a consequence of early birth.

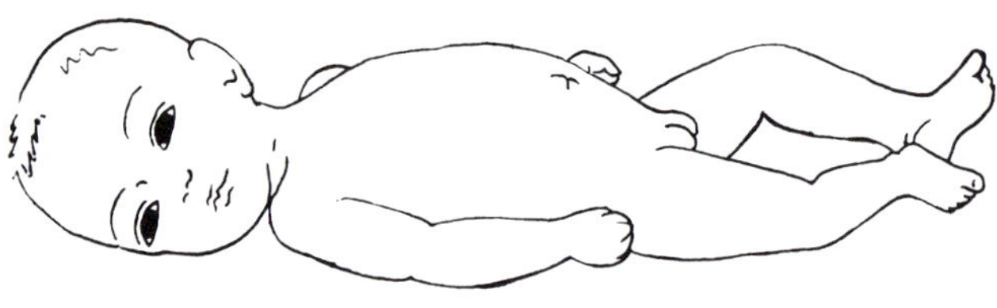

**Imposed posture at 30–33 PMW. Semiflexion of the legs, extended position of the arms and marked lateral position of the head.**

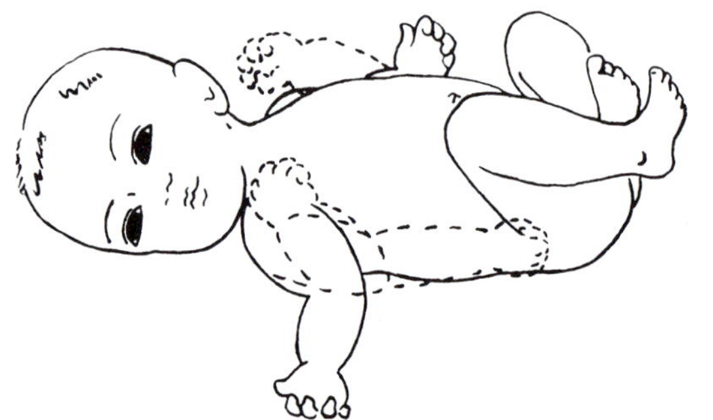

**Imposed posture at 34–36 PMW. Mostly flexion of the legs and intermediate position of the arms between flexion and extension.**

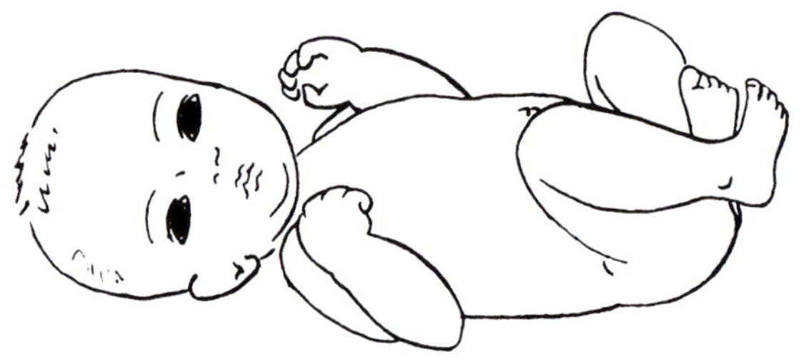

**Imposed posture from 37 PMW to about 2 months after term. Mostly flexion of the legs and flexion of the arms, head in a less marked lateral position.**

Environmental conditions, provided they are not adverse, seem to have very little influence on the development of various neural functions before 40 post-menstrual weeks (PMW). However, the immaturity of the brain limits the influence of prematurely acting environmental stimuli. In the Bonn Longitudinal Study (BLS), the extra-uterine development of preterm infants at term (40 post-menstrual weeks) and thereafter does not differ significantly from that of full-term control infants. With the better neonatal care currently available, and early as well as high-energy feeding, preterm infants are discharged from hospital sooner than in earlier decades, often before their expected date of delivery. In many cases their post-menstrual age is only 36–37 weeks. As muscle tone increases until 40 weeks (the expected date of delivery), there is the risk that this phenomenon could be misinterpreted as the beginning of spasticity (overdiagnosis). Unnecessary measures such as physiotherapy may be initiated and, for the parents, additional anxiety is produced. On the other hand, preterm infants with underlying pathology and persisting hypotonia (i.e., when the muscle tone does not increase physiologically) may be considered normal, and the diagnosis is missed (underdiagnosis). Here, the knowledge of normal neurological development in the last weeks before term is important; not only for the neonatologist, but also for all health professionals.

## Main aspects of neurological development between 27 and 40 post-menstrual weeks

Muscle tone, defined as resistance against passive movements, increases in a caudocephalic direction from about 30 PMW and reaches a peak at 40 PMW, and can be used as a maturational criterion. It can be assessed by *imposed posture*, the *popliteal angle*, the *scarf manoeuvre*.

### *Imposed posture*

There is a close age relationship with imposed posture in the preterm infant; i.e., posture after a slow passive extension of the extremities and after putting the head in the mid-line of the trunk. The posture that the baby assumes after this manipulation for at least 2–3 minutes – before being changed by spontaneous movements – can be recorded. The typical changes in imposed posture with increasing PMA are due to increasing resistance against gradual stretching of the extremities, caused by tonic myotatic reflexes.

Between 30 and 33 PMW the legs are semiflexed whereas the arms remain in an extended position. There is a marked lateral position of the head.

Between 34 and 36 PMW the legs are mostly flexed and posture of the arms is intermediate between flexion and extension.

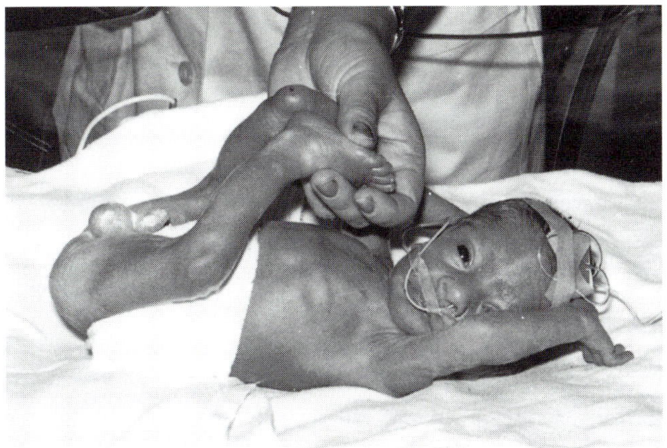

**Popliteal angle about 140° at 32 PMW in a preterm infant of 4 days. The arms – in a reflex movement – are also moved upwards.**

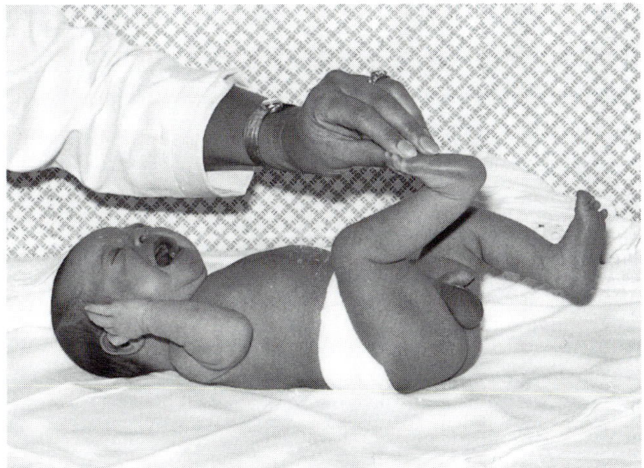

**Popliteal angle 90° at 40 PMW (term) in a full-term infant of 4 days.**

From 37 to 40 PMW and thereafter up to about 2 months after term the legs are mostly flexed, as are the arms. The head is now in a less marked lateral position with the chin in mid-line between the acromion and the sternum.

*The popliteal angle*

This decreases between 28 and 32 PMW from about 170° to about 140°, and between 35 and 37 PMW to 90°, and remains so until about 2 months after term.

*The scarf manoeuvre*

This involves moving the hand to the opposite shoulder, and seeing if there is any resistance to this. At 28–33 PMW there is mostly no resistance. The elbows cross the mid-line widely. The resistance in the upper extremities increases with increasing post-menstrual age (PMA). At 34–39 PMW the elbows reach the mid-line. At about 40 PMW the resistance against this passive movement increases significantly, and the elbows only reach the axillary line. There is a highly significant correlation of 0.72 between the resistance score and the PMA.

There are no easily quantifiable age-dependent changes, in spontaneous motor activity, except that the number of spontaneous head rotations increases. Furthermore, reflexes

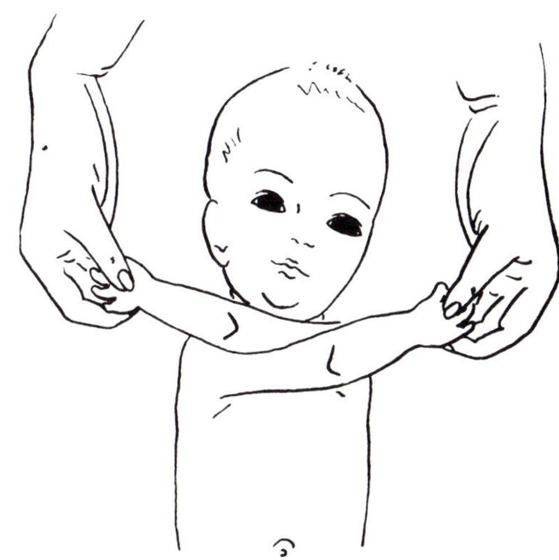

**Scarf manoeuvre at 28–33 PMW. Elbows crossing widely mid-line and nearly reaching opposite axillary line.**

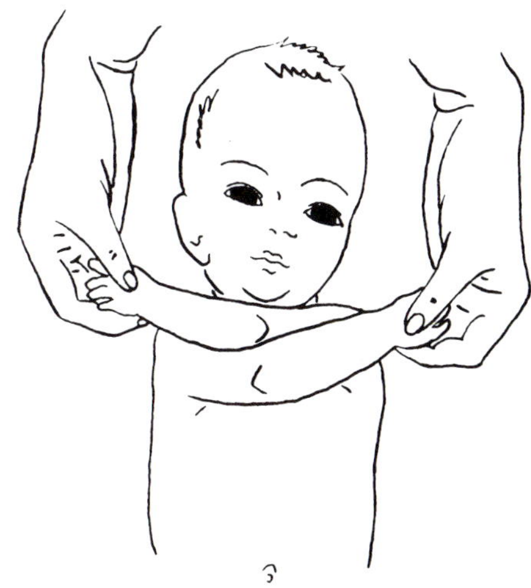

**Scarf manoeuvre at 34–39 PMW. Elbows reaching mid-line (moderate resistance).**

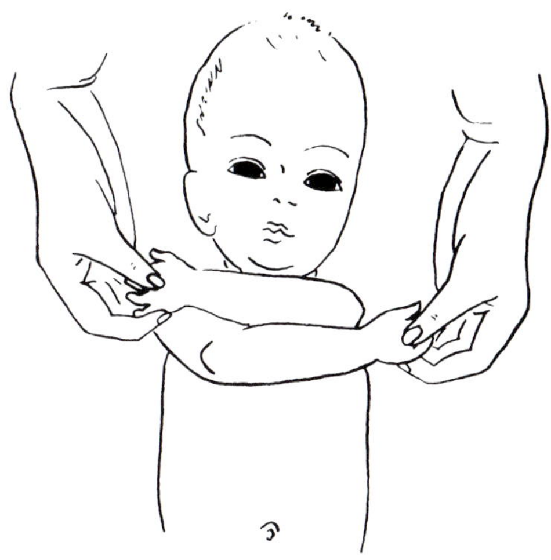

**Scarf manoeuvre at 40 PMW (term). Elbows only reaching axillary line (strong resistance).**

and reactions are completed, among those the Moro reflex and the righting reaction of the arms and head.

## Continuity of neural function from prenatal to post-natal life

There is a striking similarity between prenatal and early post-natal motor activity. For many years the human infant was considered to be born preterm. However, it is still difficult to determine the extent to which an infant born at term (40 PMW) is also premature. A major reorganization in motor output occurs around 9 to 12 weeks after term. Before 9 weeks most infants are not strikingly different from the newborn in their movement quality. This can create difficulties in the early diagnosis of central nervous system (CNS) damage.

**Grasps ring when given, median 3.5 months, variability 2–5 months.**

The first major transition in post-natal neurological development occurs at about two or three months after term birth. Then the infant proceeds from a state of 'is moved' to one of 'moving voluntarily'.

The exact time of emergence of voluntary motor activity (**6.2**) is difficult to delineate; however, this is an important step in brain development. It could be defined, for example, as the beginning of voluntary grasping, represented by the item of the Griffiths Test 'grasps ring when given', where the median is 3.5 months and the variability 2–5 months.

Knowledge of the first major transition in post-natal neurological development can be important for early diagnosis of brain dysfunction. Special attention has to be paid to hand function. The occurrence of voluntary grasping is a valuable diagnostic tool because of its small variability of 2–5 months. For example, in infants with congenital hemiplegia due to a cerebral infarction, in the first weeks motor activity of both arms and hands, as well as muscle tone, i.e. resistance against passive movements, is symmetrical in most cases. The same holds for the palmar grasp reflex and the Moro reaction. As soon as the infant starts grasping voluntarily, an asymmetry may become obvious in cases of hemiplegia. Later, severity of hemiplegia can be graded primarily on hand function since, in most cases, the arms are more affected than the legs.

The small variability of the item 'grasps ring when given', further allows an early diagnosis of a developmental delay and also a differentiation from a normal development. Here, again, it is very important to correct the age for prematurity.

*Ingeborg Brandt*

See also 'Neurological development' (**4.6**), 'Standards and references for the assessment of fetal growth and development' (**4.9**) and 'Motor development and performance' (**6.2**)

# The development of patterns of deep body temperature in infancy

Normal adults have a marked daily rhythm of deep body temperature. Core temperature falls to a minimum of about 36.2 °C in the early hours of the morning, and rises to a maximum of around 37.2 °C in the afternoon. Newborn babies show no such fluctuation, though daily variations in body temperature may occur in the fetus before birth in association with the mother's daily rhythm. By 6 months old however all infants show a temperature pattern like that of the adult. Early studies suggested that a distinct daily rhythm appeared at about 3 months of age, but more recent work has revealed that there are marked individual differences in this development associated with aspects of the baby's history and the way in which it is cared for.

Data loggers may be attached to rectal probes in order continuously to monitor deep body temperature in babies at home. There is now a large data base of several thousand recordings from infants, mostly recorded at weekly or two weekly intervals from 6 to about 20 weeks of age. In the neonatal period the deep body temperature of infants varies remarkably little over the day, remaining around 37 °C with small elevations (around 0.1–0.2 °C) associated with feeds, and small falls associated with sleeps or exposure to cool environments. From about 3 weeks old, however, rectal temperature begins to fall significantly with sleep, reaching a minimum of about 36.7 °C (± 0.2) after about one hour. The pattern of temperature change is the same for sleeps taken during the day and by night, even when night-time sleeps become very much longer.

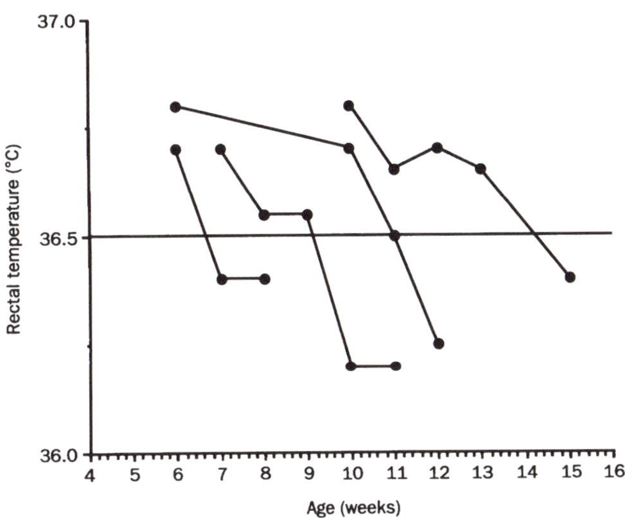


**The minimum night-time sleeping rectal temperature of a number of infants plotted versus age. Each line represents a single baby. All babies show, at some stage, a sudden fall in minimum night-time temperature which marks the onset of an adult-like temperature pattern. This occurs at different ages in different babies, ranging from 8–20 weeks.**

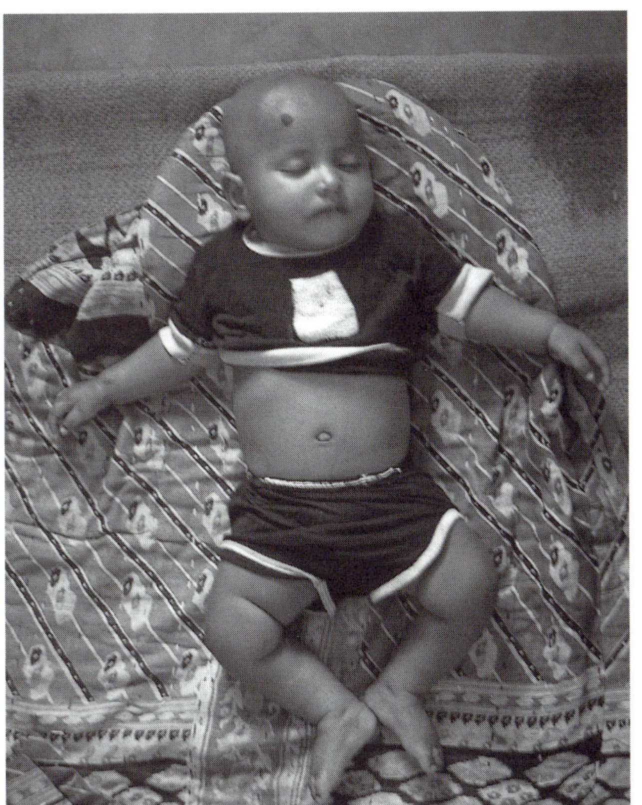

**Daysleep in a Bangladeshi infant. Photograph by Stanley Ulijaszek.**

This pattern of deep body temperature being associated with events and being unaffected by time of day, persists for different lengths of time in different babies. At some stage between 8 and 20 weeks of age however, the rectal temperature of a baby begins to fall more when it sleeps at night than during sleeps by day. At night, rectal temperature falls to about 36.3 °C within an hour or so. Even in long day-time sleeps, temperature still does not fall below 36.5 °C.

Typically, the babies who develop the adult-like pattern early do so most abruptly, with the minimum night-time rectal temperature falling by 0.5 °C or more between one week's recording and the next. Babies who develop the adult-like pattern later do so more slowly, occasionally showing a gradual decline in minimum sleeping rectal temperature over 2 or 3 weeks.

Later development of an adult-like pattern is associated with a number of factors. Babies of low birth-weight (**8.1**), and those with low Apgar scores at birth develop the adult-like pattern significantly later, as do infants who are bottle- rather than breast-fed and whose parents smoke. All babies, however, develop the adult-like pattern eventually. But in both younger and older infants the minimum sleeping rectal temperature is significantly and independently affected by a number of factors. Low birth-weight, bottle-feeding and parental smoking (**8.3**) and non-supine sleeping all significantly elevate sleeping rectal temperature. Factors such as birth-weight and type of feeding therefore affect both the rate of development of adult-like temperature patterns and the precise nature of those patterns. The mechanism of these effects is unknown, but may be related to changes in metabolic rate.

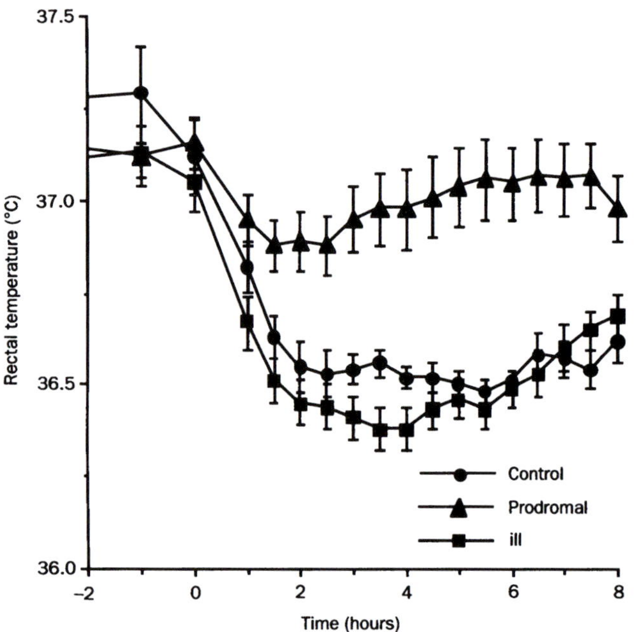


**Changes in rectal temperature pattern in babies incubating minor illness, such as a respiratory tract infection. Sleeping rectal temperature is significantly elevated during the prodromal phase when it is not obvious that the baby is ill.**

The normal daily pattern of deep body temperature is disturbed if a baby is either ill, or more significantly, incubating an illness. Babies in the prodromal phase, particularly of respiratory tract infections, show a significant elevation of sleeping rectal temperature. The temperature does not, however, rise enough to be called a fever, with a typical maximum of around 37.5 °C This temperature would not be abnormal during the day. The extent of the temperature disturbance is affected by a similar range of factors to sleeping body temperature, so that, for example, babies whose parents smoke have a greater temperature disturbance when incubating an illness than those whose parents do not smoke.

The development of an adult-like daily deep body temperature pattern is associated with changes in the daily pattern of heart-rate, which falls more with sleep at night, and also with a marked increase in the night/day fluctuation in the secretion of adrenal cortical hormones. The fact that this occurs much later in some infants than others, and is affected by infant care practice, is potentially of great significance.

As far as deep body temperature patterns are concerned, therefore, human development is completed within 6 months of birth, but the changes in that time are considerable and associated with massive re-organizations in physiology. The 3-month-old baby is very different physiologically from the neonate, and care should be taken in drawing conclusions about infant physiology from the study of newborns.

*Stewart Petersen and Mike Wailoo*

See also 'Standards and references for the assessment of fetal growth and development' (**4.9**), 'Smoking' (**8.3**), 'Infant-feeding and growth' (**9.1**) and 'Body-size at birth' (**10.3**)

# Morphology

*Morphology* refers to form and structure. In studies of the developing fetus this usually refers to the genesis of biological structures and the emergence of the neonate from the simple organism consisting of only a few cells. In studies of post-natal growth, morphology includes a broader range of subjects but, again, refers most often to the sequential emergence of the adult form from that of the infant. While morphological changes can be described qualitatively, among auxologists they are usually presented and analysed by means of anthropometric dimensions (**1.4**).

The progressive development of adult proportions, from the fetus to the adult, may be viewed as a series of shapes characteristic of various ages and developmental stages. It may also be seen as a set of differing rates of growth. It is clear that, relative to other body segments, the lower extremity grows more rapidly and the height of the head less rapidly. The differences in growth-rates are responsible for chang-

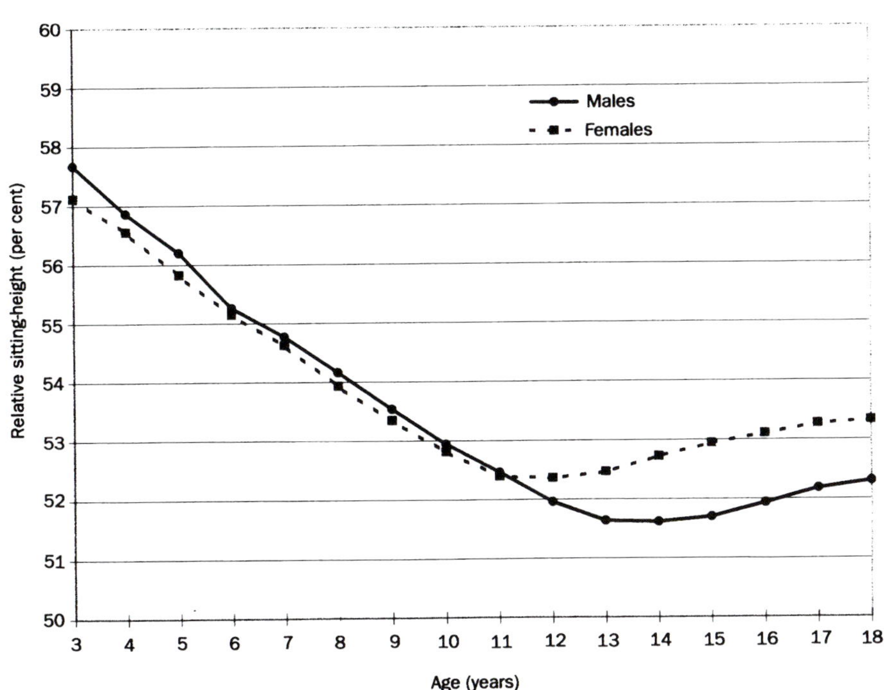


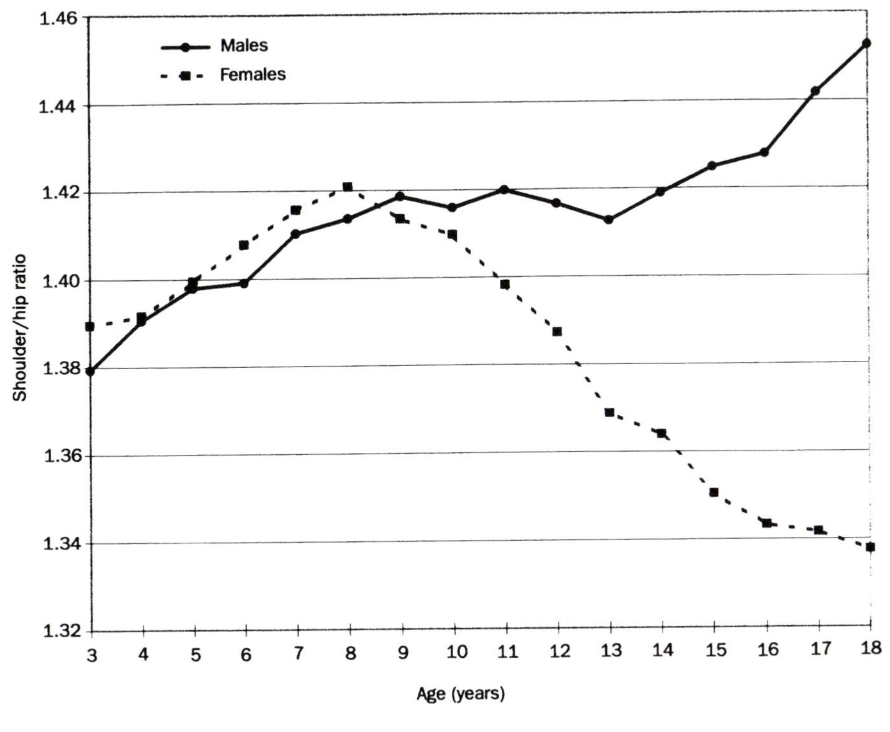


**Means of the relative sitting-height and the shoulder/hip ratio by age, from 3 through 18 years, for males and females. The data come from the Hungarian national growth study, a regionally-stratified sample of 39,035 children and youth measured between 1982 and 1985. The changing relationships are evident: in both sexes, there is a steady linear decrease in the relative sitting-height from 3 to 11 years as the legs grow more rapidly than the trunk. There is also a relative increase in shoulder breadth relative to hip over the same period.**

**From 11–18 years the curves must be described by sex. In males there is a slight but significant shift in the earlier downward trend in relative sitting-height, with an increase in the means through age 18 years. With regard to the shoulder/hip ratio, there is a levelling-off of the means until age 13 years, indicating similar rates of growth of these two diameters. This is followed by another period of increase from 13–18 years of age.**

**The shape of the curve for relative sitting-height of females is generally similar to that of males: a levelling-off followed by a gradual rise. On the other hand, the shoulder/hip ratio declines steadily from age 10 through 18 years, as the pelvis widens at an apparently greater rate than does the shoulder.**

**The sex differences in the changes of these ratios with age are clear. Prior to the circum-pubertal years there is but little if any difference. The male relative sitting-height means are consistently smaller than those of the females, the difference generally being less than 1 percentage point. Relative skeletal breadths, indicated by the shoulder/hip ratio, show no real difference.**

ing shapes and proportions. The most commonly used dimensions to reflect the changing shapes of growth are:

- height (crown–heel length below 2 years of age)
- sitting-height (crown–rump length below 2 years of age)
- leg length (height-sitting height)
- bicromial (bony shoulder) breadth
- bicristal (bony hip) breadth
- relative sitting-height (100 ×[sitting-height/height])
- shoulder/hip ratio (bicromial/bicristal breadth)

These dimensions describe the relative growth of trunk and extremities, and of the shoulder and pelvic girdles, in a sense depicting the outline of the post-cranial shape of the body. There are clear changing relationships with age in both sexes. There is a steady linear decrease in the relative sitting-height from 3 to 11 years as the legs grow more rapidly than the trunk. There is also a relative increase in shoulder breadth relative to hip over the same period. Sex difference emerge between the ages of 11 and 18 years. In males there is a slight but significant shift in the earlier downward trend in relative sitting-height, with an increase in the means through to age 18. With regard to the shoulder–hip ratio, there is a levelling-off of the means until age 13, indicating similar rates of growth of these two diameters. This is followed by another period of increase, from 13 to 18 years. The shape of the curve of relative sitting-height-by-age of females is generally similar to that of males: a levelling off followed by a gradual rise. On the other hand, the shoulder–hip ratio declines steadily from age 10 through 18 years, as the pelvis widens at an apparently greater rate than does the shoulder.

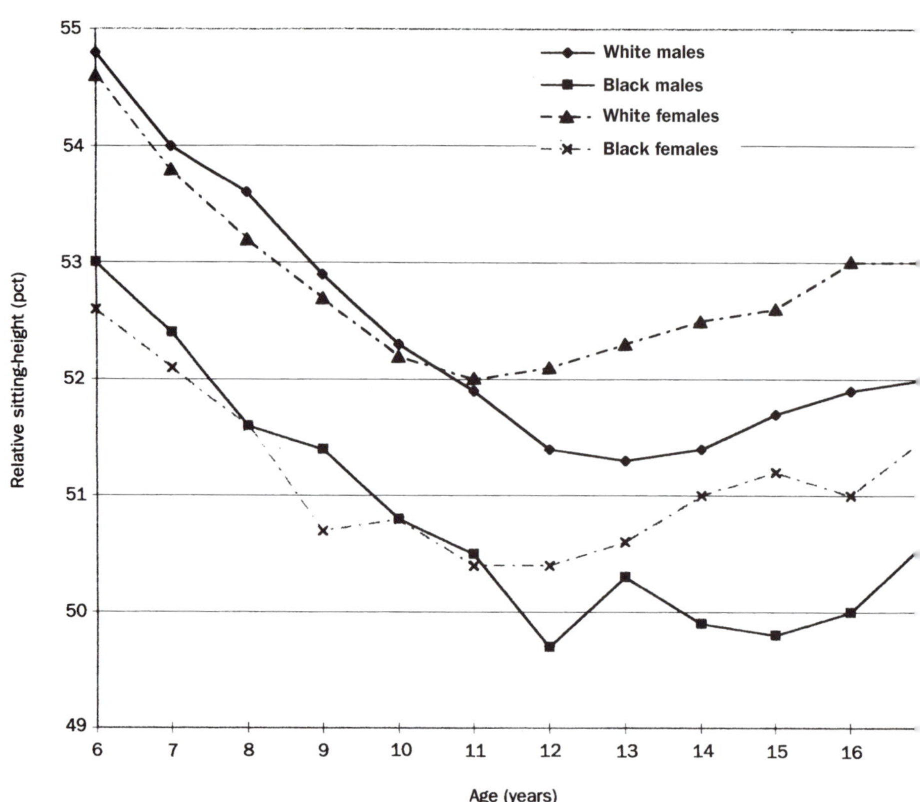


Curves of relative sitting-height by age and sex in United States Black and White children and adolescents aged 6–17 years. The data come from cycles II and III of the U. S. Health Examination Survey, which examined national probability samples of children and youth in the 1960s (Hamill, Johnston, and Lemeshow, 1973; Malina, Hamill, and Lemeshow, 1974). The general shape of the curves is similar across the four samples. However, the populatio differences are greater than are the sex differences. That is, concomitant with the growth spurt and within either ethnic/racial group, females have greater relative sitting-heights that do males. However, the same cannot be sa across ethnicity/race. Though White males have lower relative sitting-heights (i.e., longer legs) than do white females, they have higher values at every age than Black females.

With the onset of the growth spurt in adolescence, the external body morphology of the sexes displays clear differences. Females become relatively wider-hipped and shorter-legged than their male age peers, with the difference especially marked in the should/hip ratio. In other words, with respect to sex, the rather ambiguous skeletal morphology of the child becomes clearly differentiated in association with the broad range of other physical and physiological characteristics that are associated with sexual dimorphism.

Population differences in body proportions of adults have been recognized for decades, and there is some consensus that they may reflect adaptive responses to temperature. In particular, a lower relative sitting-height (i.e., relatively longer legs) is hypothesized to facilitate heat-loss in a hot environment, just as higher values are thought to help the body maintain a constant temperature in a cold climate.

The growth patterns of relative sitting-heights also vary among population. For example, although white males in the United States have lower relative sitting-heights (i.e., longer legs) than do white females, they have higher values at every age than black females. The genesis of population differences in body proportions has most often been attributed to genetic variability, with natural selection shaping morphology during the growing years to enhance adaptation to the climate. While accepting a strong environmental component for body-size, most authors have opted for genetic control of shape. However, it is now clear that environmental factors influence body-shape as well. Eveleth and Tanner (1990), in their summary of variability in the growth patterns of different populations, have shown that the secular trend (**11.3**) toward increased body-size among Japanese groups is more marked in leg length than in sitting-height; that is, Japanese are becoming longer-legged. This is in association with better nutritional and health status. Furthermore, it is in agreement with differential growth of body segments, as described above. The normally greater growth of the legs relative to the trunk which occurs during childhood would be even more if the childhood environment were more conducive to increased increments of growth.

Thus, growth is accompanied not only by an increase in size, but by proportional changes which move the individual toward an adult morphology. The rate of change in proportion is markedly increased during the growth spurt, and sexual dimorphism in body-shape is also heightened. And finally, the population differences in the morphology of adults may be seen in different patterns of growth.

*Francis E. Johnston*

See also 'Anthropometry' (**1.4**)

## SEXUAL DIMORPHISM OF PHYSIQUE

### *HEIGHT*

Until about 10 years of age, boys and girls are very close in height, with boys being very slightly taller and heavier. Consistent with this, the rate of growth, or growth velocity, is also nearly identical in boys and girls from birth until the onset of the adolescent growth spurt. These height velocity curves are derived from samples of normal boys and girls measured regularly in Edinburgh. At the onset of puberty (**5.15**), girls maturing at an average time show an acceleration in height velocity from about 5–6 centimetres per year at a mean age of 10 years to 8 centimetres per year by 12 years. In contrast, boys generally do not commence their growth spurt until about 12 years reaching a peak of somewhat greater intensity (about 9 cm/year) by 14 years.

Thus, for two years between the ages of 10 and 12 years, the average girl is somewhat taller than the average boy of the same chronological age (**5.17**). Nearly all of the adult height difference between the sexes is developed during this period, with 10–12 centimetres accruing to boys from 2 extra years of growth at 5–6 centimetres per year and 2–4 centimetres from the more intense male growth spurt. This makes a total sex difference in the United Kingdom of 14 centimetres in favour of boys.

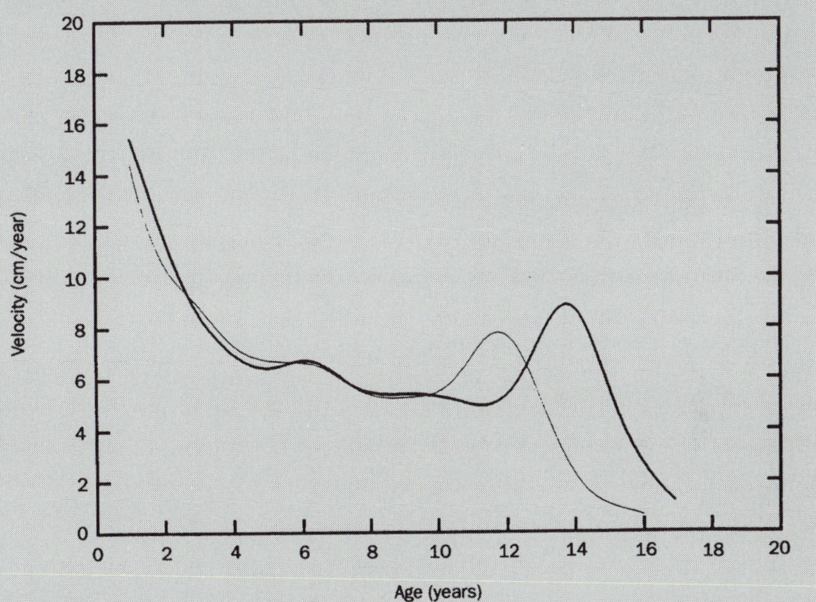


**Average height velocity curves for girls (fine line) and boys (heavy line) from 1 year of age to adult height.**

### *BODY PROPORTIONS*

Measurements of various body segments and diameters generally show the same pattern of sexual dimorphism as described above. There is consistently greater pubertal contribution to growth in stature, sitting-height, leg-length and shoulder width in boys relative to girls, while for pelvic breadth there is a slightly larger proportion of adult size developed in female puberty. This is particularly noticeable when compared to shoulder width. Thus the characteristic sexual dimorphism in torso silhouette is established.

| | Pubertal contribution (%) | |
|---|---|---|
| | Male | Female |
| Stature | 5 | 3 |
| Sitting-height | 7 | 4 |
| Leg-length | 4 | 3 |
| Shoulder width | 9 | 5 |
| Pelvic width | 7 | 8 |

### *BODY COMPOSITION*

The skeleton (**5.6**) is a major component of physique, but the final contribution comes from the soft tissues that invest the bones. Here there are also sex differences that develop at puberty. Before puberty muscle mass (**5.8**) is very similar in girls and boys, assuming we are not discussing children who take unusual amounts of physical exercise. At the onset of puberty both sexes show an increased accretion of muscle but this is much more dramatic in boys. In addition, subcutaneous fat is deposited in very different ways. Even before puberty girls tend to have thicker layers of fat particularly over the limbs, but at puberty they gain fat whereas boys actually lose fat from the subcutaneous layers. This leads to the generally more angular physique of the male with more marked muscle definition. Of course all these soft tissue effects can be modified by diet and exercise and there will be many exceptions to the rule.

*Michael Preece*

## BODY PROPORTIONS

At birth, different parts of the body are closer to the adult proportions than others, and so have different maturity gradients. For example, the upper parts of the body, particularly the head, are closer to adult proportions than the lower parts. Similarly, upper limbs are more developed than lower limbs; and distal segments, such as the hands and feet, more than proximal segments. This is known as the craniocaudal 'maturity gradient'. At puberty (**5.15**), the adolescent growth spurt has different timings in different parts of the body. On average, distal segments precede the more proximal segments in the age of peak growth velocity. The legs, hands and feet have an earlier peak velocity than do sitting-height, the arms and stature. Many of the changes show what can be called developmental overshoot. Relative sitting-height reaches adult values prepubertally but continues to fall to a minimum at around peak height velocity before increasing to adult proportions. All these changes make the adolescent look and feel gangly and big-footed. Adult stature (**5.17**) may be attained by 18 years but breadths such as the bi-acromial may increase until the mid-20s.

*N.G. Norgan*

5.6

# Skeletal development

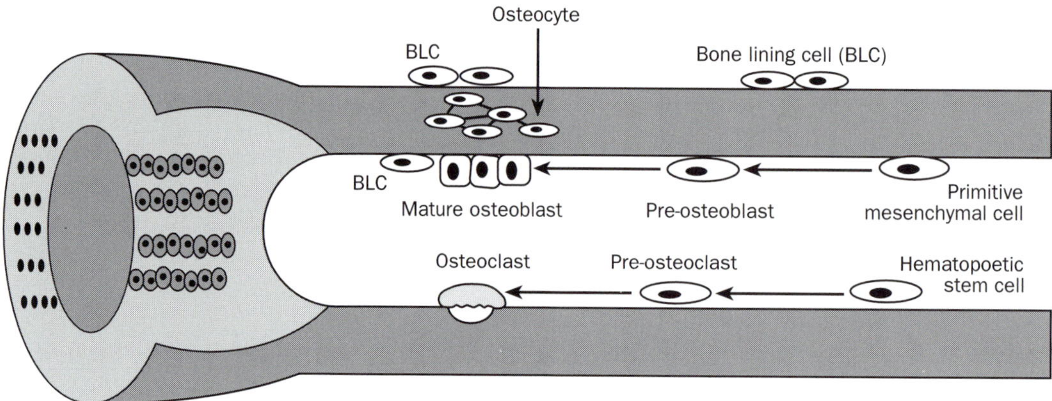


**Schematic diagram of a developing long bone illustrating the cells of bone.**

The skeleton is formed from tissues that are all included within the category of connective tissue. As such, they share their mesodermal origin with muscle and vascular tissue. The shared origins of muscle and bone are consistent with their close functional relationship in the musculoskeletal system. The importance of marrow as the site of most of the body's erythropoietic and immunogenic activity further underlines the close functional affinities of these mesodermal derivatives.

The earliest stages of osteogenesis involve the origin of cartilaginous models sheathed in a highly vascularized connective tissue envelope. This envelope, the perichondrium, is made up of two functional layers. The outer, fibrogenic, layer gives it both strength and flexibility. The inner, chondrogenic, layer is a loose matrix which includes chondroblasts, the cells that secrete the collagen filaments that will form the framework of hyaline cartilage. In addition, chondroblasts secrete a ground substance that includes chondroitin sulphate and hyaluronic acid which adheres to collagen fibres. The combination of collagen fibres and ground substance gives hyaline cartilage its tough, rubbery consistency, which is resistant to the penetration of blood vessels. Some chondroblasts remain in lacunae in the cartilage that they form. These cells are referred to as chondrocytes. For a time they retain the potential to divide and resume secretion of collagen. When this occurs, their continuing activity underlies the process of interstitial growth, by which cartilage expands as bone grows in length and diameter. However, the ability to sustain interstitial growth declines as the matrix becomes more rigid, and subsequent growth occurs at the outer edge of the cartilaginous mass and under the perichondrium. This is referred to as *appositional growth*. Because cartilage resists vascularization, the metabolic requirements of chondrocytes must be supplied through diffusion. As cartilage becomes more rigid, diffusion of nutrients is reduced, chondrocytes reduce their metabolic activity and eventually die.

The original cartilaginous models are in most instances strikingly similar to the mature bones which will eventually replace them. However, the process of development by which bones grow and are remodelled involves intermediate stages during which the shape and composition of each bone differs markedly from both its original and its final form. A critical event in the replacement of cartilaginous models by osseous tissue is the appearance of ossification centres. This process begins when dormant mesodermal cells within the perichondrium begin to divide and form new mesenchyme. The formation of a periosteal bud by this mesenchyme is followed by the invasion of adjacent cartilage which has already begun to degenerate as chondrocytes die.

Certain bones, namely those that form the cranial vault, parts of the mandible, and the clavicle, arise from the mesenchyme of membranous tissue in close

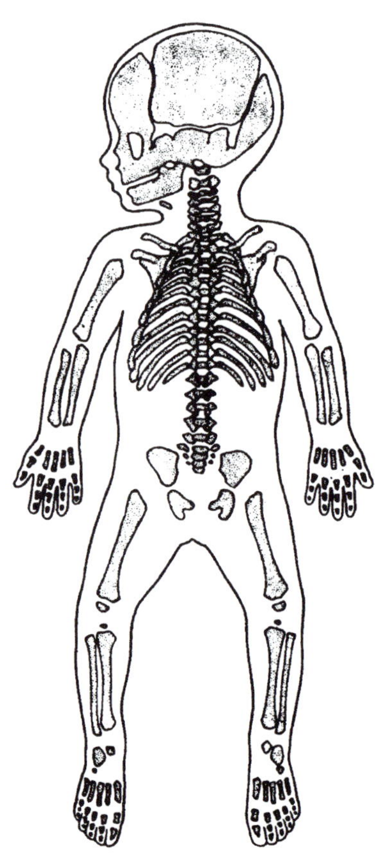

**Ossification of the skeleton in a new-born baby. Note the absence of any centres of ossification in the wrists, the separate centres for the components of hip bones, and the lack of any secondary centres of ossification for the long bones except the lower end of the femur and the upper end of the tibia. From Sinclair, D, *Human Growth after Birth*. Oxford: Oxford University Press, 1985.**

proximity to the integument and are therefore often referred to as 'dermal' bones. The ontogeny of these bones does not include the initial appearance of cartilaginous models, but in all other respects proceeds in much the same manner as in cartilaginous bone. In both types of bone, the process involves the secretion of tropocollagen by osteoblasts to form an organic matrix (osteoid) that will eventually be sheathed in crystalline mineral. The predominant constituents of bone mineral are carbonate and phosphate salts of calcium which combine to form calcium hydroxyapatite. Under certain circumstances, a variety of other constituents can be incorporated in bone, with the result that bone mineral may vary in composition to reflect exposure to such substances as fluoride, lead, and strontium, that have a high affinity for bone matrix.

The appearance of centres of ossification can be detected by an increase in radio-opacity where deposition of bone mineral occurs. In the long bone the first, or primary ossification centres appear near the centre of the diaphysis. Expansion of ossified tissue from its centre of origin produces a collar of periosteal bone that spreads toward the more distal areas of the diaphysis. Calcification of cartilage underlying the newly formed bone hastens its degeneration and facilitates its replacement by the infiltration of osseous tissue. Later, additional ossification centres will appear at what will become the epiphyses. Other centres may occur in the vicinity of major muscle attachments on the surface of large bones. The first bone formed at the ossification centres is trabeculated and will eventually be replaced by cortical bone or through remodelling of trabeculae to accommodate the stresses and strains associated with use.

Despite the degeneration and replacement of cartilage as ossification occurs, much of the increase in length occurring in long bones is the result of cartilaginous growth. New cartilage is laid down distal to areas of spreading ossification, both in the diaphysis and the epiphysis. Environmental factors that limit this aspect of growth have the potential to influence the extent to which the genetic programme for linear growth is attained. In time, however, the rate of ossification exceeds that of cartilaginous growth. Eventually, the only cartilage remaining is that forming the articular cartilage, located at the point where bones make contact with each other, and that found in the cartilaginous growth plate, located between the rapidly growing metaphyseal end of the diaphysis and the expanding osseous tissue of the epiphysis. The cartilaginous growth plate will ultimately he replaced by bone when epiphyseal closure occurs at the end of the adolescent growth spurt.

Throughout growth and maturation, the skeleton is subject to continual remodelling. The formation of a bone with all of the structural attributes required by its function necessitates destruction of bone by osteoclasts concurrent with the formation of new bone by osteoblasts. This process, following a well-defined sequence of resorption and apposition, involves the activities of bone-remodelling units, teams of osteoclasts and osteoblasts that reshape bones throughout life. The turnover of bone associated with this process varies from one part of the skeleton to another and tends to be most active in trabecular bone, which has much surface area. However, the structure of the denser cortical bone being composed of concentric layers (lamellae) of Haversion systems, each organized around blood vessels, permits ready access to nutrients as well as removal of metabolic by-products. This accessibility facilitates turnover of bone mineral essential for the important role that bone plays in the maintenance of calcium homeostasis.

*William A. Stini*

See also 'Biochemical markers' (**1.7**)

*Local mediators in skeletal tissues*

| Factor | Expression of mRNA or protein in bone and cartilage cells |
|---|---|
| *Growth factors* | |
| Insulin-like growth factors (IGF-I & II) | Osteoblasts (OB) & chondrocytes (C) |
| Transforming growth factors (TGFβs 1–3) | OB & C |
| Fibroblast growth factors acidic and basic (αFGF & βFGF) | OB & C |
| Platelet derived growth factor (PDGF) | OB |
| Bone morphogenetic proteins BMP 1–7 | OB |
| *Interleukins (IL)* | |
| IL-1β | OB & C |
| IL-3 (Multi CSF) | OB |
| IL-4 | |
| IL-6 | OB & C |
| IL-8 | OB & C |
| *Tumour necrosis factors* | |
| TNFα | OB |
| TNFβ | |
| *Interferons* | |
| IFNγ | OB |
| *Colony stimulating factors* | |
| GM-CSF | OB & C |
| M-CSF | OB & C |
| *Others* | |
| Prostaglandins | OB & C |
| PTH-RP | OB & C |
| CGRP | |

*Models for the study of skeletal development*

*Embryo:*
*Xenopus*, chick, human, rodent, bovine

*Post-natal:*
Rodent, chick, bovine, rabbit, deer antler

*Models of induced bone formation in vivo:*
Subcutaneous implantation of demineralized bone matrix.
Bone and cartilage formation in intraperitoneal diffusion chambers.

*In vitro systems:*
Cell culture: stromal cells, osteoblasts, oesteoclasts, growth plate chonrocytes, periosteal fibroblasts etc.
Organ culture: developing mandible and limb bud, rodent long bone, fetal or post-natal rodent calvariae etc.

*Genetic manipulations:*
Gene disruption: *in vivo* and *in vitro* models.

# The human growth plate

The growth plate is a cartilaginous disc that extends across the full cross-section of a bone. It is the structure that enables bones to increase in length while the bone itself retains the necessary mechanical strength for its functions within the skeletal system. Since the cells and matrix of the plate have little mechanical strength, it is kept in position by a 'corset' of fibrous tissue, the perichondrium. In many bones the disc of the growth plate is corrugated – these undulations prevent damage by shear forces.

During the years of active growth the plate is held between two bony structures – the epiphyseal bone plate and the trabecular bone that makes up the metaphysis. The nutrition of the plate is provided by blood vessels from the epiphysis, which supply sinuses between the bone plate and the cartilage. Blood vessels within the metaphyseal bone are responsible for breaking down the cartilage and for supplying nutrients to the osteoblasts and other cells engaged in replacing calcified matrix with osteoid and true bone. During the first few years of growth blood vessels are found within canals in the cartilage.

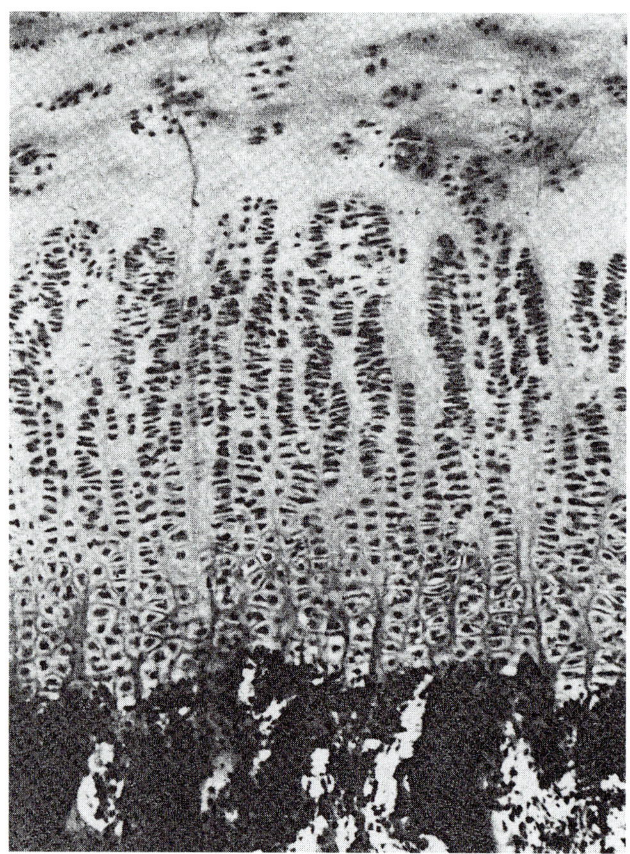

**Section through growth cartilage from a male aged 8 years, showing the inert zone and the columnar zone. Within each column the change from disc-like proliferating to expanded hypertrophic cells can be seen. (Hematoxylin and eosin).**

## Endochondral ossification

The process whereby new cartilage, produced by cell division within the plate, is converted into bone is termed endochondral ossification. In addition to the cartilage cells, the cells of the metaphysis – endothelial cells, chondroclasts, osteoblasts and osteoclasts – are involved in the growth

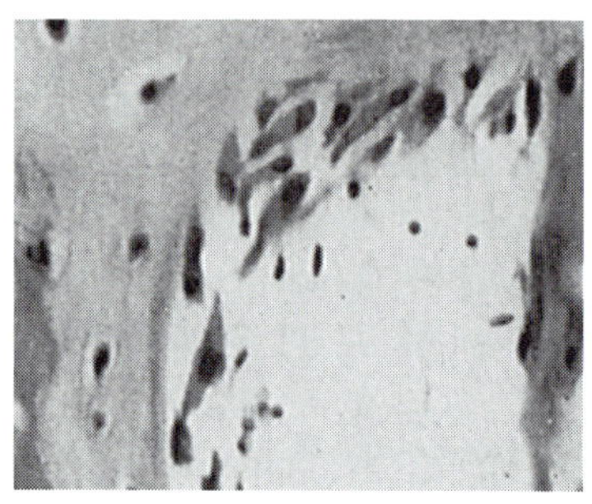

**Osteoblasts. Small and irregularly shaped with dense nuclei, the osteoblasts form a layer on the surface of developing bone, to which they are adding new material. From Sinclair, D. *Human Growth After Birth*. Oxford: Oxford University Press, 1975.**

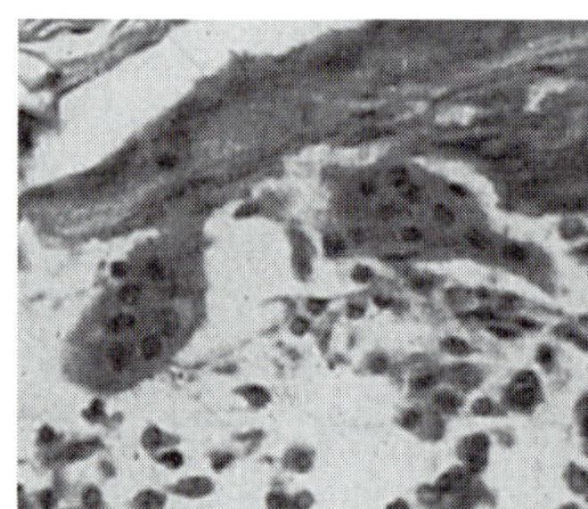

**Osteoclasts. Two of these large multinucleated cells are shown alongside a portion of bone which is undergoing absorption. From Sinclair, D. *Human Growth After Birth*. Oxford: Oxford University Press, 1975.**

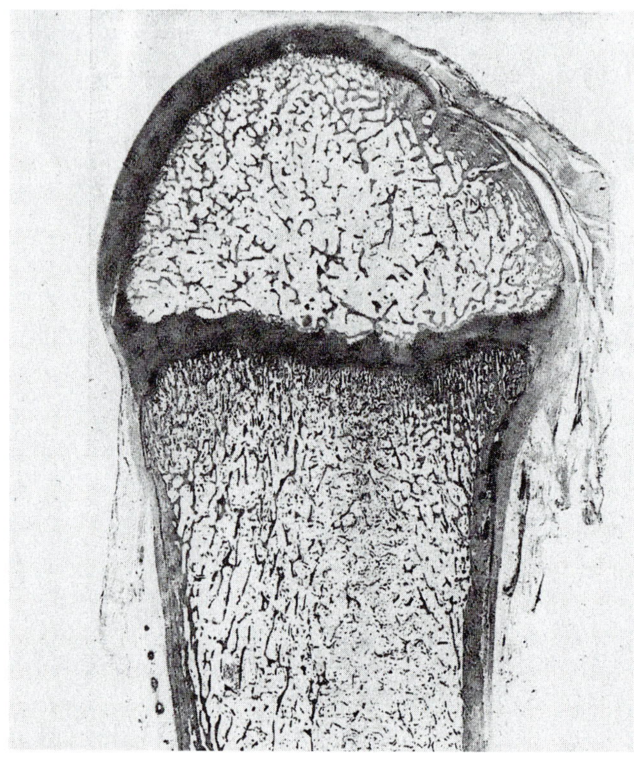

**Longitudinal section through the distal end of the femur. Male aged 7 years. The cartilage plate is seen between the coarse bony trabeculae of the epyphysis (the bone plate is too thin to be clearly visible) and the fine trabeculae of the metaphysis. (Hematoxylin and eosin).**

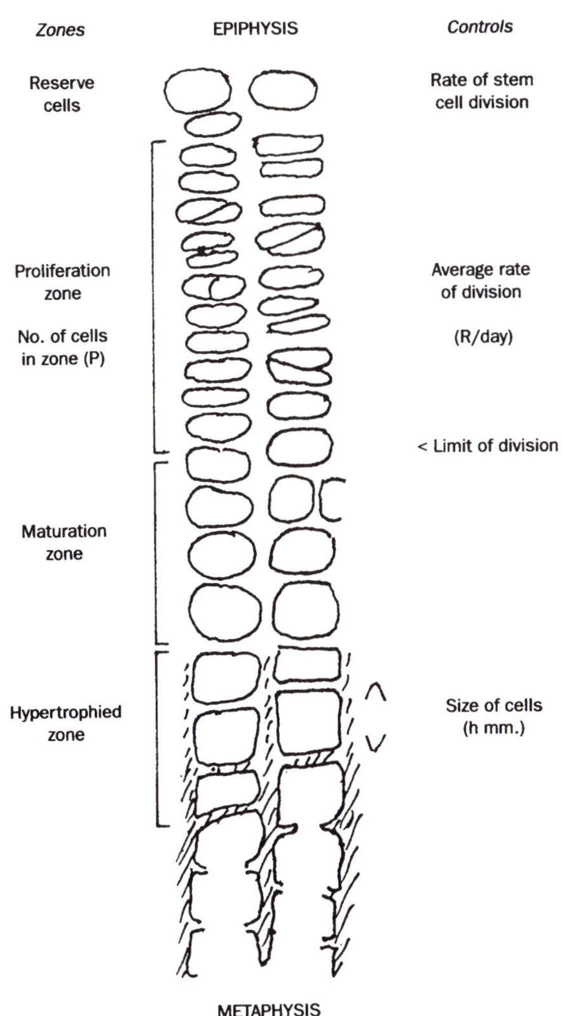


**The drawing of the two columns from the cartilage plate is labelled (to the left) to indicate the zones within each column and (to the right) the levels of controls acting on the growth mechanisms. The growth-rate is equal to P × R × H mm/day.**

process, which is simplest to describe in the long bones of the appendicular skeleton.

The familiar microscopic structure of the growth plate shows regular columns of cartilage cells within the cartilage matrix. The cells within each of the columns of the growth plate may be divided into zones which are defined partly by function and partly by structure. They do not have distinct limits as active cells move from zone to zone during the growth process. However, the human growth plate shares with the larger mammals a wide zone of inert cartilage between the bone plate and the true cartilage plate. The function of this inert zone is not clear.

The first active zone is the reserve cell zone in which the cells seldom divide. The reserve cells can be considered the stem cells for the columns and are called on to divide when the division capacity of proliferating cells is exhausted. The next zone is the proliferation zone. The cells are generally seen as flattened discs in a sagittal section. The zone is defined by the presence of mitotic figures or, more usually, by the presence of labelled cells when markers, such as tritiated thymidine, are used that are specific to cells in the division cycle. As cells are displaced along the columns by division of cells nearer the epiphysis, they are found to expand in size. This increase in size continues through the maturation zone until the cells that were originally disc-shaped approximate to spheres. No cell division occurs in the maturation zone. In the hypertrophic, or calcified, zone calcium is laid down in the matrix between the columns so that the cartilage structure is transformed into a rigid scaffolding and no further expansion of the cells is possible. In this zone the cell nuclei are resorbed and, as capillaries invade the transverse walls, the primary trabeculae of the metaphysis are formed. The rate of conversion of hypertrophied cells into bone matches the rate of cartilage-cell production, so that the width of the cartilage plate does not change perceptably over periods measured in months.

## The relationship between cell division and overall bone growth

The mathematics of growth is simple: bone growth equals the rate of production of new cells per column multiplied by the average height of hypertrophic cells. In turn: the rate of production of new cells is equal to the number of cells in the proliferation zone multiplied by their average rate of cell division multiplied by the growth fraction.

Cell size may be measured on microscopic sections, but the determination of the other parameters of cell division requires the techniques of cell kinetics. These parameters are not known with any precision. For a 10-year-old girl the distal growth plate of the femur had 28 cells in the proliferation zone, nine in the maturation zone and six hypertrophic cells with average size 20.5 microns. The cells of the proliferation zone were dividing approximately once every 15 days (assuming all cells are 'in cycle', i.e. growth factor = 1) so that 1.9 new cells were produced each day. The contribution of this cartilage plate to the overall growth was therefore 38 microns per day or about 1.4 centimetres per year.

Variations between different growth plates occur largely in the number of proliferating cells. Differences in the average size of hypertrophic cells are small. Growth in diameter of the plate is produced by additional divisions of cells to produce new columns at the periphery.

Although the kinetics of each column follows the simple mathematical relationship above, bone growth is achieved through co-ordinated cell division and maturation throughout the approximately 300,000 columns that are found across the disc of a typical human growth plate.

## Controls on growth

Growth may be affected by physical, nutritional, hormonal and local controls like growth factors (**4.2**). All such effectors must act at one or more of the following levels:

## GROWTH ARREST LINES

'Growth arrest lines', also called 'transverse lines' or 'Harris lines', have been described by a number of researchers over the past 80 years. These lines are often seen in radiographs of the long limb bones of children, particularly near the growing ends of the bone. Although the growth arrest lines form during childhood, they can persist into adulthood, and have been observed in contemporary people as well as in skeletons recovered from archaeological sites (**11.1**). They are highly mineralized trabeculae that roughly parallel the contour of the cartilage growth plate and are observable in radiographs or in a longitudinal cross-section of a bone.

Interest in growth arrest lines intensified during the late 1950s–70s during which time a number of experimental studies were undertaken, primarily using rodents, to determine precisely how growth arrest lines formed and under what circumstances. Results of those studies showed that nutritional deprivation (**9.2**) or infectious pathogens (**9.3**) such as septicemia could result in growth-arrest-line formation. These studies also demonstrated the mechanisms that govern line formation, concluding that a growth arrest line forms as a result of an imbalance between the rates of cartilage-cell and bone-cell activity where the former slows or stops while the latter continues.

During this same time-period, human biologists, anthropologists, and clinicians conducting studies of growth and development also observed growth arrest lines in their study populations. Most informative were data from longitudinal studies of growth and development of children conducted in England and the United States. Similar to the results of the animal studies, poor nutrition and disease, as well as inoculation and illness were associated with growth-arrest-line formation in children. With the apparent association between nutritional status or illness and the occurrence of growth arrest lines in animals and in humans, it seemed logical to study that same relationship in past human populations. During the past three decades numerous researchers have used growth arrest lines as a non-specific indicator of nutritional insufficiency or disease episodes to investigate a number of research problems in archaeological populations.

At first, growth arrest lines were embraced as a relatively clear and tractable indicator of morbidity experienced during an individual's growing years. Many researchers would agree now, however, that interpreting growth arrest lines (in past populations at any rate) has a number of problems. Among the most important is that the occurrence of growth arrest lines in a skeletal sample infrequently corresponds with the patterning of other skeletal pathologies indicative of nutritional and disease insults. As a consequence, many have questioned whether growth arrest lines are an accurate reflection of the morbidity experienced by any given individual or the population. This poor correspondence between growth arrest lines and other skeletal pathologies is reinforced by the studies of contemporary children mentioned above. In those studies it was shown that there is not a one-to-one correspondence between the formation of a growth arrest line and an illness episode or inoculation, and that growth arrest lines frequently form in the absence of poor nutrition, illness, or inoculation. In addition, studies also failed to show a difference in the prevalence of growth arrest lines when comparing healthy children to those with chronic undernutrition and growth failure.

The assumption that poor nutrition or illness is the primary underlying cause of growth arrest lines has been questioned recently. A study of the relationship between rate of growth and growth-arrest-line formation in generally healthy contemporary children shows that line formation is associated with rapid growth. It is argued that growth arrest lines are a byproduct of local, tissue-level hormonal control of the rate of cartilage cell proliferation and maturation. These controls are most likely to be exerted when rate of growth is rapid, but could also exert themselves under conditions of inadequate nutrition. This hypothesis remains to be tested under experimental conditions. Given that the underlying etiology of growth arrest lines is poorly understood and that it is not clear whether their etiology differs from that of other skeletal pathologies indicative of poor nutrition or disease, the general consensus among most researchers is that growth arrest lines are not a particularly sensitive nor accurate reflection of nutrition and disease insults and that caution should be exercised when attempting to interpret their patterning in skeletal populations.

*Ann L. Magennis*

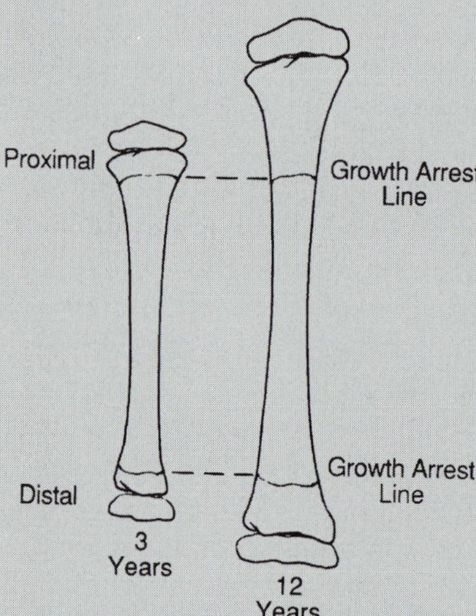


**A diagram to illustrate the use of growth arrest lines near the proximal and distal ends of the tibia at 3 and 12 years in the measurement of elongation at the proximal and distal epiphyseal zones. Since the lines are fixed, the distances from them to the ends of the bone increase due to elongation at the epiphyseal zones.**

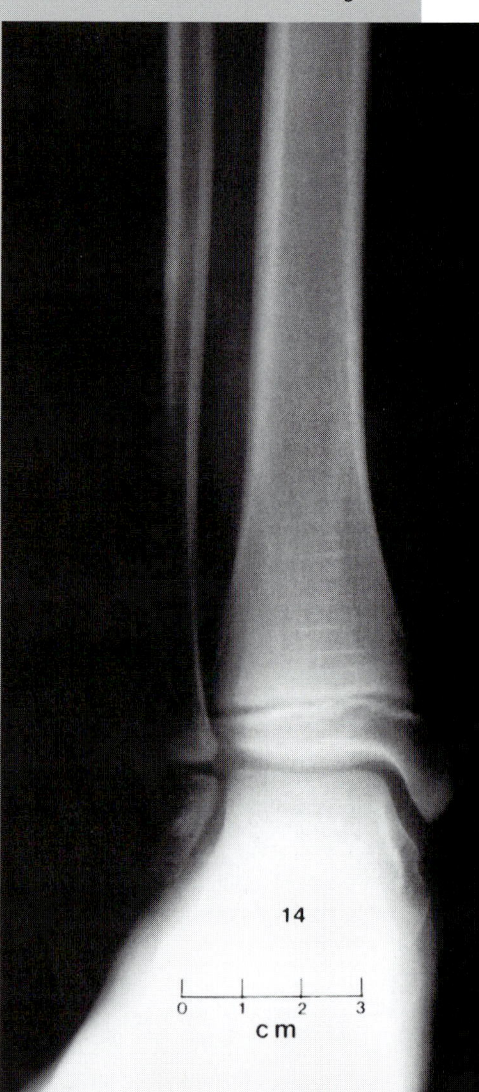


**Growth arrest lines are visible in radiographs of the distal tibia of a 10-year-old girl. Also observable are lines that formed at younger ages. This individual was a participant in the Denver Child Research Council study of growth and development, a long-term study which began in the 1930s and ended in 1967. Bone continually remodels, where old bone is removed, and new bone replaces it. In that remodelling process, some growth arrest lines may be resorbed and remodelled away, whereas others may persist well into adulthood.**

1) rate of division of stem cells
2) rate of division of proliferation zone cells
3) the command that calls a cell out of division cycle and so determines the limit between the proliferation and maturation zones – this command also sets the size of the proliferation zone
4) rate of maturation of cells – i.e. limits on their expansion in size

## Development of the growth plate

In the embryo (**4.5**), each bone is first defined as a cartilage analage, and bone forms initially at the primary centre of ossification in the diaphysis where the cartilage is hypertrophic. As the bone takes shape, the growth plates become differentiated at each end. The cells originally have no particular orientation but the distinctive columnar structure slowly emerges. This structure remains throughout the period of active growth but after the growth spurt the plate becomes thinner and eventually bridges of bone cross the plate which is gradually resorbed. This process of the closure of a plate takes place at different times, for example, in the tibia the distal-growth plate ceases activity and is resorbed before the proximal plate. Growth is still possible in some bones where the articular cartilage may be induced to act as a growth plate, for example, as in acromegaly. *N.F. Kember*

See also 'Biochemical markers' (**1.7**), 'Radiographic assessment' (**1.8**), 'Skeletal development' (**5.6**), 'Adulthood and developmental maturity' (**5.17**) and 'Bone dysplasias' (**7.5**)

## CELL KINETICS

The processes of growth, repair and the maintenance of renewing populations all depend on cell division. Cell kinetics is the quantitative study of cell division and asks the questions: which cells divide?, how many?, how frequently?, what might control their division? In a system such as erythropoeisis cell kinetics will also study the relationships between the various precursors.

Cell kinetics is limited by its methods. Originally the mitotic figure was the chief evidence of cell division. In 1957 tritiated thymidine became available. This radioactively labelled compound is taken up exclusively by cells during a period called the synthesis phase – the period during which the DNA content of a cell doubles prior to division. If thymidine is made available to a cell population only those cells in synthesis will incorporate it into their DNA. The labelling is generally stable so that the fate of a cell labelled at a given time may be followed for weeks or even years. Thus, in bone, a labelled osteogenic precursor cell can be followed through the osteoblast stage into the osteocyte stage, where it may be detected within bone a year or more later.

The detection of the tritium label requires skill at preparing reliable high resolution autoradiographs of tissue sections or other cell preparations. The method has been applied to electron microscopy. More recently, techniques using Bromodeoxyuridine (BuDr) as a DNA label have been developed that rely on the detection of immunochemical reactions (**1.7**).

### *THE CELL-CYCLE*

Inherent to cell kinetics is the concept of the cell-cycle. Each proliferating cell enters the DNA synthesis phase S. This phase lasts about 6 to 16 hours, depending on cell and species. There is then observed a time interval G2 (gap 2) before labelled cells are seen in mitosis. After mitosis the label is shared between the daughter cells. If these remain 'in cycle' there is a longer gap G1 before the cells enter phase S again. The thyxidine labelling technique is unable to follow what is occurring within cells during the G1 and G2 phases.

### *CELL KINETIC PARAMETERS*

The technique of tissue autoradiography following application of tritiated thymidine (or the BuDr technique) reveals a number of things. Firstly the types and proportions of cells in DNA synthesis can be identified. The proportion of cells labelled at, say, 1 hour after a single application, is called the labelling index. Because the S phase lasts 6 to 12 times longer than mitosis, the labelling index is correspondingly higher than the mitotic index. Secondly, the duration of S phase may be estimated but this measurement requires either multiple sampling of the labelled population or repeated applications of a labelled substance. Third, the duration of the cell-cycle may be measured for cells in continuous proliferation. This is related to the division rate, measured in cells per day, for a given population. In addition, the proportion of the cells 'in cycle' within a given population may be measured. This is called the growth fraction. Specific antibodies such as Ki67 have recently been used to investigate growth fractions. Finally, the fate of labelled cells may be followed by sequential sampling.

### *LIMITATIONS OF CELL KINETICS*

The quantitative results of cell kinetics have proved attractive to mathematicians and computer modellers. Their studies must be supported by thorough knowledge of the biology and by common sense. The accuracy of cell kinetics is low – resulting from the low precision of the techniques and the wide biological variability in the parameters being measured. For example, differences much below 30 per cent in labelling index between populations are not detectable with any reliability. It must also be recognized that cell kinetics has been developed from work with rodents, using techniques which are not always applicable to human cell populations.

If these limitations are kept in mind, cell kinetics gives valuable insights into the relationships between cell division and overall tissue growth. The study of the controls on the growth of a tissue must include explanations of the mechanism by which those controls affect the population sizes and rates of division of the relevant proliferating cells. *N.F. Kember*

# Human skeletal muscle across the lifespan

Skeletal muscle develops from mesodermal cells which, around the sixth week of gestation, differentiate to form myoblasts which aggregate and fuse to form myotubes attached at each end to the tendons and the developing skeleton (**5.6**). Some myoblasts remain as single cells with mitotic potential and these will form the satellite cells. During growth it is the satellite cells that divide and are incorporated into the muscle fibres. Likewise, if the muscle fibre is damaged the satellite cells are activated to divide and begin the process of regeneration. Around 10 to 11 weeks of gestation, developing motor axons from the spinal cord invade the fetal muscle where they make contact with the myotubes. At first, a number of axons form synapses with each embryonic fibre but as the muscle matures all but one of the synapses are lost. In a mature muscle one motor neurone will, through its axonal branches, supply many muscle fibres scattered throughout the muscle. All the muscle fibres supplied by one motor neurone form a motor unit. If a motor neurone fires, all the muscle fibres in that unit will contract at the same time, producing a twitch. Within a single muscle there is a range of motor unit size, determined by the numbers of fibres and, in addition, the type of muscle fibre varies between motor units.

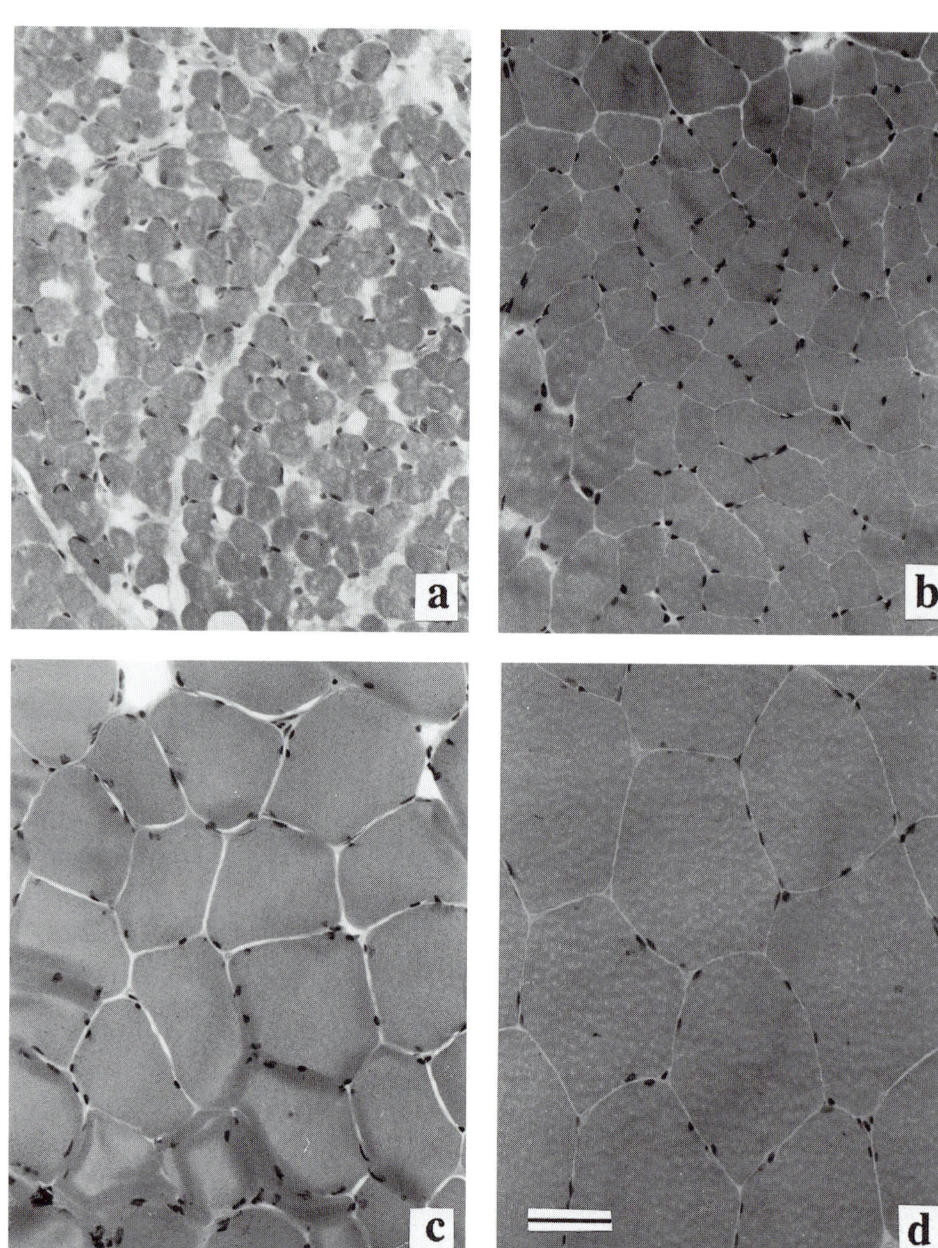

**Growth of muscle-fibre size of the human quadriceps: a = baby aged 8 months; b = child of 5 years; c = boy, aged 14 years; d = large male, aged 23 years. All at the same magnification and stained with hemaoxylin and eosin. Bar in d = 5μm.**

Muscle fibres can be divided broadly into three types, each with a different content of glycolytic and mitochondrial enzymes: type 1 fibres have high oxidative and low glycolytic enzymes; type 2b have low mitichondrial content and high glycolytic enzymes. Type 2a are similar to 2b but with more mitochondria.

The soleus muscle in most mammals is composed almost entirely of type 1 fibres, while the extensor digitorum longus consists mainly of type 2a and 2b fibres. Most other skeletal muscles contain mixtures of fibre types. The high content of myoglobin and cytochrome C in mitochondria of the type 1 fibres gives the slow muscles their red colour.

As early as the 17th century a number of authors had commented on the fact that muscles differ in their appearance but it was not until 1873 that the French physician and physiologist Ranvier recognized that skeletal muscles not only differ in colour, some being dark red, others almost white, but that they also have different contractile properties. For example, the soleus muscle is red in appearance and is slow, while the extensor digitorum longus, a white muscle, contracts and

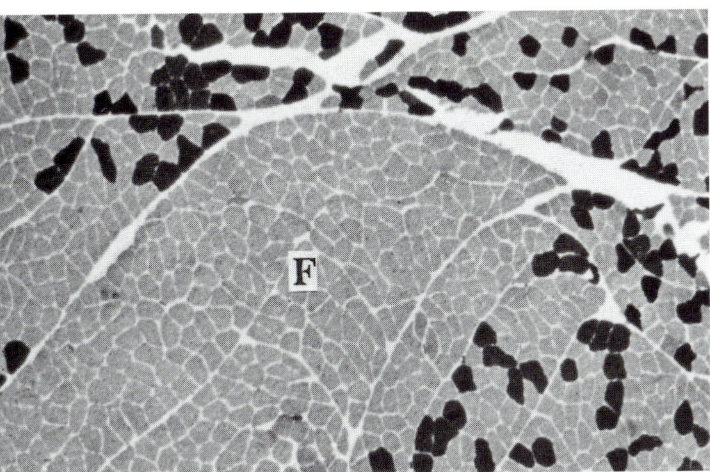


**Adductor pollicis muscle of the hand (autopsy sample, transverse section) from an elderly woman with no clinical neurological signs. With this stain (ATPase pH 9.4) the fast type 2 fibres are dark while the slow type 1 fibres stain less intensly. F = fascicle containing only type 1 fibres, indicating dennervation and re-innervation.**

relaxes rapidly. The reason for these differences lies in the nature of the fibres making up the muscle or motor unit.

On the basis of size, speed and fatiguability, motor units fall between two extremes: large, fast and fatiguable, or small slow and fatigue resistant. All the fibres in one motor unit have the same histochemical properties and the large fast units tend to be made up of type 2b fibres while the small, slow units are predominantly composed of type 1 fibres. Motor units consisting of type 2a fibres span a range of size and fatigue resistance which is reflected in the broad spectrum of their mitochondrial enzyme activities.

Differences in fibre composition determine the functional capability of a motor unit and it is important to understand how the differences arise. The first clue came in a series of experiments by Buller and colleagues (1960). Motor nerves from the slow soleus and the fast extensor digitalis longus were transplanted into the opposite muscle, and after several weeks it was noticed that the re-innervated muscles had changed colour and contractile characteristics. What was formerly a slow muscle had become fast, and the fast muscle had become slow. Imposing an artificial pattern of activity by electrical stimulation can change a fast muscle to slow, and it is apparent that the pattern of activity imposed on the muscle by repeated usage determines both the speed and the fatiguability of a muscle. The more a muscle is used the more fatigue resistant it becomes as a result of increased blood supply, mitochondrial content and a slower rate of energy turnover.

Many top endurance athletes have over 80 per cent type 1 fibres in their leg muscles, and there is considerable debate as to whether they were born this way or have achieved the fibre-type disproportion as a result of their prolonged training. While training is undoubtedly important, there is a strong genetic influence over fibre-type proportions as there is for maximum aerobic exercise capacity which is determined, in part, by the fibre-type composition.

The number of fibres in each human muscle is probably set by 24 weeks of gestation, so that growth occurs by an increase in size without a change in fibre numbers (hypertrophy without hyperplasia). In an adult man about 40–45 per cent of the body-weight is muscle and this figure is slightly lower in females. Reliable measurements of strength can be made from the age of about 5 years and is found to increase steadily in both sexes up to puberty when, during the growth spurt and sexual maturation, a more rapid increase occurs. In the years before puberty, muscles of the lower limbs grow in proportion to the body-weight (or height cubed). Teleologically this might be expected as muscles in the leg clearly have the function of bearing the body-weight. Muscles in the upper body, however, grow less rapidly, in proportion to height squared, the muscles maintaining the same proportions as the long bones increase in length. During puberty and the adolescent growth spurt (**5.15**, **5.17**), the skeletal muscles grow more rapidly in relation to height and body-weight.

For adult women the proportions of muscle strength to body-weight and height remain much the same as those for children. In adolescent boys, however, there appears to be an additional stimulus for muscle growth that is particularly noticeable in the muscles of the upper limb girdle. The relationship between strength of the biceps and height differs from that seen in young boys. This additional muscle growth probably represents the direct action of testosterone on muscle.

Most men can manage at least one press-up or chin-up, while these exercises are very difficult for the majority of women. Throwing, pole-vaulting and gymnastic exercises involving strength of the upper body are other activities where men tend to have the advantage over women. It is not clear, however, what evolutionary advantage is gained by one sex having particularly strong arms. The most likely explanation is that it is of little practical advantage but that upper-body muscle development is another male secondary sexual characteristic. Heavy musculature may originally have served the function of either attracting females or of impressing and dominating other males, the latter being the more likely explanation.

Although physical strength is an important factor in many athletic and everyday activities, it must be remembered that nervous control of muscle actions is a prerequisite for normal movement and these motor functions develop over the first 10 years of life. Many of the benefits of training are due to the acquisition or honing of skills as much as to physiological changes in the muscle itself.

During and after the sixth decade there is a marked loss of muscle mass and a decrease in strength that is particularly notable in women when the hormonal balance is changed after the menopause. The loss of muscle in elderly subjects appears to be progressive and by the age of 90 years muscle mass can be reduced by 30 per cent. The loss of muscle bulk is reported to be greater than can be accounted for by atrophy of the muscle fibres, indicating that fibre numbers may be reduced possibly as a consequence of loss of motor neurones from the anterior horn.

Histochemical examination of muscle specimens taken at autopsy from elderly subjects with no reported neurological symptoms frequently show areas of fibre-type grouping, suggesting some neurogenic disturbance has occurred. These appearances could also be due to peripheral damage to motor nerves. It is of interest that the small muscles in the foot can show clear evidence of neuropathic change with increasing age, which is thought to be a consequence of wearing shoes. Increased collagen cross-linking and lipofuchsin deposits in the muscle are also common findings. Whatever its cause, the loss of muscle may eventually

## PHYSICAL WORK CAPACITY DURING GROWTH

The functional capacity of the individual to do physical work is best measured by the maximal oxygen consumption ($\dot{V}O_2$ max; peak $\dot{V}O_2$), and is an indicator of their physical condition, or fitness. It is expressed as the total $\dot{V}O_2$ in litres per minute or relative to body-weight (ml/min/kg). The former is strongly related to body size and the latter has been widely used as indicative of physical fitness. Cross-country comparisons of $\dot{V}O_2$ max (l/min) show it to be similar in both sexes, but with girls tending towards lower values than boys. $\dot{V}O_2$ max in relation to body-weight (BW) shows some separation of the various nationalities with British boys and girls and Colombian girls tending to lower values than the others, and the Swedish children having higher values. There is a difference in pattern between the sexes with boys increasing or levelling off with age and the girls declining after 8 to 12 years of age. This decline is thought to be related to decreased physical activity in adolescent girls.

There are difficulties of normalizing $\dot{V}O_2$ max; comparing values for girls and women, the decrease in $\dot{V}O_2$ max (ml/min/kg BW) with age is

**Children playing basketball. Photograph by Stanley Ulijaszek.**

**Nepali boy carrying buffalo fodder back to the village. In many rural communities in the developing world, children are expected to work to some degree and help the household economy. Photograph by Stanley Ulijaszek.**

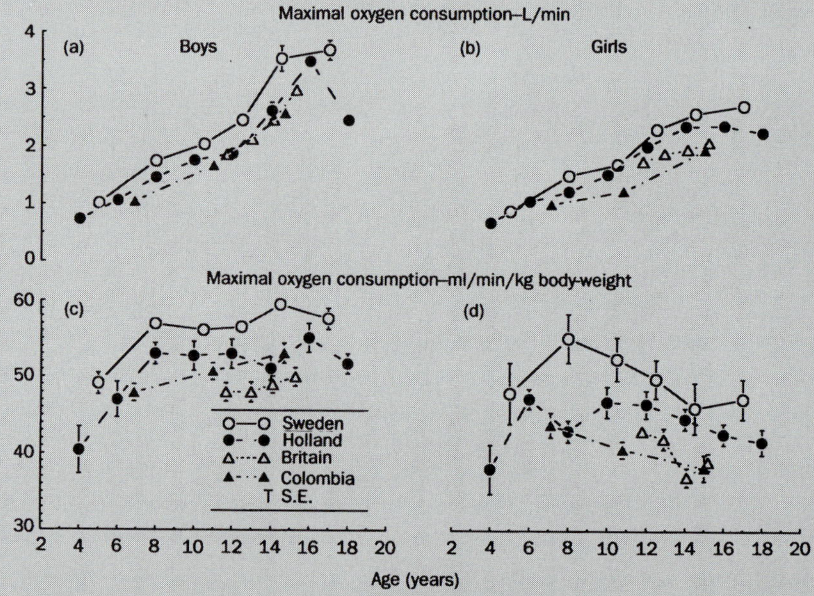


**Mean ± SE of $\dot{V}O_2$ max as L/min and ml/min/kg body weight in boys and girls from four countries; Sweden (Åstrand, 1952), Holland (Saris, 1982), Britain (Armstrong et al., 1991), Colombia (Spurr and Reina, 1989). Studies were selected from treadmill measurements on boys and girls in the same laboratory with sufficient age spread to allow for detection of growth trends.**

become disabling so that subjects cannot rise from a chair or climb stairs unaided. A decrease in mobility is an important cause of loss of independence and postponing its onset will become increasingly important as the number of elderly people surviving in the population increases. There is no evidence that habitual exercise can help to prevent the loss of muscle bulk although, by preserving cardiac and respiratory function, regular exercise will help the elderly make the best use of the muscle that remains.

*D.A. Jones and Joan M. Round*

See also 'Physical activity and training for sport as factors affecting growth and maturation' (**5.12**) and 'Motor development and performance' (**6.2**)

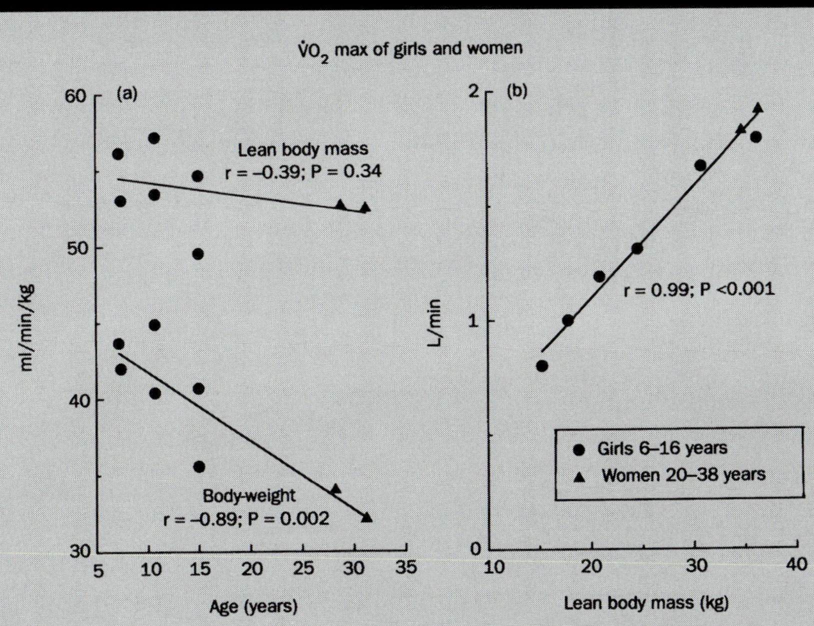


**Comparison of $\dot{V}O_2$ max of 6 groups (n=11–28) of girls 6–16 y with 2 groups (n= 22–27) of women 20–38 y. In panel A, the upper plot is $\dot{V}O_2$ max as ml/min/kg LBM and the lower ml/min/kg BW. Data from Spurr et al., 1994.**

not seen when expressed in terms of lean body mass (LBM). There is a close relationship between $\dot{V}O_2$ max (l/min) and body size expressed as LBM. The difference in aerobic capacity relative to BW in girls compared to boys may be due to differences in body composition. Certainly, in Colombian children the $\dot{V}O_2$ max in each sex is a function of the body size (LBM). However, there is a statistically significant difference ($P < 0.01$) in the slopes of the two lines, without significant difference in the intercepts, indicating an increasing separation between boys and girls with age. For each unit of LBM the girls have lower values for $\dot{V}O_2$ max than boys.

Functional capacity to do physical work as measured by $\dot{V}O_2$ max (l/min) is closely related to body size during growth. Apparently higher values (ml/min/kg BW) in girls than women disappear when normalized for LBM. Girls have lower values than boys, probably as a result of lower levels of physical activity during adolescence.

*G.B. Spurr*

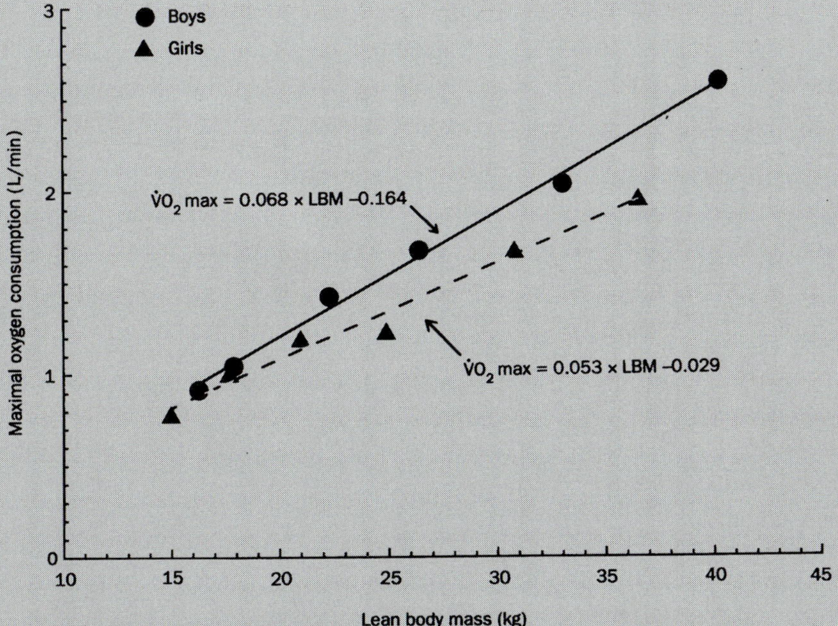


**$\dot{V}O_2$ max (L/min) in 6 groups of Colombian boys and girls 6 to 16 years of age plotted as a function of their lean body mass (Spurr and Reina, 1989).**

# Post-natal craniofacial growth

Post-natal growth in the human craniofacial complex resembles a pattern characteristic of all mammals. At birth, the human skull has an inflated appearance, the large bulbous neurocranium dominating the small face and toothless jaws. The neurocranium and orbital cavities, housing the brain and eyes, follow the growth pattern typical of neural tissues, characterized by rapid early growth which gently decelerates as the infant ages. The major portion of the human face, or viscerocranium, is associated with the respiratory and alimentary tracts and follows the general growth pattern of somatic tissues, increasing throughout the growth period, accelerating during adolescence and ceasing only at early adulthood. The viscerocranium is 'grafted onto' the neurocranium at the basicranium or cranial base; this region exhibits an overall growth pattern intermediate between the somatic and neural patterns. This cursory overview, of course, disguises the tremendous complexity and regional specificity characteristic of human post-natal craniofacial growth. A more detailed consideration requires a focus on growth patterns, growth sites and mechanisms, and growth controls.

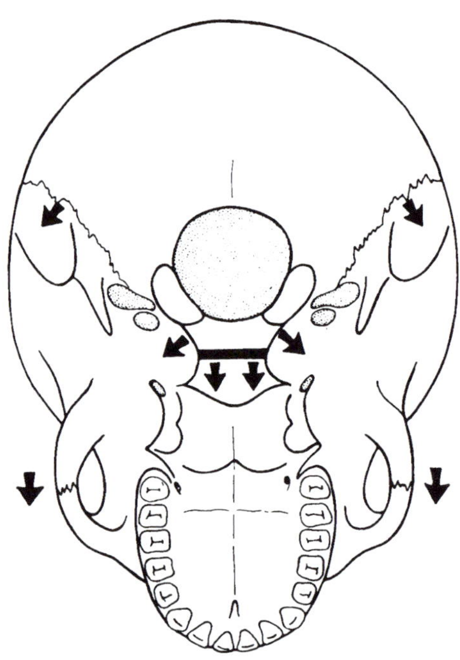

Growth of base of skull. The main points at which growth in length and breadth take place are indicated by arrows. From Sinclair, D. *Human Growth after Birth*. Oxford: Oxford University Press, 1985.

## Patterns

The quantitative description of growth patterns in the skull utilizes traditional suites of craniometric landmarks and planes. Two-dimensional radiographic cephalometry has supplemented quantitative studies of growth of solely the external skull and, more recently, promising new investigative approaches, such as three-dimensional craniometrics and computer assisted tomography (CAT) scans, have been incorporated. These and other techniques have revealed that the bony vault, or calvarium, expands outward considerably during early post-natal growth, achieving almost 90 per cent of its total growth by the age of 2 years. The regions of the calvarium located farthest from the cranial base, i.e., those associated with the cerebral hemispheres of the brain, exhibit the greatest expansion. Closely associated with the growth of the calvarium is the bony basicranium; its anterior component is made up of portions of the frontal, ethmoid and sphenoid bones, while the posterior component includes portions of the sphenoid and occipital bones. The basicranium is associated with the anterior and middle-brain segments, as well as the pituitary gland, which sits within the sella turcica of the sphenoid bone. In terms of post-natal growth patterns, radiographic studies have revealed that the basicranium consists of four fairly distinctive subregions. The span from the anterior to posterior border of the foramen magnum and the area between the foramen caecum, and the pituitary fossa both follow the neural pattern, being relatively slow-growing and completing growth early. The spans from nasion to foramen caecum and from the pituitary fossa to basion both track the general somatic growth pattern, continuing expansion until much later in ontogeny. Many detailed studies have described the complexities of angular changes in the various components of the basicranium, but for our purposes here it is sufficient to point out that the human cranial base remains strongly flexed throughout growth and into adulthood, in contrast to the situation seen in most other mammals.

In humans, the circumorbital region, nasal cavity, and the maxillae (including the dental arch and teeth) comprise the bulk of the 'mid-face'. This region undergoes dramatic growth post-natally, particularly relative to the calvarium and basicranium. Height dimensions exhibit the greatest increase, though depth and width dimensions also grow considerably. Interesting differences in degrees and timing of growth emerge in particular regions, such that the interorbital distance grows relatively little and mostly early, while the bizygomatic distance exhibits its greatest increase later in ontogeny, such that the cheekbones expand and add considerably to facial breadth. Growth in mid-facial depth displaces the maxillae forward, accommodating the erupting dentition.

The mandible exhibits a growth pattern where the condyles grow backward, upward and laterally, the coronoid processes grow upward and laterally, and the bodies and rami grow simultaneously backwards and laterally, keeping pace with the expanding cranial base. Thus, as the mid-face grows or is shifted downward and forward, the mandible or lower face exhibits dramatic increase in height and swings forward (while expanding laterally) to maintain its alignment and occlusion with the growing mid-face.

A number of craniometric studies have clearly demonstrated the presence of an adolescent growth spurt (**5.14**, **5.15**) in various facial and mandibular dimensions, including the earlier initiation and cessation in females as compared to males. This association of craniofacial growth and overall somatic growth has also been established through multivariate craniometric studies which reveal covariation among various dimensions. In addition to this general somatic effect, significant associations during growth have been identified between measures in adjoining regions (such as the posterior maxillae and the basicranium) and within more localized areas (such as the erupting dentition and the surrounding maxillary and mandibular alveolar bone).

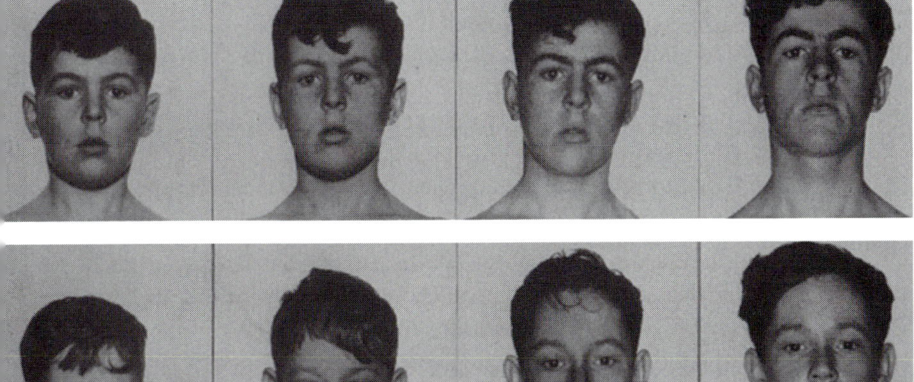

**:e changes in two boys at puberty. From Tanner, J.M. *Growth at Adolescence*. Oxford: Blackwell Scientific,1962.**

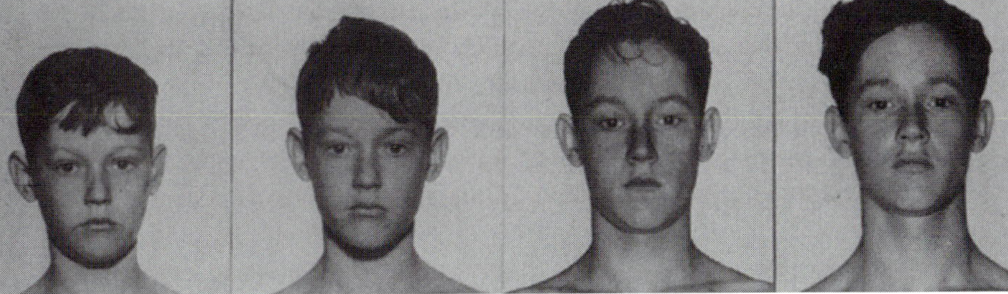

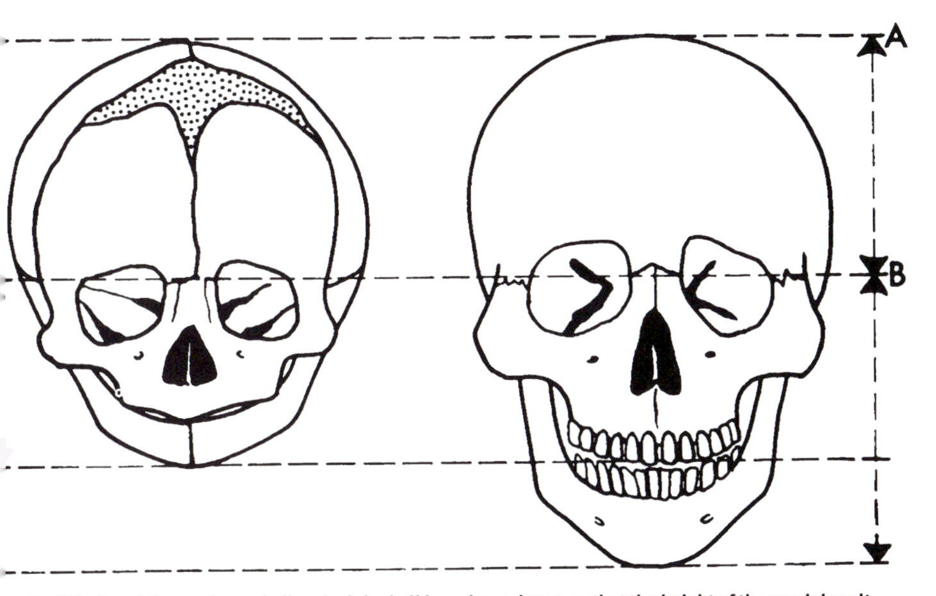


**owth of the face. The newborn skull and adult skull have been drawn so that the height of the cranial vault e distance between the planes A and B) is the same. Notice the great relative increase in the facial skeleton in e adult. From Sinclair, D. *Human Growth after Birth*. Oxford: Oxford University Press, 1985.**

## Mechanisms and sites

Changing proportions during post-natal ontogeny are mediated by different types of bone growth throughout the entire craniofacial complex. Bones grow, remodel, drift and relocate through a combined process of deposition, or the secretion by osteoblasts of a calcifiable matrix on pre-existing bony tissue, and resorption, or the removal of bone by osteoclasts. The interplay between processes of resorption and deposition are complex. All bony surfaces throughout the craniofacial complex can be mapped with remodelling profiles of depository and resorptive fields. In addition, considerable growth occurs at the various sutures of the face and cranium, at the spheno-occipital synchondrosis, at the mandibular condyle, surrounding the nasal septum in the mid-face, and in the alveolar regions. The spheno-occipital synchondrosis in particular is an active growth site contributing to antero-posterior growth in the cranial base and displacement of the mid-face until early adulthood. Other key sutures coursing across the cranial base and calvarium converge on the spheno-occipital synchondrosis, suggesting complex interactions among these sutural growth sites.

## Controls

The factors controlling craniofacial growth continue to be researched and debated. Many believe that the regulation of post-natal bone growth in the skull occurs predominantly through the associated matrix of soft tissues (e.g., muscles), capsular elements (neural, orbital), and functional spaces (mouth, nose, pharynx). The functional matrix model of Moss and colleagues holds that intrinsic genetic control of growth lies not within bones and their associated sutures, et cetera, but rather within the adjacent soft tissues. Evidence for this model comes from a variety of extirpation experiments and clinical cases (such as hydrocephaly), though the precise biomechanical and biochemical control inputs remain unknown. Others suggest that at least for certain areas of the skull, particularly the cartilaginous regions of the cranial base and nasal septum, growth may be quite strongly autonomous and even co-ordinate responses in adjacent regions. In all likelihood, both influences occur, perhaps varying by skull region. Finally, various studies on individuals of known genetic relatedness have demonstrated a moderately strong degree of heritability in facial growth, whatever the precise general or local controls involved.

*Brian T. Shea*

See also 'Dental maturation' (**1.9**), 'Human growth from an evolutionary perspective' (**2.3**), 'Genetics of child growth' (**3.1**), 'Embryonic development of teeth' (**4.4**), 'Morphology' (**5.5**), 'Variation in time of tooth formation and eruption' (**5.10**), 'Hormonal regulation of growth in childhood and puberty' (**5.14**), 'Adulthood and developmental maturity' (**5.17**), and 'Orthodontic disorders' (**7.2**)

# Variation in time of tooth formation and eruption

Studies of the time of tooth formation and eruption include efforts to develop local standards (**1.16**) and to understand the influence of various genetic and environmental factors on dental developmental timing. Standard development is important for a number of reasons. When birth dates have not been recorded, as is frequently the case in international situations and is nearly always the case in forensic and prehistoric research, dental developmental age is often used as a best proxy for chronological age. Additionally, in some situations dental age is a more appropriate yardstick for determining the time of medical and dental treatment. Last, understanding genetic and environmental influences on dental formation is a fundamental research question. Such studies may provide insights into evolutionary processes and how genes and environment interact during development. The following is a brief summary of recent findings on factors influencing tooth formation and calcification.

### Methods and processes

Dental development studies include: 1) the time of emergence of teeth through the alveolar bone and gums, called tooth emergence or eruption studies, and 2) the relative degree of formation of the dental hard tissues. The former type of study is the more common. Such studies include cross-sectional oral examinations of erupted teeth. A few longitudinal studies of dental eruption have also been completed; these involve repeated oral examination of the same individuals. Tooth formation studies are less common because they require either radiographic examination of the developing teeth in their bone crypts or, in a few earlier studies, direct anatomical observation of the developing tooth from dissection of autopsies. A few well-designed longitudinal radiographic studies have been completed since the 1950s. Unfortunately, however, these tend to miss earlier developmental stages.

Although the observations of development are relatively simple, they belie a complex and multifaceted process. Dental formation includes both matrix formation and calcification (**5.6**). Calcification may involve up to four waves. Methods of assessment of degree of formation, most commonly from estimation of opacity of a radiographic image, are based on the relative degrees of calcification that are observable on panoramic or other types of intra-oral radiographs. Factors governing control of the commencement of matrix formation are most likely very different from those controlling the rate of enamel extension and of the timing and completeness of calcification.

The control of eruption is even more multifaceted. Eruption is influenced in part by the degree of dental development. In general, teeth erupt when their roots are about two-thirds to three-quarters developed. However, little is known of what controls the movement of the unerupted tooth through bone (to erupt eventually), and eruption is clearly tied to bone size and the availability of alveolar space to accommodate the erupted tooth. With dental development it appears that the more simple the observation, the more complex the underlying process.

### Genetic control of tooth formation

Dental hard-tissue formation, the first stage of development, is considered to be under strong genetic control. It has been suggested that tooth eruption is about one-third as affected by environmental variation as bone ossification. Tooth formation, in comparison, appears to be under even tighter genetic control. Siblings display similarities in timing, whereas endocrinopathologies and nutritional variations do not greatly affect timing. Except for differences between males and females, there is very little evidence to suggest great variation in tooth formation timing among groups or individuals.

Among teeth and dental stages, the least variation occurs in earlier developing teeth of both the deciduous and permanent dentition and earlier developing stages for all teeth. Also, polar teeth, such as first molars, appear to exhibit less variation in developmental timing than non-polar teeth. Polar teeth appear to be more tightly canalized.

Whereas there is little variation in early stages of development, later tooth-formation times do vary significantly among standards. For example, over two years of difference is reported in the mean age of completion of crown formation for canine teeth among standards. These differences could be interpreted as due to genetic and/or environmental factors. However, no clear pattern emerges. The most parsimonious explanation is that the variation is mostly attributable to differences in methodology among studies.

Tooth-formation timing does consistently vary between men and women. Most studies have shown that females tend to be slightly advanced compared to males in both the time of tooth formation stages and eruption.

### Dental emergence

The available data suggest that tooth emergence, while being genetically controlled, is moderately affected by environmental and other non-genetic factors. Again, greater variation is seen in later-developing teeth, permanent versus deciduous teeth, and in teeth that are 'non-polar'. Permanent-tooth emergence is generally delayed in European populations, for example, and data suggest associations in emergence time of individuals with shared genetic make-ups.

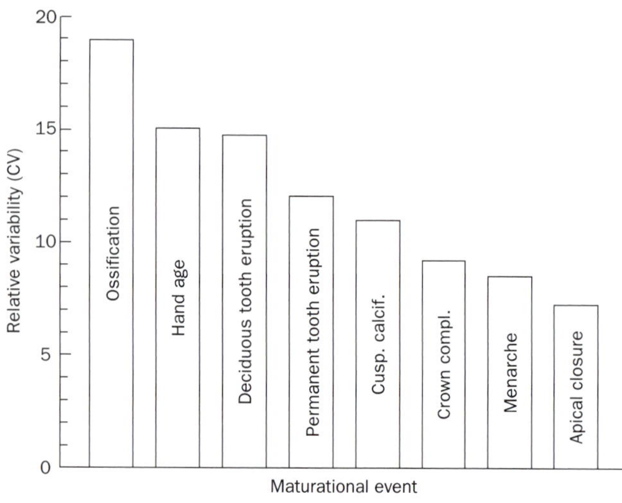


**Illustration of the relative amount of variability for various processes of maturation, demonstrating that dental calcification has very low variability compared with other criteria. From Lewis and Garn (1960).**

Because many human groups or populations do not record birth dates, tooth emergence has frequently been used as an estimator of chronological age. For this reason, many studies have attempted to better understand how much emergence might be affected by nutritional status and other environmental factors. Most studies have demonstrated that tooth emergence is affected by nutritional status, dietary intake, and economic status. For example, in Guatemala, nutritional supplementation of about 70 kilocalories per day slightly advanced tooth-emergence times, and tooth emergence was earlier in individuals with more adipose tissue. Infants who were small-for-date tend to have delayed dental development. Interestingly, it is not certain how much prenatal as opposed to post-natal factors affect tooth eruption times. Generally, the delays in emergence times with decreased nutritional status are in the order of less than a month or two for the deciduous dentition and only slightly greater for the permanent dentition. Correlations between past or current nutritional status and dental maturation tend

## DEVELOPMENTAL DEFECTS OF DENTAL ENAMEL

Tooth-formation timing is under strong genetic control (3.3). Perhaps because of this, the developing tooth is limited in means by which it can respond to environmental perturbations and suboptimal conditions. One frequent method of response is to continue the pace of development while forming thinner or less fully calcified tissues. Thinner or altogether missing enamel is referred to as an enamel hypoplastic defect, and less calcified enamel as a hypocalcification. The most frequent types of hypoplastic defects are somewhat linear (called linear enamel hypoplasia) and transverse in orientation, reflecting the relative development of the tooth crown at the time of disruption. These hypoplastic enamel defects occur because of a disruption during enamel matrix formation (a disruption during the secretory phase of amelogenesis). Decreased calcification, hypocalcification defects, can occur at almost any stage from initial matrix formation to full maturation. Both enamel pathologies are classes of developmental defects of dental enamel.

Enamel hypoplastic defects may be caused by local trauma, hereditary conditions, and systemic stress. Defects due to local trauma affect only one or a few adjacent teeth; hereditary defects are very severe and tend to involve complete crowns. Defects due to systemic physiological disruption (stress) tend to affect only the teeth developing at the time of stress, and their location on tooth crowns reflects the relative development of the teeth at the time of stress. Over 100 causes of enamel hypoplastic defects have been isolated. For this reason, it is probably more fruitful to consider these defects to be non-specific indicators of stress during development.

The location of enamel defects on tooth crowns reflects the relative completeness of crown formation at the time of the defect-forming event. Thus, the location of enamel defects on different teeth provides a record of the relative development of each tooth at the time of 'stress'. Interestingly, it is common that not all teeth developing at the time of stress stand an equal chance of developing a

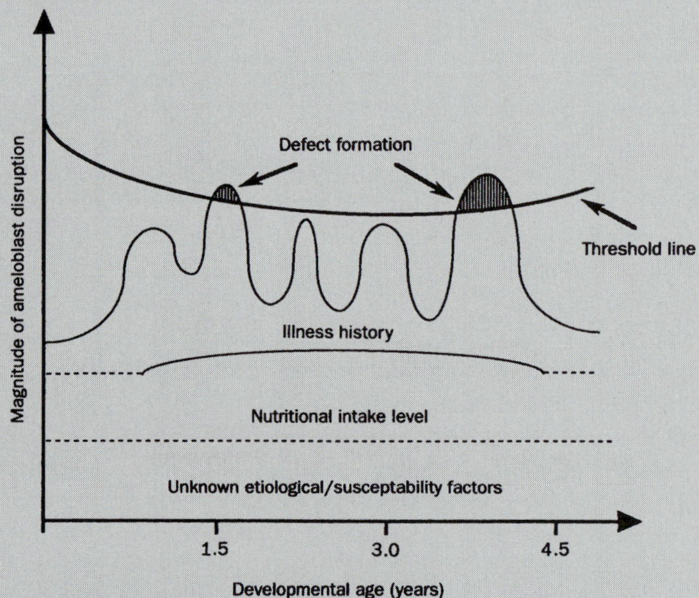


**Threshold model for the formation of a linear enamel hypoplasia (LEH). Enamel apposition and LEH formation occurs when ameloblasts temporarily cease secretion of enamel matrix. This growth stoppage has been associated with a wide variety of stressors. It is suggested that there is a threshold at which ameloblasts will cease functioning. This threshold may be reached by a combination of unknown etiological factors, chronic or acute undernutrition, and acute morbidity.**

to be very low, in the order of r = 0.1–0.2. Thus, when chronological ages are not available, tooth emergence may be used as a somewhat conservative proxy.

Questions concerning the pattern of dental development, that is the order and relative timing of dental events and stages among different teeth, have emerged in the last decade as an important research topic. The pattern of development is a phenotypic characteristic that may be used to distinguish among related taxa. One of the difficulties in such studies, however, is the lack of agreement among standards for the same species. Very little information is available at this time on dental developmental patterns for non-human primates, and great variation exists for humans among standards, due to the aforementioned methodological variations. However, with the possible development of new, low risk methods for detection of hard-tissue development, newer standards, especially for early stages of development, might help to resolve these and other questions.

*Alan H. Goodman*

See also 'Dental maturation' (1.9) and 'Embryonic development of teeth' (4.4)

hypoplastic defect. Anterior teeth and teeth that are under the strongest genetic control of developmental timing are generally the most susceptible to ameloblastic disruption.

In archaeological and paleontological studies, teeth are often found, and for this reason, enamel hypoplasias have frequently been used as an indicator of relative degrees of stress during tooth crown formation. Studies of populations undergoing the transition from hunting and gathering to agriculture have frequently shown that this transition results in an increase in the rate of enamel hypoplasias. Other studies have shown that enamel hypoplasia rates are very high in poor-house populations and among slaves, thus confirming that these were highly stressed groups. Similarly, studies of contemporary populations have shown that enamel hypoplasias are inversely related to nutritional and socio-economic status.

A number of studies have shown lower rates of enamel hypoplasias in the deciduous dentition compared to the permanent dentition. Apparently this is because of protection to the growing individuals during prenatal and early post-natal times (when the crowns of deciduous teeth are developing). Permanent-tooth enamel hypoplasias tend to develop around the second and third years of age. This has been associated with the change to a post-weaning diet and increased infectious disease at this time.

Because enamel hypoplastic defects are easily observed and can provide information on the time of developmental disruption, they may be useful adjunct measures of nutritional status. For example, enamel hypoplasias on deciduous teeth may provide a unique window into the prenatal period, and enamel hypoplasias in permanent teeth may provide a window into early periods of development in adolescents and adults. *Alan H. Goodman*

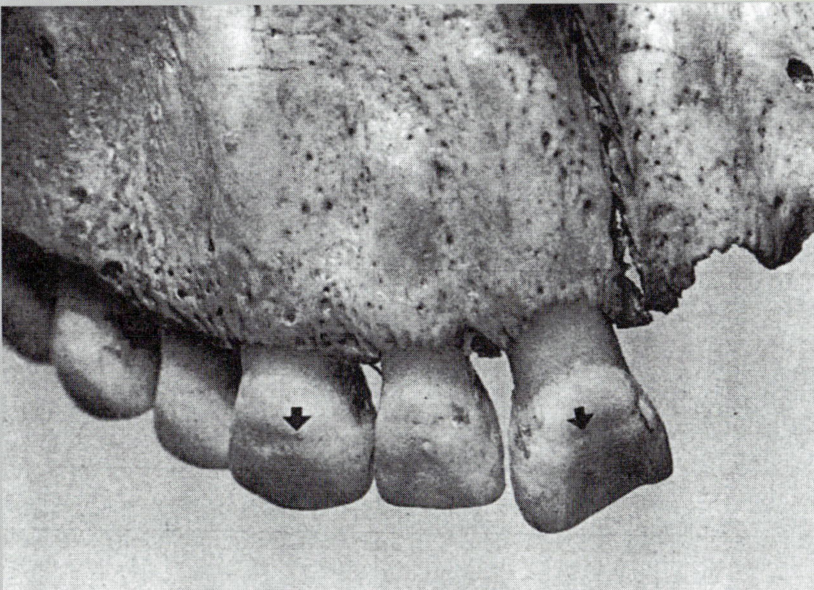

**Chronologic enamel hypoplasias (stress-hypoplasias) on right maximally central incisor and canine (arrows). Both of these shallow hypoplastic bands occurred around 3.5 year development age, based on the degree of crown completion (enamel apposition begins at the occlusal tips, which are slightly worn). The common estimated age at formation suggests that these defects were the result of the same systemic physiological perturbation (stress).**

# Body composition

Humans begin life lean and watery. By birth, they are nearer adult levels of hydration, but before adulthood is reached they are likely to have gained and lost and gained again in relative fatness, and demonstrated large differences in the sexes from puberty onwards. Body-composition changes throughout growth are complex but, on the whole, orderly. It is convenient to regard the body as made up anatomically of a lean body mass (LBM) plus adipose tissue (AT). Individuals differ in the relative proportions of these and the proportions change over the lifespan. There are also characteristic changes in the composition of the lean body mass during growth, particularly in the early years. The growth of many organs and tissues mirror the growth of height. Others, such as the brain, the reproductive organs and adipose tissue have very different growth patterns, particularly in regard to timing.

## The development of the lean body mass

Marked compositional changes take place in the fetus as it increases in size in the last trimester and in the neonate in the first year of life. The fetus is initially of high water (94%) and low fat, protein and mineral content. By the time of birth, the water content of LBM is some 82 per cent, and other constituents, such as protein and mineral, have increased proportionately. The fall in water content is a mammalian development characteristic, but it has been known for sometime that Moulton's generally accepted theory that lean body mass is chemically mature when 4 per cent of the lifespan has been reached does not apply to humans with their longer growth period. Instead chemical maturation, particularly for water and nitrogen, was

## ORGAN DEVELOPMENT AND BODY-WEIGHT

***PATTERNS OF GROWTH***

The growth of different human tissues is so complex that it is difficult to illustrate the changes in a single way. Different organs not only grow at different times and rates, but different constituents within the same tissue may undergo different developmental changes. Furthermore, some tissues increase in size or cellularity whilst others do the opposite. There are two aspects of organ development that are particularly important. First, the weight of organs at birth, expressed as a percentage of adult weight, varies considerably. At birth, when body-weight is about 5 per cent of adult weight, the brain is 25 per cent of adult weight, the thymus is 65 per cent, many other organs (reproductive organs, ovaries, testes, thyroid, pancreas, liver, kidneys, lungs and heart) are 3–9 per cent, and muscle and the skeleton of males are 3.0–3.5 per cent. Second, the growth spurt of individual tissues or organs may occur earlier (e.g. brain) or later (e.g. reproductive tissues) than that of the body. The brain begins its growth spurt in the last trimester of intra-uterine life (**4.6**) (neurones develop in the first trimester, but glial cell formation occurs predominantly in the third trimester) and continues after birth. Most of its adult size and DNA content is achieved by 2–3 years of age when body-weight is only about a quarter of its adult size.

The head and some of its structures follow a similar pattern of growth as the brain (**5.9**). For example, the eyeball at birth is 35 per cent of its adult size, and it approaches its adult size in a similar way as the brain. Many tissues and major organs (e.g. lung, heart, kidney, liver) approach their adult weight in a manner that is not very different from that of the whole body. The reproductive organs are an exception since during early childhood they grow much slower than the body as a whole, but during puberty they approach their adult size at a rate that is faster than the body as a whole (cf. brain). The thymus is unusual in at least three ways. First, its weight at birth is already two-thirds of that of a young adult. Second, between

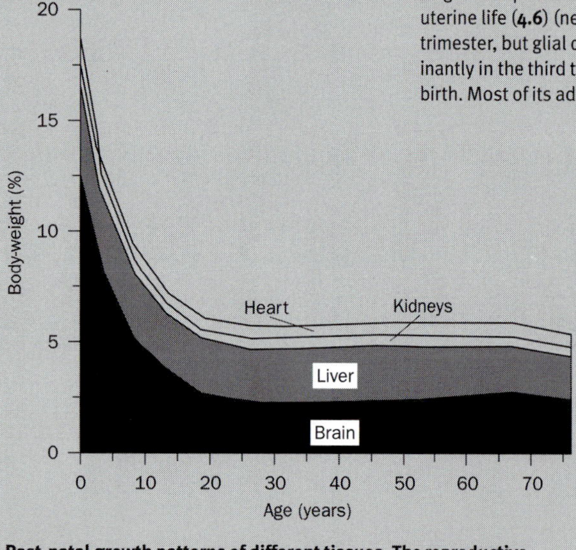


**Post-natal growth patterns of different tissues. The reproductive tissues (individual or combined) include testes, ova, prostate and uterus. The uterus is unusual in that its size decreases between birth (0.4g or ≈ 8% of adult weight of ovary) and the first 2 years of life (≈ 3% of adult weight). This is not shown.**

*Water and protein content of tissues according to developmental age (g/100g tissue)*

| | Fetus | | Baby | | Adult |
|---|---|---|---|---|---|
| | 13–14 weeks | 20–22 weeks | Full-term | 6 months | |
| *Water* | | | | | |
| Muscle | 91 | 89 | 80 | 79 | 79 |
| Liver | 83 | 81 | 79 | 76 | 75 |
| Kidneys | 92 | 88 | 84 | – | 81 |
| Brain | 92 | 92 | 90 | – | 77 |
| Skin | 92 | 90 | 83 | 68 | 68 |
| *Protein* | | | | | |
| Muscle | 6 | 8 | 13 | 16 | 17 |
| Liver | 12 | 12 | 13 | – | 15 |
| Kidneys | 8 | 9 | 12 | – | 15 |
| Brain | 6 | 5 | 6 | – | 11 |
| Skin | 7.5 | 7.9 | 14 | 34 | 34 |

thought to be delayed until 4–6 years. However, current thinking is that such maturation is much later as relative protein and mineral contents of LBM are still increasing in late adolescence in boys. This variation in LBM composition during growth is such that assuming adult values results in the overestimation of body fatness by most of the *in vivo* whole-body composition methods readily available. This has to be borne in mind in examining the early literature. It has necessitated the development and implementation of multi-component methods of body composition analysis and of age and sex-specific estimation procedures.

Skeletal and muscle mass show growth curves similar to those of height and weight. Muscle comprises 25 per cent of the body-weight of the neonate but 40 per cent of the adult male. This is achieved by hypertrophy alone. Muscle growth is more or less constant during childhood but there is a rapid increase in rate at adolescence. Relative muscle mass increases from 42 to 54 per cent in boys from 5 to 17 years. In girls, it increases from 40 to 45 per cent between 5–13 years and then falls slightly as relatively more adipose tissue is gained in adolescence. There is also a change in composition of the skeleton as mineral is deposited in cartilage.

## Changing masses and proportions of lean body mass and adipose tissue

### *Lean body mass*

In absolute terms, growth in lean body mass has a similar pattern and timing to that of height and weight, although longitudinal data are scarce. Before adolescence, boys have a slightly larger LBM than girls. The adolescent spurt in LBM is later in boys than in girls and much more intense and prolonged. Peak velocity of LBM growth in boys is around 17 years, but the maximum LBM value is not reached until 20 years. In Czech boys peak LBM velocity coincided with peak height velocity (PHV), but other data report no difference in annual increments between 10–18 years. In girls PHV and peak LBM velocity occur at 15 and 16 years, on average, but there is wide variability in all these ages. By late adolescence and young adulthood males have 50 per cent more LBM than

3 months and 19 years its weight is greater than that of the adult of 20–50 years, and in early childhood exceeds its adult weight by 50 per cent. Third, it decreases in size during adolescence, when most other tissues are growing rapidly.

#### *DIFFUSE CELLULAR SYSTEMS*

Lymphocytes form part of the immune system (**4.7**), which is widely distributed within the body – for example, in the thymus, lymph nodes, other tissues and the circulation. The mass of lymphocytes in a 'reference man' is 1.5 kilograms (1.2 kg in a 'reference woman'), but only less than 0.5 per cent is present in the circulation. The lymphocyte mass at 1 month is about 10 per cent of the adult mass but between 0.5–1 year it is as much as 43 per cent of the adult mass. The reticuloendothelial system, which has a mass of 1.8 kilograms in reference man is another example of a diffuse cellular system.

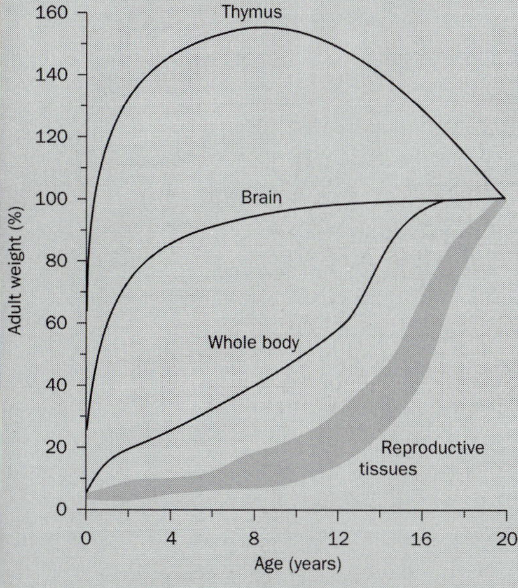


**Contribution of organ-weight (heart, kidneys, liver and brain) to body-weight according to age. The reduction is most striking for the brain.**

#### *ORGAN SIZE IN RELATION TO BODY-WEIGHT*

The contribution of most major organs to body-weight decreases between birth and adulthood. In contrast, the contribution of muscle increases (22% at birth to 40–45% in young adults, 30–35% in women). Adipose tissue is mainly accreted in the third trimester of pregnancy so that infants born prematurely (26 weeks gestation) have a small fat mass, which constitutes only 1–2 per cent body-weight. In both childhood and adulthood, adipose tissue mass is present in variable amounts but, in Western societies, its contribution to body-weight increases during adult life.

#### *CHEMICAL COMPOSITION OF TISSUES*

Although some of the developmental changes in structure are specific to individual tissues, during growth *in utero* and early childhood there is a general reduction in tissue hydration especially in extracellular water (**1.6**). This results in an increase in the concentration of other constituents, such as protein, and fat in the case of adipose tissue.

#### *DIFFERENT PATTERNS OF DEVELOPMENT IN THE SAME TISSUE*

Tissues such as the brain exhibit complex developmental changes. For example, compared to the rest of the brain, the growth of the cerebellum starts later, finishes earlier, and grows much faster during its shorter growth spurt. As a result, the cellularity of the cerebellum (and DNA content/g/tissue) rapidly increases shortly before and after birth, whilst in the cerebral hemisphere opposite changes occur, at least before birth.

#### *EFFECT OF PATHO-PHYSIOLOGICAL CONDITION*

During senescence (**12.3**) many tissues decrease in size. Muscle mass decreases, at least partly because of reduced physical activity, and bone mass also decreases, as calcium is lost from the skeleton, especially in post-menopausal women. Reproductive tissues also decrease in size. The increase in body-weight that occurs during pregnancy (e.g. 12.5 kg) occurs as a result of an increase in the size of multiple tissues (fetus 3.3 kg, placenta 0.65 kg, amniotic fluid 0.8 kg, uterus 0.4 kg, blood 0.9 kg and others, 5.2 kg). The reduction in body-weight that occurs during starvation in animals and humans is associated with preferential preservation of the brain and skeleton. In animal models of injury the central organs (e.g. liver, lung) are preferably preserved at the expense of muscle.

#### *FUNCTIONAL CHANGES*

The different patterns of growth and development of different tissues have important functional associations. For example, muscle strength is related to muscle mass, and loss of calcium from the skeleton, especially when severe (osteoporosis), is associated with increased risk of fractures. Furthermore, four organs (heart, liver, brain and kidneys), which account for about 5 per cent of body-weight, are responsible for almost 60 per cent of resting-energy expenditure. The twofold decrease in metabolic rate that occurs between infancy (e.g. 6 months) and adulthood is associated with a twofold reduction in the contribution of these organs to body-weight. *Marinos Elia*

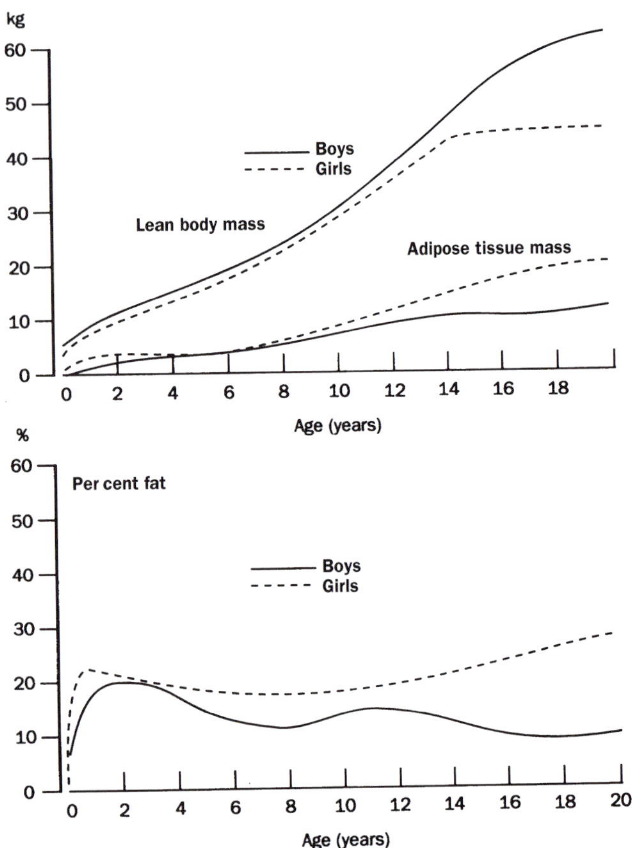


Typical distance curves for lean body mass, adipose tissue and for relative fatness (% fat). Redrawn from R. M. Malina and C. Bouchard, *Growth, Maturation and Physical Activity*, (1991).

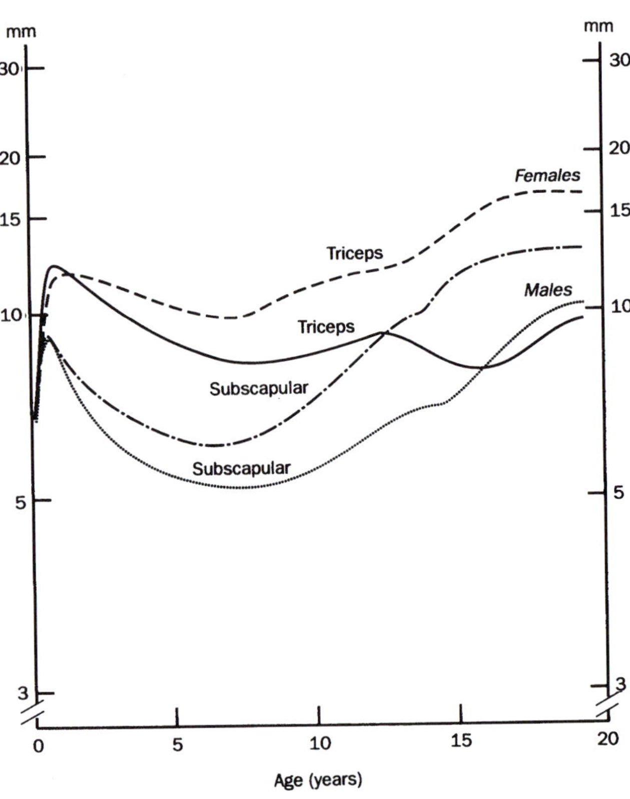


Distance curves for median triceps and subscapular skinfold thicknesses in British children. Redrawn from Tanner, J. M, and Whitehouse, R. H. (1975) *Archives of Diseases of Childhood* 50: 142–145, with permission.

females. Radiography provides absolute measures of regional body composition less compromised by uncertainties of methodology. Gain of arm and calf muscle and bone widths resembles that of LBM as a whole. Bony breadths such as bi-acromial and bi-illiac also increase maximally at PHV in boys and, a year later, in girls.

*Adipose tissue*

In mid-fetal life, the proportion of fat in the body is 1 per cent of the body-weight. This increases to some 15 per cent at birth, 25 per cent at 6 months, and 30 per cent at 1 year by both hyperplasia and hypertrophy. Little adipose tissue may be gained in early childhood and through adolescence in boys. Whether there are negative velocities is disputed. There are slight sex differences in favour of girls at all ages but the differences become striking at adolescence. Females have 50 per cent more adipose tissue mass than males in late adolescence and young adulthood, and have twice the percentage fat as males.

Regional fatness data on subcutaneous adipose tissue thickness can be obtained by measuring the thickness of a double compressed fold, called a skinfold thickness, or by measurements taken off radiographs. During childhood, skinfold thicknesses show a flow, an ebb and a flow, what is called the *adiposity rebound* (**1.11**). Peak values are reached at 9–12 months and then adiposity decreases until 6–8 years then rises again. The rebound occurs earlier in children at higher percentiles for fatness and there is a tendency for children with an early rebound to have higher fatness at the end of growth. Differences in skinfold thicknesses in the sexes are small at birth but, on average, girls always have higher skinfolds than boys. The divergence becomes marked by 8 years. Both sexes experience a prepubescent adipose tissue gain, more so in girls than boys. Girls' skinfolds continue to increase throughout adolescence, and a mid-pubescent loss occurs at the triceps in boys and to a lesser extent at the subscapular site. From radiographs, the nadir for arm adipose tissue width velocity is at PHV and for calf 6 months later. In boys the velocities will remain negative until adulthood but in girls they soon become positive.

The ratio of trunk to extremity skinfolds, a measure of fat patterning, is stable during childhood but at adolescence both sexes gain proportionally more on the trunk than the extremities. The gain is less in girls, and boys may experience a decrease at extremity sites with the result that after adolescence boys have a relatively greater amount of subcutaneous adipose tissue on the trunk than the extremities – a centripetal fat distribution. An association of advanced maturation and centripetal fat distribution has been reported in some studies but not others. Fat patterning is related to later health outcomes, but this is more so for internal-external patterning than for subcutaneous patterning. The proportion of subcutaneous adipose tissue to total adipose tissue falls between 9–19 years in girls and until 13–14 years in boys, after which it increases. There is some evidence that fat patterning is more under genetic control than relative fatness (percentage fat).

## Variations on the average pattern

A number of factors bring about differences in the proportions of the various tissues and the timing of the changes described as a general pattern above. The type and level of nutrition has a considerable influence on body composition and maturity. Breast-fed infants have been found to lighter in weight and leaner than formula-fed infants (**9.1**), but of equal length. Fat infants have double the risk of becoming fat adults than the lean infant. However, most fat infants and children do not stay fat. Fatness is associated with early maturity and leanness with late. Early maturers, on the basis of skeletal age, have larger bone muscle and adipose tissue masses, primarily because of their larger body-size.

Children and adolescents with regular involvement in physical activity tend to show more LBM and less adipose tissue than those who do not. Somatic and skeletal maturation, and the body-composition changes that accompany them, do not appear to be delayed by regular physical activity although intense physical training has been associated with delayed sexual maturation.

There are differences in body composition during growth between various ethnic groups. Adult Africans have more muscle and heavier bones per unit weight, at least in males, and less limb fat. African-American children have lower triceps skinfold thicknesses than European-Americans in the same environments and Mexican and Guatemalan children may have higher subscapular and lower triceps skinfold thicknesses. Children in Asian countries have relatively larger subscapular skinfold thicknesses than London children. However, the primary cause of body composition differences within and between populations arises from the environment in which the children are raised. *N.G. Norgan*

See also 'Assessment of body composition' (**1.6**), 'Developmental morphology of the embryo and fetus' (**4.5**), 'Morphology' (**5.5**), 'Skeletal development' (**5.6**), 'Human skeletal muscle across the lifespan' (**5.8**), 'Physical activity and training for sport as factors affecting growth and maturation' (**5.12**), 'Fat and fat patterning' (**5.16**), 'Nutrition' (**9.2**), 'Catch-up weight-gain' (**9.7**), 'A functional outcome in adulthood of undernutrition during growth' (**9.10**), 'Modernization and growth' (**11.5**), 'Obesity, fatness and modernization' (**11.8**), and 'Ageing as part of the development process' (**12.3**)

# Physical activity and training for sport as factors affecting growth and maturation

Regular physical activity is generally viewed as having a favourable influence on the growth and maturation of children and youth. Nevertheless, recently concern has been expressed about potential negative consequences of intensive training for sport on growth and maturation, specifically girls participating in sports requiring small body-size, slender builds and/or low body-weight (gymnastics, figure-skating, ballet). These contrasting views highlight some of the difficulties in evaluating the role of physical activity in growth and maturation, for example, physical activity and training are variably defined constructs, activity and training are not synonymous, generalizations based upon highly select athletes may not apply to the general population, and so on.

## Physical activity and training

In its broadest sense, physical activity includes '... any body movement produced by the skeletal muscles and resulting in a substantial increase over the resting-energy expenditure'. Most discussions refer to a child's estimated level of habitual physical activity, for example, in hours per week or by an activity score, which is ordinarily derived from questionnaires, interviews, diaries and heart-rate integrators. Physical activity is not the same as regular training. Although activity is integral to training, the latter refers to systematic, specialized practice for a specific sport or sport discipline for most of the year or to specific short-term experimental programmes, for example, 15 weeks of endurance training in running or resistance training.

The measurement and quantification of habitual physical activity and its major correlate, energy expenditure, during childhood and youth are a difficult task and need further attention. Training programmes are ordinarily specific (e.g., endurance-running, strength-training, sport-skill training, etc.), and vary in intensity and duration. As with physical activity, it is imperative that training programmes be better documented. In addition, many of the changes attributed to regular physical activity, though not all, are in the same direction as those that accompany normal growth and maturation.

## Limitations

Inferences about the influence of regular physical activity and training on growth and maturation are based largely on short-term experimental programmes, several longitudinal comparisons of active and inactive children, and comparisons of young and adult athletes and non-athletes, respectively. Longitudinal studies that span childhood and adolescence and that control for physical activity are few. Criteria used to classify children as active or inactive also vary.

Although many talented young athletes train regularly for several years, especially in sports like gymnastics and swimming, the selective nature of sport cannot be overlooked. Athletes as a group are highly select. They are selected for specific skills and, in many sports, for size and physique characteristics. Among female gymnasts and swimmers, for example, the former are already shorter than average and the latter are already taller than average prior to the start of the respective training programmes in childhood. Further, parents of gymnasts are also shorter than those of swimmers, which would imply a genotypic factor in stature. Successful young ballet dancers tend to have the thinness and proportional features of élite ballerinas. On the other hand, males who are successful in many sports at relatively young ages tend to be advanced in biological maturity status and are thus generally taller and heavier than their chronological-age peers. The selectivity of sport thus limits the utility of comparisons of athletes and non-athletes in the context of detailing potential training effects. It cannot be assumed *a priori* that differences in growth and maturation between young athletes and non-athletes are due to regular training. The same applies to studies showing differences between adult athletes and non-athletes, which are often interpreted as reflecting the influence of training during childhood and youth.

The subsequent discussion is based on healthy, adequately nourished children and youth. Intensive physical activity may be contra-indicated in chronically undernourished individuals.

## Activity and growth in body-size

### *Stature*

Regular physical activity, sport participation, and training for sport have no apparent effect on attained stature and rate of growth in stature. Longitudinal data on active and inactive boys followed from late childhood through adolescence indicate no differences in stature. Data for boys and girls regularly active in sport during childhood and adolescence indicate, with few exceptions, mean statures that either approximate or exceed reference medians. Gymnasts of both sexes, female figure-skaters, and male divers present, on average, statures shorter than reference medians. Several short-term longitudinal studies of athletes in varying sports, for example, volleyball, diving, distance-running, basketball, indicate growth-rates similar to those of reference data for non-athletes.

## Body-weight and gross composition

In contrast to stature, body-weight can be influenced by regular activity and training, resulting in changes in body composition. The latter is most often viewed in the context of the two compartment model: body-weight equals fat-free mass plus fat mass. Activity and training are associated with a decrease in fatness in both sexes and possibly with an increase in fat-free mass in boys. Changes in fatness depend on continued, regular activity or training (or caloric restriction, which often occurs in sports like gymnastics and ballet in girls, and wrestling in boys) for their maintenance. When activity or training are significantly reduced, fatness tends to accumulate. On the other hand, it is difficult to partition effects of training on fat-free mass from expected changes associated with growth and sexual maturation.

## Activity and specific tissues

Skeletal, muscular and adipose tissues are the three primary components of body composition, in addition to the viscera. The skeleton is the framework of the body and the main reservoir of mineral. Skeletal muscle is the major work-producing and oxygen-consuming tissue, while adipose tissue represents energy in stored form.

### *Skeletal tissue*

Regular physical activity and training during childhood and youth are associated with increased bone mineral content and mass. The beneficial effects are more apparent in weight-bearing than non-weight-bearing activities. Of particular importance for childhood activity and the integrity of skeletal tissue is the observation that bone mineral established during childhood and youth may be a determinant of adult bone-mineral status.

In contrast to the positive influence of physical activity and training on bone mineralization, excessive training associated with altered menstrual function in some post-menarcheal athletes is associated with loss of bone mineral. Restrictive diets and/or disordered eating are contributory factors. The interaction of disordered eating, cessation of regular menstrual cycles, and osteoporosis in high-performance athletes is of major concern for some adolescent athletes and may impact the accretion of skeletal mineral during adolescence. Thus, under conditions of altered menstrual function (secondary amenorrhea) and a deficient diet, physical activity may have a negative influence on the integrity of skeletal tissue.

### *Skeletal muscle*

Information on the effects of physical activity on skeletal muscle tissue is derived largely from short-term, specific training studies of small samples. Muscular hypertrophy is associated with high-resistance programmes such as weight- or strength-training in adolescent boys, and may not occur, or may occur to a much lesser extent, in pre-adolescent boys and girls, and in other forms of training. There is no strong evidence to suggest that fibre-type distribution in youth can be changed as a result of training.

Progressive strength training is associated with an increase in the relative area of type 2 (fast twitch) fibres, while endurance training is associated with an increase in the relative area of type 1 (slow twitch) fibres in young adults. Corresponding data for youth are variable. In 16-year-old boys, 3 months of endurance training were associated with an increase in the areas of both type 1 and 2 fibres, while three months of sprint training did not affect fibre areas.

Information on changes in the metabolic properties of skeletal muscle with training in children and youth are limited. Endurance training is associated with increased activities of both succinate dehydrogenase (SDH, oxidative) and phosphofructokinase (PFK, glycolytic) in 11-year-old boys. Among 16-year-old boys, in contrast, endurance training results in an increase in SDH but not PFK, while sprint training results in an increase in PFK but not SDH. Differences in training protocols may account for the variable results. There is also the possibility of age-associated variation in response to training. Corresponding data for young females are not available. The limited data thus suggest that regular training has the potential to modify the metabolic capacity of muscle in children and youth. However, after 6 months of no supervised training, SDH and PFK activities return to pretraining levels in the 16-year-old boys. This observation indicates an important feature of training studies. Changes in response to short-term programmes are generally not permanent and depend upon regular activity for their maintenance. An important question is: How much activity is needed to maintain the beneficial changes associated with training?

## Adipose tissue

In studies of children and youth, adipose tissue is most often measured in the form of skinfold thicknesses, which provide an estimate of subcutaneous adipose tissue. Cross-sectional data indicate thinner skinfolds in active children and young athletes compared to reference samples. However, longitudinal data for active and inactive boys and girls followed from 6 to 12 years, and adolescent boys followed from 13 to 18 years, do not differ from each other in skinfold thicknesses. Similarly, boys and girls active in sport and followed longitudinally from 8 to 18 years have skinfold thicknesses that do not differ from reference data. The discrepancies between cross-sectional and longitudinal observations should be noted. It is likely that more intensive physical activity is essential to modify skinfold thicknesses in growing children and youth. In addition, skinfolds change differentially during growth in boys, i.e., extremity and not trunk skinfolds generally decline during adolescence. Data dealing with potential effects of physical activity and training on subcutaneous fat distribution during growth are presently not available.

In young adult males, intensive training for 15 and 20 weeks is associated with a greater reduction in trunk than in extremity skinfolds, while corresponding changes in young adult females are evenly distributed between trunk and extremity sites.

Information on the effects of regular physical activity and training on adipose tissue cellularity and metabolism in children and youth is also lacking. Adipose tissue cellularity increases gradually during childhood and then more rapidly with the onset of puberty. The decrease in fatness associated with training in adults is attributable solely to a reduction in estimated adipocyte size. Trained adults also have increased ability to mobilize and oxidize fat, which is associated with lipolysis. Increases in lipolysis also occur in sedentary adults exposed to aerobic training, and the increase in lipolysis is greater in males than in females.

## Activity and maturation

### *Skeletal maturation*

Although regular activity functions to enhance bone mineralization, it does not influence skeletal maturation of the hand and wrist as assessed in growth studies, i.e., skeletal age (SA) relative to chronological age (CA). Active and inactive boys followed longitudinally from 13 to 18 years do not differ in SA. Young athletes in several sports differ in skeletal maturity status. Boys successful in a variety of sports tend to have, on average, SAs in advance of CAs. Exceptions are ice-hockey players, among whom late childhood and early adolescent athletes do not differ in SA and CA, while older adolescent athletes are advanced in SA. Among girls, those in ballet, gymnastics and track tend to have SAs that lag behind CAs, while those in swimming tend to have SAs in advance of CAs. However, longitudinal observations on young athletes indicate corresponding gains in CAs and SAs. In later adolescence, differences between individuals of contrasting maturity status are reduced and eventually eliminated as skeletal maturity is attained.

### *Somatic maturation*

Age at peak height velocity (PHV) requires longitudinal data (**1.14**) for its estimation. Age at PHV is not affected by regular physical activity and training for sport. The data are limited largely to boys, with only few observations for girls. The limited longitudinal observations for male athletes are consistent with the data for SA, i.e., age at PHV tends to be earlier in male athletes. The magnitude of PHV is also not affected by regular activity and training.

### *Sexual maturation*

Longitudinal data on the sexual maturation of either girls or boys regularly active or training for sport are not extensive. Cross-sectional observations on breast, genital and pubic hair development in young athletes are consistent with those for SA, while the limited longitudinal data indicate no effect of activity or training on the timing and progress of breast, genital and pubic hair development in boys and girls. An exception is a study of élite ballet dancers, in whom there appears to be normal progress of pubic hair, but very slow progress in breast development. The latter may be influenced by the extreme linearity of build and thinness of élite ballet dancers, i.e., a deficiency in subcutaneous fat which may influence assessment of breast development. Although evidence for élite adolescent ballet dancers is not available, observations for elite adult dancers indicate rigorous weight control and a high prevalence of disordered eating.

Much of the discussion of the potential influence of activity and training on menarche focuses on later mean ages at menarche which are characteristic of late-adolescent and adult athletes. These data, however, are primarily retrospective. Prospective and status-quo-estimated ages at menarche for gymnasts, divers and ballet dancers are reasonably consistent with retrospective estimates, while prospective and status-quo-estimated ages for swimmers, track athletes and rowers are earlier than retrospective estimates for late-adolescent and adult athletes. Training for sport is commonly indicated as the causative factor in the later mean ages at menarche with the inference that training 'delays' the onset of this late maturational event. Retrospective and cross-sectional data, of course, do not permit cause–effect statements, and results of a consensus discussion indicate no presently available evidence to support the conclusion that physical activity and training influence the age at menarche.

Studies of athletes ordinarily do not consider other factors which are known to influence menarche. For example, sister–sister and mother–daughter correlations for age at menarche in families of athletes are similar to those for the general population, and number of siblings in the family has a similar effect on the age at menarche in athletes as in the general population.

If training for sport is related to the age at menarche, it probably interacts with, or is confounded by, other factors so that the specific effect of training *per se* may be difficult to extract. Potential confounding factors include extreme linearity of physique (which is associated with late maturation in both sexes), dietary practices and/or restrictions associated with the maintenance of an optimal body-weight or thinness, disordered eating behaviours, and high levels of coach–athlete stress and/or competitive stress.

### *Summary*

Although regular physical activity is often viewed as having a favourable influence on growth and maturation, adequately nourished children and youth will grow and mature whether or not they are active. Evidence for active and inactive children and youth and for young athletes indicates that regular activity and training for sport have no effect on statural growth and on indices of biological maturation. Menarche occurs later in athletes in many sports, but the presently available data do not warrant the conclusion that training delays menarche. Other factors known to be associated with menarche also need to be considered.

In contrast, regular activity and training are important in the regulation of body-weight, particularly fatness. It is difficult to partition training effects from expected age- and maturity-associated changes in body composition, especially fat-free mass in adolescent boys. Regular activity, especially of weight-bearing type, is associated with enhanced skeletal mineralization, but intensive training in association with altered menstrual function and marginal or inadequate diets can contribute to bone loss in post-menarcheal athletes. Changes in skeletal muscle metabolism are specific to the type of training programme. Information on the influence of regular activity on adipose tissue metabolism and cellularity in children is lacking.

*Robert M. Malina*

See also 'Human skeletal muscle across the lifespan' (**5.8**), 'Body composition' (**5.11**), 'Sexual maturation' (**5.15**), 'Adulthood and developmental maturity' (**5.17**) and 'Motor development and performance' (**6.2**)

## 5.13

# Growth cyclicities and pulsatilities

The human life-cycle is continuous, passing through fixed stages in the individual person and in generation after generation. Arbitrarily, we may consider the life-cycle of the individual to begin with fertilization and then proceed through the stages of prenatal (see Part four) and post-natal growth and development (see Part five) and then senescence (**12.3**). Within each of these stages are pulses of developmental change, such as gastrulation, menarche (**1.15**), or death that are not again repeated during the individual's life-cycle of growth, but are repeated in the life-cycle of other people. This entry limits the discussion of growth-cycles and pulses to the individual person and to the stages of life from birth to maturity. Cycles and pulses are considered from the macro-level of the basic stages of post-natal growth through finer levels of analysis, including seasonal, weekly, and even daily or subdaily growth events.

At the macro-level, there are five post-natal stages of human growth: infancy, childhood, juvenile, adolescence, and adulthood. Growth and development are not smooth and continuous within or between these five stages. The major trend of decreasing growth velocity from birth to maturity is punctuated by two notable pulses. The first is the mid-childhood growth spurt and the second is the adolescent growth spurt. Both spurts are produced by the action of specific hormones which themselves are produced and released within the body in a cyclic or episodic fashion.

The maturation of the adrenal cortex and the sudden increase of the production and release of its hormones causes the mid-childhood growth spurt. Pulsatile secretion of gonadotropin-releasing hormone (GnRH) from the hypothalamus triggers the onset of puberty and the adolescent growth spurt. The pulsatile secretion of GnRH is present during early life but is inhibited by late infancy. It is reactivated just before puberty, thus exhibiting a system of cycles and pulses in terms of both its own activity and also in terms of its effects on growth and development.

At a finer level of analysis are seasonal cycles of growth. At temperate latitudes, for example, healthy well-nourished children grow more quickly in height during the spring and summer than they do during the fall and winter. To account for the seasonal cycle several investigators suggest that seasonal periodicity in the amount or intensity of sunlight may act on the human endocrine system to synchronize changes in growth-regulating hormone activity.

Experimental support for the 'sunlight hypothesis' was provided by a Swedish experiment in 1929. A group of 45 boys received 'sun-lamp' treatments (using a lamp that produced both visible and ultraviolet light) during the winter months in Stockholm. During the period of treatment the experimental group averaged 1.5 centimetres more growth in height than the control group of 292 individuals. During the summer, the control group grew at a faster rate than the experimental group, so that over the entire year there was no difference between groups in total height gain. Only the time of year when the maximum gain occurred differed between groups. This indicated that normal children may have an endogenous rhythm for growth, following approximately a sine curve of one year's duration. Some factor associated with sunlight seems to entrain this endogenous rhythm into a seasonal pattern.

Light can enter the body and influence the neuroendocrine system and growth in two ways: through the eyes and through the skin. There is some evidence that both photic pathways can influence growth and development, however, the strongest support for the control of seasonal growth comes from skin pathway. Cholecalciferol, vitamin $D_3$, is synthesized by the skin (and modified in the liver and kidneys to its active form) when people are exposed to ultraviolet light. The physiological action of vitamin $D_3$ is to increase the intestinal absorption of calcium and to control the rate of skeletal remodelling and the mineralization of new bone tissue. Vitamin $D_3$ is essential for normal bone growth, and thus growth in height. Seasonal variation in the intensity and duration of sunlight is known to correlate strongly with the level of vitamin $D_3$ production and the level of vitamin $D_3$ metabolites in the bloodstream. Research shows that growth in height is fastest during the season of

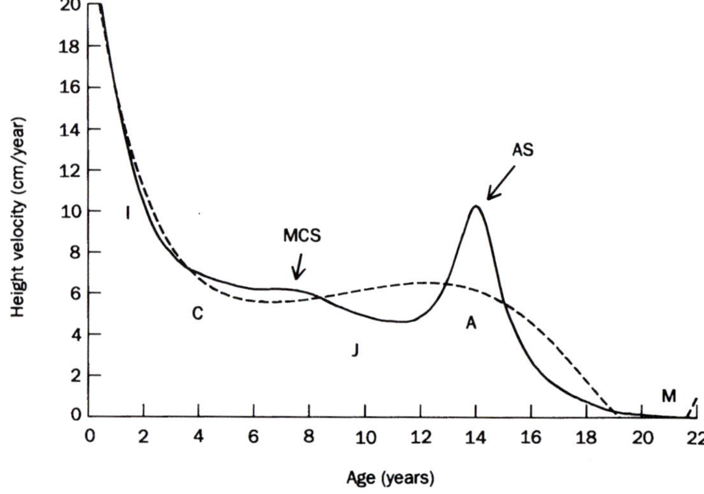


**Idealized velocity curve of human growth for boys (solid line): I = infancy, C = childhood, J = juvenile, A = adolescence, M = mature adult. The dashed line is a sixth degree polynomial curve fit to the velocity curve data. The polynomial curve does not fit well to real growth data due to the pulses of the mid-childhood spurt (MCS) and the adolescent spurt (AS). The human velocity curve cannot be fit adequately by a single continuous mathematical function. Two or more functions are required.**

maximum sunlight availability. At the temperate latitudes this is the spring–summer season, while at tropical latitudes this is the dry season.

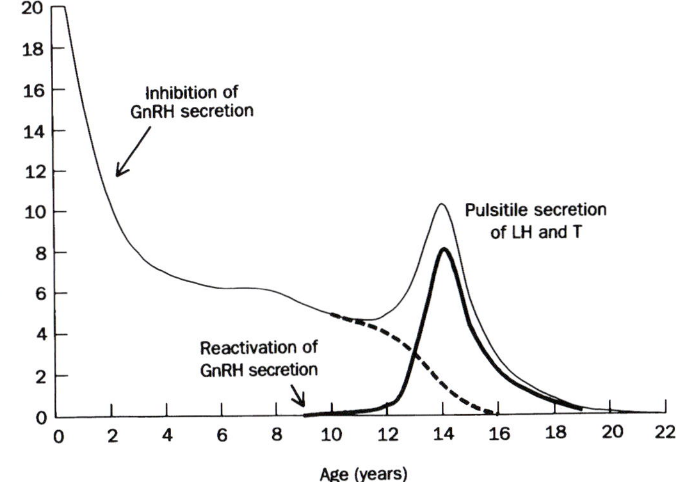


'dual function' of human growth, showing cycles and pulses of hormone inhibition and ctivation associated with changes in height velocity. The thin solid line is an idealized elocity curve of human growth for boys. This curve can be described mathematically by wo linked functions: a prepubertal and a pubertal function. The prepubertal function is ssociated with the phase of quiescent gonadal hormone activity due to inhibition of ypothalamic gonadotropin-releasing-hormone (GnRH) secretion by late infancy. The repubertal rate of growth reaches zero and is 'switched-off' by 16 years of age, as indiated by the heavy broken line. The pubertal function (heavy solid line) is 'switched n' at about 9 years of age, when GnRH secretion is reactivated. As the frequency and mplitude of GnRH pulses increase, there is a rise in the pulsitile secretion of pituitary uteinizing hormone (LH) and gonadal testosterone (T). Spermatogenesis and an ncrease in rate of growth take place. The size and shape of the adolescent growth spurt s the product of interaction between the prepubertal and pubertal functions (modified om the concept of W. Stutzle).

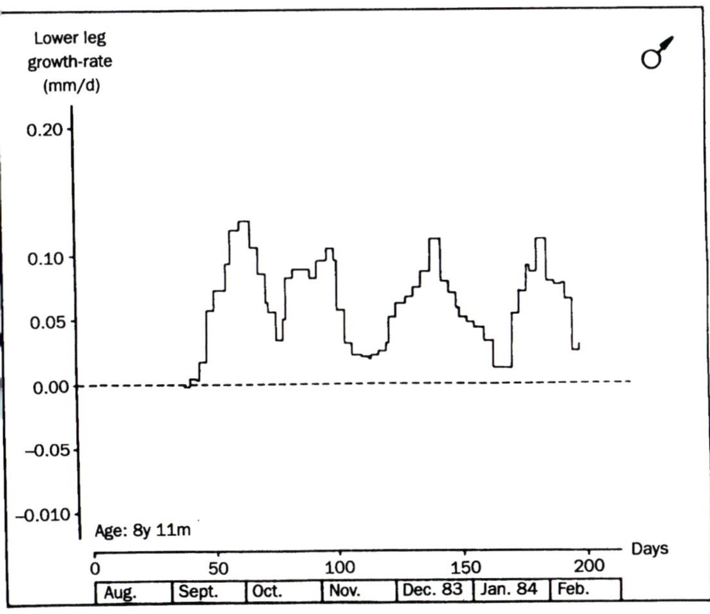


rowth velocity data for the lower leg ('knee height') of an 8.9-year-old boy, measured nce weekly. 'Mini growth spurts' occur from week to week, and a longer term cycle f increases and decreases in the size of the 'spurts' is also apparent. Data from I. Hermanussen (1988). *Annals of Human Biology* **15**: 103–109.

Studies of seasonal variation in weight-gain and -loss find that for healthy individuals, fall/winter or the rainy season are the times of maximum weight-gain. The high-sunshine season tends to be the time of minimum weight-gain for children, and the time of maximum weight-loss for adults. Several researchers also find minimum weight-gains, or even weight-losses, occur simultaneously with maximum height-gains. A study of monthly increments of weight change of 43 boys and 42 girls 5–7 years old, all healthy well-nourished children living in Guatemala, found that minimum weight gains occurred for 90 per cent of the boys and 80 per cent of the girls during the dry season, the time of maximum increments in height. In the last 3 months of the dry season, between 30 and 60 per cent of the children did not gain weight, some even lost weight, although they were growing in height. The pattern of weight change could not be explained by observable behaviour related to diet, exercise, or disease. Perhaps the change in metabolism that produced the maximum increments in height growth resulted in a loss of soft tissue or body water.

Seasonality in the onset of puberty, especially menarche, and in fertility have also been reported in the literature. Biological, meteorological, and social-behavioural explanations have been proposed, but no consensus exists as there is much geographic and cultural variation in the timing of this seasonality.

Growth cycles or pulses have been found when growth in length is measured weekly, daily, and even twice daily (morning and evening). The latter type of cycle is due to compression of the intervertebral disks. Stature loss may begin immediately after arising from bed and may exceed 1.0 centimetre within an hour and 2.0 centimetres by evening. On a weekly or daily basis, periods of little or no growth are followed by bursts of rapid growth. The discovery of this type of variation has replaced the traditional belief, based on annual or quarterly measurements, that growth is a smooth and linear process. The discovery also showed that short-term rates of growth cannot be extrapolated to annual rates, an important finding for medical intervention. Also important was the discovery that different segments of the body may be growing in cyclical or pulsatile fashion at different times. A Japanese study of two children (ages 7.5 and 6.6 years) measured daily for 1 year found that all growth in stature was confined to the lower limbs, except for a seasonal pulse of trunk growth in the spring (April–May).

How these patterns are controlled is not known. They lend support to the 'dual-effector' hypothesis, which posits that the growth of many tissues, including bone, occurs in two phases. The first is the differentiation of specific cell types from stem cells. The second is expansion of the tissue by hyperplasia and hypertrophy. Slow growth or stasis would occur during the differentiation stage, while rapid growth, or saltation would occur during the expansion stage.

*Barry Bogin*

See also 'Longitudinal analysis' (1.14)

## MINI GROWTH SPURTS

Oscillatory phenomena are ubiquitously distributed throughout nature, many examples of annual, near-monthly, circadian and shorter periodicities having been identified. Thus, it remains peculiar that oscillatory phenomena have been discovered so late in the study of human growth, and this may only be explained by the difficulty of measuring short-term length increments. Yet growth is also oscillatory, with repetitive periods of fast growth alternating with periods of decreased or even inverted (negative) growth.

The circadian (once every 24 hours) rhythm of human body-stature was discovered about 100 years ago. However, measuring techniques had to be improved for the investigation of the more minute undulations of day-to-day length increments. Instead of total body stature, lower leg length is now commonly used for the study of short-term growth: knemometry for children and adolescents, mini-knemometry for neonates and prematures, and mikro-knemometry for rabbits and rats. Now that accuracy of measurement (**1.4**) has increased, with a technical error ranging between 160 microns (knemometry) and 50 microns (mikro-knemometry), daily measurements of the lower leg length have provided evidence for circaseptan (approximately once every 7 days) periodicity in growth. Growth velocity (**1.16**) oscillates with periods of acceleration named *mini growth spurts* interposed by periods of deceleration.

Short-term growth has not always been measured daily. The first description of mini growth spurts was based on weekly determinations of lower leg length, and suggested a much longer periodicity of once every 30 to 55 days. This was explained and reconciled with circaseptan rhthmicity in the following way: weekly measurements of a rhythm that appears approximately once within 7 days produce interferences at a much slower rate than the basic circaseptan rhythm. The mechanism of mini growth spurts is not known. *Michael Hermanussen*

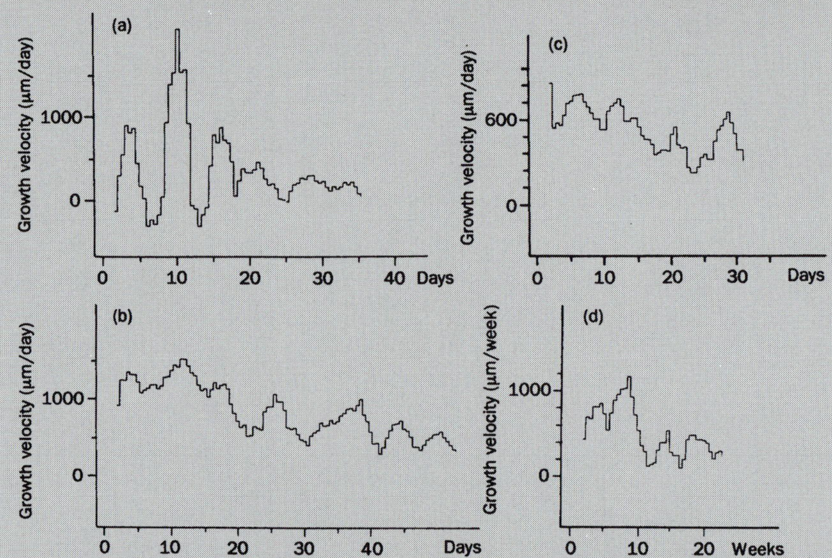


**(a) Lower leg growth velocity of a healthy male infant initially aged 28 days. Growth velocity was calculated from daily measurements smoothened within overlapping 4-day intervals. Periodic changes of growth velocity are evident with a peak-to-peak distance of approximately 6 days.**

**(b) Lower leg growth velocity of a healthy rabbit, initially aged 2 months. Growth velocity was calculated in the same way as above.**

**(c) Lower leg growth velocity of a healthy rat, aged 21 days at the beginning of the study.**

**(d) Lower leg growth velocity of a healthy boy aged 14.7 years. Growth velocity was calculated from weekly measurements smoothened within overlapping 28-day intervals. Periodic changes of growth velocity are similar to those calculated from daily measurements, but a peak-to-peak distance with 30 to 55 days clearly outranges a circaseptan period.**

## SALTATORY GROWTH

Daily growth measurements have identified human growth as a saltatory process characterized by variable amplitude pulsatile growth events, or saltations, followed by refractory intervals of stasis during which no significant growth takes place. Pulsatile growth in body dimensions implies a two-phase mechanism in which sequential discrete pulses, or saltations, are followed by stationary, quiescent intervals. The amplitude of the pulses, or amount of growth at each saltation, is variable within and between individuals, as are the time intervals between the growth saltations. The precise mechanisms underlying saltatory growth are unknown at the present time.

Beginning in fetal life, serial ultrasound assessments (**1.2**) document significant growth of long bones, the skull and abdominal dimensions during measurements taken only 2 days apart and separated by prolonged intervals, up to 25 days, when no growth in these body parts is detectable. Post-natal studies of infants aged between 2 days and 21 months document growth in total body-length of up to 1.6 centimetres during 24-hour measurement intervals, separated by variable durations of stasis that can be as long as 60 days in the second half of the first year of life in normal healthy infants. Total body-height assessed in childhood and adolescence document the same saltatory pattern, with growth pulses of up to 1.5 centimetres in 24 hours separated by variable intervals of no growth ranging up to 100 days.

Head-circumference growth follows the same pulsatile patterns in infancy and while weight is a more problematic research variable, due to the inherent daily oscillations, a study of weight-gain patterns in premature infants (**5.2**) also documents growth pulsatility. In general, in the present studies, increases in weight precede pulses in total body-length.

These observations identify adjustments in growth pulse amplitude and frequency as the underlying mechanism for changing developmental growth-rates, with high infant and adolescent growth velocities associated with more frequent growth pulses and/or pulses of higher amplitude compared to growth during the intervening childhood years. Likewise, individual variability in size and growth-rate reflects variability in saltatory growth patterns. It is likely that genetic and environmental effects are expressed and modulated through flexibility in pulsatile dynamics. Present analyses hypothesize saltatory growth to be an example of a deterministic non-linear dynamical biological system.

*Michelle Lampl*

## SEASONALITY OF GROWTH

Growth as represented on the centile chart is perfectly smooth (**1.16**), yet in reality it is an irregular process. On several different time scales, weekly, monthly or yearly, growth varies appreciably from one time period to the next. On the scale of months the variation is cyclical, determined largely by the seasons of the year, and this occurs in both temperate and tropical regions of the world.

In temperate zones, seasonal variation in diet and morbidity is relatively small compared to the tropics, and so its effect on growth is correspondingly modest. Despite this, clear and consistent seasonal changes in growth-rate during childhood have been documented in Europe and North America over the last 75 years. In summary, they show that the greatest rate of gain in height occurs during the spring, between March and July, while the pattern for weight is exactly the opposite, with the slowest growth-rate in the spring and the fastest in the autumn. Gain in lower leg length, as measured by knemometer, varies seasonally in step with height.

The fact that height and weight increase out of phase with each other is perhaps surprising. It may be related to the same phenomenon seen in infancy, when individual children grow first in length and then in weight, so that they appear to grow out of phase. Alternatively it may simply mean that different factors underlie the two patterns of growth.

The studies investigating seasonality are fairly consistent in the period of the year they identify when height growth is at its greatest, yet the effect is relatively small and only about one-third of children show a seasonal component of growth. The effect on height is also smaller in magnitude than for weight. A unique study of five Japanese siblings measured monthly for 10 years shows how each individual has his or her own characteristic seasonal pattern of growth, which reappears year after year and differs subtly from the patterns of the other siblings.

There are two reasons why the seasonal pattern of growth is of interest. Firstly, it provides evidence about the physiology of growth and the factors that might affect it; the impact of the environment on growth in temperate zones is only minor, and this limits the number of factors that might be involved. The second reason is that it is important for accurate measurement for height velocity.

W.A. Marshall carried out a series of experiments during the 1970s to test the effects of different aspects of seasonality on growth. He provided what is in effect both a confirmation and a summary of earlier papers on the size of the effect, with his studies of normal children in London and the Orkneys. He also measured blind children, and showed that like sighted children they grow cyclically, but not in synchrony with the seasons.

Height-gain in children receiving synthetic growth hormone for growth hormone deficiency (**7.9**) is known to be seasonal, even though levels of circulating growth hormone in such children do not vary seasonally. However interactions with other seasonally varying hormones like androgens, particularly in boys, could be a factor. Overall, the weakness of the association between growth-rate and season, coupled with the fact that a majority of individuals fail to show it, suggests that season itself (or proxies like day-length or temperature) are not the causative agent, and that instead children may have their own internal cycle of growth, and in a minority this becomes synchronized with the seasons.

The practical significance of a seasonal pattern of growth is that height velocity measured over a 6-month period may vary by as much as a factor of two, depending on when the measurements are made. For this reason height velocity should be measured over a period of whole years, so that the seasonal variation cancels out. The Tanner–Whitehouse velocity charts are based on a 12-month measurement interval for this reason. *T.J. Cole*

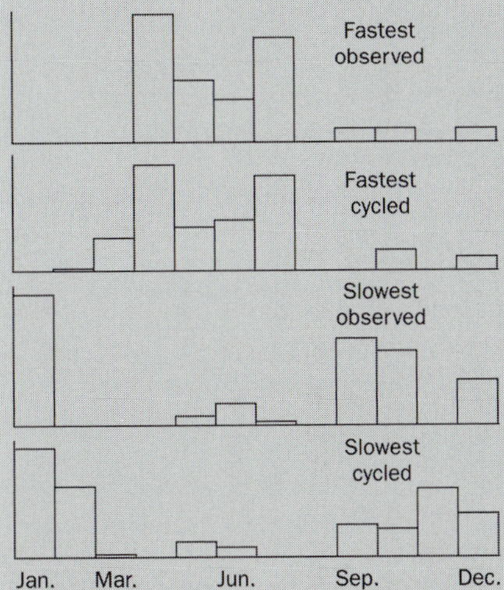


**Distribution of months in which individual boys completed their fastest and slowest periods of growth over 3 months; London schoolboys.**

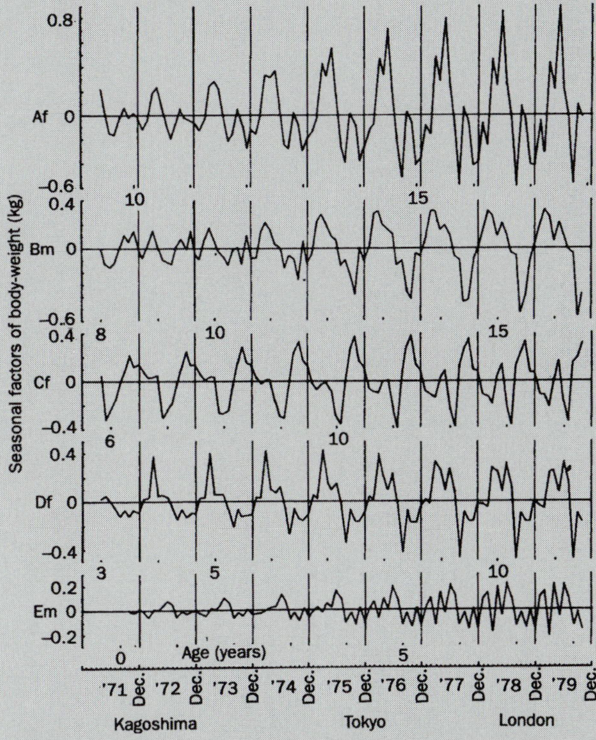


**Seasonal factors of body-weight distances in five Japanese siblings with age scales below the curves and living places at extreme bottom. Data from M. Togo and T. Togo (1982).** ***Annals of Human Biology*** **9:425–440.**

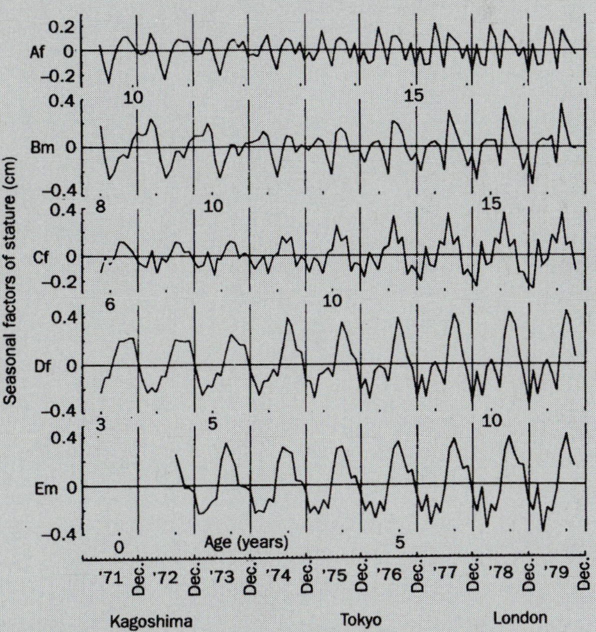


**Seasonal factors of stature distances in five Japanese siblings with age scales below the curves and living places at extreme bottom. Data from M. Togo and T. Togo (1982).** ***Annals of Human Biology*** **9:425–440.**

## PULSATILE GROWTH HORMONE RELEASE

Growth hormone (GH) (**5.1**) is a large protein molecule secreted by a distinct subpopulation of specialized cells within the anterior portion of the pituitary gland. The pituitary gland resides at the base of the brain from which it receives directive input in the form of specific chemical signals. The two primary brain signals that supervise GH secretion are growth hormone-releasing hormone (GHRH) and somatostatin. GHRH stimulates GH synthesis by turning on the gene that encodes this protein hormone(**3.2**, **7.6**). GHRH also provokes release of intracellular stores of GH. In contrast, somatostatin suppresses GH release from pituitary cells, but does not block GH synthesis. The opposing actions of these two unique brain peptide signals serve to co-ordinate brief bursts or explosions of GH secretion by the pituitary gland.

The brain's release of GHRH and concomitant interruption of somatostatin secretion serve to trigger the outpouring of GH. Indeed, when blood samples are obtained frequently (e.g. every 5 minutes over 24 hours in the human), a clearly pulsatile or episodic pattern of GH release is observed. This consists of punctuated bursts of GH release, followed by a gradual decline as GH is removed from the circulation. The liver and kidneys participate in removing GH from the blood, and impose a half-life of about 10–25 minutes in the rat and human.

Mathematical models have been used to estimate the rate of secretion (which determines the rapid upstroke of the release episode) and the GH half-life (which influences how slowly the blood concentrations fall). These methods allow quantification of the brain's regulation of pulsatile GH secretion. Normally, there is a strong 24-hour rhythm in GH pulsatility, which in the human involves 30–100-fold variations in serum GH concentrations. During the fed and the awake period of the day, GH is secreted as low amplitude events that are relatively infrequent, for example, occurring approximately every 2–3 hours. In contrast, at night-time when an individual is fasted and asleep, GH is released in bursts that occur as often as every 45 minutes, and are also of greater amplitude. Indeed, sleep and nutrients are primary regulators of GH secretion.

Recent studies using highly sensitive GH assays such as those based on chemiluminescence principles show some basal or non-pulsatile GH release between discrete bursts as well. In healthy young adults, this basal release constitutes less than 10 per cent of the total GH secreted per 24 hours.

The GH axis is among the most dynamic endocrine systems known, since multiple internal and external cues either suppress (e.g. food ingestion) or facilitate (e.g. sex steroids) sleep pulsatile GH secretion. This allows for a wide range of adaptive responses to physical activity, obesity, age, reproductive needs, varying nutritional requirements, catabolism, stress, puberty, growth, and senescence. Finally, the remarkable fall in blood GH concentrations in early adult life and with healthy ageing can be viewed on the one hand as detrimental with ill consequences to bone and muscle mass, mood and vigour, and fat deposition, and on the other hand as favourable by limiting otherwise continual stimulation of tissue growth and metabolism.

*Johannes D. Veldhuis*

*Physiological factors and disease states that regulate the amplitude and frequency of GH pulses and the metabolic clearance of growth hormone GH in humans. From Hartman et al., 1993.*

| GH pulse amplitude | Detectable GH pulse frequency | GH half-life |
|---|---|---|
| Increase | Increase | Increase |
| Estrogen | Estrogen | Uremia |
| Fasting | Fasting | Chronic liver disease |
| Type I diabetes | Type I diabetes | Turner syndrome |
| Thyrotoxicosis | Thyrotoxicosis | |
| Puberty | Glucorticoids (low dose) | |
| Exercise training | Acromegaly | |
| Testosterone | Uremia | |
| Decrease | Decrease | Decrease |
| Adiposity | Adiposity | Adiposity |
| Ageing | Ageing | |
| Food ingestion | Food ingestion | |

# Hormonal regulation of growth in childhood and puberty

Childhood is in general a period of relative growth stability, in that velocities of various dimensions tend to be regular with little change. The hypothalamus and pituitary regulate the changes that are observed through a series of negative feedback loops with other endocrine glands. Hormones secreted by the anterior pituitary (e.g. TSH, LH) stimulate target glands whose products in turn inhibit the production of hypopthalamic releasing factors, maintaining a generally steady output of regulatory hormones to the various end organs (**5.1**). The interaction of pituitary growth hormone and the insulin-like growth factors regulate the patterns in individual children of increases in dimensions measured along the axes of their long bones. Despite the pulsatilities in growth known to exist over the short term (**5.12**), these patterns are so regular that over intervals of a few months, simple linear mathematical functions of age may be fitted to various dimensions with residuals that typically do not fall outside of the range of measurement error.

While the effects of growth hormone are pervasive throughout the developing child, and while growth-hormone deficiency is manifest (albeit to varying degrees) in most developmental parameters, other hormones also exert significant effects on size and rate of development. Adequate thyroid output is a prerequisite for normal rates of skeletal maturation, while insulin, derived from the pancreas, may be a significant regulator of changes in cell size.

While much is known about the hormonal regulation of childhood growth, virtually all of it is derived from studies of children with various deficiencies and other dysfunctions of the endocrine system. There is relatively little knowledge about the relationships between variation in the output of particular hormones and variation in the size, shape, composition, or maturation of normal, healthy individuals.

## Puberty

The auxological events of puberty are striking and significant. They are likewise highly variable, among individuals as well as groups. And while puberty tends to be conceptualized as a stage of development in which a myriad of systems undergo marked changes and accelerations, it is a stage in which these systems display a considerable degree of independence from each other, making generalization and prediction difficult and uncertain.

The most evident growth event associated with puberty is the growth spurt, when the majority of anthropometric dimensions display a marked increase in velocity, followed by a (somewhat less) rapid decline that is usually referred to as the end of the growth period. As are the other events of puberty, the growth spurt is one response to the decline in the sensitivity of the hypothalamus to the negative feedback from the various endocrine glands. This is seen in a gradual increase in circulating levels of testosterone which trigger the acceleration in the growth-rate, especially at the metaphyses of long bones. Since testosterone in males is derived both from the testis and adrenals, their spurt is greater than that of females, whose testosterone is derived from the adrenal cortex. This several-fold increase in testosterone drives the growth spurt in skeletal dimensions and in lean body mass. Secondarily, in females, an increase in estrogen accompanying maturation of the ovaries may play a role in the spurt. Of greater importance, however, is the role of estrogen on receptors in the bony pelvis and the accompanying changes in morphology. It is at puberty, in response to estrogens, that the female pelvis matures to its adult form. Due to the later increase in levels of estrogen relative to testosterone in normal females, growth and maturation of the pelvis continues even after increase in height has virtually ceased. Consequently, the cessation of height growth in younger adolescent females is not a reliable indicator of a mature pelvic morphology.

Other skeletal dimensions show shifts and inflections in their growth-rates with puberty. The mandible moves both downward and forward due to increased growth at the condyle. In addition, the mandible also becomes more prominent through the deposition of bone at the chin. Both of these phenomena are more pronounced in males, reflecting their higher levels of testosterone.

Circum-pubertal changes in body composition are also mediated by hormonal agents. The spurt in muscle mass, as well as its accompanying sex difference, is associated with increased testosterone levels and the intensification of pre-adolescent patterns of fat deposition in females is correlated with heightened estrogen secretion and the maturation of the menstrual cycle. Quite possibly, insulin plays an important role in the normal adolescent, though this is much less defined.

While there is a tendency for greater muscle mass in males relative to females, even when expressed as a percentage of body-weight, at all ages, the differences are greater at each age group from 11 years on. This is due not only to greater percentages in males resulting from increased muscle growth associated with testosterone, but also in decreasing percentages in females, associated with estrogen levels.

Finally, the cessation of growth, at least as measured by conventional anthropometric techniques, is mediated by hormones. This is brought about by the closure of the epiphyses as they unite to their corresponding diaphyses. Since virtually all of the increase in the length of long bones occurs in the metaphysis, epiphyseal union signals the 'end' of the growth period. Evidence shows that the fusion of epiphysis and diaphysis is associated with gonadal maturation: in children with gonadal dysgenesis, epiphyseal fusion is markedly delayed and continued linear growth leads to extremes of height and limb lengths. Likewise, administration of steroid hormones to normal children judged clinically to be undesirably tall leads to epiphyseal union and cessation of growth. In this respect it is of interest that testosterone and estrogen, the hormones that produce the growth spurt, are also responsible for its end.

*Francis E. Johnston*

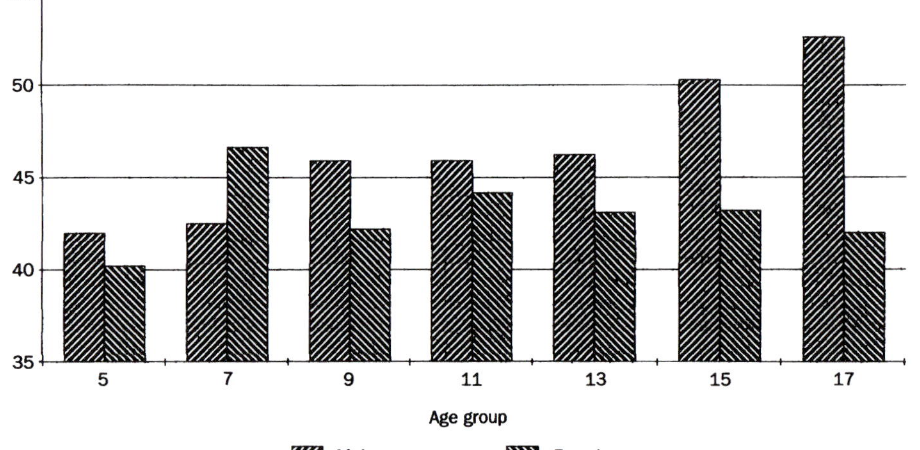


**Estimated average muscle-weight as a percentage of body-weight at various ages. Data from Malina, 1986.**

See also 'Physical examination' (**1.10**), 'Assessment of age at menarche' (**1.15**), 'Sexual maturation' (**5.15**), 'Adulthood and developmental maturity' (**5.17**), 'Endocrine growth disorders' (**7.7**), 'Between-population differences in adolescent growth' (**10.5**), 'The secular trend' (**11.3**) and 'Treatment of growth disorders: the future' (**13.4**)

## THE MID-CHILDHOOD GROWTH SPURT

One feature of the childhood phase of growth is the modest acceleration in growth velocity at about 6–8 years. Tanner (1947) called it the mid-growth spurt. Some studies note the presence of the mid-growth spurt in the velocity curve of boys, but not girls. Others find that up to two-thirds of boys and girls have mid-growth spurts.

The mid-growth spurt is linked with an endocrine event called adrenarche, the progressive increase in the secretion of adrenal androgen hormones. Adrenal androgens produce the mid-growth spurt in height, a transient acceleration of bone maturation, the appearance of axillary and pubic hair (**1.10**), and seem to regulate the development of body fatness and fat distribution (**5.16**). All of these events suggest a connection with gonadarche, the maturation of testes or ovaries that takes place during adolescence. Bolk (1926) speculated that for our early human ancestors sexual maturation took place at about 6–8 years of age, so perhaps adrenarche is a vestige of our evolutionary past. Much research, however, from clinical medicine to anthropological fieldwork, shows that there is little or no connection – adrenarche and gonadarche are independently controlled events.

The mechanism controlling adrenarche is not understood as no known hormone appears to cause it. Animal models of adrenarche and the mid-growth spurt are lacking, for these seem to be evolutionarily novel events. Adrenarche is found only in chimpanzees and humans, and the mid-growth spurt is unique to people, as is the childhood stage of growth (**2.3**). The primate data show that there is variability between species in the relation of age and adrenal activity. Rhesus monkeys and baboons have relatively high plasma concentrations of adrenal androgens at all ages after birth and, compared with chimpanzees and people, these monkeys grow rapidly and mature sexually at a relatively early age. Chimpanzees and humans have low serum levels of adrenal androgens after infancy and prior to adrenarche. Moreover, chimpanzees and humans have a relatively slow growth and a long delay in the onset of sexual maturation (**2.2, 2.3**). In the human being, the delay is so protracted that childhood and juvenile stages of growth, and growth deceleration, become major features of the total pattern of development.

Adrenarche and the human mid-growth spurt may function as life-history events, marking the transition from childhood to the juvenile stages. In addition to the physical changes induced by adrenarche there is also a change in cognitive function, called the '5–7-year-old shift' by psychologists, that leads to new learning and work capabilities in the juvenile. Perhaps the evolution of adrenarche and the human mid-growth spurt may be explained as mechanisms that maintain slow epiphyseal maturation and skeletal growth (**5.17**) in the face of the prolongation of the prepubertal stages of growth, and serve as biocultural markers of cognitive maturation. The independent control of adrenarche and gonadarche certainly points to an independent evolution for both adrenarche and the mid-growth spurt.

*Barry Bogin*

# Sexual maturation

Adolescence marks a rather abrupt transition in the human life-course from an immature and actively growing organism to a physically mature and potentially reproductive organism. The onset of reproductive maturation is closely co-ordinated with the completion of physical maturation in both sexes. The process of sexual maturation itself involves both the maturation of primary reproductive traits (the activation of the hypothalamic-pituitary-gonadal [HPG] axis and the onset of gamete production) and the maturation of secondary reproductive traits (including external genitalia, breast development in girls, and changes in body habitus). Secondary sexual maturation is dependent on primary sexual maturation and driven for the most part by the production of gonadal steroid hormones in conjunction with other hormonal determinants of pubertal growth.

Primary sexual maturation involves changes in the hormonal axis controlling gonadal function in both sexes. The mediobasal hypothalamus, anterior pituitary, and gonad are linked by a series of hormonal signals involving the peptide hormone gonadotropin-releasing-hormone (GnRH) released by the hypothalamus into the minute portal vessel system linking it to the anterior pituitary; the glycoprotein hormones follicle-stimulating hormone (FSH) and luteinizing hormone (LH) produced by the anterior pituitary; primary steroid hormone products of the gonads; estradiol and progesterone on the part of the ovary; testosterone on the part of the testis, together with secondary steroid products and the protein hormone inhibin produced by the gonads of both sexes. Hypothalamic production of GnRH in a circhoral pulsatile pattern causes release of FSH and LH by the anterior pituitary, which in turn stimulate gamete production and hormone production on the part of the gonads. Gonadal steroids direct endometrial maturation in females and contribute to the development and maintenance of secondary sexual characteristics in both sexes. Gonadal steroids and inhibin feedback on the hypothalamus and pituitary to modulate gonadotropin production, cyclically in females, tonically in males.

Experimental manipulation in primates and clinical evidence in humans indicate that primary sexual maturation is initiated by the establishment of quasi-hourly pulses of GnRH release by the hypothalamus. Levels of GnRH are not ordinarily observable in humans, so that the first indirect evidence of such hypothalamic activity occurs in the form of pulses of LH in the systemic circulation of pre-adolescent children during sleep. Subsequently, ordinarily within a year, continuous gonadotropin pulses are established throughout the day. Testicular enlargement in boys and increases in both sexes in gonadal steroids from the basal levels typical of childhood follow. Increasing steroid levels are associated with secondary sexual development, pubic hair, breast development, voice and body habitus changes. Eventually, steroid levels reach sufficient levels to support cyclical ovarian and endometrial activity in girls, resulting in menstruation, and the initiation of semen production and ejaculation in boys. In both humans and primate experimental models the process of primary sexual maturation can be initiated by proper administration of exogenous GnRH in hourly pulses and can be arrested or prevented by administration of GnRH analogues that block the activity of endogenous GnRH. Pituitary and gonadal maturation do not appear to be limiting on the process. Hence the natural initiation of primary sexual maturation is thought to involve functional changes in hypothalamic activity.

There are currently two leading hypotheses regarding the nature of the changes in hypothalamic function that initiate primary sexual maturation; the gonadostat hypothesis, and the direct drive hypothesis. According to the gonadostat hypothesis, hypothalamic sensitivity to the negative feedback of gonadal steroids, exquisitely high during childhood, is gradually eroded, either by an endogenous programme of desensitization or by exposure to increasing levels of steroids secreted

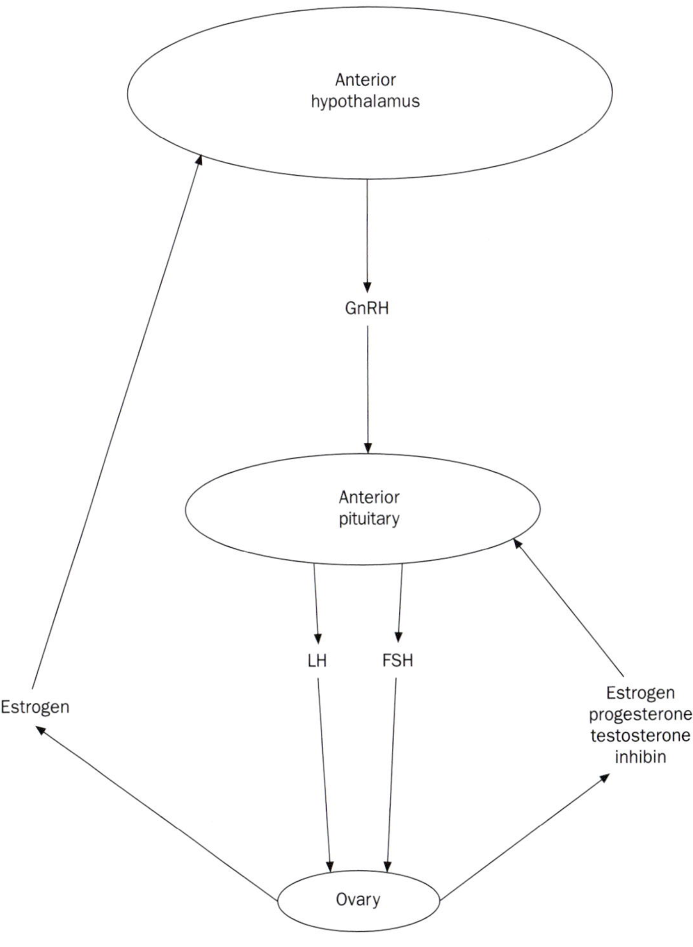


**Hypothalamic-pituitary-ovarian axis (figure drawn by Lynette Leidy).**

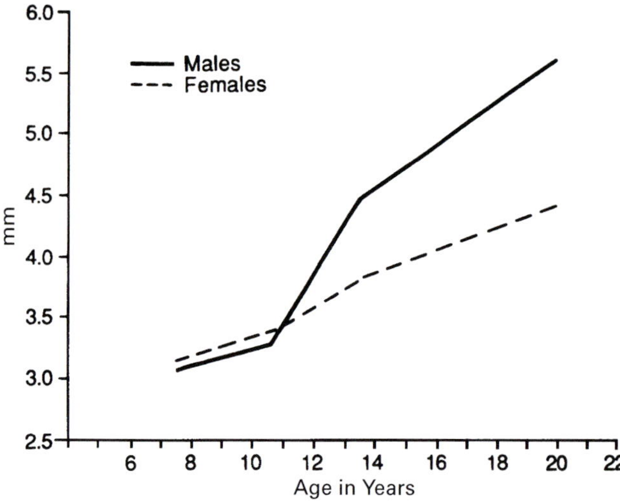


**Means for the cortical thickness of the mandible at the inferior margin of the body near the premolar teeth. Israel, H (1969). Pubertal influence upon the growth and sexual differentiation of the human mandible. *Archives of Oral Biology* 14:583–90.**

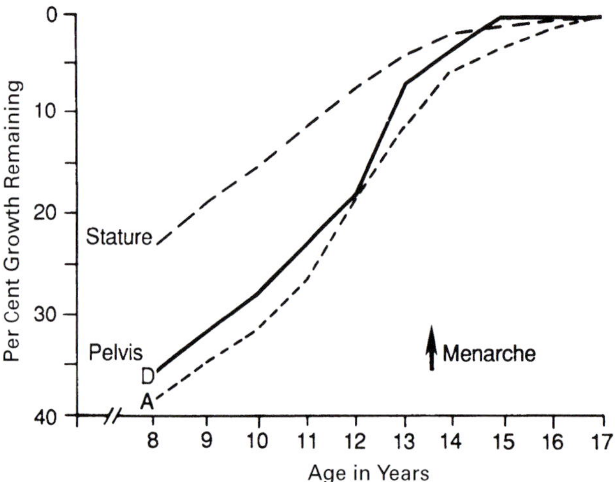


**The percentage of growth remaining for stature and pelvic breadths in girls at ages 8 to 17 years (A = inlet breadth; D = interischial breadth).**

by the adrenal gland prior to puberty. As the hypothalamus becomes desensitized to steroid feedback, a restraint on hypothalamic activity is relaxed, leading to greater stimulation of pituitary gonadotropin release and greater stimulation of gonadal activity. The resetting of hypothalamic sensitivity is compared to resetting the thermostat in a dwelling, hence the term 'gonadostat'.

In contrast, the direct drive hypothesis posits that hypothalamic-pituitary hormone production increases in its baseline activity (i.e. 'open loop' or unopposed gonadotropin production) without losing sensitivity to negative feedback by steroids. Endogenous maturational changes in the central nervous system are thought to result in this increase to direct gonadotropin drive.

The relationship of primary sexual maturation to physical and somatic maturation has also been an active area of scientific inquiry. Longitudinal research in the early and middle part of the century established a close relationship between reproductive and skeletal maturation and led to the use of indices of skeletal maturation as measures of developmental age. Scoring the status of epiphyseal growth plates observable in hand–wrist radiographs has become a standard for quantifying developmental status. In the 1970s, however, Rose Frisch and her collaborators introduced the hypothesis that menstrual onset in girls was conditional upon achievement of a minimum percentage of body fat, shifting attention to changes in body composition as key elements of developmental age. The hypothesis also included a functional explanation of the timing of primary sexual maturation as dependent on a sufficient energetic reserve to support reproduction. This hypothesis was shown to be based on flawed interpretation of data and to have little statistical validity or predictive value. Changes in female body composition themselves are consequences of primary sexual maturation, not antecedents, nor does it make sense that energetic preparation for reproduction (accumulation of fat) would be more important than physical preparation (achievement of an

## MENARCHE

Menarche, or the onset of menstruation, is the clearest indicator of sexual maturation in females (**1.15**). Brought on by heightened activity of the sex hormones, it does not indicate the beginning of puberty (**1.10**). In 99–100 per cent of girls menarche occurs after the spurt in height growth, that is, when height growth is slowing down. Nor does menarche mean that the reproductive system is functionally mature since many of the first menstrual cycles may occur without ova or with abnormal ova.

Individual differences in age of menarche may be influenced by physical training for sports or dance. Female athletes and ballet dancers consistently are reported to be late maturers. Causative factors are unclear but may be the extreme energy expenditure required during training. Individual differences also are a reflection of heredity (see Part three). Monozygotic twin girls arrived at menarche within 2 to 3 months of each other while dizygotic twins differed by an average of 9 months. Possibly some differences among populations may be linked to genetics as well, such as the north–south cline in Europe and early menarche among Asiatics.

The mean age of menarche has been decreasing for the last 150 years in industrialized countries (**11.3**). The general trend has been about 0.3 years per decade. But it should not be assumed that there has been a steady rate of decline since 1840. In Olso, for example, there was a rapid decline from 1860 to 1890 followed by a period of little or no decline from then until around 1920. At that time a steep decline began, lasting until 1960, after which there has been little change. In the 1970s and '80s other countries noted a marked slowing down of the trend. Post-war Japan has experienced the steepest trend of all; from 1950 to 1975 the decline was approximately 1.0 years per decade in the general population and about half that in the well-off. Here also the trend has levelled off.

*Phyllis B. Eveleth*

appropriate size for successful parturition). An alternative hypothesis has proposed that such size constraints are more likely to be important in the co-ordination of physical and reproductive maturation. Remodelling of the female pelvis to enlarge the birth canal is among the last elements of skeletal maturation before skeletal growth ceases. Longitudinal radiographic data indicate that first menstruation in girls is closely correlated with the attainment of adult external pelvic widths. Functionally, it would make sense for primary sexual maturation to be deferred until an appropriate physical scale for successful reproduction is attained.

The timing of male sexual maturation is not addressed by either of these hypotheses. Functionally, male maturation presumably reflects opposing selective pressures favouring early initiation of reproduction on the one hand and the increased competitive ability that comes with greater size and maturity on the other.

P.T. Ellison

See also 'Physical examination' (**1.10**), 'Assessment of age at menarche' (**1.15**), 'Genetics of child growth' (**3.1**), 'Morphology' (**5.5**), 'Physical activity and training for sport as factors affecting growth and maturation' (**5.12**), 'Fat and fat patterning' (**5.16**), 'Adulthood and developmental maturity' (**5.17**), 'The development of sexuality' (**6.6**), 'Chromosome aberrations and growth' (**7.4**), 'Endocrine growth disorders' (**7.7**), 'Teenage pregnancy' (**8.6**), 'Growth in high-altitude populations' (**9.9**), 'Growth and natural selection' (**10.1**), 'Between-population differences in adolescent growth' (**10.5**) and 'Menopause' (**12.2**)

# 5.16

# Fat and fat patterning

Adipose tissue is made up of adipocytes and their supporting structures and is mostly located in the skin and subcutaneous tissue. In normal individuals 65–70 per cent of adipose tissue is fat. Adipose tissue either stores or releases fat as fatty acids and glycerol, depending on whether dietary energy exceeds expenditure or vice versa. Fat is one of the most labile tissues in the body and alters according to both genetic (**3.1**) and environmental factors (**9.2**).

*Direct* analyses of body composition, carried out by dissection, suffer from a wide age range of the subjects, small sample sizes, causes of death that may have altered 'normal' body composition, and embalming procedures that alter body fluid contents. Using data from 42 adult cadavers dissected in the 19th and 20th centuries adipose tissue or fat accounts for 19.2 per cent (SD = 10.2%) and 38.2 per cent (SD = 9.5%) of the total body-weight of males and females respectively. Similar data that exists for babies provides values between 5 per cent and 15 per cent. Thus considerable changes in the fat content of the body occur between birth and adulthood.

The majority of our information on body composition and changes in total body fat and fat pattering is based on *indirect* methods of assessment, thus our knowledge is dependent on the accuracy of these indirect methods. In addition, indirect measures of body fat mean that we only have estimates of the relationship between superficial (subcutaneous) fat and deep body fat, and these estimates are based on the accuracy of the assumptions about the relationship between different aspects of total body composition, such as the proportion of the lean body mass that is water.

Data from skinfold measurements at the triceps and subscapular sites reflects the fact that subcutaneous fat gradually increases during the course of growth and demonstrates considerable sexual dimorphism at the time of adolescent growth and pubertal development.

Subcutaneous fat begins to be laid down in the fetus during the last trimester of pregnancy, at about 34 post-menstrual weeks, and reaches a peak about 9 months after birth. During the second year of life, as the infant begins to walk, there is a decline in skinfold values, which continue through childhood (5–10 years), but at both sites females demonstrate greater subcutaneous fat values than males. From the beginning of adolescence males and females diverge dramatically, so that females tend to be gaining fat throughout adolescence whilst males gain only in body fat and appear to lose limb fat. In terms of total body fat, pre-adolescent values are about 15 per cent whilst post-adolescent males have about 12 per cent body fat and females about 20 per cent body fat.

In addition to an increase in the absolute amount of fat during adolescence there is clear sexual dimorphism in fat patterning. Fat patterning is a relatively new concept in human growth and relates to the anatomical placement of adipose tissue and the relationship between central and peripheral fat stores. It is, therefore, distinct from differences amongst individuals in total adiposity. It can be assessed in a number of ways, for example, with anthropometry (**1.4**), ultrasound, CAT scanning, and MRIs, but current information is derived mainly from skinfolds and waist–hip ratios. Research has demonstrated that a 'centripetal' patterning of fat, in which deposits of subcutaneous fat are greater on the trunk than on the extremities, is a risk factor in adults for non-insulin-dependent diabetes and cardiovascular disease. There has thus been considerable interest in how fat patterning develops during growth.

Research strategies to elucidate fat patterning have used four approaches: 1) comparison of absolute measures, such as the skinfolds described above; 2) comparisons of the ratios of the *natural* log values of individual skinfolds to the *natural* log of the sum of individual skinfolds, for example, in

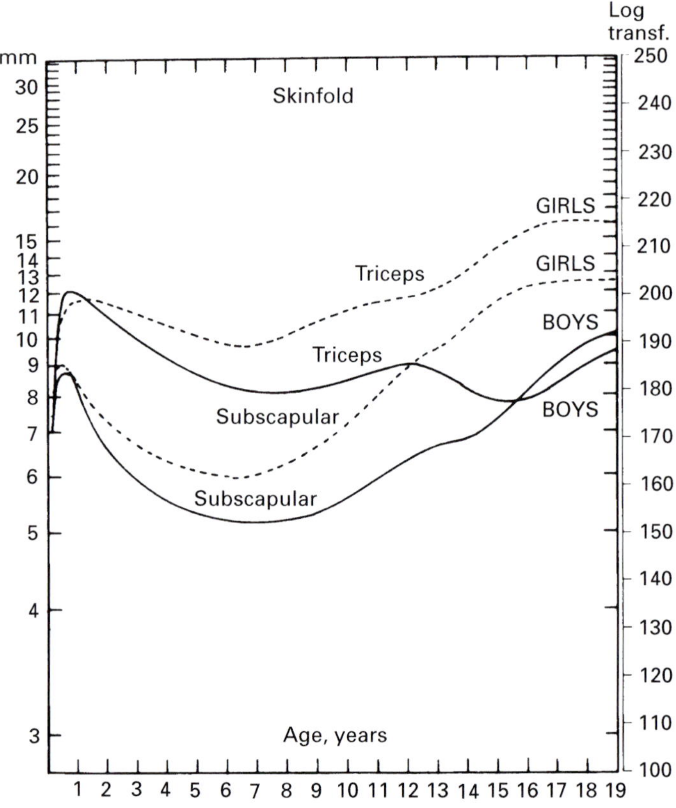


**Distance curve of subcutaneous tissue measured by Harpenden skinfold callipers over triceps (back of upper arm) and under scapula (shoulder blade). Scale is mm on the left and logarithmic transformation units on the right-hand side. British children, 50th centiles. Tanner, J.M. and Whitehouse, R.H. (1975). Revised standards for triceps and scapular skinfolds in British children. *Archives of Disease in Childhood*, 50:142–145**

(triceps)/ ln (triceps + biceps + subscapular + suprailiac), to demonstrate the contribution of each skinfold to the sum of the skinfolds; 3) comparisons of the centripetal fat ratio (CFR = subscapular/[subscapular + triceps]), and 4) principal components analysis.

All samples so far studied in North and South America, Europe and Africa demonstrate that during adolescence males develop greater centralization than females. This centralization appears to come about by changes in the relative contributions of limb and truncal fat. In males, measures of truncal fat, for example, subscapular skinfold, assume greater importance than measures of limb fat, for example, triceps skinfold. In females, subscapular fat increases over time, but relatively large increases in limb fat mitigate against centralization being demonstrated. Longitudinal analyses, in which children have been monitored throughout adolescence, demonstrate that centralization occurs in the latter part of the adolescent growth spurt *after* peak height velocity. It also appears that the level of fatness plays a role in fat patterning during adolescence. Prior to adolescence there is a negative relationship between absolute level of fatness and degree of centralization; the lower the fat content the greater the centralization. During adolescence the relationship to level of fatness disappears, i.e., centralization occurs *regardless* of the initial level of fatness. The pre-adolescent negative relationship of centralization to total fatness suggests that the body centralizes fat stores when total body fat is minimal, or that there is a depletion of peripheral fat stores. Such a phenomenon would have a good physiological basis in that the body's vital organs would be those that would benefit from the advantage of what little fat was available. In addition, measures of nutritional status and socio-economic status also have a negative association with the degree of centralization, adding emphasis to the effect of stress being associated with less body fat and greater centralization.

Principal components analyses of skinfold measurements at both limb and truncal sites have provided a greater insight into the mechanism of fat patterning. For such an analysis to be successful the effect of *total* body fat must be removed. This is accomplished by regressing each log-transformed skinfold

## FAT PATTERNING AND NON-INSULIN-DEPENDENT DIABETES MELLITUS IN MODERNIZING POPULATIONS

Patterns of adipose tissue, or fat, distribution are not only important descriptors of growth and development in children, but key predictors of chronic disease risk among adults. Regional adiposity, or fat patterning, or fat distribution directly influences metabolic processes and disease initiation and progression, independent of total adiposity. For example, the relationship between central-body depositions of adipose tissue and increased cardiovascular disease risk factors and mortality is well established in adults from industrial societies. Visceral fat deposited intra-abdominally has been implicated in particular due to differences in lipolytic activity and lipoprotein lipase activity and their effect on lipid and lipoprotein regulation and insulin and, thus, glucose metabolism.

Among modernizing populations increases in adult weight, proportions of overweight, and total adiposity are ubiquitous (**11.5**). The rapid decrease in physical activity and increase in dietary intake of calories, especially fats and simple carbohydrates, directly influence energetic balance and lead to an accumulation of adipose tissue. This fat accumulation is not equal across all body sites and regions, but tends to become more centrally deposited with increasing adiposity and with adult age.

Concomitant with the increase in adult adiposity, non-insulin-dependent diabetes mellitus (NIDDM) incidence and prevalence tends to increase among adults in modernizing populations. These temporal trends in adiposity, central adiposity and NIDDM have been observed among several Amerindian groups who experienced profound social and economic changes in the late 20th century, notably the Pima of Arizona, United States, and the Dogrib of North-west Territories, Canada. Detailed longitudinal studies of adiposity and NIDDM conducted in the Pima Indians have contributed much physiologic and epidemiologic understanding of the development of NIDDM and the role of adipose tissue in glucose metabolism.

Within the United States migrant minority groups who are experiencing increases in material life style, such as Mexican-Americans in Texas and California, have increasing levels of total adiposity, central adiposity and NIDDM, and other metabolic maladies. Within other industrialized nations, other migrant minority groups such as South Asians in Britain, Singapore and South Africa, show increases in adiposity and NIDDM.

The Aborigines of Australia provide dramatic evidence for increasing adiposity, centrality of fat distribution and NIDDM. The changes from very rural dispersed-foraging subsistence patterns to a sedentary way of life and modern diets are directly implicated for these health changes.

The rapid social and material life changes in all Pacific Island peoples have led to a pandemic of NIDDM, which is attributed to increasing adiposity. The island nation of Nauru, populated by a Micronesian population, has provided clear evidence for the relationship between rapid modernization, overweight, fat distribution and NIDDM. Among modernizing Polynesians increasing adiposity, central fat distribution and NIDDM have been linked temporally, although few detailed longitudinal studies of fat distribution and NIDDM exist. These trends have been described for Cook Islanders, Hawaiians, Maoris, Tokelauans, Tongans, and Samoans. Patterns of central adipose distribution during growth and development are also becoming more common among modernizing Polynesian children. This indicates a continuing and perhaps worsening scenario for NIDDM and cardiovascular diseases for Polynesian adults.

The importance of fat distribution on NIDDM, relative to the role of total adiposity, is controversial and remains an active area for research. Longitudinal studies of communities experiencing modernization combining epidemiologic with physiologic and genetic studies may offer unique study opportunities to disentangle the effects of fat accumulation from those of fat patterning on NIDDM and other chronic disease.

Although the associations in modernizing populations among adiposity, central fat distribution and disturbed insulin and glucose regulation are well established, there are questions about why the pathophysiologic sequence leading to NIDDM occurs so quickly, and whether these groups are at greater NIDDM risk at lower levels of adiposity. A consensus is developing that some populations may be genetically susceptible to NIDDM when there is rapid social and economic change, and positive energetic balance. Evolutionary and adaptive scenarios, such as the thrifty genotype, speculate that individuals with hyperinsulinemic responses had selective advantages during times of nutritional deprivation. With modern patterns of relative inactivity and increased caloric intakes, those with relative hyperinsulinemic responses might accumulate more fat, including centrally deposited fat, develop insulin resistance and eventually decreased insulin secretion, glucose intolerance, hyperglycemia and NIDDM.

This adaptive and evolutionary interpretation for the associations of adiposity, its patterning and NIDDM need not be restricted to modernizing populations but may represent human biological variability to food shortages through time. Those modernizing groups with rapid rises in adiposity and NIDDM may be more susceptible due to an interaction between adaptations to energetic stresses in the past and contemporary patterns of overnutrition, adult fatness and its central distribution. *Stephen T. McGarvey*

on the mean log skinfold thickness for each individual. The resulting *residual* log skinfolds are then subjected to a principal components analysis. The first principal component that emerges from such an analysis, accounting for almost 50 per cent of the total variance, reflects the differential distribution of fat on the trunk and extremities; the limb skinfolds have negative loadings and the truncal skinfolds have positive loadings. Underlying the effects of total body fat and fat centralization are components that relate to ventral–dorsal patterning and upper–lower truncal patterning. These components are not consistent between studies; in some the upper–lower component is dominant. In addition, the relationship of these components with age is not like that of the central–peripheral component that moves from high negative values at young ages to high positive values at older ages. Instead these components undulate around zero and demonstrate no clear directional change with age. It has been suggested that this lack of consistent pattern is a statistical artifact but others have argued that the consistent demonstration of ventral–dorsal and upper–lower components suggests that these may be individual components that have a genetic basis and may be associated with systematic disease in adulthood.

It is relatively clear that there are ethnic differences in subcutaneous fat thickness, suggesting genetic variation in fat patterning amongst populations. Consistent results demonstrate that those of European ancestry have a more peripheral than central pattern compared to those of African or Asian ancestry.

*Noel Cameron*

See also 'Anthropometry' (**1.4**), 'Adiposity rebound and prediction of adult fatness' (**1.11**), 'Body composition' (**5.11**), 'Sexual maturation' (**5.15**), 'Nutrition' (**9.2**), 'Modernization and growth' (**11.5**) and 'Obesity, fatness and modernization' (**11.8**)

# Adulthood and developmental maturity

Human growth and development involves both structure and function. Whilst the body is changing in size and shape (**5.5**) it is also developing to achieve functional maturation. The techniques we use to assess growth and maturation have end-points that imply a cessation of the process of both growth and maturation. 'Adulthood' in terms of somatic growth is usually taken as the time at which growth increments for height are less than 1 centimetre a year, i.e., there is no definitive end-point in somatic growth which is defined as 'adult' because somatic growth continues into the late 20s and the adult height of each individual is different because of their genetic background and the environment they have experienced during the process of growth. Height and other linear variables that are used to describe somatic growth cannot, of course, reduce in size and thus do not demonstrate growth *decrements* in response to environmental stress. Other variables pertaining to somatic growth have less definable 'end-points' because of their inherent variability and eco-sensitivity. Weight, for instance, has a wide range in adults, of more than 40 kilograms between the 5th and 95th centiles, and can either increase or decrease during the process of ageing in response to nutrition, illness, exercise and other factors.

Functional maturity is reflected in the adult status of secondary sexual development (breast/genitalia, pubic hair, axillary hair, testicular volume, menarche, spermaturia), skeletal, and dental maturity. Techniques to assess maturation do not usually involve a quantitative value but are qualitative relative to 'adulthood'. Secondary sexual development, is assessed against the 'Tanner Scale', or using an adapted scale of 'pubertal' values based on combined Tanner ratings, and have end-points that are defined as 'adult'. Of particular importance is that adult sexual maturity is normally reached prior to the cessation of somatic growth, thus an individual can be 'adult' in terms of maturity but still 'adolescent' or 'pre-adult' in terms of somatic growth. Dental maturity is reached relatively early, when the permanent dentition is completely erupted, and skeletal maturity when the epiphyses of the long bones have fused with the metaphyses.

Because the descriptions of 'adulthood' in terms of secondary sexual development are relative to a pre-adult stage they appear incomplete out of this context and use relative terms such as 'adult in appearance'. The following definitions of the Tanner Scale thus give both the pre-adult and adult descriptions.

## FERTILITY

Fertility is defined as production of live births, and shows complex relations to growth from the proximate level of physiology to the ultimate level of life-history strategy. Human life-history organizes growth and reproduction as largely mutually exclusive processes: available energy goes first into the former, and then, after puberty, into the latter (**10.1**). The switch from growth to reproduction is enforced by induction of skeletal maturation and growth cessation with gonadal activation. Because pelvic, specifically birth-canal, width is crucial to successful childbirth, ovarian maturation may be co-ordinated with skeletal maturation. Nearly all girls reach menarche after peak height velocity, as growth-rates decline rapidly. A biiliac diameter of 24 centimetres may represent a critical threshold for menarche or advanced reproductive function in girls, implying that undertaking the risk of pregnancy is predicated on adequate skeletal development. Moreover, the energy demands of pregnancy compete with those for growth; even well-nourished girls who undergo pregnancy before their own growth is complete incur measurable growth deficits. While nutrition affects timing of puberty through its general influence on maturation rates, the suggestion that attainment of menarche (**1.15**) requires a critical mass or percentage of body fat has been largely refuted. By contrast with girls, boys attain spermatogenesis well before peak height velocity, but the visible signs of androgenization (e.g. peak growth, voice change) appear late in puberty. Boys, then, develop capacity for early 'sneak fertilizations', but delay achievement of full adult characters that will lead to increased workload and high-risk adult tasks such as warfare.

Attainment by girls of reproductive capacity occurs gradually. Full co-ordination of endocrine and other physiologic processes involved in competent ovarian cycles requires some years to achieve (**5.15**). At menarche, few ovarian (menstrual) cycles are ovulatory and fully fertile, and subfecundity (reduced ability to reproduce) characterizes the post-menarcheal period. Post-menarcheal subfecundity arises from low ovulatory frequency, and long and poorly organized ovarian cycles, which diminish the likelihood of successful pregnancy. Girls with late menarche tend to have concomitantly late first intercourse, marriage, and first birth, but timing of menarche appears unrelated to age at menopause. Slow attainment of full fecundity partly explains the U-shaped distribution of pregnancy loss by maternal age, with a nadir in the late 20s. Very young mothers are also more likely to bear low-birth-weight infants than are mothers in their later teens and 20s, after which risk for low birth-weight increases with maternal age. The importance of maternal developmental history for successful reproductive outcomes is suggested by persistent correlations of birth-weight with maternal height and prepregnancy weight, but a role for ongoing effects of class or resource availability cannot be ruled out.

Recent comparative studies have revealed variation among populations in levels of gonadal hormones in both sexes; adults in post-industrial societies appear to have markedly greater circulating levels of gonadotropins and gonadal hormones than do adults in some less privileged, more slowly maturing groups. Resource availability and stability experienced in development are thought to influence not only growth-rates and timing of puberty, but also set-points for endocrine regulation of gonadal function and responsiveness to environmentally mediated stressors such as weight-loss.

Finally, fertility patterns themselves influence child growth and health. Close birth spacing diminishes infant birth-weight and survival, due partly to erosion of maternal nutritional reserves (maternal depletion), and partly to early weaning and intensification of sibling resource competition as birth intervals shorten. Low birth-weight (**10.3**) correlates with reduced adult body-size and increased lifetime morbidity and mortality risks.

*Carol M. Worthman*

## BONE MORPHOLOGY AND MINERALIZATION ACROSS ADULT LIFE

Being connective tissue, bone is characterized by the presence of specific types of cells and ground substance (**5.6**). Bone cells originate from stem cells that differentiate to become either osteoblasts or osteoclasts. The primary function of osteoblasts is the synthesis and release of collagen and the initiation of its mineralization. Some osteoblasts become trapped in the matrix they have produced, becoming osteocytes, whose primary function is the exchange of the constituents of bone mineral between bone and the vascular system. Osteoclasts are multinuclear cells whose primary function is the resorption of bone mineral either beneath the periosteum, on the external surface of the bone or the endosteum within the medullary cavity.

Some bones, such as those forming the cranial vault (**5.9**), the mandible, and the clavicles originate in the dermis and are referred to as of dermal, or intramembranous origin. Most of the human skeleton originates in cartilage and is therefore of cartilaginous origin. Functionally and histologically, the two forms are indistinguishable in the adult. In all areas of the skeleton, the outer surface of bone is made up of cortical bone which is characterized by the presence of osteons, or Haversion systems. Osteons are made up of concentric layers or lamellae of bone matrix containing a number of lacunae wherein osteocytes reside. The lacunae are connected by a network of canaliculae which ultimately provide access to the vascular system at the central canal around which the osteon is organized. It is this access that permits the vital exchange of bone mineral that insures the maintenance of blood calcium concentration within physiologically tolerable limits. Deviations from normal blood calcium levels stimulate release of hormones that will in turn stimulate the activity of osteoclasts and or osteoblasts to bring calcium levels back to normal.

Skeletal growth during adolescence involves both interstitial and appositional growth. Interstitial growth occurs in the epiphyseal cartilage, while appositional growth occurs at the metaphyses, or growing ends of the diaphysis, or shaft. Simultaneously, appositional bone growth is also occurring at the subperiosteal surface, increasing its diameter. The end of the adolescent growth spurt occurs when the epiphyseal cartilage is calcified, cartilage cells are deprived of nutrients and die, and osteoblasts invade the cartilage plate, replacing it with bone. The closure of epiphyses occurs latest in the annular epiphyses of the spine. Therefore, the last increment in the growth of stature usually is in trunk length. Consequently, relative sitting-height increases for a time before growth ceases entirely.

Although bone remodelling occurs to varying degrees throughout life, the density of adult bone is maintained at a relatively stable level from young adulthood until the decline of gonadal hormone production associated with old age. However, bone turnover, involving the synthesis of new bone by osteoblasts and the resorption of old bone by osteoclasts, is continuous. The rate of turnover varies from one region of the body to another, and even from one part of a single bone to another, and is more rapid in trabecular than in cortical bone. *William A. Stini*

## EPIPHYSEAL FUSION

The fusion of the epiphyses with the metaphyses marks the achievement of adult skeletal maturity in the 'long bones'. Both the *Atlas* and bone-specific scoring techniques for the assessment of skeletal maturity (**1.8**) concentrate on the *left* hand and wrist as the skeletal site most useful in assessing maturity. This site contains both 'long', 'short' and 'round' bones. The latter are the bones of the carpus and include the capitate, hamate, triquetral, lunate, schaphoid, trapezium (greater multangular) and trapezoid (lesser multangular). These bones ossify within a cartilaginous model and contain no epiphyseal plates. The long bones include the radius and ulna, and the short bones include the metacarpals and phalanges.

The long and short bones all have epiphyses and metaphyses with cartilaginous *growth plates* separating them. These epiphyses are at the distal ends of the radius, ulna, and metacarpals II to IV, and at the proximal ends of the phalanges and the first metacarpal. The radiographic appearance of epiphyseal fusion follows a stage of development in which the epiphyses are said to 'cap' the shaft. *Capping* is the process during which the epiphyses mould to the shape of the metaphyses and in radiographic cross-section appear to form pointed ends that wrap around the metaphyses. These pointed ends look similar to the peak on a cap, which is why the process is called capping. Capping is immediately followed by epiphyseal fusion.

Fusion is first recognized radiographically by a loss of the continuous dark line representing the growth cartilage. This loss normally begins in the centre of the epiphysis/metaphysis and gradually extends medially and laterally until the growth cartilage is totally ossified. There is a subsequent period, during which ossification is incomplete in the peripheral part of the growth cartilage, when a radiographically visible groove is present on the surface of the bone (**1.8**). The transverse densely radio-opaque layer that joins the epiphysis to the metaphysis is resorbed after fusion is complete, but this resorption may be delayed for an extended period of time.

Once fusion is complete no more maturity indicators are present in these bones to provide further assessments and the child is said to have 'adult maturity'. The chronological timing of *complete* epiphyseal fusion is variable and is earlier in females than in males at 16.0 and 18.2 years respectively for the British children that formed the source sample for the bone-specific scoring technique.

Certain bones present with fusion prior to others. Initial fusion of the first metacarpal and distal phalanges may be seen at 13.5 years in the female standard of the Greulich and Pyle *Hand–Wrist Atlas*. In males of the same source sample that initial fusion is not seen until the standard representing 15.5 years. Complete fusion is not seen until 17.0 years in the girls and over 18.0 years in the boys of this sample. Thus the process of fusion takes approximately 3 years and because all the long and short bones are used in the estimation of skeletal maturity full *adult* maturity is not apparent until the final complete fusion of these bones. *Noel Cameron*

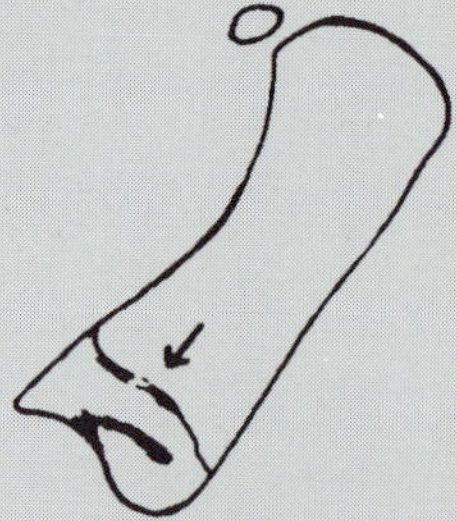

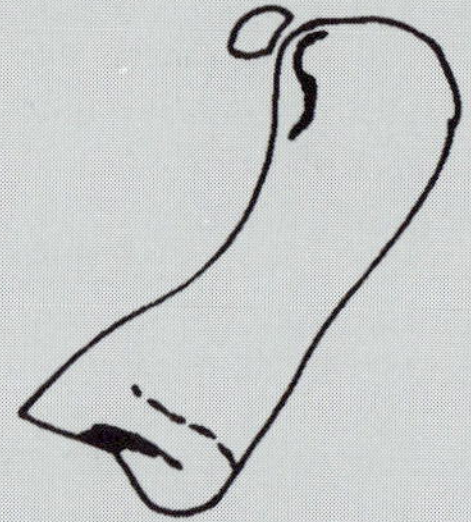

**Stage H**

**Fusion of epiphysis and metaphysis has begun. A line is still visible, composed partly of black areas where the epiphyseal cartilage remains, and partly of dense white areas where fusion has occurred.**

**Stage I**

**Fusion of epiphysis and metaphysis is completed. (Over the majority of its length the line of fusion has entirely disappeared, but some thickened remnant of it may still be visible.)**

*Pubic hair: (same for males and females)*

Pre-adult : Hair now resembles adult in type, but the area covered by it is still considerably smaller than in the adult. No spread to the medial surface of the thighs.
Adult : Adult in quantity and type, distributed as an inverse triangle of the classically feminine pattern. Spread to the medial surface of the thighs but not up the linea alba or elsewhere above the base of the inverse triangle.

*Genitalia*

Pre-adult: Increased size of the penis with growth in breadth and development of glans. Further enlargement of testes and scrotum; increased darkening of the scrotal skin.
Adult: Adult in size and shape. No further enlargement after Stage 5 is reached.

*Breasts*

Pre-adult: projection of aerola and papilla to form a secondary mound above the level of the breast.
Adult: Projection of the papilla only, due to recession of the general contour of the breast.

*Values for the mean ages of attainment of various indicators of sexual, dental, and skeletal maturity. Where figures for standard deviations have been reported these are provided in brackets.*

| Indicator | Males | Females |
|---|---|---|
| Breast stage 5 | – | 15.33 (1.20) |
| Genitalia stage 5 | 14.92 (1.10) | – |
| Pubic hair stage 5 | 15.18 (1.07) | 14.14 (1.12) |
| Testicular volume 20ml | 16.0 | – |
| Menarche | – | 13.47 (1.02) |
| Skeletal maturity RUS | 18.2 | 16.0 |
| Skeletal maturity CARP | 15.0 | 13.0 |
| Skeletal maturity TWII(20) | 18.0 | 16.0 |

See also 'Biochemical markers' (**1.7**), 'Creation of growth references' (**1.16**), 'Comparative growth and development of mammals' (**2.1**), 'Growth in non-human primates' (**2.2**), 'Genetics of child growth' (**3.1**), 'Skeletal development' (**5.6**), 'The human growth plate' (**5.7**), 'Sexual maturation' (**5.15**) and 'Endocrine growth disorders' (**7.7**)

*Testicular Volume*

Mid-puberty is usually taken when testicular volume is 12 millilitres. Adult values range between 16 and 25 millilitres from the 10th to the 90th centile of European norms.

Naturally these descriptions of 'adulthood' relate to physical appearance rather than function. Indicators of functional maturity relate to menstrual cycles in girls (menarche) and the production of spermatozoa in boys (spermarche).

*Menarche*

Menarche is said to be reached when regular menstrual cycles are experienced. This event is usually in mid-puberty, following peak height velocity, when the girl exhibits stage 3, 4, or 5 for breast and pubic hair development. Menarcheal age is a very sensitive indicator of environmental stress and is also under significant genetic control. It thus varies considerably depending on the socio-economic circumstances of the girl. Information on menarcheal age may be obtained by status quo, prospective, and retrospective or recall methods (**1.15**).

*Spermarche*

The detection of spermatozoa in urine has been proposed as a quick, non-invasive method to assess the functional state of the maturing gonad but its use may be limited because both longitudinal and cross-sectional studies have shown that spermaturia is a discontinuous phenomena.

*Dental maturity*

Maturity is said to be reached when the permanent dentition has completed erupted. Prior to complete eruption radiographic techniques can be employed to assess the level of maturation but concern over exposure of the child to radiation has resulted in tooth eruption as the most commonly used method to assess dental maturation.

*Skeletal maturation*

Two techniques, the *Atlas* method and the bone-specific scoring method, are used to assess skeletal maturation. Both use the complete fusion of the epiphyses as the adult, fully mature, stage.

Noel Cameron

## Part Six

# Behavioural and cognitive development

In human biology, the standard approach to growth and development has usually been to address all kinds of changes in the physical body, from zygote to adult, but pay comparatively little attention to mental development. This area is seen as being somewhat separate from the rest and is not easy to deal with, on account of the great diversity of concepts which can be used to study and understand mental functioning.

The obvious tendency is to rely only on clear-cut biological data, such as those provided by neurobiology, or at least on measures of behaviours, as provided by the psychometric approach of behaviourism that still characterizes mainstream contemporary psychology (**6.1**). Yet many object that such approaches, for all the fundamental contributions they make, still only brush the surface. With respect to the deeper dimensions of mental functioning assumed to underlie observed behaviours and biological correlates, this work may reveal little more than the tip of the iceberg.

Clearly, in the growth process from single cell to a some $10^{14}$-cell adult human organism, the many variables involved increasingly give rise to an interactional network in which all structures, functions or behaviours come to express themselves in a context that is essentially process-based, whatever its manifold material expressions. To explain mental processes and functioning only in terms of observed behaviours and biological correlates may grossly underestimate such complexity. Even a brief overview should, therefore, attempt to at least mention the variety of possible approaches to an area of study that ultimately deals with the very nature of the human being.

### Neurophysiology

**Opposite: Edvard Munch 'Puberty' (1895). Courtesy of the Nasjongalleriet, Oslo.**

Information processing by the central nervous system seems to involve a number of hierarchically organized centres governing the various behaviours a species is capable of expressing, and the individual acts of which they consist. The fundamental behaviours related to the maintenance of the organism, such as heart-beat, breathing, and so on, are obviously inborn and seem to be governed by the phylogenetically older portions of the central nervous system (CNS) (spinal cord, medulla and hypothalamus), while behaviours related to experience are gradually acquired and seem to be governed by the newer portions characteristic of mammals, particularly the cerebellum.

The about $10^{11}$ neurones constituting the adult human brain are virtually all formed prenatally, mostly in the first 3 months, when practically all structures are formed, and the timing of ontogenesis appears to reflect phylogenesis: subcortical structures emerge first, followed by the development of the centres underlying instinct, emotional response and postural behaviour, and eventually by that of the cortical areas associated with higher relational functions (attention, awareness, and so on). Movement behaviour can be detected by the end of the embryonic period. Even sensory functions develop to some extent prenatally, and in a sequence that also reflects phylogenetic timing (sensitivity to touch, sound, light). However, the brain remains functionally very plastic through to the neonatal period and infancy, allowing it to develop neurophysiologically and psychologically, and increasingly adjust to the environment. This is a condition that is progressively reduced, though not totally lost, in later stages of life. The brain develops much earlier and faster than any other part of the body: at birth it is already 25 per cent, at 1 year of age, over 50 per cent, and at age 5 years about 90 per cent of its adult weight. The psychological achievements and the various behaviours that gradually come to characterize the human person rest largely on the neurophysiological development and maturation accomplished prenatally and during infancy.

At birth the neonate displays a whole repertoire of simple stimulus-response sequences that represent behaviours of great adaptive significance. Approach reflexes, such as sucking or swallowing fluids, are too important for the survival of the newborn to be acquired with experience. And so are avoidance reflexes such as sneezing, coughing or blinking.

The constant unfolding of the biological programme of maturation of nervous, muscular and other structures is manifested in increasing competence in a variety of fundamental movement behaviours. Along with growth in size, motor development (**6.2**) represents the most impressive characteristic of infancy. In the course of a few months, a helpless neonate, initially unable even to roll over, gradually attains control of the head, upper trunk and upper extremities, and becomes capable of assuming a sitting position, moving around by prone progression, and eventually conquering the upright position and walking independently. The acquisition of this gross motor skill, largely a function of biological maturation more than experience, is accompanied by that of fine motor skills, such as the manipulative movements of the hands involved in the development of prehension, which also requires eye–hand coordination, and therefore a significant perceptual component.

*Perception* develops as a natural outgrowth of the sensory system and is largely conditioned biologically, although individual experiences and variability play a relevant role. Contrary to previous beliefs, infants appear to be perceptually very alert, open to the world with all their senses operating in an integrated way. They perceive odours, tastes and sound in the first days of life, engage very early in mouthing and fingering exploratory activities, discriminate colours by the age of 6 months, and develop other forms of sensitivity such as to space, forms, movements, and so on, well before 1 year of age.

## Cognition

These fundamental achievements lay the ground for cognitive development, that essentially consists in the unfolding of mental processes by which knowledge is built up, and which result from perception and judgment and involve learning and behaviour.

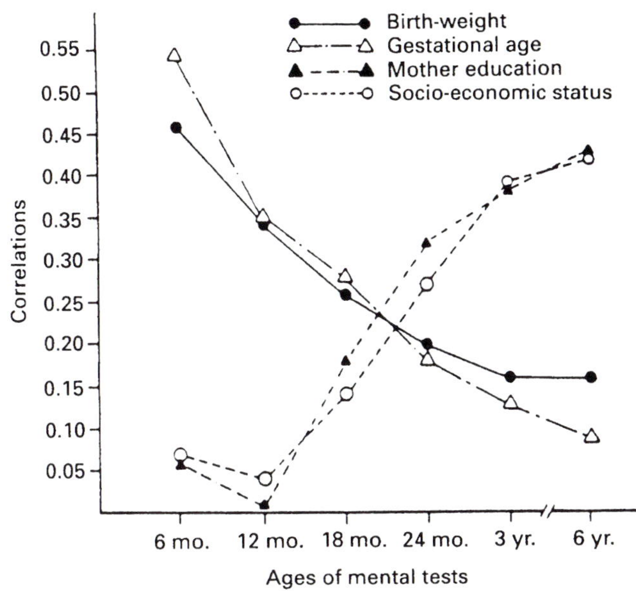


**Relation of biological and social factors to mental test scores from 6 months to 6 years of age (Wilson, 1985).**

*Learning* involves a change in behaviour, as well as in perception and judgment, based on perceptual experience and evaluation. This may in some circumstances be chronologically delayed in undernourished individuals (**6.5**). A basic mechanism is provided by *association*, whereby the occurrence of one event brings another to mind, both events having previously combined in a subjective experience. Association plays a very relevant role in cognitive and emotional development throughout infancy and childhood, starting with very early experiences, possibly even prenatally. Other mechanisms of learning include discrimination, habituation (whereby a subject learns *not* to respond to repeated stimuli), imitation, concept formation, and perceptual or psychomotor learning.

Learning necessarily implies some form of information encoding and retrieval, or *memory*. This apparently involves neural changes that are held to occur first at the level of synapses and to be temporary and reversible (short-term memory), and later to effect the structure of neurones and involve a more permanent storage (long-term memory). It appears that a rapid development of storage skills occurs in early infancy, while long-term memory progresses throughout childhood.

These variables have been classic objects of behavioural studies, focussed on the notion of *intelligence*. This term designates a complex human faculty whose definition and analysis has been highly controversial, on account of methodological and conceptual difficulties and of its ideological and social implications. Basically, intelligence can be defined as a set of *abilities* or potentialities, allowing an individual to learn and adapt, and to face novel situations. These have been alternatively unified into a single, though complex trait, or broken down into up to 120 specific abilities, and variously considered to be essentially genetic in nature, or conversely, to be largely the product of individual experiences and other environmental forces.

Measures are provided by intelligence tests, consisting of a series of tasks addressing specific mental abilities, such as verbal, mathematical or other logical skills, memory, and so on (as they are expressed at the time the test is administered, so that possible differences in development should then be accounted for). Scores are combined in a synthetic measure, the *Intelligence Quotient* (IQ), originally based on the ratio of mental age (MA) to chronological age (CA) usually multiplied by 100 (IQ = 100 [MA/CA]). This typically results in an average of 100, corresponding to the equivalence of the two (MA = CA), while lower or higher intelligence is expressed by IQ values correspondingly lower (MA < CA) or higher (MA > CA) than 100. Measures have then come to be referred to the frequency distribution of scores in the general population. The IQ distribution is typically continuous, with a mean of 100 and a standard deviation of approximately 15. By and large, about 70 per cent of individuals in the general population score within one standard deviation from the mean, that is, between 85 and 115 (and are considered to be ‘normal’). Of the remaining 30 per cent, about 12.5 per cent each fall two standard deviations below or above the mean – that is, between 70 and 85, or between 115 and 130 – and 2.5 per cent each fall in either of the two tails of the distribution – that is, below 70 (‘retarded’) or above 130 (‘very superior’). Twin, family and adoption studies, including studies of identical twins reared apart, concur in indicating fairly high genetic conditioning, accounting for about 70 per cent of the variance. However, whereas it is widely accepted that mental retardation (frequently associated with chromosomal damage) or very superior intelligence are almost totally genetically conditioned, disagreements about definitions, testing methods, the continuous distribution of scores, and the multifactorial nature of whatever is measured under the term 'intelligence', make this area still controversial.

A different, developmental approach to cognition has been proposed by Jean Piaget and referred to as *genetic epistemology*, thereby stressing the process of continuous adaptation of the individual's genetic potentialities to the environment in the acquisition of knowledge. The structure or organization of intelligence is here seen to be changing across the course of development, passing through a series of qualitatively different stages, each of which develops from and incorporates the preceding one. At each stage in this hierarchical process new skills are developed and new kinds of concepts attained. Piaget's model distinguishes four universal stages: 1) *sensorimotor operations* (age 0–2 years), in which an infant experiences its physical self through perception, response and movement, and differentiates it from the environment; 2) *pre-operational representations* (age 2–7 years), in which the child develops language and the first forms of inner abstract representation, learning to manipulate the environment in different and better ways; 3) *concrete operations* (age 7–12 years), in which the child starts developing logical reasoning in the form of classification, with a better understanding of basic categories and constructs such as causation, time, space, quantity, and so on; 4) *formal operations* (age 12 years to adulthood), in which the adult modes of thinking are acquired, with a deeper assimilation and manipulation of abstract concepts and reasoning.

Piaget's developmental model, which has received empirical support from various sources and has been very influential in educational policy, is frequently held to represent an alternative to the psychometric approach. However, the differences between the two systems might well be overestimated. Both view intelligence as a successful adaptation of biological potentialities to the environment, essentially resulting from a genetic endowment open to individual experiences and therefore to some extent modified in its expression by the action of, and interactions with, environmental factors. Also, attempts to compare the results of Piagetian tasks with those of typical psychometric tests have indicated substantial agreement.

An important extension of Piaget's model is in the area of *moral development*. Piaget made a distinction between the response to a situation and the reasoning behind it, and proposed two stages of moral reasoning, the first based on submission to social constraints, the second on the understanding and interpretation of the abstract principles of morality. This has been extended by Lawrence Kohlberg into a more complex progression, essentially including: 1) a *preconventional* stage, in which the child bases its reasonings and actions on what gives pleasure or pain; 2) a *conventional* stage, in which decisions are based on social models and rules; and 3) a *post-conventional* stage, in which the attainment of formal operational thinking allows the person to reason in terms of values and to account for the subjective and arbitrary elements of social rules.

*Language* development occupies a very central role in cognition (**6.3**). This is a complex process which is far from consisting of simple stimulus-response mechanisms, and involves the unfolding of a biological programme whereby the infant's inherent predispositions are stimulated and moulded in the specific form dictated by the environment. The human infant seems to express a natural motivation to discover and experiment vocal or gestural meaning, and passes through a sequence of developmental stages that is nearly identical in all languages – evidence that led the linguist Noam Chomsky to introduce the concept of a 'universal grammar'.

Infants of all cultures exhibit a sensitivity to sound soon after birth and an early ability to detect individual differences in the caring adult's voice and its emotional tone. Signals of pain or hunger are soon identified in their vocalization. They then pass through babbling at about 6 months, to uttering single words in a meaningful way, then two words, and to understanding prepositions by the end of their second year. By the age of 3 years, a child may build at a speed of a couple of words per day upon a productive vocabulary that may have already reached some 3000 words. These times and speeds by no means represent a rule, language development as a whole being characterized by very considerable differences among individuals.

Although it is generally agreed that language acquisition is genetically determined, appropriate stimulation in the early years of life is absolutely fundamental. Reduced language exposure, and more generally lack of appropriate relational interaction may not only hamper language development, but also seriously affect the growth and survival of a child altogether. A well-known anecdote reports Frederick II, at the time king of Sicily, to have performed an experiment to find out what kind of speech children would develop if they were reared by foster mothers who would suckle and carefully take care of them, but neither speak nor play with them at all. Apparently the experiment failed, for the children, deprived of the petting, loving words and emotional stimulation, all proved unable to survive.

## Personality

*Emotion* is an important component of perception and thus plays a fundamental role in cognition as well as in personality development (**6.4**), of which it is usually considered to be the basic aspect. Although cognition and personality are usually considered separately, their various components, and particularly perception and emotion, all interact to a fairly large extent in the course of mental development and in any behavioural expression.

Emotion represents a subjective experience, rooted in the individual perception of a given situation, involving a physiological response and related approach/withdrawal behaviours, possibly associated with a dynamic process of stimulation/inhibition of given functions or bodily movements controlled by the hypothalamus. Some basic emotions, such as interest, disgust, sorrow and joy, are expressed by the infant soon after birth, while other ones, such as sadness, anger, surprise, and eventually shame, emerge as cognitive development proceeds. The early ability to discriminate the emotional content, hidden or unconscious as it may be, in the attitude of the caring adults, plays a very important role in the infant's perception and development, and an actual or perceived lack of sufficient positive stimulation can have critical consequences for cognition and on the structuring of personality, and even possibly hamper physical growth and survival, as witnessed by conditions such as psychosocial dwarfism and anorexia nervosa, or similar ones.

Emotions play a very relevant role in regulating internal psychological processes and interpersonal relations. They express themselves in a characteristic pattern, known as *temperament*, that research on variation in infants, and particularly in twins, has shown to be already defined very early in life and largely to reflect genetic influences – although in close interaction with environmental forces, whereby temperamental modifications may occur in the course of development. Emotions and temperament are the core around which the subsequent enlargements and changes resulting from interaction with the environment build up the overall pattern of mental functioning and behaviour known as personality.

In experimental psychology, the notion of personality refers to a construct whereby individual behaviour is no longer regarded as somewhat simplistically determined by the various events and situations of life in some sort of stimulus-response process (situationalism), but is rather seen to show some degree of consistency and predictability across time and place, and therefore to express a constitutional specificity of the individual, though widely open to environmental influences. Specific behavioural characteristics showing substantial consistency and predictability – the so-called *personality traits* or *factors* – are evaluated and usually quantified through apposite rating scales. Evaluation may be based on observational methods, and include unstructured or structured interviews, or on self-report questionnaires (such as the MMPI, Minnesota Multiphasic Personality Inventory, or the EPQ, Eysenck Personality Questionnaire), or on indirect, projective techniques such as the Rorschach Inkblot Test, based on the assumption that the individual's interpretation of an ambiguous inkblot reflects cognitive, emotional and symbolic aspects of the subject's personality.

Self-report inventories are held to be fairly reliable and to provide the best approach to personality measurement; they are simple to use, easy to score, and allow large-sample testing. Extensive studies have thus been carried out on family relatives – comparing, in particular, identical versus non-identical twins, both reared together and apart, as well as adoptive versus biological relatives – and sophisticated psychometric genetic approaches have been recently applied. By and large, these studies indicate that about 40–50 per cent of the total variance in the various personality traits is due to genetic factors (essentially the *additive* genetic component with possible minor *dominance* effects) and over 50 per cent to environmental factors. The latter, however, are consistently restricted to variation *within the family*, that is, to unique personal experiences. Interestingly, no trace has been found of any considerable effect of *shared environment*, meaning that parental attitude, education and socio-economic status are not seen to play the significant role that is usually attributed to them. In fact, the correlations between subjects adopted early in life and their adoptive parents or siblings have been found to be consistently lower than 0.10.

## Personality theory, human needs and motivation

The psychometric approach is frequently criticized on the grounds that it reduces the complexity of the human mind to a number of measurable traits that represent at best the tip of an underlying iceberg of which they ultimately say little, if anything. In fact, different approaches have long characterized the history of psychology, and increasingly do so, although mainstream scientific thinking still tends to consider them as being too theoretically oriented and difficult to be verified through scientific enquiry. This results in a dichotomy between the 'scientific', biologically-oriented approach, based on experimentation and/or the observation of well-defined structural, functional or behavioural variables, and a more phenomenological approach, based on inner experience, subject-centred and much wider in scope, leaning towards human sciences and philosophical speculation.

The most influential personality theory by far is that of *psychoanalysis*, established by Sigmund Freud in the early part of the 20th century and still widely applied, either in its classic formulation or in any of the variants developed by Freud's former associates or later theorists. In essence, the theory assumes human personality to result from the continuous, dynamic interaction of three components: the *id*, representing the basic structure, where primitive, inherited drives, or instincts, are contained; the *ego*, developing later as the centre of realistic, logical thinking – that is, of the higher cognitive, logical and perceptual functions whereby these basic instinctual needs can be realistically satisfied; and the *super-ego*, a superimposed structure representing the stringencies of introjected social values and moral codes.

The drives originating from the id determine an inborn tendency requiring satisfaction, and express themselves as basic needs aimed at achieving pleasure or avoiding pain, whereby the tension is released. They are associated with a psychic energy, the *libido*, that throughout development is progressively centred in different body areas, determining which stimulation is to be gratified. The child is seen as passing through stages that involve the oral, the anal, and eventually the genital area, and as needing to gratify progressively the stimuli associated with each of these. Frustration of a need may produce a fixation of libido, that is then partially blocked at the related stage, resulting in emotional troubles and continuous attempts to obtain some form of the missed gratification throughout life. In this way the experiences of the first years are critical for the development of personality and behaviour.

All behaviour is thus seen as serving to satisfy a set of basic instincts, whether directly or indirectly. In the course of development, primary biological drives come to be associated with secondary drives, and gratification of the latter helps to satisfy the former (for example, the drive toward power and prestige is thought to be secondary to the drive for sex). The association can involve third-order drives, and so on, so that the immediate motivation of an act can become increasingly distant from its real source. The theory's greatest contribution is perhaps the notion of *unconscious motivation*, meaning that humans are often unaware of the real reasons for what they do and frequently adopt irrational or contradictory behaviours, being driven by elements that are largely excluded from ordinary awareness, being removed in the unconscious.

Flexible and elegant as it can be, the theory has been challenged on various grounds. In particular, the notion that the motives of human behaviour are all to be traced back to physiological requirements has been either rejected or smoothed, and important variants have branched out of the original theory. Perhaps the main one is that developed by Carl Gustav Jung, who introduced the notion of *collective unconscious*, a sort of storehouse of humanity shared by everyone, containing patterns of universal character (*archetypes*) that play a fundamental role in human symbolic life and motives, particularly toward spirituality. They represent frequently opposed aspects of the human experience (Good/Evil, Masculine/Feminine, and so on) that are partially hidden in the unconscious, from where they influence psychic life and motivation through symbolic messages and unconscious drives. To become aware of their existence and effects, that is, to bring them back and integrate them into consciousness – a process known as *individuation* – is the essence of personality development. Rather than being limited to the first years of existence, growth is thus seen as a potentially continuous process towards a widening of awareness.

Later trends in psychology have increasingly stressed this notion of continuous growth and the developmental character of *needs and motivation*, that humanistic psychology has extended from the purely physiological to progressively higher levels. Fundamental to this is the scheme proposed by Abraham Maslow, with five interrelated levels arranged in a ladder: 1) *physiological needs*; 2) *safety needs*; 3) *belongingness and love needs*; 4) *esteem needs*; and 5) *need for self-actualization* or *realization*, representing the most fundamental human tendency toward making actual what is potential in the self. The scheme is hierarchical and developmental in nature, in that higher needs do not fully emerge, nor can they be attended to, until adequate satisfaction is given to the lower needs (those that appear earlier in devel-

**A 2-year-old child. Photograph by Stanley Ulijaszek.**

opment and are more closely related to biological aspects, including the development of sexuality **6.6**).

At a later time, Maslow extended his scheme, adding still higher needs, meant to go even beyond the fullest expression of a mature ego, or individual self, toward the realization of *ultimate* human potentialities and states of consciousness, aiming at *self-transcendence* and the experience of the sacred. That was the basis for the emergence of *transpersonal psychology*, a new movement that is increasingly promoting the encounter of psychology and spirituality, largely based on the practice of meditation and the related widening of consciousness, and that is proving increasingly influential in science, culture and society.

Far from rejecting the approaches of psychoanalysis or behaviourism, these new trends of contemporary psychology tend to integrate their views and findings in a wider frame of reference, as has been widely theorized by Ken Wilber and others over the past 20 years. They are fundamentally distinct, however, in that they remind us – as already noted in the citation of the award presented in 1964 by the American Psychological Association to Gordon W. Allport – 'that man is neither a beast nor a statistic, except as we choose to regard him so, and that human personality finds its greatest measure in the reaches of time'.

*Paolo Parisi*

# Cognitive development

Cognition is a collective term for the psychological processes involved in the acquisition, organization and use of knowledge. It includes perception, memory, attention, problem-solving, language, thinking and imagery. The perceptual system must be able to provide a description of the structure of the world, but provides essentially selections of reality. However, selection is not uncritical, but involves descriptions or hypotheses of the objective world. The view that infant perception is piecemeal and disorganized has changed across the course of the 20th century to one where it is seen as active, highly organized and coherent. In this view, infants are born with an initial idea about the world of objects, which through learning becomes elaborated, creating an elementary understanding of the external world.

Associated with this is memory. Memory is not merely a record of experiences, but is essential to adaptive behaviour because it is organized in ways which make information gained from past experience applicable to present situations. Interaction between the child and its environment changes not only its behaviour and/or physiology, but also its experience of the environment. Memory allows experience to become knowledge. The essence of memory organization is classification, and although only individual events are experienced, they are remembered and recurrences are identified as instances of categories.

The learning of concepts and categories does not have a straightforward explanation. The 'hypothesis testing' view has it that concept formation is a form of problem-solving. The 'category-learning' position suggests that there are perhaps two ways of organizing information: on the basis of a small number of discrete categories, and on the basis of shared, or common traits. Language, thinking and imagery allow information to be integrated and used in different ways. Often when presented with information, humans not only encode its meaning, but also elaborate or explain it. Such elaborations allow the new information to be related to existing knowledge. However, the ways in which cognition develops are still controversial; certainly, theories of language acquisition which have included ideas of imitation and reinforcement have been shown to be inadequate in explaining the majority of observed measurable cognitive developmental traits.

As the understanding of cognitive and language development has expanded, the idea of an overall developmental level as represented by one score has been made obsolete. Rather, with the awareness of the existence of many different types of cognitive skills, the use of different measures representing

*Differences in correlations between non-adoptive and adoptive siblings, according to the test used*

| | | Non-adoptive | | Adoptive | |
|---|---|---|---|---|---|
| **Measure** | **Year** | r | **N (pairs)** | r | **N (pairs)** |
| Bayley MDI | 1 | .37 | 82 | .05 | 67 |
| | 2 | .42 | 70 | .12 | 61 |
| Stanford-Binet | 3 | .35 | 54 | .32 | 50 |
| | 4 | .24 | 43 | .23 | 43 |
| CAP general factor | 3 | .44 | 52 | .18 | 44 |
| | 4 | .10 | 37 | .20 | 39 |
| Bayley verbal | 1 | .22 | 82 | .21 | 66 |
| Bayley lexical | 2 | .27 | 70 | .03 | 68 |
| Bayley verbal | 2 | .31 | 70 | .05 | 60 |
| CAP verbal | 3 | .11 | 55 | −.05 | 50 |
| | 4 | .42 | 43 | .17 | 42 |
| Bayley means-end | 1 | .24 | 82 | −.16 | 66 |
| Bayley spatial | 2 | .14 | 70 | .06 | 60 |
| CAP spatial | 3 | .33 | 55 | .24 | 51 |
| | 4 | .12 | 43 | .27 | 42 |
| CAP perceptual speed | 3 | .18 | 55 | .08 | 49 |
| | 4 | .12 | 43 | −.08 | 39 |
| CAP memory | 3 | .14 | 54 | .19 | 51 |
| | 4 | .05 | 41 | .09 | 41 |
| SICD | 2 | .29 | 69 | .08 | 56 |
| | 3 | .21 | 53 | .10 | 50 |

Plomin, R., Defries, J. and Fulker, D.W. *Nature and Nurture during Infancy and Early Childhood*. Cambridge: Cambridge University Press, 1989.

*Sources of genetic influences on family environmental measures*

| **Probands** | **Objective measures** | **Subjective measures** |
|---|---|---|
| Parents | Genetically influenced characteristics of parents related to the environmental measures | Parental perceptions of the environment |
| | Parental responses to genetically influenced characteristics of offspring | Parental perceptions of their response to genetically influenced characteristics of offspring |
| Children | Parental responses to genetically influenced characteristics of children | Parental ratings: parental perceptions of their response to genetically influenced characteristics of children<br>Children's ratings: (a) parental responses to genetically influenced characteristics of children; (b) effect of genetically influenced perceptions of children on their perceptions of parental behaviour. |

Plomin, R., Defries, J. and Fulker, D.W. *Nature and Nurture during Infancy and Early Childhood*. Cambridge: Cambridge University Press, 1989.

different skills has become the norm. Global scores, such as intelligence quotient (IQ), do not provide any information about specific abilities. However, a variety of scales exist for assessing cognitive development at different ages.

From birth to 2 years of age, the Bayley Scales are most commonly used. These have two components: the Mental Development Index (MDI) that measures cognitive and perceptual development, and the Psychomotor Development Index, which measures motor development. The MDI can be further subdivided by the Kohen–Raz analysis into five subscales: eye–hand co-ordination, manipulation, conceptual relationships, imitation-comprehension, and vocalization social scales. Although the Bayley Scales are still useful at 2 years of age, the child's understanding of words, phrases and sentences can be tested in addition by the Reynell Development Language Scales (revised). This test is useful through to the age of 4 years. At the age of 3 years, other tests of cognitive and language abilities become useful. These include the McCarthy Scales of Children's Ability, which can be used to assess verbal skills, perceptual performance, knowledge of numbers and quantitative terminology, and verbal and non-verbal short-term memory. Other measures of verbal skills at this age are the Stanford–Binet test, which can be used up to the age of 5 years, and the Illinois Test of Psycholinguistic Ability Grammatic Closure subtest is useful for 3- to 7-year-olds. Between 6 and 12 years of age, the Wechsler Intelligence Scale for Children (revised) gives a composite score of cognitive function, but does not give disaggregated scores of specific abilities, nor can it be regarded as a clean measure of cognitive ability independent of educational influences.

Most tests of cognitive development give scores which show a continuous distribution, with the assumption that the magnitude of difference between scores at all points on the scale is equal. Accepting this assumption, it is possible to carry out cross-sectional parametric analyses of different aspects of cognitive function, and make comparisons of cognitive function scores with physical and motor parameters of growth and development. However, the variety of tests across different age groups makes it difficult to compare the results longitudinally, or across childhood, because it cannot be assumed that two different tests appropriate for different age groups measure the same thing, even if the index has the same name. Furthermore, the assumption that the scoring scale of tests of cognition is calibrated equally across the whole range of scores may not hold. An alternative approach involving categorization can be used, which does not depend on equal calibration of the scoring scale. There is a loss of information as a consequence of reducing a distribution into a small number of categories, but this is compensated for by greater analytical robustness. In addition, there is less confusion in interpreting longitudinal studies, since children classified into one category of development or another in early childhood have been shown to remain there into later childhood. Thus, categorization of cognitive test data, with its good prediction of future cognitive outcome, makes it appropriate for integrated longitudinal studies of growth and development.

*Stanley Ulijaszek*

**Jumping on a bouncy castle: play and motor development. Photograph by Stanley Ulijaszek.**

See also 'Language development' (**6.3**), 'Psychosocial factors in growth and development' (**6.4**) and 'Nutrition and cognitive development' (**6.5**)

# Motor development and performance

Movement activities are an integral part of children's behavioural repertoire and provide the medium through which children experience many aspects of their environments. Thus, the acquisition and refinement of skillful performance in a variety of motor activities are important developmental tasks of childhood. The terms *motor development* and *motor performance* are used to distinguish between, respectively, the acquisition or development of competence in fundamental movement behaviours, and the demonstrated level of proficiency in motor tasks. The distinction between development and performance is, of course, arbitrary.

## Motor development

Motor development is the process through which an individual acquires competence in a variety of fundamental movement tasks. The process involves the interaction of several factors including: 1) rate of neuromuscular maturation, which is probably genetically mediated; 2) the residual effects of prior movement experiences; and 3) current movement experiences. Additional factors are the physical characteristics of the child and environmental conditions which influence the opportunity to move and social interactions.

Motor development begins prenatally and continues postnatally. With advances in technology for fetal monitoring, early movement behaviours can be accurately documented. Earliest movements occur towards the end of the embryonic period. The movement repertoire of the fetus is quite diverse (**4.6**). During the second half of gestation, fetal movements increase, on average, in number through the 34th week and then decrease, indicating an inverted-U pattern. There are, however, substantial individual differences which are relatively stable. Continuity of prenatal and post-natal movement activity is

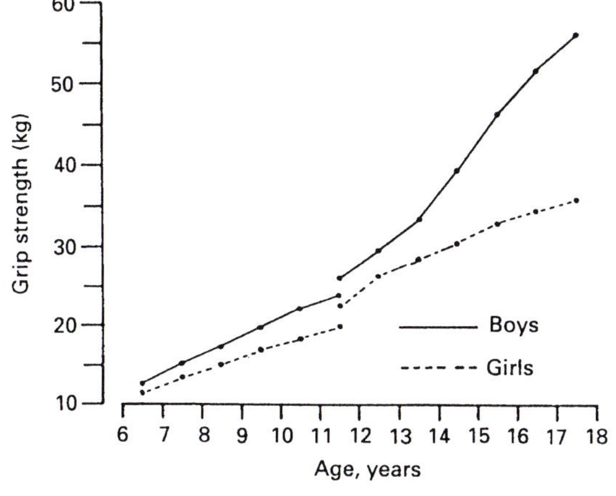


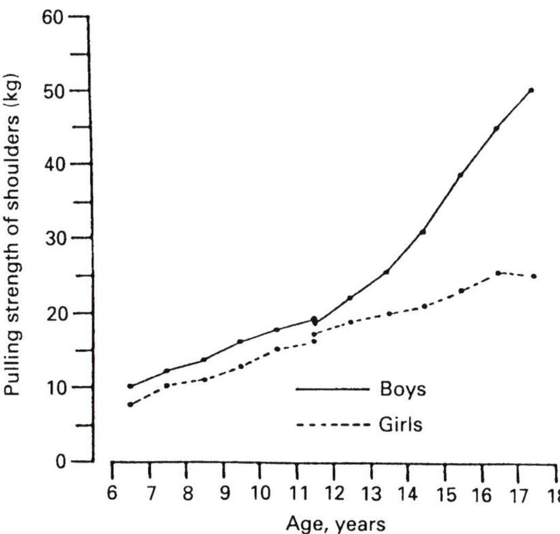


**Mean grip strength (top) and pulling strength (bottom) between 6 and 18 years of age. Mixed longitudinal data for 6 to 11 years are drawn from those of Malina for Philadelphia children, reported in Malina and Roche (1983); the longitudinal data for 11 to 18 years are drawn from those reported by Jones (1949).**

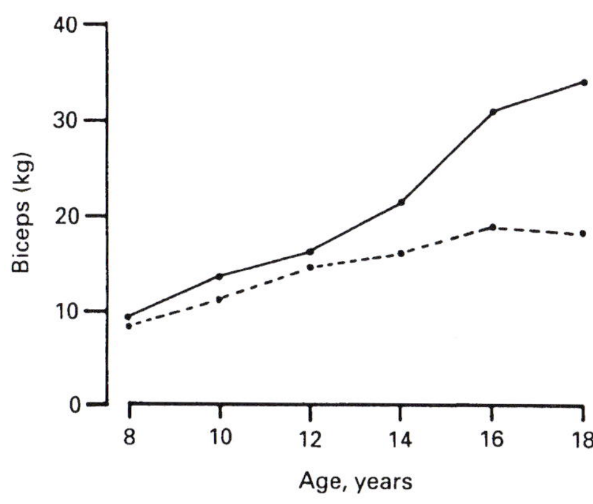


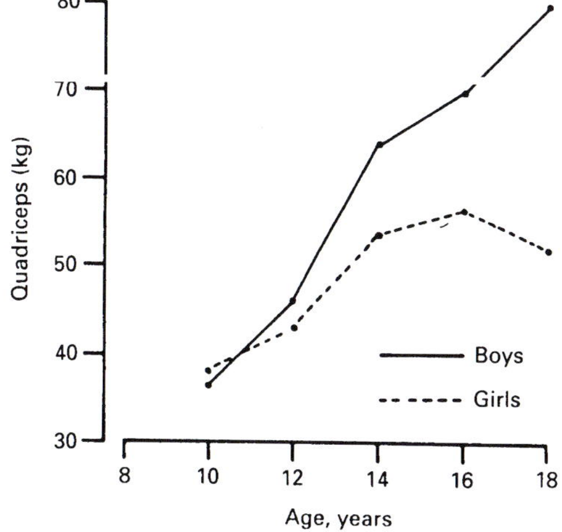


**Mean isometric strength in biceps flexion of the upper extremity (top) and in quadriceps extension of the lower extremity (bottom) between 8 and 18 years. Drawn from cross-sectional data for the biceps reported by Rodahl et al. (1961) and from unpublished cross-sectional data of Bouchard for the quadriceps.**

often assumed, but studies of the relationship between prenatal movement activity and subsequent post-natal movement activity and motor development are not available.

Motor development during the first 2 years post-natally is aimed largely at the development of upright posture and independent locomotion, i.e., walking, and of prehension. The former is a gross motor skill – movements of the entire body, while the latter is a fine motor skill – manipulative movements of the hands requiring a significant perceptual component, i.e., eye–hand co-ordination. Commonly used scales for the assessment of early motor development include a variety of gross and fine motor items. The focus of the subsequent discussion is primarily on gross motor development and achievements of children and youth, although most motor tasks incorporate both fine and gross motor elements.

The general sequence of developmental changes leading to walking can be summarized as follows. The infant gradually attains control of the head, upper trunk and upper extremities. Control of the entire trunk follows with the assumption of sitting-posture, first with support and then alone. This is followed by active efforts at locomotion by means of prone progression. Active efforts at upright posture follow, the child standing first with and then without support. Finally, independent walking follows, on average, between 11 and 14 months of age. Initial efforts at walking are rather 'stiff-legged' and 'flat-footed', with a wide base of support and the arms outstretched for balance. Once walking is initiated, proficiency in this basic motor skill develops gradually so that the adult walking pattern is established by about 4 years of age in most children.

With improvement in walking, the acquisition of proficiency in a variety of fundamental movement patterns is a major developmental task of the next 6 to 7 years. These include running, jumping, hopping, skipping, throwing, kicking, striking, catching, and others. Development of these basic movement patterns progresses rapidly during early childhood and continues into middle childhood. Ages at which 60 per cent of boys and girls are able to perform at specific developmental levels of several fundamental motor tasks vary from task to task. Note that stages *per se* are arbitrary designations superimposed for convenience of study on a continuous developmental process. Sex differences are small, with the exception of throwing and kicking (earlier in boys) and hopping and skipping (earlier in girls). On average, mature movement patterns of most basic motor skills are developed by 6 to 7 years of age. Subsequently, they are refined through practice and instruction (e.g. physical education, youth sports programmes), and integrated into more complex movement sequences such as those required for specific games and sports. The quality and quantity of performance thus improve.

The distinction between movement pattern and skill should be made. The *pattern* is the basic movements involved in the performance of a particular task, for example, arm, trunk and leg actions in overhand throwing or vertical jumping. *Skill* emphasizes the accuracy, precision and economy of the performance. Pattern is a more general concept, while skill is

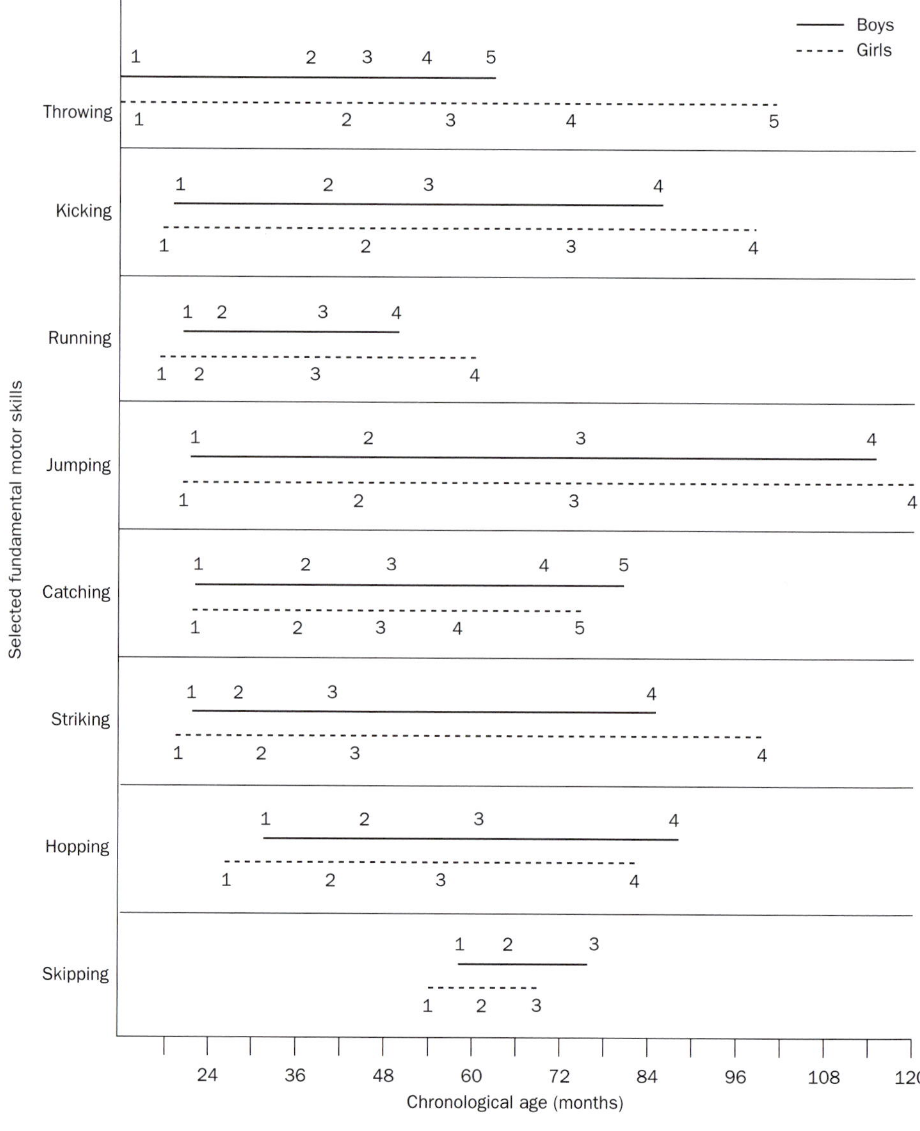


**Ages at which 60% of children were able to perform at specific developmental levels for several fundamental motor skills. Stage 1 indicates an immature movement pattern: stages 4/5 indicate the mature pattern. Stages 2 and 3/4 are intermediate levels in the development of the specific movement pattern. Redrawn after Seefeldt and Haubenstricker (1982).**

more specific. Thus, almost all young children can perform basic movement patterns, but levels of skill vary considerably.

## Motor performance

Variation in skill is evident, in part, in the outcomes of performance, i.e., the distance jumped, time elapsed in a dash, the distance that a ball is thrown, and other similar measures of the attained levels of performance. In addition to skill, motor performances are influenced by growth and maturation *per se*, i.e., changes in body-size, build and composition and differential timing of the growth spurt and sexual maturation; experience, instruction and practice; and motivation to perform, among others.

The outcome of performance is generally emphasized from 5 or 6 years of age through adolescence, and occasionally into young adulthood. A variety of specific tasks, many of which incorporate fundamental movement patterns, are used to document levels of motor performance. The motor tests are done in a standardized manner and under standardized conditions. Many test batteries include tests of jumping, throwing, running and strength. Tests of balance and flexibility are also often included.

Jumping tasks require motor co-ordination and power to project the body horizontally forward in the standing long jump or vertically in the vertical jump and reach. Throwing tasks require co-ordination and power in projecting an object, most often in the form of a ball thrown for distance. Running tests are more variable. The dashes or sprints are tests of speed that require the power and co-ordination to move the body as rapidly as possible over a specified distance. Shuttle runs include an element of agility in addition to speed, i.e., the ability to rapidly change the direction while running.

Muscular strength (**5.8**) is an essential component of most motor tasks. Strength is an expression of muscular force, or the individual's capacity to develop tension against an external resistance. There are several types of strength. *Isometric* or static strength is the force exerted against an external resistance without any change in muscle length, for example, grip-strength, and pulling-strength of the arm. *Explosive* strength or power is the ability of the muscles to release maximal force in the shortest possible time, for example, standing-, long- and vertical jumps. Dynamic strength is the force generated by repetitive contractions of muscles, for example, pull-ups, push-ups. In contrast, muscular endurance is the ability to repeat or maintain muscular contractions over time, for example, flexed arm hang.

The ages 5 to 8 years appear to be a transitional period in the development of motor performance and strength. The basic movement patterns reach mature form at this time, but there is considerable variation among children. In addition, the application of these movement patterns to specific tests must be learned and/or practised. Hence, variation in the performance of specific motor and strength tests at these ages is expected.

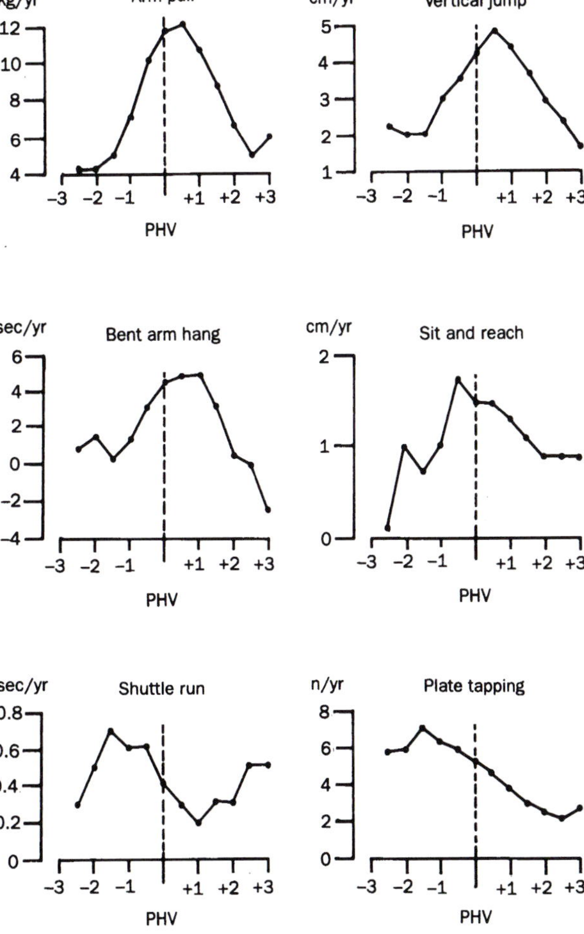


**Median velocities for several strength and performance tests aligned on peak height velocity (PHV) in adolescent boys. Drawn from data reported by Beunen et al. (1988).**

Motor performances, on average, increase with age. Some skills increase considerably between 5 and 8 years and then improve more gradually through childhood (e.g. running speed), while others show gradual, steady improvement from 5 years through childhood (e.g. jumping, throwing, static strength). Sex differences in performance are relatively small. Boys tend to perform better, on average, in tasks requiring strength, speed and power, while girls tend to perform better, on average, in tasks requiring balance.

Motor performances of boys continue to increase in a linear manner through adolescence, while those of girls are more variable. Boys show well-defined adolescent growth spurts in strength, power and speed. Spurts in strength and power tend to occur, on average, after peak height velocity (PHV), while spurts in speed tasks tend to occur, on average, before PHV. Corresponding longitudinal data for girls are not extensive. With the exception of static strength, the performances of girls do not show clearly defined adolescent spurts. When viewed cross-sectionally, strength increases gradually with age, but not to the same extent as for boys, while other performances reach a plateau, on average, at 13 to 15 years of age, or even decline. Thus, sex differences in motor performance become marked during adolescence, such that few girls perform as well as average boys in many strength, power and speed tasks in later adolescence.

An exception to the preceding trends is flexibility. A commonly used test is the sit-and-reach, which measures the flexibility of the lower back, hip and upper thighs. Mean performances of boys on this test are rather stable from 5 to 8 years, then gradually decline with age reaching a nadir at 12–13 years, and then increase through late adolescence. In girls, mean performances are stable from 5 to 11 years, increase to about 14–15 years, and then appear to reach a plateau. At all ages, girls are more flexible than boys, and it appears that the sex difference is greatest during the adolescent spurt. Boys show an adolescent spurt in the sit-and-reach which occurs, on average, before PHV, while corresponding data are not available for girls.

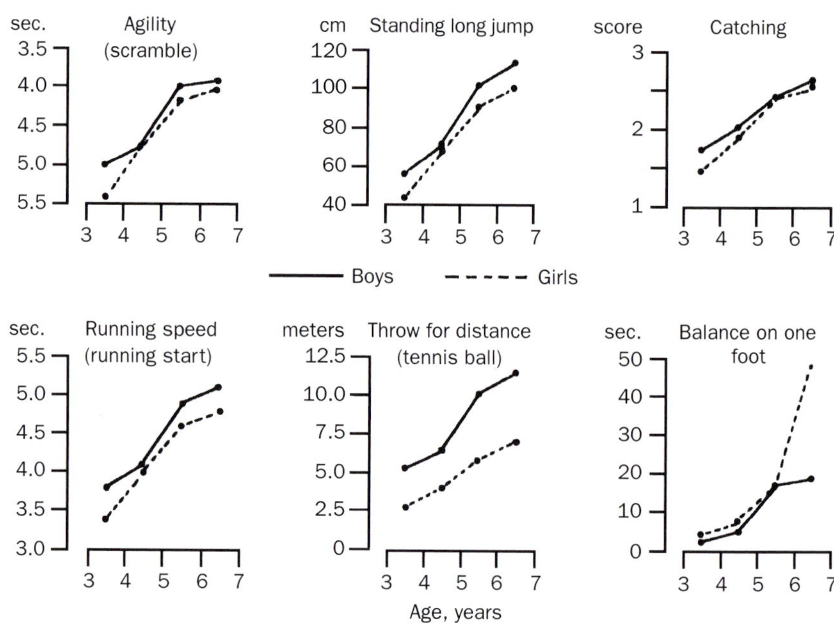


**Mean levels of motor performance for several tasks in children 3 through 6 years of age. Drawn from cross-sectional data reported by Morris et al. (1982).**

## Influencing factors

Biologically and environmentally related factors interact to influence motor development during infancy and childhood in a dynamic manner. In addition to genotype, body-size, build and composition *per se*, and changes associated with growth influence development. The influence of growth *per se* on motor development, however, has not been systematically evaluated.

Environmental conditions, many of which are themselves interrelated, include rearing style, birth order, number of children in the family, sibling and peer interactions, opportunity for movement and play, socio-economic status, and nutritional status, among others. The interaction of biological and environmental factors is cumulative. As the child grows, size and proportions change, and as the child develops in the motor sphere, levels of skill change. These in turn influence the nature of subsequent interactions between the child and his/her environments.

Planned instructional programmes can enhance the development of the fundamental motor skills of preschool children. Guided instruction by specialists or trained parents, appropriate task sequences, and adequate time and opportunity for practice are essential components of successful programmes at young ages. Body-size, build and composition are important factors that affect motor performance and strength during childhood and adolescence. The extent of relationships, however, vary with the tasks and age. Static strength is related to body-size and especially fat-free mass. For other measures, there is a need to distinguish between tasks in which the body is projected (jumps, dashes) or supported (flexed arm hang), and tasks in which an object is projected (throws). Body-weight and fatness tend to exert a negative influence upon the former, while they generally do not significantly affect the latter. Individual variation in the timing, duration and intensity of the adolescent growth spurt adds another dimension to the interrelationships between size, build and composition on one hand, and motor performance on the other.

Habitual physical activity may influence proficiency in motor tasks, but results vary among performance items and between cross-sectional and longitudinal studies. Social and motivational factors are additional considerations, but these have not, in general, received extensive, systematic study during childhood and adolescence. Severe malnutrition early in infancy and early childhood is associated with motor deficits later in childhood and adolescence, while the small body-size and reduced muscle mass associated with chronic mild-to-moderate protein-energy malnutrition is related to reduced levels of strength and motor performance during childhood and adolescence, and probably into adulthood.

*Robert M. Malina*

See also 'Neurological development' (4.6), 'Morphology' (5.5), 'Human skeletal muscle across the lifespan' (5.8), 'Physical activity and training for sport as factors affecting growth and development' (5.12) and 'Adulthood and developmental maturity' (5.17)

# Language development

Development of language is a complex process involving phenotypic initiation and environmental enhancement of auditory processing, neuromuscular co-ordination (vocal or gestural), abstract concept formation, syntactic rule formulation, and lexicon acquisition. This task of synthesis and co-ordination is accomplished by the age of 3 to 4 years in most humans, and the sequence of developmental stages is nearly identical in all languages including gestural languages used by hearing-impaired people. Chomsky introduced the concept of a 'universal grammar' in the 1960s to explain how language is acquired early in life without formal instruction or conscious effort; a 'mental organ' unique to humans was hypothesized. Although Chomsky's views are controversial, the 'innateness' hypothesis is now generally accepted; i.e., the human brain is endowed with the capacity, through the mechanism of genetic inheritance, to develop grammar-based language if the individual is exposed to a communicative model before the 'critical age' cut-off of about 6 years.

The human sensory modality with the most highly myelinated (matured) nerves at birth is the sense of hearing; this is an indication of how important reception of sound is to the well-being of the neonate. Research has demonstrated that infants can detect differences such as mother's voice versus another female voice, and infant-directed versus adult-directed speech, in addition to at least four patterns of acoustic signatures associated with varieties of emotional tone (prohibition, approval, attention, and comfort) in care-giver vocalizations. In addition, newborn infants can imitate facial expressions they observe, indicating the importance of socially relevant stimuli. Studies of the interactive nature of mother/infant communications show complex contributions from both individuals, indicating that rapid turn-taking is intrinsic and probably related to bonding. Thus, the pattern of contingent and responsive signalling is established by at least the second month of life, setting the stage for conversational, or 'give-and-take', communicative patterns throughout life.

The first formal stage of language acquisition, called *babbling*, is evident at about 6 months. The infant utters a variety of phonemic sounds, some of which are not in the language of its caretakers; these gradually conform to the phonemes and intonation contours to which the child is regularly exposed. During this same period infants, who could at first detect auditory phonemic differences that range through all languages, gradually lose that flexibility and become sensitive only to the sound segments of their native language. Deaf infants make unsystematic, random sounds during this stage, but begin to use simple *gestures* in a repetitive, systematic manner if exposed to caretaker sign language. Babbling thus appears to be a human's first consciously controlled, potentially communicative segments produced in the absence of a strong limbic/emotional (e.g. pain-cry) need-satisfaction motivation. It seems likely that human infants are born with the ability and motivation to discover segments underlying linguistic meaning, whether vocal or gestural, and to experiment with their production as an intrinsically satisfying act.

At some time between the first and second year of life, the child begins using single words in apparently referential or meaningful ways. At first the stimulus must be immediate and simple (for example, '*dada*' when greeting), but soon words take on command ('*up!*') and descriptive ('*up*' while pointing) and interrogative ('*up?*') functions. This has been termed the *holophrastic* stage. The child's lexicon grows and then, during the second year, two-word 'sentences' begin to appear, but no syntax or morphological markers are used yet. This stage is superficially similar to the signed or symbol/token-based communications produced by chimpanzees, orang-utans and gorillas in the primate language-learning studies conducted since the 1930s. However, the difference between human toddler and non-human primate sentence production is quite apparent during the next stage of human language development. In the *telegraphic speech stage*, children begin linking several content words (without small function words such as *a*, *to*, *is*, *the*) in a manner which suggests that they have acquired rules of syntax such as hierarchy and structure. They produce fully meaningful and complex sentences which are not copies of anything they have heard in adult speech; examples include, 'Where block?', 'Tommy got car!', and 'Me going now.' One very talented pygmy chimpanzee named Kanzi mastered over 100 word meanings (lexicon keyboard symbols) and showed considerable ability to comprehend spoken communication by humans, but showed extremely limited grammatical rule production after several years of very enriched environment during his early life. Contrast that with the human child's acquisition of an extensive mental dictionary and sentence structure by the age of 3 years without formal training, or even much assistance in deciphering word or phrase boundaries in their communicative environment (where disconnected phrases, inconsistent stops and starts, and long runs of conversation are the norm).

Children have been shown to form sentences on the basis of *syntactic* categories and relations (verb, noun, etc.) rather than by *semantic* roles (agent, theme, etc.). This is evident in languages such as Russian or Italian in which verbs are inflected for number and person in order to agree with the subject. In those cases, more so than in English, it is clear that syntactic relationships must be mastered in order to produce even simple sentences. By the age of 3 to 4 years, children have learned the rules (but not all the exceptions) which govern the use of inflectional morphemes ('-ed', '-ing' etc.), grammatical morphemes (function words like 'to' or 'at'), prepositions, pronouns, possessive and plural formations; i.e., they communicate very much like adults in regard to grammatical sophistication.

Theories of language acquisition have included *imitation* and *reinforcement* paradigms, but these have been shown to be inadequate in explaining the mistakes children make such as:

1. Over-generalization of categories ('*dog*' for all four-legged creatures).
2. Under-generalization of categories ('*bird*' for the family parakeet only).
3. Failure to notice irregular forms of nouns or verbs used by adults regularly (use of 'go-ed', rather than 'went', for the past tense of 'go').
4. Inability to perceive subtle corrections enough to improve their communications (reciting '*She go in*' in response to adult's prompt of '*She goes in*').

These mistakes are eventually corrected, not by tutelage but by the passage of time and neurological maturation. Language development is not a simple stimulus/response learning style, but a complex cognitive process in which inherent predispositions are activated by 'good enough' environmental conditions.

Cognitive science has contributed a great deal to this linguistics subject by hypothesizing how abstract concepts contribute to the language skill. Multimodal sensory input (hearing, seeing, etc.) must be converted into modality-neutral abstractions (state of being, condition, judgment, etc.), before the lexicon and grammar capacities can be activated prior to actual speech/gesture production. For example, an assertive toddler who sees and feels her favourite toy grabbed away by another child may exclaim 'Mine!'; her visual and tactile data are converted to the concepts of 'possession lost' and 'deprivation', whereupon she is *compelled to communicate* with both grammatical word and prosodic emphasis expression (rather than observing with no reaction, or reacting with no impulse to communicate). In addition, language development studies have shown that children do not use past tense forms (such as *worked* or *played* ) until they have mastered the abstract concept of time in which there is past, present, and future. The 'innateness' and 'critical age' hypotheses appear valid but are far from being adequately explained in terms of neurological, cognitive, or anatomical formulations. Observations and testing of children and adults who do not exhibit adequate language skills have formed the basis for acceptance of the ideas that language acquisition ability is genetically determined but that an appropriate environment of language examples in early life is necessary to stimulate this biological predisposition.

*Marilee Monnot*

See also 'Neurological development' (**4.6**), 'Post-natal craniofacial growth' (**5.9**), 'Cognitive development' (**6.1**) and 'Psychosocial factors in growth and development' (**6.4**)

# Psychosocial factors in growth and development

Any living being is part of a complex relational network involving each and all of the physical, chemical and biological agencies, factors and processes constituting its outer environment and interacting at all times with its own internal constituents. Because this implies a continuous modification of living conditions, the basic properties of life include the capability to perceive relevant stimuli, that is, modifications likely to affect an organism's internal equilibrium, survival or reproductive potential, whether positively or negatively, and to react appropriately, and in such a way to maintain the variability of the internal environment within tolerable limits (*homeostasis*).

The amount of stimuli that an organism is able to generate and perceive is clearly related to the complexity of its nervous system and relational processes, and is thus highest in social mammals. In the human species, the interaction of an individual with other individuals and with the specific structures and factors of his or her social environment results in an immense variety of stimuli, the processing of which is influenced by the nature of the stimulus, as well as by the individual's external conditioning and genetic constitution, and usually involves an emotional response.

Emotion is a concept that has long been ignored in human biology, essentially because it has to deal with feelings that are highly individual and defy any attempt at objective definition. Emotion has only been studied in terms of *behaviours* through which it is expressed in humans or other higher mammals. Yet, emotions have an experiential component that is an essential part of human nature and plays a very relevant role in growth and development, in social relations, and more generally in mental and physical well-being.

## Emotional development

In general biological terms, and as first pointed out by Darwin (1872) in a classic study, emotions are central to communication and a vital asset for the homeostasis and survival of many animal species. Some basic emotions have a high adaptive relevance, as shown by the fact that their expression – representing a fundamental *non-verbal language* – may follow

**Facial expressions associated with emotions, such as joy, are recognized in people of different cultures: (right) a European infant; (above) children in Papua New Guinea. Photographs by Stanley Ulijaszek.**

universal patterns across species and populations. Facial expressions associated with emotions such as joy, surprise, sadness, disgust, anger, contempt, and fear, are recognized by people of different cultures, literate and non-literate alike, and are most probably innate, as confirmed by observations on born-blind subjects.

Because of their adaptive value, emotions must have evolved relatively early. In fact, their expression is rooted in old parts of the brain: stem, mid-brain, thalamus, amygdala, hypothalamus. These lie between the spinal cord, the oldest part, mainly related to movement, and the neocortex, the most recent part, mainly related to higher cognitive functions, such as perception and thought. As a result, an *emotional activation* can travel on subcortical routes and, therefore, much faster, and produce an almost immediate response to the stimulus. Also, because it does not necessarily require cognitive mediation, the activation can elude cognitive recognition and processing. This implies that the feelings involved, let alone their actual origins, are sometimes not easily identified nor described by the subject. Emotional states are usually recognized from their expressions, being associated with an arousal of the autonomic nervous system and a consequent activation of visceral organs, as may be expressed by such well-known symptoms as palpitation, flushing or paleness, heavy breathing, freezing, nausea, sweating, and salivation.

The threshold of emotional activation and the extent and form of its expressions vary to some degree among individuals. The perception and processing of relevant stimuli is in fact highly influenced by the genetic make-up of the individual as well as by specific cultural background and experiences. *Cognitive processes*, largely a function of cultural conditioning and resulting perceptive filters, clearly play a fundamental, though not exclusive role. Their conditioning impact is mediated by genetic factors, thereby resulting in individually specific patterns of response to stimuli. Identical twins exposed to highly differing environmental pressures in early infancy can diverge in their emotional development and response patterns. On the other hand, siblings (or for that reason unrelated individuals) exposed to highly similar environmental pressures can also differ strikingly in their response patterns.

The ability to express the various emotions appears gradually in the course of infant development, reflecting mainly genetic effects (**3.1**) in the early stages, and the growing influence of psychosocial conditioning and maturation in later ones. The basic emotion of interest is already expressed immediately after birth, and so is the ability to execute the muscular movements involved in the expression of other emotions. Disgust and early forms of sorrow and joy are also present or appear soon after, while sadness, anger and surprise appear by the age of 3 months, when the child starts to develop early discriminative abilities. As cognitive development proceeds, new emotions appear, such as fear (7–9 months) and then shame, providing the basis for the emergence of the sense of guilt and personal responsibility.

The emotional expressions shown by the mother or other caretaking figure represent a major source of information and guidance for the infant, who is able to discriminate expressive or expressionless faces or attitudes, as well as positive or negative emotions, by the age of 3 months. This plays a fundamental role in mother–infant interactions and more generally in the infant's relation to others, in its temperament, attitude to life and social interaction, and adult personality development. Various lines of thought and research concur in indicating that an insufficient amount of emotional stimulation and/or an excessive amount of negative stimulation in the early stages of life are likely to result in a difficult temperament and personality development, with a higher risk of mental health troubles.

In recent years, as the role of emotional factors in mental as well as physical health (*psychosomatics*) started to be more overtly recognized, the term *psychosocial* appeared in the medical literature, thereby pointing to the relevance of both social stimuli and the individual mediation of their mental processing in eliciting the emotional response and its behavioural, physiological, or pathological correlates. The notion of psychosocial factors was specifically introduced in growth studies in relation to the influence that the emotional responses they elicit may have on *physical* growth processes (**9.4**), as well as mental ones. This is still a somewhat controversial issue.

*A comparison of mean scores of control, high-risk and child abuse groups on Parent/Partner Empathy Measure and its factorial components*

| | Total empathy | Cognitive F1 | Affect expression F2 | Partner empathy F3 | Empathic distress F4 |
|---|---|---|---|---|---|
| *Mean scores* | | | | | |
| Control | 131.7 | 42.3 | 35.2 | 30.2 | 21.8 |
| Clinic | 124.8 | 38.8 | 34.2 | 28.8 | 21.6 |
| Abuse | 122.8 | 39.2 | 34.1 | 27.1 | 20.2 |
| | $p < .001$ | $p < .001$ | $p < .05$ | $p < .05$ | $p < .05$ |
| *Planned comparisons* | | | | | |
| Control versus Clinic | $t = 3.46$ | $t = 4.4$ | $t = 1.99$ | $t = 3.38$ | NS |
| | $p < .01$ | $p < .0001$ | $p < .05$ | $p < .001$ | |
| Control versus Abuse | $t = 4.17$ | $t = 3.8$ | $t = 1.97$ | $t = 4.48$ | $t = 2.18$ |
| | $p < .0001$ | $p < .0001$ | $p < .05$ | $p < .0001$ | $p < .05$ |
| Clinic versus Abuse | NS | NS | NS | NS | $t = 1.5$ |
| | | | | | $p = .14$ |

From Feshbach N.D. (1989). The construct of empathy and the phenomenon of physical maltreatment of children. In *Child Abuse and Neglect* (eds D. Cicchetti & V. Carlson) pp 349–73. Cambridge: Cambridge University Press.

## Psychosocial deprivation and growth problems

Although the psychological needs of an infant have always been the object of some concern, it is only in recent decades that scientifically oriented minds have gradually come to recognize the importance that relational contact and stimulation may have on the infant's growth and survival. The first line of evidence came from infants confined in institutions. Their very high morbidity and mortality had been known for centuries, and even acknowledged at the 1915 meeting of the American Pediatric Society, but became a subject for scientific inquiry only after the 1940s, when the pediatrician Bakwin first suggested that the failure to thrive observed in infants hospitalized for relatively long periods of time might be due to loneliness and emotional deprivation.

The notion of what became known as 'hospitalism' was then extended by psychiatrist Spitz to other forms of confinement. In a classic long-term investigation, he compared infants kept in a foundling home to infants kept in the nursery of a penal institution and to controls. Although housing, hygienic and nutritional conditions were strictly comparable and 'impeccable' in the two institutions, the foundling home children exhibited a continuous decrease in their Developmental Quotient and extremely high susceptibility to infections, which Spitz attributed to their being much less exposed to social contacts than nursery children, who were cared for by their mothers and developed normally. Follow-up examination indicated a mortality-rate of over 37 per cent in 2 years and serious retardation of both mental and physical development in survivors. In another classic study, nutritionist Elsie Widdowson compared the children of two German war orphanages who, although receiving exactly the same measured amount of food, were found to show a highly different weight-gain, apparently as a result of different relational and emotional conditions.

Despite criticism levelled at the details of these studies, their findings received confirmation from various sources, including studies of children living in their own families, so that the notion of confinement was extended into that of deprivation. Non-organic failure to thrive came to be recognized as a real syndrome, variously called 'hospitalism', 'environmental retardation', 'maternal' or more generally 'emotional' or 'psychosocial deprivation', and later 'psychosocial short stature', 'non-organic failure to thrive', or other such names. Scepticism remains, but since the late 1950s and well into the 1990s, thousands of children with this condition have been reported in clinical case-reports, or in follow-up, retrospective and other studies on hospital records, providing substantial evidence that physical growth retardation can frequently not be attributed to any known organic factor and is instead associated with poor relational stimulation and emotional troubles. The condition has been found to occur in practically all industrialized countries, in all social classes, and at all ages through adolescence, perhaps more frequently in males than females.

The analysis of family environment usually discloses a history of serious emotional troubles, stemming from child neglect and abuse, parental immaturity, alcoholism, and the like. Clearly, these situations are likely to imply child malnutrition, thus providing the most obvious organic explanation for the growth retardation. This has indeed been found in a number of cases, and some growth-retarded children have been seen to recover when adequately nourished. But in many other cases the condition cannot be attributed to undereating, although eating behaviour is often seriously disturbed. These children can do such things as eat from garbage cans, drink from toilet bowls, or gorge to the point of vomiting as a way of getting the attention they lack, and their frequent histories of steatorrhea and large intakes with no excess weight-gain may rather reflect malabsorption. Interestingly, somewhat similar symptoms have been described in monkeys kept in isolation.

A deficiency in growth hormone (GH) (**4.2**, **7.6**) serum levels, first described by Powell and co-workers in the late 1960s, can be observed in perhaps 50 per cent of cases, all other laboratory findings being either normal or markedly variable from patient to patient. It has been suggested that the syndrome may consist of at least two subtypes, one mainly characterized by malnourishment and occurring in children under 2 years of age, and one mainly characterized by GH deficiency (**7.7**) and occurring in older children. The situation is, however, far from clear-cut, since normal growth does not necessarily follow when either correction of dietary inadequacies has been made (in the former case), or following GH administration, in the latter case.

There is now general consensus that the primary etiology of this condition consists of the child's lack of stimulation or disturbed relationship with the mother or other caretaker, and that the only effective therapy is to remove the child from the emotionally disturbed environment. When this is done, GH levels, if altered, rapidly return to normal, other troubles tend to disappear, and regular growth is resumed.

These observations stress the importance of psychosocial factors in child growth and development, indicating that emotional troubles may reach a point where they translate into overt physical manifestations, let alone mental and behavioural ones. This concurs with similar findings from the area of psychosomatics in indicating that the distinction between 'physical' and 'mental' appears to be much less clear-cut than previously believed; the two can interact to some extent, possibly via psychoendocrine, immunological or other agencies. Close inspection, however, frequently indicates these biological correlates to be much less the cause of the trouble than an effect of it.

*Paolo Parisi*

See also 'Cognitive development' (**6.1**) and 'Growth and psychosocial stress' (**9.4**)

# Nutrition and cognitive development

Eating is too important for the basic ingestive responses to be acquired. Therefore, newborn infants will suck and swallow fluids brought to their lips. Although such sucking and swallowing responses seem deceptively simple they require a complex set of controls to ensure the scaling of lips around the nipple, the maintenance of breathing and the separation of fluid from air to ensure that milk goes down the esophagus to the stomach and only air goes to the lungs. All these responses are innate but they can be fine-tuned during maturation.

Young school-children in south India perform better if they are adequately fed. Photograph by Stanley Ulijaszek.

As the baby develops it needs to learn to recognize appropriate foodstuffs. Such learning goes on for many years but against a background of the child's instinctively conservative approach to novelty in food tastes and textures. This is a sound evolutionary strategy that also is seen in the infants of other omnivores, such as rats. Of the tastes that are acquired in a child's early years two are of special importance in light of current views on healthy eating. Babies do not have a specific liking for salt (in fact the taste receptors that are sensitive to sodium ions mature quite late) but if fed exclusively on breast milk, which contains low levels of salt, until 6 months they develop a liking for saltiness in this kind of range. By 12 months of age a child shows dislike for unsalted food if she or he is used to eating foods with added salt. But diets with high levels of salt are physiologically unnecessary – excess salt is excreted in urine – and can lead to increased risk of cardiovascular problems in later life. A liking for sweetness may be innate or acquired by babies through exposure to foods that contain sugar. Once a baby develops a liking for sweeter foods this preference persists through infancy and probably to adulthood.

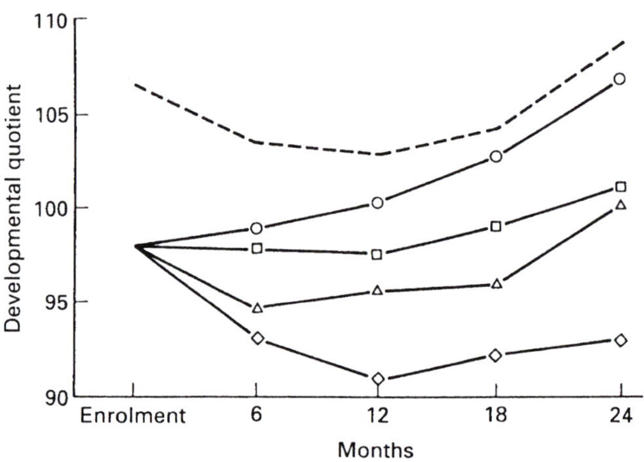


**Mean developmental quotients of stunted children adjusted for initial age and score, compared with non-stunted children adjusted for age only. Results of intervention over 2 years. ——, non-stunted; stunted: □——□, stimulated; △——△, supplemented; ○——○, stimulated and supplemented; ◇——◇, non-stimulated and unsupplemented controls. From Grantham-McGregor et al. Nutritional supplementation, psychosocial stimulation and development of stunted children: the Jamaican study. *Lancet* 338:1–5. (1991)**

Brain development can be affected by chronic effects of diet. However, the brain is remarkably robust nutritionally and it obtains preferential access to nutrients even in conditions of semi-starvation. There are of course some rare metabolic disorders where inappropriate metabolic breakdown products can accumulate and damage the developing brain. This can lead to neurological disorders and learning difficulties. Also, severe energy-protein deficiency in a mother (during gestation and lactation) can harm brain development in the fetus or breast-fed baby. This may manifest itself in slow intellectual development. However, it is difficult to disentangle the effects of dietary deficiency from the effects of economic and social disadvantage.

A contemporary issue has arisen with the ability of neonatal units to keep premature or seriously low-birth-weight babies alive. It seems that some fatty acids that are present in breast milk may be needed in the infant milk formula fed to such babies for the brain to develop to allow full intellectual potential.

Glucose is normally the brain's source of energy, although under starvation conditions it can adapt to using ketone bodies derived from the partial breakdown of fats. Brain cells, unlike most other body cells, do not need insulin to take up glucose from the blood and they always have preferential access to the available glucose. Although there is some positive evidence that giving glucose to elderly people might improve their memory it is more likely that the high sugar intake of some children has an acute sedative effect – this could lead to difficulty in clear thinking so that the child becomes restless when faced with an intellectual problem.

Are some foods or diets good or bad for brain development? Although there are persistent popular fads and fancies about this there is little evidence to support such views. For instance some people believe that eating fish aids brain development. This is based on the argument that both the brain and fish are rich in phosphorus. But the compounds in fish that contain phosphorus – phospholipids, phosphoproteins and nucleic acids – will be broken down by digestive processes and will be transported to most cells of the body not just the brain. What the brain sees of such breakdown products and what it might do with them is quite unpredictable.

Another widespread belief is that some children develop behavioural problems or mental distress because they are allergic to specific natural proteins or are intolerant to some food additives such as tartrazine. The evidence strongly suggests that intolerance to food additives has a much lower incidence than the behavioural problems attributed to them. Also, allergies to natural foods are uncommon and may not last more than a few months.

Most parents want to ensure that their child will develop their full intellectual potential. Their hopes of an easy way to benefit their children seemed to be answered by some recent studies that appeared to show that dietary supplements with one or more unspecified vitamins and minerals (micronutrients) helped some children gain better non-verbal intelligence test scores (IQ). The children who benefited were alleged to be deficient in these nutrients, which suggests that lack of these micronutrients led to impaired intellectual development. There are several problems with these studies since the evidence that the children were nutritionally deficient came from children's dietary records and these are likely to be confounded by factors that influence verbal IQ scores. Moreover, the children who consumed fewer micronutrients also had diets with lower energy intakes. There is no plausible mechanism yet proposed that could explain how moderate micronutrient deficiency could affect non-verbal IQ.

Childhood obesity (**1.11**) is the most important dietary problem, both numerically and because it predisposes the individual to develop body-weight problems in adulthood. Its impact on a child's self-esteem can be profound because fatness is usually denigrated by other children. Of less importance than obesity, but much more widely discussed in the popular press, is the misnamed disorder anorexia nervosa. This disorder does not usually involve loss of appetite, which *anorexia* implies, but does involve self-starvation. The incidence of this disorder is low; only 1 to 2 per cent of girls develop it and many fewer boys do. Because anorexia nervosa does not usually occur until puberty (when brain development has already taken place) the effects of starvation on the brain are less serious than they would be if the disorder developed earlier. There is no consensus either on the origins of the disorder or its treatment.

*Robin Stevens*

See also 'Infant-feeding and growth' (**9.1**) and 'Nutrition' (**9.2**)

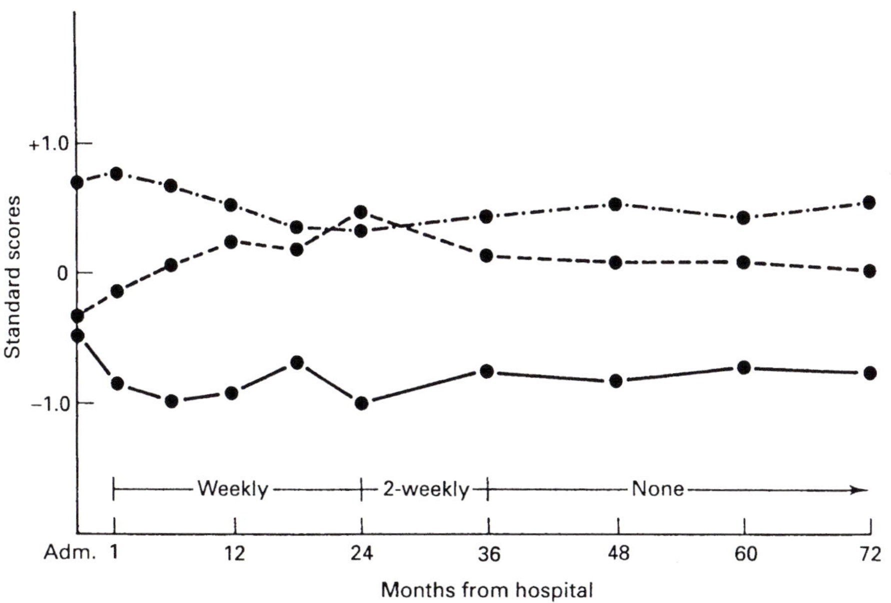


**Developmental quotient and IQ scores transformed to standard scores for each test session for Jamaican children, from admission to hospital to 72 months following discharge. —·●·—, 20 adequately nourished; – – ● – –, 16 non-intervened malnourished; — ● —, 18 intervened malnourished children. From Grantham-McGregor et al., 1987.**

6.6

# The development of sexuality

A common everyday view of male sexuality is that sexual feelings are powerful urges which arise during adolescence and that require conscious control. The everyday view of female sexual feelings is rather different. It is often held that adolescent girls seek relationships rather than sex and that female sexual desire requires awakening through male interest. For them, sex comes as a consequence of their relationships. These common sense views neither fit what systematic enquiry shows is the usual pattern of sexual development nor its explanation in terms of physiological changes and social conduct. The origins of human sexual development occur much earlier than puberty.

In the first years of life, children begin to identify themselves as either male or female. They view the world from one or other of these identities and model their behaviour on what they see as appropriate to their gender. They perceive that bodies come in two kinds, and develop a keen interest in both the variety within their own kind and the differences with the other. Games in which children explore each other's bodies are often seen as sexual by adults. However, it may be more accurate to see adult reactions to the ways in which children touch and expose parts of their bodies as one of the ways in which children begin to learn what is sexual. Children learn that special rules apply to their genitals, and that they should be hidden in most situations. These rules are somewhat different for boys and girls. In early childhood, mothers may be more punitive in their treatment of girls who touch their genitals than they are of boys. A boy's genitals are more likely to be named by parents than are girl's genitals. By adolescence, boys are likely to have many more words for their genitals than girls. Long before they reach puberty, boys and girls have a wide knowledge of conventions in their families and the wider social world that govern attitudes toward their bodies. They also understand intergenerational boundaries – that some things are better not done in front of adults but may be shared with their peers. Inappropriate or abusive behaviour from adults breaches these boundaries. Adults forcing children to take part in overtly adult sexual activities may have lasting effects on the development of their sexuality.

Both sexes are acutely aware of the changes in their bodies and those of their peers at puberty (**1.10**). Many boys say that they would like to be taller, more muscular and heavier (**5.8**). Typically, girls are more dissatisfied with their bodies than boys and would like to be thinner. A majority of girls diet in attempts to reduce their weight. For girls there are links between self-esteem and perceptions of physical attractiveness, and for boys with body fitness and strength. However, there are some indications that physical development in girls is related to making and sustaining relationships with boys, at least in Western societies, though age and parental control may be much more important.

Puberty, either in terms of bodily changes or hormonal levels, is neither necessary nor sufficient for the initiation of sexual intercourse. While in some populations links have been found in boys between post-puberty increases in the level of the hormone testosterone and sexual attractiveness and the beginning of sexual intercourse, in other populations a significant number of boys were found to have had intercourse *before* they had reached puberty. Varied results have also been found in different female populations. In some, social control seemed to be the key issue related to sexual initiation while, in others, the physical changes seem more important. Overall, studies of the beginning of heterosexual sex show a complex process in which the social context, individual difference, family influences, young people's attitudes, physical development and hormonal changes all have at least some part to play, but the ways in which the var-

**Edvard Munch 'Puberty' (1895). Courtesy of the Nasjongalleriet, Oslo.**

ious factors mix together may vary considerably from culture to culture even within America and Western Europe where most of the research has been carried out.

For young women, adolescence is the time when the knowledge, attitudes and feelings about their own bodies are beginning to be used in social encounters. Young girls are expected to seek love and not sex, but what they do may not always follow expectations. A girl must tread a narrow path (at least in English culture) between being seen as a 'slag' or a 'drag'. A woman's feelings are supposed to be a spontaneous response to the situation, so preparing herself by arranging contraception beforehand may indicate that she is 'easy'. As a result, the practice of safe sex may involve risking losing her reputation. In early teenage years confiding relationships are largely confined to female friends and mothers, but by the late teens male partners often become more important.

Studies suggest that, though fewer young women masturbate than young men, women may start at younger ages. Most young women do not find their early experiences of sexual intercourse very pleasurable. But those who do are much more likely to have had experience of masturbation, as well as having had sexuality represented to them by their mothers as pleasurable. But experience is very variable; for some the first experience of orgasm may be well before puberty while for others only after years of heterosexual intercourse. For some, masturbation, too, may not begin until after years of varied sex with partners. Sexual desire, too, is very variable in its development in relation to age and puberty.

The great majority of boys masturbate to orgasm long before they first have sexual intercourse. When boys begin to make relationships with girls the cultural expectations that shape these are rather different. They often perceive sexual desire as something to be satisfied whenever opportunity allows. Young men, unlike young women, are more likely to first experience sexual intercourse outside an established relationship and often with a partner with whom they do not have sex again. While reported experiences of early sexual intercourse is varied, more men than women describe it as being pleasurable. For most men, as with some women, an important process in the early experience of intercourse is to bring together the early erotic experience of masturbation with that of sex with a partner.

A considerable minority of both young women and men engage in sexual activities with others of the same sex. Others may fantasize about this but not do it. A person's self-identification as homosexual or bisexual does not automatically follow from sexual activity with a same-sex partner. Early in sexual development same-sex activity may be quite frequent, and it does not necessarily lead to homosexual self-identification or interest in same-sex partners later on in life. In cultures which are strongly heterosexual there may be some delay before individuals see themselves as homosexual, and bisexual identity may form even more slowly.

*Martin Richards*

See also 'Assessment of age at menarche' (**1.15**), 'Sexual maturation' (**5.15**), 'Adulthood and developmental maturity' (**5.17**), and 'Teenage pregnancy' (**8.6**)

PART SEVEN

# Clinical growth abnormalities

As auxology is a central theme to the study of normal childhood development, then the study of clinical growth abnormalities forms a central part of the study of pediatric medicine. Most childhood illnesses of any severity or length will affect growth, usually by impairing it. These changes can happen in many different ways. For instance, this may occur by specifically interfering with the normal control of growth (**7.1**), as in many endocrine disorders (**7.7**), or by a more general intrusion onto widespread body functions as may happen in chronic childhood disease. Very specific genetic disorders may only affect a single receptor or enzyme system and yet have profound affects on growth. The growth defect may affect the whole body or be very specific and only involve quite limited regions of the body, such as the head or spine(**7.2**, **7.3**).

## Regional growth disorders

Later in this section we discuss craniofacial and spinal disorders and their management (**7.2**). The underlying causes may be many and varied, ranging from genetic disorders with very precise genetic mutations as may occur leading to distinctly abnormal growth. This is particularly true of craniofacial growth with a group of disorders including Crouzon, Pfeiffer and Apert syndromes. These disorders, collectively termed craniosynostoses, are due to abnormal fusion of certain cranial sutures leading to distortion of the skull and face. Their classification has historically depended on clinical delineation of the phenotypes, but now we are in a position to identify the responsible gene mutation in most cases. In the case of these three conditions different mutations in the gene for Fibroblast Growth Factor Receptor 2 (*FGFR2*) have been found, but in some cases the same mutation is seen in the phenotypically quite distinct Crouzon and Pfeiffer syndromes. Furthermore, some cases of the latter syndrome have shown mutations in the related but distinct *FGFR1* gene. Thus, even with these powerful genetic tools, there are still some areas of uncertainty where phenotype and genotype do not map exactly.

Although not strictly regional disorders, the bone dysplasias (**7.5**) share many aspects in common with the craniosynostoses. They predominantly affect bone growth but in a much more general way. They are also very many in number and have generally depended on a phenotypic classification, taking into account both clinical, auxological and radiological assessments. In some cases we are beginning to identify the genetic abnormalities that

**Opposite: In the dominantly inherited type of achondroplasia, growth at the ends of the long bones and at the base of the skull is stunted. Somerset (aged 11), who has the condition and is the smaller of the two boys, is 2 years older than his brother Sebastian. Courtesy of Douglas Oram.**

underlie the clinical syndromes (**7.6**). Two of the more common bone dysplasias are achondroplasia and hypochondroplasia, in both of which there is predominantly impaired growth of the limbs although the spine is not entirely normal and the skull and facial bones are abnormal in achondroplasia. These two conditions are also due to a mutation in a Fibroblast Growth Factor Receptor gene, but this time *FGFR3*.

No such detailed genetic data relate to the problems of spinal growth and asymmetry that are discussed in (**7.3**). Here we are dependent on clinical assessment and management and, in addition, require the tools of growth assessment discussed by Charles Brook. When seen against the background of the rather more glamorous molecular biology of today's medicine it is easy to undervalue these skills. However, without them the clinical disorders would not be identified or carefully defined, which would make the genetic analyses impossible. In the practice of pediatrics the skills of good growth assessment can not be overstated, and this depends not only on obtaining appropriate and precise measurements, but also on their correct interpretation. The latter requires understanding of the measurements, but also needs appropriate and up-to-date reference data with which comparison may be made.

## Chromosomal disorders

In many ways we can think of the chromosomal disorders as being more severe genetic abnormalities than the single mutations discussed above, and in some cases that is true. In the Prader–Willi syndrome (**7.4**) there may be an associated small chromosomal deletion of part of chromosome 15 which is visible by appropriate cytogenetic techniques, but it seems as if only a single gene is affected. In contrast, in the Turner syndrome (**7.4**), and many other chromosomal disorders such as Down syndrome, there is loss or gain of a whole chromosome or, in some cases of the former, major rearrangements of one X-chromosome. In these circumstances many genes are involved, often in different combinations in different individuals. This makes for a more varied phenotypic outcome. This is particularly exemplified in Turner syndrome where affected girls may show almost no features at one end of the clinical spectrum and yet others may suffer many different abnormalities at the other.

## Endocrine disorders

These disorders may also be genetic or acquired. Disruption of growth hormone synthesis or release may occur as a result of specific gene mutations (**7.6, 3.1**) which by their nature start to exert their effects soon after birth. As fetal growth hormone has little part to play in pregnancy these children are of normal, or near normal, size at birth. Further down the growth hormone axis there is a further genetic disorder that disrupts the actions of growth hormone at its target cells. This condition, which is often referred to as Laron-type dwarfism after the physician who first described it, is due to one of several mutations in the growth hormone receptor gene.

Other endocrine disorders are acquired, and therefore tend not to exert their effects until later in childhood. One of the most insidious is growth hormone insufficiency which may be associated with a difficult birth, often with a breech delivery. It is believed that there may be temporary interruption of the pituitary blood supply leading to death of the growth-hormone-secreting cells and sometimes other cells responsible for secretion of other pituitary hormones. A newer cause of acquired growth hormone deficiency is the use of cranial irradiation as part of the treatment of childhood leukemia and cancer. The growth-hormone-producing cells of the pituitary sensitive to this and growth hormone insufficiency may become apparent after 4 or more years have elapsed.

## Growth and chronic disease

From time to time we have alluded to the effect of chronic childhood disease on growth (**7.8**). In many ways this is becoming an increasing constituency for the growth specialist as more children survive from severe childhood disease, but at a price. The price that they often have to pay in terms of impaired growth is sometimes due to the nature of their original disease. For example, in the case of inflammatory bowel disease (**7.8**) the actual disease process impairs growth, but if it is treated effectively and while there is still growth potential, complete catch-up is likely. On the other hand, growth may be affected more by the treatment than the disease. For example, over-enthusiastic use of steroids in treating asthma may have profound effects on a child's growth. In these situations there is always a need to balance the benefit of the treatment of the primary disorder with potential deleterious effects of the treatment on the child as a whole, including growth.

Much of the treatment of growth disorders is endocrine in nature (**7.9**) and therefore hinges on the use of various hormones or agents that modify their actions. However, these are not the only modalities of treatment. For example, there is great interest at present in methods of lengthening limbs by surgical means. These techniques were largely pioneered in managing children or adults with severe limb inequality. In the past the only option was to shorten the long limb, which would usually mean an undesirable loss of total height. The approach that is now used is to stretch the short limb by applying a carefully stabilized traction force across a region of new bone formation created by dividing the shaft of a long-bone. In this way, as new bone is laid down it is stretched and the overall effect is to lengthen that bone and hence the limb. Many centimetres may be added in this way. The technique is now being used in children with symmetrically short limbs, such as in achondroplasia. It is particularly suitable in this disorder as, because the trunk is of relatively normal size, stretching the legs not only increases total stature, but restores normal body proportions.

Growth disorders are a major contribution to pediatric disease. However, in many instances they are not accorded the level of importance that they deserve. It is true that nobody dies of being short, but it is equally true that many short children are made to be very unhappy by their situation. In looking after children where impaired growth forms part of their disease complex it is very important to remember this and give it appropriate weight. It is also important to remember that growth disorders are not restricted to one particular organ system and that most areas of pediatrics will include some impact on growth. Along with psychological development it is an aspect of a child's life which continually changes throughout childhood and which may be disrupted by a great variety of often unexpected life events.

*Michael Preece*

See also 'Post-natal craniofacial growth' (**5.9**)

# Identification of abnormal growth

The measurement of stature is not difficult and data collected by a trained individual are highly reproducible. If measurements are made at different times by different measurers at poor precision, the data acquired become meaningless.

## Use of height attained

Height measurement is highly sensitive in identifying a case of abnormal growth. In other words, in an extremely tall or short person there is more likelihood of discovering an abnormality of growth than in less extreme cases. The specificity (the likelihood of obtaining a negative test result in a person without disease or disorder) is very poor; in other words, most tall people and short people have nothing wrong with them.

## Screening for growth problems by growth-rate

The calculation of growth-rate requires measurement of the same individual on two different occasions separated by a period of time. The time interval between the two measurements is dictated by the increment in stature which might be expected. This needs considerably to exceed the errors on each of the two measurements: in practice intervals of 3 months may be satisfactory in patients expected to grow more than 7.5 centimetres in one year, but 4 or 6 months are required for patients growing at rates seen in middle childhood.

Screening by height velocity requires the services of the same trained observers at different intervals and a recall system for the subject. The value of height velocity obtained (increment/time) is much more difficult to interpret than the value of a height attained, but the value is much more specific (a person growing at a normal rate is likely to be normal, regardless of height obtained) and it is also highly sensitive (a person growing excessively quickly or slowly needs a diagnosis and treatment).

## Screening for growth disorders: auxology versus biochemistry

The rate at which children grow is determined principally by the amplitude of the growth hormone concentration which is achieved by pituitary secretion. Surrogate markers of growth and growth hormone secretion (indices of bone and collagen turnover, such as osteocalcin or hydroxyproline and insulin-like growth factors and their binding proteins (**4.2**)) correlate well at the extremes of stature or abnormality, but they fail to assist in the marginal situation. Since growth velocity is what defines normality, screening for growth abnormalities is best achieved by this means.

## Screening for growth abnormalities in children of normal stature

The acquisition of a normal stature for age does not exclude the possibility of a growth abnormality. For example, a poor growth-rate is characteristic of patients with untreated brain tumours, of patients with brain tumours which have been treated by radiotherapy, of patients with asthma, renal failure, anorexia and so on. The identification of these abnormalities can only be achieved through the recording of repeated height measurements.

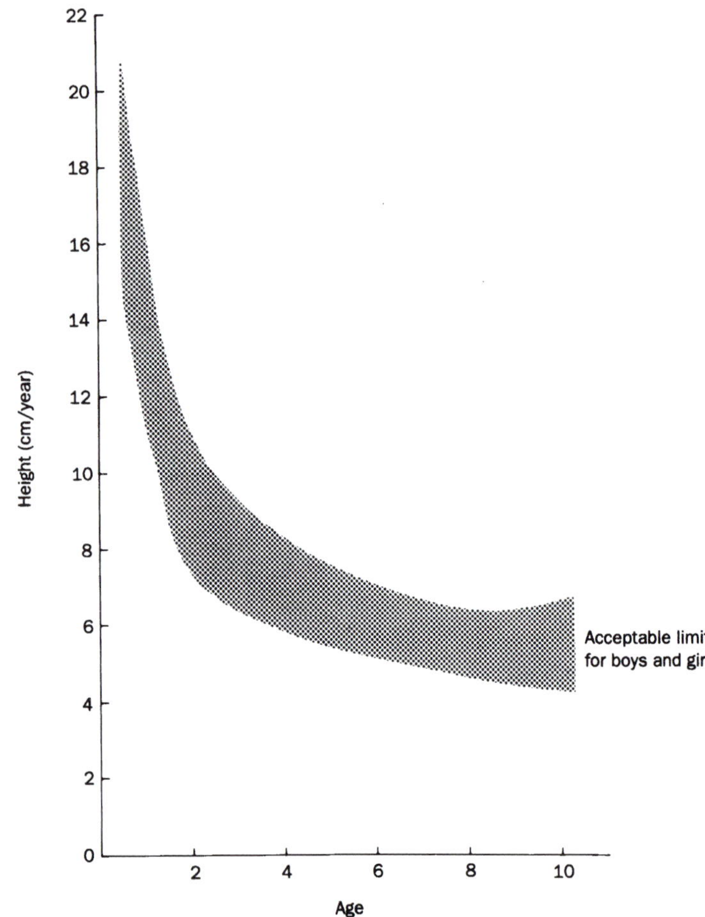


**How to use this chart:**

1. **Measure height on two occasions.**
2. **Plot the rate of growth (cm year) against the age of the child midway between the two measurements, e.g. heights measured at ages 4 and 6 years show a difference of 12cm. Rate is 6cm/year, which should be plotted at age 5**
3. **Seek advice/explanation for rates falling outside the shaded area.**

## The practical way forward

Growth monitoring is desirable within a population to define the health of children. Repeated measurements are useful for defining the normality of growth in individual cases. Screening for growth abnormalities by height attained is cheap and sensitive but not specific. Screening by height velocity requires trained personnel and is thus more expensive, but it has both high sensitivity (a positive test result in a person with disease) and specificity (a negative result excludes disease) (**1.13**).

Screening for growth disorders in individual cases requires repeated measurements of growth and the use of charts.

C.G.D. Brook

See also 'Anthropometry' (**1.4**), 'Biochemical markers' (**1.7**), 'Physical examination' (**1.10**), 'Creation of growth references' (**1.16**), 'The use of growth references' (**1.17**), 'The human growth curve' (**2.4**), 'Endocrine growth disorders' (**7.7**), 'Treatment of growth disorders' (**7.9**), 'Growth patterns associated with new problem complexes' (**13.3**) and 'Treatment of growth disorders: the future' (**13.4**)

# Orthodontic disorders

Orthodontic therapy is often a valuable adjunct to surgery in the treatment of craniofacial anomalies, including those related to disorders of the growth process. The objective of orthodontic treatment is first and foremost to provide the patient with a dentition that functions well and will be stable for a lifetime. Secondarily (but no less important), the goal is to improve facial aesthetics by restoring a culturally acceptable balance to the physiognomy.

The proximate causes of malocclusion may be discrepancies in tooth relationships within the bony alveolus, in the dental occlusion, in the relationship of maxillary to mandibular basal bone or in some combination of these factors. Effective treatment depends upon correct assessment of the etiology. The most frequently used diagnostic aids are casts of the dentition which are mounted in an articulator (simulating the relationship of the jaws) and the cephalometric X-ray. A series of planes and angles is traced over the radiograph and, by reference to normal standards for these planes and angles, the clinician may determine the locus of the problem.

The most enduring improvement is obtained through the application of fixed orthodontic appliances. In this technique, a wire – bent to approximate the desired arch shape – is fixed via ligatures to brackets bonded to the labial (or in some cases, lingual) surfaces of the dentition or to metal bands fitted to individual teeth. The material used for these arch wires has a 'memory', such that, after being deformed when affixed to the teeth, the wire returns to its original shape, moving the teeth with it. A variety of bends and twists may be incorporated in the arch wire in order to bring upright, extrude or rotate individual teeth. This treatment is only sufficient when the discrepancy involves malposition of the teeth alone. When there is not enough room within the alveolus to house the teeth in acceptable alignment, extraction is sometimes the only alternative. Usually, the first premolars are extracted, because their loss represents the most minor obstacle to achieving optimal occlusion.

The natural course of tooth eruption usually is compensatory so that, in spite of the degree of intermaxillary imbalance, some form of occlusion will be produced. When there is a discrepancy in skeletal development (**5.6**), however, treatments to stimulate or inhibit growth (dental orthopedics) may be required. The timing of treatment is extremely important, for those treatment modalities which either enhance or diminish naturally occurring growth, while promoting optimal occlusion, improve the chances for success. Skeletal imbalance is seen in three dimensions: transverse, antero-posterior and vertical, and a variety of appliances is available to correct each.

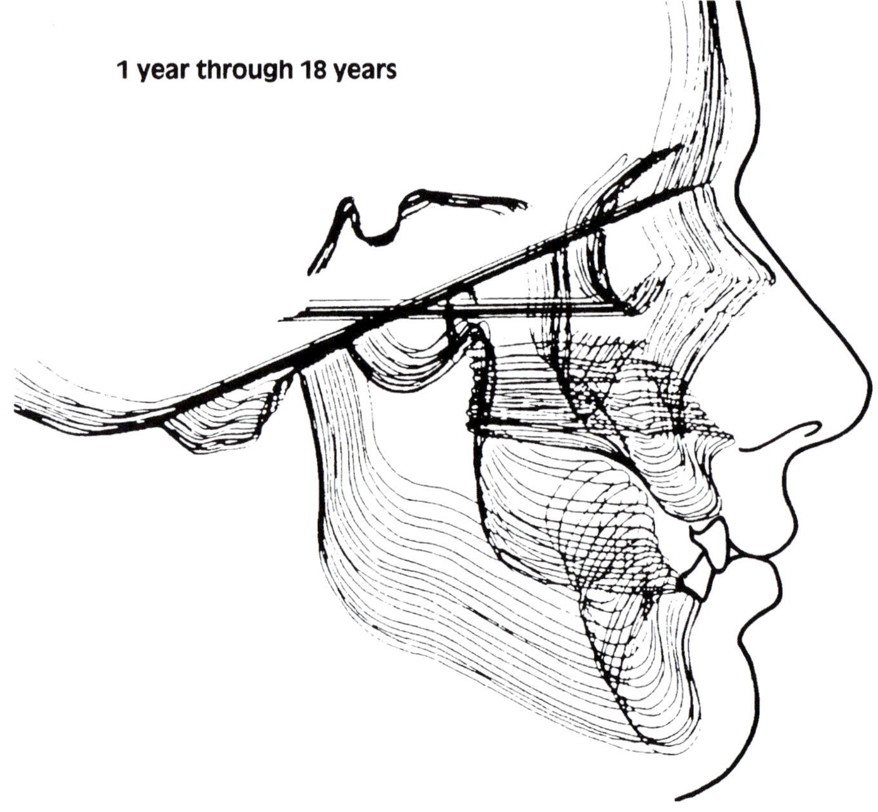


**Composite tracing of annual stages of facial growth for the first 18 years. Taken from lateral radiographs. Adapted from P. Sullivan (1994). *Growth Matters.***

Insufficient palatal width (sagittal dimension), for example, results in unilateral or bilateral crossbite: a form of occlusion which is unstable. This is correctable through the use of a fixed or removable maxillary appliance with either a coiled lingual arch wire or a screw device which the caregiver opens a small amount at regular intervals. Thus, the mid-palatal suture is separated and the teeth are moved laterally until occlusion is optimal. Compensatory growth at the suture eventually stabilizes the result. This is often a valuable option in the treatment of cleft palate as well.

Antero-posterior discrepancies may be influenced through retraction or pressure applied in the direction opposite to the vector of growth. Malocclusions in which the maxilla is adjudged to be growing too rapidly may be treated with headgear, in which posterior force is applied to the maxillary first molars, restraining the maxilla while the mandible undergoes catch-up growth. Vertical discrepancies are correctable by changing the angle of force applied by the headgear to intrude or extrude posterior teeth. The same principle is employed to apply pressure to a mandible growing too rapidly: a chin-cup is worn for restraint, permitting the maxilla to catch up.

Several varieties of *activator* appliance treat mandibular insufficiency by repositioning the mandible forward during the period of maximum mandibular growth. The success of this appliance and others like it depends on compensatory growth in the condyle to achieve a stable result. The perioral musculature also may sometimes constrain growth and so most appliances, both fixed and removable, incorporate devices to eliminate muscle pressure on the dentition. In most cases, some combination of all of these treatments is required. Dysplasias so extreme as not to be correctable by orthodontic appliances alone require orthognathic surgery in combination with orthodontics, but this treatment frequently is postponed until most growth has ceased.

*Diane Markowitz*

See also 'Dental maturation' (**1.9**), 'Creation of growth references' (**1.16**), 'Embryonic development of teeth' (**4.4**), 'Post-natal craniofacial growth' (**5.9**) and 'Variation in time of tooth formation and eruption' (**5.10**)

7·3

# Regional growth disorders

A regional growth disorder is defined as a condition where tissues within a part of the body undergo some form of abnormal growth process. This can vary from a particular area such as a limb to the whole half of the body as, for example, in hemihypertrophy. Such growth may be in the form of overgrowth or undergrowth but generally results in a departure from normal shape.

Growth disorders are usually primary, affecting the growth and development of the skeleton (**5.6**), and are attributed to an intracellular developmental failure. Secondary growth disorders may be due to a problem outwith the skeletal system, such as abnormalities of hormonal supply, and consequently the pattern of local malformation may not be manifest until months or years after birth. This should normally exclude local manifestations of a general underlying pathological process leading to classification problems as examples of apparently local growth manifestations subsequently may be found to a part of a general growth disorder. A good example of this is Perthes disease of the hip, where there is a localized process of necrotic destruction of the head of the femur in children around the age of 5 years, which subsequently repairs itself. Research has, however, demonstrated local growth effects in these subjects affecting their general skeletal maturity, and local growth retardation affecting their feet. Similarly, the overgrowth of the hands associated with acromegaly is a local manifestation of a known general body disorder.

Local growth disorders may also follow damage to nerve supply to the region resulting in subsequent under- or overgrowth on a regional basis. These changes are also encountered in a wide range of conditions, including those due to congenital causes which collectively form an important group of syndromes.

Regional growth problems are not common. One group of rare disorders, often easily detected by the lay person, are the syndromes called the hemihypertrophies or hemiatrophies. Hemiatrophy is undergrowth and hemihypertrophy is overgrowth of the body. The results of these conditions disturb the normal symmetry of the body and produce one side that is a different size to the other. Although the causes of such growth disturbances are still not clear, there is undoubtedly a disturbance to the control mechanism for local growth present, such as an abnormality of part of the nervous system. Hemihypertrophies can be generalized, where they involve half of the body, or localized on a more regional basis – for example, with the involvement of a single affected limb or in unilateral facial asymmetry. When considering such regional disorders, the abnormality can involve either soft tissue or bone or both types of tissue. For example, facial asymmetry may be due to local abnormal growth of the mandibular condyle (jaw) alone, whereas regional hemihypertrophy of the face involves bone and soft tissue.

Current research also suggests that a previously unsuspected localized growth disturbance might be associated with pediatric conditions such as club foot. Here the foot is subject to abnormal developmental pressures of as yet unknown causes and present at birth as a distortion in the shape of the foot. Treatment, while seeking to correct the deformity, often is difficult to achieve due to underlying growth disturbances limited to the foot affected.

Abnormal morphogenesis may result in malformations. These processes may be associated with growth abnormalities at a specific time during embryonic development, resulting in the lack of development of a structure or organ, such as an absent nostril or renal agenesis; abnormal development of a structure, such as the septal defects in the heart; or the development of redundant structures, such as in polydactyly, where an extra digit is present on the hand or foot. In this context, there is also some support for the radical hypothesis that a range of later-onset non-growth-related adult diseases may also be attributed to embryonic growth-programming errors.

Regional growth disorders also form part of a wide range of rare congenital and late-developing syndromes. These often have a regional component as well as a more general body disorder, as, for example, in syndromes such as Sotos syndrome, Wiedemann–Beckwith syndrome and the Proteus syndrome. In the Proteus syndrome, many localized growth abnormalities are present, including hemihypertrophy, scoliosis, cavernous hemiangiomas, macrodactyly and rugated soles of the feet. Similarly, there is a clear embryological role in the causation of regional abnormalities affecting the trunk in Poland syndrome, resulting in hypoplasia and absence of the pectoralis major muscle on one side of the body. This particular problem is also often associated with developmental disturbances to the upper limb itself. Other localized disorders with probably a similar embryological cause include Ollier disease, dysplasia epiphysealis hemimelica and reduplication of a limb. Although a clear mesenchymal origin of some of the tissues involved in these conditions is apparent, involvement of other layers of tissue underlines the complex embryological processes involved, with the regional growth disorder representing a local expression of a general body growth disorder. The close developmental link between the central nervous system and hormonal controls of growth must not be overlooked.

As already described, abnormal morphogenesis can result in specific regional growth disorders which affect organs within the body, resulting in their over- or underdevelopment. Often an isolated incident, such manifestations can include agenesis of an organ, such as the kidney, on one side of the body, or local gigantism of a structure. Local developmental abnormalities also may become a cluster of other defects, such as the VATER associations.

The etiology and pathogenesis of these regional disorders is not known, although in a number of cases the abnormality is clearly embryological in origin. The activities of protein-signalling molecules present in early embryogenesis could generate cellular hyperplasia rather than cellular hypertrophy, resulting in the range of localized regional growth problems. However, if a syndrome is recognized, clearly a genetic defect is likely to be behind the embryological problem.

There are different types of intra-uterine growth retardation (**4.8**) recognized, each depending on the stage of the pregnancy when growth is restricted. Early growth restriction results in a proportionately retarded infant who is uniformly small, but growth restriction in the third trimester results in a baby with local disproportionate growth restriction such as a normal head-size but reduced body-length or weight. Malnutrition late in pregnancy results in the fetus being underweight but normal in all other respects. For a regional disorder to occur, a trigger such as a chemical or infectious insult to the embryo must occur at a critical period. For

## ASYMMETRY

A minor degree of asymmetry is always present in animals, and the human is no exception. These asymmetries can express themselves as, for example, an obvious difference in facial symmetry where there might be a slight difference between the shape of the right and left cheeks, mouth or the outline of the nose, or perhaps the slight difference in size of the feet that makes fitting shoes sometimes a problem. These minor asymmetries are consistent right–left differences within the normal individual and are usually of no major biological or medical significance. More obvious body asymmetry does occur in nature in some species which have evolved to meet particular ecological situations. This is fairly rare. Examples include fiddler crab males who have a large right claw, or the male narwhal and its elongated overgrown tooth on the left which develops into a tusk. However, consistent right–left differences relating to the siting of the body organs, termed *handed asymmetry*, are present in many vertebrates. Examples include the dextral looping of the heart during embryological development, the siting of the liver on the right side of the abdomen, or variations in the lobes of the lungs between right and left. If the study of biological or medical asymmetry is undertaken, the asymmetry can be classified into three different categories of either *directional asymmetry*, *antisymmetry* or *fluctuating asymmetry*. Each asymmetry is characterized by differing combinations of the right-minus-left differences.

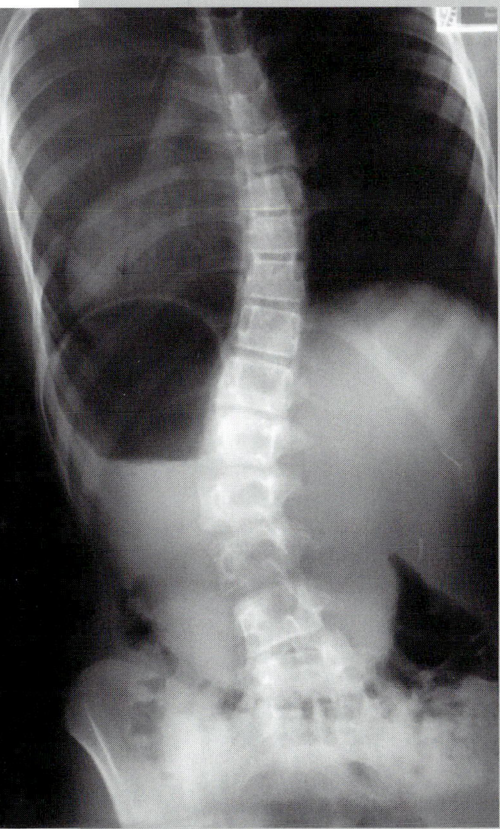

**Radiograph of an vertebral column of a patient with scoliosis. The normal straight spine is displaced laterally, creating directional asymmetry of the trunk. This condition has been found to be associated with asymmetry of the skull and upper and lower limbs. Untreated medically, it can lead to severe deformity of the trunk.**

Directional asymmetry occurs when there is a greater development of a body segment or character on one side of the axis of the body. This can include asymmetry of appendicular structures, such as the limbs, where an arm and a leg might be longer on the right side than on the left side – as in the clinical condition called hemihypertrophy.

Antisymmetry occurs when asymmetry is a normal situation in development of a part of the body. The variable between one side or the other is such that it produces a bimodal distribution, or broad peak, of the right-minus-left values about a mean of zero. There are many examples of this in the animal world, such as the claw size asymmetry of the fiddler crab and the tooth asymmetry of the narwhal.

Fluctuating asymmetry is a random deviation from perfect bilateral symmetry which relates to developmental perturbation resulting from the inability of individuals to undergo identical development of a bilateral trait on both sides of the body. It occurs when small, random deviations from a perfect bilateral symmetry in a body character can be identified, in which the normal distribution produces a right-minus-left value around a mean of zero. An example of fluctuating asymmetry is human breast-size variation during the menstrual cycle. Fluctuating asymmetry is a measure of biological quality and could well provide important biological signalling mechanisms between individuals in a population for sexual selection and reproduction or as a result of external stresses. Unfortunately, scientific debate still has not resolved the most appropriate method of quantifying and analysing fluctuating asymmetry.

Asymmetry becomes significant if it is associated with an abnormality of development, or disease process, which might, for example, occur where there is abnormality in the mirror symmetry of morphogenesis for the right and left sides of the body. The asymmetrical visceral organs, such as the liver, heart and pancreas, probably arise out of the intrinsic 'handedness' of the embryo. Recent discoveries have revealed a gene that determines laterality and can reverse the direction of post-implantation turning, one of the earliest manifestations of left–right asymmetry during early embryonic development. Other genes have also been identified that determine the side on which the body organs develop in the embryo. A pair of protein-signalling molecules called activin receptor IIa and Sonic Hedgehog have been shown to be produced asymmetrically, suggesting that the left–right asymmetry of the body is caused at a very specific time during development of the embryo (**4.5**) in part by the transient asymmetrical activation of the genes *cAct-RIIa* and *Shh* controlling these two proteins (**3.3**). This process might explain both the occurrence of unilateral malformations of paired structures which consistently appear to favour one side of the body more than the other, and the range of abnormal asymmetries, such as spinal curvature and limb length inequalities, encountered in medical situations. *Peter H. Dangerfield*

example, drugs are known to act asymmetrically on the brain. Research has demonstrated that acetazolamide will cause a unilateral forelimb on the right side of treated embryos, and rubella, or German measles, virus infection will affect the development of the eye, heart and ear.

Clearly, in the field of growth much further research is required to clarify processes involved in the complexity of development both on a general and regional basis. What is almost certain is that local manifestations of disorder are probably due to underlying centrally controlled mechanisms which involve genetically controlled intercellular messenger exchanges.

*Peter H. Dangerfield*

See also 'Body-imaging by three-dimensional surface anthropometry' (**1.5**)

## SCOLIOSIS

Scoliosis is a common disorder of the spine which occurs more frequently in humans than animals and almost exclusively during the growing period of life. The normal human spine when viewed from behind in the standing position is generally straight. If it is curved to one side, i.e., laterally towards an arm or a leg, it is termed a *scoliosis*. The child is then asked to bend forwards and the examiner looks for a hump on one or other side of the back. In healthy adolescents, posterior trunk asymmetry in a forward-bending position is common and has a pattern (mainly right thoracic or left lumbar humps) and is termed *constitutional back asymmetry*. If a relatively large hump is found it is attributed to the axial rotation of vertebrae, which accompanies most forms of the scoliosis; an X-ray examination confirms the diagnosis. Scolioses may be physiological or pathological, but most involve some local deformity of vertebrae and intervertebral discs in three-dimensions.

Small scoliosis spinal curves are commonly seen in normal adolescents. Some result from one leg being slightly shorter than the other (*pelvic tilt scoliosis*). Other lateral spinal curves found without known cause are termed *idiopathic scoliosis* – not all scolioses are idiopathic: about 10 per cent are either congenital with vertebral abnormality or caused by known disease of nervous system, muscles, ligaments or bones which are known as *secondary scolioses*. Idiopathic scoliosis (IS) develops at different periods of human skeletal growth (**2.4**) and three types are identified – *infantile* (IIS), *juvenile* (JIS) and *adolescent* (AIS), the latter being the most common. IIS, rare in North America, is slightly more common in boys than girls; it is usually a thoracic curve convex to the left which resolves without treatment. Of those that progress (as some do if not treated), deformation of the trunk can interfere with lung development and cause heart failure and death in early adult life. JIS has recently been shown, by a special imaging technique (MRI), to be associated with either abnormalities of the nervous system (such as syringomyelia) or intraspinal tumours in 20 per cent of children, leaving 80 per cent as idiopathic. AIS is 10 times more common in girls than boys, and is usually either a right thoracic curve convex to the right, or a left lumbar curve (or both), the enantiomorph pattern of which is related to handedness. These scolioses can be detected by screening in the school by an experienced person. Fortunately, most such spinal curves do not progress to need treatment by brace and/or surgery. The cause of idiopathic scoliosis in all its forms is still unknown, but all types are thought to involve a lengthening of the front of the spine relative to its back.

There is evidence of a hereditary predisposition to the AIS, which is probably multifactorial. Girls with AIS, having moderate to severe scolioses, are taller than average (after correcting for the loss of height due to the lateral spinal curve), and they show an increase in most linear dimensions of the body. Those with a right thoracic scoliosis have a longer right upper arm, which may be an expression of predisposition to AIS, or be secondary. Anomalies have been found in the nervous system, muscles and connective tissues (joint laxity) – in particular, changes in vestibular function, proprioception, vibratory response and posterior column function have been detected. A major problem in such research is whether the changes either result from, or cause, the deformity. Scoliosis causation is best viewed in the perspective of: 1) human evolution; 2) human development and growth; 3) spinal movements, some of which result from bipedal gait (these include rotations of the pelvis and lumbar spine with counter-rotations of the thoracic and cervical spines above, to ensure that such spinal axial rotations are eliminated before they reach the skull, so enabling the eyes to maintain their focus on an object in question); and 4) proprioceptive input (the learning by repetition of neuromuscular mechanisms which continually adjust to maintain the balance of the growing human body in its upright posture during gait).

According to this view, IS can result when genetic abnormalities, unilaterally expressed, impair proprioceptive input; this putatively leads to neuromuscular asymmetry in trunk muscles, which causes eccentric loading of vertebral growth plates. Such loading, by altering spinal growth, causes a secondary growth-induced torsion of the spine (scoliosis), with the vertebral bodies growing more than the vertebral arches. A similar neuromusculoskeletal perspective during development and growth explains both the physiological changes and the abnormalities of torsion in the femora and tibiae of humans some of which need surgical correction.

*R.G. Burwell*

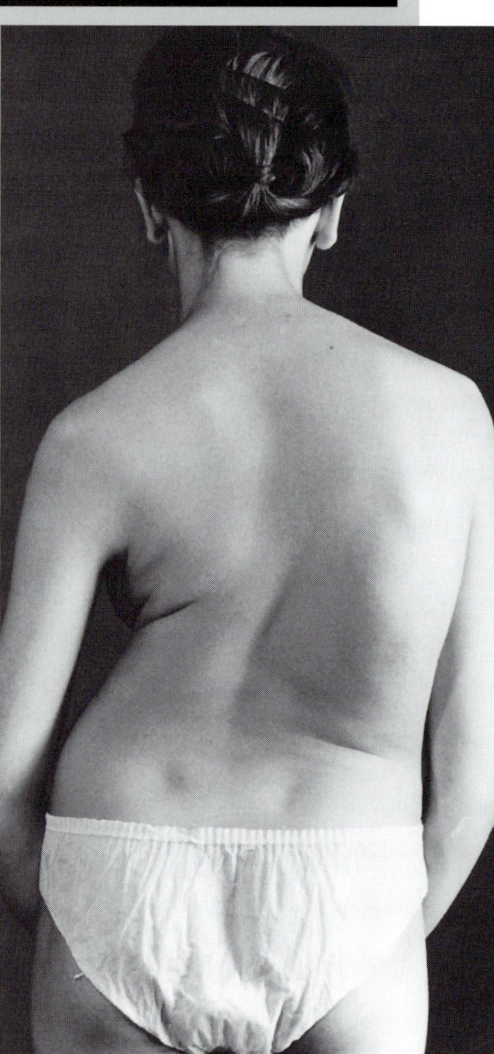

**A girl with adolescent idiopathic scoliosis. Note: (1) the lateral curvature of the spine to the right and the prominent rib hump; (2) the hip asymmetry, the left hip being more prominent than the right (girl's hip on the left, boy's hip on the right); and (3) left shoulder at a lower level than the right.**

# Chromosome aberrations and growth

Humans have 46 chromosomes consisting of 22 pairs of autosomes (non-sex chromosomes) and one pair of sex chromosomes. The autosomes are numbered 1–22 and sex chromosomes are labelled XX in females and XY in males (the Y chromosome is much smaller than the X). Individual genes are short sequences of DNA, and in each chromosome an immensely long strand of DNA is coiled to make a compact, string-like structure. Therefore, each chromosome carries many thousands of genes. A chromosome has a short arm and a long arm, separated by the centromere.

Chromosome abnormalities occur in about 1 in 200 newborn babies. A variety of abnormalities can occur. Some individuals have a whole extra chromosome in a pair, giving *trisomy* for that chromosome. *Monosomy* for a whole chromosome is less common and is mainly seen in Turner syndrome, where a female has only one X, giving 45 in all. Other individuals have the right number of chromosomes but there may be a *deletion* or *duplication* of chromosomal material. About 1 in 500 normal people carries a *translocation*, where two chromosomes from different pairs have exchanged material. In many cases no genetic material is lost, no clinical abnormalities are caused and the translocation is said to be balanced. However, carriers of balanced translocations have a risk producing children with a combination of deleted and duplicated chromosome material.

Height is affected in many chromosome abnormalities. A reduction in height is most commonly seen, but increased stature is characteristic of some abnormalities. In general, the affect on height cannot yet be explained by the involvement of known genes. Chromosome defects involve a large number of genes, and the cause of alteration in growth in individual cases is likely to be complex. The most common group of chromosome abnormalities affecting height involves the sex chromosomes. The presence of an extra sex chromosome to give an XXX female or XXY or XYY male occurs in approximately 1 in 1000 individuals for each type. In general, an extra X chromosome increases stature to give average adult height of 4–5 centimetres greater than XY or XX individuals. There must also be genes for height on the Y chromosome, as XYY males are on average taller than XY males.

Turner syndrome, where there is only one sex chromosome, occurs in around 1 in 3000 females. Adult height is reduced, at 135–150 centimetres. Some females have only part of one X chromosome missing. This provides clues as to where genes important in stature are situated. The short arm seems to be a particularly important region.

Many deletions or duplications of the autosomes result in reduced stature. This indicates that there are a large number of genes involved in growth scattered on different chromosomes. In the most common autosomal chromosomal defect, Down syndrome, where there is an extra chromosome 21 in every cell, average adult height in females is 145 centimetres and in males 155 centimetres. In a few cases the precise cause for short stature can be found. For example, individuals with a deletion of the short arm of one chromosome (18) frequently have measurable growth hormone deficiency. This is not due to a deletion of the growth hormone gene, which is situated on another chromosome. It may be caused by deletion of a gene important in pituitary development, which is thought to be located on chromosome 18.

In most people, one member of each chromosome pair comes from the mother and one from the father (both the

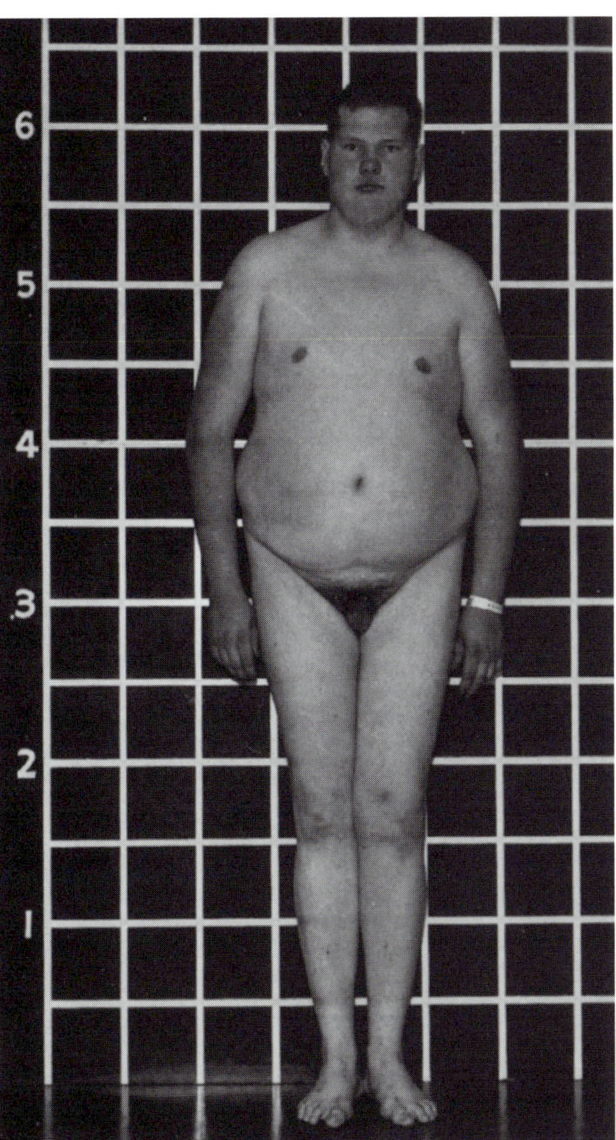


**Adult XXY Klinefelter syndrome patient whose height of 193 cm is 2.7 SD above the mean. Note the incomplete virilization, the moderate obesity, and the immaturity of facial bone structure.**

sperm and the egg contain 23 chromosomes, one from each pair). However, rarely it is possible for two chromosomes of a pair to come from one parent and none from the other. This situation is known as *maternal* or *paternal disomy* for the chromosome concerned. In some cases stature is affected, for example, individuals with maternal disomy for chromosome 7 (two chromosome 7s from mother and none from father) have a low birth-weight and short stature. The number of chromosomes and genes in this case is normal, and yet there are clinical effects. The reason is thought to be that some of the genes on chromosome 7 must be marked to indicate that they come from the mother or father. This 'marking' process is known as *imprinting*. Normal function of the genes is dependent on there being both a maternal copy and a paternal copy. Two maternal copies do not function completely normally. Imprinting is also thought to explain the overgrowth seen in children with Beckwith–Wiedemann syndrome. Affected children usually have a high birth-weight and can show hypertrophy of the limbs, tongue and internal organs. In some cases one arm or leg is significantly larger

## TURNER SYNDROME

The main clinical features of this condition were first described by Ulrich (1930) and Turner (1938); it is caused by the absence or structural abnormality of one X chromosome. Typical karyotypes are 45 X, where one X is completely absent, accounting for about 50 per cent of cases; 46i(Xq); or 46 X rX. In the second of these (46 i[Xq]) one X chromosome is an isochromosome, which means that there is a loss of only the short (p) arm and a duplication of the remaining long (q) arms. In the third abnormality (46X rX), one X has been converted into a ring chromosome, usually with considerable loss of chromatin material at the breakpoints. Normal and abnormal cell lines may be combined in an individual (e.g. 45 X/46 XX), a condition which is termed *mosaicism*.

The incidence in the population is approximately 1 in 2000 live-born females. The main features of the syndrome are: short stature; ovarian dysgenesis, and a variable collection of dysmorphic features and internal organ malformations. The most common dysmorphic features are ptosis, high arched palate, low-set ears, low posterior hair-line, neck webbing, a shield-shaped chest with widely spaced nipples, cubitus valgus, cutaneous naevi and dysplasia of the nails.

Abnormalities of the internal organs include congenital heart or aortic disease and horseshoe kidney. Many of these structural abnormalities are probably secondary to impairment in intra-uterine development of the lymphatic system. At least part of the short stature is due to a form of bone dysplasia. In later life hypothyroidism, hypertension and non-insulin-dependent diabetes mellitus occur with greater frequency than in the general population. In general, intelligence is normal although there may be selective learning difficulties often related to impaired visuo-spatial appreciation. Ovarian dysgenesis means that the ovaries are usually rudimentary and non-functional, leading to a failure to enter puberty (**5.15**) unaided, and infertility. Rarely, a girl with the Turner syndrome will have fully functioning ovaries (about 15%) and may even be fertile, although the menopause (**12.2**) is usually early. Short stature seems to be the most consistent feature. It is noteworthy that there is very little correlation between the chromosomal make-up and the clinical severity of the condition.

The growth pattern is very specific. Size at birth is reduced (mean 2800 g) and height growth is consistently at a lower than average velocity so that girls with the condition gradually fall further behind the general population of girls. Adult height is on average 20 centimetres below the

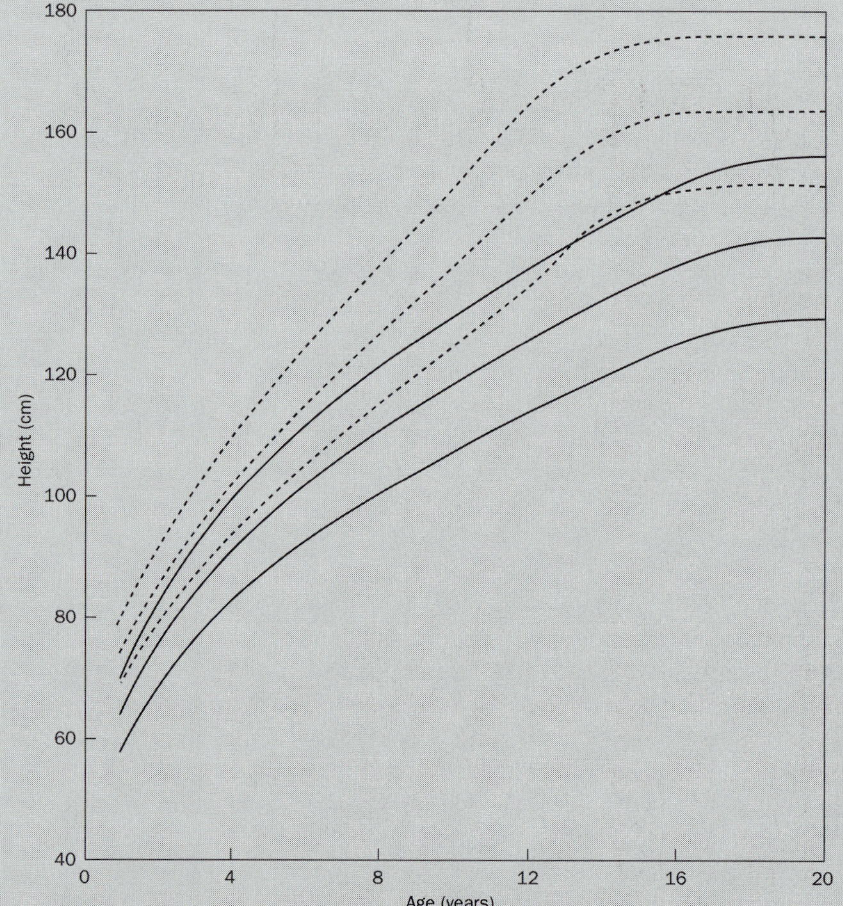


**Growth of girls with Turner syndrome (solid lines; 2nd, 50th and 98th centile) compared to normal girls (broken lines; 2nd, 50th and 98th centiles).**

than the other (hemihypertrophy). Imprinted genes on the short arm of chromosome 11 are thought to cause the syndrome. The insulin-like growth-factor II (IGF-II) gene may be particularly important in the growth abnormalities seen. One maternal and one paternal copy seem to be essential. A deletion of the maternal IGF-II gene region, or a duplication of the paternal gene region, results in features of the condition. Some children may be *mosaic*. This means that some cells have normal genes but in others there may be a duplication of the paternal IGF-II gene region. If the abnormal cells are present more on one side of the body than the other then hemihypertrophy occurs. A further mechanism, rather than deletion or duplication, may be a change in the imprinting pattern to make maternal genes look like paternal genes or vice versa.

*Robin Winter*

See also 'Physical examination' (**1.10**), 'Assessment of age at menarche' (**1.15**), 'Genetics of child growth' (**3.1**), 'Genetic and environmental influences on fetal growth' (**3.2**), 'Skeletal development' (**5.6**), 'Sexual maturation' (**5.15**), 'Treatment of growth disorders' (**7.9**) and 'Treatment of growth disorders: the future' (**13.4**)

normal female mean, although the height of the parents still exerts the same effect as in normal children. Thus, a girl with the Turner syndrome who has very tall parents will be tall for the diagnosis and may even fall just within the normal range. Skeletal maturation is usually delayed.

Multidisciplinary long-term care is required to ensure normal development and a socially integrated life. Malformations are corrected to restore normal organ function. In most individuals life-long replacement of female sex steroids is required from about 12–14 years of age in order to ensure normal feminization and sexual functioning. The short stature may be partially corrected by treatment with growth hormone during childhood. There is no evidence of a major premature mortality in the syndrome.

*Michael B. Ranke*

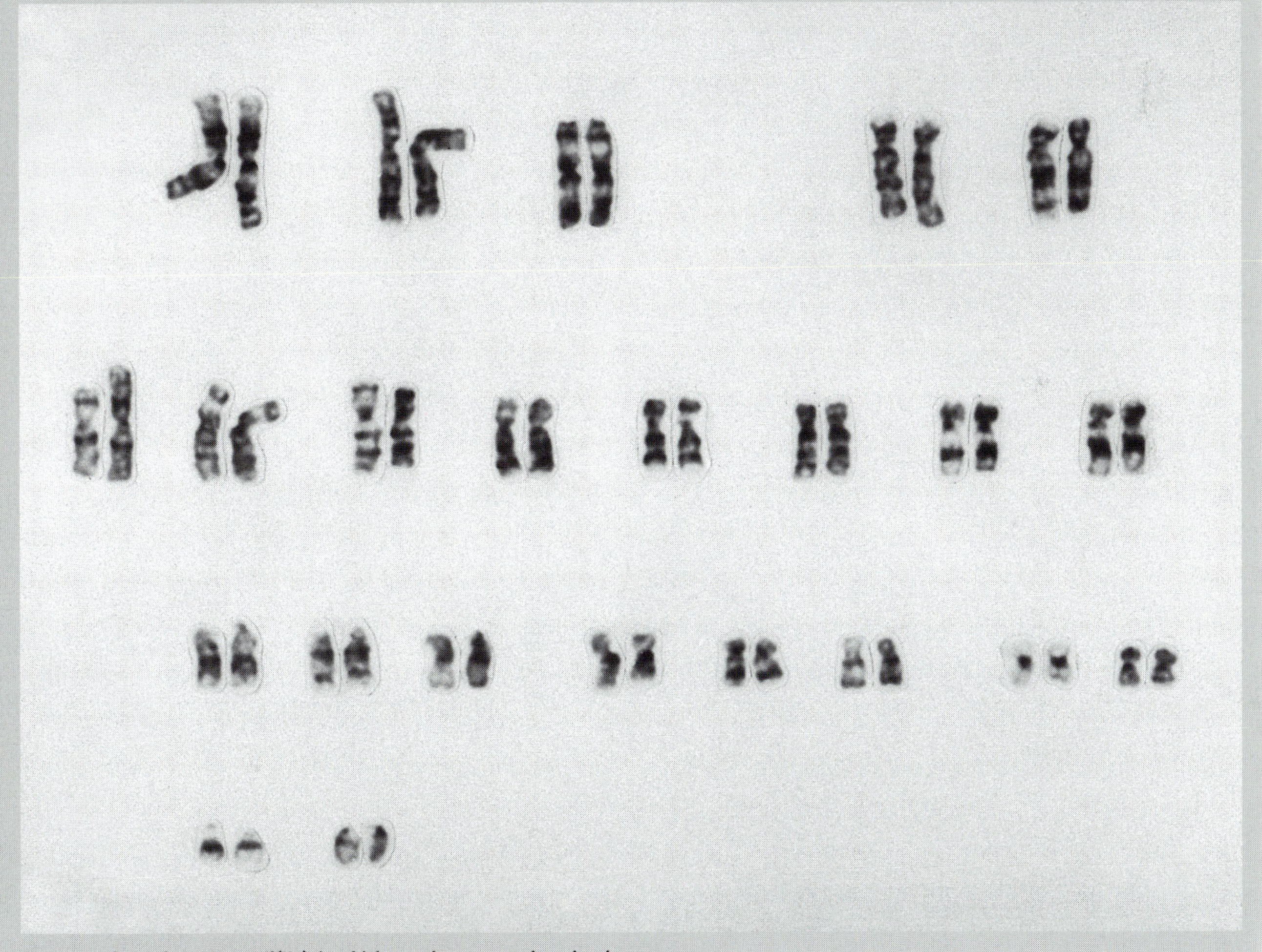

**Turner syndrome karyotype 46i(Xq), in which one chromosome is an isochromosome.**

## PRADER–WILLI SYNDROME

First described in 1956, Prader–Willi syndrome (PWS) is the most common syndromal cause of obesity. It has an incidence of approximately 1 in 15,000 individuals and affects males and females equally. The disorder is associated with a loss of a paternally derived contribution to a specific region within 15q11-13. Loss of the same region from the maternally derived chromosome 15 gives rise to the clinically distinct Angelman syndrome. Together, these two syndromes demonstrate the non-Mendelian form of inheritance known as *imprinting*.

### *Clinical characteristics*

Even before birth, reduction in fetal movements have been recorded, and the neonate with Prader–Willi syndrome often first presents with pronounced hypotonia with poor reflexes, which, as they include sucking, can lead to feeding difficulties and failure to thrive. In the neonatal period, diagnosis is often assisted in males by the presence of hypogonadism as half of them have bilateral and a further quarter have unilateral cryptorchidism. Hypopigmentation is also common, the PWS infant having paler eyes, skin and hair than the rest of the family. In infancy, sticky saliva is another frequent observation.

Gradually, over the first months of life, the hypotonia improves and muscle tone increases. The feeding difficulties experienced early on are replaced by an excessive appetite and the onset of obesity, with weight reaching more than 2 standard deviations above the normal by about 4 years of age. Weight has been noticed to increase before appetite.

Although the hands and feet have been recorded as normally sized in younger children with PWS, by the early teens the probands tend to be of short stature and to have small hands and feet. Growth retardation is characteristic of PWS, with the average height of adult males reaching about 155 centimetres and of adult females only about 145 centimetres. A deficiency in the secretion of growth hormone may contribute to the abnormal growth pattern, and in some cases treatment with growth hormone has led to an increase in height. Low growth hormone levels are not always present in PWS probands. Faces take on a typical PWS appearance, with small head, narrow bifrontal diameter and almond-shaped eyes.

The growth spurt associated with puberty (**1.10**) is lacking, and the appearance of secondary sex characteristics is either delayed or they do not appear at all. As a result, males have a small penis with undescended testes or cryptorchidism, and females have hypoplasia of the labia minora and either late or no menarche, with 70 per cent of females having primary amenorrhea, and a further 30 per cent oligomenorrhea. The primary gonadal defect is accompanied by a reduction in hypothalamic function, leading to lowered gonadotropin levels.

Obesity is the most prominent physical characteristic of probands with PWS and is the major contributor to the health problems they experience. These include cardiovascular compromise, hypertension, and diabetes mellitus. Scoliosis (**7.3**), dental problems, eye problems and severe skin-picking are also common. Death, due to the complications of obesity, is often between 20 and 40 years of age.

Most affected people have significantly delayed motor and language development(**6.2**, **6.3**). Cognitive impairment is associated with an IQ which generally lies between 60 and 90, although some studies have placed the average IQ as low as 62 with a range of 39–96. Seventy-five per cent of patients require special education but 5 per cent have an IQ within the normal range.

In early childhood, generally between 3 and 5 years of age, probands become subject to temper tantrums and stubbornness. Children tend not to interact well with their peer group and, combined with inappropriate social behaviour, this leads to anxiety and depression. Even those with a normal or borderline IQ can have specific cognitive difficulties (**6.1**), tending to fare better on visual rather than verbal performance, and to be poorer at arithmetic than at reading. In Britain, about half of all probands work in sheltered workshops, but over a third never work at all.

Two behavioural problems endemic to PWS are known to cause stress in families. Gross hyperphagia, accompanied by persistent hunger (which is the cause of the obesity) leads to food foraging, hoarding, stealing and even scavenging, while disturbed sleeping patterns lead to obstructive sleep apnea. Medication aimed at controlling the appetite in PWS patients has not been successful, so that dietary control is essential. Apart from the hyperphagia, it has been suggested that in PWS the basal metabolic rate (**9.2**) is reduced, so that only 60 per cent of the calories needed by control individuals are required to maintain normal body-weight. Children with problems of growth who are treated with human growth hormone (hGH) (**7.9**) show an increased growth-rate and those with a growth hormone deficiency secondary to a hypothalamic dysfunction sometimes may even lose weight. Cryptorchidism may be treated with gonadotrophin while testosterone may be effective in stimulating growth of the small penis.

PWS had already been associated with abnormalities involving chromosome 15 when, in 1981, Ledbetter and colleagues observed a microscopic deletion of 15q11-13 in a girl with the syndrome. It was quickly established, with high resolution cytogenetics, that between 60–70 per cent of probands demonstrated the same deletion, while a further 5 per cent had other cytogenetic abnormalities, such as translocations, which always involved proximal chromosome 15. In 1989, Kaplan et al. observed an apparently identical 15q11-13 deletion in three patients who patently did not have PWS, but one of whom had Angelman's syndrome (AS). Angelman's syndrome, first described in 1965, is clinically distinct from PWS yet was associated with an apparently identical chromosomal abnormality. The dilemma appeared to be resolved when it was discovered that in PWS all the deletions in 15q11-13 were located on the paternally derived chromosome 15, while those in AS were on the maternally derived homologue. So Prader–Willi and Angelman syndromes are both associated with a loss of chromatin from the same region of the genome, but derived from opposite parents, indicating that this region is imprinted. Parent-specific imprinting results in the differential expression of genes, depending upon the parent of origin. In PWS, as the deletion is always paternally derived, it is maternal imprinting which causes lack of gene expression. In AS the opposite occurs; the deletion is of maternal origin, implying that it is the paternal homologue which is imprinted. Despite this, approximately 90 per cent of patients with either PWS or AS and deletions in 15q11-13 have the same chromosomal breakpoints.

That PWS is due to lack of paternal expression of gene(s) in 15q11-13 was confirmed when it was reported that a boy with a 13q; 15q maternally inherited Robertsonian translocation had also inherited his non-translocated 'normal' homologue of chromosome 15 from his mother. He thus had two copies of maternally derived chromosome 15 but no copy from his father, a phenomenon known as maternal uniparental disomy or UPD. It quickly became clear, using molecular methods, that nearly all of the 30 per cent of PWS patients who did not have a deletion in 15q11-13 had maternal UPD. Similarly, a number of AS cases were found to have paternal UPD, with both of their chromosomes 15 derived from their fathers. In AS, however, the probands with paternal UPD represent only about 5 per cent and there remains a further 20–25 per cent of cases which apparently have a maternally inherited autosomal dominant form of the disease, with an accompanying high recurrence risk. In PWS, the recurrence risk is low (estimated at $< 1\%$), and if it does recur it is usually the consequence of an unbalanced familial chromosomal translocation. Such families can present with cousins who have either PWS or AS, depending upon the mode of inheritance. An unusual family in which three siblings with AS were found to carry a submicroscopic deletion inherited from their unaffected mother who, in her turn, had inherited it from her father, served to demonstrate that the PWS and AS critical regions were not allelic but contiguous.

The rapid advances made in the field of molecular genetics, together with the availability firstly of probes detecting RFLP (restriction fragment length polymorphisms) in loci from 15q11-13 and then of STR (short tandem repeat) polymorphisms detectable by PCR (polymerase chain reaction) confirmed that the critical regions for PWS and AS were probably contiguous, with that for PWS lying proximal or closer to the centromere, and that the deletions in either syndrome covered a region of several megabases.

Methylation differences had long been associated with the control of gene expression, particularly in the inactivation of the X-chromosome, and these were soon discovered at different loci within the PWS critical region. Two such loci were found to have parent-of-origin methylation differences determined by the restriction of DNA with methylation-sensitive enzymes. As both of these loci lie within the Prader–Willi syndrome critical region rather than the more distal Angelman syndrome region, it is the maternally derived homologue which is imprinted, and therefore not active, in both cases. As expected, the imprinted maternal band was found to be methylated and uncut, while the paternal band, which is not imprinted, is also not methylated and was therefore cut by the enzyme, resulting in a smaller band after electrophoretic separation. Thus, normal individuals have two bands, whereas those with PWS have only the maternal band, and those with AS have only the paternal band.

Intensive study of a region of mouse chromosome 7 which is homologous to human 15q11q13 permitted the isolation of an expressed gene *SNRPN* (small nuclear riboprotein N), coding for a ubiquitous ribonucleoprotein subunit N (SmN), which is involved in the splicing of pre-mRNA to mRNA. *SNRPN* lies within the PWS critical region rather than the AS, is maternally imprinted, has parental-specific methylation and may be upregulated in the brain, thus suggesting that SmN may have neural involvement. The *SNRPN* gene fulfils many of the criteria necessary for a candidate gene for PWS.

Besides *SNRPN*, comparatively few expressed genes have been isolated from the deleted region, and the majority of those which have, such as those for *ZNF127*, *GABRβ3* and *GABRα5* have gradually been eliminated as candidate genes for either syndrome after study of rare individuals with small deletions which do not encompass the whole of the critical region that becomes deleted in the majority of probands.

Study of such rare individuals with unusual small submicroscopic deletions has permitted the critical region for PWS to become progressively better defined and allowed the isolation of further expressed sequences from within it. Some of these expressed sequences carry a parental imprint, so may lead to the isolation of candidate gene(s) for PWS. Recently, probands have been described who have neither a detectable deletion in 15q11-13 nor maternal UPD and appear to have biparental inheritance at all loci within the region, but who appear to carry a maternal imprint on both chromosomes 15. This suggests the presence of an imprinting control region which can exert its influence throughout the whole of the PWS critical region, and which, if altered, can cause the clinical manifestation of the disease.

In PWS the clinical phenotype is therefore associated with abnormalities in 15q11-13 caused by a paternal deletion, maternal uniparental disomy or an alteration in an imprinting control element, which causes a maternal imprint to be present in both maternally and paternally derived homologues of chromosome 15.

*Tessa Webb*

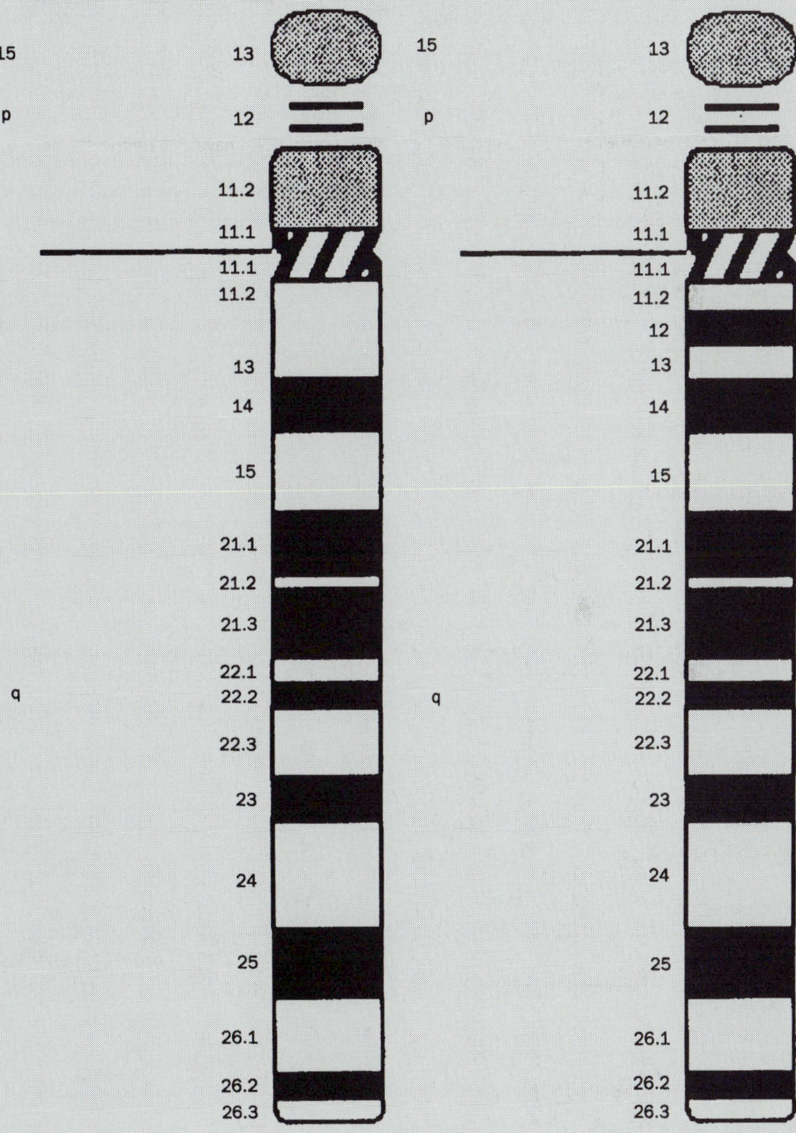


**A normal chromosome 15.**

**A chromosome 15 with band 15q 12 deleted.**

# Bone dysplasias

The bone dysplasias represent a large group of disorders involving short stature caused by a defect intrinsic to bone or cartilage (**5.6**). Other terms used to describe this group include *skeletal dysplasias*, *osteochondrodysplasias* or *chondrodysplasias*. All of the conditions are caused by abnormal genes. It has been estimated that about 1 in 5000 individuals suffers from a bone dysplasia. In addition, about 1 in 10,000 babies is still-born with a lethal form of bone dysplasia. Short stature in the bone dysplasias is usually disproportionate and the pattern is characteristic for each condition. Thus, in some cases the long bones are primarily involved and the spine spared, giving a short-limbed appearance. Particular segments of the limbs may be preferentially involved to give *rhizomelic*, *mesomelic* or *acromelic* shortening, depending upon whether there is proximal (humerus/femur), medial (radius/ulna/tibia/fibula) or distal (hands/feet) shortening. In other cases the spine might be primarily involved to give a short trunk.

Bone dysplasias are diagnosed by several different means. Most have a characteristic radiographic appearance of the bones (**1.8**), and until recently this has been the main way of achieving a diagnosis. In some cases the microscopic appearance of the bone and cartilage is diagnostic, particularly around the growth plates of the long bones. As the genes responsible for different bone dysplasias are identified, it is increasingly possible to identify specific conditions by DNA tests on a simple blood sample.

Some bone dysplasias are caused by abnormalities of the genes coding for the main structural protein in bone – *collagen*. A typical collagen molecule is made up of three chains wound as a triple helix. Type I collagen is the main collagen in bone. Each molecule is made up of two identical type I chains and one type II chain. Mutations in the genes coding for either chain can cause *osteogenesis imperfecta*, or brittle bone disease, causing the bones to fracture easily, sometimes causing progressive deformity. The commonest type of osteogenesis imperfecta gives a tendency to easy fracture of the long bones, but these usually heal without major deformity. Other features of the condition are blue sclerae (the normally white part of the eye) and abnormal dentin in the teeth to give a grey or opalescent appearance. About 1 in 50,000 babies has a lethal form of osteogenesis where there are multiple fractures of the long bones and ribs even in the womb and death occurs early because of breathing difficulties.

Type II collagen is found in cartilage. Mutations in type II collagen genes cause abnormalities of the joints and the growing points of the long bones (the *epiphyses*) (**5.7**) and the spine (which causes flattened vertebrae, or *platyspondyly*). These disorders are classified as spondyloepiphyeal dysplasias and are characterized by a short spine and limbs with progressive early osteoarthritis in the joints. Because type II collagen is also present in the eye, abnormalities such as extreme short-sightedness and detachment of the retina may also be present. Although type I and II collagens are the most common forms in bones and cartilage, other collagen molecules and enzymes help to stabilize or cross-link the helices, and mutations in the genes coding for these have been found to cause rarer bone dysplasias.

One of the more common bone dysplasias causing short stature is *achondroplasia*. There is proximal shortening of the limbs, short digits, a pronounced curve of the lower back and a large head with a prominent forehead. The condition occurs in about 1 in 40,000 newborn babies and is caused by a dominant gene (affected individuals have one normal gene and one abnormal gene). There is a 50 per cent chance of passing on the abnormal gene to a child. Nevertheless, many affected children have normal parents. This is because they receive a new mutation where a normal gene in the parents is mis-copied before it is passed on in the sperm or the egg. Achondroplasia has been found to be caused by a mutation in the gene coding for the Fibroblast Growth Factor Receptor type 3 (FGFR3). Fibroblast growth factors (**4.1**, **4.2**) are proteins, important in development, that stimulate cells to grow and to differentiate into specific tissues. The growth factor receptor sits on the surface of the cell and recognizes and binds to a growth factor outside the cell. When this happens, a message is passed to the nucleus giving instructions to alter the growth pattern of the cell. One of the functions of FGFR3 may be to suppress growth of the long bones – mice in which the gene has been knocked out are longer than normal. FGFR3 mutations in achondroplasia are very specific (they almost always change the same amino acid in the protein) and they may result in *gain of function* of the receptor to inhibit growth more than normal. Different mutations in the FGFR3 gene cause *thanatophoric dysplasia*. The name means 'death-seeking', indicating that these children all die around birth of respiratory failure, because of the severely shortened ribs and to very short limbs.

A subgroup of skeletal dysplasias is made up of conditions caused by abnormal lysosomal storage. The lysosomes are the small vesicles inside the cell where complex macromolecules are broken down. Individuals with one of the *mucopolysaccharidoses*, have a specific enzyme deficiency which means that long mucopolysaccharide chains cannot be degraded. This material accumulates with the bones, skin, liver and other tissue and disrupts normal development. Six well-defined syndromes are known, all caused by separate enzyme deficiencies. The skeletal abnormalities in each condition involve the spine and long bones and have overall similarities. The term *dysostosis multiplex* is use to describe the pattern of skeletal dysplasia seen.

*Robin Winter*

# Genetic disorders of the growth hormone axis

Clinical growth abnormalities other than chromosomal alterations, bone dysplasias, endocrinological disturbances, dysmorphic syndromes or chronic disease, mainly include genetic disorders involved with abnormalities, either deletions or mutations, in the growth hormone (GH) 1 gene (*GH1*), the GH receptor gene (*GHR*), the pituitary transcription factor 1 (*Pit1*) and, possibly, the growth hormone-releasing hormone (GHRH) receptor (*GHRHR*). So far, the GHRH gene has been excluded as a candidate to explain growth disorders. Potential candidates in the future will include: the pituitary adenylate cyclase activating protein (PACAP) and its receptor, prohormone convertases and co-activators for *Pit1* and several loci on the X-chromosome possibly involved in GH control.

Short stature associated with growth hormone (GH) deficiency has been estimated to occur in about 1 every 4000 to 10,000 live births. It is believed that 5–30 per cent of cases have an affected first-degree relative suggesting a genetic etiology. Evidence of environmental factors or anatomic anomalies by hypothalamic-pituitary magnetic resonance imaging studies is found in less than 20 per cent of sporadic cases of GH deficiency.

At least four Mendelian types of isolated GH deficiency (IGHD) have been delineated based on the mode of inheritance and the degree of GH deficiency: IGHD type IA, autosomal recessive with endogenous GH absent; type IB, autosomal recessive with diminished GH; type II, autosomal dominant with diminished GH; and type III, X-linked with diminished GH. Panhypopituitary Dwarfism (PD), a condition characterized by deficiency of at least one pituitary trophic hormone in addition to GH deficiency, can also have autosomal and X-linked modes of inheritance. The table summarizes data on GH gene abnormalities, including their GH secretion and response to treatment.

## IGHD IA

IGHD IA is the most severe of the Mendelian forms of IGHD. The incidence is unknown. Normal length and hypoglycemia are common at birth. All exhibit severe dwarfism by 6 months of age. The circulating levels of GH are undetectable. Following treatment with exogenous GH, most but not all patients develop anti-GH antibodies in sufficient titre to cause arrest of response to GH replacement. The molecular basis of this autosomal recessive disorder in most cases has been deletion of both alleles of the gene encoding pituitary GH, *GH1* (**3.1**).

## IGHD IB

IGHD IB is a second autosomal recessive form of IGHD associated with deficient but detectable amounts of GH after provocative stimuli and a less severe phenotype than IGHD IA. Affected individuals usually do not develop immune intolerance to GH replacement therapy. Criteria for diagnosis of IGHD IB include: two siblings with IGHD and parents of normal height, stature greater than 2 standard deviation scores below the mean for age and sex with delayed bone age and deficient growth velocity, peak GH levels under 10 nanograms per millilitre after provocative stimuli, otherwise normal endocrine function and absence of a history of immunodeficiency in cases from pedigrees with only affected males. The loci and derangements responsible for this genetic condition remain unclear. Linkage studies, how-

*Genetic abnormalities of human growth hormone*

| Type | Inheritance | Endogenous GH | Response to GH | Molecular basis |
|---|---|---|---|---|
| ***Isolated GH deficiency*** | | | | |
| IA | AR | Absent | Variable | GH1 gene deletions nonsense, frameshift and splice GH1 gene mutations |
| IB | AR | Decreased | + | ? |
| II | AD | Decreased | + | Intron III GH1 gene mutations |
| III | X-linked | Decreased | + | ? |
| ***Panhypopituitarism*** | | | | |
| I | AR | Decreased | + | ? |
| IB | AR or AD | Absent | + | Pit1 gene mutations |
| II | X-linked | Decreased | + | ? |
| ***Embryological abnormalities*** | | | | |
| Absence of pituitary | AR? | Absent | Temporal? | |
| Holoprosencephaly | AD or AR | Decreased | +? | |
| Rieger syndrome | AD | Decreased | +? | |
| EEC syndrome | AR | Decreased | + | |
| Fanconi anemia | AR | Decreased | + | |

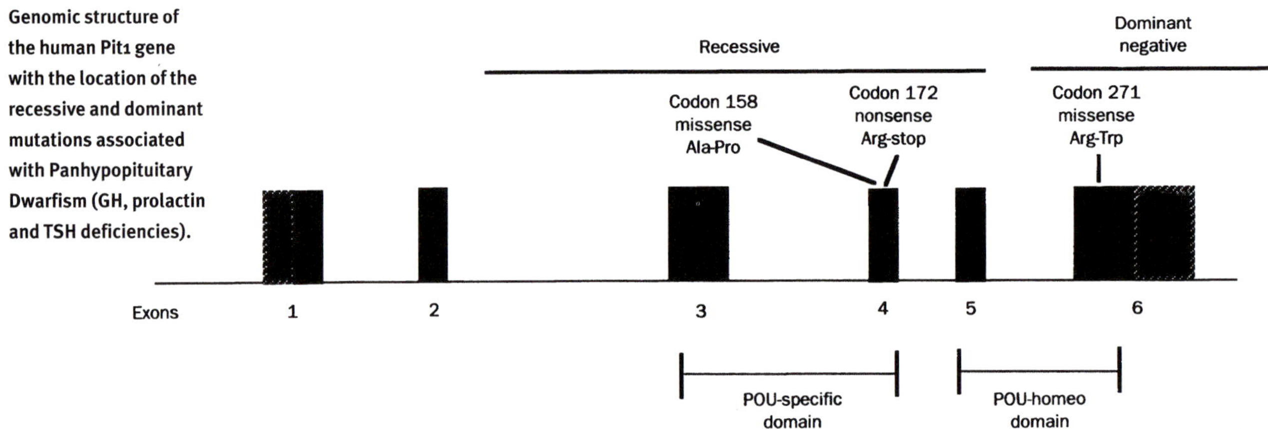


**Genomic structure of the human Pit1 gene with the location of the recessive and dominant mutations associated with Panhypopituitary Dwarfism (GH, prolactin and TSH deficiencies).**

ever, have excluded the *GH1* locus in most cases of IGHD IB. The GHRH gene is not mutated in the IGHD IB families studied. The human GHRH receptor is a possible candidate to explain the molecular basis of patients with IGHD IB.

## IGHD II

Patients with IGHD II have the same clinical characteristics and similar diagnostic criteria to those of IGHD IB subjects. The mode of inheritance in IGHD II is autosomal dominant, as shown by the presence of both an affected parent and an affected child. In contrast to IGHD IB, linkage studies are consistent with co-segregation of GH1 gene alleles and the IGHD II phenotype in most of the studied families. The GHRH gene has been excluded by linkage analysis in all studied families. The molecular defect causing IGHD II could be an intron III *GH1* mutation. The predicted mutant protein would be the described 17.5 GH variant that lacks amino acids 32 to 71 including one cysteine residue. Nevertheless, there is also molecular heterogeneity since some of the IGHD II families show recombination with GH gene markers.

## IGHD III

Three non-related families with X-linked recessive IGHD have been reported in which all the affected males had hypogammaglobulinemia. Linkage analysis in two of these families suggested that the combination of X-linked agammaglobulinemia (XLA) and IGHD could be due to a genomic alteration that perturbs the recently isolated gene for XLA on Xq21.3-q22 and a probable contiguous locus necessary for GH expression. The existence of other X-linked forms of IGHD is probable.

## Panhypopituitary dwarfism (PD)

PD is a condition characterized by deficiency of one or more of the pituitary trophic hormones (TSH, ACTH, FSH, LH) in addition to GH deficiency. Most cases are sporadic; however, several families have been reported suggesting an autosomal recessive mode of inheritance (type 1) as well as an X-linked form (type II). The loci responsible for most of these hereditary disorders have not been established except for a subtype of PD type 1 that associates GH, prolactin and TSH due to abnormalities in the human *Pit1* gene. The human gene encoding *Pit1* has been cloned and mapped to chromosome 3p11. A nonsense mutation in codon 172 (Arg to stop) was described in a Japanese patient with a phenotype remarkable for TSH deficiency leading to congenital cretinism. A missense mutation (Ala to Pro) in the POU-specific domain at amino acid 158 has been found in two Dutch families either in the homozygous state or associated with a deletion of the other *Pit1* allele. A *de novo* missense mutation just beyond the DNA-binding domain (Arg to Trp) has been reported in a heterozygous patient with the combined PD phenotype.

*Manuel Hernández and Jesús Argente*

See also 'Genetics of child growth' (**3.1**), 'Genetic and environmental influences on fetal growth' (**3.2**), 'Genetic regulation of growth-promoting factors by nutrition' (**3.4**), 'Chromosome aberrations and growth' (**7.4**) and 'Treatment of growth disorders: the future' (**13.4**)

# Endocrine growth disorders

Children who are tall, small, too fat, too thin, have early puberty or late puberty are generally those at the end of the distributions of normality. Identifying those who have an endocrine disorder of growth principally requires clinical skills and confirmation of a suspected diagnosis by appropriate biochemical tests. Most endocrine diagnoses are missed because observers do not think about them.

## Short stature

An algorithm is helpful for screening patients who are found to be short. The key to making a diagnosis of an endocrine disorder lies in identifying an abnormal growth-rate (**7.1**). Once this has been identified and other causes of abnormality have been excluded, endocrine screening tests and detailed

**An algorithm for the screening of patients found to be short.**

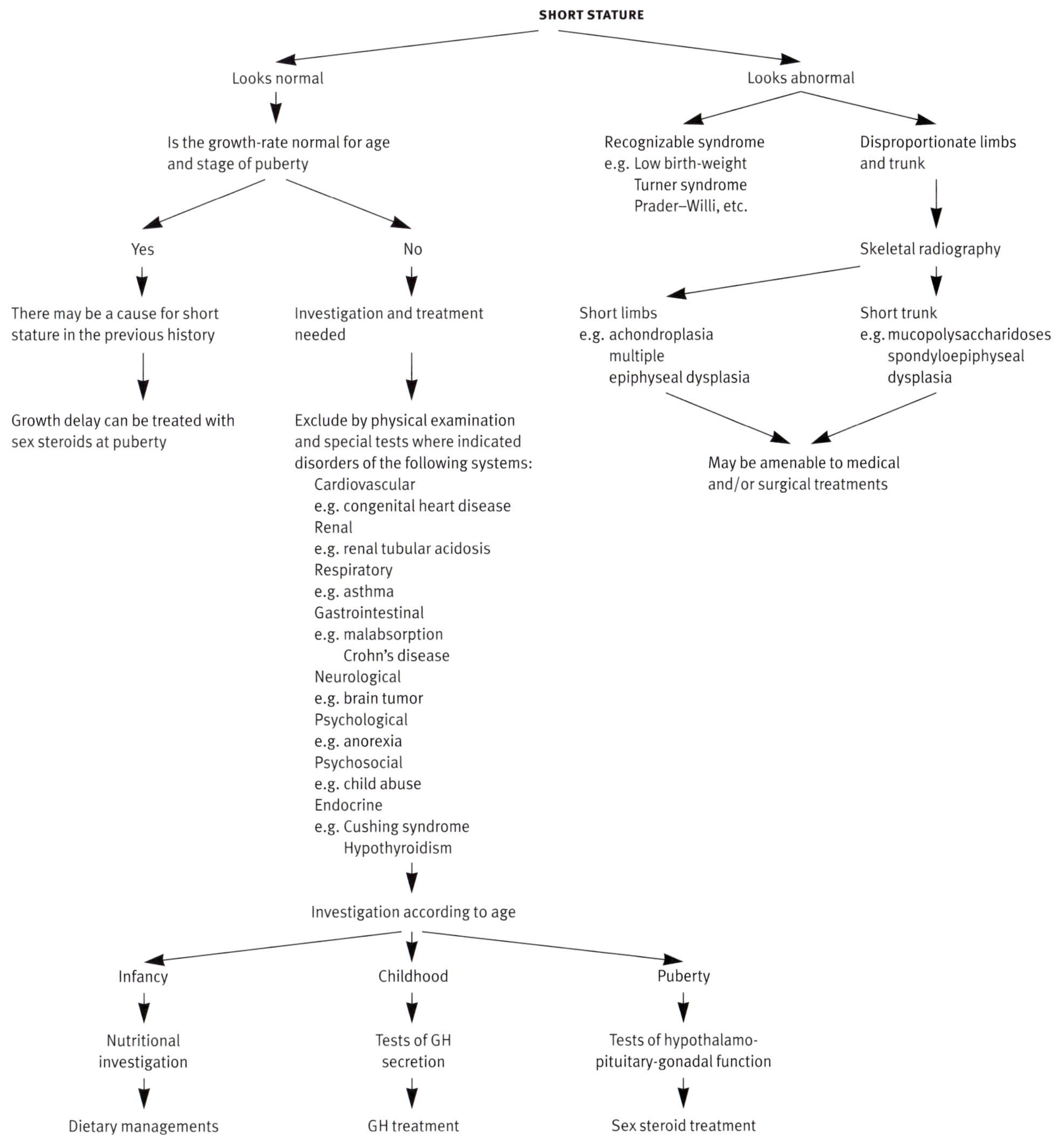

**Algorithm for the diagnosis of tall stature.**

**TALL STATURE**

- Looks normal
  - Is the growth-rate normal for age and stage of puberty
    - Yes
      - There may be a cause for tall stature in the previous history e.g. parents obesity from infancy
        - Is the height prediction acceptable? Is it amenable to treatment?
    - No
      - Are there signs of puberty
        - Yes
          - Investigate if appropriate → Is the height prediction acceptable? Is it amenable to treatment?
        - No
          - Why is the velocity increased? e.g. GH excess thyrotoxicosis What treatment is needed?
- Looks abnormal
  - Recognizable syndrome e.g. Cerebral gigantism
  - Disproportionate growth e.g. Marfan, Klinefelter
    - Is the height prediction acceptable?

      Is it amenable to treatment?

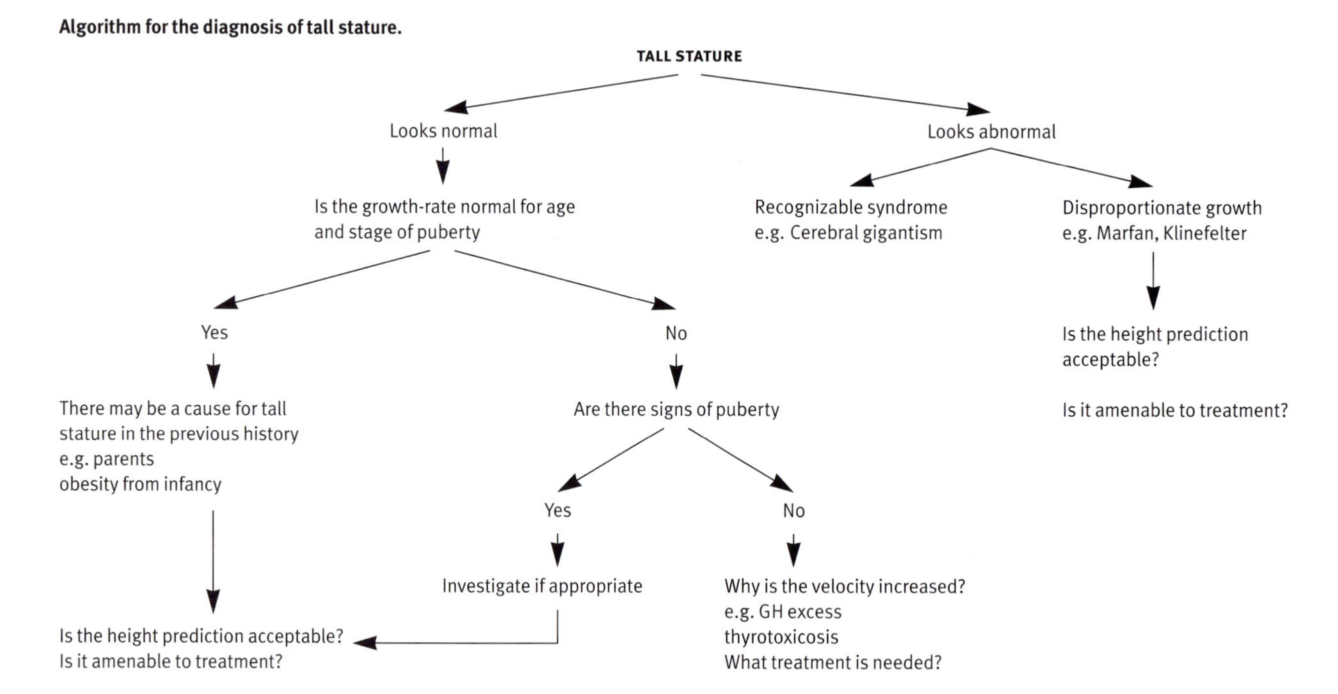

hypothalamo-pituitary function tests are indicated. Endocrine treatments for short stature are highly specific and effective – but they depend upon getting the diagnosis right.

## Tall stature

The use of an algorithm for the diagnosis of tall stature and the identification of an abnormal growth velocity in the absence of stimulation by sex hormones is the key to alerting the possibility of a diagnosis of an endocrine disorder. Growth hormone excess is usually treated by surgical removal of a pituitary adenoma, and thyrotoxicosis can also by treated surgically but medications are available to suppress both GH and thyroid hormones.

## Disorders of shape

Most children with excess fatness do not have an endocrine disorder. If they have become obese during the phase of infant growth which is determined by nutrition (**9.2**), they will not only be fat but also tall for the family. All of the endocrine causes of obesity (hypothalamopituitary problems, hypogonadism, Cushing syndrome, hypothyroidism and pseudohypoparathyroidism) are associated with a diminished growth-rate and short stature. The child who is short and fat or the child who has mental retardation and

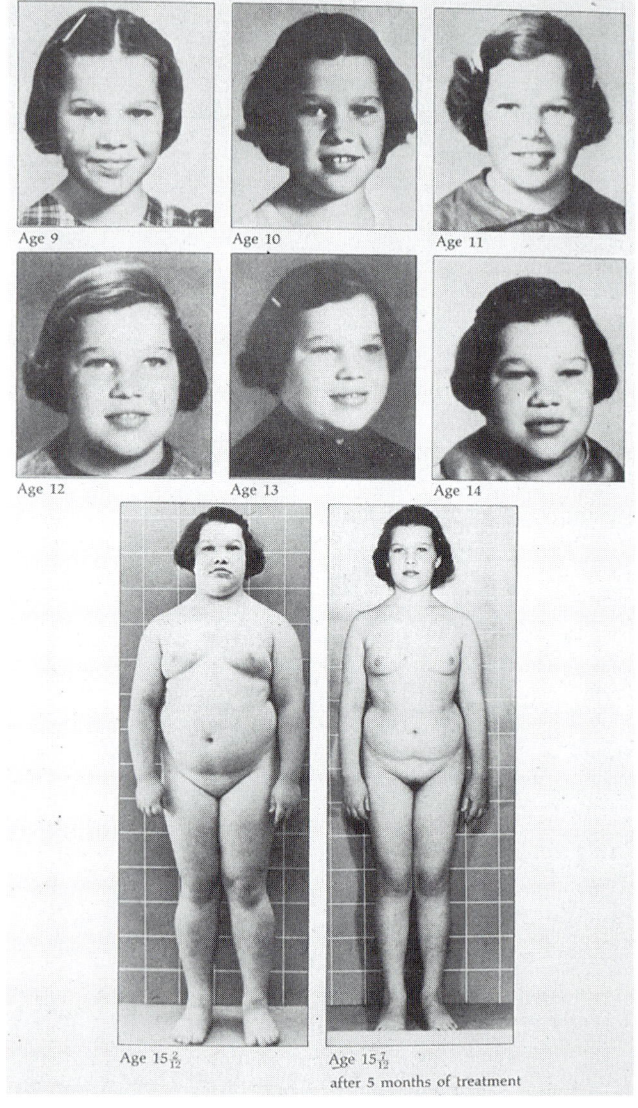


**The progressive effects of hypothyroidism. While height measurements from the school records of this girl show that thyroid malfunction began between age 9 and 10, the full clinical picture required several more years to emerge. From H. L. Barnett and A. H. Einhorne, eds. *Pediatrics*, 15th ed., New York: Prentice-Hall, 1972.**

**Algorithm for the management of early puberty.**

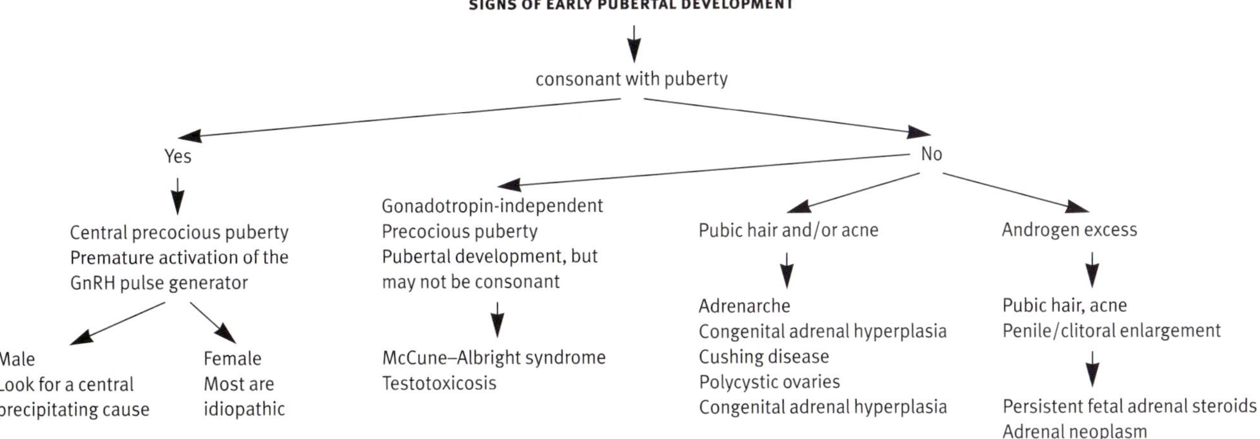


**Algorithm for the management of late puberty.**

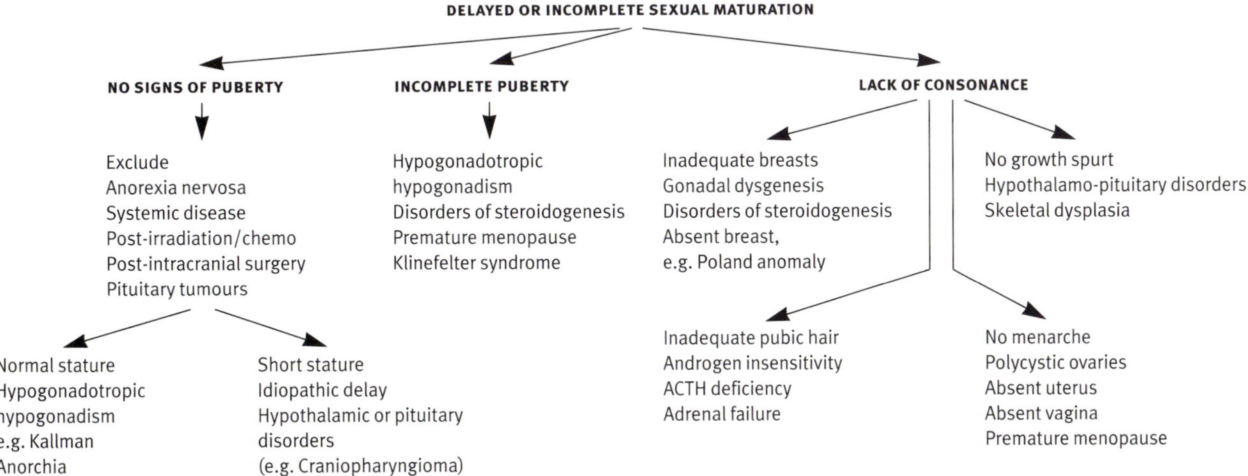


fatness may well have an endocrine disorder and may need detailed investigation to define it.

Primary endocrine disorders are not causes of thinness but there are causes (e.g. anorexia nervosa, bulimia, emotional deprivation and calorie restriction) in which endocrine abnormalities may be a secondary consequence, such as delayed or absent puberty in anorexia nervosa. It is rare that the imposition of an endocrine treatment ameliorates causes of thinness, although sometimes the manipulation of puberty may assist in the treatment of the emotional disorder.

## Puberty

The appearance of secondary sexual characteristics does not necessarily mean puberty has started. For this, all the signs of puberty have to be present. The endocrine events which lead

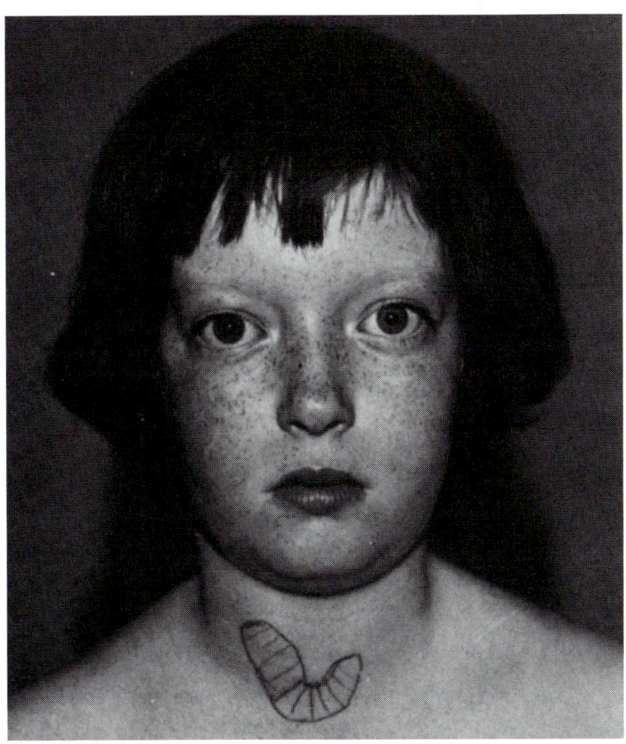

**Eleven-year-old girl with hyperthyroidism, whose diagnosis and management were instituted before there had been an appreciable impact on linear growth. Note the mild exophthalmos and the lines drawn over the enlarged thyroid gland. Medical suppressive therapy was instituted.**

to the development of secondary sexual characteristics antedate their appearance by a considerable amount of time and so isolated signs of puberty may be seen.

Early puberty is particularly common in girls and late puberty is particularly characteristic of boys because the pulse amplitude of gonadotropin secretion has to be greater to induce testicular secretion of testosterone than ovarian secretion of estradiol. Organic causes of early and late puberty are equally distributed between boys and girls, but with girls with early puberty outnumbering boys by 20 to 1, and the reverse for boys outnumbering girls with delayed puberty, the diagnostic search for a cause of early puberty in boys (usually a brain tumour) has to be more diligent and the same applies to girls with delayed puberty.

Fifty per cent of boys and girls show signs of puberty development by their 11th birthday. Children who show signs of pubertal development before the age of 8 years need consideration about causation as do children of 14 years who have failed to show any signs of puberty.

*C.G.D. Brook*

See also 'Anthropometry' (**1.4**), 'Biochemical markers' (**1.7**), 'Radiographic assessment' (**1.8**), 'Physical examination' (**1.10**), 'The use of growth references' (**1.17**), 'Standards and references for the assessment of fetal growth and development' (**4.9**), 'Sexual maturation' (**5.15**), 'Adulthood and developmental maturity' (**5.17**), 'Identification of abnormal growth' (**7.1**), 'Bone dysplasias' (**7.5**), 'Growth in chronic diseases' (**7.8**) and 'Treatment of growth disorders: the future' (**13.4**)

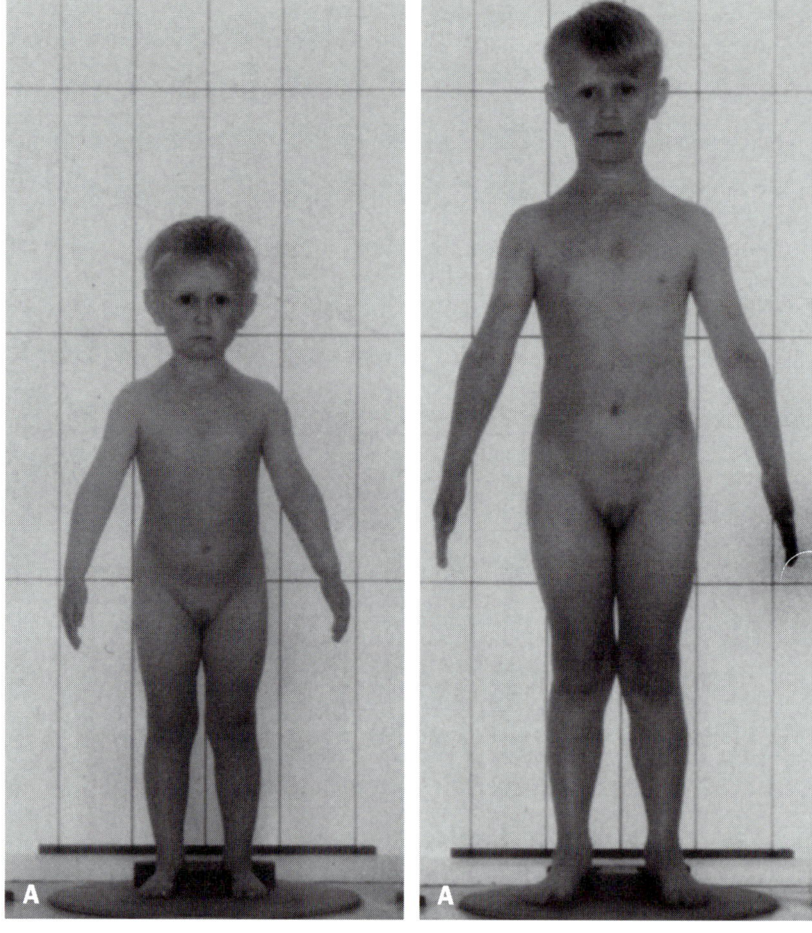


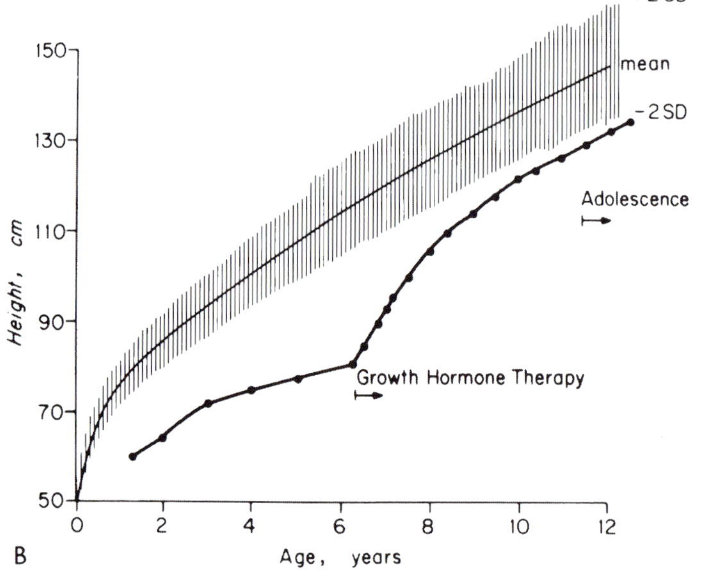


**Boy with isolated idiopathic pituitary growth hormone deficiency before and after 2 years of replacement therapy with intramuscular human growth hormone (A). Note the pretreatment mild adiposity, small facial bone structure, and relatively small penis. The pre- and post-treatment linear growth is shown below (B). Photos of same patient previously published in Tanner, J. M.: Human growth hormone. *Nature* 237:433, 1972.**

# Growth in chronic diseases

All of those who are responsible for the care of children with chronic illnesses are well aware that growth is often impaired. In some cases this effect is relatively trivial and certainly far less important than the problems of the primary disease. However, in others it forms a major problem of increasing importance. For some children this is becoming less of a clinical problem as the primary diseases are treated early, and the full impact on growth is avoided, but in others the reverse is true. This is particularly the case with a number of diseases including childhood malignancy and congenital heart disease, where previously lethal conditions are now being cured but where the impact on growth remains a significant problem. There is thus a new generation of children who have survived such conditions and are now reaching adolescence, but who are failing to grow and develop in a normal manner. These children present a completely new therapeutic challenge. As this is a phenomenon of many chronic diseases, it is clearly impractical to provide an exhaustive review of the growth patterns seen in each severe disease.

## General points

There are two distinct dimensions to growth which may be affected by any disease process. Firstly there is growth delay with implied reversibility (**9.6**). This may, of course, occur as part of the exceedingly common constitutional delay of growth in adolescence and in many endocrine and other specific growth disorders. It does, however, nearly always accompany the poor growth seen with most of the chronic diseases. Secondly, and in principle more seriously, there is absolute stunting of growth with an implication of irreversibility. The major problem is to discover where these two processes blur into each other, as it is often very difficult to tell how much of the growth failure is still in the reversible delay phase and how much is due to irreversible stunting.

Generally, delayed growth is the first effect and stunting only becomes manifest when the disease process has been chronic for a long period of time. In either case it is still very unclear as to the exact mechanisms by which these two effects are realized. There are of course diseases where there is a direct effect on the growth regulatory processes and then it is relatively easy to understand the impact on growth. One of the simpler examples in this area would be the effect of cranial irradiation given for, say, treatment of an intracranial tumour, with a subsequent effect on hypothalamic function leading to disordered secretion of pituitary growth hormone. Usually, however, the situation is nothing like as clear, but it is possible to identify five particular mechanisms which probably play a major role. These are by no means mutually exclusive or exhaustive and it is probable that there is an element of each of them in many conditions.

## Nutritional factors

Probably the most important factor in the pathogenesis of growth failure in chronic disease is poor nutrition. There are really three different mechanisms which need to be taken into account: 1) malabsorption; 2) intake limited by volume or content; and 3) anorexia. Clearly, in conditions such as coeliac disease, where malabsorption forms a part of the disease process, it is easy to understand how a reduced nutrient intake can occur with its subsequent effects upon growth. Reduced dietary intake may also occur due to severe dietary restriction as a consequence of treatment of multiple food allergies. Alternatively, the volume of food intake may be physically restricted, despite a reasonably normal appetite. The latter may occur in children with, for example, inadequately controlled disorders of fluid regulation who are drinking large volumes of water. Finally, and probably of great importance in many chronic diseases, the children may be anorexic. Many sufferers of prolonged and unpleasant illnesses have a relatively poor intake simply due to a very reduced appetite.

The manner in which undernutrition leads to poor growth is in itself very interesting. Chronic malnutrition leads to elevation of plasma growth hormone levels in the presence of a fall in insulin-like growth factor I (IGF-I). This peptide is known to stimulate cartilage activity and hence promote growth in long bones, and is probably the principal mediator of the growth-promoting actions of growth hormone.

The reduced IGF-I levels rise when malnutrition is corrected. The effect of the high growth hormone secretion with lowered IGF-I levels allows the separation of the growth-promoting actions of growth hormone (IGF-I mediated) from the metabolic properties of fat mobilization. It is possible that this mechanism, which may operate in a number of conditions where there is impaired endogenous glucose production, may in fact be protective by limiting energy-expensive growth while preserving cerebral and other more important functions.

## Protein-loss

Chronic protein-loss as part of a disease may also occur in a number of ways. It may have specific effects leading to relative protein malnutrition, or it may just contribute to a generally poor caloric balance. Such protein-loss can occur rather specifically, as in protein-losing bowel disorders, or in a less specific way in such conditions as severe eczema, where protein is lost in exudate from the skin lesions.

## Chronic inflammation

In a more obscure way it seems that diseases associated with chronic severe inflammation of one or more organs may lead to poor growth. The classic example is juvenile rheumatoid arthritis particularly if there is severe involvement of visceral organs. It is, however, notable that in this condition there may be locally increased bone growth in or around specifically inflamed joints. This argues for a double effect with locally increased growth, possibly secondary to increased blood flow, associated with more generalized reduced skeletal growth. Other conditions which may be cited are the inflammatory bowel diseases, although in these cases there may be significant malabsorption as well. There is also a possible role for the cytokines, a family of peptides that mediate the inflammatory process, but which may interfere with the actions of local growth factors such as IGF-I.

## Tissue anoxia or acid-base disturbance

This is a rather less certain area, although it has been long believed that the poor growth which is often a feature of severe congenital heart disease is due at least in part to tissue anoxia. In a similar manner, the poor growth found with some renal disorders may be due to chronic acidemia due to impaired excretion of acidic metabolites. As in both these cases there is probably significant nutritional deficiency as well, it is very difficult to define the exact contributions of these various pathogenetic mechanisms.

### GROWTH IN RENAL DISEASE

Growth in chronic renal failure (CRF) is adversely affected by inadequate energy intake due to anorexia, nausea and vomiting; salt and water abnormalities; acidosis; metabolic bone disease; and endocrine disturbances (**7.7**). Growth post-transplant depends on the success of the graft, and is also influenced by steroid therapy. The mean height for children with end-stage renal failure (ESRF) and post-transplant is less than 2 standard deviations below the mean for the normal population.

#### *THE PATTERN OF GROWTH IN CRF*

Growth in the first year of life is most severely affected: 70 per cent of infants presenting at this age are below the third centile for height. Bone age is retarded, the more so the earlier the onset of CRF. Thereafter, increase in height and bone age continues parallel to the percentiles, but catch-up is rarely seen. Subsequently, appearance of secondary sexual characteristics (**1.10**) occurs on average 2.5 years later than the normal population. The growth spurt is shortened, with a pubertal height-gain of less than 50 per cent of normal and there is rapid bone maturation, resulting in loss of height potential. Deterioration of CRF is usually associated with a declining height velocity that is rarely improved by dialysis.

#### *THE PATTERN OF GROWTH POST-TRANSPLANT*

Following successful transplantation in early childhood using standard immunosuppression with steroids, cyclosporin A, and azathioprine, there is rapid catch-up growth during the prepubertal years, but subsequently there is delayed bone age and appearance of secondary sexual characteristics and a delayed, attenuated, prolonged pubertal growth spurt, probably due to steroid therapy.

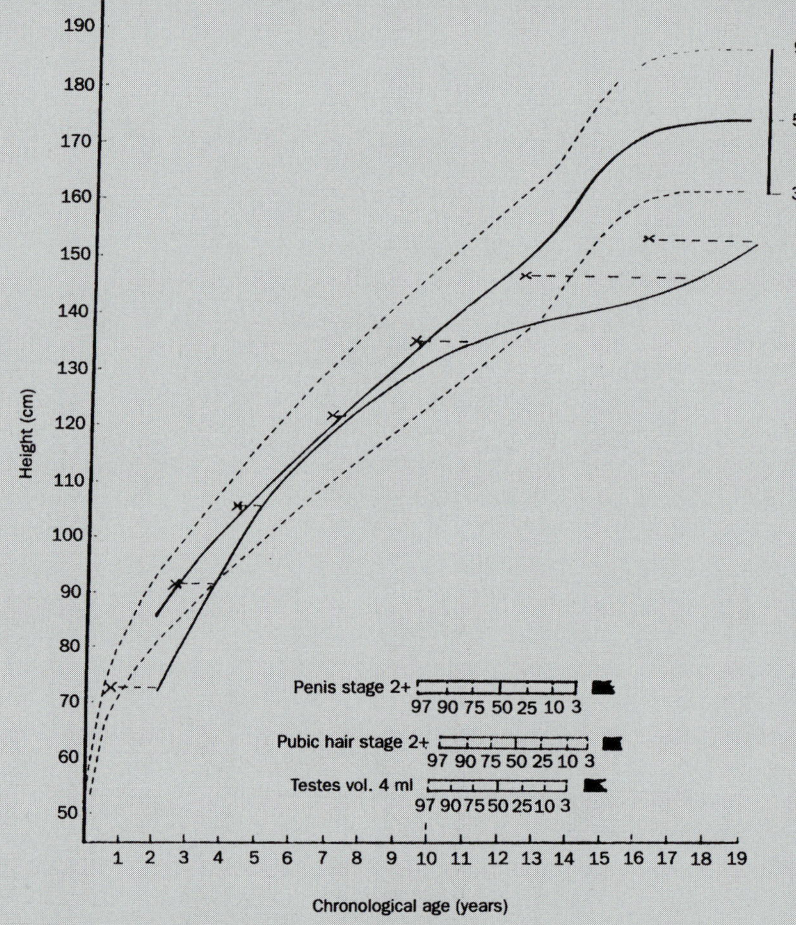


**Typical growth chart of a child with chronic renal failure.**

#### *CAUSES OF POOR GROWTH IN CRF*

*Nutrition*

Provision of adequate nutrition (**9.2**) is a particular problem in infancy. Growth is at its most rapid at this time, and is principally dependent on nutrition rather than growth hormone (GH). Energy and protein requirements are 2 to 3 times higher per kilogram than in the adult. Infants with CRF are difficult to feed, and vomit frequently. Nasogastric or gastrostomy tube feeding is often necessary to provide adequate energy. If vomiting is severe, gastric fundal plication may help. The principal cause of CRF in infancy is structural abnormality of the urinary tract, so that multiple surgical operations, requiring periods of fasting, may be required. Loss of height has been reported to be as great as 0.6 standard deviations per month at this age. Anorexia and vomiting may continue into later childhood, worsening as renal failure progresses.

*Abnormalities of salt and water balance and acidosis*

Many of the structural and congenital renal abnormalities disproportionately affect the renal tubules, resulting in salt and water loss so that

## Therapy

Without doubt there are a number of potent therapies, essential for halting or curing life-threatening diseases, which are themselves toxic to the growth process. Most striking among these are the corticosteroids and in particular the more potent fluorinated analogues. It is now clear that these compounds have some effect on the growth process on at least three levels as they tend to reduce pituitary growth hormone production, interfere with hepatic IGF production and have profound effects on protein synthesis in growing cartilage. It seems likely that the last mechanism is the most important in terms of the degree of clinical impact.

Over the last two decades the growth problems that may accompany chronic childhood disease have assumed greater importance. This is due to the increasing number of children who survive previously lethal conditions, but also because of the recognition that stunting of growth as a penalty for such survival is unacceptable. At present many of the growth sequelae are not readily amenable to treatment or prevention, but hopefully this will not always be the case. Attention to this complication is already bringing improvements as, for example, in modern schedules for the use of corticosteroids and the greater emphasis placed on nutrition in any long-standing pediatric illness.

*Michael Preece*

See also 'Nutrition' (**9.2**), 'Infection' (**9.3**) and 'Treatment of growth disorders: the future' (**13.4**)

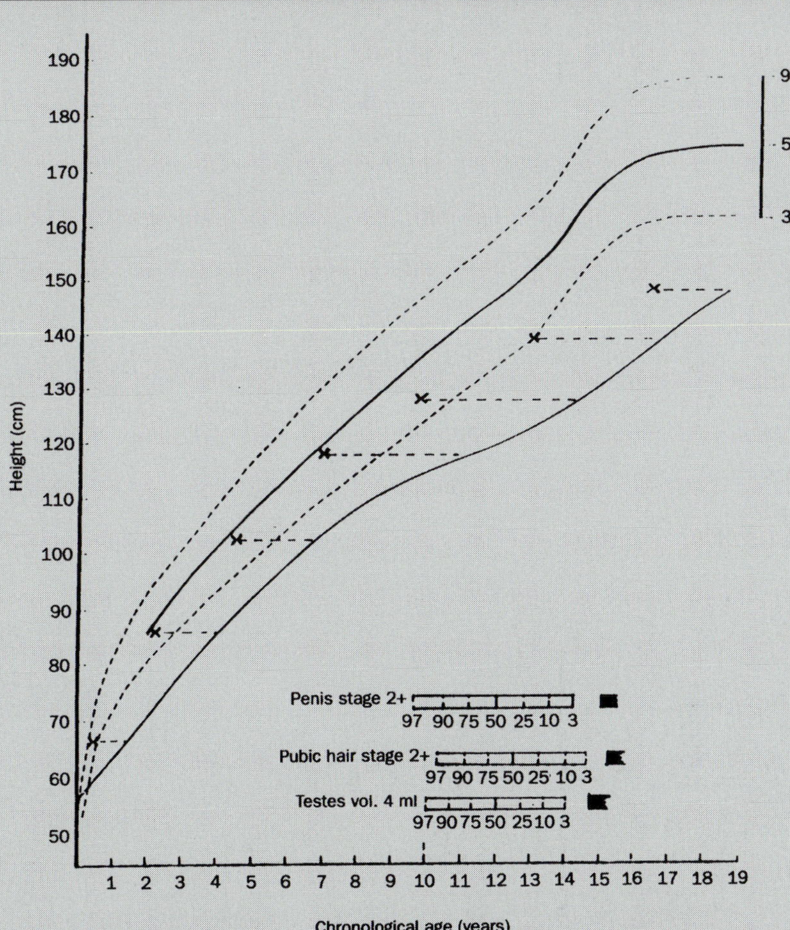


**Typical growth chart of a child with chronic renal failure following successful transplantation in early childhood, using steroids, cyclosporin A and azathisprine as immunosuppressives.**

the children may be in a chronic state of salt and water depletion. This, along with metabolic acidosis, may adversely affect growth.

*Metabolic bone disease*

Renal osteodystrophy can result in destruction of the metaphysis, and bone softening and deformity. However, such abnormalities can be prevented with phosphate binders, and activated vitamin D, and are now rarely seen.

***CAUSES OF POOR GROWTH, POST-TRANSPLANT***

The children who grow the best post-transplant are the youngest ones, with the greatest growth delay and with the best graft function. Chronic graft failure and high steroid dose adversely affect growth.

***ENDOCRINE ABNORMALITIES IN CRF***

GH levels are normal or high in CRF due to decreased clearance. GH binding protein is low, suggesting that there is deficient GH receptor expression. Insulin-like growth factor I (IGF-I) bioactivity is reduced, but by radioimmunoassay, levels are normal, suggesting the presence of circulating IGF-I inhibitors. These may be insulin-like growth factor binding proteins (IGFBPs), or their fragments, which are poorly cleared from the circulation in CRF. The increased binding capacity results in decreased free, bioactive IGF-I.

Response of testosterone to HCG is reduced in boys, and basal levels of estrogen and progesterone are low in girls. Pulses of luteinizing hormone (LH) are irregular and bioactivity is reduced. Failure of GH and the sex steroids to act in concert cause delayed puberty and an abnormal pubertal growth spurt.

Steroids have both a local effect on collagen synthesis that may be direct or as a result of interference with the action of IGF-I, and a central depressant effect on the pulsatility of GH secretion, particularly during puberty. Growth in adolescents with pubertal delay may continue into their 20s.

***EFFECT OF RECOMBINANT HUMAN GROWTH HORMONE (RHGH) IN CRF AND POST-TRANSPLANT***

The GH resistance of CRF and potentially blood levels of GH in children taking steroid therapy post-transplant provide a rationale for the use of rhGH. In the short term, rhGH increases the rate of growth in CRF, with maximal effects in those on conservative management, and the least effect in dialysed children. Post-transplant, the maximum effect is seen in the youngest children with the best renal function and the lowest steroid dose. Results of 5 years of treatment are becoming available and suggest that the improvement in growth is maintained. Side effects of rhGH appear to be uncommon.

*L. Rees*

## GROWTH OF CHILDREN WITH GASTROINTESTINAL DISEASE

A number of gastrointestinal disorders are associated with delayed growth and malnutrition (9.2, 9.3). Despite the fact that in many of these disorders there are absorptive and digestive defects as well as excess nutrient loss from the gastrointestinal tract these alone cannot account for the growth problems which occur in children with gastroenterological disorders. In fact, a major factor in virtually all gastrointestinal disorders associated with poor growth is associated anorexia. Thus, a vital part of the assessment of the child with gastrointestinal diseases associated with impaired growth is a detailed assessment of the energy intake in each child.

As the child is a developing and growing organism, when gastrointestinal disease causing growth impairment is unrecognized and untreated, this has an adverse prognosis for the fulfilment of their growth potential. In general, it is important to appreciate that the capacity of the human body to respond to changes in the availability of nutrients by slowing or halting growth is generally a reversible process. In this way adaptation to change in nutrient supply can occur. However, the ability to resume normal growth after the correction of metabolic insults is influenced by how long the insult lasts, as well as the phase of development during which it occurs. When damage occurs during susceptible periods of growth, the effect on growth may be irreversible. However, before puberty there is usually the possibility of catch-up growth and the pubertal growth spurt. So when disorders associated with growth impairment can be effectively treated before puberty, there is a real chance of catch-up growth occurring.

### INADEQUATE INGESTION

Especially in infancy, there are a number of disorders causing considerable problems for the child to secure adequate dietary energy related to mechanical reasons involving the upper gastrointestinal tract. When surgical correction is possible the prognosis may be good but very often this is not possible. Then there is the need for a detailed team approach to co-ordinate the feeding care of these children.

### ANOREXIA

Whenever there is chronic inflammation of the gastrointestinal tract there is usually associated anorexia. Furthermore, in disorders such as coeliac disease, anorexia plays a major part in the development of impaired growth. Various iatrogenic causes may be associated with anorexia, for example, when the child is placed on a therapeutic diet which is essentially unpalatable, or a diet that is deficient in basic nutrients.

### DEFECTIVE DIGESTION OR ABSORPTION

These disorders, while usually accompanied by anorexia, may have significant malabsorption which plays a part in growth impairment. A significant example of this is cystic fibrosis. However, modern treatment with a high energy intake associated with appropriate pancreatic enzyme supplementation usually leads to resolution of the growth problem. In disorders such as coeliac disease, once the disease is specifically treated with a gluten-free diet there is good prognosis for growth.

### EXCESSIVE LOSSES

When there is severe vomiting, for example in gastroesophageal reflux, this can be associated with considerable growth impairment. Equally, when there is excess energy loss from the gastrointestinal tract, as in protein-losing enteropathy, there is also impaired growth.

### OTHER FACTORS

When there is increased metabolic demands there may also be impaired growth failure. Drug therapy such as steroids can lead to impaired growth.

### EXAMPLES OF GASTROINTESTINAL DISORDERS ASSOCIATED WITH POOR GROWTH

*Coeliac disease*

Severe growth retardation, delayed development, sometimes with delayed onset of puberty were well known to the early observers of coeliac disease. Samuel Gee, writing in 1888, remarked 'while the disease is active children cease to grow; even when it tends slowly to recover they are left frail and stunted'. Growth retardation and late onset of puberty continue to be important complications of coeliac disease, albeit less common now owing to earlier diagnosis and treatment. Screening for anti-endomysial antibodies is valuable but small intestinal biopsy remains the gold standard for diagnosis. Catch-up growth may occur once there has been a response for gluten-free diet.

Coeliac disease may present with a growth failure as the sole manifestation, without any gastrointestinal symptoms. In a study of 93 newly diagnosed children with coeliac disease, the majority were under 50th height centile and 27 were below third height centile. Late diagnosis of coeliac disease in these circumstances carries a poor prognosis for final height.

*Chronic inflammatory bowel disease*

Short stature and delayed puberty are frequent complications of childhood inflammatory bowel disease, both for Crohn's disease and ulcerative colitis. A survey of 96 patients with Crohn's disease attending an outpatient clinic showed that 23 per cent were below the third centile for height. Short stature may be a presenting feature of Crohn's disease. Clinical pointers to the diagnosis of Crohn's disease include mouth ulcers and anal tags. These features should always be sought in children presenting with unexplained growth failure. Appropriate therapy with drugs, diet or surgery offer a good prognosis for height in most circumstances.

*J.A. Walker-Smith*

*Gastrointestinal causes of poor growth*

**Inadequate Ingestion**
Transfer dysphagia
  Neuromuscular co-ordination
Dysphagia
  Neuromuscular inco-ordination
  Post-operative repair of esophageal atresia
  Esophagitis and stricture/corrosive acid/peptic

**Anorexia**
Inflammatory lessions
  Gastritis
  Post-enteritis enteropathy
  Crohn's disease
  Ulcerative colitis
Coeliac disease
Stagnant loop syndrome
Therapy
  Unpalatable diet
  Deficient diet
  Drugs – Sedatives
    – Gastric irritants

**Defective digestion or absorption (usually accompanied by anorexia)**
Pancreatic insufficiency
  Cystic fibrosis
  Others
Intestinal
  Coeliac disease
  Stagnant loop syndrome
  Cow's-milk-sensitive enteropathy

**Excessive losses**
Gastro-esophageal reflux
Pyloric stenosis
Incomplete, recurrent bowel obstruction
Protein-losing enteropathy

**Other protein**
Increased metabolic demands
Corticosteroid therapy

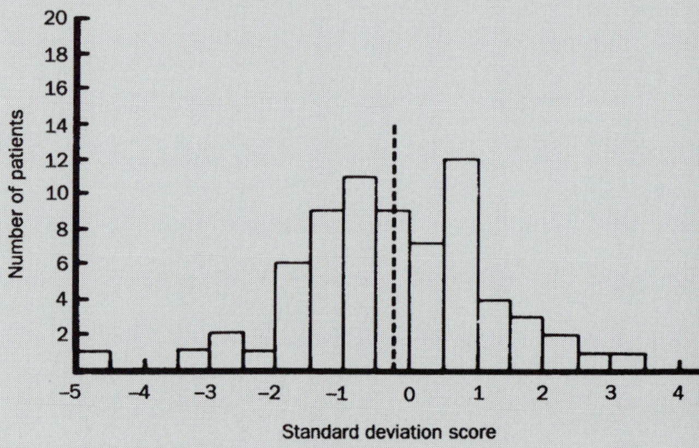


**Height centiles at transfer to adult clinic in 70 patients. The dotted line is the median.**

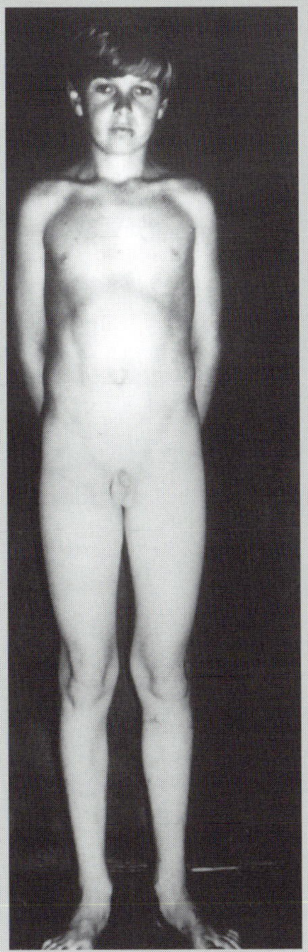

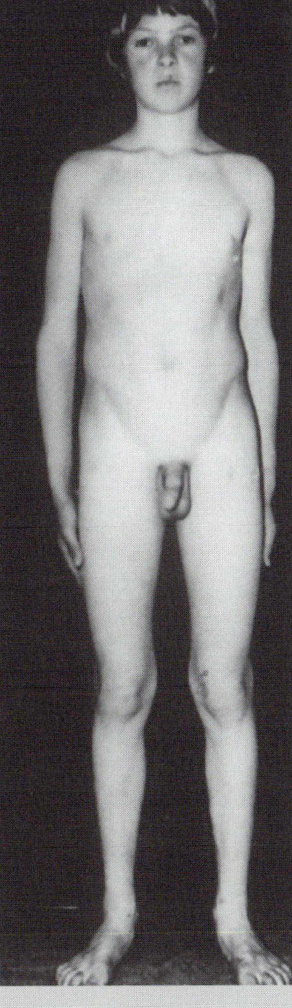

**Boy with coeliac disease**
**(above) undiagnosed with delayed puberty and retarded growth (above right) entering puberty and showing 6 months after gluten-free diet.**

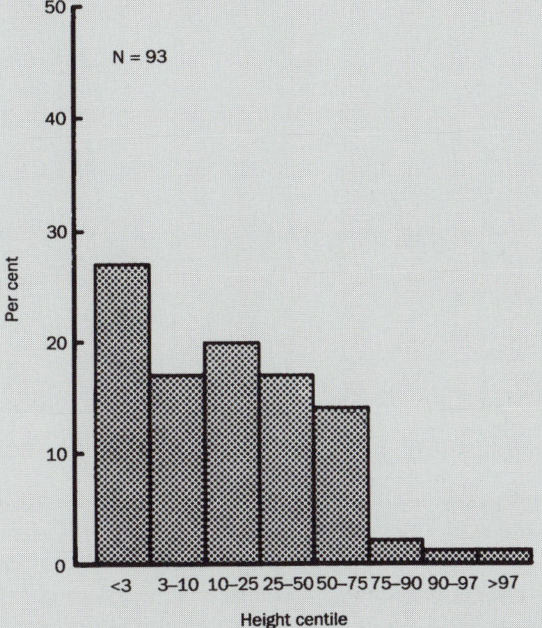


**Height centiles of children with coeliac disease at time of diagnosis.**

## GROWTH OF CHILDREN WITH ASTHMA

Over 50 years, both cross-sectional and longitudinal studies have shown that asthmatic children are of shorter stature than the general population. Both asthma clinic-based studies and investigations of whole communities have consistently demonstrated that independent of treatment, significant growth retardation with delayed bone age occurs in children with asthma, relating directly to the severity of disease. However, the pattern of growth is that of healthy late developers, in that final adult height (**5.17**) is not compromised. Indeed, one study of Israeli conscripts suggested that those with mild childhood asthma were marginally taller than their non-asthmatic peers, whilst those with severe disease were of the same height as non-asthmatic peers.

Many mechanisms have been suggested to explain the poor growth associated with asthma. Tissue hypoxia, particularly in relation to nocturnal desaturations, might interrupt biological rhythms of the growth-related hormones, but there is no evidence of this. Aberrant nutrition, perhaps because of poor appetite or impaired absorption, might occur in some asthmatics, particularly in those with associated eczema and food intolerance, where there is a direct effect on the gut. One recent study has indicated that resting energy expenditure (**9.2**) is increased in asthmatic children compared with matched controls. However, other factors such as a metabolic effect of beta-adrenergic agonists might have contributed to the observation. Chronic stress, both physical and psychological, is present in asthmatic children associated with increased severity of disease. Poor growth, secondary to emotional deprivation, accompanied by disruption of hypothalamic pituitary function, could conceivably explain the associations. However, this is difficult to quantify and being related to the severity of disease, is hard to dissociate as a causative factor. Finally the use of corticosteroids, with their well-established effect on growth, has a progressively greater impact, as they are used more frequently in more severe disease.

Oral corticosteroids have a dose-dependent effect in reducing linear growth in asthmatic children. While alternate-day steroids might have some degree of growth-sparing which has even been demonstrated using day-to-day measurements of leg length by knemometry, a cumulative dose which is greater than 6 milligrams per metre$^2$ per day will lead to evident growth retardation in asthmatics. Recent attempts to give simultaneous growth hormone treatment have demonstrated very limited success with effects only evident when relatively low doses of prednisolone were used (equivalent to < 0.35 mg/kg/day of prednisone). There is limited evidence that substitution of glucocorticoid therapy with adrenocorticotrophic hormone might decrease the effects on growth, though it seems improbable, at equivalent doses to control disease, that there would be any difference.

Inhaled corticosteroids (ICS) have now been in common usage for the treatment of childhood asthma for over 25 years. Prolonged follow-up studies using low doses of beclomethasone, up to 400 micrograms per day, suggested that there were few, if any, side-effects and no impairment of growth. This led to the concept that ICS were totally safe and devoid of systemic effects, which is clearly not the case. Dose-dependent adrenal suppression has been shown, using 24-hour urinary-free cortisol and overnight cortisol profiles. Beclomethasone, at a dose of 400 micrograms per day, consistently suppresses the small pulses of cortisol release overnight, though having little impact on the morning peak. Other studies have demonstrated a dose-dependent effect on short-term leg length growth using knemometry (**5.13**). There is only one double blind controlled clinical trial which set out to elucidate effects of ICS on growth. In children with mild recurrent wheeze, beclomethasone at 200 micrograms per day produced a mean decrement in growth over 7 months' use, relative to a placebo group, of 0.9 centimetres with no catch-up growth after stopping treatment over a further 5-month period. The effect may be different in patients with more severe disease, but with increasing doses, growth suppression has been clearly described. It is possible that newer ICS with lower absorption and high first-pass clearance will have fewer effects on growth but controlled comparative studies are lacking.

While physicians might feel that it is more important to treat the asthma than worry about effects on stature, there must be a continual assessment of the risks and benefits of particular treatments. Thus, ICS should be used when the disease warrants it, but the minimum dose necessary to control the problem should be established by appropriate titration and if alternative and safer prophylactic medications are effective, then they should be used in preference. This has prompted a judicious step-wise approach to the management of chronic asthma as proposed by the International Pediatric Asthma Consensus Group.

*J.O. Warner*

# Treatment of growth disorders

The treatment of growth disorders depends on the underlying cause but can broadly be classified into whether the growth failure takes place during infancy, childhood or puberty. In infancy the major component of growth is nutrition and disorders of nutrition, for example, malabsorption, should be looked for and managed appropriately. During childhood most of growth is growth hormone dependent and as such growth-hormone deficiency and its treatment comes to the fore. The important principles in the management of growth hormone treatment relate to the dose of hormone used, the frequency of administration of the treatment and the condition for which treatment is prescribed.

There are a series of dose-response curves which relate growth response to dose and the pretreatment growth-rate of the individual. Children growing extremely poorly with severe growth-hormone deficiency (upper line on the top figure) have the best response on a dose-for-dose basis compared to children who are growing with a near normal growth-rate prior to intervention (lower line on the same figure).

The frequency of administration should ideally mimic the physiological pulse pattern, but this is rarely achieved because of the pharmacokinetics of a subcutaneous growth hormone injection. Frequency is important as the more frequently the dose is administered the better is the effect. The condition being treated needs to be considered also. Although the classic indication for growth hormone therapy is growth-hormone deficiency, a number of other conditions will respond to treatment, but the doses required will be greater. For example, the growth response for growth-hormone deficiency and skeletal dysplasia hypochondroplasia differs, the latter showing a much reduced growth response on a dose-for-dose basis than that seen in the straightforward deficient state.

Using the hypothalamic peptides growth-hormone-releasing hormone and a novel growth-hormone-releasing peptide (GHRP) opens up a new possibility for therapy which might more closely mimic the physiological pulsatile pattern of secretion. These peptides have very short half-lives, so a growth hormone profile more approaching that seen in physiology can be achieved. A particular advantage of the GHRPs is that they can be given orally and, with their particular pharmacokinetics, a more physiological replacement regimen might be achieved.

Treatment of growth disorder in puberty largely centres on the management of delayed puberty. This is an exaggeration of the normal deceleration in growth which takes place prior to the pubertal growth spurt. In these individuals pubertal growth is delayed in time so that in the short term the individual becomes shorter compared to their peers, although ultimately, once the growth spurt gets under way, their final height will be the same as their peers. In this situation management centres on mimicking the pubertal growth

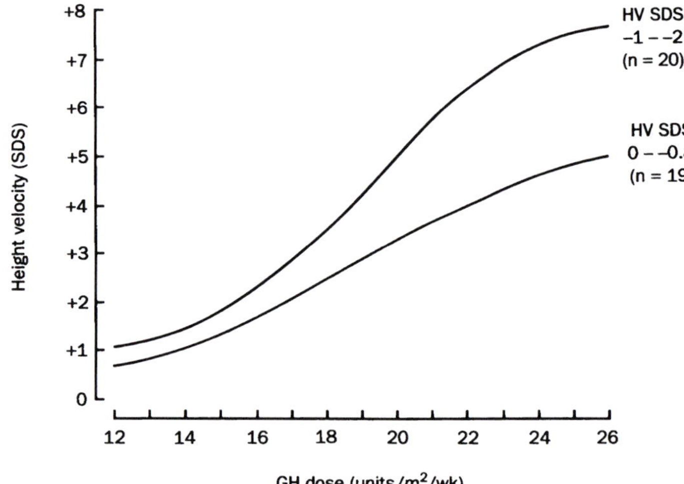


**Dose reponse curves for two different height velocity groups.**

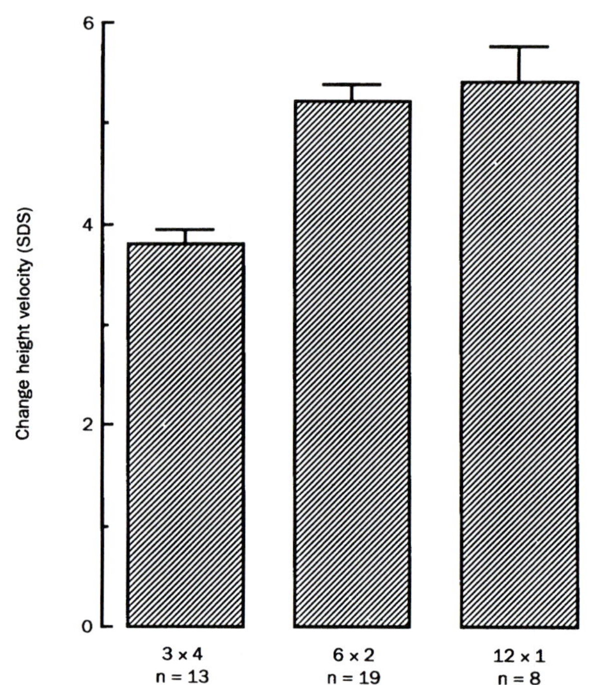


**Response following 12 months of growth hormone administration at three dose regimens. Here the same total dose per week (12 units) is given three times per week (4 units), six times per week (2 units) or twice daily (1 unit). There is a limit to the frequency of administration which is determined ultimately by patient acceptability of frequent subcutaneous injections. A once-daily regimen is considered optimal.**

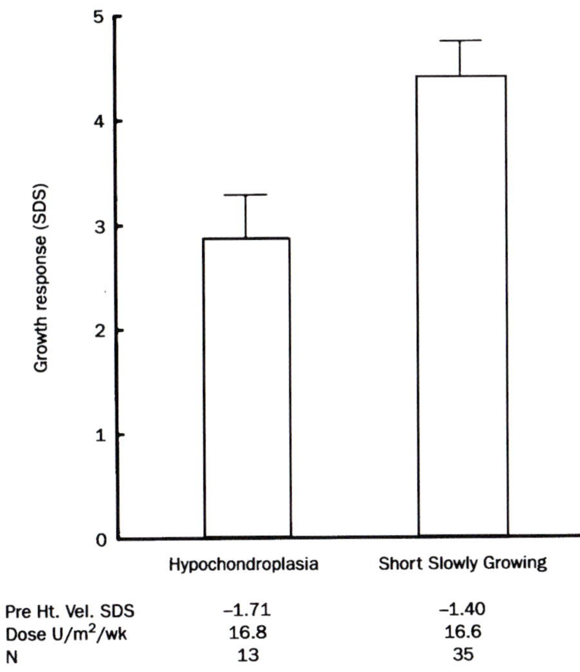


| | Hypochondroplasia | Short Slowly Growing |
|---|---|---|
| Pre Ht. Vel. SDS | −1.71 | −1.40 |
| Dose U/$m^2$/wk | 16.8 | 16.6 |
| N | 13 | 35 |

**Comparison of growth response to biosynthetic growth hormone in 12 hypochondroplastic and 35 short, slowly growing children.**

spurt at an earlier stage and, in effect, relies on using low doses of sex steroid to achieve this aim. High doses of sex steroid lead to growth acceleration but are associated with an unacceptable advance in skeletal maturation which is not seen with the low dose approach. *Peter Hindmarsh*

See also 'Physical examination' (**1.10**), 'Growth cyclicities and pulsatilities' (**5.13**), 'Adulthood and developmental maturity' (**5.17**), 'Identification of abnormal growth' (**7.1**), 'Genetic disorders of the growth hormone axis' (**7.6**), 'Endocrine growth disorders' (**7.7**) and 'Treatment of growth disorders: the future' (**13.4**)

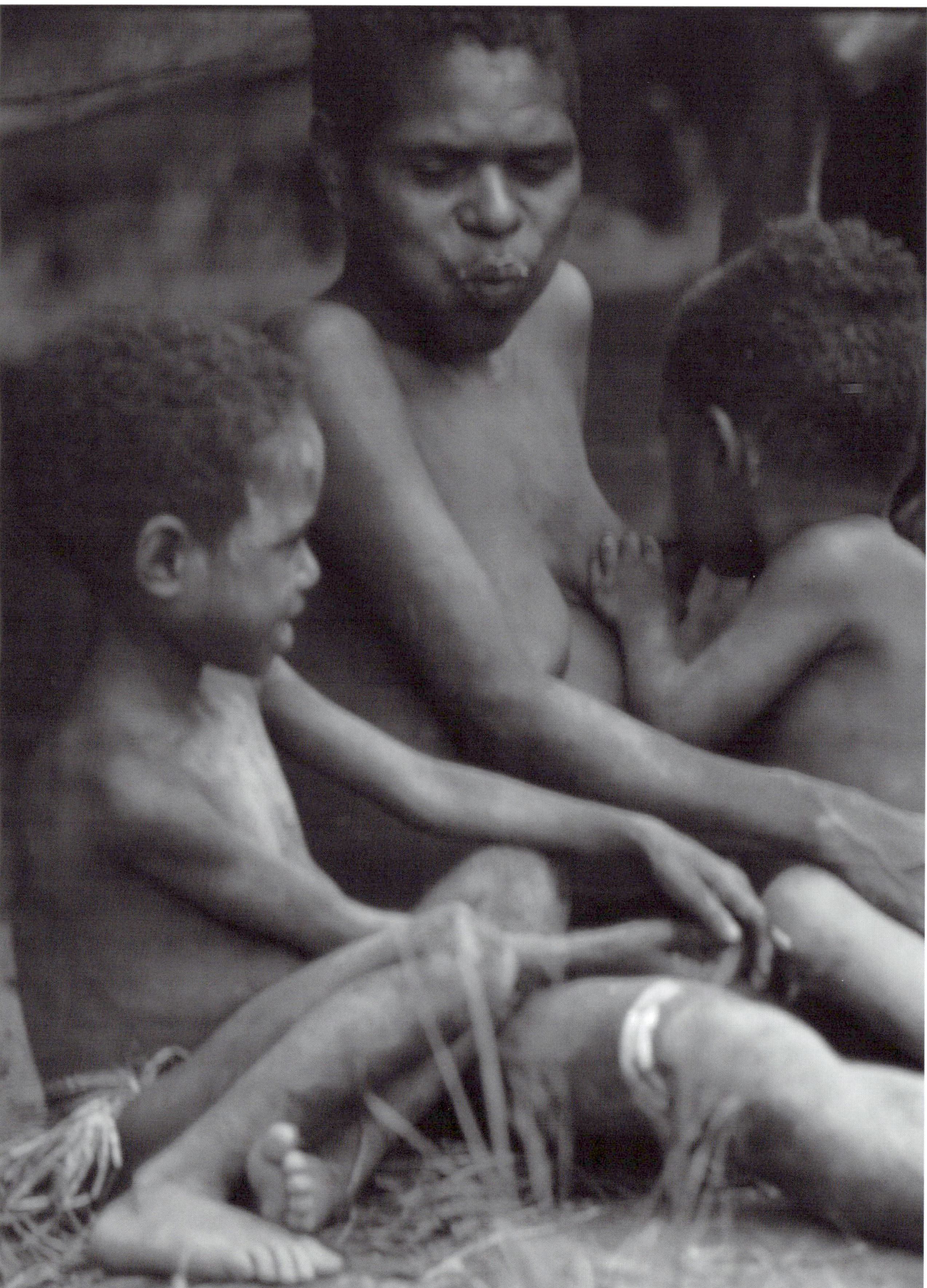

## Part Eight

# Environmental factors influencing birth-weight

Birth-weight is an easily measured integrated result of prenatal growth and is regularly used by clinicians and public health workers as an indicator of newborn viability. Considerable variation exists within and among populations in newborn size. Indeed, so great is this variation among populations that the birth-weight mean in one population could be considered in another population to be low and indicative of a poor prognosis for the infant. Birth-weight is multifactorial and sensitive to a large variety of environmental influences, the largest of these, after the length of gestation itself, are mother's size and weight-gain (**8.1**), reflecting both maternal genetic endowment for her infant's size, her past nutritional status and her current energy balance (**8.2**), and cigarette smoking (**8.3**).

The causes of variation in size vary among populations. It is useful to identify the causes, particularly the causes of low birth-weight, in order to design effective interventions to prevent the condition and its attendant risks of greater subsequent morbidity and mortality. Decreasing adverse influences on prenatal growth to prevent low birth-weight is a major goal of public health programmes in both the developed and developing world.

Birth-weight is a product of the rate of weight-gain and the duration of prenatal growth. Small size at birth can result from either shortened gestation or slowed rate of weight-gain, or a combination of the two. The length of gestation is 40 weeks from the onset of the last menstrual period, and in most women, is approximately 38 weeks from conception. Definitions of norm term for the gestation of a fetus must be observed if rates of prematurity, and weight-gain are to be compared across studies of different populations. Preterm birth is defined as one occurring before the completion of the 37th week of pregnancy, while a full-term birth occurs between the completed 37th week and before the completion of the 42nd week of pregnancy. Births occurring thereafter are termed post-term. The rate of weight-gain changes immensely from conception to delivery, with the greatest relative weight-gain occurring in the ovular period immediately after conception. However, the greatest absolute weight-gains occur much later, between the 30th and 37th weeks of pregnancy in most samples, weight-gain slowing to less than 100 grams per week during the final weeks of pregnancy.

Fetal weights can be estimated by using ultrasonographic measurements of size *in utero* at specific times, while the end-product of gestation can be obtained from weighing infants at birth. The birth-weights of preterm neonates have been used to estimate patterns of

**Opposite: A mother and her two children in Papua New Guinea. Photograph by Stanley Ulijaszek. Among the many things that affect birth-weight are socio-economic status, and maternal weight and nutritional status.**

weight growth earlier in gestation. Since most viable fetuses are born after 27 weeks of pregnancy, the rate of weight-gain during the final 13 weeks of pregnancy is most important for understanding the interrelationship between the duration of gestation, weight-growth and birth-weight.

## Typical or normal birth-weight

The average birth-weights of populations in Europe and North America are approximately 3400 grams. Birth-weight is not a normally distributed variable because there is a long tail of weights less than the mean that decreases to the limit of viability, and the tail of heavy birth-weights is shorter due partly to the obstetric practice of inducing labour at 42 weeks gestation, owing ultimately to the risk of mortality associated with post-term delivery after this time. The standard deviation of birth-weight is typically slightly less than 500 grams.

Average birth-weight varies considerably around the world. The average birth-weight of poor Delhi newborn girls being on average 2.75 kilograms, and male Lumi newborns of New Guinea weighing even less, 2.4 kilograms. Among well-off populations around the world, birth-weight averages are usually close to 3.4 kilograms. Most populations have mean birth-weights somewhere within this range (**8.1**).

## Birth-weight pathology

Low birth-weight is defined by the World Health Organization as a birth below 2500 grams. VLB (Very Low Birth-weight) usually refers to births of less than 1500 grams. A distinction is made between newborns appropriate in weight for their gestation age, whether that age is 40 weeks or far less, and newborns weighing less than is appropriate for their gestational age. The latter group can be termed SGA (small for gestational age), or SFD (small for dates), or IUGR (intra-uterine growth retardation). Other forms of prenatal growth retardation, and their diagnostic titles, are recognized as departures from normal allometric relationships. Departures from normal weight-to-length relationships, or the weight-to-chest-circumference relationship are sometimes termed asymmetrical growth. For example, 'chronic' IUGR, refers to small for dates newborns with normal weight-to-length proportions reflecting, in theory, chronic reductions in sustenance beginning early in gestation; 'subacute' IUGR refers to reductions in weight more than in length occurring between the 27th and 30th weeks of gestation; and 'acute' IUGR refers to reductions in weight occurring in the last month of gestation.

## What does birth-weight mean?

In general, there are usually greater risks to health for neonates and infants weighing less than 2500 grams at birth, particularly when size is inappropriately small for gestational age. These adverse health risks include greater mortality, morbidity, developmental delay and congenital anomaly, and slower post-natal growth. Birth-weight itself may not be the cause of the greater health risk, but the causes of the lower birth-weight may be responsible also for the greater health risk subsequently. Researchers have sought to refine the association between size and subsequent health risk by delineating different diagnostic categories based on relationships of, for example, size-to-age, and weight-to-length to different types or magnitudes of subsequent health risk.

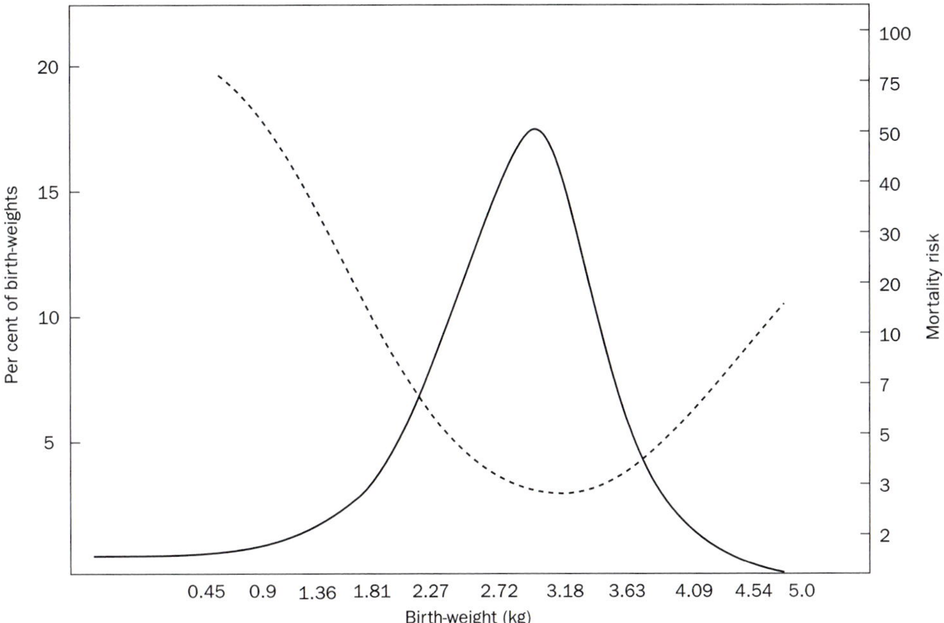


**Relationship of birth-weight (solid line) to mortality risk (dashed line): mortality is highest at the smallest birth-weights, reaches a minimum and then increases as weight increases.**

The general shape of the relationship of neonatal mortality (mortality in the first month of life) to birth-weight is that of a reverse 'J'. At the left of the birth-weight distribution are the lowest birth-weights, and the risk of neonatal death is highest. The risk decreases steadily as weights increase up to the weight of lowest mortality risk (about 3.1 kg in the figure above). This is usually slightly greater than the mean birth-weight. From this point begins the smaller rise of neonatal mortality associated with larger birth-weights. There is also a greater risk of mortality as birth-weight increases beyond the same optimum birth-weight. The risk may be 30–40 times greater in low birth-weight neonates than in ones weighing more than 2500 grams. Very low birth-weight babies may have 200 times the mortality risk of normal birth-weight babies.

The exact shape of the mortality curve depends on many factors, including the availability of care in the prenatal and immediate post-natal periods. The relationship of birth-weight to post-neonatal mortality (mortality from the beginning of the second month until the end of the first year) is not reverse 'J' shaped: infants with lower birth-weights have larger risks of mortality but infants with larger ones do not. It is thought that neonatal mortality reflects prenatal conditions and events at delivery, while post-neonatal mortality reflects the quality of the post-natal environment. Although it is customary to dichotomize the distribution of birth-weight at 2500 grams into low and normal birth-weight categories, that point originally chosen because of greater mortality risk below it, most studies show a continuum of increasing risk as birth-weight decreases. Low birth-weight and very low birth-weight categories are arbitrary, but serve the clinician in determining prognosis and treatment.

The frequency of other health risks usually increase as birth-weight decreases. Low birth-weight infants are slightly more likely than normal birth-weight infants to experience developmental delay or congenital anomaly. Very low birth-weight infants are three to four times more likely to do so, and five to seven times more likely to have a severe impairment. The risk of significant illness is also slightly higher among low birth-weight infants, again, largely due to substantial increase in risk associated with very low birth-weight.

The pattern of growth during infancy also is related to birth-weight. There is a negative correlation between size at birth and weight-growth during the first 6 months post-natally. Small neonates usually grow faster than neonates with median birth-weights, and large ones usually grow more slowly. Among healthy small neonates this may be a form of catch-up growth, compensating for any late occuring limitations on fetal growth imposed by a relatively small or insufficient placenta and/or growth-constraining uterus. After the immediate post-term compensation in growth-rate, larger neonates tend to grow more during childhood. Weight and height at 7 years of age is positively correlated with size at birth. However, birth-weight is not a good predictor of adult size, partly because of the post-natal compensation, because of differences in adolescent growth and the influence of environmental factors during post-natal life.

## Sources of variation in normal birth-weight

Since weight at birth depends on the rate of prenatal growth and the duration of gestation, sources of normal variation may be grouped as to whether they influence one or the other outcome. Birth-weight is known to have a multifactorial determination. It is desirable to identify modifiable causes of variation in birth-weight, particularly of low birth-weight, so as to prevent decreased birth-weight and the attendant morbidity and mortality risks. The World Health Organization has published a review and meta-analysis of research published since 1970 on intra-uterine growth and gestational duration determinants. The causal determinants of intra-uterine growth are: infant sex, racial/ethnic origin, maternal height, prepregnancy weight, paternal height and weight, parity, previous low birth-weight infant, gestational weight-gain, energy intake, general maternal morbidity, malaria, cigarette-smoking, alcohol consumption and tobacco-chewing. Maternal age and socio-economic status also influence birth-weight, but through one or more of the causal factors. Since the importance of a determinant depends on its prevalence in a population, and the prevalence of determinants varies substantially among populations (i.e., the use of cigarettes varies considerably), the review considered the relative contribution of the determinants in rural developing populations and developed populations separately.

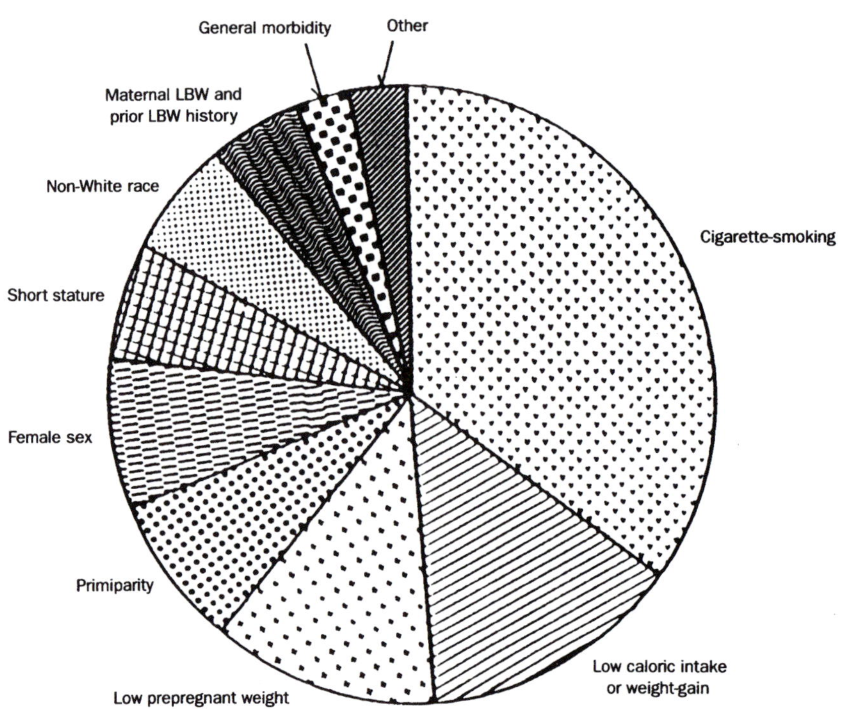


**Relative importance of causal influences on low birth-weight in developed countries. From Kramer, 1987.**

In developed countries, the largest direct causal factor on birth-weight is cigarette-smoking (**8.3**), with more than double the influence of the next most influential factor, low energy intake during gestation. Next in importance is low prepregnant weight, primiparity, female sex, short stature, and non-European race/ethnicity. The three most influential factors are all potentially modifiable. Among rural populations in developing countries the two most important factors are non-European race/ethnicity and low energy intake and low weight-gain during gestation (**8.2**). Low prepregnant weight, short stature and malaria are next. These five factors account for approximately three-fourths of the causal influences on birth-weight. The remaining 25 per cent is attributed to female sex, primiparity, maternal low birth-weight and prior low birth-weight history, general morbidity, small paternal size and other factors.

Other factors associated with low birth-weight, usually of low prevalence, include twinning, congenital rubella syndrome, uterine malformation, and medical complications of pregnancy such as toxemia, and abruptio placentae. Altitude (**9.9**) and cocaine use (**8.5**) are also important for certain populations. It is also important to determine the extent to which more general substance abuse (**8.5**), HIV infection (**8.4**) and teenage pregnancy (**8.6**) influence birth-weight.

An attempt to define the determinants of gestation length and their relative importance has been relatively unsuccessful. Known factors account for less than 25 per cent of the relative contribution to variation in birth-weight in rural populations in developing countries, while in developed countries, known factors account for just over 25 per cent of variation. More research on the determinants of gestation length is necessary before public health programmes can be designed to intervene and improve prenatal growth.

## Biological sources of variation

Sex and multiple birth are two normal sources of variation in birth-weight. The male fetus is heavier than the female one from the 28th week of gestation onwards, and at term, boys

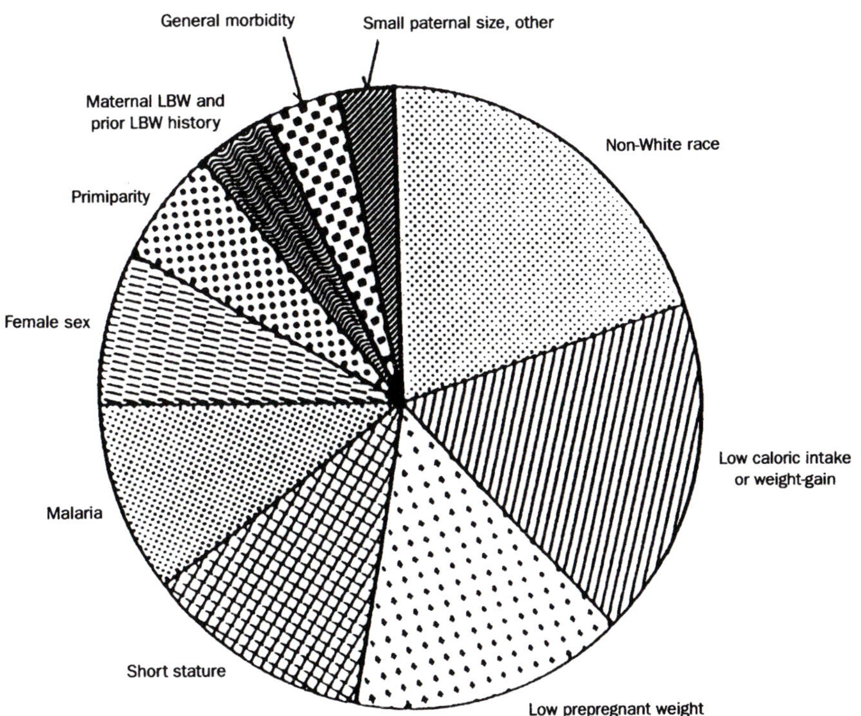


**Relative importance of causal influences on low birth-weight in rural populations of developing countries. From Kramer, 1987.**

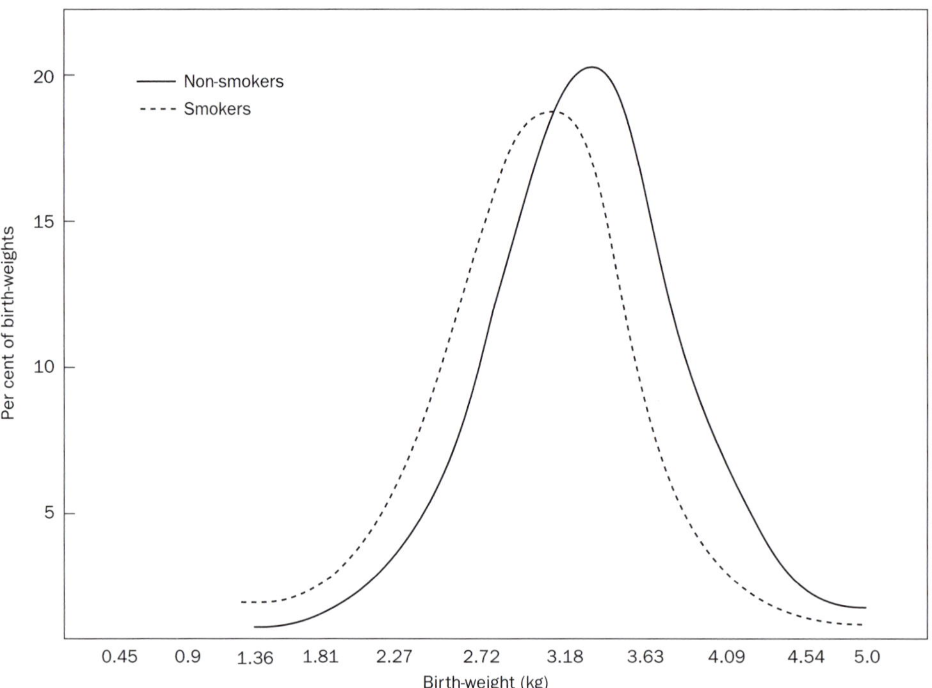


**The influence of smoking (8.3) on the distribution of birth-weights within a population.**

average 150 grams more than girls. However, girls have better survival risks than boys of the same weight, probably because once matched for weight, girls are more mature than boys. This points to maturity, more than weight itself, as the more important factor in neonatal viability. The weight-growth of twins slows prenatally, usually before the 34th week, and when the combined weight of the twins approximates the singleton's 36th week weight. At birth, twins may weigh 0.9 kilogram less than a singleton of the same gestational age; length is also affected. Usually one twin is larger than the other, but catch-up growth occurs in the first few months of life and the difference is narrowed considerably by 3 years of age.

There is considerable controversy surrounding the influence of race/ethnicity on birth-weight. In the United States, Black neonates average less weight at birth than Whites. The difference can easily be attributed to differences in socio-economic status between the two groups. However, when Black and White neonates are matched for weight, Black neonates have a lower mortality risk, suggesting to some researchers that the means of the distributions, and the distributions themselves, are intrinsically different. Since it is exceedingly difficult to distinguish the effects of socio-economic status and race/ethnicity in the United States, the true cause of the difference cannot be discerned at present.

## Public-health use of birth-weight statistics

The percentage of neonates with low birth-weight is used as a basic indicator of public health by many governments and agencies including the World Health Organization (WHO). In their *World Health Report, Bridging the Gaps* (1995), the WHO notes the frequency of LBW for many countries as of 1990. The following countries have the lowest rates of LBW (4% or less): Finland, Iceland, Ireland, Luxembourg, Malta, Netherlands, Norway, Spain, and Tonga. African countries have generally higher frequencies (mostly 10–19%) while European countries have lower ones (all between 3–9%). Countries in the Americas are intermediate with frequencies between 7 and 15 per cent. It is important to remember that frequencies can vary considerably within a country as they do in America where the overall rate is 7 per cent but the rate among Black Americans is nearly twice that.

*Lawrence M. Schell*

# Maternal anthropometry and birth outcome

Since 1906, when Newmann suggested that infants weighing less than 2500 or 3000 grams would be considered as premature, there has been no overall agreement as to what constitutes low birth-weight (LBW). Whereas prematurity was at one time defined as a birth-weight of 2500 grams or less, today the term *prematurity* means a gestational age less than 37 weeks, and WHO defines low birth-weight as a weight less than 2500 grams, regardless of gestational age. The definition is complicated further by the use of several other terms, such as *intra-uterine growth retardation* (IUGR), which refers to term babies (born at more than 37 weeks of gestation) with low birth-weight (< 2500 g), and *small for gestational age* or *small for dates* babies, denoting preterm or full-term newborns with weight significantly less than the expected mean for gestational age. The use of a birth-weight cut-off point, such as less than 2500 grams, to assess fetal undernutrition has limitations as it fails to distinguish prematurity from IUGR. In spite of this, the cut-off of 2500 grams for LBW has been widely used in epidemiological studies as it has proved to be a good indicator of risk.

In addition to the impact of socio-economic and environmental characteristics of the mother there is increasing interest in the use of maternal anthropometric measurements as predictors of birth-weight. An important component of maternal nutritional status that bears significantly on pregnancy outcome is energy balance, which is reflected in the size of the body stores. Anthropometric assessment of nutritional status is valuable because it provides information about fat stores as well as other aspects of maternal body-size and composition that affect pregnancy outcomes. Assessment of maternal nutritional status relies on measurements of stature, prepregnancy weight (PPW), pregnancy weight at different trimesters, weight-gain during pregnancy (PWG), skinfold thicknesses and limb circumferences. Some measures reflect a woman's nutritional status or energy stores as she enters pregnancy (height, PPW), while others reflect changes in her nutritional status over the course of pregnancy (skinfold thickness, limb circumference, PWG, gestational weight). The latter are attributable to changes in maternal energy stores, development of maternal reproductive tissue, increased blood volume and extracellular fluid and fetal growth. Maternal size is well recognized as being associated with birth-weight.

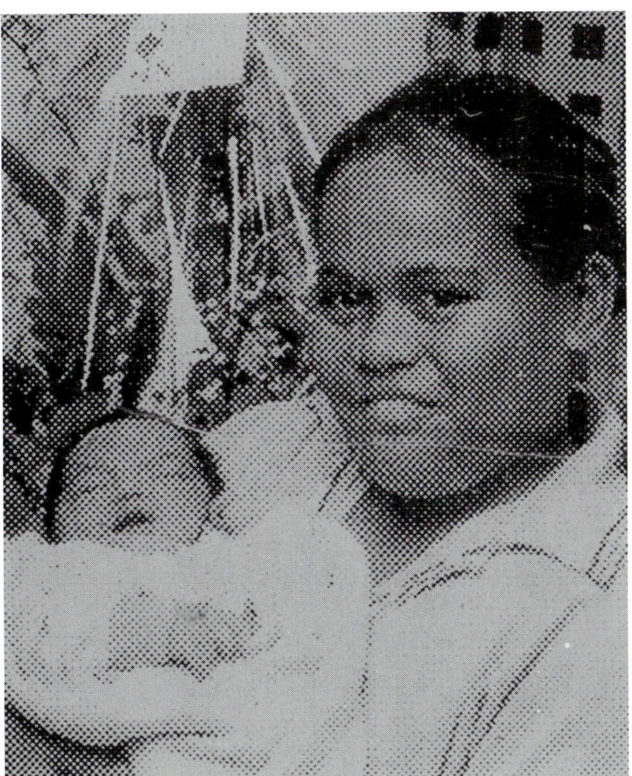

**Overweight women with excessive weight-gain during pregnancy have more large-for-dates babies. Photograph with permission from the *Cook Islands News*.**

Studies assessing the relationship between mother's height and birth-weight are contradictory. Several researchers have quantified the relationship between maternal height and birth-weight, with regression coefficients ranging from 12.1 to 22.1 grams per centimetre when height alone is considered, and from 12.5 to 20.0 grams per centimetre when weight is entered independently into the regression equation. Although height appears to have a direct relationship with birth-weight and infant survival, it is necessary to assess whether it is height *per se*, or the association of height with weight, that affects pregnancy outcome, since taller women are generally heavier and have more lean body mass than shorter women.

Measurement of prepregnancy weight is used to assess initial risk of poor pregnancy outcome, to assign weight-gain recommendations appropriate for women of different risk status, and to target nutritional interventions for those most in need. The prepregnancy weight and weight-gain in pregnancy are generally additive (or subtractive) in their effect on birth-weight. The combination of low prepregnancy weight and low weight-gain during pregnancy greatly increases the risk of low birth-weight, and perinatal, neonatal, and infant mortality. Ideally, prepregnancy weight should be taken before the woman is diagnosed as pregnant. However, this is very difficult, particularly in developing countries, and most studies use measurements early in pregnancy as a proxy for prepregnancy status.

Several researchers have shown prepregnancy weight to be a significant determinant of birth-weight in both developed and developing countries. Kramer (1987), in his meta-analysis, found the causal effect of prepregnancy weight on intrauterine growth retardation to be well-established, important, and sensitive to modification. Taffel (1980) reported that women in the United States who weighed less than 51 kilograms prior to pregnancy were more than twice as likely to have low birth-weight infants than those above this weight.

A cut-off of 40 kilograms for prepregnancy weight is the most commonly cited figure in the developing countries used to assess risk of various pregnancy outcomes. Studies in India and East Java have confirmed that a cut-off of 40 kilograms was important in identifying women at risk of delivering low birth-weight babies. Anderson (1989), in her study in India, found that women with weights of less than 40 kilograms in the first 6 months post-partum had 2.1 times the risk of severely low weight-for-age infants (< 60% of median) and 3.5 times the risk of severe stunting in their infants (< 85% of median) compared with women weighing more than this threshold. A slightly lower value of under 38 kilograms was used as a risk indicator in the Kasa Project in India. Women weighing less than this value prior to pregnancy delivered infants with an average birth-weight of 2467 grams, compared to 2595 grams for women with prepregnancy weights of 41 kilograms or more.

A number of indices have been used to define weight-for-height and the most common are the body mass index (BMI) or Quetelet index ($wt/ht^2$), and the ponderal index ($\sqrt[3]{wt}/ht$). Ponderal index, prepregnancy weight, height and weight-gain during pregnancy in studies in England, America and Chile were examined in relation to small-for-date, average-for-date and large-for-date babies born to them. Women with lower ponderal index had lighter infants, while women who gained less weight during pregnancy had lighter infants after controlling for ponderal index.

*Prevalence of LBW and mean birth-weight for different countries*

| Country | Prevalence of LBW (% of live births) | Mean (±SD) birth-weight (g) |
|---|---|---|
| **Asia:** | | |
| China | 4 | 3164 (535) |
| India | | |
| Low SES | 28 | 2633 (417) |
| High SES | 19 | 2782 (414) |
| Nepal | 14.4 | 2782 (416) |
| Sri Lanka | 18.4 | 2843 (458) |
| Indonesia | 10.5 | 2936 (415) |
| Myanmar | 17.8 | 2852 (469) |
| Thailand | 9.6 | 3004 (462) |
| Vietnam | 5.2 | 2997 (369) |
| **Africa:** | | |
| Botswana | 12.5 | 3031 (543) |
| Gambia | 12.1 | 2937 (410) |
| Lesotho | 10.3 | 3078 (543) |
| Nigeria | 12.4 | 3052 (584) |
| **Europe:** | | |
| Ireland | 5.6 | 3436 (612) |
| England | 6.2 | 3239 (505) |
| **Americas:** | | |
| Argentina | 6.3 | 3239 (535) |
| Colombia | 16.1 | 3015 (658) |
| Cuba | 8.1 | 3174 (544) |
| Guatemala | 12.5 | 2996 (492) |
| United States | | |
| Black | 10.6 | 3144 (570) |
| Hispanic | 4.8 | 3347 (519) |
| White | 6.0 | 3355 (566) |

Studies in developing countries show significant relationships between BMI or other weight-for-height indices and birth-weight. In a study in Colombia, it has been shown that when maternal weight divided by height at 6 months gestation was less than 360 grams per centimetre, mean birth-weight was 200 grams less than women with greater weight.

Several studies have quantified the effect of pregnancy weight-gain to birth-weight. Kramer (1987) in his meta-analysis showed that for an average woman with adequate prenatal nutrition, the effect on birth-weight is 20.3 grams per kilogram pregnancy-weight-gain. However, optimal weight-gains are different for women who begin pregnancy at a different nutritional status. However, the combination of low prepregnancy weight and low pregnancy weight-gain puts women at the greatest risk of delivering a low birth-weight baby.

Low pregnancy weight-gain is also related to fetal and neonatal mortality. A significant positive influence of prenatal weight-gain on birth-weight after controlling for gestational age, maternal social class, ethnicity, cigarette consumption, marital status, age, parity, education and weight-for-height has been demonstrated.

The need for an optimal range of weight-gain is also important as studies have shown significant associations of unfavourable birth outcomes (including macrosomia, labour abnormalities, meconium staining, and unscheduled caesarean section) with both increased maternal pregnancy BMI and increased maternal gestational weight-gain. Findings from a Finnish study showed increased incidence of induced labour and a tendency for emergency caesarean section in obese women with excessive weight-gain during pregnancy, who also had more large-for-dates babies.

Weight-gain varies across pregnancy, but is complicated by age, gravida, smoking status (**8.3**) and other morbidity status. As a consequence of the difficulty of measuring maternal weight-gain during pregnancy, several investigators have looked at the relationship between maternal post-partum weight with birth-weight. Luke and Petrie (1980) found increased birth-weight to be associated with heavier maternal post-partum weight in underweight and normal weight women. Several Indian studies have also shown a positive relationship between maternal post-partum weight and infant birth-weight.

There have been attempts to relate foot length to birth outcome. In a study of 351 British women in 1985, a relative risk of caesarean section was found to be 8.6 in a woman with shoe size under 4.5 compared with a woman with a

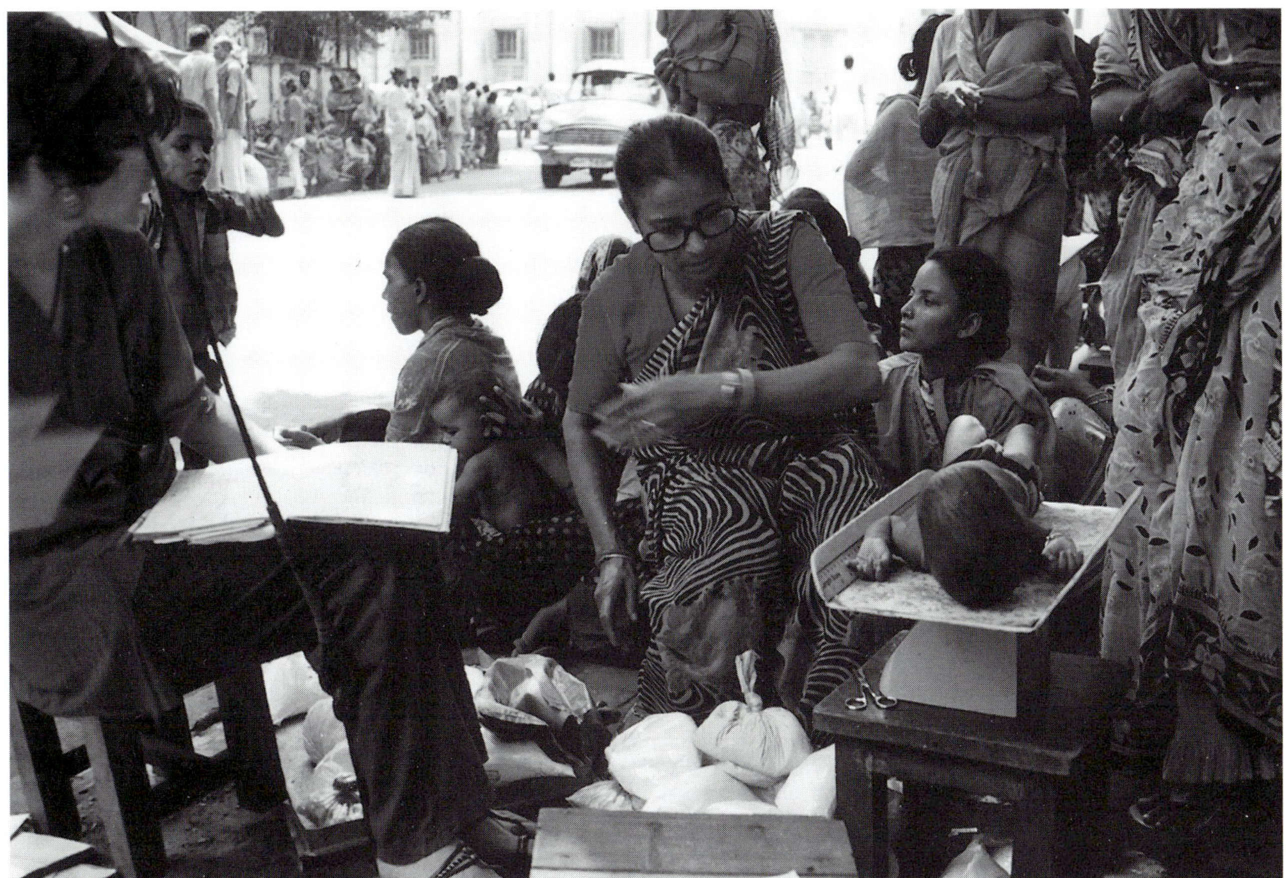

**Weighing infants in India. Photograph by Patricia van Someren.**

shoe size larger than 6. However, other studies found an increased caesarean section rate in women of short stature, but no association has been found between mode of delivery and shoe size. A more recent study in Bangladesh found only a very slight relationship between foot length and LBW.

Limb circumference and skinfold thicknesses are well recognized as important indicators of nutritional status. Due to its practical advantages compared to other anthropometric measurements, the use of arm circumference has been recommended by various researchers in predicting pregnancy outcomes. For example, Lechtig examined the advantages and disadvantages of various maternal anthropometric indicators, namely arm circumference, (with a 23.5 cm cut-off point), weight-gain (< 10th centile for gestational age, equivalent to 8 kg at term), uterine height (< 10th centile for gestational age) and weight-gain expressed as a percentage of standard weight-for-height (≤ 90% at the start of a pregnancy) in predicting risk of low birth-weight have been compared. Arm circumference at any time in pregnancy was found to be comparable to weight-gain during pregnancy for gestational age for predicting birth-weight, while weight-gain for standard height, and uterine height were slightly better predictors.

In a study of pregnant women in rural Bangladesh, maternal arm circumference was strongly related to other anthropometric measures (such as prepregnancy weight and weight-for-height, weight and weight-for-height at different stages of pregnancy, weight-gain per trimester and height), except for the last trimester when it was not related to these anthropometric indices. In explaining fetal and infant mortality, maternal weight and weight changes during pregnancy were found to be only slightly better than arm circumference. However, sensitivity/specificity analyses across a range of cut-off points revealed that arm circumference was slightly better than height, weight or pregnancy weight-gain in predicting fetal and infant mortality.

*Enamul Karim*

See also 'Anthropometry' (**1.4**), 'Cross-sectional studies' (**1.13**), 'Comparative growth and development of mammals' (**2.1**), 'Genetic and environmental influences on fetal growth' (**3.2**), 'Hormonal regulation of fetal growth' (**4.1**), Fetal growth retardation' (**4.8**), 'Standards and references for the assessment of fetal growth and development' (**4.9**), 'Nutrition' (**8.2**), 'Body-size at birth' (**10.3**)

# Nutrition

A satisfactory birth-weight is one of the best markers of a favourable outcome of pregnancy and a good prognosis with respect to infant mortality and morbidity. Many of the determinants of birth-weight are related to maternal nutritional status in some way. These include rate of fetal growth and maternal food intake, maternal weight-gain and prepregnant body mass index. Maternal weight-gain during gestation and nutritional status prior to and during pregnancy are closely linked and each affects the other. Weight-gain is influenced by the mother's body-size (weight, height, body mass index, percentage body fat) prior to pregnancy, the quality and quantity of dietary intake and physical activity during pregnancy. Overall maternal nutritional status is influenced by dietary intake, socio-economic status and level of physical activity. Two other major determinants of fetal growth, which are themselves closely associated and both of which interact with, and may have an adverse impact on, maternal nutritional status, are alcohol intake (**8.5**) and tobacco-smoking (**8.3**). Generally, birth-weight is lower, and the incidence of low birth-weight higher, in babies born to women who during pregnancy were underweight, had a low total energy intake, or poor quality diet or who drank or smoked heavily.

Body adiposity is one indicator of fitness for reproduction and so, at extremes of over- or undernutrition, fertility is suppressed, but once pregnancy is established nutritional requirements have to be met from dietary intake and body stores. A good overall nutritional status prior to and during pregnancy is necessary both to cover the extra requirements of pregnancy and be sufficient to support the lifestyle of the mother. Pregnancy is therefore a period of considerable nutritional risk to both mother and infant since it often occurs under marginal, and in some cases severely inadequate, nutritional conditions.

'Nutrition' encompasses total energy intake and intake of energy-yielding macronutrients (protein, fat, carbohydrate and alcohol), vitamins and minerals, and it is difficult to separate out the possible effects on birth-weight of these different constituents. However, the energy requirements of pregnancy (maintenance energy expenditure and fat deposition) vary over a wide range according to maternal energy status prior to and during pregnancy (body fat and energy intake) and this may well be a proxy for overall nutritional status. This

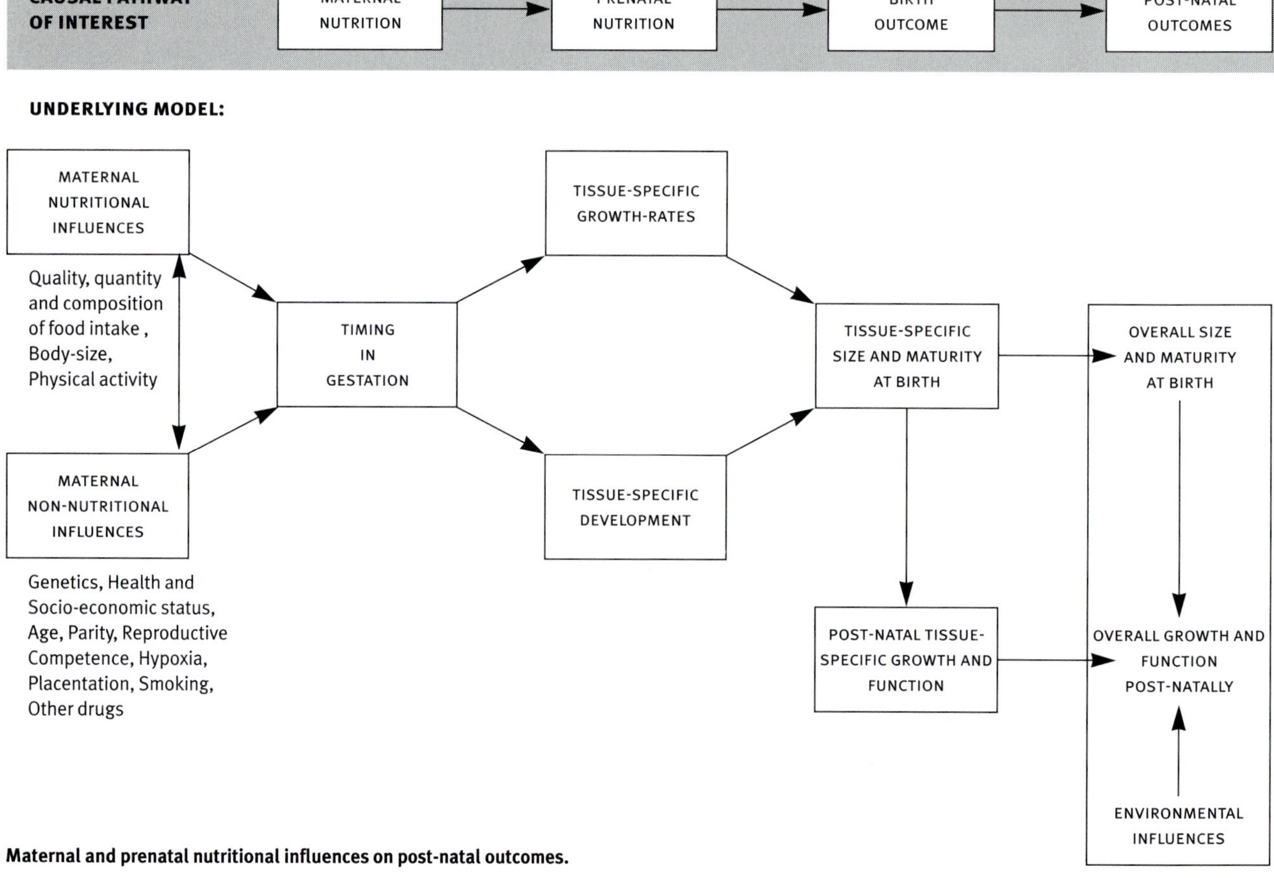


**Maternal and prenatal nutritional influences on post-natal outcomes.**

variation seems to be an important way of protecting fetal growth in an attempt to ensure that the fetus is appropriate for maternal body-size and therefore has the best chances of survival. However, this is probably not without cost. The possible effects on the mother of successfully sustaining a pregnancy under adverse nutritional conditions have yet to be established and subtle nutritional influences on the fetal environment might have long term consequences for the offspring (**12.1**).

The relatively slow rate of fetal growth and large maternal fat stores in humans, together with the potential for adaptive changes in metabolism, may help to counter some of the adverse effects of undernutrition. Although adequate fetal growth appears to be sustained at different levels of nutrition, in extreme situations of chronic negative energy balance due to, for example, famine, or very high levels of physical activity in combination with low energy intake, birth-weight is substantially reduced. Since weight is not a wholly adequate measure of a baby's overall condition at birth (**12.1**) and

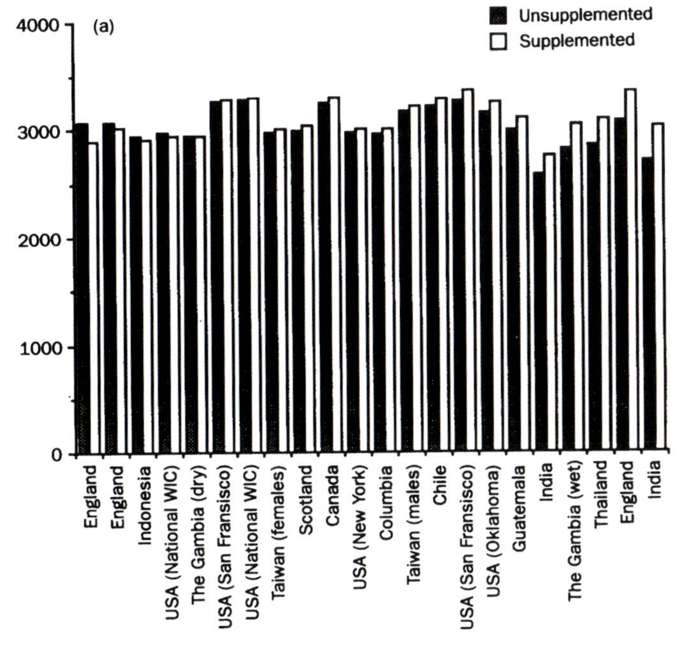


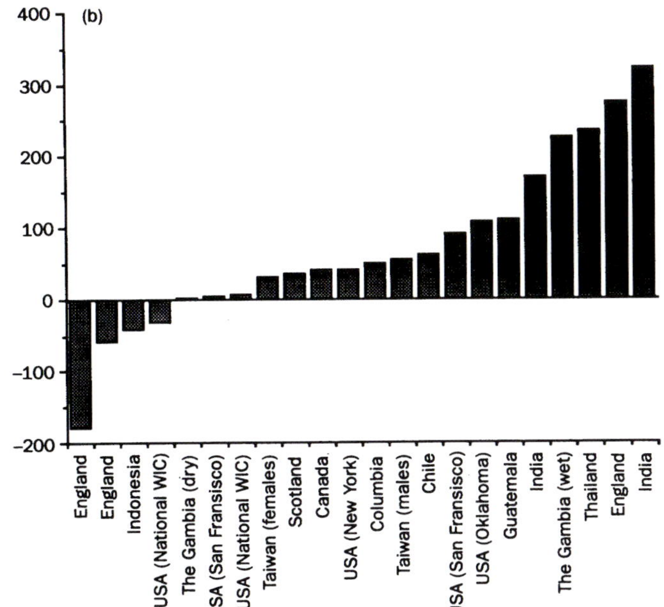


**Influence of dietary supplementation on (a) mean birth-weight, and (b) birth-weight difference.**

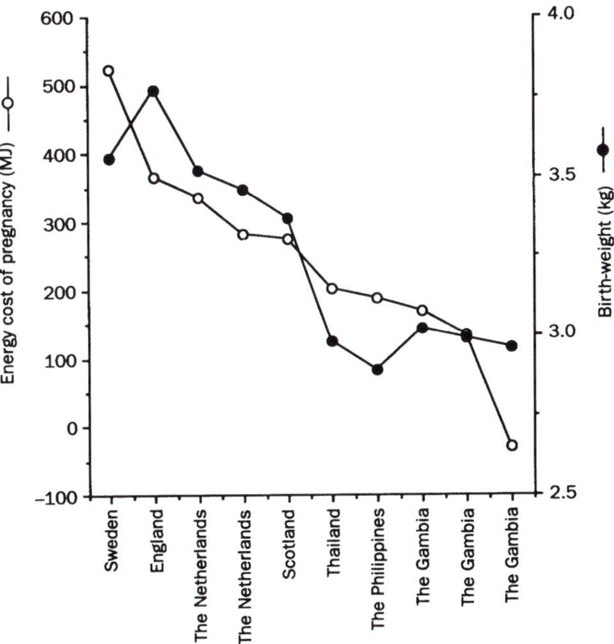


**Average absolute birth-weights from women from diverse nutritional backgrounds in 10 longitudinal studies of the metabolic (energy) costs of pregnancy. Values are ranked according to the total metabolic costs (maintenance plus fat deposition). The data are metabolic costs and do not include energy expended in physical activity.**

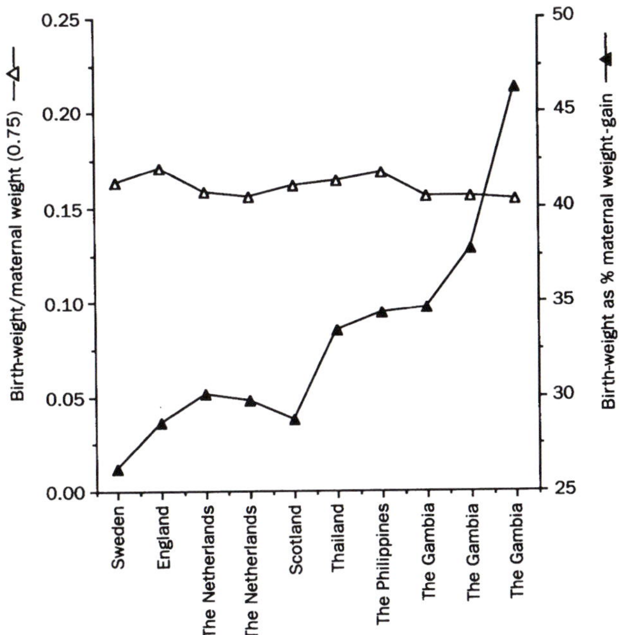


**Data as from top figure, but with birth-weight expressed relative to maternal metabolic size, and as a proportion of total maternal weight-gain during pregnancy.**

because of the higher rates of morbidity and mortality in low birth-weight infants, pregnant women, especially in developing countries, need to be considered as nutritionally at risk.

## Dietary supplementation

Many dietary supplementation studies have been conducted in pregnant women in order to increase birth-weight and particularly to reduce the incidence of low birth-weight. Most studies have focused on overall energy and protein intake (**9.2**) and supplements have usually included extra vitamins and micronutrients. The primary outcome measure of studies in which supplements have specifically been given as vitamins and micronutrients only has not been birth-weight.

Dietary supplementation studies have been conducted in the field and under experimental conditions in both developed and developing countries. While some supplementation studies have led to significant improvements in birth-weight and/or in reducing the incidence of low birth-weight, others have not. However, differences in experimental design, amount, timing and frequency of supplementation, choice of target and control groups, and compliance of subjects all make comparisons very difficult. Where no significant effect was found, supplements tended to replace, rather than add to, usual intake, or they were given to women who were probably not undernourished in the index pregnancy. Studies which have demonstrated that nutritional intervention can significantly improve the outcome of pregnancy with respect to birth-weight tend to be those in which undernutrition was identified in the target population, or in a given season where birth-weights were less than predicted, or where the supplement was given directly to the pregnant women herself with no dilution by family members. There is a large potential improvement in birth-weight that appropriately targeted supplementation can have and it is important that the results of the relatively small-scale trials which have been conducted under experimental conditions can be replicated in real-life conditions.

*Gail Goldberg*

See also 'Genetic and environmental influences on fetal growth' (**3.2**), 'Hormonal regulation of fetal growth' (**4.1**), 'Fetal growth retardation' (**4.8**), 'Standards and references for the assessment of fetal growth and development' (**4.9**), 'Maternal anthropometry and birth outcome' (**8.1**), 'Substance abuse' (**8.5**) and 'Body-size at birth' (**10.3**)

# Smoking

Women who smoke cigarettes during pregnancy deliver babies weighing 200 grams less, on average, than babies of non-smokers. This finding has been confirmed in over 50 well-designed studies conducted largely in developed countries. The frequency of low birth-weight (LBW) (**10.3**) is approximately doubled among smokers. The 1983 report by the United States Surgeon General surveyed studies of over 113,000 births in America, Canada and Wales and found that maternal cigarette-smoking could account for between 21 per cent and 39 per cent of the incidence of low birth-weight.

Maternal cigarette-smoking during pregnancy is associated with birth-weight in a dose-response fashion: the more the mother smokes during pregnancy, the greater is the reduction in birth-weight. Among smokers, the entire distribution of birth-weights is shifted downward; the mean reduction due to smoking is not due only to an increase in very small neonates or a decrease in very heavy ones. Thus, the amount of average reduction due to smoking and the increase in LBW depends on the prevalence of cigarette-smoking in the study sample and in the general population of pregnant women. According to the Centers for Disease Control's National Health Interview Survey, (NHIS) in 1992, 14.3 million American women aged 18–44 years smoked cigarettes, resulting in an overall prevalence rate of 26.9 per cent. In 1965, the first year that prevalence was monitored nationally by NHIS, 33 per cent of American women smoked cigarettes. The prevalence varies considerably by several social and demographic variables. For example, among African-American women the rate in 1992 was 22.6 per cent. A review of studies of women in 14 countries from Asia, Europe and Latin America, where there is wide variation in smoking prevalence, found increases in the frequency of low birth-weight and a reduction of birth-weight due to maternal smoking that are similar to effects seen in developed countries.

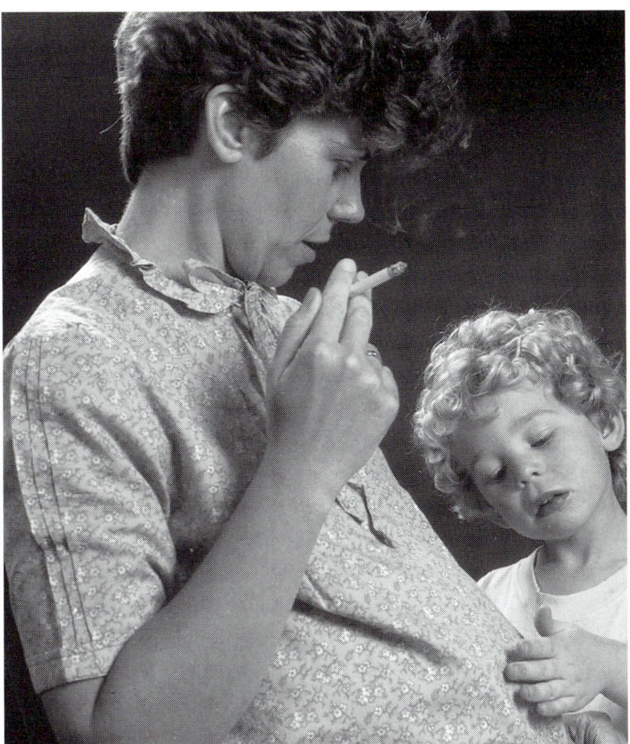

**A pregnant woman smoking. As well as the effects of smoking on the birth-weight of the unborn child, the effects of passive smoking on the other child could result in respiratory problems, such as asthma and bronchitis, which may affect his growth adversely. Courtesy of James King-Holmes/Science Photo Library, London.**

Maternal smoking causes reductions in gestation length of 2 days or less, this reduction being insufficiently large to account for the accompanying birth-weight decrement. When birth-weights of smokers' and non-smokers' infants are compared at each week of gestation from weeks 36–43, infants of smokers consistently have lower mean birth-weights. Likewise, the frequency of low birth-weight among smokers is greater than among non-smokers for term births whether the delivery occurred early, average or late.

Smoking is one of the strongest influences on birth-weight. In one well-designed study in which the effects of multiple factors were controlled, smoking had a larger effect than previous pregnancy history, hospital pay status, mother's prepregnant weight, height, age-parity, or the sex of the child. A recent World Health Organization review of all influences on birth-weight concluded that in developed countries cigarette-smoking is the largest influence on birth-weight relative to other factors, while in developing countries its effect is negligible; the difference in effect between the types of countries being due to different prevalences of smoking and variation in nutritional levels.

Women who stop smoking before conception have babies similar in average size and risk of low birth-weight to non-smokers. The effect of stopping smoking after conception depends on when cessation occurred and on how much was smoked. Stopping before the fourth month of pregnancy is thought to restore the risk of low birth-weight to that of a non-smoker and result in birth-weights of average size for non-smokers. Some of this effect is because stopping is more common among light smokers than heavy smokers. When the amount of smoking and the effect of stopping smoking are considered independently, heavy smokers do not achieve the same risk of LBW or have birth-weights equal to those of non-smokers, although this analysis should be repeated with a sample with measured thiocyanate levels to ensure that stoppers did in fact stop. The fact that stopping smoking reduces or erases the effects of early smoking provides some evidence for a causal relationship of smoking to fetal growth.

Maternal smoking is regularly associated with reduced growth in length and arm circumference as well as birth-weight. A reduction in head circumference is sometimes reported as well. Decrement sizes depend on the population prevalence of smoking and the dose to the individual. A longitudinal ultrasonographic study of prenatal biparietal diameter (BPD) growth indicated that among newborns delivered between 266 and 294 days, BPD increases faster from the 28th week of gestation onwards among non-smokers. This indicates an effect of smoking on prenatal growth well before the last weeks of pregnancy.

Placental ratios are larger among smokers largely due to the reduction in birth-weight. Some studies in which the placentas are carefully trimmed according to a standard protocol have noted that the placentas of light smokers are lighter than those of non-smokers, while placentas of heavy smokers are slightly heavier than those of non-smokers. Other studies have found a decrease in placenta size associated with heavy smoking and the nicotine content of cigarettes smoked. There is some evidence that placenta ratios are higher for African-American smokers. Smokers' placentas are also thinner with larger minimum diameters. These changes in placental morphology may be an adaptation to the increased carboxyhemoglobin and reduced oxygen-carrying capacity of the blood associated with maternal smoking. Other changes consistent with chronic ischemia in the placenta do not appear to be adaptive.

There is a small increase in perinatal mortality associated with smoking that is not clearly evident unless sample sizes are large. However, because smoking prevalences range around 30 per cent in developed countries, the increase in deaths due to smoking on a population-wide basis is quite large. Smoking is associated with greater risk of infant death from Sudden Infant Death Syndrome (SIDS), or cot death. Many studies have found lower Apgar scores in newborns of smokers; however there is no clear evidence for an increase in birth defects. Smoking may interact with other potential birth-weight-lowering factors: for example alcohol may have a greater effect on prenatal growth among smokers than non-smokers.

The biological cause of smoking-associated growth retardation is not certain; both direct biological effects and indirect ones are posited. Direct biological effects are likely because tobacco smoke contains thousands of compounds; carbon monoxide is its primary constituent. Cigarette-smoking exacerbates fetal hypoxia. Carbon monoxide, with an affinity for hemoglobin 200 times that of oxygen, has an even greater affinity for fetal hemoglobin. It is estimated that if a mother smokes 40 cigarettes per day there is a 10 per cent concentration of carboxyhemoglobin equivalent to a 60 per cent reduction in blood flow to the fetus. Cigarette-smoking provides a physiological dose of nicotine; one puff of a cigarette equals the exposure of an intravenous injection of 0.1 milligram of nicotine. Nicotine stimulates adrenal production of noradrenalin (norepinephrine), adrenalin (epinephrine) and acetylcholine, and results in less uteroplacental perfusion. It also crosses the placenta where it acts directly to increase fetal blood pressure and respiratory rate. A negative association between the nicotine content of cigarettes smoked and birth-weight has been observed.

Growth inhibiting poisons such as cyanide, lead and cadmium are found in cigarette smoke. Cyanide and thiocyanate levels are increased in smokers, and serum levels of vitamin B12 are decreased. This may reflect a disorder of cyanide detoxification. Smoking also may affect fetal growth by damaging the placenta.

An indirect biologic effect is possible also. It is clear from the very small effect of smoking on gestation length, however, that the reduction in size is not due to shortened gestations. Birth-weight could be affected by way of a smoking-induced reduction in maternal appetite, lower intake of food and lower maternal weight-gain, but most evidence indicates that weight-gains are similar among smokers and non-smokers, or that controlling, by matching for weight or through statistical procedures, does not erase the influence of smoking. Complete control over this variable is difficult because weight-gain is associated with fetal growth (fetal weight is a portion of weight-gain), and with gestational age.

A persistent theory used to explain reduced growth in the offspring of smokers is that it is the 'smoker' not the 'smoking' that influences growth. Since nearly all studies of the effects of smoking involve comparisons of self-selected groups, the possibility of bias is present. Only randomization would completely eliminate this problem. A randomized trial of smoking cessation treatment (literature and counselling) was conducted to eliminate bias due to self-selection of smoking status. The treatment group experienced a 43 per cent reduction in smoking and the control group a 20 per cent reduction (salivary thiocyanate levels were measured in both groups). Keeping in mind that control and treatment groups contained smokers and non-smokers though in different proportions, newborns were 92 grams heavier and 0.6 centimetres longer in the treatment group than in the control group. When smokers from either group were compared to all non-smokers, birth-weights differed by the amount found in most studies without randomization of treatment. This study demonstrates a causal effect of a specific anti-smoking intervention on size at birth, strengthening the evidence for a direct biological effect of maternal smoking on fetal growth.

The existence of a dose-response relationship between the amount smoked and the reduction in fetal weight, the consistency of results from innumerable studies, the replication across levels of weight-gain, economic well-being, geography, race/ethnicity, and age, and the knowledge that there are several biological pathways for the components of smoking to affect fetal growth, justifies the firm conclusion that maternal smoking reduces fetal growth. *Lawrence M. Schell*

# Maternal HIV infection

By 1998, over 12,000 children had been reported with AIDS in the United States and Europe, the majority infected via mother-to-child (vertical) transmission. The incidence of HIV infection among women of child-bearing age in developing countries, particularly in sub-Saharan Africa and South-east Asia, is higher than in developed countries, and it is estimated that 90 per cent of future pediatric HIV infections worldwide will be among children born to HIV-infected women in these countries. The issues associated with the impact of maternal HIV infection on infant and childhood morbidity and mortality are complex.

## Timing of vertical transmission

The risk of vertical transmission ranges between 15 and 35 per cent; the lowest rates are reported in Europe, the highest in Africa where most women breast-feed which approximately doubles the risk of transmission. Timing of vertical transmission may be relevant to pregnancy outcome since early exposure of the fetus to HIV could induce reduced birth-weight, prematurity and congenital malformations, as observed in other congenital infections. Vertical HIV transmission may take place in the intra-uterine, intrapartum or post-natal (via breast-feeding) period, but the relative contribution of each of these modes of transmission remains unclear. Although there is evidence based on examination of placental and fetal tissues that HIV infection may occur *in utero*, increasing indirect evidence suggests that a substantial proportion of infants acquire their infection during the peripartum period. Interventions initiated in late pregnancy and or during delivery offer scope for reducing the risk of vertical HIV infection.

*Evidence for* in-utero *and peripartum transmisson of HIV from mother to child*

| In utero |
|---|
| Presence of HIV in aborted fetuses from HIV-infected women |
| HIV isolation from peripheral blood in only about 50% of HIV infected infants in the first days of life |
| Rapid progression of HIV disease in approximately 25% of HIV infected infants |
| **Peripartum** |
| Increased HIV transmission rate in first-born compared with second-born twins |
| No detectable circulating virus in about 50% of HIV-infected infants in the first week of life |
| Suggestion from some studies of decreased transmission with Caesarian section |
| Administration of the anti-HIV drug zidovudine (AZT) in late pregnancy, peripartum and in the neonatal period decreases HIV transmission by about two thirds |
| The presence of a slow pattern of HIV disease progression in the majority of children |
| Absence of congenital malformations and symptoms and signs at birth |

## Birth-weight

There is conflicting data from developed and developing countries regarding the relationship between maternal HIV infection, birth-weight and the HIV-infection status of the child. This may be due in part to differences in study design, variation in the clinical and immunological stage of HIV disease in the women studied, and other factors such as drug use, other infectious diseases, and nutritional status.

Prospective European and American studies have not established an association between maternal HIV infection and adverse pregnancy outcome. Similar birth-weights have been reported in the infants born to HIV-infected women, regardless of the eventual HIV-infection status of the infant. These studies have been relatively small in number, and maternal drug use during pregnancy (**8.5**) has been a confounding factor associated with low birth-weight. In addition, none of these studies included a comparison group of infants born to HIV-negative women. However, in a recent retrospective study from Edinburgh, the birth-weights and growth of uninfected children up to 3 years of age born to HIV-infected women were found to be similar to those of children born to HIV-negative women. HIV-infected children were not compared in this study. The majority of the women in the European and American studies had asymptomatic or mildly symptomatic HIV infection with relatively good immune function.

Most studies from Africa (Zaire, Zambia, Congo, Kenya and Rwanda) and Haiti have shown a small but significant difference of approximately 100–400 grams between the birth-weights of children born to HIV-infected women compared with infants born to similar uninfected controls. Infants born to women with AIDS had the lowest birth-weights, suggesting an association between the clinical and immunological disease status of the woman and birth-weight.

The majority of studies have attributed the low birth-weight among infants born to HIV-infected women to either intra-uterine growth retardation or prematurity. Most have been relatively large and have included a control group of uninfected women.

Although there is potential for bias due to differences in associated maternal characteristics between the two groups, attempts were made to control for these in most studies. It is also plausible that women with more advanced HIV disease may suffer from other factors which might affect pregnancy outcome, such as other infectious diseases (malaria, tuberculosis), poor nutritional status and inadequate weight-gain in pregnancy.

Other adverse outcomes reported among HIV-infected women in the African studies include higher rates of intrauterine and intrapartum death and chorioamnionitis. However, as in European and American studies, no increase in congenital abnormalities has been reported among infants of HIV-infected women compared with those of uninfected women.

The majority of African studies have not reported birth-weight by the HIV-infection status of the infants born to infected women. This is in part due to the difficulties in diagnosing HIV infection in very young children in Africa, some of whom will die before infection status can be resolved. It is, therefore, difficult to determine whether the low birth-weight and adverse pregnancy outcome described in the African studies are solely a result of maternal HIV infection or other related factors. In one African study, where the HIV-infection status of the infants was classified, HIV-infected infants were reported to be on average 290 grams lighter than the uninfected infants born to both HIV-infected and uninfected women. However, the numbers of children in this study were relatively small.

In summary, discrepancies regarding the impact of maternal HIV infection on birth-weight between developed and developing countries are probably multi-factorial. Studies in the developed world have not shown an association between maternal HIV infection and adverse pregnancy outcome. However, African studies consistently report lower birth-weights in children born to HIV-infected mothers as well as an increase in the incidence of adverse pregnancy outcome, compared with infants of uninfected mothers.

*Kristina Formica and Diana M. Gibb*

See also 'Fetal growth retardation' (**4.8**), 'Standards and references for the assessment of fetal growth and development' (**4.9**), 'Extra-uterine growth after premature birth' (**5.2**), 'Maternal anthropometry and birth outcome' (**8.1**), 'Nutrition' (**8.2**), 'Substance abuse' (**8.5**), 'Infection' (**9.3**) and 'Growth patterns associated with new problem complexes' (**13.3**)

# Substance abuse

Substance abuse during pregnancy is an escalating public health problem. Gestational exposure to legal or illicit chemicals has been implicated in fetal growth retardation and reduced post-natal growth velocities, congenital malformations, impaired neurobehavioural development, neonatal withdrawal symptoms and health complications that contribute to infant morbidity and mortality. The specific effects of prenatal drug exposure on the newborn depend on a variety of factors including: the type of drug and its mechanisms of action; the frequency, amount, and timing of exposure *in utero*; and the status of maternal health in a potentially high-risk, drug-abusing environment.

Intra-uterine substances may manifest primary effects on the developing fetus through direct actions of the drug and its metabolites on placental blood flow or on fetal tissues or cells after crossing the placenta. Drug quantity, mode of administration, and trimester of exposure are key variables for determining locus and severity of impact, and for identifying possible teratogenic properties of the drug compound. Prenatal substance abuse may also subject the fetus to negative secondary effects mediated through the influences of drugs on the mother. Such indirect effects have been shown to suppress appetite, heighten vulnerability to infection, impair decision-making processes, and otherwise undermine maternal physical and mental well-being and adaptation.

Polydrug-use, poor maternal health, lack of prenatal medical care, and low socio-economic status are foremost among confounding variables that hinder clinical efforts directed at identifying the effects of specific drugs. A host of substantive methodological issues make epidemiological data and clinical comparisons problematic. These include: lack of reliable estimates, reporting errors, difficulties with quantification and testing procedures, differences in sample populations, lack of appropriate control groups, and varying study designs. Experimental animal models provide for greater precision and maximum control of variables (e.g., dose-response, drug purity, duration, timing and amount of exposure), but remain limited in their applicability to humans.

The adverse effects of gestational exposure to alcohol and tobacco have been well documented, while the effects of marijuana, opiates, and cocaine are less conclusively established. Evidence for impaired growth parameters related to illicit drugs *in utero* is compelling; however, the evidence for associated teratogenic, cognitive or behavioural effects is tenuous. The most convincing finding concerning prenatal exposure to illicit substances is that of intra-uterine growth retardation (IUGR) (**4.8**, **4.9**). Impaired fetal growth, especially of the brain, may indirectly mediate drug effects on cognitive and physical development.

## Alcohol and tobacco

Fetal alcohol exposure has been shown to cause abnormal growth in length, weight, and head circumference; as well as particular teratogenic effects on the developing brain. The pattern of compromise shows length to be compromised more than weight, although dose-response effects on birth-weight have been described for as little as 100 grams per week (roughly 1.5 drinks per day). In severe cases, alcohol-related post-natal growth deficiencies have been known to persist, despite documented adequate caloric intake, resulting in stunted growth and microcephaly.

Like those of alcohol, the effects of prenatal tobacco-use on low birth-weight manifest in a dose-dependent manner, such that reduced size has been correlated directly with number of cigarettes smoked (**8.3**). In general, infants of mothers who smoke tobacco throughout pregnancy are born at full-term, but weigh less and tend to be shorter than infants of non-smokers, although length differences usually disappear when matched for birth-weight. Neonates gestationally exposed to tobacco show diminished lean body mass, with deposition of body fat relatively unaffected. This proportional decrease in weight and length results in a symmetric pattern of intra-uterine growth retardation, and is indicative of impaired fetal oxygenation.

## Marijuana

After alcohol and tobacco, marijuana is the most popular psychoactive substance used during pregnancy in the United States. However, the results of research on its prenatal effects are equivocal. Some studies have reported correlations between marijuana-use during pregnancy and smaller size at birth; others, however, have found no association after controlling for confounding variables. A large prospective study found a 79-gram decrease in birth-weight and a 0.5-centimetre decrease in length in the offspring of women who tested positive for marijuana at the time of delivery. Another prospective study which examined expectant mothers at regular and frequent intervals found no relationship between marijuana-use and birth-weight, yet reported a small effect on birth length. It has been claimed that gestational marijuana-exposure exhibits differential effects on neonatal body proportionality, such that body fat is unaffected, while arm circumference measures indicate significant decreases in lean body mass. Compared with infants of normal weight or asymmetric IUGR, symmetric growth retardation of lean body mass in general carries a worse prognosis for later growth and neurodevelopment, especially if the head-size is also small. The few studies of growth in marijuana-exposed infants beyond the neonatal period yield inconsistent findings.

Despite several reports of marijuana-related birth malformations, the teratogenicity of marijuana is unproven, and its mechanism of action unknown. The sparing of fat deposition despite concomitant decrements in lean body mass suggests prolonged hypoxia *in utero*, and not nutritional deficiency. However, as with nicotine, marijuana-use during pregnancy suppresses weight-gain, presumably through the direct influence of smoking on maternal metabolism. Animal experiments indicate that the last trimester is the gestational period most vulnerable to marijuana-related growth effects.

## Narcotics

Heroin and methadone are the chief opiates linked to fetal exposure, and both have been correlated with harmful effects on infant growth and development. Opiate-exposed infants show high rates of IUGR, which tends to follow a symmetric pattern. Among heroin-exposed infants, preterm delivery also contributes to low birth-weight, with one report noting a twofold increase in the incidence of preterm births. In contrast, the prolonged use of low-dose methadone has been associated with a normal gestational calendar. Reduced length and head circumference measures have also been reported in methadone-exposed infants.

Intra-uterine exposure to opiates appears to have a prolonged effect on infant growth. In one study, mean growth values among methadone-exposed infants remained in the 10th percentile for weight and fifth percentile for head-size for up to 4 months of age, after which growth accelerated to the 40th and 25th percentiles respectively, accompanied by abatement of neurobehavioural symptoms. In another study, methadone-exposed children who had suffered withdrawal syndromes were shorter than comparable infants who did not. The risk of post-natal environmental injury added to prenatal biologic insult appears to be particularly acute for low birth-weight or preterm infants who have been exposed to narcotics *in utero*.

There is little information with regard to the mechanisms by which fetal opiate exposure affects growth, but results from laboratory studies suggest that opiates interrupt the normal growth process by causing an absolute reduction in the total number of cells in various organs.

## Cocaine

Infants exposed to cocaine *in utero* have been deemed at greater risk for low birth-weight and preterm delivery. Early reports of LBW and high rates of IUGR following gestational cocaine exposure have been challenged on methodological grounds. However, many subsequent studies controlling for covariates have corroborated these findings. Clinical comparisons of cocaine-exposed with drug-free infants reveal exposed infants to have reduced overall body-size, with significant numbers classified in the small-for-gestational-age category for weight. A number of studies have declared intra-uterine cocaine a major risk factor for reduced head circumference with increased incidence of microcephaly, wherein circumferential values fall below the 10th percentile for gestational age. The selective diminution of fetal brain growth and head-size appears to be independent of effects on somatic growth, and implies direct effects of cocaine on the developing brain surpassing those expected on the basis of reduced somatic growth. Follow-up studies of the long-term effects of gestational cocaine indicate normal catch-up growth in all parameters, although reduced head circumference may be a chronic condition for severely compromised infants. Cocaine exerts a vasoconstrictive effect on fetal vasculature, and decreases fetal cerebral blood flow.

## Other drugs

Limited availability of research data make it difficult to verify the effects of numerous other street drugs. Claims have been made for a serious impact of PCP (phencyclidine) on reduced intra-uterine growth, leading to increased likelihood for birth-weight, length and head circumference parameters falling below the 25th percentile for gestational age. Maternal use of amphetamines also has been associated with decreased birth-weight, length and head circumference, but not for hallucinogens. In a study of Ts and blues (pentazocine and tripelennamine), infants born to abusers had significantly reduced birth-weight, length, and head-circumference values, and an increased frequency of major congenital anomalies.

Interpreting outcome studies of drug-use in pregnancy is complicated by the frequent presence of numerous confounding variables associated with the drug-abusing lifestyle. Interrelated and multiple adverse maternal health and lifestyle conditions make it difficult to isolate and characterize independent drug effects in clinical and epidemiological studies. With respect to polydrug-use, known *in vivo* additive or interactive effects and the formation of co-metabolites with disruptive potential may complicate pharmacokinetic and physiological responses. In this manner, the interaction of cocaine and alcohol has been found to yield an active compound – cocaethylene – with possible adverse, long-term effects.

The synergistic nature of confounding variables associated with substance abuse make it difficult to arrive at any definitive conclusions about the short and long-term effects of gestational drug exposure on the developing fetus. None the less, it is increasingly clear that the multifactorial clustering of risk factors within a single individual and/or population and geographic area identify substance abuse as a systemic problem (**13.3**). Thus, attention must be directed to the contextual risk-factors that predispose to compromised infant growth and development.

*Kyra Marie Landzelius*

## CRACK COCAINE

Cocaine has become the illicit drug of choice among certain groups of pregnant women in the United States, resulting in a growing epidemic of stimulant abuse with high prevalence rates for urban women of lower socio-economic status. An estimated 1 million of the total 103 million women in the United States use the drug habitually, with a correspondent 100,000 cocaine-related births reported annually in America since the late 1980s.

'Cocaine-exposed' infants are so-designated on the basis of intra-uterine exposure to the toxic compound of cocaine (benzoylmethylecgonine, $C_{17}H_{21}NO_4$) and its secondary and tertiary metabolites, which enter the fetal circulatory system via the maternal bloodstream and the direct venous flow of the placenta. Gestational exposure is generally substantiated by a positive drug toxicology screening of either member of the mother–infant dyad at the time of delivery, or may be indicated indirectly through maternal drug history. Infants born to mothers who used crack-cocaine 1–2 days before delivery have been found to excrete the drug's metabolites for 12 to 24 hours post-natally, with no withdrawal symptoms reported.

### *DEVELOPMENTAL EFFECTS*

Intra-uterine cocaine poses an environmental stress to the developing fetus through a host of adverse health and physical effects. However, a well-defined 'fetal cocaine syndrome' has not been established (**13.3**). Contradictory data and the existence of confounding covariables have increased problems of understanding the impact of prenatal cocaine on the fetus and maturing child.

Nevertheless, a distinctive pattern of cocaine-associated risk factors is evident. The neonatal outcomes commonly related to maternal cocaine-use involve the complications of low birth-weight, preterm delivery, and intra-uterine growth retardation in length and head circumference. Other suggested health compromises include: structural defects in organ systems, particularly those of the gastrointestinal tract; neurological and behavioural abnormalities; an overall greater dependency on neonatal intensive care resources; and possible increased risks of spontaneous abortion and sudden infant death syndrome.

A survey of research published between 1981 and 1995 found gestational cocaine to be highly correlated with intra-uterine growth retardation (IUGR) and low birth-weight (characterized by neonatal weight under 2500 grams for a full-term infant). Epidemiological studies with clinical experimental and control populations deem exposed newborns at higher risk for reduced body-weight and decreased linear and head circumference measures at birth; and more likely rank them below the 50th percentile of reference values, with frequent reporting of small-for-gestational age (SGA) status. Prenatal cocaine exposure appears to induce a symmetric pattern of IUGR, in which indicators of both body fat and body leanness are reduced, and overall body composition resembles that following maternal malnutrition. In a large-scale survey of 17,466 singleton births including 408 who were cocaine-exposed, infants with a history of intra-uterine cocaine exposure manifested significantly increased risks of SGA and low birth-weight (LBW). SGA rankings constituted a higher percentage of group total for cocaine-positive than for drug-negative groups, ranging between 15 and 35 per cent in the drug-exposed group, versus a 7 per cent high in the drug-free group. Another study showed LBW in 43 per cent of exposed infants and SGA in 36 per cent, values significantly above those of the general hospital population. The odds ratios of LBW occurring in cocaine-exposed newborns are consistently elevated, ranging between 3.0 and 6.6. On average, full-term exposed infants may experience approximately 300 grams decrement in mean birth-weight than newborns of non-users.

Complicating the effects of LBW and IUGR is the association of gestational cocaine with an increased likelihood of preterm birth (parturition at less than 37 weeks of gestational age, compared to full-term gestation of 38–40 weeks). Discriminant function analyses conducted on nine variables known to contribute to early labour identified cocaine as the strongest predictor of preterm status in one prospective cohort of 425 pregnant women. Of expectant mothers who lacked prenatal care and were admitted for active labour before the 34th week of pregnancy, twice as many (38% versus 19%) tested positive for cocaine metabolites. In studies controlling for confounders, the odds of preterm birth in exposed neonates ranged from 2.8 to 20.4 times that of unexposed neonates.

Preterm infants are especially vulnerable to the negative health implications of low birth-weight. Even attenuated use of cocaine can pose complications, as evidenced for one population wherein 17 per cent of first trimester-only users delivered early, as did 31 per cent of women who used cocaine throughout pregnancy, compared to 3 per cent of non-drug using controls. Data suggest that cocaine-exposure episodes can directly precipitate onset of labour.

There appears to be a direct relationship between severity of birth-weight compromise and quantity of drug used, as demonstrated by measures of cocaine from the hair shaft wherein the level of exposure was found to be significantly and inversely correlated with birth-weight and head-circumference values. Reduced fetal growth may result from the vasoconstrictive ability of cocaine to decrease uterine blood flow and restrict nutrient delivery and uptake in the fetus and/or may represent an effect of cocaine crossing the placenta and directly producing vascular disruptions in the fetus, leading to reduced fetal growth and compensatory diminished circulation to non-vital organs.

Conclusions about the short- and long-term effects of prenatal cocaine have been controversial and equivocal. The methodological difficulties of clinical research with a drug-abusing population and the synergism of confounding socio-demographic and behavioural variables create obvious impediments to isolating the effects of cocaine from those of other potentially high-risk determinants. Lack of prenatal health care, maternal tobacco use, and polydrug exposure are frequently cited confounders. Problems of drug measurement, user classification, and reporting bias thwart comparative interpretations. Discrepancies may also result from small epidemiological sample sizes that preclude thorough control for covariates, and may be statistically inadequate for assessing risk.

A review of available data suggests that prenatal cocaine-use predicts negative birth outcomes directly, as well as through obstetric risk factors of maternal health status and life-style practices. Since the 1980s, the United States has witnessed an acute crack-cocaine epidemic that has disproportionately compromised the section of the population which is of reproductive age, and consequently has placed at risk a sizable cohort of cocaine-exposed infants who are more likely to experience adverse birth and growth complications, and whose long-term physical and cognitive health prognoses remain undetermined.

*Kyra Marie Landzelius*

## ALCOHOL CONSUMPTION AND PREGNANCY OUTCOME

Alcohol is a toxic compound and a potent teratogen and may influence fetal growth and development in a number of ways. Alcohol may affect general maternal nutritional status by changing appetite or by substituting for the other energy-yielding macronutrients. Fetal growth may be affected by alcohol independently of nutritional status, by reduction of placental uptake and transport of nutrients, decreased fetal glycogen and blood glucose, reduced blood flow to the placenta, hypoxia and hormonal changes, and interference with protein synthesis in fetal tissues. Pre-existing maternal liver disease, for example, due to alcohol abuse, will increase circulating amounts of toxic metabolites which may affect the developing fetal brain and may also disturb the balance of amino acids in the maternal and fetal circulation, contributing to impaired fetal growth. It is not yet clear whether there are differential effects of alcohol at various stages of gestation with respect to growth, congenital abnormalities and mental development, whether there is a threshold of total alcohol intake, or if there is a dose-response effect. The consequences of binge-drinking episodes also remain to be established.

*Fetal alcohol syndrome* (FAS) describes a variety of features common to babies born to mothers who were alcohol abusers and/or who drank heavily during pregnancy (> 21 units, 210 g per week). Criteria for FAS include evidence of 1) prenatal and/or post-natal growth retardation, i.e. weight, length, and/or head circumference below the 10th percentile when corrected for gestational age; 2) signs of neurological abnormality or developmental delay; 3) characteristic facial dysmorphology for example microcephaly (head circumference below the third percentile), short palpebral fissures; poorly developed philtrum, thin upper lip and/or flattening of the maxillary area. Other birth defects including cardiac, genito-urinary and musculoskeletal abnormalities may also be present. There is some evidence of longer term effects such as growth and mental retardation in childhood.

The fact that large amounts of alcohol have adverse consequences for the fetus, has raised the possibility that FAS is the extreme end of a

*Overview of major behavioural findings of the Seattle Longitudinal Study on alcohol and pregnancy from birth to 7 years of age*

| Time and finding | Procedure | Reference |
|---|---|---|
| ***Day 1*** | | |
| Poorer habitation | Neonatal Behavioural Assesment Scale | Streissguth, Barr, & Martin, (1983) |
| Lowered arousal level | | |
| More body tremors; more head-turns to left; more open-eye time; less high-level body activity; more hand-to-face | Naturalistic observations of neonatal movement | Landesman-Dwyer et al. (1978) |
| ***Day 2*** | | |
| Decreased sucking pressure (whole trial) decreased latency to suck | Sucking pressure transducer | Martin et al. (1979) |
| ***8 months*** | | |
| Decreased performance on Mental Developmental Index; decreased performance on Psychomotor Developmental Index | Bayley Scales of Infant Development (MDI and PDI) | Streissguth, Barr, Martin, & Herman (1980) |
| ***4 years*** | | |
| Lowered full-scale IQ; trend toward lowered Verbal IQ; lowered performance IQ | WPPSI (intelligence test) | Streissguth, Barr, Sampson, Darby, & Martin (1980) |
| More motor errors; longer latency to correct errors; more total time on the pegboard | Wisconsin Fine Motor Steadiness Battery | Barr, et al. (1990) |
| Decreased finger-tapping performance | Halstead-Reitan Battery (portions) | Barr, et al. (1990) |
| More total time on Tactual Performance Test for young children | Ontario Battery (portions) | |
| Poorer balance | Gross motor tasks | Barr, et al. (1990) |
| More errors of omission; more errors of commission; lower ratio of correct responses; longer reaction time (funal trial) | Vigilance task on microcomputer (preschoool version); Motion detector | Streissguth, Martin, et al. (1984) |
| ***7 years*** | | |
| Lowered full-scale IQ; lowered verbal IQ; lowered performance IQ | WISC-R (intelligence test) | Streissguth, Barr, & Sampson (1990) |
| Lowered reading scores; lowered arithmetic scores; [spelling scores not lowered] | WRAT-R (achievement test) | Streissguth, Barr, & Sampson (1990) |
| Longer word reading performance time; longer colour-naming time | Stroop (example of a neuro-psychological task) | Streissguth, Bookstein, et al. (1989) |
| More errors of omission and commission on X and AX Tasks; longer average reaction time | Vigilance task on microcomputer (7-year version) | Streissguth, Barr, et al. (1986) |

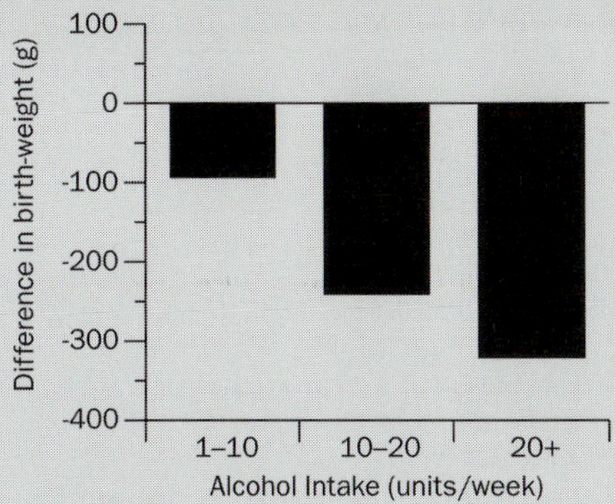


**The influence of alcohol consumption on birth-weight.**

spectrum and that more subtle effects occur at more moderate levels of intake. In contrast to heavy and chronic alcohol intakes, there is no consensus about the effects of normal light–moderate 'social' drinking prior to and during pregnancy. Across studies of alcohol consumption at non-extreme levels, there are many conflicting results, and the evidence for and against effects is limited and inconsistent. In some studies, decreased neonatal weight, length and head circumference and increased neurological abnormalities, malformations, prematurity, miscarriage and still-births, have been observed even at low levels of alcohol consumption, which suggests that there is no safe level of alcohol intake during pregnancy.

Overall, studies suggest that women who consume 1–2 drinks per day (14 units, 140 g per week) are more likely to have an infant with growth retardation and possibly other complications. In women who drink less than 7 units per week the situation is much less clear. Authorities advocate sensible drinking habits (no more than 1–2 units per week) and recommend that women should abstain from all alcohol during pregnancy because even occasional drinking may have minimal, but preventable risks. Because the first trimester, especially weeks 2–8 of gestation when embryogenesis occurs, is a critical period for the induction of anatomical abnormalities, non-pregnant women who are planning to conceive are also recommended to abstain from alcohol or at least avoid binge-drinking and consume no more than 10 units of alcohol per week.

*Gail Goldberg*

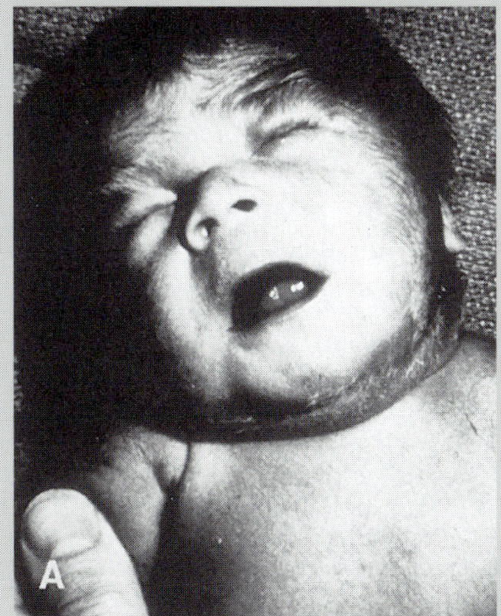


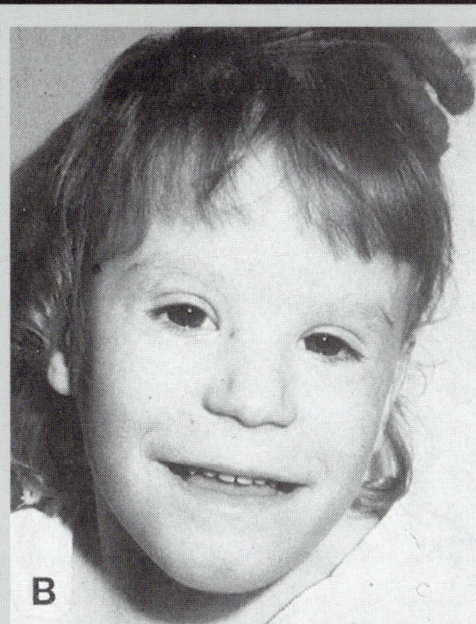


**Children with fetal alcohol syndrome: A shows a severe case; B shows a slightly affected child. Both children have short palpebral fissures and hypoplasia of the maxilla. Courtesy of Dr David Smith.**

**A full-term pregnant woman drinking beer. Courtesy of Science Photo Library, London.**

# 8.6

# Teenage pregnancy

The increased risk of infant low birth-weight (< 2500 g) associated with adolescent pregnancy is well known. For young primiparas age 16 years and under, risk is increased approximately twofold when compared to mature women aged 19–29 years, while for teenage multiparas risk is increased nearly twofold when compared to same aged primiparas or to mature multiparous women of the same ethnic group. Low birth-weight is a leading cause of neonatal and infant mortality, associated with childhood morbidity, with problems in growth and cognition as well as with an increased risk of chronic disease (diabetes and hypertension) in later life. For the young mother, a history of having borne a low birth-weight infant, whether the infant is preterm or small for gestational age, is one of the strongest risk factors for poor outcome in subsequent pregnancies.

It is well recognized that women raised under favourable social and economic conditions are likely to terminate a pregnancy which is unintended or occurs at an early age. However, women who grow up under suboptimal circumstances are more likely to give birth as teenagers. In the United States, for example, teenage childbearing is particularly prevalent among disadvantaged minorities – Blacks (116/1000) and Hispanics (107/1000) where the rate is 2 to 3 times higher than among young white women (42/1000). Conversely, more than 80 per cent of American teenagers who give birth come from families that are poor or low income.

Suboptimal life circumstances are likely to give rise to both early childbearing and the associated risk of poor outcome. Social and cultural factors which underlie the decision to maintain or terminate a pregnancy result in different rates of teenage childbearing between richer and poorer women. Consequently, the same constellation of social risk factors also will be present when infant birth-weights of young and older mothers from different socio-economic circumstances are compared. In this instance, the risk associated with young maternal age often is 'explained' by variables which are surrogates for social class differences that underlie early childbearing (maternal school or poverty status, grandparent education and occupation). While chronological age *per se* may not be an independent predictor of pregnancy outcome, adolescents are at risk because certain factors are more common among them: being part of an ethnic minority group; having a short interval between pregnancies; inadequate prenatal care; a prior history of poor outcome; and poor maternal nutritional status (low pregravid body mass, inadequate gestational weight-gain, maternal anemia). Smoking, drinking and drug use also affect the outcome of pregnancy although, as a rule, these risk factors are usually not found to be in excess when teenagers are compared to mature women from the same life circumstances.

The biological immaturity of the adolescent mother will also influence the outcome of her pregnancy. At adolescence the tempo of growth and development depends more upon underlying biological maturation than upon chronological age. In young women, rate and amount of adolescent growth vary considerably in age at onset (age 10–14 on the average), in the size of the increments following peak height velocity (10.8–22.3 cm) and menarche (74–10.6 cm), and the amount of time post-menarche that growth continues (1.8–6.7 years). At age 17 years, approximately 50 per cent of teenage women have achieved mature stature and the remaining 50 per cent are still growing. When aligned by chronological age, inter-individual variation can be so great that, at adolescence, no effect of age on growth can be demonstrated. Whereas when realigned by an index of maturation, such as years since menarche (gynecological age), differences are apparent. This also is evidenced during teenage pregnancy. For example, after controlling for confounding variables associated with the duration of gestation, one of the most important causes of infant low birth-weight, young maternal age, is weakly associated with an increased risk of preterm birth (< 37 weeks gestation) and amounts to approximately a 3 per cent decline in risk per year of (increasing) age. Apart from meta-analysis, where individual studies are combined, this effect usually is not statistically significant. Whereas low gynecological age, an indicator of biological maturity practically unique to the teenage years, shows a much stronger relationship with risk of preterm birth (20% decline in risk/year) even after chronological age is controlled for.

Adolescent growth and teenage pregnancy often coincide. However, when serial measures of stature are taken, growth is not apparent because of a diminution in maternal stature over the course of pregnancy. Vertebral compression in association with gestational weight-gain and postural lordosis contribute to the mistaken impression that little growth of consequence occurs. Using sensitive methods (Knee Height Measurement Device or KHMD) and a body segment less susceptible to 'shrinkage' (lower leg) to index growth, about 50 per cent of young primiparas and multiparas grow while pregnant. Compared with non-growing teenagers and with mature women, growing gravidas, both primiparas and multiparas, give birth to infants weighing 150–200 grams less. This occurs despite the fact that growing teenagers gain more weight during pregnancy and that greater gestational weight-gain is associated with increased infant growth and thus larger size at birth. One reason for larger weight-gains is that growing and pregnant adolescents continue to accrue fat during the third trimester when fat stores should be mobilized to support fetal growth. Infant birth-weight is substantially diminished when late pregnancy fat stores increase.

Additional weight and fat accrued during pregnancy is retained by young still-growing mothers in the post-partum period as part of their own continued growth and development. On the average, growing teenagers retain more than 40 per cent of their gestational weight-gain at 4–6 weeks post-partum, compared to 14 per cent in mature women and same age non-growing teenagers.

A decade or more ago, observations made under conditions of famine suggested that mother and fetus competed for available nutrients and that maternal needs took precedence over fetal growth. Later, data from pregnant teenagers were used as another illustration with the presumption that maternal growth triggered competition. The situation in the developing world, where childhood malnutrition delays biological maturation and thus extends the period of adolescent growth, was particularly compelling because of the increased incidence of intra-uterine growth restriction that might reasonably be secondary to maternal–fetal competition. Consistent with this hypothesis, decreased transmission of nutrients to the fetus has recently been documented in inner-city American teenagers by measurement of nutrients in cord blood and by Doppler ultrasound. In addition, the use of Doppler velicometry has identified greater risk of abnormal maternal-fetal placental vascular resistance, an index of reduced blood flow. Each indicator of reduced nutrient transmission between mother and fetus was more prevalent when the young mother continued to grow during an adolescent pregnancy.

*Theresa O. Scholl*

See also 'Ultrasound measurement of prenatal growth' (**1.2**), 'Physical examination' (**1.10**), 'Assessment of age at menarche' (**1.15**), 'Fetal growth retardation' (**4.8**), 'Sexual maturation' (**5.15**), 'Adulthood and developmental maturity' (**5.17**), 'The development of sexuality' (**6.6**), 'Maternal anthropometry and birth outcome' (**8.1**), 'Nutrition' (**8.2**) 'Body-size at birth' (**10.3**) and 'Long-term consequences of early environmental influences' (**12.1**)

## Part Nine

# The ecology of post-natal growth

Ecology is a part of the earth sciences and is the study of the interaction of organisms with their physical, biological, and chemical environments. However, because of the breadth of its subject matter and its methodology, ecology overlaps considerably with the biological sciences. Human ecology, in which the focus is on human populations, is based on principles of ecology and the premise that our species is subject to the same evolutionary and ecological laws as are others. However, its boundaries are broadened as a result of the particular complexities of human behaviour, and especially of human culture. It is impossible to study human/environment relationships without a major emphasis on technology, social behaviour, and the worlds created by the varieties of human culture. Consequently, an ecological approach to growth means the interpretation of patterns of development within the broadest possible environmental framework.

An ecosystem is a concept, formulated to allow researchers to operationalize an ecological approach for specific purposes and to test specific hypotheses. To analyse an ecosystem is to determine the nature of the interactions which are observed, as much as possible to quantify them, and to assess their significance as part of the total system being studied. It is important to understand that an ecosystem is an operational model, defined in the context of the research at hand, and operationalized as a set of variables which become proxies for the components of interest.

**Opposite: Nepali children. One of the factors influencing post-natal growth is altitude. Photograph by Stanley Ulijaszek.**

An ecosystem is often studied as the transfer of energy along some chain which, with other chains, form a network, or web – for example, the food chain. An ecological analysis of human growth takes as its central focus those variables which through their interactions direct, channel, and sometimes constrain, the growth process and its outcome. The number of variables and the nature of the interactions specified are limited only by an investigator's ability to conceive and study a situation, and the availability of appropriate methods of analysis. In some instances, certain variables may only be described qualitatively while, in others, the interactions may be analysed quantitatively, resulting in a formal model capable of being generalized broadly.

**Illustration of types of relationships that may occur between variables A and B; 1 = causality; 2 association; 3 = interaction; 4 = feedback (synergism).**

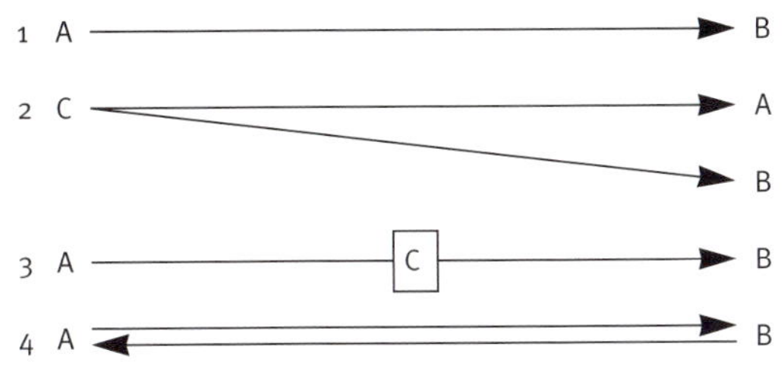


The range of interactions among variables is broad. The figure shows schematically the kinds of interactions between variables that may occur. In type 1 the relationship is causal, such that a change in A causes a

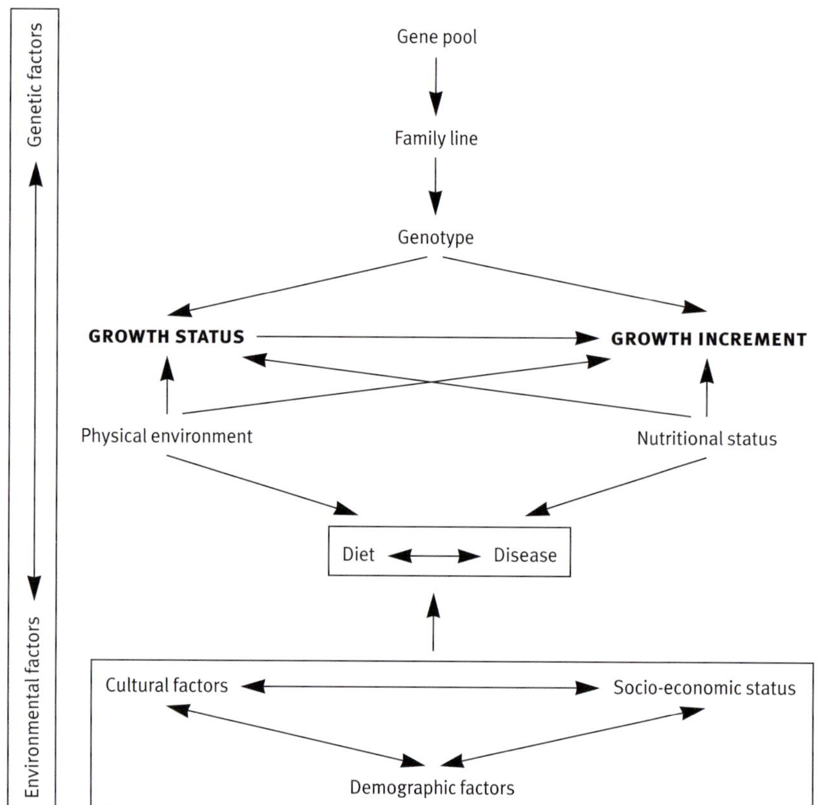


**Schematic diagram of one ecosystem of human growth.**

change in B, for example, when an experimental animal is deprived of a particular essential nutrient and growth failure results. A type 2 interaction is one of covariance in which two variables respond to changes in a third, variable C. For example, improved economic circumstances may result in increased height and improved cognitive development, with no causal relationship between height and cognition. Type 3 illustrates the case in which the effect on B of A is dependent upon the state of C. This could occur if the effects of energy restriction on the weight-loss of obese children depended on their individual genotypes. And type 4 is a feedback, or synergistic, relationship; a change in A will cause a change in B which will feed back and change A. Feedback relationships may be positive or negative. For example, malnutrition increases the risk of infectious disease which in turn increases the risk of malnutrition – a positive feedback. The regulation of the hypothalamic-pituitary-target gland system is an example of negative feedback.

The next figure presents one conceptualization of an ecosystem of growth. This model is an heuristic one, designed to illustrate clusters of variables that influence growth outcome, and to suggest pathways by which they act. The decomposition of such a model requires the specification of a set of operational variables (those that will actually be measured) which will be valid representations of each conceptual variable in the model (for example, nitrogen balance as a proxy for nutritional status). Studies may then be designed which can measure the extent and significance of the effects hypothesized.

At the top of the second figure are genetic components and at the bottom, environmental ones. The bar at the left indicates that genetic and environmental components interact with each other, but any attempt to represent the interactions would render the figure unreadable. Illustrations of the types of gene–environment interactions are modelled in the third figure, which illustrates the responses of four hypothetical genotypes, along the y-axis, to changes in the environment, along the x-axis.

Genotypes A and C demonstrate additive relationships, in that environmental effects are added to genetic ones. Environmental effects are clear but are independent of the genotype. Genotype B displays an interaction in that, though its initial phenotypic value was lower, its response to any environmental change is greater than that of either A or C. Genotype C shows another type of interaction characterized by a threshold that limits further increases in the phenotype beyond a particular environmental level.

Since an interaction means that genes and environments are not independent, even more complex interactions are possible, especially so in the case of behaviour. For example, individ-

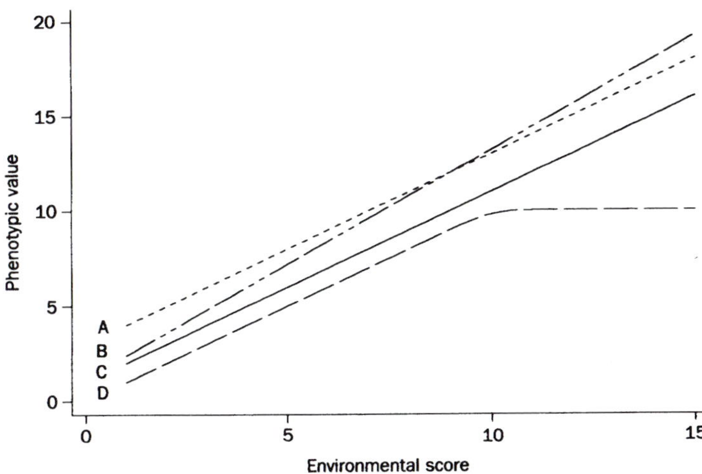


uals who are genetically more intelligent are more likely to seek out environments which are more enriching and which will stimulate increased mental development. Unfortunately, present levels of scientific knowledge preclude, except in particular and uncommon situations, the identification of specific genotypes for normal growth and development as well as their delineation from other genotypes. As a consequence, an ecological analysis of growth is currently restricted to measurement of the environment, the genotype being treated either as a 'black box' or in simple descriptive terms.

**Responses of four hypothetical genotypes to environmental change; A & B = additive; C & B = genotype/environment interaction.**

## Perspectives on the ecology of post-natal growth

*Evolutionary/adaptive perspectives*

The basis for all life and for all of its varieties is evolution. Central to evolution is adaptation, the process by which genetic variability has been shaped to match the environment and, since the ecosystem is the unit within which populations interact with their environments, adaptation is central to any ecological consideration. Growth, of course, is the process by which the newly fertilized zygote becomes an adult, and achieves the body-size, shape, and composition, as well as the functional capacity (**9.10**) capable of interacting successfully with the surrounding environment. An ecology of post-natal growth must hence begin with an adaptive perspective.

The evolution of the human growth curve has been characterized primarily by an attenuation of childhood, followed by a relatively brief, intense adolescent spurt. While one can never be certain of the primary selective pressure underlying this evolutionary trend, it is clear that the lengthened period of biological immaturity, has heightened sensitivity to the environment resulting in a plasticity of growth as a consequence of environmental interactions. The growth of children is so sensitive to the environment that it serves as a reliable indicator of the quality of that environment. Environmental factors which can influence growth include nutrition (**9.2**), infection (**9.3**), psychosocial stress (**9.4**), environmental toxins (**9.5**) and hypoxia (**9.9**). Furthermore, given the somewhat dominant position of nutrition and infection in influencing growth and development, cultural and behavioural factors associated with infant feeding (**9.1**) also affect growth patterns, as do climatic and cultural factors that create seasonality in nutritional availability and disease transmission (**9.8**).

Adult body-size and morphology vary systematically among populations and the pattern of the variation is in accordance with what is predicted by the 'ecological rules' of Bergmann and Allen. The result are particular physiques which are adapted to the thermoregulatory demands of the different ecosystems. Since variation in growth patterns is responsible for variation in adult morphology, it seems clear that the interaction between the growth process and the ecosystem is an adaptive one. The ways in which genetic and environmental determinants of growth interact is still uncertain. There is presumptive evidence of the role of differences among gene-pools which may have evolved as part of the adaptive process. But there is also clear experimental evidence from animal studies of the effects of temperature on growth which result in regulatory adaptations among adults of the species.

The sensitivity of human growth to the environment is further demonstrated by the ability of the process to recover, or to catch-up, following episodes of disease or malnutrition (**9.6, 9.7**). Human growth is a target-seeking process and, insofar as is possible, seeks to move back to its individual trajectory when it is driven off-course. The ability to recover fully is of course a function of the timing, the duration, and the intensity of the insult. When the mechanisms responsible for growth faltering or failure are ubiquitous, the process will be significantly constrained and individuals will fail to reach their innate potential. None the less, at its most basic, growth is an adaptive process acting within specific ecosystems to produce healthy and functional adults.

*Biomedical perspectives*

While the heightened sensitivity of the individual to the environment is a basic feature of human growth, and crucial to the adaptive capacity of populations in different settings, it increases the impact of the environment upon development. When the interactions among ecological components is a positive one, the environment supports the genetic template and the result is positive. This is the evolutionary model in its ideal form. In the evolutionary past,

*A proposed list of risk and opportunity factors influencing human development and functioning*

| Variables | Risk factors | Opportunity factors |
|---|---|---|
| Mother's age | Younger or older than normal childbearing years | Within optimal childbearing years |
| Parent education | Low educational attainment | High educational attainment |
| Income | Inadequate income | Adequate income |
| Occupation status | Low occupation status of head of household | High occupation status of head of household |
| Socio-economic status (SES) | Low SES | High SES |
| Job stability | Repeated job changes or unemployment | Stable job |
| Pregnancy | Unplanned | Planned |
| Number of siblings | More than four children | One or two children |
| Residential stability | Repeated relocations | None or few relocations |
| Marital status | Absence of spouse or partner | Supportive spouse or partner present |
| Marital relationship | Conflictive | Harmonious |
| Marital stability | Repeated changes in a conjugal relationship | Stable conjugal relationship |
| Child temperament | Avoidant, difficult | Warm, responsive |
| Infant separation | Prolonged separation in first year | Limited separation in first year |
| Parental health | Poor physical health | Excellent physical health |
| Parental mental health | Repeated occurances of mental health related problem | Stable emotional well-being |
| Parental self-esteem | Low self-esteem | High self-esteem |
| Parental locus of control | External | Internal |
| Parental social skills | Poor | Good |
| Coping strategies | Reactive | Proactive |
| Quality of primary caregiver/ child interaction | Controlling and emotionally unavailable | Stimulating and warm |
| Parenting style | Authoritarian/Directive | Responsive/Facilitative |
| Toxic substances | High exposure | No exposure |
| Nutritional intake | Inadequate | Adequate |
| Accidents | Frequent | Infrequent |
| Infections/illnesses | Frequent | Infrequent |
| Alternative caregivers | No | One or more |
| Presence of extended family | None or few available | Many and supportive |
| Extrafamily support | Poor/Unsupportive | Good/Supportive |
| Life events | Negative life events | Positive life events |

From Dunst,C.M. (1994). Methodological considerations and strategies for studying the long-term effects of early intervention. In *Developmental Follow-up Concepts, Domains and Methods*, ed. S.L. Friedman and H.C. Haywood. San Diego: Academic Press.

demographic parameters were appropriate to the conditions of the environment and, when disease and malnutrition occurred, though acute they tended to be short-term. But in more recent times, many populations exist under conditions of great social and economic disadvantage, with overcrowding, poverty, malnutrition and disease characteristic of the environments of millions of children. Under these conditions, environmental stress and deprivation, while ranging from moderate to acute, are chronic and intractable in the short term. The situation is exacerbated by the forces of acculturation in which cultural values alter in response to changing ideas. Tradition infant-feeding practices, for example, may be abandoned and replaced by more 'modern' ones which cannot be supported by the traditional economy and/or value system.

Thus the evolutionary adaptation for increased environmental sensitivity has also become a burden upon human populations. Rather than providing an ecosystem within which individuals may reach their genetic potential, the environment becomes a systematic force which constrains human development.

## Integrating the adaptive and the biomedical

The integration of an adaptive approach to growth, with its emphasis upon the ecosystem as a force shaping growth in a positive way, and the biomedical, with its emphasis upon the ecosystem as a barrier to the achievement of potential, is not easy. In the extreme, one approach focuses on homeostasis and self-regulation and the other on improvement through change. In the final analysis, an ecological approach to growth must combine both views of human ecosystems. An ecosystem is not a monolithic structure, but rather is a dynamic set of interactions. Understanding its role both in channelling and in constraining growth is essential to understanding the process of human growth in all of its variety.

An ecological approach to growth is one that begins with a model of an ecosystem and successively disaggregates it to analytically manageable units. The results obtained are then combined as the model is re-aggregated and synthesized into a picture of the interactions among components and of the relationships among those interactions. In so doing, individual interactions are subsumed and replaced by the ecosystem itself. Depending upon the purpose of the investigation and the nature of the population under study, both adaptive and mal-adaptive outcomes are included. If an intervention is deemed advisable, quantitative analysis can suggest pathways and nodes where specific programmes may be applied.

What then is the unit of comparison from one system to another? It is not the interaction but the system. The impact upon subsequent growth of abandoning breast-feeding will be different in a poor community from a lesser-developed nation that in an affluent sector of industrialized society (**9.1**). Or the capacity for catch-up growth will not be the same in societies with an enormous load of intestinal parasites (**9.3**) as in others with high levels of medical care. The unit of comparison must be the ecosystem itself and growth outcomes within its parameters. Though this is not an easy task, either in terms of study design or analytic strategy, it is the only way to view post-natal growth from a truly ecological perspective.

*Francis E. Johnston*

# Infant-feeding and growth

Growth is a function of genetic potential and environmental influences, of which one of the most important determinants in infants and children is nutrition. The World Health Organization has recommended the use of the United States National Center for Health Statistics (NCHS) growth references for the assessment and monitoring of all ethnic groups. While useful as a reference, the charts do not represent the actual growth-rates for most children of the world. The mean weight of formula-fed infants in Western countries is similar to the 50th percentile throughout their first year of life, but infants who are breast-fed fall below the 50th percentile during their second 6 months of life.

During the 1950s and '60s there was a substantial decline in breast-feeding in industrialized countries, and in the United Kingdom in 1975 only 30 per cent of mothers were breast-feeding their infants at 3 months of age. The use of infant formula became very widespread in industrialized countries and for a period in the '60s and '70s larger infants became the norm. The mean growth-rates of infants in the United States and the Britain was greater than the reference values and has since declined slightly. Breast-feeding is able to sustain adequate growth of children of almost any population in the first 4 months of life. In the developing world (and in some disadvantaged groups in developed countries) the growth-rates of infants usually begins to flatten at between about 4 and 6 months of age. While this is a result of nutritional deficiencies, the immediate antecedent causes may include delayed or inadequate weaning and the effect of infections (**9.3**). Each year 14 million children die in the developing world, and most of these deaths have nutritional deficiency as a contributing cause at least. For those children who survive, 39 per cent of them remain stunted in their growth. For these children, even if catch-up growth occurs later, it is highly unlikely that they will reach their genetic potential for stature. In many of these populations, puberty is delayed by as much as 1–2 years.

## Breast-feeding

Breast-feeding is the most important factor in the nutrition and growth of the infant. Until the 20th century, breast-feeding, by mother or surrogate, was the only way to feed an infant. It remains the preferred method and numerous studies have shown that breast-feeding has advantages for both infants and mothers. Breast-feeding provides nutritional, immunological and psychological benefits to the infant and also offers economic benefits to both family and society. While the digestive and excretory systems of the infant are still immature, the infant is in a period of rapid growth and development and has high nutritional demands relative to other phases of life. Breast-milk provides for the nutritional needs of infants from birth until at least 4–6 months of age, during which time the birth-weight will double. Most recommended dietary allowances for infants are based on the proposition that breast-milk provides optimum nutrition for an appropriate growth-rate, body composition (**1.6**), and

*Composition of mature breast-milk (per 100 ml)*

| | | | | | |
|---|---|---|---|---|---|
| Energy (kcal; kJ) | 70;293 | Thiamin (µg) | 16 | Calcium (mg) | 35 |
| Water (g) | 89.97 | Riboflavin (µg) | 30 | Phosphorous (mg) | 15 |
| Protein (g) | 1.3 | Nicotinic Acid (µg) | 230 | Magnesium (mg) | 2.8 |
| Fat (g) | 4.2 | Vitamin B6 (µg) | 6 | Iron (mg) | 76 |
| Carbohydrate (g) | 7.4 | Vitamin B12 (µg) | 0.01 | Copper (µg) | 39 |
| Cholesterol (mg) | 16 | Total Folate (µg) | 5.2 | Zinc (µg) | 295 |
| Vitamin A (µg) | 60 | Pantothenic acid (µg) | 260 | Manganese (µg) | 1.2 |
| Vitamin D (µg) | 0.01 | Biotin (µg) | 0.76 | Chromium (µg) | 0.6 |
| Vitamin E (µg) | 0.35 | Sodium (mg) | 15 | Selenium (µg) | 1.4 |
| Vitamin K (µg) | 0.21 | Potassium (mg) | 60 | Iodine (µg) | 7 |
| Vitamin C (mg) | 3.8 | Chloride (mg) | 43 | Flourine (µg) | 7.7 |

Dept of Health and Social Security. Report on Health and Social Subjects 32. Present day practice in infant-feeding: Third report. London: HMSO, 1988.

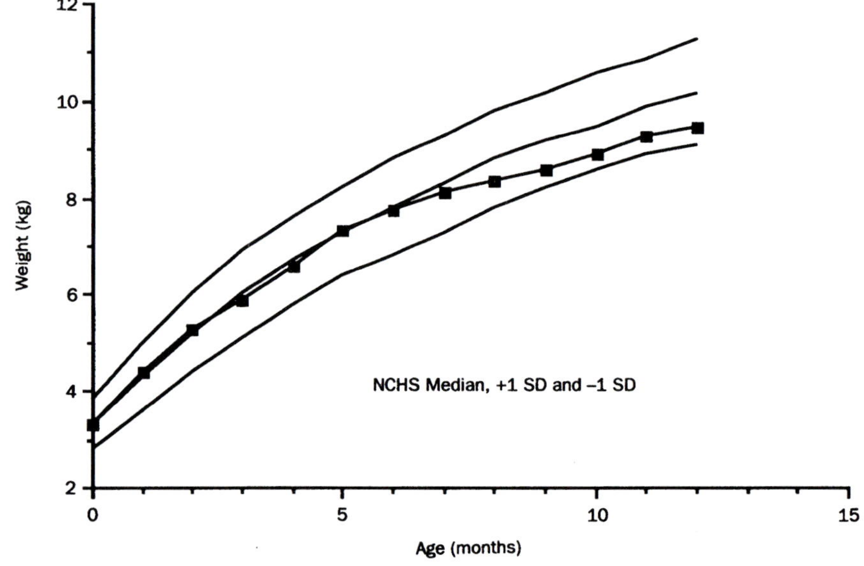


**Growth of breast-fed Australian aboriginal infant boys.**

hildren in Guatemala. Photograph by Barry Bogin.

level of physical activity which are conducive to optimal health and neurobehavioural development (**4.6**, **6.2**).

In recent years there has been substantial research into the growth factors and hormones which are found in human milk, often in concentrations higher than maternal plasma. Many of these survive the neonatal gastrointestinal tract and become absorbed into the neonatal circulation. Effects can be readily demonstrated in animal models on the growth of gastrointestinal epithelium and it is likely that there are other effects on the growth and development of tissues yet to be documented.

A number studies have shown an association between breast-feeding and certain pathological conditions, suggesting a pivotal role for breast-milk in specific organ development. These include: physiological reflux; pyloric stenosis; juvenile diabetes; inflammatory bowel disease; some childhood cancers; and delayed onset of coeliac disease. Breast-milk contains a number of growth-promoting agents that are particularly important in the maturation of the gastrointestinal tract. There are also some benefits for the mother's health, with lactation reducing the incidence of osteoporosis and premenopausal breast cancer. In many cultures it is customary to delay breast-feeding until the colostrum is replaced by milk. This carries a risk to the infant as infection may be transmitted by any complementary feed given in the interim.

Evidence concerning the influence of breast-feeding on growth after the age of 6 months is conflicting. The limited evidence from industrialized countries suggests that prolonged breast-feeding in the 1990s has little influence on child health and growth, providing of course that weaning foods provide adequate energy and nutrients. By way of contrast in the developing world, the continuation of breast-feeding for 1–2 years has potential benefits from the delivery of nutrients and protective factors. The potential value of breast-feeding in birth spacing, a critical factor in maternal and infant health, should not be ignored. For these reasons it is important in poor areas of the developing world to encourage breast-feeding until 2 years, providing that energy intake is maintained from adequate amounts of weaning foods.

There remains debate about the optimum nutrition of infants who are born prematurely. However, breast-milk feeds have the advantage of more effective utilization of proteins, fat, minerals and trace elements. It also has passive immunologic factors, provides active immunostimulation and reduced rates of necrotizing enterocolitis. The one disadvantage is the potential for the transmission of viral disease, although technology for eliminating this risk is now becoming available.

The importance of breast-feeding in the health and growth of infants has meant that efforts are being made for its protection, support and promotion. Efforts are now being made to control the inappropriate marketing of infant formulas and to promote the benefits of breast-feeding. For those who choose not to breast-feed, modern infant formulas provide adequate nutrient supply for growth, provided the preparation is sterile and prepared in the correct concentration. In many developing countries this is not the case, and contaminated and diluted infant feeds are the cause of much growth retardation and morbidity. The composition of infant formulas has been modified considerably in recent years as research has refined knowledge of the composition of breast-milk.

At the age of 4–6 months, breast-milk or infant formula no longer provides sufficient energy for growth, and at this time a weaning food of sufficient energy density is required. For the breast-fed infant this introduces the new hazard of microbiological contamination. It is estimated that a high proportion of the infants and children under 5 years of age who die each year is due to the interaction of infections and poor nutrition. Diarrheal diseases have traditionally been ascribed to poor water supply and sanitation. However, many weaning foods prepared under unhygienic conditions or using contaminated water supplies are frequently heavily contaminated and are a major factor in the cause of diarrheal diseases and associated malnutrition.

## Protein

The protein content of human milk is relatively low compared to other mammals, and this is probably related to the slow rate of growth of human infants relative to other species. The protein content of human milk was previously overestimated because methods of protein estimation relied on the measurement of nitrogen content, and milk contains non-protein nitrogen-containing compounds. Human milk proteins are readily digested, contain appropriate mixtures of amino acids and their low concentrations give low solute loads for immature kidneys. The protein content of infant formula has been higher than breast-milk and it is now thought that a reduced level of 4.5 grams protein per 100 kilocalories or as low as 3.5 grams would be more appropriate for growth.

## Lipids

A relatively high proportion of fat (> 50% energy) is required in the diets of infants and young children because of their high energy needs and the limited volumetric capacity of their gastrointestinal tract. Soon after birth and the start of oral-feeding serum, total and low-density lipoprotein (LDL) cholesterol rise significantly. After 7 days the levels stabilize and the concentration after this time is highly dependent upon the cholesterol and polyunsaturated fat content of the diet. By 12 months of age there is no residual effect of the type of milk fed earlier in infancy on serum cholesterol and lipoproteins. Because of its high energy value, the amount of fat in the infant and child's diet determines the energy availability and hence is a major determinant of growth-rates. In addition, the level of fat in the diet is correlated with the levels of vitamins in the diet, and infants placed on low-fat diets often have growth retardation and multiple deficiencies.

In addition to the role of fat as a provider of energy and a substrate for vitamins, there are important requirements for specific fatty acids. It is important for early human growth and development that there are sufficient long-chain polyunsaturated fatty acids (LCP), such as arachidonic (C20:4n-6) and docosahexanoic (C22:6n-3) acids, available. The fetus receives supplies of these fatty acids by placental transfer. However, the newborn baby has limited, if any, capacity for the synthesis of LCP from the precursor fatty acids, linoleic (C18:2n-6) and alpha-linolenic (C18:3n-3) acid. Babies who are breast-fed are supplied with LCP in human milk, but at the present time most infant formulas are deficient in LCP. Rapid depletion of LCP levels occurs in premature infants fed on standard formulas and this causes reduced visual acuity during the first post-natal months. Breast-fed infants have significantly greater amounts of the long-chain polyunsaturated fatty acids, such as docosahexanoic acid (DHA) in their cerebral cortex than infants being given formulas.

When infants on formula are supplemented with DHA their functional performance as shown on the Bayley Scale shows improvement. The quality of the lipid content of an infant's diet thus has a limiting effect on the growth and maturation of its central nervous system. It is not known whether the differences in intelligence detected in breast-fed infants when compared to those who are formula-fed persist into adulthood, or even whether infant diet is a factor in the adult onset of neurodegenerative disease.

## Iron

On a worldwide basis iron deficiency is probably the most common of all nutritional deficiencies after deficiencies of energy and somtimes protein. While breast-milk contains only around 1 milligram of iron per litre in the form of lactoferrin, it is extremely well absorbed. Infant formula generally contains a minimum of 4–6 milligrams per litre to have the same physiological effect. During the first 6 months of life, the iron stores of a breast-fed infant become depleted and at that age an additional source of iron is required if deficiency is to be avoided. Generally, iron deficiency is unlikely to cause a growth deficit under the age of 12 months.

**Rural children in the former Soviet Union. Photograph by Elena Godina.**

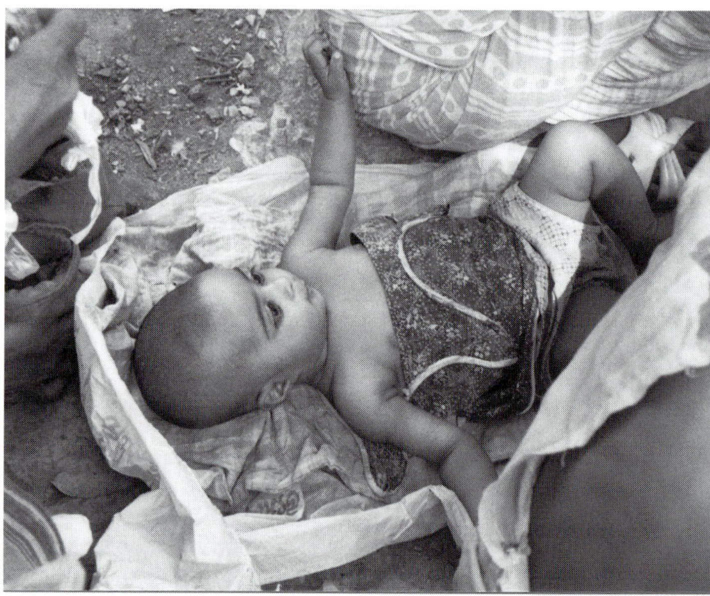

**Young Indian child. Photograph by Patricia van Someren.**

## Sodium

Sodium (Na) is an important growth factor which stimulates cell proliferation. It also has a role in protein synthesis. In animal studies, sodium chloride deficiency results in reduced

## GROWTH FACTORS IN MATERNAL MILK AND DEVELOPMENT OF THE GASTROINTESTINAL TRACT

In 1976, Widdowson and colleagues published findings of a study in which elevated growth of the gastrointestinal tract was shown to be associated with colostrum-feeding in neonatal pigs. Since then, a number of studies have identified mitogenic properties for different mammalian milk, including that of human beings. Growth factors found in human milk include the insulin-like growth factors and three insulin-like growth-factor-binding globulins (**4.2**). Milk-based growth factors have direct trophic effects on the intestine, as well as effects on the autocrine regulation of intestinal growth. However, studies of human-milk-stimulated growth of cultured gut cells show that only a fraction of the increased growth can be explained by the addition of epidermal growth factor, or the insulin-like growth factors. This has lead to the search for synergistically operating factors, and other independently operating growth-promoting substances. While the apparently normal growth and development of healthy infants receiving formula feeds devoid of growth factors suggests that their absence is not vital, there may be long-term consequences of gut growth which has not involved human-milk growth-factor-regulated growth, such as greater susceptibility to gastrointestinal disease through childhood and into adult life.

*Stanley Ulijaszek*

*Growth factors found in human milk*

- Epidermal growth factor
- Transforming growth factor $\alpha$
- Insulin
- Insulin-like growth factors I and II
- Insulin-like growth-factor-binding globulins 1, 2 and 4

## THE PROMOTION OF BREAST-FEEDING

The importance of breast-feeding for the health and growth of infants has meant that efforts are being made for its protection, support and promotion worldwide. The encouragement of breast-feeding is an important aspect of primary health care, since it is an unequalled way of providing nutrients for the growth and development of infants, while the anti-infective properties of breast-milk help to protect against disease. Inappropriate feeding practices can lead to infant malnutrition, morbidity and mortality and this is particularly marked in countries with poorer environmental standards.

During the decades of the 1960s and 1970s there was a general decline in breast-feeding in both developing and developed countries. In many countries infant formulas were marketed aggressively, often with the support of health workers. In poor communities, infant formula was often heavily contaminated with bacteria (**9.3**) and for economic reasons was over-diluted by mothers. This meant that many infants were being denied the benefits of breast-feeding with often disastrous effects on their nutrition and survival.

To counteract this trend, in 1981 the World Health Assembly almost unanimously endorsed the World Health Organization International Code of Marketing of Breast-Milk Substitutes. One country opposing the code was the United States of America on the grounds of impairment of free trade (perhaps more cynically seen as the right of multinational companies to kill innocent infants in countries far away from their own shores for the sake of economic gain). It was amended in 1994 to eliminate loopholes. Many countries have adopted the principles of the code and have developed policies for monitoring and controlling infant formulas. One of the earliest countries to implement a policy was Papua New Guinea, which took the simple step of banning the sale of feeding bottles and teats.

The WHO Code applies to the marketing and promotion of all breast-milk substitutes. It also has provisions related to complementary foods which are bottle-fed and which could be marketed for use in addition to breast-feeding. It points out the pivotal role health care systems and their health workers play in guiding infant-feeding practices and in encouraging and facilitating breast-feeding. In view of the vulnerability of infants in the early months of life, the marketing of breast-milk substitutes requires special restrictions. Breast-feeding does not have the financial or marketing resources to compete on an equal basis with the manufactured alternatives. For this reason manufacturers and distributors should not provide samples of products such as infant formula, bottles, teats or dummies, directly or indirectly, to the general public. Nor should any health workers do this distribution for them by giving samples of infant formula to pregnant women, parents of infants and young children, or members of their families. Representatives of companies are not permitted to educate or promote their products in any way.

The Code has a lot to say about the role of the health system in promoting healthy infant-feeding. The employees of infant formulas companies are not permitted to work in maternity hospitals. While health workers require some knowledge of the safe use of formulas in specific situations, their education needs to be balanced and include information on the benefits of breast-feeding and the disadvantages of using formulas.

The original version of the Code could be interpreted as only applying to maternity hospitals. So in May 1994 the 47th World Health Assembly adopted resolution 47.5 on infant and young-child nutrition to extend coverage to all health care facilities. This resolution urges Member States to ensure that no donations of free or subsidized supplies of breast-milk substitutes and other products covered by the WHO Code be accepted in any part of the health-care system. This applies to the following institutions: 1) maternity, pediatric and other hospitals; 2) medical practitioners and other health workers; 3) early childhood centres/maternal and child health centres; and 4) pharmacies.

*C.W. Binns*

*In 1989 WHO and UNICEF adopted the following guide to the promotion of breast-feeding in health care facilities, commonly known as the 'Baby Friendly Hospital Initiative'.*

**Ten steps to successful breast-feeding**

Every facility providing maternity services and care for new-born infants should:

1. Have a written breast-feeding policy that is routinely communicated to all health care staff.
2. Train all health care staff in skills necessary to implement this policy.
3. Inform all pregnant women about the benefits and management of breast-feeding.
4. Help mothers initiate breast-feeding within half an hour of birth.
5. Show mothers how to breast-feed, and how to maintain lactation even if they should be separated from their infants.
6. Give new-born infants no food or drink other than breast-milk, unless medically indicated.
7. Practise rooming-in (allow mothers and infants to remain together) 24 hours a day.
8. Encourage breast-feeding on demand.
9. Give no artificial teats or pacifiers (also called dummies or soothers) to breast-feeding infants.
10. Foster the establishment of breast-feeding support groups and refer mothers to them on discharge from hospital or clinic.

growth-rates, as shown by reduced body- and brain-weight. A restoration sodium intake restores the growth velocity to normal, but catch-up growth does not occur.

This situation is sometimes found in premature infants who can be in negative sodium balance in the first few years of life and are vulnerable to hyponatremia. This can be prevented by increasing sodium intake, which also produces accelerated weight-gain that persists beyond the period of supplementation. Since the early nutrition in preterm infants can affect subsequent growth and cognitive function, an adequate intake of salt (NaCl) is important in this group. The mechanism whereby sodium promotes cell growth is not understood, but probably involves cell membrane transportation mechanisms. After infancy it is important to maintain salt intakes at recommended low levels to avoid the development of an early tendency towards hypertension.

## Trace elements

Numerous trace elements are important in human growth and development. However, for many trace elements there is still a lack of detailed information about their dietary requirements during infancy. In many cases current dietary recommendations are based on the composition of human milk, and

### GROWTH AND WEANING IN PAPUA NEW GUINEA

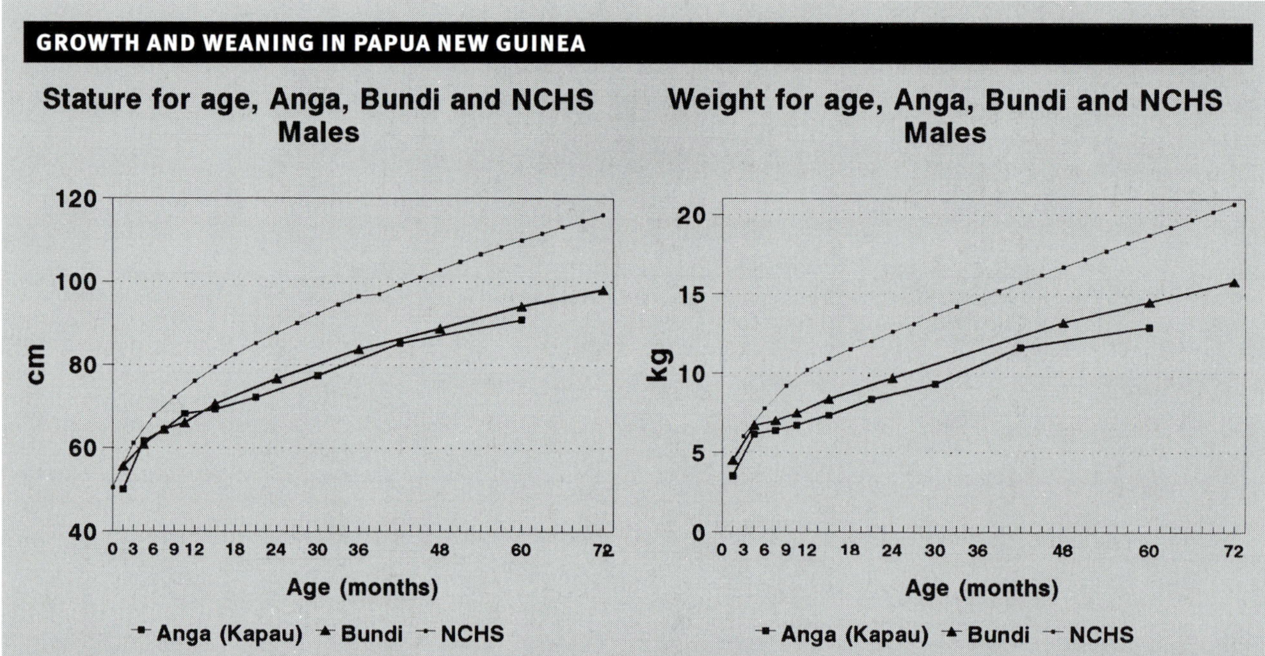


**Height and weight by age of young Anga and Bundi male children relative to the National Center for Health Statistics 50th centile. The decline relative to international reference values is typical of many populations in Papua New Guinea.**

Research on growth and nutrition in Papua New Guinea was carried out for many decades until the 1990s. In all studies, conclusions pointed to the interaction of prolonged breast-feeding and early feeding with foods of inadequate nutrient density as the basis of stunted growth throughout childhood. The cultural and ecological conditions which foster poor weaning diets and strengthen reliance on breast-milk have been described. The problem derives from root-crop-based subsistence strategies. The nation has no indigenous staple grains, and where root crops are not the primary staple, as in about one-third of the nation, the main source of energy is sago, a starch derived from the palm *Metroxylon sagu*. All of these staples are bulky and low in energy density.

In many Papua New Guinea societies it is not customary to feed children solids until they are walking or at least sitting. If motor development is retarded, then the feeding of solids is also likely to be delayed. This practice is buttressed by post-partum sex taboos which require abstinence until the child walks. When feeding begins, infants are fed on the society's major staple and bananas. In some cases, as among the Amele of lowland Madang Province, beliefs about the importance of watery foods for 'soft' infants encourage the feeding of cooked papaya and other juices or soups. In other cases, the lack of cooking pots requires that food is roasted over the fire or steamed in leaf packets or bamboo tubes. In such societies, many mothers traditionally premasticated food for infants. With the advent of modern health services, health workers discouraged this practice on the grounds it would contribute to the transmission of disease.

In the energy-deficient societies of Papua New Guinea, extra cooking effort for infants who can more easily be breast-fed is considered wasteful. Foods cooked for adults are rarely appropriate unless mashed or strained. With few utensils, this extra effort is difficult. With modernization, tea strainers, spoons and forks are available and mothers have been taught in some areas to use these. Occasionally some improvement has been evident; however, the most positive influence on the growth of weaning-age children occurs when the household has access to a steady cash income. Unfortunately, because of the nation's rugged terrain, undeveloped infrastructure, and high unemployment in urban areas, many households do not have access to cash. It is not likely that this will improve in the foreseeable future.

Documented growth patterns of Papua New Guinean children are consistent with adequate breast-feeding for 6 months, after which growth falters. An immense burden of disease from repeated respiratory infections, malaria, diarrhoea, and other infections contributes to cumulative stunting throughout childhood. Where sustained economic improvement occurs at the household level, as in many urban families, the weaning-age child benefits from increased fats and protein in the diet. *Carol Jenkins*

## PROLONGED BREAST-FEEDING AND GROWTH

While there is little debate about the advantages of breast-feeding for the first 4–6 months of life, the benefits beyond that time depend on the family and community environment. In the developing world, the continuation of breast-feeding for 1–2 years has potential benefits from the delivery of nutrients and protective factors (**9.3**). However, the association of between prolonged breast-feeding and growth has been a matter of some controversy. While a number of previous studies suggested that prolonged breast-feeding was associated with poor growth performance, others have shown that breast-fed children who are higher on anthropometric indices tend to be weaned earlier. Infants who are not growing as well are kept at the breast longer, suggesting that prolonged breast-feeding is confounded by poverty and poor environment.

A well-controlled study from Hubei province, China, has confirmed the positive relationship between growth and duration of breast-feeding. A cross-sectional nutrition survey was undertaken of 2148 children who were recruited in four economically disadvantaged counties. The response rate to the survey was in excess of 95 per cent of those recruited. Information was collected on the duration of breast-feeding and the introduction of other foods and a number of other variables, including details of the socio-economic status of the family. The anthropometric data was collected using standardized methods and analysed by the calculation of Z-scores using the National Center for Health Statistics growth references. Using multivariate analysis, the length of breast-feeding was shown to be correlated with weight-for-age, height-for-age and weight-for-height.

The Hubei study is important because of the lengths to which the authors went to control for important socio-economic and environmental variables. Rural Chinese children who were breast-fed for longer periods grew better than their peers who were breast-fed for shorter periods. These findings strongly suggest that in the developing world children would benefit from breast-feeding for 2 years, while benefiting from good-quality solid foods from around 4–6 months of age. *C.W. Binns*

*Duration of breast-feeding and anthropometric status among Chinese children (Z-scores relative to NCHS references)*

| Duration (months) | Height for age | Weight for age | Weight for height |
|---|---|---|---|
| <12 | −1.84 | −1.62 | −0.62 |
| 12–17 | −1.63 | −1.44 | −0.55 |
| 18–23 | −1.55 | −1.30 | −0.43 |
| = or > 24 | −1.52 | −1.29 | −0.44 |

All results significant. Results were adjusted for a number of factors related to weaning foods and socio-economic status.

detailed physiological studies are still awaited. Of particular interest in human growth are iodine, zinc, copper, manganese, molybdenum, chromium, fluoride, and selenium.

Infants who are born to iodine-deficient mothers may develop cretinism, a condition which is fairly rare. However, children who are born in many parts of the world where subclinical iodine deficiency is widespread show growth retardation and delayed mental development. Prevention is possible through the use of public health interventions such as the addition of iodine to a suitable carrier, such as salt or bread, or the antenatal use of iodized oil injections. Iodine is almost unique amongst nutritional deficiencies in that it can occur in populations whose food supplies are adequate in all other respects.

## Other nutrients

There are several other nutrients which may be essential for growth, including choline, taurine and carnitine. Breast-milk contains small amounts of nucleotides which are biologically active as anti-infective agents. Indirectly, these contribute to growth. C.W. Binns

See also 'Anthropometry' (**1.4**), 'The use of growth references' (**1.17**), 'Growth in non-human primates' (**2.2**), 'The human growth curve' (**2.4**), 'Genetic regulation of growth-promoting factors by nutrition' (**3.4**), 'The development of patterns of deep body temperature in infancy' (**5.4**), 'Growth cyclicities and pulsatilities' (**5.13**), 'Nutrition and cognitive development' (**6.5**), 'Nutrition' (**9.2**), 'Infection' (**9.3**), 'Growth and natural selection' (**10.1**), 'Growth in infancy and pre-adolescence' (**10.4**) and 'Long-term consequences of early environmental influences' (**12.1**)

# 9.2

# Nutrition

It has proved difficult to study *in vivo* the effects on growth of deficiencies of individual nutrients because of the nature of the food supply in most communities. More often than not a community in which malnutrition is severe enough to show growth retardation, will have diets which are likely to be deficient in total energy as well as a number of macro- and micronutrients.

The time at which a nutrient deficiency occurs may modify its effect on growth. There are periods of growth spurt or of organ development when a deficiency may have a substantial impact. There appear to be two or three periods of childhood which are critical for the development of subsequent obesity: the last trimester of pregnancy, the periods of adiposity rebound (ages 5–7 years) and adolescence. All may be critical for adult obesity.

There are approximately 40 nutrients which are known to be important in human nutrition. A deficiency in many of these will result in growth retardation. In many situations a diet which provides adequate energy will provide most nutrients. A malnourished child will often have multiple nutrient deficiencies, making the study of growth under field conditions to be quite difficult. As the rate of growth changes with time, particularly during infancy and adolescence, nutrient requirements and feeding practices obviously undergo considerable change. In good environmental circumstances, growth in height ceases at around 16 years of age in girls and 18 years in boys and at this time nutrient and energy requirements are those of active adults.

**Mother and child in Papua New Guinea. Photograph by Stanley Ulijaszek.**

During later childhood it becomes important for energy intake to be in balance with energy expenditure. In malnourished children of 2–4 years of age, increased physical activity may lead to better linear growth when compared to a control group. In Western countries there is a secular trend of declining energy expenditure. Without adequate exercise there is a risk of obesity in many children.

## Dietary energy

It is difficult to study the impact of energy deficiency on growth since low energy intake under field conditions is usually associated with other nutritional deficiencies. However, evidence suggests that limited energy intake diminishes growth-rates, but there have been conflicting reports on the growth response of malnourished children who are given high-energy dietary supplements. In some reports, linear growth responds to energy supplements, but other studies suggest more complete nutritional supplements are needed.

## Protein

Protein and its constituent amino acids are essential for adequate growth. In most cultures, if the diet is of sufficient energy density, protein will be sufficient for growth, provided that there are no additional needs, such as the demands of illness or tissue repair. Some studies have now shown that linear growth of stunted children can be improved if given a protein supplement.

## Lipids

Because of its high energy density, the amount of fat in the child's diet largely determines the energy availability and hence is a major influence on growth-rates. Children living in areas of the world where diets are predominantly bulky carbohydrate staple, such as the sago or sweet potato consuming societies of Papua New Guinea, are low in fat content and high in bulk, which limits energy intake. In addition, the level of fat in the diet is correlated with the levels of vitamin intake. Young children placed on very low-fat diets often have growth retardation and can develop multiple nutritional deficiencies. There are a number of reports in the literature of mothers who were themselves placed on low-fat diets, who fed their children in a similar fashion with the consequence of growth delay.

## Iron

Growth and behavioural development during infancy and early childhood is dependent on adequate bodily iron stores and a dietary source of iron which is reasonably well-absorbed. A number of studies have shown that children with iron deficiency and who are also anemic will respond with linear growth and weight-gain when given iron supplementation. The response in iron-deficient children who are small in stature but who are not anemic is less consistent.

The first major study suggesting that iron deficiency caused behavioural abnormalities was published in 1978. Since then, there have been a number of studies which have confirmed that iron deficiency in the early years of life can impair psychomotor development and cognitive function. The more severe and prolonged the iron deficiency, the greater the developmental delays. In three recent studies, the treatment of iron-deficiency did not produce an improvement in test scores, suggesting that iron deficiency during a critical period of brain development and differentiation may produce permanent abnormalities. These observations are consistent with the theory that iron is important to the normal development and functioning of dopaminergic neurons and that early changes could lead to permanent damage.

## Calcium and phosphorus

Calcium and phosphorus are essential for the development of bone tissue and have major roles in mineral homeostasis. Adequate calcium, phosphorus and protein intakes, together with adequate exercise, are essential for optimum bone development and mineral deposition. The importance of adequate calcium nutrition lies in achieving peak bone density by late adolescence as insurance against subsequent osteoporosis. Breast-fed infants are able to maintain calcium–phosphorus homeostasis and bone development, provided that vitamin D synthesis or intake is adequate. However after 6 months of age, additional sources of calcium are required or the phosphorus content of breast-milk may limit mineralization. In recent years soy-based infant formulas have been quite widely used. The issue of adequate mineralization of infants fed exclusively on soy-based formulas requires further study.

## Zinc

There have been many studies of the role of zinc in human growth, since the initial descriptions of deficiency in the Middle East in the 1960s. Initial studies of supplementation with zinc were not particularly successful, probably because of binding with high phytate diets. More recent studies have shown an improvement in the linear growth of children who are of short stature with confirmed zinc deficiency in Western countries when they are given zinc supplements. However, the response of more severely malnourished children in developing countries to zinc has been less consistent.

Recently there has been speculation that the lower growth-rates observed in breast-fed infants after the age of 5–6 months may be due to zinc deficiency. There is a decline in breast-milk zinc concentrations during the first 6 months of lactation. Children who are on vegetarian diets are more at risk of zinc deficiency.

## Copper

Children who are recovering from severe malnutrition and who have low serum copper and ceruloplasmin often benefit if copper supplementation is given in addition to their other nutritional therapy. Children who are of low birth-weight or who have low copper intakes, such as those who consume high levels of cow's milk, are particularly at risk of growth retardation due to copper deficiency.

## Vitamins

Children who are suffering from vitamin A deficiency and who have exophthalmia or night-blindness have growth retardation. When children recover spontaneously from vitamin A deficiency their growth-rates improve. Supplementation studies with vitamin A have had varying effects on

*Nutrient intakes for growth. A comparison of four sets of recommendations.*

| Nutrient | UK (0–3 months) | USA (0–6 months) | EU (6–11 months) | Australia (0–6 months) |
|---|---|---|---|---|
| Protein | 12.5 | 13 | 15 | |
| Energy (Male) | 2.28 | | | |
| Energy (Female) | 2.16 | | | |
| Thiamine | 0.2 | 0.3 | 0.3 | 0.15 |
| Riboflavin | 0.4 | 0.4 | 0.4 | 0.4 |
| Niacin | 3 | 5 | 5 | 4 |
| Vitamin $B_6$ | 0.2 | 0.3 | 0.4 | 0.25 |
| Folate | 50 | 25 | 50 | 50 |
| Vitamin C | 25 | 30 | 20 | 25 |
| Vitamin A | 350 | 375 | 350 | 425 |
| Vitamin E | | 3 | | 2.5 |
| Vitamin D | 8.5 | 7.5 | | |
| Vitamin $B_{12}$ | | | 0.5 | 0.3 |
| Calcium | 13.1 (524) | 400 | 400 | 300 |
| Phosphorous | 13.1 (406) | 300 | 300 | 150 |
| Magnesium | 2.2 (52.8) | 40 | | 40 |
| Sodium | 9 (207) | | | 140–280 |
| Potassium | 20 (780) | | 800 | 390–580 |
| Iron | 0.03 (1.68) | 6 | 6 | 0.5 |
| Zinc | 0.06 (3.9) | 5 | 4 | 3 |
| Copper | 0.005 (0.32) | | 0.3 | |
| Selenium | 0.1 (7.9) | 10 | 8 | 10 |
| Iodine | 0.4 (50) | 40 | 50 | |

**Malnourished 2 year old. Photograph by Patricia van Someren.**

weight-gain and on linear growth. This may be related to the severity of other coincidental nutritional deficiencies or to the frequency of administration of the vitamin A supplement.

## Other nutrients

Carnitine is important as a long-chain fatty acid carrier through mitochondrial membranes, which is essential for the oxidation of fatty acids and energy production. While carnitine is normally synthesized in the adult, it is a required nutrient in infancy and also during pregnancy and lactation. Carnitine has many physiological functions in metabolic pathways related to acyl transfer for long-, medium- and short-chain fatty acids which are important for growth and development. Children suffering from severe protein malnutrition have lowered serum carnitine levels, but further research is required to elucidate the relationship between carnitine deficiency and growth.

Taurine is an amino acid that is found in human milk. Taurine has been shown to be important in animal studies for the development of the retina and other neural tube tissues. In recent years it has been added to infant formulas in approximately the same amounts as found in breast-milk.

Choline is another 'vitamin-like' molecule which has an important role as a precursor in the biosynthesis of the neurotransmitter acetyl choline, as a source of methyl groups and

### LONG-TERM EFFECT OF IRON SUPPLEMENTATION ON GROWTH IN INDONESIAN PRESCHOOL CHILDREN

Iron deficiency anemia and decreased growth-rate are two major nutritional problems among preschool children in developing countries. Prevalence of anemia among preschoolers in developing countries is high but the prevalence reduces again when children grow older. In preschoolers a chronic lack of nutrients in general can reduce growth-rate, potentially leading to stunting. Anemia and reduced growth-rate are often associated among preschool children. In a study of Indonesian children in Jakarta aged 2–5 years with low hemoglobin status, subjects were randomly allocated to a treatment group (n = 40) receiving 30 milligrams iron and 20 milligrams vitamin C per day, and a placebo group (n = 40) receiving 20 milligrams vitamin C per day over an 8-week period. Both groups received deworming treatment at the beginning of the study. Initial hemoglobin values were 102 and 103 grams per litre for the iron and placebo groups respectively.

At the end of the study, the hemoglobin status of the iron-supplemented group had increased. In addition, height-for-age Z-score of the iron group showed a significantly larger increase relative to the control group. Of the original 76 subjects, 40 were followed up 2 years after the end of the study (the placebo group was given iron supplements for two months, after the end of the study, for ethical reasons). At the 2-year follow-up, mean hemoglobin level was 122 grams per litre for the iron group and 111 grams per litre for the placebo group, while mean height-for-age Z-score increased significantly in both groups. An improved iron status persisted 2 years after the intervention, although a between-group difference in hematological status persisted. The significant increase in height which occurred in the iron-supplemented (experimental) group was visible 2 years after children received iron supplements. Although the improved growth-rate over 2 years might be due factors other than iron supplementation, it is unlikely that spontaneous improvements in height-for-age would occur in the poor environment in which the children live. Thus, iron supplementation in anemic young children may influence longer-term, as well as short-term growth.

*Imelda T. Angeles, Werner Schultink, Rainer Gross, Soemilah Sastroamidjojo*

*Mean weight-for-age, height-for-age and weight-for-height Z-scores before and after 2 months supplementation of the complete sample of subjects*

| | Before | After | Difference |
|---|---|---|---|
| **Fe + vit C (n = 39)** | | | |
| Weight-for-age | −2.53 | −2.37 | 0.17 |
| Height-for-age | −2.33 | −1.96 | 0.37 |
| Weight-for-height | −1.48 | −1.57 | −0.06 |
| **Vit C (n = 37)** | | | |
| Weight-for-age | −2.54 | −2.33 | 0.21 |
| Height-for-age | −2.18 | −2.10 | 0.07 |
| Weight-for-height | −1.56 | −1.36 | 0.21 |

*Mean weight-for-age, height-for-age and weight-for-height Z-scores in a sub-group of subjects before supplementation, after supplementation (2 months), and 2 years after the supplementation ended*

| | Baseline | 2 months | 2 years |
|---|---|---|---|
| **Fe + vit C (n = 19)** | | | |
| Weight-for-age | −2.66 | −2.45 | −1.94 |
| Height-for-age | −2.44 | −2.05 | −1.73 |
| Weight-for-height | −1.57 | −1.58 | −1.21 |
| **Vit C (n = 21)** | | | |
| Weight-for-age | −2.49 | −2.30 | −2.02 |
| Height-for-age | −2.07 | −1.96 | −1.59 |
| Weight-for-height | −1.59 | −1.43 | −1.45 |

as precursor of the phospholipids found in cell membranes. In animal models, choline deficiency produces severe growth retardation. In practice, choline deficiency is more likely to occur as part of a more general malnutrition syndrome.

Nutrition has a profound influence on growth and development. Children are not just small adults, but have periods during growth which are critical for organ development (**12.1**). Nutritional deficiencies during these critical phases often leads to growth which is lost and which may not be caught up subsequently. For these reasons it is important that children in the developing world have access to a variety of nutrient-dense foods which can supply all of their requirements for growth. Their cousins in the more affluent world also need adequate nutrition which is not inappropriately restricted by placing them on diets meant for the prevention of chronic disease in adults. *C.W. Binns*

See also 'Adiposity rebound and prediction of adult fatness' (**1.11**) and 'Infant feeding and growth' (**9.1**)

## BODY-SIZE AND ENERGY NEEDS

Body-size is a major determinant of the energy and protein needs of animals and humans. Daily energy needs are estimated from measures of total daily energy expenditure or heat production. The basal heat production at rest, termed basal metabolic rate (BMR), represents the major proportion of daily energy expenditure required for the maintenance of vital functions of the body at physical and mental rest. The proportional contribution of BMR to the total daily energy expenditure varies considerably with the level of physical activity (**5.12**) of the individual and may vary from just over 50 per cent in extremely physically active individuals, to accounting for over 70 per cent of the total energy expended in individuals with sedentary life styles (where much of the time is spent in activities at or near basal energy expenditure).

In any individual, BMR is determined principally by body-size and body composition. This implies that the energy needs of an individual are determined largely by these two factors and by the level of physical activity of the individual. The relationship between body-size and energy expenditure has been examined both at the inter- and intraspecies level. For 26 species of animals studied by Kleiber, the correlation coefficient between body-weight and BMR was of the order of 0.98. In order to enable comparisons of basal energy expenditure or BMR between or within species, a common base of expression is needed which is free from the variations introduced by the large differences in body-size or composition. Inter-species comparison shows that BMR increases with the body-size of an animal; however, it is much greater per kilogram of mouse than per kilogram of man, the latter much greater than per kilogram of elephant.

BMR per unit of body surface area, on the other hand, is much more consistent both across variations in body-sizes between species and within the same species. This relationship between basal heat production and body surface area can be accounted for in terms of heat transfer, since it would be advantageous for larger animals with a low surface-area to body-mass ratio to have lower BMRs per kilogram than smaller animals who have a relatively high surface-area to body-mass ratio. This is not to imply that heat transfer is a prime determinant of BMR since in a neutral thermal environment with an animal at rest, heat loss is not a driving force.

Body-weight is a more accurate indicator of body-size than surface area. Kleiber demonstrated that the logarithms of BMR and body-weight are linearly related and concluded that BMR is directly proportional to body-weight raised to a power 0.73; body-weight and surface area predict basal heat production of mature animals of homeothermic species as well as Kleiber's power equation. It thus appears that Kleiber's power equation may form the inter-species basis to estimate heat production from the body-weights of homeothermic animals. Few species show large enough differences in mature body-weight to test the applicability of the power function of weight as an index of metabolic body-size. Within species, the BMR in a given individual varies with sex and age. These variations are imposed by the differences in the body composition of the tissues. However, animal studies do not show differences in BMR expressed per unit metabolic body-size (0.73 kg) even when animals of the same species with similar body-weights vary in the degrees of fatness. In humans, over the normal range of body-weight, both for men and women, metabolic rates correlate directly with body-weight as closely as with body-weight to the power 0.73. Outside the normal range of body-weights, changes in body composition (**5.11**) supervene to distort this direct linear relationship of BMR to body-weight.

For the prediction of BMR of human subjects of both sexes and different age ranges, body-weight is the most useful index. The inclusion of other anthropometric indices, such as stature or the type of predictive equation tested, i.e. linear, quadratic, logarithmic, does not add to the accuracy of the prediction based solely on body-weight. It is now generally accepted that body-weight is a good proxy for body-size to estimate the BMR of adults. BMR of an individual is relatively constant, despite wide fluctuations in levels of daily physical activities and habitual energy intakes; the within-individual variability no greater than that for other physiological data. Assessment of energy requirements or energy needs can thus be made by specifying the energy needs for BMR and then considering all physical activity costs as effectively occurring in the fed state and the energy cost incurred by this physical activity expressed as a multiple of the BMR. Energy needs of humans can hence be easily determined by knowing the energy equivalent of BMR and adding to it the multiple of this BMR which is equivalent to the energy needed for habitual levels of physical activity. *P.S. Shetty*

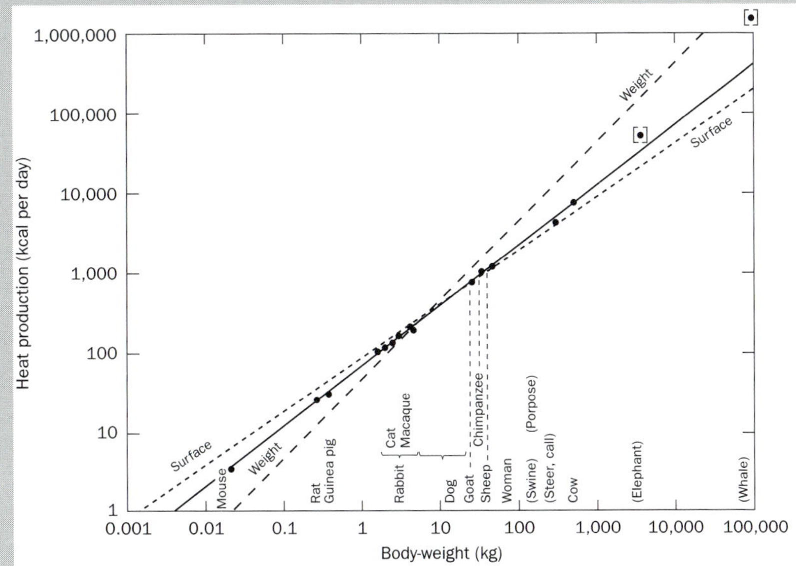


**Interspecific comparison of mean maintenance metabolism and mean body-weight, showing that BMR is proportional to body-weight raised to a power of 0.73. From S. Brody *Bioenergetics and growth*, New York: Reinhold, 1945.**

## THE INFLUENCE OF NUTRITION IN EARLY LIFE ON STATURE AND RISK OF OBESITY

In animals as in humans, early feeding experience has a profound influence on body-size and fatness later in life. Studies have shown that well-fed animals are heavier, longer and fatter at weaning, and that these differences persist after weaning under ad libitum feeding. These observations stress the importance of early feeding practices (**12.1**).

Children's energy needs are related to the energy needed for growth, which is comprised of the specific requirements for the different tissues which grow at different rates. During the first years of life, height velocity is high and the brain grows rapidly. During this period, energy needs (expressed as kcal/kg body-weight) are high. By the age of 6 years, brain growth is nearly completed (when body-weight is less than one third of its mature size) and height velocity is low. These important changes suggest that the desirable nutrient balance should change with age. Indeed, in rats, spontaneous choices of food are age-dependent as the young animal chooses a high-fat, low-protein diet, while the older one prefers protein at the expense of fat.

Studies conducted in industrialized countries generally report low fat (< 30% of energy intake) and high protein (> 15% of energy intake) intakes during early childhood and, paradoxically, the proportion of fat increasing with age. The high-fat, low-protein content of human milk is adapted for the high energy needs for growth and for the myelinization of the nervous system. A recent study of nutrition and growth has shown that a high protein diet early in life is positively associated with increased fatness via an early adiposity rebound (**1.11**). In young children consuming a high protein diet, fatness steeply decreases, but over a shorter period, resulting in increased fatness development. This pattern of growth reflects advanced bone age and accelerated but shortened maturation. The early adiposity rebound is likely to be promoted by early production of insulin-like growth factor I (IGF-I) triggering precocious cell multiplication in all tissues. This is consistent with the well-known observation of accelerated maturation in obese children (**1.11**). They tend to be taller and have a larger skeletal and muscular mass. These characteristics cannot be explained by fat-intake excess alone.

Tall stature without being overweight can be another response to protein-energy excess. It is sometimes interpreted as achievement of full genetic potential. However, increased stature is associated with various diseases, including cardiovascular diseases and cancer.

The pattern of growth promoted by chronic protein-energy excess is opposite to that observed in chronic protein-energy deficiency. Inappropriate intakes can increase (bolting) or decrease (stunting) stature, but are not necessarily accompanied by changes in weight-for-

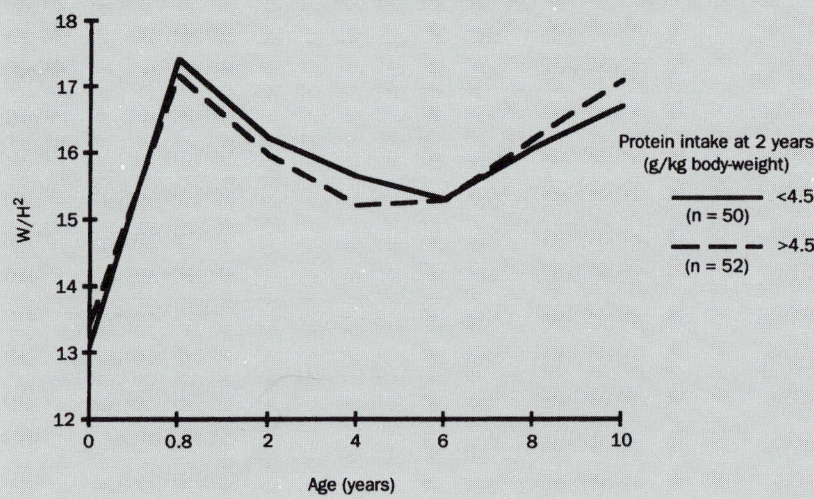


**BMI pattern according to protein intake at the age of 2 years. A high protein intake at the age of 2 years is associated with a steep but shortened fatness decrease, promoting subsequent enhanced fatness development.**

## PROTEIN–BODY-SIZE RELATIONSHIPS

The allometry of protein metabolism, universally studied, has been that of obligatory or endogenous nitrogen loss which is the output in animals on a protein-free diet. Such measurements were made following the proposal by Folin in 1905 that a dichotomy existed in protein metabolism. Folin observed that the nitrogenous compounds excreted in the urine could be partitioned between creatinine and creatine, the excretion of which was (on a creatine-free diet) independent of nitrogen intake, and urea and ammonia, the amount of which excreted was related to the protein intake. Folin's postulated duality in protein metabolism was that there existed a constant endogenous metabolism of body tissue and a variable exogenous metabolism reflecting protein intake. Following this, much work was done in assessing the magnitude of the endogenous nitrogen loss. Folin used the term 'endogenous nitrogen loss' to define the *qualitative* nature of the urinary nitrogen excretion products, and he and his immediate successors tended to measure endogenous nitrogen losses as urinary losses alone, since that was the way it was defined. On the other hand, as interest shifted towards quantification of the minimal rates of nitrogen loss, later workers used the term 'endogenous nitrogen loss' to cover all nitrogen losses from the body on a protein-free diet – that is, nitrogen losses in the urine, feces and by other routes, which is the same as the term 'obligatory nitrogen loss'.

The term endogenous nitrogen loss (ENL) was then systematically used by Mitchell (1923–4), who equated it with nitrogen loss in both urine and feces on a protein-free diet; that is to say *not* the way Folin had initially used the term to mean urinary losses alone.

By concentrating on two types of different investigation, the quantification of total minimal nitrogen loss, and the quantification of that part of nitrogen loss due to the metabolism of body substance rather than diet has led to the use of subtly different conditions of measurement by researchers in this field. The extent to which these have made their results incompatible has been the source of much of the confusion and controversy. Mitchell showed for the first time that ENL could be approximated for a wide range of mammalian species as 2 milligrams of nitrogen per basal kilocalorie expended.

Brody (1945) extended these studies by showing that for mammals the allometry of ENL was $146W^{0.72}$ milligrams per day. Despite Brody's clear description that in his work ENL referred to *urinary losses alone*, later investigators took his allometry as referring to *total* ENL. Using this height proportions. Similar oppositing situations are observed for hormonal status. During periods of nutritional deficit, plasma concentration of Growth Hormone (GH) is increased while IGF-I is reduced. By contrast, GH deficiency is associated with established obesity, and IGF-I is increased in obese children. Subsequently, decreased IGF-I is reported in adult obesity.



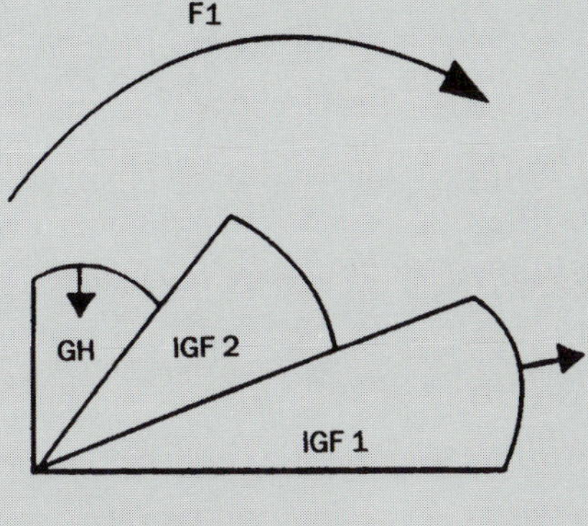


A sedentary lifestyle may also play a role in promoting positive energy balance. As exercise and cold decrease IGF-I levels, active open-air life styles may regulate IGF-I release, serving to limit growth-factor production and the avoidance of an early adiposity rebound. Since human milk has a low protein content compared to formula, feeding practices at birth may also be involved in the pattern of growth, breast-feeding avoiding accelerated growth and early adiposity rebound.

The secular trend of an increasing prevalence of obesity cannot be explained by increased energy and fat consumption, as most studies of trends in food consumption over the last one or two decades demonstrate a decrease in average intake. Increasingly sedentary life styles and increased protein intake are more likely explanations. They probably also contribute significantly to the secular trend of increased stature, at least in industrialized countries.

*Marie-Françoise Rolland-Cachera*

**The figure illustrates the concept of a balance between two opposite situations controlled by various hormones, growth factors, and cycles. Equilibrated alternative F2 and F1 situations may be necessary for good health, while chronic F2 or chronic F1 situations may promote abnormal growth and various health disorders (HPA axis: Hypothalomo-pituitary-adrenal axis) (after Rolland-Cachera, 1995).**

allometry, together with the allometry of BMR, $70W^{0.73}$, they arrived at yet another confirmation of the relationship that ENL was 2 milligrams of nitrogen per basal kilocalorie (i.e. $146W^{0.72}$ / $70W^{0.73}$ = 2mg/N/basal kcal). The minimal level of urinary nitrogen loss is achieved when two conditions are met. First, the protein intake should be as close to zero as possible and second, there should be no restriction of energy intake.

Several years ago Brody (1945), using allometry, showed obligatory nitrogen loss (ONL) in a range of homeotherms to be of the form:

$$ONL = 272W^{0.75}$$

It has been shown that fasting urinary nitrogen loss (FUNL), over a wide range of homeotherm species varies with metabolic body-size ($W^{0.75}$). Although the FUNL changes systematically from roughly 1150 milligrams per kilogram in a mouse to 100 milligrams per kilogram in a steer, the allometric study indicates that the mean FUNL $W^{0.75}$ from mouse to steer is roughly 428 milligrams per day Thus:

$$FUNL = 428W^{0.75}$$

Since the mass exponents are the same in the two equations, using the allometric cancellation technique, and dividing equation 2 by 1, we have:

$$FUNL = 1.5\ ONL$$

The factorial division of nutrients into components needed to match obligatory losses and those required for growth and activity has proved to be of great practical value in estimating protein requirements. Having noted that ONL contributes significantly to the estimation of human protein requirements, we can next explore the relationship between FUNL and ONL. The former is the nitrogen loss during total energy and protein restriction (fasting) and the latter is the nitrogen loss during the feeding of a non-protein diet adequate in energy.

It is known that the amount of body fat in an individual significantly influences protein mobilization: i.e., the greater the adiposity, the smaller the FUNL. Now, linking the above relationship between FUNL and ONL, this suggests a smaller ONL in those with a smaller FUNL. It may, therefore, be concluded that subjects with increasing adiposity will have a lower FUNL, and ONL, and hence lower protein requirements.

*C.J.K. Henry*

## CHILDHOOD OBESITY AND GROWTH

The proportion of fat in the body varies with age. Changes with age anticipate some of the nutritional stresses of growth: weaning; adolescence; pregnancy and lactation. Obesity results from an excess of body fat-for-age, deposited either subcutaneously or internally, predominantly intra-abdominally. Direct methods of measuring fatness, such as measurement of fatfold thicknesses, have tended to record subcutaneous fat only (**1.4**). Computerized tomography and magnetic resonance imaging show intra-abdominal fat (**1.6**) in childhood, but this is proportionally much less in children than in adults, irrespective of nutritional state.

Measuring total body fat directly and non-invasively is not practical in most circumstances. Recognition of obesity is thus commonly made by relating weight-for-age and height-for-age; average weight-for-height in childhood is not independent of height-for-age. Children who are tall-for-age are, on average, heavier than those of the same height but greater age who are short-for-age. Lean body mass for height in tall-for-age children may be greater irrespective of any increase in fat mass. Weight-for-height estimations are thus biased towards defining normal tall children as overweight. Since otherwise normal obese children are usually above average stature for age, this bias also overestimates the extra fat of obese children. The relative body mass index (BMI) expressed as a percentage, centile position or standard deviation (Z-score) of the BMI for age, is a practical clinical measure of overweight. Severe overweight (>120% expected BMI) usually indicates excessive fatness – 'obesity'. Relative BMI can be used to assess overweight in populations, or changes in relative fatness over time in individuals (**1.16**).

### *Causes of obesity*

Obesity results from environmental factors acting on a susceptible genotype. Where the genetic predisposition to obesity is not great, or the environmental factors minor, obesity may occur transiently, perhaps as over-exuberant physiological fattening at those ages when fat is being deposited rapidly. Such obesity, which resolves spontaneously, may be common but largely unnoticed. Where there are strong genetic predispositions (**3.1**) and significant environmental influences, obesity may persist and increase in severity. Obesity is more likely to persist if it develops at ages when children would usually be expected to slim than at ages of physiological fattening. Children showing the onset of obesity, or progressive obesity, after the first 6 months of life and before 6 years (early adiposity rebound), are likely to remain obese into adult life. Eighty per cent of overweight and obese infants are likely to return to normal fatness by the age of 5 years, although changes in the causes and prevalence of obesity in infancy may be affecting the relation of fatness in infancy and later childhood. Correlations between weight-for-height indices at different ages are still more dependent on correlations in lean body mass-for-age than correlations in relative fatness-for-age.

The relative tallness of obese children may be associated with advanced bone age, early pubertal growth spurt and normal, or even short, adult stature. This is not invariable. Bone age assessed from radiographs of the short arm bones (radius and ulna) (**1.8**) is commonly more advanced than that of the carpal bones in the obese child. Some girls who have early growth spurt seem particularly at risk of obesity post-menarche. Boys with pubertal delay but stature within the normal range for age may become very obese with the prolonged period of prepubertal fat deposition but achieve normal weight and stature in adult life. Boys with early puberty have greater fatfolds in adolescence than those with average or late onset of puberty but, at the age of 30, only truncal fatfolds remain significantly higher in the early developers (**5.16**).

*Estimated percentage of body-weight as fat at different ages (from various sources)*

| | Body adipose tissue % | |
|---|---|---|
| **Age** | **Boys** | **Girls** |
| Birth | 13.7 | 14.9 |
| 4 months | 25.4 | 26.0 |
| 12 months | 22.5 | 23.7 |
| 5 years | 16.5 | 15.4 |
| 10 years | 17.5 | 16.0 |
| 14 years | 18.4 | 27.0 |
| 15 years | 11.4 | 23.3 |
| Young adults | 18.3 | 23.7 |

*Formula for percentage Relative Body Mass Index*

$$\frac{\text{Actual weight}^* \text{ of child}}{(\text{height}^* \text{ of child})^2} \div \frac{50^{\text{th}} \text{ centile weight}^* \text{ for age from reference centiles}}{(50^{\text{th}} \text{ centile height}^* \text{ for age})^2} \times 100\%$$

< 80% = underweight
80–110% = normal weight
>110–120% = overweight
>120% = 'obese'

*weight in kg, height in m, from reference centiles

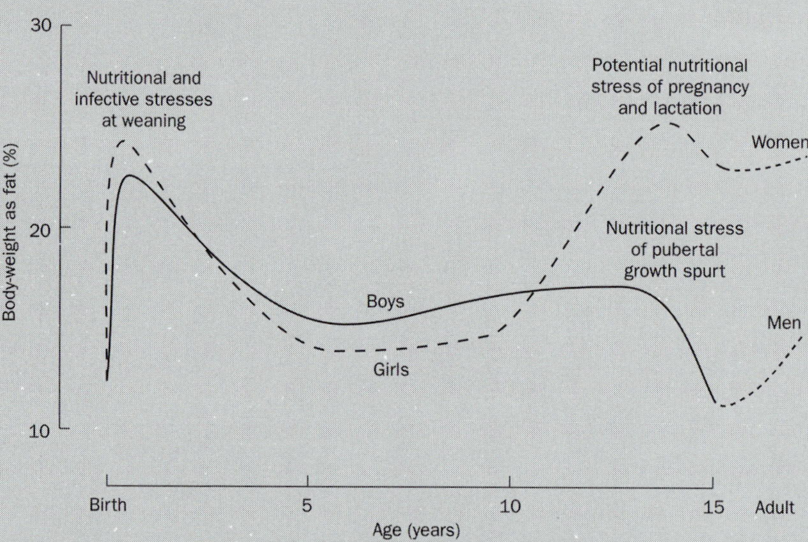


**Diagrammatic representation of changing percentage body fat with age and physiological change in childhood.**

*Risk factors for obesity in childhood*

**Genetic?**
Obesity in one or both parents
Adiposity rebound before age 6 years

**Environmental**
Low social class in Westernized societies
Affluence in developing countries
Single child in family

**Activity related?**
Low levels of energy expenditure
High levels of television-watching

**Diet related**
Energy dense, high-fat diet
Disorganized eating patterns

**Medical**
Specific chromosomal, genetic and endocrine disorders

Obesity is multifactorial in origin. Thus it is not surprising that just as growth patterns vary between individuals, so do the risk factors for obesity. To become obese, energy intakes must, at some stage, be in excess of energy expenditures. But falling mean energy intakes throughout childhood and rising prevalence of obesity over recent decades in Britain at least, suggest that falling energy expenditure may be as important as excessive energy intake in the precipitation of obesity. Some obese children do have very high energy intakes, but this is not true for all. Obese children may be 'faddy' eaters with few vegetables, fruit or unrefined cereals in their diets. Meals take little time to eat and satiety from meals may thus be reduced, leading to excessive consumption. A high proportion of fat energy in the diet may contribute through the preferential deposition of fat energy as body fat. Eating may be disorganized, with meals providing little satisfaction through social and family interaction. Obese children are often intense snackers. Deprived home environments prevent children taking informal exercise easily and may result in the over-use of television as entertainment for the children. Treatment and prevention of obesity therefore require dramatic lifestyle as well as dietary changes which, given the disadvantaged background from which many obese children come and the importance of peer group pressures in childhood, are virtually impossible to effect, at least in the United Kingdom and United States.

*E.M.E. Poskitt*

9·3

# Infection

Malnutrition and infections are so frequently associated that the existence of a vicious circle of infections leading to malnutrition, and malnutrition predisposing to infection, has been long postulated. This concept implies that the prevention of infectious diseases could lead to an improvement of growth of children, and has major public-health implications. Improving the diet of children in deprived communities is notoriously difficult and requires political decisions regarding resource allocation which are highly sensitive. Perhaps for that reason, the alternative route of improving the growth of children by preventing infections has often been favoured by international organizations. In 1990, UNICEF stated that 'Most cases of malnutrition occur in families where there is enough food and the cause of malnutrition are likely to be repeated infections'. In the report of the 1992 WHO/FAO Rome Conference on Nutrition, there are similar statements about the necessity of breaking the infection–malnutrition cycle to improve the nutritional situation of the world.

Diarrhea is the most frequent infection seen in malnourished children, and received particular attention. The World Health Organization (WHO) programme of control of diarrheal diseases was created in 1980 with the specific objective, among others, 'to reduce malnutrition among young children in developing countries'. Yet, the interaction between infection and malnutrition is complex, and a careful examination of previous studies is needed to assess the effect of infections on growth.

## Lack of direct evidence of the effect of infections on growth

There is no doubt that in some cases, a severe infection can lead to malnutrition by the combined effect of reduced appetite, fever and increased digestive losses. In poor communities, however, growth retardation is a general phenomenon which cannot be explained by individual observations. Ideally, to prove that infections are a major cause of growth retardation, one should show that a programme reducing their prevalence can also improve the average growth of children. This type of direct evidence, however, is lacking. Indeed, growth retardation may persist at high levels after implementation of effective immunization programmes eliminating measles and whooping cough, the most debilitating infectious diseases among children.

In the Mirzapur area of Bangladesh, a carefully monitored water and sanitation programme, was successful in reducing the prevalence of diarrhea in children under-5 compared to a control area. This gave an opportunity to measure directly to what extent growth would improve after reduction of the prevalence of diarrhea, the type of infection supposed to have the greatest effect on growth. There was no measurable improvement of growth in children living in the intervention area. This is perhaps the only controlled attempt to improve growth of children in a community study by reducing bacterial contamination.

## Effects of infections on growth: what is the evidence?

The debate concerning the role of infections on growth relies mostly on indirect evidence. The hypothesis that infections might be an important cause of growth retardation in poor communities was spelled out for the first time in a monograph by Scrimshaw, Taylor and Gordon published by the World Health Organization in 1968. At that time, the frequency of infections in poor communities had only recently been recognized. It was also realized that many supplementary feeding programmes, then based on the distribution of high-protein food to children, had little effect on growth. Also, the farming industry had just discovered that antibiotics promoted growth in animals living in unhygienic conditions. Yet, the report gave no real measure of the possible importance of infections as a cause of growth retardation.

The first attempt to measure the effects of infections on growth was made in 1977 in a longitudinal study carried out in the Gambia. This was estimated by a multiple regression model, with monthly weight-gain as dependent variable and prevalence of different infections as independent variables. This study showed that growth was depressed in months of

*Keneba study (The Gambia). Estimation of the effect of different types of infections on weight-gain of children.*

| | Effects on weight-gain estimated for 100% prevalence[1] (g/months) | Prevalence (%) | Calculated effect (in grams per month) |
|---|---|---|---|
| **Type of infections** | | | |
| Upper respiratory tract infections | −81 | | |
| Lower respiratory tract infections | −53 | | |
| Gastroenteritis | −746*** | 13.7 | −102 g/m |
| Infectious fevers | 79 | | |
| Malaria | −1072*** | 1.4 | −15 g/m |
| Giardiasis | −131 | | |

From Rowland et al. *British Journal of Nutrition* 1977 37: 441–50

[1]Estimated from linear multiple regression with weight-gain measured on monthly intervals as dependent variables and prevalence of infections as independent variables. This type of analysis does not take into account the possibility of catch-up growth after each infection. Major effects of infection on growth, as reported in this study, are not found in long-term studies, which suggest that some degree of catch-up growth does take place.

the year with seasonally high prevalence of infections, and indicated that diarrhea had the most important effect on growth. The apparent effect of diarrhea was due to its high prevalence. Other infections associated with fever had a more pronounced effect on growth by day of prevalence, but were not frequent enough to have a comparable theoretical effect on growth. In the following years, similar findings were reported from other settings, giving support to the idea that infections, and in particular diarrhea, might be a major cause of growth retardation in developing countries.

## Evidence against a major effect of infection on growth

Failure of the Mirzapur study to produce any effect on growth after reduction of diarrhea led to a re-analysis of the relationship between diarrhea and growth. In Mirzapur, as in previous studies, growth was reduced across the period when diarrhea was present, but this effect was shown to be transient, and to disappear when examined over longer time intervals. This seemed to be due to the occurrence of catch-up growth in the weeks following a diarrhea episode, a phenomenon which was overlooked in other studies. This result suggested that the effect of infection on growth had been previously overestimated. A similar discrepancy between short-term and long-term effects of infections on growth was later confirmed in other studies in Zimbabwe and in Jamaica.

Catch-up growth of severely malnourished children with a 30–40 per cent body-weight deficit has a large energy demand, and requires approximately double the usual energy and protein intakes to be sustained. Epidemiological studies suggest that the reduction in weight-gain in months when children have diarrhea is less than 100 grams, representing a body-weight deficit of around 1 per cent. Children recovering from minor infections have much lower nutritional requirements for catch-up growth than children recovering from severe malnutrition: it is estimated that if growth stops for a few days as a result of infection in children over 6 months of age, a increase of 1–2 per cent of energy intake for a few days is enough to promote catch-up growth. It seems quite plausible that some degree of catch-up growth can often take place in poor communities between different infectious episodes. Studies of food intake of children suffering from different types of infections in Bangladesh and in Nigeria also failed to show a major effect of infections on food intake. Infections associated with fever depress appetite, but this effect seems too small to have a lasting effect on food intake and hence on growth. Minor acute infections not associated with fever are not associated with a significant decrease of food intake and are unlikely to have any influence on growth.

Energy and protein losses in the stools in cases of diarrhea are usually small and transient and are unlikely to have a lasting effect on growth. Even in cases of severe diarrhea, the effect on nutritional balance seems to be minor in comparison to the reduction of food intake. Micronutrient losses in cases of diarrhea are more difficult to assess: they have been estimated only in a few hospital studies, and these have dubious relevance to the community.

None of the mechanisms proposed to explain an effect of infections on growth, such as increased energy requirements for repeated episodes of catch-up growth, reduction of food intake or nutrient losses in the stools, have been shown to be prevalent or important enough in the community to have a sustained effect on the growth of young children. This suggests that the main effect of diarrhea, and presumably of other acute infections, is to induce a transient cessation of growth or transient weight-loss with no major long-term effects, and so is unlikely to explain the high prevalence of growth retardation observed among children in poor environments.

*Matlab community study (Bangladesh). Multiple regression showing the effect of different diseases on food intake of children after adjustment on age, sex and season.*

| **Dependent variable: energy intake in kcal/kg/jour** | | | | |
|---|---|---|---|---|
| | **Regression coefficient** | **se** | **F** | **P** |
| Fever | −10.8 | 3.15 | 10.24 | 0.001 |
| Diarrhea (with or without fever) | −2.68 | 2.71 | 0.98 | ns |
| Respiratory infections (without fever) | −3.85 | 2.06 | 3.50 | ns |
| Other diseases | −3.36 | 3.23 | 1.08 | ns |

From Brown et al., *American Journal of Clinical Nutrition* 1985 41: 343–55

$r^2 = 0.10$ The regression coefficients give an estimate of the energy intake reduction for each day of disease. For instance, fever is associated with a reduction of 10.8 kcal/day (compared to requirements of about 100 kcal/day in this age group). Prevalence of different types of infection is not stated in the paper. If fever were present 10% of the time, this would lead to a global reduction of food intake of 1%, not taking into account the possibility of increased intake in the convalescence period. In contrast to hospital studies, community studies do not support the hypothesis that food intake is sharply reduced during infections.

## Unanswered questions

### *Infections and growth*

Some children in poor communities have biological markers of infections, such as high erythrocyte sedimentation-rate or high white-cell counts, even in the absence of apparent clinical infection. It is conceivable that unapparent infections, overlooked by previous studies, are sufficient to produce a chronic overproduction of the interleukins IL-1, IL-6 or TNF-$\alpha$, which could in turn chronically depress appetite and growth. If these low-key, unapparent infections are not transient and are prevalent enough, this could be a plausible

*Major cytokine effects on bone growth and remodelling*

| **Cytokine** | **Effect on: Resorption** | **Formation** | **Osteoblast maturation** | **Production in inflammation** |
|---|---|---|---|---|
| IL-1$\alpha$ | ↑ | ↑ | ↓ | ↓ |
| TNF$\alpha$ | ↑ | ↑ | ↑+↓ | ↑ |
| IFN$\gamma$ | ↓ | ↓ | ↑+↓ | ↓ |

mechanism for growth depression by infections. Bacterial contamination of the upper gut seems to be quite prevalent, and in conjunction with a depressed appetite could induce some degree of malabsorption of some nutrients. If this is the case, longitudinal studies on the interaction of growth and infections need to be undertaken again with parallel biological explorations. They should focus on the effect of persistent infections more likely to have a sustained effect on growth than on acute infections. However, interpretation of results may prove difficult, since, for example, depressed immunity due to micronutrient deficiencies may lead to non-causal associations between subclinical infections and growth faltering in the absence of a direct effect of infection on growth.

**Children in the developing world (here, Venezuela) often live in an unclean environment, making it difficult to avoid infection for any length of time. Photograph by Patricia van Someren.**

*Nutrition and growth*

Nearly 20 years ago, research on the effect of infection on growth was stimulated by the observation that nutritional supplementation programmes were ineffective in improving growth. Failure of infection control programmes to substantially improve growth should now in turn stimulate research on different types of nutritional interventions. Present views on nutritional requirements of malnourished children are more complex than 20 years ago, when it was considered that increasing protein intake was enough to promote growth. We know that nutrients needed to promote growth include various minerals and vitamins to which no attention was given 20 years ago. These issues should be re-examined.

In a study from The Netherlands, children consuming macrobiotic diets with adequate protein and energy intakes, and protected from bacterial contamination, had a growth pattern similar to that of children from poor developing countries. Their diet contained no animal food, and some nutrients, perhaps needed for optimal growth, were shown to be absent. The same type of nutritional deficiency may affect growth of children from poor communities consuming a diet with little or no animal food. While nutritional deficiencies associated with poor diets are not clearly identified, it may be unwise to ascribe all unexplained growth retardation to associated infections. *André Briend*

See also 'Ontogeny of the immune system' (**4.7**), 'Growth in chronic diseases' (**7.8**), 'Maternal HIV infection' (**8.4**), 'Infant-feeding and growth' (**9.1**) and 'Growth and natural selection' (**10.1**)

## INTESTINAL PARASITISM

Parasitic worm infections affect over one-third of the world's population at any time, and the majority of those infected are children and adolescents residing in developing countries. Lack of sanitary human waste disposal, vector control and regular anti-parasitic treatments create conditions leading to endemic infection levels in many regions.

The major intestinal worm parasites which influence growth and development during the first two decades of life are ascaris, trichuris, hookworm, and schistosomes. These four parasites are helminths, for which humans are the definitive host where male and female adult worms mate and produce eggs. However, they do not actually reproduce offspring worms in their human hosts, but require developmental life-cycle events outside humans in other species and in the external environment. The first three parasites are geohelminths whose excreted eggs require contact with soil to develop the infective form. Schistosomes require contact with fresh water and specific snail species for their development. Infection with these four helminth species is usually assessed by measuring the presence and number of their respective eggs in stool specimens or, for *Schistosoma hematobium*, in urine.

Regardless of their complex life-cycles, these helminths are important in human nutrition, growth and development. Growing children are disproportionately burdened by these worm infections due to their exposure to infective soil and water through play and domestic activities, such as washing and fetching water. In addition, resistance to re-infection is age- and maturity-dependent and may take repeated infections. This results in older children and adolescents having the highest age-specific prevalence and incidence of infection. Without regular treatment, children and adolescents also have the highest intensity of infections.

Although it was recognized many years ago that these helminths affected growth and nutritional status, it has been difficult to estimate their impact in poor developing world communities. The environmental context of rural poverty includes other infections, limited dietary intake of protein, energy and micronutrients such as vitamin A and iron, and limited access to primary health care. Until recently there were few detailed community studies which combined both quantitative parasite methods and reliable assessments of child growth and nutritional status. Large samples and follow-up designs are also necessary to detect the independent effects of parasitic infections on childhood growth in such communities.

A number of studies in the last 10 years have demonstrated strong and significant insults to childhood growth and nutritional status due to helminth infections. These include studies of ascaris, trichuris and hookworm, as well as all three schistosome infections, *S. mansoni*, *S. hematobium*, and *S. japonicum*. Children with any one of these helminth infections have reduced weight-for-height, skinfold thickness, and in some cases, height-for-age compared to uninfected peers. In short-term randomized studies, infected children treated with anti-helminthic drugs show increases in skinfold thickness and weight relative to the untreated children. It appears that intestinal parasitic infections play an important role in producing the characteristic shorter stature and reduced weight in growing children and adolescents in developing countries, as well as the reduced body-size of adults. These may have direct implications for work capacity and economic productivity, although such effects are difficult to study.

The mechanisms responsible for reduced nutritional, growth and developmental status due to helminths include blood-loss leading to iron-deficiency anemia, decreased appetite and food intake, putative immune and cytokine responses to infection, and their interactions. There are also specific parasite infection effects such as the hepatosplenomegaly of schistosomiasis or the large blood-loss due to hookworm infections from *Ancylostoma duodenale* which exacerbate nutrient loss and mobilization.

Finally, helminth infections, especially hookworm but also trichuris and schistosomiasis, cause reduced hemoglobin levels and anemia. The effects of these parasitic infections on childhood cognition is becoming an important area for study because of the greater appreciation for the interactions among parasitic infections, iron-loss, dietary iron deficiencies and cognitive development and performance (**6.5**). These may have implications for educational performance, school attendance and ultimately economic productivity (**9.10**).

*Stephen T. McGarvey*

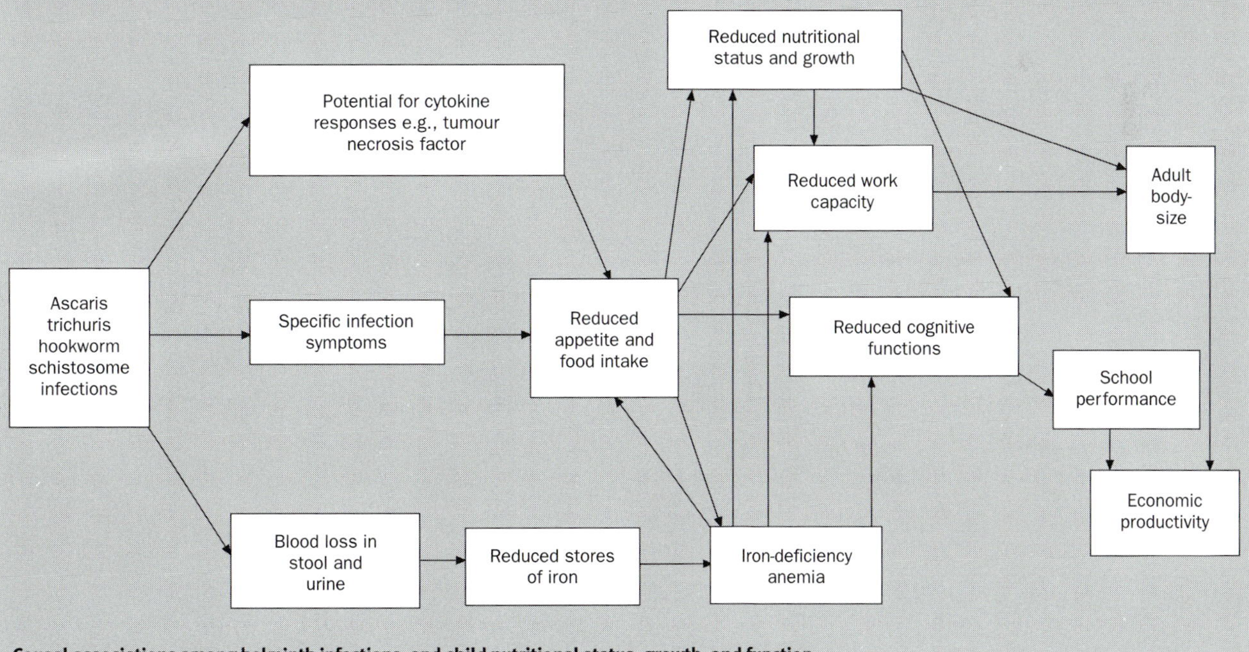


**Causal associations among helminth infections, and child nutritional status, growth, and function.**

## HIV AND GROWTH

*Prevalence of nutrient malabsorption in children with symptomatic HIV infection*

| | |
|---|---|
| Sugar malabsorption | 32% |
| Steatorrhea | 30% |
| Protein loss | 17% |
| Iron malabsorption | 62% |

The growth of a child with symptomatic infection is frequently impaired. So-called 'failure to thrive', which can be defined as an abnormally slow rate of weight-gain and poor growth in height, is included in the case definition of AIDS by the World Health Organization. The prevalence of failure to thrive ranges from 20–80 per cent of children with symptomatic HIV infection. The typical presentation is reduced body-weight associated with a proportionally reduced height compared with the reference values obtained from age-matched healthy children. Therefore, a child with HIV infection often appears 'normal' but younger than he or she really is. However, there are three distinct growth patterns in HIV infection: 1) a progressive delay of growth-rate; 2) a stable delay of

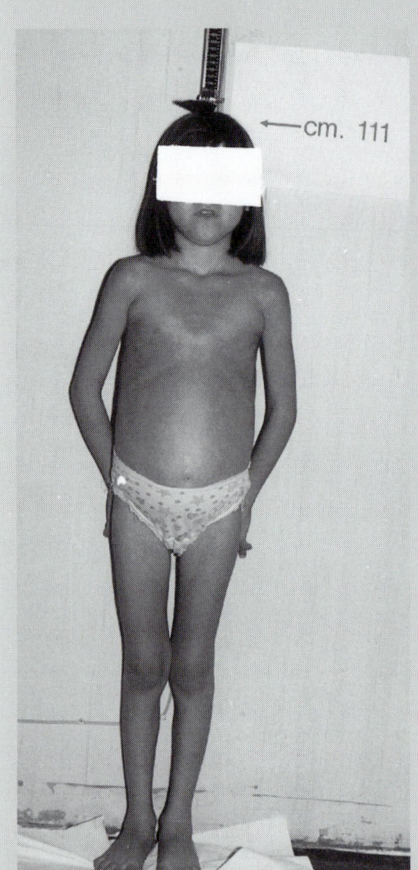


**A 9-year-old girl with symptomatic HIV infection. Her height is appropriate for that of a 6-year-old girl. The prominent abdomen is a typical sign of malnutrition.**

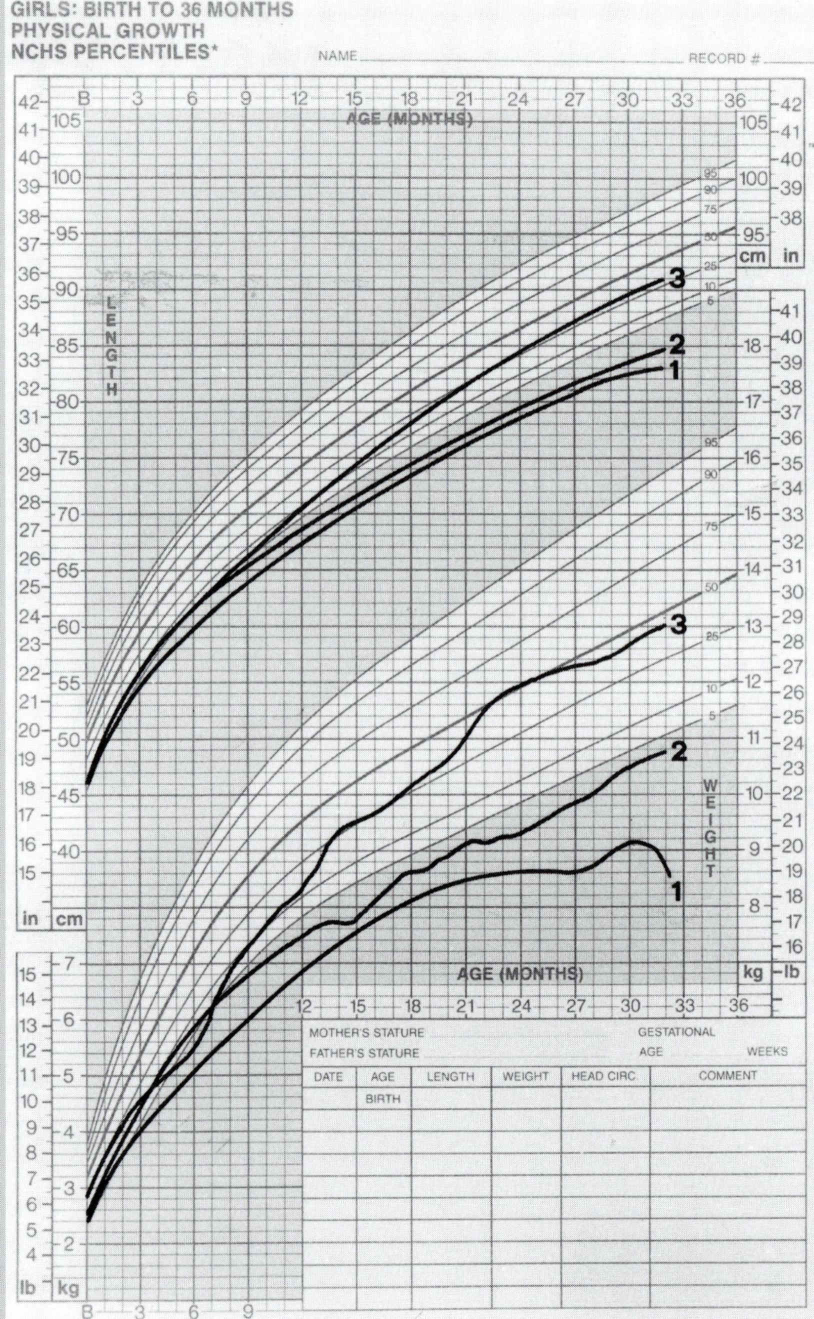


**The weight and height rates of three young females with HIV infection have been plotted on the reference anthropometric chart. Three different patterns may be seen: 1) a progressive growth delay with a curve which tends to be increasingly flatter; 2) a stable delay of growth curve, whose pattern is parallel to the lowest normal centile and 3) a normal growth curve, following a period of catch-up growth. The catch-up growth coincided with the onset of specific antiretroviral therapy associated with nutritional support.**

growth-rate and 3) a normal growth-rate, often following a catch-up growth (**9.6, 9.7**) phase. With the progression of the disease, the growth curve becomes irreversibly impaired and the loss of body-weight leads to the wasting syndrome or cachexia which is a terminal event of AIDS.

There are three possible pathways for failure to thrive: 1) reduced energy intake, 2) intestinal malabsorption and 3) increased metabolism leading to catabolic state. All three mechanisms may play their part in a patient with immune impairment. In a HIV-infected child, the energy intake may be reduced, either by oral or esophageal infections, neurogenic swallowing problems, or inadequate feeding due to the poor socio-economic conditions that HIV-infected parents usually live in.

The role of intestinal dysfunction in the growth failure is well established. The intestine is a target organ in AIDS. Between 20 and 50 per cent of infected children have abnormalities of the intestinal digestive-absorptive processes including carbohydrate malabsorption, abnormal steatorrhea, fecal protein loss and malabsorption of other nutrients such as iron or vitamins. One typical feature of intestinal dysfunction in children is the lack of evident symptoms, such as diarrhea, and loss of body-weight may therefore be the only sign of intestinal disease.

The etiology of intestinal dysfunction in AIDS is unknown. Classical or opportunistic enteric pathogens are seldom detected in children with intestinal malabsorption, and a direct enteropathogenic role has been suggested for the human immunodeficiency virus itself.

A normal response to weight loss is the reduction of metabolism which can be estimated by measurement of the resting energy expenditure. Most children with HIV infection show a paradoxical response to the loss of body-weight by increasing their metabolism. A catabolic state may be associated with fever or with evolving infections, but also with neuroendocrine abnormalities or with a chronic inflammatory state. A complex interplay of infections, inflammation, malabsorption and neuroendocrine dysfunction, together with a reduced energy intake, is responsible for growth failure in children with HIV infection, and a central triggering role is played by infectious agents. Several therapeutic strategies are used, often in combination, to treat children with poor growth but the results are variable.

Currently, the use of growth factors, such as the genetically engineered human-derived recombinant growth hormone (GH), in treatment is being investigated. GH possesses a well-established anabolic effect and a more recently discovered anti-diarrheal effect through direct interaction with the enterocyte. GH has proved effective in improving body composition and function and in reversing the loss of body-weight in HIV-infected adults. Progress is being made in other related fields and together with the availability of new anti-retroviral drugs will certainly reduce the impact of growth failure in children with AIDS, allowing them a longer lifetime and a better quality of life.

*A. Guarino*

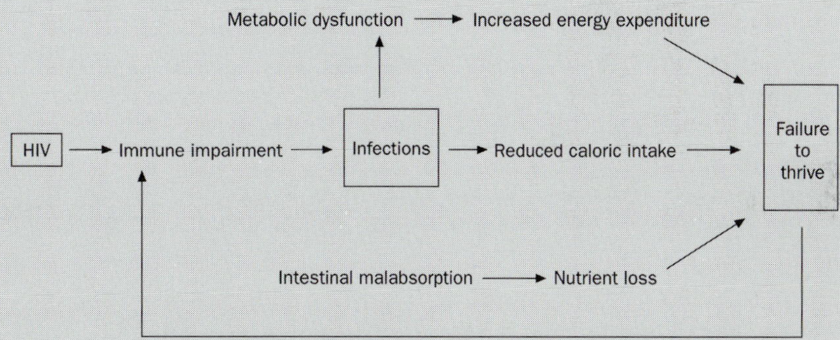


**Growth may be impaired through 3 main pathways: 1) reduced caloric intake; 2) metabolic dysfunction which increases energy expenditure and 3) intestinal malabsorption which is responsible for nutrient loss. A central role is played by recurrent bacterial, viral and protozoal infections, namely those caused by the so-called opportunistic agents. The ultimate result is growth failure. This, in turn, contributes to the immune derangement initially induced directly by retrovirus.**

*Therapeutic approaches to HIV-infected children with failure to thrive*

| | | | | | | |
|---|---|---|---|---|---|---|
| **Goals** | Reduce HIV burdens | Reduce opportunistic infections | Optimize nutrition | Optimize absorption | Counteract hormonal deficiencies | Psychosocial factors |
| **Tools** | Antiretroviral therapy | Antimicrobial agents | High caloric feeding | Elemental diet | Replacement therapy | Parental counselling, Foster care |
| | | | Enteral tube feeding | | | Psychological support |
| | | | Vitamin and oligoelement support | | | |
| | | | Parenteral nutrition | | | |

The therapeutic approach to a child with HIV-infection and failure to thrive requires multiple competence, including those of pediatric infection specialists, immunologists, nutritionists, endocrinologists and psychiatrists.

## IMMUNOCOMPETENCE

One aspect of growth and development which mediates the overall growth response to undernutrition and infection is the development of the immune system (**4.7**). B lymphocytes of the newborn child are functionally immature, and this is reflected in the low levels, relative to adult values, of circulating immunoglobulins G, A, and M. However, complete B cell maturity (as assessed by the proportion of adult circulating immunoglobulin values attained) is not approached until the pubertal period. Attained IgA maturity is lowest at all ages.

IgA in the mucous secretions of the lungs and gastrointestinal tract is protective against respiratory and diarrheal infection, but may not be adequate in the infant. At birth, the neonate is largely protected from pathogens in the birth canal by the IgG antibody from the mother. The breast-fed infant receives maternal IgA which is protective, with a broad spectrum of antipathogenic activity. There is little evidence to suggest that diarrhea has any effect on the growth of exclusively breast-fed infants in developing countries. Dietary supplementation of infants increases their exposure to pathogenic agents, while cessation of breast-feeding both increases pathogenic exposure and removes the maternal antibody contribution to the infants' immune system. Thus it is little surprise that in the developing world diarrheal and respiratory infections are most common at this stage of a child's life. Furthermore, undernutrition at the time of weaning (**9.1**) may well lead to further maturational delays in immune system development, as well as growth faltering. This is supported by measurements of immunoglobulin containing cells in the jejunal mucosa of undernourished children with gastroenteritis, which have shown reduced numbers of IgA-containing cells, but normal numbers of the IgM-containing cells which are the precursors of IgA-producing cells. What is not clear, however, is the extent to which infants are able to mount an adequate immune response despite low levels of immune system maturity.

*Stanley Ulijaszek*

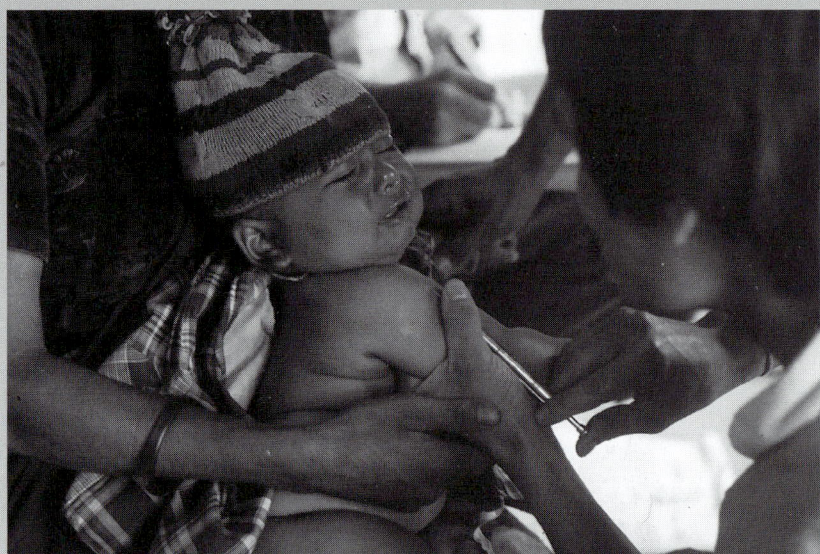

**Infant in Nepal receiving immunization. Immunization can go some way to reducing undernutrition-infection interactions, and their effects on growth. Photograph by Stanley Ulijaszek.**

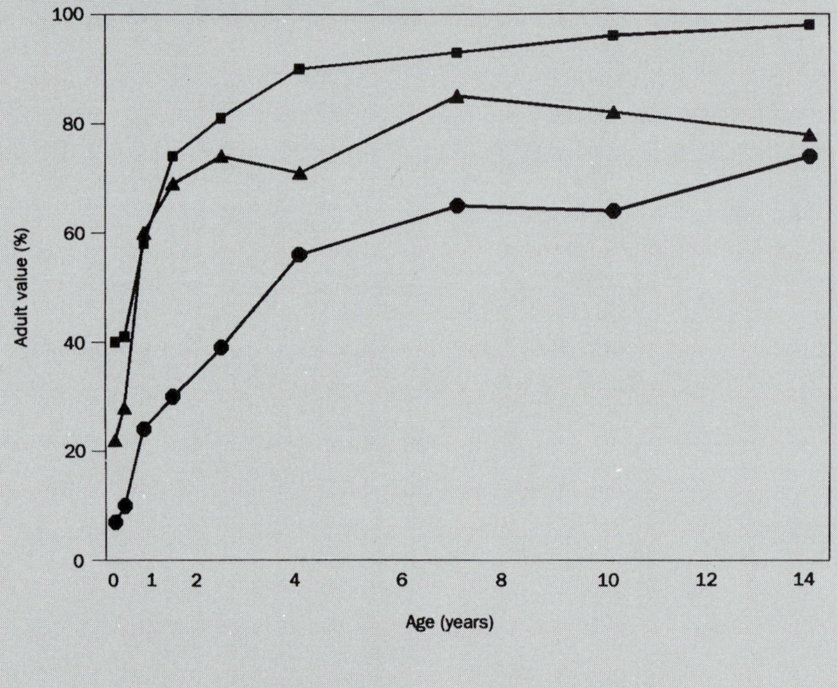


**Immune system maturity as determined by the proportion of adult levels of immunoglobulins G, A, and M during growth and development in healthy, well-nourished children. After Johnson, Moore and Jeffries, 1978.**

# Growth and psychosocial stress

It is now widely accepted that an adverse family and social environment can retard children's physical development, but there is still uncertainty about the mechanisms involved. Evidence of an association between social deprivation and impaired physical growth has accumulated mainly from cross-sectional epidemiological studies (**1.12**, **11.4**). When growth failure occurs during infancy it is usually described in terms of poor gain in weight (which is easier to measure accurately than length), and is known as 'failure to thrive'. Failure to thrive in the absence of physical disease or disability is termed 'non-organic'. Non-organic failure to thrive used to be considered indicative of neglect, but it is not usually associated with frankly abusive or neglectful parenting. In most cases the etiology is unknown. Recent evidence suggests failure to thrive in developed countries is often the outcome of a maladaptive interaction between specific child characteristics (including failure to signal hunger unambiguously, or dysfunctional oral-motor skills) and specific parental characteristics, such as a failure appropriately to interpret feeding cues. The child is consequently undernourished.

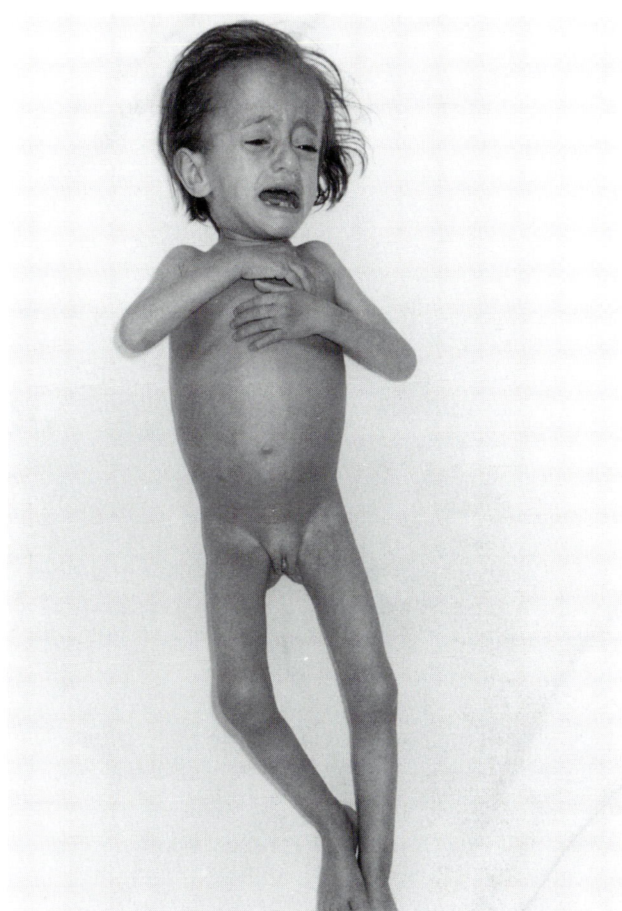

**Child with failure to thrive due to neglect.**

If failure to thrive persists long enough it will eventually lead to stunting, in which height is low in relation to genetic potential (**3.1**) (as assessed by mid-parental height). Adverse psychosocial circumstances can delay the rate of skeletal maturation by other mechanisms too. A few studies have suggested that boys are more vulnerable to the effects of psychosocial stress than girls.

## Epidemiological evidence

Height is normally distributed in the general population, but among English school children the distribution is 'spaced out' at the lower extremes of social class, so that elementary school-aged children of manual workers are about 2 centimetres shorter than those of non-manual workers at the 50th centile, but nearly 4 centimetres shorter at the third centile (**11.4**). Differences in height between children of the employed and the unemployed have been found within each social class.

Can we be more specific about the nature of the risks for impaired growth that are associated with low socio-economic status? A cluster of factors that affect families seems to be important. They include single parenthood, overcrowding, low disposable income, paternal ill health, dependence on social welfare, and parental abuse of alcohol and drugs (**8.5**). A number of studies have also found a broader association between the quality of maternal care, as assessed by health visitors, and children's stature. In both the developed and the developing world a similar association is found between stature and socio-economic status (**1.12**), height correlating positively with wealth.

## Causal mechanisms

How do poor home conditions 'cause' short stature? Deprivation and stress are not synonymous, although there is a tendency for co-occurrence, especially among socio-economically disadvantaged families. Few attempts have been made to disentangle their influence. By far the greater part of the growth deficit in disadvantaged groups, relative to national norms, is evident by the time the child enters elementary school. Thereafter, children from deprived or psychosocially stressed homes develop at much the same rate as those from more advantaged backgrounds, both in terms of height-for-age and in terms of bone age (skeletal maturation) (**1.8**). Thus, the deficit seen at school age was usually first apparent during the preschool period. A number of lines of evidence converge to suggest that deprivation or stress during a 'sensitive period' of infancy is the crucial factor. In other words, a given insult that affects growth at that time in a child's life, especially within the first post-natal year, may have more severe and persistent impact than an equivalent insult during a later period of development.

Epidemiological surveys using longitudinal data have shown many short children at school-entry failed to thrive during infancy. Could early failure to thrive have led to their poor growth in height? It is certainly possible, but the mechanisms of failure to grow well in infancy and in later life are rather different. The initial phase of physical development is primarily under nutritional control (**9.2**). Consequently, 'growth faltering' (poor weight-gain) during infancy is not a necessary condition for later stunting, which seems due to a lag in the onset of the hormonally mediated 'childhood phase' of linear growth (**2.4**). Further research in this area is needed.

## Opportunities for catch-up growth

During the first year, children are especially vulnerable to environmental insults because their rate of growth is greater than at any other time, even puberty. Diminution in that normal velocity, for even a relatively brief time, may have long-lasting consequences. If the growth trajectory diverges substantially and persistently from the population norm due to environmental adversity during this period there is probably limited opportunity for full catch-up growth in later childhood, even if circumstances improve. How much catch-up growth is achieved depends on the degree to which biological maturation is delayed. If maturation is not delayed substantially, there will be less opportunity to catch up through a prolonged growth period. Some evidence suggests accelerated growth-rates are possible when deprived children are moved into a substantially more advantaged environment (**9.6**). Unfortunately, this may be at the expense of accelerated maturation, the consequence being an early puberty and short adult stature. The mechanism by which accelerated maturation occurs is unknown.

## Psychosocially induced short stature

Evidence is accumulating to suggest that some children are genetically vulnerable to stress, and respond to it in an idiosyncratic way, resulting eventually in excessively short stature. The usual mechanism is believed to be loss of appetite, and anorexia, with a correspondingly chronically deficient intake, hence nutritional stunting. However, just over 25 years ago a condition was reported which simulated primary pituitary deficiency of growth hormone, but was potentially reversible. It occurred in association with a variety of unusual behaviours, including hyperphagia (insatiable appetite) and polydipsia (insatiable thirst) among children who were emotionally or physically abused. When these children were removed from their abusive homes they usually started to grow at an accelerated rate. Food restriction does not seem to be the cause of these children's obsession with eating, and the growth-hormone deficiency (**7.7**) has nothing to do with malnutrition, in which levels are characteristically higher than normal. Parents of hyperphagic short stature (HSS) children often go to

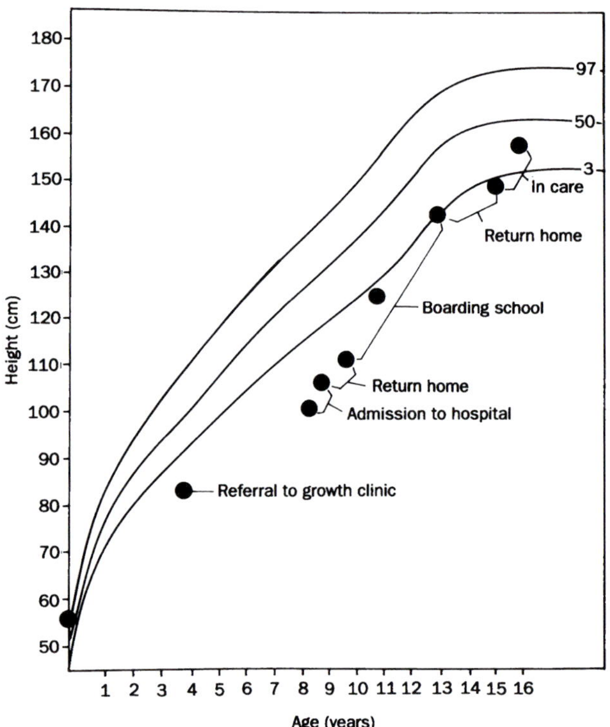


**Growth trajectory of child with short stature due to emotional abuse, demonstrating catch-up height when living away from home.**

extraordinary lengths to stop uncontrolled eating behaviour at home. They lock food cupboards, refrigerators and freezers, and put alarms on kitchen doors.

Phenotypically HSS children are very similar to those with the Prader–Willi syndrome (**7.4**), which is caused by a small deletion on the paternal chromosome 15q 11-13. Prader–Willi children are typically overweight and also hyperphagic; their parents usually attempt to restrict food intake by similar means. Strangely, although on average overweight for their height, HSS children are never obese. Perhaps this is because, in stark contrast to the lethargic Prader–Willi child, they are almost invariably hyperactive.

The endocrinology of the condition is intriguing. Investigations of growth hormone dynamics show that HSS children secrete very small quantities of growth hormone while living under conditions of high stress, typically at home, but they may not respond to therapy with exogenous growth hormone (**7.9**). Yet if they are removed from the abusive parent or stressful home environment, for example into hospital, there is a rapid and spontaneous increase in their endogenous growth hormone output to above normal levels within a few weeks. If the reduction in stress can be sustained for a period of several months there will be catch-up growth (**9.6**, **9.7**) to a degree that is often spectacular. This ceases rapidly if the child is once more returned to the stressful home environment. Other behaviours associated with the syndrome, including the hyperphagia, wax and wane in severity in parallel with the rate of linear growth. We still have much to learn about the subtle and elusive relationship between the mind and growth during childhood.

*David H. Skuse*

# Environmental toxicants

Human physical growth may be adversely affected by some pollutants, even at low to moderate doses. Evidence is strongest for effects of lead, mercury, noise, air pollution, and a group of dioxin-like compounds including the polychlorinated biphenyls and dibenzofurans. However, many pollutants have not yet been examined for effects on human growth. Effects of pollutants on growth would be consistent with one of the traditionally accepted indicators of toxicity: weight-loss in mature mammals and lack of weight-gain in immature ones. The ability to define precisely the relationship between mundane exposure to a pollutant and human growth depends greatly on the methods employed.

A pollutant may be defined as a material or energy that is unwanted to some degree, is thought to interfere with health or well-being, and is produced by human activity either in part or entirely. However, it should be noted that some pollutants, such as sulphur dioxide and methane, are created by natural processes. Pollutants are ubiquitous, and probably no human population is completely unaffected. Thus, it is unrealistic to seek unexposed populations that can be used as a control in studying the effects of pollution. Levels of pollutants vary markedly, depending partly on the degree of industrialization or proximity to an industrialized society (**11.7**). Samples of arctic ice contain lead that dates to the onset of the industrial revolution, while lead has been found in some neolithic skeletons. Polychlorinated biphenyls, as well as DDT and its metabolites, are found in breast-milk in many populations around the world.

The main problem in the study of pollution and growth is ascertaining the level and extent of exposure (dose) to the pollutant in question. Studies vary greatly in how well the dose is measured, and whether it is measured in individuals or determined for a group and then estimated for individuals. Group attributions of exposure are weaker because it is likely that exposure varies within the group and some individuals can be misclassified. The United States National Research Council has identified a hierarchy of exposure classifications that range from the biomarkers (quantified individual measurements) as the best measure, to residence or employment in an area where a pollutant is found as the worst, with five intermediate levels of dose/exposure quality.

*Lead* is retained in the body (90% in the adult skeleton) after exposure, and can be measured precisely in individuals, allowing quantification of a dose-response relationship. Measurements in teeth and skeleton, hair and blood can provide estimates of long-term exposure, exposure over the past few months, and contemporary exposure, respectively. Lead enters the body primarily by ingestion and to a far lesser extent by respiration. In children, lead is absorbed easily through the gastrointestinal tract, and a greater percentage is absorbed than in adults. A toxic burden can be reached with the consumption of only a few chips of lead-based paint, but with the decline in use of such paint, there has been a reduction in the contribution from this source to the lead burden of the population. Ubiquitous lead-containing dust is the main environmental source, although food and water also contribute.

Lead poisoning, once commonly defined as a level of lead above 60 micrograms per decilitre and infrequently observed today, is known to reduce post-natal physical growth. However, more recently the low levels of lead that are now commonplace in industrialized countries (for example, below 25 micrograms per decilitre), have also been shown to affect post-natal physical growth and development. However, it is difficult to compare earlier studies with more recent ones because of between-study differences in dose levels, techniques for measuring lead content, control over extraneous factors, and changing definitions of low versus high doses.

Among a national probability sample of United States children aged 6 months to 7 years (National Health and Nutrition Examination Survey II) significant negative relationships were observed between lead level and height, weight and chest circumference. At the mean blood-lead level, each of these dimensions was reduced by 1.5 per cent compared to a blood-lead level of zero. Among 5–13 year old children in the American Hispanic Health and Nutrition Examination Survey, those with lead levels between 11–40 micrograms per decilitre had a 1 centimetre deficit in stature compared to children with lower lead levels.

Prenatal exposure may influence prenatal growth and size at birth as well, since lead crosses the placenta freely. Several longitudinal studies begun in the 1970s and 1980s, when lead levels in the environment were decreasing, have examined this relationship and discovered that levels of lead below 15 micrograms per decilitre may reduce birth-weight with each log unit of lead associated with a 150–200 gram decrement in birth-weight. Head-circumference and body-length are not affected, or are affected very little, while skinfolds may be affected most. Early effects on growth may extend into the post-natal period. It is difficult to separate the effects of prenatal exposure and growth reduction from the effects of post-natal exposure because there are few individuals who differ markedly in prenatal and post-natal exposures. The Cincinnati cohort study carefully examined the contributions of prenatal and post-natal lead levels to post-natal growth, and found that high exposures in both periods were related to depressed post-natal growth in the second and third year of life, but catch-up growth was evident among children whose post-natal lead levels declined to the sample median or below it. Currently, the level of blood-lead at which no affects on physical growth occur is not known, and there is considerable concern that neurobehavioural effects may be evident at very low levels. For the United States, the Centers for Disease Control have set 10 micrograms per decilitre as the level above which a health intervention should occur to protect human health and child development.

*Polychlorinated biphenyls* (PCB) are a class of 209 structurally similar organochlorines that have come to be considered common pollutants. Analysis of adipose-tissue samples from Canada and America indicate that few samples in the early 1980s were entirely free of PCBs. The frequency of heavy burdens have fallen following bans of PCB-use in manufacturing. Dramatic evidence for an effect of PCBs on growth comes from two incidents of PCB poisoning, one in Japan in 1968, and the other in Taiwan in 1979, when rice-oil contaminated with PCBs and closely related compounds (dibenzodioxins and dibenzofurans) was consumed. The resulting disease was termed *Yusho* and *Yu-cheng*, respectively. Children exposed post-natally showed depressed growth, and though the size of increments 4 years later did not differ between exposed and unexposed children, there was no catch-up growth among the exposed children. Babies exposed *in utero* were characteristically smaller at birth. Since a variety of halogenated hydrocarbon compounds were present in the contaminated oil, the contribution of PCBs alone is not certain. In other studies where PCBs may be acting alone, childhood post-natal growth has been shown either to be related to levels of PCBs measured near the time of birth, or to be unrelated. Several studies of size at birth and PCB exposure found significant reduction in birth-weight associated with measured PCBs in cord blood or with a surrogate measure of maternal prenatal exposure. Thus, evidence for an effect of PCBs on prenatal growth is strong, while evidence for an effect on post-natal exposure is weaker. There is evidence that low levels of PCB exposure may interfere with neurobehavioural development.

High levels of *air pollution* are associated with reduced physical growth, both prenatal and post-natal, as well as with slowed skeletal development(**1.8**, **5.17**). Since air pollution is a very heterogeneous entity, it is difficult to know which component of it contributes to growth impairment. Furthermore, all studies of air pollution have compared entire towns, or parts of towns differing in air pollution levels without measuring pollutants in individuals. Other differences between comparison groups are likely and the possibility of unmeasured confounding variables undercuts the internal validity of virtually all studies of air pollution and growth. Studies involving measurements on individuals may clarify the effect of air pollution on prenatal and post-natal growth.

Pollution from *hazardous waste* sites may impair growth and development. Studies of children born and raised in Love Canal, New York, where a leaking hazardous waste site was discovered in 1978, indicated that children born and raised there were significantly shorter than children from a community similar in social characteristics but located far from the dump-site. Two studies of birth-weights at Love Canal,

**A view of Widnes, Liverpool, taken in 1895. Coal-smoke pollution, produced by factories and domestic heating, remained a major cause of respiratory problems in Britain until the 1960s, when smoke-abatement legislation and the introduction of smokeless fuel proved effective in producing cleaner air. Courtesy of the Wellcome Institute, London.**

**Noise pollution, such as that associated with large airports, may affect fetal and child growth by way of stress responses from the adrenal cortex and the autonomic nervous system. Photograph by Stanley Ulijaszek.**

**Commuters in Tokyo sometimes wear masks in an attempt to reduce their exposure to air pollution. Photograph by Stanley Ulijaszek.**

one by the New York State Department of Health, also indicated that reduced birth-weight, or the frequency of low birth-weight, was related to residence at Love Canal, especially in areas of the community where exposure may have been greater, and particularly to residence during past periods of active waste-dumping. However, it is difficult to determine the exposure to pollutants from hazardous waste sites because many of the pollutants are not stored in the skeleton or adipose tissue and cannot be measured after the exposure when the studies are conducted.

*Energy*, such as *noise* and *electromagnetic fields*, may be also considered as pollutants. Studies of exposure during pregnancy to airport noise on size at birth have been conducted in Japan, The Netherlands, and the United States. The studies differ in sample size, how well socio-economic factors were controlled for, and how accurately the noise exposure was measured. All have reported a negative relationship between size at birth and noise exposure. One study noted a marked increase in the frequency of small birth-size exactly at the time when jet plane service began at a large airport. A study of occupational noise exposure and size at birth found that the frequency of small-for-gestational-age births was increased among women with greater exposure to noise. Studies of large numbers of Japanese children measured at 3 years of age during compulsory health checks found a consistent association between noise levels from a nearby airport, and the frequency of short stature. Another study conducted in America with a far smaller number of subjects but strong control over other influences on growth, found evidence for an effect only among the most highly exposed (noise levels above 100 decibels at residence). Noise may affect growth through hormonal pathways since it activates the typical stress response, including reactions from the adrenal cortex and the autonomic nervous system. The fetus may be affected by the mother's reaction to noise stress, and the child could be affected directly.

In sum, studies show that at very high doses, a variety of pollutants can affect growth. Other pollutants that can be measured precisely in individuals (such as lead), can affect growth at moderate to low levels. The impact of many pollutants at low doses is not clear.

*Lawrence M. Schell*

See also 'Fetal growth retardation' (**4.8**), 'Body-size at birth' (**10.3**) and 'Urbanism and growth' (**11.7**)

## 9.6

# Catch-up growth in height

About one third of the world's children are less than 2 standard deviations below North American standards in height; this stunting, which starts in infancy or earlier, usually persists to give rise to small adults. Children treated in hospital rarely have a measurable catch-up in height over the course of their stay while their diet is under control, no-matter whether there is or is not a real catch-up measured by sophisticated techniques. The main question is whether an adverse early nutritional environment permanently affects the individual to give a 'growth scar' or whether catch-up can occur. This is quite unlike wasting where catch-up in weight is predictable and reproducible.

Martorell and colleagues examined the increment in height of stunted and non-stunted Guatemalan children between 5 and 18 years of age. The absolute gain in height did not decrease their deficit over this time span – there was no evidence of catch-up; this is the usual state of affairs in most of the impoverished world. Martorell concluded that a period of malnutrition in the first years of childhood irrevocably changes the child so that he or she is 'locked into' a growth trajectory with a lower potential for final adult height and that catch-up growth does not occur.

There is experimental data to support this view. Malnourished rats very rarely achieve the length of rats that have never been malnourished. The classical experiments of cross-breeding shire horses with Shetland ponies by Walton and Hammond (1938) dramatically demonstrate the effect of the intra-uterine environment in determining later growth. The human equivalent is the fate of monozygotic twins of different size; the smaller twin remains small and fails to catch up as an adult.

However, compelling data to show that remarkable catch-up is possible, even under conditions of extreme privation, is presented by Steckel. Following the abolition of the American slave-trade from Africa in 1807, all slaves transported by sea from one American port to another had to have their ages, sex and heights recorded. Steckel analysed the heights of 50,000 such slaves. The recorded heights presented as centiles of present day (NCHS) standards show that they are tiny by modern standards. However, between 15 and 17 years of age they quite suddenly undergo a spectacular catch-up in height to above the modern 25th centile. At this age they received extra food, in the form of salt-pork, although they also started exhausting work in the fields. These data clearly show that adolescents can have an almost complete catch-up in height without any change in hygiene or exposure to disease.

In Chile, Peru and Cape Town, the follow-up of previously malnourished children shows clear evidence of spontaneous catch-up; although these children, who are among the poorest, do not achieve normality. In Kenya, Kulin and colleagues (1982) conducted a cross-sectional survey comparing girls from three privileged schools in Nairobi against an impoverished rural district with a very high prevalence of malnutrition. By the age of 18 years, the previously malnourished rural children had completely caught up with the affluent girls. Another source of evidence comes from the study of refugee children; they virtually all gain height at an accelerated rate when the environment is enriched.

Affluent children with stunting secondary to disease have the advantage that they are less likely to have had early developmental problems and malnourished parents, although on-going disease and treatment may affect their growth. There are many reports of impressive catch-up under these circumstances, including children with coeliac disease, Crohn's disease and trichuris dysentery syndrome.

It is quite clear that very substantial catch-up in height to totally eradicate a deficit is possible, even long after the early stages of growth should be complete – although the animal and human data suggest that may not occur with intra-uterine insult or with imprinting of genomic regions associated with growth. It is equally clear that this is very unusual – it is unreasonable to expect that those who developed stunting associated with poverty will catch up spontaneously without a major change in their circumstances, and yet this seems to be what many investigators look for. Because growth in height is sufficiently slow, experiments to demonstrate catch-up are expensive, tedious, difficult to control and commensurately uncommon. Furthermore, until recently, stunting was not seen to be a major problem, although it was known to be associated closely with mental ability.

There have been no studies conducted where optimum nutrition for good growth has been provided to chronically stunted children, so we just do not know the potential for catch-up in most groups of stunted children. Unsuspected growth-limiting nutrients are probably deficient in many populations. For example, in America, addition of zinc to head-start children's diets led to dramatic increases in height growth in girls. The problem with these 'growth nutrients', or type II nutrients, is that they are not associated with any specific clinical signs or symp-

*Height (cm ±SD) of school-children from rich (urban) and poor (rural) schools in Kenya*

| | **Male** | | **Female** | |
|---|---|---|---|---|
| | **age 10** | **age 18** | **age 10** | **age 18** |
| Well-nourished | 142.9 ±7.4 | 164.4 ±6.0 | 144.3 ±5.9 | 156.9 ±7.3 |
| Malnourished | 130.4 ±5.8 | 164.8 ±5.0 | 123.7 ±8.0 | 157.3 ±5.9 |
| Difference (cm) | −12.5 | +0.4 | −20.6 | +0.4 |

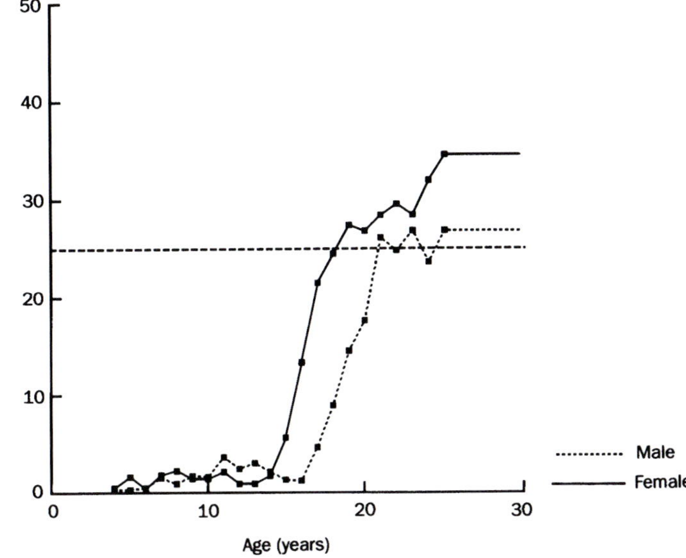


**Heights of slaves transported by sea from one part of America to another between 1820 and 1860, by age. The data are plotted as centiles of the NCHS standard, the horizontal line is at the 25th centile. Data from Steckel, R.H. (1987). 'Growth depression and recovery: the remarkable case of American slaves'. *Annals of Human Biology*, 14: 111–132.**

toms except for growth failure, and they are exceptionally difficult to diagnose without an intervention trial.

Unfortunately we do not know the dietary requirements for either normal or catch-up growth in height, although these are now being defined more closely. When we ask if stunting is or is not reversible in an individual or a society, at the moment we cannot make any prediction, because the pathogenic mechanisms and gene–nutrient interactions that control longitudinal growth in humans are still under early investigation (**9.3**, **3.4**); indeed, none of the determinants we examined to see if we could predict whether a particular stunted child will or will not have a height 'spurt' during treatment were effective.

*Michael H. N. Golden*

See also 'Longitudinal analysis' (**1.14**), 'Creation of growth references' (**1.16**), 'The use of growth references' (**1.17**), 'The human growth curve' (**2.4**), 'Within-population variation in growth patterns' (**2.5**), 'Genetics of child growth' (**3.1**), 'Genetic regulation of growth-promoting factors by nutrition' (**3.4**), 'Fetal growth retardation (**4.8**), 'Extra-uterine growth after premature birth' (**5.2**), 'Growth cyclicities and pulsatilities' (**5.13**), 'Growth in chronic diseases' (**7.8**), 'Nutrition' (**9.2**), 'Catch-up weight-gain' (**9.7**), 'Growth in infancy and pre-adolescence' (**10.4**), 'Physical growth during industrialization' (**11.2**), 'Migration and changing population characteristics' (**11.6**) and 'Long-term consequences of early environmental influences' (**12.1**)

## 9·7

# Catch-up weight-gain

Severely wasted children gain weight very rapidly when they are given access to abundant, wholesome food. It is not unusual for them to gain weight at 20 times the normal rate. When they achieve normal body proportions they spontaneously decrease their intake to slightly above normal.

The first figure shows the cumulative weight-gain of 1503 refugee children recovering from malnutrition in several centres in Rwanda, for each day after their minimum weight. After 1 and 3 weeks they had gained about 12.5 per cent and 25 per cent of their body-weight. This phenomenal rate of weight-gain was not related to the magnitude of the initial deficit, although all the patients were less than 85 per cent of the expected weight. The second figure shows that the rate of weight-gain was also not related to the age of the children. Indeed, 54 adults that were treated in the centres gained weight at the same rate as the children (13.4 vs. 13.8 g/kg/day) when they were given the same dietary regimen as the children. This is equivalent to a 70-kilogram adult putting on over 13 kilograms in two weeks.

The rate of weight-gain is not steady, rather it declines continuously as the patient recovers. Although this has been assumed to be an exponential decline, in fact the curve fits a power function much more closely; the equation for the line in the figure is weight-gain = $(36.6 \pm 0.7) \times \text{Time}^{(0.640 \pm 0.007)}$ ($r^2 = 0.999$). From this equation the average length of time that it takes for a wasted person to recover from a particular deficit if they are treated adequately, can be readily calculated. This function is almost exactly the same as that obtained from 437 6–18-month-old children treated for severe malnutrition in the Tropical Metabolism Research Unit in Jamaica.

The rate of weight-gain (recovery) is linearly related to the energy intake and not to the content of protein, provided it is at least 9 per cent of energy intake. There does not appear to be an upper limit to this relationship in *ad libitum*-fed patients. For this reason, fat has been added to the diets of these children to levels up to 1.35 kilocalories per gram. The data shown in the first figure were obtained with a diet containing 1.0 kilocalories per gram so that an energy density above this does not seem to give an additional increment in weight-gain. Indeed, with very high energy densities, water becomes a limiting nutrient, particularly in small infants with a large surface-area-to-weight ratio. However, when dilute porridge is given with an energy density below about 0.7 kilocalories per gram, it is impossible to ingest sufficient volume to achieve energy intakes necessary for rapid rates of weight-gain, no matter how much of the other nutrients are supplied.

Energy balance during recovery from malnutrition shows that there is a clear relationship between the type of tissue that is being synthesized and the amount of energy that needs to

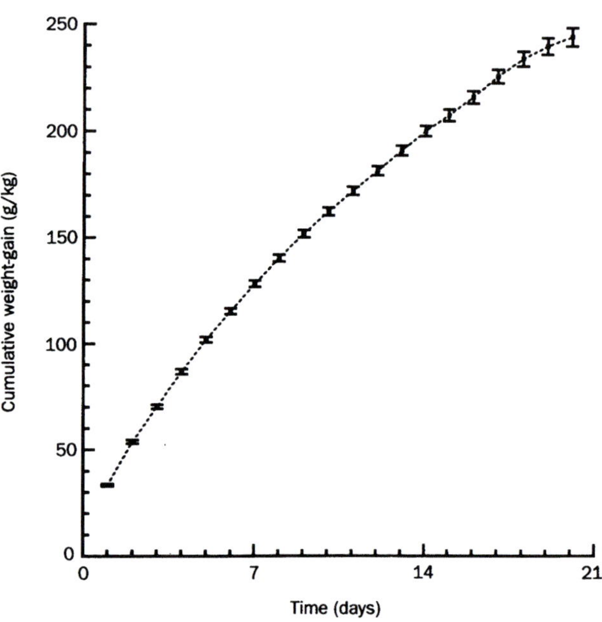


**Cumulative weight-gain in 1503 patients treated for malnutrition in five therapeutic centres in Rwanda in 1994/1995. Commencing on the day of minimum weight (when edema and infections were under control), daily weights were divided by the minimum weight and expressed as the increment over unity in grams. Mean ± standard error of the mean.**

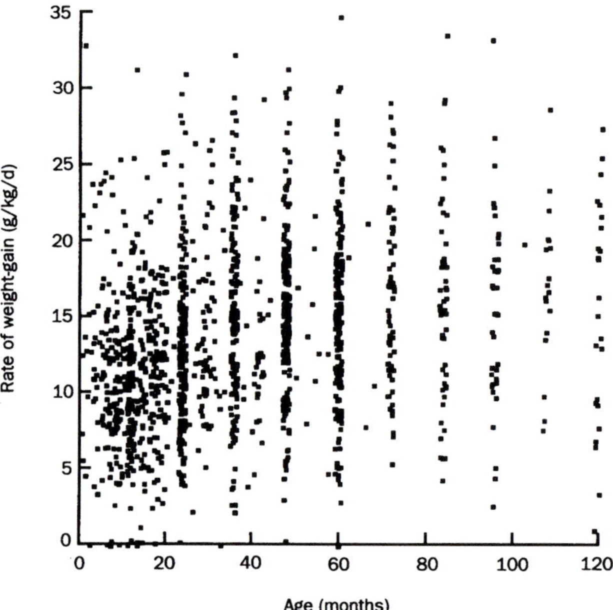


**Mean rate of weight-gain between time of minimum weight and discharge expressed as g/kg/d plotted against age of subject. Each point represents one patient. The plot has been 'jittered' to reduce overlapping data. Most ages are only recorded to the nearest 1 year.**

be ingested to gain one gram of new tissue. Theoretically, one gram of adipose tissue will store 7 kilocalories and one gram of lean tissue about 1 kilocalorie if the tissues are hydrated at 20 per cent and 80 per cent respectively. The energy cost of actually synthesizing the tissue is about 1 kilocalorie per gram for both the lean and fat tissue; direct measurements give total requirements of 2.2 and 11.1 kilocalories per gram for fat and fat-free tissue deposition respectively. These figures can be used to determine the composition of the tissue being deposited if there are careful measurements of dietary intake, as well as the weight, and the nutrients are absorbed from the intestine. Such calculations have been validated with elegant isotopic measurements of muscle mass, so that the 'energy cost of tissue deposition' is a non-invasive way of determining changing body composition.

The energy cost of tissue deposition during recovery from severe malnutrition varies continuously with the stage of recovery. In early recovery the patient synthesizes mainly lean tissue, whereas, as recovery progresses a higher proportion of fat tissue is laid down until, in late recovery, only adipose tissue is being synthesized; so that as recovery progresses there is less weight gained for each unit of energy ingested. This continuous change in the 'efficiency' of growth and thus the type of tissue being synthesized, may account for the shape of the curve of weight-gain; however, one would have expected a different curve for those with different deficits – a situation which has not been demonstrated in practice.

Although it would seem that the treatment of wasting is simply a matter of giving abundant energy, this is not the case. The children that have recovered to the 50th centile weight-for-height, and therefore should then be 'normal', in fact have a very abnormal body composition; their muscle is still atrophic, their thymus has not re-grown, they have increased fat-fold thickness, and balance studies show that they are not retaining nitrogen and therefore synthesizing protein. Clearly, although the children may gain weight at a greatly accelerated rate with the simple addition of energy, this does not return the children to normal. Indeed, with their 'attempt' to synthesize lean tissues, rapid weight-gain can itself lead to an acute nutritional exhaustion with secondary faltering in growth, anemia and even acute clinical illness. Not only is the chemical composition of their bodies and their anatomy abnormal, but they also have disordered electrolyte physiology and immune function.

In many diets used for malnourished children zinc is the limiting nutrient; this was the case in the early 1980s in Jamaica. We supplemented routinely with zinc during catch-up weight-gain and found that not only did additional zinc

## RECOVERY FROM KWASHIORKOR

Animal studies in the 1960s suggested that the earlier in life a severe nutritional insult occurred, the greater and more protracted the eventual retardation and the less likelihood of reaching full genetic potential in adult size. In other words, any catch-up growth that was likely to occur would be incomplete as a result of early and severe nutritional insult. Studies on malnourished children in both developed and developing countries provided conflicting results of 1) ex-patients being stunted in relation to 'local' unaffected children; 2) ex-patients having subnormal mental capacities; 3) little or no evidence of stunting 2 to 8 years after nutritional insult, and 4) a relationship to the timing of malnutrition such that severe and continuous malnutrition in the first year is likely to cause permanent retardation, but a single episode of protein deficiency is less likely to have such consequences.

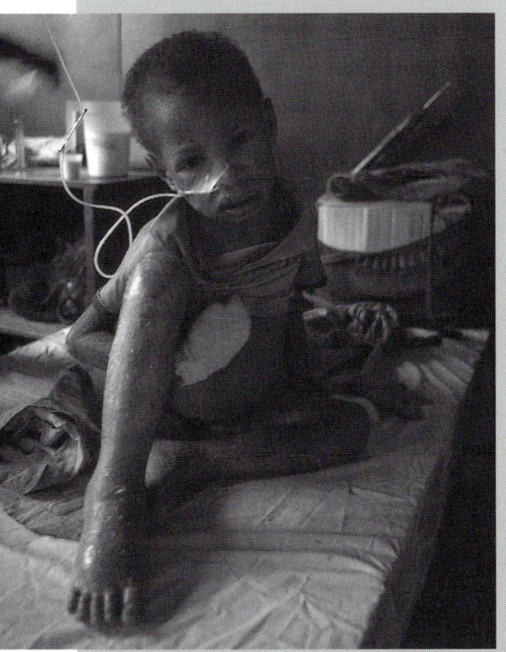

Child with kwashiorkor. Photograph by Stanley Ulijaszek.

First recognized in Ghana, the Ga word *kwashiorkor* literally means 'first–second', identifying that it is prevalent in the post-weaning stages of infancy and early childhood when the first child is deposed from breast-feeding by the second. Kwashiorkor, the most severe form of protein-energy malnutrition, is often precipitated by a series of infections (**9.3**) occurring successively or concurrently in the presence of a diet with a low protein content relative to energy. Often this is a result of the child being fed starchy 'paps' which do not provide an adequate source of protein to substitute for the protein of breast-milk which has been the child's only protein source before weaning – perhaps for the first 18 to 24 months of life. The symptoms of the disease are retarded growth, apathy, anorexia, hypoalbuminemia with edema, and characteristic hair and skin changes. The normally dark hair becomes fine, depigmented or reddish-yellow, and may fall out in patches. The normally dark skin may become flaky, may peel, and leave a reticulated, patchy surface. In addition there are changes to the liver and pancreas. Growth retardation is related to retardation in skeletal development with subnormal mineralization, thinning of the cortices of the long bones, diminished trabecular pattern, and growth arrest.

Few long-term studies have followed malnourished infants to adulthood to investigate the eventual affects on the adult. Those that have been completed indicate that, given an equal nutritional chance and home environment, the ex-kwashiorkor patient is little different from his/her siblings in terms of the timing and magnitude of adolescent growth or secondary sexual development, but that females tend to be less affected than males. Two major studies in the West Indies and South Africa suggest that ex-patients may actually surpass their non-affected siblings in terms of adult height and weight. The latter study reported greater adult heights and weights in ex-patients of both sexes that were significant for males. Ex-patients were delayed at ages of entry to some aspects of secondary sexual development, notably genitalia development in males, but in both sexes the sequence of pubertal events (**1.10**) was the same.

Thus catch-up growth following kwashiorkor appears to be complete in the absence of further nutritional insult. In addition, adolescent development, although slightly delayed, is not significantly affected by an early episode of severe malnutrition.

*Noel Cameron*

increase weight-gain but it also decreased the energy cost of tissue deposition. In a second group of children we confirmed that additional zinc led to an increased proportion of lean tissue, and less adipose tissue deposited, by measuring body-water and protein synthesis sequentially.

The diets used in these experiments turned out to have zinc as their limiting nutrient. However, any of the type II nutrients may similarly limit lean tissue synthesis and may lead to illness if attempts are made to re-nourish patients with the simple addition of energy to standard diets. It is distressing that many of the regimens that are still in use have insufficient potassium, magnesium or available (non-phytate) phosphorus, as well as insufficient zinc. During the very intense metabolism that accompanies catch-up in weight, where the child gains weight and synthesizes new tissue with each meal, there is a danger of disequilibrium syndromes being precipitated. Clearly we wish to return the patient to a state with normal physiology, anatomy and body composition. Surprisingly, it seems that the mere achievement of an astonishingly fast rate of weight-gain is neither evidence of good health nor normal physiological function. Much more attention needs to be paid to the quality of the catch-up growth.

*Michael H.N. Golden*

See also 'Assessment of body composition' (**1.6**), 'Body composition' (**5.11**), 'Growth in chronic diseases' (**7.8**), 'Nutrition' (**9.2**), 'Infection' (**9.3**), 'Catch-up growth in height' (**9.6**), 'Seasonality of subsistence and disease ecology' (**9.8**), 'Growth and natural selection' (**10.1**) and 'Growth in infancy and pre-adolescence' (**10.4**)

# Seasonality of subsistence and disease ecology

In populations living in tropical regions, environmental factors which can influence growth and development, such as infectious disease, nutrition, energy intake, expenditure and balance, often vary seasonally. However, they may not vary in the same way, or at the same time. There are societies in which individual seasonal stresses operate, for example in urban Mexico, where the major seasonal stress is infection, and there are societies in which multiple stresses operate. Urban communities in the developing world are more likely to fall into the first category, while rural communities are more likely to fall into the second. Of societies where multiple stresses have been shown to operate, these may or may not be coincidental. In general, agricultural communities are more likely to experience multiple, coincidental seasonal stress than other types of group, including pastoralists.

The most important disease categories associated with poor growth in tropical agricultural communities are those associated with high population densities, such as respiratory and diarrheal infections. Most respiratory infections peak in the dry season, with factors such as a dry, dusty climate, and closely packed sleeping arrangements, being important. Incidence of diarrhea is also seasonal, varying with the disease agent concerned. Furthermore, seasonality of disease incidence is likely to be related to patterns of land-use and differences in household size and population density at different stages of the subsistence cycle. Thus, both hunter-gatherers and pastoralists experience less seasonality of infectious disease than more settled agricultural groups living at higher population densities.

However, even small differences in population size and disease ecology can have important effects on seasonal patterns of growth and development. For example, the Wodaabe and Ferlo pastoralists of West Africa share similar subsistence practices, but the peak of seasonal stress in young children

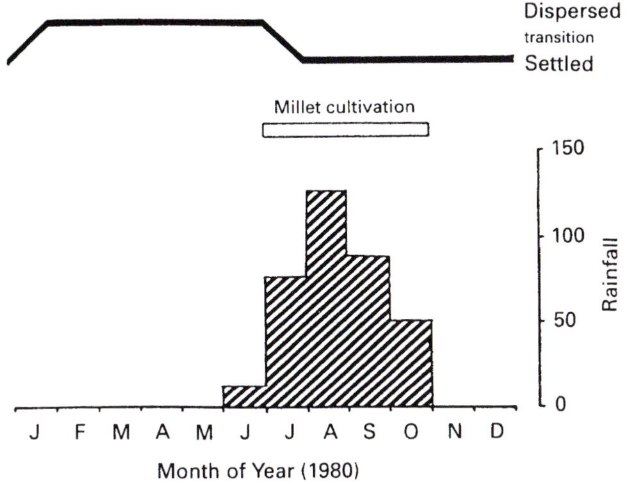


**The seasonal cycle of agropastoralism practised by the Ferlo of Senegal.**

*Effects of climatic seasonality on infant and maternal nutritional status according to stage in pregnancy and lactation at which the wet season is experienced*

| Period | Age of child | Mother | Child |
|---|---|---|---|
| 1 | −3 to 0 | Poor pregnancy outcome: proximate variables: poor nutrition, high work load, infection | Survival, birth-weight |
| 2 | 0 to 3 | Reduced lactation on demand: proximate variable: high work load | Growth |
| 3 | 3 to 6 | Constraints to lactation on demand and/or inadequate supplementary feeding | Infection, growth |
| 4 | 6 to 9 | Constraints to lactation on demand and/or inadequate supplementary feeding | Infection, growth |

*Seasonal variation in the nutritional status of children below the age of 5 years*

| Country | Group | Subsistence type | Age group | Anthropometric | Malnourished (%), by season | | | | Source |
|---|---|---|---|---|---|---|---|---|---|
| | | | | | dry | dry | dry | wet | |
| Niger | Wodaabe | Pastoralism | 0–5 | < 80% W/A | 5 | 6 | 15 | 8 | Loutan & Lamotte, 1984 |
| Senegal | Ferlo | Agropastoralism | 1–5 | % of standard W/H | 95 | 94 | 94 | 91 | Benefice et al., 1984 |
| Bangladesh | | Rice cultivation | 1–5 | Weight as Z-scores below | male female | | 2 | 3 | Becker et al., 1986 |
| Mexico | | Periurban | 0–2 | % of standard W/H | | | 88 | 84 | Sepulveda et al., 1988 |
| Gambia | | Urban | 0.5–3 | < −2 SD W/H | | | 4.5 | 8.5 | Tomkins et al., 1986 |

W/A: weight for age; W/H: weight for height; SD: standard deviation

*Seasonality of infection in West Africa (From Tomkins, 1993)*

| Dry season | Wet season |
|---|---|
| Measles | Malaria |
| Meningitis | Typhoid |
| Pneumonia | Diarrhea |
| Tetanus | Guinea-worm |
| Scabies | PEM/infection syndromes |
| Vitamin A deficiency | Anemia |

*Factors influencing seasonality of infection – dysentery (From Tomkins, 1993)*

| Variable | Host/Pathogen | Vector |
|---|---|---|
| Climate | + | + |
| Land pressure | – | + |
| Urbanization | – | + |
| Chemotherapy | + | – |
| Vaccines | + | – |
| HIV | + | – |
| Vector avoidance | + | + |
| Nutrition | + | – |

*Factors influencing seasonality of infection – malaria (From Tomkins, 1993)*

| Variable | Host/Pathogen | Vector |
|---|---|---|
| Climate | + | + |
| Land pressure | – | + |
| Urbanization | – | + |
| Chemotherapy | + | – |
| Vaccines | – | – |
| HIV | – | – |
| Vector avoidance | – | + |
| Nutrition | + | – |

occurs in different seasons. Both groups suffer food shortages toward the end of the dry season and throughout the wet, and the incidence of disease is high for both groups during the wet season. Malaria is hyperendemic among the Ferlo at this time, while the frequency of episodes of diarrhea amongst the Wodaabe rises from 13 per cent in the dry season to 47 per cent in the wet. These different seasonal patterns may be related to small differences in subsistence practice. The Ferlo live in social units of about 50 individuals, and they are sedentary during the rainy season as they cultivate millet at this time. During the dry season they pursue their nomadic activities, either together or in household units of 8 to 10 people. In contrast, the Wodaabe pursue their nomadic activities throughout the year in family groups of about seven individuals. Aggregation of the Ferlo during the wet season may allow seasonal transmission of malaria, precisely at the time of year when food shortages arise. Associated with this is growth performance, which is poorest in the wet season. For the Wodaabe, the poorer growth experienced by children during the dry season may be due more to seasonal food shortages, and less to infectious disease.

In the Gambia, both urban and rural children show a decline in weight velocity in the wet season. The decline in weight velocity is greater in the rural children, however, and is associated with food shortages, diarrheal and malarial morbidity. In the urban children, the effect of diarrhea on growth does not vary seasonally, but the age-adjusted impact of lower respiratory tract infection does. The reduced seasonal impact of diarrheal disease on growth has been attributed to the better nutritional status of the urban children. In older children, seasonality of nutritional status is more likely to be due to food shortages than to infectious disease. Billewicz and McGregor (1982) have shown that the effect of season on growth in height in Gambian children extends into adolescence.

Women in rural Gambia show differences in energy balance, as estimated from measures of body composition, at all stages of pregnancy, between wet and dry seasons. One result of this difference in energy balance is that birth-weights are about 200–300 grams lower in the wet season than in the dry. High seasonal mortality in early infancy may be related to low birth-weight (less than 2.5 kg) (**8.1**). Although the majority of low birth-weight births may be attributable to the lower energy intakes experienced by women passing their third trimester of pregnancy during the wet season. Infection by

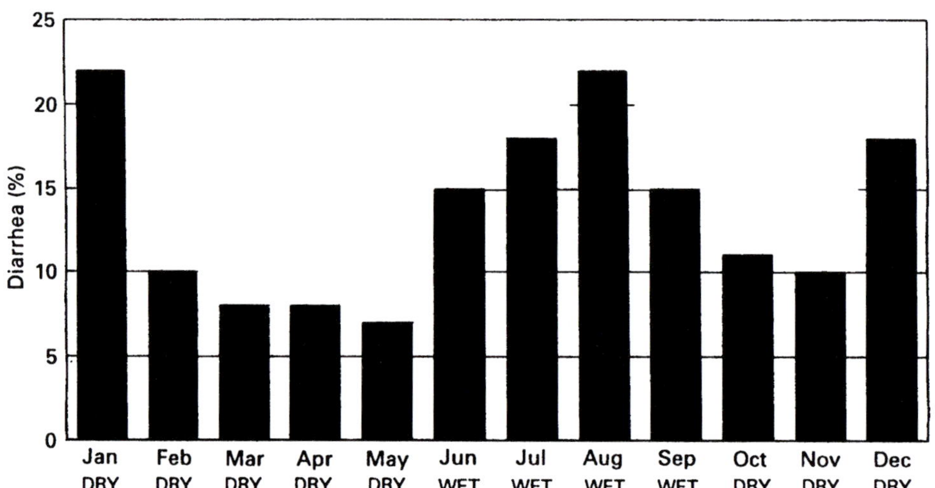


**Diarrhea prevalence among children aged 6 to 12 months of age in the Gambia, West Africa (From Tomkins, 1993).**

malaria is also important, particularly among primiparous women. In rural Gambia, primiparas represent 18 per cent of all births, but about 30 per cent of them are affected by malarial infection *in utero*, as assessed by placental pigmentation. First-born infants with pigmented placentas due to malarial infection have mean birth-weights of 2.6 kilograms, compared to 3.15 kilograms for unaffected babies. The influence of maternal undernutrition on fetal growth may have long-term consequences for the growth and development of the child by way of imprinting. In particular, the potential for catch-up growth associated with improved nutritional conditions and/or reduced exposure to infectious disease typical of seasonal growth in young children may be reduced. Although growth-rates of Gambian children born at different times of year have been compared by various authors, these are not able to determine the catch-up potential (**9.7**) of different seasonal birth cohorts, since the measurements are made in the suboptimal environment of rural Gambia.

The observed seasonal pattern of growth does not reveal seasonal patterns of survivorship associated with different seasonal birth cohorts. For example, Gambian children born late in the wet season have the lowest birth-weights and the highest mortality rates in the first 3 months of life of all rural Gambian children. Conversely, children born at the end of the dry season have the highest birth-weights and greatest survivorship. Breast-feeding (**9.1**) is likely to shield this birth group from excessive exposure to diarrheal diseases for most of the wet season, the introduction of weaning foods taking place when the worst of the wet season is over. Thus the ways in which the combined influences of undernutrition and infection on growth and development are played out in seasonal environments are complex.

*Stanley Ulijaszek*

See also 'Growth cyclicities and pulsatilities' (**5.13**), 'Infection' (**9.3**), 'Catch-up growth in height' (**9.6**) and 'Catch-up weight-gain' (**9.7**)

## GROWTH AMONG NEPALI AGRICULTURALISTS

Nepal is a difficult and highly seasonal environment where most of the population derives subsistence from agriculture. It is one of the poorest countries of the world, with very high child mortality rates. During the monsoon season, food supplies are at their lowest, work loads of adults are at their highest, and the transmission of infectious diseases (**9.3**), diarrhea in particular, is at its height. In a self-sufficient population of central Nepal, children aged 0 to 4 years show stunting after the age of one year, but little wasting, a profile typical of many developing countries. Monthly increments in weight are highest in the winter months and lowest during the monsoon (June–August). Height increments are low after the monsoon.

The prevalence of infection, especially diarrhea, is sixfold greater in the monsoon than in winter. Gains in weight and height were lower for children having between two and five illnesses at the time of measurement, versus 0–1 illnesses. Two- and three-year-olds bear the brunt of ill-health. Death-rates also peak in the monsoon.

The agricultural workloads of village women significantly increase during the monsoon. But because 0–2 year olds invariably accompany the mother, nursing times show no seasonality. Older children are increasingly left behind, but by then are in a good position to survive. Women's work *per se* does not seem to prejudice small children. Rather, growth-rates are affected by recurrent illnesses in the monsoon and poor quality weaning foods. In turn, extensive and early stunting leads to very short stature among Nepali adults.

*C. Panter-Brick*

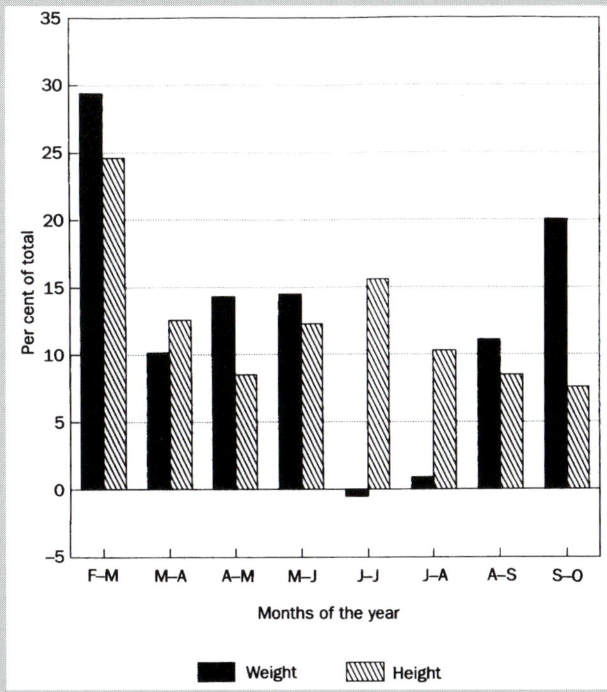


**Monthly increments in weight and height as a percentage of total observed gain (Repeated measures over 9 months for 70 children).**

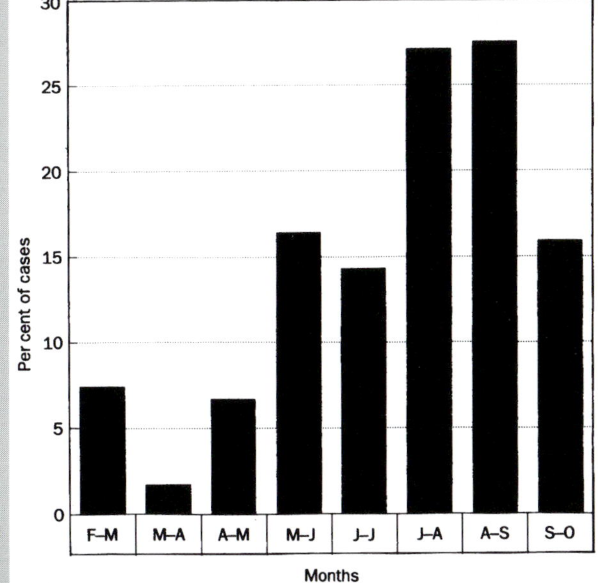


**Prevalence of illnesses reported by mothers at the time of weighing children (% of 572 cases).**

## GROWTH OF AFRICAN PASTORALIST CHILDREN

Pastoralism is a pattern of subsistence in which livestock are kept and managed and where their products (milk, meat, blood, hides) are used for food, materials, and trade. African savanna lands are inhabited largely by pastoralists whose history is obscure but who are identified often as of Nilotic origin. Pastoral patterns of subsistence are variable, but there is some similarity from the Bedouin, Berber, and Tuareg from North Africa to the Fulani people in West Africa to the multiple ethnic groups of East Africa. Savanna lands are less congenial for cultivation, but these will support livestock, albeit with considerable risk to their survival. The risks arise from patterns of savanna drought, including seasonal and longer-term drought, and the limited vegetation for livestock, particularly during the long annual dry seasons. This leads to a need for mobility in pastoralists in order to move their animals in the search for available vegetation and water.

There are several interesting research problems that arise from these pastoral patterns. First, food supply tends to be highly seasonal as rainfall produces vegetation that feeds livestock and consequent livestock products feed people. Food (particularly milk) is more abundant following the annual wet season, and scarce during the usually long dry season. Famine may ensue if a rainy season fails or long-term drought occurs. Fluctuating food availability can have a profound influence on child growth, disease status (**9.3**), and reproduction (**5.15**), and produce corresponding cycles of body composition and health status in adults and children alike. Second, because animal products make up a substantial portion of the diets of pastoralists, most of these peoples have both high-protein and high-fat food intakes, but comparatively low energy intakes (**9.2**). These nutritional conditions are likely to influence growth processes but in different ways than seasonality. Third, a special concern for nomadic pastoralists is the balance struck between livestock numbers and the human labour (**9.10**) needed to manage the herds. One of the solutions to this problem for African pastoralists is polygyny. When herd owners have several wives, fertility is maximized and labour resources are derived from the children of these polygynous family units. Polygyny influences both male and female fertility and is almost certain to influence child growth through complex behavioural patterns associated with polygynous households. Fourth, interactions between nutritional status and disease status tend to be different in nomadic pastoralists in contrast to cultivators. Breast-feeding, supplementation, and weaning practices in infants tend to be unusual because pastoralists have access to animal milk for supplementation. These practices (early or late supplementation; early or late weaning) will influence both infant growth and maternal fecundity, the latter will feed back on the growth of later-parity infants. Finally, other questions have been raised about the growth of children raised under hot, dry conditions and why African pastoralist physiques are markedly linear. An explanation for this phenomenon probably lies with study of the holistic picture of nutrition, infection, growth, and climate in the context of infant, child, and adolescent development.

*Michael A. Little*

**Nomadic Turkana children of north-west Kenya queuing up to be measured for a growth study. Photograph by Michael Little.**

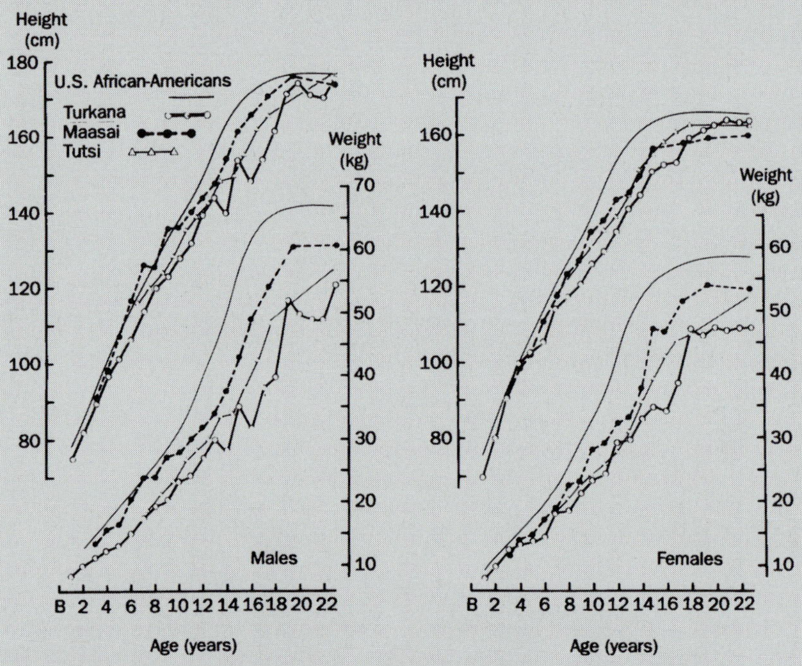


**The growth of children, adolescents, and young adults from three East African populations compared with the growth of African-Americans from the United States.**

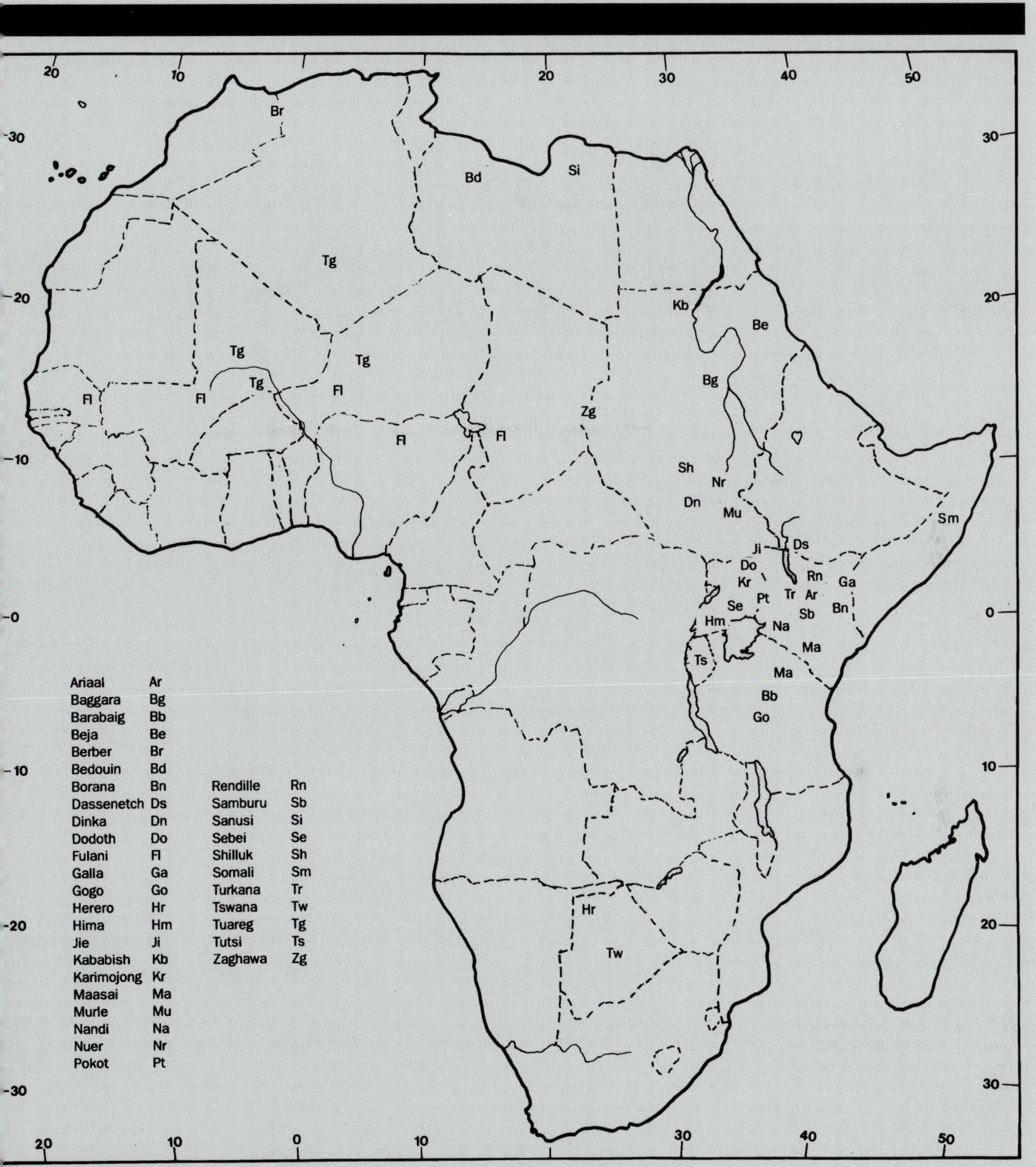


A map of Africa with identification of ethnic pastoral populations. The list is not comprehensive, but includes those peoples who have been studied by ethnographers. Studies of child growth have been conducted on only a handful of these populations

# Growth in high-altitude populations

High-altitude populations are characterized on the basis of their adaptations primarily to hypoxic stress due to exposure to low partial pressure of oxygen relative to sea-level values. Such populations are to be found predominantly in the Andes of South America, the Himalayas, the Tibetan plateau, and the high plains of Ethiopia. However, hypoxia is not the only stress, others being poor nutritional availability, and the cold, respectively. Slower growth and smaller adult body-size relative to lowland populations and Western norms are among the various adaptations to high altitude observed.

Growth faltering starts *in utero*. Mean birth-weight is negatively correlated with altitude, while the lower birth-weight of Andean infants is associated with high placental weight, suggesting that at least in part, this is due to poor oxygen delivery to the fetus. Mean birth-weight varies inversely with altitude, with mean birth-weights on average 400 grams lower at 3500 metres than at sea-level. Fetal growth retardation at high altitude is believed to take place in the final 6–8 weeks of gestation. Since growth retardation due to nutritional stress takes place during the final 13 weeks of gestation, and many high-altitude populations are exposed to nutritional stress, the influence of maternal undernutrition on birth-weight cannot be completely discounted.

Relative to lowland populations, post-natal growth is slow and prolonged in stature, but with disproportionately faster growth in chest-size. This pattern of growth is associated with a constant increase in total lung capacity and other measures of pulmonary function relative to lowlanders. Although this pattern of slow and prolonged growth has been interpreted as an adaptive response to atmospheric hypoxia, the magnitude of this effect varies considerably from study to study. Such variability is probably due to confounding of comparisons by factors other than the partial concentration of oxygen, especially socio-economic conditions (**11.4**). That such factors can mask the effects of hypoxia on growth has been demonstrated in both Ethiopian and Peruvian populations, where high-altitude children were shown to be taller and less healthy than low-altitude children. Studies of high socio-economic status children living at high altitude in Bolivia support the view that the limiting effect of hypoxia on linear growth is small relative to the potential effect of other factors, especially low dietary availability and poor nutritional status.

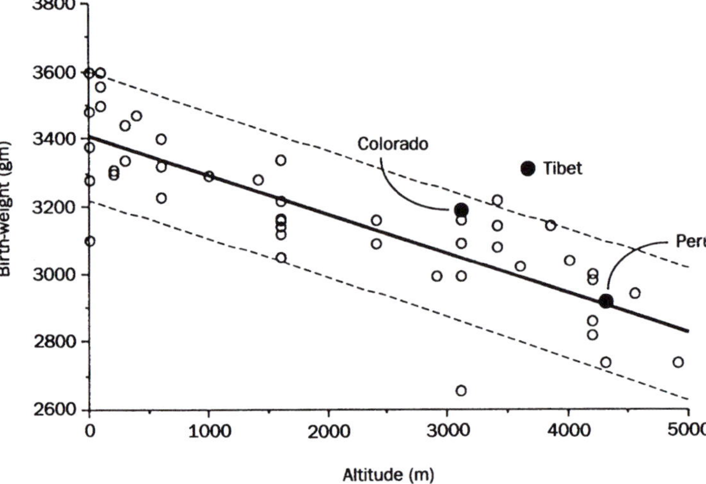


**Mean birth-weights reported from previously published studies in North and South America (open circles) are shown with the upper and lower 90% confidence limits (dotted lines). The average birth-weights from the Colorado and Peru samples (filled circles) fall within the 90% confidence limits, whereas the Tibet birth-weights are higher than previously reported values.**

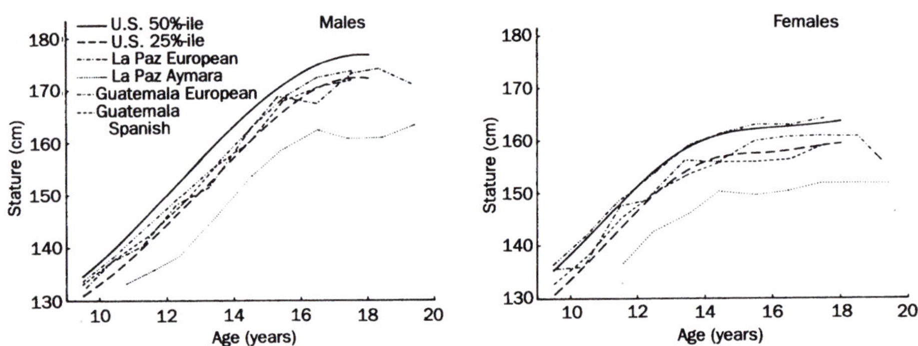


**Stature for age in selected high- and low-altitude populations.**

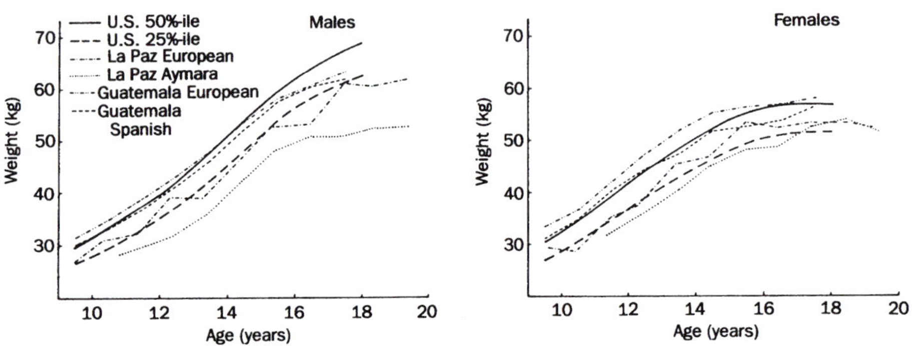


**Weight-for-age in selected high- and low- altitude populations.**

The distinctive pattern of slow growth in stature, faster growth in chest-size is probably established by the age of 9 years and then maintained, rather than accentuated to any great degree, in the later developmental period.

Regardless of the primary source of growth-retarding stress, slow growth is associated with delayed skeletal maturation and a delay in the age at which epiphyseal fusion takes place. Growth in height continues until the age of 20 years, and sometimes a little beyond that, compared with 18 years among healthy well-nourished low-altitude populations. Sexual maturation is also delayed. In Himalayan populations, age at menarche varies between 15.2 and 18.4 years at altitudes between 1500 and 4050 metres, and 14.4 and 16.1 years at altitudes between 330 and 1400 metres. In Andean populations living between 3600 and 4300 metres, age at menarche varies between 13.6 and 13.7 years, compared to 11.6 years among low-altitude populations. Although later menarche is observed in Himalayan relative to Andean populations, high-altitude groups show delay relative to lowland groups. But again, the importance of hypoxia to delayed skeletal and sexual maturation relative to nutritional stress is probably low.

*Stanley Ulijaszek*

See also 'Fetal growth retardation' (**4.8**), 'Sexual maturation' (**5.15**), 'Adulthood and developmental maturity' (**5.17**), 'Nutrition' (**9.2**), 'Catch-up growth in height' (**9.6**), 'Growth in infancy and pre-adolescence' (**10.4**), 'Between-population differences in adolescent growth' (**10.5**) and 'Body-proportion differences' (**10.6**)

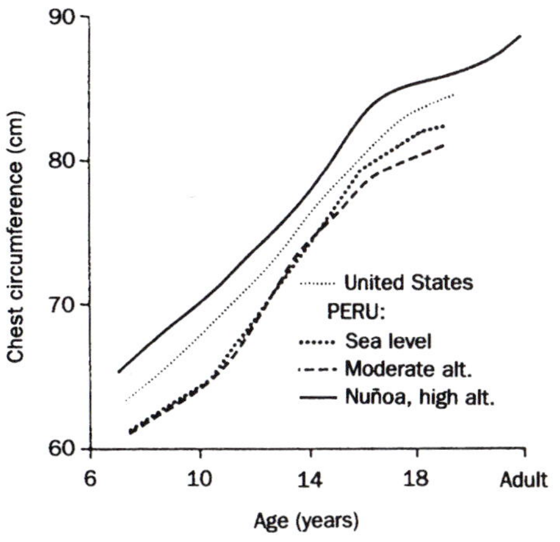


**Comparison of chest circumference among Peruvian children and adults from sea-level, moderate altitude, and high altitude. Highland Quechuas from Nunoa exhibit accelerated growth in chest-size. (From A. R. Frisancho and P. T. Baker. 1970. *American Journal of Physical Anthropology* 32:279–92.)**

# 9.10

# A functional outcome in adulthood of undernutrition during growth

Productivity in heavy physical work of individuals (e.g., sugar-cane-cutting or loading, logging industry) has been shown to be positively correlated with their maximum oxygen consumption ($\dot{V}O_2$max). In the first figure, the $\dot{V}O_2$max (l/min) and aerobic capacity (ml/min/kg body-weight) are shown in three groups of sugar-cane cutters of differing productivity (tons cut per day) together with their correlations with productivity. Both expressions of $\dot{V}O_2$max were significantly correlated with the tonnage of sugar-cane cut per day. In relating various anthropometric measures and age to productivity, a stepwise multiple regression demonstrated that $\dot{V}O_2$max (l/min), percentage body fat and height (cm) contributed significantly to the variation in productivity such that:

$$\text{Productivity} = 0.81 \times \dot{V}O_2\text{max} \times 0.14\%\ \text{fat} + 0.03 \times \text{height} - 1.962$$

$r = 0.685;\ p < 0.001$

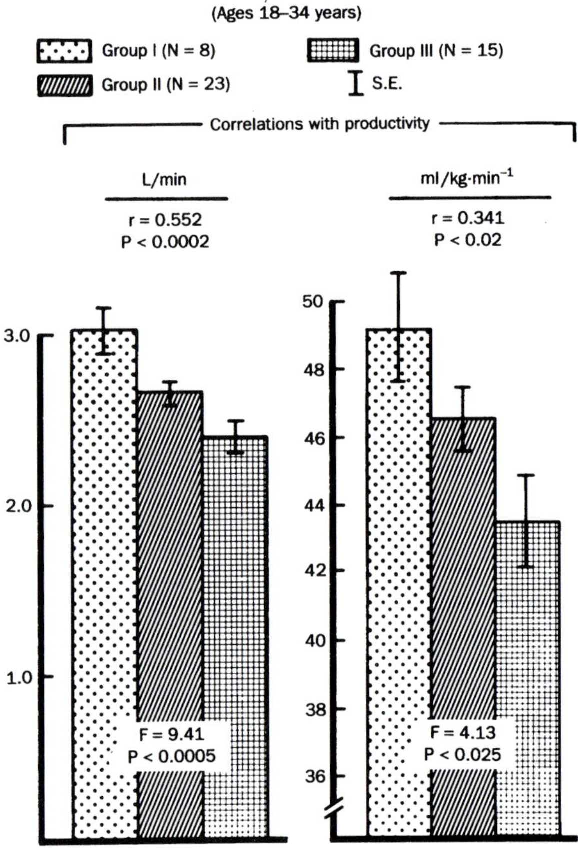


**$\dot{V}O_2$ max (l/min) and maximum aerobic power (ml/min/kg body-weight) of good (Group I), average (Group II) and poor (Group III) sugar-cane cutters classified according to their daily productivity. F-Ratios are from one-way analysis of variance.**

This equation states simply that those who are in poor physical condition (low $\dot{V}O_2$max) or are stunted because of past undernutrition (**9.2**) are at a disadvantage in terms of ability to produce in cutting sugar-cane. The negative coefficient for fat indicates some advantage for having low percentage body-fat content.

The $\dot{V}O_2$max, in turn, is related to body-size as represented by height and lean body mass (LBM) (**5.11**). Similar relationships exist for girls and women. Higher correlations in children than in adults is probably due to the wider range of body-sizes in children.

Nutritional status, body composition and $\dot{V}O_2$max are also positively correlated due to the reduced muscle mass which occurs in malnourished adults. The second figure shows the expression of $\dot{V}O_2$max in terms of body composition data for three groups of malnourished men classified as mild (M), intermediate (I) and severe (S) in the degree of their nutritional deprivation and the dietary repletion of the latter group during about 4 months. Over 80 per cent of the difference in $\dot{V}O_2$max between M and S subjects is accounted for by the difference in muscle cell mass (MCM). The remaining difference might be ascribed to reduced capacity for oxygen transport either because of low blood hemoglobin or reduced maximum cardiac output. Consequently, it can be deduced that poor nutritional status will be accompanied by reduced

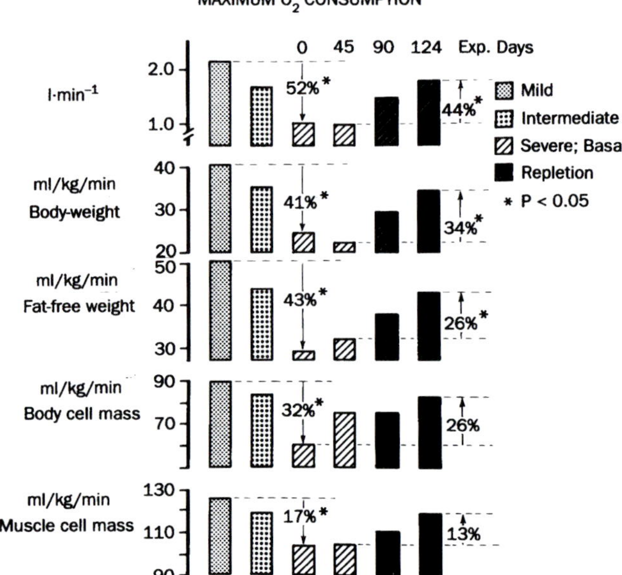


**$\dot{V}O_2$ max (l/min) expressed in terms of various body compartments in three groups of undernourished subjects with mild (M), intermediate (I) and severe (S) nutritional deficits, and during the dietary repletion of the severely malnourished group.**

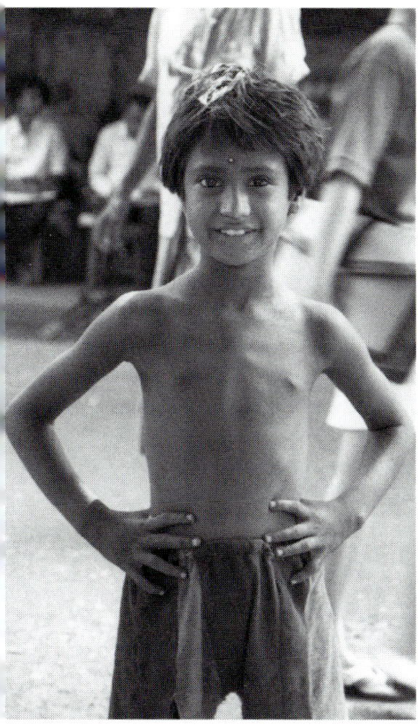

**Children in India are often required to work quite hard – this fit individual is small for age, but willing to work hard. Photograph by Patricia van Someren.**

productivity in heavy physical work. This relationship has not been measured directly in malnourished men because of physical debilitation and the fact that unemployment is usually one of the primary reasons for the poor nutritional status.

The effects of chronic undernutrition on the growth of nutritionally normal and marginally undernourished Colombian boys 6 to 8, 10 to 12 and 14 to 16 years of age are presented in panels A to C of the third figure and compared with the values for nutritionally normal Colombian men (C) and those with mild (M), intermediate (I) and severe (S) malnutrition described in the second figure. Also shown are data for 10 nutritionally normal North American men and the 50th percentile values for height and weight of the United States National Center for Health Statistics. Panels D to F show values for $\dot{V}O_2$max of these same subjects (expressed as l/min, ml/min/kg of body-weight and of LBM). All of the data in this figure are from various studies in the same laboratory. It can be seen that growth in height, body-weight and LBM are progressively and significantly attenuated in the undernourished children and that this is reflected in the significant depression of $\dot{V}O_2$max throughout the age range studied when expressed per kilogram body-weight. With the exception of the youngest age group, the undernourished boys had higher aerobic capacities than normal boys, which was thought to be the result of higher LBM per unit of body-weight due to lower body-fat content in the nutritionally deprived boys. However, subsequent studies demonstrated that even when expressed in terms of LBM, the aerobic capacity of the undernourished children is higher than normal boys in the older age groups, indicating better physical condition.

All the Colombian subjects in the third figure were recruited from lower socio-economic groups and, when the adults are compared with North American men, show the results in adulthood of chronic undernutrition during growth and of various grades of malnutrition on the $\dot{V}O_2$max of men. Since the sugar-cane cutters of the first figure were working at about 35 per cent of their $\dot{V}O_2$max during the 8 hour work day, which is close to the maximum that can be sustained for this period of time, the conclusion is inescapable that undernourished children become smaller adults who will have lower $\dot{V}O_2$max and therefore, reduced productivity in heavy physical work.

G. B. Spurr

**Body-size, composition and maximum work capacity (measured by the $\dot{V}O_2$ max) of nutritionally normal and marginally malnourished Colombian boys 6–16 years of age, and adult agricultural workers classified as nutritionally normal control subjects (C) or mild (M), intermediate (I) and severe (S) in the degree of their nutritional deprivation. Data from 10 North American men and the NCHS 50th percentiles for height and weight are also shown.**

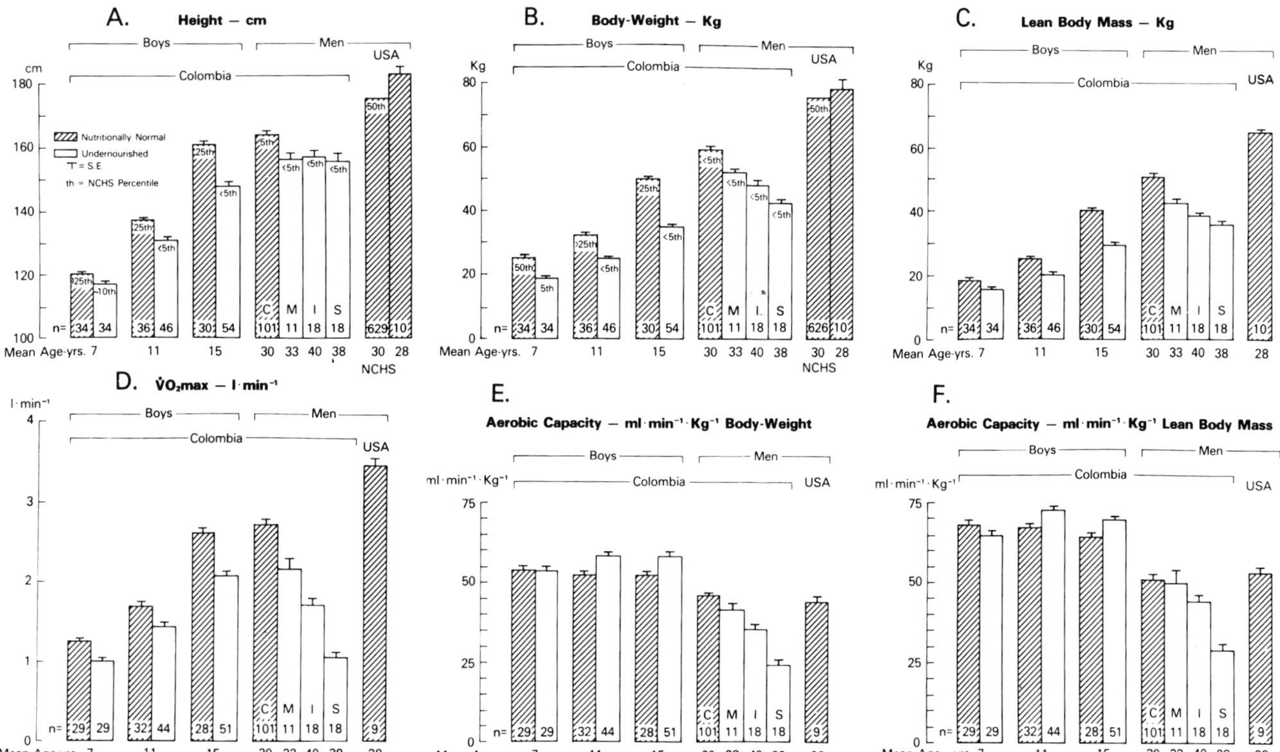

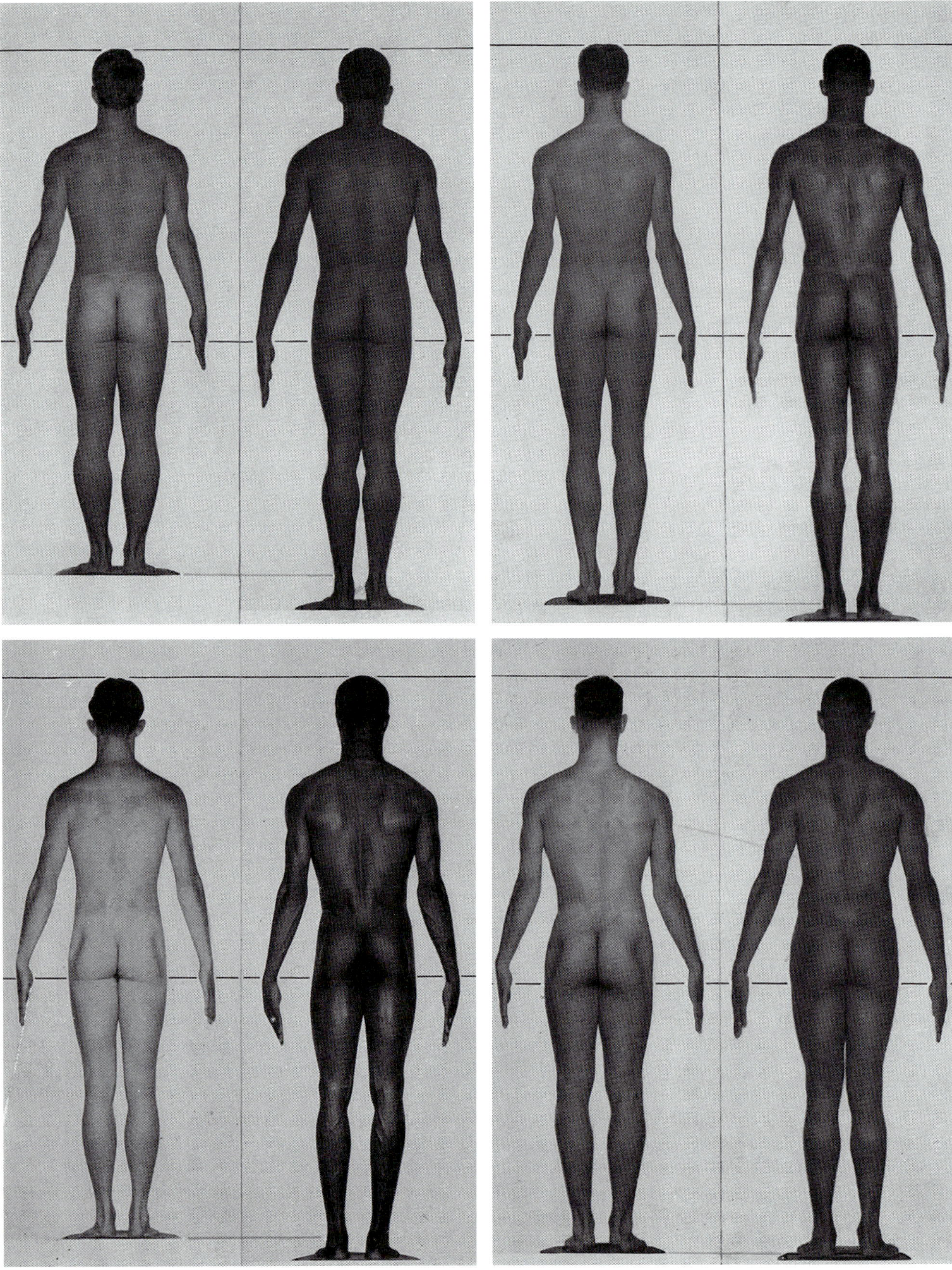

**Part Ten**

# Between-population differences in human growth

**Opposite: A comparison of European and African physiques, showing the comparatively longer legs of the Africans. Courtesy of James Tanner.**

Although between-population differences in aspects of adult human body-size, shape and form were documented prior to the 20th century, systematic compilations of growth data from different parts of the world have only been made more recently. The most notable publications in this field are Eveleth and Tanner's *Worldwide Variation in Human Growth*, volumes 1 and 2. The efforts that gave rise to these rich sources of human growth data came with the International Biological Programme Human Adaptability studies (IBP/HA), instigated in 1966.

The aims of the IBP/HA were several, but included the classification and documentation of human variation in a variety of traits, and the interactions of human populations with their

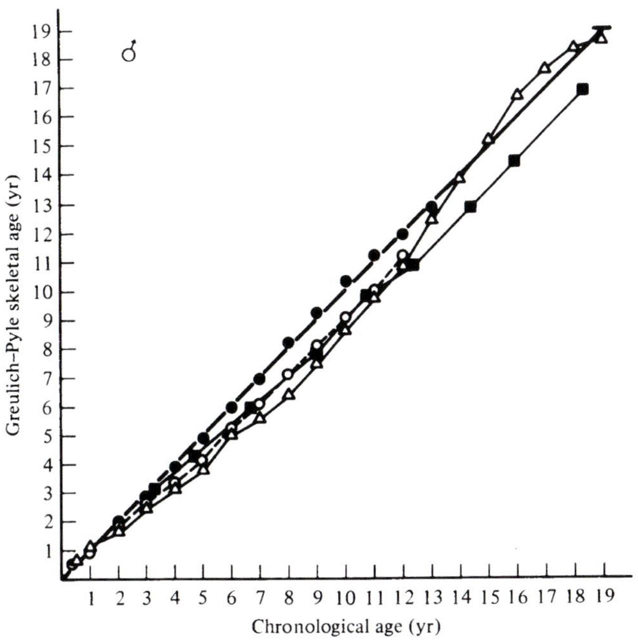


**Mean skeletal age of boys in reference to Greulich–Pyle standards (as assessed by the Greulich–Pyle method). Samples are Melbourne Australians (●, Roche, 1967), Hong Kong Chinese (△, Hang et al., 1967; Low et al., 1964), Zürich Swiss (○, Budliger & Prader, 1972) and Nuñoa Quechua Amerindians (■, Frisancho, 1969). The bold line represents skeletal age equal to chronological age.**

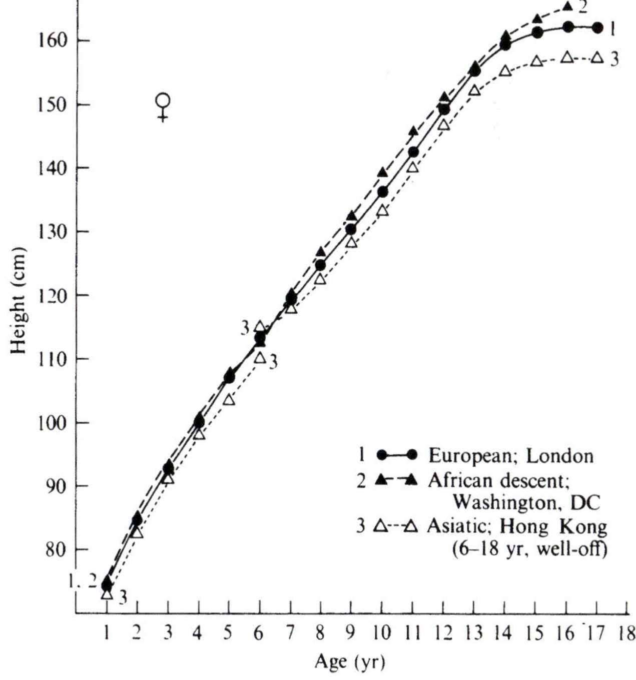


**Height means of European girls (London), Asiatic girls (Hong Kong) and girls of African origin (Washington, DC). From Eveleth, P.B. and Tanner, J.M. *Worldwide Variation in Human Growth*. Cambridge: Cambridge University Press, 1990.**

environments. This work created baseline measures for the study of more complex human–environment interactions. The methods used included genetic, morphometric, physiological and nutritional techniques among others. Although subsequent work was not as globally comprehensive as had been initially intended, there is little doubt that the IBP/HA was a beginning rather than an end; if one of the current concerns in the study of growth and development is with adaptation and change in contemporary populations, then much of the data collected at that time provide baselines from which to measure such change.

The study of body-size and shape is classically part of the remit of anthropology, featuring Franz Boas and Francis Galton among the earliest practitioners. Awareness of the plasticity of human size and shape as a consequence of phenotypic responses in the growth process to environmental factors came with the studies of European migrants to the United States made by Boas after the turn of the 20th century. By the time of the IBP/HA, it was appreciated that the study and understanding of variation in body-size and growth patterns provides important information about the nature and extent of human adaptability. Arising from this was the acceptance that modelling the human growth curve was an important technique for improving understanding of variation in growth patterns of populations both locally and internationally. This type of research has become a subdiscipline of human biology, as researchers now use this method to improve their understanding of human growth patterns, short-term fluctuations in growth-rates and the relative importance of environmental quality on variation in growth-rates. Such work has implications for human adaptability study since an understanding of growth plasticity in response to environmental stresses is reliant upon an understanding of growth patterns among individuals and populations in relatively unstressed environments.

It is difficult to determine the extent to which between-population differences can be attributed to genetic and environmental factors (**10.2**). The greatest differences are to be found between populations in industrialized and non-industrialized nations, and between well-off and poorer groups within countries. Although genetic factors cannot be discounted, such differences can largely be attributed to differences in environmental quality experienced, influencing growth largely through differentials in nutritional well-being and exposure

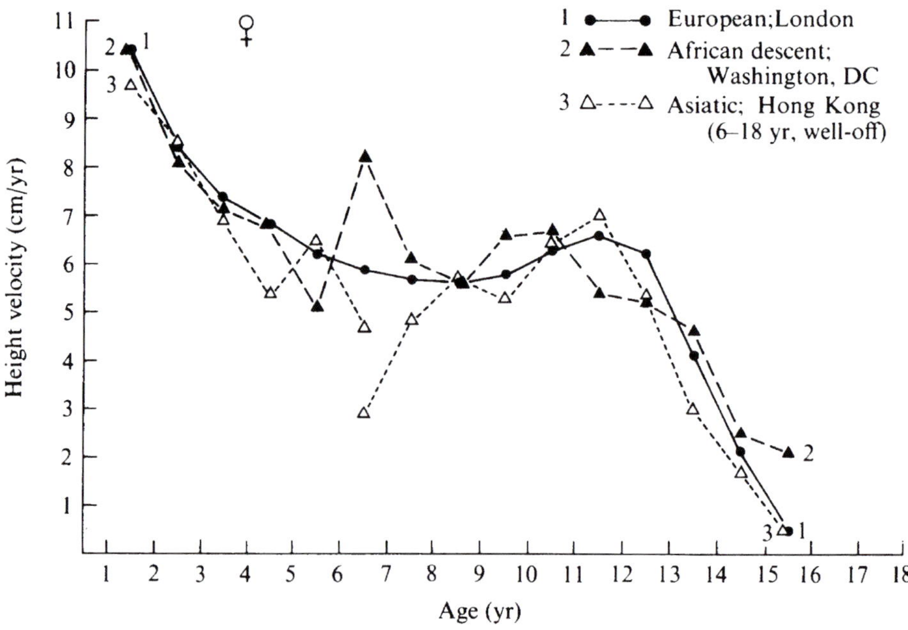


**Height velocities (derived from cross-sectional means) of European girls (London), Asiatic girls (Hong Kong) and girls of African origin (Washington, DC). From Eveleth, P.B. and Tanner, J.M. *Worldwide Variation in Human Growth*. Cambridge: Cambridge University Press, 1990.**

to, and treatment of, infectious disease. Growth patterns in well-off populations and groups of high socio-economic status are less heterogeneous than those of lower socio-economic status. However, differences between major global population groupings may still exist after socio-economic factors have been controlled for, possibly bringing into question the validity of the concept of an international reference for the growth of children.

Processes which have taken place in the past have shaped the human populations found in the world today. The most fundamental process has been genetic, namely the shaping of human gene-pools through natural selection (**10.1**). In addition, human population migration has served to isolate different groups across sufficiently long time-spans to allow between-population genetic variation. Knowledge of the pattern of human dispersal gives us clues about how human populations might best be classified: 1) since evidence points to humans having a common point of origin, with dispersion only coming late in evolutionary time, human populations, regardless of where they live, have tremendous genetic similarity; 2) the migratory route suggests that variation in any trait is likely to be a continuum across populations, and that any classification is to some extent arbitrary; 3) prior to the onset of agriculture, some 12,000 years ago, humans lived as hunter-gatherers at low population densities often in isolation from each other, leading to the possibility of the development of regional population genetic differences.

Migrations taking place after the onset of agriculture served to create larger, more genetically homogeneous populations across wide areas, with genetically isolated populations left in less hospitable ecological niches. Genetically isolated populations to be found in the world today include tribal groups of hunter-gatherers in Africa, Latin America, and Asia. Large-scale colonization of the Pacific Islands took place at this time. Later migrations, during colonial times, include the migrations of 1) Europeans to the Americas, Australasia and parts of Africa; 2) Africans, mostly of Bantu origin, to the Americas and the Caribbean; and 3) Asiatics, Chinese in particular, and Indo-Mediterraneans, largely South Asians, to most parts of the tropical world and to parts of the New World. Migrations in the post-colonial period are largely related to economics and urbanization. Examples include the migrations of Mexicans and Hispanics to

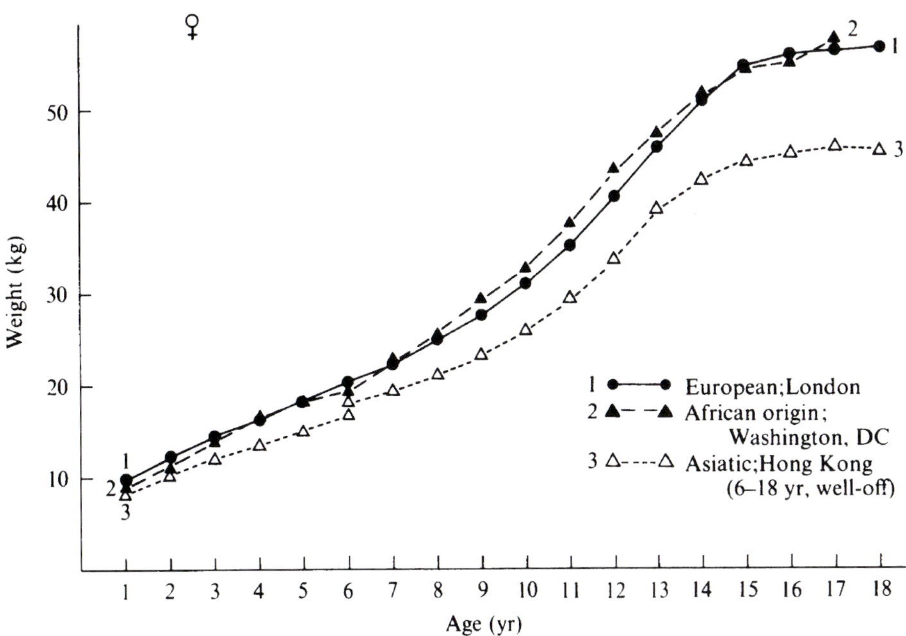


**Height velocities (derived from cross-sectional means) of European boys (London), Asiatic boys (Hong Kong) and boys of African origin (Washington, DC). From Eveleth, P.B. and Tanner, J.M. *Worldwide Variation in Human Growth*. Cambridge: Cambridge University Press, 1990.**

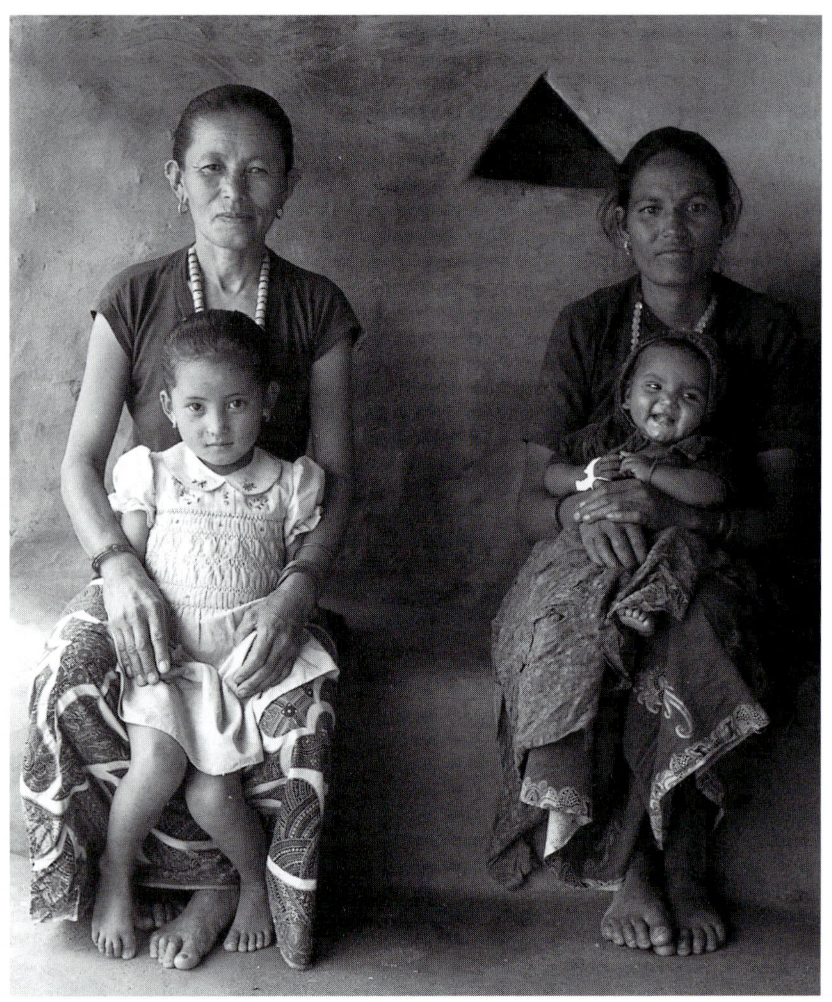

**There are clear morphological differences in different populations as, for example, in the two mother-child pairs above, one of Tibeto-Burman origin (left) and the other, Indo-Aryan (right), both from rural Nepal. Photograph by Stanley Ulijaszek.**

America, and of South Asians and Afro-Caribbeans to Britain. In addition, there is the global trend of rural-to-urban migration in post-colonial times. Thus, human populations have great homogeneity for most genetically determined characteristics, between-population variation having taken place in the recent evolutionary past. Migrations and population expansions have created several larger population groupings within which genetically isolated groups may be found.

Relative to other species, humans grow very slowly and reach maturity. The greater body-size across taxa and evolutionary time has been taken to mean that greater size confers advantages which have been selected for. With respect to humans, the childhood phase of growth, which is slow, has been interpreted to be a reproductive strategy that distributes parental investment over time, and a survival strategy that reduces the levels of nutrient intakes needed for growth. Although the selective effects on human body-size are difficult to evaluate, there are clear limits to human body-size which are easy to observe. However, the importance of genetic factors in the different growth patterns of various populations is difficult to establish in all but the most extreme cases, including the African pygmies.

Most of the between-population variation in body-size at birth (**10.3**) can be attributed to maternal environmental factors, although studies attempting to control confounding environmental differences still find residual differences between different populations. Notably, there are differences in birth-weight between European and African-American infants. Post-natal growth-rates then vary, with populations with lower mean birth-weights showing greater growth-rates in the first months of life than populations with higher mean birth-weights. In developing countries, intra-uterine reduction in growth leading to lower birth-weight is probably compensated for by a burst of catch-up growth once in the robust phase of early breast-feeding. Although between-population differences in growth in childhood (**10.4**) can largely be accounted for by differences in environmental quality, the extent to which this is the case for adolescent growth is not clear, although there are strong hereditary effects on maturation (**10.5**). Conversely, the genetic contribution to population differences in body proportion is probably less than was previously thought to be the case, although there are clear differences, for example in limb length between European and African-origin populations, that are likely to be under strong genetic influence (**10.6**).

*Stanley Ulijaszek*

# Growth and natural selection

Growth patterns in humans reflect evolved responses to adaptive problems, adaptations produced through natural selection, or differential reproductive success (fitness). Recent advances in life-history theory (**2.2**), or the comparative analysis of life-cycle parameters, have illuminated the role of growth in life-history evolution. Growth is a function of productivity, or a species' ability to achieve food intakes and nutrient availability above the maintenance demands of the body. Maintenance includes ordinary energy costs of metabolic turnover, as well as those for body repair. Productivity among primates averages only 40 per cent that of other mammals, possibly due to higher maintenance costs because they have larger, more energetically expensive brains, are long-lived and necessarily spend more energy on repair, and incur higher behavioural costs for sociality. Unsurprisingly, primates grow relatively slowly and mature late, humans especially so. Growth-rates interact with other decisive factors in life-history: adult mortality rates determine adult life expectancy, and life expectancy along with selective pressure for fitness maximization determine the age at reproductive maturation. In turn, age at maturation compounded by productivity-determined growth-rates, control adult size. A pervasive cross-taxonomic trend to increased body-size with evolutionary time, a trend represented among primates by humans and great apes, suggests that size confers advantage. The distinctive 'childhood' phase in human life-history (**2.3**), with its characteristic slow, and slowing, growth-rates may be both a reproductive strategy that distributes parental investment over time, and a juvenile survival strategy that reduces rate of energy intake required for growth.

The juvenile period is subject to particularly intense selective pressure because pre-reproductive mortality yields null individual fitness, barring inclusive fitness; hence, traits that influence survival to reproduction are strongly selected. Facultative adaptation, or fitness-enhancing phenotypic plasticity expressed as a function of conditions encountered by the individual, is an important component of ontogeny. Indeed, the insight that developmental biological processes are to a significant extent designed to rely on inputs from the environment, has stimulated many recent advances in developmental biology. A norm of reaction is the range of phenotypic outcomes from a specific (species) genotype manifested under different circumstances, but it can be difficult to discriminate between an adaptive response and tolerable impairment. Whether variation in human growth-rates and adult height in relation to environmental quality (especially nutrition) constitutes environmental insult or facultative adaptation is subject to intense debate. On the one hand, growth-slowing and short adult stature reduce caloric needs and allow toleration of nutritional shortfalls. On the other, stunting, or growth deficits, associate with long-term impairments in work capacity (**9.10**), health, and psychosocial function, and the inequities that cause stunting are uniformly considered undesirable. Adaptations carry costs by definition and their desirability is not guaranteed by virtue of their being adaptive.

The close relationship of growth and nutrient availability, or nutrition, is 'hard-wired' in physiology: circulating levels of growth hormone and other growth factors closely follow blood-sugar levels, and several hormones regulating growth are also involved in metabolic process and regulation throughout life. Nutritional stress, psychological stress, and illness each result in acute growth-rate reduction, followed by rebound (catch-up growth) if conditions improve (**9.6**, **9.7**). Moreover, some data suggest that growth-rate correlates with mortality risk, and that rapidly growing, large-for-age children are more likely to suffer and die from acute malnutrition. Mortality rates, and thus the stringency of post-natal selection, are greatest in the newborn period, then in those under 2 years, then in the under-5s. It is perhaps significant that the ranking of mortality risk by age parallels that of post-natal growth-rates, which peak shortly after birth and fall briskly during the first 2 years.

Adaptive design integral to human life-history strategy is represented not only in overall growth, but also in growth trajectories of specific organ systems. Contrast the rapid growth of the brain and early hypertrophy of the immune system with the slow development of reproductive structures. Infants attain 60 per cent of adult brain-size by age 2 years, and over 80 per cent by 5 years of age, at only 30 and 40 per cent of adult stature, respectively. Early brain development supplies cognitive capacity to support the childhood tasks of learning and socialization. That immunological (lymphoid) structures in prepubertal children attain twice the mass present in adults reflects both the intense selective pressure to acquire immune competence to deal with survival-threatening infective agents, and the post-pubertal pattern of thymic reduction with increased decentralization of immune function. Virtual absence of post-natal growth of the reproductive organs until the second decade reflects biology of the human life-history strategy of prolonged childhood and delayed reproductive maturation. In life-history terms, during puberty children switch from using productivity for growth to its adult use for reproduction. The transition is biologically built in to puberty, in that rising hormone levels resulting from gonadal activity initially stimulate bone growth and propel the pubertal growth spurt, but then higher concentrations overstimulate and produce epiphyseal closure and the cessation of bone growth. Populations with strong developmental delay, reflected in mean menarcheal

## BODY-SIZE AND NATURAL SELECTION

Natural selection on body-size operates at all levels of biological organization. Multicellularity that gives the potential for larger body-size emerged on several independent occasions during evolution. At those simple levels the advantages are clear – a group of cells can feed more effectively than isolated single cells, it takes longer to starve to death and is more difficult for a predator to consume. A disadvantage is that as size increases reproductive-rate decreases.

Body-size in higher organisms is a more complex concept. To say that someone is a big man may mean that he is very tall, very fat or very heavy. In other words human size can be measured in terms of stature (a linear measurement), in terms of girth or surface area (a quadratic), or in terms of volume or mass (a cubic measurement). It is likely that selection will operate in different ways on each of these three measures. Sex dimorphism (**5.5**) means that there may be different selection pressures in the two sexes. Size in adulthood is not an isolated entity, but is last in a whole series of stages of size (i.e. growth) on each of which selection may be acting, and the nature of such selection may be quite different from that operating on final size. Selective effects on human size, therefore, are difficult to appraise.

The small or large size that occurs in some genetic disorders(**7.4**, **7.6**) is selected against because it is part of the deleterious clinical condition, not because of the size itself. Human size is generally regarded as a multifactorial character determined by both genetic and environmental influences. Its environmental components include nutrition, disease, diurnal rhythm and activities, postural habits as well as other less obvious factors. Its genetic component, controlling timing and rate of growth processes, is polygenic; there are many genes, each of small effect at loci widely distributed throughout the chromosomal array, acting with or against each other, so that selection cannot operate on individual genes but only on the combinations in which they occur. For stature, heritability (the proportion of the phenotypic variance due to the genotype) is high, the best estimates clustering round 80 per cent in Europe and 60 per cent in West Africa. The pattern of inheritance and the high heritability of stature mean that while there is ample genetic control which can be modified by evolutionary processes, the effectiveness of natural selection is greatly reduced by the number of loci and the buffering effect of the environmental factors. The genetic effectiveness of natural selection on adult weight is yet more attenuated.

Evidence of natural selection acting on size in humans is mostly indirect. Skeletal material over the last 3 million years shows that the earliest ancestor or near-ancestor of human beings, *Australopithecus* was a small animal. Different species varied widely in weight, the gracile forms averaging 36 kilograms, the robust 59 kilograms.

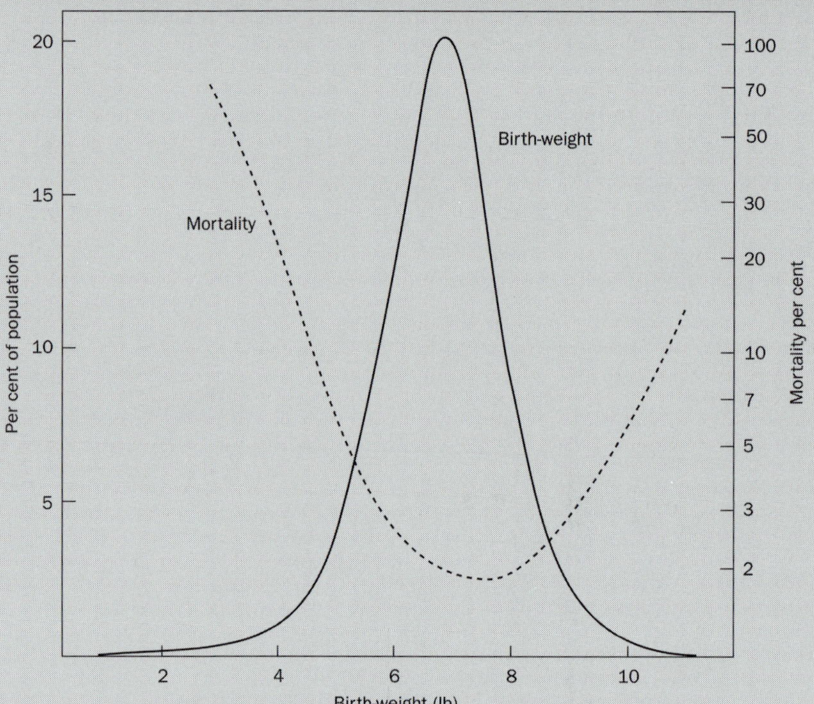


**Distribution of birth-weights correlated with mortality.**

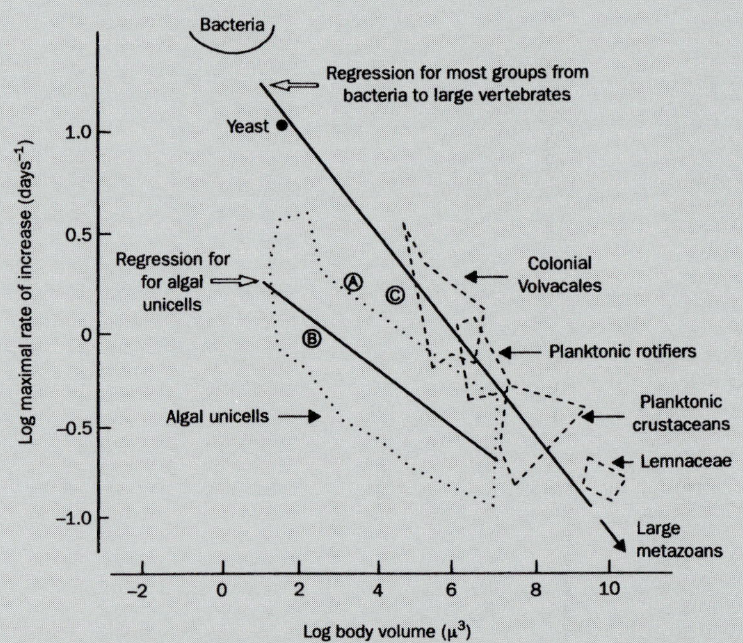


**The effect of increase in size on reproductive rate. The dotted outlines surround data points for large numbers of species of the different groups indicated. A marks the average for amoebas, B for Chlamydomonas, C for ciliates.**

Had the body proportions been those of modern man, stature (estimated from femur length) would have been some 122 centimetres, if that. The *Sinanthropus* material suggests a mean height of approximately 150 centimetres, and Neanderthal stature averaged 152.4 centimetres. By the time of the Upper Paleolithic, average stature seems to have been very similar to that of today. This clear evolutionary trend of a long-term increase in stature is likely to have been selective in origin (**2.3**). It seems, however, that selection on size is no longer directional. Size at birth, measured by birth-weight, appears to be maintained by balancing selection, as shown by comparison of birth-weight and neonatal survival. Survival in the first few weeks after birth is greatest in infants of medium size and steadily diminishes away from the middle of the distribution until it is worst in very small infants, who are least prepared for extra-uterine existence, and in the very large, who are liable to more trauma at birth. The disadvantage of, and the selection against, the infants at the extremes tend to produce a balance, maintaining a birth-weight distribution around its present average. Similarly, balancing selection appears to be operating on adult size, though less intensely, since fertility of very tall and very short males tends to be lower than in those of medium size.

However, there is also empirical and experimental evidence that low weight may be selectively advantageous in tropical climates, not only in human beings but also in other warm-blooded animals, the selective factor being thermoregulatory advantage. In all continental groups of man there is a clear negative correlation of mean body-weight and the temperature of their habitats. Weight and stature are closely correlated, and together they provide a measure of shape (**10.6**); the lower the weight-per-unit stature, the greater the linearity of form, and the greater the surface area. The greater the ratio of surface area to body mass, the greater the ratio of heat-loss to heat-production, and the greater the advantage in situations of heat-stress. Thus there is no conflict in regarding both the short stature of African pygmies and the tallness of the long, lanky Nilotics as adaptations to high temperatures. The pygmy reduction of stature reduces the body mass and therefore the potential heat-production. In the linear Nilotic, the tall stature, achieved especially by the greatly increased slender limb length, increases the body surface area and the potential heat-loss. Both have achieved the same overall effect, a reduction of the ratio of weight to surface area.

Despite the difficulties of showing natural selection in action, and the consequent need to use indirect evidence instead of proof, it is clear that natural selection on body-size in human beings has been appreciable, operating through a variety of different mechanisms. *Derek Roberts*

ages over 15 years, show a distinct prolongation of growth into the late teens and twenties that allows some degree of catch up through continued growth. Such extended growth may be facilitated by slow gonadal maturation with lower endocrine regulatory set points for reproductive function.

Human populations vary substantially in rates of growth and in timing of puberty, and again the degree to which this variation reflects evolved capacities to adjust rates of development to availability of necessary resources remains disputed. Some have argued that growth-slowing and deficits reflect biological impairment of ability to achieve full growth potential, rather than a positive adaptive response to resource limitation. On the other hand, the pervasiveness of food shortfalls and pathogen pressure across human history argue for strong selective pressures for developmental plasticity. Nutrition and health are the key factors linked to growth variation, although many other environmental and even psychosocial factors have been implicated.

*Carol M.Worthman*

See also 'Comparative growth and development of mammals' (**2.1**), 'Growth in non-human primates' (**2.2**), 'Human growth from an evolutionary perspective' (**2.3**), 'Genetic regulation of growth-promoting factors by nutrition' (**3.4**), 'Ontogeny of the immune system' (**4.7**), 'Sexual maturation' (**5.15**), 'Nutrition and cognitive development' (**6.5**), 'Infant-feeding and growth' (**9.1**), 'Infection' (**9.3**), 'Catch-up growth in height' (**9.6**) and 'Long-term consequences of early environmental influences' (**12.1**)

## GROWTH IN AFRICAN PYGMIES

African pygmies occupy heavily forested regions of the narrow equatorial band between the 4° North and 4° South parallels, from the Atlantic coast to Lake Victoria. Linguistic groups such as The Aka and Mbuti are identified by their hunter-gatherer lifestyle and their short stature, which is taken to be an adaptation to the hot and humid forest climate. Hiernaux distinguished five main groups of people in sub-Saharan Africa, of which pygmies were one. Pygmies show similar genetic distance from the more numerous Bantu and Sudanic populations practising agriculture, as do the San Bushmen of the Kalahari Desert.

Adult pygmies are on average 87 per cent as long as the average sub-Saharan African values for most linear body measurements. Although their adult short stature is extreme in relation to all other populations on earth, body proportions do not deviate from the norm of larger populations. The growth pattern of pygmy populations is remarkably similar to that of other, taller African groups prior to puberty. However, pygmies show little if any growth spurt. Where it takes place, it is smaller than in non-pygmy Africans. The mean birth-weight of pygmies, at 2.61 kilograms, is much lower than that of non-pygmy Africans (2.93 kg). This smaller size at birth may be a consequence of the smaller size of the mother. Although there is no information on catch-up growth in early post-natal life of pygmies, there is plentiful evidence to show that, on average, infants in developing countries born at lower birth-weight than those in industrialized nations, show greater weight velocities in the first 3 months of life than do infants in the United States or Britain. Thus it is easy to square the lower birth-weight with similar post-natal growth, at least post-infancy, if early catch-up growth is assumed.

The growth endocrinology of pygmies is as striking as their pubertal growth pattern (**5.17**). Prior to puberty, their serum insulin-like growth factor I (IGF-I) levels are slightly lower than reference values. However, serum IGF-I (**4.2**) is much lower than reference norms during the pubertal period and beyond. Although this could be due to either growth-hormone (GH) or growth-hormone-receptor (GH-R) deficiencies, GH treatment does not increase IGF-I, suggesting that the deficiency is primarily one of GH-R. Supporting this is the observation of good GH responsiveness to a standard provocative stimulus. It has been postulated that the genetic basis of this may be a defect in promotor activity for the GH-R gene (**7.6**).

*Stanley Ulijaszek*

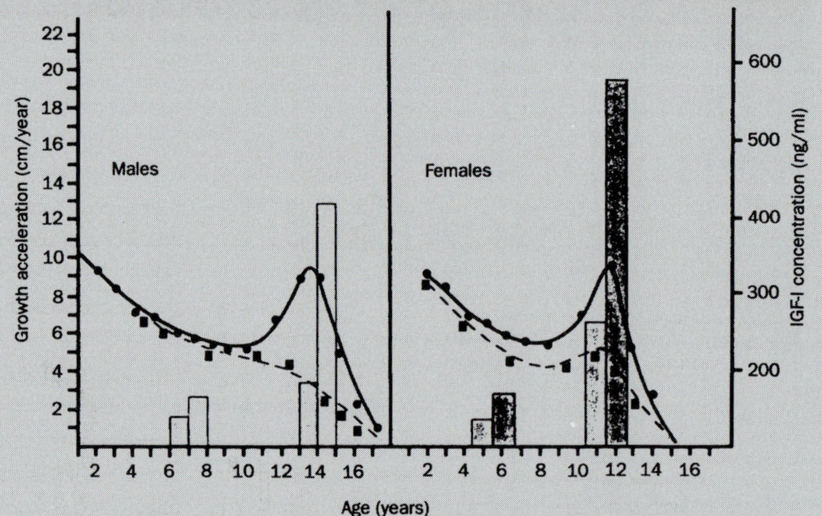


**Growth acceleration (cm/year) and IGF I concentrations in serum are shown for controls (-●-●-) and pygmies (-■-■-). There is a slight increase in growth-rate in females at puberty, but this is not discernible with mean data from males.**

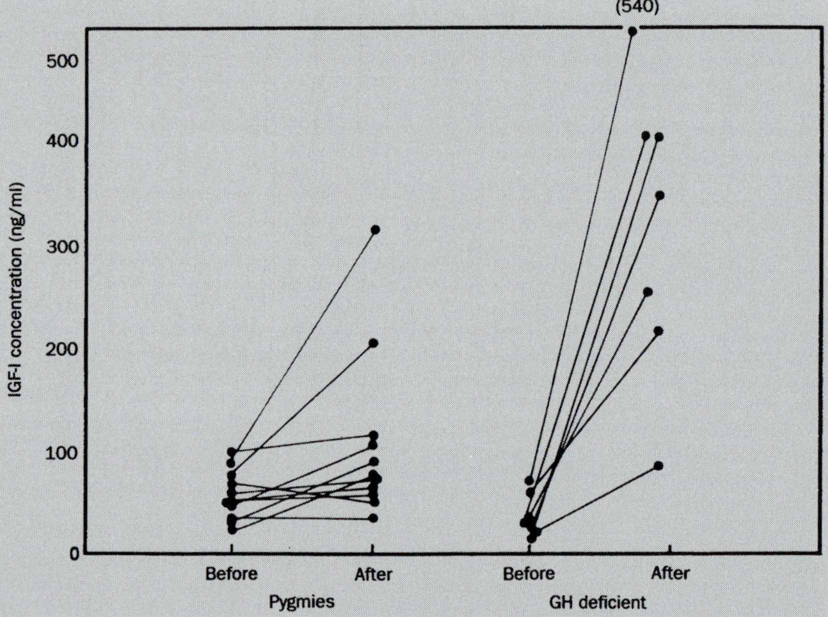


**The effect of GH treatment (5 mg twice daily for 5 days) on serum IGF I of individual pygmy subjects and GH-deficient patients. The GH-deficient patient with the smallest response was treated for only 3 days.**

# Gene–environment interactions

There are few examples of environmental influence on the genetic programme of development as clear as in the case of the axolotl. For years, this rare tadpole-like creature, living and reproducing in the waters of Lake Xochimilco in Mexico, was regarded as a distinct species until the discovery by experiment that treatment with thyroxin caused it to metamorphose into a terrestrial salamander. Many salamander species occur in Central America, but this new creature, *Ambystoma mexicanum*, was an archaic form now seen by humans for the first time.

This example is important in several ways. First, from the evolutionary and population point of view it means that, presumably in response to some environmental pressure, the ancestral array of genes controlling rates of growth and development (**3.3**) had changed by natural selection to produce a slowing down of the maturation process and eventually the total disappearance of axolotl metamorphosis and the transition to the adult stage. The effect of the modern array of genes is the retention of infantile characters in later, reproductive life stages; the state of neoteny (**2.3**). From the point of view of ontogeny and the individual, experimental evidence shows that the action of genes controlling growth and development can be modified by external factors, in this case the thyroxin administration. Thirdly, the example shows that there are two types of environment. There is the external environment in which the organism, the phenotype, functions and is exposed to natural selection, and to which it responds by physiological adjustment and selective change in the genetic constitution. Then there is the internal environment, inside the organism and inside the cell, in which the genes in the nucleus operate and respond to the messages conveyed to them as intracellular physicochemical changes.

One direction of interaction, effect of environment on gene action and genetic constitution, is clearly of great importance in respect to both the internal and external environments. Indeed, the influence of the internal environment is an essential mechanism in the control of gene action (**3.4**), perhaps especially in genes controlling growth and development, while the influence of the external environment has been critical in evolution, for it is a principal factor in the dynamics of gene-frequency change in populations. For the other direction of interaction, there is much less potential for gene influence on the external environment other than for genes controlling behaviour and habits of environment utilization; the role of genes in influencing the internal environment is of much greater significance.

The importance of the internal environment in humans, how it affects and is affected by gene action, is well illustrated from the rapidly increasing knowledge of the physiology of growth processes, the molecular basis of physiological substances, and experimental techniques of manipulation of the animal genome, notably in the mouse. The fundamental controller of growth is growth hormone (GH) (**5.1**), its structure being determined by a gene located in man at chromosome 17q22-24. Produced by the anterior pituitary and circulating bound to growth-hormone-binding protein (GHBP), its synthesis and secretion are affected by many circulating hormones and metabolic substances. Growth-hormone-releasing hormone (GHRH) and somatostatin (SS) released into the hypophyseal-portal system interact to produce the pulses of GH secretion (**5.13**), the surges in which result from a reciprocal increase in GHRH neuronal activity and a decrease in SS release, while the converse produces the troughs. Overproduction of GH produces gigantism and underproduction dwarfism; it can be produced artificially and applied to ameliorate some disorders of growth in children. The gene for the GH receptor is located at chromosome 5p13-12 and, for the controlling peptides, the gene for GHRH is located at 20p11-23. But these peptides are affected by many neurotransmitters and neuropeptides, such as creatinine, gamma-aminoisobutyric acid, and dopamine. The adrenergic system affects release of GHRH, and the cholinergic system influences SS release. Also other neuropeptides affect GH secretion; thyroid-releasing hormone, vasoactive intestinal peptide, gastrin, and neurotensin are thought to stimulate it, and calcitonin, neuropeptide Y, and corticotropic-releasing hormone (CRH) to diminish it. Growth hormone acts mainly through its induction of the two insulin-like growth factors (IGF-I and IGF-II) of which IGF-I is the most important, at least in postnatal life; their genes are at 12q and 11p, respectively. These IGFs circulate in the serum bound to binding proteins (IGF-BPs), which both inhibit and enhance the bioactivity of the IGFs by controlling their clearance from the blood, and delivery to the target cell. Like the GH binding proteins, the IGFBPs not only transport but also serve as a reservoir, inhibit the degradation, and consequently prolong the half-life of the IGFs.

A number of other substances in the internal environment are also involved in the control of growth. Chromosomal locations and molecular structures are known for the genes that determine several of these substances, their receptors, and the proteins to which they bind. The picture that is emerging is of a highly complex interaction. There are numerous genes, widely separated on different chromosomes. These are switched on and off in different tissues at different times in response to the messages they receive in the internal environment. There is a complex interaction between these genes, the rates at which the proteins for which they code are produced, the internal environment in which these gene products circulate and which they thereby modify, all of these factors in combination control and regulate growth balanced throughout the body of the developing individual.

Rendering this interaction more complex still are other mechanisms that bring about selective expression of gene products, in addition to sequential activation by switching. For in the normal process by which a cell acts on genetic instructions there may occur gene amplification by differential replication, giving multiple copies so that the number of RNA molecules produced per unit time is increased. There may be different rates of transcription, giving altered rates of synthesis of RNA. There may be changes in the rate of passage of the RNA from the nucleus to the cytoplasm. The messenger RNA, once synthesized, may function a number of times before being degraded, and so enhance the synthetic activity of the cell, or it may be stored in an inactive form, accumulating in preparation for periods when bursts of activity are required.

It is through this interaction that influences from the external environment penetrate. The adverse effect of deficiency of an essential substance in a habitat, causing a nutritional deficit, is well illustrated by the cretinous dwarfism and other less extreme thyroid disorders in individuals in areas where iodine is deficient but who would have grown normally elsewhere. Such environmentally caused disorders can be reduced by remedying the deficit at the population level. Less extreme effects in normal human populations are shown in the numerous empirical studies where change in body form and size is associated with change of environment, implying modification of growth, and in similar studies of maturation. There are experimental studies monitoring the effects of supplementary feeding on children's growth. And it is intervention at various points in this interacting complex that brings benefits, not only where the deficit is environmental but also where it is inborn, due to some genetic error inhibiting utilization of the substance. Women genetically at high risk of producing a baby with spina bifida may have that high risk reduced by prepregnancy folic acid administration; while others, where the risk is for a baby with a cleft lip and palate, may benefit from similar treatment. Administration of growth hormone may help a child whose small stature is due to a deficiency in its secretion, but not where it is the genes controlling the growth hormone receptor that are defective. Not only does understanding of this interaction throw light on many disorders of growth in individuals, it also promises to help understanding of population differences in growth. The mechanism is not yet clear

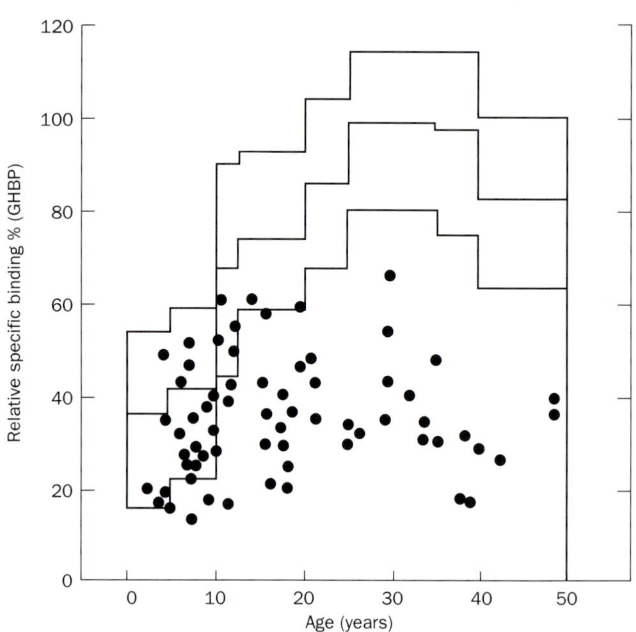


**The relative specific binding by serum of growth hormone is shown for individual pygmy subjects (n = 62). The mean ±2SD for European and Black African controls are shown by the lines. Distribution of values for pygmies older than 10.5 years differed from those for both controls and younger pygmies with $P < 0.001$.**

by which the slightly greater maturity of African infants at birth, by comparison with White, is brought about. But the small stature of African pygmies is due to a difference in genetic control, the reduction in number of GH receptors. The GHBP levels are only slightly reduced in infancy but then do not show the expected rise during childhood, and at puberty they are only 30 per cent of those of Africans of normal stature. There appears to be a progressive limitation in GH action due to GH receptor activity, and this limits the IGF-I response, particularly during adolescence.

The concept of gene–environmental interaction, especially in relation to the internal environment, has been considerably developed in the last few years. It is proving to be one whose implications and applications hold great promise in future exploration.

*Derek Roberts*

See also 'Genetics of child growth' (**3.1**) and 'Genetic disorders of the growth hormone axis' (**7.6**)

# Body-size at birth

In many Western cultures, when the birth of a new baby is welcomed with best wishes for future happiness, greetings often are accompanied by a standard question: 'How much did the baby weigh?' This simple inquiry by friends signals to the mother additional interest in the well-being of her new infant, and it demonstrates beautifully a culturally engrained recognition of an association between the newborn's health and size at birth. Scientific findings bear out these cultural values. Even though many body dimensions may be measured on a newborn, birth-weight in particular is the single most important determinant of the child's survival and subsequent healthy growth and development.

Size at birth is the cumulative result of growth and developmental processes in the embryo and fetus following conception. These intra-uterine changes during gestation occur within the larger physiological environment of the mother, and she carries out her daily activities in an external environment that may vary dramatically from place to place in the world. Size at birth is affected by differences among mothers and their environments, and these differences are reflected in variation among populations. Because of its importance, most research on body-size at birth has focused on birth-weight. A number of technical terms and abbreviations that are customarily used to classify and study birth-weight are presented in the table below.

*Mean birth-weights and percentage of low birth-weight (LBW) in selected countries*

| Country | Mean birth-weight (grams) | LBW (%) |
|---|---|---|
| ***Africa*** | | |
| Egypt | 3200–3240 | 7.0 |
| Kenya | 3143 | 12.8 |
| Nigeria | 2880–3117 | 18.0 |
| Zaire | 3163 | 15.9 |
| ***Asia*** | | |
| China | 3215–3285 | 6.0 |
| India | 2493–2970 | 30.0 |
| Indonesia | 2760–3027 | 14.0 |
| Japan | 3200–3208 | 5.2 |
| Malasia | 3027–3065 | 10.6 |
| Pakistan | 2770 | 27.0 |
| ***Europe*** | | |
| France | 3240–3335 | 5.6 |
| Hungary | 3144–3162 | 11.8 |
| Italy | 3445 | 4.2 |
| Sweden | 3490 | 4.0 |
| United Kingdom | 3310 | 7.0 |
| ***Latin America*** | | |
| Brazil | 3170–3298 | 9.0 |
| Chile | 3340 | 9.0 |
| Colombia | 2912–3115 | 10.0 |
| Guatemala | 3050 | 17.9 |
| Mexico | 3019–3025 | 11.7 |
| ***North America*** | | |
| Canada | 3327 | 6.0 |
| USA | 3299 | 6.9 |

*Glossary of selected terms and abbreviations relating to birth-weight*

***Gestational age*** – Duration of pregnancy, usually expressed in weeks

***Intra-uterine growth retardation (IUGR)*** – Birth-weight below a given cut-off (usually 10th or 5th percentile) for gestational age

***Low birth-weight (LBW)*** – Birth-weight < 2,500g

***Preterm (or gestationally premature)*** – Birth at < 37 weeks gestation

***Small-for-gestational-age (SGA) or small-for-dates*** – Same as intra-uterine growth retardation

***Term infant*** – Birth between 37 and 42 completed weeks of gestation

***Very low birth-weight (VLBW)*** – Birth-weight < 1,500g

## Significance of birth-weight

Birth-weight is an important indicator of the health and well-being of the infant because size at birth is accompanied by biological and developmental processes that reflect the maturity of function required for survival, and for healthy compatibility with the extra-uterine environment. Of greatest concern for populations are infants considered to be of low birth-weight (LBW), premature, and intra-uterine growth retarded (IUGR). Small babies are many times more likely to die early in life than larger babies. VLBW and LBW infants who survive are at increased risks of cerebral palsy, seizure disorders and other neurodevelopmental problems, including poor school performance. Also, increased risks for lower respiratory tract conditions and chronic pulmonary disease early in life

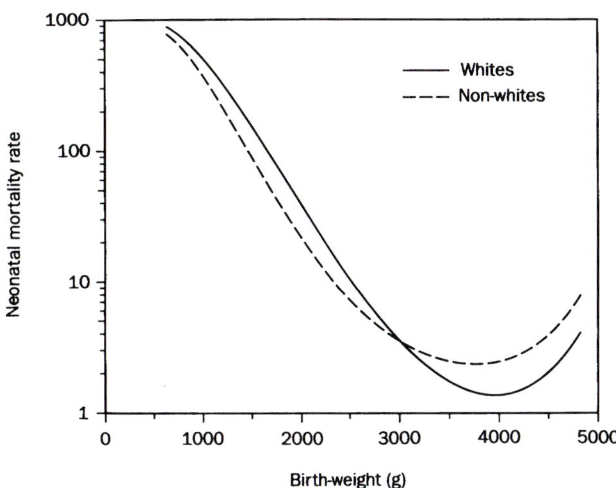


**Death-rates during the first month of life (per 1,000 live births) for singleton births in South Carolina, USA, 1975–1980 (n = 282, 366). (Data redrawn from Alexander et al. *Public Health Reports* 1985: 100: 539).**

are associated with LBW. Impaired fetal growth may be related to impaired post-natal growth in height and weight.

The increased requirements for medical care to attend LBW and VLBW babies have significant implications for the sophistication and costs of health services. Having a high-risk infant within a family may add additional emotional and economic stress or disrupt family function.

## Chief determinants of population differences

Most research on birth-weight has been conducted to identify risk factors for individuals within populations. This approach is appropriate so that mothers at most risk may be identified for intervention, infant needs anticipated, appropriate treatment and management prescribed, and prognoses defined. Nevertheless, unless risk factors are sufficiently prevalent within a population *and* they differ systematically among populations, they will not be determinants of differences in birth-weight among populations. For example, being a female newborn is a significant risk factor for LBW within a population. Nevertheless, because the sex ratio of births does not differ appreciably across populations, gender of the newborn is not a chief determinant of population differences in LBW.

Relatively strong risk factors for LBW within populations may not contribute appreciably to observed population differences because the contribution of the risk factor is not very large relative to other existing risk factors. For example, maternal smoking during pregnancy is a well known risk factor for LBW within populations and is associated with decreased mean birth-weights of between 75–150 grams, and relative risks for LBW of 1.1–2.5, when compared with mothers who did not smoke during pregnancy. Even so, because of the fairly low prevalence of smoking in absolute terms and the size of the effects on birth-weight, smoking *per se* is not a chief determinant of the differences between most populations in birth-weight.

Many of the risk factors for IUGR between populations are associated with poverty. Insufficient dietary energy and accompanying low maternal weight-gains during pregnancy increase the risk for IUGR, especially if the mother is not adequately nourished prior to pregnancy. Low prepregnancy body-weight is an indicator of poor nutritional status immediately preceding the pregnancy relative to requirements for energy expenditure and daily activities. Many women in developing countries often engage in strenuous work regularly and have a considerably less body-weight relative to their height compared with better-off women in developed countries. Short maternal stature is associated with increased risk for IUGR. Most of the short stature in women in developing countries reflects poor nutritional status much earlier in the lives of the women, perhaps even during their own infancy. Genetic factors in mothers with short stature may contribute to IUGR as well.

Illness during pregnancy, especially malaria, probably contributes to differences between populations in IUGR. These illnesses may affect maternal nutritional status through anorexia and hypermetabolic periods, or infections may act more directly on the placenta and fetus.

Effects of race or ethnic origin are difficult to interpret precisely because of common accompanying differences in other factors that may be related to LBW or IUGR, for example, nutrition, maternal size, education, birth interval, antenatal care, et cetera. Nevertheless, in careful studies attempting to control for confounding factors, consistent differences apparently due to race persist. Relative to Whites of European extraction, North American Blacks, Pakistanis and Indians all have smaller mean birth-weights and greater risks for IUGR. Other factors that may contribute to risks of IUGR in developing countries are short paternal stature and poor maternal education. North American Indians, Samoans and North African Sephardic Jews have reduced risks of IUGR. Whether the relatively large birth-weights in these groups reflect genetic or environmental factors is unknown.

Another factor that is an important determinant of LBW and IUGR in some populations is the altitude at which the mothers reside during pregnancy. Altitude effects probably result from maternal hypoxia. The lowest part of the distribution of birth-weights starts to shift downward at altitudes as low as 1500 metres above sea level, mostly due to IUGR. The Andes mountains of South America, the highlands of northern Ethiopia and the Tibetan plateau are areas of the world in excess of 2500 metres that support sizeable permanent populations. In these areas mean birth-weights may differ by 200–300 grams compared with similar ethnic groups living near sea level (**9.9**).

*John H. Himes*

See also 'Genetic and environmental influences on fetal growth' (**3.2**), 'Fetal growth retardation' (**4.8**), 'Standards and references for the assessment of fetal growth and development' (**4.9**), 'Smoking' (**8.3**), 'Nutrition' (**9.2**) and 'Long-term consequences of early environmental influences' (**12.1**)

# Growth in infancy and pre-adolescence

Different human populations vary from each other in a large and ever-growing list of genetic markers; growth patterns, body-size and composition are polygenic in character, and the genetic component to human growth (see Part three) in different populations is still a debated but poorly resolved issue. It is possible to classify populations around the world in a number of different ways, the Eveleth and Tanner (1990) classification dividing them into European, African Asiatic, and Indo-Mediterranean. This typology is rather rigid and simplistic, since it does not comfortably include mixed populations such as Spanish-Indians in the Americas, and European-Africans in the Caribbean, United States and Britain. Furthermore, it aggregates populations that have been shown to have clear differences in growth pattern. For example, the term 'African' includes the majority Bantu-descended populations of Africa, as well as distinctively short-statured hunter-gatherer groups such as the Mbuti of Zaire and the !Kung bushmen of Namibia. Furthermore, the term 'Australian Aborigines and Pacific Islanders' covers populations with considerable genetic heterogeneity. However, clumsy though the classification is, it may be the best that is currently available.

The number of studies of human growth and body-size carried out worldwide during the latter part of this century runs to three figures. However, many of them are of groups and populations living in poor environmental circumstances. Possible differences in growth patterns which might be attributed to genetic factors can be examined in various ways, but among the most popular is the comparison of child body-size measures such as height and weight of children across populations and socio-economic status.

There is considerable variation in mean birth-weight across populations, but in general, the lower birth-weights of children in the developing countries can be largely attributed to intra-uterine stress such as undernutrition. At birth, mean body-weights of different populations reflect the size and sometimes the nutritional state of the mother. Infants in populations of the developing world have weight velocities which are often in excess of those shown by children in Western industrialized nations. This is best explained in terms of catch-up growth as part of the process of recovery from intra-uterine growth retardation. Between 1 and 3 months of age, weight velocities are similar for infants from both developing and economically developed countries. However, after this time, growth faltering as a consequence of undernutrition and infection can take place in children from developing countries anywhere from the third to the sixth month of post-natal life, and for a length of time according to the degree of environmental stress, but is usually finished before the age of 3 years. In extreme environments, growth faltering may be prolonged and carry on at some level into puberty and beyond, sometimes resulting in delayed maturation and achievement of full physical maturity.

A comparison of mean heights of 7-year-old boys of high and low socio-economic status has been used to demonstrate that the largest differences in mean values are a function of poverty, with low socio-economic status children having much lower values than children of higher socio-economic status. Differences in mean stature between high socio-economic status children from different populations is comparatively small. In a comparison of 49 well-off populations around the world, the range of means for 28 European and European-origin populations was found to be 119.1 to 126.5 centimetres, similar to those for African and African-origin

*Classification of population types for the purposes of growth studies*

| Classification | Definition | Other terms used |
|---|---|---|
| European and European origin | Living in Europe or elsewhere, of European ancestry | Caucasian, White, Anglo-American, European-American |
| African and African origin | Living in Africa or elsewhere, of African ancestry | Negro, Black, Black British, Afro-Caribbean, Afro-American, African-American |
| Asiatic | Living in Asia or elsewhere, of Asian ancestry | Asian, Indian (American) |
| Indo-Mediterranean | Peoples of the Near East, North Africa, the Indian subcontinent and their descendents | Indian, Asian, Indo-Pakistani, Arab, South Asian |
| Australian Aborigines Pacific Island peoples | | Melanesian, Polynesians, Micronesians |

*Birth-weight and weight velocity of infants*

| Country | Birth-weight (kg) | Weight velocity by age group (months) (g/kg body-weight/day) | | | | | |
|---|---|---|---|---|---|---|---|
| | | 0–1 | 1–2 | 2–3 | 3–4 | 4–5 | 5–6 |
| Guatemala | 2.55 | 11.9 | 5.8 | 4.6 | 1.6 | 3.0 | 2.0 |
| Gambia | 2.8 | 8.6 | 7.3 | 5.9 | 2.4 | 1.5 | 2.2 |
| Tanzania | 3.03 | 11.7 | 6.1 | 5.4 | 2.8 | 1.8 | 2.1 |
| Nigeria | 3.05 | 10.1 | 6.2 | 4.3 | 3.2 | 1.9 | 1.5 |
| Kenya | 3.05 | 9.2 | 6.8 | 4.5 | 2.8 | 3.1 | 1.9 |
| Singapore | 3.12 | 9.1 | 7.1 | 4.6 | 3.2 | 2.1 | 1.5 |
| USA | 3.25 | 8.1 | 5.8 | 4.7 | 3.6 | 3.3 | 2.3 |
| UK | 3.31 | 6.1 | 6.8 | 5.3 | 4.2 | 3.4 | 2.6 |

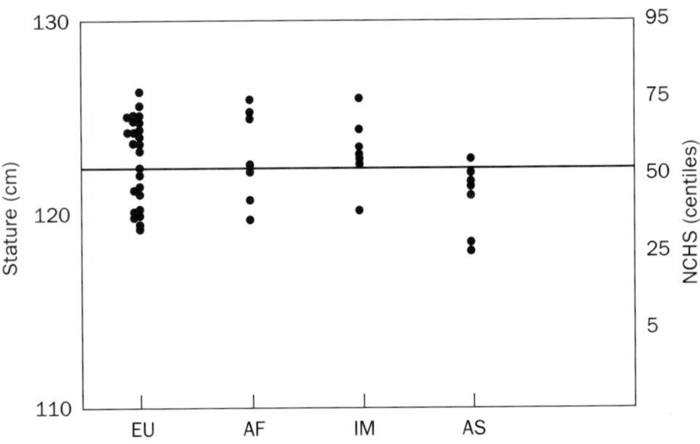


**Mean statures of 7-year-old males from industrialized countries, and from well-off populations in developing countries European (EU), African (AF), Indo-Mediterranean (IM) and Asiatic (AS). Each dot represents a mean value for a population.**

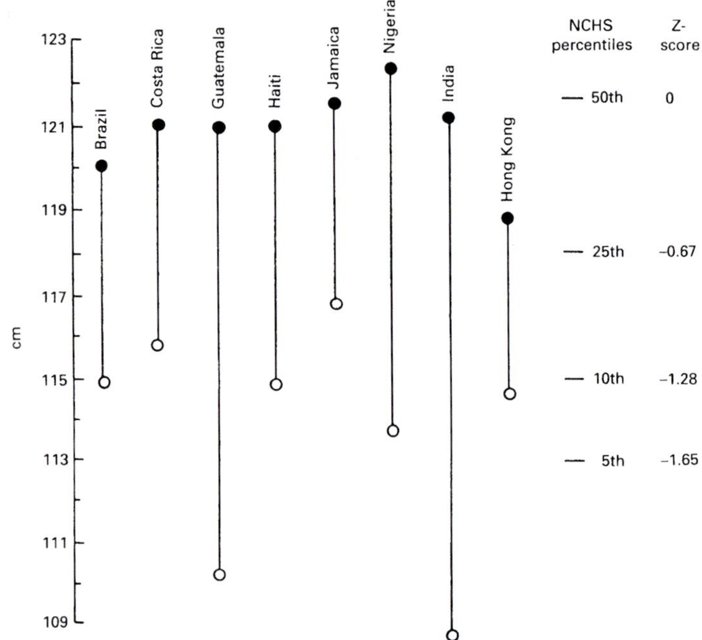


**Mean heights of 7-year-old boys of high (●) and low (○) socio-economic status.**

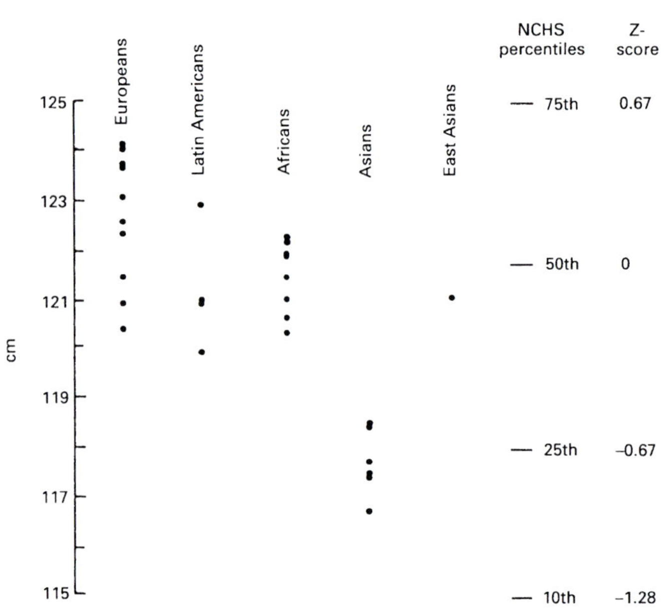


**Mean heights of 7-year-old boys of high socio-economic status, of differing ethnicities.**

populations (119.6–126.0 cm), and Indo-Mediterranean populations (120.2–126.0 cm), but higher than that for Asiatic populations (118.1–122.6 cm). This supports the suggestion that genetic potential for growth is similar for all groups examined in this way, apart from Asiatic populations. It is not clear, however, whether Asiatic populations have achieved their maximal stature as a consequence of the secular trend yet. In addition there is no data available on high socio-economic status children who are Australian Aborigines or Pacific Islanders, and it is not known whether more genetically isolated groups such as those of Africa, can be included within the broader population typologies put forward.

When considering differences in the growth characteristics of well-off populations, the following generalizations are striking. On average, African-origin children and European-origin children have very similar mean heights before adolescence, with Indo-Mediterranean children showing slightly slower growth, leading to lower attained height across the ages 3 to 10 years. Asiatic children, however, are shorter on average throughout the pre-adolescent period. However, African-origin children are lighter at birth than European-origin children, but show greater weight-gains in the first 3 months of life. Indo-Mediterranean and Asiatic populations have lower mean birth-weights and lengths than European-origin or African-origin children. For the Asiatic populations, this difference is maintained throughout childhood. Whether or not this is due primarily to genetic factors is not clear. However, for the Indo-Mediterranean populations, such relative smallness is unlikely to be genetic, since the mean birth-weights and lengths of Indo-Mediterranean children in migrant families (**11.6**) in the United Kingdom are increasing.

*Stanley Ulijaszek*

See also 'Growth as an indicator of social inequalities' (**1.12**), 'Longitudinal analysis' (**1.14**), 'Within-population variation in growth patterns' (**2.5**), 'Sexual maturation' (**5.15**), 'Infant-feeding and growth' (**9.1**), 'Growth in high-altitude populations' (**9.9**), 'Body-size at birth' (**10.3**), 'Between-population differences in adolescent growth' (**10.5**), 'The secular trend' (**11.3**) and 'Social and economic class' (**11.4**)

# Between-population differences in adolescent growth

As a general rule the adolescent period is one of heightened variability in growth. Differences in the rate of maturation (**5.17**) will increase the age-associated variance in measures of status – the 'out-of-phase effect'. The spurt in growth is both intense and rapid, and greater variability will follow because of differences in velocity along with enhanced sexual dimorphism. And the process of attainment of adult body-size, shape, and composition (**5.11**), which occurs during this phase of development will introduce a component related to characteristics of the adult.

For example, the coefficients of variation (CVs) of height among 9-year-old Dutch children are 4.2 per cent for males and 4.3 per cent for females. At 12 years of age, these CVs are 4.8 and 4.6, increases of approximately 10 per cent. Among Guatemalan children of low socio-economic status, the CV in males increases from 4.2 to 4.7 per cent, and in females from 4.5 to 4.9 across the same ages. Other anthropometric dimensions show even greater increases in their coefficients of variation. In the Guatemalan sample, the changes ranged from a 10.5 per cent increase in the CV of the triceps skinfold to a 42.3 per cent increase in that of head circumference.

The increase in the variance during adolescence of almost any measure of growth is a phenomenon which may be observed in any population as a result of individual variability. However, different populations may display different mean rates of maturation. This can result in populations differing in measurements of growth, in their being out of phase with each other. Groups maturing on average more rapidly will be larger during the early stages of adolescence than groups maturing more slowly. Then, as the individuals in the slow-maturing population moves through their adolescent period (**5.15**, **5.17**), the differences between the two groups will diminish. This is sometimes wrongly interpreted as growth catch-up(**9.6**).

This is illustrated in the figure, which shows the curves of height growth for three samples of males, age 12–18 years. The American sample is drawn from a community of upper socio-economic status (SES) (**11.4**) in the north-east part of the United States (n = 173), the India high SES sample is from the northern part of the country (n = 210) as is the low SES group (n = 108). In each sample the heights were fitted to a logarithmic curve.

The curve of the high SES Indian sample steadily moves closer to that of the American boys throughout the age range, with the differences decreasing from 9 centimetres at 12 years to 5 at 18. This suggests that the maturation rate of this Indian sample is slower, with the curves moving toward convergence as the Indian group moves into their own adolescent spurt.

Just as steadily, the curve of the low SES Indian sample moves away from that of the United States, with differences increasing from 10 centimetres at 12 years to 15 centimetres at 18. This suggests a slower rate of maturation.

However, the maturation rate is not the only mechanism affecting population differences in growth during adolescence. Given the environmental constraints on the growth of the low SES Indian sample, it is highly likely that they will never recover in mean height to that of the other two samples, ending up as short adults. Likewise, from what we know about adults from India and the United States, it is also likely that the high SES Indian sample will be significantly shorter as adults than will the American boys. In other words, the analysis and interpretation of population differences in growth

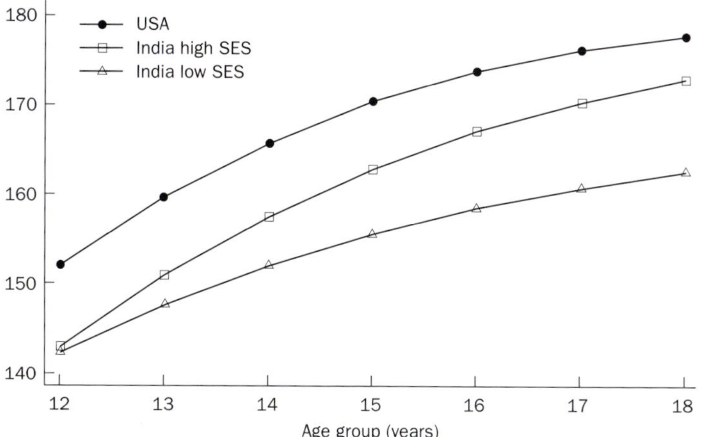


**Logarithmic fits of height on age in three samples of males (n = 491). See text for description of samples.**

*Estimated mean values and standard deviations (SDS) for the pre-adolescent and adolescent components of longituinal growth in height (cm) of Guatemalan and European children living in Guatemala and European children living in the United States*

| | Pre-adolescents | | Adolescents | |
|---|---|---|---|---|
| | **mean** | **SD** | **mean** | **SD** |
| *Guatemalans in Guatemala* | | | | |
| Males (20)[1] | 147.1 | 6.0 | 26.8 | 3.0 |
| Females (15) | 141.2 | 6.3 | 20.8 | 6.0 |
| *Europeans in Guatemala* | | | | |
| Males (14) | 145.4 | 9.0 | 29.9 | 6.7 |
| Females (9) | 138.7 | 7.0 | 25.7 | 4.2 |
| *Europeans in California* | | | | |
| Males (65) | 151.4 | 6.7 | 29.5 | 5.3 |
| Females (64) | 143.0 | 6.7 | 25.1 | 4.4 |

[1] Sample size given in parentheses.
Source: Johnston et al. (1976)

during adolescence requires attention paid both to the likely effects of maturation rate and final size.

## Population differences, genes, and the environment

Inter-population differences in growth result from genes, the environment, and their interaction. During childhood, it is generally accepted that environmental factors (see Part eight) are the major determinants of differences between populations. This conclusion is based on different kinds of evidence: the comparison of poor and well-off samples from the same country and ethnic background; the analysis of medical and nutritional interventions; and secular change in specific populations in response to environmental improvements.

Whether growth differences among populations during adolescence are as overwhelmingly regulated by the environment is still open to question. It is clear that environmental changes can affect adolescent growth, especially through an acceleration of the process of maturation. Likewise, the sensitivity of body composition, especially adiposity, to the environment is well-established. But there may also be a still-unquantified component of adolescent growth attributable to differences between gene-pools. Using different research designs, Johnston et al. (1976) and Frisancho et al. (1980) have presented data to this effect. Their analyses suggest that during childhood, samples of children grow in a fashion similar to other samples who share the same environment, even if they differ in ethnic background. However, during adolescence, growth patterns are similar to those who are ethnically similar, even if the environments are not the same.

Insofar as the rate of maturation is concerned there is ample evidence of hereditary influences (**3.1**), whether one looks at the age at menarche, bone age, or the development of secondary sex characteristics. However, it is also clear that environmental stresses and constraints tend to overpower genetic influences such that the most important component of the between-population variance in the maturation rate is environmental.

*Francis E. Johnston*

### VARIATION IN AGE AT MENARCHE

There is a wide age-range for reaching menarche (**1.15**) in present-day populations. Within any population there is considerable individual variation of age when girls begin their menstrual periods, and among populations the range is equally as great. Different European populations are actually quite similar in mean age of menarche, although Mediterranean populations tend to be earlier (12.1 to 12.8 years) than English, Dutch, or Scandinavians (13.0–13.5 years). Populations in underdeveloped rural areas experience menarche later, as did all European populations 100 years ago. American and Canadian girls are fractionally earlier than their root European populations (12.8–12.9). Latin-American girls also are similar to their parent populations or slightly earlier (12.0–12.6). African girls, even apparently well-off, are relatively late (13.1–13.4), although African-Americans experience menarche earlier (12.4–12.5) than Americans of European descent. Both Chinese and Japanese have an early menarche (12.4–12.5). Well-off girls from Hong Kong and Singapore are as early or earlier than the majority of Mediterranean populations. Well-off Indians in the Punjab (12.5) and Madras (12.9) are not unlike Mediterranean populations, although Pakistani girls in London are later (13.1) and poor, rural Indian groups are considerably later (13.7–14.6). More traditional Melanesian populations from New Guinea have the latest recorded menarche, ranging from 15.5 years for the Megiar to 18.0 years for the Bundi. Other groups reporting late menarche are in poorer areas of sub-Saharan Africa (15.0–17.0).

Environmental conditions influence the mean age of menarche: altitude, nutrition, disease, family size, social conditions. Menarche occurs later in populations living at high altitude, such as La Paz, Bolivia (13.4), and Nepal (16.2). Chronic undernutrition (**9.2**) delays menarche; likewise, improvement in nutrition and health-care appears to result in earlier maturation. Family size has been reported to influence timing of menarche; girls from smaller families in north-east England matured earlier than those from large families. Well-off girls in all populations for which data are available reach menarche earlier than the poorly-off, and urban girls earlier than rural.

The mean age of menarche has been decreasing for the last 150 years in industrialized countries (**11.2**). The general trend has been about 0.3 years per decade. But it should not be assumed that there has been a steady rate of decline since 1840. In Olso, for example, there was a rapid decline from 1860–90 followed by a period of little or no decline from then until around 1920. At that time a steep decline began, lasting until 1960, after which there has been little change. In the 1970s and '80s other countries noted a marked slowing down of the trend. Post-war Japan has experienced the steepest trend of all; from 1950 to 1975 the decline was approximately 1.0 years per decade in the general population and about half that in the well-off. In the West also the trend has levelled off.

*Phyllis B. Eveleth*

*Age at menarche in some populations*

| Country | Year | Mean | S.E. |
|---|---|---|---|
| Belgium: Brussels | 1980 | 13.1 | 0.06 |
| Netherlands, all | 1980 | 13.3 | 0.04 |
| Spain: Madrid, middle-class | 1980 | 12.1 | 0.04 |
| Italy: Turin | 1984 | 12.6 | 0.05 |
| Yugoslavia: Zagreb | 1982 | 12.7 | 0.02 |
| United States, | | | |
| European origin | 1968 | 12.8 | 0.04 |
| African origin | 1968 | 12.5 | 0.11 |
| Brazil: Sto. Andre, well-off | 1978 | 12.2 | 0.03 |
| Venezuela: Caracas, middle-class | 1976 | 12.0 | 0.04 |
| Cuba: Havanna | 1973 | 12.8 | 0.01 |
| Mexico: Mexico City, middle-class | 1985 | 12.3 | 0.05 |
| Nigeria: Ibadan, university educated | 1974 | 13.3 | 0.06 |
| Somalia: Mogadish, better-off | 1975 | 13.1 | 0.18 |
| South Africa, | | | |
| African origin, urban | 1988 | 13.2 | — |
| rural | 1989 | 14.0 | — |
| India: Punjab, well-off | 1986 | 12.5 | 0.13 |
| poor | 1986 | 13.7 | 0.18 |
| Japan: Tokyo | 1988 | 12.5 | 0.01 |
| Hong Kong, better-off | 1978 | 12.4 | 0.18 |
| New Guinea | | | |
| Bundi, highlands | 1967 | 18.0 | 0.19 |
| Kaipit, lowlands | 1967 | 15.6 | 0.25 |
| Fiji, Melanesians | 1985 | 13.6 | — |

## AGE AT MENARCHE IN THE FORMER SOVIET UNION

Although there have been many reviews of menarcheal age for separate countries and continents, the former Soviet Union was a unique 'laboratory' for bioanthropologists, having populations of different ethnicity that inhabit various geographical and climatic zones. This population marker of sexual development has never been reviewed on a 'whole-country' basis, though the evaluation of menarcheal age (**1.15**) for different populations has been produced in many publications. Here, published material as well as unpublished data collected at the Institute of Anthropology, Moscow State University are compiled. Although there is no geographical gradient north, south, east or west across this huge continent, there is a clear rural–urban gradient, with urban girls almost always being earlier maturers. This is true regardless of ethnicity in all but one case. While Russians, Byelorussians, Armenians, Abkhazians, Turkmens, Tadjiks, Buryats, and Komi show the expected urban–rural difference, girls from western Ukraine show the opposite, with village girls maturing earlier. Populations in the big cities have almost the same age at menarche, regardless of location, while for rural population of the same ethnicity the difference may be as great as 0.8 years. The most likely reason for differences in age at menarche in populations across the former Soviet Union is socio-economic (**1.12**, **11.4**), despite the claim that such differences were minimal in the earlier political climate of state socialism. *Elena Godina*

Key to the map
Nos. Republic, Region (oblast), city/town/village, Ethnicity

1. Russia, Arkhangel'sk city, Russians
2. Russia, Arkhangel'skaya region (village), Russians
3. Russia, Karelia, Olonets town, Karels
4. Russia, Syktyvkar city, Russians
5. Russia, Syktyvkar city, Komi
6. Russia, Nenetskiy National district, Nar'yan Mar town, Russians
7. Russia, Nenetskiy National district, Nar'yan Mar town, Nenets
8. Russia, Permskaya region (village), Komi-Permyaks
9. Russia, Moscow city, Russians
10. Russia, Polotnyaniy Zavod town, Russians
11. Russia, Lipetsk city, Russians
12. Russia, Voronezh city, Russians
13. Russia, Smolensk city, Russians
14. Russia, Samara city, Russians
15. Russia, Nizhniy Novgorod city, Russians
16. Russia, Tomsk city, Russians
17. Russia, Omsk city, Russians
18. Russia, Tyumen' city, Russians
19. Russia, Tyumenskaya region (village), Russians
20. Russia, Yamalo-Nenetskiy National district (village) Komi-Izhems
21. Russia, Khanty-Mansiyskit National district (village) Khanty
22. Russia, Khanty-Mansiyskit National district (village) Mansy
23. Russia, Yakutia (Sakha Republic), (village) Yakuts
24. Russia, Ulan Ude city, Russians
25. Russia, Ulan Ude region (village), Buryats
26. Russia, Irkutsk city, Russians
27. Russia, Abakan and Krasnoyarsk cities, Khakass
28. Russia, Norilsk city, Russians
29. Russia, Yuzhno-Sakhalinsk city, Russians
30. Russia, Petropavlovsk-Kamchatsky city, Russians
31. Russia, Kamchatskaya region (village), Russians
32. Russia, Kamchatskaya region (village), Chukchi, Koryaks, Itelmens
33. Estonia, Tartu city, Estonians
34. Latvia, Riga city, Latvians
35. Lithuania, Vilnus city, Lithuanians
36. Byelorussia, Vitebsk city, Byelerussians
37. Byelorussia, Vitebskaya region (village), Byelerussians
38. Byelorussia, Pinsk city, Byelerussians
39. Ukraine, Zakarpatskaya region (village), Ukrainians
40. Ukraine, Ivano-Frankovsk city, Ukrainians
41. Ukraine, Odessa city, Ukrainians
42. Azerbaydzhan, Baku city, Azerbaydzhans
43. Armenia, Yereven city, Armenians
44. Armenia, Yereven district (villages), Armenians
45. Georgia, Tbilisi city, Georgians
46. Georgia, Abkhazia, Sukhumi city, Abkhazians
47. Georgia, Abkhazia (village), Abkhazians
48. Kazkhstan, Alma Ata city, Kazakhs
49. Uzbekistan, Andizhan city, Uzbeks
50. Tadzhikistan, Dushanbe city, Tadzhiks
51. Tadzhikistan, Khorog town, Tadzhiks
52. Turkmenistan, Ashkhabad city, Turkmens
53. Turkmenistan, Ashkhabad region (villages), Turkmens
54. Kirgizstan, Kyzyl-Dzhar region (village), Kirgizs
55. Kirgizstan, Kirovskoye region (village), Kirghizs

# Body-proportion differences

Differences in body proportions are observed in various ethnic groups. The largest differences between ethnic groups, when all are growing up in good environments, are those of shape rather than size. Sub-Saharan Africans and African-Americans have narrower shoulders and hips compared with European origin populations, and narrower hips relative to shoulders than European and Asiatic children. Africans and African-Americans are on average proportionally longer-limbed than Europeans. The South-eastern and Far-Eastern Asians are proportionally shorter-limbed than Europeans and have proportionally broader hips to shoulders than all other major populations groups. The differences in body proportions between Africans and African-Americans compared to Europeans are evident from birth onwards, but the Asiatic–European differences arise during growth.

Secular trends in body proportions illustrate the interplay of genetic and environmental factors. Some of the ethnic differences in body proportions have tended to decrease with improved living standards, others have tended to increase. The Japanese and other Asian populations have traditionally been regarded as being of low proportional leg-length. Secular trends of increased stature (3–4 cm) in Japanese adults between 1957–77, and in many other populations throughout the world, have been almost all due to increases in subischial leg-length rather than sitting-height. Nowadays, leg-lengths in prepubertal Japanese and Chinese children are only just below those of European children for a given sitting-height (1–2 cm) but because puberty is about a year earlier, the phase of rapid prepubertal leg growth is shorter than in Europeans. This results in the persistence of an adult form of relatively short leg-length, but the Japanese are much more similar to Europeans than previously. Relative arm-lengths have been less affected by secular changes.

Powerful genetic influences remain (**3.1**). As the Japanese leg-length proportions have become more like those of Europeans, their head shape has become more divergent, towards marked brachycephalization. Africans, African-Americans and in particular Australian Aborigines all have proportionally longer legs than Europeans or European-origin populations, even when growing up in inferior environments. African-Americans have experienced many generations of interbreeding but their shape is not intermediate between African and Europeans but rather a heightening of the African trait of long-leggedness. This heightening may reflect a more favourable environment in America, but there is little evidence of changes in relative sitting height in American young men over the last 60 years.

Observations on Mexican and Mexican-American children in America show that poverty status is related to length measurements in pre-adolescent children, but not to relative proportions. These observations support the view that, in response to malnutrition, size is sacrificed and shape is preserved. Animal studies suggest this arises because those dimensions furthest along the road to maturity are held up most during undernutrition, with the effect that shape is more typically adult than is size.

Migration studies further illustrate the genetic–environment interplay. The first migrant studies by Boas showed that both stature and head shape of first and second generation migrants in New York City differed from those adults remaining in Europe. These and other migrant studies lead to the conclusion that proportions were relatively constant, suggesting a

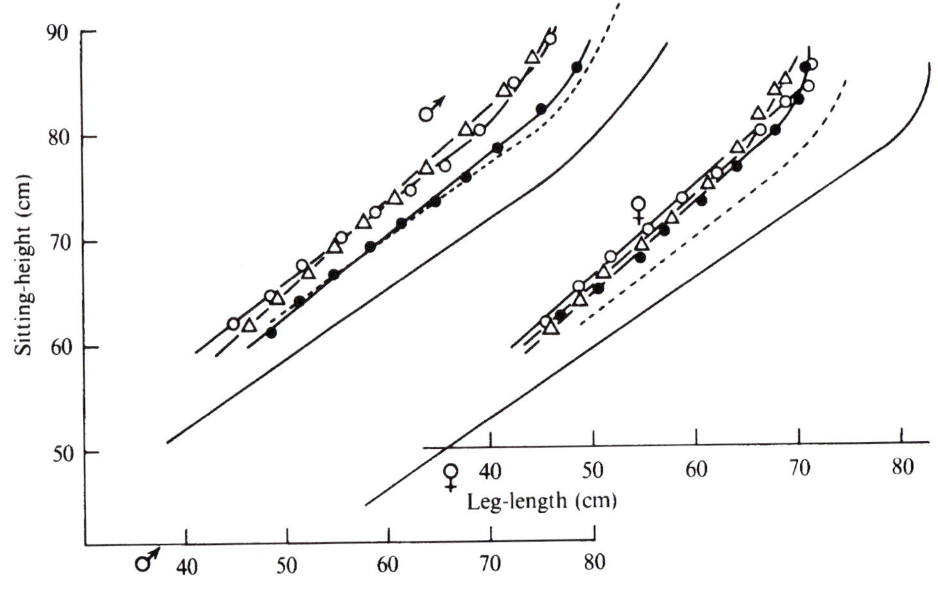


**Regression of sitting-height means (cm) on leg-length means (cm) at successive ages in Japanese (△; sample 1953–62), African-Americans in Washington, DC (solid line), Euro-Americans in Philadelphia (dashed line), and in children resulting from the matings of Japanese women with Euro-American (○) and with African-American (●) men. From Eveleth P.B. and Tanner, J.M. *World-wide Variation in Human Growth.* Cambridge: Cambridge University Press, 1990.**

strong genetic component and that size was more labile. This was not true for the cephalic index (head-breadth to head-depth), once considered a definitive index of race, or for those proportions showing recent rapid secular changes.

*N G Norgan*

See also 'Morphology' (**5.5**), 'Body composition' (**5.11**), 'Adulthood and developmental maturity' (**5.17**), 'Nutrition' (**9.2**), 'Growth in high-altitude populations' (**9.9**), 'Growth and natural selection' (**10.1**), 'Growth in infancy and pre-adolescence' (**10.4**), 'Between-population differences in adolescent growth' (**10.5**), 'Physical growth during industrialization' (**11.2**) and 'Migration and changing population characteristics' (**11.6**)

HARROW

## Part Eleven

# Changing human growth patterns

The growth pattern of an individual is the outcome of a continuous interaction process between genes and environment, which includes the psychosocial micro-environment of family and school as well as the ecological environment of district and country. Changes in pattern therefore reflect changes in one or more of these factors. Changes in the gene-pool of a population, or even in patterns of inbreeding, produce their effects so slowly that in the recent historical period it is environmental changes which must occupy us.

In 1978 Tanner coined the phrase '*growth as a mirror of conditions in Society*' to refer to the way socio-economic conditions influence the physical growth of children and youth in populations or in subgroups within these populations. The changing pattern of physical growth in populations going far back in history has been studied through skeletal remains, especially long bones and teeth, and then in the modern era through data on the height of conscripts, children and adults. These data go back as far as the 1720s, and provide crucial evidence for theories mainly concerned in the end with economic history.

Anthropologists and archaeologists have debated, during the last 30 years, whether the nutritional and health status of hunters and gatherers was better or worse than that of later agriculturists (**11.1**). A number of studies have pointed to a decline in growth-rate and adult bone lengths with agricultural intensification; at the same time other studies have not found this change of pattern. The causes of the change, in instances where it occurs, are far from certain, but change in the quality and quantity of the diet is certainly involved.

Economic historians – following the pioneer in this field, Robert Fogel – have been interested in how the average heights of the populations of different countries (for example, the United States, Great Britain, Sweden and Austro-Hungary) relate to the changing general socio-economic situation. Height, in their terminology, is a proxy for socio-economic conditions at a specific time, and changes in height over time thus reflect the changes of economic conditions in the society. Of special interest to these historians has been what happened to adult height during the period of industrialization (**11.2**). From about 1800 to 1950 the overall increase in height of European populations – adults as well as children – was over 10 centimetres. This trend was not a smooth one; in Britain, Sweden and Hungary there were decades in the middle of the nineteenth century when average height decreased, as it did also in the

**Opposite: An engraving from 1870 by Gustave Doré depicting Houndsditch, on a foggy evening, showing street traders in a polluted slum area of London. Courtesy of Mary Evans Picture Library, London.**

United States during the same period. Growth in height has been more marked and also more consistent in the twentieth century than in the nineteenth, in spite of wars and economic depressions.

In this context the term 'secular trend' is often used – to describe a slowly continuing change in growth and development over successive generations living in the same territories (**11.3**). Since, however, the changes over shorter periods of time can be either positive, negative or nil, van Wieringen, in 1986, wrote that it is preferable not to use 'secular trend' or 'the positive trend in industrialized countries' as synonyms for 'secular changes'. During the last 200 years the phenomenon of young people maturing earlier and becoming taller has been rather well researched, first in the industrial countries and then also in the developing ones. As time has gone by, not only height, weight and weight-for-height (or body mass index) have been studied, but more and more, other physical and physiological growth characteristics have been taken into consideration, like birth-weight, head-breadth and circumference, and maturational events such as menarcheal age, age at pubertal stages, peak height velocity age (PHV-age) and peak weight velocity age (PWV-age); all of these characteristics mostly show positive secular growth changes (**11.3**).

However, a change of one physical characteristic in one direction is not always accompanied by a change in the same direction by another characteristic. The net increase in height of European children of school age during the last 100 years is about 10 centimetres. However, that increase has not always been accompanied by a corresponding weight increase, and an earlier PHV-age does not seem automatically to be associated with an earlier PWV-age or earlier menarche. Indications that the secular growth changes in height, PWV-velocity or menarche have stopped for the moment have been reported from a few European countries, including Hungary, Norway and Sweden. At the same time, weight-for-height has increased in some countries and decreased in others. These positive and negative changes are largely a function of modernization processes (**11.5**).

Modernization is not always beneficial for growing individuals; consider the pollution of industrial areas. So far, little research has been done within this area, but in Poland, where menarcheal age was investigated in a polluted versus an unpolluted area, the result was that menarcheal age occurred earlier in the polluted area, contrary to expectation. However, polluted areas are often urban, and urban girls generally mature earlier than rural ones, so the difference in menarcheal age between polluted and unpolluted areas might well indicate that the difference between urban/polluted and rural/unpolluted actually has become smaller because of the urban area being polluted. Certainly, where confounding effects have been controlled for, negative effects of high levels of pollution on growth have been identified (**9.5**).

**Children in a British inner city in the 1970s, a time when slum-clearance was already well-advanced. Photograph by Stanley Ulijaszek.**

Sometimes, when secular growth changes have been studied in population surveys, socio-economic differences and changes in social class have also been investigated (**11.4**). The purpose of these studies has often been of a political nature; to propose or evaluate social reforms. Surprisingly perhaps, to the modern eye, the original hypothesis on the effects of social class (suggested by Virey and others in France in the late 18th century) was actually that people from the upper social class were the ones that would have shorter lives because of their overeating, overdrinking and excessive sexual behaviour – something shown by Villermé in 1828 to be totally wrong. Social-class distinctions in growth (adults and children) have uniformly been present from that time until fairly recently. Children in 'upper' socio-economic groups have in general been reported to be taller, to have a faster tempo of growth and also to be taller as grown-ups than children of 'lower' socio-economic groups. In most industrialized countries – both West and East – the main difference is nowadays between children whose fathers have manual and those who have non-manual professions – or between workers and intellectuals/managers. In developing countries the differences in growth are as much or even more related to the father's and especially the mother's education, as, for example, in Hanoi, Vietnam. In Sweden and in Oslo, socio-economic differences disappeared for children born in the 1950s (**11.3**), and the social class gradients in height are almost non-existent in British children of the 1990s (**11.4**). Swedish studies have shown, however, that social-class differences were slowly coming back in the 1960s. The levelling out of the differences came from a comparatively higher secular growth increase in the lower social class than in the higher one. The reason why differences between social classes were coming back has been claimed to be a worsening of the general economic situation of the country, which has greatest impact on the lower social groups.

Changes in physical growth characteristics have also been observed when people migrate from one country to another or move from a rural to an urban area within the same country. However, there is always the question of whether people who migrate are a sample of selected individuals usually bigger and more powerful – rather than random members of the population (**11.6**). Migration is rarely random. For example, migration in and from Europe from the sixteenth century onwards involved selected groups. In his diary, the Swede, Johan Svensson, who migrated to America in 1891 to become the Swedish-American John Swanson, wrote of Chicago with pardonable hyperbole: 'And such policemen they have in the city, who are so very big, that I who am a 6-feet-and-2-inches-(188-cm)-tall passenger need a ladder to climb in order to look such a man straight into his eyes and so thick as both you and I and another two of us'. Johan was an unusually tall Swede for his time. In some instances – for example Japanese moving to the United States – the immigrants have become taller and have matured earlier than Japanese remaining in Japan. Also, children of Mayan refugees living in the United States are taller and heavier than Mayan children living in the village in Guatemala (**11.4**). In other cases – for instance Italian immigrants in Belgium – there have been no overall differences in height and weight. However, when the Italian migrants' age was controlled for, differences were found, in that the younger the migrants were when they migrated, the more their height and weight increased compared to coevals in their native country (**11.6**). As migration has become more and more usual, migrants may become less select and if the migrants in this case become taller, fatter or mature earlier, it might be because they have moved from a poorer environment to a richer and/or healthier one.

The same reasoning can be applied to migration from urban to rural areas and/or vice versa (**11.7**). However, there will always be some individuals, who will not fit into this generalization but exemplify Darwin's 'survival of the fittest', like the homeless street-boys in Nepal (**11.7**).

So why is it important to study the change of human growth patterns over time, and why does taller seem better and fatter seem worse? Being taller or fatter has implications regarding the ability to be socially mobile, get a longer formal education, a better income and especially a longer life, which is one criterion of health. A relationship between height and mortality as well as body mass index and mortality has been found in recent investigations, for example, by Waaler in Norway. During the period 1963–75 an extensive chest X-ray study was conducted in Norway, and associated with it was the measurement of heights and weights of 1.8 million Norwegian adult females and males. Between 1975 and 1983 considerable numbers of these people died. After having analysed height and body mass index in relation to age of death, Waaler concluded that there is a strong relationship between height and age of death or, put another way, tall people had a lower overall mortality than short people at all ages up to about 80 years. This relationship is probably the result of socio-economic conditions in the infant years affecting growth and predisposing, in some unknown way, to earlier death. Much contemporary work centres around this theme, particularly championed by Barker and his colleagues (**12.1**).

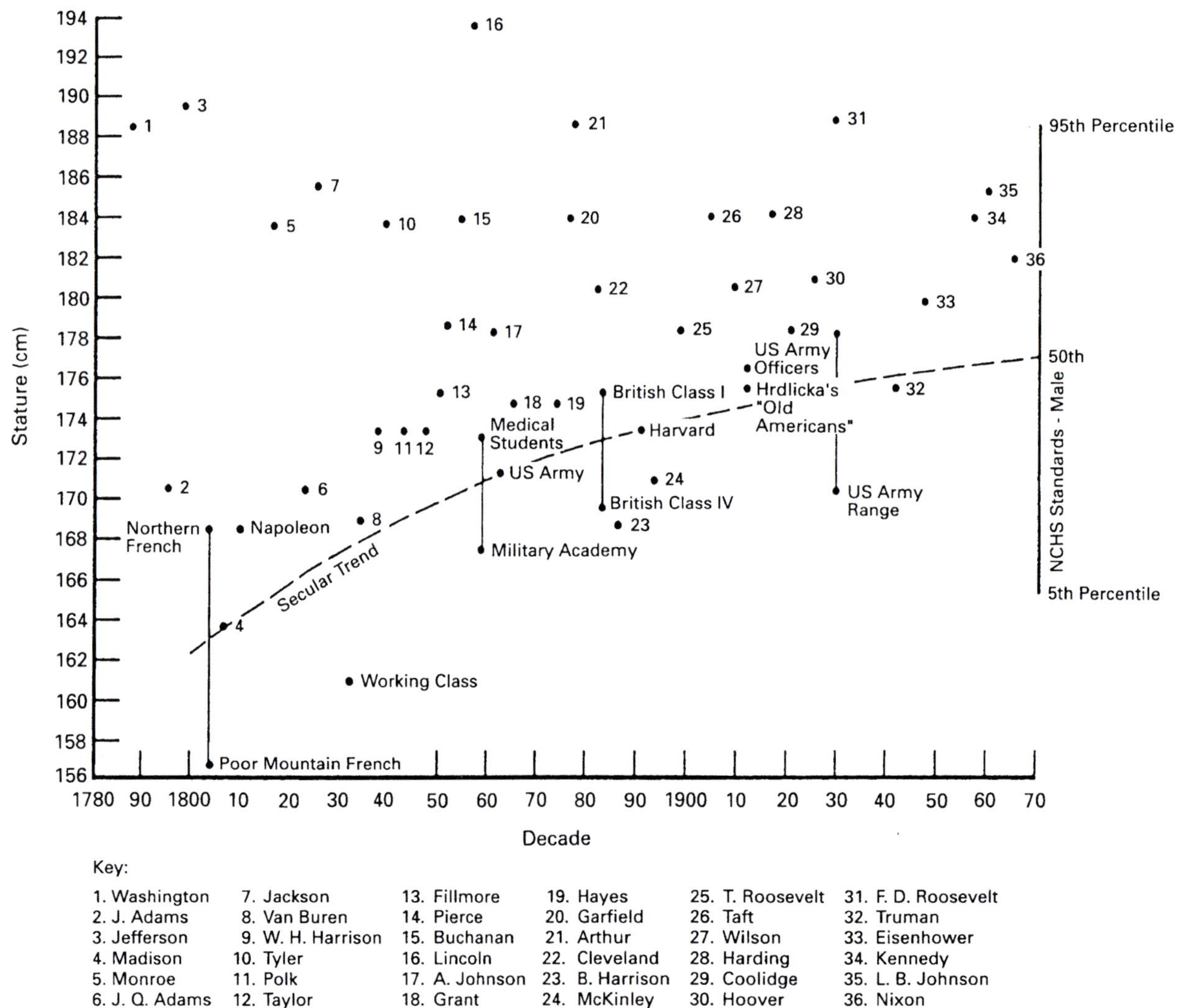


**Height of the presidents of the United States across time, showing the majority of them to be above average for the time in which they lived. This reflects above average socio-economic status. From Cassidy, C.M. (1991). The good body: when big is better. *Medical Anthropology* 13:181–213.**

In many European countries – including Sweden and the United Kingdom – the gap between rich and poor has widened, so also has inequality of health, as mirrored in life expectancies and prevalence of different diseases in the different socio-economic strata. For this reason the study of changes in socio-economic differences in growth as well as mortality rates is more interesting than ever now. Marmot, in his report *Health Inequalities in the British Civil Service*, from 1994, showed that the differences in mortality rates between socio-economic classes were only diminished by a quarter by controlling for age, height, blood pressure, blood cholesterol, blood sugar and smoking. Marmot concludes that it is something to do with the social position itself from early childhood throughout life – and not its material attributes – that most affects the health of humans. Wilkinson has taken this one step further. Comparing different countries he found that life expectancy at birth was clearly related to gross national product per capita amongst countries that were poor, but that among the rich countries, with per capita GNP above 5000 dollars, it was the equality of the income distribution within the country that related to its life expectancy. In the United Kingdom recently, for example, the bottom 60 per cent of households had a declining share of the total income and the increase of life expectancy was low compared with other industrialized countries. The main influence in rich countries seems to be relative not absolute deprivation.

So, how to explain these relationships? Studies have shown that tempo of physical growth can be delayed and height stunted by psychosocial/ emotional stress during childhood (**9.4**) and also that physical growth and above all socio-economic background are related to measured cognitive growth and ability (**6.4** and **6.5**). Cognitive ability during childhood in turn is related to further education and ultimate profession. Low cognitive ability and short formal education seem to contribute to shorter life expectancy and this seems to be more true for males than for females. However, it is probably not the cognitive ability and the shorter formal education *per se* that leads to a shorter life, but rather the psychosocial/emotional stress these individuals are trying to cope with when they are growing up, confronted with the value systems of the family, school and society in combination with the living conditions they are actually experiencing. Bogin, in this section, presents some relevant models (**11.4**).

The modern thought about short stature in the socially deprived is that it is a product of the interaction between episodes of infection and relative undernutrition (**9.3**). Especially in early childhood, the poor, living in overcrowded houses, get more infections which last longer because of inefficient use or inadequate provision of medical services. During these infections growth may slow or even stop; after recovery catch-up occurs. But to compensate perfectly this needs very good nutritional intake (**9.6**), which is perhaps unavailable. Thus it is the interaction which produces the effect. The post-modernist view incorporates these elements but also adds another: the sense of self-worth with which the child confronts the illness/undernutrition. The child may accept what society seems intent to inflict on him or her: or may deny the usual niche and aim for something felt to be more suitable for him or her. So a triple interaction seems to occur: infection–undernutrition–psychological deprivation (or betrayal), and together, uncorrected, these spell the symptom short stature and associated early death. The way the individual perceives and values her/himself (as well as how a social class or a country perceives itself) in relation to what is regarded as valuable in the family, school, society, country and even internationally is a very much overlooked variable.

The higher the evaluation, the higher the self-esteem and the will to produce and survive. Thus we could speak of a general psychosocial stress, that is more pronounced in the lower socio-economic groups within a country, as well as between countries in an international comparison of economic success, which in this case is the prime value-system. The problem of obesity fits well in this hypothetical framework (**11.8**). There will always be a time-lag to changes in human growth patterns. Thus people who are starved (for instance the working class during industrialization and up to and including the turn of the century, or malnourished children in developing countries) will try to get more food; and if they get it, they will overeat, get obese and still have shorter lives. On the other hand people who, while growing up, always have had plenty of food, would not find it very important to overeat – instead their minds are focused not on survival but on attractiveness. That is why fashion trends play such an important role in every society. Compare Japan in an historical perspective (**11.8**), and the European countries during the so called 'Twiggy-period'. In the countries where we now have quite obese populations; like America, Australia, the United Kingdom, France and Sweden, we might, in the future, develop another body-ideal which will cause human growth changes in the future. It does not seem enough just to study human growth changes over time, it is equally important to discuss and understand why these changes are occurring in each particular instance; 'to achieve a clearer picture we need to discern the details within the generalization'.

*Gunilla Lindgren*

# Skeletal growth and time of agricultural intensification

One of the more intensely debated questions in anthropology and archaeology concerns the nutritional and health status of hunters and gatherers versus later agriculturalists. The effect of agricultural development and intensification on human biological well-being occasions more than an isolated academic debate. It strikes at the core of main historical themes, has great popular appeal, and may even have significance for understanding determinants of quality of life for humans in industrial societies. At its core, this question is about progress. Have we, with evolutionary changes from foraging to farming and post-industrial economies, done better or worse?

Over the last 30 years many paleoepidemiological studies have attempted to shed light on this question quantitatively. These studies involve comparisons of the skeletal remains (bones and teeth) of early agriculturalists with those of prior hunter-gatherer groups, and comparisons of variables such as estimated mean ages at death and signs of infectious and other disease processes. Additionally, evidence has been gathered on subadult growth patterns, and lengths and widths of adult long bones. The assumption underlying these studies is that a change in nutritional status would be reflected by a change in growth and adult size.

## Methodological issues in anthropometric studies of skeletal populations' growth

*Subadults*

Studies of growth in prehistoric populations involve the measurement of long bones and, less frequently, of widths and sizes of other bones and dimensions (vertebral canal, skull-base height). These are typically plotted against dental age. From here, the data are managed in much the same way that anthropometric data from contemporary cross-sectional studies are managed. However, the following qualifications and special characteristics of growth studies from skeletal remains should be kept in mind:

- Because archaeological samples are cross sectional 'death assemblage', results can be used to infer periods of peak stress only when conditions are relatively stable over time.
- Cemetery-based studies represent not the healthy or even the 'average' child, but just those who died. This property has great potential for introducing a directional bias in comparison to the growth and size characteristics of the live population. For example, if children who suffered from chronic malnutrition (and were shorter) more often died in childhood, then the death assemblage would be shorter.
- Prehistoric series are frequently limited by small sample-size, particularly after 5 years of age. This is because the probability of death is typically low in late childhood and early adolescence. Thus, most estimations of growth parameters for ages after 5 are based on small numbers.
- There is a technical problem of measuring long bones with and without epiphyses (the unattached growing ends of long bones), which are frequently destroyed or otherwise lost to archaeological recovery. This irregularity has the potential of adding considerable measurement error.
- Dental age is usually used as a best proxy for chronological age. This limits power to find true changes because dental age is also likely to be somewhat affected by environment. (Unfortunately, the differential effects of environment on growth in long bones versus dental age are not precisely known.)

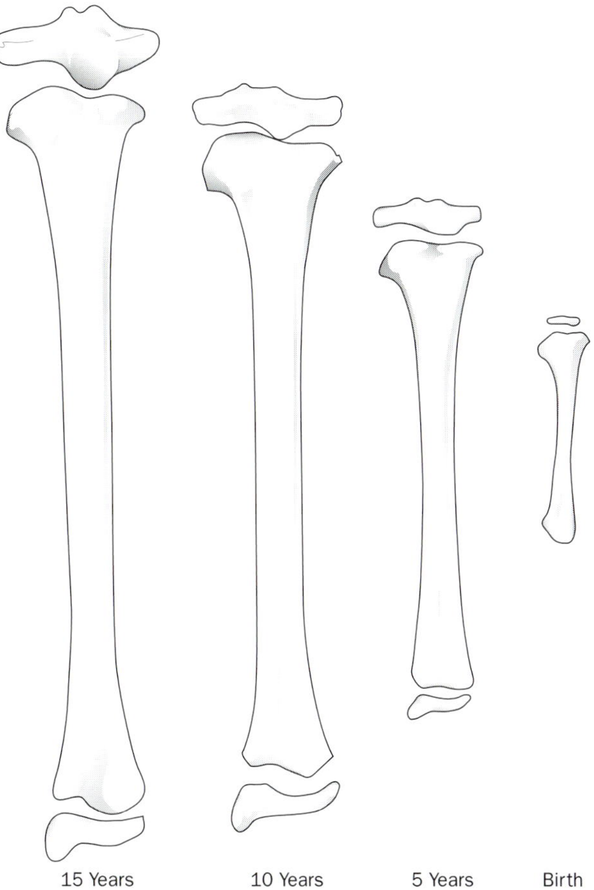


**Anterior view of the tibia at various ages showing development of proximal and distal epiphyses.**

- Poor ability to distinguish sex of subadult skeletons disallows comparisons between boys and girls.
- There is limited ability to 'standardize' growth of prehistoric long bones to that of contemporary groups. The only sample from which longitudinal growth of long bones is well established is from the Denver Longitudinal Growth Study.

*Adults*

Studies of adults are not constrained to as great a degree by problems of assignment of age and sex, nor are small sample-sizes as frequently a limiting factor. As in studies of adult anthropometry of living populations, the drawbacks to studies in prehistory revolve around the loss of sensitivity for clarifying underlying processes affecting growth and ultimate size at adulthood. The loss of the most stressed segment of the population (subadults) due to death before adulthood, coupled with the ability to catch up in growth, renders adult morphology less sensitive to environmental variation when compared to growth and development of subadults.

Adult anthropometric studies of skeletal remains are most often of lengths of major long bones. A number of equations have been generated to estimate stature based on lengths of one or more long bones. These sometimes include lengths of other bones, such as the talus and calcaneus. Finally, 'robusticity' estimates are often derived as measures of, for example, length versus width of a long bone; and ratios of sexual dimorphism in height have been used as measures of environmental stress.

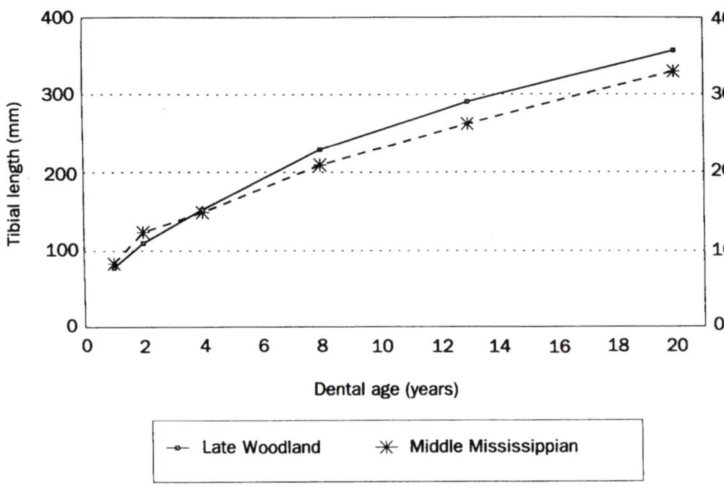


**Comparison of the growth in length of the tibia for individuals at Dickson Mounds, Illinois during the late Woodland (circa 950–1100 AD) period of hunting and gathering and Middle Mississippian (circa 1150–1300 AD) period of maize horticulture. Mean growth appears to slow for those who lived during the Middle Mississippian period around the age of 2 years.**

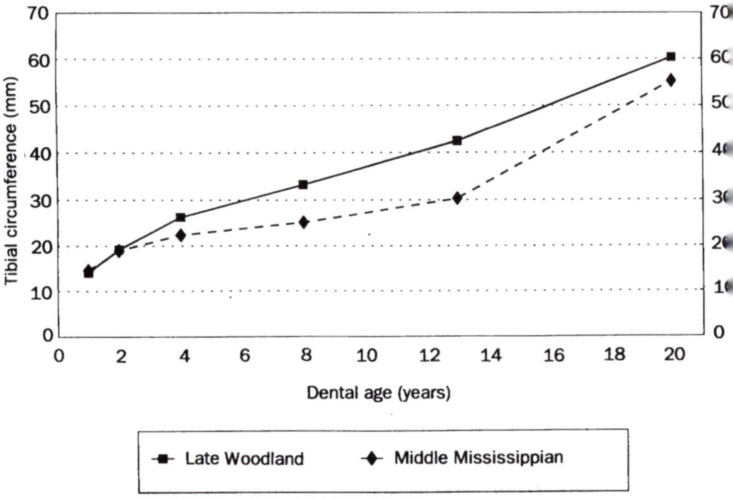


**Comparison of the growth in circumference of the tibia for individuals at Dickson Mounds, Illinois during the Late Woodland (circa 950–1100 AD) period of hunting and gathering and Middle Mississippian (circa 1150–1300 AD) period of maize horticulture. Mean growth slows dramatically for those who lived during the Middle Mississippian period around the age of 2 years. There is some evidence for 'catch up' growth in the second decade.**

## Agriculture and skeletal anthropometry

A handful of paleoepidemiological studies include sufficient sample sizes to compare growth of subadults before and after agricultural development and intensification. Results from North America, where the majority of such studies have been undertaken, show that children of early agricultural groups tend to have shorter and less robust long bones. In a few instances, evidence is available to suggest a reduction in growth velocities around the ages of 2 to 5 years. This decrease in velocity has often been interpreted as being due to the adoption of a weaning diet that is reduced in variety and is low in the availability of key nutrients. Interestingly, shorter children tend to have more evidence of disease on their bones and teeth. Thus, the decrease in growth velocity is consistent with evidence on bones of infectious disease (periosteal inflammations) and porotic hyperostosis (a thickening of bone associated with iron deficiency). For example, in North America the agricultural staple of maize (corn) is low in essential amino acids, and high phytate levels reduce the bioavailability of key nutritional elements such as iron and zinc.

Comparisons of adult stature also suggest a trend of declining nutrition and health with agricultural intensification. Data from skeletal series around Greece and the Mediterranean show a profound decrease in estimated mean adult stature in males from 177 centimetres during the Paleolithic to 172 centimetres during the Mesolithic and 169 centimetres during the Neolithic periods. The decline in mean female stature is equally profound. Declines in stature are also observed in other parts of Europe and India as well as in the Americas. Conversely, a number of studies have not found a similar pattern.

In summary, there are a number of limitations to the study of growth and size in past populations. None the less, when these studies are applied to the important question of changes in nutritional status with agricultural development,

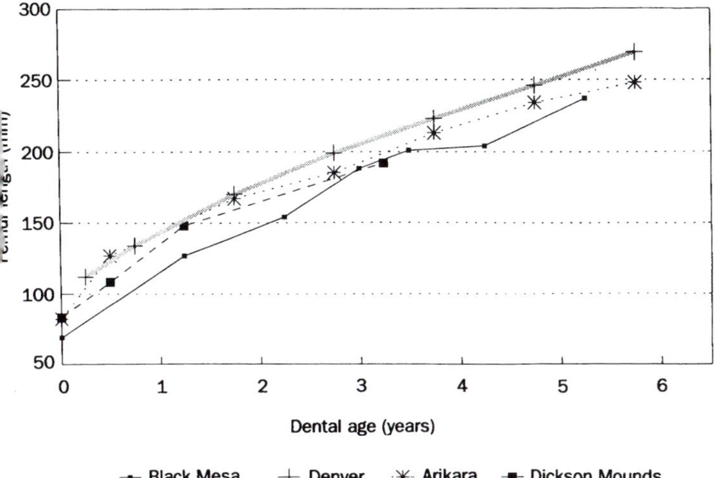


Comparison of the growth in length of the femur of various archaeological series compared to radiographic data from the Denver Growth Study (Maresh, 1955).

some intriguing results have been obtained. Although far from invariable, a number of studies point to a decline in growth velocity and adult long-bone lengths with agricultural intensification. Whereas this generalization is of great interest, understanding of the cause of this pattern is far from certain. Changes in dietary quality and quantity are almost certainly involved as proximate causes. The next question concerns whether these dietary changes are due to economic change alone (intensification of agriculture) or are the result of more complex changes between economy, political developments, and local ecologies. Whatever the answer, these data call into question unilinear notions of progress.

*Alan H. Goodman*

See also 'Dental maturation' (**1.9**), 'Growth as an indicator of social inequalities' (**1.12**), 'Creation of growth references' (**1.16**), 'Morphology' (**5.5**), 'Skeletal development' (**5.6**), 'The human growth plate' (**5.7**), 'Variation in time of tooth formation and eruption' (**5.10**), 'Adulthood and developmental maturity' (**5.17**), 'Growth in chronic diseases' (**7.8**), 'Infant-feeding and growth' (**9.1**), 'Nutrition' (**9.2**), 'Infection' (**9.3**), 'The secular trend' (**11.3**) and 'Social and economic class' (**11.4**)

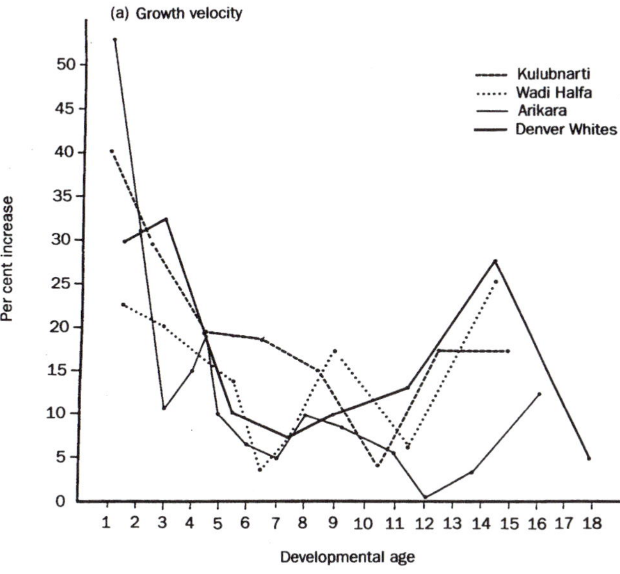


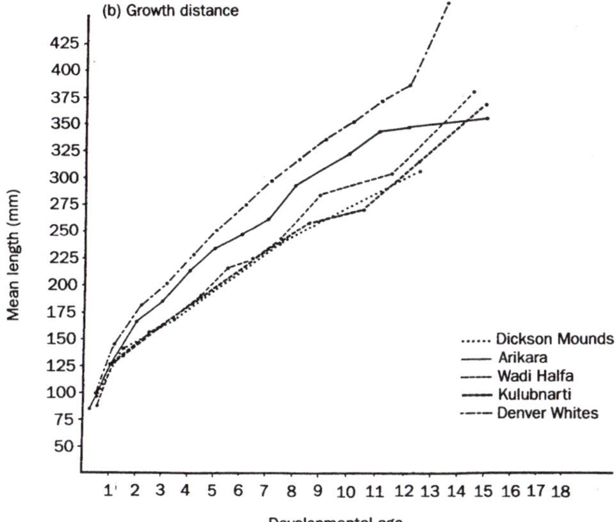


**Comparison of cross-sectional femoral growth curves of modern Europeans with those derived from archaeologicallly recovered skeletons from Dickson Mounds, Illinois, Arikara Indians from North Dakota, and the Mediterranean (Wadi Halfa, Kulubnarti), (a) velocity, (b) distance.**

## PATAGONIAN GIANTS: MYTHS AND POSSIBILITIES

When Magellan first arrived near San Julian in Patagonia, he was supposed to have cried '*ha, patagon!*' but it is not clear what he saw or what he meant by this statement. However, this part of the earth has since been called Patagonia, and legends have arisen about its past inhabitants. In 1511, Francisco Vazquez published *Palmeria de oliva*, and in 1512 the second volume appeared at Salamanca. It was called *Primaleon*, a fantastic story about the knight Primaleon which was very popular at that time. Primaleon sails to a far-off island where he meets a wild population of 'Patagonians'. He fights against a horrible giant beast, named the 'Big Patagon', overwhelms it, brings it to his ship and sails back to Polonia to present it to Queen Gridonia and her daughter Zerfira.

Magellan might well have known that story. When he discovered Patagonia, he found it populated with Tehuelche Indians, who are extinct now, but used to be of enormous size. According to Oviedo in 1520, Magellan estimated their stature '*doce a trece palmos* (12–13 times 21 cm) *de altura*', and Pigafetta wrote that 'our heads just reached their hips'. In 1579 Sarmiento described '*colosos de tres varas* (three times 83 cm)', and in 1592, Knivet estimated their stature even 15–16 spans of the hand, whereas Hawkins did not measure at all, but simply called them giants (1593). Lemaire and Schouten measured them at 10 to 11 feet (1615), while Carman described them at 9 to 10 feet (1704). Byron found the tallest to be 7 English feet and more, the smallest at least 6 feet, 6 inches. Bougainville (1767) measured them at between 5 French feet, 8 inches, to 6 feet, 4 inches, and Wallis (1767) found statures between 6 English feet, 7 inches, and 5 feet, 10 inches. Despite the great variation in reported height, all agree in that they were a tall population.

The current view is that mean adult male height may have been 173 centimetres, with a few individuals as tall as 192 centimetres, and a mean female height of 162 centimetres. This makes them taller than any contemporary Asiatic population, however wealthy, but shorter than many European, and European-origin populations, including those of Belgium, the Czech Republic, the Slovak Republic, Denmark, France, Poland, Ireland, The Netherlands, Norway, Sweden and the United States (**10.4, 10.5**). The Tehuelche Indians disappeared in the middle of the 19th century, and what is left are single bones of impressive size in the Museo Regional Provincial 'Padre Manuel Jesus Molina' at Rio Gallegos. Magellan's chronicle told that one of these Patagonians was caught and brought into Magellan's ship, just as the knight Primaleon did with the Big Patagon. It may be that Magellan had this story in mind when he cried '*ha, patagon!*'.

*Michael Hermanussen*

**E. Garnier's drawing of Captain Byron's encounter with giants in Patagonia, from *Nains et géants* (1964–5). The caption reads: 'An English sailor giving a biscuit to a Patagonian woman.' Courtesy of Mary Evans Picture Library, London.**

**Giants have been commented upon throughout history. A caricature of Patrick Byrne, the Irish giant, by Thomas Rowlandson. From Sinclair, D. *Human Growth After Birth*. Oxford: Oxford University Press, 1975.**

# Physical growth during industrialization

Historians have recently explored in great detail the changes which occurred in the heights of European populations since 1700 because of what can be revealed about the nature of improvements in the standard of living of the whole population and of groups within it (**11.3**, **11.4**). Mean height stands as a proxy for welfare – perhaps the best that is available – in the historical study of periods when such modern measures as average wages or per capita income can be estimated only imperfectly.

The welfare of a population is also, however, indicative of its ability to produce both the agricultural goods on which it can feed and the industrial goods and services which it can sell. Recent studies have therefore exploited the fact that the demand and supply of energy must be in balance in the long- as well as the short-run to explore the historical interrelationships between nutritional status, productivity and mortality.

## Data sources

These interests have influenced the search for data. What is needed is much information on very large numbers of people over long periods of time. Military records have been most useful. From the middle of the 18th century onwards, just as industrialization was beginning to happen in a number of European countries, the armies of those countries became bureaucratized. Proof was needed of height – always seen as a sign of strength – and of other physical details, if only to identify deserters. In some countries, though not in Britain, universal conscription was introduced and every young man had to be examined for fitness for the military life, yielding invaluable evidence.

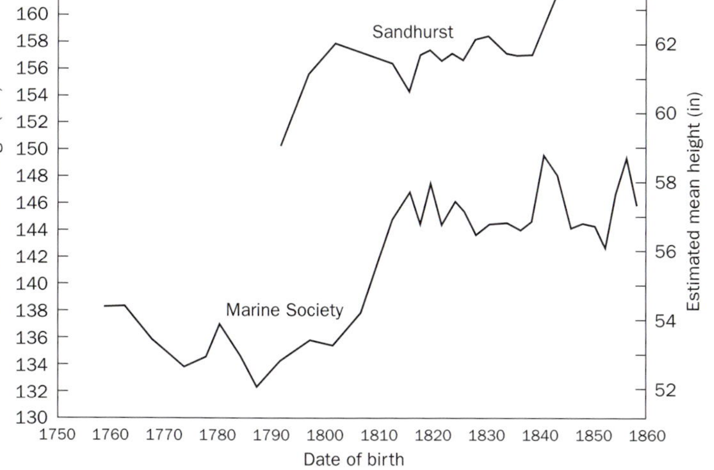


**Height by social class: a comparison of the heights of recruits aged 14 to the Marine Society and Sandhurst, England, 1760–1860.**

Volunteer armies, such as those in Britain and the United States, cause historians more difficulty. First, volunteers came mainly from the working classes; second, recruiters preferred men who were tall and therefore strong. Little can be done about the first bias, except to remember its existence whenever comparisons are made between such data for other countries or for modern periods which are based on samples of the whole population. The second bias can be tackled statistically; height is the classic example of the normal distribution so that a decision by recruiters to reject short men leads to the truncation of, or absence of, observations in the lower tail of a normal distribution. It is possible, however, to use a number of different methods, including Reduced Sample Maximum Likelihood Estimation, to determine the parameters of the underlying distribution and thus the mean height of the population from which recruits were drawn. Armies did not recruit women and children, creating large gaps in knowledge which are filled only by records of women criminals, such as those transported from Britain to Australia, and by the increasing numbers of medical inspections of school children.

## The economic and social context

In every European country, the average person earns at least 12 times as much today as he or she did in 1700. This is the legacy of industrialization or, more precisely, of modern economic growth, since it entailed much more than the growth of industry. Food output expanded while agricultural labour shrank from the overwhelming occupation to one of the smallest. Manufacturing industry, inside and outside the factory, grew rapidly, as did the service sector – from banking to nursing – which now predominates in most developed countries. This process of economic change was neither continuous nor untroubled in any country. Regions, industries, occupations, rose and fell and it is this constant process of change that is illuminated by data on human growth.

## Starting points

In 1800, after some decades of industrialization in Britain and France, but when most European countries were still largely agricultural, average adult male heights ranged between 164 centimetres (some parts of Germany, France, The Netherlands) and 167 centimetres (England, Sweden). Other countries, for which data is only available for later periods, were certainly shorter: Italians and Spaniards were only 163 centimetres as late as 1880. Perhaps most surprising, because they are now so tall, Norwegians are said by one author to have been below 160 centimetres on average in the 1760s. These average heights imply that significant numbers of the populations of these countries were severely malnourished, so short that they would not even reach the 5th centile of a modern height distribution. Moreover, the averages conceal wide variation within these societies between different groups. In Britain, for example, it has been shown that male children aged 13–15 from the London slums of the early 19th century were on average as much as 20 centimetres shorter than children from the British upper classes. Although such differences would have diminished as final height was reached, it is no exaggeration, in this period, to say that the upper class could 'look down' on the working class. Regional differences were marked in Britain and in other countries.

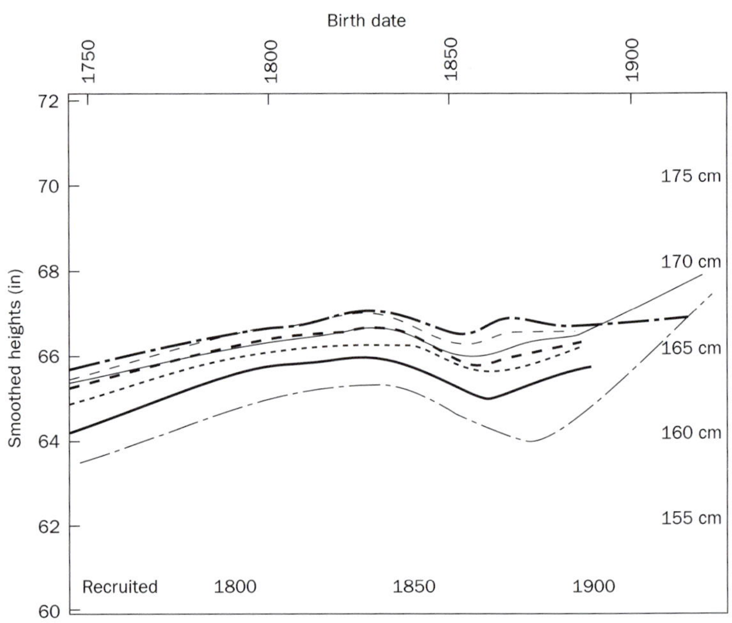


**Smoothed series for the heights of British men aged 18, 19, 20, 21, 22, 23 and 24–30, 1747–1916.**

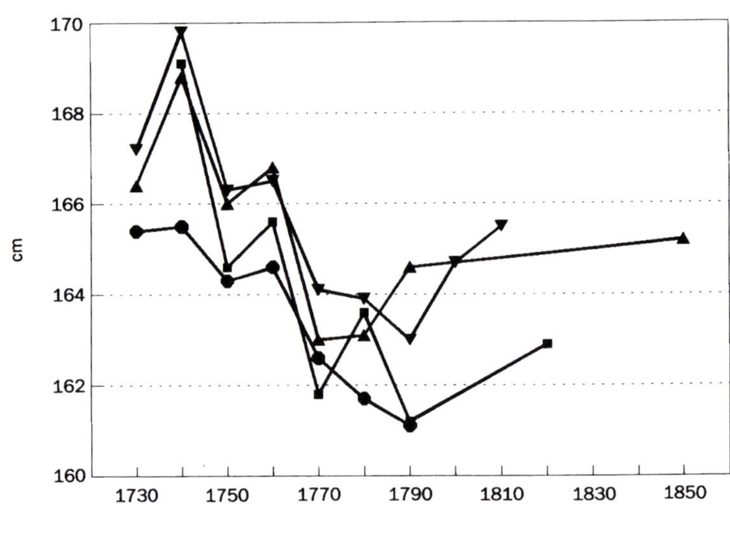


**Heights of European soldiers aged 23–45 years, 1730–1850.**

## Changes over time

The overall impact of the onset of modern economic growth was to make European populations healthier and wealthier and thus taller and heavier. By 1900, the peoples of all the industrializing countries were several centimetres taller than they had been in 1800. Britain, at 169 centimetres, had been surpassed by The Netherlands (170 cm) and Sweden (172.5 cm). The process was not, however, one of smooth growth, either for whole populations or for groups within them. In Britain, Sweden and Hungary, for example, there were periods of several decades in the middle of the 19th century in which average height actually fell, an event paralleled at the same time in the United States. Research into this phenomenon continues, but it seems likely that its origins lie in the growth of the cities and the unhealthy conditions of urban life at the time, which cancelled out the benefits of rising incomes.

As industrialization progressed, there were significant changes in the relative heights of different population groups. In every country, the populations of the towns, which were the focus of industrialization, grew in height more rapidly than did rural groups so that, over time, the rural advantage which was marked in the 18th and early 19th centuries was first eroded and then reversed. In Britain, for example, the rural Scots populations were the tallest subgroup, despite their apparently restricted diet, at the onset of industrialization but, by the 20th century, they were among the shortest. Londoners, the shortest and renowned for their stunted appearance, had become among the tallest.

## End points

Growth in average height has been much more marked and consistent in the 20th century than in the 19th, despite the intervention of wars and economic depressions which temporarily moderated the rate of increase. By 1950 the population of The Netherlands already had an average adult male height of 178 centimetres, well on the way to its current height of 181 centimetres, making them the tallest population in the world. Sweden, another country with relative income equality and a strong welfare state, was close behind, with Britain, France and Germany lagging. The overall improvement during the period of industrialization since 1800 was of over 10 centimetres; modern economic growth had changed the shape of the European peoples, as it is now also doing for peoples throughout the world. *Roderick Floud*

## GROWTH PATTERNS FROM RECRUITING DATA

Officers of the British 54th Regiment recorded recruit stature in either of two columns, 'Size when enlisted' or 'Size after growing'. This regimental record was started in 1762, such usage being widespread in the mid-18th century. A practical side of using height and height standards was in keeping with the early military belief that short men could not tolerate the 'fatigues' of a soldier's life. Furthermore, the record's structure also shows that young men were recruited below minimum height limits, recognizing that many 16 to 18 year-olds might still grow a few more centimetres.

How much growth potential was realized at different times is an interesting puzzle, particularly given the shortage of published information on height by age in mid-teen years. Cross-sectional stature data (**1.16**) for European and American groups shows that slowest growth and shortest heights at maturity are those from the poorest circumstances(**1.12, 11.4**). While 18th- and 19th-century boys did not reach modern heights, even with longer growth periods, relative higher economic status always brought better growth; for example, élite British cadets of 1844 and American soldiers of 1760 were taller than European immigrant soldiers of 1760 and Scots prisoners of the 1840s. Remarkably, while these four groups span the course of the Industrial Revolution, there is little variation in mature mean statures (between 167–168 cm).

There were nevertheless large differences in growth-rate, and fluctuations in mean height across the age of industrialization. For example, fluctuation in heights of 16-year-old Marine Society recruits – London boys 'rescued' from poverty for naval service, and among the poorest of the poor – shows great variation between 1760 and 1850. Between 1760 and 1800 the Industrial Revolution began but growth did not improve. The early 19th century saw industry expand and food fall in price, with dramatic growth benefits. By the mid-1830s economic stratification intensified and the poorest in London fell back to early industrial growth patterns. In America, in 1864, whereas native-born soldiers were taller than contemporary Europeans, they were only slightly taller than men from those same states 100 years earlier. Americans had better food but negative effects of industrialization including those of overcrowding and poor sanitation may have limited any possible gains in growth.

*A. T. Steegmann, Jr.*

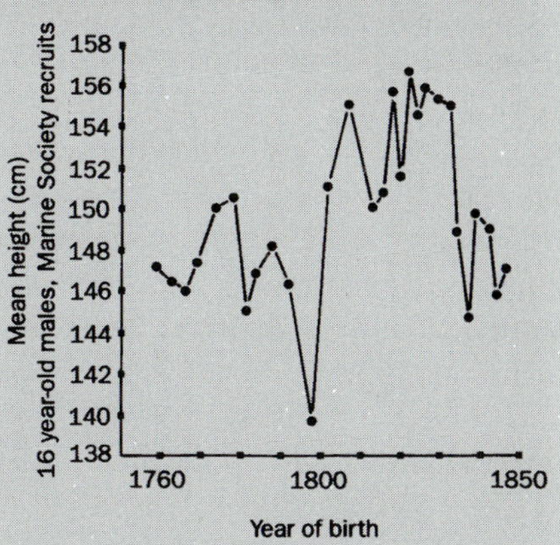


**Secular trend in height of 16-year-old male recruits of the Marine Society. These boys were recruited from the poorest areas of London, England, for maritime service. Note that the time axis is year of birth rather than year of measurement.**

*Cross-sectional data on growth and fully-attained male stature, primarily from military recruiting records. Columns headed 16 through 25 show age in years.*

| 16 | 17 | 18 | 19 | 20 | 21 | 22 | 23 | 24 | 25 | Population and dates |
|---|---|---|---|---|---|---|---|---|---|---|
| 173.1 | 176.1 | 177.6 | 178.0 | 178.0 | | | | | 177.5 | Dutch, general population, 1965. van Wieringen et al. 1971:18 |
| 174.5 | 175.5 | 176.6 | | | | | | | 176.7 | American, general population, 1970s. Frisancho 1990:59 |
| 170.0 | 172.1 | 173.7 | 173.8 | | | | | | | British, élite military cadets, 1844. Floud et al. 1990:77–88 |
| 169.3 | 170.5 | 171.6 | 174.3 | 173.9 | 173.3 | 172.9 | | | | British, élite military cadets, 1852. Floud et al. 1990:77–88 |
| 166.6 | 168.3 | 169.6 | 171.2 | 171.3 | 173.9 | 172.0 | 171.9 | 172.7 | 171.9 | American, soldiers, 1790. Steegmann: original data |
| 166.1 | 167.7 | 168.3 | 167.6 | 167.3 | 167.8 | 168.1 | 168.1 | | | Englih, soldiers (54th Regt.) 1762–99 Steegmann: 1985:86 |
| 162.6 | 165.4 | 167.6 | 169.4 | 170.8 | 171.7 | 172.0 | 172.2 | 172.4 | 172.5 | American, soldiers (NY, NJ, PA), 1864. Gould 1869:113–144 |
| | 164.7 | 165.0 | 168.4 | 168.0 | 169.8 | 169.9 | 168.4 | | | Irish, soldiers (54th Regt.), 1762–99. Steegmann: 1985:80 |
| 162.1 | 164.1 | 165.7 | 167.0 | 167.9 | 168.7 | 169.2 | 169.5 | 169.6 | 169.6 | Europeans, soldier immigrants to US, 1864. Gould 1869:113–114 |
| 162.1 | 163.6 | 164.4 | 168.0 | 168.6 | 167.0 | 167.8 | 168.6 | 167.1 | 167.4 | Europeans, soldier immigrants to US, 1760. Steegmann: original data |
| | | 163.8 | 166.1 | 167.1 | 167.4 | 167.1 | 168.1 | 167.9 | 167.9 | British/Irish, soldiers, 1800–14 Mokyr & O'Grada 1994:4 |
| 158.5 | 160.1 | 164.5 | 167.1 | 167.6 | 168.1 | 167.9 | 167.9 | | | Scots, urban prisoners, 1840s–1850s. Riggs 1994:67 |

## GROWTH AND DEVELOPMENT IN ART HISTORY

European historic paintings and sculpture can offer insight into normal and abnormal growth (see Part seven), sometimes supplementing information gleaned from written records and archaeological finds. In 1502 the physician Gabrielo de Zerbis had described anatomical differences between adults and children, while later in that century, Leonardo da Vinci dissected and sketched aborted fetuses and still-born neonates in addition to doing similar work on adult cadavers. While anatomic drawings of the era used a paradigm adapted from architecture in their focus on spatial relationships, the fact that Leonardo included both infants and adults in his studies suggests an awareness that not just size, but structure changed during maturation (**5.17**). Methods for the accurate rendition of the human figure were worked out by Albrecht Dürer in the late 15th century, contributing to significant changes in the portrayal of children.

The Spanish painter Velázquez in the first half of the 17th century is well-known for his numerous portrayals of adults with dwarfism, both alone and in groups. In one, a male achondroplastic dwarf is shown alongside the 2-year-old Prince Baltasar Carlos; there is no apparent difference in stature. A 19th-century lithograph by an anonymous artist, entitled 'Shakers', includes in a row of normal-sized dancers a man with achondroplasia; he is dancing towards the back along with African-American members of this religious sect. These pictorial representations of growth pathology are of particular interest in conjunction with the contemporaneous emergence of interest in the 'normal' development of the human body. The art of the grotesque plays a role in constructing the ideal body (see Introduction) through its opposite; in this case, depictions of growth abnormalities emphasize the existence of a normal course for human development.

In Western painting and portraiture, children are rarely the central subjects unless they are of noble birth, or belong to the artist, but when they appear, they may do so numerous times over years, often accompanied by other data of interest to the auxologist, such as records of birthdays. Velázquez' series of paintings of Prince Baltasar Carlos began in 1631, when the prince was 2 years old, and continue about every other year until he was about 7. According to Gombrich, 'Interpretation on the part of the image-maker must always be matched by the interpretation of the viewer'. Thus, one problem with using information derived from historic paintings is that of accounting for changing cultural and stylistic contexts. Contemporary esthetics as well as technical expertise determine the form of representations; da Vinci demonstrates an ability equalling and surpassing that of modern medical illustrators, and yet his masterpieces are fully cast within Renaissance ideals for beauty and form.

Standardized methods, currently lacking, are needed to allow longitudinal information to be gleaned from series of paintings of the same child or children over many years. Such information might thus supplement that obtained from, for example, local burials of a contemporary sample. Also, art and artifacts of non-Western origin depicting children may prove to be a rich source of knowledge concerning growth in a variety of populations (see Part ten).

*Leslie Carlin and C.M. Hill*

**Diego Velázquez 'Las Meninas' (1656). Courtesy of Museo del Prado, Madrid. Note the dwarf standing before the dog on the right, represented with great realism.**

**Rodrigo de Villandro's painting 'Philip IV as a Prince and the Dwarf Soplillo' (*c.*1616). Courtesy of Museo del Prado, Madrid. Here the dwarf could be a young boy. The painter has not realistically captured the physical characteristics of achondroplasia. Both de Villandro and Velázquez painted in the court of Philip IV.**

# The secular trend

The term 'secular trend' is used to describe marked changes in growth and development of successive generations of human populations living in the same territories. This has been predominantly documented in European and European-origin populations during the last 100–200 years. In a number of groups, average heights and weights have been shown to be greater with each generation, while the onset of puberty, especially the adolescent growth spurt and menarche, have been taking place at progressively younger ages. It has been shown to plateau in some European populations. However, studies of non-European populations have identified a lack of change, and even reductions in body-size and delays in sexual maturation across generations in some but not all groups studied. Documented secular effects relate either to an increase in the rate of human growth and development, or to an increase in adult body-size.

Most evidence adduced for the presence of a secular trend involves measures of height and/or weight. Secular changes have been demonstrated in physical characteristics of newborn babies, growing children, and adults. Under improved environmental conditions, the mean birth-weight, head-breadth and -circumference of newborn babies all tend to increase, while the proportion of babies born at low birth-weight (less than 2.5 kg) or prematurely (gestational age less than 37 weeks) declines. In growing children, the growth-rate for height, weight and other bodily measures increases, while the mean age of onset of puberty decreases. In adults, the secular trend has been identified by increased mean height, weight, head-breadth, facial width, head circumference, mesiodistal and buccolingual diameters of tooth crowns, and absolute and relative anteroposterior diameter of the pelvis.

Mean heights and weights of children in The Netherlands, Belgium, Britain, Austria, Norway and Sweden have all shown positive secular trends, having reached a plateau for Dutch, Swedish and Norwegian girls, mean heights for all three lying between the 50th and 75th centiles of the National Center for Health Statistics (NCHS) (1977) references. For populations of European origin in the New World, positive secular trends have been shown for Canadian, United States and Australian children. It has plateaued for the former two groups, the United States population reaching a maximum average stature which is lower than that of northern European populations, of which the Swedish and Dutch populations are currently tallest. A possible explanation for this difference between northern Europe and North America is the likelihood that the

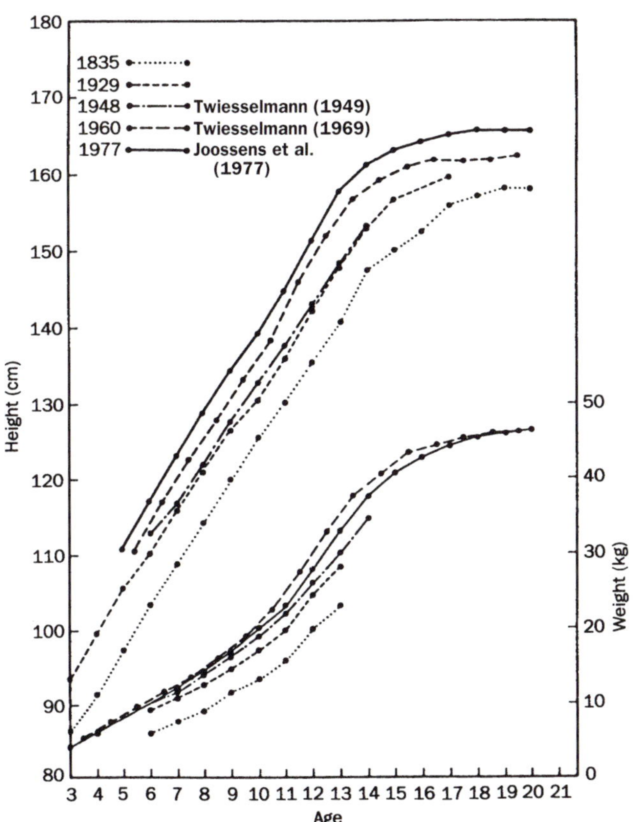


**Secular trend of height and weight of Belgian girls.**

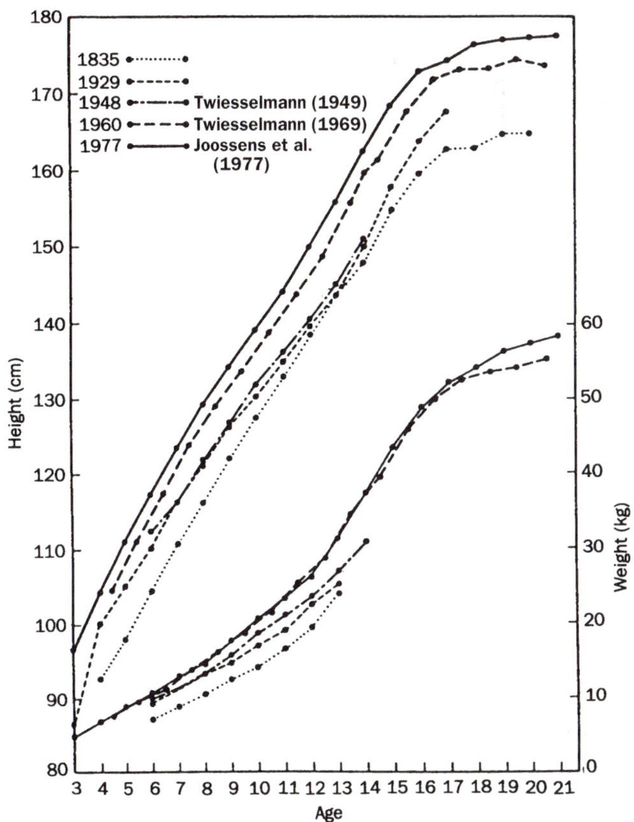


**Secular trend of height and weight of Belgian boys.**

United States growth patterns represent a mixture of north and south European growth patterns. South Europeans, including Spanish and Italian, are on average shorter in childhood than their north European counterparts.

The positive secular trend in height is not necessarily accompanied by a proportional weight increment. Although relatively greater increase in height than in weight has taken place in Belgium and Poland, greater increases in weight relative to height have been observed in Sweden, England, Australia, Canada and the United States. Although the prime determinants of the secular increase are claimed to be increased availability of public health services and improved nutrition, a number of other possibilities have been put forward by various authors. These include the expression of heterosis resulting from ethnic and social class migration and intermarriage; effects of assortative and selective mating without sizeable immigration; increased formula-feeding of infants; reduction in child labour and other forms of child abuse; and improvements in prenatal care. The latter two explanations are likely to play out through improved health and

## SECULAR CHANGES AND CLASS DISTINCTIONS IN GROWTH IN SWEDEN

Secular growth changes and class differences of the Swedish population have been studied through conscript data on height dating as far back as the 1750s. Conclusions from a study of three cohorts of conscripts in the Swedish county of Uppland during the 19th century were that a height increase had occurred during that century, but that this increase was unevenly distributed over different regions and different socio-economic strata. The socio-economic differences (**1.12, 11.4**) were more pronounced in the latest registered cohort, 1873–74, which the investigators claimed reflected the ongoing proletarianization process at that time. Students, in contrast to peasants, workers and farm labourers, were, in their height increase, more than two generations ahead of the population as a whole. A further suggestion was that it was environmental influences during childhood which caused these adult differences between population subgroups. As evidence for this they found a close relationship between the height increase and the declining mortality figures during the same time period.

The changing pattern of growth of Swedish children and youth, however, can only be studied over the last 100 years. The first study on children's growth was reported in 1885 by the professor of medicine Axel Key. Not only height at different ages but also weight, weight for height and maturational rate have since been studied, both regarding secular growth changes and in relation to socio-economic differences. Swedish children have grown taller by about 10 centimetres during the last 100 years. This trend seems now to have stopped, at least for the moment. A small increase seems to have taken place around puberty (**5.17**) between school-children born in 1955 and those born in 1967, simply indicating that those born in 1967 were maturing a little earlier. Their heights after puberty, however, did not differ from those of the children born in 1955.

Weight in relation to height at the school-age years decreased during the period from 1883 to 1964–73. This tendency, however, seems subsequently to have changed. School-children born in 1967 compared to school-children born in 1955 were considerably heavier at the ages 7–15 years – especially the girls at the ages 13–15 years. Comparable results have been reported for Swedish conscripts. At the same time it was found that the dietary energy intake was about the same for the school-children born in 1955

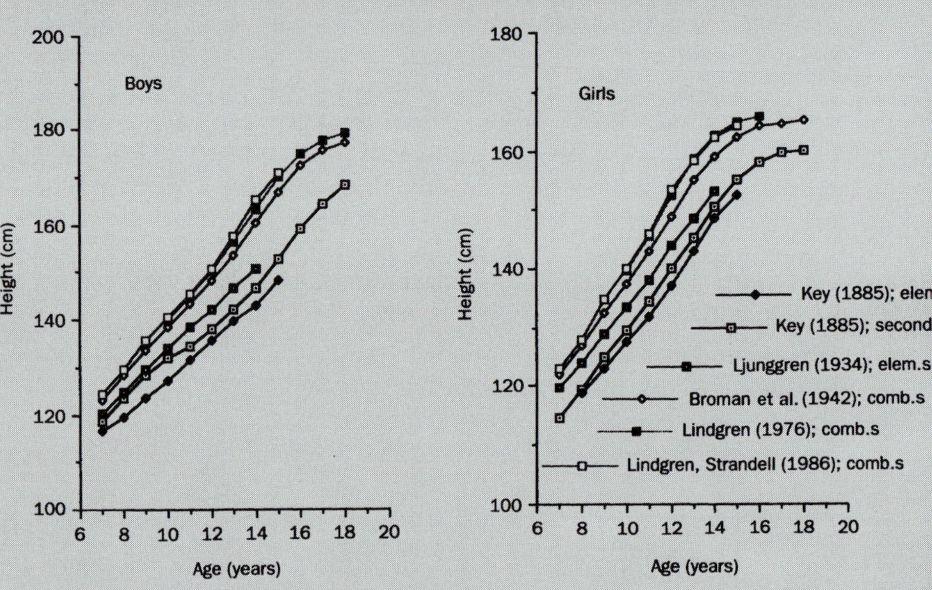


**Secular growth changes in height in Swedish school-children from 1883 to 1982.**

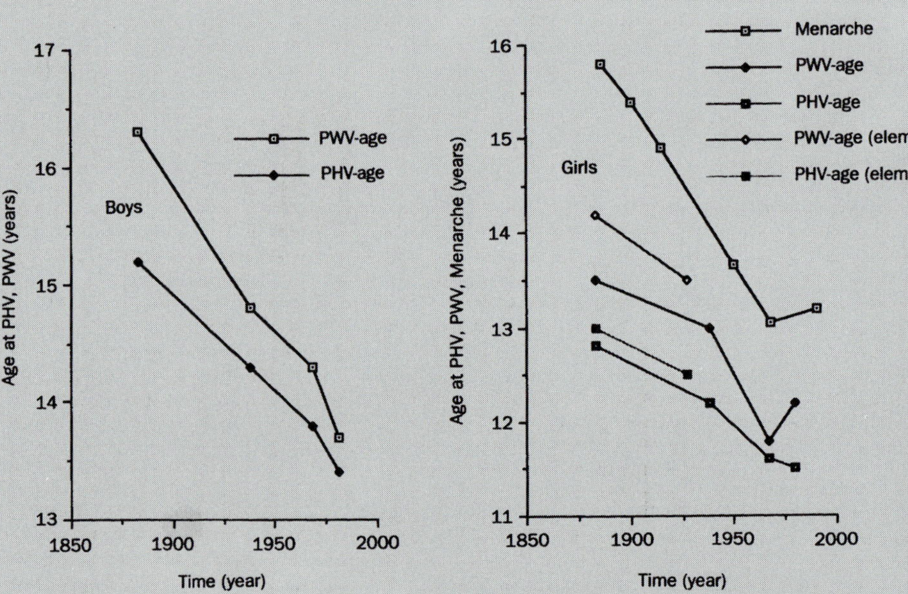


**Secular growth changes in maturational rate (PHV-age, PWV-age and menarcheal age) in Swedish youth from 1883 to 1990.**

nutrition, while all are likely to be related to socio-economic circumstances. Changes in weight-for-height relationships across the secular trend suggest changes in body-weight perceptions and the manipulation of body-size, for example, in slimming regimes; or may reflect levels of dietary availability and dietary quality. Higher weight-for-height in England and the United States may reflect greater dietary energy availability and/or formula-feeding across the period of the secular trend, while the 'taller and slimmer' pattern of secular increase may reflect other effects, such as assortative mating or heterosis, although this possibility has not been rigorously tested. Another possibility is that the greater skeletal growth relative to soft tissue growth may be a function of greater intake of non-energy dietary sources which might have been growth-limiting at some stage. Deficiencies of calcium and zinc have been associated with growth faltering, and it is plausible that increased intake of these micronutrients could have an influence on the secular trend independently of energy intake.

Evidence for a positive secular trend in non-European origin populations has been shown for African-Americans, San

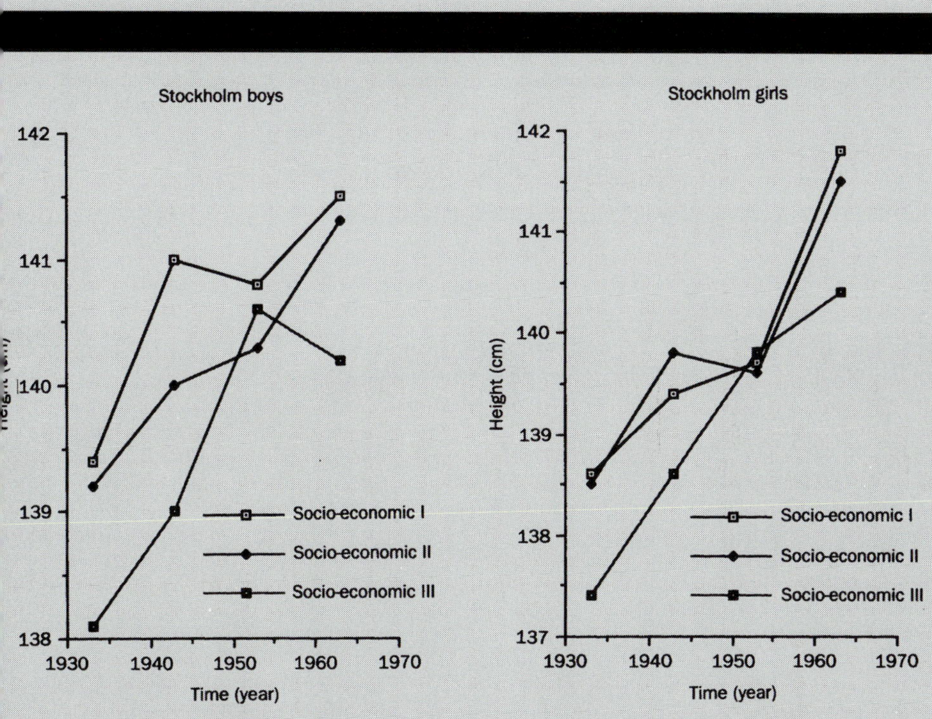


**Social class differences in height at the age of 10 in Stockholm school-children born in 1933, 1943, 1953 and 1963.**

and 1967, so the conclusion was that children born in 1967 were less physically active (**5.12**).

The pattern of maturational rate of Swedish youth during the last 100 years shows that girls' menarcheal age has gone down from 15.8 years in 1886 to about 13.0 years in 1990. The trends for peak-height-velocity (PHV) age and peak-weight-velocity (PWV) age have not been as rapid as that for menarcheal age (**1.15**), probably because of the greater genetic control of PHV-age. Thus, not only do Swedish boys and girls enter puberty at an earlier age than formerly, but the passage from PHV to PWV and (in girls) menarche is faster nowadays than 100 years ago.

Social class differences in height and maturational rate, which existed from 1883 up to the 1940s when comparing elementary (low social class) and secondary (high social class) school-children, seemed to have disappeared in a nationwide sample of urban children born in 1955. There were no differences between social groups I, II and III (I is high, III is low) in height at the ages 10–18 years, or in PHV-age, PWV-age or menarche. The only social group difference that was found was weight for given height – boys and, especially, girls in social group III weighed more in relation to height than those in the other social groups. A later, also nationwide, study concerning school-children born in 1967 gave similar results, in that there were no socio-economic differences in height at the ages 7, 11 and 15 years, with the exception that girls from social group III at the age of 7 years were about 1 centimetre shorter than the girls from the other social groups. The children of social group III born in 1967, like these in 1955, weighed more in relation to height than those in the other social groups.

Despite these nationwide studies of children and youth born in 1955 and after, indicating that former socio-economic differences were levelled out at a national level, analyses of school-children in the city of Stockholm born in 1933–63 showed a different pattern. There were socio-economic differences in height at ages 7, 10 and 13 years in Stockholm children born in 1933 and 1943. These differences were, however, not present in Stockholm children born in 1953 (thus confirming the results for the country as a whole). However, in Stockholm boys born in 1963 differences appeared again at ages 7 and 13 years, and the same tendency, although not reaching statistical significance, was found in Stockholm girls. Thus, results on Stockholm children born subsequent to 1963 were awaited with some trepidation. In effect, in the 1980-born cohort there were differences in height of border-line significance between social classes I and III. These averaged only 0.5 centimetre in the preschool years, but the value increased with age and for girls became statistically significant at the age of 5.5 years. There were no apparent weight differences, which implies that, as expected, weight for height was slightly greater in the lower class. In head circumference, however, there were clear differences, significant at least in girls, the lower class being smaller by 1 per cent.

Thus the secular trend towards increasing tallness seems to have come to a halt in Sweden, and so also has the trend towards an earlier maturational rate as indicated by menarche. The trend that still seems to be going on is a gradually higher weight for height during childhood and especially during puberty. Regarding social-class differences, there seems to have been a golden period for mainly the lower class born in the 1950s, which for a short period levelled out the social-class differences. However, as worsening socio-economic conditions overtook the Swedish population, these differences seem to have returned. *Gunilla Lindgren*

bushmen of the Kalahari desert, southern Africa, Indians in south India and the United Kingdom, and among Japanese, Brazilian, Australian Aboriginal, Inuit and some Papua New Guinea populations. In none of these cases is there clear evidence that the secular trend has reached a plateau. The absence of secular changes has also been observed, notably among African populations in Ivory Coast, Senegal, Sudan and Burkina Faso, and in the majority of Indian and Papua New Guinea populations observed. Negative secular trends have been identified in a small number of groups in India, Africa, and Papua New Guinea. Since human growth and body-size responds with considerable sensitivity to environmental quality, the positive secular trend in populations of the developing world has largely been attributed to improved nutritional and health conditions, while the lack of any such trend has been taken as being suggestive of a lack of change in environmental quality. The negative secular trend is taken to be a response to environmental deterioration, and reversals of positive trends have been observed in historical sequences of population stature, as well as in contemporary populations.

In summary, population-level growth and development characteristics can change across generations in either a positive or negative way. When historical data is in question, it is usually very difficult to ascribe precise causes to such changes. In more recent sequences, causes for secular changes are easier to identify. *Stanley Ulijaszek*

See also 'Growth as an indicator of social inequalities' (**1.12**), 'Skeletal growth and time of agricultural intensification' (**11.1**), 'Physical growth during industrialization' (**11.2**), 'The secular trend' (**11.3**), 'Social and economic class' (**11.4**), 'Modernization and growth' (**11.5**) and 'Defining the growth characteristics of new populations' (**13.2**)

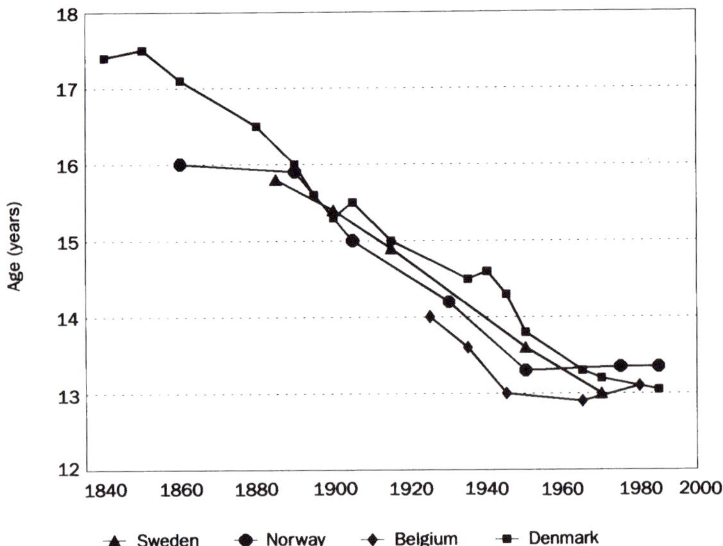


**Secular trend in age at menarche, Northern Europe.**

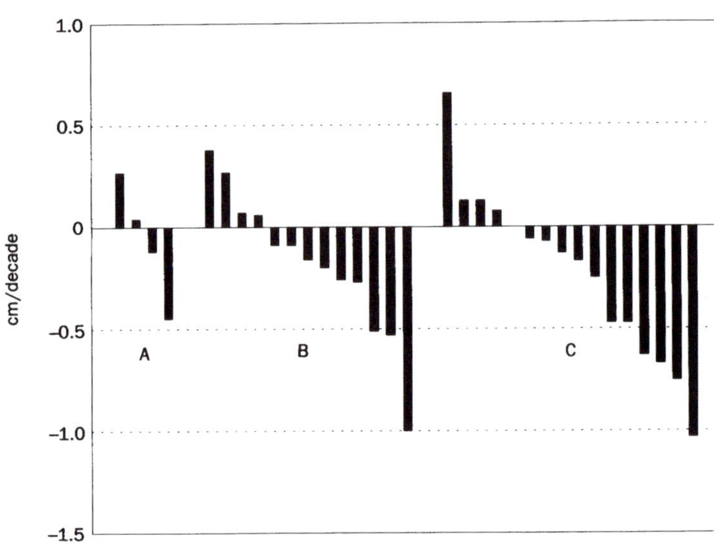


**Secular trend in adult male stature in India across the 20th century (calculated from Ganguly, 1972).**

# Social and economic class

Social and economic class are powerful influences on human growth and development. Conversely, stature, body composition (fatness and muscularity), and rate of development influence the social, emotional, and economic status of children, youth, and adults. An appreciation of these biosocial interactions between growth and socio-economic status (SES) has been realized only in the past 200 years. In the late 18th century it was shown that during the years of growth, sons of the European nobility were, on average, taller than sons of bourgeoisie families. A century later in Italy Luigi Pagliani (See Appendix 1, 11) noted that children from the higher social classes were taller, heavier, and had larger vital capacities (the maximum volume of air that can be inspired in one breath) than poverty-stricken children. About the same time (1875), H.P. Bowditch (See Appendix 1, 10) gathered measurements of height and weight of 24,500 schoolchildren from the Boston, Massachusetts, area. He found that children of the labouring classes were smaller than children from the non-labouring classes. Bowditch said the non-labouring classes were taller because of the '... greater average comfort in which [they] live and grow up...'

The SES effect is often considered to be a proxy for more direct influences on growth, such as nutrition, health care, physical labour among children, and physical/psychological stimulation. The relationships of the first three to growth have been abundantly described in the literature, and there is more limited evidence for the effects of physical/psychological stimulation. Inadequate physical and emotional contact during infancy is known to inhibit physical growth (**9.4**). Conversely, growth stimulation may occur in cultures requiring children and youth to undergo a painful or emotionally difficult rite of passage, such as circumcision or scarification. Many studies have shown that urbanization is associated with increased stature (**11.7**). Of course, cities provide many of the nutrition and health benefits that are positive influences on growth, but urbanization also may increase social/psychological stimulation. Studies in Japan and Poland have associated increased stature with the rate of urban population growth itself, independent of nutrition and health care, and with Engel's coefficient (the amount of family income not needed for shelter and food). Larger urban populations lead to an increase in the number and variety of public and private educational and social services, as well as more variety in occupations, entertainment, and social life. An urban life style and a higher Engel's coefficient (i.e., a higher SES), allows people to avail themselves of these physical/psychological stimulants that may promote growth.

The urbanization-higher SES-increased growth relationship is particularly strong in the higher-income nations of the world. In The Netherlands, Norway, and Sweden, social class differences in child growth and development have largely been eliminated, although there is evidence that it is creeping in again in Sweden. SES differences, as measured by education, occupation and income, do exist in these nations, but federal systems of guaranteed health care and social support services attempt to provide people of all SES levels with an equal share in the environmental opportunities for growth and development. In other high-income nations the SES stratification is still an important determinant of human growth. In the United States, for example, much research establishes a direct link between poverty and a high prevalence of low birth-weight, chronic illness, lead-poisoning, low quantity and quality of health care, short stature, obesity, and increased absenteeism and school failure. Thus, economic, political and cultural systems act to differentially allocate the benefits and risks for growth between SES groups. Epidemiologists refer to this process as 'risk focusing'.

In the lower-income nations the risk-focusing effect of SES is even more pronounced. Poor living conditions, including high rates of undernutrition, infectious disease, childhood mortality and heavy workloads are common. Consequently, small average stature of low SES children and adults, and the slow rate of maturation of children, are consistent features of these populations. Studies in the low-income nations of Central America, Africa, and South Asia show that small body-size limits the possibility for SES advancement. This is due to the effects of undernutrition, which leads to reduced physical work capacity (**9.10**) due to a loss of muscle mass and altered body proportions (for example, short legs in proportion to length of the trunk and head). Mild-to-moderate undernutrition also effects the aptitudes and cognitive abilities of preschool children, and determines, in part, the likelihood that

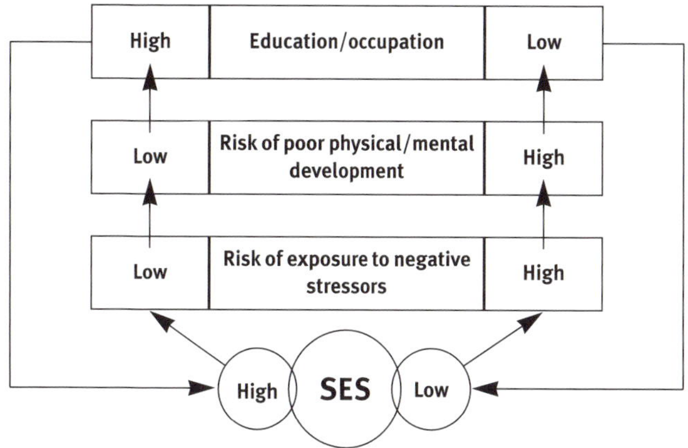


**Model of risk-focusing developed by Lawrence Schell. In this model, socio-economic status (SES) factors allocate and focus risk of exposure to negative stressors leading to disability and poor qualifications for employment, lower occupational status, and low SES.**

the child will enter school. Short stature, reduced work capacity, and lower educational attainment work synergistically to recycle low SES and poverty from generation to generation.

There is another side to the issue of socio-economic factors and growth, which is that within all social classes taller individuals tend to move up in SES and shorter individuals tend to move down or remain stable in SES. Stature-related social mobility even occurs within families, in that taller sons and daughters tend to have higher SES than their shorter siblings. In these studies SES is usually measured by education and occupation. Research in several European nations and Japan finds that within families the taller siblings were better educated than shorter siblings. Better education, of course, is likely to lead to more skilled and higher-paying employment and, consequently, higher SES.

Stature, by itself, does not determine occupation, education, or socio-economic attainment, but there is a strong social bias in favour of the tall, and the non-obese, which may help facilitate their SES climb. Throughout the world there is a general bias in favour of taller men by women in courtship and mating. Numerous studies, from a diversity of disciplines, find that the taller, non-obese man or woman is given preference in many arenas linked with SES, such as the perception of intelligence, academic performance, and social skills, as well as in finding jobs and perception of both current and future job performance. The empirical data show that height and weight have a small but statistically significant effect on measured academic performance, and both an initial and long-term effect on job earnings in the higher income countries. The cognitive and school effect is almost entirely due to rate of maturation – faster maturation during the growing years results in both greater stature and cognitive performance. Educational attainment is the major predictor of occupation and job earnings, and the tall are often better educated. Yet, stature

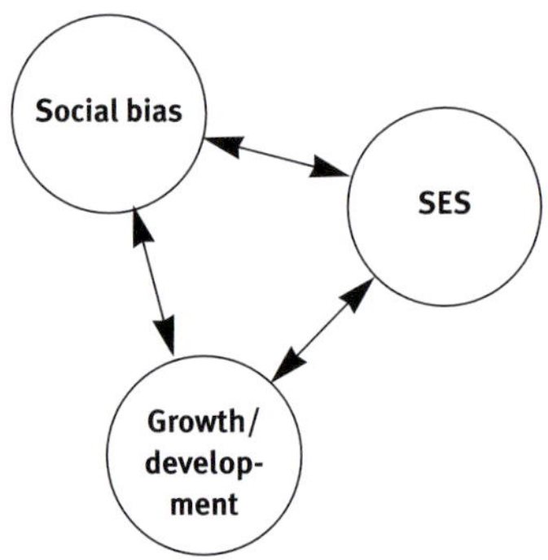


**Growth and development as both a cause and a consequence of social and economic effects. Double-headed arrows indicate two-way interactions within a cyclical process. SES = socio-economic status.**

Poverty

Ethnicity
■ social class
■ behaviours
■ customs

Location
■ urban
■ rural

Dietary risk
■ quantity
■ quality
■ hunger
■ micronutrients

Housing risk
■ crowding
■ safety
■ contamination
■ allergens

Healthcare risk
■ quantity
■ availability
■ usage
■ cultural barriers

Parental factors
■ maternal health
■ education
■ behaviours
■ beliefs

Health
■ low birth-weight
■ chronic illness
■ lead levels
■ other

Growth
■ low stature
■ low weight
■ overweight/obesity

Cognitve development/ function

Behaviour/ attention

School-loss days

Motivation

School achievement
■ grades/tests
■ drop-out/promotion

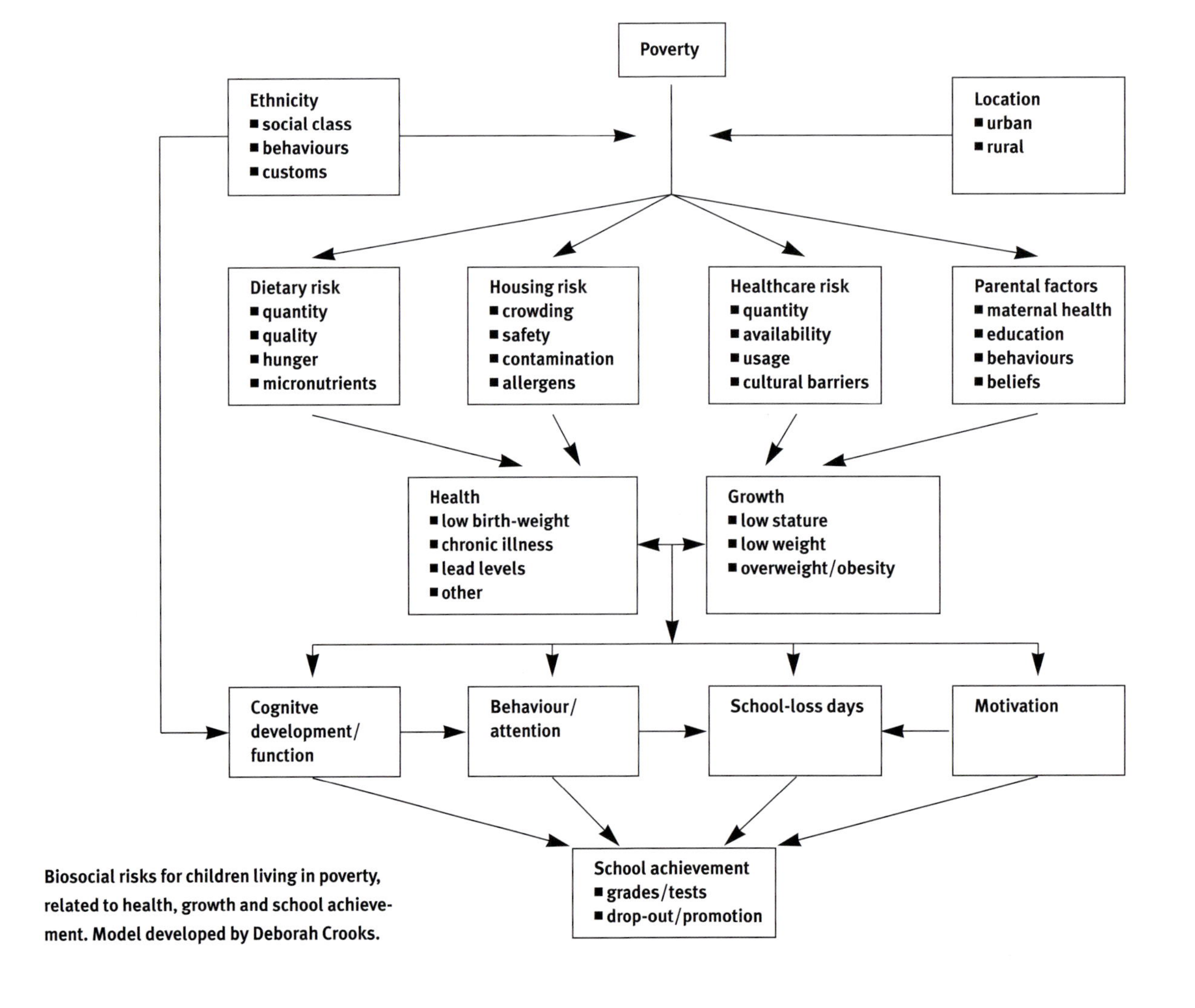

**Biosocial risks for children living in poverty, related to health, growth and school achievement. Model developed by Deborah Crooks.**

has an independent effect on earnings and job status, as height is a strong predictor of social dominance.

A growth bias for desirable types, in terms of stature and body composition, exists. The concrete result of this bias is similar to racial discrimination: individuals or groups of the accepted 'type' are more likely to receive better care, in the widest sense, than individuals or groups of the undesired type. A positive feedback relationship between growth and socio-economic status results from the social bias, better environmental conditions lead to larger size, taller individuals tend to rise in SES, and higher SES leads to better environmental conditions. An opposing cycle exists for those from lower SES, or shorter individuals from any social class. The result is that differences in physical size between individuals are both a consequence and a cause of socio-economic effects on growth.

*Barry Bogin*

## SOCIAL CLASS AND HEIGHT IN BRITAIN

Socio-economic indicators are used in the assessment of the impact of social disadvantage on an individual's physical and mental development (**1.12**). A socio-economic gradient in the population related to anthropometric measurements and health can be assessed using a large number of variables. In Britain social class, based on the occupation of individuals or the breadwinner in a family, has been commonly used in the assessment of differences in height in the population. The Office of National Statistics (ONS) publishes an updated version of the 'Classification of Occupations' every 10 years. Social class is based on this classification of occupations and is in many ways an imperfect measure of social disparity in stature.

*Social classification based on occupations*

| Social class | |
|---|---|
| I | Professional occupations |
| II | Managerial and technical occupations |
| III | Skilled occupations<br>(N) non-manual<br>(M) manual |
| IV | Semi-skilled manual occupations |
| V | Unskilled manual occupations |

*Some alternatives to using social class to assess the impact of socio-economic factors on health and physical characteristics*

**Occupational classifications**
Job security
Employment status (whether employed or unemployed)
Other classifcations using occupational information (e.g. Goldthorpe class scheme)
Income

**Non-occupational classification**
Characteristics of neighbourhood
Level of household overcrowding
Housing tenure
Education level and qualifications
Household resources (car, central heating etc.)
Receipt of welfare benefits
Indices based on a range of independent variables
Demographic indicators associated with hardship

Recently a committee of the Economic and Social Research Council (ESRC) identified that the principal advantages of the social class classification were that it provides an authoritative standard for the country, that it allows time series analysis and hence historical continuity of socio-economic trends, that its use has demonstrated a social-class gradient of health variables, and the small number of categories make it a manageable system. The use of social class for assessing changes over time is not necessarily consistent because the classification is modified every 10 years as some occupations disappear, others are created and some are reclassified: occupations are reclassified to maximize social gradients in relation to health indicators. The classification has been criticized also for lack of conceptual rationale in that it is not clear what the classification is supposed to measure. In addition, the classification has limited coverage because the population who are frequently the most disadvantaged are not included in the classification, i.e. the long-term unemployed and those who have never worked. There are, however, alternatives to the social class classification.

### SOCIAL CLASS AND HEIGHT IN CHILDHOOD

Most of the available national data are from school-children. Using information from the 1946, 1958 and 1970 British birth cohort studies, the differences in height between children from non-manual and manual-worker social classes in the period 1953 to 1981 in 7 and 11 year olds has slightly decreased from 2.4 centimetres to little less than 2 centimetres (between 1.9–1.3 cm). A height difference of 1 centimetre has been reported in a study of British children carried out between 1979 and 1983 contrasting children from manual-work and non-manual-work backgrounds. Adding the 1988 data from the National Study of Health and Growth, a monitoring system of growth, to the above information it can be shown more convincingly that the unadjusted difference in height between manual and non-manual social classes is now less than 1 centimetre.

The National Study of Health and Growth provides the most appropriate current data set for assessing the possible association between social class and children's height. Based on recent data, unadjusted differences between children whose fathers have a non-manual occupation and those whose fathers have a semi-skilled or unskilled manual occupation, in a representative sample, is of the order of 0.75 centimetre in England and Scotland. The difference in a sample restricted to inner-city-area children is heterogeneous over ethnic groups. The association between child's height and father's social class is not statistically significant after adjustment for parental heights, child's birth-weight, size of sibship, birth order and ethnicity. The most recent national data of social class differences in height in children under the age of 5 were collected in the 1970s. The difference between non-manual social classes and social class IV and V was approximately 1 centimetre.

### SOCIAL CLASS AND HEIGHT IN ADULTS

In the 1991 Health Survey for England, a difference of approximately 3 centimetres was reported between men and between women in social classes I and II and those in social classes IV and V. After adjusting for age, education level and geographic region, a significant association persisted between manual and non-manual occupations of approximately 1.5 centimetres in both sexes. Using data from a previous cross-sectional study carried out by ONS in 1980 a persistent difference of 2–3 centimetres was reported between manual and non-manual workers born between 1910 and 1960. This difference was also found in the British Regional Heart Study and 1946 birth cohort study. However, in the lifestyle survey carried out in the 1980s the differences in height between the manual- and non-manual-labouring social classes are smaller in the younger than the older group. Several research workers have demonstrated, using social class of origin and own or husband's social class, that tall men and women tend to move socially upward; conversely short men and women tend to move downward.

Social class gradients in height based on father's occupation are almost non-existent in British children of the 1990s. In the adult population recent cross-sectional surveys contradict each other. Two surveys showed a persistent difference in height between manual and non-manual social classes, but another survey reported a flattening of the social-class gradient. The most likely scenario is that a social class-gradient in height will disappear in the young adult groups at the beginning of the 20th century. If this does not occur, possible explanations to the persistence of a gradient are social mobility favouring the taller or social-class gradient related to the growth spurt during adolescence. The lack of social class differences in height will not necessarily mean that socio-economic factors do not affect the height of the British people; this should be assessed using a wider range of socio-economic measures.

*Roberto Rona*

# Modernization and growth

Modernization encompasses all the developments that have taken place in the wake of industrialization (**11.2**) and mechanization, including the loosening of boundaries between social classes and increased social mobility. In addition, it includes the growth of widespread education, the evolution of procedures of industrial negotiation and the development of social welfare systems at the national and international level. These things have happened in most Western nations and many Asian nations as their populations underwent the transition from a predominantly agricultural mode of subsistence to one of wage-earning within an industrialized cash-based economy. In Britain, as elsewhere in Europe, social class differences in average weights and heights of (**1.12**, **11.4**) children were great, with working-class children being more than 10 centimetres shorter than upper-class children, between 1800 and 1850. Modernization has resulted in reduced height differences between upper and lower classes, both groups showing secular increases in growth-rates and body-size, but with working-class children showing greater secular increases in the 50 years or so spanning the turn of the 20th century. The secular trends (**11.3**) reported in Western countries are largely a function of modernization processes, and there is no clear evidence that the secular trend in these countries is of necessity associated with increased levels of obesity.

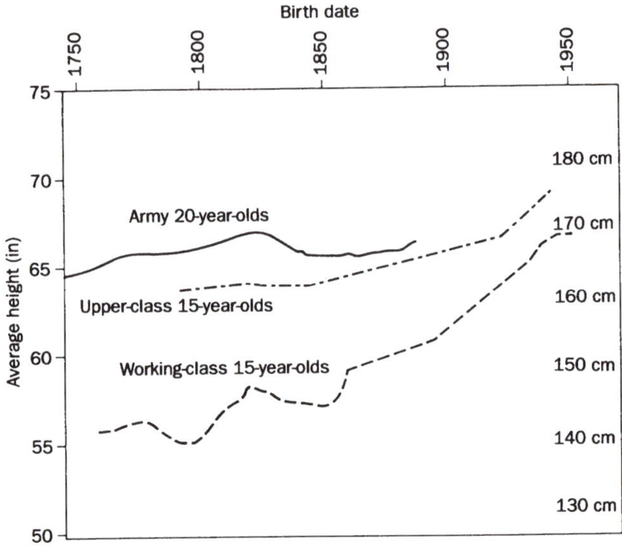


**Heights in Britain, 1750–1950.**

The experiences of newer nation states becoming independent in the 20th century are different from those of the now-industrialized nations which underwent their industrial revolutions in the 18th and 19th centuries. In this arena, modernization has been defined as the interaction of less complex energetic, technological and socio-economic systems characterized by regional production and consumption with contemporary economic systems of industrial technology influenced by the national and international market, as well as by social and political factors. Thus, modernization in the non-Western nations takes place within a framework of colonial and neocolonial dominance in which growth- and health-related outcomes cannot be predicted from what is known of the modernization of the Western nations. The growth and development of children in developing countries is influenced by many factors, most importantly food intake and infectious and invasive diseases. These two factors are modulated by household food availability, the caring capacity of household members, housing and environmental conditions, and the existence and utility of health services. Modernization includes the process of urbanization, and people's experience of modernization will vary according to whether they are located rurally, or in urban or peri-urban settings.

**High-rise public housing in modernizing Seoul, Korea. Photograph by Stanley Ulijaszek.**

**The modernization of Kuala Lumpur, 1985. The redevelopment of city-centres involves changing environmental conditions for all, including the urban poor, whose low-cost housing disappears. Photograph by Stanley Ulijaszek.**

In rural Sarawak, cash income due to logging and cash-cropping activities has increased the possibility of fatness and disease risks associated with obesity. Photograph by Stanley Ulijaszek.

*Selected structural and functional characteristics of rural and urban areas (from Gross, 1994)*

| Characteristics | Rural | Urban |
|---|---|---|
| Number of population | Low | High |
| Population density | Low | High |
| Participation in monetary economy | Lower | High |
| Main income source | Agriculture | Heterogeneous |
| Diversity of information | Low | High |
| Cultural influence | Monocultural | Monocultural/Multicultural |
| Literacy | Lower | Higher |
| Food sources | Monotonous | Heterogeneous |
| Diseases | Communicable | Communicable + non-communicable |
| Social mobility | Lower | Higher |
| Nutritional problems | Undernutrition | Undernutrition + overnutrition |
| Complexity | Low | High |

In modernizing societies, per capita food availability is usually higher in rural than in urban areas since. However, although food may be more abundant and diverse in the latter setting, availability is limited by purchasing power and access (often, markets are located far from newly emerging squatter settlements, and transport is often difficult). Perhaps paradoxically, growth status is often marginally better in urban areas. It has been suggested that this may be due the greater availability and uptake of health services among urban populations, rich and poor alike, resulting in lower overall impact of infectious disease on growth (**9.3**).

Among rural populations, the process of modernization results in generalized improvements which are reflected in improved growth. However, the introduction and adoption of newer technologies and the mechanization of agriculture serves to amplify social and economic disparities between sectors of the rural community. Such disparities are likely to be reflected in differences in growth performance between children from richer and poorer families (**11.4**). In urban populations, the far greater complexity of economic and health factors makes it more difficult to generalize. Although the trend among urban populations has been to reduce breast-feeding duration, there is evidence to suggest a reversal of such patterns, at least in Brazil and Malaysia. There is greater variety of food available in urban centres, with a much greater preponderance of Western-style foods, especially highly processed foods such as sugar, soft drinks, and energy-dense 'fast foods'. As a consequence of increased consumption of such foods, there is also widespread increase in the prevalence of obesity. This is common in low-income communities and families as well as among middle- and high-income families. In higher-income groups, greater growth velocities resulting in higher stature across generations is likely to be associated with increased levels of obesity. In poorer groups, undernutrition and slow growth in some individuals can co-exist with overnutrition and faster growth in others. Furthermore, since many societies in developing countries are undergoing dietary and nutritional transitions, undernutrition and overnutrition can occur in the same individual at different stages of life. For example, individuals who have undergone growth faltering across childhood may carry a higher risk of obesity in adult life, as well as cardiovascular disease, hypertension, and non-insulin dependent diabetes (**5.15**).

Increased social mobility and exposure to a wider range of cultural influences relative to village life can also influence growth patterns, through changing infant-feeding practices (**9.1**), health behaviours, and housing and local environmental conditions. The growth patterns of modernizing human populations do not easily fit into rural, urban, and peri-urban categories, even when differentials between higher and lower socio-economic status are taken into consideration. This is because environmental factors which influence growth are affected by linked behavioural and economic relationships which in turn are mediated by the exchange of people, information and resources between these three economic contexts. For example, each category will include migrants (**11.6**), either temporary or permanent, who possess characteristics of their place of origin and of the place they have migrated to. Thus, an urban population will have mixed characteristics composed of those individuals born there, of permanent migrants from a rural area, and of temporary migrants, as well as settlers whose parents (and sometime grandparents) were born in the urban setting. Conversely, rural populations may include temporary and permanent back-migrants. The growth characteristics of modernizing populations may be conveniently labelled 'urban', 'rural', or 'peri-urban', but these are overlapping categories, inasmuch as urban characteristics flow into the rural context, and vice-versa.

*Stanley Ulijaszek*

# Migration and changing population characteristics

Isolates are very rare in human populations, with total isolation being exceptional and most populations exchanging individuals and mates by migration. Migration is the geographical movement of individuals, and is highly variable in: 1) orientation of movement; 2) the distances involved; 3) in the number of migrants; and 4) their eventual differences to their populations of origin. In addition, motivations to migrate are numerous and diverse: people migrate to improve their economic position, to obtain a better education, to escape war, because they are in trouble with the law, as well as for political reasons. Most recently, the category of 'environmental migrant' was created to take account of those

## GROWTH OF GUATEMALAN MIGRANTS IN THE UNITED STATES OF AMERICA

The Maya of Guatemala are characterized by many biocultural features. One of them is short stature. It is sometimes asserted that Mayans are 'genetically short' due to generations of adaptation to an environment of poor health and nutrition. Alternatively, the Maya are described as 'small but healthy'. These notions imply that no intervention or economic development are needed for the Maya. Recent migration of Mayan refugees to the United States has afforded the opportunity to study the consequences of life in a new environment on the growth of Mayan children.

Mayan children living in Indiantown, Florida and Los Angeles, California, between 4 and 12 years old (n = 240), were measured for height, weight, fatness, and muscularity. These refugee children are taller, heavier, and carry more fat and muscle mass than Mayan children living in the village of San Pedro in Guatemala. The San Pedro Maya are representative in height, weight and other physical dimensions of children of low socio-economic status (SES) (**11.4**) in Guatemala. They live in a village with no safe supply for drinking-water, an irregular supply of any water, and unsanitary means for waste disposal. The parents of these children are employed, predominately, in very low-wage occupations. There is one public health clinic in the village, and treatment of infants and preschool children with clinical undernutrition is common. Most children exhibit signs of chronic mild-to-moderate undernutrition. In contrast, the Mayan refugees in America, while still poor, have access to safe drinking-water, better health care, nutrition supplementation programmes, and more remunerative wage-labour.

The Maya refugee children are similar to United

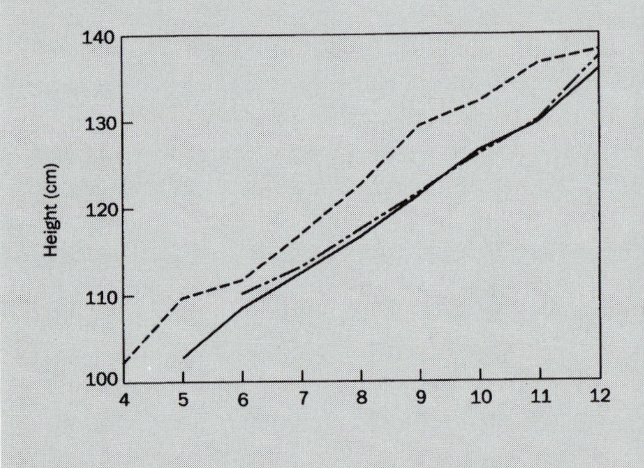

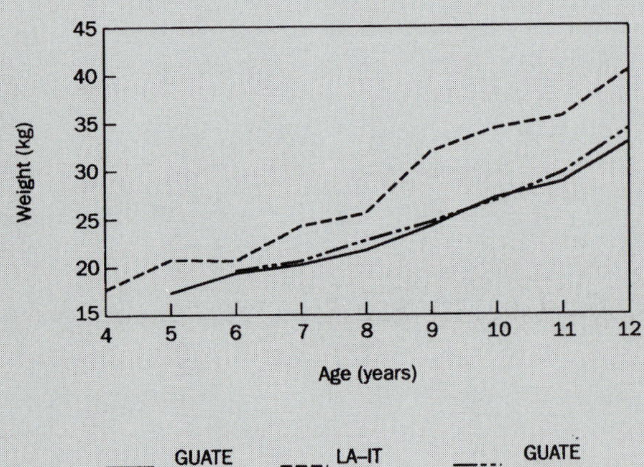


**Above. Guatemala Maya children in Guatemala. Photograph by Barry Bogin.**

**Left. Comparisons of mean height or weight of Los Angeles and Indiantown Maya (LA-IT) with San Pedro, Guatemala Maya (GUATE) children by age. The San Pedro children were measured in two time periods. Note the lack of secular change between the earlier and later periods; evidence for environmental continuity between the two times. Data for boys and girls within samples are combined.**

**Opposite left. Comparison of mean arm muscle area or arm fat area of Los Angeles and Indiantown Maya (LA-IT) and San Pedro, Guatemala Maya (GUATE) children by age. Data for boys and girls within samples are combined.**

**Opposite right. Mean height or weight of Indiantown children by ethnic group and age. NCHS data (a reference sample for United States children) are shown for height. Data for fatness and muscularity are not shown, but are analogous to weight in that the Maya refugees are similar to the other Indiantown ethnic groups.**

leaving the sites of human-made environmental disasters such as the one at Chernobyl. In the past few centuries, migration has taken place on a scale previously unknown in human history or prehistory. For example, more than 50 million Europeans have left their continent since the 16th century and, on a worldwide scale, more than 50 million people have migrated or have become refugees since the end of the last global war, 1945.

In population genetics, migration is used in models where the migrants are considered to be a random sample of their population of origin. If the number of migrants is low and that they remain isolated, genetic drift and founder effect could occur. In the absence of different selection pressures and random drift, migration can reduce gene-pool differences across populations. In such models immigrants are considered as a random sample of their population of origin. However, some authors have suggested that the physical characteristics of more mobile individuals is likely to be different from the less mobile ones, and with selective migration, populations and their gene-pools can diverge.

## Methodological problems

Migration differentials between populations, characterized by factors such as race, age, sex, marital status, occupation, and family status are well documented. Migrants are not

States-born children in terms of body-weight, fatness and muscularity, indicating rapid improvements in health and nutritional status. However, they are shorter, on average, than children of Black, Mexican-American, and White ethnicity living in the same Florida town. It is likely that the present generation of Mayan refugee children are in the first stage of a process of increasing stature from generation to generation. This process, the secular trend in growth, is often associated with migration from a worse to a better environment. Within a few generations, United States-born children of Mayan descent, if allowed equal access to health, economic, and social opportunities, should have an average stature similar to that of the general North American population.

These findings are typical of migrant adaptation to new environments with improved SES conditions, first described by Franz Boas (see Appendix 1, 13) in 1912. In 1969, Gabriel Lasker used the migration and growth research to develop a formal model of human biological plasticity. The term *plasticity* refers to a physical or behavioural change in the phenotype of an organism which results from a change in the environment. Much of the variation in adult human phenotype is the result of developmental plasticity occurring from conception to maturity. Using Lasker's plasticity model it can be shown that most of the variation in adult Mayan stature observed in living populations, and in the archaeological record, is a response to environmental change, mostly to change in the quality of health and nutrition, and not a genetic adaptation.

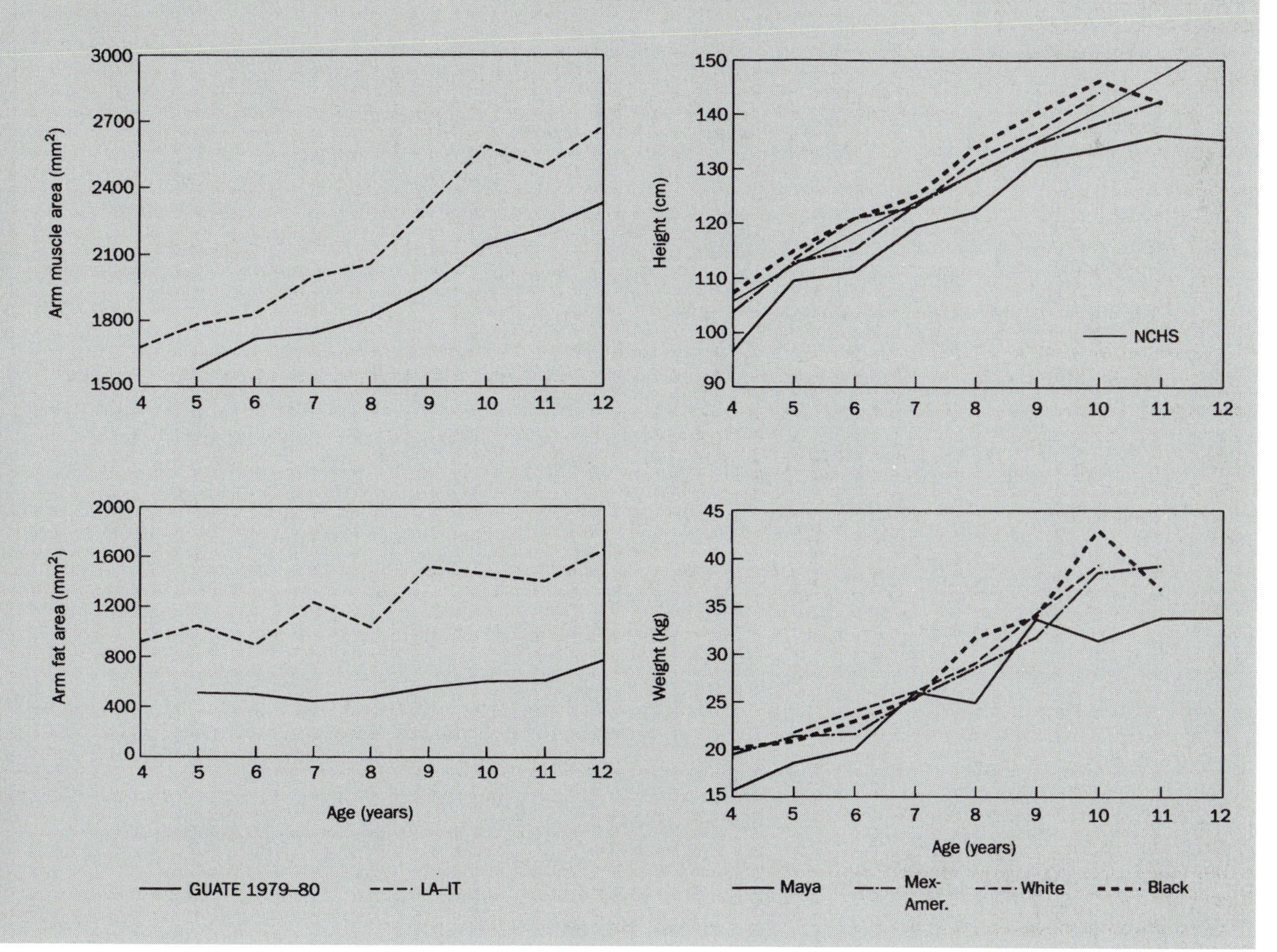

*Barry Bogin*

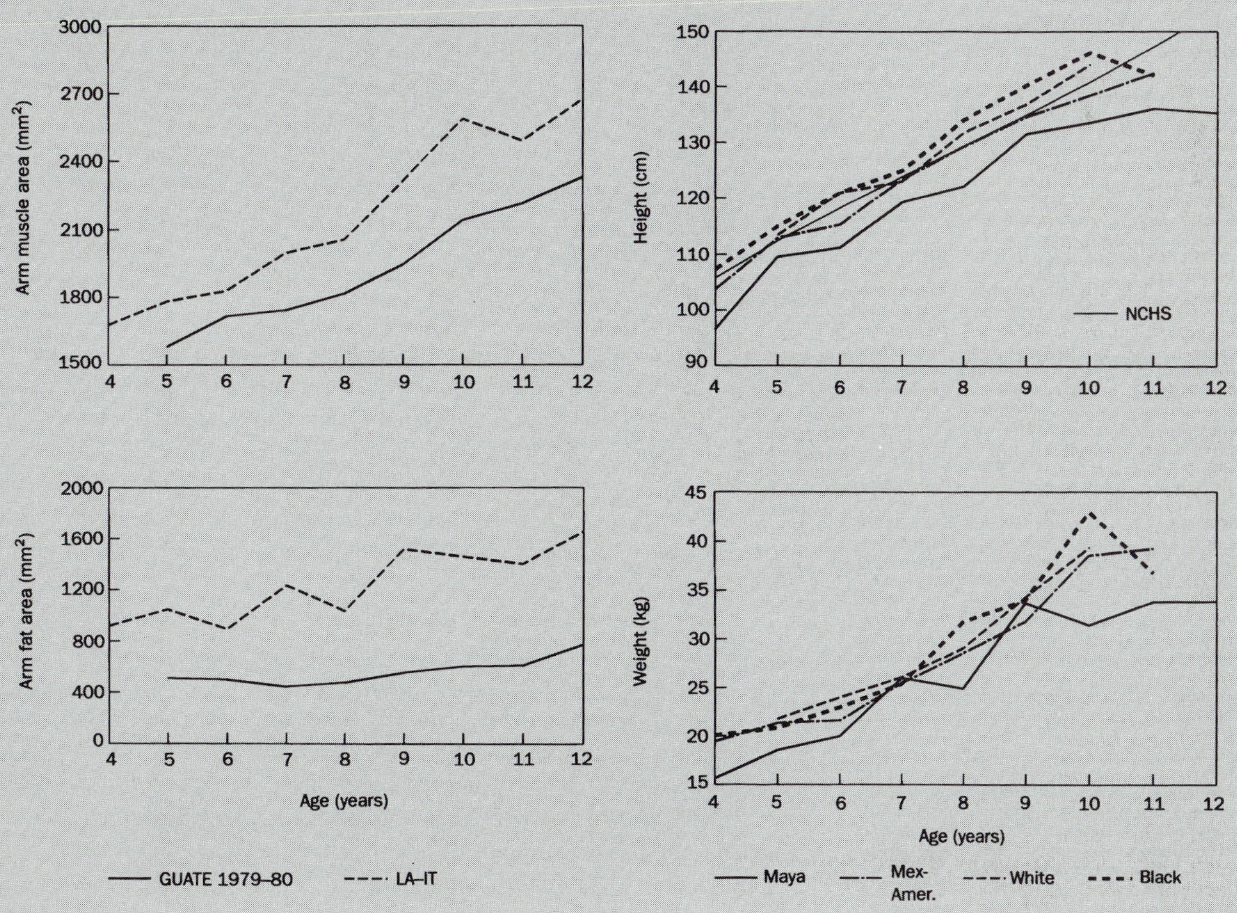

necessarily socially and demographically representative of their population of origin: for example, young adults and especially young adult males are usually over-represented, and more first-born boys are likely to leave than later-born ones. In addition, economic conditions greatly affect the spatial pattern, volume and type of migration. Selective migration also takes place between rural and urban settings: this has become particularly common in the past few centuries, but does not mean that migrants are of necessity biologically different from those that do not migrate.

In anthropological studies on selective migration, the characteristics of interest are multifactorial and are determined by genetic and environmental factors, and it is difficult or impossible to separate these two components with respect to the proportion of variation in the phenotypic measurement that they contribute. Selective migration can be tested only by comparison of newly arrived immigrants with their population of origin. To separate genetic and environmental factors, it might be possible to observe premigrants, when they are still living on the country of origin, with the same subject a few years after migration.

Heterosis can happen as a consequence of migration when there is an increase in the number of exogamous marriages, a decrease of consanguinity and an increase in the frequency of heterozygotes. In this case, accelerated growth, larger stature and earlier puberty would be observed. However, no studies in which social factors have been adequately controlled have been able to demonstrate an effect of heterosis in human populations.

In any comparison of migrants with their population of origin, the age of migrants has to be taken into account; often young male adults are over-represented in migrant populations. In addition, the age at migration needs to be considered when judging the environmental influences on the biology of migrants. Furthermore, comparisons of migrants with their parents can be misleading if the populations under consideration are undergoing secular trends.

The socio-economic origin of the sample is also of great importance. For example, a significant increase in stature of migrants from Ireland has been shown by Hulse (1969) to be due to an over-representation of farmer's sons in the sample of migrants and of migrants who were at the same time civic guards. The average stature of migrants who were not civic guards did not differ significantly from that of sedentes. Sometimes migrants are of upper social class, but most often people are motivated to migrate to improve the economic position of their families. However, migrants from warfare might display the opposite social mix if only wealthier people are able to 'buy themselves out'.

The length of residence has an effect on many growth-related parameters. In migrant studies, the number of years between the migration of the parents and the birth of the child is often not recorded. Sometimes, the number of generations between the migration and the studied individuals is unknown or not recorded. If there is a favourable environmental influence with migration, this is likely to increase with the number of years the children have lived in the new environment.

## Migrant studies involving growth and body-size

In 1912, Franz Boas (see Appendix 1, 13) demonstrated physical changes in the American-born progeny of immigrants from several European countries, including significant differences in stature, weight, head shape, and facial width. Boas ascribed the heavier and taller features to the better health care and nutrition received by the children in the United States. Virtually all of the differences which Boas had shown in the first American-born generation continued in subsequent generations, and in some cases the rates of change were the same. This indicates that the limits of environmental plasticity were not reached in one generation after migration.

Studies taking into account the length of residence in a foreign country during youth have shown no evidence for selective migration on the basis of physical characteristics. A study of Italian migrants in Belgium carried out by Charles Susanne (1979) lent no support to the hypothesis of selective factors for height and weight; indeed the means calculated for Italian immigrants in Belgium were no higher than the means of the sedentee reference population.

Evidence for environmental influences on the physical characteristics of migrants becomes apparent when the age at which migration occurred is taken into consideration. Italians born in Belgium, or individuals who migrated before the age of 25 years, are taller and heavier than those who migrated after the age of 25 years, after controlling for differences in age, level of study, and professional status. In addition, mean heights and weights of Italians born in Belgium and of individuals who migrated before 5 years of age are significantly greater than the means of the individuals who migrated after 5 years of age. A recent study of Chinese people born in Paris confirms these results and seems to show that only improved living conditions (and especially nutrition) have been of great importance in the changes observed.

Globally, in the majority of comparisons, the height or the weight of migrant children are greater than those of sedentes. In some studies, migrant women are lighter than sedentes, and here the influence of modern fashion may be important.

Although environmental conditions of migrants are usually better than those of the sedente population, this is not always so. For example, urban slums in some African and Asian countries have very poor environmental conditions and migrants do not necessarily experience the positive influences of the urban environment.

Some physical changes upon migration may reflect the plasticity of human development to some particular environment. For instance, greater breadths and girth measurements, especially of the thorax, are observed in high-altitude compared to low-altitude migrants; in this case, adaptation to hypoxia may be part of the explanation (**9.9**). This adaptation is directly related to the age at migration and to the length of residency during childhood, but not during adulthood.

Other environmental influences have been demonstrated at an epidemiological level; people migrating in childhood acquire the prevalence of characteristic diseases of the country to which they emigrate. This frequently leads to an increase in overweight and obesity. Investigators have reported that Mexican and Japanese migrants to the United States have increasing prevalence-rates of many cancers, hypertension, coronary heart disease and other chronic disease over time and even over generations.

Despite some methodological problems, studies of growth, development and adult body-size show that physical change is most usually due to changed environmental quality. At least in studies of rural–urban migration, factors including improved living conditions, better medical care, better nutrition during childhood, increased stature and weight and decreased age at menarche have been observed.

*Mean height and weight of Italian migrants in Belgium by age at migration (values corrected for age)*

| Migration status | n | Height (cm) | Weight (kg) |
|---|---|---|---|
| Born in Belgium | 54 | 169.6 | 63.4 |
| Before 5 years | 55 | 169.1 | 65.9 |
| Between 5 and 15 years | 88 | 167.3 | 62.8 |
| Between 15 and 25 years | 140 | 167.4 | 62.9 |
| After 25 years | 79 | 166.5 | 62.1 |

This confirms that the environment, as mediated by such factors as socio-economic status, food availability, and health status, is the primary determinant of biological change in growth and development following migration.

*C. Susanne, M. Vercauteren and M. Zavattaro*

See also 'Growth as an indicator of social inequalities' (**1.12**), 'The secular trend' (**11.3**), 'Modernization and growth' (**11.5**) and 'Defining the growth characteristics of new populations' (**13.2**)

## URBAN–RURAL DIFFERENCES IN GROWTH PATTERNS OF NEPALI CHILDREN

A study of pre-adolescents, 307 boys and 124 girls, 6 to 14 years of age, was designed to compare health status and life styles in contrasting environments. Poor rural children in a remote but self-sufficient village of central Nepal (1870 m) were contrasted to three urban groups in Kathmandu (1300 m): poor children in squatter settlements, children of the street (all homeless boys) and middle-class school-children. The rural and urban poor (**11.7**) but not schoolchildren, are stunted, yet not wasted. Unexpectedly, homeless boys are taller than either village or squatter boys, whether recent arrivals or living on the streets for up to 10 years; this sample (23% of a total census of homeless boys) includes those from both low- and high-caste backgrounds, and those receiving little or frequent help from non-governmental organizations. Ethnic composition differed between samples, but differences are consistent for both the Tibeto-Burmese and Indo-Nepalese. Elsewhere in Nepal, studies confirm the short stature of rural populations.

A number of life-style variables contribute to such differences in prepubertal growth status. First, the village diet is a highly monotonous regime of cereals with low energy-density sauces, excessive fibre and limited nutrients, such as protein and vitamins A and C. By contrast, the urban diet is more varied; in particular, the homeless spend most of their earnings on restaurant foods, and may also obtain nutritious meals from benevolent tourists or institutions.

The village life style is also the most arduous: children carry up to 130 per cent of their body weight up and down mountain slopes. Rag-picking and begging for the urban homeless and squatters are physically less demanding.

Villagers have no access to medical care. While few rural boys (13%) complain about ill-health, intestinal ailments and diarrhea (known to affect growth-rates (**9.3**)) are the main burden of disease and prevalent from early childhood. As expected, school-children report fewer complaints than squatters or homeless children. Serum levels of acute phase proteins, indicating inflammatory responses to infection, are very high among villagers.

The general finding of rural–urban differences in growth status echoes those of other studies and also emphasizes the considerable heterogeneity of urban environments. The specific finding that homeless street-boys are less stunted than poor villagers and squatters in stable households is more surprising; while there is evidence for selective migration of taller boys amongst recent arrivals on the street, the fact that duration of up to 10 years of street-life has no impact on levels of wasting or stunting indicates that boys adopt relatively successful strategies to cope with urban homelessness. *C. Panter-Brick*

*Growth status of Nepali children (Mean NCHS (1977) Z-scores)*

| | Village | Squatter | Homeless | School |
|---|---|---|---|---|
| Boys (n) | 52 | 62 | 111 | 82 |
| Stunting | −2.91 | −2.68 | −2.39 | −1.86 |
| Wasting | 0.10 | −0.27 | −0.22 | −0.17 |
| Girls (n) | 23 | 45 | – | 56 |
| Stunting | −2.90 | −2.61 | – | −1.68 |
| Wasting | 0.49 | −0.21 | – | −0.28 |

*Number of health complaints (data from boys only) self-reports during a health survey lasting 10 days*

| | Village | Squatter | Homeless | School |
|---|---|---|---|---|
| Number of reports | 364 | 153 | 444 | 237 |
| Number of complaints (n) | 47 | 92 | 317 | 107 |
| Complaints | | | | |
| Diarrhea | 4% | 11% | 10% | 6% |
| Stomach-ache & worms | 6% | 9% | 12% | 15% |
| Respiratory | 1% | 13% | 16% | 6% |
| Fever | 0% | 3% | 5% | 1% |
| Headache | 1% | 9% | 7% | 10% |
| Skin/injury | 1% | 11% | 14% | 3% |
| Eye/ear infection | 0% | 2% | 3% | 0% |
| Other | 1% | 3% | 6% | 5% |

# Urbanism and growth

Urban–rural differences in child-growth patterns have been known for more than a century. While urban–rural differences are well known, the extent of the difference between urban and rural samples varies according to specific features that characterize the urban and rural communities that are compared. Thus, urban–rural differences may be non-existent, as they often are in contemporary comparisons involving children from long-industrialized countries, or they may be substantial as they frequently were in American and European countries 100 years ago, or are today in developing countries where there may be large differences in nutrition, medical care, morbidity, pollutant exposure, parental characteristics and energy expenditure. With this in mind, generalization about urban rural differences is difficult and laden with qualification, although conclusions about differences in growth between specific urban and rural samples are possible.

Recent comparisons of urban and rural communities in many European countries and North America show no urban advantage in children's heights. By the 1960s, urban and rural children of similar socio-economic status in the United States did not differ in height, while in Canada, differences in height were small and often favoured rural children. In other parts of the world, urban children are generally taller. In Poland the average height of army conscripts in 1976 decreased consistently with decreasing population size of the conscripts' place of residence. In the 1960s, rural Finnish 8-year-old boys were about 2.5 centimetres shorter than their peers from Helsinki. In Greece, the difference between rural boys and boys in Athens was nearly 5 centimetres. When a growth advantage exists, it usually involves faster maturation as well, such that advantaged urban girls may reach menarche as much as 1.5 years earlier than disadvantaged rural girls. Thus, the advantage in size of either urban or rural samples depends on the age and maturational status of the children being compared. The urban-rural difference is generally larger during early adolescence since whomever reaches adolescence first will add extra height from the earlier occurrence of the adolescent growth spurt.

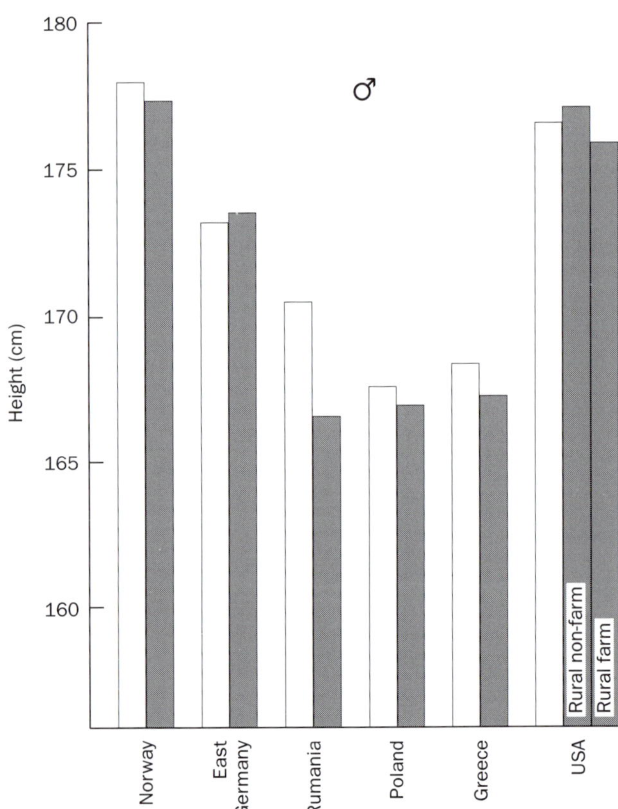


**Mean adult male height in urban (open columns) and rural (shaded columns) areas in Europe and the USA. From Eveleth, P.B. and Tanner, J.M. *Worldwide Variation in Human Growth*. Cambridge: Cambridge University Press, 1990.**

There may be a pattern to urban–rural differences in child growth. They are non-existent or small in long-industrialized countries. In developing countries, well-off urban children are larger than poor rural children. In South Africa, well-off urban children from Soweto had heights and weights close to the National Center for Child Health Statistics' 50th percentile while poor Soweto children and rural children grew less well. Furthermore, the poor Soweto children were smaller in more age groups than the rural children. This is consistent with studies in other developing countries, including Guatemala, Mexico, Nigeria, and Zaire. In general, studies of urban slum children show the depressing effect of poverty regardless of the degree of urbanization. Urban slum children suffer high rates of morbidity and growth retardation which may be related to crowding, increased person-to-person contacts and more infections, poor sanitary facilities and feeding patterns. However, the alignment of environmental conditions such as these are hardly the same across all urban slums, and some urban children, such as infants in Cebu, Philippines, experience less stunting than their rural counterparts, despite more infection. Furthermore, even within a poor urban community such as in Guatemala City, variation in household socio-economic status is related to growth. Thus, when urban environments include better nutrition, less infection, better medical care, better educated parents, and larger parents whose own growth benefited from previous urban environments, urban children will grow larger than rural ones having fewer advantages for growth, but when conditions in urban environments are poor, as in urban slums, urban children may grow poorly in comparison to rural ones.

Urban–rural differences in child growth can change over time as well as vary among countries. In many of the

**Urban children in Bangladesh. Photograph by Stanley Ulijaszek.**

developed countries where there is now little difference between urban and rural child growth, there were once substantial differences. Fifty years ago in the United States, the average urban child was generally larger than the average rural one, the difference being as much as 6 centimetres in pre- and post-pubertal boys, and even larger during puberty. Cohorts from the United States born 100 years ago also show an urban advantage of generally less than 6 centimetres. Analysis of American military records from the late-18th through the mid-19th centuries showed that rural and urban native-born men had similar final heights until the second decade of the 19th century after which the rural men had the height advantage for some time. These fluctuations in height can be used to suggest changing economic well-being for the population (**11.2**).

While the urban-rural comparison can summarize differences in health and growth, it is not very specific in that it is rarely able to isolate the specific factors responsible for differences in growth patterns. Furthermore, each environment may contain growth-promoting and detracting factors that may offset one another. Since there is remarkable heterogeneity among urban and among rural environments over time and space, the promoting and detracting forces in each environment may be different though the growth patterns that result may sometimes be similar. The urban–rural comparison is capable of masking the effects of significant specific influences on growth. Thus, the beneficial effects of early prenatal care, and education, for example, may sometimes be offset by detrimental influences such as greater exposure to human-borne pathogens, stress, and pollution. If additional pollution, crowding and stress cause urban environments to deteriorate further in well-off countries, the physical growth of children may become poorer in urban areas in comparison to rural ones.

*Laurence M. Schell*

See also 'Growth as an indicator of social inequalities' (**1.12**), 'Environmental toxicants' (**9.5**), 'Physical growth during industrialization' (**11.2**), 'Social and economic class' (**11.4**), 'Migration and changing population characteristics' (**11.6**) and 'Defining the growth characteristics of new populations' (**13.2**)

11.8

# Obesity, fatness and modernization

Obesity, fatness and large body-size are terms which vary in the degree to which they describe a desirable or undesirable, healthy or unhealthy state, while health and perceptions of healthiness vary across cultures and time. In England and The Netherlands, the positive sense with which body fatness was perceived in the past has been displaced by a negative one. This is probably true of other industrialized countries too. The same change in perception of body fatness is currently taking place in Malaysia, Samoa, and Nauru, as well as in many other modernizing populations.

A social definition of obesity is one of fatness beyond the socially accepted norms for a given society, while a medical definition relates to individuals who weigh more than the upper acceptable limit for their height and frame. The medical definition may rely on actuarial information such as United States life assurance statistics, or reference data, such as body mass index, for adults and children. However defined, health risks associated with obesity are quite clear: although high blood pressure, raised concentration of plasma low-density cholesterol and a low concentration of high-density cholesterol fractions are all important risk factors for heart disease, weight-gain makes these factors worse, and weight-loss makes them better, in both males and females. However, these relationships differ across populations.

The social definition of obesity also varies across societies and time since, although fatness is and has been symbolically linked to psychological dimensions such as self-worth and sexuality, the nature of that symbolic association is not constant. Traditional positive perceptions of obesity and fatness are easy to understand: societies that express this experience food shortages and uncertainty of food supply, and individuals with larger, fatter body-size represent variously success, better reproductive performance, higher social status, and better survivability in times of shortage. The advantages of body fatness in buffering adversity and promoting reproductive success are considerable; indeed, the energy store represented by body fat in adequately nourished women is usually about equivalent to the energy cost of a pregnancy.

Cultural elaboration of the desirability of fatness in the form of ritual fattening practices is still present in many traditional societies. This may have symbolic importance for communities in coping with a highly seasonal and often marginal subsistence environment. In past Nauru, food supplies were irregular, and the traditional fattening of young females was related to reproductive performance: the strength of the young woman was nurtured at the time when she became an adult, and she was supported in her role as the creator of new life, in a community which perceived itself to be under demographic threat. Elsewhere, ritual fattening of Annang women in Nigeria is perceived to enhance fertility, while among the Azande of Central Africa, obesity and fatness is associated with greater fertility as well as higher social status.

Comparisons of perceptions of obesity across cultures are made difficult by differences in the process and rate of modernization (**11.5**) across societies. Thus, the changes in body-size, shape, health and disease risk must be considered in relation to changes in life style and perception associated with the particular set of social and economic changes in any group, community or population. In Nauru society, puberty ceremonies and fattening processes diminished in importance with new forms of social differentiation, which included cash income and education. However, food as a mark of prestige persisted through a period of introduction of new foods that were energy-dense and high in refined carbohydrates. With epidemiological evidence of very high levels of non-insulin dependent diabetes mellitus (NIDDM), the process of modernization seems to have created a contradiction for the Nauruans, who maintain an identity in which food sharing and consumption are very important, but are faced with 'the slur of obesity', which they are told, makes

**Mechanisms for the influence of socio-economic status upon obesity. Model of Sobal, 1991.**

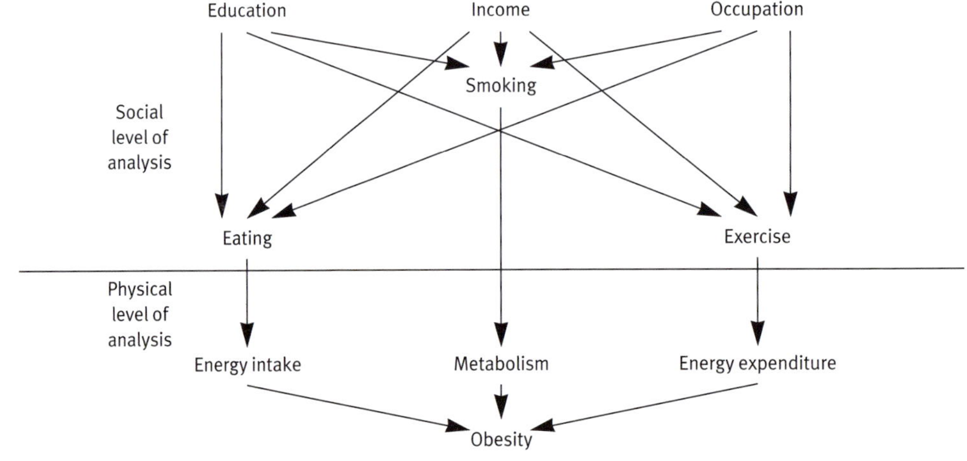

them unhealthy. Obesity and overweight have been described as biological adjustments to modernization among Samoan populations, with women becoming more obese than males in the course of this process. This gender difference has been largely attributed to gender differences in changing work patterns, women initially becoming less active, but men retaining higher levels of activity despite changes in occupation, although the role of increased dietary energy intake cannot be denied. In Malaysia, changes in food resources of villagers between 1968/9 and 1984 toward greater availability of commercial foodstuffs, increased income, and reduced physical activity led to an increase in the number of overweight individuals, and greater concern about obesity as a consequence of exposure to Western views about obesity, fatness and health.

Modernization has taken place in the industrialized nations, and is underway in most populations of the developing world. However, the former second-world countries of the former Eastern European bloc are in a rather ambiguous position in relation to theory about modernization, obesity and disease. The course of modernization in countries like Czechoslovakia and Poland has been very different from elsewhere in the world. In the late 1970s, there were higher levels of obesity in the higher socio-economic groupings of the Polish population than in the lower ones. This pattern seemed to reflect positive cultural and social attributes associated with fatness, and thus it is interesting that the social distribution of obesity in Poland prior to the economic transition of the early 1980s was similar to that found in countries prior to, or early in the process of modernization elsewhere.

Although associations between body fatness, body composition, fat patterning and risk of coronary heart disease and NIDDM have been clearly demonstrated, these relationships vary across populations, and are made complicated by environmental factors associated with different levels of modernization. A further complication is the role of stress in this relationship. For example a neuroendocrine link between stress and day-to-day levels of adrenaline may play an intermediary role in fat distribution and the morbidity risk associated with it. Furthermore, in addition to the stress associated with modernization-driven change, changes in the social perception of obesity could also be stressful, and carry a health risk. Thus, modernizing populations, facing contradictions between their traditional perceptions of obesity and the 'modern' medically-oriented view of obesity as an unhealthy condition, may have additional health burdens associated with this and other social incongruities associated with dietary and health transitions.

Obesity and fatness are usually desirable in the context of general food shortage or uncertainty. In some communities this desirability is elaborated in fattening rituals which may have both biological and psychological importance for the whole community in their coping with low food availability. However, the social desirability of fatness or obesity is usually reversed in the course of modernization, when large body-size and fatness become available to all, and the negative health effects of obesity become apparent.

*Stanley Ulijaszek*

See also 'Fat and fat patterning' (**5.16**), 'Nutrition' (**9.2**), 'Social and economic class' (**11.4**) and 'Modernization and growth' (**11.5**)

**Lady's costume in medieval period (right) and one-piece Kimono in the Edo period. Changing perceptions of female beauty in Japan. (From Ishige, 1995).**

*Characteristics traditionally associated with body-shapes of women in Japan after the 18th century (from Ishige, 1995)*

| Figure | Associated characteristics | | | | | |
|---|---|---|---|---|---|---|
| Plump | pigeon breast & protruding rump | barbarousness | ugliness | toil | rural | healthy |
| | ↕ | ↕ | ↕ | ↕ | ↕ | ↕ |
| Slender | willow waist | elegance | beauty | enjoyment | urban | unhealthy |

## Part Twelve

# The human lifespan

If the human lifespan is considered in an evolutionary context, links between environmental influences on growth and development, and biological and behavioural characteristics in adult life are easier to understand. Although nothing is known about the lifespans of the fossil hominids, some inferences can be made about the evolution of the human lifespan by drawing analogies with extant primate species whose life-history characteristics have been defined. On the basis of available DNA sequence homology, humans appear to be most closely related, evolutionarily, to chimpanzees, gorillas, and orang-utan; with gibbons being the most closely related of the non-ape primates.

Excluding industrialized humans, an important feature of human evolution has been the increase in the time between birth and sexual maturity, or of extended childhood. This has been accompanied by a less pronounced increase in longevity, as reflected in the difference in total lifespan between gibbons and apes and humans. There is little difference in longevity between human hunter-gatherers and the apes. However, when longevity of human hunter-gatherer populations is compared with industrialized humans in the second half of the 20th century, there is a vast and clear difference between the two. Notably the period between sexual maturity and death is nearly three times longer for industrialized humans than for hunter-gatherers. Thus a striking feature of contemporary human populations is the greatly extended period of adult life. From an evolutionary perspective, the disorders of late onset are a new phenomenon, since they are unlikely to have presented in any proportion in past populations. Furthermore, they are unlikely to have been selected against, since many disorders associated with industrialized populations reveal themselves at post-reproductive ages. In this sense, the extended human post-reproductive maturity lifespan presents us with a new analytical challenge, since many of the disorders seen are likely to represent genetic–environmental outcomes which are outside of the human design specifications moulded, at population level, in the course of human evolution.

**Opposite: A 100-year-old man holding a photograph of himself as a young man. Courtesy of Getty Images, London.**

*Maximum lifespan recorded for a range of mammals*

| Species | Maximum lifespan (years) |
|---|---|
| Humans | 115 |
| Indian elephant | 70 |
| Horse | 46 |
| Chimpanzee | 44 |
| Brown Bear | 36 |
| Rhesus monkey | 29 |
| Domestic cat | 28 |
| Sheep | 20 |
| Domestic dog | 20 |
| Grey squirrel | 15 |
| Black rat | 5 |
| House mouse | 3 |

Greater population longevity has significance for more than the present industrialized world, since the global trend has been toward increased life expectancy at birth, even in countries with very high levels of under-5s mortality rates. This may be due largely to reduction in infant mortality rates associated with primary health care, medical intervention and economic development of one sort or another. If this trend continues, greater numbers of people in both the developed and developing world will be susceptible to diseases associated with early environmental influences on human growth and development.

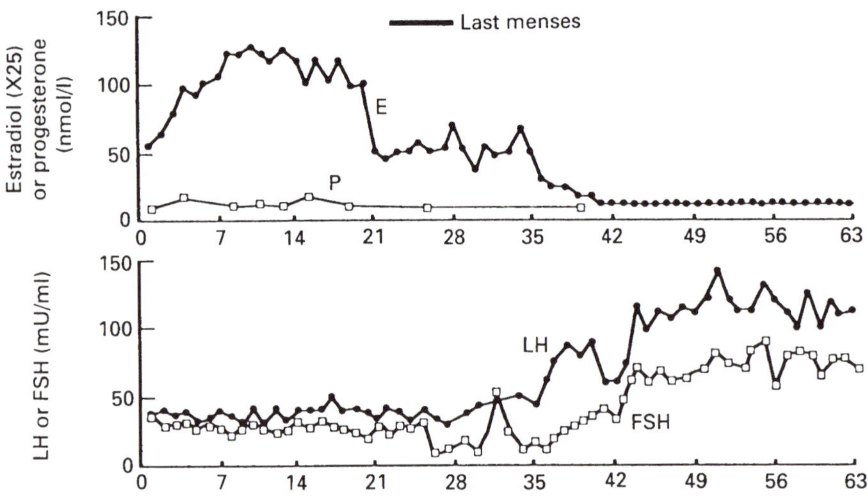


**Changes in the plasma concentrations of ovarian steroids (top) and gonadotropins (bottom) during the menopausal transition in a woman aged 48 years. Note that the fall in estradiol concentration results in the last menses and is followed by a marked rise in the concentrations of LH and FSH. E = estradiol-17β; P = progesterone; LH = luteinizing hormone; FSH = follicle-stimulating hormone. (From Wood, 1994).**

In relation to other species, the human growth pattern is characterized by a prolonged period of infant dependency, an extended childhood, and a rapid and large acceleration in growth velocity at adolescence leading to physical and sexual maturation. The chronology of growth and differentiation of specific organs and tissues varies, and it has been suggested that there may be critical periods when environmental factors such as nutritional stress or exposure to infection can significantly alter the developmental process from its genetic trajectory. There are many ways in which environmental influences at different stages of development can influence morphology, and have consequences for the adult phenotype (**12.1**). These include disease risk, fatness and fat patterning, skeletal muscular morphology, the development of sexuality, neurodegenerative disorders, body composition, physiological set-points for nutritional needs, and immune system ageing.

One biological correlate of the extended human lifespan is the menopause (**12.2**). This is the stage of life in women characterized by follicular depletion and the hormonal changes associated with it. The number of eggs in the ovary is set prior to birth, and a woman's reproductive lifespan is partly defined by this, and partly by the rate of loss, which may be due to ovulation, or a degeneration called *atresia*, which begins before birth. There is a loss of feedback regulation by ovarian steroids when the pool of viable follicles remaining in the ovary becomes too small to produce the amount of estrodiol needed to support further follicular development and the feedback control of the gonadotrophin-releasing hormone pulse generator. Furthermore, there may be some deterioration of hypothalamo-pituitary function with age, and high levels of luteinizing hormone and follicle stimulating hormone may be produced, even in this presence of measurable estrodiol levels. Thus the hormonal profile associated with menopause involves the removal of the control of the hypothalamo-pituitary axis by the ovarian steroids estrogen and progesterone, with a consequent elevation of levels of the gonadotrophins luteinizing hormone and follicle-stimulating hormone (FSH). These levels are high and often variable, and ephemeral peaks of these hormones coincide with the hot flashes of menopause. Although the menopause signals the end of reproductive life for a woman, there is clear evidence that reproductive function, as assessed by reduced levels of estrodiol and progesterone output and increased levels of FSH, declines with age prior to the menopause, after peaking between the ages of 20 and 30 years. Consequences of the hormonal changes associated with the menopause include changing patterns of fatness and fat patterning, and increased risk of cardiovascular disease and osteoporosis.

Ageing and senescence are inseparable from the processes of growth and development, given the increased life expectancies of contemporary populations (**12.3**). Mammalian species differ enormously in their maximum recorded lifespans, but for each there is a fixed maximum lifespan beyond which the probability of survival is effectively zero, and which cannot be extended. In general, among the mammals, lifespan is positively related to body-size. Although

**The seven stages of man. A 15th-century illustration from a French translation of Bartholomew's *De proprietatis rerum* (1491). Reproduced from Tanner, J.M. (1981). A *History of the Study of Human Growth*. Cambridge: Cambridge University Press.**

there may be genetic factors which influence longevity, the idea that ageing is programmed, in the sense that active processes count the passage of time and bring about senescence after an allotted period, is difficult to sustain. Observation of a variety of biological traits suggests that senescence is not under tight genetic control, but that antagonistic pleiotropy may be an underlying mechanism. That is, natural selection may have taken place for traits which increase fertility and reproductive success, some of which may be associated with greater risk of disease and death in later life. Another mechanism may be the expression of specific, detrimental genes later in life. Such genes will not have been selected against, because the intensity of natural selection decreases with age.

The study of ageing has become a large and important discipline, in both industrialized, and more recently, industrializing nations. This is largely because the epidemiology of nations changes with changing population structure and the diseases associated with it. In the industrialized nations, the phenotypes associated with ageing are reasonably well defined, while in the developing world these are less clear, since many populations are currently in the process of epidemiological and demographic change. More clearly defined are the metabolic processes associated with ageing at the cellular level. Of particular interest is the functional decline of mitochondria as a consequence of free-radical damage. Free radicals are molecules or molecular fragments containing a single unpaired electron, making them highly reactive, and damaging to cellular and subcellular membranes. Although free radicals are implicated in many disease processes, they are not always harmful. They are generated in various metabolic reactions, such as those involving dehydrogenases, and may be important in stimulating the bactericidal action of neutrophils of the immune system when activated by bacteria. In the mitochondria, damage by free radicals produced in association with the various dehydrogenase enzymes of the tricarboxylic acid cycle is usually kept in check by free-radical scavenging by the enzymes glutathione peroxidase, which is generally distributed, and manganese superoxide dismutase, which is specific to the mitochondria. When such scavenging becomes less efficient, damage to membranes can take place.

Cell death takes place across the human lifespan and may be either programmed, or not. Programmed death is more common than non-programmed death, and is a mechanism whereby damaged, malfunctioning or unnecessary cells can be removed from the body. It is important in growth and development, and in the repair and maintenance of the mature body. Programmed cell death is not a cause of ageing, but defects in this system may contribute to age-related processes.

The ageing process involves changing function of different physiological systems. Age-related changes in immune function include a reduction in the ability of stem cells to expand

*Changes in cognition with normal ageing (from Bonder and Wagner, 1994)*

| Cognitive abilities | Changes |
|---|---|
| Problem-solving | Delayed until late in sixth decade<br>Many changes remedial through instruction and practise |
| Memory | |
| Sensory | Little if any decline |
| Short-term | None |
| Long-term (secondary) | Some decline, deficits in encoding processes |
| Very long-term (remote) | Little decline |
| Psychomotor skills | Decline may begin in early 50s<br>Not altered by intervention |
| Information processing | Decline may begin in early 50s<br>Not altered by intervention |
| Verbal skills | Decline not until 80 years |
| Abstract reasoning | Older adults may be less proficient on laboratory tests |

clonally and to produce B cells. Osteoclasts, which reabsorb bone, are derived from hemopoitic stem cells, and the bone-loss, or osteopenia, of the aged may be in part a function of defective functions of such cells, as well as other, immunologically related (interleukin 1, tumour necrosis factor $\alpha$, tumour necrosis factor $\beta$, transforming growth factor $\alpha$, transforming growth factor $\beta$, $\gamma$ interferon) and non-related (parathyroid hormone, calcitonin) endocrinological stimuli. Macrophage number, and their handling of antigens, does not appear to change with age, but circulating serum antibody levels and T-cell-dependent cell-mediated functions do. The thymus has been implicated as the ageing 'clock' of T cells. Furthermore, since T cells are able to regulate hemopoiesis, some of the age-related changes in stem cells and B cells may also be due to the thymus.

Changes in neuromuscular structure and function with age vary enormously from person to person. Changes include declines in muscle strength, decrease in muscle mass, change in posture, impairments in balance, co-ordination and gait, decreases in the speed of movement and in the initiation of responses to stimuli, and increases in the threshold for vibration sensation. With respect to muscle structure, the most important change is the loss of both number and size of muscle fibres and increase in connective tissue and fat within the cells.

Cognitive changes associated with ageing were first investigated in the 1920s. The tempo of decline in cognitive function varies from person to person, while the approximate age of onset of different function categories differs according to type. Verbal skills show no decline until the eighth decade of life, while the speed of information processing begins to decline in the fifth decade. In general, memory and attention spans show little decline until extreme old age. However, age-associated memory impairment, which describes long-term memory-loss in conjunction with complaints of memory impairment in everyday activities, is a condition which may point to depression or neurological disorders.

Increasingly, it is important to consider processes associated with human function from a lifespan perspective. Broad though this framework may be, there are biological phenomena that link the fetus to the child, the child to the adult, and the younger adult to the older one. The linkages between different realms of growth and development and long-term form and function are being forged, and along with them our understanding of growth and development and its implications for the species deepen.

*Stanley Ulijaszek*

# Long-term consequences of early environmental influences

A number of early environmental factors influencing human growth and development have been shown to have long-term biological or behavioural consequences. These include relationships between: 1) the intra-uterine environment and adult cardiovascular disease, chronic bronchitis and hypertension; 2) infant diet and cholesterol metabolism; 3) respiratory infection in infancy and chronic lung disease in adult life; 4) adverse experiences in childhood and adult psychosocial functioning; and 5) the intra-uterine environment and non-insulin-dependent diabetes mellitus (NIDDM), and schizophrenia. Although often describing disparate phenomena, these observations are linked by the notion that human developmental processes are environmentally sensitive in a variety of ways, and the outcomes of these processes only become manifest in adult life, or alternatively, appear in childhood and persist into later life. Of relevance to this framework is the lifespan perspective of human development, first used to describe processes and states in psychological development but since broadened to encompass all adaptive phenomena in human biology.

In the present context, the lifespan perspective is limited to the influence of factors during growth and development on different stages in adult life, including ageing and senescence (**12.3**). The human lifespan is much greater than it was even in the recent past, and certainly more than it was in the course of human evolution. Thus it is reasonable to assume that the manifestation of long-term consequences of early environmental influences are greater, partly as a consequence of this.

The chronology of the growth and differentiation of specific organs and tissues varies, and it has been suggested that there may be critical periods when environmental factors such as nutritional stress or exposure to infection (**9.3**) can significantly alter the developmental process from its genetic trajectory. McCance (1962) identified the importance of chronological time, or age, on the pattern of development in relation to nutritional stress in humans and other mammalian species. Since that time, work on the developmental biology of invertebrate species has revealed various and subtle ways in which environmental changes or cues at different stages of development can influence morphology (**4.5**). Following the rich literature on this topic, Lucas (1991) used the term 'programming' to describe such phenomena when associated with some health-related outcome.

There are two ways in which programming might happen: 1) by the induction, deletion, or impaired development of a permanent somatic structure as the result of a stimulus or insult operating at a critical time; and 2) by physiological 'setting' by an early stimulus or insult at a 'sensitive' period, resulting in long-term consequences for function, the effects being either immediate or deferred.

It has been argued that 'early environment' is intra-uterine, or ceases by the age of 1 year, on the basis of the observation that most critical periods in development may have been passed by this age. However, while this may be true for the development of physiological systems which are associated with cardiovascular disease risk, it is not true of critical periods in behavioural development. Behavioural development is closely related to physical and physiological growth and development, set in social contexts. There are various pathways from childhood experiences to adult psychosocial functioning (**6.4**, **6.6**), but a common feature is that long-term effects depend for their occurrence on several, sometimes many, intervening links in a chain of indirect connections. When such links are all present, these long-term effects can be strong, but in their absence, there may be no enduring consequences of even severe early adversities. However, epidemiological associations have been found between stress in early childhood and adult behaviour and parenting, and between parents' divorce and children's experience of depression, particularly if divorce took place before the child was of school age.

An epidemiological relationship between adult NIDDM or glucose tolerance and low birth-weight or low weight at 1 year of age has led to the proposal of a mechanism relating impaired development of the pancreas in the undernourished fetus to the development of NIDDM in adults. This is in opposition to the genetic explanation first put forward by Neel (1962). The postulate is that maternal malnutrition and/or other maternal placental abnormalities lead to fetal malnutrition. The consequences of a generalized undernutrition include reduced fetal growth (**4.8**), and a postulated reduced β-cell mass and islet function of the pancreas. Deficiencies of protein and amino acids are believed to be of particular importance in creating these defects of structure and function. Of critical importance to this framework is the view that impaired β-cell function persists into adulthood, predisposing the individual to NIDDM. This disease then expresses itself in association with environmental risk factors, obesity (**9.2**) and physical inactivity in particular, as well as increasing age, and possibly other processes leading to insulin resistance.

Recent evidence suggests that the increased prevalence of diabetes in adults who were growth-retarded during fetal life is mediated by increased insulin resistance, rather than by abnormalities of insulin secretion. However, genetic explanations appear to be ascendant again, despite many obstacles hindering the study of this field, with claims of various candidate genes for NIDDM. Furthermore, U-shaped relationships between the prevalence of NIDDM and birth-weight among Pima Indians in Arizona have lead to the suggestion that high mortality of low-birth-weight infants may be associated with selective survival in infancy of individuals who are genetically predisposed to insulin resistance and diabetes. This debate on the causation of NIDDM is yet to be resolved, but is an important one, given the rapid increase in the global prevalence of NIDDM, particularly among many modernizing populations of the developing world.

Other examples of possible long-term consequences of early environmental triggers include: infant-feeding and cholesterol metabolism; and the timing of onset of puberty in relation to fecundity and breast cancer. Both of these speculations are based on an understanding of the hormonal imprinting of a variety of metabolic processes which take place in the course of growth and development, and that such physiological setting can be modified at critical periods by exogenous factors. For example, cholesterol intake from breast-milk has an important role in the development of cholesterol regulatory mechanisms, and preweaning cholesterol intake may prevent diet-induced hypercholesterolemia later in life. *Stanley Ulijaszek*

See also 'Sexual maturation' (**5.15**) and 'Infant-feeding and growth' (**9.1**)

## FETAL DEVELOPMENT AND DISEASE IN ADULT LIFE

We have become accustomed to the idea that the major diseases in the Western world, which include coronary heart disease, stroke and diabetes, arise through an inappropriate life style – a high-fat diet, obesity, smoking – combined with genetically determined susceptibility. Recent research, however, suggests that failure of development *in utero* (**4.5**, **4.8**) may play an important role. Animal studies long ago showed that undernutrition before birth leads to persisting changes which 'programme' the body's function and structure. It may, for example, change the way food is used or hormones are released. Precisely what happens depends on when in intra-uterine life undernutrition occurs. Different tissues and systems in the body undergo so-called 'critical' periods of development at different times; during these periods they seem most vulnerable to permanent change.

The great majority of human development is accomplished before birth (see Part four), in as much as the fertilized egg goes through 42 rounds of cell division before birth, and only a further five cycles after birth. As in other species, the growth of the human fetus depends on nutrients and oxygen from the mother.

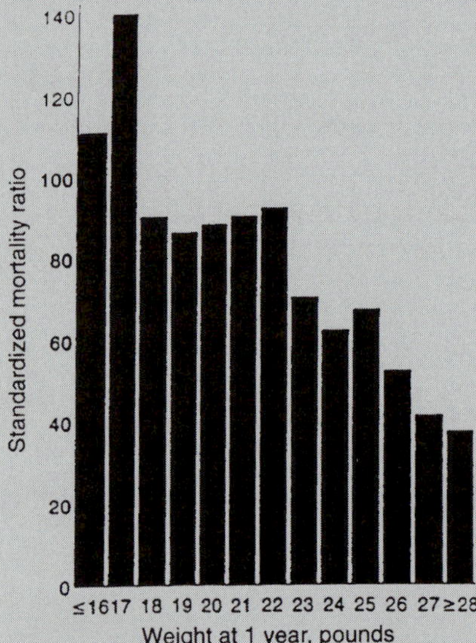


**Standardized mortality ratios for coronary heart disease in 8175 British men according to their weight at 1 year of age.**

*Prevalence of Syndrome X (type 2 diabetes, hypertension and hyperlipidaemia) in British men according to birth-weight*

| Birth-weight pounds (kg) | Total number of men | % with Syndrome X | Odds ratio adjusted for body mass index (95% confidence interval) |
|---|---|---|---|
| ≤5.5(2.50) | 20 | 30 | 18 (2.6 to118) |
| –6.5(2.95) | 54 | 19 | 8.4(1.5 to 49) |
| –7.5(3.41) | 114 | 17 | 8.5(1.5 to 46) |
| –8.5(3.86) | 123 | 12 | 4.9(0.9 to 27) |
| –9.5(4.31) | 64 | 6 | 2.2(0.3 to 14) |
| >9.5(4.31) | 32 | 6 | 1.0 |
| Total | 407 | 14 | |

*Mean serum lipid concentration according to abdominal circumference at birth in British men and women aged 50–53 years*

| Abdominal circumference (inches) | Number of people | Total cholesterol (mmol/l) | Low density lipoprotein cholesterol (mmol/l) |
|---|---|---|---|
| –11.5 | 53 | 6.7 | 4.5 |
| –12.0 | 43 | 6.9 | 4.6 |
| –12.5 | 31 | 6.8 | 4.4 |
| –13.0 | 45 | 6.2 | 4.0 |
| >13.0 | 45 | 6.1 | 4.0 |
| Total | 217 | 6.5 | 4.3 |

Restriction in the supply of these leads to small size at birth and disproportion, such that a baby is thin or short. Evidence that programming may be important in human disease has come from studies in which people have been followed up from birth to middle- or old age. They have shown that men and women who were below average birth-weight, or were thin or short at birth, are at increased risk of developing coronary heart disease as adults together with the conditions associated with the disease – raised blood pressure, diabetes and disturbances of cholesterol metabolism. The increases in risk are large. For example, people who weighed 2.5 kilograms or less at birth are six times more likely to develop diabetes than those who weighed more than 4.3 kilograms.

We are beginning to understand how undernutrition before birth can have such profound long-term effects. It seems, for example, that muscle may be altered during its critical period of development in such a way that it becomes permanently less sensitive in the action of insulin, thereby predisposing to non-insulin-dependent diabetes mellitus. The way the liver handles cholesterol may similarly be permanently changed. Remarkably, disease in adult life is linked not only to the size of the newborn baby but to the size of the placenta through which nutrients were extracted from the mother. Either a disproportionately small or large placenta is associated with later coronary heart disease. In animals, expansion of the placenta is one of the mechanisms by which an undernourished fetus may attempt to extract more nutrients from the mother. Similarly, in high altitude populations, large placentas are associated with the need to scavenge oxygen in a hypoxic environment.

We know surprisingly little about what regulates fetal and placental growth in humans. Research in domestic animals points clearly to the importance of the nutrition and physiology of the mother. Fetal adaptations to undernutrition are becoming an important area of medical research, and could hold the key to preventing disease in the next generation. *D.J.P. Barker*

## MUSCULAR DEVELOPMENT

Physiological processes which take place across childhood and which result in modified muscle morphology may have long-term consequences for the health and human biology of adults. Childhood undernutrition has negative consequences for health and growth, but positive ones for the energetic efficiency of muscular contraction and the performance of physical work (**9.10**). Mechanical efficiency is the ratio between work done and the energy expended in doing it, and Shetty (1993) has summarized reports of this variable in adults experiencing chronic energy deficiency, compared with well-nourished controls. Although the comparison of these studies, which were carried out under a variety of conditions with disparate experimental designs, is not straightforward, in general, undernourished subjects use less energy in performing some standardized task than do their well-nourished controls. Waterlow (1990) proposed a physiological explanation for this phenomenon, relating undernutrition, the selective development of slow and fast twitch muscle fibres, their relative energetic efficiencies in muscular contraction, and overall work performance.

The relation between the force developed or the amount of mechanical work done to the amount of adenosine triphosphate (ATP) used is called the contraction coupling effiency. Slow twitch fibres have a higher contraction coupling efficiency than fast twitch fibres. Muscles with a high proportion of slow twitch fibres use less ATP per unit of isometric tension than muscles containing a high proportion of fast twitch fibres. In conditions of low energy intake and in hypothyroidism there is a reduction in the proportion and diameter of fast twitch fibres, with increased contraction coupling efficiency in hypothyroidism.

Studies of low-intensity, long-duration training in healthy adults showed a reduction in fast twitch fibres in the triceps brachii and quadriceps muscles, while slow twitch fibres were not affected. Similarly, investigations of muscle size and composition in malnourished patients showed that the size of the slow twitch fibres in the calf muscle was better preserved than that of the fast twitch fibres. In the developing world, low dietary-energy availability and prolonged low intensity work output may lead to such changes in muscular development, with the associated lower energy expenditure made possible by them.

Differential development of slow twitch over fast twitch fibres in the course of the growth faltering has not been demonstrated. However, muscle phenotype characteristics may be modified or determined by thyroid-hormone status. Notably, clinically hypothyroid patients show significantly lower energy expenditure compared with euthyroid controls during quadriceps muscle function tests. A possible mechanism for this effect might be that thyroid hormones influence the sequestration and release of calcium ions by the sarcoplasmic reticulum, and ATP consumption in calcium pumping during the contraction cycle, thereby increasing overall efficiency. It is not clear whether the disappearance of the low thyroid status associated with growth faltering leads to reversion of the muscle phenotype to one with a greater proportion of fast twitch fibres. If nutritional stress persists through life, then low T3 (triiodothyronine) will persist, and with it, greater muscular efficiency associated with selective persistence of slow twitch fibres over fast twitch fibres. *Stanley Ulijaszek*

Undernutrition in childhood
↓
↓ Thyroid hormone status
↓
↓ Proportion and diameter of fast twitch muscle fibres during growth and development
|
(? ↑ Contraction coupling efficiency)
↓
↓ Energy expenditure in aerobic muscular activity in adult

**Mechanism whereby undernutrition in childhood may influence the efficiency of muscular contraction and energy cost of physical activity in adulthood (from Waterlow, 1990).**

## FAT PATTERNING

The issue of whether there are early influences on adult fat patterning require three separate lines of research for resolution:

1) does the environment influence fat patterning in young children?
2) is there tracking of fat patterns in children and youth?
3) assuming the first two, are there later consequences of early fat patterns?

Among adults there are significant differences in fat patterning by age, sex, ethnicity, and environmental variables, such as socio-economic, nutritional and health status. We know that fat patterns are present at birth and that, while they do not attain statistical significance, they suggest the well-known sex differences which are seen from childhood on, with greater centripetal patterning in males (**5.16**).

*Centripetal fat ratios*[1] *in 100 male and 96 female neonates*

| | |
|---|---|
| Males | 0.485 |
| Females | 0.475 |

[1] (Subscapular skinfold)/(Subscapular + Triceps)

The tracking of fat patterns refers to their persistence across time – for example, do young children who have a peripheral distribution of fat maintain that distribution, relative to their peers, through their growing years and into adulthood? Roche (1988) has summarized the evidence from the relatively few longitudinal studies which have collected adequate data. He finds that there is little tracking of the basic patterns during infancy or from infancy into the pubertal years. The lack of significant associations for indices of relative fat distribution across more than a few years is likely to be related to changes in distribution that occur normally with age. Males, for example, display increasing centralization and females increasing peripheralization of fatness during growth, with an intensification of the patterns during the adolescent years. Differences in the pathways by which dual patterns change vary considerably, as a result of hereditary, endocrinological, and maturational factors. It is important to realize that even while cross-sectional studies may indicate a persistence of differences with growth, longitudinal data may very well show something quite different.

Nutritional status (**9.2**) seems to affect the distribution of fatness in children, independent of the amount. Bogin and Sullivan (1986) reported a more centripetal distribution of subcutaneous fat in urban Guatemalan children of low socio-economic status, compared to those from a more privileged environment.

The extent to which the fat patterns of infants and young children are indicators of risk is also unclear from the available data. While it is well known that these patterns are significant predictors of the risk of chronic disease among adults, there is little basis at present for suggesting similar associations among children, except perhaps among extreme cases.

In summary, the available data fail to support an hypothesis that fat patterns early in life: 1) are associated with the environment; 2) persist in individuals into the later years; and 3) are associated with an elevated risk of chronic disease as adults. It is important to recognize that this conclusion is flawed to an extent by a lack of data sufficient to test properly the proposition. However, it is also important to point out that it can only be tested with data on individuals and not from cross-sectional studies. Clearly there is much to be learned. *Francis E. Johnston*

## SENSITIVE PERIODS IN BRAIN GROWTH AND DEVELOPMENT

The term *sensitive period* is defined here as a stage at which some aspect of development occurs with greatest ease, and is preferred to the terms 'critical' and 'vulnerable' period. 'Critical period' is probably best restricted to the most dramatic cases in which a process can only occur within a sharply defined period of development. 'Vulnerable period' focuses on harmful effects and ignores the positive.

The growth of the brain after organogenesis (**4.3**, **4.5**) can be described as occurring in two somewhat overlapping phases: the first being one of neuronal multiplication followed by the second one of glial-cell proliferation, during and after which axons grow, dendrites branch extravagantly and synapses are formed. Once a particular class of glial cell, the oligodendroglia, are present, the process of myelination can proceed. Within this gross progression there is a finer-grained chronology. For instance, different neuronal populations divide at different times. In particular, the microneurons in the hippocampus, cerebellum and olfactory bulbs arrive very late, well into the phase of glial-cell proliferation.

The timing of brain growth is interesting in two ways. First, it occurs in advance of most of the growth of the rest of the body; babies have relatively big heads. Second, its timing in relation to birth differs markedly between species. Precocial species, like guinea pigs, sheep and monkeys, grow most of their brain prenatally, whereas altricial species, like rats and humans, show most brain growth post-natally. The growth in weight of the human brain is only a quarter complete by birth, but spurts thereafter to attain about three-quarters of its adult weight by 2 years.

The course of brain development can be affected by a variety of factors, the nature and extent of the perturbation dependent on the timing of the treatment with respect to stage of brain development, and also on its duration and severity. For there to be a sensitive period, there must be some growth and/or development occurring. Growth processes completed before treatment will be unaffected. Only those processes occurring at the time of the treatment will be affected, plus some later-occurring processes, as a result of a cascade of effects.

Examples follow of factors with effects that depend on when they are applied; which, in other words, exemplify sensitive periods. Ionizing radiation is one such factor. Undernutrition too can have negative effects on brain development. Its principal effect is to slow down the rate of brain-growth processes concurrent with the undernutrition, causing deficits in growth attainment. Distortion of brain growth can result from undernutrition which is not uniform in severity over the whole period of brain growth, or which covers only a part of it.

Numerous hormones influence brain development in a variety of ways. Gonadal hormones provide a nice example of a sensitive period (**4.1**, **5.1**), indeed a critical period, for the development of sexual dimorphism in sexual behaviour (**6.6**). For instance, castration of male rat pups within the first few days after birth, but not later, has a lasting feminizing effect, whereas administration of androgen to female pups at the same stage has a masculinizing effect. Corresponding changes described as differential neuronal growth in the preoptic area and hypothalamus, parts of the brain mediating sexual behaviour, appear to be the morphological basis for these effects.

A final example demonstrates subtle but far-reaching consequences of disturbances of visual experience on the development of vision. A variety of mechanisms are involved in establishing the final pattern of connectivity within the visual pathway from retina to visual cortex: cell proliferation and migration, axon outgrowth, overconnection, competitive interaction, and the regulation of connectivity by cell death and axon withdrawal. The visual cortex undergoes a distinct period of plasticity in early post-natal life, during which the synaptic input to cortical cells is being regulated on the basis of activity reaching the cortex from the eyes. This would appear to be the basis of amblyopia, a deficiency of vision (usually impaired visual acuity) resulting from reduced or altered input from one eye (as from a 'squint') early in life. For complete recovery in such cases, treatment needs to be instituted early.

*James L. Smart*

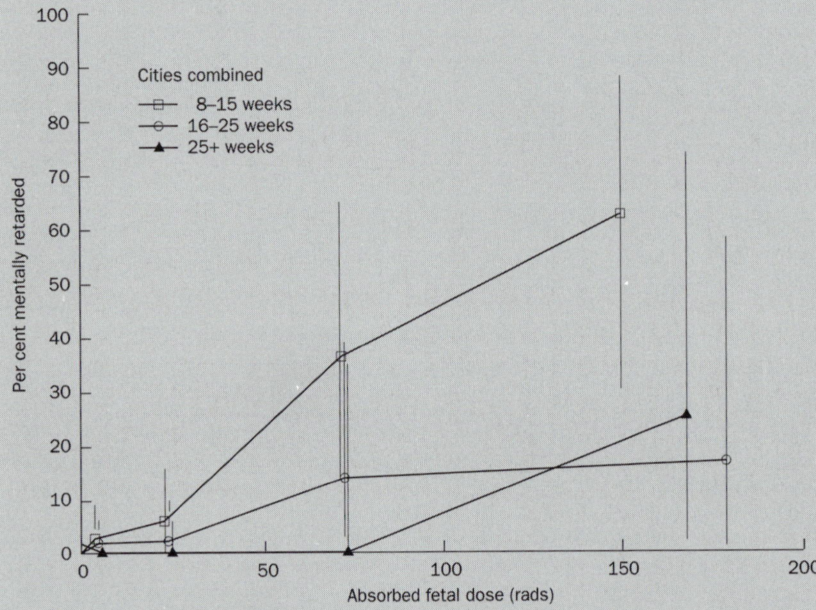


**Ionizing radiation, such as is released by a nuclear bomb, kills dividing cells only. The survivors of the atomic bomb attacks on Japan in 1945 provide epidemiological evidence about the occurrence of mental retardation which suggests a sensitive period in brain development. The incidence of mental retardation corresponds both to the radiation dose and to the stage of development of fetuses at the time they were exposed. The greatest incidence is highly correlated with exposure between 8 and 15 weeks of gestation (uppermost line), which corresponds closely to the period of neuronal multiplication in the forebrain. (Redrawn with the permission of the publisher and authors, from Otake and Schull, 1984.)**

## VIRAL INFECTION OF THE PREGNANT WOMAN AND RISK OF SCHIZOPHRENIA IN HER OFFSPRING

Infection by the influenza virus in mid-pregnancy is believed to adversely affect the developing brain of the fetus (**4.5**), predisposing it to later schizophrenia, a view consistent with both neurodevelopmental and genetic theories of the disorder. The association between influenza and schizophrenia was first alluded to by Menninger (1928) who described 67 cases of dementia praecox in a group of 175 patients with mental illness following the 1918 European influenza pandemic. In Denmark, the winter-birth excess in schizophrenics has been associated with higher than expected rates of influenza in the sixth month of gestation, while individuals *in utero* during the second trimester of pregnancy during the 1957 influenza epidemic had a higher than expected susceptibility to subsequent schizophrenia. Furthermore, influenza epidemics have been consistently correlated with higher rates of subsequent schizophrenic births in England and Wales between 1939 and 1960. The striking decline in the incidence of schizophrenia in Britain, Denmark, Australia and New Zealand has been attributed, at least in part, to general improvements in living conditions in recent decades, resulting in a decline in influenza morbidity. However, epidemiological studies do not explain why only some mothers exposed to influenza during the second trimester of pregnancy produce offspring who develop schizophrenia. It has been proposed that this pattern of schizophrenia could be initiated by auto-immune disorders arising in genetically at-risk mothers and fetuses exposed to a viral trigger.

Since the influenza virus does not cross the placenta, it has been suggested that the process is most likely to be mediated immunologically, possibly by the induction of maternal antibodies by the influenza virus, which unlike the virus, are able to cross the placenta and immature blood–brain barrier and cross-react with fetal tissues, interfering with neuronal migration.

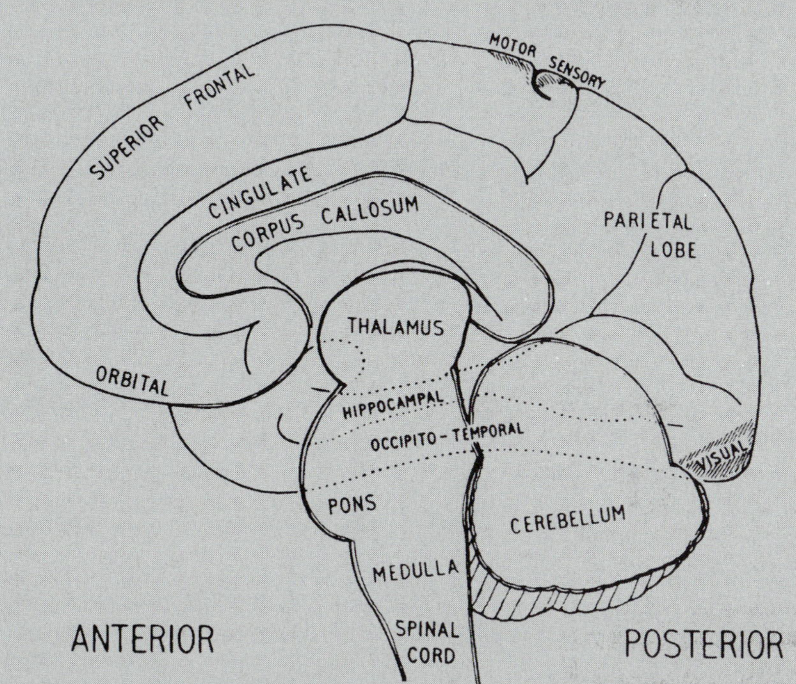


**Cross-section of the brain, showing the hippocampus.**

The cytotoxic response to influenza virus depends on Class I human leucocyte antigens (HLA), while antibody response depends on Class II HLA. Paranoid schizophrenia has been associated with HLA Class I A9, and HLA Class II DR8 in Japanese schizophrenics and the Class II HLA DRw6 in African-Americans. Thus the extent of the potentially neuronally damaging immune response (**4.7**) may be influenced by genetic variation in HLA both within and between populations, and that patterns of schizophrenia morbidity may have some basis in the impaired development of permanent somatic structures in the brain of fetuses of genetically susceptible mothers. The structural modification most characteristic in schizophrenics is a smaller than usual volume and cell number in the hippocampus, possibly as a consequence of impaired proliferation in the ventricular zone in mid-gestation. *Stanley Ulijaszek*

# Menopause

Menopause marks the end of menstruation and the onset of post-reproductive life. In contrast to menarche (the first menstrual period), women identify the timing of menopause by counting monthly absences of menstrual blood. Most investigators now agree that menopause is recognized in retrospect following 12 months of amenorrhea (without menstruation).

Humans experience menopause, in part, because we are mammals. Mammalian ovaries (with few exceptions) are gradually depleted of eggs over the course of the lifespan. In the female fetus, germ cells replicate through mitosis to form oocytes (undeveloped eggs). Unlike males, who are able to produce sperm throughout life, human females never produce another egg after the fetal period. Instead, during the fifth month of gestation, the store of oocytes peaks at 7 million before plummeting down to only 2 million oocytes at birth. Oocytes are lost through a process called atresia that continues on, past birth, through childhood and adulthood, until menstruation ceases. Atresia is a process of cellular degeneration. Some investigators attempt to account for atresia as a type of apoptosis, or programmed cell death (**12.3**). The rate of atresia appears to be modified by the environment of the ovary, including hormonal changes and toxins.

The timing of menopause, then, is determined by two factors: the peak number of oocytes present in the ovary during the fifth month of gestation, and the rate of loss of oocytes across the lifespan. While some oocytes are ovulated and become capable of fertilization, 99 per cent are lost through the process of atresia.

Among humans the average age at menopause varies from about 45 years of age in undeveloped countries to 50 or 51 years in developed countries. Within any one population, age at menopause ranges from 40 to 60 years of age. Variables associated with age at menopause include smoking habits, body-size, income, and marital status. Clinically, the cessa-

## AGE CHANGE IN REPRODUCTIVE FUNCTION

While the cessation of natural female reproductive ability at menopause is obvious, the trajectory of female fecundity in the years and decades preceding menopause continues to be a matter of debate. Demographers have long observed a steady decline in age-specific female fertility rates after an age of 30 years that is extremely consistent in its pattern across diverse populations. The cause of this decline, however, was generally thought to represent behavioural changes, especially reduced frequency of sexual intercourse, rather than any physiological change in fecundity, the biological capacity for procreation. This impression was reinforced by a large compilation of data on menstrual patterns in American women representing over 25,000 person-years of observations on more than 2000 women. These data indicated that menstrual patterns are very consistent and do not change appreciably with age from a few years after menarche (**1.15**) to a few years before menopause. A different, contemporary set of data on basal-body-temperature recordings representing 3264 person-months of observations on 481 German women, however, indicated that the frequency of ovulation and luteal-phase sufficiency begins to decline after age 30 years, despite regular menstruation.

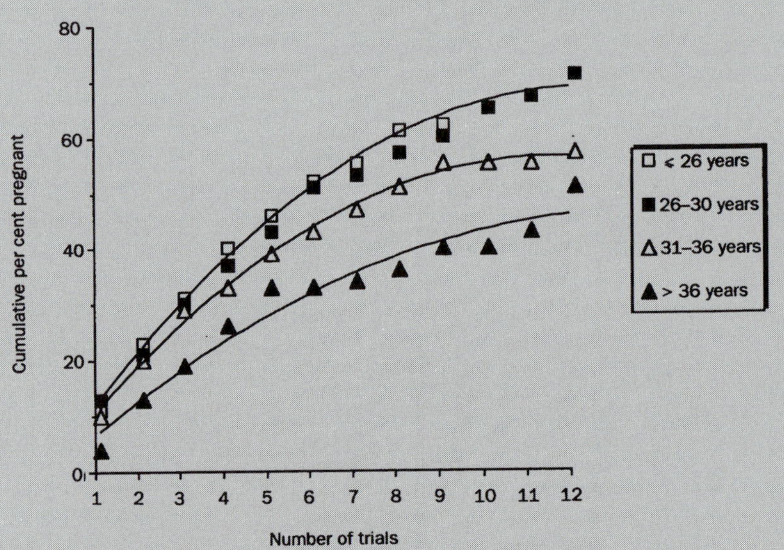


**Cumulative success rates in artificial insemination by donor in France by woman's age. Success rates are significantly lower in women over age 30 than in younger women.**

More recently, clinical evidence from the area of assisted reproduction technology has challenged the notion of constant female fecundity in advance of the menopausal transition. Declining success rates in artificial insemination with donor semen as female age increases beyond 30 years drew both popular and scientific attention to the issue in the 1980s. Similar age-related declines in success rates have since been reported for ovulation induction, *in vitro* fertilization, and ovum donation. Ovum donation results indicate that uterine function is not ultimately limiting on female fecundity, but rather implicate either the age of the ovum itself or the hormonal milieu produced by the ovary as the immediate causes of declining success rates.

Direct evidence is now available documenting declining ovarian hormonal production with age after a peak in the mid- to late-20s. Both estradiol and progesterone levels decline with increasing age with potential effects on oocyte fertilizability, embryo quality, endometrial thickness, and hormonal support for implantation. These data confirm the earlier German data based on observations of indirect, metabolic consequences of ovarian hormone production. Oocyte age independent of ovarian hormone production may also reduce the probability of conception and increase the probability of early embryonic loss, since ova from younger donors more often result in pregnancy than those from

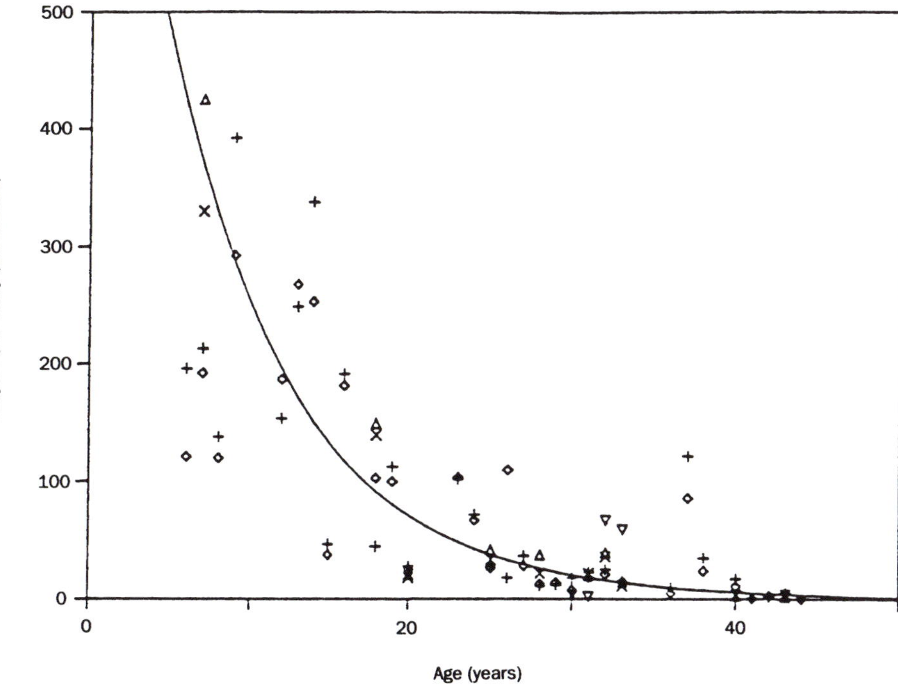


**An exponential model of oocyte number per ovary versus age, with the best fit line.**

tion of menstruation prior to the age of 40 years is called 'premature ovarian failure' (POF). The age of 40 years is an arbitrary and changeable cut-off, however, and many of the same factors that contribute to POF, both genetic and environmental, also contribute to the onset of menopause.

## The symptoms of menopause

After the first great wave of atresia between the fifth month *in utero* and birth, each ovarian oocyte is eventually surrounded by a follicle. The follicle is composed of specialized cells that nourish the oocyte and produce the ovarian hormones of estrogen, progesterone, and inhibin. Each of these hormones targets particular tissues, including the pituitary and hypothalamus in the brain. The relationships

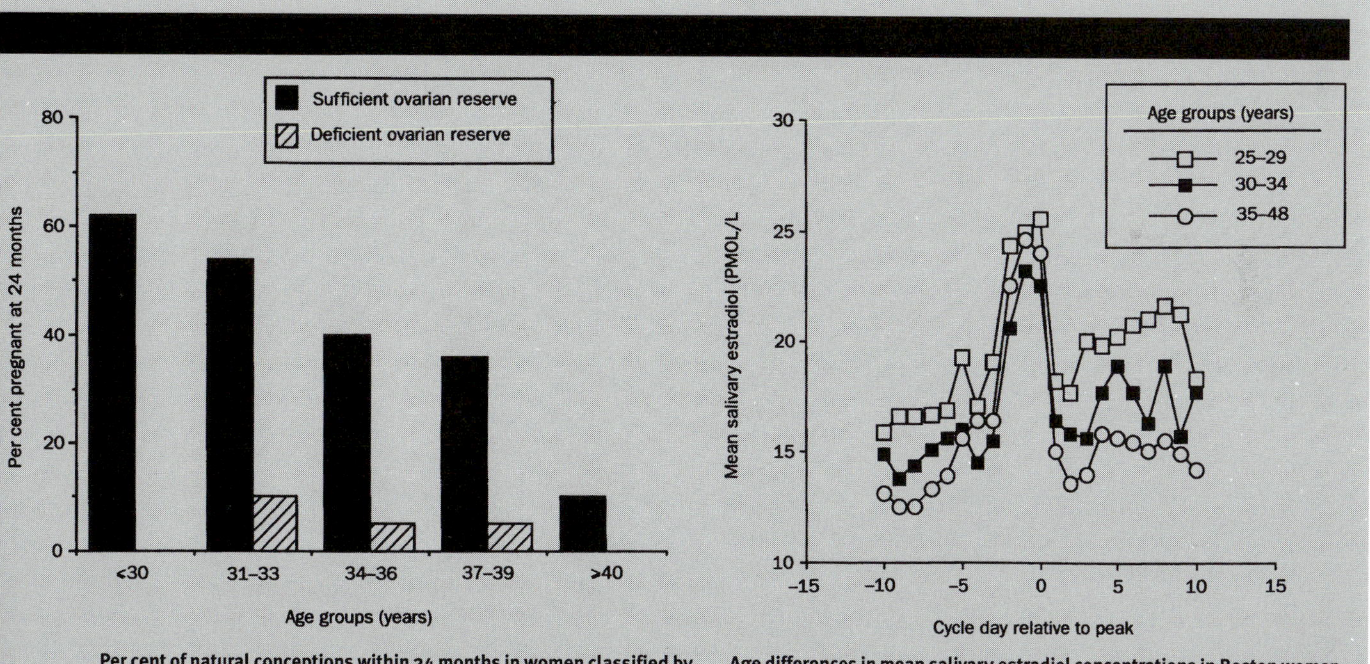


**Per cent of natural conceptions within 24 months in women classified by age and ovarian-reserve status. Probability of conception is low in women with diminished ovarian reserve regardless of chronological age. Probability of conception declines steadily with age after 30 years in women with sufficient ovarian reserve.**

**Age differences in mean salivary estradiol concentrations in Boston women. Age group differences in the follicular phase (days −10 to −2 relative to the mid-cycle drop in estradiol) and in the luteal phase (days + 1 to + 10) are statistically significant ($p < 0.05$), while those at mid-cycle (days −1 and 0) are not, reflecting a constancy in the level of estradiol necessary to trigger ovulation.**

older donors in recipients undergoing similar hormonal preparation for implantation.

Since most of the data to date are cross-sectional in nature, it still remains a question whether the fecundity of individual women declines steadily with age over one to two decades or more precipitously at different ages in different women. Elevated levels of follicle-stimulating hormone (FSH) early in the follicular phase of the menstrual cycle have been interpreted as a sign of impending menopause, indicating a diminished follicular reserve. Elevated FSH levels are associated with markedly reduced fecundity regardless of age. Yet conception rates decline with age among women in the absence of diminished ovarian reserve as well. Longitudinal studies of changing reproductive function with age will be needed to thoroughly resolve these issues. *P.T. Ellison*

between the ovary, pituitary, and hypothalamus can be visualized as interrelating feedback loops that, ultimately, regulate the monthly menstrual cycles.

Ovarian inhibin, for example, regulates levels of pituitary FSH (follicle-stimulating hormone). As menopause approaches, the store of ovarian follicles is reduced until inhibin levels begin to fall. When this happens, pituitary FSH levels rise. This rise, measured in women as young as 37 or 38 years, may explain why menstrual periods become irregular or heavier in some, but not all, women prior to menopause.

Like inhibin levels, estrogen levels also fall as ovarian follicles are depleted. This decline affects tissues rich in estrogen receptors, such as the lining of the vagina and breast tissue. Estrogen is also involved in bone metabolism and the regulation of blood cholesterol levels. As estrogen levels fall, some women experience hot flashes (flushes, in the United Kingdom) and sweating, particularly at night. These are called vasomotor symptoms and are relieved by the widely prescribed 'hormone replacement therapy' (HRT). HRT appears to reduce many of the acute and chronic risks attributed to menopause, but HRT carries side effects of its own, including a possible increased risk of breast cancer.

Studies of menopause can be divided into those that emphasize biomedical aspects and those that give more weight to socio-cultural factors. While biomedicine emphasizes the effects of declining estrogen levels, cross-cultural investigations suggest that the frequency of acute and chronic menopausal symptoms, such as hot flashes and hip-fracture risk, vary in relation to activity levels, smoking habits, reproductive history, diet, and cultural expectations.

Marcha Flint was one of the first to carry out a cross-cultural, community-based investigation of menopause. She showed that menopause was experienced differently in India compared with the United States. Her work (published in 1975) helped move the study of menopause away from a biomedical emphasis and opened the way for other cross-cultural studies of variation in age at menopause and symptom experience. Investigators who continue to advance a holistic view of menopause as a natural process, rather than a biomedical event, include anthropologists Margaret Lock and Yewoubdar Beyene, and sociologists Patricia Kaufert and Sonya McKinlay.

**THE WESTERN BIO-MEDICAL MODEL; MENOPAUSE AS AN ENDOCRINOPATHY**

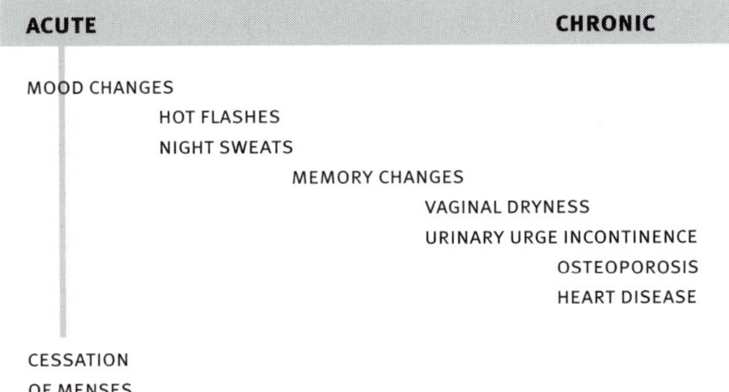


**Acute and chronic symptoms attributed to menopause by biomedical investigators in relation to age.**

## Menopause as a human phenomenon

Many mammals experience a reduction in oocyte number and a decline in fertility with advancing age; however, only humans experience such a long post-reproductive life. At some point during hominid evolution, our somatic lifespan, that is the function of our hearts, lungs, kidneys and so forth, exceeded the functional capability of our ovaries. The best primate model to date for understanding the hormonal and histologic changes of menopause in humans is the macaque, an Old World monkey. However, the macaque does not spend a potential one-half of its lifespan in a post-reproductive period.

The evolutionary biologists who have pondered this problem have generally agreed that the increased length of post-menopausal life paralleled an increase in hominid brain development and a concomitant increase in the length of gestation, infancy, and childhood (**2.3**, **10.1**). The increased lifespan could have been adaptive (for grandmothering), or simply an architectural happenstance.

*Lynnette Leidy*

See also 'Comparative growth and development of mammals' (**2.1**), 'Growth in non-human primates' (**2.2**) and 'Sexual maturation' (**5.15**)

# Ageing as part of the developmental process

Growth, development, and ageing are inseparable co-evolved processes in all animal species. Gerontological research has revealed extensive connections among growth, development, maturity, reproductive output, and senescence. Intimate connections of individual growth and development with variable patterns of individual senescence flow naturally from antagonistic pleiotropy and age-specific action of genes, two evolutionary models of senescence. Growth and development are necessary processes for attaining maximum reproductive potential (MRP), which occurs during the early phases of mature adulthood (**5.17**). MRP is followed by declining reproductive potential throughout adulthood. This decline with increasing age after attainment of MRP seems to be an inevitable consequence of sexual reproduction. Even with a constant force of mortality, the probability of reproducing during each additional year is a decreasing function of time. Thus, any investment in somatic maintenance at the expense of investment in reproduction after attainment of MRP is necessarily selected against. Eventual loss of somatic function is the ultimate outcome of such evolutionary pressures.

Senescence is the process of physiological decline in all body systems with age that occurs concurrent with increased risk of death. Unlike growth and development, senescence does not appear to be under tight genetic control. Rather, wide variation is observed in almost all biological, physiological, and socio-cultural measures with increasing age. A pattern to be expected if numerous intrinsic systems that regulate homeostatic processes are composed of numerous, but not infinite, redundant and mortal cells and functional units (for example, nephrons in kidney, eggs in ovaries, neurons).

Humans are a relatively long-lived species, showing average and maximum lifespans that few other animals attain. It is likely that human long-livedness evolved in concert with the evolution of altrical infants and extended prereproductive immaturity (infancy and childhood) (**2.3**, **10.1**). As attainment of MRP has been postponed, selection either favoured long-life or mechanisms favouring long-life came along through pleiotropy or linkage disequilibrium. Since reproduction was postponed and investment in individual offspring increased, selection pressures favouring longer life and longer-term reproduction probably existed in the past and may still operate today.

Human growth and development are characterized by an extended period of childhood growth, rapid maturation, gradual decline in function, and post-reproductive senescence. Extended growth and development and the existence of a post-reproductive period are uniquely human; however, the latter appears to be a more recent phenomenon. Human maximum lifespan potential (120 years) has not changed since *Homo sapiens sapiens* evolved. Life expectancy at birth, however, has changed dramatically throughout our species' history, particularly in the last century. Since the mid-19th century, life expectancy at birth has increased from 40 to almost 80 years in the United States. This was more due to decreased childhood and young adult mortality than to any change in adult life-expectancy, processes that now allow more people to live well past their maximum reproductive potential than ever before.

**Grandmothers on the park-bench, a common sight throughout the world – in this case, Bulgaria. Photograph by Stanley Ulijaszek.**

Most human physiological systems change continually over the lifespan. Patterns of change differ across organs, cell types, and metabolic systems. For age-related variation to be considered senescence, however, it must meet specific criteria, it must be *cumulative, universal, progressive, irreversible,* and *degenerative* (CUPID). Thus, much of what we see as daily examples of ageing in cosmopolitan societies: hypertension, presbyopia, hearing-loss, cardiovascular diseases, osteoporosis, osteoarthritis, may be culturally and environmentally specific examples of age-related disease. Traits more elusive in their etiology, lipofuscin deposits, greying of hair, loss of skin elasticity, decreased physiological response to and recognition of environmental stressors, appear to be true markers of senescence. An ever increasing risk of death is an additional commonly accepted aspect of senescence. Mortality rate doubling time for humans is about 7 years; for every 7 years after attainment of MRP the mortality rate doubles, producing a geometric curve of mortality.

With increased lifespan, more humans live sufficiently long to experience various chronic diseases and age-related frailty. This is partially because all elders experience declines in physiological reserve capacity with increased age, and partly because ageing is accompanied by the onset and progression of co-morbid disease. Numerous aspects of failing physiology compromise an elder's health, including poor stress reactions, lower immune function, and a weakened cardiovascular system. Thus, elderly persons suffer a higher burden of disease and disability compared to younger people. Although many older individuals experience disability due to failing physiological function with age, other older adults maintain physiological vigour into their ninth decade and beyond.

Over the last century there has been a gradual decrease in death-rates from infectious diseases with a concurrent increase in deaths from cancer, cardiovascular disease, and other chronic diseases. For example, deaths due to diarrhea, a common condition in infants and children, decreased from 194 per 100,000 in 1900 to 1 per 100,000 in 1986 in the United States. During the same period, mortality due to cancer increased from 64 per 100,000 in 1900 to 195 per 100,000 in 1986. Thus, life expectancy has increased at almost all ages. In addition, throughout the lifespan, men experience greater mortality than women. Older women, however, have more chronic diseases; therefore, although women are less likely to die from chronic disease than are men, they suffer greater disability related to chronic diseases.

Possible limits to human longevity currently are hotly debated in gerontology. Although some chronic diseases appear to be decreasing, there seems little hope of any significant increase in human longevity beyond the age of 120 years. Two opposing views on human lifespan are current: 1) lifespan is a genetically determined species-specific characteristic and human longevity will not increase beyond 120 years; 2) lifespan is an environmentally and genetically labile characteristic and therefore there need not be a species-specific lifespan. According to the latter perspective, there is no reason to believe that humans have met their limit to longevity.

Two evolutionary models have strong research backing. Antagonistic pleiotropy builds on the fact that the force of natural selection necessarily decreases with increasing age. In this model, traits that increase fertility at young ages are favoured by natural selection, even if they are associated with

## AGEING IN INDUSTRIALIZED SOCIETIES

Life-expectancies have increased steadily in industrialized societies (**11.2**) for more than 100 years, this being in large part attributable to reduced mortalities from infectious disease. However, when the focus is shifted to the conditional probability of survival at ages beyond 60 years, remaining years of life projected for residents of non-industrialized societies may actually exceed those for residents of industrialized ones. This apparent paradox arises from differences in the prevalence of degenerative conditions associated with ageing and the life styles predominant in industrialized as opposed to non-industrialized circumstances.

Cardiovascular diseases including hypertension, ischemic heart disease and stroke remain the leading causes of mortality in the industrialized nations even though the mortality rate associated with these conditions has been declining for 40 years or more in the United States. Meanwhile, cancer mortalities continue to increase and conditions such as late-onset diabetes, emphysema, Alzheimer's disease and osteoporosis have become increasingly prevalent. While all of these conditions can also be found in developing countries, especially among the upper socio-economic strata, their characteristically late onset makes their occurrence less common in populations where the demographic profile favours the younger age groups. Still, demographic factors alone may be insufficient to explain differences in these age-associated mortality patterns.

There is a general tendency for body composition to change with age in both sexes (**5.11**), with the major factors being loss of muscle mass and increase in adipose fat mass. Associated with this trend is a decline in energy requirements. This decline is, in most cases, amplified by reduced activity levels, often without a compensatory reduction in food intake. The result is an enhanced tendency toward obesity in middle-age. There is a clear association between obesity and diabetes which is thought to potentiate degenerative changes in tissues throughout the body through disruption of peripheral circulation.

Late onset insulin-resistant diabetes mellitus (NIDDM) (**5.16**) is increasingly common after middle age. Although the cause of this condition is still only poorly understood, there is a growing body of evidence that the function of the insulin-secreting B-cells of the pancreas is impaired by the presence of the protein amylin which also induces insulin resistance in non-pancreatic tissues. The presence of this aberrant protein with its widespread effects raises the question of whether other degenerative changes associated with ageing are associated with abnormal proteins. It has long been known that Alzheimer's disease is associated with the accumulation of amyloid plaque that is associated with neurofibrillary tangles.

However the question of whether the association implies a cause-and effect relationship remains controversial. The presence of two other aberrant proteins, BAP, and Tau, the former the result of faulty cleavage of a normal cellular protein and the latter the product of deviant aggregations of the cytoskeletal protein tubulin may be evidence of a general tendency for irregularities in the process of protein folding to accumulate with increasing age. Since such irregularities often enhance the tendency of such proteins to form insoluble aggregates that may disrupt normal cellular function, they are now viewed as potential agents of the ageing process.

It remains to be seen whether environmental factors peculiar to industrialized societies play a role in potentiating errors in protein folding that may underlie some of the pathologies of ageing. The presence of industrial pollutants in air and water, use of pesticides and herbicides, and exposure to sources of ionizing radiation (**11.7**) all increase the risk of accumulation of free radicals and damaged DNA. These factors may potentiate transcriptional and translational errors that underlie pathological ageing and such conditions as cancer. *William A. Stini*

early mortality in middle or later adult years(**12.1**). Traits beneficial at young ages that may be strongly associated with disease or death in middle-age will be preserved by natural selection if their early advantage leads to improved reproduction. Several genes in humans have been identified as candidates for antagonistic pleiotropy, APO E4 (which increases risk of cardiovascular disease, Alzheimer's Disease, and disability), genes controlling testosterone level (which may contribute to risk for some cancers among men), and those controlling cholesterol extraction and retention. The second model is age-specific gene action, some loci may include alleles that have detrimental effects at later points in the lifespan. Huntington's disease and amyotropic lateral sclerosis are possible examples of age-specific gene action.

*Douglas E. Crews and Gillian J. Harper*

See also 'Comparative growth and development of mammals' (**2.1**), 'Body composition' (**5.11**), 'Physical growth during industrialization' (**11.2**) and 'Menopause' (**12.2**)

## AGEING AND METABOLIC CHANGE

In evolutionary terms it has been proposed that senescence is the end-result of an energy conservation strategy operating in somatic cells. During the course of an organism's lifespan, total available energy (**9.2**) has to be differentially allocated to a variety of functions, including macromolecular synthesis and degradation, cell and organ maintenance and reproduction of the species. Since the energy supply is finite, and in order to ensure propagation of the species by the successful transmission of genes to future generations, a compromise has to be reached between the energy made available for each function. This accommodation in energy-saving is achieved by the maintenance of absolute or near-total accuracy in germ-cell replication but less rigorous correction of mutations which occur in somatic cells. If maintenance of the energy supply is critical, then in the event of a reduction in energy production or supply all energy-dependent processes in somatic cells will be adversely affected in proportion to their particular energy requirements.

The area of ageing research which is currently attracting greatest interest links functional decline in the mitochondria, sub-cellular organelles responsible for aerobic adenosine triphosphate (ATP) production, to the disorganizing effects of free-radical-induced damage. Large numbers of free radicals are formed during normal cellular respiration, and both the inner mitochondrial membrane and mitochondrial DNA (mtDNA) have been identified as potential targets for their destructive action. As the inner membrane is the site of the electron transport chain, and enzymes of the tricarboxylic acid-cycle are located within the mitochondrial matrix, oxidative damage by free radicals generated as by-products of cell respiration could severely compromise the ability of an organism to produce ATP and so meet its essential energy requirements.

Experimental studies have indicated that human mitochondria show a decrease in number, an increase in size and the occurrence of structural abnormalities with ageing, and mtDNA analysis has revealed a spectrum of deletions and point mutations in the 16 569 nucleotides of the human mitochondrial genome with advancing age. The best described of these mutations is a 4977 base pair (bp) deletion in a region of the mitochondrial genome which encodes components of the respiratory system. The deletion has been confirmed in a wide variety of tissues which, to varying degrees, are dependent on aerobic respiration, including heart, brain, liver, kidney, spleen and skeletal muscle (**5.11**). In ovarian tissue the 4977 bp deletion has been observed only after onset of the menopause (**12.2**).

This evidence provides strong support for the hypothesis that the accumulation of somatic mtDNA deletions and other forms of mutation may be an important determinant of ageing, with cytoplasmic segregation of the mutant mitochondrial genomes (heteroplasmy) leading to cells possessing variable bio-energetic capacities. Studies have indicated that, even at the end of their *in vitro* lifespan, the mitochondria of human diploid fibroblasts remain potentially functional. However, there is a large body of information which shows an age-related decline in the activity of the cellular respiratory chain and its constituent enzymes, resulting in a significant decrease in available energy. Whether these changes can be ascribed primarily to functional declines in the mitochondrial or nuclear genomes has yet to be clarified, as has the nature of the co-operative mechanisms utilized by the two genomes in the regulation of cellular energy production.

*Alan H. Bittles*

Factors influencing longevity

| | |
|---|---|
| **Longer lifespan** | |
| Increased rates of UV-induced DNA repair | Hart and Setlow (1974)<br>Hall et al. (1984) |
| Enhanced superoxide dismutase activity | Tolmasoff et al. (1980) |
| Longer red cell lifespan *in vivo* | Rohme (1981) |
| Longer fibroblast lifespan *in vitro* | Rohme (1981) |
| Elevated level of carotenoids | Cutler (1984) |
| **Shorter lifespan** | |
| Increased number of chromosomal abnormalities | Curtis and Miller (1971) |
| Telomere shortening | Harley et al. (1990) |
| Reduced number of functional mitochondria | Lipetz and Cristofalo (1979)<br>Johnson (1979) |
| Increased numbers and size of lysomes | Lipetz and Cristofalo (1972)<br>Johnson (1979) |
| Increased cytochrome P-488 content | Pashko and Schwartz (1982) |
| Increased DNA binding of activated 7.12 dimethyl benz anthracene | Schwartz and Moore (1977) |
| Increased benzopyrene conversion to water-soluble metabolites | Moore and Schwartz (1978) |
| Increased rates of auto-oxidation | Cutler (1985) |
| Elevated free radical production | Harman (1992) |
| Elevated rates of superoxide and hydrogen peroxide generation | Ku et al. (1993) |

## AGEING IN NON-WESTERN POPULATIONS

*Demographic and biological characteristics of different populations*

| Country | Population in 1986 (×$10^6$) | % aged 15–64 years | Expectation of life at birth | Proportional CVD mortality for those > 30 years | Mean systolic BP (mm/Hg) men aged >30–35 years | Mean diastolic BP (mm/Hg) men aged > 30–35 years | Mean body mass index (kg/m$^2$) men aged >30–35 years | Mean total cholesterol (mg/dl) men aged > 30–35 years |
|---|---|---|---|---|---|---|---|---|
| Cuba | 10 | 65.7 | 72 | 44 | 125 | 80 | 22.0 | 200 |
| Ghana | 14 | 50.5 | 49 | 22 | 134 | 82 | 20.6 | 180 |
| Mauritius | 1 | 65.0 | 65 | 45 | 121 | 76 | 21.5 | 215 |
| Sri Lanka | 17 | 61.0 | 67 | 21 | 120 | 75 | 20.6 | 211 |
| Tanzania | 23 | 49.0 | 50 | 22 | 124 | 75 | 21.0 | 135 |
| Thailand | 52 | 60.0 | 63 | 30 | 125 | 75 | 21.1 | 189 |
| USA with Finland | – | 67.0 | 71 | 50 | 125 | 80 | 26.0 | 215 |

Western and non-Western populations differ in ethnic diversity and demographic structure. To what extent do they also differ in the biological properties of ageing? 'Ageing' can be described at the individual level in terms of the morphological and functional changes associated with increasing age, but is perhaps best defined at a population level as the net effect of such changes on the organism's ability to survive to a given age. For it is this which results in progressively increasing age-specific mortality with age.

Contemporary non-Western populations represent various points along the continuum of epidemiological, demographic and nutritional characteristics which change with varying economic and social circumstances. Life-expectancy at birth is usually shorter, and rates of chronic disease mortality lower, than for Western populations. But it cannot simply be claimed that non-Western populations show slower rates of biological ageing and comparatively few problems of ageing when judged by Western criteria.

Environmental conditions influence growth and development, resulting in differential survivorship, secular trends and cohort-associated effects which make it difficult to describe biological ageing adequately from cross-sectional data. The need for fully longitudinal studies of ageing in developing countries has not yet been met, so it is necessary to rely on cross-sectional data for groups showing little evidence of secular trends. Long-bone length, as represented by tibial height, iliac height, or arm span, does not change during adulthood (**5.17**) and so comparison between age cohorts will suggest the presence or absence of cohort or secular trend effects on the morphology, size and composition of the body. It is then possible to construct ageing profiles by first correcting age-associated variables for chronological age, expressing these as standard deviation scores, and then plotting the values for the subpopulations of interest.

Both Western and non-Western populations show a loss of standing height at a rate of about 1 centimetre per decade in both sexes. This results from degenerative changes in vertebral and intervertebral bodies, leading to progressive spinal curvature, and from the postural consequences of arthritic joints. This is a universal property of normal ageing. It is associated, especially in those over the age of 40, with a loss of lean body mass varying between 3 and 6 per cent per decade. This also appears to be universal. Non-Western populations tend to show declining fatness (**5.16**) and fat content with increasing age, in marked contrast to Western populations. Trends in blood

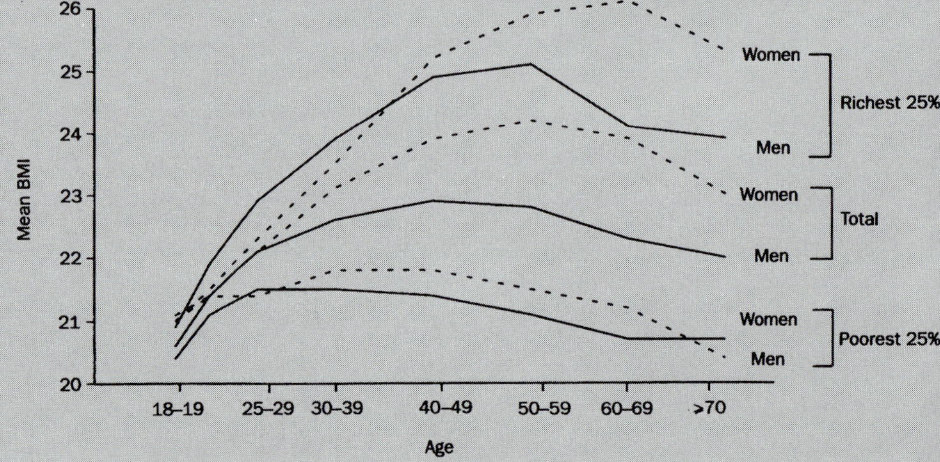


**Body mass index (kg/m²), age and income in Brazil, 1974-75 (data from Vasconcellos, 1994).**

pressure follow this contrasting pattern, though largely unacculturated Inuits living on diets rich in fat and protein resemble Westerners in showing a moderate increase with age. However, the direction and magnitude of change in body-weight, when corrected for height, varies along social economic gradients (**1.12, 11.4**) within non-Western populations, as for example among Brazilians. Thus, these are not necessary properties of ageing.

The causes of these differences in age-related fatness and body-weight are obscure. They may be related to the general absence of a 'retirement threshold' in the physical activity level of people living in developing countries and its dependence on occupation (**9.10**). However, the anabolic hormone insulin stimulates fat synthesis by enhancing glucose uptake into adipocytes, and fatness is often associated with blood pressure and levels of glucose and insulin. Thus there may be underlying population and subpopulation differences in insulin/glucose metabolism. Non-Westernized highlanders of Papua New Guinea showed a slight age-related increase in postprandial blood glucose, and as with Inuits on traditional diets, these were markedly lower values than in Americans and affluent Pacific islanders.

Forced expiratory volume (FEV-1) and Forced Vital Capacity (FVC) are measures of pulmonary function which also show characteristic declines with age. These are correlated with anthropometric dimensions (weight, height) and indices of muscularity, so observed declines are partly a function of changing body composition. In Tibetan and American adults aged 20–60 years, the FVC fell by about 24–26 millilitres per year in men and 21–22 millilitres per year in women, and the FEV-1 by about 30–33 millilitres per year in men and 22–26 millilitres per year in women. Values of peak expiratory flow.in highlanders of Papua New Guinea fell by about 3.4 litres per minute per year. These values were intermediate between those of Australian Whites and Welsh farmers, but were probably confounded by pulmonary disease. In this respect further studies of normal and pathological ageing are needed.

Women in non-Western populations have tended to show comparatively late age at menarche (**5.15**) and it has been supposed that menopause would be delayed or even accelerated to a similar extent. Menopausal ages of 45 years in highlanders of Papua New Guinea, 47 in Tibetans, 47–49 in Indian Rajputs (depending on altitude), and 49 in Greenland Inuits, suggest that late menarche can sometimes be followed by slightly more rapid completion of female reproductive lifespan (**12.2**). *Simon S. Strickland*

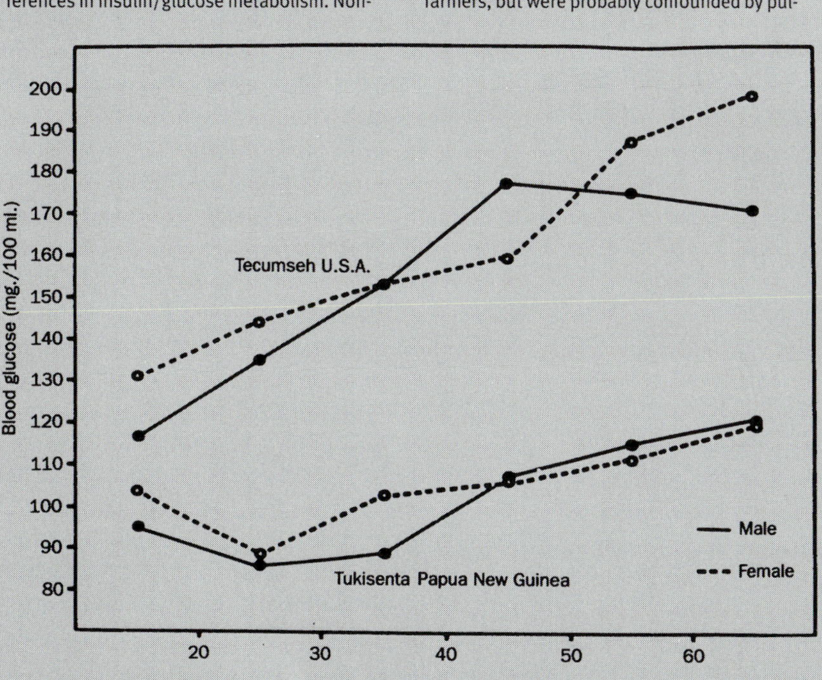


**Postprandial blood glucose concentrations in highlanders of Papua New Guinea (Tukisenta), and in the United States of America (Tecumseh, USA).**

**An older woman in rural Sarawak, with low body mass index (BMI).**

## SARCOPENIA: MUSCLE-LOSS

Sarcopenia is the deficiency of muscle mass associated with wasting in whatever context, including during the course of ageing. In European populations, there is marked loss of muscle mass and strength decrease after about the age of 50 years, especially in women following the menopause. The decline in males is less severe, although by the age of 90 years, muscle mass can be reduced by up to 30 per cent in both sexes. The loss of muscle mass is greater than can be accounted for on the basis of muscle-fibre atrophy (**5.8**), suggesting that the number of fibres is also reduced. It has been suggested that there is a loss of motorneurones from the anterior horn causing a loss of motor units. Although it is not clear why motor unit loss should happen, it has been suggested that impaired neural function or peripheral damage to motor nerves may be the initiating factor.

In the absence of disease, sarcopenia is reversible by physical activity regimens even in subjects up to the age of 90 years. However, decreases in muscle strength increase the likelihood of falls, while low values for muscle mass, especially in the extremities where most skeletal muscle is located, is associated with higher than expected mortality rates. This might be as a consequence of higher infectious disease risk through reduced immunocompetence (**4.7**), and/or greater risk of cardiovascular disease associated with low high density lipoprotein-C levels. In both cases the effect would be mediated by low somatic protein nutritional status.

Roche (1994) has argued that associations between sarcopenia and mortality rates require accurate measures of total muscle mass by such methods as dual X-ray absorptiometry or high frequency energy absorption (**1.6**), since biochemical measures such as urinary creatinine and 3-methyl histidine are deemed to show too much day-to-day variability to be of use in assessing small changes. Since the physical methods cannot be widely used in epidemiological study, Roche recommends the use of body mass index (BMI) in studies of sarcopenia and mortality. *Stanley Ulijaszek*

*Characteristics of five major studies that relate sarcopenia to mortality rates (from Roche, 1994)*

| **Author** | **Tayback et al. (1990)** | **Tao et al. (1991)** | **Wienpahl et al. (1990)** | **Cornoni-Huntley et al. (1991)** | **Lindsted et al. (1991)** |
|---|---|---|---|---|---|
| Sample | 4710 aged 55–74 years | 3043 aged 40–59 years | 5184 aged 30–79 (men) and 40–79 years (women) | 1472 aged 65–74 | 8828 aged from less than 40 years (22.8%) to 80 years or older(5.6%) |
| Gender | Male | Male | Male and female | Male and female | Male |
| Exclusion of early follow-up deaths | 1 year | 5 and 10 years | 5 years | 7 years | No |
| Exclusion of those with diseases recognized at physical examinations | Yes | Yes | Yes | Yes | No |
| Exclusion of current smokers | Yes | Yes | Yes | Yes | Yes |
| Exclusion of ex-smokers | No | No | Yes | Yes | No |
| BMI data[1] | Measured | Measured | Measured | Measured | Self-reported |
| Follow-up | 9 years | 17–20 years | 15 years | 7–13 years | 26 years |
| Major finding *re* mortality rate | *Increased* if BMI < 22.0 $kg/m^2$ | *Increased* if BMI < 23.3 $kg/m^2$ | *Increased* if BMI < 24.1 $kg/m^2$ (men) or < 23.5 $kg/m^2$ (women) | *Increased* if BMI < 21.4 $kg/m^2$ and sum of triceps and skinfold thicknesses < 16.0 m (men) or BMI <25.0 $kg/m^2$ (women) | *Decreased* if BMI ≤ 22.3 $kg/m^2$ or < 20.0 $kg/m^2$ |

[1]Calculated from measured or self-reported weight and stature.

## APOPTOSIS: PROGRAMMED CELL DEATH

Animal cells carry an intrinsic genetic 'death' programme; one that has been conserved by evolution from fruit flies and roundworms through to humans. Originally defined by Kerr, Wyllie and Currie in 1972, apoptosis (pronounced apo-*to*-sis) refers to the programmed series of morphological changes that occur during physiological cell death. Apoptosis is more common than necrosis, another major form of cell death, and appears to be a property of all vertebrate somatic cells. During necrosis, there are early changes in mitochondria and other organelles; the cell becomes unable to maintain homeostasis and the plasma membrane fails to regulate osmotic pressure, finally the cell swells and bursts, releasing cytoplasmic fluids and structures into the matrix, and triggering an inflammatory response. No energy is needed for necrosis.

In contrast to necrosis, apoptosis requires energy in the form of ATP and is characterized by a regular and controlled series of morphological changes. Apoptosis removes damaged or unwanted cells without triggering an inflammatory response. Apoptosis can be quite rapid, sometimes completing its course in minutes. The plasma membrane of the apoptotic cell loses its normal contacts with surrounding cells. The cell then emits processes which may break off and become intact cellular fragments called apoptotic bodies. Since the intracellular contents are not spilled, there is no inflammatory response (**9.3**). Early in the apoptotic process, phagocytic cells recognize apoptotic cells and apoptotic fragments. As apoptosis continues, the cell shrinks by up to a third of its volume, decreases its RNA and protein synthesis, and increases protein and RNA degradation. The nucleus then shrinks and chromatin condenses against the nuclear membrane.

Nuclear DNA is cleaved between nucleosomal sites, thereby producing a characteristic ladder of regular DNA subunits in multiples of 180–200 base pairs. Although transcription ceases, other processes must be involved in programmed cell death because apoptosis kills the cell more rapidly than could be accounted for by loss of transcription alone. There are no noticeable changes in mitochondria or other organelles during apoptosis. Genetic regulation of apoptosis currently is poorly understood. However, it is believed that *bcl-2*, *bax*, *Fas*, *p53*, *c-myc*, *ced-4*, *ced-3*, *ced-9*, reaper and other oncogenes and tumour suppressor genes play a role in the regulating apoptosis.

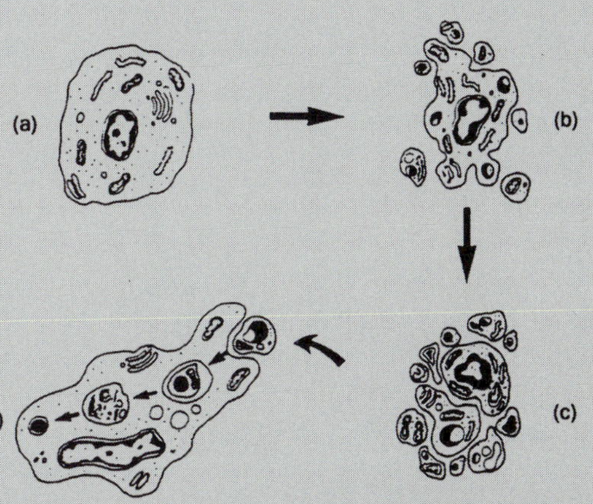

**Schematic representation of apoptotic cell death. Upon being triggered to undergo apoptosis (generally a physiological or mild pathological stimulus), a normal cell (a) condenses its cytoplasm and DNA (b), and proceeds to fragment into many intact vesicles (apoptotic bodies), many containing fragments of condensed chromatin and morphologically normal organelles (c), that are then recognized and engulfed by neighbouring phagocytes (d). Apoptotic cells generally do not provoke an inflammatory response.**

Although apoptosis has a characteristic series of steps, there is no specific and invariable set of metabolic processes that lead to or necessarily occur during apoptosis. Rather, apoptosis may be stimulated in a multitude of different ways, with some factors both enhancing and depressing apoptosis. Induction mechanisms depend on gene expression following exposure to an apoptotic stimulus. Release mechanisms behave as if the apoptotic programme is always present, but is inhibited by molecular factors with short half-lives. Transduction mechanisms also behave as if the apoptotic programme is present at all times, but is activated by an external signal delivered by cytotoxic T cells or other mechanisms.

Apoptosis is a common and continuous process throughout the lifespan. During growth and development many more cells are produced than are needed; during differentiation and specialization these unnecessary cells are eliminated through apoptosis. For example, digits are formed embryonically (**4.5**) by the apoptotic death of cells in the interdigital zone. Throughout our lifetime, as cells malfunction or become damaged, they are eliminated by apoptosis. In adult humans this amounts to approximately 25 million cells every second. Many populations of cells must undergo constant loss and renewal (for example, lymphocytes, cells lining the gut, epidermal cells of the skin). Thus, apoptosis is an important process both in development and in repair and maintenance of the mature soma.

It is important to recognize that the existence of a mechanism for programmed cell death is not equivalent to a 'death programme' for organisms. Evolution has produced apoptosis as an exquisite mechanism for the elimination of malfunctioning, damaged, or unnecessary cells. Although to date there is no convincing evidence that variations in apoptotic processes become uncontrolled with ageing, some evidence that apoptosis may be associated with longevity and degenerative disease is available. For example, peripheral blood mononuclear cells from healthy centenarians may be less susceptible to induction of apoptosis from multiple agents than those of young healthy individuals. In addition, apoptosis is seen in cardiac cells following ischemic events that lead to necrosis of nearby cells and apoptosis also may play a role in some neurodegenerative disease processes. Thus, apoptosis is an important biological phenomenon during growth, development, disease progression, and even ageing. However, it does not appear to be the ultimate cause of ageing. Rather, as with many age-related phenomena, defects in apoptosis, whether leading to early cell death or extended longevity of cells, may increase due to pleiotropic or age-specific action of multiple genetic loci.

*Jay D. Pearson and Douglas E. Crews*

## AGEING AND BONE-LOSS

Turnover of bone mineral is a normal element of the physiology of bone. In a normal, healthy young adult, the rates of resorption and formation of bone are roughly equal, so that the density of the skeleton remains quite constant despite the continual remodelling of bone (**5.6**). Heavy occupational use of a specific part of the body, as in the case of professional tennis players, may lead to asymmetrical development of limb bones through stimulation of bone formation in areas of stress. On the other hand, lack of use or reduction in weight-bearing stress, such as occurs in the microgravity of space flight will lead to loss of bone density as the balance between bone resorption and formation is disturbed by inadequate stimulation of osteoblasts while osteoclasts continue their work of resorption.

Because osteoblasts and osteoclasts respond to a number of endocrinological stimuli, many factors, both local and systemic, can alter the balance of bone turnover. Vitamin D and its metabolites affect calcium homeostasis through mechanisms involving calcium absorption by intestinal epithelium, calcium reabsorption in the kidneys and resorption of bone mineral by osteoclasts. Calcitonin, produced by specialized cells in the thyroid, inhibits osteoclast activity, slowing bone resorption while simultaneously reducing reabsorption of calcium in the kidney tubules, increasing calcium loss in urine. Parathyroid hormone, produced by the parathyroid glands, stimulates osteoclasts to resorb bone while stimulating resorption of calcium in the kidney tubules, thereby reducing urinary calcium loss. Estrogen enhances bone formation by stimulating osteoblasts which possess estrogen receptors. Likewise, the anabolic properties of testosterone appear to affect bone turnover primarily by stimulating bone formation.

Many local factors modify the effects of the aforementioned systemic ones in the regulation of bone turnover. Lymphocytes, macrophages, and monocytes can stimulate osteoclast activity through release of interleukin-1, transforming growth factor alpha tumour necrosis factor alpha, and tumour necrosis factor beta. Prostaglandin $E_2$, produced by many cells of the monocyte-macrophage line, as well as by osteoblasts, also stimulates resorption of bone by osteoclasts. It is thought that the effects of most of these local factors are indirect, and that osteoblasts may serve as intermediaries in the modulation of osteoclast activity. Both gamma interferon (INF-γ) and transforming growth factor beta (TGF-β) are thought to inhibit the process of osteoclast formation.

The complex system of interacting factors that control calcium homeostasis is subject to perturbations arising from alterations of a wide range of systemic and local factors. The most widely experienced form of bone-loss associated with disturbance of the balance of these factors is the accelerated bone loss associated with the decline in estrogen production occurring at menopause. The balance of bone turnover is tipped in the direction of bone resorption in the years immediately following menopause (**12.2**) with the result that the risk of fractures increases dramatically. As a rule, the rate of bone-loss declines 4 or 5 years after menopause, but loss of bone continues for the rest of a woman's life. Men experience a later onset of bone-loss associated with declining testosterone production (**5.15**), and the rate of loss is not as rapid as that experienced by women in the first years after menopause. None the less, virtually anyone who lives beyond the age of 90 can anticipate sufficient loss of bone density to be classified as osteoporotic. *William A. Stini*

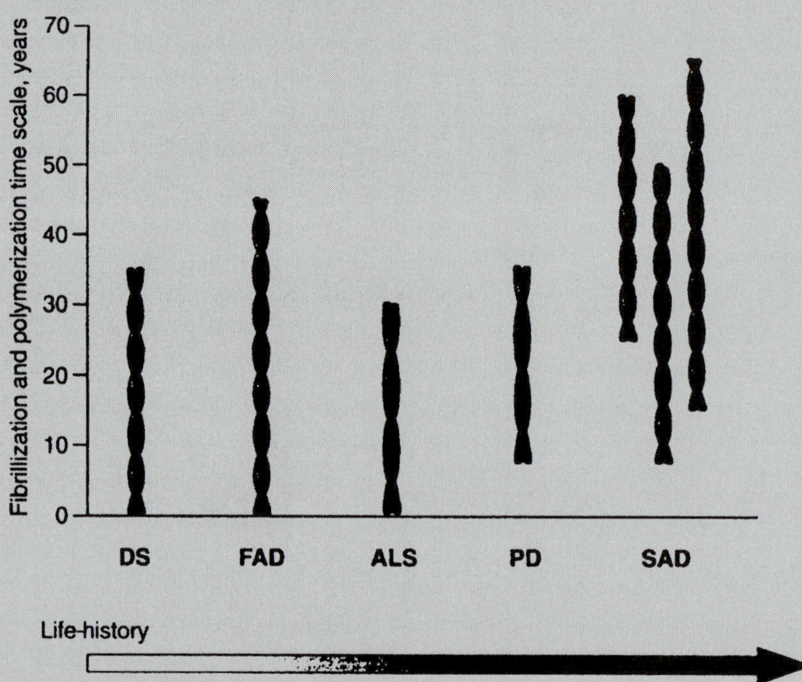


**The process of protein fibrillization and polymerization in neurodegenerative disorders of long-latency and slow progression. The initial insult can be environmental or genetic and occur anytime during the life-cycle. The number of years necessary for the process of fibrillization to occur sufficiently enough to produce clinical disease in measured in decades and can be accelerated or decelerated by the magnitude of the insult, the chronicity of the insult and the age at which the insult begins. If the process is turned off or delayed or the insult removed, clinical expression of disease may not occur. DS = Down syndrome; FAD = familial Alzheimer's disease; ALS = amyotrophic lateral sclerosis; PD = Parkinsonism-dementia; SAD = sporadic Alzheimer's disease.**

'The Seven Ages of Man' by William Mulready, taken from Shakespeare's *As You Like It*.

... At first the infant,
Mewling and puking in the nurse's arms.
And then the whining schoolboy, with his satchel,
And shining morning face, creeping like snail
Unwillingly to school. And then the lover,
Sighing like furnace, with a woeful ballad
Made to his mistress' eyebrow. Then a soldier,
Full of strange oaths, and bearded like the pard,
Jealous in honour, sudden and quick in quarrel,
Seeking the bubble reputation
Even in the cannon's mouth. And then the justice,
In fair round belly with good capon lined,
With eyes severe and beard of formal cut,
Full of wise saws and modern instances;
And so he plays his part. The sixth age shifts
Into the lean and slippered pantaloon,
With spectacles on nose and pouch on side,
His youthful hose well saved a world too wide
For his shrunk shank; and his big manly voice,
Turning again towards childish treble, pipes
And whistles in his sound. Last scene of all,
That ends this strange eventful history,
Is second childishness, and mere oblivion,
Sans teeth, sans eyes, sans taste, sans everything.

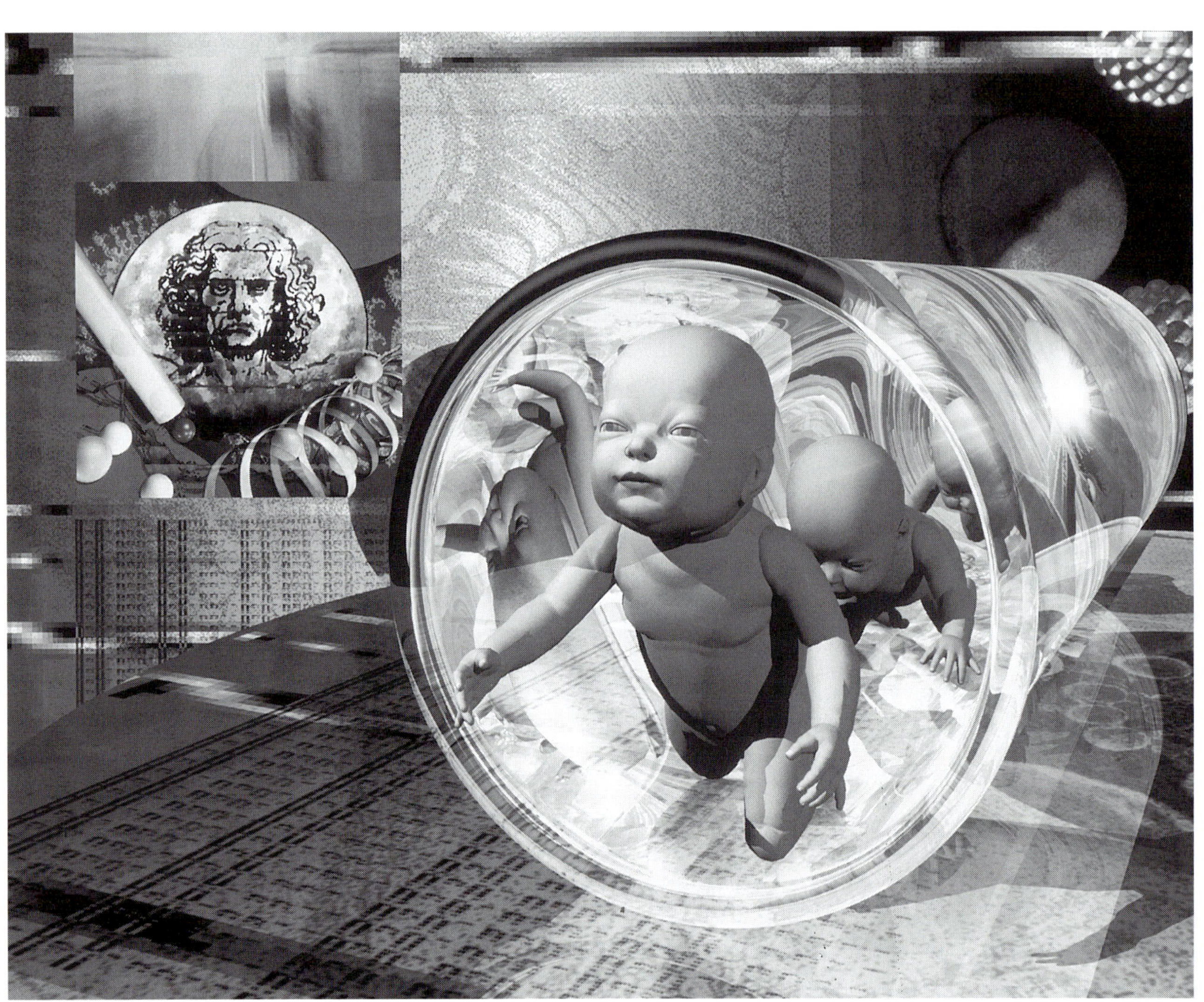

## Part Thirteen

# The future

The children are the future; but who can predict the future? With respect to human populations, one thing is certain. In the near future at least, there will as many, if not more children alive than at any time in the history and prehistory of humanity. With respect to the study of growth and development there is another certainty: that research has barely moved from the descriptive to the analytic. While many of the descriptions are adequate for certain types of public health interventions, there are many uncertainties, largely because mechanisms of disease processes associated with growth and development are in most cases still poorly understood. Thus, one strand with which the tapestry of the future will be woven is the pursuit of understanding of human growth and development mechanisms, at all levels: endocrinological, immunological and genetic.

However, the old questions will not go away. Anthropologists are as interested in the interaction of nature and nurture on the physiological, morphological and developmental characters of human populations now as they were at the time of Franz Boas (see Appendix 1, 13). The methods used to study these processes and traits span the range of human biological research. Understanding of the nature of human plasticity is still poor, despite myriad descriptive studies. For example, it is common knowledge that poor growth relative to Western reference values at the age of weaning is widespread in the developing world; the primary reason traditionally given for this is undernutrition. However, there is evidence from epidemiological, immunological and nutritional studies that there may be many instances where growth faltering may be primarily due to various metabolic disturbances associated with infection. Thus, diverse research into the mechanisms of growth-related processes can shed new light on the ecology of growth in a way that has tremendous public health significance: children's lives depend on this knowledge. Similarly, it is reasonable to expect that future insights into mechanisms of growth will change understanding of human plasticity, adaptation and growth pathology, perhaps in ways currently unknown.

Human populations are experiencing change on an unprecedented scale. Population numbers are greater than ever before, and this is causing ecological stress due to the scale of numbers in relation to resources in some parts of the world. Diseases like coronary heart disease and non-insulin-dependent diabetes, previously dubbed 'diseases of civilization', are increasing in prevalence in the less-developed world; whether these prevalences reflect a transition from generalized undernutrition to overnutrition at the population level is not clear. However, hot pursuit of the epidemiology and mechanisms of diseases and disorders possibly associated with early environmental factors (**12.1**) is both intellectually stimulating and, again, of vital public health importance.

**Opposite: The face of the future? Courtesy of Laguna Design/ Science Photo Library, London. The ability to clone genetic material has opened the door to many previously unthinkable medical prospects.**

New stresses and new problem complexes (**13.3**) have arisen across the course of the present century. New ecological stresses include emerging diseases, of which HIV infection is one. This disease will have growing impact on child growth, and because it is also a disease of adults, will continue to change the rural ecologies of many parts of Africa and elsewhere by virtue of the socio-economic implications of disease and death among productive adults. This will be in addition to changing disease ecologies as a consequence of the natural history of HIV infection itself; death by opportunistic infections means that the spread of diarrheal and respiratory infections (common killers of the very young; shapers of the human growth curves of the survivors) may increase. One respiratory infection in particular, tuberculosis, has become a disease of the late 20th century just as it was a disease of the 19th, and is likely to be a disease of the 21st century too.

Migration, urbanization and exposure to and control of new technologies both change human populations (**13.2**) and expose them to new stresses. Although human biologists and anthropologists in the 1960s were aware that most of the societies they were studying were undergoing rapid change, the pace of change has accelerated since the 1960s, with no sign of abatement. The impact of modernization on the human biology of traditional societies since that time has been documented for populations in most parts of the developing world.

New questions have arisen from the study of populations undergoing change as a consequence of both local and international environmental change. The developing world has experienced and continues to experience dietary, nutritional, fertility and mortality transitions taking place, which will fundamentally affect the population structure and their biology.

Developments in the use of analytical methods and models, and in study design have become all-important in auxology. The latter has focused on the choice of population, the type and size of sample, and the type of statistical design (**1.13**) for the question in hand and the nature of the biological data being collected. Statistical methods appropriate for the handling of dietary, disease frequency, anthropometric, and nutritional data have all been published, as has work on multilevel modelling for use with hierarchically organized data.

The study of body-size and shape is classically part of the remit of anthropology, featuring Franz Boas and Francis Galton among the earliest practitioners. The modelling of the human growth curve has been recognized as being important as a way in which the variation in growth patterns of populations both locally and internationally could be understood. Modelling has been used to improve understanding of: 1) human growth patterns; 2) short-term fluctuations in growth-rates, and 3) the relative importance of environmental quality on variation in growth-rates. Such work has implications for the understanding of human population biology, nutrition, and public health since an understanding of growth plasticity in response to environmental stressors is reliant upon an understanding of growth patterns among individuals and populations in relatively unstressed environments. The description and use of human growth patterns in predicting and screening for abnormality depends on what is considered normal. Apart from the general agreement that growth is a sensitive measure of environmental quality, there has been constant debate about the modelling of growth, and associated with this, the nature of normal or acceptable growth. The near future is likely to see the greater use of computing in growth screening, and a greater use of meta-analyses of human growth studies in better characterizing human growth patterns (**13.1**).

The linking of human growth patterns with physiological processes calls for greater accuracy and sophistication of measurement of a wide variety of parameters, anthropometric and otherwise. Thus, the creation of the knemometer, and more recently the mikro-knemometer will allow short-term growth to be examined in greater detail, as will the use of biochemical markers of short-term growth such as collagen metabolites (**1.7**). Imaging and scanning devices also provide much more information about morphology and composition than would ever have been thought possible 50 years ago.

Most importantly, the future lies also with genetics. The study of the genetic basis of normal and abnormal growth, the mediation of genetic control by growth factors, and the regulation of genetics by environmental factors is still in its infancy. However, such understanding as is gleaned will underpin our understanding of the growth process, growth patterns, ecology and plasticity of growth, as well potentially providing the basis for future treatment of abnormal growth (**13.4**).

*Stanley Ulijaszek*

13.1

# Growth modelling and growth references: the future

Two contrasting influences are likely to direct the use and construction of growth references in the future – one is the increasing availability of computers, and the other is a changing perception of what constitutes 'healthy' growth. Computers have already revolutionized the collection of anthropometric measurements. Large national research studies such as the Health and Nutrition Examination Surveys in America, or the National Diet and Nutrition Survey in Britain are now relatively common. The collection, storage and dissemination of such data is so easy that increasingly large samples are becoming available for analysis.

In addition there are vast, accumulating and relatively untapped sources of computerized anthropometry collected routinely in primary care. There are two important needs here – like all such data they need to go through a stage of monitoring for accuracy and data cleaning, otherwise their value will be unproven; and they should also be made available to general practitioners, linked to other clinical information, to justify the data collection.

The proliferation of body measurement data highlights the need for specialist software to handle them. This need is apparent in two distinct areas – to simplify the clinical interpretation of anthropometry, and to construct different (and possibly new) types of growth reference.

There are a few programs that are designed to display growth data – the Castlemead Growth Program, for example, is aimed at general practitioners or growth specialists, and some drug companies have produced software targeted at growth hormone specialists. However, none of the programs (so far) interface with primary-care databases, which greatly reduces their value to general practitioners. This is one area where things are likely to improve.

The Castlemead Growth Program gives a few clues as to the future of such software. Including a database, spreadsheet and graphics, it allows data to be input manually, saved and plotted on a suitable growth reference, either as the measurement itself or as a standard deviation score (SDS) or Z-score. However, the program would be more valuable if it could update its database from a central database automatically, say, at night, and subsequently flag children whose growth pattern gave cause for concern. A degree of intelligent data cleaning would also be needed, to distinguish between gross measurement error and genuine growth failure.

The availability of large quantities of anthropometry (**1.4**) should greatly simplify the assembly of reference samples to construct or update growth references. The numbers available may even make ethnic minority references possible, although this will raise other issues of definition. Already birth-weight centile charts have been published based on databases containing hundreds of thousands of measurements. Yet large numbers do not necessarily produce a high quality reference – the key issues of data quality and sample selection do not disappear just because the sample is large.

The greatest benefit of such data would be in allowing national references to be updated regularly, every 15 years or so. Few countries (The Netherlands is one) can afford national surveys on a regular basis, due to the high costs involved, but a sample based on data collected routinely would be both cheap and timely.

The dual pressures of more data and better software will lead to improved forms of growth reference. Most growth charts are very simple, involving just two variables, because more complex references, for example conditioning on other relevant information, cannot easily fit on a chart. However, on a computer screen customization is very easy. The principles of conditional references (as described for example by Berkey) will provide tailor-made charts on screen, with correspondingly greater relevance to the individual patient. Birth-weight centile charts can be customized for women of differing body-size and ethnic group, showing how these factors affect the interpretation of the infant's birth-weight(**4.9**, **10.3**).

In the developing world the emphasis on computers will be less noticeable in the immediate future, though not absent. Instead, growth charts will continue to be valuable in

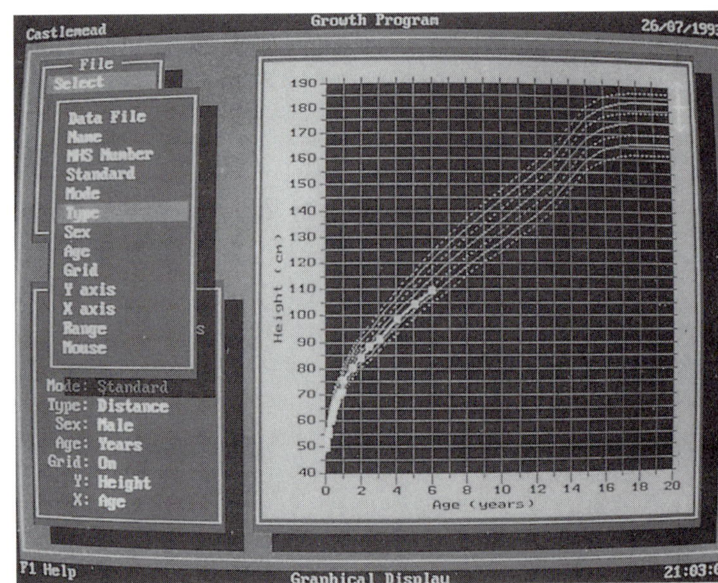


**The Castlemead Program graphical display window showing 'Select'. A typical graph with growth standard centiles and patient measurements plotted automatically.**

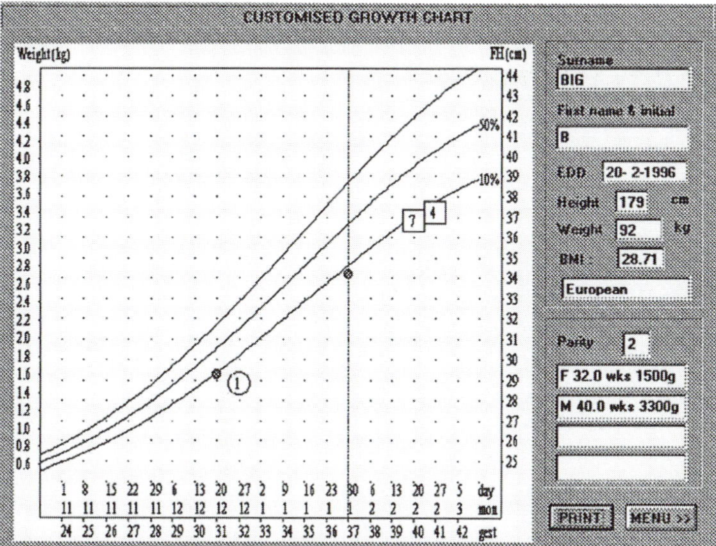


Customized growth chart for 'Mrs Big', showing the weight percentiles of two previous babies and a 3400-g boy born at 41 weeks in the current pregnancy. The fetal weights, based on two antenatal scans at 31 and 37 weeks, are also shown.

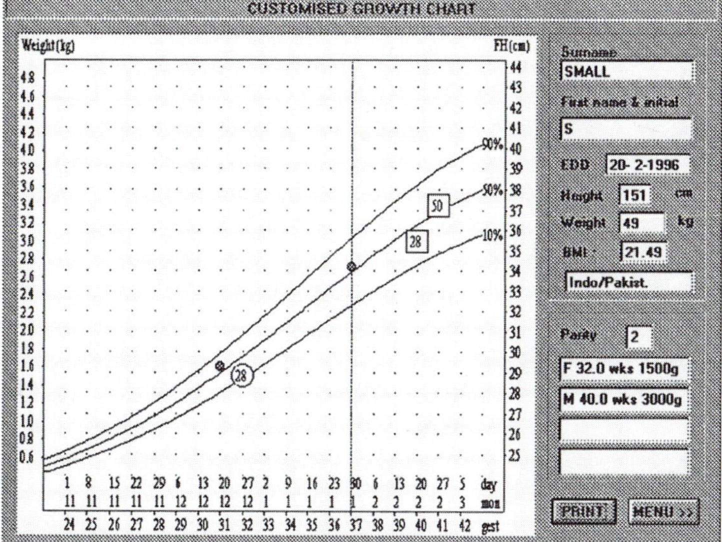


Customized growth chart for 'Mrs Small', showing the same weight data but different percentiles as the norm of fetal growth has been adjusted to the different pregnancy characteristics.

the form of 'Road to Health' charts, and low-tech alternatives to computers, such as nomograms and slide-rules, will also be useful. However, the most important change in the developing world will involve the World Health Organization (WHO) international growth reference. This is currently based on the United States National Center for Health Statistics standard, but there are pressures, both political and health-related, to change it. Politically it is felt to be increasingly inappropriate to impose the United States reference on other countries, but more importantly the old idea that American growth is 'healthy' growth is becoming less and less credible. North American adults are far from the tallest in the developed world (The Netherlands probably takes that title), and obesity (**1.11**, **9.2**) is a serious and increasing problem in the United States (and to a lesser extent throughout the developed world). Thus the idea that the international growth reference represents good growth, and hence doubles as a growth standard, is no longer viable. A new form of reference is needed, representative of several countries rather than just one, and its relationship (if any) to 'good' growth will be less critical.

*T.J. Cole*

See also 'Longitudinal analysis' (**1.14**), 'Creation of growth references' (**1.16**) and 'The use of growth references' (**1.17**)

# Defining the growth characteristics of new populations

Between-population differences in rates of physical growth and development and attained body-size are well documented. Although the extent to which these differences can be attributed to genetic and environmental factors varies from group to group and across time, the stature growth patterns of all major population groupings are likely to have similar genetic potential, with the possible exception of Asiatic populations. Deviations from these patterns are largely due to any of a variety of environmental insults. However, the characterization of growth patterns must continue, since human populations continue to change with respect to: 1) the type and degree of environmental insult exposed to; 2) patterns of marriage and procreation; and 3) structure, as a consequence of migration (**11.6**).

Changes in environmental stress produce at the population level the secular trends which have been observed initially in the industrialized nations, and subsequently in many places elsewhere. Although predominantly associated with increased body-size, negative trends showing decreased body-size have also been demonstrated. Where positive, trends in increased stature are not necessarily accompanied by proportionate increase in weight, such that increased body-size may be associated with increased linearity, or fatness, or neither of these. Where negative, decreased stature may be associated with increased young-child mortality, especially in the developing countries. If, as is likely, the mortality is differentially greater among the smaller individuals of the young child population, then this is likely to contribute to changes in the genetic constitution of the population in the future.

With the increased connectedness and urbanism of human populations especially during the second part of the 20th century, marriage patterns reflect patterns of association within and between culturally defined groups in different places. For example, in San Francisco, although like ethnicity generally marries like, Jewish people are quite likely to marry Europeans of various sorts; and, while Philippinos and people from the Middle East are easy candidates for intermarriage, African-Americans do not intermarry with either Philippinos and are more likely to marry Latin-Americans than Europeans or other Asians, while African-Americans, when they intermarry, may do so across a wide range of ethnicities. In this case study, the greatest influence on ethnic intermarriage frequency is the living distance between the two spouses before they married. This effect tails off beyond a distance of 5 miles. In the case of Philippinos, closer residency to Latino populations than Chinese or Japanese groups may explain the observed pattern. The second greatest effect on ethnic intermarriage is education. Marriage between major groupings is greater among individuals with college education or beyond. This is especially true of African-Americans, who, once educated beyond the poor urban environment in which the

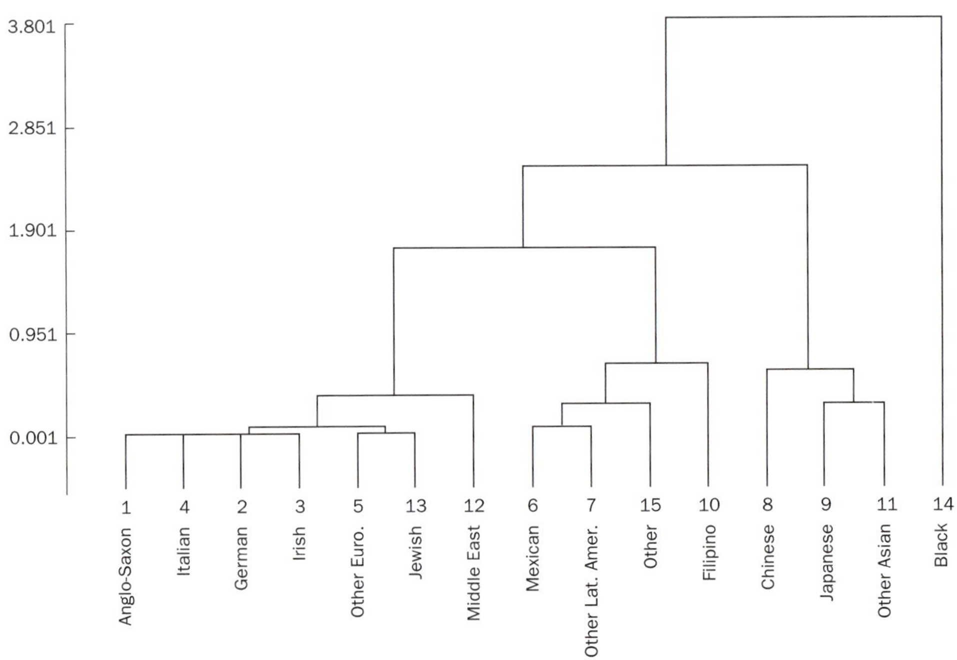


**San Francisco groom's ethnic group, against bride's ethnic group, Ward's method. From Peach & Mitchell, 1988.**

majority of them find themselves, may intermarry with most other ethnicities. Similar patterns of intermarriage, involving differing ethnic groups, are played out in most of the big cities of the world. The urban environment and the human groupings within it thus shapes the population structure of the next generation, with increasing numbers of growing children who do not fall into functionally straightforward categories such as 'European', 'Asian', or 'African-origin'. The implications of this for growth are not clear, other than it becoming more difficult to use established growth norms for screening and surveillance.

Human migrations taking place after the onset of agriculture (**11.1**) in various parts of the world served to create larger, more genetically homogeneous populations across wide areas, with genetically isolated populations left in less-hospitable ecological niches. Genetically isolated populations to be found in the world today include tribal groups of hunter-gatherers in Africa, Latin America, and Asia. Later migrations, during colonial times, included the migrations of Europeans to the Americas, Australasia and parts of Africa; of Africans, mostly of Bantu origin, to the Americas and the Caribbean; and of Asiatic and South Asian peoples to most parts of the tropical world and to parts of the New World. For the purposes of the study of growth, these have been adequately classified by Eveleth and Tanner (1990). However, migrations in the post-colonial period are largely related to economics and urbanization, and are usually highly selective.

Migrant populations in towns and cities of the industrialized world are usually not typical of the populations from whence they came. Thus, the overwhelmingly urban South Asian population of Britain is both heterogeneous within Britain, because South Asians form groups according to place of origin (be it Pakistan, the Punjab, Gujarat, East Africa, or elsewhere), and different from the population in their place of origin. In the developing world, the growth and development of urban populations also differs from those of rural populations, largely because of differences in environmental quality and access to resources. In both industrialized and developing nations, urban life provides a nexus for ethnic mixing and intermarriage, which can alter the growth characteristics of populations. While the genetic potential for growth in height may be small in comparison with the extent to which growth potential is denied by environmental insult in most parts of the developing world, there may be important differences in other aspects of growth which might have health consequences under different environmental conditions.

Currently, many populations of the developing world are undergoing various transitions: in mortality rates, fertility, health, and diet. As conditions change, so do growth patterns. The validity of measures of growth and body-size in assessing health risk and outcome depend heavily on the understanding of normal or acceptable growth patterns of the population under consideration. Thus, despite the vast array of growth data accumulated during this century, it is unlikely that auxologists will be asked to put down their stadiometers and callipers in the foreseeable future. However, should it become feasible to predict accurately the growth curves of different populations from large numbers of cross-sectional measurements at a small number of key ages (eg. 0, 1, 7 and 20 years of age), the nature of future population-based data collection would change.

*Stanley Ulijaszek*

See also 'The secular trend' (**11.3**), 'Modernization and growth' (**11.5**), 'Migration and changing population characteristics' (**11.6**) and 'Urbanism and growth' (**11.7**)

13.3

# Growth patterns associated with new problem complexes

The context within which growth occurs is a continuously novel one, made so by the changes which occur in environments, in societal structures, and in technology. In particular, the social and demographic upheavals which have characterized human history have impacted on children and the milieu that channels their development. It is the responsibility of society to provide the conditions which will allow children to attain their potential, by appropriate responses to the problems created by change.

The Industrial Revolution of the 19th century fundamentally altered the environments of children (**11.2**). At the same time, it accelerated the development of a technology that could respond to the new challenges to growing children, and to maintain, as much as possible, conditions which would foster child growth. However, the context has continued to change and new responses have become necessary – an interplay between the new problems that arise and society's response to them.

What are the new problem complexes that challenge the growth of children? They arise from changes in human demography, in human technology, and in human political economy. These changes do not arise anew, but reflect the course of human history. For example, the response to an increasing human population, and the consequent pressures exerted on the environment, led some 12,000 years ago to settled village life characterized by farming and animal hus-

## GROWTH IN HIV-POSITIVE INFANTS AND CHILDREN

HIV infection has become one of the most important health problems facing children worldwide. It is estimated that 5 to 10 million were infected in the 1990s, almost exclusively due to transmission from HIV-infected mothers. Many more children born to infected mothers will remain uninfected but will suffer privation as a consequence of their mother's HIV disease (**8.4**). Growth failure is recognized as one of the significant manifestations of HIV disease in children. In the United Kingdom failure to thrive was an AIDS indicator in 38 of the 212 infected children reported with AIDS by the end of July 1995. However, the relationship between HIV infection in the child, the mother and background factors to growth in weight and height is as yet poorly described. Their interaction is likely to be complex and the contribution of individual components difficult to distinguish.

In studies in Africa and elsewhere, infants born to HIV-positive mothers are often smaller, with a lower birth-weight (**10.3**) and length, than those born to uninfected mothers. The majority are uninfected and most of these will show catch-up growth, attaining the NCHS median standard weight and height by 3 to 4 years of age, such that the anthropometric characteristics become similar to those of children born to seronegative mothers. When such catch-up does not occur it probably reflects the influence of maternal illness and perhaps orphanhood of the child. Infected children have been shown not only to be significantly shorter and lighter for their age than uninfected children, but also to have reduced linear growth- and weight-gain, although these differences may be small. Ill-health, as indicated by HIV related morbidity (**9.3**) and not HIV infection *per se*, is associated with poor growth.

The relationship between growth and symptoms is unclear below the age of 18 months, after that symptomatic HIV infected children show less gain in height and weight than other HIV infected asymptomatic children or those who are uninfected. It is as yet unclear as to what occurs beyond the age of 5 years. Despite lower than normal height- and weight-for-age in HIV infected children, differences in weight are largely accounted for by stunting, and wasting seems to be not as common in children as it is in adults. Other observations of normal gross nutritional status, of more frequent stunting in infected than in uninfected children and of a more pronounced decrease in linear growth than in weight-for-age may reflect a chronic state of malnutrition.

It is unclear whether, or to what extent, growth failure is a direct consequence of HIV infection, acquired opportunistic infections or as an indirect consequence of adverse environmental conditions. The fact that children born to HIV positive mothers may have low birth-weight and length and subnormal growth in the first few months of life regardless of their own infection status has been taken to reflect secondary effects of ill-health in mothers, and of factors such as malnutrition (**9.2**), stress and other maternal risk factors such as injecting-drug use. However the observed differences between infected and uninfected children born to positive mothers of similar backgrounds suggest involvement of the infants' HIV infection in growth failure. Early impaired weight-gain has been apparent in both symptomatic and asymptomatic infected infants. This may be related to effects of primary HIV infection on growth. Slower growth of infected children between years 1 and 6 is more closely related and attributable to the occurrence of HIV-related disease. In HIV-infected children poor growth can be strongly associated with loss of appetite, diarrhea, frequent vomiting, fever and bacterial infections although indirect involvement of the endocrine system cannot be excluded. The role of gastrointestinal dysfunction is unclear. It cannot account for all growth impairment given the short-term and often rare nature of the complaint, the chronic nature of clinical outcome, and the fact that many infected children show no direct evidence of malnutrition. Hence it is more likely that observed differences between infected and uninfected children reflect a multifactorial cause of growth failure related to HIV-infection and the environment in which that infection was transmitted.

Continued follow-up is required to determine the importance of growth to the long-term prognosis of infected infants and children. It has been suggested that growth hormone therapy (**7.9**) should be employed for those children who are at risk of endocrinological abnormality. However, the use of this costly intervention has yet to be convincingly argued in light of an infected child's total health and social needs. While clinical markers are needed to identify those at risk of infection at an early age, differences in growth between infected and uninfected children are too small and their predictive value too weak to use growth as an indicant of infection. However, interpretation of growth studies must be made cautiously because of their non-prospective nature, the methodology used (which may not be optimal for the study of growth), and potential bias towards relatively 'healthy' subsets of the HIV-infected population, those alive and responding to clinical attention.

*Anna Molesworth and Angus Nicoll*

bandry. While this new subsistence strategy permitted the existence of larger populations, it resulted also in an increase in problems associated with poor sanitation, in the intensity of infectious disease, and in malnutrition. The Industrial Revolution further concentrated people into cities and exacerbated the problems of crowding, disease, and nutrition (**9.2**). These changes inevitably affected the patterns of child growth.

The spread of human groups into increasingly remote areas of the world has brought them into contact with new disease organisms, especially viruses, and the social stratification of large complex societies, with the rise of an underclass, has created new vectors for the spread of those diseases. HIV is a classic example of the introduction of a virus into the human population and its spread through particular groups rendered at increased risk because of their social status and role. While the behavioural basis of the AIDS epidemic is incontrovertible, and the differences in its epidemiology among populations a function of cultural and social variables, the impact on the growth of children is clear and has become a major concern of both public-health and clinical medicine. Research into the effects of HIV on child growth is still in the early stages.

Societal changes have led to a general change in sexual behaviour. While there is variation among societies and regions of the world, the overall trend is one of decreasing age at first pregnancy. The effects of adolescent pregnancy on child growth are covered elsewhere in this volume (**8.6**), but as social patterns continue to change so will social behaviour, which will inevitably lead to new challenges to growth.

One of the outcomes of pregnancy among young teenagers is an increase in low-weight births. Biomedical technology now permits the survival of neonates who would in the past have died at birth, and has led to a new category – very low birth-weight – as distinct from simply low birth-weight. The effects on growth of children whose birth-weight was in the range of 1500 grams are still to be determined, as are the effects of altered growth on adult function. Technological advancements create other challenges to normal growth. Not too long ago space travel and gene therapy were still topics for authors writing science fiction. Today both are upon us.

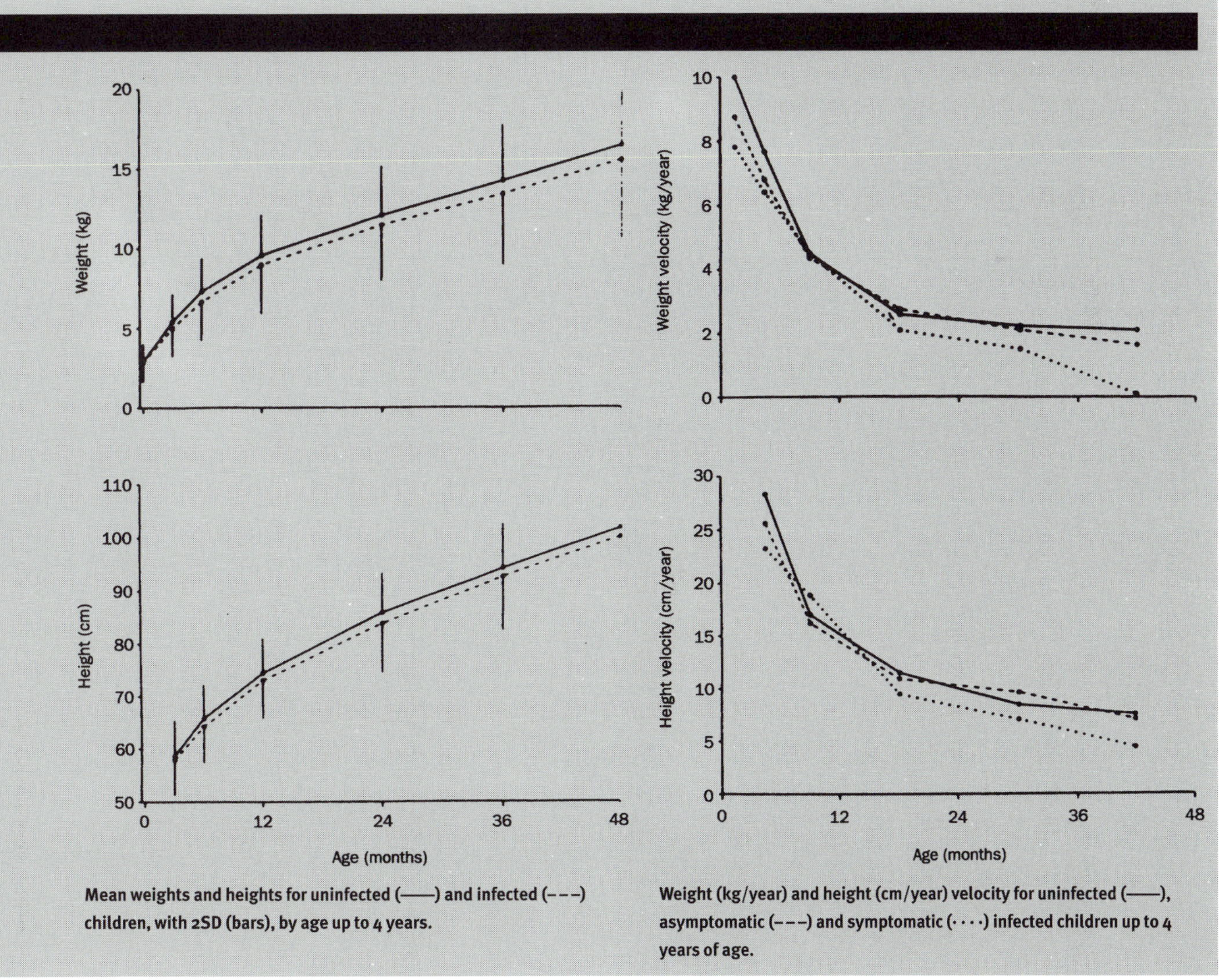


**Mean weights and heights for uninfected (——) and infected (– – –) children, with 2SD (bars), by age up to 4 years.**

**Weight (kg/year) and height (cm/year) velocity for uninfected (——), asymptomatic (– – –) and symptomatic (· · · ·) infected children up to 4 years of age.**

Human growth has evolved within a milieu that includes gravity and one of the dangers of weightlessness is a failure to maintain bone mass within normal limits. The existence of long-term space stations involving large numbers of people is still only a matter for speculation, but at one time so were cities of 10 million inhabitants, as was hormone therapy for altering growth and changing sex.

Today's environments are as much a creation of human technology as a reflection of geological evolution. As children are placed increasingly in these new environments, patterns of growth will surely be affected and responses required to maintain normal channels of development. The new growth-problem complexes that emerge as humans, their accomplishments, and their environments continue to change cannot be predicted with any certainty. However, it is certain that they will emerge, just as it is certain that any responses will increasingly be social and technological. It is also certain that bio-ethical issues will come even more to the front as technology and the values underlying it interact.

*Francis E. Johnston*

See also 'Identification of abnormal growth' (**7.1**) and 'Maternal HIV infection' (**8.4**)

## GROWTH OF CHILDREN BORN TO COCAINE USERS

Cocaine represents a potential compromise to child growth and development via placental exposure during gestation and through passive exposure in the post-natal environment. Preliminary data on the longitudinal effects of intra-uterine cocaine indicate normal growth in weight, but suggest inadequate growth of head circumference (**4.9**).

Cocaine exposure *in utero* has been associated with pre- and post-natal growth retardations in weight, length, head and abdominal circumferences. Delayed growth in exposed infants need not be chronic, however. Given adequate post-natal care and few health complications, exposed newborns characteristically exhibit average or accelerated growth velocities. In one outreach study, mean age- and sex-adjusted percentile rankings for birth-weight, length, and head circumference increased from 23 per cent, 29 per cent and 18 per cent, respectively, to 43 per cent, 49 per cent, and 54 per cent in children of age 12 months. For full-term cocaine-positive infants receiving proper nutrition (**9.2**), weight appropriate for age is often achieved by 12 months of age, despite initially lower mean birth-weights than controls. The deleterious effects of gestational cocaine on neonatal body mass thus appear reversible, with weight being the most responsive parameter of rehabilitation; prenatally exposed infants in one foster-care programme with nutrition, health, and parental surveillance showed substantial catch-up in weight (**9.7**) by 12 months of age yet failed to achieve comparable catch-up in linear growth.

Intra-uterine cranial growth retardation has been declared the most common brain abnormality of cocaine-exposed infants, and the best single predictor of head-growth velocity. The gaseous form of smoked crack-cocaine may pose particularly high risk, causing 2.8 times the likelihood of microcephaly (defined as head circumference ranking more than 2 standard deviations below the mean for gestational age) in one crack-cocaine exposed cohort compared with a cohort exposed to powdered cocaine through maternal insufflation. One study found microcephaly in 21 per cent of the exposed group; and, in another study, 16 per cent of cocaine-exposed newborns had decreased head circumferences, compared with 6 per cent of non-exposed neonates. Although catch-up growth for weight and linear measures (**9.6**) has been demonstrated by the first year of post-natal life for cocaine-exposed infants in optimal care-giving environments, smaller head circumference values for this population frequently persist. Research with adopted children found that, despite comparable values in mean weight and stature by 34 months follow-up, an eight-fold increase in the risk for microcephaly at birth failed to be fully attenuated, with exposed infants still exhibiting significantly smaller occipito-frontal head circumferences than their drug-free peers. A longitudinal evaluation of exposed children in home environments with their biological mothers showed weight and length measures standardized by age 1 year, whereas head size remained smaller through to 2 years of age. These reports suggest that cocaine exerts a direct effect on fetal brain growth (**12.1**) beyond values attributable to restricted somatic growth.

Previous research has shown that small head-size and poor head growth are associated with delayed cognitive developmental scores (**6.1**). Intensive remedial programmes give a favourable prognosis for somatic catch-up growth in cocaine-exposed infants, while cranial growth appears more sensitive to persistent effects of the drug. Data are necessarily incomplete, however, as research indicates that exposed infants lost-to-follow-up represent a more severely compromised group. In addition, potential inadequacies of the post-natal environment may impair the processes and timetable of maturation (**1.10**). And while the short-term growth characteristics of 'cocaine babies' are increasingly known, the long-term characteristics are not.

*Kyra Marie Landzelius*

# Treatment of growth disorders: the future

Improved treatments for growth disorders are going to depend on a variety of actual or potential advances in medical and biological technology and the relevance of such advances will clearly depend on the underlying cause of the growth problem in any particular patient group. One way of approaching the issue is to take some rather arbitrary etiological groups and then see what future benefits advancing science might bring.

## Endocrine deficiencies

This group is typified by abnormalities in the growth hormone axis. The most common of these is isolated growth hormone (GH) insufficiency, which in turn has a number of underlying etiological factors, both genetic (**7.6**) and acquired. At present the treatment of this condition involves the simple replacement of the GH, but this is dependent on a relatively recent advance – the development of recombinant DNA technology to allow the production of GH by bacteria or eukaryotic cells programmed to produce human GH by the insertion of the human GH gene into their DNA makeup. This ensures a virtually limitless supply of GH, which might be seen as the end of the problem. However, there remain at least two problems: the GH requires daily injection, which is relatively invasive and can lead to compliance problems; and the daily injection poorly imitates the physiological pattern of GH secretion which is characterized by pulses of GH at approximately 3-hourly intervals, mostly at night (**5.13**).

The issue of frequent injections would be overcome if it were possible for GH to be given by mouth, but this is unlikely to be possible. Alternatives are to use agents that stimulate the child's own pituitary (as the problem is mostly hypothalamic) and a number of potential substances are currently under investigation, of which the most hopeful are the GH-releasing peptides (**7.9**). The problem of pulsatility is more difficult and it is not clear how important this is. If the releasing peptides become a reality it may be possible to use these in a manner that would better imitate nature.

Further down the GH axis are the insulin-like growth factors (IGFs) which appear to mediate most of the growth promoting actions of GH (**4.2**). Recombinant IGF-I has been produced in limited amounts and tested in some GH-resistant conditions that have been totally unresponsive to treatment, with promising benefits that certainly deserve further development.

Among the endocrine deficiencies we must also consider those that arise as secondary phenomena in the treatment of other disease. The most striking example is the GH insufficiency that will often follow the use of cranial irradiation in the treatment of brain tumours and childhood leukemia. The hypothalamus and pituitary are vulnerable to this irradiation and GH insufficiency is a common sequel. At present this is managed by the use of GH after the situation has developed, but it would be much better if the situation could be prevented. Valuable advances would be the development of other forms of cancer treatment that would remove the need for irradiation, or methods of protecting the endocrine tissues during irradiation.

## Skeletal dysplasias

In these conditions there is a structural bone abnormality which currently does not lend itself to medical treatment, although some small benefit may be gained by the use of GH in high dosage. At present, therapeutic advance is mostly surgical with the development of methods of limb-lengthening. These can be strikingly effective in some cases, but there is still much refinement required, particularly in the maintenance of acceptable body proportions and good mechanical functioning.

It is noteworthy that these conditions are all genetic in etiology and that inevitably the thought of gene therapy comes to mind. Sadly, this is likely to be a very long time away as it would require the introduction of the normal gene into certain specific tissues, such as the growth plates, and the technical problems are overpowering at the moment.

## Glucocorticoid excess

The glucocorticoids are a group of steroid hormones normally secreted by the adrenal glands, and which are normally involved in many aspects of the body's response to stress, but exposure of a child to excessive quantities of the glucocorticoids is a very effective way of stunting its growth. This may occur as the result of an endogenous problem, for example, an adrenal tumour that secretes excessive hydrocortisone, which is the principal glucocorticoid in humans. This is treated by removal of the tumour. More difficult is the situation where the glucocorticoid excess arises from treatment of some other condition. Some synthetic glucocorticoids are very valuable in the management of certain quite common and often serious childhood illnesses, including asthma, arthritis and many forms of kidney disease. Unfortunately, they also have the major side effect of stunting growth, mostly by direct effects on the growth-plate cartilage. Generally this effect is dose dependent and much of the harm can be avoided by reducing the dose to the minimum that effectively controls the primary disease. As these disorders are relatively common there are quite a large number of children who are at risk of stunted growth through this mechanism. Major benefits would accrue from the development of alternative treatments that would avoid the use of these powerful agents, or possibly newer glucocorticoids that do not have this growth-inhibiting side-effect.

## Chronic pediatric disease

We have already discussed the importance of chronic pediatric disease and its effect on growth (**7.8**). In many instances the secondary growth problems will be most helped by improved management of the underlying illness. For example, chronic renal failure is a major cause of growth failure and improvement in the availability of renal transplants will not only improve the renal well being of the children but also improve their growth. On a more worldwide basis the most important single advance would be reduction in the prevalence of malnutrition (**9.2**) and infection (**9.3**) as these remain the most important interacting causes of growth failure in the developing world.

*Michael Preece*

See also 'Identification of abnormal growth' (**7.1**), 'Chromosome aberrations and growth' (**7.4**), 'Bone dysplasias' (**7.5**) and 'Endocrine growth disorders' (**7.7**)

# Brief biographies James Tanner

## 1 Hippolyt Guarinoni (1571–1654)

Guarinoni, an exact contemporary of Francis Bacon, was the son of Emperor Rudolf II's personal physician in Prague. He studied medicine in Padua, where Galileo and Sanctorius were professors, and William Harvey a student. He lived in Hall, near Innsbruck in the Tyrol, from 1598 onwards, doctor to the large convent there and to the townspeople. Long forgotten, his name was resurrected in a short article in a local paper in 1903; by 1927 he was hailed as the forerunner of preventive and social medicine in Germany, and the tercentenary of his death was marked by a full-scale commemorative symposium at the University of Innsbruck.

He emerges as a straightforward and good man of immense energy and enthusiasm, a people's doctor 'ready even in his old age to visit the lowliest hut at any hour of the night', and a staunch campaigner for clean air, pure water and belief in an old-fashioned God. He wrote an immense book, published in 1610, dedicated to Rudolf II together with the Virgin Mary, which boasts 1330 folio pages, and is bound and locked like an old family bible. It is entitled *Die Grewel der Verwüstung menschlichen Geschlechts*, which we may roughly translate as 'The Horrors (Faults, Outrages) which Desolate Humankind'. It is full of stories, personal anecdotes and harangues about clean air, good nutrition, exercise, sleep and God, and is now recognized as a major source of knowledge of peasant life in 17th-century Germany. In it there are two passages relevant to our subject.

Guarinoni was the first to describe retardation of growth caused by emotional stress at school. 'Many children do not grow properly despite good food, because coming home from school they feel still the pain of rough blows and anticipate their renewal with anxiety and fear, so they are never happy or light-hearted'. He also wrote 'The peasant girls of this Landschaft in general menstruate much later than the daughters of the townsfolk or the aristocracy, and seldom before the 17th, 18th or even 20th year .... the townsfolk have usually borne several children before the peasant girls have yet menstruated. The cause seems to be that the inhabitants of the town consume more fat foods and drink and so their bodies become soft, weak and fat and come early to menstruation, in the same way as a tree which is watered too early produces earlier but less well-formed fruit than another'. Guarinoni was nothing if not a country-man.

## 2 Johann Sigismund Elsholtz (1623–88)

Elsholtz, a German physician, graduated at Padua with a thesis entitled *Anthropometria* (1654). This is the first use of the word, meaning 'the measurement of man'. The instrument he used for his measurements he called the *Anthropometron*; it was a vertical-scaled rod, with a travelling horizontal piece. It was clearly derived from an instrument devised two centuries earlier by the architect artist and sculptor Leon Battista Alberti (1404–72) for measuring the proportions of statues, so as to make correct copies, and figures conforming to the accepted Greek and Renaissance canons of proportion.

## 3 Johann Augustin Stöller (1703–80)

Johann Stöller was the author of a remarkable work, the first textbook on human growth. It was published in Magdeburg in 1729, mentioned in Haller's great *Elementa Physiologiae* in 1778 and then wholly lost sight of until Tanner was shown one of the very few remaining copies in the Countway Library in Boston in 1978. It runs to 224 pages and has a long 18th-century title, beginning *Historisch-Medizinische Untersuchung des Wachstums der Menschen in die Länge* ('A Medical-historical Investigation of Human Growth in Length').

Stöller was born in Windsheim, near Nuremburg. His brother, Georg Wilhelm, became famous as the surgeon botanist on von Behring's expedition to Alaska and the pioneer of Alaskan natural history. Johann Augustin graduated at Halle, and worked in Barby and Köthen, Saxony, before going (in 1738) to Eisenach where he remained as physician to the Duke of Sachsen-Eisenach.

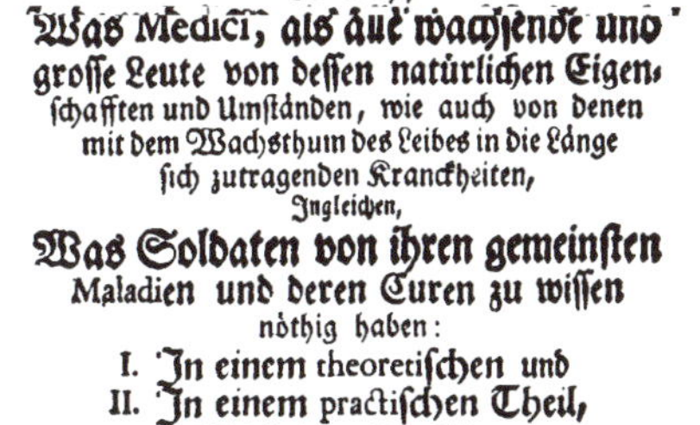

Historisch-Medicinische Untersuchung des Wachsthums Der Menschen In die Länge, So wohl, Was Medici, als auch wachsende und grosse Leute von dessen natürlichen Eigenschafften und Umständen, wie auch von denen mit dem Wachsthum des Leibes in die Länge sich zutragenden Kranckheiten, Ingleichen, Was Soldaten von ihren gemeinsten Maladien und deren Curen zu wissen nöthig haben: I. In einem theoretischen und II. In einem practischen Theil, Nebst einem Vorschlag Von einer Medicinischen Revuë, Denen vielen auserlesensten Königl. Preußis. Trouppen zum Besten also abgefasset von Johann Augustin Stöller, Medicinæ Licentiato & Practico in Barby.
MAGDEBURG, Bey Christoph Seidels sel. Wittwe, und George Ernst Scheidhauer. 1729.

'I have in mind', writes Stöller, 'to present human growth in length somewhat more rigorously than has yet been done in our country, using as my basis *Natur-lehre* or physiology ... thus I take the opportunity presented by the differences in men's heights to write, from a medical point of view, a somewhat detailed account of growth and its bodily benefits and injuries, which so far as I can discover no one has done before'.

The book presents no actual data on growth: the first were to come just 25 years later. It retails the still-current views about growth slowing down because of the decrease in elasticity of the body's fibres. But there is much that is very sensible, especially concerning conditions governing growth. Stöller gave the first clear description of catch-up growth. 'Lastly', he writes, 'I note that people grow really visibly, especially in length, when they have overcome a severe illness, provided they behaved appropriately during it .... Frequently illnesses stop people growing ... but if a feverish or not very long-standing malady is properly overcome then people grow very much: so as a rule those persons shoot up in height, who particularly in their childhood have been held in cheek by hot or cold fevers. This is the basis of the well-known proverb: "Illness laid him low and stretched him out"'.

## 4 Christian Friedrich Jampert (d. 1758)

Jampert graduated at Halle in 1754, 'promoted' by Michael Alberti, who 30 years earlier had promoted Stöller's thesis. In Jampert's thesis, entitled *Causas incrementum corporis animalis limitantes* (factors which control the growth of the animal body) was the first real table of measurements of the human. This thesis was also forgotten for 200 years, until discovered in a comprehensive search in the British Library. Contemporary accounts of Jampert's teaching activity in Halle and the dissertations he sponsored indicate the presence of an outstanding man, of broad views and extensive knowledge. Alas, he died young, probably of tuberculosis.

Jampert's thesis is the first that has an entirely modern ring. He understood the problems of variation and sampling. He was the first to realize the difference between a cross-sectional study (which his was) and a longitudinal study (which he thought would be better). In Jampert's table successive years of age are listed, and values for length, weight (he stresses without clothes) and nine other measurements are given. At each age a single individual only is measured, the one selected from the children at the Berlin Friedrich's orphanage as typical of that age (a living average, as it were). When plotted on modern standards these 18th-century orphan children are very small indeed – far below the current third centile.

## 5 George LeClerc de Buffon (1707–88) and Philip Gueneau de Montbeillard (1720–85)

George LeClerc de Buffon was one of most magisterial figures of 18th-century Europe. He laid the foundations of modern geology; he wrote a book on literary style which was a model for its time; with Rousseau, Diderot and Voltaire, he formed the core of the Encyclopédists, that most potent of all educational time-bombs. He lived at Montbard, near Dijon. In 1739 he was appointed director of the Royal Medical Herb Garden in Paris and asked to prepare a catalogue of its collection. Instead, he set out to prepare a catalogue of all Nature; comparable to but surpassing that of Aristotle. This was the *Histoire naturelle, générale et particulière*, in 15 volumes (1749–67). The natural history of man occupies parts of volumes two and three (1749).

Buffon and his colleagues measured aborted fetuses and newborns, being probably the first persons to do so. He also gives lengths of 1, 2 and 3 year-olds, consistent with the modern third centile. He also wrote 'There is a quite remarkable thing about growth of the human. The fetus grows more and more rapidly up to the moment of birth; in contrast the child grows less and less rapidly up to the age of puberty, when he grows, one might say, in a bound, and arrives in very little time at the height he has thence forth.'

One of Buffon's assistant colleagues, Philip Gueneau de Montbeillard, was also a close family friend, and Buffon persuaded him to make measurements of the height of his son from birth all the way to adulthood. This was the first longitudinal study ever made, so far as we know, and it remains one of the best, and certainly the most

**George LeClerc de Buffon. Courtesy of Wellcome Institute Library, London.**

**Louis-René Villermé. Courtesy of Wellcome Institute Library, London.**

famous (see Introduction). The table of measurements was published in 1777 in one of the volumes supplementary to the great *Natural History*. Montbeillard's growth study was well known to Quetelet and others in the 19th century, but then fell from view until revived by Richard Scammon (1883–1952), Professor of Anatomy in the University of Minnesota, who converted the old-style values to modern centimetres and published the results as a graph in 1927. D'Arcy Thompson added velocity curves in the revised edition of his *On Growth and Form* (1942) and Tanner used the combined distance and velocity as the opening figure in his *Growth at Adolescence* (1955).

## 6 Louis-René Villermé (1782–1863)

Louis-René Villermé grew up in a small village outside Paris and served for 10 years as a surgical assistant in the army of the Napoleonic Wars, working under atrocious conditions. His thesis for qualification as a doctor, entitled 'Effects of famine on health in places which are the theatres of war' described his experiences in the Estremadura Spanish campaign precisely in the terms of Goya's *Horrors of the War*. The experience affected him deeply; after only 4 years of medical practice he mobilized his small family fortune to support him while he examined the plight of the unfortunates in post-Napoleonic France – the weavers and cotton-spinners, the silk-workers, the prison population.

Villermé became the chief architect of public health and the public concern for health in France. He was one of the founders of the *Annales d'Hygiène Publique* and probably wrote its prospectus. 'Medicine', this said, 'does not only have as its object the study and treatment of sick people; it has close relationships with social organization; sometimes it helps the legislature in drawing up laws, often it clarifies their application and always ... it watches over the public health'.

In 1828 Villermé published a paper on the relative mortality of rich and poor, a disputed subject at that time. He showed that in Paris the mortality rate at all ages combined was nearly twice as great in the poorest district of the city as in the richest. In 1829 he followed this with a memoire in the first number of the *Annales* in which he showed that the height of conscripts to the French army of 1812–13 varied according to the wealth of the region from which they came. Among the different districts of Paris height was associated with percentage of persons owning their own houses. Poverty, Villermé said, was much more important than climate in influencing growth. 'Human height', he wrote, 'becomes greater and growth takes place more rapidly ... in proportion as the country is richer, comfort more general, houses, clothes and nourishment better, and labour, fatigue and privation during infancy and youth less. In other words, the circumstances which accompany poverty delay the age at which complete stature is reached, and stunt adult height'.

Villermé's views were strongly opposed by Paul Broca (1824–80), the leading physical anthropologist in France; while agreeing that poor nutrition could delay growth, he averred that only 'ethnic heredity' governed final height. The clash of their views echoed round the academies for 100 years.

Villermé himself never made any studies of children's heights, but he was the prime investigator of the enormously influential work of Quetelet on this topic.

## 7 Adolphe Quetelet (1796–1874)

Adolphe Quetelet was one of the major figures in European science and public health in the first half of the 19th century. Born in Ghent, he was Professor of Mathematics at the Royal Brussels Atheneum at an early age, and in 1823 was given the job of creating and directing the Brussels Observatory. Political events postponed the Observatory's opening till 1832, and in the meantime Quetelet trav-

**Adolph Quetelet. Courtesy of Museé de l'Observatoire Royale de Belgique and Roland Hauspie.**

elled widely, meeting Laplace and Gauss, Herschel, the British Astronomer Royal, Fourier, who introduced him to Villermé in Paris, and Goethe, in whose garden in Weimar he demonstrated his newly constructed instrument for measuring magnetic fields. Villermé became a life-long friend and was the influence behind his celebrated study of children's growth. From 1834 until his death, Quetelet was permanent secretary of the Brussels, later Belgian, Royal Academy of Science, Letters and Fine Arts. From this position he dominated for two decades the cultural and scientific life of Brussels, shining with a brilliance scarcely equalled since. It was Quetelet who introduced into practical statistics the Normal curve, discovered in the context of astronomical error by Laplace and Gauss. Quetelet was, if not the father of modern work on growth (for which see Boas) at least its prophet.

Quetelet made two series of measurements of children's heights; the first in 1831, the second, in which weights were also measured, in 1832. His tables, given in his most famous book *Sur l'homme et le développement de ses facultées* (published in Paris in 1835, and renamed *Physique Sociale*, 'Social Physics', in the second enlarged edition published in Brussels in 1869), cover the age range from birth to 25, but the actual numbers of children are nowhere stated. Consideration of Quetelet's later writings makes it likely there were only about 10 boys and 10 girls per year; and at age 7 a value for boys is missing. In addition, there is a peculiarity in both these series which dogged subsequent measurers for decades – such was Quetelet's authority. At no age is the girls' mean height greater than the boys'; this is a unique finding and inevitably puts a question mark against the objectivity of selection.

**Edwin Chadwick. Courtesy of Wellcome Institute Library, London.**

Quetelet's aim in measuring the children was to determine the ideal Curve of Growth. Gauss and Goethe were Quetelet's intellectual parents and Goethe predominated. His later anthropometric work was a quest for canons of Beauty and enshrined his belief that Beauty was to be found in the harmony of numbers; so he fitted a curve to the succession of his children's mean heights (being the first person so to do), using an elegant expression which gives a continuously falling velocity of growth from birth to maturity, with no trace of a pubertal spurt, though his empirical means did show one. The fitted values were published as a Table of Growth, which confused later workers for years. For Quetelet, this beautiful curve was the *homme moyen* in the making, that average man whom Quetelet regarded as the ideal type.

Later, Quetelet measured numbers of adults in his search for perfect harmony (*Anthropométrie, ou mésures des différentes faultés de l'homme* 1870). In this context he introduced the index weight (kg)/height (m)$^2$, now usually called Body Mass Index (BMI). This measure is now widely used as an epidemiological characterization of undernutrition (low BMI) and obesity (high BMI) in both children and adults.

## 8 Edwin Chadwick (1800–90) and Leonard Horner (1785–1864)

Edwin Chadwick, who played a role in England similar to that of Villermé in France, was a civil servant and social reformer who, in his enquiries, deployed the most up-to-date statistical methods; his emphasis on proper sampling and suitable controls is revolutionary in its impact. In his youth he had been secretary to the Utilitarian philosopher Jeremy Bentham. Amongst the many government reports of which he was co-author the *Report of the Commissioners on the Employment of Children in Factories* (1833) and the *Report on the Sanitary Condition of the Labouring Population of Great Britain* (1842), made for the Poor Law Commissioners, are the most relevant to our subject, and perhaps the most famous. The 1833 report resulted in a Factories Regulation Act which prohibited children under 9 years of age from working in various types of textile factories, and stipulated that children aged 9 to 13 should have 1.5 hours each day for meals and rest (out of the 10 to 12 hours a day then worked). As part of the Enquiry the Medical Commissioners were required to measure the heights of girls and boys working in textile factories in the north of England, to see 'whether there be any difference at any age, and what age, and in either sex, between persons brought up from an early period in a Factory and persons of the same age and sex and station not brought up in a Factory'. Only two of the Commissioners actually measured children, however, and they did not provide the requested controls. But the data, buried deep in a Parliamentary Report, represent the first effort to use children's heights as a measure of their well-being.

Leonard Horner, son of a well-to-do wholesale linen merchant in Edinburgh, was for 25 years one of the four Government Inspectors of Factories. He was a passionate, if amateur, geologist, a friend of T. H. Huxley, Charles Lyell (his son-in-law) and Charles Darwin. He was also a passionate educator of working men and children, founder of the first of the numerous Mechanics Institutions which sprang up in Great Britain in the 1820s and 1830s and he was the first Principal of the University of London (1826–30). In 1833 he joined the Factory Inspectorate and in 1837 organized measurements of height on a far larger sample of children (16,000) than Chadwick's. Horner campaigned continually for better factory conditions and, in 1840, reported on the condition of children working in the mines to such good effect that a total ban was placed on the employment there of women and children.

The heights of the 1833 and 1837 children are very similar and the means are well below the present third centile. Horner's survey, it has to said, was made in the interests of checking the ages attributed to children by medical practitioners anxious to please parents by alleging a higher age than reality and thus allowing the child to obtain part or full-time factory employment.

## 9 Charles Roberts (d. 1901) and Francis Galton (1822–1911)

In 1872 a Parliamentary Commission was set up to enquire again into the conditions of women and children in textile factories in the north of England. Five doctors examined and measured nearly 10,000 'factory and other children of the working classes'. One of these doctors, Charles Roberts, was an admirer, albeit a critical one, of Quetelet, and examined with exemplary care the mass of data this survey generated. His place in

**Francis Galton. Courtesy of Wellcome Institute Library, London.**

the history of growth rests on just two papers, published 1874 and 1876, and his book *A Manual of Anthropometry* published in 1878. In 1875, on the initiative of Francis Galton, the British Association for the Advancement of Science had set up an Anthropometric Committee which organized, amongst other things, measurements of children in schools and colleges. From 1878 Roberts was secretary of this Committee and wrote its Final Report in 1883.

Roberts' reports contrast the heights of working boys with those of boys educated in private schools (in England called 'Public Schools', for example, Marlborough, Wellington and Eton Colleges). Thus the first clear indication of social class differences in growth was obtained. Roberts, in contrast to Quetelet and all his predecessors, laid great emphasis on variation about the mean. Charged, for Factory Act purposes, with using height measurements to establish a boy's age, he gave a frequency distribution of the heights of 771 boys aged 14 and pointed out that their huge variation precluded any such estimate. Roberts was the first author to give frequency distributions of children's heights at successive years of age. He conducted a lively trans-Atlantic correspondence with Henry Bowditch, to whom he transmitted Galton's method of describing a distribution by percentiles. As for his relations with Galton, history is strangely silent. They certainly worked together in the 1880s as chairman and secretary of the Anthropometric Committee. Roberts put together his and Galton's surveys to produce statistics on some 85,000 boys, the largest survey of its time. But Roberts and Galton came from very different backgrounds; Roberts was a retiring, scholarly man, born in Yorkshire and qualified in medicine at St George's Hospital, London. Galton, in contrast, was a rich gentleman-amateur of science, with extensive social connections. In Karl Pearson's three-volume biography of Galton, Roberts, the public health doctor, is never mentioned. But if Roberts and Galton did not work hand in hand, at least they laboured back to back.

Quetelet had thought that the variation shown by human height was analogous to the Gaussian Curve of Error, with deviation from the modal 'type' somehow to be deplored. Galton was one of the first to grasp that the curve of distribution, though it resembled in shape the Gaussian Curve of Error, had altogether different biological implications. It was not a curve of God's errors, but of evolution's possibilities. Besides introducing the system of percentiles to describe the curve and an individual's place in it, he also originated the concepts and words regression and correlation, in relating, for the first time, heights of parents and offspring in bivariate diagrams.

## 10 Henry Pickering Bowditch (1840–1911)

Henry Pickering Bowditch was a member of a prominent Bostonian family, a family which every Christmas measured the heights of the many children then assembled. In 1872 Henry Pickering, returned from a European training in physiology, read a paper to the Boston Society of Medical Sciences which was the beginning of all North American work on human growth. He described the individual growth curves of 13 girls and 12 boys followed from between 1 year and 7 years to between 13 and 19 years. (The measurements themselves only appeared in the British Anthropometric Committee's Report of 1880, through the agency of Roberts). Bowditch found that girls had their pubertal spurt before boys, and were taller than boys at certain ages, which contradicted Quetelet, and gave rise to anxiety about the 'alleged inferiority in physique of American women'. In the eastern United States about this time there was a rather hysterical literature, some by Harvard professors of medicine, concerning the allegedly ill effects on the bodies of girls brought about by educating their minds.

**Henry Pickering Bowditch. Courtesy of the Francis Countway Library, Boston.**

Bowditch, now the first Professor of Physiology in America and soon to be Dean of Harvard Medical School, persuaded the Boston School authorities to survey the growth of all pupils in the school system, to see what light could be thrown on this matter. In 1875 24,500 children were measured. His analyses of these data, in the 1877 Report of the Board of Health of the State of Massachusetts, is a classic of the international growth literature. Besides presenting charts of the mean height and its variation at each age, Bowditch gives a table of the successive differences between yearly means. This confirmed the girls' earlier and girls' taller result of the longitudinal family study. In a subsequent Report, for 1891, he applied the method of percentiles communicated to him by Roberts, giving the first centile reference values, or standards of growth, by which the status of a given child could be judged in relation to the population. These were the first practical growth charts: they showed the 5th, 10th, 20th, 30th, 40th, 50th, 60th, 70th, 80th, 90th and 95th centiles. Unlike modern charts, however, age is represented on the vertical axis, height on the horizontal.

## 11 Luigi Pagliani (1847–1932)

Luigi Pagliani played much the same public health role in continental Europe as Bowditch did in the United States. He was an assistant in the Physiological Institute of the University of Turin when he published the two papers (1875, 1879) which established his reputation as the foremost European auxologist and exerted an overriding influence on the many educational auxologists who appeared in the 1880s and 1890s. A few years later, in 1881, he was made Professor of Hygiene in Turin and charged with organizing the Italian Public Health Service.

Pagliani was principally interested in what a change of circumstances would do to growth, and in June 1872 (3 months before Bowditch read his paper to the Boston Society of Medical Sciences, and 4 months before Roberts began his factory survey) he established the first short-term mixed longitudinal study, at the Institute Bonafous, a boarding school for orphaned and abandoned boys. He compared the growth of these 60 boys, in the age range 10 to 16 years, with the growth of 120 boys from wealthy families attending one of the best-appointed institutions in Turin. He also measured a similar number of girls at a school for officers' daughters.

Pagliani was one of the first to understand precisely the relative advantages of what he called the 'individualizing' and the 'generalizing' method (our longitudinal and cross-sectional). The individualizing method is the more natural way to study the growth process, he said, but it takes a very long time. He was the first to realize that just calculating means cross-sectionally at each year of age in a mixed longitudinal study is wrong; he calculated the increments for each individual, averaged them and constructed a growth curve of height by adding these mean increments

**Luigi Pagliani. Courtesy of Lodovico Benso.**

successively onto the 10 year-old value.

Pagliani realized at once Quetelet's error concerning puberty, and, just like Bowditch, persuaded the school authorities to measure large numbers of children over a greater age range to confirm his new growth curve. The results were made public in 1878 at the Paris Anthropological Congress, just as news of Bowditch's great 1877 Report was reaching Europe. Pagliani was delighted with Bowditch's '*bellisimo lavoro*', made, he carefully points out, quite independently. He and Bowditch obtained exactly similar results; they were the first to obtain a true population curve of children's growth. Bowditch sadly developed Parkinsonism and retired in 1911, but Pagliani, luckier, retired from his chair in 1924 aged 77 and, like Boas after him, returned to his first love, writing a little monograph on growth 'with its applications for education and social hygiene' (1925). It still makes up-to-date reading today.

## 12 Paul Godin (1860–1942)

Paul Godin, a French army doctor, was a pioneer and an enthusiast of serially made growth measurements as a guide to a child's health. In 1895 he was appointed physician to a school for non-commissioned officers' sons and, over the next 5 years, took personally, without missing a single one, he says, 129 measurements on each of 230 boys every 6 months for, mostly, 4 years. This is a total of $1.3 \times 10^5$ measurements, a number only equalled by R.H. Whitehouse, who took 15 measurements on 9000 child-occasions at the Harpenden Growth Study.

Godin was the first to use ratings for development of the secondary sex characters in the way that is now standard. It was Godin, also, who introduced the word *Auxology*, in an article entitled '*La méthode auxologique*' (1919). Being a convinced longitudinalist, he restricted the word to refer to longitudinal researches; others, perhaps, scarcely merited a *Méthode*.

Godin's horrendous number of measurements, which was standard physical anthropology of the time, may raise a smile nowadays, but his basic approach is very much alive. He advocated preventive pedagogical medicine: in discussing a boy with retardation of one of his indices, he writes 'The boy is not sick, and that removes the usefulness of a clinical doctor: but ... not of an educational doctor' (1919).

**Paul Godin. Courtesy of Archives Institut Jean-Jacques Rousseau, Université de Genève.**

## 13 Franz Boas (1858–1942)

If any single person can be called the father and mother of auxology, that person is Franz Boas. Born and educated in Germany he studied physical anthropology in Berlin with Virchow, a major influence as regards both his scientific empiricism and his lifelong stance against prejudice and chauvinism. In 1886 he visited the Pacific Northwest of America for ethnological work and stayed on from 1888 with Stanley Hall (1844–1924) at Clark University in Worcester, Massachusetts, where he initiated the first, though ill-fated, longitudinal study in America.

In 1891 Boas took charge of the Physical Anthropology section of the World's Columbian Exposition. He collected the school records from six cities in North America, and produced composite North American growth standards, based on 90,000 subjects aged 5 to 18 years (1898). From 1896 he worked in New York, at the American Museum of Natural History and Columbia University.

Boas' work on growth divides into two periods, 1890 until 1912, and 1930 till 1940. It was he who first emphasized that some individuals are throughout their childhood further along the road to maturity than others. He introduced the concept of physiological or developmental age, and the phrase 'tempo of growth' to describe how fast the process of childhood was played out, an analogy with the markings of classical music. Examining Bowditch's statistics, he noted that the variability of height increased at the time of puberty, and also that the distribution of heights became first positively, then negatively skewed as puberty progressed. Both these things resulted, he realized, from some of the children having their pubertal spurt early, others late.

**Franz Boas. Courtesy of Columbiana Collection, Columbia University.**

Secondly, Boas realized that variation in tempo of growth coupled with the fact of a pubertal spurt meant that most children departed at puberty from Bowditch's cross-sectionally derived channels of growth. He also realized that some or all of the social-class and ethnic differences during the growth period are due to acceleration/retardation.

In his second period of growth research, starting when he was aged 72, Boas wrote three of the most important papers in the whole field. In them he gave height curves from longitudinal individual data, classifying the curves by age at peak height velocity or age at menarche, introducing diagrams made classic by Shuttleworth and later writers. Boas distinguished clearly individual growth curves from population-derived statistics, thus laying the foundation for modern course-of-growth reference standards.

Boas had an extraordinary width of interest, as the tributes in his memorial volume show (Goldschmidt, 1959). Folklorists and linguists claimed him as pre-eminent in their fields too. He transformed traditional physical anthropology and gave the death-blow to the current version of racism. He was in the forefront of the application of statistical method to the study of humankind. But his writings in the small field of auxology show no trace of dispersion of talent. On the contrary, most of his work makes the other writers of his time look like dilettantes; he it is who penetrates below the surface.

## 14 D'Arcy Wentworth Thompson (1860–1948)

No account of the history of growth studies could ignore the author of *On Growth and Form* (1917, 1942). D'Arcy Thompson's enduring fame rests on a single, huge book and the trans-

**D'Arcy Wentworth Thompson. Courtesy of the University of St Andrews.**

mitted memory of a scholar and teacher of Renaissance proportions. D'Arcy, as he was universally called, was brought up on the Classics, wrote *A Glossary of Greek Birds* and *A Glossary of Greek Fishes*, and translated Aristotle's *Historia Animalium* with an understanding only possible to a professional zoologist who was also president of the Classical Associations of Scotland, England and Wales. He was Professor of Natural History, first in Dundee, Scotland (1884–1917) then St Andrews (1917–48).

*On Growth and Form* is above all a Renaissance book, crammed with all sorts of considerations and written in a prose of such clarity and elegance that one reviewer compared its writer with the author of the *Anatomy of Melancholy*. It has running page-headings which include 'Of modes of flight' and 'The comparative anatomy of bridges' as well as 'Of growth in infancy'. Two chapters only, out of 17, deal with growth in the strict sense. One introduces the theory of transformations, showing how the method of transformed co-ordinates depicts the evolution of a species or the relations between different genera. The second is entitled 'The rate of growth'. Form is produced by growth of varying rates in different directions. D'Arcy emphasized growth rate, or as he called it, growth velocity (whence the introduction of that term to human auxology by Tanner in the 1950s). 'To say that children of a given age vary in the rate at which they are growing would seem to be a more fundamental statement than that they vary in the size to which they have grown', he wrote.

D'Arcy is a link between the human auxologist and the more general auxologist, concerned with theories of change in form, and the comparative evolution of growth curves. The 1942 *Festschrift* for D'Arcy edited by Clark and Medawar (whose passion was general auxology even if he was honoured for his role in founding immunology) is a collection of essays of a very high level indeed. Though D'Arcy made no explicit contribution to the facts of human growth, he elevated the whole intellectual level of the field and broadened the outlook of all who worked in it.

## 15 Bird T. Baldwin (1875–1928) and Howard Meredith (1903–85)

In North America a powerful child-welfare movement arose in the 1920s and provided the soil for a crop of longitudinal studies, whose harvesting shaped the whole pattern of human auxology in the years 1935–55. The first centre for these studies was the University of Iowa Child Welfare Research Station, established in 1917, with Bird T. Baldwin as its director. Baldwin was an East-Coast educationist, whose first substantial paper was on the relation between height, weight, vital capacity and school marks. He understood entirely that 'the composite curve of the average heights based on simple means of different individuals at different ages does not represent the growth of any individual' (1914) and gave 28 charts on which were plotted the individual growth curves of 170 children, collected from the records of schools attached to the University of Chicago and Columbia University.

As soon as he was called to Iowa, Baldwin set up a longitudinal study programme. Ales Hrdlicka, the legendary founder of physical anthropology in North America, secured the measuring instruments and trained the measurers. In 1921 Baldwin published a classic monograph *The Physical Growth of Children from Birth to Maturity* as Volume 1 in University of Iowa Studies of Child Welfare. In it he reanalysed the longitudinal data he had previously collected, and broke new ground by calculating the intercorrelations between different measurements at each age and between the same measurements at successive ages. He also drew what was probably the first graph of velocity of growth in height for different age-at-menarche groups, stimulating the later, more trenchant analyses of Boas.

The remainder of Baldwin's monograph is a magnificent review of previous studies of growth. There are 80 pages of mean heights and weights of children from all over the world, followed by an annotated bibliography of 911 papers, the prime source for any historian of growth.

Baldwin died young, of erysipelas, in 1928, and the physical growth section of the Welfare Station was directed first by Charles McCloy (1886–1959) and then Howard Meredith. Meredith was born and educated in England, then took a postgraduate degree at Iowa and stayed there for practically all his professional life. Neither pediatrician nor educational psychologist, he was one of the small band of professional human auxologists. Meredith's data had the deserved reputation of being the most accurate on the American continent. Tanner described him as the Tycho Brahe of his subject. On his retirement in 1973 the Iowa Orthodontic Society republished 44 of his papers concerning head and face growth as a book, which remains a major landmark in this field. In 1948, in the course of visiting all the major centres of American longitudinal growth studies, Tanner, then teaching at Oxford, spent some days with Meredith learning his anthropometric techniques for use in the Harpenden Growth Study. Subsequently the International Children's Centre Longitudinal Studies followed Harpenden. Thus, Howard Meredith's methods of measurement spread around the world.

## 16 Frank Shuttleworth (1899–1958)

Frank Shuttleworth was a psychologist who obtained his Ph.D at the University of Iowa, where he was a contemporary of Nancy Bayley. From 1928 to 1939 he worked in the Department of

**Bird T. Baldwin. Courtesy of University of Iowa Archives.**

**Howard Meredith. Courtesy of University of Iowa Archives.**

**Frank Shuttleworth. Courtesy of the City College of the City University of New York archives.**

Education and the Institute of Human Relations at Yale University, and later he counselled students at the City College of New York. Like Boas and Baldwin, Shuttleworth was impressed and dismayed by the lack of understanding amongst psychologists and others of the difference between longitudinal and cross-sectional studies. The initial impetus for his studies, he wrote, 'was supplied by the many reports which, though based on longitudinal data, have yielded only cross-sectional findings. The implications of the so called *adolescent spurt* in growth for physical care, diet and mental hygiene have been elaborated in many text books but upon an insecure foundation of fact derived from cross-sectional studies' (1937).

To counter this, Shuttleworth published four monographs in the Society for Research in Child Development series in the period 1937–38 with two atlases in 1949. All concern aspects of growth in the adolescent period. (He also published a short paper urging that standards of growth should be couched in terms of increments, 'progress rather than status' as he put it. In this he was 30 years ahead of his time and preaching to a totally unconvertible audience).

Shuttleworth used for his analyses data accumulated in the Harvard Growth Study, the second major American longitudinal study, set up in 1922 by the Harvard Department of Education in the schools of three suburbs of Boston. About 3600 children aged 5 or 6 were recruited, and at the end of 12 years, when the study closed, nearly 1000 were still being measured annually. This was a massive archive, not equalled till the 1970s. There were 10 physical measurements, X-rays of the hand and wrist, records of the number of teeth erupted, and tests of intelligence and school performance. All the raw measurements were actually published by Dearborn and Rothney, the originators, together with Shuttleworth, in 1938.

Shuttleworth took age at peak height velocity (PHV) as a measure of tempo of growth applicable to both girls and boys, and plotting individuals' growth velocity curves in terms of PHV age he showed that each physical measurement had its characteristic curve, with a peak at a defined position with respect to the others. Thus the spurt in leg length preceded the spurt in sitting height and was flatter. Those children who spurted early had higher peaks than those who spurted late. On average children who were relatively tall at age 8 were nearer their spurts than those who were still small; but this did not mean they would end up taller as adults. All Shuttleworth's analyses were done very simply, using annual individual increments, never measurements attained. Yet few of the elaborate and often beautiful curve-fitting procedures done by later generations of auxologists have produced generalizations which Shuttleworth failed to make, or proved any of Shuttleworth's generalizations wrong. Shuttleworth's work of the late 1930s moulded a whole new generation of auxologists and began to exert its effect on what was much later to be the practice of adolescent medicine and psychology.

**Nancy Bayley. Courtesy of University of California Institute of Human Development, and Frank Falkner.**

**Wilton Marion Krogman. Courtesy of the University of Pennsylvania and Mary Krogman.**

## 17 Nancy Bayley (1899–1994)

Nancy Bayley was a psychologist, whose contributions to human growth, like those of Frank Shuttleworth, are classic in their penetration and simplicity. She took her Ph.D in 1926 at Iowa, attending Baldwin's seminar on physical growth as well as studying psychology. In 1928 she was recruited to run the Berkeley Growth Study at the University of California Institute of Child Welfare. She remained there until 1954. Parallel to her work on physical growth, she had a distinguished career in psychology, becoming President of two Divisions of the American Psychological Association.

The Berkeley Growth Study was a longitudinal study of 31 boys and 30 girls of whom 47 were followed to maturity. Bayley took 22 physical measurements (all herself until the boys reached puberty and preferred a male physician) as well as photographs and radiographs of the hand and wrist. There were a number of psychometric tests, whose analyses provided, in Bayley's hands (1955, 1957) critically important information on the development of mental performance.

In 1954 Bayley published the correlations between children's heights and parents' heights and, for the first time, correlations of children's heights at successive ages with their heights at maturity. This showed the correlation of adult height with length at birth to be about 0.3 rising swiftly to a level of 0.7 at about 18 months and 0.8 at age 3, before falling again during puberty. Bayley was also able to calculate, for the first time, the percentage of his/her own adult stature that a child had reached at each age. She thus had a valid measure of tempo other than bone age.

In 1956 Bayley made the first effort to produce standards for heights which took into account an individual's tempo. Developing Shuttleworth's ideas and using the Berkeley Guidance Study data in addition to her own, she took all children whose skeletal ages and percentages of mature height indicated a near-average tempo (± 1 year of chronological age). The averages of their successive heights were plotted as the central line – the average height of the average-tempo children. Another line characterized the average growth of the early maturers, and a third the average growth of the late maturers. The three lines began to diverge at about age 2, were greatly separated at puberty and ended together again at maturity, confirming Shuttleworth's generalization. Bayley's graphs showed also the lines for very tall, very early maturing and very short, very late-maturing.

She also put in the curve of velocity for the average maturers. This was a radical new departure in the whole approach to standards of growth. It was before its time, and published in the *Journal of Pediatrics*, whose readers had little idea of auxology and less of tempo. But things were moving and just 10 years later the first practical, clinically-oriented tempo-conditional standards appeared (Tanner et al., 1966).

Bayley also, with S.R. Pineau, gave tables for predicting mature height from height and skeletal age at earlier ages (1954), essentially using skeletal age to compensate for the diminution of the adult–child correlation during puberty. The method is still used, as is Bayley's scale for infant development. Nancy Bayley, the most charming and courteous of persons, left an indelible mark on auxology.

## 18 Wilton Marion Krogman (1903–87)

Wilton Marion Krogman was a Chicago-trained physical anthropologist who worked with T. Wingate Todd (1885–1938), Director of the Cleveland Brush Foundation Growth Study and originator of the famous *Atlas of Skeletal Maturation*, 1937 (revised as the *Greulich–Pyle Atlas* in 1944).

He moved back to Chicago in 1938 and then went to Philadelphia in 1948 as professor at the University of Pennsylvania Graduate School of Medicine, with a laboratory at the Children's Hospital as well as the title Professor of Anthropology and Curator of the famous University Museum. As outgoing and friendly as he was physically large, Krogman exercised a huge influence on American auxology in the 1940s and 1950s. Most of the American human biologists currently engaged in auxological work are his first-generation or second-generation students. Like Baldwin and Meredith, Krogman was a great bibliographer, and his collection of studies of growth, published in *Tabulae Biologicae* (1941) and his *Bibliography of Human Morphology* (1941), are worthy successors to Baldwin's 1921 monograph. He had a special interest in orthodontics, which he acquired while working with Holly Broadbent (1894–1977) in Todd's laboratory. His laboratory was later named the W.M. Krogman Center for Research in Child Growth, and continues to exemplify in practice Krogman's insistence that physical anthropology and medicine should be the closest of partners.

## 19 Andrea Prader (1919– )

Among the European longitudinal growth studies co-ordinated under Natalie Mass (1919–75) of the International Children's Centre (ICC) in Paris, the Zürich element was perhaps the most successful in terms of its final data analysis, and the most long-lasting, with a second generation series now continuing. The director of this study was Andrea Prader, who succeeded the famous Fanconi as Professor of Pediatrics in Zürich. Prader's main clinical interest was in endocrinology and he was the founder of the first pediatric endocrinology association in the world, the European Society for Pediatric Endocrinology (1962). Involved in a longitudinal study, and soon a life-long colleague and friend of the professional auxologist James Tanner, Prader rapidly became himself a proficient auxologist. It was he who, with Masse, was chiefly responsible for turning the ICC longitudinal studies in the direction of clinical application and for bringing Tanner at the Institute of Child Health in London back from pure research to the practice of medicine. Prader and Tanner, together with von Harnack, a German pediatrician, were authors of a much-quoted paper on so-called *Catch-up Growth* (1963), demonstrating in clinical practice the truth of Waddington's theory of canalization, or homeorrhesis of development.

**Andrea Prader. Courtesy of Theo Gasser.**

Above all, Prader was responsible for introducing auxological principles to pediatricians and making the subject respectable, as a century before Claude Bernard had made physiology respectable to clinicians. No physical anthropologist could do this, and not even a clinician like Tanner, trained in adult medicine rather than pediatrics, could succeed. Prader's influence raised greatly in Europe the pediatric endocrinologists' level of sophistication about growth. North America suffered, perhaps, from the lack of a comparable authority. When human growth hormone first became available, Prader, together with Tanner and a few others, became the pioneers of the treatment in Europe. This application again strengthened the bonds between auxology and medicine, so desired by Krogman.

## 20 Michael R.J. Healy (1923– )

Though Boas, Shuttleworth and Bayley found ways of handling auxological data without invoking much statistical theory, subsequent advances have been increasingly dependent on the collaboration of professional statisticians and the use of often quite complicated mathematical methods. Needless to say, the successful statistical collaborators have been those who were also immersed in auxology.

**Michael R.J. Healy. Courtesy of Gunilla Tanner.**

First amongst these was Michael Healy, a Cambridge mathematical graduate, whose father was a general medical practitioner in Paignton, Devon. Healy never lost his childhood interest in medical research and after spending many years in that holy of holies, the Statistical Department of Rothamsted Experimental Station at Harpenden with Frank Yates, Ronald Fisher's successor there, he chose to pass the latter part of his career not there but as professor at the London School of Hygiene and Tropical Medicine and the Medical Research Council's Clinical Research Centre.

Healy's particular interest in auxology stemmed from his meeting Tanner at a Biometric Society meeting held at Rothamsted in 1949. The two became lifelong collaborators and friends. Beginning with a paper in 1952 on *Some statistical aspects of anthropometry*, Healy produced a steady flow of articles and advice on the construction of growth standards, statistics of reliability, design and analysis of population surveys and the fitting of curves to longitudinal growth data. An extraordinarily clear writer of English, he taught a whole generation of statisticians interested in growth and development; one, Harvey Goldstein (1939– ), worked in Tanner's department for a number of years and was statistical adviser and tutor to all the International Children's Centre Longitudinal Studies, statistician to the immense National Child Development Study and subsequently professor at the London University Institute of Education. Healy's influence spread widely. In the 1970s and 1980s auxological analysis progressed to a quite new level of sophistication; Healy and his students and collaborators were the main group who pushed it there.

**Professor Tanner deliberately excluded contemporary auxologists from his biographies, citing only two living persons, both working outside the field, though impacting upon it. The editors have therefore added brief biographies of James Tanner himself, and his contemporaries Frank Falkner, Stanley Garn and Alex Roche.**

Frank Falkner (1918–) Stanley Ulijaszek
James M. Tanner (1920–) Francis E. Johnston
Alex Roche (1921–) Roger M. Siervogel
Stanley Garn (1922–) Stanley Ulijaszek

## 21 Frank Falkner (1918– )

Frank Falkner has been one of the great initiators in the study of human growth and development during the second half of the 20th century. He was born and educated in Britain, where he graduated in 1945. At an early stage in his career as a pediatrician he was involved in the International Children's Centre studies of longitudinal growth. The first of these European studies was initiated in London in 1949 by Alan Moncrieff, Professor of Child Health at the University of London, and Falkner was put in charge of this project. The International Children's Centre (ICC) was founded in 1950 and supported by UNICEF and the French government, and in 1953 Falkner was sent to Paris for a year to initiate the French study. Subsequently, longitudinal growth studies of the ICC were initiated in Zürich (1954), Stockholm and Brussels (both 1955). In all surveys, common assessments and measurements were made. Anthropometric techniques used were those adopted in the Harpenden Growth study, ratings of pubertal characteristics were made, as were radiographs of the hand and wrist. In this way, the somatic growth characteristics of European populations came to be defined.

In 1956, Falkner joined the Department of Pediatrics in Louisville, Kentucky, where, uniquely, he initiated a study of the longitudinal growth of twins from birth. Falkner left before substantive findings were published, but his work was carried on under Ronald Wilson, who showed that while at birth the differences in weight and length between monozygotic (MZ) pairs was greater than differences between dizygotic (DZ) pairs, MZ twins came to resemble each other more with increasing age as a consequence of catch-up growth after uterine restriction in the smaller twin. Falkner was appointed Director of the Fels Research Institute in 1971, where he remained for 8 years. During this time, he established the Fels Division of Pediatric Research at the University of Cincinnati, initiated a multinational study of infant mortality, and became interested in the assessment of nutritional status. Falkner's contribution to the study of growth and development was pivotal to the establishment and understanding of physical growth indicators of nutritional status. Subsequently, Falkner moved to San Francisco, where his research interests moved to the characterization of intra-uterine growth, and the factors that influence it.

## 22 James M. Tanner (1920–)

Unquestionably the leading figure in the field of human growth of the 20th century, Tanner's contributions to our knowledge of auxology are so numerous that it is impossible to do them justice in the space allotted here. In this time of reductionism, Tanner has been persistent in stressing the necessity of viewing the child holistically, as a biobehavioural organism, as part of a complex social network, and as a member of an evolving population that interacts with the ecosystem of which it is a part.

Tanner was trained both as a physician and a human biologist (as well as a mathematician) and holds MD and Ph.D degrees. This undoubtedly has moulded his approach to the study of growth, an approach which emphasizes the integration of the clinical and the basic, the one learning from the other in a true synergism. In his Growth Disorders Clinic at London's Hospitals for Sick Children, more than anyone else he brought the breadth of auxological knowledge about the normal to the diagnosis and treatment of children suspected of abnormal growth. And the knowledge he gained from dealing with abnormality was used to elucidate the basic processes of growth and maturation.

Tanner is probably known best for moving auxology from the study of growth as a discrete process and of children as separate from their worlds into those disciplines which deal with biological phenomena and their interactions with social forces. The phrase 'growth as a mirror of the conditions of society' was coined by Tanner. The importance of variability among individuals and groups as an indicator of the environments in which growth is expressed gave rise to a set of subdisciplines within fields such as epidemiology, economic history and ecology.

In terms of research, Tanner's major contributions derived from the Harpenden Growth Study, a longitudinal study of children at a residential home in Hertfordshire that began in 1948 and ended in 1971. This study, which examined children every 6 months (every 3 months during the years of the growth spurt) provided the raw data from which were developed the first modern growth reference data for height, weight, and skinfolds (mid-1960s), the radiographs which led to the Tanner–Whitehouse method of assessing skeletal age (now the world standard), and the tables for predicting the adult height of still-growing children.

Important theoretical advances came from the Harpenden Study. The anthropometric reference data led to a new conceptualization of how to develop and apply standards. Data were smoothed and presented systematically as percentiles. The assessment of growth status was done with respect to other variables – bone age

James M. Tanner. Courtesy of Gunilla Tanner.

and parental size, for example – and the term 'conditional standards' was born. And the availability of precise reference standards for the United Kingdom has led to a major concern with the theory of a 'reference' and its applicability to populations around the world.

Tanner revolutionized the assessment of skeletal age and maturity status. T. Wingate Todd had formulated the concept of the (skeletal) maturity indicator, and he and his students had produced a series of atlases of different body regions (especially the hand–wrist) which were used in clinical medicine and in human biological research. In the 1950s, Roy Acheson, at Oxford, questioned some of the bases of Todd's inspectional method and introduced the point system, in which bones were rated individually and the ratings summed. Tanner built on this notion, but refined and expanded it greatly by developing a biologically based system which reduced the number of centres of ossification that were assessed and which weighted them mathematically, producing a maturity rating (in points) which was then converted to a bone age.

As Professor of Child Health and Growth at the Institute of Child Health, University of London, Tanner has had an enormous influence on the biological and social sciences, and on medicine. This has come about by teaching, by writing, and by a range of other professional activities. His example has set the tone for the future.

## 23 Alex Roche (1921–)

Alex Roche graduated in medicine (M.B., B.S.) from the University of Melbourne in 1946 and subsequently obtained Ph.D., D.Sc., and M.D. degrees from the same University before becoming a Fellow of the Royal Australasian College of Physicians (1978). After completing a residency, he joined the faculty of the Anatomy Department

**Alex Roche. Courtesy of Roger Siervogel.**

of the University of Melbourne in 1948 and became a Reader in 1962. Alex spent 1952 in America as a Fulbright Scholar with Idell Pyle (skeletal maturation) and Holly Broadbent (cephalometric radiography) at Western Reserve University in Cleveland. In 1968 he became Chairman of the Department of Growth and Genetics of the Fels Research Institute in Yellow Springs, Ohio, and held this position, albeit with some changes in titles, until 1992. After the Fels Research Institute merged with Wright State University in 1977, he became Fels Professor of Pediatrics and of Community Health and was chosen as University Professor (1990–95). The latter is a prestigious title that had been granted on only three occasions during the history of Wright State University. In addition, he has served as President of the Human Biology Council and of the American Dermatoglyphics Association.

While at Melbourne, Alex began a longitudinal study of child growth that included more than 1200 children. The major groups studied, other than normal children, were those with trisomy 21, anisomelia, or unusual statures. The data collected related to size and maturity, particularly of the craniofacial region and the teeth, dietary intakes, and data from physical examinations. The major published papers concerned maturation and growth in normal children and in children with trisomy 21, and the treatment of excessively tall girls and short children.

At the Fels Research Institute, the most important achievements of Alex Roche are the continuation of the Fels Longitudinal Study (now in its 67th year with four generations from birth in some families), and the extension of this study to body composition, risk factors for selected cardiovascular and metabolic diseases, and genetic mechanisms. His published works number over 750, including more than 40 books and monographs. His major scientific contributions are the development of new methods for the prediction of adult stature and for the assessment of maturity, the standardization of anthropometric procedures, and analyses of the validity of new and old body-composition methods, and of body mass index, blood pressures, and plasma lipids and lipoproteins during middle-age in relation to childhood values. Some of his other activities include the analysis of data from national surveys and the development and revision of the 1977 National Center for Health Statistics Growth Charts.

## 24 Stanley Garn (1922– )

**Stanley Garn. Photograph by David W. Smith.**

Stanley Garn is a physical anthropologist who graduated from Harvard in 1948, and who has been prolific in many areas of growth and development. Indeed, his breadth of research interests is difficult to match. He joined the Department of Growth and Genetics at the Fels Research Institute in 1952, where his early interests included the study of tissue growth, especially that of fat. Subsequent work included studies of tooth eruption and calcification, skeletal maturation, and the gain and loss of bone tissue. Garn's work is characterized by his ability to ask new questions and to analyse data in different ways. In the 1950s, Garn had studied fat patterning, the applications of pattern analysis to anthropometric data, and the use of X-ray methods for assessing body composition. Through the 1960s, Garn's interests in growth broadened to include the influence of malnutrition on growth and development, the use of anthropometry in assessing nutritional status and human body-size and its variation, and the genetics of maturation.

In 1968, Garn moved to the University of Michigan, where he has maintained his broad interests in many areas of growth and development to the present day. Throughout his life his work has been informed by the evolutionary perspective he gained in his earlier years at Harvard.

# Glossary

## A

**absorptiometry** any technique for measuring absorption of waves or particles

**accuracy** indication of the amount of error associated with a measurement

**acetylcholine** mammalian neurotransmitter common in many parts of the brain

**achondroplasia** genetic disorder and the most common form of short-limbed dwarfism, characterized by a failure of normal development of endochondral bone

**acromegaly** a disease of adults due to a **growth-hormone**-secreting **pituitary** tumour

**acromelic shortening** limb shortening affecting the most distal parts (hand or foot)

**adaptation** the process by which change leads to increased survival value of an individual or a **population**; also refers to the genetic, physiological or morphological feature that has resulted from the process

**adipocyte** fat cell

**adipose tissue** the fat tissue of the body made up of large numbers of **adipocytes**

**adjusted means** the average values of a variable in a set of groups after removing the effects of confounding factors

**adolescent growth spurt** a rapid increase in the rate of growth which accompanies the entry into the pubertal stage of development

**adrenal cortex** the outer layer of the adrenal gland which is responsible for the secretion of a number of steroid **hormones** which are responsible for maintenance of many bodily functions

**adrenaline** an alternative name for **epinephrine**

**adrenarche** the onset of secretion of **androgens** by the **adrenal cortex** which occurs in mid-childhood

**adrenergic system** part of the nervous system associated with noradrenaline-mediated nerve-impulse transmission: the **sympathetic nervous system**

**adulthood** the period of life after full physical **maturation** has been achieved

**agenesis** failure of development; usually refers to an organ in the body

**AIDS** acquired immune deficiency syndrome

**alleles** alternative forms of a particular gene

**allometry** the differential growth-rate of one part of the body relative to the body as a whole

**altricial mammals** mammalians species producing offspring which are very dependent on their parents (cf. **precocial mammals**)

**Alzheimer's disease** a chronic dementia often occurring at an unusually young age which has a characteristic neuropathological appearance; it may be familial

**amblyopia** blindness due to disuse of an eye, usually because of an untreated squint

**amenorrhea** the absence of normal menstrual cycles

**amyotrophic lateral sclerosis** a progressive degeneration of motor neurones leading to widespread paralysis, usually but not exclusively affecting those over 50 years of age

**anabolism** a state of positive protein synthesis; the opposite of **catabolism**

**anaemia** *see* **anemia**

**analysis of variance** a statistical technique for evaluating differences among the means of several discrete groups

**anastomosis** a surgical joining of two ducts, blood vessels or bowel segments to allow flow from one to the other

**androgenization** the process of virilization or acquisition of masculine characteristics

**androgens** steroid **hormones** that are related to **testosterone** and which induce masculinization

**anemia** deficiency in circulating hemoglobin, red blood cells, or packed-cell volume

**aneuploidy** a state of an abnormal number of **chromosomes**

**Angelman syndrome** a rare genetic disorder involving mental retardation due to a gene on the long arm of **chromosome** 15

**anorexia** the absence of appetite for food

**anthropometer** an instrument used to measure lengths and breadths of the body

**anthropometry** the process of making careful measurements on the body using precisely defined **landmarks** as points of reference and comparison

**antibody** a protein produced as a result of the introduction of an **antigen** which has the ability to combine with the antigen that stimulated its production

**antigen** a substance that can induce a detectable immune response

**Apgar score** a score from 0–10 (higher being better) used to classify the physical state of a newborn baby

**apoptosis** programmed cell death as part of the normal development of organs or tissues

**appendicular skeleton** the skeleton of the limbs

**aromatization** strictly applies to the creation of an aromatic 6-carbon atom ring containing three double bonds; in the context of this, volume refers to the more specific conversion of **androgens** to **estrogens** with the latter containing an aromatic ring in the A position

**atresia** the failure of development of a tubular structure, such as bowel; this creates a blind-ending tube

***Australopithecus afarensis*** an early **hominid** species from East Africa, **accuracy** dated to 3.0–3.7 million years

**autoantigen** an endogenous body constituent that stimulates the production of an auto-antibody and a resulting auto-immune reaction; also known as self-antigen

**autocrine** a mechanism of **hormone** action where the hormone acts directly on the cell that produces it

**autoimmune disease** a disease where an organ or tissue is attacked by the body's own immune system

**autonomic nervous system** subsystem of the motor branch of the nervous system which regulates the smooth and cardiac muscles, and the **endocrine system**

**autoradiographs** the result of exposing X-ray film to radioactive or light-emitting compounds so that there position can be determined; commonly used to locate biological substances in tissues or after separation by chromatography

**autosomal recessive/dominant** modes of inheritance; a recessive gene only creates a clinical effect when present on each of a pair of **chromosomes** and so has to be inherited from both parents; a dominant gene produces an effect when present on only one of the chromosomes

**autosome** any **chromosome** other than a sex chromosome

**auxology** the study of physical growth and development

**axillary hair** body hair in the armpit that appears at **puberty**

**axon** long fibrous portion of a nerve cell along which nerve impulses are transmitted

## B

**B cells** cells in the immune system responsible for producing circulating antibodies

**basal metabolic rate** the body's metabolic rate measured in the preprandial state, at rest, and in a thermoneutral environment

**Bayley scale** a test of infant psychomotor development developed by Nancy Bayley

**biacromial diameter** the straight-line distance between the most lateral projections of the acromial processes of the right and left clavicles

**biiliac diameter** the straight-line distance between the most lateral projections of the crests of the right and left ilia of the pelvis

**binding proteins** proteins that bind to active substances such as **hormones** and usually carry them in the circulation

**biparietal diameter** the maximum straight-line distance between the right and left parietal bones of the skull, also know as head breadth

**bipedal gait** a gait pattern characterized by habitual locomotion using the hind, or lower, limbs

**bivariate analysis** an analysis involving two variables

**blastocyst** an embryonic stage when the cell mass consists of 16–32 cells and becomes hollow (about 4 days after conception)

**body composition** refers to the quantification of the components of body mass, e.g. fat, lean tissue, bone mineral

**body mass index (BMI)** an index of weight-for-height, calculated as the height in meters divided by the square of the weight in kg, also known as **Quetelet's Index**

**bone dysplasia** one of many genetic disorders of bone which lead to abnormal growth of the bone often with striking deformity and shortening

**breast-milk substitute** any food marketed as a replacement for milk, either total or partial, whether or not it is suitable for that purpose

**buccal** refers to the cheek

**bulimia** an eating disorder characterized by the consumption of copious amounts of food, followed by forced purging

## C

**cachexia** a state of advanced **wasting** and **catabolism** usually resulting from profound malnutrition or advanced malignant disease

**calcification** hardening of tissue by the process of deposition of calcium (and magnesium) salts into bone cells

**calvarium** the skull cap

**canalization** the process by which the pattern of an individual's growth becomes more regular and predictable with the passage of time

**capitate** one of the bones of the wrist

**carpus** the group of 7 small round bones that make up the wrist

**catabolism** a state of net protein loss, the opposite of **anabolism**

**catch-down growth** a regulatory slowing of growth after a period of abnormally accelerated growth

**catch-up growth** an acceleration of the growth-rate following medical treatment such that normal size is attained

**central limit theorem** a mathematical theorem which states that the distribution of means of a variable of samples of a given size is normally distributed, even though the variable itself may not be

**centripetal fat distribution** a relatively greater concentration of fatness in the body's more centralized trunk depots

**cephalic index** an index of head shape, calculated as head length divided by head breadth

**ceruloplasmin** also known as ferroxidase, an enzyme of the oxoreductase class; principal means of transport and maintenance of tissue levels of copper

**chemical maturity** level of bone **mineralization** and body potassium relative to adult values

**chemiluminescence** the emission of light due to a chemical reaction

**cholecalciferol** also known as vitamin D3, a steroid **hormone** related to cholesterol that is important in regulation of calcium and in bone formation

**chondroclast** a multinucleate giant cell that is considered to be responsible in part for resorption of cartilaginous tissue

**chondrocyte** a cell found in cartilage that is responsible for the secretion of the cartilage matrix

**chondrogenic** giving rise to or forming cartilage

**chorionic somatomammotrophin** also known as placental lactogen, a **hormone** related to **growth hormone** that is produced by the placenta

**chromatin** collectively the **DNA** of the **chromosomes**, molecular complex made of DNA and associated specific proteins

**chromosome** chain-like assembly of nucleic acids and proteins in the nucleus of the cell; the site of the vast majority of the genetic material and hence inheritable properties of the cell

**chronic hemolytic anemia** a form of **anemia** caused by chronic destruction of the red corpuscles

**chronological age** age in days, months, or years, since birth

**circadian rhythm** biological rhythm that recurs every 24 hours

**circaseptan** about 7 days (in relation to growth-rate periodicities)

**circumpubertal years** the age period which encompasses the development of the reproductive system

**cleft palate** a congenital condition characterized by a failure of the palatine bones to fuse in the midline

**cleidocranial dystosis** a genetic disorder, which is rather variable but includes an absence of the clavicles, short stature and characteristic cranial and facial abnormalities

**clonal expansion** the generation of several generations of cells that all share certain characteristics

**codon** in **DNA** a group of three **nucleotides** that together code for a particular amino acid in the gene product; e.g. AAA = Lysine

**coefficient of determination** the proportion of **variance** in a dependent variable attributable to variance in one or more dependent variables; also known as r-squared

**coefficient of reliability** a quantitative indicator of the error associated with a particular measurement

**coeliac disease** a bowel disorder in which the patient is sensitive to gluten, a protein component of wheat, leading to partial atrophy of nutrient-absorbing structures in the small bowel and a subsequent failure of food absorption

**collagen** a **glycoprotein** which forms the major component of connective tissues, bones and teeth

**colostrum** breast-milk produced in the earliest days of breast-feeding after the infant is born

**computer assisted tomography** an X-ray technique that produces images that have the appearance of thin slices through regions of the body, allowing visualization of internal organs

**continuous variable** a variable which, conceptually, can take on any value

**correlation** a statistical measure of the relationship between a set of variables

**cortex** outer layers of an organ, for example, the **adrenal cortex**

**cranial base** the inferior aspect of the brain case, formed by the articulation of several bones that separate it from the facial and respiratory components of the skull

**creatine** a nitrogenous constituent of muscle, the phosphorylated form of which is essential for muscle contraction

**creatinine** a nitrogen-containing substance derived from the **catabolism** of **creatine**, and excreted in the urine

**Crohn's disease** also known as inflammatory bowel disease, this is a condition of unknown cause that leads to chronic inflammation of variable lengths of bowel

**cross-sectional study** a type of growth study in which each subject is examined once

**cryptorchidism** failure of normal descent of the **testes** from the abdomen to the scrotum

**Cushing syndrome** a collection of signs and symptoms due to excessive production of **glucocorticosteroids**; it may also be due to the excessive use of glucocorticosteroids as medical treatment

**cystic fibrosis** the commonest inherited disorder due to a **mutation** in a gene on chromosome 7; in its most severe form it leads to chronic lung disease and intestinal malabsorption

**cytokine** hormone-like chemical messenger

**cytotoxic T cells** part of the immune system, these are cells dependent on the **thymus** and responsible for the killing of cells infected with viruses or bacteria

## D

**demography** the vital statistics of a **population**

**developmental quotient** a measure of an individual's mental development

**diabetes mellitus** effectively a lack of insulin action leading to elevated blood **glucose** concentrations and an inability to process the glucose; may be due to insulin deficiency (Type-I or insulin dependent) or insulin resistance (Type-II or **non-insulin dependent**)

**diaphysis** the end of the shaft of a long bone, where growth in length occurs

**dimerization** the process when two molecules (usually proteins) are physically associated; they may be identical (homodimerization) or different (heterodimerization)

**discrete variable** a variable which can take on only specific values (e.g. sex)

**discriminant analysis** a statistical analysis which calculates that linear function of a combination of variables that best separates two groups; it has been largely replaced by approaches which utilize logistic functions

**dizygotic** arising from two **zygotes**, as in dizygotic **twins** which are non-identical

**DNA** deoxyribonucleic acid, the fundamental basis of inheritance and the chemical structure of genes; consists of two interlocking strands arranged as a double helix, with each strand being exactly complementary to the other; during cell division the strands separate and a daughter strand is laid down alongside each with a constitution determined by the base sequence of its parent strand

**dopamine** a neurotransmitter and intermediate in the synthesis of catecholamine **hormones**

**dopaminergic neurons** nerve cells responsive to the neurotransmitter **dopamine**

**dorsal** back view

**Down syndrome** one of the more common chromosomal disorders due to an extra chromosome 21; affected individuals are mentally handicapped and have other serious problems, with short stature being universal; serious congenital heart defects may occur

**downstream** in a **chromosome**, the **nucleotide** sequences 3' to the end of the last **exon** of a gene

**dysmorphic** a term used to describe any congenital abnormally formed features, such as a cleft lip or extra finger

## E

**ectoderm** the embryological cell layer that lies on the outside surface of the embryo and ultimately gives rise to surface tissues but also the nervous system

**ectodermal dysplasia** one of a group of disorders which lead to a weakness in the layers of the skin leading to easy blistering, often with severe scarring

**edema** presence of abnormal amounts of fluid in intercellular spaces

**encephalization** increase in brain-size relative to body-size, across evolutionary time; the greater role of the cerebral **cortex** in various physiological functions

**endochondral ossification** new bone formation within cartilage

**endocrine system** system of glands under neural control responsible for the secretion of chemical messenger molecules into the blood for the widespread regulation of metabolic processes

**endocrine** the **endocrine system** is classically described as a collection of glands that release chemical messengers, or **hormones**, into the blood circulation that then have actions on other glands or organs elsewhere in the body; it is now realized that this is too restricted a definition and should also include **autocrine** and **paracrine** modes of action

**endogenous nitrogen loss** represents the loss of protein nitrogen from the tissues of the body

**endonuclease digestion** the process of digestion of nucleic acids by specific **enzymes** that cut within the nucleic acid chain

**Engel's coefficient** index of poverty based on the proportion of income spent on food; the higher the proportion, the poorer the individual, group, or **population**

**enteritis** inflammation of the intestine

**enteropathy** any disease of the intestine

**enzyme** protein chemicals that are able to modify other chemicals, usually in a highly specific way; e.g. the digestive enzymes initiate the breakdown of food in the gut prior to absorption

**epidermal growth factor** an important factor in growth and development of many tissues, particularly in the developing fetus; its actions are more widespread than its name would suggest and are not limited to epidermal development

**epinephrine** an alternative name for **adrenaline**; a major neurotransmitter substance released by nerve cells of the sympathetic system and also the inner part of the adrenal gland (adrenal medulla)

**epiphysis** the end of the long bones where the **growth plates** are situated; the epiphysis is distal to the growth plate, whereas the **metaphysis** is **proximal**; a secondary centre of **ossification** which unites to the primary centre during the course of development

**epistasis** a type of interaction between genes at different loci on a **chromosome** in which one is able to suppress or mask the expression of the other

**epithelium** a layer of cells that surround body organs and cavities

**eruption** the emergence of a tooth, most often with reference to the gum (alveolus)

**erythropoesis** red blood cell formation

**estrogen** the family of female **sex steroids** which all contain an aromatic A-ring

**ethology** the study of behaviour by means of observation

***Eutheria*** the placental mammals

**exon** many genes are composed of regions that code for the protein product of the gene (exons) separated by regions that carry no meaningful code for the protein (**introns**); during gene expression the introns are excised and the exons are spliced together to code for the messenger **RNA** that carries the full code for the protein

**exopthalmia** a condition in which the eyes protrude; often associated with an overactive thyroid gland

## F

**Fanconi anemia** a rare, inherited form of **anemia** in which there is a defect in red blood cell formation

**fat patterning** the tendency for fat to be deposited preferentially in certain regions of the body, resulting in characteristic patterns, e.g. by age, sex, life style

**fecundity** refers to reproductive ability due solely to biological factors

**fertility** a measure of reproductive performance resulting from the interaction of biological, behavioural, and environmental factors

**fetal alcohol syndrome** a congenital syndrome of mental retardation, poor growth and a number of other physical abnormalities; it is caused by excess alcohol consumption by the mother during the early stages of pregnancy

**fibroblast** a flat, elongated, undifferentiated cell in the connective tissue, which gives rise to a variety of precursor cells such as the chondroblast, collagenoblast and **osteoblast**, that form the fibrous, binding and supporting tissue of the body

**fibroblast growth factor** a family of **growth factors**, originally identified because of their actions on **fibroblasts**, but which have much more widespread functions in growth and development

**fibroblast growth factor receptor** the tissue **receptor** for the **fibroblast growth factors**; there are at least three types; **mutations** in these receptors have been implicated in a number of cranial and long-bone growth disorders

**folivorous animals** leaf/foliage-eating animals

**follicle stimulating hormone** one of two **hormones** secreted by the **pituitary** gland which are involved in pubertal development and subsequent reproductive function

**follicular phase** the first half of the menstrual cycle, immediately following **menstruation** and prior to ovulation

**forced expiratory volume** standardized measure of lung capacity

**forced vital capacity** greatest amount of air that can be expired after maximal inspiration of air; a physiological measure of lung function

**founder effect** characteristic patterns of gene frequency in a **population** that can arise when it has only a limited number of ancestors

**frugivorous** refers to an habitual diet characterized by a dependence of fruits

## G

**gastrulation** this occurs in the embryo when a hollow ball of cells starts to fold in on itself, thus creating a two-cell layered structure

**Gaussian distribution** a distribution that conforms to the normal function

**gene therapy** a still experimental treatment for certain gene defect disorders where a normal gene is introduced into the affected individual to replace or supplement their own non-functioning gene

**genetic drift** stochastic changes through time in the gene frequencies of a **population**, resulting from a range of non-deterministic factors

**genetic potential** the maximum expression of a variable that can be attained by a genotype

**genome** the complete set of genes in the **chromosome** of each cell of a particular organism

**gestational age** the age, counted from the time of the last menstrual period, of a fetus or delivered baby, usually expressed in weeks

**gestosis** any toxemic manifestation of pregnancy

**gingiva** the gums around the teeth

**glucocorticosteroids** steroid **hormones** secreted by the adrenal gland whose predominant action was originally thought to be related to synthesis of **glucose**, but whose actions are now known to be more varied

**glucose** the most common simple sugar

**glycerol growth** increase in body mass due to cellular **hypertrophy**, cellular **hyperplasia**, or the accretion of tissue; may also refer to directional changes in body dimensions, e.g. height

**glycoprotein** a protein that has sugar molecules covalently bonded to the **polypeptide** backbone

**gonad** the testis in the male and the ovary in the female

**gonadal steroids** steroids secreted by the **gonads**

**gonadarche** the onset of gonadal development prior to **puberty**

**gonadotrop(h)in** **follicle stimulating hormone** and **luteinizing hormone**, secreted by the **pituitary** gland, and responsible for the induction of **puberty** and maintenance of reproductive capacity

**gonadotrop(h)in releasing hormone** a small peptide **hormone** secreted by the **hypothalamus** that stimulates the release of the **gonadotrop(h)ins** by the **pituitary** gland

**growth arrest lines** fine transverse lines sometimes seen towards the ends of long bones on X-rays; they are thought to represent a brief period of diminished or absent growth, during which time extra calcium is laid down, making the line denser to X-rays

**growth curve** a line which describes age changes in some tissue, organ, or dimension; it may be formed by joining individual data points, or more formally by quantitative methods

**growth factors** collectively a group of **hormones** or other chemical molecules that stimulate or regulate growth of cells and tissues; some are relatively specific and only act on certain tissues, whereas most have a more widespread function on multiple cell types and tissues

**growth hormone** also known as somatotropin, a major stimulant of **somatic growth** which is secreted by the **pituitary** gland under the control of the **hypothalamus**; its growth-promoting activities are mediated through the **insulin-like growth factors**, but it also has important metabolic effects that are probably achieved directly

**growth plate** an area of cartilage near the ends of long bones where growth occurs, sandwiched between the **epiphysis** and **metaphysis**; the growth plates eventually disappear when growth is complete

**growth reference** a generally accepted set of data which serves to facilitate comparisons across samples or across time

**growth standard** sometimes synonymous with **growth reference**; or it may refer to a well-defined, usually healthy, dataset used in evaluating satisfactory growth

**growth velocity** the rate of growth, most often expressed as change per year

## H

**haemo-** *see* **hemo-**

**half-life** a characteristic attributed to many substances and chemicals that by their nature disappear over a period of time; most commonly used in the case of radioactivity where the half-life is the length of time taken for the level of radioactive emission to fall by half; also commonly used of **hormones** and other chemicals in the circulation where the half-life is the length of time taken for the substance's concentration to fall by half

**hamate** a carpal bone, or short bone of the wrist; other carpal bones are the scaphoid, lunate, triquetrum, pisiform, **trapezium**, trapezoid and **capitate**

**Harris lines** lines of increased density seen in the radiographs of long bones of children; also called transverse lines, they may indicate periods of environmental stress

**hemiatrophy** in essence the reverse of **hemihypertrophy** where one side of the body is reduced in size compared to the other

**hemihypertrophy** a condition where one side of the body (often only partial) is enlarged compared to the other

**hemolytic disease** actually a group of diseases of various etiologies that share the accelerated destruction of red blood cells leading to **anemia**

**hemopoeisis** the formation of blood

**hepatocytes** the main functioning cells of the liver

**heritability** a measure of the genetic contribution to the variation in a physical trait

**heterodont** the characteristic dentition of mammals, in which there are different types of teeth

**heterosis** the phenomenon of increased vigour or size resulting from greater heterozygosity

**heterozygote** a genotype which manifests different **alleles** at a particular locus

**heterozygous** possessing dissimilar pairs of genes for any hereditary trait

**hippocampus** part of the brain associated with encoding short-term memory

**histology** the study of the minute structures of the tissues of organisms

**HIV** human immunodeficiency virus

**holoprosencephaly** a rather variable group of conditions that share mid-line defects of cranial and intracranial structures

**homeobox genes** an important group of genes that have a fundamental role in early embryonic development; they share a highly conserved region referred to as the **homeodomain**

**homeodomain proteins** proteins associated with the homeobox region of the **chromosome**

**homeodomain** the highly conserved region shared by all **homeobox genes**; it encodes a group of approximately 60 amino acids in the specific gene products

**homeosis** formation of a body part which displays characteristics normally found in a related body part in a different place

**homeostasis** maintenance of physiological equilibrium

**homeothermic** warm-blooded

**hominid** a member of the primate family consisting of the genera *Australopithecus* and *Homo*

**hominoids** a member of the primate superfamily consisting of the *Hominidae*, the living apes, and related fossil groups

***Homo erectus*** a fossil species from Africa, Asia, and Europe that appears first in Africa approximately 1.8 million years ago

***Homo habilis*** the oldest known species of the genus *Homo*, appearing first in Africa over 2 million years ago

**homodont** refers to a dentition in which there are not different kinds of teeth

**homozygote** a genotype where both **alleles** at a particular locus are the same

**homozygous** having identical pairs of genes for any given pair of hereditary traits

**hormone** substance produced in an organ to produce a specific effect in another organ

**Hox genes** *Hox*, with an appended number (e.g. *Hox-1*, -2 etc.) is used to identify the locus (position on a **chromosome**) of a particular homeobox gene

**hydrocephaly** a pathological condition where there is excess fluid surrounding or within the cavities of the brain

**hydrodensitometry** determination of body density by underwater weighing

**hydroxyapatite** major component of bone crystals

**hydroxyproline** a non-indispensable amino acid found in abundance in **collagen**

**hypercholesterolemia** abnormally high **serum** cholesterol levels

**hyperglycemia** a raised concentration of **glucose** in the blood

**hyperinsulinemic** abnormally high **serum** insulin levels

**hyperphagia** excessive appetite for food

**hyperplasia** abnormal multiplication of normal cells

**hypertension** a clinical condition defined as an abnormally high blood pressure

**hypertrophy** enlargement of an organ due to an increase in size of its constituent cells

**hypocalcification** low level of calcium deposition

**hypoglycemia** a low concentration of **glucose** in the blood

**hypogonadism** reduced activity of the **gonads** (ovaries or **testes**) leading to inadequate sexual development or function; may be due to any of several causes

**hyponatremia** low **serum** sodium level

**hypophyseal portal system** the highly specialized system of blood vessels running from the **hypothalamus** down into the **pituitary** gland, which conveys regulatory **hormones**

**hypothalamus** the region of brain immediately above the **pituitary** gland which is responsible for regulation of many basic bodily functions; it produces the **hormones** that control **pituitary** gland function (such as the release of **growth hormone**) but also has a role in **thermoregulation**, appetite and other more basic functions

**hypotonia** resistance to stretch in muscle

**hypoxia** low oxygen availability at tissue level

## I

**immunofluorescent staining** a technique of visualizing particular cells or tissue types under the microscope; antibodies to the substance of interest are linked to chemicals that fluoresce a particular colour under ultraviolet light and these antibodies are then applied to the tissue; when exposed to ultraviolet light the antibodies fluoresce and identify the location of the target substance

**immunoglobulins** immune system **glycoproteins** that function as antibodies

**imprinting** the fixing of the affections and interests of a newborn animal on another animal or object

**indirect calorimetry** measurement of heat produced by the body from determination of the consumption of oxygen and sometimes carbon dioxide

**inflammatory bowel disease** *see* **Crohn's disease**

**inhibin** a **hormone** produced by the **gonad** that suppresses the secretion of **follicle stimulating hormone** by the **pituitary** gland

**insulin-like growth factors** two closely related **growth factors** that show much chemical and structural similarity to insulin; they have widespread effects on multiple tissues and have a fundamental role in growth and development

**intelligence quotient** a measure of mental development obtained by the application of a standardized test

**interferons** class of lymphokines produced by immune cells and **fibroblasts** which have antiviral actions

**interorbital distance** the distance between the internal epicanthic folds of the eye

**intra-uterine growth retardation** inadequate fetal growth which leads to a baby that is born abnormally small for the length of the pregnancy; it should be distinguished from prematurity, where the baby is born small, but early and is of appropriate size given the length of the pregnancy

**intron** variable length sequences of **DNA** that are interspersed between **exons** but are not expressed and therefore do not contribute to the code for the protein product of the gene

**ischemia** decrease in tissue oxygenation

**isometric strength** an estimate of strength based on measures of static muscular contractions that result from pushing or pulling against a stationary object

**isotopes** atoms of the same element having the same atomic numbers and chemical properties, but differing in mass

## K

**knemometry** a highly precise method of measuring the lower leg which can also be used to measure small changes over short periods of time

**knockout mutation** an artificial mutation in a gene that completely removes its function as if the gene were completely deleted

**kurtosis** the third moment about the mean of a distribution, referring to the frequency of values near the mean and further from it

## L

**landmark** a precisely defined anatomical point used as a reference in **anthropometry**

**lean body mass** an *in vivo* concept, defined as body-weight minus body fat

**least squares method** one method of quantifying the relationship between a dependent variable and one or more independent variables

**life-history** the length and timing of key events, including gestation length, onset of **puberty**, reproductive function and, in humans and some primates, **menopause**, across a species lifespan

**ligand** a molecule, ion or group bound to the central atom of a chemical compound

**linear enamel hypoplasia** under-development of dental enamel during the course of growth and development as a consequence of any of a number of stresses, including nutritional stress

**logarithmic transformation** the use of the logarithm of a value instead of the actual value in an attempt to reduce **skewness**

**longitudinal study** a growth study in which individuals are seen at one or more regular intervals

**low birth-weight** birth-weight below some standard of weight for **gestational age**; most commonly, this is taken to be lower than 2.5 kg

**luteinizing hormone** a hormone synthesized in the **pituitary** which stimulates the synthesis of sexual hormones from the ovary in women, and from the testis in men

**lymphokine** one of the chemical factors produced and released by T lymphocytes that attract macrophages to the site of infection and prepares them for attack

## M

**macronutrients** potentially energy-supplying nutrients, including carbohydrate, fat, and protein

**macrosomia** large organ size; often found in infants born to diabetic mothers

**mass exponent** a measure of body-size that controls for proportionality, usually in the form of weight to a fractional power between 0.5 and 1

**maternal disomy** a chromosomal abnormality where the embryo inherits both of its copies of a particular **chromosome** from the mother and none from father

**maturation** the development of functional capacity of organs, physiological systems and of the whole individual; the process of achievement of adult status in structure or function

**maxillae** large paired bones that comprise most of the anterior surface of the midfacial region and hold the upped teeth

**meiosis** the process whereby germ line cells with 23 pairs of **chromosomes** undergo division and reduction to 23 single chromosomes in the sperm or ovum

**menarche** the attainment of the first menstrual period

**menopause** cessation of ovulation and **menstruation** in women

**menstruation** periodic vaginal bleeding occurring with shedding of the uterine mucosa

**mesenchyme** a diffuse network of tissue derived from the embryonic **mesoderm**

**mesoderm** the middle of the 3 cell layers of the developing embryo

**Mesolithic** a stage in the development of human culture, falling between the **Paleolithic** and the Neolithic

**mesomelic shortening** limb shortening affecting the middle part of the limb such as the fore-arm or lower leg

**metabolism** the sum of all chemical modifications taking place to meet energy requirements, and to synthesize and maintain levels of **enzymes**, **hormones** and antibodies, and to repair tissue damage

**metacarpals** the long finger bones

**metaphysis** the part of a long bone, furthest from the mid-point of the bone and adjacent to the **growth plate**

**micronutrients** trace elements with nutritional function, such as iron, calcium, zinc and vitamins

**mid-growth spurt** a modest increase in the growth-rate, which can be detected in many children at about 7–8 years of age

**mineralization** deposition of calcium and other mineral matter in skeletal development

**missense mutation** a genetic mutation where a single **nucleotide** is changed; a missense mutation leads to a **codon** which codes for a different amino acid; the resultant protein is thus of normal length and may retain many of the normal functions but usually will vary in some important characteristics *see also* **nonsense mutation**

**mitogen** a substance that promotes cell division

**mitosis** the process of cell division when the **DNA** is shared equally between the daughter cells which are thus genetically identical

**modernization** the process by which a society develops a more cosmopolitan set of values

**monosomy** an abnormal situation when a **chromosome** has no partner, as opposed to the usual situation when every chromosome is one of a pair; the exception is the **Y chromosome** in males, which is partnered by an **X chromosome**; monosomy may be partial in that one of the pair of chromosomes has lost a part of its length so that the partner is monosomic for that region

**monotreme** a member of the lowest order of *Mammalia, Monotremata*, which has a single opening for the genital and digestive organs

**monozygotic twins** who arise from the division of a fertilized ovum

**morbidity** refers to the disease process, e.g. the morbidity rate

**morphogen** molecules that determine the form of the body during development

**morphogenesis** the generation of shape in the embryo

**morphology** the study of physical size and shape, of humans and other species

**mosaic** a situation where two or more cell lines co-exist with different chromosomal complements; an example may occur in **Turner syndrome** where, for example, 45 X and 46 XX cell lines may co-exist

**motor development** development of bodily functions such as posture, locomotion, tool use and handwriting

**mucopolysaccharidoses** a rare group of genetic disorders where the mucopolysaccharide chemicals are not metabolized correctly leading to a build-up in the tissues with various physical and functional consequences

**muscular hypertrophy** enlargement of muscles due to an increase in the size of their constituent cells

**mutation** a change in a particular gene

**myoblast** an embryonic cell which becomes a cell of the muscle fibre

**myoglobin** the oxygen-transporting pigment of muscle

**myotatic** performed or induced by stretching or extending a muscle

**myotube** a developing muscle fibre with a centrally, rather than peripherally, located nucleus; also known as myotubule

## N

**natural selection** the primary mechanism of evolutionary **adaptation**, in which **alleles** that confer greater survivability are transmitted preferentially to the next generation

**Neanderthal** a variety of humans who lived in Europe about 150,000 years ago; paleoanthropologists are divided as to whether they were *Homo sapiens* or a separate species

**necrosis** death of cells (cf. **apoptosis**), or a portion of tissue

**necrotizing enterocolitis** bacterial infection of the gut leading to gut cell death

**neocortex** part of the brain controlling some of the higher functions of the nervous system, including language

**neonatal** pertaining to the newborn

**neoteny** showing infantile physical features at a more advanced stage of development

**neurotransmitters** a group of chemical mediators of electrical impulses across nerve cells

**non-additivity** refers to an interaction between genetic loci, such that the action of genes at one locus is influenced by genes at another

**non-insulin dependent diabetes mellitus** multifactorial disorder characterized by insulin resistance and **glucose** intolerance; insulin production is not impaired, but more is needed for glucose levels to be managed; advanced stages are associated with functional impairments involving, primarily, the vascular system

**nonsense mutation** a genetic mutation where a single **nucleotide** is changed; in a nonsense mutation the change creates a chain termination **codon** that stops gene expression and leads to a shortened protein that will usually have abnormal or absent function *see also* **missense mutation**

**noradrenaline** *see* **norepinephrine**

**norepinephrine** an alternative name for noradrenaline; a major neurotransmitter substance released by nerve cells of the sympathetic system and also the inner part of the adrenal gland (adrenal medulla)

**nucleotide** basic building blocks of **DNA** composed of a purine or pyramidine base combined with a sugar, deoxyribose; in practical terms they are usually identified by the base as adenine (A), guanine (G), cytosine (C) and thymine (T); A and G are purines whilst C and T are pyrimidines

## O

**obesity** a medical condition characterized by an excessive level of fatness

**occlusion** a blood vessel obstruction in the brain that can lead to a stroke

**occlusion** the interdigitation of the upper and lower dentition

**oedema** *see* **edema**

**oestrogen** *see* **estrogen**

**oligomenorrhea** irregular menstrual cyclicity

**oligonucleotide** short sequence of nucleic acids

**omnivorous** refers to species that exploit the nutritional resources of their ecosystem widely; humans are omnivorous

**oncogenes** genes found in the **chromosomes** of tumour cells whose activation is associated with the initial and continuing conversion of normal cells into cancer cells

**ontogeny** change through development, as opposed to **phylogeny**, or change through evolution

**organogenesis** formation of organ systems during fetal growth and development

**ossification** the process of bone formation, in which it is laid down bone in a pre-existing matrix of cartilage or fibrous tissue

**osteoblasts** bone-forming cells developing from stromal cell precursors in the bone marrow

**osteocalcin** protein associated with the process of bone **calcification**

**osteoclasts** multinuclear cells that erode and resorb previously formed bone

**osteogenesis imperfecta** a group of genetic diseases of bone which lead to weakened bones with increased liability to fracture

**osteomalacia** softening of the bone; occurs mainly in adults

**osteoporosis** reduction in bone quantity, occurring mainly in women after the **menopause**

## P

**paleoepidemiology** the study of disease processes in ancient **populations**

***Pan*** the genus of apes which comprises chimpanzees

**paracrine** a method of hormonal transmission where the **hormone** acts on cells in close proximity to those secreting the hormone; *see also* **autocrine** and **endocrine**

**parametric statistics** a set of analytic statistical procedures that makes certain assumptions about the distributions of the variables

**parturition** giving birth to a child

**pastoralist** a society in which the primary subsistence base is herding

**pathogens** agents of infection

**peak height velocity** the maximum rate of growth achieved by an individual

**peptide growth factors** usually short proteins with any or all of **endocrine**, **paracrine** and **autocrine** activity; includes **insulin-like growth factors**

**perichondrium** the layer of dense, fibrous connective tissue which invests all cartilage except the articular cartilage of the synovial joints

**Perthe's disease** a disease of the **proximal epiphysis** of the femur where there is disturbance of the blood supply and the bone dies and is partially destroyed; healing takes place spontaneously, but there may be permanent deformity unless appropriate action is taken

**phenotype** the result of the interaction between a genotype and the environment

**phospholipids** major fat components of cell membranes, critical to their structure and function

**phosphorylation** the metabolic process of introducing a phosphate group into an organic molecule

**phylogeny** change through evolution as opposed to **ontogeny**, or change through development

**physical activity level** a way of expressing energy expended in physical activity; defined as total daily energy expenditure divided by **basal metabolic rate**

**physiological age** the mean **chronological age** associated with individuals of a specified level of physiological development

**physique** a term which encompasses the whole body

**pituitary endocrine** gland releasing **hormones** directly in blood, which act on other endocrine glands in the individual

**plasticity** the ability of an individual to be altered by the environment in which he or she develops

**platelet derived growth factor** a **cytokine** stimulating vascular smooth muscle cell division

**platyspondyly** flattening of the vertebral bodies

**pleitropic effect** a single (usually genetic) event that triggers a series of different events that affect several functions; a single gene influencing more than one phenotypic character

**pluripotential stem cells** cells that have several potential lines of development and specialization that form the origins of other more specialized and differentiated cells after subsequent divisions

**Poland syndrome** a rare association of partial absence of the pectoral muscles, abnormal breast or nipple development and hand deformities

**polydipsia** excessive drinking

**polygenic** a trait whose expression results from the effects of more than one genetic locus

**Polymerase chain reaction (PCR)** a fundamental technique of molecular biology in which minute quantities of **DNA** are amplified to produce vastly increased quantities for further analysis

**polynomials** algebraic expressions containing multiple terms of increasing powers, e.g. $x + x^2 + x^3$

**polypeptide** a molecule made up of a number of amino acids; proteins are large polypeptides

**ponderal index** an index expressing weight-for-height and calculated as height divided by the cube root of weight

**population** a group of individuals who form a cluster because of some shared activity; commonly it refers to a group which shares a common gene pool

**porotic hyperostosis** porous overdevelopment of bone tissue, often associated with **anemia**, in past **populations**

**portal vessel system** a system of blood vessels that has a capillary bed at both ends, as opposed to a larger artery or vein at one end; there are two such systems: the **hypophyseal portal system** and the hepatic

**Prader–Willi syndrome** a genetic syndrome with various combinations of low birth-weight, poor feeding and floppiness as an infant, subsequent **obesity** and characteristic **dysmorphic** features

**precision** an indication of the degree of exactness with which a measurement is taken

**precocial mammals** mammalian species producing offspring which are reasonably independent from birth (cf. **altricial mammals**)

**presbyopia** the tendency for increasing long-sight with increasing age

**primiparous** referring to women experiencing their first birth

**principle components analysis** a method of statistical analysis which reduces a set of individual, and intercorrelated, variables to a smaller number of constructs, or components

**probit analysis** a technique for fitting observations to a model to the likelihood of an outcome

**prodromal phase** the second phase of an infection characterized by non-specific symptoms, making diagnosis difficult

**progesterone** a steroid **hormone** produced by the ovary in the second half of the menstrual cycle (the luteal phase) and which prepares the uterine lining for implantation if an ovum is fertilized

**prohormone convertases** **enzymes** responsible for the conversion of the inactive form of a **hormone** to the active one

**prosimians** a collective term for 'lower' primates, encompassing lemurs, lorises, and tarsiers

**prostaglandins** a family of fatty acids that all contain a 5-carbon atom ring and two side chains; there are many members of the family which have a variety of regulatory functions in many tissues

**proteinases** **enzymes** responsible for the breakdown of proteins of different kinds

**proteolysis** the break down of protein into constituent chemicals

**Proteus syndrome** an exceedingly rare syndrome, of unknown cause, in which different tissues, organs and regions of the body overgrow in an unco-ordinated manner, leading to severe deformity

**proximal** nearest to the head, or point of attachment

**pseudogene** a sequence of **nucleotides** that resembles a gene, and may be derived from one, but lacks a genetic function

**puberty** physical changes during growth and development resulting in the development of sexual function and maturity

## Q

**quadrapedal** four-legged

**Quetelet's Index** *see* **body mass index (BMI)**

## R

**radius** one of the two bones of the forearm

**ramus, rami** a bony process as in the mandibular ramus

**receptor** a protein molecule which is inserted through the cell membrane and which binds certain chemicals such as **hormones** and then mediates their action within the cell

**receptor kinases** a large family of **receptors** have kinase **enzymes** associated with their intracellular extension; these enzymes initiate a cascade of intracellular actions when the relevant specific substance binds to the extracellular component of the receptor

**recombination** an event that occurs in **meiosis** when **chromatin** is exchanged between the members of a **chromosome** pair and which leads to the appearance in the offspring of traits that were not found together in either parent

**recumbent** lying down

**reticuloendothelial system** tissue macrophage component of the immune system

**retromolar** situated behind the most posterior upper and lower molar tooth

**rhizomelic shortening** limb shortening affecting the most **proximal** part of the limb, as in the upper arm or thigh

**Rieger syndrome** a structural defect of the anterior chamber of the eye, which may be associated with increased intraocular pressure

**RNA** ribonucleic acid is the single stranded nucleic acid which is synthesized during gene **transcription** when the **DNA** is decoded and a complementary strand of RNA is produced; any **introns** in the gene are cut out so that the RNA reflects just the **exons** of the DNA; the RNA is subsequently translated to programme the synthesis of the protein product

## S

**sagittal axis** the axis or plane that goes from anterior to posterior

**saltation** a mutation causing a significant difference in appearance between parent and offspring, or an abrupt variation in the characteristics of the species

**sarcopenia** muscle **wasting** and loss

**schizophrenia** a form of psychosis in which an individual suffers from disturbed thinking, moods and behaviour, often involving hallucinations and delusions

**scoliosis** abnormal curvature of the spine

**secondary sexual characteristics** physical changes in **puberty** other than ovarian development in females, and testicular development in males; includes breast development in females, and the development of pubertal and **axillary hair** in both males and females

**secular trend** systematic changes in population growth characteristics across time, usually generations

**sella turcica** a small bony pocket in the base of the skull in which the **pituitary** gland lies

**senescence** old age or the process of ageing

**sensitivity** in epidemiology, a measure of the success of an indicator at identifying correctly individuals who have some disease or condition

**serum** fluid component of the blood

**sex steroids** steroid **hormones** including the **estrogens, androgens** and **progesterone** that are involved in **sexual maturation** and reproduction

**sexual dimorphism** differences between the sexes in certain traits, such as height, body-size and shape

**sexual maturation** full development of sexual function

**sickle cell disease** a disease most common in those of African descent where there is a **mutation** in the hemoglobin gene which leads to a tendency for red blood cells to become sickle shaped in response to reduced oxygen tension; this change of shape leads to poor flow through small blood vessels with subsequent clot formation

***Sinanthropus*** not a taxonomic category but the collective name of a group of ***Homo erectus*** fossils from China

**skeletal age** a measure of growth tempo in which the level of skeletal **maturation** is expressed as a **chronological age** equivalent of a reference sample

**skewness** a measure of the symmetry of the distribution of a variable about its mean

**skinfold** a 'fold' of skin and subcutaneous tissue which is pulled away from the underlying tissue and measured; its thickness is used as an indicator of fatness

**small-for-gestational age** synonymous with **intrauterine growth retardation**

**small-but-healthy hypothesis** an interpretation of small body-size in lesser developed countries as a healthy **adaptation** to nutritional deficiency, arising through slowed rates of growth

**somatic growth** growth of the body

**somatostatin** also known as **growth-hormone**-release inhibiting **hormone** this is a small peptide secreted by the **hypothalamus** and pancreatic islet cells; in the **pituitary** gland it inhibits **growth hormone** release but its function in the pancreatic islets is less clear

**somatotrophs** cells in the **pituitary** gland that secrete **growth hormone** (somatotropin)

**Sotos syndrome** a syndrome where children are very tall, moderately mentally retarded and tend to be poorly co-ordinated; because they also have advanced skeletal **maturation** they stop growing relatively early and therefore are not as tall as adults as it might seem during childhood

**Southern blot analysis** a basic technique of molecular biology where **DNA** fragments are separated by gel electrophoresis and then transferred to a nylon membrane where the DNA is fixed to the membrane and can be further studied

**specificity** in epidemiology, a measure of the success of an indicator at identifying correctly individuals who do not have some disease or condition

**spermatogenesis** the production of sperm in the **testes**

**spermaturia** the passage of sperm in the urine

**spline method** a method of modelling complex relationships between variables by fitting pieces of the data and then joining these fits into an overall model

**spondyloepiphyseal dysplasias** a family of **bone dysplasias** which affect the vertebral bodies and the epiphyses, usually causing severe short stature

**stable isotope** in contrast to a radioactive isotope, a stable isotope does not undergo radioactive decay; it is detected by sophisticated methods of measuring atomic mass

**stagnant loop syndrome** a bowel disorder in which a region of the upper bowel, which is normally sterile, becomes colonized by bacteria usually from the lower bowel; this is usually due to an anatomical or functional abnormality of the bowel leading to stagnation of the contents; it leads to disordered absorption of nutrients

**stepwise regression** a statistical technique which successively adds (or removes) independent variable to an analysis until the most economical model is achieved

**steroid hormones** a family of **hormones** that are all chemically related to cholesterol; they have a variety of functions, including the maintenance of **glucose** concentration in the blood (**glucocorticosteroids**), salt and water balance (mineralocorticosteroids) and sexual differentiation (**gonadal steroids**)

**stratified random sample** a sampling technique which defines strata on some external criterion, and then samples randomly within each stratum

**strength** basic muscular force required for movement

**stunting** a retardation of growth processes resulting in an individual who is small but normal in body proportions

**subcutaneous fat** adipose tissue deposited in cells located under the skin

**subfecundity** abnormally low biologically ability to conceive a child

**substrate** the chemical substance(s) on which **enzymes** act

**supine** lying down

**suprailiac** in **body composition**, refers to a **skinfold** which is located immediately superior to the superior and lateral aspects of the iliac crest

**sympathetic nervous system** branch of the **autonomic nervous system** which carries signals to activate internal organs including the lungs, heart, liver and smooth muscle

**synpolydactyly** having more than the usual number of fingers or toes which are fused

**systemic** pertaining to the body as a whole

## T

**T cells** lymphocyctes, dependent on the **thymus**, that are involved in **antigen**-recognition pathogen killing

**technical error of measurement** an estimate measurement error, calculated as the sum of the squared differences between replicates, divided by twice the number of pairs

**tempo of development** the speed at which an organism matures, relative to its peers; an early developer has fast tempo and a late developer has slow tempo

**testes** male sex organs responsible for production of sperm and **testosterone**

**testosterone** male sex **hormone** produced by the Leydig cells of the **testes**

**thalamus** part of the brain adjacent to the cerebral **cortex** associated with complex integrative functions

**thanatophoric dysplasia** a very severe **bone dysplasia** which is usually fatal soon after birth

**thermic effect of food** the increased **thermogenesis** following a meal

**thermogenesis** production of heat in the body by physiological processes

**thermoregulation** maintenance of constant body temperature

**thymus** gland producing **T cells** of the immune system, and **hormones** that stimulate T cells after they leave the thymus

**thyrotoxicosis** an overactive thyroid gland

**thyroxin** with tri-iodothyronine (T3), one of the principal thyroid **hormones**

**tissue hydration** degree of water content in bodily tissues; this differs with age, and between types of tissue

**transcription** a process whereby the genetic information contained in **DNA** is used to produce a complementary sequence of bases in an **RNA** chain

**transcription factors** a group of peptide factors that act directly on genes and modify their **transcription**

**transforming growth factors** a family of at least two factors (a and b) which have a fundamental role in early embryonic and fetal development and seem particularly important in cell differentiation

**transgenic** a term used to describe experimental animals which have artificially inserted genes from other species or strains

**translocation** the rearrangement of genetic material within the same chromosome; the transfer of a part of one **chromosome** to another different one, involving a single break in only one of the chromosomes

**transmembrane domain** that part of a **receptor** that actually crosses the cell membrane

**trapezium** short bone of the wrist

**triceps skinfold** in **body composition**, a **skinfold** located over the triceps muscle and oriented vertically

**triquetral** short bone of the wrist

**trisomy** the state of having three copies of a particular **chromosome** rather than the usual two; **Down syndrome** is an example where there is trisomy of chromosome 21

**tritiated thymidine** radioactively labelled **nucleotide** base

**Turner syndrome** a condition affecting females where one **X chromosome** is either completely or partially deleted or structurally abnormal; this leads to a variable number of **dysmorphic** features among which short stature is invariably present

**twins monozygotic** twins originate from the same fertilized egg and have an identical genetic composition; **dizygotic** twins originate from two independent fertilized eggs and have a different genetic composition

**Type I collagen** the most abundant **collagen** found in skin, bone, tendons and ligaments and other tissues; composed of two $\alpha 1$ chains and one $\alpha 2$ chain

**Type II collagen** **collagen** found predominantly in the vitreous body of the eye and ageing human cartilage; composed of three $\alpha 1$ chains

**type 1 error** the error that occurs when an investigator concludes that two samples differ when in fact they do not

**type 2 error** the error that occurs when an investigator concludes that two samples do not differ when in fact they do

## U

**ulcerative colitis** a chronic inflammatory condition of the large bowel

**ulna** one of the two bones of the forearm

**ultrasound** imaging device based on the detection of density differences using very high frequency sound waves

**urea** main nitrogenous constituent of the urine, formed in the liver when amino acids are deaminated

## V

**$\dot{V}O_2$ max** oxygen consumption at exercise to exhaustion, reflecting individual capacity for aerobic work

**variance** a measure of the dispersion of a distribution: the sum of squared differences from the mean divided by the number of observations

**VATER** acronym for *v*ertebral defects, imperforate *a*nus, *t*racheo*e*sophageal fistula, and *r*adial and *r*enal dysplagia, which together form a non-random association of congenital defects

**vertical transmission** the passage of an infectious agent from mother to child

**visceral fat** fat deposited within the body's subperitoneal deposits; also called deep fat

## W

**wasting** a disproportionate loss of fat and **lean body mass** relative to height or length

**work capacity** *see* **$\dot{V}O_2$ max**

## X

**X chromosome** one of the sex **chromosomes**; in the female there are two X chromosomes, although one is inactivated for most of the time

## Y

**Y chromosome** the male has one Y **chromosome** and one **X chromosome**; in the absence of the Y chromosome the individual will develop as a female, with the one exception of the situation where the vital male determining gene *SRY* is translocated on an X chromosome

## Z

**Z score** a transformation of a value to a standardized deviate, calculates as the value minus the mean, divided by the standard deviation

**zygote** the fertilized egg cell

# Further reading

## Introduction and brief history of the study of human growth

*All references prior to 1980 are to be found in the bibliography of Tanner (1981).*


Bouchalová M (1987). *Vyvoj během détstvi a jeho ovlivnéni. Brnénská rustová studie*. [Development in childhood and influences acting upon it. The Brno longitudinal study]. Praha: Avicenum.

Boyd E & Scammon R E (1980). *Origins of the study of human growth*. Portland University of Oregon: Health Sciences Center Foundation.

Butler G E, McKie M & Ratcliffe S G (1990). The cyclical nature of prepubertal growth. *Annals of Human Biology* 17: 177–198.

Eiben O G (1995). The Budapest longitudinal growth study 1970–1988. In *Essays on Auxology Presented to James Mourilyan Tanner by Former Colleagues and Fellows* (eds R Hauspie, G Lindgren & F Falkner) pp 211–223. Welwyn Garden City: Castlemead Publications.

Eiben O G, Barabás A & Panto E (1991). The Hungarian National Growth Study. *Humanbiologia Budapestinensis* 21: 1–121.

Eveleth P & Tanner J M (1990). *Worldwide Variation in Human Growth* 2nd edn. Cambridge: Cambridge University Press.

Falkner F & Tanner J M (eds) (1986). *Human Growth: a Comprehensive Treatise* 2nd edn. New York: Plenum.

Fogel R (1994). Economic growth, population theory and physiology: the bearing of long-term processes on the making of economic policy. Nobel Prize Lecture in Economic Sciences 1993. *American Economic Review* 84: 369–395.

Fogel R (1995). Anthropometric history: notes on two decades of a new field of research. In *Essays on Auxology Presented to James Mourilyan Tanner by Former Colleagues and Fellows* (eds R Hauspie, G Lindgren & F Falkner) pp 271–284. Welwyn Garden City: Castlemead Publications.

Gasser T, Kneip A, Ziegler P, Largo R & Prader A (1990). A method for determining the dynamics and intensity of average growth. *Annals Human Biology* 17: 459–474.

Hauspie R, Lindgren G & Falkner F (eds) (1995). *Essays on Auxology Presented to James Mourilyan Tanner by Former Colleagues and Fellows*. Welwyn Garden City: Castlemead Publications.

Keynes M (ed) (1993). *Sir Francis Galton FRS: the legacy of his ideas*. London: Macmillan.

Komlos J (1989). *Nutrition and Economic Development in the Eighteenth-Century Habsburg Monarchy: an anthropometric history*. Princeton: Princeton University Press.

Komlos J (ed) (1994). *Stature, Living Standards and Economic Development: Essays in Anthropometric History*. Chicago: University of Chicago Press.

Komlos J (ed) (1995). *The Biological Standard of Living on Three Continents: Further Explorations in Economic History*. Boulder: Westview Press.

Prokopec M (1989). Growth surveys and growth surveillance in Czechoslovakia. In *Auxology 88: Perspectives in the Science of Growth and Development* (ed J M Tanner) pp 121–131. London: Smith-Gordon.

Roche A F (1992). *Growth, Maturation and Body Composition. The Fels Longitudinal Study 1929–91*. Cambridge: Cambridge University Press

Tanner J M (1981). *A History of the Study of Human Growth*. Cambridge: Cambridge University Press.

Tanner J M (1982). The potential of auxological data for monitoring economic and social well-being. In 'Trends in Nutrition, Labor, Welfare and Labor Productivity'(eds R W Fogel & S L Engerman). *Social Science History* 6: 571–81.

Tanner J M (1986). Use and abuse of growth standards. In *Human Growth: A Comprehensive Treatise* (eds F Falkner & J M Tanner) 2nd edn, vol 3 pp 95–109. New York: Plenum.

Tanner J M (1989). The first study of human growth; Christian Friedrich Jampert. *International Journal of Anthropology* 4: 19–26.

Tanner J M (ed) (1989). *Auxology 88: Perspectives in the Science of Growth and Development*. London: Smith-Gordon.

Tanner J M, Hayashi T, Preece M A & Cameron N (1982). Increase in length of leg relative to trunk in Japanese children and adults from 1957 to 1977: comparison with British and with Japanese Americans. *Annals of Human Biology* 9: 411–424.

Tanner J M, Whitehouse R H, Cameron N, Marshall W A, Healy M J R & Goldstein H (1983). *Assessment of Skeletal Maturity and Prediction of Adult Height (TW2 Method)* 2nd edn. London: Academic Press.


## Part one: Measurement and assessment

### Embryonic staging


Butler H & Juurlink B H J (1987). *An Atlas for Staging Mammalian and Chick Embryos*. Boca Raton, Florida: CRC Press.

O'Rahilly R (1988). One hundred years of human embryology. *Issues & Reviews in Teratology* 4: 81–128.

O'Rahilly R & Müller F (1987). *Developmental Stages in Human Embryos*. Washington, DC: Carnegie Institution of Washington.

O'Rahilly R & Müller F (1994). *The Embryonic Human Brain. An Atlas of Developmental Stages*. New York: Wiley-Liss.


### Prenatal age


Cadkin A V (1992). Crown–rump length dating of pregnancy at less than nine weeks' gestation. *American Journal of Obstectrics and Gynecology* 166: 269.

Dickey R P & Gasser R F (1993). Computer analysis of the human embryo growth curve. *Anatomical Record* 237: 400–407.

Dickey R P & Gasser R F (1993). Ultrasound evidence for variability in the size and development of normal human embryos. *Human Reproduction* 8: 331–337.

Dickey R P, Gasser R F, Olar T T et al. (1994). The relationship of initial embryo crown–rump length to pregnancy outcome and abortus karyotype. *Human Reproduction* 9: 366–373.

O'Rahilly R (1997). Gestational age? *Clinical Anatomy* 10: 367.

O'Rahilly R & Müller F (1984). Embryonic length and cerebral landmarks in staged human embryos. *Anatomical Record* 209: 265–271.

Saito M et al. (1972). Time of ovulation and prolonged pregnancy. *American Journal of Obstetrics and Gynecology* 112: 31–38.

Tercanli S & Holzgreve W (1995). Das erste Trimenon. In *Ultrashall in Gynäkologie und Geburtshilfe* (eds C Sohn & W Holzgreve). Stuttgart: Thieme.

Wisser J, Dirschedl P & Krone S (1994). Estimation of gestational age by transvaginal sonographic measurement of greatest embryonic length in dated human embryos. *Ultrasound Obstet.Gynecol.* 4: 457–462.


### Anthropometry


Cameron N (1984). *The Measurement of Human Growth*. London: Croom-Helm.

Cameron N, Hiernaux J, Jarman S, Marshall W A, Tanner J M & Whitehouse R H (1981). Anthropometry. In *Practical Human Biology* (eds J S Weiner & J A Lourie) pp 27–52. London: Academic Press.

Jones P R M (1995). From tape to technology. In *Essays on Auxology Presented to James Mourilyan Tanner by Former Colleagues and Fellows* (eds R Hauspie, G Lindgren & F Falkner) pp 3–17. Welwyn Garden City: Castlemead Publications.

Lohman T G, Roche A F & Martorell R (1988). *Anthropometric Standardization Reference Manual*. Champaign, Illinois: Human Kinetics Books.

Martin R (1914). *Lehrbuch der Antropologie*. Jena: Fischer.

Tanner J M (1981). *A History of the Study of Human Growth*. Cambridge: Cambridge University Press.

Ulijaszek S J & Mascie-Taylor C G N (1994). *Anthropometry: the Individual and the Population*. Cambridge: Cambridge University Press.


### *Measurement error*


Frisancho A R (1990). *Anthropometric Standards for the Assessment of Growth and Nutritional Status*. Ann Arbor: University of Michigan Press.

Heymsfield S B, McManus C B, Seitz S B, Nixon D W & Andrews J S (1984). Anthropometric assessment of adult protein-energy malnutrition. In *Nutritional Assessment* (eds R A Wright, S Heymsfield & C B McManus) pp 27–81. Boston: Blackwell Scientific Publications.

Mueller W H & Martorell R (1988). Reliability and accuracy of measurement. In *Anthropometric Standardization Reference Manual* (eds T G Lohman, A F Roche & R Martorell) pp 83–86. Champaign, Illinois: Human Kinetics Books.

Ulijaszek S J (1997). Anthropometric measures. In *Design Concepts in Nutritional Epidemiology* (eds B M Margetts & M Nelson) pp 289–311. Oxford: Oxford University Press.

Ulijaszek S J & Lourie J A (1994). In *Anthropometry: the Individual and the Population* (eds S J Ulijaszek & C G N Mascie-Taylor) pp 30–55. Cambridge: Cambridge University Press.

Zerfas, A J (1985). *Checking Continuous Measures: Manual for Anthropometry*. Los Angeles: Division of Epidemiology, School of Public Health, University of California.


### Body imaging by three-dimensional surface anthropometry


Addelman D & Addelman L (1985). Rapid 3–13 digitizing. *Computer Graphics World* November: 42–44.

Jones P R M, West G M, Harris D H & Read J B (1989). The Loughborough anthropometric shadow scanner (LASS). *Endeavour* 13: 162–168.

Jones P R M (1997). Three-dimensional surface anthropometry: applications to the human body. *Optics and Lasers in Engineering* 28: 89–117.

Uesigi M (1991). Three dimensional curved shape measuring system using image encoder. *Journal of Robotics and Mechatronics* 3: 190–195.


### Assessment of body composition


Davies P S W & Cole T J (1995). *Body Composition Techniques in Health and Disease*. Cambridge: Cambridge University Press.

Forbes G B (1978). Body Composition in Adolescence. In *Human Growth : A Comprehensive Treatise* (eds F Falkner & J M Tanner) vol 2. pp 239–272 New York: Plenum.

Lohman T G (1992). *Advances in Body Composition Assessment*. Champaign, Illinois: Human Kinetics Publishers.


### *Chemical immaturity*


Jebb S A & Elia M (1995). Multi-compartment models for the assessment of body composition in health and disease. In *Body Composition Techniques in Health and Disease* (eds P S W Davies & T J Cole) pp 240–254. Cambridge: Cambridge University Press.

Lohman T G (1992). *Advances in Body Composition Assessment*. Champaign, Illinois: Human Kinetics Publishers.
Lohman T G, Boileau R A & Slaughter M H (1985). Body composition of children and youth. In *Advances in Pediatric Sport Sciences* (ed R A Boileau) pp 29–57. Champaign, Illinois: Human Kinetics Publishers.
Lohman T G, Gonig S B, Slaughter M H & Boileau R A (1989). Concept of chemical immaturity in body composition estimates: implications for estimating the prevalence of obesity in childhood and youth. *American Journal of Human Biology* 1: 201–204.
Siri W E (1956). The gross composition of the body. *Advances in Biological & Medical Physics* 56: 680–688.

### Biochemical markers

Blumsohn A, Hannon R A, Wrate R, Barton J, Al-Dehaimi A W, Colwell A & Eastell R (1994). Biochemical markers of bone turnover in girls during puberty. *Clinical Endocrinology* 40: 663–670.
Bollen A-M & Eyre D R (1994). Bone resorption rates in children monitored by the urinary assay of collagen type I cross-linked peptides. *Bone* 15: 31–34.
Fujimoto S, Kubo T, Tanaka H, Miura M & Seino Y (1995). Urinary pyridinoline and deoxypyridinoline in healthy children and in children with growth hormone deficiency. *Journal of Clinical Endocrinology and Metabolism* 80: 1922–1928.
Rauch F, Schoenau E, Woitge H, Remer T & Seibel M J (1994). Urinary excretion of hydroxy-pyridinium cross-links of collagen reflects skeletal growth velocity in normal children. *Experimental and Clinical Endocrinology* 102: 94–97.
Robins S P (1994). Biochemical markers for assessing skeletal growth. *European Journal of Clinical Nutrition* 48 (Suppl 1): S199–S209.
Saggese G, Bertelloni S, Baroncelli G I & Di Nero G (1992). Serum levels of carboxyterminal propeptide of type I procollagen in healthy children from 1st year of life to adulthood and in metabolic bone diseases. *European Journal of Pediatrics* 151: 764–768.
Trivedi P, Hindmarsh P, Risteli J, Risteli L, Mowat A P & Brook C G D (1989). Growth velocity, growth hormone thereapy and serum concentrations of the amino-terminal propeptide of type III procollagen. *Journal of Pediatrics* 114: 225–230.

### Radiographic assessment

Acheson R M (1966). Maturation of the skeleton. In *Human Development* (ed F Falkner) pp 465–502. Philadelphia & London: W B Saunders.
Greulich W W & Pyle S I (1950). *Radiographic Atlas of the Skeletal Development of the Hand and Wrist*. Stanford: Stanford University Press.
Tanner J M, Whitehouse R H, Cameron N, Marshall W A, Healy M J R & Goldstein H (1983). *Assessment of Skeletal Maturity and Prediction of Adult Height* 2nd edn. London: Academic Press.

### Dental maturation

Demirjian A, Goldstein H & Tanner J M (1978). A new system of dental age assessment. *Human Biology* 45: 211–227.
Eveleth P B & Tanner J M (1990). *Worldwide Variation in Human Growth* 2nd edn. Cambridge: Cambridge University Press.
Harris E F & McKee J H (1990). Tooth mineralization standard for blacks and whites from the Middle Southern United States. *Journal of Forensic Sciences* 35: 859–872.
Moorrees C F A, Fanning E A & Hunt E E Jr (1963). Age variation of formation stages for ten permanent teeth. *Journal of Dental Research* 42: 1490–1502.
Ranta R A (1986). Review of tooth formation in children with cleft lip/palate. *American Journal of Orthodontics and Dentofacial Orthopedics* 90: 11–18.
Tanguay R, Demirjian A & Thibault H W (1984). Sexual dimorphism in the emergence of the deciduous teeth. *Journal of Dental Research* 63: 65–68.

### Physical examination

Brook C G D & Stanhope R (1995). Normal puberty: physical characteristics and endocrinology. In *Clinical Paediatric Endocrinology* (ed C G D Brook) 3rd edn. Oxford: Blackwell Scientific Publications.
Buckler J M H (1990). *A Longitudinal Study of Adolescent Growth*. London: Springer-Verlag.
Tanner J M (1962). *Growth at Adolescence* 2nd edn. Oxford: Blackwell Scientific Publications.

### Adiposity rebound and prediction of adult fatness

Forbes G B (1978). Body composition in adolescence In *Human Growth: A Comprehensive Treatise* (eds F Falkner & J M Tanner) 2nd edn, vol 2 pp 119–139. New York: Plenum.
Garn S M & Clark D C (1975). Nutrition, growth, development and maturation *Pediatrics* 56: 306–319.
Knittle J L, Timmers K, Ginsberg-Fellner F, Brown R E & Katz D P (1979). The growth of adipose tissue in children and adolescents. Cross sectional and longitudinal studies of adipose cell number and size. *Journal of Clinical Investigations* 63: 239–246.
Nouguès J, Reyne Y, Barenton B, Chery T, Garandel V & Soriano J (1993). Differentiation of adipocyte precursors in a serum-free medium is influenced by glucocorticoids and endogenously produced insulin-like growth factor-1. *International Journal of Obesity* 17: 159–167.
Poskitt E M E & Cole T J (1977). Do fat babies stay fat? *British Medical Journal* 7–9.
Rolland-Cachera M F (1993). Body composition during adolescence: methods, limitations and determinants. *Hormone Research* 39 (Suppl 3): 25–40.
Rolland-Cachera M F, Bellisle F & Sempé M (1989). The prediction in boys and girls of the Weight/Height$^2$ index and various skinfolds measurements in adults: a two-decade follow-up study. *International Journal of Obesity* 13: 305–311.
Rolland-Cachera M F, Deheeger M, Avons P, Guilloud-Bataille M, Patois E & Sempé M (1987). Tracking adiposity patterns from 1 month to adulthood. *Annals of Human Biology* 14: 219–222.
Rolland-Cachera M F, Deheeger M, Bellisle F, Sempé M, Guilloud-Bataille M & Patois E (1984). Adiposity rebound in children: a simple indicator for predicting obesity. *American Journal of Clinical Nutrition* 39: 129–135.
Sempé M, Pédron G & Roy-Pernot MP (1979). *Auxologie, Méthode et Séquences*. Paris: Théraplix.
Siervogel R M, Roche A F, Guo S, Mukherjee D & Chumlea W C (1991). Patterns of change in weight/stature$^2$ from 2 to 18 years: findings from long-term serials data for children in the Fels longitudinal growth study. *International Journal of Obesity* 15: 479–485.
Zapf J, Walter H & Froesch E R (1981). Radioimmunological determination of insulin-like growth factors I and II in normal subjects and in patients with growth disorders and extrapancreatic tumor hypoglycemia. *Journal of Clinical Investigation* 68: 1321–1330.

### Growth as an indicator of social inequalites

Bielicki T (1986). Physical growth as a measure of the economic well-being of populations: the twentieth century. In *Human Growth: A Comprehensive Treatise* (eds F Falkner & J M Tanner) 2nd edn, vol 3. New York: Plenum.
Bielicki T, Malina R M & Waliszko, H (1992). Monitoring the dynamics of social stratification: statural variation among Polish conscripts in 1976 and 1986. *American Journal of Human Biology* 4: 345–352.
Brundtland G H, Liestol K & Walloe L (1980). Height, weight and menarcheal age of Oslo schoolchildren during the last 60 years. *Annals of Human Biology* 7: 307–322.
Eveleth P B & Tanner J M (1990). *Worldwide Variation in Human Growth* 2nd edn. Cambridge: Cambridge University Press.
Fogel R W (1986). Physical growth as a measure of the economic well-being of populations: the eighteenth and nineteenth centuries. In *Human Growth: A Comprehensive Treatise* (eds F Falkner & J M Tanner) 2nd edn, vol 3. New York: Plenum.
Tanner J M (1981). *A History of the Study of Human Growth*. Cambridge: Cambridge University Press.
Tanner J M (1986). Growth as a mirror of the condition of society: secular trend and class distinctions. In *Human Growth: A Multidisciplinary Review* (ed A Demirjian). London: Taylor & Francis.
Tanner J M (1982). The potential of auxological data for monitoring economic and social well-being. *Social Sciences History* 6: 571–581.
Ulijaszek S J & Strickland S S (1993). Nutritional studies in biological anthropology. In *Research Strategies in Human Biology* (eds G W Lasker & C G N Mascie-Taylor). Cambridge: Cambridge University Press.
Volkova T V (1979). The secular trend of stature in workers of Central Russia. *Voprosy Antropologii* 60: 97–103 (in Russian).

### Cross-sectional studies

Lemeshow S, Hosmer D W Jr, Klar J & Lwanga S K (1990). *Adequacy of Sample Size in Health Studies*. Chichester: World Health Organization and John Wiley & Sons.
Mascie-Taylor C G N (1994). Statistical issues in anthropometry. In *Anthropometry: the Individual and the Population* (eds S J Ulijaszek & C G N Mascie-Taylor) pp 56–77. Cambridge: Cambridge University Press.
Neave H R (1981). *Elementary Statistics Tables*. London: Unwin Hyman.
Zar J H (1984). *Biostatistical Analyses* 2nd edn. New Jersey: Prentice Hall.

### Longitudinal analysis

Gasser T, Kneip A, Ziegler P, Largo R & Prader A (1990). A method for determining the dynamics and intensity of average growth. *Annals of Human Biology* 13: 129–141.
Karlberg J (1987). On the modelling of human growth. *Statistics in Medicine* 6: 185–192.
Preece M A & Baines M J (1978). A new family of mathematical models describing the human growth curve. *Annals of Human Biology* 5: 1–24.
Togo M & Togo T (1982). Time series analysis of stature and body weight in five siblings. *Annals of Human Biology* 9: 425–440.

### Creation of growth references

Cole T J (1988). Fitting smoothed centile curves to reference data. *Journal of the Royal Statistical Society Series A: Statistics in Society* 151: 385–418.
Cole T J, Freeman J V & Preece M A. British 1990 reference centiles for height, weight, body mass index and head circumference fitted by maximum penalized likelihood. *Statistics in Medicine 1996* (submitted).
Healy M J R, Rasbash J & Yang M (1988). Distribution-free estimation of age-related centiles. *Annals of Human Biology* 15: 17–22.

### *Body mass index references*

Cole T J (1990). Weight–Stature indices to measure underweight, overweight and obesity in anthropometric classification of nutritional status. In *Anthropometric Assessment of Nutritional Status* (ed J Himes) pp 83–111. New York: Wiley-Liss Inc.
Cronk C E, Roche A F et al. (1982). Longitudinal trends and continuity in Wt/Stature$^2$ from 3 months to 18 years. *Human Biology* 54: 729–749.
Garrow J S & Webster J (1985). Quetelet's index (W/H$^2$) as a measure of fatness. *International Journal of Obesity* 9: 147–153.
Hammer L D (1991). Standardized percentile curves of Body Mass Index for children and adolescents. *American Journal of Diseases in Childhood* 145: 259–263.
Keys A, Fidanza F, Karvonen M J, Kimura N & Taylor H L (1972). Indices of relative weight and obesity. *Journal of Chronic Disease* 25: 329–343.
Must A, Dallal G E & Dietz W H (1991). Reference data for obesity: 85th and 95th percentiles of body mass index (wt/ht$^2$). *American Journal of Clinical Nutrition* 54: 773.
Roche A F, Siervogel R M, Chumlea W B & Webb P (1981). Grading body fatness from limited anthropometric data. *American Journal of Clinical Nutrition* 34: 2831–2838.
Rolland–Cachera M F, Cole T J, Sempé M, Tichet J, Rossignol C & Charraud A (1991). Body mass index

variations: centiles from birth to 87 years. *European Journal of Clinical Nutrition* 45: 13–21.

Rolland-Cachera M F, Sempé M, Guilloud-Bataille M, Patois E, Péquignot- Guggenbuhl F, Fautrad V (1982). Adiposity indices in children. *American Journal of Clinical Nutrition* 36: 178–184.

***Skinfolds***

Hammond W H (1955). Measurement and interpretion of subcutaneous fat, with norms for children and young adult males. *British Journal of Preventive and Social Medicine* 9: 201–211.

Harrison G G, Buskirk E R, Carter J E L, Johnston F E, Lohman T G, Pollock M L, Roche A F & Wilmore J (1988). Skinfold thicknesses and measurement technique. In *Anthropometric Standardization Reference Manual* (eds T G Lohman, A F Roche & R Martorell) pp 55–70. Champaign, Illinois: Human Kinetics Publishers.

***Skeletal development***

Acheson R M (1966). Maturation of the skeleton. In *Human Development* (ed F Falkner) pp 465–502. Philadelphia & London: W B Saunders.

Greulich W W & Pyle S I (1950). *Radiographic Atlas of Skeletal Development of the Hand and Wrist*. Stanford: Stanford University Press.

Tanner J M, Whitehouse R H, Cameron N, Marshall W A, Ilealy M J R & Goldstein H (1983). *Assessment of Skeletal Maturity and Prediction of Adult Height (TW2 Method)* 2nd edn. London: Academic Press.

***Growth software***

Healy M J R, Rasbash J, Yang M (1988). Distribution-free estimation of age-related centiles. *Annals of Human Biology* 15: 17–22.

Cole T J & Green P J (1992). Smoothing reference centile curves: the LMS method and penalized likelihood. *Statistics in Medicine* 11: 1305–1319.

Jolicoeur P, Pontier J & Abidi H (1992). Asymptotic models for the longitudinal growth of human stature. *American Journal of Human Biology* 4:461–468.

***Alternative measures for assessing linear growth***

Snyder R G, Schneider L W, Owings C L, Reynolds H M, Golomb D H & Schork M A (1977). Anthropometry of infants, children and youths to age 18 for product safety design. *Highway Safety Research Institute Report* LIM-HSRI-77-17. Michigan: Ann Arbor.

***Height and weight references***

Hamill P V V, Drizd T, Johnson C L, Reed R B & Roche A F (1977). *NCHS Growth Curves for Children, Birth–18 Years, United States*. DHEW Pub (PHS) 78–1650. Hyattsville, Maryland: National Center For Health Statistics.

Waterlow J C, Buzina A, Keller W, Lane J M, Nichaman M Z & Tanner J M (1977). The presentation and use of height and weight data for comparing the nutritional status of groups of children under the age of 10 years. *Bulletin of the World Health Organization* 55: 489–498.

***Weights defining small-for-gestational-age in twins***

Blair E & Stanley F (1985). *Intra-uterine Growth Charts*. Fyshwick, ACT: Canberra Publishing & Printing Co.

Henderson-Smart D J (1995). Post-natal consequences of chronic intra-uterine compromise. *Reproductive Fertility Development* 7: 559–565.

Levy F, Hay D, McLaughlin M, Wood C & Waldman I (1996).Twin sibling differences in parental reports of ADHD, speech, reading and behavioural problems. *Journal of Child Psychology and Psychiatry and Allied Disciplines* 37: 569–578.

Powers W F, Kiely JL & Fowler M G (1995). The role of birth weight, gestational age, race and other infant characteristics in twin intra-uterine growth and infant mortality. In *Multiple Pregnancy* (eds L G Keith, E Papiernik, D M Keith & B Luke) pp 163–174. New York: Parthenon Publishing.

**The use of growth references**

Cole T J (1993). The use and construction of anthropometric growth reference standards. *Nutrition Research Reviews* 6: 19–50.

Cole T J (1990). *UK 1990 Boys Height and Weight Centiles*. London: Child Growth Foundation.

Preece M A, Law C M & Davies P S W (1986). The growth of children with chronic paediatric disease. *Clinics in Endocrinology and Metabolism* 15: 453–477.

Tremlett G (1991). Making growth monitoring more effective. In *Diseases of Children in the Subtropics and Tropics* (eds P Stanfield, M Brueton, M Chan, M Parkin & T Waterston) p 274. London: Edward Arnold.

***Screening for growth failure in industrialized countries***

Ahmed M L, Allen A D, Sharma A, Macfarlane J A & Dunger D B (1993). Evaluation of a district growth screening programme: the Oxford growth study. *Archives of Disease in Childhood* 69: 361–365.

Cernerud L & Edding E (1994). The value of measuring height and weight of schoolchildren. *Paediatric and Perinatal Epidemiology* 8: 365–372.

Greco L, Power C & Peckharn C (1995). Adult outcome of normal children who are short or underweight at age 7 years. *British Medical Journal* 310: 696–700.

Grumbach M M (1988). Growth hormone therapy and the short end of the stick. *New England Journal of Medicine* 319: 238–241.

Hakama M (1991). Screening. In *Oxford Textbook of Public Health* (eds W W Holland, R Detels & C Knox) vol 3 pp 91–106. Oxford: Oxford University Press.

Voss L D, Mulligan J, Betts P R & Wilkin T J (1992). Poor growth in school entrants as an index of organic disease: the Wessex growth study. *British Medical Journal* 305: 1400–1402.

Wilson J & Jungner C (1968) Principles and practice of screening for disease.
*World Health Organization Public Health Paper* 34. Geneva: WHO.

***Screening for growth failure in developing countries***

Victoria C G (1992). The association between wasting and stunting: an international perspective. *Journal of Nutrition* 122: 1105–1110.

Waterlow J C (ed) (1988). *Linear Growth Retardation in Less Developed Countries*. New York: Raven Press.

Waterlow J C & Schürch B (eds) (1994). Causes and mechanisms of linear growth retardation. *European Journal of Clinical Nutrition* 48, Suppl 1.

***Screening and surveillance during famine***

ACC/SCN (1989). *Suggested Approaches to Nutritional Surveillance with Particular Reference to Structural Adjustment. A Report to the Management Committee of the Inter-Agency Food and Nutritional Surveillance Programme from the Chairman of the ACC/SCN*. Geneva.

Beaton G et al. (1990). Appropriate uses of anthropometric indices in children. A report based on an ACC/SCN workshop. *ACC/SCN state of the art series, nutrition policy discussion paper* 7. Geneva: UN, ACC/SCN.

Bern C & Nathanail L (1995). Is mid-upper-arm circumference a useful tool for screening in emergency situations? An evaluation among Rwandan refugee children in Zaire. *Lancet* 345 (8950): 631–633.

Habicht J-P (1980). Some characteristics of indicators of nutritional status for use in screening and surveillance. *American Journal of Clinical Nutrition* 33 (3): 531–535.

Ismail S J (1991). Nutritional surveillance: experiences in developing countries. *Proceedings of the Nutrition Society* 50: 673–679.

Lawrence M et al. (1994). Nutritional status and early warning of mortality in Southern Ethiopia 1988–1991. *European Journal of Clinical Nutrition* 48: 38–45.

Médecins sans Frontières (1995). *Nutrition Guidelines* 1st edn. Paris: MSF.

Seaman J (1991). Management of nutrition relief for famine-affected and displaced populations. *Tropical Doctor* 21 (suppl 1): 38–42.

Shoham J (1987). Does nutritional surveillance have a role to play in early warning of food crisis and in the management of relief operations? *Disasters* 11/4/1987.

UN (1975). *Report of the World Food Conference, Rome 5–6 November 1974*. Publication E/Conf.65/20. New York: UN.

de Ville de Goyet C (1978). *The Management of Nutritional Emergencies in Large Populations*. Geneva: WHO.

WHO (1976). Methodology of nutritional surveillance. Report of a joint FAO/UNICEF/WHO Expert Committee. *Technical Report Series* No. 593. Geneva: WHO.

WHO (1986). Use and interpretation of anthropometric indicators of nutritional status. Report of a WHO working group. *Bulletin of the World Health Organization* 64(6):929–941.

***Screening and surveillance during warfare***

Robertson A, Fronczak N, Jaganjac N, Hailey P, Copeland P & Duprat M. Nutrition and immunization survey of Bosnian women and children. *International Journal of Epidemiology* (in press).

Robertson A & James W P T. War in the former Yugoslavia: coping with nutritional issues. In *Essentials of Human Nutrition* (eds J I Mann & A S Truswell).

Vespa J & Watson F. Who are the nutritionally vulnerable in Bosnia–Hercegovina? *British Medical Journal* (in press).

Watson F & Vespa J. The impact of a reduced and uncertain food supply in three besieged cities of Bosnia–Hercegovina. *Disasters* (in press).

Watson F & Vespa J. Selected papers from a seminar in Sarajevo. *European Journal of Clinical Nutrition*. (Suppl.)

WHO Annual Report 1994. WHO Mission in the Former Yugoslavia. Regional Office for Europe. Zagreb Area Office. 1995.

## Part two: Patterns of human growth

Bogin B (1995). Growth and development: recent evolutionary and biocultural research. In *Biological Anthropology: The State of The Science* (eds N T Boaz & L D Wolfe) pp 49–70. Bend, Oregon: International Institute for Human Evolutionary Research.

Bogin B & Smith B H (1996). Evolution of the human life cycle. *American Journal of Human Biology* (in press).

Bonner J T (1965). *Size and Cycle*. Princeton: Princeton University Press.

Galdikas B M & Wood J W (1990). Birth spacing patterns in humans and apes. *American Journal of Physical Anthropology* 83: 185–191.

Goodall J (1983). Population dynamics during a 15-year period in one community of free-living chimpanzees in the Gombe National Park, Tanzania. *Zeitschrift für Tierpsychologie* 61: 1–60.

Halpern C T, Udry R J, Campbell B & Suchinddran C (1993). Testosterone and pubertal development as predictors of sexual activity: a panel analysis of adolescent males. *Psychosomatic Medicine* 55: 436–447.

Lancaster J B & Lancaster CS (1983). Parental investment: the hominid adaptation. In *How Humans Adapt* (ed D J Ortner) pp 33–65. Washington, DC: Smithsonian Institution Press.

Pereira M E & Altmann J (1985). Development of social behavior in free-living nonhuman primates. In *Nonhuman Primate Models for Human Growth and Development* (ed A S Watts) pp 217–309. New York: Alan R Liss.

Shea B T (1990). Dynamic morphology: growth, life history, and ecology in primate evolution. In *Primate Life History and Evolution* (ed C J DeRousseau) pp 325–352. New York: Wiley-Liss.

Smith B H (1992). Life history and the evolution of human maturation. *Evolutionary Anthropology* 1: 134–142.

Watts E S & Gavan J A (1982). Postnatal growth of non-human primates: the problem of the adolescent spurt. *Human Biology* 54: 53–70.

Weisner T S (1987). Socialization for parenthood in sibling caretaking societies. In *Parenting Across the Life Span: Biosocial Dimensions* (eds J B Lancaster, J Altmann, A S Rossi & L R Sherrod) pp 237–270. New York: Aldine de Gruyter.

Worthman C (1986). Developmental dyssynchrony as normative experience: Kikuyu adolescents. In *School-age Pregnancy and Parenthood: Biosocial Dimensions* (eds J Lancaster & B A Hamburg) pp 95–112. New York: Aldine de Gruyter.

**Comparative growth and development of mammals**
Brody S (1964). *Bioenergetics and Growth*. New York: Hafner Publishing Company.
Grether W F & Yerkes R M (1940). Weight norms and relations for chimpanzee. *American Journal of Physical Anthropology* 27: 181–197.
Reeds P J & Fiorotto M L (1990). Growth in perspective. *Proceedings of the Nutrition Society* 49: 411–420.
Richards A F (1985). *Primates in Nature*. New York: W H Freeman.

***The allometry of basal metabolic rate***
Brody S (1945). *Bioenergetics and Growth*. New York: Reinhold.
Huxley J S (1932). *Problems of Relative Growth*. London: Methuen.
Kleiber M (1975). *The Fire of Life*. Revised edn. New York: Kleiber.

***Birth-weight, gestation length and relative maturity***
Peters R H (1988). Allometric comparisons of growth, reproduction and nutrition. In *Comparative Nutrition* (eds K Blaxter & I Macdonald) pp 1–19. London: John Libbey.
Reeds P J & Fiorotto M L (1990). Growth in perspective. *Proceedings of the Nutrition Society* 49: 411–420.

***Locomotion***
Alexander R McN (1992). *The Human Machine*. London: Natural History Museum and New York: Columbia University Press.

***Mammalian life-history***
Harvey P H, Martin R D & Clutton-Brock T H (1987). Life histories in comparative perspective. In *Primate Societies* (eds B B Smuts et al.) pp 181–196. Chicago: University of Chicago Press.
Miller J S (1981). Pre-partum reproductive characteristics of eutherian mammals. *Evolution* 35: 1149–1163.

***Allometry: a brief history***
Brody S (1945). *Bioenergetics and Growth*. New York: Reinhold.
Huxley J S (1932). *Problems of Relative Growth*. London: Methuen.
Kleiber M (1975). *The Fire of Life*. Revised edn. New York: Kleiber.

**Growth of non-human primates**
Charnov E L & Berrigan D (1993). Why do female primates have such long lifespans and so few babies? Or life in the slow lane. *Evolutionary Anthropology* 1: 191–194.
DeRousseau C J (ed) (1990). *Primate Life History and Evolution*. New York: Wiley-Liss.
Janson C R & van Schaik C P (1993). Ecological risk aversion in juvenile primates: slow and steady wins the race. In *Juvenile Primates* (eds M E Pereira & L A Fairbanks) pp 57–74. New York: Oxford University Press.
Kirkwood L K (1985). Patterns of growth in primates. *Journal of Zoology* 205: 123–136.
Leigh S K (1994). Ontogenetic correlates of diet in anthropoid primates. *American Journal of Physical Anthropology* 94: 499–522.

***Primate life-history***
DeRousseau C J (ed) (1988). *Primate Life History and Evolution*. New York: Wiley-Liss.
Martin R D (1993). *Primate Origins*. London: Chapman & Hall.
Lee P C, Majluf P & Gordon I J (1991). Growth, weaning and maternal investment from a comparative perspective. *Journal of Zoology* 225: 99–114.
Pianka E R (1983). *Evolutionary Ecology*. New York: Harper & Row.
Ross C & MacLamon A M (1995). Ecological and social correlates of maternal expenditure on infant growth in primates. In *Motherhood in Human and Nonhuman Primates* (eds C K Pryce, R D Martin & D Skuse) pp 37–46. Basel: Karger.
Ross C (1992) The intrinsic rate of natural increase and reproductive effort in primates. *Oecologia* 90: 383–390.
Stearns S C (1993). *The Evolution of Life Histories*. Oxford: Oxford University Press.

**Human growth from an evolutionary perspective**
Bogin B (1993). Why must I be a teenager at all? *New Scientist* 137: 34–38.
Bogin B (1995). Growth and development: recent evolutionary and biocultural research. In *Biological Anthropology: the State of the Science* (eds N T Boaz & L D Wolfe) pp 49–70. Bend, Oregon: International Institute for Human Evolutionary Research.
Deacon T W (1992). The human brain. In *The Cambridge Encyclopedia of Human Evolution* (eds S Jones et al.) pp 115–123. Cambridge: Cambridge University Press.
Goodall J (1983). Population dynamics during a 15-year period in one community of free-living chimpanzees in the Gombe National Park, Tanzania. *Zeitschrift für Tierpsychologie* 61: 1–60.
Harvey P et al. (1987). Life histories in comparative perspective. In *Primate Societies* (eds B Smuts et al.) pp 181–196. Chicago: University of Chicago Press.
Martin R D (1983). *Human Brain Evolution in an Ecological Context. Fifty-second James Arthur Lecture*. New York: American Museum of Natural History.
Nishida T et al. (1990). Demography and reproductive profiles. In *The Chimpanzees of the Mahale Mountains: Sexual and Life History Strategies* (ed T Nishida) pp 63–97. Tokyo: University of Tokyo Press.
Smith B H (1993). Physiological age of KMN-WT 15000 and its significance for growth and development of early Homo. In *The Nariokotome Homo Erectus Skeleton* (eds A C Walker & R F Leakey) pp 195–220. Cambridge, Massachusetts: Belknap Press.

***Neanderthal growth and development***
Aiello L & Dean M C (1990). *An Introduction to Human Evolutionary Anatomy*. London: Academic Press.
Anemone R L, Mooney M P & Siegel M I (1996). Longitudinal study of dental development in chimpanzees of known chronological age: implications for understanding the age at death of Plio-Pleistocene Hominids. *American Journal of Physical Anthropology* 99: 119–133.
Bilsborough A & Thompson J L (1996). Dentition of the Le Moustier 1 Neanderthal. *American Journal of Physical Anthropology* (Suppl) 21: 69.
Creed-Miles M, Rosas A, & Kruszynski R (1996). Issues in the identification of Neanderthal derivative traits at early post-natal stages. *Journal of Human Evolution* 30: 147–153.
Dettwyler K A (1995). A time to wean: the hominid blueprint for the natural age of weaning in modern human populations. In *Breastfeeding: Biocultural Perspectives* (eds P Stuart-Macadam & K A Detwyller) pp 39–73. New York: Aldine de Gruyter.
Lee P C, Majluf P & Gordon I J (1991). Growth, weaning and maternal investment from a comparative perspective. *Journal of Zoology* 225: 99–114.
Makiko K (1995) Reconstructing the growth pattern of the Dederiyeh Neanderthal. In *Dederiyeh: Neanderthals and Modern Humans* pp 32–35. Tokyo: University Museum, University of Tokyo.
Mann A, Lampl M, & Monge J M (1996). The evolution of childhood: dental evidence for the appearance of human maturation patterns. *American Journal of Physical Anthropology* (Suppl) 21: 156.
Minugh-Purvis N (1988). Patterns of craniofacial growth and development in Upper Pleistocene hominids. Ph. D. dissertation, University of Pennsylvania.
Rak Y, Kimbel W H, & Hovers E (1994). A Neanderthal infant from Amud Cave, Israel. *Journal of Human Evolution* 26: 313–324.
Rosenberg K R (1992). The evolution of modern human childbirth. *Yearbook of Physical Anthropology* 35: 89–124.
Smith B H (1991). Dental development and the evolution of Life History in Hominidae. *American Journal of Physical Anthropology* 86: 157–174.
Smith B H & Tompkins R L (1995). Toward a Life History of the Hominidae. *Annual Review of Anthropology* 24: 257–279.
Stringer C B, Dean M C & Martin R D (1990). A comparative study of cranial and dental development within a recent British sample and among Neandertals. In *Primate Life History and Evolution* (ed C J DeRousseau) pp 115–152. New York: Wiley-Liss.
Thompson J L (1995). Terrible teens: the use of adolescent morphology in the interpretation of Upper Pleistocene human evolution. *American Journal of Physical Anthropology* (Suppl) 20: 210.
Thompson J L & Bilsborough A (1996). Le Moustier 1: characteristics of a late western European Neanderthal. *American Journal of Physical Anthropology* (Suppl) 21: 229.
Thompson J L & Bilsborough A (1997). The current state of the Le Moustier 1 skull. *Acta Praehistorica et Archaeologica* 29.
Thompson J L & Nelson A J (1997). Relative postcranial development of Upper Pleistocene hominids compared to human population variation. *Journal of Human Evolution* 32: A23–24.
Tillier A-M (1986). Quelques aspects de l'ontogense du squelette cranien des Neandertaliens. *Anthropos (Brno)* 23: 207–216.
Tompkins RL (1996). Relative dental development of Upper Pleistocene hominids compared to human population variation. *American Journal of Physical Anthropology* 99: 103–118.
Trinkaus E (1983). *The Shanidar Neanderthals*. New York: Academic Press.
Trinkaus E & Tompkins R L (1990). The Neanderthal life cycle: the possibility, probability, and perceptibility of contrasts with recent humans. In *Primate Life History and Evolution* (ed C J DeRousseau) pp 153–180. New York: Wiley-Liss.
Vlcek E (1964). Einige in der Ontogenese des modernen Menschen untersuchte Neandertalmermale. *Zeitschrift für Morphologie und Anthropologie* 56: 63–83.
Wolpoff M H (1978). The dental remains from Krapina. In *Krapinski Pracvojek i Evolucija Hominida* (ed M Malez) pp 119–144. Zagreb: JAZU.

**The human growth curve**
Preece M A & Henrich I (1981). Mathematical modelling of individual growth curves. *British Medical Bulletin* 37: 247–252.
Karlberg J (1987). On the modelling of human growth. *Statistics in Medicine* 6: 185–192.
Karlberg J (1989). A biologically-oriented mathematical model (ICP) for human growth. *Acta Paediatrica Scandinavica* 350: 70–94.
Steutzle W, Gasser T H, Molinari L, Largo R H, Prader A & Huber P J (1980). Shape-invariant modelling of human growth. *Annals of Human Biology* 7: 507–528.
Karlberg J, Engström I, Karlberg P & Fryer J G (1987). Analysis of linear growth using a mathematical model I. From birth to three years. *Acta Paediatrica Scandinavica* 76: 478–488.
Karlberg J, Fryer J G, Engström I, Karlberg P (1987). Analysis of linear growth using a mathematical model II. From three to twenty-one years of age. *Acta Paediatrica Scandinavica* 337: 12–29.
Gasser T, Mueller H G, Köhler W, Prader A, Largo R & Molinari L (1985). An analysis of the mid-growth and adolescent spurts of height based on acceleration. *Annals of Human Biology* 12: 129–148.
Burt C (1937). *The Backward Child*. New York: Appleton-Century.
Karlberg J (1990). The infancy-childhood growth spurt. *Acta Paediatrica Scandinavica* 367: 111–118.
Karlberg J, Jalil F, Lam B, Low L & Yeung C Y (1994). Linear growth retardation in relation to the three phases of growth. *European Journal of Clinical Nutrition* 48 (Suppl 1): S25–S44.

***Curve-fitting***
Gasser T, Köhler W, Müller H-G, Kneip A, Largo R, Molinari L & Prader A (1984). Velocity and acceleration of height growth using kernel estimation. *Annals of Human Biology* 11: 397–411.
Hauspie R C (1989). Mathematical models for the study of individual growth patterns. *Rev. Epidém. et Santé Publ.* 37: 461–476.
Hauspie R C & Chrzastek-Spruch H (1993). In *Kinanthropometry IV* (eds W Duquet & J A P Day) pp 68–83. London: E & F N Spon.
Jolicoeur P, Pontier J, & Abidi H (1992). Asymptotic models for the longitudinal growth of human stature. *American Journal of Human Biology* 4: 461–468.

Largo R H, Gasser T, Prader A, Stuetzle W & Huber P J (1978). Analysis of the adolescent growth spurt using smoothing spline functions. *Annals of Human Biology* 5: 421–434.

Marubini E & Milani S (1986). In *Human Growth: A Comprehensive Treatise* (eds F Falkner & J M Tanner) 2nd edn, vol 3 pp 79–94. New York: Plenum.

Preece M A & Baines M K (1978). A new family of mathematical models describing the human growth curve. *Annals of Human Biology* 5: 1–24.

***A technique for analysing the growth of population sub-groups***

Brush G (1995). Growth variation and comparative growth homeostasis. In *Human Populations, Diversity and Adaptations* (eds A J Boyce & V Reynolds) pp 122–137. Oxford: Oxford University Press.

Brush G, Harrison G A & Zumrawi F Y (1993). A path analysis of some determinants of infant growth in Khartoum. *Annals of Human Biology* 20: 381–387.

Goldstein H (1995). The multilevel analysis of growth data. In *Essays on Auxology* (eds R Hauspie, G Lindgren & F Falkner) pp 89–99. Welwyn Garden City: Castlemead Publications.

Harrison G A. & Brush G (1990). On correlations between adjacent velocities and acclerations in longitudinal growth data. *Annals of Human Biology* 17: 55–57.

Harrison G A, Brush G & Zurnrawi F Y (1993). Motherhood and infant health in Khartoum. *Bulletin of the World Health Organization* 71: 529–533.

Smith D W, Truog W, Rogers J E, Greitzer L J, Skinner A L, McCann J J & Harvey M A S (1976). Shifting linear growth during infancy – illustration of genetic factors in growth from fetal life through infancy. *Journal of Pediatrics* 89: 225–230.

Tanner J M (1986). Growth as a target-seeking function. In *Human Growth: A Comprehensive Treatise* (eds F Falkner & J M Tanner) 2nd edn, vol 1 pp 167–179. New York: Plenum.

Zurmawi F Y (1991). Effects of the relative importance of different factors and their degree of interactions on child growth. *Journal of Tropical Pediatrics* 37: 131–135.

## Part three: Genetics of growth

Bailey S M & Garn S M (1986). The genetics of maturation. In *Human Growth: A Comprehensive Treatise* (eds F Falkner & J M Tanner) 2nd edn, vol 3 pp 169–195. New York: Plenum.

Bouchard C, Despres J P & Tremblay A (1991). Genetics of obesity and human energy metabolism. *Proceedings of the Nutrition Society* 50: 139–147.

Dauncey M J (1995). From whole body to molecule: an integrated approach to the regulation of metabolism and growth. *Thermochimica Acta* 250: 305–318.

Dauncey M I, Burton K A, Vffite P, Harrison A P, Gilmour R S, Duchamp C, Cattaneo D & Straus D S (1994). Nutritional regulation of hormones and growth factors that control mammalian growth. *FASEB Journal* 8: 6–12.

Lewis E B (1978). A gene complex controlling segmentation in Drosophila. *Nature* 276: 565–570.

Little R E & Sing C F (1987). Genetic and environment influences on human birthweight. *American Journal of Human Genetics* 40: 512–526.

Maffei M, Halaas J, Ravussin E, Pratley R E, Lee G H, Zhang Y, Fei H, Kim S, Lallone R, Ranganathan S, Kern P A & Friedman J M (1995). Leptin levels in human and rodent: measurement of plasma leptin and of RNA in obese and weight-reduced subjects. *Nature Medicine* 1: 1155–1161.

Muragaki Y, Mundlos S, Upton J & Olsen B R (1996). Altered growth and branching patterns in synpolydactyly caused by mutations in HOX13. *Science* 272: 548–550.

Ratcliffe S G (1981). The effect of chromosome abnormalities on human growth. *British Medical Bulletin* 37: 291–295.

Roberts D F (1986). The genetics of human fetal growth. In *Human Growth: A Comprehensive Treatise* (eds F Falkner & J M Tanner) 2nd edn, vol 3 pp 113–143. New York: Plenum.

Sharpe P (1996). HOX gene mutations – the wait is over. *Nature Medicine*.

Thissen J-P, Ketelslegers J-M & Underwood L E (1994). Nutritional regulation of the insulin-like growth factors. *Endocrinology* 15: 80–101.

**Genetics of child growth**

Bailey S M & Garn S M (1986). The genetics of maturation. In *Human Growth: A Comprehensive Treatise* (eds F Falkner & J M Tanner) 2nd edn, vol 3 pp 169–195. New York: Plenum.

Byard P L, Guo S & Roche A F (1993). Family resemblance for Preece–Baines growth curve parameters in the Fels Longitudinal Growth Study. *American Journal of Human Biology* 5: 151–157.

Eveleth P B (1986). Population differences in growth. In *Human Growth: A Comprehensive Treatise* (eds F Falkner & J M Tanner) 2nd edn, vol 3 pp 221–239. New York: Plenum.

Hauspie R C, Bergman R, Bielicki T & Susanne C (1994). Genetic variance in the pattern of the growth curve for height: a longitudinal analysis of male twins. *Annals of Human Biology* 21: 347–362.

Hauspie R C, Das S K, Preece M A & Tanner J M (1982). Degree of resemblance of the pattern of growth among siblings in families of West Bengal (India). *Annals of Human Biology* 9: 171–174.

Mueller W H (1986). The genetics of size and shape in children and adults. In *Human Growth: A Comprehensive Treatise* (eds F Falkner & J M Tanner) 2nd edn, vol 2 pp 145–168. New York: Plenum.

Ratcliffe S G (1981). The effect of chromosome abnormalities on human growth. *British Medical Bulletin* 37: 291–295.

Ratcliffe S G (1995). The ontogenesis of sex chromosomal effects on growth. In *Essays on Auxology Presented to James Mourilyan Tanner* (eds R Hauspie, G Lindgren & F Falkner) pp 480–488. Welwyn Garden City: Castlemead Publications.

Roberts D F (1986). The genetics of human foetal growth. In *Human Growth: A Comprehensive Treatise* (eds F Falkner & J M Tanner) 2nd edn, vol 3 pp 113–143. New York: Plenum.

Roberts DF (1994). Genetics of growth. In *Auxology '94* (ed O G Eiben). *Humanbiol. Budapest*. 25: 23–30.

Susanne C (1986). Bases génétiques. In *L'Homme, son évolution, sa diversité* (eds D Ferembach, C Susanne & M-C Chamnla) pp 3–16. Paris: Doin Editeurs.

Tanner J M (1978). Interaction of hereditary and environmental factors in controlling growth. In *Education and Physical Growth* (ed J M Tanner) 2nd edn pp 90–114. London: Hodder & Stoughton.

***Genetics and energy metabolism***

Bogardus C, Lilloja E, Ravussin E, Abbott W, Zawadzki J K, Young A, Knowler W C, Jacobowitz R & Moll P (1986). Familial dependence of the resting metabolic rate. *New England Journal of Medicine* 315: 96–100.

Bouchard C, Despres J P & Tremblay A (1991). Genetics of obesity and human energy metabolism. *Proceedings of the Nutrition Society* 50: 139–147.

Bouchard C, Tremblay A, Nadeau A, Despres J P, Theriault G, Boulay M R, Lortie G, Leblanc C & Fournier G (1989). Genetic effect in resting and exercise metabolic rates. *Metabolism* 38: 364–370.

Fontaine E, Savard R, Tremblay A & Bouchard C (1985). Resting metabolic rate in monozygotic and dizygotic twins. *Acta Geneticae Medicae et Gemellologiae (Roma)* 34: 41–47.

Griffiths M & Payne P R (1976). Energy expenditure in small children of obese and non-obese parents. *Nature* 260: 698–699.

Henry C J K, Piggott S M, Rees D G, Priestley L & Sykes B (1990). Basal metabolic rate in monozygotic and dizygotic twins. *European Journal of Clinical Nutrition* 44: 717–723.

Perusse L, Tremblay A, Leblanc C & Bouchard C (1989). Genetic and environmental influences on level of habitual physical activity and exercise participation. *American Journal of Epidemiology* 129: 1012–1022.

***Genetics and body fatness***

Bodurtha J N, Mosteller M, Hewitt J K, Nance W E, Eaves U, Moskowitz W B, Katz S & Schieken R M (1990). Genetic analysis of anthropometric measures in 11-year-old twins: the Medical College of Virginia Twin Study. *Pediatric Research* 28: 1–4.

Bouchard C (ed) (1994). *The Genetics of Obesity*. Boca Raton, Florida: CRC Press.

Brook C G D, Huntley R M C & Slack J (1975). Influence of heredity and environment in determination of skinfold thicknesses in children. *British Medical Journal* 2: 719–721.

Cardon L R (1994). Height, weight, and obesity. In *Nature and Nurture During Middle Childhood*, (eds J C DeFries, R Plomin & D W Fulker). Oxford: Blackwell Publishers.

Harsha D W, Voors A W & Berenson G S (1980). Racial differences in subcutaneous fat patterns in children aged 7–15 years. *American Journal of Physical Anthropology* 53: 333–337.

Meyer J M, Bodurtha J N, Eaves U, Hewitt J K, Mosteller M, Nance W E, Neale M C & Schieken R M (1996). Tracking genetic and environmental effects on adolescent BMI: the MCV Twin Study (MS).

Neale M C & Cardon L R (1992). *Methodology for Genetic Studies of Twins and Families*. Dordrect: Kluwer Academic Publishers.

Price R A & Gottesman H (1991). Body fat in identical twins reared apart: roles for genes and environment. *Behav. Genet.* 21: 1–7.

Sørensen T I A, Holst C & Stunkard A J (1992). Childhood body mass index – genetic and familial environmental influences assessed in a longitudinal adoption study. *International Journal of Obesity* 16: 705–714.

Stunkard A J, Sørenson T I A, Hannis C, Teasdale T W, Chakraborty R, Schull W J & Schulsinger F (1986). An adoption study of human obesity. *New England Journal of Medicine* 314: 193–198.

Towne B, Roche A F, Chumlea W C & Siervogel R M (1996). Genetics of blood pressure and body mass index before and after puberty. *Circulation* (in press).

***Sex steroids and growth hormone-releasing hormone gene expression***

Argente J, Evain-Brion D, Gilmour R S (1994). Regulation of porcine insulin-like growth factor I and growth hormone receptor mRNA. *Journal of Clinical Endocrinology and Metabolism* 63 (3): 680–682.

Argente J, Chowen J A, Zeitler P, Clifton D K & Steiner RA (1991). Sexual dimorphism of growth hormone-releasing hormone and somatostatin messenger RNA levels in the rat hypothalamus over the course of development. *Endocrinology* 128: 2369–2376.

Argente J & Chowen J A (1993). Control of the transcription of growth hormone-releasing hormone and somatostatin genes by sex steroids. *Hormone Research* 40: 48–53.

Chowen J A, Argente J, González-Parra S & García-Segura L M (1993). Differential effects of the neonatal and adult sex steroid environments on the organization and activation of hypothalamic growth hormone-releasing hormone and somatostatin neurons. *Endocrinology* 133: 2792–2802.

Chowen J A, Argente J, González-Parra S, Torres-Aleman I & García-Segura L M (1993). Effects of the neonatal sex steroid environment on growth hormone-releasing hormone and somatostatin gene expression. *Journal of Pediatric Endocrinology* 6: 211–218.

Evain-Brion D, Donnadieu M, Liapi C, Argente J, Tonon M C, Garnier P & Job J C (1986). Plasma growth hormone releasing factor levels in children: physiological and pharmacological induced variations. *Hormone Research* 24: 116–120.

Gaylinn B D, von Kap-Herr C, Golden W & Thorner M O (1994). Assignment of the human growth hormone-releasing hormone receptor gene (GHRHR) to 7p14 by in situ hybridization. *Genomics* 19: 193–195.

Guillemin R, Brazeau P, Bohlen P, Esch F, Ling N & Wehrenberg W B (1982). Growth hormone releasing factor from a pancreatic tumor that caused acromegaly. *Science* 218: 585–587.

Jansson J O, Edén S & Isaksson O (1985). Sexual dimorphism in the control of growth hormone secretion. *Endocrine Reviews* 6: 128–150.

Jansson J O, Ishikawa K, Katakami H & Frohman L A (1987). Pre- and post-natal developmental changes in hypothalamic content of rat growth hormone releasing factor. *Endocrinology* 120: 525–530.

Mayo K E, Cerelli G M, Rosenfeld M G & Evans R M (1983). Characterization of cDNA and genomic clones encoding the precursor to rat hypothalamic growth-hormone releasing factor. *Nature* 314: 464–467 .

Mayo K E (1992). Molecular cloning and expression of a pituitary-specific receptor for the growth hormone-releasing hormone. *Molecular Endocrinology* 6: 1733–1744.

Meister B, Hökfelt T, Johansson O & Hulting A-L (1987). Distribution of growth hormone releasing factor, somatostatin and coexisting messengers in the brain. In *Growth Hormone: Basic and Clinical Aspects* (eds O Isaksson, C Binder, K Hall & T Hökfelt) pp 29–52. Amsterdam: Excerpta Medica.

Pezzolo A, Gimelli G, Sposito M, Giussani U, Rossi E & Suffardi O (1994). Definitive assignment of the growth hormone releasing factor gene to 20q11.2. *Human Genetics* 93: 213–214.

Rivier J, Spiess J, Thorner M & Vale W (1982). Characterization of a growth hormone releasing factor from a human pancreatic islet tumour. *Nature* 300: 276–278.

Shirasu K, Stumpf W E & Sar M (1990). Evidence for direct action of estradiol on growth hormone releasing factor (GRIF) in rat hypothalamus: localization of [$^3$H] estradiol in GRF neurons. *Endocrinology* 127:344–349.

Zeitier P, Argente J, Chowen-Breed J A, Clifton D K & Steiner R A (1990). Testosterone regulation of growth hormone-releasing hormone messenger RNA levels in hypothalamic neurons of the male rat. *Endocrinology* 127:1362–1368.

#### Genetic and environmental influences on fetal growth

Kline J et al. (1989). *Conception to Birth. Epidemiology of Pre-natal Development*. New York: Oxford University Press.

Kramer M S (1987). Intra-uterine growth and gestational age determinants. *Pediatrics* 80: 502–511.

Little R E & Sing C F (1987). Genetic and environment influences on human birthweight. *American Journal of Human Genetics* 40: 512–526.

Roberts D F (1986). The genetics of human fetal growth. In *Human Growth: A Comprehensive Treatise* (eds F Falkner & J M Tanner) 2nd edn, vol 3 pp 113–143. New York: Plenum.

#### Homeobox genes

Bateson W (1894). *Materials for the Study of Variation, Treated with Especial Regard to Discontinuity in the Origin of Species*. London: Macmillan.

Burke A C, Nelson C E, Morgan B A & Tabin C (1995). Hox genes and the evolution of vertebrate axial morphology. *Development* 121: 333–346.

Carroll S (1995). Homeotic genes and the evolution of arthropods and chordates. *Nature* 376: 479–485.

Duboule D (ed) (1994). *Guidebook to the Homeobox Genes*. Oxford: Oxford University Press.

Hall B K (ed) (1994). *Homology: The Hierarchical Basis of Comparative Biology*. San Diego: Academic Press.

Hanson I, von Heyningen V (1995). Pax6: more than meets the eye. *Trends in Genetics* 11: 268–272.

Kappen C, Schughart K, Ruddle F H (1993). Early evolutionary origin of major homeodomain sequence classes. *Genomics* 18: 54–70.

Krumlauf R (1994). Hox genes in vertebrate development. *Cell* 78: 191–201.

Krumlauf R (1993). Hox genes and pattern formation in the bronchial region of the vertebrate head. *Trends in Genetics* 9: 106–112.

Manak J R, Scott M P (1994). A class act: conservation of homeodomain protein functions. *Development* (Suppl): 61–71.

Read A P (1995). Pax genes: Paired feet in three camps. *Nature Genetics* 9: 333–334.

Scott M P, Tamkun J W & Hartzell G W (1989). The structure and function of the homeodomain. *Biochimica et Biophysica Acta* 989: 25–48.

Slack J M W, Holland P W H & Graham (1993). The zootype and the phylotypic stage. *Nature* 361: 490–491.

Sordino P, van der Hoeven F & Duboule D (1995). Hox gene expression in teleost fins and the origin of vertebrate digits. *Nature* 375: 678–681.

Stuart E D, Kioussi C & Gruss P (1993). Mammalian Pax genes. *Annual Review of Genetics* 48: 219–236.

#### *Growth hormone gene deletions*

Chen E Y et al. (1989). The human growth hormone locus : nucleotide sequence, biology, and evolution. *Genomics* 4: 479–497.

Cogan J D et al. (1993). Heterogeneous growth hormone (GH) gene mutations in familial GH deficiency. *Journal of Clinical Endocrinology and Metabolism* 76: 1224–1228.

Duquesnoy P et al. (1990). A frameshift mutation causing isolated growth hormone deficiency type IA. *American Journal of Human Genetics* 47: 110A.

Goossens M et al. (1986). Isolated growth hormone (GH) deficiency type IA associated with a double deletion in the human GH gene cluster. *Journal of Clinical Endocrinology and Metabolism* 62: 712–716.

Igarashi Y et al. (1993). A new mutation causing inherited growth hormone deficiency: a compound heterozygote of a 6.7 kb deletion and a two base deletion in the third exon of the GH-1 gene. *Human Molecular Genetics* 2: 1073–1074.

Illig R et al. (1971). Hereditary pre-natal growth hormone deficiency with increased tendency to growth hormone antibody formation ('A-type' of isolated growth hormone deficiency). *Acta Paediatrica Scandinavica* 60: 607.

Kamijo T et al. (1992). Detection of molecular heterogeneiety in GH-1 gene deletions by analysis of polymerase chain reaction amplification products. *Journal of Clinical Endocrinology and Metabolism* 74: 786–789.

Mullis P E et al. (1992). Prevalence of human growth hormone-1 gene deletions among patients with isolated growth hormone deficiency from different populations. *Pediatric Research* 31: 532–534.

Nishi Y et al. (1993). Treatment of isolated growth hormone deficiency type IA due to GH-1 gene deletion with recombinant human insulin-like growth factor I. *Acta Paediatrica* 82: 983–986.

Phillips J A III (1995). Inherited defects in growth hormone synthesis and action. *The Metabolic and Molecular Bases of Inherited Disease* 7: 3023–44.

Phillips J A III et al. (1981). Molecular basis for familial isolated growth hormone deficiency. *Proceedings of the National Academy of Sciences USA* 78: 6372–6375.

Phillips J A III et al. (1994). Molecular basis of familial human growth hormone deficiency. *Journal of Clinical Endocrinology and Metabolism* 78: 11–16.

Vnencak-Jones C L et al. (1988). Molecular basis of human growth hormone gene deletions. *Proceedings of the National Academy of Sciences USA* 85: 5615–5619.

Vnencak-Jones C L et al. (1990). Hot spots for growth hormone gene deletions in homologous regions outside of Alu repeats. *Science* 250: 1745–1748.

Walker J L et al. (1992). Stimulation of statural growth by recombinant insulin-like growth factor I in a child with growth hormone insensitivity syndrome (Laron type). *Journal of Pediatrics* 121: 641–646.

#### Genetic regulation of growth-promoting factors by nutrition

Dauncey M J (1995). From whole body to molecule: an integrated approach to the regulation of metabolism and growth. *Thermochimica Acta* 250: 305–318.

Dauncey M I, Burton K A, Vffite P, Harrison A P, Gilmour R S, Duchamp C & Cattaneo D (1994). Nutritional regulation of growth hormone receptor gene expression. *FASEB Journal* 8: 81–88.

Duchamp C, Burton K A, Herpin P & Dauncey M I (1996). Perinatal ontogeny of porcine growth hormone receptor gene expression is modulated by thyroid status. *European Journal of Endocrinology* (in press).

Pell J M (1996). Regulation of post-natal growth by insulin-like growth factor I and its binding proteins. In *Molecular Biology of Growth SEB Symposium* (eds PT Loughan & L M Pell). Cambridge: Cambridge University Press.

Pell J M, Saunders J C & Gilmour R S (1993). Differential regulation of transcription initiation from insulin-like growth factor I (IGF-I) leader exons and of tissue IGF-I expression in response to changed growth hormone and nutritional status in sheep. *Endocrinology* 132: 1797–1807.

Stewart C E H, Bates P C, Calder T A, Woodall S M & Pell J M (1993). Potentiation of insulin-like growth factor I (IGF-I) activity by an antibody: supportive evidence for enhancement of IGF-I bioavailability *in vivo* by IGF binding proteins. *Endocrinology* 133: 1462–1465.

Strauss D S (1994). Nutritional regulation of hormones and growth factors that control mammalian growth. *FASEB Journal* 8: 6–12.

Thissen J-P, Ketelslegers J-M & Underwood L E (1994). Nutritional regulation of the insulin-like growth factors. *Endocrinology* 15: 80–101.

Weller P A, Dauncey M J, Bates P C, Brameld L M, Buttery P L & Gilmour R S (1994). Regulation of porcine insulin-like growth factor I and growth hormone receptor mRNA expression by energy status. *American Journal of Physiology* 266: E776–E785.

## Part four: Fetal growth

Baker J, Lin J P, Robertson E J & Efstratiadis A (1993). Role of insulin-like growth factors in embryonic and post-natal growth. *Cell* 75: 73–82.

Gluckman P D (1995). The endocrine regulation of fetal growth in late gestation – the role of insulin-like growth factors. *Journal of Clinical Endocrinology and Metabolism* 80: 1047–1050.

Milner R D G & Gluckman P D (1996). Regulation of intra-uterine growth. In *Pediatrics and Perinatology: The Scientific Basis* (eds P D Gluckman & M A Heymann) 2nd edn pp 284–289. London: Arnold.

Woods K A, Camacho-Hubner C, Savage M O & Clark A J (1996). Intra-uterine growth retardation and post-natal growth failure associated with deletion of the insulin-like growth factor I gene. *New England Journal of Medicine* 335: 1363–1367.

#### Hormonal regulation of fetal growth

Baker J, Liu J-P, Robertson E J & Efstratiadis A (1993). Role of insulin-like growth factors in embryonic and post-natal growth. *Cell* 75: 73–82.

Chard T (1989). Hormonal control of growth in the human fetus. *Journal of Endocrinology* 23: 3–9.

D'Ercole A (1992). The insulin-like growth factors and *in utero* growth. *Growth, Genetics & Hormones* 8: 1–5.

Gluckman P D & Harding J E (1994). Nutritional and hormonal regulation of fetal growth: evolving concepts. *Acta Paediatrica* (Suppl) 399: 60–63.

Gluckman P D (1995). The endocrine regulation of fetal growth in late gestation: the role of insulin-like growth factors. *Journal of Clinical Endocrinology and Metabolism* 4: 1047–1049.

Lassarre C, Hardouin S, Daffos F et al. (1991). Serum insulin-like growth factors and insulin-like growth factors binding proteins in the human fetus. Relationships with growth in normal subjects and in subjects with intra-uterine growth retardation. *Pediatric Research* 29: 219–225.

Massa G, de Zegher F & Vanderschueren-Lodeweyckx M (1992). Serum growth hormone binding proteins in the human fetus and infant. *Pediatric Research* 32: 69–72.

Milner R D G & Hill DJ (1987). Interaction between endocrine and paracrine peptides in prenatal growth control. *European Journal of Pediatrics* 146: 113–122.

Tapanainen P J, Bang P, Wilson K et al. (1994). Maternal hypoxia as a model for intra-uterine growth retardation: effects on insulin-like growth factors and their binding proteins. *Pediatric Research* 36: 152–158.

**Growth factors and development**

Centrella M et al. (1994). Transforming growth factor-ß gene family members and bone. *Endocrine Reviews* 15: 27–39.

Hogan B L M et al. (1994). Growth factors in development: the role of TGF-beta related polypeptide signaling molecules in embryogenesis. *Development* 1994 (Suppl): 53–60.

Jessel T M & Melton D A (1992). Diffusible factors in vertebrate embryonic induction. *Cell* 68: 257–270.

Nislen-Hamilton M (ed) (1994). *Growth Factors and Signal Transduction in Development*. New York: Wiley-Liss.

Massagué J & Pandiella A (1993). Membrane-anchored growth factors. *Annual Review of Biochemistry* 62: 515–541.

Mayo K E (1994). Inhibin and activin: molecular aspects of regulation and function. *Trends in Endocrinology and Metabolism* 5: 407–415.

Muenke M & Schell U (1995). Fibroblast-growth-factor receptor mutations in human skeletal disorders. *Trends in Genetics* 11: 308–313.

Pimentel E (ed) (1994). *Handbook of Growth Factors*. Boca Raton, Florida: CRC Press.

Sporn M B & Roberts A B (eds) (1991). *Peptide Growth Factors and Their Receptors*. New York: Springer-Verlag.

Wilkie A O M et al. (1995). Functions of fibroblast growth factors and their receptors. *Current Biology* 5: 500–507.

**Growth of the human embryo**

Bossy J & Katz J M (1964). Croissance sémentaire d'une série de 100 foetus humains. *An. Desarrollo* 12: 181–213.

O'Rahilly R & Müller F (1984). Embryonic length and cerebral landmarks in staged human embryos. *Anatomical Record* 209: 265–271.

O'Rahilly R & Müller F (1986). Human growth during the embryonic period proper. In *Human Growth: A Comprehensive Treatise* (eds F Falkner & J M Tanner) 2nd edn, vol 1 pp 245–253. New York: Plenum.

O'Rahilly R & Müller F (1996). *Human Embryology and Teratology* 2nd edn. New York: Wiley-Liss.

**Embryonic development of teeth**

Berkovitz B K, Boyde A, Frank R M, Höhling H J, Moxham B J, Nalbandian J & Tonge C H (1989) *Teeth*. Berlin: Springer-Verlag.

Cohn S A (1957). Development of the molar teeth in the albino mouse. *American Journal of Anatomy* 101: 295–320.

Ferguson M (1990). The dentition through life. In *Dentition and Dental Care* (ed R J Elderton) pp 1–29. Oxford: Heinemann.

Lumsden A G S (1988). Spatial organization of the epithelium and the role of neural crest cells in the initiation of the mammalian tooth germ. *Development* 103S: 155–169.

Peyer B (1968). *Comparative Odontology*. Chicago: University of Chicago Press.

Ruch J V (1988). Determinisms of odontogenesis. *Cell Biology Reviews* RBC 14: 1–81.

Ruch J V, Lesot H & Bögue-Kirn C (1995). Odontoblast differentiation. *International Journal of Developmental Biology* 39: 51–68.

Slavkin H C (1974). Embryonic tooth formation. A tool for developmental biology. *Oral Sciences Reviews* Vol 4 (eds A H Melcher & G A Zarb). Copenhagen: Munksgaard.

Smith M M & Hall B K (1990). Development and evolutionary origins of vertebrate skeletogenic and odontogenic tissues. *Biological Reviews* 65: 277–373.

Smith M M & Hall B K (1993). A developmental model for evolution of the vertebrate exoskeleton and teeth. *Evolutionary Biology* 27: 387–448.

Ten Cate A R (1995). The experimental investigation of odontogenesis. *International Journal of Developmental Biology* 39: 511.

Thesleff I & Hurmerinta K (1981). Tissue interactions in tooth development. *Differentiation* 18: 75–88.

Thesleff I, Vaahtokari A & Partanen A-M (1995). Regulation of organogenesis. Common molecular mechanisms regulating the development of teeth and other organs. *International Journal of Developmental Biology* 39: 35–50.

Weiss K M (1990). Duplication with variation: metameric logic in evolution from genes to morphology. *Yearbook of Physical Anthropology* 33: 1–23.

Weiss K M (1993). A tooth, a toe, and a vertebra: the genetic dimensions of complex morphological traits. *Evolutionary Anthropology* 2: 121–134.

Zeichner-David M, Diekwisch T, Fincham A, Lau E, MacDougall M, Moradian-Oldak J, Simmer J, Snead M & Slavkin H C (1995). Control of ameloblast differentiation. *International Journal of Developmental Biology* 39: 69–92.

**Developmental morphology of the embryo and fetus**

Blechschmidt E (1963). *Der menschliche Embryo*. Stuttgart: Schattauer.

Blechschmidt E (1973). *Die pränatalen Organsysteme des Menschen*. Stuttgart: Schattauer.

Hinrichsen K V (ed) (1990). *Humanembryologie*. Berlin: Springer.

Müller F & O'Rahilly R (1997). The timing and sequence of appearance of neuromeres in staged human embryos. *Acta anatomica* 158: 83–99.

O'Rahilly R & Müller F (1987). *Developmental Stages in Human Embryos*. Washington, DC: Carnegie Institute of Washington.

O'Rahilly R & Müller F (1994). *The Embryonic Human Brain. An Atlas of Developmental Stages*. New York: Wiley-Liss.

O'Rahilly R & Müller F (1996). *Human Embryology and Teratology*. 2nd edn. New York: Wiley-Liss.

***Fetal body composition***

Metcoff J (1986). Association of fetal growth with maternal nutrition. In *Human Growth: A Comprehensive Treatise* (eds F Falkner & J M Tanner) 2nd edn pp 333–388. New York: Plenum.

Prentice A & Bates C J (1994). Adequacy of dietary supply for human bone growth and mineralisation. *European Journal of Clinical Nutrition* 48 Suppl 1: S161–S177.

Schwartz R (1990). Magnesium metabolism. In *Nutrition and Bone Development* (ed D J Simmons) pp 148–163. Oxford: Oxford University Press.

Southgate D A T (1986). Fetal measurements. In *Human Growth: A Comprehensive Treatise* (eds F Falkner & J M Tanner) 2nd edn, vol 1 pp 379–395. New York: Plenum.

Southgate D A T & Hey E N (1976). Chemical and biochemical development of the human fetus. In *The Biology of Human Fetal Growth* (eds D F Roberts & A M Thomson) pp 195–209. London: Taylor & Francis.

Widdowson E M (1968). Growth and composition of the fetus and newborn. In *The Biology of Gestation* (ed N S Assali) vol 2 pp 1–49. New York: Academic Press.

Widdowson E M & Dickerson J W (1981). Composition of the body. In *Geigy Scientific Tables* (ed C Lentner) vol 1 pp 217–25. Basle: Ciba-Geigy Ltd.

Widdowson E M & Spray C M (1951). Chemical development *in utero*. *Archives of Disease in Childhood* 26: 205–214.

**Neurological development**

Ballantyne J W (1902). *Manual of Antenatal Pathology and Hygiene*. Edinburgh: W Green & Son.

Brandt I (1986). Patterns of early neurological development. In *Human Growth: A Comprehensive Treatise* (eds F Falkner & J M Tanner) 2nd edn, vol 2 pp 469–518. New York: Plenum.

Prechtl H F R (ed) (1984). Continuity of neural functions from prenatal to postnatal life. *Clinics in Developmental Medicine* 94. London: Spastics International Medical Publications.

Robinson H P & Fleming J E E (1975). A critical evaluation of sonar crown–rump length measurements. *British Journal of Obstetrics and Gynaecology* 82: 702.

de Vries J I P, Visser G H A & Prechtl H F R (1982). The emergence of fetal behaviour. I Qualitative aspects. *Early Human Development* 7: 301–322.

de Vries J I P, Visser G H A & Prechtl H F R (1984). Fetal motility in the first half of pregnancy. In *Continuity of Neural Functions from Prenatal to Postnatal Life* (ed H F R Prechtl). *Clinics in Developmental Medicine* 94: 46–64. London: Spastics International Medical Publications.

**Ontogeny of the immune system**

Berardi A C, Wang A, Levine J D, Lopez P & Scadden D T (1995). Functional isolation and characterization of human hematopoietic stem cells. *Science* 267: 104–108.

Byun D G, Demeure C E, Yany Y P, Shu U, Ishihara H, Vezzio N, Gately M K & Delespesse G (1994). *In vitro* maturation of neonatal human CD8 T lymphocytes into IL4 and IL5 producing cells. *Journal of Immunology* 153: 4862–4871.

Demeure C, Wu C Y, Shu U, Schneider P V, Heuser C, Yssel H & Delespesse G (1994). *In vitro* maturation of human neonatal CD4 T lymphocytes. II. Cytokines present at priming modulate the development of lymphokine production. *Journal of Immunology* 152: 4775.

Ehlers S & Smith K A (1991). Differentiation of T cell lymphokine expression: the *in vitro* acquisition of T cell memory. *Journal of Experimental Medicine* 173: 25–36.

Fowlkes B J & Schweighoffer E (1995). Positive selection of T cells. *Current Opinion in Immunology* 7: 188–195.

Freedman A R, Zhu H, Levine J D, Kalams S & Scadden D T (1996). Generation of human T lymphocytes from bone marrow CD34+ cells *in vitro*. *Nature Medicine* 2: 46.

Hayes B F et al. (1988). Early events in human T cell ontogeny. Phenotypic characterization and immunohistologic localization of T cell precursors in early human fetal tissues. *Journal of Experimental Medicine* 168: 1061.

Hunt D W, Huppertz H I, Jiang W & Petty R E (1994). Studies of human cord blood dendritic cells: evidence for functional immaturity. *Blood* 84: 4333–4343

Lukens J N (1993). Blood formation in the embryo, fetus and newborn. In J Wintrobels *Clinical Hematology* 9th edn pp 79–100. Philadelphia: Lea & Febiger.

Nunez C, Nishomoto N, Gartland G L, Billips L G, Burrows P D, Kubagawa H & Cooper M D (1996). B cells are generated throughout life in humans. *Journal of Immunology* 156: 866–872.

Oettinger M A, Schatz D G, Gorka C & Baltimore D (1990). RAG-1 and RAG-2, adjacent genes that synergistically activate V(D)J recombination. *Science* 248: 1517–1523.

Phillips J H, Hori T, Nagler A, Bhat N, Spits H & Lanier L L (1992). Ontogeny of human natural killer (NK) cells: fetal NK cells mediate cytolytic function and express cytoplasmic CD3 epsilon,delta proteins. *Journal of Experimental Medicine* 175: 1055–1066.

Ryan D H, Nuccie B L, Ritterman I, Liesveld J L & Abboud C N (1994). Cytokine regulation of early human lymphopoiesis. *Journal of Immunology* 152: 5250.

Splawski J B, Nishioka J, Nishioka Y & Lipsky P E (1996). CD40 ligand is expressed and functional on activated neonatal T cells. *Journal of Immunology* 156: 119–127.

**Fetal growth retardation**

Barker D J P (1992). The fetal origins of diseases of old age. *European Journal of Clinical Nutrition* 46 (Suppl 3): S3–S9.

Doubilet P M & Benson C B (1995). Sonographic evaluation of intra-uterine growth retardation. *American Journal of Radiology* 164: 709–717.

Kramer M S (1987). Determinants of low birth weight: methodological assessment and meta-analysis. *Bulletin of the World Health Organization* 65: 663–737.

McCormick M C (1985).The contribution of low birth weight to infant mortality and childhood morbidity. *New England Journal of Medicine* 312: 82–90.

World Health Organization Expert Committee on Physical Status (1995). *Physical Status: The Use and Interpretation of Anthropometry*. Chapter 4: The newborn infant. World Health Organization Technical Report Series No. 854 pp 121–160.

***Intra-uterine growth retardation: perspectives from the Bonn study***

Albertsson-Wikland K, Wennergren G, Wennergren M, Vilbergsson G & Rosberg S (1993). Longitudinal follow up of growth in children born small for gestational age. *Acta Paediatrica* 82: 438–443.

Brandt I (1981). Kopfumfang und Gehirnentwicklung – Wachstumsretardierung bei intrauteriner Mangelversorgung und ihre Aufholmechanismen. *Klin. Wschr.* 59: 995–1007.

Brandt I (1985). Growth dynamics of low-birth-weight infants. *Acta Paediatrica Scandinavica* (Suppl) 319: 38–47.

Brandt I (1986). Growth dynamics of low-birth-weight infants with emphasis on the perinatal period.In *Human Growth: A Comprehensive Treatise* (eds F Falkner & J M Tanner) 2nd edn, vol 1 pp 415–475. New York: Plenum.

Brandt I (1989). Growth of head circumference: relationship to brain development, parent–child correlations and secular trend.In *Auxology 88: Perspectives in the Science of Growth and Development* (ed J M Tanner) pp 221–226. London: Smith-Gordon.

Brenner W E, Edelmann D A & Hendricks C H (1976). A standard of fetal growth for the United States of America. *American Journal of Obstetrics and Gynecology* 126: 555–564.

Hansmann M (1976). Ultraschall-Biometrie im II. und III. Trimester der Schwangerschaft. *Gynaekologe* 9: 133–155.

Karlberg P, Engström L & Selstam U (1979). Birth weight and height of newborn Swedish infants with relation to gestational age. Personal communication.

Lubchenco L O, Hansman C, Dressler M & Boyd E (1963). Intra-uterine growth as estimated from live born birth-weight data at 24 to 42 weeks of gestation. *Pediatrics* 32: 793–800.

Lubchenco L O, Hansman C & Boyd E (1966). Intra-uterine growth in length and head circumference as estimated from live births at gestational ages from 26 to 42 weeks. *Pediatrics* 37: 403–408.

Meredith H V (1971). Human head circumference from birth to early adulthood: racial, regional, and sex comparisons. *Growth* 35: 233–252.

Tanner J M, Whitehouse R H & Takaishi M (1966). Standards from birth to maturity for height, weight, height velocity, and weight velocity: British children, 1965-I and -II. *Archives of Disease in Childhood* 41: 454–471, 613–635.

Usher R H & McLean F H (1969). Intra-uterine growth of live-born Caucasian infants at sea level: standards obtained from measurements in 7 dimensions of infants born between 25 and 44 weeks of gestation. *Journal of Pediatrics* 74: 901–910.

**Standards and references for the assessment of fetal growth and development**

Brandt I (1981). Brain growth, fetal malnutrition & clinical consequences. *Journal of Perinatal Medicine* 9: 3–26.

Brandt I (1986). Growth dynamics of low-birth-weight infants with emphasis on the perinatal period. In *Human Growth: A Comprehensive Treatise* (eds F Falkner & J M Tanner) 2nd edn, vol 1 pp 415–475. New York: Plenum.

Brenner W E, Edelmann D A & Hendricks C H (1976). A standard of fetal growth for the United States of America. *American Journal of Obstetrics and Gynecology* 126: 555–564.

Dunn P M (1979). Perinatal terminology definitions & statistics. In *Perinatal Medicine, Sixth European Congress, Vienna.* (eds O Thalhammer, K Baumgarten & A Pollack). Stuttgart: Georg Thieme Publishers.

Hansmann M, Hackelöer B-J & Staudach A (eds) (1986). *Ultrasound Diagnosis in Obstetics and Gynecology*. Berlin: Springer-Verlag.

Hohenauer L (1980). Intrauterine Wachstumskurven für den Deutschen Sprachraum. *Zeitscrift für Geburtshilfe und Perinatologie* 184: 167.

Jeanty P & Romero R (1986). Critical reading of the biometry literature. In *Ultrasound Diagnosis in Obstetics and Gynecology* (eds M Hansmann, B-J Hackelöer & A Staudach) pp 161–188. Berlin: Springer-Verlag.

Keen & Pearce (1988). Weight, length, and head circumference curves for boys and girls of between 20 and 42 weeks' gestation. *Archives of Disease in Childhood* 63: 1170.

Lubchenco L O, Hansman C, Dressler M & Boyd E (1963). Intra-uterine growth as estimated from live born birth-weight data at 24 to 42 weeks of gestation. *Pediatrics* 32 793–800.

Lubchenco L O, Hansman C & Boyd E (1966). Intra-uterine growth in length and head circumference as estimated from live births at gestational ages from 26 to 42 weeks. *Pediatrics* 37 403–408.

Robinson H P & Fleming J E E (1975). A critical evaluation of sonar crown–rump length measurements. *British Journal of Obstetrics and Gynaecology* 82: 702.

Streeter G L (1920). Weight, sitting height, head size, foot length, and menstrual age of the human embryo. *Contrib. Embryol. Caneg. Inst.* 11: 143.

Thomson A M, Billewicz W Z & Hytten F E (1968). The assessment of fetal growth. *Journal of Obstetrics and Gynaecology of the British Commonwealth* 75.

Usher R H & McLean F H (1969). Normal fetal growth and the significance of fetal growth retardation.In *Scientific Foundations of Pediatrics* (eds J A Davis & J Dobbing). London: Heinemann Medical Books.

## Part five: Post-natal growth and Maturation

Malina R M (1995). Issues in normal growth and maturation. *Current Opinion in Endocrinology and Diabetes* 2: 83–90.

Malina R M & Bouchard C (1991). *Growth, Maturation, and Physical Activity*. Champaign, Illinois: Human Kinetics Publishers.

Oklund S & Prolo D J (1990). Human skeletal growth and development. In *Handbook of Human Growth and Developmental Biology* (eds E Meisami & P S Timiras) vol II pp 53–67. Boca Raton, Florida: CRC Press.

Roche A F, Heymsfield S B & Lohman T G (eds) (1996). *Human Body Composition*. Champaign, Illinois: Human Kinetics Publishers.

Tanner J M (1949). Fallacy of per-weight and per-surface area standards, and their relation to spurious correlation. *Journal of Applied Physiology* 2: 1–15.

Tanner J M (1962). *Growth at Adolescence* 2nd edn. Oxford: Blackwell Scientific Publications.

**Endocrinological regulation of post-natal growth**

Brook C G D (1995). *Clinical Paediatric Endocrinology* 3rd edn. Oxford: Blackwell Science.

Isaksson O G P, Jansson J-O & Gause I A M (1982). Growth hormone stimulates longitudinal bone growth directly. *Science* 364: 1237–1239.

Kelly P A, Dijane J, Postel-Vinay M C & Edery M (1991). The prolactin/growth hormone receptor family. *Endocrine Reviews* 12: 235–251.

Lin S C, Lin C R, Gukovsky I, Lusis A J, Sawchenko P E & Rosenfeld M G (1993). Molecular basis of the little mouse phenotype and implications for cell-specific growth. *Nature* 364: 208–213.

Savage M O, Bourgnignon J-P & Grossman A B (eds) (1994). *Frontiers in Paediatric Neuroendocrinology*. Oxford: Blackwell Science.

**Extra-uterine development after premature birth**

Brandt I (1983). *Griffiths-Entwicklungsskalen (GES) zur Beurteilung der Entwicklung in den ersten beiden Lebensjahren*. Weinheim: Beltz Verlag.

Brandt I (1986). Patterns of early neurological development. In *Human Growth: A Comprehensive Treatise* (eds F Falkner & J M Tanner) 2nd edn, vol 2 pp 469–518. New York: Plenum.

Gesell A & Amatruda C S (1947). *Developmental Diagnosis, Normal and Abnormal Child Development* 2nd edn. New York: Harper & Row.

Hopkins B & Prechtl H F R (1984). A qualitative approach to the development of movements during early infancy. In *Continuity of Neural Functions from Prenatal to Postnatal Life* (ed H F R Prechtl) pp 189, 192, 193. *Clinics in Developmental Medicine* 94. London: Spastics International Medical Publications.

Prechtl H F R, Fargel J W, Weinmann H M, & Bakker H H (1975). Development of motor function and body posture in pre-term infants. In *Aspects of Neural Plasticity/Plasticité nerveuse* (eds F Vital-Durand & M Jeannerod) vol 43 p. 55–66. Paris: INSERM.

**Morphology**

Eiben O G et al.*The Hungarian National Growth Study*. Budapest: Humanbiologia Budapestinensis.

Eveleth P B & Tanner J M (1990). *Worldwide Variation in Human Growth*. Cambridge: Cambridge University Press.

Harrison G A, Tanner J M, Pilbeam D R & Baker P T (1988). *Human Biology*. Oxford: Oxford University Press.

Hamill P V V, Johnston F E & Lemeshow S (1973). Body weight, stature, and sitting height: white and Negro youths 12–17 years of age. *DHEW Pub (HRA)* 74–1608, Washington: US Government Printing Office.

Malina R M, Hamill P V V & Lemeshow S (1974). Body dimensions and proportions, white and Negro hildren 6–11 years, United States. *DHEW Pub (HRA)* 76–1625, Washington: US Government Printing Office.

Stratz C H (1909). Wachstum und proportionem des menschen vor und nacj der geburt. *Archiv für Anthropologie* 8: 287–297.

***Body proportions***

Eveleth P B & Tanner J M (1990). *Worldwide Variation in Human Growth*. Cambridge: Cambridge University Press.

Tanner J M (1989). *Foetus into Man* 2nd edn. Ware: Castlemead Publications.

**Skeletal development**

Price J S, Oyajobi B O & Russell R G G (1994). The cell biology of bone growth. *European Journal of Clinical Nutrition* 48 Suppl 1: S131–S149.

**The human growth plate**

Apte S S (1988). Application of monoclonal antibody to bromodexyuridine to detect chondrocyte proliferation in growth plate cartilage *in vivo*. *Med Sci. Res.* 16: 405–406.

Apte S S (1990). Expression of the cell proliferation-associated nuclear antigen reactive with the Ki-67 monoclonal antibody by cells of the skeletal system in humans and other species. *Bone and Mineral* 10: 37–50

Kember N F (1993). Cell kinetics and the control of bone growth. *Acta Paediatrica* (Suppl) 391: 61–65.

Kember N F (1979). Proliferation controls in a linear growth system: theoretical studies of cell division in the cartilage growth plate. *Journal of Theoretical Biology* 78: 365–374.

Kember N F (1983). Cell kinetics of cartilage. In *Cartilage* (ed B K Hall) pp 149–180. New York: Academic Press.

Kember N F & Sissons H A (1976). Quantitative histology of the human growth plate. *Journal of Bone & Joint Surgery* 58: 426–435.

Moss-Salentijn L, Kember N F, Shinazuka M, Wu W F & Bose A (1991). Computer simulations of chondrocytic clone behaviour in rabbit growth plates. *Journal of Anatomy* 175: 7–17.

Thurston M N & Kember N F (1985). *In vitro* labelling in human and porcine growth plates. *Cell and Tissue Kinetics* 18: 575–582.

***Growth arrest lines***

Acheson R M (1950). Effects of starvation, septicaemia and chronic illness on the growth cartilage plate and metaphysis of the immature rat. *Journal of Anatomy* 93: 123–130.

Blanco R A et al. (1974). Height, weight, and lines of arrest growth in young Guatemalan children. *American Journal of Physical Anthropology* 40: 39–48.

Dreizen S et al. (1956). Observations on the association between nutritive failure, skeletal maturation rate and radiopaque transverse lines in the distal end of the radius in children. *American Journal of Roentgenology* 68: 709–723.

Dreizen S et al. (1964). The influence of age and nutrition status on 'bone scar' formation in the distal end of the growing radius. *American Journal of Physical Anthropology* 22: 295–306.

Follis R H Jr & Park E A (1952). Some observations on bone growth, with particular respect to zones and transverse lines of increased density in the metaphysis. *American Journal of Roentgenology* 68: 709–723.

Garn S M et al. (1968). Lines and bands of increased density: their implications to growth and development. *Medical Radiography and Photography* 44: 58–89.

Gindhart P S (1969). The frequency of appearance of transverse line in the tibia in relation to childhood illnesses. *American Journal of Physical Anthropology* 31: 17–22.

Goodman A H et al. (1984). Indicators of stress from bone and teeth. In *Paleopathology at the Origins of Agriculture* (eds M N Cohen & G J Armelagos) pp 23–49. Orlando: Academic Press.

Hummert J R & VanGerven D P (1985). Observations on the formation and persistence of radiopaque transverse lines. *American Journal of Physical Anthropology* 66: 297–306.

Huss-Ashmore R et al. (1982). Nutritional inference from paleopathology. *Advances in Archaeological Method and Theory* 5: 395–475.

Maat G J R (1984). Dating and rating of Harris's lines. *American Journal of Physical Anthropology* 63: 291–299.

Marshall W A (1968). Problems in relating the presence of transverse lines in the radius to the occurrence of disease. In *The Skeletal Biology of Earlier Human Populations* (ed D R Brothwell) pp 245–261. London: Pergamon Press.

Mays S A (1985). The relationship between Harris lines formation and bone growth and development. *Journal of Archaeological Science* 12: 207–220.

Murchison M A et al. (1984). Transverse line formation in protein-deprived rhesus monkeys. *Human Biology* 56: 173–182.

Park E A & Richter C P (1953). Transverse lines in bone: the mechanism of their development. *Johns Hopkins Hospital Bulletin* 93: 234–248.

Park E A (1964). The imprinting of nutritional disturbances on the growing bone. *Pediatrics* 33 (Suppl): 815–862.

Platt B S & Stewart R J C (1962). Transverse trabeculae and osteoporosis in bones in experimental protein-calorie deficiency. *British Journal of Nutrition* 16: 483–495.

**Human skeletal muscle across the lifespan**

Buller A J, Eccles J C & Eccles R M (1960). Interactions between motor neurones and muscles in respect of the characteristic speeds of their responses. *Journal of Physiology* 150: 417–439.

Burke R E, Levine D M, Tsairis P & Aajai F E (1973). Physiological types and histological profiles in motor units of cat gastrocnemius. *Journal of Physiology* 234: 723–748.

Garnett R A F, O'Donovan M J, Stephens J A & Taylor A (1979). Motor unit organization of human medial gastrocnemius. *Journal of Physiology* 287: 33–43.

Grimby G & Saltin B (1983). The ageing muscle: a mini review. *Clinical Physiology* 3: 209–218.

Jones D A & Round J M (1990). *Skeletal Muscle in Health and Disease: A Textbook of Muscle Physiology*. Manchester: Manchester University Press.

Merton P A (1972). How we control the contraction of our muscles. *Scientific American* 226: 30–37.

Parker D F, Round J M, Sacco P & Jones D A (1990). A cross-sectional survey of upper and lower limb strength in boys and girls during childhood and adolescence. *Annals of Human Biology* 17: 199–211.

Peter J B, Barnard V R, Edgerton V R, Gillespie C A & Stempel K E (1972). Metabolic profiles of three fibre types of skeletal muscles in guinea pigs and rabbits. *Biochemistry* 11: 2627–2633.

Ranvier M L (1873). Propriétés et structures différentes des muscles rouges et des muscles blancs, chez les Lapins et chez les Raies. *Comptes Rendus des Académies de Sciences* 77: 1030–1034.

Rowe R W D & Goldspink G (1969). Muscle fibre growth in five different muscles in both sexes of mice: I, normal mice. *Journal of Anatomy* 104: 519–530.

Rutherford O M & Jones D A (1992). The relationship of muscle and bone loss and activity levels with age in women. *Age and Ageing* 21: 286–293.

Salmons S & Henriksson K (1981). The adaptive response of skeletal muscle to increased use. *Muscle & Nerve* 4: 94–105.

Salmons S & Vrbova G (1969). The influence of activity on some contractile characteristics of mammalian fast and slow muscles. *Journal of Physiology* 201: 535–549.

Vrbova G & Lowrie M (1989). Role of activity in developing synapses, search for molecular mechanisms. *News in Physiological Sciences* 4: 75–78.

***Physical work capacity during growth***

Åstrand P O (1952). *Experimental Studies of Physical Working Capacity in Relation to Age and Sex*. Copenhagen: Ejnar Munksgaard.

Åstrand P O & Rodahl K (1986). *Textbook of Work Physiology*. New York: McGraw-Hill.

Armstrong N, Williams J, Balding J, Gentle P & Kirby B (1991). The peak oxygen uptake of British children with reference to age, sex and sexual activity. *European Journal of Applied Physiology* 62: 369–375.

Sallis J F (1993). Epidemiology of physical activity and fitness in children and adolescents. *Critical Reviews in Food Science and Nutrition* 33: 403–408.

Saris W H M (1982). *Aerobic Power and Daily Physical Activity in Children*. The Netherlands: Meppel Kripps Repro.

Spurr G B & Reina J C (1989). Maximum oxygen consumption in marginally malnourished Colombian boys and girls 6–16 years of age. *American Journal of Human Biology* 1: 11–19.

Spurr G B, Reina J C, Dufour D L & Naruaez J V (1994). $\dot{V}O_2$ max and nutritional status in urban Colombian girls and women. *American Journal of Human Biology* 6: 641–649.

**Post-natal craniofacial growth**

Enlow D H (1982). *Handbook of Facial Growth* 3rd edn. Philadelphia: W B Saunders.

Goose D H & Appleton J (1982). *Human Dentofacial Growth*. Oxford: Pergamon Press.

Moore W J & Lavelle C L B (1974). *Growth of the Facial Skeleton in the Hominoidea*. London: Academic Press.

Ranly D M (1980). *A Synopsis of Craniofacial Growth*. New York: Appleton-Century-Crofts.

**Variation in tooth formation and eruption**

Eveleth P B & Tanner J M (1976). Worldwide variation in human growth. *International Biological Programme 8*. Cambridge: Cambridge University Press.

Garn S M, Rohmann C G & Guzman M A (1963). Genetic, nutritional and maturational correlates of dental development. *Journal of Dental Research* 44: 228–242.

Gustafson G & Koch G (1974). Age estimation up to 16 years of age based on dental development. *Odontologisk Revy* 25: 297–306.

Macho G A & Wood B A (1995). The role of time and timing in Hominid dental evolution. *Evolutionary Anthropology* 17–31.

Smith B H (1991). Standards of human tooth formation and dental age assessment. In *Advances in Dental Anthropology* (eds M A Kelley & C S Larsen) pp 143–168. New York: Wiley-Liss.

Ubelaker D H (1981). *Human Skeletal Remains: Excavation, Analysis, Interpretation*. Chicago: Aldine.

***Developmental defects of dental enamel***

Corruccini R S, Handler J S & Jacobi K P (1985). Chronological distribution of enamel hypoplasias and weaning in a Caribbean slave population. *Human Biology* 57: 699–711.

Goodman A H & Capasso L L (eds) (1992). Recent contributions to the study of enamel developmental defects. *Journal of Paleopathology Monographic Publication* 2.Termano, Italy: Edigrafital.

Goodman A H & Rose J C (1990). Assessment of systemic physiological perturbations from dental enamel hypoplasias and associated histological structures. *Yearbook of Physical Anthropology* 33: 59–110.

Goodman A & Rose J C (1991). Dental enamel hypoplasias as indicators of nutritional status. In *Advances in Dental Anthropology* (eds M Kelley & C Larsen) pp 279–293. New York: Wiley–Liss.

Sarnat B G & Schour I (1941). Enamel hypoplasias (chronic enamel aplasia) in relationship to systemic diseases: a chronological morphological and etiological classification. *Journal of the American Dental Association* 28: 1989–2000.

Suckling G. Developmental defects of enamel – historical and present-day perspectives on their pathogenesis. *Advances in Dental Anthropology* 3 (2):87–94.

**Body composition**

Boileau R A, Lohman T G, Slaughter M H, Ball T E, Going S B & Hendrix M K (1984). Hydration of the fat-free body in children during maturation. *Human Biology* 56: 651–666.

Eveleth P B, Tanner J M (1990). *Worldwide Variation in Human Growth* 2nd edn pp 215–219. Cambridge: Cambridge University Press.

Fomon S J, Haschke F, Ziegler E E & Nelson S E (1982). Body composition of reference children from birth to age 10 years. *American Journal of Clinical Nutrition* 35: 1169–1175.

Forbes G B (1987). *Human Body Composition: growth, ageing, nutrition and activity* pp 125–168. New York: Springer-Verlag.

Lohman T G (1986). Applicability of body composition techniques and constants for children and youth. *Exercise and Sports Science Reviews* 14: 325–357.

Malina R M, Bouchard C (1991). *Growth, Maturation and Physical Activity*. Champaign, Illinois: Human Kinetics Books.

Roche A F (ed) (1992). *Growth, Maturation and Body Composition: the Fels Longitudinal Study, 1929–91* pp 199–235. Cambridge: Cambridge University Press.

***Organ development and body-weight***

Dobbing L & Sands J (1973). Quantitative growth and development of the human brain. *Archives of Disease in Childhood* 48: 757–767.

Elia M (1991). Organ and tissue contribution to metabolic rate. In: *Energy Metabolism Tissue Determinants and Cellular Corrolaries* (ed J M Kinney & H N Tucker) pp 61–80. New York: Raven Press.

Snyder W S, Cook M J, Nasset E S, Karhausen L R, Parry-Howells G & Tipton I H (1974). Report of the Task Group on Reference Man. *International Commission on Radiological Protection No.23*. Oxford: Pergamon Press.

Widdowson E M & Dickerson J W T (1960). The effect of growth and function on the chemical composition of soft tissues. *Biochemical Journal* 77: 30–43.

Widdowson E M & Dickerson J W T (1964). Chemical composition of the body. In *Mineral Metabolism* (eds C L Comar & F Bronner) vol 2 pp 2–247. New York: Academic Press.

**Physical activity and training for sport as factors affecting growth and maturation**

Baxter-Jones A D G, Helms P, Baines-Preece J C & Preece M (1994). Menarche in intensely trained gymnasts, swimmers and tennis players. *Annals of Human Biology* 21: 407–415.

Baxter-Jones A D G, Helms P, Maffulli N, Baines-Preece J C & Preece M (1995). Growth and development of male gymnasts, swimmers, soccer and tennis players: a longitudinal study. *Annals of Human Biology* 22: 381–394.

Beunen G P, Malina R M, Renson R, Simons J, Ostyn M & Lefevre J (1992). Physical activity and growth, maturation and performance: a longitudinal study. *Medicine and Science in Sports and Exercise* 24: 576–585.

Bouchard C, Shephard RJ (1994). Physical activity, fitness and health: the model and key concepts. In *Physical Activity, Fitness, and Health: International Proceedings and Consensus Statement* (eds C Bouchard, R J Shephard & T Stephens) pp 77–88. Champaign, Illinois: Human Kinetics Publishers.

Eriksson B O, Gollnick P D & Saltin B (1974). The effect of physical training on muscle enzyme activities and fiber composition in 11-year-old boys. *Acta Paediatrica Belgica* 28 (Suppl): 245–252.

Fournier G B, Ricci J, Taylor A W, Ferguson R J, Montpetit R R & Chaitman B R (1982). Skeletal muscle adaptation in adolescent boys: sprint and endurance training and detraining. *Medicine and Science in Sports and Exercise* 14: 453–456.

Loucks A B, Vaitukaitis J, Cameron J L, Rogol A D, Skrinar G, Warren M P, Kendrick J & Limacker M C (1992). The reproductive system and exercise in women. *Medicine and Science in Sports and Exercise* 24 (Suppl): S288–S293.

Malina R M (1983). Menarche in athletes: a synthesis and hypothesis. *Annals of Human Biology* 10: 1–24.

Malina R M (1991). Darwinian fitness, physical fitness and physical activity. In *Applications of Biological Anthropology to Human Affairs* (eds C G N Mascie-Taylor & G W Lasker) pp 143–184. Cambridge: Cambridge University Press.

Malina R M (1994). Physical activity and training: effects on stature and the adolescent growth spurt. *Medicine and Science in Sports and Exercise* 26: 759–766.

Malina R M (1994). Growth and maturation of young athletes. *Exercise and Sport Sciences Reviews* 22: 389–433.

Malina R M (1994). Physical activity: relationship to growth, maturation, and physical fitness. In *Physical Activity, Fitness, and Health: International Proceedings and Consensus Statement* (eds C Bouchard, R J Shephard & T Stephens) pp 918–930. Champaign, Illinois: Human Kinetics.

Malina R M & Bielicki T (1996). Retrospective longitudinal growth study of boys and girls active in sport. *Acta Paediatrica* (in press).

Malina R M & Bouchard C (1991). *Growth, Maturation, and Physical Activity*. Champaign, Illinois: Human Kinetics Publishers.

Warren M P (1980). The effects of exercise on pubertal progression and reproductive function in girls. *Journal of Clinical Endocrinology and Metabolism* 51: 1150–1157.

**Growth cyclicities and pulsalities**

Ashizawa K & Kawabata M (1990). Daily measurements of the heights of two children from June 1984 to May 1985. *Annals of Human Biology* 17: 437–443.

Bogin B A (1978). Seasonal pattern in the rate of growth in height of children living in Guatemala. *American Journal of Physical Anthropology* 49: 205–210.

Bogin B (1979). Monthly changes in the gain and loss of growth in weight of children living in Guatemala. *American Journal of Physical Anthropology* 51: 287–292.

Bogin B (1988). *Patterns of Human Growth*. Cambridge: Cambridge University Press.

Hermanussen M et al. (1988). Periodical changes of short term growth velocity ('mini growth spurts') in human growth. *Annals of Human Biology* 15: 103–109.

Kobayashi M & Togo M (1993). Twice-daily measurements of stature and body weight in two children and one adult. *American Journal of Human Biology* 5: 193–201.

Lampl M et al. (1992). Saltation and stasis: a model of human growth. *Science* 258: 801–803.

Perera A D & Plant T M (1992). The neurobiology of primate puberty. *Ciba Foundation Symposium* 168: 252–262

***Saltatory growth***

Bernstein I M, Blake K & Badger G (1995). Evidence that normal fetal growth is not continuous. *American Journal of Obstetrics and Gynecology* 172: 323.

Hermanussen M, Geiger-Benoit K, Burmeister J & Sippell W G (1988). Periodical changes of short term growth velocity ('mini growth spurts') in human growth. *Annals of Human Biology* 15: 103–109.

Lampl M, Veldhuis J D & Johnson M L (1992). Saltation and stasis: a model of human growth. *Science* 258: 801–803.

***Mini growth spurts***

Burmeister J & Hermanussen M (1989). The measurement of short-term growth. Addendum. In *Auxology 88: Perspectives in the Science of Growth and Development* (ed J M Tanner) pp 60–61. London: Smith-Gordon.

Camerer W (1880). Versuche über den Stoffwechsel, angestellt mit 5 Kindern im Alter von 2 bis 11 Jahren. *Zeitschrift für Biologie* 16: 24–41.

Curtiss F H (1898). Some investigations regarding loss in weight and gain in height during sleep. *Amer Phys Ed Rev* 3: 270–273.

Hermanussen M (1998). The analysis of short-term growth. *Horm Res* 49: 53–64.

Hermanussen M, Geiger-Benoit K, Burmeister J & Sippell W G (1988a). Knemometry in childhood: accuracy and standardization of a new technique of lower leg length measurement. *Annals of Human Biology* 15: 1–16.

Hermanussen M, Geiger-Benoit K, Burmeister J & Sippell W G (1988b). Periodical changes of short term growth velocity ('mini growth spurts') in human growth. *Annals of Human Biology* 15: 103–109.

Hermanussen M, Bugiel S, Aronson S & Moell C (1992). A non-invasive technique for the accurate measurement of leg length in animals. *Growth Development and Aging* 56: 129–140.

Hermanussen M (1993). Rhythms in growth. *Acta Med Auxologica* 25: 75–79.

Perez-Romero A, Rol de Lama M A, Grenados B, Tresguerres J A F, Hermanussen M & Ariznavarreta M C (1994). Daily measurements of tibial growth in a GH deficient rat model. *Neuroendocrinology* 60 (S1): P2.64.

Rol de Lama M A, Perez-Romero A, Hermanussen M, Tresguerres J A F & Ariznavarreta R C (1994). Daily measurement of rat growth velocity. *European Journal of Clinical Investigation* 24 (Suppl 12): P241

Valk I M, Langhout Chabloz A M E, Smals A G H, Kloppenborg P W C, Cassorla P G, Schutte E A S T (1983). Accurate measurement of the lower leg length and the ulnar length and its application in short term growth measurements. *Growth* 47: 53–66.

***Pulsatile growth hormone release***

Hartman M L, Iranmanesh A, Thorner M O & Veldhuis J D (1993). Evaluation of pulsatile patterns of growth-hormone release in humans: a brief review. *American Journal of Human Biology* 5: 603–614.

Iranmanesh A, Grisso B & Veldhuis J D (1994). Low basal and persistent pulsatile growth hormone secretion are revealed in normal and hyposomatotropic men studied with a new ultrasensitive chemiluminescence assay. *Journal of Clinical Endocrinology and Metabolism* 78: 526–535.

Martha P M Jr, Goorman K M, Blizzard R M, Rogol A D & J D Veldhuis (1992). Endogenous growth hormone secretion and clearance rates in normal boys as determined by deconvolution analysis: relationship to age, pubertal status and body mass. *Journal of Clinical Endocrinology and Metabolism* 74: 336–344.

Veldhuis J D, Carlson M L & Johnson M L (1987). The pituitary gland secretes in bursts: appraising the nature of glandular secretory impulses by simultaneous multiple-parameter deconvolution of plasma hormone concentrations. *Proceedings of the National Academy of Science, USA* 84: 7686–7690.

Veldhuis J D, Iranmanesh A, Lizarralde G, & Urban R J (1994). Combined deficits in the somatotropic and gonadotropic axes in healthy older men: an appraisal of neuroendocrine mechanisms by deconvolution analysis. *Neurobiology of Aging* 15: 509–517.

**Hormonal regulation of growth in childhood and puberty**

Hernández M & Argente J (1992). *Human Growth: Basic and Clinical Aspects*. Amsterdam: Excerpta Medica.

Malina R M (1986). Growth of muscle tissue and muscle mass. In *Human Growth: A Comprehensive Treatise* (eds F Falkner & J M Tanner) 2nd edn, vol 2 pp 77–99. New York: Plenum.

Preece M A (1986). Prepubertal and pubertal endocrinology. In *Human Growth: A Comprehensive Treatise* (eds F Falkner & J M Tanner) 2nd edn, vol 2 pp 211–224. New York: Plenum.

***The mid-childhood growth spurt***

Bogin B *Patterns of Human Growth*. Cambridge: Cambridge University Press.

Bolk L (1926). *Das Problem der Menschwerdung*. Jena: Gustav Fischer

Cutler G B Jr et al. (1978). Adrenarche: a survey of rodents, domestic animals, and primates. *Endocrinology* 103: 2112–2118.

Katz S H et al. (1985). Adrenal androgens, body fat and advanced skeletal age in puberty: new evidence for the relations of adrenarche and gonadarche in males. *Human Biology* 57: 401–413.

Parker, L N (1991). Adrenarche. *Endocrinology and Metabolism Clinics of North America* 20: 71–83.

Smail P J et al. (1982). Further studies on adrenarche in non-human primates. *Endocrinology* 111: 844–848.

Tanner J M (1947). The morphological level of personality. *Proceedings of the Royal Society of Medicine* 40: 301–303.

Weirman M E & Crowley W F Jr (1986). Neuroendocrine control of the onset of puberty. In *Human Growth: A Comprehensive Treatise* (eds F Falkner & J M Tanner) 2nd edn, vol 2 pp 225–241. New York: Plenum.

Worthman, C M (1986). Later-maturing populations and control of the onset of puberty. *American Journal of Physical Anthropology* 69: 282.

**Sexual maturation**

Boyar R M, Rosenfeld R S, Kapen S, Finkelstein J W, Roffwarg H P, Weitzman E D & Hellmann L (1974). Human puberty. Simultaneous augmented secretion of luteinizing hormone and testosterone during sleep. *Journal of Clinical Investigation* 54: 609–618.

Ellison P T (1982). Skeletal growth, fatness and menarcheal age: a comparison of two hypotheses. *Human Biology* 54: 269–281.

Frisch R E (1990). Body fat, menarche, fitness and fertility. *Progress in Reproductive Biology and Medicine* 14: 1–26.

Plant T M (1988). Puberty in primates. In *The Physiology of Reproduction* (eds E Knobil & J Neill) vol 2 pp 1763–1788. New York: Raven Press.

***Menarche***

Eveleth P B & Tanner J M (1990). *Worldwide Variation in Human Growth* 2nd edn. Cambridge: Cambridge University Press.

**Fat and fat patterning**

Bogin B & Sullivan T (1986). Socio-economic status, sex, age, and ethnicity as determinants of body fat distribution for Guatemalan children. *American Journal of Physical Anthropology* 69: 527–535.

Bouchard C (1988). Inheritance of human fat distribution. In *Fat Distribution During Growth and Later Health Outcomes* (eds C Bouchard and F E Johnston) pp 103–125. New York: Liss.

Cameron N, Johnston F E, Kgamphe J S & Lunz R (1992). Body fat pattering in rural South African black children. *American Journal of Human Biology* 4: 353–364.

Cameron N, Gordon-Larsen P & Wrchota E M (1994). Longitudinal analysis of adolescent growth in height, fatness, and fat pattering in rural South African black children. *American Journal of Physical Anthropology* 93: 307–321.

Healy M J R & Tanner J M (1981). Size and shape in relation to growth and form. *Symposia of the Zoological Society of London* 46: 19–35.

Malina R M & Bouchard C (1988). Subcutaneous fat distribution during growth. *In Fat Distribution During Growth and Later Health Outcomes* (eds C Bouchard & F E Johnston) pp 63–84. New York: Liss.

Mueller W H (1982). The changes with age of the anatomical distribution of fat. *Social Science and Medicine* 16: 191–196.

Mueller W H (1985). The biology of human fat pattering. In *Human Body Composition and Fat Distribution* (ed N G Norgan) pp 159–174. Wageningen: Stichting Nederlands Instituut voor de Voeding. Euro-Nut report 8.

Roche A F (ed) (1985). *Body Composition Assessment in Youth and Adults. Report of the Sixth Ross Conference on Medical Research*. Columbus Ohio: Ross laboratories.

***Fat patterning and non-insulin-dependent diabetes mellitus in modernizing populations***

Bouchard C & Johnston F E (eds) (1988). *Fat Distribution During Growth and Later Health Outcomes*. New York: Liss.

McGarvey S T, Bindon J R, Crews D E & Schendel D E (1989). Modernization and adiposity: causes and consequences. In *Human Population Biology*. (eds M A Little & J D Haas) pp 263–279. New York: Oxford University Press.

Zimmet P Dowse G, Serjeantson S, King H & Finch C (1990).The epidemiology and natural history of NIDDM – lessons from the South Pacific. *Diabetes/Metabolism Reviews* 6: 91–124.

**Adulthood and developmental maturity**

Cameron N (1993). Assessment of growth and maturation during adolescence. *Hormone Research* 39 (Suppl 3): 9–17.

Marshall W A & J M Tanner (1969). Variations in the pattern of pubertal changes in girls. *Archives of Diseases in Childhood* 44: 291–303.

Marshall W A & J M Tanner (1970). Variations in the pattern of pubertal changes in boys. *Archives of Diseases in Childhood* 45: 13–23.

Prader A (1966). Testicular size: assessment and clinical importance. *Triangle* 7: 240.

Tanner J M, Whitehouse R H, Cameron N, Marshall W A, Healy M J R & H Goldstein (1983). *Assessment of Skeletal Maturity and Prediction of Adult Height* 2nd edn. London: Academic Press.

***Fertility***

Wood J W (1994). *Dynamics of Human Reproduction*. New York: Aldine de Gruyter.

***Epiphyseal fusion***

Roche A F, Wainer H & Thissen D (1975). *Skeletal Maturity: The Knee Joint as a Biological Indicator*. New York: Plenum.

Tanner J M, Whitehouse R H, Cameron N, Marshall W A, Healy M J R & Goldstein H (1983). *Assessment of Skeletal Maturity and Prediction of Adult Height* 2nd edn. London: Academic Press.

## Part six: Behavioural and cognitive development

Eaves L J, Eysenck H J & Martin N G (1989). *Genes, Culture and Personality. An Empirical Approach*. London & New York: Academic Press.

Eysenck H J & Fulker D W (1979). *The Structure and Measurement of Intelligence*. Berlin, Heidelberg & New York: Academic Press.

Krech D, Crutchfield R S, Livson N, Rollin A R & Wilson W A (1969). *Elements of Psychology* 2nd edn. New York: Alfred A Knopf.

Mussen P H, Conger J J & Kagan J (1969). *Child Development and Personality* 3rd edn. New York: Harper & Row.

Maslow A H (1970). *Motivation and Personality* 2nd edn. New York: Harper & Row.

Plomin R (1986). *Genetics, Development and Psychology*.

Scarr S, Webber P L, Weinberg R A & Wittig M A (1981). Personality resemblance among adolescents and their parents in biologically related and adoptive families. In *Twin Research 3. Part B, Intelligence, Personality and Development* (eds L Gedda, P Parisi & W E Nance). New York: Liss.

Wilber K (1993). *The Spectrum of Consciousness* 2nd edn. Wheaton, Illinois: Theosophical Publishing House.

Wilson R S (1978). Synchronies in mental development: an epigenetic perspective. *Science* 202: 939.

**Cognitive development**

Bayley N (1969). *Bayley Scales of Infant Development*. New York: Psychological Corporation.

Estes W K (1994). *Classification and Cognition*. Oxford: Oxford University Press.

Illingworth R S (1987). *The Development of the Infant and Young Child. Normal and Abnormal*. Edinburgh: Churchill Livingstone.

Kirk S, McCarthy J & Kirk W D (1968). *Illinois Test of Psycholinguistic Abilities*. Urbana: University of Illinois Press.

Kohen-Raz R (1967). Scalogram analysis of some developmental sequences of infant behaviour as measured by the Bayley Infant Scale of Mental Development. *Genetic Psychology Monographs* 76: 3–21.

Largo R H, Graf S, Kundu S, Hunziker U & Molinari L (1990). Predicting developmental outcome at school age from infant tests of normal, at-risk and retarded infants. *Developmental Medicine and Child Neurology* 32: 30–45.

McCarthy D (1972). *McCarthy Scales of Children's Abilities*. New York: Psychological Corporation.

Medin D L & Ross B H (1990). *Cognitive Psychology*. New York: Harcourt Brace Jovanovich.

Reynell J K (1985). *Reynell Developmental Language Scales-Second Edition*. Windsor: NFER-Nelson.

Siegel L S (1994). Assessment of cognitive and language functioning: a developmental perspective. In *Developmental Follow-up* (eds S L Friedman & H C Haywood) pp 217–233. San Diego: Academic Press.

Thorndike R L, Hagen E P & Sattler J M (1986). *The Stanford-Binet Intelligence Scale*. Chicago: Riverside Publications.

Wechsler D (1974). *Manual for the Wechsler Intelligence Scale for Children-Revised*. New York: Psychological Corporation.

**Motor development and performance**

Beunen G & Malina R M (1988). Growth and physical performance relative to the timing of the adolescent spurt. *Exercise & Sport Sciences Reviews* 16: 503–540.

Beunen G, Malina R M, Van's Hof M A, Simons J, Ostyn M, Renson R & Van Gerven D (1988) *Adolescent Growth and Motor Performance: A Longitudinal Study of Belgian Boys*. Champaign, Illinois: Human Kinetics Publishers.

Eaton W O & Saudino K J (1992). Prenatal activity level as a temperament dimension? Individual differences and developmental functions in fetal movement. *Infant Behavior and Development* 15: 57–70.

Espenschade A (1960). Motor development. In *Science and Medicine of Exercise and Sports* (ed W R Johnson) pp 419–439. New York: Harper.

Haubenstricker J & Seefeldt V (1986). Acquisition of motor skills during childhood. In *Physical Activity and Well-Being* (ed V Seefeldt) pp 41–102. Reston, Virginia: American Alliance for Health, Physical Education, Recreation and Dance.

Malina R M (1975) Anthropometric correlates of strength and motor performance. *Exercise and Sport Sciences Reviews* 3: 249–274.

Malina R M (1980). Biosocial correlates of during infancy and early childhood. In *Social and Biological Predictors of Nutritional Status, Physical Growth, and Neurological Development* (eds L S Greene & F E Johnston) pp 143–171. New York: Academic Press.

Malina R M (1994) Anthropometry, strength and motor fitness. In *Anthropometry: The Individual and the Population* (eds S J Ulijaszek & C G N Mascie-Taylor) pp 160–177. Cambridge: Cambridge University Press.

Malina R M & Bouchard C (1991). *Growth, Maturation, and Physical Activity*. Champaign, Illinois: Human Kinetics Publishers.

Prechtl H F R (1986). Pre-natal motor development. In *Motor Development of Children: Aspects of Co-ordination and Control* (eds M G Wade & H T A Whiting) pp 53–64. Dordecht, The Netherlands: Martinus Nijhoff.

Seefeldt V & Haubenstricker J (1982). Patterns, phases, or stages: an analytical model for the study of developmental movement. In *The Development of Movement Control and Co-ordination* (eds J A S Kelso & J E Clark) pp 309–319. New York: John Wiley.

**Language development**

Dunbar R (1996). *Grooming, Gossip, and the Evolution of Language*. London: Faber & Faber.

Fernald A (1992). Human maternal vocalizations to infants as biologically relevant signals: an evolutionary perspective. In *The Adapted Mind: Evolutionary Psychology and the Generation of Culture* (eds J Barkow, L Cosmides & J Tooby). New York: Oxford University Press.

Fromkin V & Rodman R (1993). *An Introduction of Language*. Fort Worth: Harcourt Brace College Publishers.

King B (1994). *The Information Continuum*. Santa Fe: School of American Research Press.

Locke J L (1994). Phases in the child's development of language. *American Scientist* 82: 436–445.

**Psychosocial factors in growth and development**

Bakwin M (1942). Loneliness in infants. *American Journal of Diseases in Children* 62: 30–40.

Blizzard R M (1990). Psychosocial short stature. *In Pediatric Endocrinology: A Clinical Guide* (ed F Lifsnitz) pp 87–107. New York: Marcel Dekker.

Darwin C E (1966). *The Expression of the Emotions in Man and Animals*. Chicago: University of Chicago Press.

Green W H (1990). A theoretical model for classical psychosocial dwarfism (psychologically determined short stature). In *Psychoneuroendocrinology: Brain, Behavior and Hormonal Interaction* (ed C S Holmes) pp 92–112. New York: Springer Verlag.

James W (1884). What is emotion? *Mind* 4: 188–204.

Laird J D (1974). Self-attribution of emotion: the effects of expressive behavior on the quality of emotion experience. *Journal of Personality and Social Psychology* 29: 475–486.

Parisi P & De Martino V (1980). Psychosocial aspects in human growth: deprivation and failure to thrive. In *Human Physical Growth & Maturation* (eds F E Johnston, A F Roche & C Susanne) pp 339–356. New York & London: Plenum.

Patton R G & Gardner L I (1963). *Growth Failure in Maternal Deprivation*. Springfield: Charles C Thomas.

Powell G F, Brasel J K & Blizzard R M (1987). Emotional deprivation and growth retardation stimulating idiopathic hypopituitarism. I. Clinical evaluation of the syndrome. *New England Journal of Medicine* 276: 1271–1278.

Powell G F et al. (1987). Emotional deprivation and growth retardation stimulating idiopathic hypopituitarism. II. Endocrinological evaluation of the syndrome. *New England Journal of Medicine* 276: 1279–1283.

Spitz R (1946). Hospitalism – a follow-up report. *Psychoanalytical Study of Children* 2: 113–117.

Spitz R (1946). Hospitalism – an inquiry into the genesis of psychiatric conditions in early childhood. *Psychoanalytical Study of Children* 1: 53–74.

Talbot N B et al. (1947). Dwarfism in healthy children: its possible relation to emotional, nutritional and endocrine disturbances. *New England Journal of Medicine* 236: 783–793.

Widdowson E M (1951). Mental contentment and physical growth. *Lancet* 260: 1316.

**Nutrition and cognitive development**

Grantham-McGregor S M, Schofield W & Powell C (1987). Development of severely malnourished children who received psychosocial stimulation: six year follow-up. *Pediatrics* 79: 247–254.

**The development of sexuality**

Gagnon J H & Simon W (1973). *Sexual Conduct: the Social Sources of Human Sexuality*. Chicago: Aldine.

Jackson S (1982). *Childhood and Sexuality*. Oxford: Blackwell.

Richards M P M (1996). The childhood environment and the development of sexuality. In *Long Term Consequences of Early Environments*. (eds C J K Henry & S J Ulijaszek). Cambridge: Cambridge University Press.

## Part seven: Clinical growth abnormalities

**Orthodontic disorders**

Bhatia S N & Leighton B C (1993). *A Manual of Facial Growth*. New York: Oxford University Press.

Enlow D H (1990). *Facial Growth* 3rd edn. Philadelphia: W B Saunders.

Moyers R E (1988). *Handbook of Orthodontics*. Chicago: Yearbook Medical Publishers.

Proffit W R, Fields H M, Ackerman J L et al. (1993). *Contemporary Orthodontics*. Philadelphia: Mosby Year Book.

Van der Linden, F P G M (1986). *Facial Growth and Facial Orthopedics*. Chicago: Quintessence Publishing.

**Regional growth disorders**

Brown N A, Hoyle C I, McCarthy A & Wolpert L (1989). The development of asymmetry: the sidedness of drug-induced limb abnormalities is reversed in situs-inversus mice. *Development* 107: 637–642.

Brown N A & Wolpert L (1990). The development of handedness in left/right asymmetry. *Development* 109: 1–9.

Burwell R G & Dangerfield P H (1992). Pathogenesis and the assessment of scoliosis. In *Surgery of the Spine: a combined neurosurgical and orthopaedic approach* (eds G F G Findley & R Owen) pp 365–408. London: Blackwell Scientific.

Dangerfield P H (1994). Asymmetry and growth. In *Anthropometry: the Individual and the Population* (eds S J Ulijaszek & C G N Mascie-Taylor) pp 7–29. Cambridge: Cambridge University Press.

Polak M & Trivers R (1994). The science of symmetry in biology. *Trends in Ecology and Systematics* 9: 122–124.

Van Valen L (1962). A study of fluctuating asymmetry. *Evolution* 16: 125–142.

Yokoyarna T, Copeland N G, Jenkins A, Montgomery C A, Elder E F & Overbeek P A (1993). Reversal of left-right asymmetry: a situs inversus mutation. *Science* 260 (5108): 679–682.

Yost H J (1995). Vertebrate left-right development. *Cell* 82: 689–692.

**Chromosome aberrations and growth**

***Turner syndrome***

Ranke M B (1992). Growth disorder in the Ullrich–Turner syndrome. *Baillière's Clinical Endocrinology and Metabolism* 6: 603–619.

Ranke M E, Pflüger H, Rosendahl W, Stubbe P, Enders H, Bierich J R & Majewski F (1983). Turner syndrome: spontaneous growth in 150 cases and review of the literature. *European Journal of Pediatrics* 141: 81–88.

**Genetic disorders**

Pérez Jurado L A & Argente J (1994). Molecular basis of familial growth hormone deficiency. *Hormone Research* 42: 189–197.

Phillips III J A (1989). Inherited defects in growth hormone synthesis and action. In *The Metabolic Basis of Inherited Disease* (eds C R Scriver, A L Beaudet, W S Sly & D Valle) 6th edn pp 1965–1983. New York: McGraw-Hill.

**Growth in chronic diseases**

***Growth of children with asthma***

Balfour-Lynn L (1986). Growth and childhood asthma. *Archives of Disease in Childhood* 61: 1049–1055.

Cogswell J J & El-Bishti M M (1982). Growth retardation in asthma: role of calorie deficiency. *Archives of Disease in Childhood* 57: 473–480.

Doull I J M, Freezer N J & Holgate S T (1995). Growth of prepubertal children with mild asthma treated with inhaled Beclomethasone dipropionate. *American Journal of Respiratory & Critical Care Medicine* 151: 1715–1719.

Hauspie R, Susanne C & Alexander F (1976). A mixed longitudinal study of the growth in height and weight in asthmatic children. *Human Biology* 48: 271–283.

Kamada A K & Zefter S J (1995). Glucocorticoids and growth in asthmatic children. *Pediatric Allergy & Immunology* 6: 145–154.

Nikolaizik W H, Marchant L L, Preece M A & Warner J O (1994). Endocrine and lung function in asthmatic children on inhaled corticosteroids. *American Journal of Respiratory & Critical Care Medicine* 150: 624–628.

Warner J O (1992). Asthma: a follow-up statement from an international paediatric asthma consensus group. *Archives of Disease in Childhood* 67: 240–248.

Wolthers O D & Pedersen S (1991). Growth of asthmatic children during treatment with budesonide: a double blind trial. *British Medical Journal* 303: 163–165.

Zeitlin S R, Bond S, Wootton S, Gregson R K & Radford M (1993). Increased resting energy expenditure in childhood asthma: does this contribute towards growth failure? *Archives of Disease in Childhood* 67: 1366–1369.

***Growth of children with gastrointestinal disease***

Walker-Smith J A, Durie P R, Hamilton J R, Walker-Smith J A & Watkins J B (1996). *Pediatric Gastrointestinal Disease. Pathophysiology – Diagnosis, Management* 2nd edn. Toronto: B C Decker.

Walker-Smith J A & MacDonald T T (1994). Chronic inflammatory bowel disease in childhood. *Baillière's Clinical Gastroenterology: International Practice and Research*. Vol 8 Number 1. London: Baillière Tindall.

**Treatment of growth disorders**

Brook C G D (1993). *Clinical Paediatric Endocrinology* 3rd edn. Oxford: Blackwell Science.

Savage M O, Bourgnignon J-P & Grossman A B (1994). *Frontiers in Paediatric Neuroendocrinology*. Oxford: Blackwell Science.

## Part eight: Environmental factors influencing birth-weight

Eveleth P B & Tanner J M (1976, 1990). *Worldwide Variation in Human Growth*. 1st and 2nd edns. Cambridge: Cambridge University Press.

Kramer M S (1987). Determinants of intrauterine growth and gestational duration: a critical assessment and meta-analysis. *Bulletin of the World Health Organization* 65(5): 663–737.

Kline J, Stein Z & Susser M (1989). *Conception to Birth*. Oxford: Oxford University Press.

**Maternal anthropometry and birth outcome**

Alberman E (1984). Low birth weight. In *Perinatal Epidemiology* (ed M B Bracken). New York: Oxford University Press.

Baaqeel H S (1989). Perinatal outcome: is young maternal age a risk factor? *Annals of Saudi Medicine* 9 (2): 144–149.

Backstrand J R (1995). Maternal anthropometry as a risk predictor of pregnancy outcome: the Nutrition CSRP in Mexico. *Bulletin of the World Health Organization* 73 (Suppl): 96–97.

Barell V, Wax Y & Ruder A (1988). Analysis of geographic differentials in infant mortality rates. *American Journal of Epidemiology* 128: 218–230.

Becerraa J E, Atrash H K, Perez N & Saliceti J A (1993). Low birth weight and infant mortality in Puerto Rico. *American Journal of Public Health* 83: 1572–1576.

Chen R, Wax Y, Lusky A, Toppelberg G & Barell V (1991). A criterion for a standardised definition of low birth weight. *International Journal of Epidemiology* 20: 180–186.

Defo B K & Partin M (1993). Determinants of low birth weight – a comparative study. *Journal of Biosocial Science* 25: 87–100.

Fox S H, Koepsell T D & Daling J R (1994). Birth weight and smoking during pregnancy – effect modification by maternal age. *American Journal of Epidemiology* 139 (10): 1008–1015.

Goldenberg R L, Cliver S P, Cutter G R, Hoffman H J, Cassady G & Davis R O (1991). Black-White differences in newborn anthropometric measurements. *Obstetrics and Gynecology* 78: 782–788.

Grunerberger W, Ledolter S & Parshalk O (1979). Maternal hypotension: fetal outcome in treated and untreated cases. *Gynecologic and Obstetric Investigation* 10: 32–38.

Institute of Medicine (1985). *Preventing Low Birth Weight*. Washington DC: National Academy Press.

Jimenez R & Bacalao J (1995). Prognostic performance of several anthropometric indicators for predicting low and insufficient birth weight. *American Journal of Human Biology* 7: 303–311.

Kramer M S (1988). *Determinants of Intrauterine Growth and Gestational Duration: A Methodological Assesment and Synthesis*. Geneva: World Health Organization.

Miller H C (1983). A model for studying the pathogenesis and incidence of low birth weight infants. *American Journal of Diseases in Childhood* 137: 323–327.

Neumann C, Ferguson L & Bwibo N O (1995). Maternal anthropometry as a risk predictor of pregnancy outcome: the Nutrition CRSP in Kenya. *Bulletin of the World Health Organization* 73 (Suppl): 91–95.

Newman G (1906). *Infant mortality*. London: Methuen.

Pettersson R et al. (1978). Birth weight distribution and socioeconomic variables. In *Birth Weight Distribution – an Indicator of Social Development*. (eds G Sterky & L Mellander). Uppsala: SAREC Report #R: 245–253.

Rasmussen K M et al. (1985). *The Biological Meaning of Low Birthweight and the Use of Data on Low Birthweight for Nutritional Surveillance*. Ithaca, New York: Cornell Nutrition Surveillance Program No. 43.

Singh G K & Yu S M (1994). Birth weight differentials among Asian-Americans. *American Journal of Public Health* 84 (9): 1444–1449.

Starfield B, Shapiro S, McCormick M & Ross D (1982). Mortality and morbidity in infants with intrauterine growth retardation. *Journal of Pediatrics* 101: 978–983.

Taha T E, Gray R H & Mohamedani A A (1993). Malaria and low birth weight in Central Sudan. *American Journal of Epidemiology* 138 (5): 318–325.

Vega J, Saez G, Smith M, Agurto M & Morris N M (1993). Risk factors for low birth weight and intrauterine growth retardation in Chile. *Revista Medica De Chile* 121: 1210–1219.

World Health Organization (1995). Maternal anthropometry and birth outcome. *Bulletin of the World Health Organization* 73 (Suppl).

World Health Organization, Division of Family Planning (1980). The incidence of low birth weight: a critical review of available information. *World Health Statistics Quarterly* 33: 197–224.

**Nutrition**

Institute of Medicine (1990). *Nutrition during pregnancy: Part 1 Weight Gain; Part 2 Nutrient Supplements*. Washington DC: National Academy Press.

Nutrition in pregnancy. *British Nutrition Foundation Briefing Paper*. August 1994. ISSN 1350 6854.

Prentice A M, Cole T J, Foord F A, Lamb W H & Whitehead R G (1987). Increased birthweight after pre-natal dietary supplementation of rural African women. *American Journal of Clinical Nutrition* 46: 912–925.

Prentice A M, Poppitt S D, Goldberg G R & Prentice A (1995). Adaptive strategies regulating energy balance in human pregnancy. *Hum Reprod Update* 1: 149–161.

**Smoking**

Abel E L (1984). *Smoking and Reproduction: An Annotated Bibliography*. Boca Raton, Florida: CRC Press.

Rosenberg M J (ed) (1987). *Smoking and Reproductive Health*. Littleton, Massachusetts: PSG Publishing.

Schell L M, Relethford J H, Madan M, Naamon P B N & Hook E B (1994). Unequal adaptive advantage of changing cigarette use during pregnancy for heavy, moderate, and light smokers. *American Journal of Human Biology* 6: 25–32.

Sexton M & Hebel J R (1984). A clinical trial of change in maternal smoking and its effect on birth weight. *Journal of the American Medical Association* 251: 911–915.

Sexton M, Fox N L & Hebel J R (1991) Pre-natal exposure to tobacco: II Effects on cognitive functioning at age three. *International Journal of Epidemiology* 19: 72–77.

US Department of Health and Human Service (1980). *The Health Consequences of Smoking for Women: A Report of the Surgeon General* pp 217–221. Rockville, Maryland: U S Department of Health and Human Services, Public Health Service.

**Maternal HIV infection**

European Collaborative Study (1994). Perinatal findings in children born to HIV-infected mothers. *British Journal of Obstetrics and Gynaecology* 101: 136–141.

Johnstone F D (1995). Issues related to pregnancy. In *HIV Infection in Children. A Guide to Practical Management* (eds J Y Q Mok & M-L Newell) pp 140–151. Cambridge: Cambridge University Press.

Johnstone F D. Pregnancy outcome and pregnancy management in HIV-infected women. In *HIV Infection in Women* (eds M A Johnson & F D Johnstone) pp 187–198. Edinburgh: Churchill Livingstone.

**Substance abuse**

Brown E R & Zuckerman B (1991). The infant of the drug-abusing mother. *Pediatric Annals* 20: 555–563.

Chiriboga C A (1993). Fetal Effects. *Neurologic Complications of Drug and Alcohol Abuse* 11 (3): 707–728.

Jones C & Lopez R E (1990). Drug abuse and pregnancy. In *New Perspectives on Prenatal Care* (eds I R Merkatz, J E Thompson, P D Mullen & R L Goldenberg) pp 273–318. New York: Elsevier.

***Crack cocaine***

Plessinger M A & Woods J R Jr (1993). Maternal, placental, and fetal pathophysiology of cocaine exposure during regnancy. *Clinical Obstetrics and Gynecology* 36 (2): 267–276.

Slutsker L (1992). Risks associated with cocaine use during pregnancy. *Obstetrics & Gynecology* 79 (5) 1: 778–789.

Young S L, Vosper H J & Phillips S A (1992). Cocaine: its effects on maternal and infant health. *Pharmacotherapy* 12 (1): 2–17.

***Alcohol consumption and pregnancy outcome***

British Nutrition Foundation (1994). Nutrition in pregnancy. *British Nutrition Foundation Briefing Paper*. August 1994.

Institute of Medicine (1990). *Nutrition during pregnancy: Part 1 Weight Gain; Part 2 Nutrient Supplements*. Washington DC: National Academy Press.

**Teenage pregnancy**

Allen Guttmacher Institute (1994). *Sex and America's Teenagers*. New York: Allen Guttmacher Institute.

Hofferth S L & Hayes C D (eds) (1987). Risking the Future. Adolescent Sexuality, Pregnancy and Childbearing. *Working Papers and Statistical Appendixes*. Washington DC: National Academy Press.

Scholl T O, Hediger M L, Schall J I, Khoo C S & Fischer R L (1994). Maternal growth during pregnancy and the competition for nutrients. *American Journal of Clinical Nutrition* 60: 183–188.

## Part nine: Ecology of post-natal growth

Frisancho A R & Greksa L P (1989). Developmental responses to the acquisition of functional adaptation to high altitude. In *Human Population Biology: A Transdisciplinary Science* (eds M A Little & J D Haas) pp 203–221. Oxford: Oxford University Press.

Roberts D F (1978). *Climate and Human Variability*. Menlo Park: Cummings.

Shea B T & Bailey R C (1996). Allometry and adaptation of body proportions and stature in African pygmies. *American Journal of Physical Anthropology* 100: 311–340.

Watts E S (1986). The evolution of the human growth curve. In *Human Growth: A Comprehensive Treatise* (eds F Falkner & J M Tanner) 2nd edn, vol 3 pp 153–166. New York: Plenum.

**Infant-feeding and growth**

Binns C W (1976). Food sickness and death in children of the highlands of Papua New Guinea. *Journal of Tropical Pediatrics and Environmental Child Health* 22: 9–13.

British Paediatric Association (1994). Is breast feeding beneficial in the UK? Statement of the standing Committee on Nutrition of the British Paediatric Association. *Archives of Disease in Childhood* 71 (4): 376–380.

Butte N F, Wong W W, Garza C, Stuff J E, Smith E O, Klein P D & Nichols B L (1991). Energy requirements of breast-fed infants. *J Am Coll Nutr* 10(3): 190–195.

Carlson S E (1991). Plasma cholesterol and lipoprotein levels during fetal development and infancy. *Annals of the New York Academy of Science* 623 (1): 81–89.

Decsi T & Koletzko B (1994). Polyunsaturated fatty acids in infant nutrition. *Acta Paediatrica* (Suppl) 83 (395): 31–37.

Gibson R A, Makrides M, Neumann M A, Simmer K, Mantzioris E & James M J (1994). Ratios of linoleic acid to alpha-linolenic acid in formulas for term infants. *Journal of Pediatrics* 125: S48–55.

Gross S J & Slagle T A (1993). Feeding the low birth weight infant. *Clin Perinatol* 20: 193–209.

Grosvenor C E, Picciano M F & Baumrucker C R (1993). Hormones and growth factors in milk. *Endocrine Reviews* 14 (6): 710–28.

Grummer-Strawn L M (1993). Does prolonged breast-feeding impair child growth? A critical review. *Pediatrics* 91 (4): 766–71.

Lawrence P B (1994). Breast milk. Best source of nutrition for term and preterm infants. *Pediatric Clinics of North America* 41 (5): 925–941.

Oski F A (1993). Infant nutrition, physical growth, breastfeeding, and general nutrition. *Current Opinion in Pediatrics* 5 (3): 385–388.

Prentice A (1991). Breast-feeding and the older infant. *Acta Paediatrica Scandinavica* (Suppl) 374 (1): 78–88.

Sheard N F (1993). Growth patterns in the first year of life: what is the norm? *Nutritional Review* 51 (2): 52–54.

Wang I Y Fraser I S (1994). Reproductive function and contraception in the postpartum period. *Obstetric Gynecological Survey* 49(1): 56–63.

***Growth factors in maternal milk and development of the gastrointstinal tract***

Donovan S M & Odle J (1994). Growth factors in milk as mediators of infant development. *Annual Reviews in Nutrition* 14: 147–167.

Ellis L A & Picciano M F (1992). Milk-borne hormones: regulators of development in neonates. *Nutrition Today* 27: 6–14.

Ichiba H, Kusuda S, Itagane Y, Fujita K & Issiki G (1992). Measurement of growth promoting activity in human milk using a fetal small intestinal cell line. *Biology of the Neonate* 61: 47–53.

Klagsbrun M (1978). Human milk stimulates DNA synthesis and cellular proliferation in cultured fibroblasts. *Proceedings of the National Academy of Sciences* 75: 5057–5061.

Kohno Y, Shiraki K & Mura T (1991). The effect of human milk on DNA synthesis of neonatal rat hepatocytes in primary culture. *Pediatric Research* 29: 251–255.

Widdowson E M, Colombo V E & Artavanis C A (1976). Changes in the organs of pigs in response to feeding for the first 24 hours after birth. II The digestive tract. *Biology of the Neonate* 23: 272–281.

***The promotion of breast-feeding***

National Health and Medical Research Council (1995). *Guidelines for Health Workers to Encourage and Support Breast Feeding*. Canberra: NHMRC.

WHO/UNICEF (1989). Protecting, promoting and supporting breast-feeding: the special role of maternity services. A joint WHO/UNICEF statement. Geneva: World Health Organization.

World Health Organization (1981). International Code of Marketing of Breast Milk Substitutes. World Health Assembly Resolution May 1981 (as amended in resolution 47.5 of the 47th World Health Assembly May 1994).

***Growth and weaning in Papua New Guinea***

Jenkins C, Orr-Ewing A & Heywood P (1984). Cultural aspects of early childhood growth and nutrition among the Amele of lowland Papua New Guinea. *Ecology of Food and Nutrition* 14: 261–275.

Smith T, Earland J, Bhatia K, Heywood P & Singleton N (1993). Linear growth of children in Papua New Guinea in relation to dietary environmental and genetic factors. *Ecology of Food and Nutrition* 31: 1–25.

***Prolonged breast-feeding and growth***

Prentice A (1994). Extended breast-feeding and growth in China. *Nutrition Reviews* 52: 144–146.

Taren D & Chen J (1993). A positive association between extended breast feeding and nutritional status in rural Hubei Province, People's Republic of China. *American Journal of Clinical Nutrition* 58: 862–867.

**Nutrition**

Argao E A & Heubi J E (1993). Fat-soluble vitamin deficiency in infants and children. *Current Opinion in Pediatrics* 5 (5): 562–566.

Fairweather-Tait S J (1992). Iron deficiency in infancy easy to prevent – or is it? *European Journal of Clinical Nutrition* 46 (1): Suppl 4 S9–14.

Fall C (1992). Nutrition in early life and later outcome. *European Journal of Clinical Nutrition* 46 (1): Suppl 4 S57–63.

Filer L J Jr (1990). Iron needs during rapid growth and mental development. *Journal of Pediatrics* 117 (2 Pt 2): S143–146.

Filer L J Jr (1992). A glimpse into the future of infant nutrition. *Pediatric Annals* 21 (10): 633–636, 639.

Giovannini M, Agostoni C & Salari P C (1991). Is carnitine essential in children? *Journal of International Medical Research* 19 (2): 88–102.

Gracey M & Falkner F (eds) (1985). *Nutrition and assessment of normal growth*. New York: Vevey-Raven Press.

Hardy S C & Kleinman R E (1994). Fat and cholesterol in the diet of infants and young children: implications for growth development, and long-term health. *Journal of Pediatrics* 125 (5 Pt 2): S69–77.

Haycock G B (1993). The influence of sodium on growth in infancy. *Pediatric Nephrology* 7 (6): 871–875.

Milner J A (1990). Trace minerals in the nutrition of children. *Journal of Pediatrics* 117 (2 Pt 2): S147–155.

Motarjemi Y, Kaferstein F, Moy G & Quevedo F (1993). Contaminated weaning food: a major risk factor for diarrhoea and associated malnutrition. *Bulletin of the World Health Organization* 71 (1): 79–92.

National Health & Medical Reasearch Council (1993). *Dietary Guidelines for Australians*. Canberra: AGPS.

Neumann C G & Harrison G G (1994). Onset and evolution of stunting in infants and children. Examples from the Human Nutrition Collaborative Research Support Program. Kenya and Egypt studies. *European Journal of Clinical Nutrition* 48 (1) Suppl 1: S90–102.

Reeds P J & Hutchens T W (1994). Protein requirements: from nitrogen balance to functional impact. *Journal of Nutrition* 124 Suppl 9: 1754S–1764S.

Shils M E & Young V R (1988). *Modern Nutrition in Health & Disease*. Philadelphia: Lea and Febiger.

Yip R, Scanlon K & Trowbridge F (1993). Trends and patterns in height and weight status of low-income US children. *Critical Review in Food Science and Nutrition* 33 (4–5): 409–421.

***Long-term effect of iron supplementation on growth in Indonesian preschool children***

Angeles I M, Schultink J W, Matulessi P, Gross R & Sastroamidjojo S (1993). Decreased rate of stunting among anaemic Indonesian pre-school children through iron supplementation. *American Journal of Clinical Nutrition* 58: 339–342.

Aukett M A, Parks Y A, Scott P H & Wharton B A (1986). Treatment with iron increases weight gain and psychomotor development. *Archives of Disease in Childhood* 61: 849–857.

Billewicz W Z & McGregor I A (1982). A birth to maturity longitudinal study of heights and weights in two West-African (Gambian) villages. *Annals of Human Biology* 9: 309–320.

United Nations ACC/SCN (1987). First report on the world nutrition situation. Geneva: United Nations Administrative Committee, Co-ordination Subcommittee on Nutrition.

***Body-size and energy needs***

FAO/WHO/UNU (1985). *Energy and Protein Requirements*. Geneva: World Health Organization.

Kleiber M (1975). *The Fire of life: An Introduction to Animal Energetics*. New York: R E Krieger.

***The influence of nutrition in early life on stature and risk of obesity***

Agostini C, Salari P & Riva E (1992). Metabolic needs, utilization and dietary sources of fatty acids in childhood. *Progress in Food and Nutrition Science* 16: 1–49.

Clemmons D R & Underwood L E (1991). Nutritional regulation of IGF-I and IGF binding proteins. *Annual Reviews of Nutrition* 11: 393–412.

Kim S, Mauron J, Gleason R & Wurtman R (1991). Selection of carbohydrate to protein ratio and correlations with weight gain and body fat in rats allowed three dietary choices. *International Journal of Vitamin Nutrition Research* 61: 166–179.

Micosi M S (1993). Functional consequences from varying patterns of growth and maturation during adolescence. *Hormone Research* 39 (Suppl 3): 49–58.

Nicklas T A, Webber L S, Srinivasan S R & Berenson G (1993). Secular trends in dietary intakes and cardiovascular risk factors of 10-year-old children: The Bogalusa Heart Study (1973–1988). *American Journal of Clinical Nutrition* 57: 930–937.

Rolland-Cachera M-F (1995). Prediction of adult body composition from infant and child measurements. In *Body Composition Techniques in Health and Diseases* (eds P S W Davies & T J Cole) pp 100–145. Cambridge: Cambridge University Press.

Rolland-Cachera M-F, Deheeger M, Akrout M & Bellisle F (1995). Influence of macronutrients on adiposity development: a follow up study of nutrition and growth from 10 months to 8 years of age. *International Journal of Obesity* 19: 1–6.

Rosskamp R, Becker M & Soetadji S (1987). Circulating somatomedin-C levels and the effect of growth realising factor on plasma levels of growth hormone and somatomedin-like immunoreactivity in obese children. *European Journal of Pediatrics* 146: 48–50.

Sorensen T I & Price R A (1990). Secular trends in body mass index among Danish young men. *International Journal of Obesity*: 14: 411–419.

Waterlow J C, Buzina R, Keller W et al. (1977). The presentation and use of height and height data for comparing nutritional status of age groups of children under the age of 10 years. *Bulletin of the World Health Organization* 55: 489–498.

***Protein–body-size relationships***

Brody S (1945). *Bioenergetics and Growth*. New York: Reinhold.

Henry C J K (1995). Influence of body composition on protein and energy requirements. In *Body Composition Techniques in Health and Disease* (eds P S W Davies & T J Cole) pp 85–99. Cambridge: Cambridge University Press.

***Childhood obesity and growth***

Bouchard C (1992). Genetic aspects of obesity. In *Obesity* (eds P Bjorntorp & B N Brodoff) pp 343–351. Philadelphia: Lippincott.

Brambilla P, Manzoni P, Sironi S, Simone P, DelMaschio A, Di Natale B & Chiumello G (1994). Peripheral and abdominal adiposity in childhood obesity. *International Journal of Obesity* 18: 795–800.

Buenen G, Malina R M, Lefevre J, Claessens A L, Rensen R, Simons J, Maes H, Vanreusel B & Lysens R (1994). Size, fatness and relative fat distribution of males of contrasting maturity status during adolescence and as adults. *International Journal of Obesity* 18: 670–678.

Dietz W H & Gortmaker S L (1985). Do we fatten our children at the television set? Obesity and television viewing in children and adolescents. *Paediatrics* 75: 807–812.

Fomon S J & Nelson S E (1993). Size and growth. In *Nutrition of Normal Infants* (ed S J Fomon) pp 36–84. St Louis: Mosby.

Garby L & Astrup A (1992). A simple hypothesis for the development of obesity. *European Journal of Clinical Nutrition* 93: 695–686.

Lissau-Lund-Sorensen I & Sorensen T I (1992). A prospective study of the influence of social factors in childhood on risk of overweight in young adulthood. *International Journal of Obesity* 16: 169–176.

Mellbin T & Vuille J-C (1973). Physical development at 7 years of age in relation to velocity of weight gain in inf ancy with special reference to incidence of overweight. *British Journal of Preventative Social Medicine* 27: 225–235.

Polito C, Di Toro A, Collini R, Cimmarute E, D'Alfonso C & Del Guidice G (1995). Advanced RUS and normal carpal bone age in childhood obesity. *International Journal of Obesity* 19: 506–507.

Poskitt E M E (1995). The fat child. In *Clinical Paediatric Endocrinology* (ed C G D Brook) 3rd edn pp 210–233. Oxford: Blackwells.

Poskitt E M E (1987). Personal practice: management of obesity. *Archive of Disease in Childhood* 62: 305–310.

Poskitt E M E (1995). Defining childhood obesity: the relative body mass index. *Acta Paediatrica* 84.

Poskitt E M E (1993). The prevention of obesity. *Obesity 93. Proceedings of 5th European Congress on Obesity.*

Poskitt E M E & Cole T J (1977). Do fat babies stay fat? *British Medical Journal* 1: 7–9.

Rolland-Cachera M-F, Deheeger M, Bellisle F, Sempé M, Guilloud-Bataille M & Patois E (1984). Adiposity rebound in children: a simple indicator for predicting obesity. *American Journal of Clinical Nutrition* 39: 129–135.

Whitehead R G, Paul A A & Cole T J (1982). Trends in food energy intakes throughout childhood from 1–18 years *Human Nutrition: Clinical Nutrition* 36A: 57–62.

**Infection**

Black R E, Brown K H & Becker S (1984). Effects of diarrhea associated with specific enteropathogens on the growth of children in rural Bangladesh. *Pediatrics* 73: 799–805.

Briend A (1990). Is diarrhoea a major cause of malnutrition among the under-fives in developing countries? A review of available evidence. *European Journal of Clinical Nutrition* 44: 611–628.

Briend A, Hasan Kh Z, Aziz K M A & Hoque B A (1989). Are diarrhoea control programmes likely to reduce childhood malnutrition? Observation from rural Bangladesh. *Lancet* 2: 319–322.

Brown K H, Black R E, Robertson A D & Becker S (1985). Effects of season and illness on the dietary intake of weanlings during longitudinal studies in rural Bangladesh. *American Journal of Clinical Nutrition* 41: 343–355.

Dagnelie P C, van Dusseldorp M, van Staveren W A & Hautvast J G A J (1994). Effect of macrobiotic diets on linear growth in infants and children until 10 years of age. *European Journal of Clinical Nutrition* 48 (Suppl 1): S103–112.

Dickin K L, Brown K H, Fagbule D, Adedoyin M, Gittelsohn J, Esrey S A & Oni G A (1990). Effect of diarrhoea on dietary intake by infants and young children in rural villages of Kwara State, Nigeria. *European Journal of Clinical Nutrition* 44: 307–317.

Hasan Kh Z, Briend A, Aziz K M A, Hoque B A, Patwary M Y & Huttly S R A (1989). Lack of impact of a water and sanitation intervention on the nutritional status of children in rural Bangladesh. *European Journal of Clinical Nutrition* 43: 837–843.

Lutter C K, Habicht J P, Rivera J A & Martorell R (1992). The relationship between energy intake and diarrhoeal disease in their effects on child growth: biological model, evidence, and implications for public health policy. *Food and Nutrition Bulletin* 14: 36–42.

Moy R J D, Choto R, Booth I W & McNeish A S (1990). Diarrhoea and catch-up growth. *Lancet* 1: 600.

Price J S, Oyajobi B O & Russell R G G (1994). The cell biology of bone growth. *European Journal of Clinical Nutrition* 48 (Suppl 1): S131–S149.

Rowland M G M, Cole T J & Whitehead R G (1977). A quantitative study into the role of infection in determining nutritional status in Gambian village children. *British Journal of Nutrition* 37: 441–50.

Scrimshaw N S, Taylor C E, Gordon J E (1968). *Interactions of Nutrition and Infections WHO Monograph Series* 57.

Solomons N W, Mazariegos M, Brown K H & Klasing K (1993). The underprivileged, developing country child: environmental contamination and growth failure revisited. *Nutrition Reviews* 51: 327–332.

***HIV and growth***

Guarino A et al. (1993). Intestinal malabsorption of HIV-infected children: relationshp to diarrhoea, failure to thrive, enteric micro-organisms and immune impairment. *AIDS* 7: 1435–1440.

Henderson R A et al. (1994). Effect of enteral tube feeding on growth in children with symptomatic human immunodeficiency virus infection. *Journal of Pediatric Gastroenterology and Nutrition* 18: 429–434.

Laue L & Cutler G B Jr (1994). Abnormalities in growth and development. In *Pediatric AIDS* (eds P A Pizzo & C M Wilfert) pp 575–590. Baltimore: William & Wilkins.

McKinney R E et al. (1993). Effect of human immunodeficiency virus infection on the growth of young children. *The Journal of Pediatrics* 123: 579–582.

Nicholas S W et al. (1991). Guidelines for nutritional support of HIV-infected children. *The Journal of Pediatrics* 119: S59–S62.

Tovo P A et al. (1992). Prognostic factors and survival in children with perinatal HIV-1 infection. *The Lancet* 339: 1249–1253.

***Intestinal parasitism***

McGarvey S T, Aligui G, Graham K K, Peters P, Olds G R & Olveda R (1996). Schistosomiasis japonica and childhood nutritional status in Northeastern Leyte, The Philippines: a randomized trial of praziquantel versus placebo. *American Journal of Tropical Medicine and Hygiene.*

Stephenson L S (1987). *Impact of Helminth Infections on Human Nutrition*. London: Taylor & Francis.

Warren K S, Bundy D A P, Anderson R M, Davis A R, Henderson D A, Jamison D T, Prescott N & Senft A (1993). Helminth infections. In *Disease Control Priorities in Developing Countries* (eds D T Jamison & W H Mosley) pp 131–160. New York: Oxford University Press.

***Immunocompetence***

Alvarado T & Luthringer D G (1971). Serum immunoglobulins in edematous protein calorie malnourished children. *Clinical Pediatrics* 10: 174.

Briend A (1990). Is diarrhoea a major cause of malnutrition among the under-fives in developing countries? A review of available evidence. *European Journal of Clinical Nutrition* 44: 611–628.

Chandra R K (1990). Nutrition and immunity. *American Journal of Clinical Nutrition* 53: 1087–1101.

Green F & Heyworth B (1980). Immunoglobulin-containing cells in jejunal mucosa of children with protein-energy malnutrition and gastroenteritis. *Archives of Disease in Childhood* 55 380–383.

Hayward A (1986). Development of Immune responsiveness. In *Human Growth: A Comprehensive Treatise* (eds F Falkner & J M Tanner) 2nd edn, vol 3 pp 593–607. New York: Plenum.

Hoffman-Goetz L (1986). Malnutrition and immunological function with special reference to cell-mediated immunity. *Yearbook of Physical Anthropology* 29: 139–159.

Johnson T R, Moore W M & Jeffries J E (1978). *Children are Different: Developmental Physiology*. Columbus, Ohio: Ross Laboratories.

Ogra P L, Fishaut M & Theodore C (1979). Immunology of breast milk: maternal neonatal interactions. In *Human Milk: Its Biological and Social Value* (eds S Freier & A I Eidelman) pp 115-124. Amsterdam: Excerpta Medica.

Solomons N W (1993). Pathways to the impairment of human nutritional status by gastrointestinal pathogens. *Parasitology* 107: S19–35.

Stiehm E R & Fudenberg H H (1966). Serum levels of immune globulins in health and disease: a survey. *Pediatrics* 37: 715.

Tomkins A M (1986). Protein-energy malnutrition and risk of infection. *Proceedings of the Nutrition Society* 45: 289–304.

Ulijaszek S J (1990). Nutritional status and susceptibility to infectious disease. In *Diet and Disease in Traditional and Developing Societies* (ed G A Harrison & J C Waterlow) pp 137–154.

Ulijaszek S J. Relationships between undernutrition, infection, and growth and development. *International Journal of Anthropology* (in press).

Zumrawi F Y, Dimond H & Waterlow J C (1987). Effects of infection on growth in Sudanese children. *Human Nutrition: Clinical Nutrition* 41: 453–461.

**Growth and psychosocial stress**

Curfs L M & Fryns LP (1992). Prader–Willi syndrome: a review with special attention to the cognitive and behavioural profile. *Birth Defects: Original Article Series* 28: 99–104.

Frasier S D & Rallinson M L (1972). Growth retardation and emotional deprivation: relative resistance to treatment with human growth hormone. *Journal of Pediatrics* 80 (4): 603–609.

Karlberg J (1989). On the construction of the infancy-childhood-puberty growth standard. *Acta Paediatrica Scandinavica* 356 (Suppl): 26–37.

Powell G F, Brasel J A, Raiti S & Blizzard R M (1967a). Emotional deprivation and growth retardation simulating idiopathic hypopituitarism: I Clinical evaluation of the syndrome. *New England Journal of Medicine* 276: 1271–1278.

Powell G F, Brasel J A, Raiti S & Blizzard R M (1967b). Emotional deprivation and growth retardation simulating idiopathic hypopituitarism: II Endocrinologic evaluation of the syndrome. *New England Journal of Medicine* 276: 1279–1283.

**Catch-up growth in height**

Alvear J, Artaza C, Vial M, Guerrero S & Muzzo S (1986). Physical growth and bone age of survivors of protein energy malnutrition. *Archives of Disease in Childhood* 61: 257–262.

Barr D G D, Shmerling D H, Prader A (1972). Catch-up growth in malnutrition, studied in coeliac disease after institution of gluten-free diet. *Pediatric Research* 6: 521–527.

Boudraa G, Touhami M, Pochart P, Soltana R, Mary J Y & Desjeux J F (1990). Effect of feeding yogurt versus milk in children with persistent diarrhoea. *Journal of Pediatric Gastroenterology and Nutrition* 11: 509–512.

Cooper E S, Bundy D A P, MacDonald T T & Golden M H (1990). Growth suppression in the *trichuris* dysentry syndrome. *European Journal of Clinical Nutrition* 44: 285–291.

Doherty J F, Golden M H & Brooks S E (1991). Peroxisomes and the fatty liver of malnutrition: an hypothesis. *American Journal of Clinical Nutrition* 54: 674–677.

Flyvbjerg A, Dorup I, Everts M E & Orskov H (1991). Evidence that potassium deficiency induces growth retardation through reduced circulating levels of growth hormone and insulin-like growth factor I *Metabolism: Clinical & Experimental* 40: 769–775.

Golden M H (1988). The role of individual nutrient deficiencies in growth retardation of children as exemplified by zinc and protein. In *Linear Growth Retardation in Less Developed Countries* (ed J C Waterlow) pp 143–163. New York: Raven Press.

Golden M H (1991). The nature of nutritional deficiency in relation to growth failure and poverty. *Acta Paediatrica Scandinavica* 374: 95–110.

Golden M H (1994). Is complete catch-up possible for stunted malnourished children? *European Journal of Clinical Nutrition* 48: S58–S71.

Graham G G, Adrianzen B, Rabold J, Mellits E D (1982). Later growth of malnourished infants and children. Comparison with 'healthy' siblings and parents. *American Journal of Diseases in Childhood* 136: 348–352.

Grantham-McGregor S M, Powell C A & Fletcher P D L (1989). The relationship between stunting, an episode of severe malnutrition and mental development in young children. *West Indian Medical Journal* (Abstract)

Grantham-McGregor S M, Powell C A, Fletcher P D L (1989). Stunting, severe malnutrition and mental development in young children. *European Journal of Clinical Nutrition* 43: 403–409.

Keet M P, Moodie A D, Wittmann W & Hansen J D L (1971). Kwashiorkor: a prospective ten-year follow-up study. *South African Medical Journal* 45: 1427–1449.

Kulin H E, Bwibo N, Mutie D & Santner S J (1982). The effect of chronic childhood malnutrition on pubertal growth and development. *American Journal of Clinical Nutrition* 36: 527–536.

Prader A, Tanner J M & von Harnack G A (1963). Catch-up growth following illness or starvation. An example of developmental canalization in man. *Journal of Pediatrics* 62: 646–659.

Scientific Committee for Food: Commission of The European Communities (1994). *Reference Nutrient Intakes for the European Community* (CS/RDA 15). Brussels: Commission of the European Communities.

Walker S P & Golden M H (1988). Growth in length of children recovering from severe malnutrition. *European Journal of Clinical Nutrition* 42: 395–404.

Walravens P A, Hambidge K M & Koepfitr D M (1989). Zinc supplementation in infants with a nutritional pattern of failure to thrive: a double-blind, controlled study. *Pediatrics* 83: 532–538.

**Catch-up weight-gain**

Ashworth A (1969). Growth rates of children recovering from protein-calorie malnutrition. *British Journal of Nutrition* 23: 835.

Ashworth A (1969). Metabolic rates during recovery from protein-calorie malnutrition: the need for a new concept of specific dynamic action. *Nature* 223: 407–409.

Ashworth A (1974). Ad lib feeding during recovery from malnutrition. *British Journal of Nutrition* 31: 109–112.

Brooke O G & Wheeler E F (1976). High energy feeding in protein-energy malnutrition. *Archives of Disease in Childhood* 51: 968–971.

Castillo-Duran C, Fisberg M, Valenzuela A, Egana J I & Uauy R (1983). Controlled trial of copper supplementation during the recovery from marasmus. *American Journal of Clinical Nutrition* 37: 898–903.

Fjeld C R, Schoeller D A & Brown K H (1989). A new model for predicting energy requirements of children during catch-up growth developed using doubly labeled water. *Pediatric Research* 25: 503–508

Forbes G B (1974). A note on the mathematics of catch-up growth. *Pediatric Research* 8: 929–931.

Golden B E & Golden M H (1981). Plasma zinc, rate of weight gain, and the energy cost of tissue deposition in children recovering from severe malnutrition on a cow's milk or soya protein based diet. *American Journal of Clinical Nutrition* 34: 892–899.

Golden B E & Golden MH (1985). Zinc, immunocompetence and catch-up growth in malnourished children. In *Zinc in Human Medicine* (eds K M Hambidge & P M Aggett) pp 95–102. Isleworth: T I L Publications.

Golden B E & Golden M H (1985). Zinc, sodium and potassium losses in the diarrhoeas of malnutrition and zinc deficiency. In *Trace Element Metabolism in Man and Animals -5*. (eds C F Mills, I Bremner & J K Chesters) pp 228–232. Aberdeen: Commonwealth Agricultural Bureau.

Golden B E & Golden M H (1992). Effect of zinc on lean tissue synthesis during recovery from malnutrition. *European Journal of Clinical Nutrition* 46: 697–706.

Golden M H (1991). The nature of nutritional deficiency in relation to growth failure and poverty. *Acta Paediatrica Scandinavica* 374: 95–110.

Golden M H & Golden B E (1981). Effect of zinc supplementation on the dietary intake, rate of weight gain, and energy cost of tissue deposition in children recovering from severe malnutrition. *American Journal of Clinical Nutrition* 34: 900–908.

Golden M H, Golden B E, Harland P S E G & Jackson A A (1978). Zinc and immunocompetence in protein-energy malnutrition. *Lancet* 1: 1226–1228.

Golden M H, Jackson A A & Golden B E (1977). Effect of zinc on thymus of recently malnourished children. *Lancet* 2: 1057–1059.

Hansen-Smith F M, Picou D I M & Golden M H (1979). Growth of muscle fibres during recovery from severe malnutrition. *British Journal of Nutrition* 41: 275–282.

Hommes F A, Drost Y M, Geraets W X M, Reijenga M A A (1975). The energy requirement for growth: an application of Atkinson's metabolic price system. *Pediatric Research* 9: 51–55.

Jackson A A, Picou D I M & Reeds W (1977). The energy cost of repleting tissue deficits during recovery from protein-energy malnutrition. *American Journal of Clinical Nutrition* 30: 1514–1517.

Kerr D S, Ashworth A, Picou D I M et al. (1973). Accelerated recovery from infant malnutrition with high calorie feeding. In *Endocrine Aspects of Malnutrition* (eds L Gardner & P Amacher) p 467. Santa Ynez, California: Kroc Foundation.

MacLean W C & Graham G G (1980). The effect of energy intake on nitrogen content of weight gained by recovering malnourished infants. *American Journal of Clinical Nutrition* 33: 903–909.

McCance R A (1936). Experimental human salt deficiency. *Lancet* 1: 823–830.

Patrick J (1977). Death during recovery from severe malnutrition and its possible relationship to sodium pump activity in the leucocyte. *British Medical Journal* 1: 1051–1054.

Patrick J (1978). Interrelations between the physiology of sodium, potassium and water, and nutrition. *Journal of Human Nutrition* 32: 405–418.

Patrick J, Golden B E & Golden M H (1980). Leucocyte sodium transport and dietary zinc in protein energy malnutrition. *American Journal of Clinical Nutrition* 33: 617–620.

Reeds P J, Jackson A A, Picou D I M & Poulter N (1978). Muscle mass and composition in malnourished infants and children and changes seen after recovery. *Pediatric Research* 12: 613–618.

Waterlow J C (1961). The rate of recovery of malnourished infants in relation to the protein and calorie levels of the diet. *Journal of Tropical Pediatrics* 7: 16–22.

Wheeler E F (1974). Changes in anthropometric measurements of children recovering from protein-energy malnutrition. *Proceedings of the Nutrition Society* 34: 3SA.

***Recovery from kwashkiorkor***

Burton B T & W R Foster (1988). *Human Nutrition* 4th edn. New York: McGraw-Hill.

Cameron N, Jones P R M, Moodie A, Mitchell J, Bowie M D, Mann M D & Hansen J D L (1986). Timing and magnitude of adolescent growth in height and weight in Cape Coloured children after kwashiorkor. *Journal of Pediatrics* 109: 548–555 (1986).

Cameron N, Mitchell J, Meyer D, Moodie A, Bowie M D, Mann M D & Hansen J D L (1988). Secondary sexual development of 'Cape Coloured' girls following kwashiorkor. *Annals of Human Biology* 15: 65–76.

Cameron R, Mitchell J, Meyer D, Moodie A, Bowie M D, Mann M D & Hansen J D L (1990). Secondary sexual development of 'Cape Coloured' boys following kwashiorkor. *Annals of Human Biology* 17: 217–228.

Gibson R S (1990). *Principles of Nutritional Assessment.* Oxford: Oxford University Press.

Golden M H N (1982). Protein deficiency, energy deficiency, and the oedema of malnutrition. *Lancet* 1, 1261–1265.

Golden M H N (1985). The consequence of protein deficiency in man and its relationship to the features of kwashiorkor. In *Nutritional Adaptation in Man*, (eds K L Blaxter & J C Waterlow) pp 169–187. London: John Libbey.

Golden M H N & Ramdath D (1987). Free radicals in the pathogenesis of kwashiorkor. *Proceedings of the Nutrition Society* 46: 53–68.

Jackson A A (1990). The aetiology of kwashiorkor. In *Diet and Disease in Traditional and Developing Societies* (eds G A Harrison & J C Waterlow) pp 76–113. Cambridge: Cambridge University Press.

Waterlow J C (1992). *Protein Energy Malnutrition.* London: Edward Arnold.

**Seasonality of subsistence and disease ecology**

Benefice E, Chevassus-Agnes S & Barral H (1984). Nutritional situation and seasonal variations for pastoralist populations of the Sahel (Senegalese Ferlo). *Ecology of Food and Nutrition* 14: 229–247.

Billewicz W Z & McGregor I A (1981). The demography of two West African (Gambian) villages 1951–75. *Journal of Biosocial Science* 13: 219–240.

Billewicz W Z & McGregor I A (1982). A birth-to-maturity longitudinal study of heights and weights in two West African (Gambian) villages 1951–1975. *Annals of Human Biology* 9: 309–320.

Chambers R, Longhurst R & Pacey A (eds) (1981). *Seasonal Dimensions to Rural Poverty*. London: Frances Pinter.

Lawrence M, Lawrence F, Cole T J, Coward W A, Singh J & Whitehead R G (1989). Seasonal pattern of activity and its nutritional consequence in Gambia. In *Seasonal*

*Variability in Third World Agriculture* (ed D E Sahn) pp 47–56. Baltimore: Johns Hopkins University Press.
Loutan L & Lamotte J M (1984). Seasonal variations in nutrition among a group of nomadic pastoralists in Niger. *Lancet* 2: 945–947.
Prentice A M, Whitehead R G, Roberts S B & Paul A A (1981). Long-term energy balance in child-bearing Gambian women. *American Journal of Clinical Nutrition* 34: 2790–2799.
Tomkins A (1993). Environment season and infection. In *Seasonality and Human Ecology* (eds S J Ulijaszek & S S Strickland) pp 123–134. Cambridge: Cambridge University Press.
Ulijaszek S J & Strickland S S (eds) (1993). *Seasonality and Human Ecology*. Cambridge: Cambridge University Press.
Ulijaszek S J & Strickland S S (1993). *Nutritional Anthropology. Prospects and Perspectives*. London: Smith-Gordon.
Watkinson M & Rushton D I (1983). Plasmodial pigmentation of placenta and outcome of pregnancy in West African mothers. *British Medical Journal* 28: 251–254.

#### *Growth among Nepali agriculturalists*

Farquharson S M (1976). Growth patterns and nutrition in Nepali children. *Archives of Disease in Childhood* 51: 3–12.
Panter-Brick C (1991). Lactation birth spacing and maternal work-loads among two castes in rural Nepal. *Journal of Biosocial Science* 23: 137–154.
Panter-Brick C (1992). Women's work and child nutrition: the food intake of 0–4 year old children in rural Nepal. *Ecology of Food and Nutrition* 29: 11–24.
Panter-Brick C (1993). Seasonality and levels of energy expenditure during pregnancy and lactation for rural Nepali women. *American Journal of Clinical Nutrition* 57: 620–8.
Panter-Brick C (1995). Child-care strategies in Nepal: responses to ecology demography and society. In *Human Populations: Diversity and Adaptation* (eds A J Boyce & V Reynolds) pp 174–188. Oxford: Oxford University Press.
United Nations (1993). *ACC/SCN Second Report on the World Nutrition Situation*. Vol II. Geneva: United Nations.

#### *Growth of African pastoralist children*

Galvin K A, Coppock D L & Leslie P W (1994). Diet nutrition and the pastoral strategy. In *African Pastoralist Systems: an Integrated Approach* (eds E Fratkin, K A Galvin & E A Roth) pp 113–131. Boulder, Colorado: Lynne Rienner.
Little M A (1989). Human biology of African pastoralists. *Yearbook of Physical Anthropology* 32 215–247.
Little M A & Johnson B R Jr (1987). Mixed-longitudinal growth of nomadic Turkana pastoralists. *Human Biology* 59 695–707.
Little M A, Gray S J & Leslie P W (1993). Growth of nomadic and settled Turkana infants of northwest Kenya. *American Journal of Physical Anthropology* 92: 273–289.
Wheeler E F (1980). Nutritional status of savanna peoples. In *Human Ecology in Savanna Environments* (ed D R Harris) pp 439–55. London: Academic Press.

#### Growth in high-altitude populations

Beall C M (1981). Growth in a population of Tibetan origin at high altitude. *Annals of Human Biology* 8: 31–38.
Beall C M (1983). Ages at menopause and menarche in a high-altitude Himalayan population. *Annals of Human Biology* 10: 365–70.
Frisancho A R (1978). Human growth and development among high-altitude populations. In *The Biology of High Altitude Peoples (IBP 14)* (ed P T Baker) pp 117–171. Cambridge: Cambridge University Press.
Frisancho A R (1993). Prenatal and postnatal growth and development at high altitude. In *Human Adaptation and Accommodation* (ed A R Frisancho) pp 281–307. Ann Arbor: University of Michigan Press.
Frisancho A R, Frisancho H G, Milotich M, Brutsaert T, Albalak R, Spielvogel H, Villena M, Vargas E & Soria R (1995). Developmental genetic and environmental components of aerobic capacity at high altitude. *American Journal of Physical Anthropology* 96: 431–442.
Greksa L P (1990). Developmental responses to high-altitude hypoxia in Bolivian children of European ancestry: a test of the developmental adaptation hypothesis. *American Journal of Human Biology* 2: 603–612.
Greksa L P & Beall C M (1989). Dvelopment of chest size and lung function at high altitude. In *Human Population Biology: A Transdisciplinary Science* (eds M A Little & J D Haas) pp 222–238. New York: Oxford University Press.
Greksa L P, Spielvogel H & Caceres E (1985). Effect of altitude on the physical growth of upper-class children of European ancestry. *Annals of Human Biology* 12: 225–232.
Greksa L P, Spielvogel H & Caceres E (1994). Total lung capacity in young highlanders of Aymara ancestry. *American Journal of Physical Anthroplogy* 94: 477–486.
Moore L G (1990). Maternal $O_2$ transport and fetal growth in Colorado Peru and Tibet high-altitude residents. *American Journal of Human Biology* 2: 627–637.
Pawson I G (1977). Growth characteristics of populations of Tibetan origin in Nepal. *American Journal of Physical Anthropology* 47: 473–482.

#### A functional outcome in adulthood of undernutrition during growth

Astrand I (1967). Degree of strain during building work as related to individual aerobic capacity. *Ergonomics* 10: 293–303.
Barac-Nieto M, Spurr G B, Maksud M G & Lotero H (1978a). Aerobic work capacity in chronically undernourished adult males. *Journal of Applied Physiology* 44: 209–215.
Barac-Nieto M, Spurr G B, Lotero H & Maksud M G (1978b). Body composition in chronic undernutrition. *American Journal of Clinical Nutrition* 31: 23–40.
Barac-Nieto M, Spurr G B, Lotero H, Maksud M G & Dahners H W (1979). Body composition during nutritional repletion of severely undernourished men. *American Journal of Clinical Nutrition* 32: 981–991.
Barac-Nieto M, Spurr G B, Dahners H W & Maksud M G (1980). Aerobic work capacity and endurance during nutritional repletion of severely undernourished men. *American Journal of Clinical Nutrition* 33: 2268–2275.
Barac-Nieto M, Spurr G B & Reina J C (1984). Marginal malnutrition in school-aged Colombian boys: body composition and maximal $o_2$ consumption. *American Journal of Clinical Nutrition* 39: 830–839.
Michael E D, Hutton K E & Horvath S M (1961). Cardiorespiratory responses during prolonged exercise. *Journal of Applied Physiology* 16: 997–1000.
Spurr G B, Barac-Nieto M & Maksud M G (1975). Energy expenditure cutting sugar cane. *Journal of Applied Physiology* 39: 990–996.
Spurr G B, Barac-Nieto M & Maksud M G (1977). Productivity and maximal oxygen consumption in sugar cane cutters. *American Journal of Clinical Nutrition* 30: 316–321.
Spurr G B (1983). Nutritional status and physical work capacity. *American Journal of Physical Anthropology* (Suppl 4) 26: 1–35.
Spurr G B, Reina J C & Barac-Nieto M (1983a). Marginal malnutrition in school-aged Colombian boys: anthropometry and maturation. *American Journal of Clinical Nutrition* 37: 119–132.
Spurr G ,B Reina J C, Dahners H W & Barac-Nieto M (1983b). Marginal malnutrition in school-aged Colombian boys: functional consequences in maximum exercise. *American Journal of Clinical Nutrition* 37: 834–847.
Spurr G B & Reina J C (1986). Marginal malnutrition in school-aged Colombian boys: body size and energy costs of walking and light load carrying. *Human Nutrition Clinical Nutrition* 40: 409–419.
Spurr G B (1988). Body size, physical work capacity, and productivity in hard work. Is bigger better? In *Linear Growth Retardation in Less Developed Countries* (ed J C Waterlow) pp 215–243. New York: Raven Press.
von Dobein W (1956). Human standard and maximal metabolic rate in relation to fat free mass. *Acta Physiologica Scandinavica* (Suppl) 126: 1–79.

### Part ten: Between-population differences in human growth

Baker P T & Weiner J S (1966). *Human Adaptability*. Oxford: Oxford University Press.
Boas F (1912). *Changes in Bodily Form of Descendents of Immigrants*. New York: Colombia University Press.
Butler G E, McKie M & Ratcliffe S G (1990). The cyclical nature of prepubertal growth. *Annals of Human Biology* 17: 177–198.
Byard P, Go S & Roche A F (1991). Model fitting to early childhood length and weight data from the Fels longitudinal study of growth. *American Journal of Human Biology* 3: 33–40.
Collins K J & Weiner J S (1977). *Human adaptability: a history and compendium of research*. London: Taylor & Francis.
Eleveth P B & Tanner J M (1976). *Worldwide Variation in Human Growth*. Cambridge: Cambridge University Press.
Eveleth P B & Tanner J M (1990). *Worldwide Variation in Human Growth* 2nd edn, vol 2. Cambridge: Cambridge University Press.
Goldstein H (1981). Measuring the stability of individual growth patterns. *Annals of Human Biology* 8: 549–557.
Harrison G A & Schmitt L G (1989). Variability in stature growth. *Annals of Human Biology* 16: 45–51.
Hermanussen M, Geiger-Benoit K, Burmeister J & Sippell W G (1988). Periodical changes of short-term growth velocity ('mini growth spurts') in human growth. *Annals of Human Biology* 15: 103–109.
Johnson ML (1993). Analysis of serial growth data. *American Journal of Human Biology* 5: 633–640.
Jolicoeur P, Pontier J & Abidi H (1992). Asymptotic models for the longitudinal growth of human stature. *American Journal of Human Biology* 4: 461–468.
Lampl M, Veldhuis J D & Johnson M L (1992). Saltation and stasis: a model of human growth. *Science* 258: 801–803.
Largo R H, Gasser T, Prader A, Stuetzle W & Huber P J (1978). Analysis of the adolescent growth spurt using smoothing spline functions. *Annals of Human Biology* 5: 421–434.
Preece M A & Baines M J (1978). A new family of mathematical models describing the human growth curve. *Annals of Human Biology* 5: 1–24.
Tanner J M (1981). *A History of the Study of Human Growth*. Cambridge: Cambridge University Press.

#### Growth and natural selection

Charnov E (1993). *Life History Invariants*. Oxford: Oxford University Press.
Eveleth P B & Tanner J M (1990). *Worldwide Variation in Human Growth* 2nd edn. Cambridge: Cambridge University Press.
Martorell R, Khan L K, Schroeder D G (1994). Reversibility of stunting: epidemiological findings in children from developing countries. *European Journal of Clinical Nutrition* 48: S45–57.
Smith-Gill S J (1983). Developmental plasticity: developmental conversion versus phenotypic modulation. *American Zoologist* 23: 47–55.
Stearns S C & Koella J C (1986). The evolution of phenotypic plasticity in life-history traits: predictions of reaction norms for age and size at maturity. *Evolution* 40: 893–913.

#### *Growth in African pygmies*

Cavalli-Sforza L L (1986). *African Pygmies*. New York: Academic Press.
Merimee T J, Zapf J & Froesch E R (1981). Dwarfism in the pygmy: an isolated deficiency of insulin-like growth factor I. *New England Journal of Medicine* 305: 965–968.
Merimee T J, Zapf J & Froesch E R (1982). Insulin-like growth factors in pygmies and subjects with the pygmy trait: characterisation of the metabolic actions of IGF I and II in man. *Journal of Clinical Endocrinology and Metabolism* 55: 1081–1087.

Merimee T J, Zapf J, Hewlett B & Cavalli-Sforza L L (1987). Insulin-like growth factors in pygmies: the role of puberty in determining final stature. *New England Journal of Medicine* 316: 906–911.

Merimee T J, Hewlett B, Wood W, Bowcock A & Cavalli-Sforza L L (1989). The GH receptor gene in the African pygmy. In *Transactions of the Association of American Physicians* (ed A S Fauci) pp 163–169. Baltimore: Waverly Press.

Merimee T J (1993). Why are pygmies short? *Growth Matters* 12: 4–6.

**Gene–environment interactions**

Falkner F & Tanner J M (1986). *Human Growth: A Comprehensive Treatise* 2nd edn. New York: Plenum.

Merimee T J, Baumann G & Daughaday W (1990). Growth hormone binding protein: II Studies in pygmies and normal statured subjects. *Journal of Clinical Endocrinology and Metabolism* 71: 1183–1188.

Pritchard D J (1986). *Foundations of Developmental Genetics*. London: Taylor & Francis.

Roberts D F (1994). Genetics of growth. In *Auxology 94: Children and Youth at the End of the Twentieth Century* (ed O G Eiben) pp 23–30. Budapest.

**Body-size at birth**

Alexander G R et al. (1996). A United States national reference for fetal growth. *Obstetrics & Gynecology* 87(2): 163–168.

David & Lucile Packard Foundation. *Low Birth Weight. The Future of Children*. 5 (1) (1995). Los Angeles: David & Lucile Packard Foundation.

Kramer M S (1987). Determinants of low birth weight: methodological assessment and meta-analysis. *Bulletin of the World Health Organization* 65: 663–737.

WHO (1995). The newborn infant. In 'Physical Status: The Use and Interpretation of Anthropometry. Report of a WHO Expert Committee.' *WHO Technical Report Series* 854: 121–160.

WHO, Division of Family Health (1980). The incidence of low birth weight: a critical review of available information. *World Health Statistics Quarterly* 33: 197–224.

Yip R (1987). Altitude and birth weight. *Journal of Pediatrics* 111: 869–876.

**Growth in infancy and pre-adolescence**

Bogin B (1988). *Patterns of Human Growth*. Cambridge: Cambridge University Press.

Bogin B (1991). Measurement of growth variability and environmental quality in Guatamalan children. *Annals of Human Biology* 18: 285–294.

Eveleth P B & Tanner J M (1990). *Worldwide Variation in Human Growth* 2nd edn. Cambridge: Cambridge University Press.

Eveleth P B & Tanner J M (1976). *Worldwide Variation in Human Growth*. Cambridge: Cambridge University Press.

Harrison G A & Schmitt L H (1989). Variability in stature growth. *Annals of Human Biology* 16: 45–51.

Martorell R (1985). Child growth retardation: a discussion of its causes and its relationship to health. In *Nutritional Adaptation in Man* (eds K L Blaxter & J C Waterlow) pp 13–29. London: John Libbey.

Martorell R, Mendoza F & Castillo R (1988). Poverty and stature in children. In *Linear Growth Retardation in Less Developed Countries* (ed J C Waterlow) pp 57–73. Nestlé Nutrition Workshop Series Volume 14. New York: Raven Press.

Offringa P J & Boersma E R (1987). Will food supplementation in pregnant women decrease neonatal morbidity? *Human Nutrition: Clinical Nutrition* 41C: 311–315.

Schmitt L H & Harrison G A (1988). Patterns of within-population variability of stature and weight. *Annals of Human Biology* 15: 353–364.

Ulijaszek S J (1994). Between-population variation in pre-adolescent growth. *European Journal of Clinical Nutrition* 48 (Suppl 1): S5–S13.

Waterlow J C (1988). Observations on the natural history of stunting. In *Linear Growth Retardation in Less Developed Countries* (ed J C Waterlow) pp 1–16. Nestlé Nutrition Workshop Series Volume 14. New York: Raven Press.

**Between-population differences in adolescent growth**

Eveleth P B & Tanner J M (1990) *Worldwide Variation in Human Growth* 2nd edn. Cambridge: Cambridge University Press.

Frisancho A R, Guire K, Babler W, Borkan G & Way A (1980). Nutritional influence on childhood development and genetic control of adolescent growth of Quechuas and Mestizos from the Peruvian lowlands. *American Journal of Physical Anthropology* 52: 367–375.

Johnston F E, Wainer B E, Thissen D, MacVean R (1976). Hereditary and environmental determinants of growth in height in a longitudinal sample of children and youth of Guatemalan and European ancestry. *American Journal of Physical Anthropology* 44: 469–476.

Martorell R (1985). Child growth retardation: a discussion of its causes and its relationship to health. In *Nutritional Adaptation in Man* (eds K Blaxter & J C Waterlow) pp 13–29. London: John Libbey.

***Age at menarche in the former Soviet Union***

Avetisyan L R (1991). Osbennosti sostoyaniya zdorovya i somaticheskogo razvitiya gorodskikh i selskikh shkol'nikov Armenil. (Peculiarities of health status and somatic development in urban and rural Armenian school children). Doctoral thesis, Yerevan. In Russian.

Balcunene I, Nainis I, Pavilonis S & Tutkuvene J (1991). Letuve antropologius natmenis. (Outlines of Lithuanian Anthropology). *Mokslas*. Vilnus. In Lithuanian.

Bazuntsev G A 1972. Morpfologitcheskie i funktsionalnye proyavleniya proyavleniya polovogo sozrevaniya i physycheskoi rabotosposobnosti detei i podrostkov uzbekskoi natsionalnosti. (Morphological and functional traits of sexual development in Uzbek children and adolescents). Doctoral thesis, Moscow. In Russian.

Berenshtein G F, Polevol D A, Nurbaeva M N & Karnoushko T P (1991). Physicheskoe razvitiye shkolnikov Vitebska za poslednie 15 let. (Physical development of Vitebsk school children for the last 15 years.) *Zdravookhranenle Beloruseii* 12: 33–37. In Russian.

Berenshtein G F, Nurbaeva M N, Polevol D A & Karnoushko T P (1992). Polovoe sozrevanie selskikh shkolnikov Vitebskoi oblasti. (Sexual maturation of rural school children of Vitebskaya region). *Zdravookhranenle Belorussii* 9: 23–25. In Russian.

Bobokhodzhayev I Ya, Usmanov Ya A, Khaidarova Kh R & Kodyrova Kh V (1990). Vremya nastupleniya menstruatsiy u devushek tadzhikskoi natsionalnosti. (Menarcheal age in Tadjik girls). *Zdravookhraneniye Tadzhikistana* 4: 53–56. In Russian.

Danker-Hopfe H (1986). Menarcheal age in Europe. *Yearbook of Physical Anthropology* 29: 81–112.

Deryaev I (1972). O pubertatnom periode sel'skikh shkol'nikov. (Pubertal development of rural school-children). *Zdravookhranenie Turkmenistana* 54–56. In Russian.

Godina E Z (1994).Some latest trends in the somitic development of Moscow school children. In *Growth and Ontogenetic Development in Man 1Y* (ed K Hajnis) pp 123–127. Prague.

Godina E Z, Danilkovich N M, Zadorozhnaya L V, Mlklashevskaya N N & Khomyakova I A (1992). Nekotorye ocobennosti protsessov rosta i razvitlya karel'skikh detei. (Some peculiarities of growth and development in Karel children). *Voprosy antropologii* 86: 70–86. In Russian.

Godina E Z, Yampol'skaya Yu.A, Gilarova O A & Zubareva V V (1995). Vozrast poyavienlya pervykh regul u zhenchshin raznykh regionov Rossii. (Meanarcheal age in women from different regions of Russia). *Gigiena i Sanitariya* 3: 30–32. In Russian.

Golotuk A I, Volosyanko R P & Nedel'ko V P (1989). Osnovnye pokazateli morphofunctsional'nogo razvitiya shkol'nikov Prikarpatya (Ivano-Frankovsk). (Main Indices of morphofunctional development of the school children from Prikarpatye). *Pediatriya* 9: 109–110. In Russian.

Kask V, Matt K & Saar K (1991). Contemporary differentiation of risk groups of girls disturbances of a reproductive system. *Acta et commentationes Universitatis Tartuensis* 920: Tartu 14–24.

Kozhanov V V & Balmuratova P F (1983). Sootnoshenle osnovnykh pokazatelei fizicheskogo razvitiya devochek shkol'nogo vozrasta (Interaction of the main indices of somitic development in schoolgirls). *Zdravookhranenie Kazakhstana* 3: 25–29. In Russian.

Kozlov A I, Chistikina G L & Vershubskaya G G (1994). Etnicheskaya izmenchivost' akusherskikh razmerov taza (Ethnic variations in the obstetric pelvic dimensions). In *Zhentchshina v aspecte phyzicheskoi antropologii* (ed G A Aksyanova) pp 110–117. Moscow. In Russian.

Kuliyev Kh I (1972). Osobennosti pubertatnogo razvitlya shkolnikov AshkhabadA (Peculiarities of pubertal development of Ashkhabad schoolchildren). *Zdravookhranenie Turkmenistana* 8: 12–14. In Russian.

Miklashevskaya N N, Godina E Z, Zadorozhnaya L V, Rusakova T V & Khomyakova I A (1992). Rostovie protsessy u russkikh detei i podrostkov Severa Evropelskoi chasti Rossiyiskoi Federtsii. (Growth processes in Russian children and adolescents of Northern European part of Russian Federation). *Voprosy antropologii* 86: 53–69. In Russian.

Miklashevskaya N N, Solovyeva V S & Godina E Z (1988). *Rostovye protsessi u detei i podrostkov*. (Growth Processes in Children and Adolescents). Moscow: Moscow University Publishers. In Russian.

Osnach A V (1992). *Physicheskoe razvitie detei sel'skikh i gorodskikh mestnostei nekotorykh regionov Ukrainy*. (Physical development of children from some rural and urban regions of the Ukraine). Doctoral thesis, Moscow. In Russian.

Pashkova V I & Tsandekov D S (1989). Kriterii vozrasta lits zhenskogo pola nekotorych korennykh narodnostei Kamchatki. (Age criteria for females of some aboriginal Kamchatka populations). *Sudebno-meditsinskaya ekspertiza* 1: 25–27. In Russian.

Salivon I I, Tegako L I & Polina N I (1989). *Okruzhaushchaya, sreda i zdorovye chelovek* (Environment and Human Health Status). Minsk: Nauka i tekhnika. In Russian.

Sukhanov S G, Gubkina Z D & Smirnov A V (1990). Sposobi otsenki reproductivnoi funktsii na Evropeiskom Severe. (Evaluation of the reproductive function in women of the European North). *Syktyvkar, Komi N Ts. Ural. otd. AN SSSR* 94: 24. In Russian.

Vlastovsky V G (1984). Sravnitel'niy analiz ocobennostei protsessov rosta i somaticheskogo razvitiya yakutskikh i russkikh detei v vozraste 8–18 let. (Comparative analysis of growth and somatic development of Yakut and Russian children from 8 to 18 years.) *Voprosy antropologii*, 3: 25–38. In Russian.

Zubareva V V. Materialy po polovomu sozrevaniu devochek raznykh regionov byvshego Sovetskogo Soyuza (Materials to sexual maturation of girls from different regions of the former Soviet Union). (Unpublished). In Russian.

**Body-proportion differences**

Boileau R A, Lohman T G, Slaughter M H, Ball T E, Going S B & Hendrix M K (1984). Hydration of the fat-free body in children during maturation. *Human Biology* 56: 651–666.

Eveleth P B & Tanner J M (1990). *Worldwide Variation in Human Growth* 2nd edn pp 185–189, 215–219. Cambridge: Cambridge University Press.

Fomon S J, Haschke F, Ziegler E E & Nelson S E (1982). Body composition of reference children from birth to age 10 years. *American Journal of Clinical Nutrition* 35: 1169–1175.

Forbes G B (1987). *Human Body Composition: Growth, Ageing, Nutrition and* Activity pp 125–168. New York: Springer-Verlag.

Lohman T G (1986). Applicability of body composition techniques and constants for children and youth. *Exercise and Sports Science Reviews* 14: 325–357.

Malina R M & Bouchard C (1991). *Growth, Maturation and Physical Activity*. Champaign, Illinois: Human Kinetics Books.

Roche A F (ed) (1992). *Growth, Maturation and Body Composition: the Fels Longitudinal Study, 1929–91*. pp 199–235. Cambridge: Cambridge University Press.

Tanner J M (1989). *Foetus into Man* (2nd edn). Ware: Castlemead Publications.

Tanner J M, Hayashi T, Preece M A & Cameron N (1982). Increase in leg length relative to trunk in Japanese children and adults from 1957 to 1977: comparison with British and with Japanese Americans. *Annals of Human Biology* 9: 411–423.

## Part eleven: Changing human growth patterns

Aurelius G, Khanh N C, Truc D B, Ha T T & Lindgren G (1996). Height, weight and body mass index (BMI) of Vietnamese (Hanoi) school children aged 7–11 years related to parents' occupation and education. *Journal of Tropical Paediatrics* 42: 21–26.

Barker D J P (ed) (1992). *Fetal and Infant Origins of Adult Disease*. London: British Medical Journal.

Barker D J, Osmond C & Golding J (1990). Height and mortality in the counties of England and Wales. *Annals of Human Biology* 17: 1–6.

Bielicki T, Szczotka H & Charzewski J (1981). The influence of three socio-economic factors on body height in Polish military conscripts. *Human Biology* 53: 543–555

Bielicki T & Welon Z (1982). Growth data as indicators of social inequalities: the case of Poland. *Yearbook of Physical Anthropology* 25: 153–167.

Bielicki T & Waliszko H (1992). Stature, upward social mobility and the nature of statural differences between social classes. *Annals of Human Biology* 19: 589–593.

Brundtland G H, Liestol K & Walloe L (1980). Height, weight and menarcheal age of Oslo school children during the last 60 years. *Annals of Human Biology* 7: 307–322.

Calnan M (1986). Maintaining health and preventing illness: a comparison of the perceptions of women from different social classes. *Health Promotion Journal* 1: 167–177.

Danker-Hopfe H & Hulanicka B (1995). Maturation of girls in lead-polluted areas. In *Essays on Auxology Presented to James Mourilyan Tanner by Former Colleagues and Fellows* (eds R Hauspie, G Lindgren & F Falkner) pp 334–342. Welwyn Garden City: Castlemead Publications.

Dornbusch S M, Carlsmith J M, Duncan P D, Gross R T, Martin J A, Ritter P L & Siegel-Gorelick B (1984). Sexual maturation, social class, and the desire to be thin among adolescent females. *Journal of Developmental and Behavioral Pediatrics* 5: 308–314.

Dubrova Y E, Kurbatova O L, Kholod O N & Prokhoroyskaya V D (1995). Secular growth trend in two generations of the Russian population. *Human Biology* 67: 755–767.

Eveleth P B & Tanner J M (1990). *Worldwide Variation in Human Growth* 2nd edn. Cambridge: Cambridge University Press.

Flegal K M, Harlan W R & Landis J R (1988). Secular trends in body mass index and skinfold thickness with socio-economic factors in young adult men. *American Journal of Clinical Nutrition* 48: 544–551.

Fogel R (1994). Economic growth, population theory, and physiology: the bearing of long-term processes on the making of economic policy. In *Les Prix Nobel, 1993* reprinted in *American Economic Review* 84: 369–395.

Furu M, Lindgärde F, Ljung B-O, Munch I & Kristenson H (1984). *Premature Death. Cognitive Ability and Socio-economic Background. A Longitudinal Study of 834 Men*. Stockholm: Stockholm Institute of Education (Department of Educational Research.)

Gulliford M C, Chinn S & Rona R J (1991). Social environment and height: England and Scotland 1987 and 1988. *Archives of Disease in Childhood* 66: 235–240.

Townsend P & Davidson N (eds) (1984). *Inequalities in Health. The Black Report* 4th edn. Harmondsworth: Penguin.

Kahn H S & Williamson D F (1990). The contributions of income, education and changing marital status to weight change among US men. *International Journal of Obesity* 14: 1057–1068.

Karp R J, Scholl T O, Decker E & Ebert E (1989). Growth of abused children: contrasted with the non-abused in an urban poor community. *Clinical Pediatrics* 28: 317–320.

Kerr G R, Lee E S, Lorimor R J, Mueller W H & Lam M M (1982). Height distributions of US children: associations with race, poverty status and parental size. *Growth* 46: 135–149.

Komlos J (ed) (1995). *The Biological Standard of Living on Three Continents: Further Explorations in Economic History*. Boulder: Westview Press.

Kuh D L, Power C & Rodgers B (1991). Secular trends in social class and sex differences in adult height. *International Journal of Epidemiology* 20: 1001–1009.

Lasker G W & Mascie-Taylor C G N (1989). Effects of social class differences and social mobility on growth in height, weight and body mass index in a British cohort. *Annals of Human Biology* 16: 1–8.

Lindgren G (1979). Peak velocities in height and mental performance. A longitudinal study of school children aged 10–14 years. *Annals of Human Biology* 6: 559–584.

Lindgren G W (1988). Psycho-social aspects of growth with special regard to the relation between physical/physiological and mental/cognitive growth. *Colloquia in Anthropology* 12: 47–66.

Lindgren G W & Hauspie R C (1989). Heights and weights of Swedish school children born in 1955 and 1967. *Annals of Human Biology* 16: 397–406.

Marmot M (1994). Social differentials in health within and between populations. *Daedalus* 123: 197–216.

Matsumoto K (1982). Secular acceleration of growth in height in Japanese and its social background. *Annals of Human Biology* 9: 399–410.

Power C & Moynihan C (1988). Social class and changes in weight-for-height between childhood and early adulthood. *International Journal of Obesity* 12: 445–453.

Proos L A, Hofvander Y & Tuvemo T (1991). Menarcheal age and growth-pattern of Indian girls adopted in Sweden. *Acta Paediatrica Scandinavica* 80: 852–858.

Roberts D F & Dann T C (1992). Social class and diachronic trends in physique in young university women. *Journal of Biosocial Science* 24: 269–279.

Rona R J & Chinn S (1991). Father's unemployment and height of primary school children in Britain. *Annals of Human Biology* 18: 441–448.

Sandberg L G & Steckel R H (1980). Soldier, soldier what made you grow so tall? *Economy and History* 23: 23–81.

Shorter E (1981). L'Age des premières règles en France, 1750–1950. (The age of menarche in France, 1750–1950.) *Annales* 36: 495–511.

Smith A M, Chinn S & Rona R J (1980). Social factors and height gain of primary school children in England and Scotland. *Annals of Human Biology* 7: 115–124.

Socialstyrelsen (1987). *Folkhälsorapport 1987* (Public Health Report 1987) Socialstyrelsen redovisar 1987: 15. Stockholm: Socialstyrelsen.

Socialstyrelsen (1994). *Folkhälsorapport 1994*. (Public Health Report 1994) SoS-rapport 1994: 9. Stockholm: Socialstyrelsen.

Steckel RH (1988). Heights and health in the United States, 1710–1950. In *Auxology 88. Perspectives in the Science of Growth and Development* (ed J M Tanner). London: Smith-Gordon.

Tanner J M (1978). *Education and Physical Growth* 2nd edn. London: Hodder and Stoughton and New York: International Universities Press.

Tanner J M (1981). *A History of the Study of Human Growth*. Cambridge: Cambridge University Press.

Tanner J M (1989). *Foetus into Man* 2nd edn. Ware: Castlemead Publications.

Walker M, Shaper A G & Wannamethee G (1988). Height and social class in middle-aged British men. *Journal of Epidemiological Community Health* 42: 299–303.

van Wieringen J C (1986). Secular Growth Changes. In *Human Growth. A Comprehensive Treatise* 2nd edn, vol 3 (eds F Falkner & J M Tanner) pp 307–331. New York: Plenum Press.

Waaler H Th (1984). Height, weight and mortality. The Norwegian experience. *Acta Medica Scandinavica* (Suppl 679).

Westin-Lindgren G (1979). *Physical and Mental Development in Swedish Urban School Children. Studies in Education and Psychology* 5. Stockholm: Stockholm Institute of Education (Department of Educational Research).

Westin-Lindgren G (1981a). Achievement and mental ability of boys in relation to physical maturity and social background. In *Human Adaptation* (eds P Russo & G Gass) pp 99–109. Sydney: Cumberland College of Health Sciences.

Westin-Lindgren G (1981b). Physical and physiological characteristics of boys 17–19 years related to social background. In *Human Adaptation* (eds P Russo & G Gass) pp 67–76. Sydney: Cumberland College of Health Sciences.

Westin-Lindgren G (1982). Achievement and mental ability of physically late and early maturing schoolchildren related to their social background. *Journal of Child Psychology and Psychiatry* 23: 407–420.

Westin-Lindgren G (1989). Auxology and education: some educational implications from the relationships between physical and mental growth in Swedish school children. *Acta Paediatrica Scandinavica* Suppl 350: 105–120.

Wilkinson R (1992). Income distribution and life expectancy. *British Medical Journal* 304: 165–168.

### Skeletal growth and agricultural intensification

Armelagos G J, Mahler P E, Owen K, Dewey J R, Mielke J & Van Gerven D P (1972). Bone growth and development in prehistoric populations from Sudanese Nubia. *Journal of Human Evolution* 1: 89–119.

Cohen M N (1989). *Health and the Rise of Civilizations*. New Haven: Yale University Press.

Cohen M N & Armelagos GJ (eds) (1984). *Paleopathology at the Origins of Agriculture*. New York: Academic Press.

Gilbert R & Mielke J (eds) (1985). *The Analysis of Prehistoric Diets*. New York: Academic Press.

Goodman A H, Thomas R B, Swedlund A C & Armelagos G J (1988). Biocultural perspectives on stress in prehistoric, historical, and contemporary population research. *Yearbook of Physical Anthropology* 31: 169–202.

Jantz R L & Owsley D W (1984). Temporal changes in limb proportionality among skeletal samples of Arikara Indians. *Annals of Human Biology* 11 (2): 157–163.

Johnston F E & Zimmer L O (1989). Assessment of growth and age in the immature skeleton. In *Reconstruction of Life from the Skeleton* (eds M Y Iscan & K A R Kennedy) pp 11–22. New York: Alan R Liss.

Larsen C S (1995). Biological changes in human populations with agriculture. *Annual Review of Anthropology* 24: 185–213.

Martin D L, Goodman A H, Armelagos G J & Magennis A L (1991). *Black Mesa Anasazi Health: Reconstructing Life from Patterns of Death and Disease*. Carbondale: Southern Illinois University Press.

### *Patagonian giants: myths and possibilities*

d'Orbigny A (1944). *L'Homme américain*, translated by Alfredo Cepeda as *El hombre Americano*. Buenos Aires, Argentina: Editorial Futuro.

Vazquez F (1512). *Palmeria de oliva* vol 2. Salamanca.

### Physical growth during industrialization

Floud R, Wachter K & Gregory A (1990). *Height, Health and History: Nutritional Status in the United Kingdom, 1750–1980*. Cambridge: Cambridge University Press.

Komlos J (1989). *Nutrition and Economic Development in the Eighteenth-Century Habsburg Monarchy. An Anthropometric History*. Princeton: Princeton University Press.

Komlos J (ed) (1994). *Stature, Living Standards and Economic Development. Essays in Anthropometric History*. Chicago: Chicago University Press.

Komlos J (ed) (1995). *The Biological Standard of Living on Three Continents: Further Explorations in Anthropometric History*. Westview Press.

### *Growth patterns from recruiting data*

Costa D L (1993). Height, weight, wartime stress, and older age mortality: evidence from the Union Army records. *Explorations in Economic History* 30: 424–429.

Floud R, Wachter K & Gregory A (1990). *Height, Health, and History: Nutritional Status in the United Kingdom, 1750–1980*. Cambridge: Cambridge University Press.

Fogel R W (1993). New sources and new techniques for the study of secular trends in nutritional status, health, mortality, and the process of aging. *Historical Methods* 26: 5–43.

Frisancho A R (1990). *Anthropometric Standards for the Assessment of Growth and Nutritional Status*. Ann Arbor: University of Michigan Press.

Gould B A (1869). *Investigations in the Military and Anthropological Statistics of American Soldiers*. New York: Cambridge Riverside Press. (Reprinted 1979 by Arno Press, New York).

Komlos J (1994). *Stature, Living Standards and Economic Development: Essays in Anthropometric History*. Chicago: University of Chicago Press.

Mokyr S & O'Grada C (1994). Heights of the British and Irish *c.* 1800–1815: Evidence from recruits to the East India Company's army. In *Stature, Living Conditions, and Economic Development* (ed J Komlos) pp 39–59. Chicago: University of Chicago Press.

Riggs P (1994). The standard of living in Scotland 1800–1850. In *Stature, Living Conditions, and Economic Development* (ed J Komlos) pp 39–59. Chicago: University of Chicago Press.

Steegmann A T Jr (1985). Eighteenth-century British military stature: growth cessation, selective recruiting, secular trends, nutrition at birth, cold and occupation. *Human Biology* 57: 77–95.

van Wieringen J C, Wafelbakker F, Verbrugge H P & de Haas J H (1971). *Growth Diagrams 1965 Netherlands: Second National Survey on 2–24 year-olds*. Groningen: Wolters Noordhoff.

***Growth and development in art history***

Bogin B (1988). *Patterns of Human Growth*. Cambridge: Cambridge University Press.

Dewhurst K C, MacDowell E & MacDowell M (1983). *Religious Folk Art in America* (plate 108). New York: E P Dutton.

Gombrich E H (1972). The visual image. *Scientific American* 227: 82–96.

**The secular trend**

Cameron N (1979). The growth of London school children 1904–1966: and analysis of secular trend and intra-county variation. *Annals of Human Biology* 6: 505–525.

Cernerud L & Lindgren G W (1991). Secular changes in height and weight of Stockholm school children born in 1933, 1943, 1953 and 1963. *Annals of Human Biology* 18: 497–505.

Chinn S, Rona R J & Price C E (1989). The secular trend in height of primary school children in England and Scotland 1972–79 and 1979–86. *Annals of Human Biology* 16: 387–395.

Helm P & Helm S (1987). Uncertainties in designation of age at menarche in the nineteenth century: revised mean for Denmark, 1835. *Annals of Human Biology* 14: 371–374.

Kikuta F & Takaishi M (1987). Studies on physical growth standards for schoolchildren in Japan. Part I Centile curves for height and weight based on cross-sectional data and a consideration of secular trend of the centile curves. *Japanese Journal of Child Health* 46: 27–33.

Liestol K & Rosenberg M (1995). Height, weight and menarcheal age of schoolgirls in Oslo – an update. *Annals of Human Biology* 22: 199–205.

Lindgren G W & Hauspie R C (1989). Heights and weights of Swedish school children born in 1955 and 1967. *Annals of Human Biology* 16: 397–406.

Ljung B O, Bergsten-Brucefors A & Lindgren G (1974). The secular trend in physical growth in Sweden. *Annals of Human Biology* 1: 245–256.

Meredith H V (1976). Findings from Asia, Australia, Europe, and North America on secular change in mean height of children, youths, and young adults. *American Journal of Physical Anthropology* 44: 315–326.

Rona R J & Chinn S (1984). The national study of health and growth: nutritional surveillance of primary school children from 1972 to 1981 with special reference to unemployment and social class. *Annals of Human Biology* 11: 17–28.

Susanne C (1985). Living conditions and secular trend. *Journal of Human Evolution* 14: 357–370.

Tanner J M, Hayashi T, Preece M A & Cameron N (1982). Increase in length of leg relative to trunk in Japanese children and adults from 1957 to 1977: comparison with British and with Japanese Americans. *Annals of Human Biology* 9: 411–423.

Tobias P V (1985). The negative secular trend. *Journal of Human Evolution* 14: 347–356.

Ulijaszek S J (1993). Evidence for a secular trend in heights and weights of adults in Papua New Guinea. *Annals of Human Biology* 20: 349–355.

Vercauteren M (1984). *Evolution seculaire et normes de croissance chez des enfants belges. Bulletin de la Société Royale Belge d'Anthropologie Préhistorique* 95: 109–123.

Wachholder A, Hauspie R C (1986). Clinical standards for growth in height of Belgian boys and girls, aged 2 to 18 years. *International Journal of Anthropology* 1: 327–338.

van Weiringen J C (1986). Secular growth changes. In *Human Growth: A Comprehensive Treatise* (eds F Falkner & J M Tanner) 2nd edn, vol 3 pp 307–331. New York: Plenum Press.

Zemel B & Jenkins C (1989). Dietary change and adolescent growth among the Bundi (Gende-speaking) people of Papua New Guinea. *American Journal of Human Biology* 1: 709–718.

***Secular changes and class distinction in growth in Sweden***

Broman B, Dahlberg G & Lichtenstein A (1942). Height and weight during growth. *Acta Paediatrica* (Uppsala) 30: 1–66.

Cernerud L (1994). Are there still inequalities in height and body mass index of Stockholm children? *Scandinavian Journal of Social Medicine* 22: 161–165.

Key A (1885). *Läroverkskomite' ns underdåniga utlåtande och förslag angående organisationen af rikets allmänna läroverk och dermed sammanhängande frågor. Bilaga E Första afdelningen text samt andra afdelningen tabeller och figurer*. (The School Committee's humble report and proposals for the organization of the Secondary Schools and thereto related questions. Supplement E Part one text, and part two: tables and figures). Stockholm: Kongliga Boktryckeriet.

Kuskowska-Wolk A & Rössner S (1990). Interrelationships between socio-demographic factors and body mass index in a representative Swedish adult population. *Diabetes Research and Clinical Practice* 10 (S1): S271–275.

Kuskowska-Wolk A & Bergström R (1993). Trends in body mass index and prevalence of obesity in Swedish women 1980–1989. *Journal of Epidemiology and Community Health* 47:195–199.

Lindgren G (1976). Height, weight and menarche in Swedish urban school children in relation to socio-economic and regional factors. *Annals of Human Biology* 3: 501–528.

Lindgren G W & Strandell A (1986). *Fysisk utveckling och hälsa. En analys av hälsokortsuppgifter för grundskoleelever födda 1967*. (Physical growth and health. An analysis of health chart data of elementary school children born in 1967.) Report No. 4. Stockholm Institute of Education (Department of Educational Research).

Lindgren G W (1988). Genetics of Growth and Development: the case of Sweden or how old was jerker? *Colloquia in Anthropology* 12: 23–45.

Lindgren G W & Hauspie R C (1989). Heights and weights of Swedish school children born in 1955 and 1967. *Annals of Human Biology* 16: 397–406.

Lindgren G W (1991). End of the secular trends in height and maturational rate of Swedish youth? *Anthropologiai Közlemenyek* 33: 17–22.

Lindgren G W et al. (1991). Menarche 1990 in Stockholm schoolgirls. *Acta Paediatrica Scandinavica* 80: 953–955.

Lindgren G W & Cernerud L (1992). Physical growth and socio-economic background in Stockholm school-children born 1933–1963. *Annals of Human Biology* 19: 1–16.

Lindgren G (1994). Aspects of research on socio-economic conditions and growth based on Swedish data. *Humanbiologia Budapestinensis* 25: 125–135.

Lindgren G, Aurelius G, Tanner J M & Healy M (1994). Socio-economic circumstances and the growth of Stockholm pre-school children: the 1980 birth cohort. *Acta Paediatrica* 83: 1209–1211.

Lindgren G, Strandell A, Cole T, Healy M & Tanner J (1995). Swedish population reference standards for height, weight and body mass index attained at 6 to 16 years (girls) or 19 years (boys). *Acta Paediatrica* 84: 1019–28.

Ljunggren C A (1933). *Skolbarnens häsouppfostran. Handbok för lärare och föräldrar.* (Schoolchildren's health education. A handbook for teachers and parents). Stockholm, Sweden: Naturoch Kultur.

Nyström-Peck A-M (1992). Childhood environment, inter-generational mobility, and adult health – evidence from Swedish data. *Journal of Epidemiology and Community Health*, 46: 71–74.

Nyström-Peck M & Lundberg O (1995). Short stature as an effect of economic and social conditions in childhood. *Social Science and Medicine* 41: 733–738.

Sandberg L G & Steckel R H (1987). Heights and economic history: the Swedish case. *Annals of Human Biology* 14: 101–110.

Sunnegårdh J (1986). Physical activity in relation to energy intake, body fat, physical work capacity and muscle strength in 8- and 13-year-old children in Sweden. *Acta Universitatis Upsaliensis* 41. Uppsala: Reprocentralen HSC.

Westin-Lindgren G (1979). Physical and mental development in Swedish urban school children. *Studies in Education and Psychology* 5. Stockholm Institute of Education (Department of Educational Research).

Westin-Lindgren G (1982). Relation between secular trends in height growth, weight increase and age at menarche in Swedish urban youth. *Tjids. Soc. Geneesk*. 60: 591–596.

Åkerman S, Högberg U & Danielsson M (1990). Height, health and nutrition in early modern Sweden. In *Society, Health and Population During the Demographic Transition* (eds A Brandstrom & L-G Tedebrand) pp 413–428. Stockholm, Sweden: Almqvist and Wiksell.

**Social and economic class**

Bogin B (1988). *Patterns of Human Growth*. Cambridge: Cambridge University Press.

Crooks D L (1995). American children at risk: poverty and its consequences for children's health, growth, and school achievement. *Yearbook of Physical Anthropology* 38. (in press.)

Ekwo E et al. (1991). The effect of height on family income. *American Journal of Human Biology* 3: 181–188.

Garn S M et al. (1984). The interaction between prenatal and socio-economic effects on growth and development in childhood. In *Human Growth and Development* (eds J Borms et al.) pp 59–70. New York: Plenum Press.

Pollitt E & Lewis N (1980). Nutrition and educational achievement. Part 1, Malnutrition and behavioral test indicators. *Food and Nutrition Bulletin* 2: 32–34.

Schell L M (1992). Risk focusing: an example of biocultural interaction. In *Health and Lifestyle Change: MASCA Research Papers in Science and Archaeology* (eds R Huss-Ashmore et al.) vol 9 pp 137–144.

Spurr G B (1983). Nutritional status and physical work capacity. *Yearbook of Physical Anthropology* 26: 1–35.

***Social class and height in Britain***

Goldstein H (1971). Factors influencing the height of seven year old children. Results from the National Child Development Study. *Human Biology* 43: 92–111.

Gulliford M C, Chinn S & Rona R J (1991). Social environment and height: England and Scotland 1987 and 1988. *Archives of Disease in Childhood* 66: 235–240.

Kuh D L, Power C & Rodgers B (1991). Secular trends in social class and sex differences in adult height. *International Journal of Epidemiology* 20: 1001–1009.

Macintyre S (1988). A review of the social patterning and significance of measures of height, weight, blood pressure and respiratory function. *Social Science and Medicine* 27: 327–337.

Power C, Manor O & Fox J (1991). *Health and Class: the Early Years*. London: Chapman & Hall.
Rona R J (1981). Genetic and environmental factors in growth in childhood. *British Medical Bulletin* 37: 265–272.
Rona R J (1995). Monitoring nutritional status in England and Scotland. In *Essays on Auxology Presented to James Mourilyan Tanner by Former Colleagues and Fellows* (eds R Hauspie, G Lindgren & F Falkner) pp 291–301. Welwyn Garden City: Castlemead Publications.
Rose D (1995). *ESRC Review of OPCS Social Classification. A Report on Phase 1 to the Office of Population Censuses and Surveys*. OPCS.
White A, Nicolaas G, Foster K, Browne F & Carey S (1993). *Health Survey for England 1991*. OPCS Social Survey Division. London: HMSO.
Walker M, Shaper A G & Wannamethee G (1988). Height and social class in middle-aged British men. *Journal of Epidemiology and Community Health* 42: 299–303.

**Modernization and growth**

Floud R, Wachter K & Gregory A (1990). *Height, Health and History*. Cambridge: Cambridge University Press.
Gross R (1994). Nutrition, growth and disease – the impact of urbanization in developing countries. In *Nutrition in a Sustainable Environment. Proceedings of the XV International Congress of Nutrition: IUNS Adelaide* (eds M L Wahlqvist, A S Truswell, R Smith & P J Nestel) pp 362–366. London: Smith-Gordon.
Ljung B O, Bergsten-Brucefors A & Lindgren G (1974). The secular trend in physical growth in Sweden. *Annals of Human Biology* 1: 245–256.
McGarvey S T. Economic modernization and human adaptability perspectives. In *Health and Lifestyle Change* (eds R Huss-Ashmore, J Schall & M Hediger) pp 105–113. Philadelphia: Museum Applied Science Center for Archaeology Research Papers in Science and Archaeology Vol 9. (The University Museum of Archaeology and Anthropology, University of Pennsylvania.)
Schell L M, Smith M T & Bilsborough A (eds) (1993). *Urban Ecology and Health in the Third World*. Cambridge: Cambridge University Press.
Susanne C (1985). Living conditions and secular trend. *Journal of Human Evolution* 14: 357–370.
Tanner J M (1981). *A History of the Study of Human Growth*. Cambridge: Cambridge University Press.
Ulijaszek S J (1995). Development, modernization and health intervention. In *Health Intervention in Less Developed Nations* (ed S J Ulijaszek) pp 82–136. Oxford: Oxford University Press.

**Migration and changing population characteristics**

Boas F (1912). Changes in the bodily form of descendants of immigrants. *American Anthropologist* 14: 530–563.
Boyce A J (ed) (1984). *Migration and Mobility. Biosocial Aspects of Human Movement* p 378. London & Philadelphia: Taylor & Francis.
Mascie-Taylor C G N, Lasker G W (1988). *Biological Aspects of Human Migration*. Cambridge: Cambridge University Press.
Susanne C (1979). Comparative biometrical study of stature and weight of Italian migrants in Belgium. *American Journal of Physical Anthropology* 50: 349–356.

***Growth of Guatemalan migrants in the United States of America***

Boas F (1912). Changes in the bodily form of descendants of immigrants. *American Anthropologist*, 14: 530–563.
Bogin B (1988). Rural-to-urban migration. In *Biological Aspects of Human Migration* (eds C G N Mascie-Taylor & G W Lasker) pp 90–129. Cambridge: Cambridge University Press.
Bogin B & MacVean R B (1984). Growth status of non-agrarian, semi-urban living Indians in Guatemala. *Human Biology* 56: 527–538.
Bogin B (1995). Plasticity in the growth of Mayan refugee children living in the United States. In *Human Variability and Plasticity* (eds C G N Mascie-Taylor & B Bogin) pp 46–74. Cambridge: Cambridge University Press.
Haviland W A (1967). Stature at Tikal Guatemala: implications for ancient Maya demography and social organization. *American Antiquity* 32: 316–325.
Lasker G W (1969). Human biological adaptability. *Science* 166: 1480–1486.

***Urban–rural differences in growth patterns of Nepali children***

Adair L S, Vanderslice J & Zohoori N (1993). Urban-rural differences in growth and diarrhoeal morbidity of Filipino infants. In *Urban Ecology and Health in the Third World* (eds L Schell, M T Smith & A Bilsborough) pp 75–98. Cambridge: Cambridge University Press.
Bogin B (1998). Rural-to-urban migration. In *Biological Aspects of Human Migration* (eds C G N Mascie-Taylor & G W Lasker) pp 90–129. Cambridge: Cambridge University Press.
Johnston F E, Low S M, de Baessa Y & MacVean R B (1985). Growth status of disadvantaged urban Guatemalan children of a resettled community. *American Journal of Physical Anthropology* 68: 215–224.
Keller W (1988). The epidemiology of stunting. In *Linear Growth Retardation in Less Developed Countries* (ed J C Waterlow) pp 17–40. New York: Raven Press.
Koppert W A (1988). *Alimentation et culture chez les Tamang, les Ghale et les Kami du Nepal*. Thèse de 3ème cycle, Faculté de Droit et de Science Politique, Aix-Marseille.
Panter-Brick C, Todd A & Baker R Growth status of homeless Nepali boys. *Social Science and Medicine*. In press.
Zemel B, Worthman C & Jenkins C (1993). Differences in endocrine status associated with urban-rural patterns of growth and maturation in Bundi (Gende-speaking) adolescents of Papua New Guinea. In *Urban Ecology and Health in the Third World* (eds L Schell, M T Smith & A Bilsborough) pp 38–60. Cambridge: Cambridge University Press.

**Obesity, fatness and modernization**

Brown P J & Konner M (1987). An anthropological perspective on obesity. In *Human Obesity* (eds R J Wurtman & J J Wurtman) pp 29–46. New York: New York Academy of Sciences.
Cassidy C M (1991). The good body: when big is better. *Medical Anthropology* 13: 181–213.
de Garine I & Pollock N (1995). *Social Aspects of Obesity*. New York: Gordon & Breach.
Garrow J S (1988). *Obesity and Related Diseases*. Edinburgh: Churchill Livingstone.
Garrow J S (1993). Obesity. In *Human Nutrition and Dietetics* (eds J S Garrow & W P T James) pp 465–479. Edinburgh: Churchill Livingstone.
Ishige N (1995). Evaluation of fatness in traditional Japanese society. In *Social Aspects of Obesity* (eds I de Garine & N Pollock) pp. 281–289. New York: Gordon & Breach.
McGarvey ST (1989). Five-year longitudinal changes in fatness and blood pressure in American Samoa. *American Journal of Physical Anthropology* 78: 269–270.
Rittenbaugh C (1982). Obesity as a culture-bound syndrome. *Culture, Medicine and Psychiatry* 6: 347–361.
Rolland-Cachera M F, Sempé M, Guilloud-Bataille M, Patois E, Péquignot-Guggenbuhl F & Fautrad V (1982). Adiposity indices in children. *American Journal of Clinical Nutrition* 36: 178–84.
Sobal J (1991). Obesity and socio-economic status: a framework for examining relationships between physical and social variables. *Medical Anthropology* 13: 231–47.

## Part twelve: The human lifespan

Barker D J P (1991). The intrauterine environment and adult cardiovascular disease. In *The Childhood Environment and Adult Disease* (eds G R Bock & J Whelan) pp 3–10. Chichester: John Wiley & Sons.
Bonder B R & Wagner M B (1994). *Functional Performance in Older Adults*. Philadelphia: F A Davis.
Clark G A, Aldwin C M, Hall N R, Spiro A & Goldstein A (1989). Is poor early growth related to adult immune aging? A follow-up study. *American Journal of Human Biology* 1: 331–337.
Clark G A, Hall N R, Aldwin C M, Harris J M, Borkan G A & Srinivasan M (1988). Measures of poor early growth are correlated with lower adult levels of thymosin-1. *Human Biology* 60: 435–451.
Kay M M B (1985). Immunobiology of aging. In *Nutrition, Immunity and Illness in the Elderly* (ed R K Chandra) pp 97–119. New York: Pergamon Press.
Kirkwood T B L (1992). Comparative lifespans of species: why do species have the lifespans they do? *American Journal of Clinical Nutrition* 55: 1191S–1195S.
Hamosh M (1988). Does infant nutrition affect adiposity and cholesterol levels in the adult? *Journal of Pediatric Gastroenterology and Nutrition* 7: 10–16.
Henry C J K & Ulijaszek S J (eds) (1994). *Long-term Consequences of Early Environment. Growth, Development and the Lifespan Developmental Perspective*. Cambridge: Cambridge University Press.
Martyn C N (1991). Childhood infection and adult disease. In *The Childhood Environment and Adult Disease* (eds G R Bock & J Whelan) pp 93–102. Chichester: John Wiley & Sons.
McCance R A (1962). Food, growth, and time. *Lancet* 2: 621–626.
Pearson J D & Crews D E (1989). Evolutionary, biosocial, and cross-cultural perspectives on the variability in human biological aging. *American Journal of Human Biology* 1: 303–306.
Waterlow J C (1992). *Protein Energy Malnutrition*. London: Edward Arnold.
Wood J W (1994). *Dynamics of Human Reproduction*. New York: Aldine de Gruyter.

**Long-term consequences of early environmental influences**

Baltes P B & Brim O G Jr (eds) (1981). *Life-span Development and Behavior* vol 4. New York: Academic Press.
Baltes P B & Schaie K W (1973). *Life-span Developmental Psychology: Personality and Socialization*. New York: Academic Press.
Barker D J P (1990). The intrauterine origins of adult hypertension. In *Fetal Autonomy and Adaptation* (ed G S Dawes). Chichester: John Wiley & Sons.
Barker D J P (1991). The intrauterine environment and adult cardiovascular disease. In *The Childhood Environment and Adult Disease* (eds G R Bock & J Whelan) pp 3–10. Chichester: John Wiley & Sons.
Datan N & Ginsberg L H (1975). *Life-span Developmental Psychology: Normative Life Crises*. New York: Academic Press.
Datan N & Reese H W (1977). *Life-span Developmental Psychology: Dialectical Perspectives on Experimental Psychology*. New York: Academic Press.
Fomon S J (1971). A pediatrician looks at early nutrition. *Bulletin of the New York Academy of Medicine* 47: 569–578.
De Fronzo R A, Bonadonna R C & Ferrannini E (1992). Pathogenesis of NIDDM. *Diabetes Care* 15: 318–358.
Gardner J (1983). Adolescent menstrual characteristics as predictors of gynaecological health. *Annals of Human Biology* 10: 31–40.
Hales C N & Barker D J P (1992). Type 2 (non-insulin-dependent) diabetes mellitus: the thrifty phenotype hypothesis. *Diabetologia* 35: 595–601.
Hamosh M (1988). Does infant nutrition affect adiposity and cholesterol levels in the adult? *Journal of Pediatric Gastroenterology and Nutrition* 7: 10–16.
Henry C J K & Ulijaszek S J (eds) (1995). *Long-term Consequences of Early Environments: Growth, Development, and the Life-span Developmental Perspective*. Cambridge: Cambridge University Press.
Hodgson P A, Ellefson R D, Elveback L R, Harris L E, Nelson R A & Weidman W H (1976). Comparison of serum cholesterol in children fed high, moderate or low cholesterol milk diets during neonatal period. *Metabolism* 25: 739–746.
Lucas A (1991). Programming by early nutrition in man. In *The Childhood Environment and Adult Disease* (eds G R Bock & J Whelan) pp 38–50. Chichester: John Wiley & Sons.

Marmot M G, Page C M, Atkins E & Douglas J W B (1980). Effect of breast feeding on plasma cholesterol and weight in young adults. *Journal of Epidemiology and Community Health* 34: 164–167.

McCance R A (1962). Food, growth, and time. *Lancet* 2: 621–626.

McCance D R, Pettitt D J, Hanson R L, Jacobsson L T H, Knowler W C & Bennett P H (1994). Birth-weight and non-insulin dependent diabetes: thrifty genotype, thrifty phenotype, or surviving small baby genotype? *British Medical Journal* 308: 942–945.

Mott G E, Lewis D S & McGill H C (1991). Programming of cholesterol metabolism by breast or formula feeding. In *The Childhood Environment and Adult Disease* (eds G R Bock & J Whelan). Chichester: John Wiley & Sons.

Neel J V (1962). Diabetes Mellitus: a 'thrifty' genotype rendered detrimental by 'progress'? *Journal of Human Genetics* 14: 353–362.

Phillips D I W, Hirst S, Clark P M S, Hales C N & Osmond C (1994a). Fetal growth and insulin secretion in adult life. *Diabetologia* 37: 592–596.

Phillips D I W, Barker D J P, Hales C N, Hirst S & Osmond C (1994b). Thinness at birth and insulin resistance in adult life. *Diabetologia* 37: 150–4.

Robins L & Rutter M (1990). *Straight and Devious Pathways from Childhood to Adulthood*. Cambridge: Cambridge University Press.

Rutter M (1991). Childhood experiences and adult psychosocial functioning. In *The Childhood Environment and Adult Disease* (eds G R Bock & J Whelan) pp 189–200. Chichester: John Wiley & Sons.

Wadsworth M E J (1984). Early stress and association with adult behaviour and parenting. In *Stress and Disability in Childhood* (eds N R Butler & B D Corner) p 100. Bristol: John Wright.

Wadsworth M E J & Maclean M (1986). Parents' divorce and children's life chances. *Children and Youth Services Review* 8: 145–159.

***Fetal development and disease in adult life***

Barker D J P (1994). *Mothers, Babies and Disease in Later Life*. London: British Medical Journal Publications.

Barker D J P & Osmond C (1986). Infant mortality, childhood nutrition, and ischaemic heart disease in England and Wales. *Lancet* 1: 1977–81.

Barker D J P, Martyn C N, Osmond C, Hales C N, Jespersen S & Fall C H D (1993). Growth *in utero* and serum cholesterol concentrations in adult life. *British Medical Journal* 307: 1524–1527.

Benediktsson R, Lindsay R S, Noble J, Seckl J R & Edwards C R W (1993). Glucocorticoid exposure *in utero*: a new model for adult hypertension. *Lancet* 1: 339–341.

McCance R A & Widdowson E M (1974). The determinants of growth and form. *Proceedings of the Royal Society of London* (Series Biology) 185: 1–17.

***Muscular development***

Awan M Z & Goldspink G (1972). Energetics of the development and maintenance of isometric tension by mammalian fast and slow muscles. *Journal of Mechanochemical Cell Motility* 1: 97–108.

Crow M T & Kushmerick M J (1982). Chemical energetics of mammalian muscle. *Journal of General Physiology* 79: 147–166.

Goldspink G (1975). Biochemical energetics for fast and slow muscles. In *Comparative Physiology – Functional Aspects of Structural Materials* (eds L Bolis, H P Maddrell & K Schmidt-Nielsen) pp 173–185. Amsterdam: North Holland Publishing Company.

Jollesz F & Sreter F A (1981). Development, innervation, and activity-pattern induced changes in skeletal muscle. *Annual Reviews in Physiology* 43: 531–552.

Kulkarni R N & Shetty P S (1992). Net mechanical efficiency during stepping in chronically energy deficient human subjects. *Annals of Human Biology* 19: 421–425.

Limas C J (1978). Calcium transport ATPase of cardiac sarcoplasmic reticulum in experimental hyperthyroidism. *American Journal of Physiology* 235: H745–751.

Nicol C J M & Bruce D S (1981). Effect of hyperthyroidism on the contractile and histochemical properties of fast and slow twitch skeletal muscle in the rat. *Pflugers Archiv* 390: 73–79.

Russell D McR, Walker P M, Leiter L A, Sima A A F, Tanner W K, Mickle D A G, Whitwell J, Marliss E B & Jeejeebhoy K N (1984). Metabolic and structural changes in skeletal muscle during hypocaloric dieting. *American Journal of Clinical Nutrition* 39: 503–513.

Schantz P, Henriksson J & Jansson E (1983). Adaptation of human skeletal muscle to endurance training of long duration. *Clinical Physiology* 3: 141–151.

Shetty P S (1993). Chronic undernutrition and metabolic adaptation. *Proceedings of the Nutrition Society* 52: 267–284.

Suko J (1973). The calcium pump of cardiac sarcoplasmic reticulum. Functional alterations at different levels of thyroid state in rabbits. *Journal of Physiology* 228: 563–582.

Waterlow J C (1990). Mechanisms of adaptation to low energy intakes. In *Diet and Disease in Traditional and Developing Societies* (eds G A Harrison & J C Waterlow) pp 5–23. Cambridge: Cambridge University Press.

Wiles C M, Young A, Jones D A & Edwards R H T (1979). Muscle relaxation rate, fibre-type composition and energy turnover in hyper- and hypo-thyroid patients. *Clinical Science* 57: 375–384.

***Fat patterning***

Bogin B & Sullivan T (1986). Socioeconomic status, sex, age, and ethnicity as determinants of body fat distribution for Guatemalan children. *American Journal of Physical Anthropology* 69: 527–535.

Bouchard C & Johnston F E (eds) (1988). *Fat Distribution During Growth and Later Health Outcomes*. New York: Liss.

Johnston F E (1988). Sex differences in fat patterning in children and youth. In *Fat Distribution During Growth and Later Health Outcomes* (eds C Bouchard & F E Johnston) pp 85–102. New York: Liss.

Roche A F & Baumgartner R N (1988). Tracking in fat distribution during growth. In *Fat Distribution During Growth and Later Health Outcomes* (eds C Bouchard & F E Johnston) pp 147–162. New York: Liss.

***Sensitive periods in brain growth and development***

Blakemore C (1991). Sensitive and vulnerable periods in the development of the visual system. In *The Childhood Environment and Adult Disease* pp 129–154. Chichester: Wiley (Ciba Foundation Symposium 156).

Dobbing J (1981). The later development of the brain and its vulnerability. In *Scientific Foundations of Paediatrics* 2nd edn (eds J A Davis & J Dobbing) pp 744–759. London: Heinemann.

Goy R W & McEwen B S (1980). *Sexual Differentiation of the Brain*. Cambridge, Massachusetts: MIT Press.

Herschkowitz N (1988). Brain development in the fetus, neonate and infant. *Biology of the Neonate* 54: 1–19.

Otake M & Schull W J (1984). *In utero* exposure to A-bomb radiation and mental retardation; a reassessment. *British Journal of Radiology* 57: 409–414.

Smart J L (1990). Vulnerability of developing brain to undernutrition. *Upsala Journal of Medical Sciences* Supplement 48: 21–41.

Smart J L (1991). Critical periods in brain development. In *The Childhood Environment and Adult Disease*, pp 109–128. Chichester: Wiley (Ciba Foundation Symposium 156).

***Viral infection of the pregnant woman and risk of schizophrenia in their offspring***

Barr C E, Mednick S A & Monk-Jorgensen P (1990). Exposure to influenza epidemics during gestation and adult schizophrenia. *Archives of General Psychiatry* 47: 869–874.

Bradbury T N & Miller G A (1985). Season of birth in schizophrenia: a review of evidence, methodology, and aetiology. *Psychology Bulletin* 98: 569–574.

Castle D J (1993). Some current controversies in the epidemiology of schizophrenia. *Current Medical Literature: Psychiatry* 4: 3–7.

Der G, Gupta S & Murray R M (1990). Is schizophrenia disappearing? *Lancet* 335: 513–516.

Eagles J M & Whalley L J (1985). Decline in the diagnosis of schizophrenia among first admissions to Scottish mental hospitals from 1969–1978. *British Journal of Psychiatry* 146: 151–154.

Joyce P R (1987). Changing trends in first admissions and re-admissions for mania and schizophrenia in New Zealand, Australia and New Zealand. *Journal of Psychiatry* 21: 82–86.

McGuffin P & Stuart E (1986). Genetic markers in schizophrenia. *Human Heredity* 36: 65–88.

Mednick S A, Machon R A, Huttenen M O & Bonet D (1988). Adult schizophrenia following prenatal exposure to an influenza epidemic. *Archives of General Psychiatry* 45: 189–192.

Menninger K A (1928). The schizophrenic syndromes as a product of acute infectious disease. *Archives of Neurological Psychiatry* 20: 464–481.

Miyanaga K, Machiyama Y & Juji T (1984). Schizophrenic disorders and HLA DR antigens. *Biological Psychiatry* 19: 121–129.

Munk-Jorgensen P & Jorgensen P (1986). Decreasing rates of first admission diagnoses of schizophrenia among females in Denmark 1970–1984. *Acta Psychiatrica Scandinavica* 74: 379–383.

Murray R M, Jones P & O'Callaghan E (1991). Fetal brain development and later schizophrenia. In *The Childhood Environment and Adult Disease* (eds G R Bock & J Whelan) pp 155–163. Chichester: John Wiley & Sons.

O'Callaghan E, Sham P, Takei N, Glover G & Murray R M (1991). Schizophrenia after pre-natal exposure to 1957 A2 influenza epidemic. *Lancet* 337; 1: 1248–1250.

Owen M J & McGuffin P (1991). DNA and classical genetic markers in schizophrenia. *European Archives in Psychiatry and Clinical Neuroscience* 240: 197–203.

Parker G, O'Donnell M & Walter S (1985). Changes in the diagnosis of the functional psychoses associated with the introduction of lithium. *British Journal of Psychiatry* 146: 377–382.

Sham P C, O'Callaghan E, Takei N et al. (1992). Schizophrenia following pre-natal exposure to influenza epidemics between 1939 and 1960. *British Journal of Psychiatry* 160: 461–466.

Wright P, Gill M & Murray R M (1993). Schizophrenia: genetics and the maternal immune response to viral infection. *American Journal of Medical Genetics* 48: 40–46.

**Menopause**

Avis N E, Kaufert P A, Lock M, McKinlay S M & Vass K (1993). The evolution of menopausal symptoms. *Baillière's Clinical Endocrinology and Metabolism* 7: 17–32.

Beyene Y (1989). *From Menarche to Menopause: Reproductive Lives of Peasant Women in Two Cultures*. Albany: State University of New York Press.

Flint M P (1975). The menopause: reward or punishment? *Psychosomatis* 16: 161–163.

Lock M (1993). *Encounters with Ageing: Mythologies of Menopause in Japan and North America*. Berkeley: University of California Press.

McKinlay S M, Brambilla D J & Posner J G (1992). The normal menopause transition. *American Journal of Human Biology* 4: 37–46.

Pavelka M S & Fedigan L M (1991). Menopause: a comparative life history perspective. *Yearbook of Physical Anthropology* 34: 13–38.

Thomford P J, Jelovsek F R & Mattison D R (1987). Effect of oocyte number and rate of atresia on the age of menopause. *Reproductive Toxicology* 1: 41–51.

Utian W H (1987). Overview on menopause. *American Journal of Obstetrics and Gynecology* 156: 1280–1283.

Whitehead M I, Whitcroft S I J & Hillard T C (1993). *An Atlas of the Menopause*. New York: Parthenon Publications.

***Age change in reproductive function***

Batista M C, Cartledge T P, Zellmer A W, Merion M J, Axiotis C, Bremner W J & Nieman L K (1995). Effects of ageing on menstrual cycle hormones and endometrial maturation. *Fertil Steril* 64: 492–499.

Ellison P T (1996). Age and developmental effects on adult ovarian function. In *Variability in Human Fertility: a Biological Anthropological Approach* (eds L Rosetta & C G N Mascie-Taylor) pp 69–90.

Cambridge: Cambridge University Press.
Federation CECOSF, Schwartz D & Mayaux M J (1982). Female fecundity as a function of age. *New England Journal of Medicine* 306: 404–406.
O'Rourke M T & Ellison PT (1993). Salivary estradiol levels decrease with age in healthy, regularly-cycling women. *Endocrine* 1: 487–494.
Toner J P & Scott RT (1995). Chronologic versus ovarian age: impact on pregnancy among infertile couples. *Semin Reprod Endocrinol* 13: 1–15.
Treloar A E, Boynton R E, Behn B G & Brown B W (1967). Variation of the human menstrual cycle through reproductive life. *International Journal of Fertility* 12: 77–126.
Weinstein M, Wood J, Stoto M A & Greenfield D D (1990). Components of age-specific fecundability. *Population Studies* 44: 447–467.

**Ageing as part of the development process**

Arking R (1991). *Biology of Ageing:Observation and Principles*. Englewood & Cliffs: Prentice Hall.
Crews D E & Garruto R A (1994). *Biological Anthropology and Ageing: Perspectives on Human Variation Over the Life Span*. Oxford: Oxford University Press.
Finch C (1990). *Longevity, Senescence, and the Genome*. Chicago: University of Chicago Press.
Rose M (1991). *Evolutionary Biology of Ageing*. Oxford: Oxford University Press.

***Ageing in industrialized societies***

Beall C M & Weitz C A (1989). The human population biology of ageing. In *Human Population Biology* (eds M A Little & J D Haas) pp 189–200.
Bittles A H & Collins K J (eds) (1986). *The Biology of Human Ageing*. Cambridge: Cambridge University Press.
Crews D E & Garruto R A *Biological Anthropology and Ageing: Perspectives on Human Variation over the Life Span*. Oxford: Oxford University Press.
Garruto R M (1994). Early environment, long latency and slow progression of late onset neurodegenerative disorders. In *Long-term Consequences of Early Environment. Growth Development and the Lifespan Developmental Perspective* (eds C J K Henry & S J Ulijaszek) pp 219–249. Cambridge: Cambridge University Press.

***Ageing and metabolic change***

Curtis J H & Miller K (1971). Chromosome aberrations in liver cells of guinea pigs. *Journal of Gerontology* 26: 292–294.
Cutler R G (1984). Carotenoids and retinol: their possible importance in determining longevity of primate species. *Proceedings of the National Academy of Sciences, USA* 81: 7627–7631.
Cutler R G (1985). Peroxide-producing potential of tissues: inverse correlation with longevity of mammalian species. *Proceedings of the National Academy of Sciences, USA* 82: 4798–4802.
Hall K Y, Hart R W, Benirschke A K & Walford R L (1984). Correlation between ultraviolet-induced DNA repair in primate lymphocytes and fibroblasts and species maximum achievable life span. *Mechanisms of Ageing and Development* 24: 163–173.
Harley C B, Futcher A B & Greider C W (1990). Telomeres shorten during ageing of human fibroblasts. *Nature* 345: 458–460.
Harman D (1992). Free radical theory of aging. *Mutation Research* 275: 257–266.
Hart R W & Setlow R B (1974). Correlation between deoxyribonucleic acid excision repair and lifespan in a number of mammalian species. *Proceedings of the National Academy of Sciences, USA* 71: 2169–2173.
Johnson J E (1979). Fine structure of IMR-90 cells in culture as examined by scanning and transmission electron microscopy. *Mechanisms of Ageing and Development* 10: 405–443.
Lipetz J & Cristofalo V J (1972). Ultrastructural changes accompanying the ageing of human diploid cells in culture. *Journal of Ultrastructure Research* 39: 43–56.
Ku H-K, Brunk U T & Sohal R S (1993). Relationship between mitochondrial superoxide and hydrogen peroxide production and longevity of mammalian species. *Free Radicals in Biology and Medicine* 15: 621–627.
Moore C J & Schwartz A G (1978). Inverse correlation between species lifespan and capacity of cultured fibroblasts to convert benzo(a)pyrene to water-soluble metabolites. *Experimental Cell Research* 116: 359–364.
Pashko L L & Schwartz A G (1982). Inverse correlation between species' lifespan and species' cytochrome P-488 content of cultured fibroblasts. *Journal of Gerontology* 37: 38–41.
Rohme D (1981). Evidence for a relationship between longevity of mammalian species and life spans of normal fibroblasts in vitro and erythrocytes *in vivo*. *Proceedings of the National Academy of Sciences, USA* 78: 5009–5013.
Schwartz A G & Moore C J (1977). Inverse correlation between species life span and capacity of cultured fibroblasts to bind 7.12-dimethyl benz(a)anthracene to DNA *Experimental Cell Research* 109: 448–450.
Tolmasoff J M, Ono T & Cutler R G (1980). Superoxide dismutase: correlation with lifespan and specific metabolic rates in primate species. *Proceedings of the National Academy of Sciences, USA* 77: 2777–2781.

***Ageing in non-Western populations***

Beall C M (1984). Ageing and growth at high altitude in the Himalayas. In *The People of South Asia: the Biological Anthropology of India, Pakistan, and Nepal* (ed J R Lukacs) pp 365–85. New York: Plenum Press.
Beall C M & Goldstein M C (1982). Biological function, activity and dependency among elderly Sherpa in the Nepal Himalayas. *Social Science and Medicine* 16: 135–140.
Borkan G A (1986). Biological age assessment in adulthood. In *The Biology of Human Ageing* (eds A H Bittles & K J Collins) pp 81–93. Cambridge: Cambridge University Press.
Dowse G K, Zinimet P Z, Finch C T & Collins V R (1991). Decline in incidence of epidemic glucose intolerance in Nauruans: implications for the 'thrifty genotype'. *American Journal of Epidemiology* 133: 11: 1093–1104.
Dowd L E & Manton K G (1990). Forecasting chronic disease risks in developing countries. *International Journal of Epidemiolgy* 19: 4: 1019–1036.
Høygaard A (1941). *Studies on the nutrition and physio-pathology of Eskimos*. Oslo: I Kommisjon Hos Jacob Dybwad.
Kirkwood T B W & Rose M R (1991). Evolution of senescence: late survival sacrificed for reproduction. *Philosophical Transactions of the Royal Society of London* B 332: 15–24.
Rowe L W & Kahn R L (1987). Human ageing: usual and successful. *Science* 237: 143–149.
Sinnett P F (1975). *The people of Murapin*. Faringdon: E W Classey.
de Vasconcellos M T L (1994). Body mass index: its relationship with food consumption and socio-economic variables in Brazil. *European Journal of Clinical Nutrition* 48 Suppl 3: S115–S123.

***Sarcopenia: muscle loss***

Cornoni-Huntley J C, Harris T B, Everett D F, Albanes D, Micozzi M S, Miles T P & Feldman J J (1991). An overview of body weight of older persons, including the impact on mortality. *Journal of Clinical Epidemiology* 44: 743–753.
Fiatarone M A, Marks E C, Ryan N D, Meredith C N, Lipsitz L A & Evans W J (1990). High-intensity strength training in nonagenarians. *Journal of the American Medical Association* 263: 3029–3034.
Grimby G & Saltin B (1983). The ageing muscle: a mini review. *Clinical Physiology* 3: 209–218.
Jones D A & Round J M (1990). *Skeletal Muscle in Health and Disease*. Manchester: University of Manchester Press.
Lindsted K, Tonstad S & Kuzma J W (1991). Body mass index and patterns of mortality among Seventh-Day Adventist men. *International Journal of Obesity* 15: 397–406.
Roche A F (1994). Sarcopenia: a critical review of its measurements and health-related significance in the middle-aged and elderly. *American Journal of Human Biology* 6: 33–42.
Tayback M, Kumanyika S & Chee E (1990). Body weight as a risk factor in the elderly. *Archives of Internal Medicine* 150: 1065–1072.
Weinphal J, Ragland D R & Sidney S (1990). Body mass index and 15-year mortality in a cohort of Black men and women. *Journal of Clinical Epidemiology* 43: 949–960.
Yao C H, Slattery M L, Jacobs D R, Folsom A R & Nelson E T (1991). Anthropometric predictors of coronary heart disease and total mortality: findings from the US Railroad Study. *American Journal of Epidemiology* 134: 1278–1289.

***Apoptosis: programmed cell death***

Carson D A & Ribeino M (1993). Apoptosis and disease. *Lancet* 341: 1251–1254.
Kerr J F K, Wyllie A H & Currie A H (1972). Apoptosis, a basic biological phenomenon with wider implications in tissue kinetics. *British Journal of Cancer* 26: 239–245.
Schwartzman R A & Cidlowski J A (1993). Apoptosis: the biochemistry and molecular biology of programmed cell death. *Endocrine Reviews* 14 (2): 133–151.
Steller H (1995). Mechanisms and genes of cellular suicide. *Science* 267: 1445–1449.

## Part thirteen: The future

Boas F (1912). *Changes in bodily form of descendants of immigrants*. New York: Colombia University Press.
Cole T J (1990). The LMS method for constructing normalized growth standards. *European Journal of Clinical Nutrition* 44: 45–60.
Gasser T, Kneip A, Ziegler P, Largo R & Prader A (1990). A method for determining the dynamics and intensity of average growth. *Annals of Human Biology* 17: 459–474.
Goldstein H (1981). Measuring the stability of individual growth patterns. *Annals of Human Biology* 8: 549–557.
Goldstein H (1986). Efficient modelling of longitudinal data. *Annals of Human Biology* 13: 129–141.
Goldstein H (1987). *Multilevel Models in Educational and Social Research*. London: Charles Griffin.
Harrison G A & Brush G (1990). On correlations between adjacent velocities and accelerations in longitudinal growth data. *Annals of Human Biology* 17: 55–57.
Hauspie R C, Wachholder A, Baron G, Cantraine F, Susanne C & Graffar M (1980). A comparative study of the fit of four different functions to longitudinal data of growth in height of Belgian girls. *Annals of Human Biology* 7: 347–358.
Hermanussen M & Brandt K. Mini-knemometry: an accurate technique for lower leg length measurements in early childhood. *Annals of Human Biology* (in press.)
Hermanussen M & Burmeister J (1993). Children do not grow continuously but in spurts. *American Journal of Human Biology*.
Hermanussen M, Brandt K & Burmeister J. Circaseptan periodicity in daily growth of seven healthy neonates, measured by mini-knemometry. *Annals of Human Biology* (in press.)
Jolicoeur P, Pontier J & Abidi H (1992). Asymptotic models for the longitudinal growth of human stature. *American Journal of Human Biology* 4: 461–468.
Lampl M, Veldhuis J D & Johnson ML (1992). Saltation and stasis: a model of human growth. *Science* 258: 801–803.
Mascie-Taylor CGN (1994). Statistical issues in anthropometry. In *Anthropometry: the Individual and the Population* (eds S J Ulijaszek & C G N Mascie-Taylor) pp 56–77. Cambridge: Cambridge University Press.
Preece M A & Baines M J (1978). A new family of mathematical models describing the human growth curve. *Annals of Human Biology* 5: 1–24.

Robins S P (1994). Biochemical markers for assessing skeletal growth. *European Journal of Clinical Nutrition* 48 (Suppl 1): S199–209.

Tanner J M (1981). *A History of the Study of Human Growth.* Cambridge: Cambridge University Press.

**Defining the growth characteristics of new populations**

Eveleth P B & Tanner J M (1990). *Worldwide Variation in Human Growth.* Cambridge: Cambridge University Press.

Hermanussen M, Theil C & von Büren E (1997). *Synthetic Growth Charts.* Abstract, Australian Society for Human Biology, Adelaide.

Peach C & Mitchell J C (1988). Marriage distance and ethnicity. In *Human Mating Patterns* (eds C G N Mascie-Taylor & A J Boyce) pp 31–45. Cambridge: Cambridge University Press.

Ulijaszek S J (1994). Between-population variation in pre-adolescent growth. *European Journal of Clinical Nutrition* 48 (Suppl 1): S5–S14.

**Growth patterns associated with new problem complexes**

***Growth in HIV-positive infants and children***

European Collaborative Study (1995). Weight, height and HIV infection in young children of HIV infected mothers. *Pediatric Infectious Disease Journal* 14: 685–690.

Lepage P et al. (1991). Clinical and endocrinal manifestations in perinatally human immunodeficiency virus type 1-infected children aged 5 years or over. *American Journal of Diseases in Childhood* 145: 1248–1251.

McKinney R et al. (1993). Effect of human immunodeficiency virus infection on the growth of young children. *Journal of Pediatrics* 123: 579–582.

Preble E et al. (1990). The impact of HIV/AIDS on African children. *Social Science and Medicine* 31: 671–680.

Ross A et al. (1995). Maternal HIV infection, drug use, and growth of uninfected children in their first 3 years. *Archives of Disease in Childhood* 73: 490–495.

Saavedra J et al. (1995). Longitudinal assessment of growth in children born to mothers with human immunodeficiency virus infection. *Archives of Pediatric and Adolescent Medicine* 149: 497–502.

***Growth of children born to cocaine users***

Harsham J, Keller J H & Disbrow D (1994). Growth patterns of infants exposed to cocaine and other drugs *in utero*. *Journal of the American Dietetics Association* 9: 999–1007.

Weathers W T, Crane M M, Sauvain K J & Blackhurst D W (1993). Cocaine use in women from a defined population: prevalence at delivery and effects on growth in infants. *Pediatrics* 91 (2): 350–354.

# Books about growth and development


Anderson M (1992). *Intelligence and Development. A Cognitive Theory*. Oxford: Blackwell.

Barker D J P (1994). *Mothers, Babies and Disease in Later Life*. London: British Medical Journal Publications.

Bayley N (1969). *Manual for the Bayley Scales of Infant Development*. Berkeley: Pyschological Corporation.

Blizzard R M (1966). *Human Pituitary Growth Hormone*. Colombus OH: Ross Laboratories.

Bock G R & Whelan J (1991). *The Childhood Environment and Adult Disease*. Chichester: John Wiley & Sons.

Bogin B (1988). *Patterns of Human Growth*. Cambridge: Cambridge University Press.

Borms J, Hauspie R, Sand A, Susanne C & Hebbelinck M (eds) (1984). *Human Growth and Development*. New York: Plenum.

Bouchard C & Johnston F E (eds) (1988). *Fat Distribution During Growth and Later Health Outcomes*. New York: Alan R Liss.

Boyd E & Scammon R E (1980). *Origins of the study of human growth*. Portland University of Oregon: Health Sciences Center Foundation.

Brauth S E, Hall W S & Dooling R J (eds) (1991). *Plasticity of Development*. Cambridge MA: The MIT Press.

Brody S (1945). *Bioenergetics and Growth*. New York: Reinhold

Cameron N (1984). *The Measurement of Human Growth*. Croom-Helm.

Cox L A (1992). *A Guide to the Measurement and Assessment of Growth in Children*. Welwyn Garden City: Castlemead Publications.

Davies P S W & Cole T J (1995). *Body Composition Techniques in Health and Disease*. Cambridge: Cambridge University Press.

Eiben O (ed) (1994). *Auxology 94. Children and Youth at the End of the 20th Century*. Budapest: Humanbiologia Budapestinensis 25.

Eveleth P B & Tanner J M (1976). *Worldwide Variation in Human Growth*. Cambridge: Cambridge University Press.

Eveleth P B & Tanner J M (1990). *Worldwide Variation in Human Growth*. Cambridge: Cambridge University Press.

Falkner F & Tanner J M (eds) (1986). *Human Growth. A Comprehensive Treatise* vols 1, 2 & 3. New York: Plenum Press.

Floud R, Wachter K & Gregory A (1990). *Height, Health and History*. Cambridge: University of Cambridge Press.

Forbes G B (1987). *Human Body Composition: Growth, Aging, Nutrition, and Activity*. New York: Springer Verlag.

Friedman S L & Haywood H C (1994). *Developmental Follow-up. Concepts, Domains, and Methods*. San Diego: Academic Press.

Frisancho A R (1990). *Anthropometric Standards for the Assessment of Growth and Nutritional Status*. Ann Arbor: University of Michigan Press.

Greulich W W & Pyle S I (1959). *Radiographic Atlas of Skeletal Development of the Hand and Wrist*. Stanford: Stanford University Press.

Hauspie R, Lindgren G & Falkner F (eds) (1995). *Essays on Auxology presented to James Mourilyan Tanner by former Colleagues and Fellows*. Welwyn Garden City: Castlemead Publications.

Henry C J K & Ulijaszek S J (1996). *Long-term Consequences of Early Environment. Growth, Development and the Lifespan Developmental Perspective*. Cambridge: Cambridge University Press.

Illingworth R S (1987). *The Development of the Infant and Young Child*. Edinburgh: Churchill Livingstone.

Johnston F E, Roche A F & Susanne C (eds) (1980). *Human Growth and Maturation: Methodologies and Factors*. New York: Plenum.

Keynes M (ed) (1993). *Sir Francis Galton FRS. The legacy of his ideas*. London: Macmillan

Kleiber M (1975). *The Fire of Life*. New York: Kleiber.

Komlos J (1989). *Nutrition and Economic Development in the Eighteenth-Century Habsburg Monarchy: an Anthropometric History*. Princeton: Princeton University Press.

Komlos J (ed) (1994). *Stature, Living Standards and Economic Development: Essays in Anthropometric History*. Chicago: University of Chicago Press.

Komlos J (ed) (1995). *The Biological Standard of Living on Three Continents: Further Explorations in Economic History*. Boulder: Westview press.

Krogman W M (1972). *Child Growth*. Ann Arbor: University of Michigan Press.

Lohman T G (1992). *Advances in Body Composition Assessment*. Champaign IL: Human Kinetics Books.

Lohman T G, Roche A F & Martorell R (ed) (1988). *Anthropometric Standardization Reference Manual*. Champaign IL: Human Kinetics Books.

Malina R M & Bouchard C (1991). *Growth, Maturation, and Physical Activity*. Champaign IL: Human Kinetics Books.

Malina R M & Roche A F (1983). *Manual of Physical Status and Performance in Childhood: Volume 2. Physical Performance*. New York: Plenum.

McCammon R B (1970). *Human Growth and Development*. Springfield IL: Charles C Thomas.

Meredith H V (1978). *Human Body Growth in the First Ten Years of Life*. Colombia SC: State Printing.

Plomin R, DeFries J C & Fulker D W (1988). *Nature and Nurture During Infancy and Early Childhood*. Cambridge: Cambridge University Press.
Publishing Corporation.

Roche A F, Heymsfield S B & Lohman T G (eds) (1996). *Human Body Composition*. Champaign IL: Human Kinetics Books.

Senterre J (ed) (1989). *Intrauterine Growth Retardation*. New York: Raven Press.

Shephard R J (1991). *Body Composition in Biological Anthropology*. Cambridge: Cambridge University Press.

Tanner J M (1981). *A History of the Study of Human Growth*. Cambridge: Cambridge University Press.

Tanner J M (1989). *Foetus into Man*. Ware: Castlemead Publications.

Tanner J M (ed) (1989). *Auxology 88. Perspectives in the Science of Growth and Development*. London: Smith-Gordon.

Tanner J M, Whitehouse R H, Cameron N, Marshall W A, Healy M J R & Goldstein H (1983). *Assessment of Skeletal Maturity and Prediction of Adult Height (TW2 Method)*. London: Academic Press.

Thompson D (1988). *On Growth and Form*. Abridged Edition, edited by J T Bonner. Cambridge: Cambridge University Press.

Waterlow J C (ed) (1988). *Linear Growth Retardation in Less Developed Countries*. New York: Raven Press.

Weiner J S & Lourie J A (1981). *Practical Human Biology*. London: Academic Press.

# Index

## A



## B

## C

## G

## H

## I



## J



## K



## L

## M



## N

## O



## P